SOME SI BASE UNITS

Physical Quantity	*Name of Unit*	*Symbol*
length	meter	m
mass	kilogram	kg
time	second	s
electric current	ampere	A
thermodynamic temperature	kelvin	K
amount of substance	mole	mol

SOME SI DERIVED UNITS

Physical Quantity	*Name of Unit*	*Symbol*	*SI Unit*
frequency	hertz	Hz	s^{-1}
energy	joule	J	$kg \cdot m^2/s^2$
force	newton	N	$kg \cdot m/s^2$
pressure	pascal	Pa	$kg/m \cdot s^2$
power	watt	W	$kg \cdot m^2/s^3$
electric charge	coulomb	C	$A \cdot s$
electric potential	volt	V	$kg \cdot m^2/A \cdot s^3$
electric resistance	ohm	Ω	$kg \cdot m^2/A^2 \cdot s^3$
capacitance	farad	F	$A^2 \cdot s^4/kg \cdot m^2$
inductance	henry	H	$kg \cdot m^2/A^2 \cdot s^2$
magnetic flux	weber	Wb	$kg \cdot m^2/A \cdot s^2$
magnetic flux density	tesla	T	$kg/A \cdot s^2$

SI UNITS OF SOME OTHER PHYSICAL QUANTITIES

Physical Quantity	*SI Unit*
speed	m/s
acceleration	m/s^2
angular speed	rad/s
angular acceleration	rad/s^2
torque	$kg \cdot m^2/s^2$, or $N \cdot m$
heat flow	J, or $kg \cdot m^2/s^2$, or $N \cdot m$
entropy	J/K, or $kg \cdot m^2/K \cdot s^2$, or $N \cdot m/K$
thermal conductivity	$W/m \cdot K$

SOME CONVERSIONS OF NON-SI UNITS TO SI UNITS

Energy:

1 electron-volt (eV) = 1.6022×10^{-19} J
1 erg = 10^{-7} J
1 British thermal unit (BTU) = 1055 J
1 calorie (cal) = 4.186 J
1 kilowatt-hour (kWh) = 3.6×10^6 J

Mass:

1 gram (g) = 10^{-3} kg
1 atomic mass unit (u) = 931.5 MeV/c^2
= 1.661×10^{-27} kg
1 MeV/c^2 = 1.783×10^{-30} kg

Force:

1 dyne = 10^{-5} N
1 pound (lb or #) = 4.448 N

Length:

1 centimeter (cm) = 10^{-2} m
1 kilometer (km) = 10^3 m
1 fermi = 10^{-15} m
1 Angstrom (Å) = 10^{-10} m
1 inch (in or ") = 0.0254 m
1 foot (ft) = 0.3048 m
1 mile (mi) = 1609.3 m
1 astronomical unit (AU) = 1.496×10^{11} m
1 light-year (ly) = 9.46×10^{15} m
1 parsec (ps) = 3.09×10^{16} m

Angle:

1 degree (°) = 1.745×10^{-2} rad
1 min (′) = 2.909×10^{-4} rad
1 second (″) = 4.848×10^{-6} rad

Volume:

1 liter (L) = 10^{-3} m^3

Power:

1 kilowatt (kW) = 10^3 W
1 horsepower (hp) = 745.7 W

Pressure:

1 bar = 10^5 Pa
1 atmosphere (atm) = 1.013×10^5 Pa
1 pound per square inch (lb/in^2) = 6.895×10^3 Pa

Time:

1 year (yr) = 3.156×10^7 s
1 day (d) = 8.640×10^4 s
1 hour (h) = 3600 s
1 minute (min) = 60 s

Speed:

1 mile per hour (mi/h) = 0.447 m/s

Magnetic field:

1 gauss = 10^{-4} T

PHYSICS

FOR SCIENTISTS AND ENGINEERS

PHYSICS

FOR SCIENTISTS AND ENGINEERS

Extended Version:

Volume II

Paul M. Fishbane

UNIVERSITY OF VIRGINIA

Stephen Gasiorowicz

UNIVERSITY OF MINNESOTA

Stephen T. Thornton

UNIVERSITY OF VIRGINIA

PRENTICE HALL, Englewood Cliffs, New Jersey 07632

About the cover: The early twentieth century witnessed the simultaneous explosion of abstract art and the birth of so-called modern physics with artists such as Picasso, Duchamp, and Klee and physicists like Einstein, Heisenberg, and de Broglie. Critics and scholars have long recognized the connection between this "new painting" and the "new physics."

Russian painter Wassily Kandinsky, as much as any of his contemporaries, sought to define a new form of art that fully considered contemporaneous discoveries in physics. Kandinsky was one of the first artists to pursue a style known as "nonobjective painting" or painting without subject matter. Kandinsky threw out traditional assumptions about the tangible world and sought to define reality from a more fundamental perspective—objectives he shared with quantum physicists.

Kandinsky's painting on the cover—*Leichtes* (Light)—was painted near the end of the artist's career and is representative of his work. The title refers not to light as in sunlight but rather to light as in weight.

Editor-in-Chief and Aquisitions Editor: Tim Bozik
Development Editor: Karen Karlin
Editorial/production supervision: Barbara DeVries
Senior Production Editor: Barbara Grasso Mack
Director of Marketing: Gary June
Copy Editors: Lynne Lackenbach/Reynold Rieger
Design Director: Florence Dara Silverman
Interior and cover design: Lee Goldstein
Photo Editor: Lori Morris-Nantz
Photo Research: Fran Antmann
Artwork by Hans & Cassady, Inc., Westerville, OH 43081
Cover art: "Leichtes,". Wassily Kandinsky, 1930. Musee National D'Art Modern, Center Georges Pompidou, Paris. Service Photographiquede Musee.
Prepress Buyer: Paula Massenaro
Manufacturing Buyer: Lori Bulwin
Editorial Assistants: JoMarie Jacobs/Nancy Bauer

A Paramount Communications Company
Englewood Cliffs, New Jersey 07632

Printed in the United States of America
10 9 8 7 6 5 4 3 2

ISBN 0-13-673021-3

Prentice-Hall International (UK) Limited, *London*
Prentice-Hall of Australia Pty. Limited, *Sydney*
Prentice-Hall Canada Inc., *Toronto*
Prentice-Hall Hispanoamericana, S.A., *Mexico*
Prentice-Hall of India Private Limited, New Delhi
Prentice-Hall of Japan, Inc., *Tokyo*
Simon & Schuster Asia Pte. Ltd., *Singapore*
Editora Prentice-Hall do Brasil, Ltda., *Rio de Janeiro*

BRIEF CONTENTS

CONTENTS

PREFACE

This text is designed to give students in a calculus-based physics course a solid foundation in the core science of all science and technology and to teach them to apply their knowledge. The students taking this course are typically in their first or second year and may be prospective engineers, scientists, physicians, or indeed any students who will use technology in their careers. These students may either be concurrently taking or have already taken a course in calculus. From our point of view, it makes little difference which of the above categories the student fits; it is the nature of physics that the few principles that unify the subject are truly fundamental and hold whatever the discipline in which they are applied.

Some of the special features that we believe distinguish this text are listed below, but the most important "feature" of the text is, we hope, the clarity and correctness with which the principles of physics, their relations, and their applications are set forth. Such an exposition must balance rigor with a more intuitive understanding of the subject, and we have tried to maintain this balance throughout.

Modern Physics

Introductory texts traditionally treat physics from Galileo to the late nineteenth century as "classical" and most of the developments in physics from the late nineteenth century to the present as "modern"—something to be treated differently from the "classical" topics. This approach seems particularly odd in light of the fact that it has been only a little over 300 years since Isaac Newton started the scientific revolution that has produced a world he would not recognize. The bulk of what we refer to as classical physics was set forth in the nineteenth century; temporally, "modern" physics comprises roughly half of our body of knowledge. Many of the ideas of modern physics are not hard in the mathematical sense. However, they can be nonintuitive, and we think that it is important that students begin to develop intuition about this material as early as possible. Without sacrificing the essential aspects of classical physics, we have introduced modern notions from the beginning. Although much of this material appears in optional sections, in many cases it is intertwined with the classical material. We conclude the text traditionally, with chapters on modern physics. We think that the preparation we have laid down for this material will make it more easily assimilated.

Mathematics

We introduce the mathematics that students need to know the first time they need to know it, in the context of the physics being presented. Even though students may be concurrently acquiring calculus skills in mathematics courses, we have tried to make the calculus and other mathematical tools used here self-contained such that the student can understand the material without having to go elsewhere for mathematical help. The mathematics thus appears in progressive degrees of difficulty. We believe that this approach fosters better understanding and less reliance on formula memorization.

We teach the correct usage of significant figures and vary the number of significant figures in examples and in problems—in keeping with the way problems arise in the real world. In this way students must maintain an awareness of significant figures and are not lulled into thinking that all problems involve the same number of significant figures.

The ability to make quantitative estimates is one of the most important skills that a scientist or engineer can have. We have made the development of that ability an important part of our approach, both in the text and in problems.

Optional Material and Boxes

An asterisk (*) indicates optional material that can be skipped without loss of continuity. Some chapter subsections or entire sections are optional; for the most part, such material is either self-contained modern physics or an in-depth discussion of material already introduced. In addition, there are three types of marked-off boxes. **Derivation** boxes set off particularly detailed derivations of important equations. **Application** boxes contain technological consequences of principles discussed in the main body of the text. **A Closer Look** boxes contain more detailed discussions of physics principles and of their place in the larger picture.

Examples, Problem-Solving Techniques, Problems, and Questions

We feel that the ability to solve quantitative problems is the best measure of whether the physics is truly understood. With this in mind, we have included certain features that will strengthen the student's problem-solving ability. **Examples** representing typical problems are numerous throughout. Their solutions illustrate sound approaches to problem-solving. We "think out loud" in the example solutions to reveal *how* and *why* we reach our answers and not just the answers themselves. We thereby deemphasize a plug-in approach. We have included throughout the text **Problem-Solving Techniques**, boxed in blue. This material appears in the key chapters that introduce new ways of looking at problems in physics.

At the end of each chapter are **Problems**, totaling some 2500, and thought **Questions**, totaling about 700. The problems are keyed to text sections, in increasing order of difficulty within each section. There are also **General Problems**, which bring together material from the entire chapter as well as from previous chapters. In addition to the ordering, problems are labeled I, II, or III. Level I problems are "easy," not going much beyond plug-ins. These problems develop student recognition of particular physics concepts and build confidence. Level II problems are typically multistep and require an increased understanding of the interconnectedness of physics; the general problems carry this requirement a step further. Level III problems are especially challenging, in some cases demanding significant synthesis of concepts in the text. Unlike the problems, the thought questions at the end of each chapter may not have simple or direct answers. They challenge students to consider concepts and to recognize the meaning of the principles taught in the text.

Other Study Aids

Effective keys and reviewing aids appear in several forms. **Important equations** are boxed, and the **Summary** at the end of each chapter reminds students of the most important results of the chapter. For chapters that introduce numerous or especially difficult concepts, we offer **Interim Summaries** to give students extra assistance. In addition, coordinated with the text are **margin notes** that play several key roles. First, important concepts or results, ones that might be "on the test," are signaled by these notes. Second, when previously developed concepts are used, margin notes recall where those concepts were covered. Finally, some margin notes point out problem-solving techniques.

Color is used in this text as a pedagogical aid, primarily where the consistent use of color to represent physical quantities and analytic elements in illustrations improves clarity. For example, color allows us to remind students visually not to confuse acceleration vectors with force vectors.

Supplements

Quality supplements make a difference. Each of the supplements that accompany this text has been carefully developed to create a unique set of high-quality, tested tools that provide professors and students the means to enliven and reinforce the teaching and learning of physics.

The Interactive Physics Player. Available only from Prentice Hall, this version of Knowledge Revolution's highly acclaimed Interactive Physics II computer program brings the text to life with real-time simulations and animations of over 100 of the examples and problems (each identified by the icon). The program allows the user to change values and view the outcomes, creating and observing "what if" scenarios. Tools showing numerical results, graphs, and vectors make it easy to analyze the simulation. For students, the Interactive Physics Player provides a way to visualize the concepts they are trying to learn. Instructors can use it as a demonstration tool in class or in lab. The program is available in a Macintosh version; a Windows version is scheduled for late 1992. See the reply card at the back of this text for additional information.

The Learning Guide. The Learning Guide, or study guide, for our text has been adapted from guides developed at Princeton University during the past 17 years. The guide features complete problem assignments with helping questions and keyed answers. Help is given in the form of increasingly detailed hints and references to the text.

Test-Item File and PH Test. The test questions that accompany this text have been developed by Professor Charles Scherr of the University of Texas and have been used at Austin by thousands of engineering and science students over the course of a decade. Approximately 500 questions have been constructed in algorithmic form with a broad range of input data, which allows each question to generate a significant and reliable number of versions. PH Test, the testing program, provides the means to create in minutes a variety of student testing devices, homework and drill sheets, and answer sheets. PH Test is available in Macintosh and DOS versions.

Instructor's Solution Manual. Written by Professor Irv Miller of Drexel University, this manual contains complete solutions to all problems in the text and separate answers to the even-numbered problems.

Transparencies. Over 150 four-color transparencies are available in the transparency set.

"Physics You Can See" Demonstration Experiments Video. This video is a collection of brief demonstrations for classroom use.

The New York Times Contemporary View Program. An annual newspaper supplement containing articles from the New York Times, it demonstrates the connection between what goes on in the classroom and in the world around us. It is available in quantity, *free* for each student.

Text Versions and Teaching Alternatives

This text provides a complete introduction to physics. It is available in a 41-chapter version, which ends with a chapter on special relativity and a chapter on quantum physics. For those who want more detailed coverage, a 46-chapter extended version containing additional coverage of quantum physics, nuclear physics, particles, and cosmology is also available. The text is published in split volumes as well. Volume I includes Chapters 1 to 21, covering mechanics, waves, and thermodynamics. Volume II includes Chapters 22 to 46 of the extended version, covering electricity and magnetism, light and optics, and modern physics.

A Lean Alternative. At many institutions, the constraints of time coupled with the degree of preparation of the students necessitate that the introductory physics course be compressed or that more time be spent on fewer topics. We propose here one possible way—out of many—in which this text could be used in a highly streamlined course. The criteria we used for suggesting what to retain and what to delete are associated with the degree to which the material is self-contained or will be studied later in a more applications-oriented course, not with how "hard" that material is. We would eliminate or deemphasize the following topics or sets of topics: relative motion and fictitious forces; multidimensional motion, including the more complicated aspects of rotational motion; statics; the more advanced aspects of harmonic motion; the Doppler effect; the more advanced aspects of wave superposition, including diffraction; fluids and solids; much of statistical physics, including probability distributions and entropy; the effects of materials in electromagnetism; applied aspects of time-dependent circuits, including alternating currents; aspects of electromagnetic radiation; mirrors and lenses; the Lorentz transformations; and the more detailed applications of quantum mechanics.

Thus our compressed course would consist of the following:

Chapter	Sections	Chapter	Sections
1	1, 2, 3, 5, 7	20	1, 2, 3, 4
2	1, 2, 3, 4, 5	22	1, 2, 3, 4
3	1, 2, 3, 4, 5	23	1, 2, 3, 4
4	1, 2, 3, 4, 5	24	1, 2, 3, 4
5	1, 2, 3, 4	25	1, 2, 3, 4, 6, 7
6	1, 2, 3, 4, 5	26	1, 2, 3
7	1, 2	27	1, 2, 3, 7
8	1, 3, 4, 6	28	1, 2, 3
9	1, 2, 3, 5, 6	29	1, 2, 3, 4
10	1, 2, 3, 4	30	1, 2, 3, 5
12	2, 3, 4	31	2, 3, 4, 5, 6
13	1, 3, 4, 5	33	1, 2, 3
14	1, 2, 3, 5, 6	35	1, 2, 3, 4
15	1, 3, 4	36	1, 2, 3
17	1, 2, 3	38	1, 2, 3
18	1, 2, 3, 4, 5, 6	40	1, 2, 3, 4, 7
19	1, 2, 3	41	1, 2, 3, 4

Acknowledgments

It is clear that a project of this magnitude cannot be accomplished by three authors alone. We are grateful to the many people who have contributed to making this a better text.

A special thanks goes to Professor Irv Miller of Drexel University. In addition to writing the Instructor's Solution Manual, Irv was an invaluable resource in creating, checking, and refining the problem set and answers. Working with the input of Professor T. S. Venkataraman, also of Drexel, Irv coordinated the independent critique and feedback of the problems and answers from the following: Ian Avruck; Roger Freedman, University of California, Santa Barbara; Joseph Hemsky, Wright State University; Joey Huston, Michigan State University; Alvin Jenkins, Jr., North Carolina State University; Rex Joyner, Indiana Institute of Technology; Alok Kumar, California State University, Long Beach; Howard Miles, Washington State University; George Miner, University of Dayton; Hans Plendl, Florida State University; and Virginia Roundy, California State University, Fullerton.

We want to thank John Malone of the teaching laboratory at the University of Virginia for his help with demonstrations and photographs.

We would also like to acknowledge and thank the following reviewers, who provided valuable feedback. We took all of their comments very seriously.

Maris Abolins, *Michigan State University*
Ricardo Alarcon, *Arizona State University*
Bradley Antanaitis, *Lafayette College*
Carl Bender, *Washington University*
Hans-Uno Bengtsson, *University of California, Los Angeles*
Robert Bowden, *Virginia Polytechnic Institute and State University*
Bennet Brabson, *Indiana University*
Edward Chang, *University of Massachusetts*
Albert Claus, *Loyola University*
Robert Coakley, *University of Southern Maine*
James Dicello, *Clarkson University*
N. John DiNardo, *Drexel University*
P. E. Eastman, *University of Waterloo*
Gabor Forgacs, *Clarkson University*
A. L. Ford, *Texas A & M University*
William Fickinger, *Case Western Reserve University*
Rex Gandy, *Auburn University*
Simon George, *California State University, Long Beach*
James Gerhart, *University of Washington*
Bruce Harmon, *Iowa State University*
Joseph Hemsky, *Wright State University*
Alvin Jenkins, *North Carolina State University*
Karen Johnston, *North Carolina State University*
Garth Jones, *University of British Columbia*
Leonard Kahn, *University of Rhode Island*
Charles Kaufman, *University of Rhode Island*
Thomas Keil, *Worcester Polytechnic University*
Carl Kocher, *Oregon State University*
Karl Ludwig, *Boston University*
Robert Marande, *Pennsylvania State University*
David Markowitz, *University of Connecticut*
Roy Middleton, *University of Pennsylvania*
George Miner, *University of Dayton*
Lorenzo Narducci, *Drexel University*
Jay Orear, *Cornell University*
Patrick Papin, *San Diego State University*
Kwangjai Park, *University of Oregon*
R. J. Peterson, *University of Colorado*
Frank Pinski, *University of Cincinnati*
Lawrence Pinsky, *University of Houston*
Stephen Pinsky, *Ohio State University*
Richard Plano, *Rutgers University*
Hans Plendl, *Florida State University*
Shafigur Rahman, *Allegheny College*
Peter Riley, *University of Texas, Austin*
L. David Roper, *Virginia Polytechnic Institute and State University*
Richard Roth, *Eastern Michigan University*
Carl Rotter, *West Virginia University*
Charles Scherr, *University of Texas*
Eric Sheldon, *University of Lowell*
Charles Shirkey, *Bowling Green State University*
Marlin Simon, *Auburn University*
Robert Simpson, *University of New Hampshire*
James Smith, *University of Illinois*
J. C. Sprott, *University of Wisconsin*
Malcolm Steuer, *University of Georgia*
Thor Stromberg, *New Mexico State University*
Smio Tani, *Marquette University*
William Walker, *University of California, Santa Barbara*
George Williams, *University of Utah*

The development of this book began under the banner of Allyn and Bacon, Inc. We particularly want to thank Jim Smith, Nancy Forsyth, and Judy Hauck of that firm for their encouragement and help.

Finally, we would like to thank the publishing team at Prentice Hall who have carried this project through. Editor-in-Chief Tim Bozik, who directed the project, has been a constant source of encouragement and material help. Karen Karlin, our developmental editor, insisted that the conceptual material be presented more clearly than we had imagined possible. We feel that she has done a splendid job. Director of Marketing Gary June has provided constant support and ideas, and we thank him. Last but not least, the production of a book such as this one is an enormous task demanding the most careful attention to detail. We especially want to thank Barbara DeVries and Barbara Grasso Mack for their support during the later and most difficult stages of production and John Morgan who handled the earlier production process. To all the individuals listed above, and to the many others at Prentice Hall who have worked to make this book a success, we extend our heartfelt thanks.

The cumulative and accelerating nature of science and technology make it more imperative than ever that our emerging scientists and engineers understand how few and how solid are the pillars of the enterprise. From this view, the distinctions between "science" and "engineering," and between "classical physics" and "modern physics," melt. We want in this book to make evident the pillars of physics as well as the highly interconnected structure that has been erected on those pillars.

PHYSICS

FOR SCIENTISTS AND ENGINEERS

CHAPTER 22

Investigations of the nature of lightning were important in the understanding of electrical phenomena. Here Benjamin Franklin is performing one of his famous kite experiments, the results of which he published between 1751 and 1753.

ELECTRIC CHARGE

Here we begin our study of electricity and magnetism, a subject with ramifications throughout the physical world. Electromagnetic forces control the structure of atoms and of all materials, and light and other electromagnetic waves are pervasive. The understanding of these forces is one of the great success stories of science. In this chapter we shall introduce electric charge, a new property carried by the constituents of atoms, and the fundamental law of the interaction of two charges at rest, Coulomb's law. This force law is as fundamental, and has the same form, as the universal law of gravitation. However, the force described by Coulomb's law can be either attractive or repulsive.

22–1 PROPERTIES OF CHARGED MATTER

Matter and Electric Charge

In most of our discussions to this point, we have characterized bulk matter, and the atoms that make up matter, by a single attribute: mass. When the structure of atoms is probed more deeply, we find that atoms are made up of electrons and nuclei. Electrons and nuclei can be characterized by another attribute, **electric charge** (usually labeled q). Electric charges exert (electric) forces on one another proportional to the product of their charges, just as masses exert gravitational forces on one another proportional to the product of their masses. Electric forces hold the atom together. However, a new element enters into electric forces that does not occur in gravitation: Charges come with two signs, and, depending on the signs of the interacting charges, the forces between charges can be repulsive *or* attractive. The set of phenomena associated with the forces between stationary charges form the subject of **electrostatics**, or **static electricity**.†

A CLOSER LOOK

A Brief History of the Study of Electricity and Magnetism

Most students have at least some acquaintance with electric charges, the forces between them, and the fact that magnetism has something to do with electricity. As obvious and simple as these things may seem, the experimental evidence and the understanding of that evidence developed only over a long time period.

The word *electricity* has its roots in the Greek word for "amber" (*electrum*), and the first written mention of the curious effects of rubbed amber dates from the fifth century B.C. Far earlier, ancient peoples surely observed the crackling and sparking of rubbed fur. It was not until the eighteenth century that the critical discovery that electric forces can be either repulsive or attractive was made. Over time, the idea developed that a quantity (which we now call electric charge) is associated with electric forces. Among the many important names associated with these discoveries are Stephen Gray, Charles Dufay, and Benjamin Franklin.

Franklin, fascinated with parlor demonstrations of electrical effects (which were fashionable in the eighteenth century), did much scientific research. He is best known for his exploitation of the existing idea that associated electrical phenomena with a kind of fluid contained in matter. Repulsion and attraction were associated with an excess or deficiency of the fluid. Implicit in this model is what we now recognize as the phenomenon of the conservation of charge: If the fluid were to flow out of a body, it would leave behind a deficiency. Franklin introduced the terms "positive" and "negative" for the two types of charge. He also set the standard sign convention, in which the electron, the actual particle that moves in conductors, has negative charge. Franklin was known, particularly to the general public, for his spectacular (and dangerous!) experiments with lightning, which he recognized as an electrical effect. Franklin and his friend Joseph Priestley, as well as Henry Cavendish, are linked with the discovery that the fundamental force between electric charges is an inverse-square law. This law was confirmed more directly first by John Robison and then by Charles Coulomb, whose name is now tied to the law, in the mid- and late eighteenth centuries, respectively.

By the first part of the nineteenth century, magnetism, which was then believed to be a phenomenon unrelated to electricity, became the object of intensive experimentation. The nature of magnetism and its relation to electricity began to be clarified around 1820, primarily through the work of Hans Christian Oersted, André-Marie Ampère, and Michael Faraday. This relation was finally understood and unified through the formulation of the theory of electromagnetism by James Clerk Maxwell in the 1860s.

The real nature of charged matter was revealed only with the experimental exploration of atoms. Quantum mechanics is an additional element needed to explain the properties of atoms. All the electrical properties of matter can now be understood within the framework of quantum theory.

†When charges move, the forces are more complicated, as we shall see in later chapters.

Atoms are **electrically neutral** (or just "neutral"); that is, an atom as a whole has no electric charge. We know this because the electric forces between atoms are small.[†] However, an atom's electrons (symbolized by e) each carry the same unit of negative charge, $q_{\text{electron}} = -e$. The electrons orbit in shell-like regions around the much heavier nucleus, consisting of neutrons (abbreviated n), which are neutral, and protons (abbreviated p), which have a positive charge with a magnitude equal but opposite to that of an electron, $q_{\text{proton}} = +e$. Although the nucleus has 99.95 percent of the mass of an atom, the nuclear radius is only about 10^{-5} of the atomic radius. In a (neutral) atom the number of electrons equals the number of protons. Chemical elements differ in the number of electrons in their atoms (or, equivalently, in the number of protons in their nuclei). The electrons that are on average closer to the nucleus are difficult to dislodge because of the strength of their attraction to it. The outermost electrons are attracted less strongly to the nucleus and are more easily dislodged, and the ease with which this happens determines to a large extent both the physical and chemical properties of the element that contains those electrons. An atom that has lost one or more electrons, and is thus left with a positive charge, is called a *positive ion*. A *negative ion* is an atom that has gained electrons.

The evidence that led to the discovery of electric charge and electric forces depended on the electrical properties of bulk matter and only indirectly on the fact that matter is made of atoms. For that reason, we want to give a brief glimpse of the electrical properties of bulk matter. If the outer electrons of atoms in bulk matter are especially easy to dislodge (are *weakly bound* to their nucleus), they behave as though they are almost free, and can move through the material almost unimpeded. Such materials are good **conductors**. Metals are good conductors; some metals, such as copper, silver, and aluminum, are better conductors than others. A certain class of materials, when cooled to low enough temperatures, contain electrons that in effect move with no inhibition. These materials, called **superconductors**, have other remarkable properties, as we shall learn. The electrons of most nonmetallic solids do not travel easily; such solids—including rubber, glass, and plastics—are **insulators**. Silicon, germanium, and an increasing number of synthetic combinations of materials are substances that we can make into insulators or conductors by controlling the electric forces on them or the temperature. Such substances are called **semiconductors**, and they play an important role in technology.

Conductors, superconductors, insulators, and semiconductors

The ease with which charges move through matter is closely related to our ability to transfer charges back and forth between different materials. When we transfer charges between different materials, we say that we have either *charged* or *discharged* them. Rubbing a material in which the outer electrons are loosely bound, such as amber, may carry those electrons elsewhere, to be ultimately deposited on some object. The original material then has an excess positive charge: It has lost some electrons. The object to which the electrons have been carried will have an excess of electrons and is now negatively charged. When charge is carried from one body to another in this way, the bodies are said to be **charged by conduction**. Note that both the original material and the object have acquired a charge.

Charging by conduction

We can gain another measure of control over the charge on an object by connecting that object to the earth with a good conductor. When a negatively charged object is connected to the earth in this way, electrons flow from the object to the earth and leave the object neutral. If, instead, an object has an excess positive charge, then electrons flow from the earth and neutralize the object. Why does the charge flow? The earth is itself a good conductor; in effect, the conducting line

[†] Not, however, precisely zero. The reasons will become apparent later.

from the object to the earth allows the charge on the object to be shared with the earth, but the earth is so large that the remaining additional charge on the object is undetectable. Such an object is said to be **grounded** (Fig. 22–1). By walking across a carpet on a dry winter day, we may slowly accumulate a charge. When we become grounded by touching a conductor connected to the earth, such as a metal pipe, we suddenly discharge our electric charge to the earth. The resulting spark can be startling.

FIGURE 22–1 Portable "lightning rods" were fashionable early in the nineteenth century. The gentleman is electrically grounded.

Evidence that Charges Are of Two Types

Some important properties of electric charge can be demonstrated by performing experiments with materials that are readily available in an elementary physics laboratory. We can transfer electric charge by rubbing a Teflon rod on a piece of fur, or by rubbing a glass rod on silk. The Teflon acquires a charge, and the fur acquires an equal but opposite charge (Fig. 22–2a); similarly, the glass acquires a charge and the silk, an equal but opposite charge (Fig. 22–2b). (Actually, the Teflon now has an excess of electrons and the fur, a deficiency of electrons, whereas the glass has a deficiency of electrons and the silk, an excess. Thus, for example, the Teflon rod becomes negative, and the glass rod becomes positive. We shall speak of these charges when we describe the experiments below. However, the particular signs of the charges acquired is irrelevant to our results. The sign of the electron is called "negative" just by convention.) The Teflon and fur seem to transfer charge more efficiently than glass and silk do and give more easily observable effects.

To study the effects of forces between charges, we want to transfer our charges to small (initially neutral) masses, because small masses react to forces most visibly. This we can do by using small cork balls coated with a conducting paint, paint that allows charge to move around easily on the surface. A cork ball is hung by a thin insulating thread (Fig. 22–3a). If we touch a negatively charged Teflon rod to a cork ball, the ball is immediately repelled by the rod (Fig. 22–3b). If we touch the negatively charged Teflon rod to *two* suspended (neutral) cork balls, the balls strongly repel each other (Fig. 22–3c). Similar behavior occurs between two cork balls that have been touched by a positively charged glass rod. However, if we

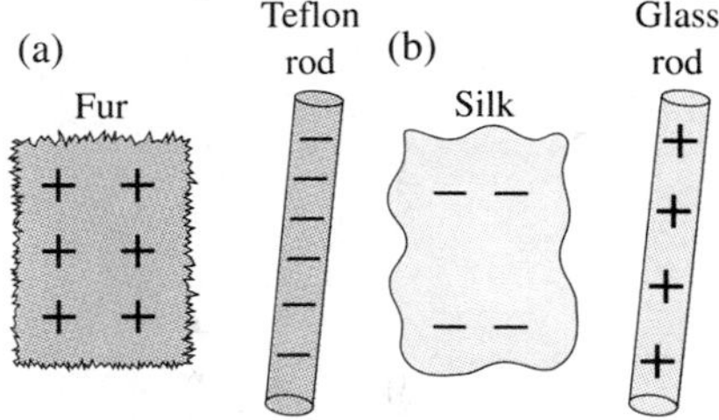

FIGURE 22–2 When (a) Teflon is rubbed against fur and (b) glass is rubbed against silk, electric charge is transferred.

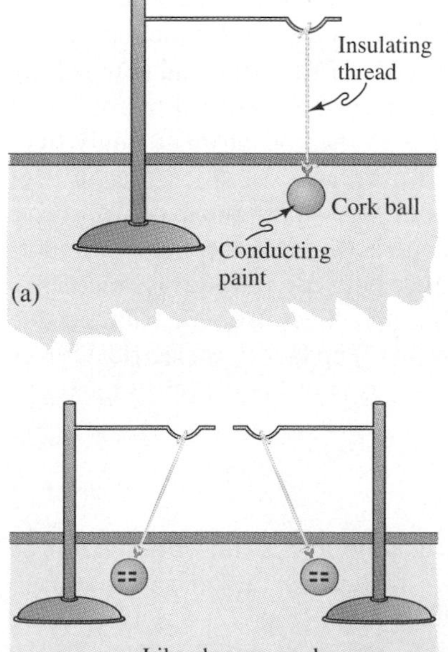

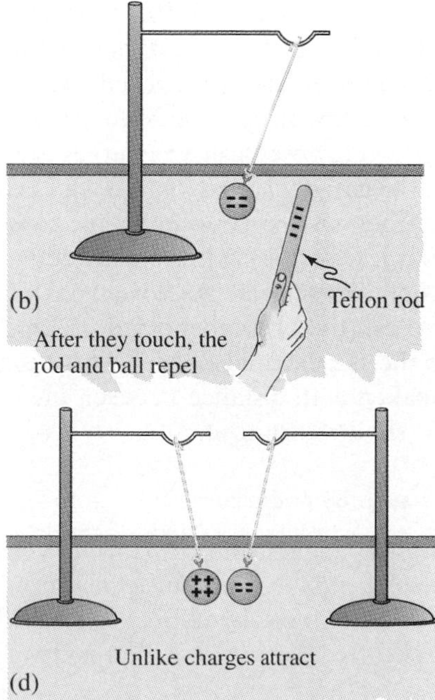

FIGURE 22–3 (a) An insulated cork ball covered with a thin layer of conducting paint can indicate the presence of small electric charges. (b) A negatively charged Teflon rod first approaches the neutral coated cork ball, which is initially attracted to the rod. After the rod touches the ball, the ball becomes charged and strongly repels the charged rod. (c) If we touch two initially neutral cork balls with a negatively charged Teflon rod, the two balls repel each other: Like charges repel. (d) If we touch one initially neutral cork ball with a negatively charged Teflon rod and a similar ball with a positively charged glass rod, the two balls attract each other: Unlike charges attract.

Fig. 5

FIGURE 22–4 An eighteenth-century experiment on static electricity by Stephen Gray. The boy, suspended in air, was charged electrostatically; bits of paper were then attracted to him.

touch the Teflon rod to one cork ball and the glass rod to one cork ball, the balls attract each other (Fig. 22–3d).

We conclude from these experiments that the electric charges on the Teflon and glass rods are different, and that

Like charges repel, unlike charges attract.

Like charges repel, and unlike charges attract.

These conclusions are the simplest explanation for what has been observed. For example, while the Teflon rod touches the cork ball, some of the rod's negative charge is transferred to the ball. Now both the ball and the rod have a negative charge. The ball, which has been charged by conduction, immediately jumps away from the rod. Our other observations are similarly explained by the rule that like charges repel and unlike charges attract (Fig. 22–4).

When the experiments above are done carefully, we can notice another effect. Before the negatively charged Teflon rod touches the neutral cork ball, the ball is *attracted* to the rod, not repelled by it. (After they touch, they strongly repel, and we have learned why.) How can we explain the initial attraction? Because we have coated the cork ball with conducting paint, there are mobile electrons on the surface of the ball. When the negatively charged Teflon rod comes near, the mobile electrons are repelled and move to the far side of the cork ball (Fig. 22–5). That leaves an equal amount of excess positive charge on the area of the ball near the rod. Those positive charges are then attracted to the rod more strongly than the negative charges on the ball are repelled. In other words, when the positive charges on the cork ball are closer to the Teflon rod than the ball's negative charges are, the net force is attractive. The initial attraction can therefore be understood if the electric force weakens when the distance between the charges increases. We call the phenomenon in which charges within an object are redistributed due to the presence of external charge **charge polarization**. The fact that electrical forces weaken with distance between the interacting charges is of great importance, and we shall return to it.

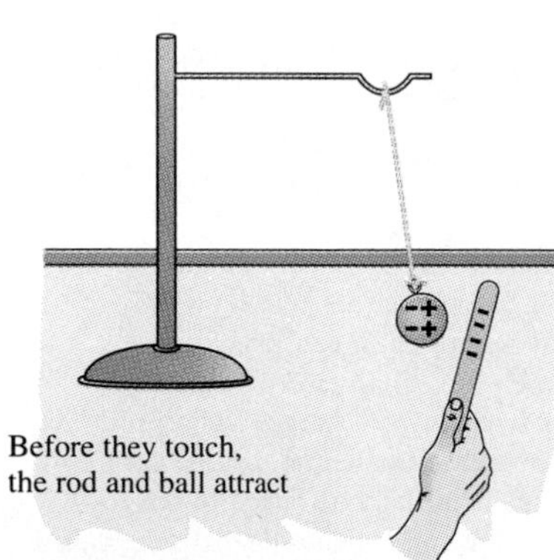
Before they touch,
the rod and ball attract

FIGURE 22–5 The neutral cork ball is initially attracted to the charged Teflon rod because some electrons on the ball move to the far side due to the repulsive force from the rod. The positive charges on the ball are on average closer to the rod, so the attractive force on them due to the rod is greater than the repulsive force on the shifted electrons.

Charge by Induction

Another experiment explains how initially neutral conductors can obtain a *charge by induction*, or an *induced charge*. Consider two neutral metal spheres that each stand on an insulated post and are in contact side by side (Fig. 22–6a). If we bring a negatively charged Teflon rod very near to one sphere, mobile electrons in the

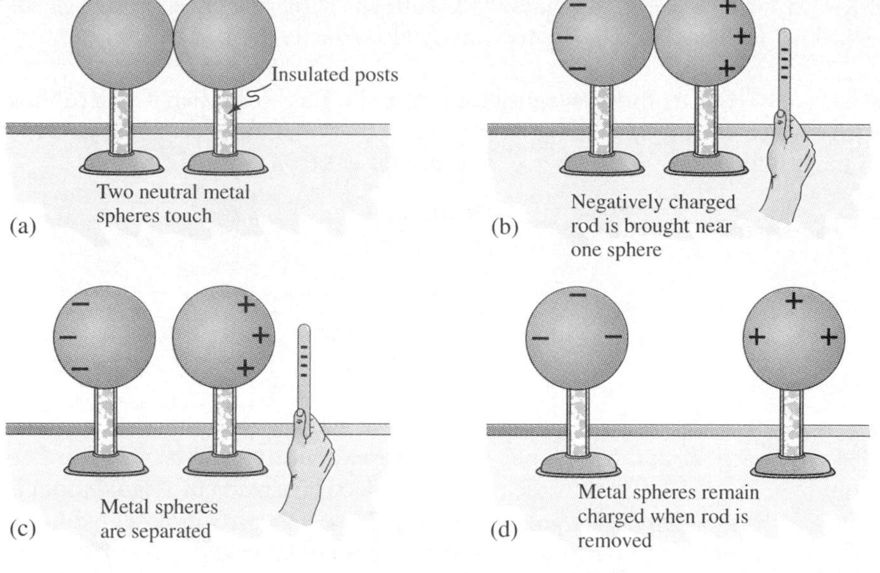
Insulated posts
Two neutral metal spheres touch
(a)
Negatively charged rod is brought near one sphere
(b)
Metal spheres are separated
(c)
Metal spheres remain charged when rod is removed
(d)

FIGURE 22–6 (a) Two neutral metal spheres on insulated posts touch. (b) A negatively charged Teflon rod polarizes the metal spheres. (c) If the metal spheres are separated while the Teflon rod is nearby, the spheres are charged oppositely. (d) When the Teflon rod is removed, the two metal spheres are still charged oppositely. Note that the total charge of the two spheres remains zero throughout.

sphere move to the opposite side of the far sphere, leaving opposite charges on the two spheres (Fig. 22–6b). The spheres have a total charge of zero, but one is positive and the other negative. While the Teflon rod is still near, we separate the two spheres, leaving them oppositely charged (Fig. 22–6c). Even when we remove the Teflon rod, the charges induced by the rod remain on the two metal spheres (Fig. 22–6d). We say that the spheres have been **charged by induction**. These charges can then be transferred to cork balls coated with conducting paint. The attractive force between the cork balls, which is easily seen because the cork balls are lightweight, demonstrates that the charges are opposite in sign. Note that only conductors can be charged by induction.

Charging by induction

Units of Charge

Just how much charge an electron carries depends on how the scale for charge is defined. The SI unit of charge is called the **coulomb** (C).† We can determine the value of the coulomb by specifying the magnitude of the force between two objects separated by a distance of 1 m, with each object carrying 1 C of charge.

The coulomb is the SI unit of charge.

The magnitude of the charge on the electron, the smallest charge found in nature, has been measured to high precision. An approximation good enough for our purposes is

$$e \simeq 1.602 \times 10^{-19}\ \text{C}. \qquad (22\text{–}1)$$

The electron charge

The mass and charge of the neutron, proton, and electron are given in Table 22–1.

TABLE 22–1

MASS AND CHARGE OF ATOMIC CONSTITUENTS

	Mass (kg)	*Charge (C)*
Neutron, n	1.675×10^{-27}	0
Proton, p	1.673×10^{-27}	1.602×10^{-19}
Electron, e^-	9.11×10^{-31}	-1.602×10^{-19}

† The coulomb is formally defined in terms of current (charge per unit time), which in the SI has units of *amperes* (A): A coulomb is the amount of charge that flows through any section of a wire in 1 s if the current in the wire is 1 A.

EXAMPLE 22–1 A glass rod rubbed with silk has a charge of $+110$ nC (110×10^{-9} C). By how many electrons is the rod deficient?

SOLUTION: Electrons were transferred from the glass rod when it was rubbed with silk, leaving an excess positive charge on the rod. Each electron has charge of magnitude e, so the number of transferred electrons must be

$$\text{transferred electrons} = \frac{\text{net charge}}{\text{charge on each electron}}$$

$$= \frac{110 \times 10^{-9}\ \cancel{\text{C}}}{1.6 \times 10^{-19}\ \cancel{\text{C}}/\text{electron}} = 6.9 \times 10^{11}\ \text{electrons.}$$

EXAMPLE 22–2 The largest American Eagle gold coin has a mass of 28.4 g. The atomic number of gold—the number of protons in the nucleus of an atom of gold—is 79, and thus the number of electrons in a neutral gold atom is also 79. The atomic mass of gold is 197, which means that 1 mol of gold has a mass $m_{Au} = 197$ g. How many electrons are contained in one pure-gold coin? What is the total negative charge contained in the coin?

SOLUTION: The number of gold atoms in a mass of 28.4 g is

$$\frac{mN}{m_{Au}} = \frac{(28.4\ \cancel{\text{g}})(6.02 \times 10^{23}\ \text{atoms}/\cancel{\text{mol}})}{197\ \cancel{\text{g}}/\cancel{\text{mol}}} = 8.68 \times 10^{22}\ \text{atoms,}$$

where $N = 6.02 \times 10^{23}$ atoms/mol is Avogadro's number, the number of atoms in 1 mol of any substance. Each neutral Au atom contains 79 electrons, so the total number of electrons is

$$\text{number of electrons} = (79\ \text{electrons}/\cancel{\text{atom}})(8.68 \times 10^{22}\ \cancel{\text{atoms}})$$
$$= 6.85 \times 10^{24}\ \text{electrons.}$$

The total charge of these electrons is

$$\text{total electron charge} = (\text{number of electrons})(\text{charge per electron})$$
$$= (6.85 \times 10^{24}\ \cancel{\text{electrons}})(-1.6 \times 10^{-19}\ \text{C}/\cancel{\text{electron}})$$
$$= -1.1 \times 10^{6}\ \text{C.}$$

The gold is neutral, so a positive net charge of the same magnitude is present due to the protons. Notice that the number of electrons transferred by rubbing the glass rod in Example 22–1 with silk is 10^{13} times smaller than the number of electrons contained in the gold coin.

22–2 CHARGE CONSERVATION AND QUANTIZATION

Charge conservation

The simple experiments described in Section 22–1 suggest strongly that *charge is conserved.* This turns out to be a fundamental physical law: All the experiments ever performed on this issue have shown that *net* charge is the same before and after any interaction, which is a statement of the **conservation of charge.** The fact that charge conservation occurs at the microscopic level means that it will occur at the macroscopic level, too.

Evidence of Charge Conservation

Let us look at some of the microscopic interactions that lead to the conclusion that charge is conserved. One of the reactions between atomic nuclei that takes place in a nuclear reactor is

$$n + {}^{235}_{92}\text{U} \rightarrow {}^{143}_{56}\text{Ba} + {}^{90}_{36}\text{Kr} + 3n + \text{energy}.$$

Here the total number of protons (92) is the same on both "sides" of the reaction.†

Even when the number of electrons or protons changes during a reaction, the total charge remains unchanged. Thus another reaction that can take place in a nucleus is *electron capture*,

$$e^- + p \rightarrow n + \nu,$$

where ν stands for a neutral particle called the *neutrino*. (The neutrino, unlike the neutron, is massless, as far as we can tell.) In this reaction, the numbers of both protons and electrons change, but charge is still conserved.

Another type of charged particle is called a *pion*, denoted by π, with a superscript to indicate the sign of the charge it carries. In the reaction

$$\gamma + p \rightarrow n + \pi^+,$$

a *photon*, γ, which is a form of very high-energy neutral radiation, impinges on a proton and produces a neutron and a pion. To the high accuracy of the experiments that investigate this reaction, the charge of the pion is *exactly* the same as that of the proton. Other particles, called *positrons*, are practically identical to electrons, except for the *sign* of the charge, and are denoted by e^+. In the reaction

$$\gamma + p \rightarrow p + e^+ + e^-,$$

an electron is produced, but then only in partnership with a positron, whose charge has exactly the magnitude. In fact, in observed reactions involving the so-called elementary particles, *no one has ever seen a single case of net charge appearing or disappearing.*

We have given a number of examples of the conservation of charge. The details of the particular examples are unimportant, as are the names of the particles involved. What is important is the principle of the conservation of charge. It applies to *any* situation in which there is a transfer of charge.

Is it possible for a little of the charge on an electron or a proton to wear off, like paint? Again, all the evidence points to the fact that the electron and the proton charges are always the same, no matter where or when they are measured. In looking at quasars, which are powerful sources of light billions of light-years away, we are looking at matter that existed billions of years ago (it took that long for the light to reach the earth). Observations of the color of the light quasars emit suggest that, to a very high accuracy, the properties of their atoms are identical to the properties of atoms we observe in the laboratory. This result implies that the charge of electrons has remained constant over billions of years.

Charge Quantization

We have already indicated that charges appear to be organized in discrete bundles. The size of such a bundle is the value of one electron charge (or one proton charge, of equal magnitude). Greater charges are always multiples of these values. The

† The superscript on the element symbol is the atomic mass, the sum of the numbers of protons and neutrons in one atom; the subscript is the number of protons.

facts that, within experimental accuracy, charge occurs in integral multiples of the electron charge, known as **charge quantization**, and that charges are never observed with values smaller than the electron charge were first established in 1909 by the pioneering and now-classic experiments of Robert Millikan. In addition, his experiments were the first in which the electron charge was measured directly and are the basis for high-precision measurements of this quantity.

Charge quantization

FIGURE 22–7 The track of charged particles (electrons) passing through a detector. It is through experiments with such detectors that the electron charge has been measured and its quantization verified.

In the 1970s and 1980s, some physicists proposed that protons and neutrons are made up of even more fundamental particles, called **quarks**, whose charges are postulated to be either $2e/3$ or $-e/3$. Despite many searches, such fractional electron charges have never been observed directly in the laboratory. Now most physicists believe that only combinations of quarks with a net charge that is an integer multiple of e can ever be isolated and independently observed (Fig. 22–7). We refer to any charge that can be isolated as **free charge**.

In summary, we can say that

Charge is conserved absolutely

and that

Free charge is quantized in positive or negative integral multiples of *e*.

22–3 COULOMB'S LAW

Encouraged by Benjamin Franklin, Joseph Priestley concluded in the mid-eighteenth century from Franklin's and his own experiments that the electric force between two charged objects varies as the inverse square of the distance between the objects. Priestley made this deduction after he observed that there is no charge on the inside surface of a closed or nearly closed metal vessel—all the charge is on the outside surface—and that the force on a charged object placed inside such a vessel is zero. This is like the phenomenon we discussed in Chapter 12: There is no gravitational force on an object inside a uniform spherical shell of matter. In gravitation, this result is a direct consequence of the $1/r^2$ nature of the force law. By analogy, Priestley argued that the electric force responsible for his observations must have a $1/r^2$ dependence.

In 1785 Charles Coulomb determined the force law for electrostatics directly. He performed the relevant experiments with a torsion balance similar to the one Henry Cavendish used in 1798 to measure the gravitational constant, G (Fig. 22–8). In Coulomb's work, small charged balls replaced the massive ones of the Cavendish apparatus. Coulomb showed that the electrostatic force is central—directed on the line between the point sources (here, balls)—and varies as

The Cavendish apparatus is described in Chapter 12.

$$F \propto \frac{1}{r^2}, \qquad (22\text{–}2)$$

where r is the distance between the centers of the charge sources. By changing the charge on the balls, possibly as we discussed in Section 22–1, Coulomb inferred that the force is proportional to the product of the charges q_1 and q_2 on the balls:

$$F \propto q_1 q_2. \qquad (22\text{–}3)$$

To demonstrate the results of Eq. (22–3), we can ground one cork ball, neutralizing it, and charge another identical ball, giving it net (unknown) charge q. After we touch the two balls together, they each have a charge of $q/2$. Then we measure the force between these two balls. Next, we ground one ball again to neutralize it, and touch the balls together once more. Each then has a charge of $q/4$, and we measure the force between them to have decreased by a factor of 4 for the same amount of separation. This set of results is consistent with Eq. (22–3): In the first case, $F \propto (q/2)(q/2) = q^2/4$, and in the second, $F \propto (q/4)(q/4) = q^2/16$.

Combining Eqs. (22–2) and (22–3) gives us a first view of **Coulomb's law**, the electrostatic force law. The magnitude of the force is

$$F = \frac{k|q_1 q_2|}{r^2}, \tag{22–4}$$

where k is a proportionality constant. The force is attractive when the charges have opposite sign and repulsive when they have the same sign.

The constant k plays the same role that the constant G plays in Newton's law of universal gravitation. The magnitude of k depends on the units used for charge. For gravity, the unit of mass had already been defined as the kilogram, and so the units of G can be determined by setting the units of force as $\text{kg}\cdot\text{m/s}^2$. Cavendish had to determine the magnitude of G by measurement. For the case of Coulomb's law, charge appears only in electromagnetic interactions and cannot be determined independently, as is true for mass. It is then possible to define the coulomb by assigning a value to k:

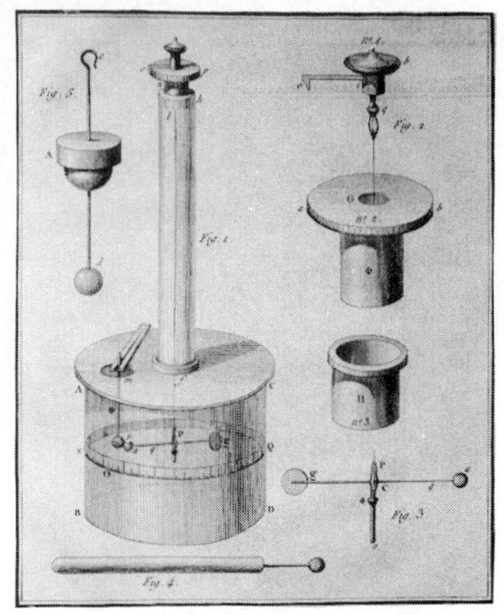

FIGURE 22–8 Coulomb's torsion balance, used to verify the inverse-square form of the force between electric charges.

$$k = \frac{1}{4\pi\epsilon_0}, \tag{22–5}$$

where ϵ_0 is known as the *permittivity of free space.* (We shall see later that the value of ϵ_0 follows directly from the defined value of the speed of light, so in this sense ϵ_0 is itself defined.) The permittivity is approximately

$$\epsilon_0 \simeq 8.854 \times 10^{-12}\ \text{C}^2/\text{N}\cdot\text{m}^2. \tag{22–6}$$

The value of k (to four significant figures) follows from Eqs. (22–5) and (22–6):

$$k = 8.988 \times 10^9\ \text{N}\cdot\text{m}^2/\text{C}^2. \tag{22–7}$$

In this book we will usually round k off to $9 \times 10^9\ \text{N}\cdot\text{m}^2/\text{C}^2$. Now that we have assigned a value to k, we can tentatively define the coulomb. From Eqs. (22–4) and (22–7), we say that *when the force between two particular charges separated by 1 m is equal to the numerical value of k in newtons* (8.988×10^9 N), *these charges are each 1 C.*

The definition of the coulomb

Note that Coulomb's law expresses the force between charged *pointlike* objects. We mentioned in Chapter 12 (and shall prove in Chapter 24) that because the gravitational force has a $1/r^2$ dependence for pointlike objects, the force between spherically symmetric distributions of mass is the same as the force between pointlike objects of the same mass placed at the centers of the spheres. The same behavior holds for electric charges, because the $1/r^2$ spatial dependences of the gravitational and electrical forces are the same. This is why Coulomb was able to measure a $1/r^2$ force even though the objects he used were not point charges. All that is required is that the charge on the balls used in the experiment be distributed in a spherically symmetric way—for example, uniformly over their surfaces. In turn, to avoid charge polarization, which would redistribute the charges on the balls, the balls should not be conducting.

The electric force, often called the **Coulomb force**, is associated with a direction and is therefore a vector. We write Coulomb's law as

$$\mathbf{F}_{12} = \frac{1}{4\pi\epsilon_0}\left(\frac{q_1 q_2}{r_{12}^2}\right)\hat{\mathbf{r}}_{12}, \tag{22–8}$$

Coulomb's law

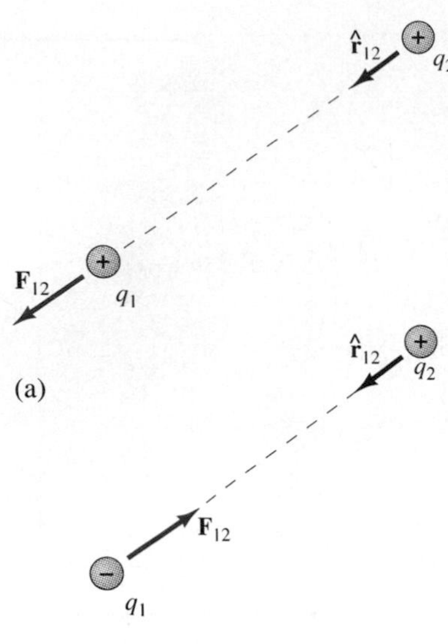

FIGURE 22–9 (a) $\mathbf{F}_{12}$ is the force on q_1 due to q_2. The force is in the direction of the unit vector $\hat{\mathbf{r}}_{12}$ for like charges and $-\hat{\mathbf{r}}_{12}$ for opposite charges.

where $\mathbf{F}_{12}$ is the force exerted on point charge q_1 by point charge q_2 when they are separated by a distance r_{12}. The unit vector $\hat{\mathbf{r}}_{12}$ is directed from q_2 to q_1 along the line between the two charges (Fig. 22–9). Note that if q_1 and q_2 have opposite signs, Eq. (22–8) indicates that the force is attractive, along $-\hat{\mathbf{r}}_{12}$. But rather than remembering the subscripts on $\mathbf{F}$ and the unit vector $\hat{\mathbf{r}}$, just remember that like charges repel and unlike charges attract.

EXAMPLE 22–3 Compare the electric force and the gravitational force for the single proton and single electron in a hydrogen atom. Assume a purely classical model of the hydrogen atom, in which the electron moves in a circular orbit around the proton, which is at the center. The radius of a hydrogen atom is about 5×10^{-11} m.

SOLUTION: First we calculate the gravitational force, obtaining the masses (m_e for the electron, m_p for the proton) from Table 22–1. Both the gravitational and electric forces are attractive in this case, so we need to calculate only the magnitudes. Using Eq. (12–4), we have

$$F_g = \frac{Gm_e m_p}{r^2}.$$

When we insert the values into this equation, we find that

$$F_g = \frac{(6.67 \times 10^{-11}\ \mathrm{N \cdot \cancel{m^2}/\cancel{kg^2}})(9.11 \times 10^{-31}\ \cancel{\mathrm{kg}})(1.67 \times 10^{-27}\ \cancel{\mathrm{kg}})}{(5 \times 10^{-11}\ \cancel{\mathrm{m}})^2}$$
$$= 4 \times 10^{-47}\ \mathrm{N}.$$

The electric force, from Eq. (22–8), is

$$F_E = \frac{(9 \times 10^9\ \mathrm{N \cdot \cancel{m^2}/\cancel{C^2}})(1.6 \times 10^{-19}\ \cancel{\mathrm{C}})(1.6 \times 10^{-19}\ \cancel{\mathrm{C}})}{(5 \times 10^{-11}\ \cancel{\mathrm{m}})^2} = 9 \times 10^{-8}\ \mathrm{N}.$$

The ratio of the two forces is

$$\frac{F_E}{F_g} = \frac{9 \times 10^{-8}\ \cancel{\mathrm{N}}}{4 \times 10^{-47}\ \cancel{\mathrm{N}}} \simeq 2 \times 10^{39}.$$

This result is independent of r. It shows that on the atomic scale, the electric force is much greater than the gravitational force, and justifies ignoring gravitation at that level.

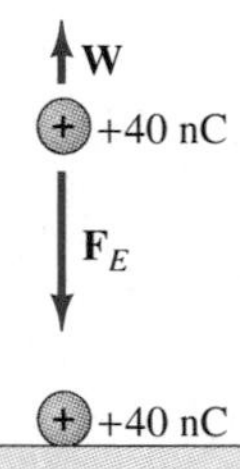

FIGURE 22–10 Example 22–4. A small object suspended in space because equal but opposite gravitational and electric forces act on it.

EXAMPLE 22–4 Two small cork balls are both charged to 40 nC and placed 4 cm apart. What is the magnitude of the electric force between them? Each cork ball has a mass of 0.4 g. Compare the electric force and the cork ball's weight.

SOLUTION: The electric force is

$$F_E = \frac{kq_1q_2}{r^2} = \frac{(9 \times 10^9\ \mathrm{N \cdot \cancel{m^2}/\cancel{C^2}})(40 \times 10^{-9}\ \cancel{\mathrm{C}})(40 \times 10^{-9}\ \cancel{\mathrm{C}})}{(4 \times 10^{-2}\ \cancel{\mathrm{m}})^2} = 0.01\ \mathrm{N}.$$

The weight of each cork ball is

$$W = mg = (0.4 \times 10^{-3}\ \mathrm{kg})(9.8\ \mathrm{m/s^2}) = 0.004\ \mathrm{N}.$$

Therefore, the (repulsive) electric force is strong enough to lift the upper cork ball if one is placed 4 cm above the other (Fig. 22–10). (A charge of 40 nC is somewhat larger than that which could realistically be placed on a ball of diameter 1 cm.)

22–4 FORCES THAT INVOLVE MULTIPLE OR CONTINUOUS CHARGES

What happens if multiple charges are present? Experiment shows that the **principle of superposition** applies: The force on any one charge due to a collection of other charges is the vector sum of the forces due to each individual charge. In this respect, the Coulomb force is again like the gravitational force, for which superposition also holds. Indeed, the only relevant difference between Coulomb's law and the law of universal gravitation is the fact that gravitational forces are always attractive, whereas Coulomb forces can be either attractive or repulsive.

The superposition principle applies to Coulomb's law.

As an example of how superposition is applied, consider four charges, numbered 1, 2, 3, 4 (Fig. 22–11). The total force on charge q_2 is the *vector sum* of the forces due to the other individual charges, q_1, q_3, and q_4:

$$\mathbf{F}_{2,\text{total}} = \mathbf{F}_{21} + \mathbf{F}_{23} + \mathbf{F}_{24}. \tag{22–9}$$

If there are N charges—$q_1, q_2, \ldots, q_N$—all acting on a charge q, the total force $\mathbf{F}$ on charge q is the vector sum of the individual forces F_i on charge q due to charge q_i:

$$\mathbf{F} = \sum_{i=1}^{N} \mathbf{F}_i = \frac{q}{4\pi\epsilon_0} \sum_{i=1}^{N} \frac{q_i}{r_i^2} \hat{\mathbf{r}}_i. \tag{22–10}$$

The vector $\hat{\mathbf{r}}_i$ is the unit vector from charge q_i to charge q. We have moved the common factor $q/4\pi\epsilon_0$ out of the sum.

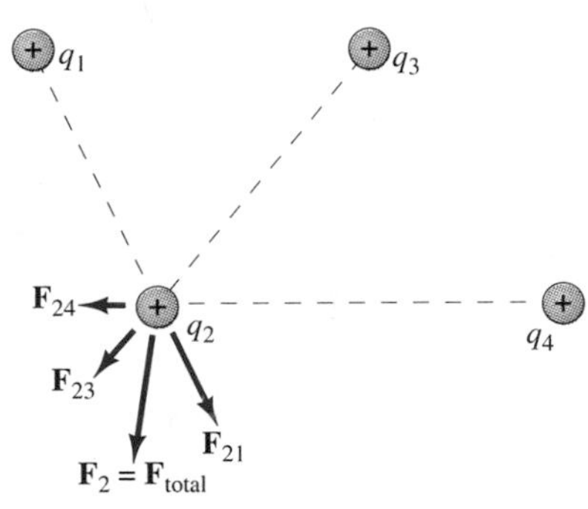

FIGURE 22–11 The superposition principle applies for multiple charges. The total force on charge q_2 is the vectorial sum of the individual forces on q_2 due to charges q_1, q_3, and q_4.

PROBLEM-SOLVING TECHNIQUES

We often need to calculate electric forces on a given charge when several other fixed charges or continuous distributions of charges are present. In these cases, keep in mind the following techniques:

1. Draw a clean diagram of the situation. Be sure to distinguish between the fixed external charges and the charges on which the forces must be found. The diagram should contain coordinate axes for reference.
2. Do not forget that the electric force that acts on a charge is a vector quantity and that when many charges are present, the net force is a vector sum. It is usually simplest to use unit vectors in a Cartesian coordinate system.
3. Search for symmetries in the distribution of charges that give rise to the electric force. When symmetries are present, the net force along certain directions will be zero. For example, if a point charge is midway between two identical charges, we know without performing any calculations that the net force on it will be zero.

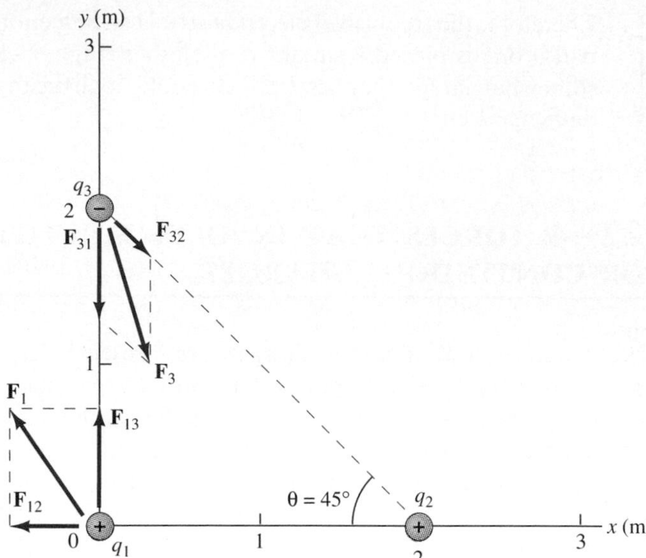

FIGURE 22–12 Example 22–5. The positions of three point charges are indicated. Charges q_1 and q_2 are positive, while q_3 is negative. Forces $\mathbf{F}_{12}$ and $\mathbf{F}_{13}$ on charge q_1, and their resultant, $\mathbf{F}_1$, as well as forces $\mathbf{F}_{31}$ and $\mathbf{F}_{32}$ on charge q_3, and their resultant, $\mathbf{F}_3$, are drawn.

EXAMPLE 22–5 Consider three point charges $q_1 = q_2 = 2.0$ nC and $q_3 = -3.0$ nC placed as in Fig. 22–12. Find the forces on q_1 and q_3.

SOLUTION: The force on q_1 is due to the presence of charges q_2 and q_3. We want to find the vector forces on q_1 due to each of the charges q_2 and q_3 separately, then add them vectorially to find the net force on q_1. Similar remarks hold for the calculation of the force on q_3.

The force on q_1 is

$$\mathbf{F}_1 = \mathbf{F}_{12} + \mathbf{F}_{13} = \frac{q_1}{4\pi\epsilon_0}\left[\left(\frac{q_2}{r_{12}^2}\right)\hat{\mathbf{r}}_{12} + \left(\frac{q_3}{r_{13}^2}\right)\hat{\mathbf{r}}_{13}\right].$$

From Fig. 22–12, we can deduce that $\hat{\mathbf{r}}_{12} = -\mathbf{i}$ and $\hat{\mathbf{r}}_{13} = -\mathbf{j}$. Thus

$$\begin{aligned}\mathbf{F}_1 &= (9.0 \times 10^9\ \mathrm{N{\cdot}\cancel{m}^2/\cancel{C}^2})(2.0 \times 10^{-9}\ \cancel{\mathrm{C}})\\ &\quad\times\left[\frac{(2.0 \times 10^{-9}\ \cancel{\mathrm{C}})}{(2.0\ \cancel{\mathrm{m}})^2}(-\mathbf{i}) + \frac{(-3 \times 10^{-9}\ \cancel{\mathrm{C}})}{(2.0\ \cancel{\mathrm{m}})^2}(-\mathbf{j})\right]\\ &= (-9.0 \times 10^{-9}\ \mathrm{N})\mathbf{i} + (13.5 \times 10^{-9}\ \mathrm{N})\mathbf{j}.\end{aligned}$$

The direction of force $\mathbf{F}_1$ is shown in Fig. 22–12.

The force on q_3 is calculated in much the same way, with the caution that the unit vector $\hat{\mathbf{r}}_{32}$, which points from q_2 to q_3, is given by $-\cos\theta\,\mathbf{i} + \sin\theta\,\mathbf{j}$:

$$\begin{aligned}\mathbf{F}_3 = \mathbf{F}_{31} + \mathbf{F}_{32} &= \frac{q_3}{4\pi\epsilon_0}\left[\left(\frac{q_1}{r_{31}^2}\right)\hat{\mathbf{r}}_{31} + \left(\frac{q_2}{r_{32}^2}\right)\hat{\mathbf{r}}_{32}\right]\\ &= (9.0 \times 10^9\ \mathrm{N{\cdot}\cancel{m}^2/\cancel{C}^2})(-3.0 \times 10^{-9}\ \cancel{\mathrm{C}})\\ &\quad\left[\frac{(2.0 \times 10^{-9}\ \cancel{\mathrm{C}})}{(2.0\ \cancel{\mathrm{m}})^2}\mathbf{j} + \frac{(2.0 \times 10^{-9}\ \cancel{\mathrm{C}})}{(\sqrt{2.0^2\ \cancel{\mathrm{m}}^2 + 2.0^2\ \cancel{\mathrm{m}}^2})^2}(-\cos\theta\,\mathbf{i} + \sin\theta\,\mathbf{j})\right].\end{aligned}$$

The angle θ is 45°, or $\pi/4$ rad, so $\mathbf{F}_3$ becomes

$$\begin{aligned}\mathbf{F}_3 &= (-13.5 \times 10^{-9}\ \mathrm{N})\mathbf{j} + (4.8 \times 10^{-9}\ \mathrm{N})\mathbf{i} - (4.8 \times 10^{-9}\ \mathrm{N})\mathbf{j}\\ &= (4.8 \times 10^{-9}\ \mathrm{N})\mathbf{i} - (18.3 \times 10^{-9}\ \mathrm{N})\mathbf{j}.\end{aligned}$$

The fact that charge is quantized will have no physical consequence when we deal with charges that are much larger than e. Such charges are actually composed of large numbers of electrons or of protons; the net charge is e(number of protons) − e(number of electrons). It may then be a good approximation to treat a large collection of point charges as a *continuous distribution* of charge. To analyze this situation, we can follow the approach suggested in Chapter 12, where continuous distributions of mass were discussed briefly. Consider first the interaction of a point charge q with a large continuous charge distribution (Fig. 22–13). The force on q due to the tiny element of volume shown, which contains charge Δq and is a distance r' from q, is

$$\Delta \mathbf{F} = \frac{q}{4\pi\epsilon_0} \frac{\Delta q}{r'^2} \hat{\mathbf{r}}'. \tag{22–11}$$

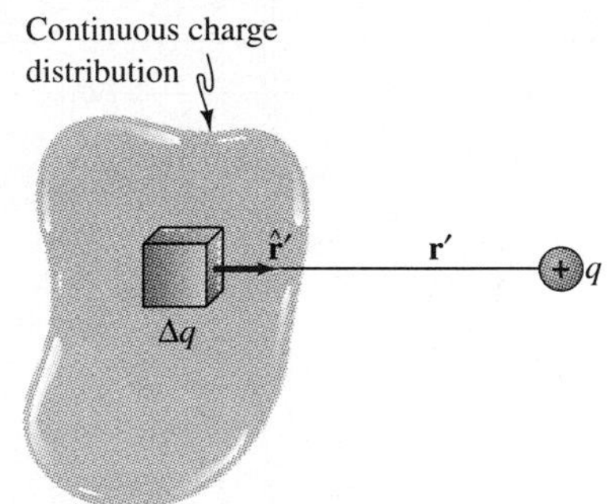

FIGURE 22–13 To find the total force on a point charge q due to a continuous charge distribution, integrate over the tiny charge elements Δq. Notice that the vector $\hat{\mathbf{r}}'$ will change as we move through the distribution.

We denote the charge density of the distribution by $\rho(\mathbf{r}')$, meaning that, at a point located a displacement $\mathbf{r}'$ from q, the charge Δq contained in the small volume $\Delta V'$ is

$$\Delta q = \rho(\mathbf{r}')\,\Delta V'.$$

In terms of the charge density of the continuous charge distribution, the force due to the volume element shown is

$$\Delta \mathbf{F} = \frac{q}{4\pi\epsilon_0} \frac{\rho(\mathbf{r}')}{r'^2} \hat{\mathbf{r}}'\,\Delta V'. \tag{22–12}$$

The *net* force on q is the sum over the terms such as Eq. (22–12). In the limit where the distribution is cut into smaller and smaller pieces, this sum becomes the integral

$$\mathbf{F} = \frac{q}{4\pi\epsilon_0} \int \frac{\rho(\mathbf{r}')}{r'^2} \hat{\mathbf{r}}'\,dV'. \tag{22–13}$$

A formal expression for the force on a point charge due to a continuous distribution of charge

The integral operates over the entire volume of the charge distribution. We often deal with *uniform* charge distributions, in which charge is distributed evenly throughout a region. In such cases, the function ρ can be removed from the integral over the region. Keep in mind that a fixed uniform charge distribution is not possible with a conductor, within or on which charges are free to move.

The charge distribution over which to integrate may be one-dimensional, for charge spread along a line; two-dimensional, for charge spread across a surface; or three-dimensional, for charge spread throughout a volume. In one dimension, the charge density (charge per unit length) is λ; in two dimensions, the charge density (charge per unit area) is σ; and in three dimensions, the charge density (charge per unit volume) is ρ. In each case the argument of the charge distribution is the vector position $\mathbf{r}'$, because what counts is the vector displacement from an element of the charge distribution to the point charge.

The integral in Eq. (22–13) that expresses the force may well have a simple answer, expressible in terms of simple functions. Symmetry is always helpful here. Conversely, the integral may be difficult to perform, particularly if there is no symmetry in the distribution. Numerical integration on a computer can always be used if the integration is too difficult to perform analytically. Example 22–6 illustrates a typical integration and how symmetry can simplify a problem.

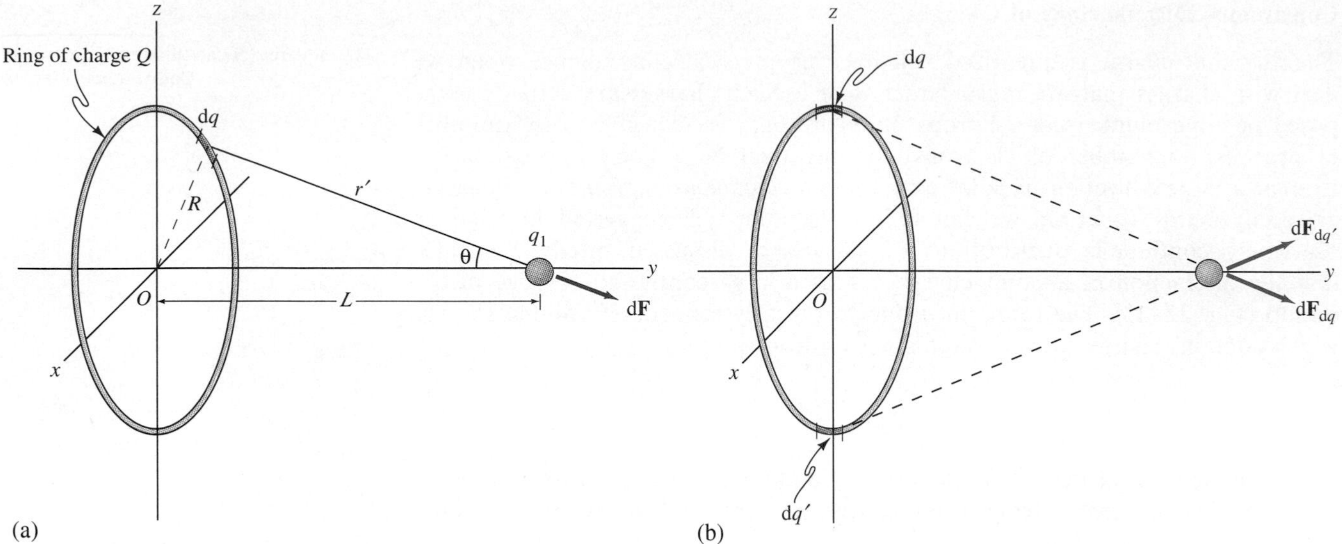

FIGURE 22–14 (a) Example 22–6. The force on a point charge q_1 due to a ring with total charge Q. First we find the force between the point charge and a tiny ring segment with charge dq. (b) Only the y-component of the force needs to be determined, because the x- and z-components will cancel due to symmetry.

EXAMPLE 22–6 Find the force on a point charge q_1 located on the axis of a uniformly charged ring of total charge Q (Fig. 22–14). The radius of the ring is R, and q_1 is located a distance L from the center of the ring.

SOLUTION: The ring has a continuous charge distribution, but it is spread along a (curved) line, so the integration must be one-dimensional. A small segment of the ring contains charge dq (Fig. 22–14a). All such segments are located a distance $r' = \sqrt{L^2 + R^2}$ from charge q_1, and the line to any segment on the ring makes the angle θ with the y-axis.

Next look at the components of the force on q_1. Because every segment of the ring is the same distance r' from q_1, the *magnitude* of the infinitesimal force from each infinitesimal slice is the same. This is not true for the direction. The force from segment dq at the top of the ring ($x = 0$, $z = R$) is $\mathbf{dF}_{dq}$, and this force has components in the $+y$-direction and the $-z$-direction (Fig. 22–14b). The force from segment dq' at the bottom of the ring ($x = 0$, $z = -R$) is $\mathbf{dF}_{dq'}$, and this force has components in the $+y$-direction and the $+z$-direction. If the magnitude dq equals the magnitude dq', the z-components of the force will cancel each other, and the y-components of the force will be additive. The z-components are the components perpendicular to the axis of the ring. This cancellation will hold for every perpendicular component of the force, because we can always consider the charge elements in pairs. Thus we need compute only the component F_y. The y-component from the element shown in Fig. 22–14a is

See the problem-solving technique about symmetry.

$$dF_y = \frac{q_1}{4\pi\epsilon_0}\frac{dq}{r'^2}\cos\theta = \frac{q_1}{4\pi\epsilon_0}\frac{\cos\theta}{R^2 + L^2}\,dq.$$

The net force has only a y-component and is the sum over the infinitesimal y-components:

$$F_y = \int dF_y = \int \frac{q_1}{4\pi\epsilon_0}\frac{\cos\theta}{R^2 + L^2}\,dq.$$

The entire coefficient of dq is a constant and can be placed outside the integral sign. Thus

$$F_y = \frac{q_1}{4\pi\epsilon_0}\frac{\cos\theta}{R^2 + L^2}\int dq = \frac{q_1 Q}{4\pi\epsilon_0}\frac{\cos\theta}{R^2 + L^2}.$$

Finally, from trigonometry we get

$$\cos\theta = \frac{L}{\sqrt{R^2 + L^2}},$$

so

$$F_y = \frac{q_1 Q}{4\pi\epsilon_0}\frac{L}{(R^2 + L^2)^{3/2}}. \tag{22–14}$$

As always, a check is desirable, and we can make two checks on this result. When point charge q_1 is very far from the ring, the ring should appear as a distant point of total charge Q, and the force should take on the Coulomb form $q_1Q/(4\pi\epsilon_0 L^2)$; this is indeed the limit of Eq. (22–14) when $L \gg R$. When $L = 0$, the charge is at the middle of the ring, and by symmetry the net force should be zero. This limit also fits our result.

EXAMPLE 22–7 A straight rod of length L is aligned along the x-axis, with the ends at $x = \pm L/2$ (Fig. 22–15). The total charge on the rod is zero but the charge density is not; it is given by $\lambda(x) = 2\lambda_0 x/L$ (positive to the right of the origin, negative to the left). Find the force on a charge q located at a point $x = R$ on the x-axis, to the right of the right-hand end of the rod.

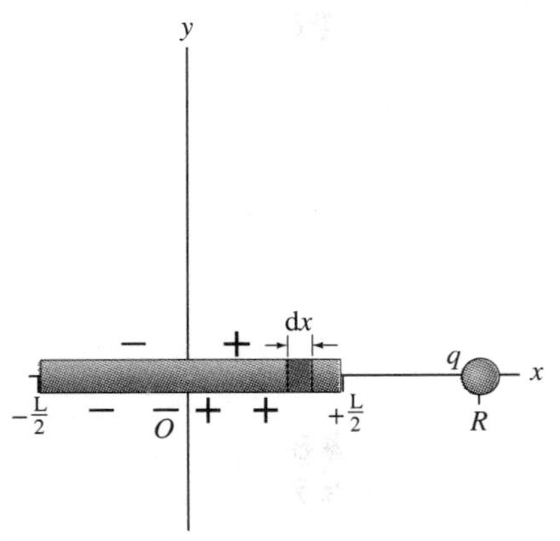

FIGURE 22–15 Example 22–7. A nonuniform charge density.

SOLUTION: Consider a thin slice of the rod located at point x, with thickness dx. The charge on that slice is given by

$$dQ = \lambda\, dx = \frac{2\lambda_0}{L} x\, dx.$$

The infinitesimal force exerted by this charge on charge q is given by

$$d\mathbf{F} = \frac{q}{4\pi\epsilon_0}\frac{2\lambda_0}{L} x\, dx \frac{1}{(R - x)^2}\mathbf{i},$$

and the total force on the charge is the integral over $d\mathbf{F}$,

$$\mathbf{F} = \int d\mathbf{F} = \int_{-L/2}^{L/2}\frac{q}{4\pi\epsilon_0}\frac{2\lambda_0}{L}\frac{x\, dx}{(R - x)^2}\mathbf{i} = \frac{2q\lambda_0}{4\pi\epsilon_0 L}\mathbf{i}\int_{-L/2}^{L/2}\frac{x\, dx}{(R - x)^2}$$

$$= \frac{2q\lambda_0}{4\pi\epsilon_0 L}\mathbf{i}\int_{-L/2}^{L/2}\left[-\frac{1}{R - x} + \frac{R}{(R - x)^2}\right]dx$$

$$= \frac{2q\lambda_0}{4\pi\epsilon_0 L}\left\{\ln\left[\frac{R - (L/2)}{R + (L/2)}\right] + R\left[\frac{1}{R - (L/2)} - \frac{1}{R + (L/2)}\right]\right\}\mathbf{i}.$$

The Force Due to a Spherically Symmetric Charge Distribution

A charge distribution that is *spherically symmetric* is both important physically and easy to treat. Such a distribution is in the form of a sphere centered at, say, point P, and the charge density has a constant value at a given distance from P. Notice

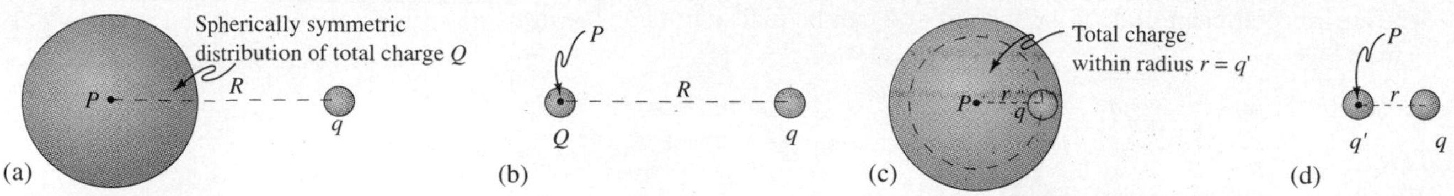

FIGURE 22–16 (a) A spherically symmetric charge distribution of total charge Q is centered on the point P. The force on a point charge q outside the distribution a distance R from P is the same as (b) the force it would experience if a point charge Q were located at P. (c) If q lies inside the distribution a distance r from P, and q' is the total charge that lies within a sphere of radius r centered on P, then it experiences the same force it would have (d) if there were a point charge q' at P.

that the charge density could vary with the distance from P, but that, when there is spherical symmetry, the charge density must look the same in any direction as seen from P. This case was discussed extensively in Chapter 12 for gravitational force. Those results depend only on the fact that the force varies inversely with the distance squared, so we can use the results here. The force of the spherically symmetric charge distribution on a point charge q outside the distribution (Fig. 22–16a) is the same as though the entire charge of the distribution were concentrated at P (Fig. 22–16b). If, as in Fig. 22–16c, the point charge q is inside any part of the distribution, then the force on q due to the part of the distribution that lies outside q is zero (Fig. 22–16d).

22–5 THE SIGNIFICANCE OF THE ELECTRICAL INTERACTION

We have now introduced a second basic force of nature; to the law of universal gravitation, we have added the electrical interaction, represented by the Coulomb force. Both the electric and gravitational forces have an inverse-square dependence on the distance between the interacting pointlike objects. Both forces are also proportional to the product of a characteristic attribute of the two objects—either mass or charge.

On the cosmic scale, the gravitational law looms large. It is the force that keeps the earth rotating around the sun and the moon rotating around the earth. The reasons why gravitation dominates electric forces on the astronomical scale are twofold. First, astronomical bodies have a great deal of mass. Second, astronomical bodies are almost exactly charge neutral, so the electric forces between them are relatively small. However, on anything less than an astronomical scale, the electric forces are normally much larger than the gravitational ones, and, apart from the direct effects of the earth's gravity, our everyday experience depends far more on the electric force than on the gravitational one.

As we have seen for the hydrogen atom, the electric force dominates the gravitational force on a microscopic scale. Even though a full explanation requires quantum physics, we can now state that the electric force is responsible for

1. electrons binding to a positive nucleus, forming a stable atom;
2. atoms binding together into molecules;
3. atoms or molecules binding together into liquids and solids;
4. all chemical reactions; and
5. all biological processes.

The electric force is behind such nonfundamental forces as friction and other contact forces. Electric energy fuels our homes, starts our cars, and runs our factories.

SUMMARY

Electric charge occurs in two forms, which we label as positive and negative charge. Charges of the same sign repel each other, and charges of unlike sign attract each other. In SI units, charge is measured in coulombs.

Much of the behavior of materials under the influence of electric forces is characterized by the ease with which electrons are dislodged from their constituent atoms and molecules and move through the material. Metals are normally good conductors of electric charge, whereas most nonmetals are not and are called insulators.

The basic electric charge is that of the electron. The electron has a charge of $-e$, and the proton has a charge of $+e$, with $e = 1.602 \times 10^{-19}$ C. Electric charge in matter is quantized in multiples of e. Charge is conserved in all interactions, meaning that the net charge before an interaction is the same as the net charge after the interaction.

The electric force between point charges q_1 and q_2 separated by a distance r_{12} is

$$\mathbf{F}_{12} = \frac{1}{4\pi\epsilon_0}\left(\frac{q_1 q_2}{r_{12}^2}\right)\hat{\mathbf{r}}_{12}, \tag{22–8}$$

where the factor $1/4\pi\epsilon_0$ characterizes the strength of the force. This equation is Coulomb's law.

The principle of superposition applies when multiple charges are present. The forces on a charge due to all other charges add together vectorially. For continuous charge distributions, we must integrate over the charges, and the force of such a distribution on a point charge q is

$$\mathbf{F} = \frac{q}{4\pi\epsilon_0}\int \frac{\rho(\mathbf{r}')}{r'^2}\,\hat{\mathbf{r}}'\,dV'. \tag{22–13}$$

The quantity $\rho(\mathbf{r}')$ is the density of charge at a point a distance r' from the point charge.

On all but the astronomical scale, electric forces tend to be much stronger than gravitational forces. The electric force is responsible for making atoms, molecules, solids, and liquids stable, and for producing all chemical reactions and biological processes.

QUESTIONS

1. When the weather is muggy (as it seems to be in all classrooms where professors do demonstrations), experiments in electrostatics work poorly. Can you explain why this would be true? What makes it easier for us to draw sparks in cold weather?

2. Why do we need to coat the cork balls in Section 22–1 with conducting paint? Would the experiment work without it? Explain.

3. Walking across a carpet in shoes often leads to a pickup of electric charge large enough to cause a spark when one touches a doorknob. In climates that are dry in winter, this phenomenon is much more common in the winter than in the summer. Why?

4. In Fig. 22–6, why can we obtain only a limited induced charge even though the number of mobile electrons is extremely large?

5. By using the apparatus discussed in Section 22–1, how could you determine what charge you accumulate by walking across a wool rug?

6. Different electrons of an atom circulate around the nearly pointlike nucleus at different distances from the nucleus.

Why might the inner electrons be more strongly bound to the nucleus than the outer ones are?

7. Neutrons and protons are believed to be made of two types of charged particles called quarks, having charge $-\frac{1}{3}e$ and $\frac{2}{3}e$, as mentioned in Section 22–2. List the possible combinations of only three quarks that make up neutrons and protons.
8. Some materials lose electrons easily by rubbing, so why are many of the objects around us not charged at all times?
9. *Earnshaw's theorem* states that a point charge cannot be in stable equilibrium while purely electrostatic forces act on the point charge. Consider a ring that is uniformly positively charged, with a positive charge at the center. It appears that the center charge suffers an identical repulsive force from every direction. How can the theorem be true?
10. How is the existence of a battery, which sends negative charges out of one of its contact points, consistent with the conservation of charge?
11. You have a cork ball with a charge of -4.8×10^{-19} C and three uncharged cork balls. Can you devise a method of touching cork balls together in sequence that will give a charge of -0.8×10^{-19} C to one of the balls?
12. We spoke of generating a spark on a winter's day when we touch a conducting line to the earth and become grounded. Automobile tires are such good insulators that a car body is not connected to the earth by a conductor. How do you explain the spark that occurs when you touch a car door after you have rubbed the car upholstery?
13. Suppose that the electric charge of a fundamental particle such as an electron depends on the speed v of the particle, so that $e = e_0[1 + (\kappa v^2/c^2)]$, where e_0 is the particle's "rest charge," c is the speed of light, and κ is some tiny number. Discuss ways in which you might measure κ. Is there any experimental reason why κ must be small, if not zero?
14. Does the modification of the electric charge proposed in Question 13 necessarily violate the principle that it should not be possible to detect the absolute velocity of a body by means of any experiment?
15. The color of the light emitted by quasars is evidence that the charge on electrons has not changed over billions of years. Is saying that the charge on electrons and protons is unchanged equivalent to the statement that charge is conserved?
16. Suppose that electrons had charge $-e$ and protons had charge $+e(1 + \delta)$, with δ very small. Would there necessarily be an additional repulsive $1/r^2$ force between the moon and the earth, for example, that could overpower the gravitational attraction between these bodies?

PROBLEMS

22–1 *Properties of Charged Matter*

1. (I) A cork ball is charged to $+1$ nC. How many fewer electrons than protons does the ball have?
2. (I) What is the total charge of all the electrons in 1 g of H_2O?
3. (II) A cork ball that is covered with conducting paint and charged to -2×10^{-11} C is touched by an identical but uncharged cork ball; the balls then separate. This second cork ball is then touched by a third uncharged cork ball, and they separate. What is the charge of each ball at the end, and how many excess electrons does each ball have?
4. (II) A cork ball covered with conducting paint is charged to -1.6×10^{-12} C. You have three similar but uncharged cork balls. Describe a method by which to produce a cork ball with a charge of -0.2×10^{-12} C. Do you need all three extra balls? Explain.
5. (II) Two cork balls of mass 0.5 g hang from the same support point by massless insulating threads of length 10 cm (Fig. 22–17). A total positive charge of 2.0×10^{-7} C is added to the system. Half this charge is taken up by each ball, and the balls spread apart to a new equilibrium position. (a) Draw a force diagram for each cork ball. (b) What is the tension in the threads before the charge is added, and what is it after? (c) What is the value of angle θ in the figure? This device is a type of *electrometer*, a meter that measures electric charge. Angle θ measures the amount of charge on the balls if we can be sure that the charge is divided between them equally. This constraint is circumvented when the electrometer is made of a single strip of conducting material draped at its midpoint over a hook; the charge is then distributed over the strip equally, and half the strip repels the other half.

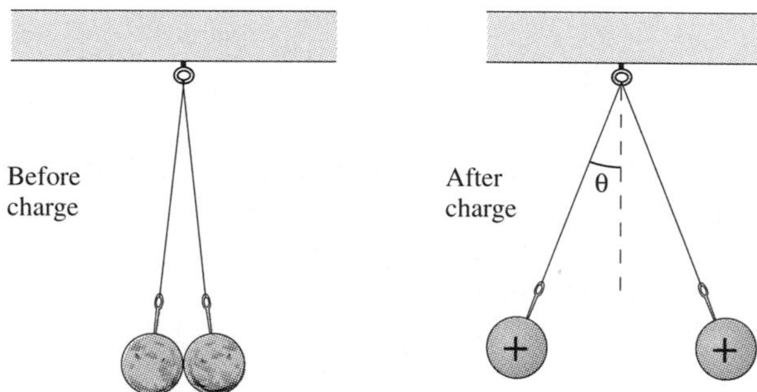

FIGURE 22–17 Problem 5.

6. (II) Silicon is the most abundant material on the surface of the earth. (a) Assume that the earth is made of silicon (28 g/mol), and calculate the total number of negative charges contained in the earth. (b) When we neutralize a cork ball that has a charge of 1 μC by grounding it to the earth, what fractional change are we making in the total negative charge contained in the earth?

22–2 *Charge Conservation and Quantization*

7. (I) *Antiparticles* have the same mass as their counterpart particles but have an opposite charge. For example, the antiparticle of an electron, e^-, is the positron, e^+. Most antiparticles are denoted by a bar over the particle, so $\bar{p}$ is the antiparticle of the proton, and it has a charge of $-e$. Which of the following reactions satisfy the conservation of charge: (a) $p + \bar{p} \to e^+ + e^- + e^+ + e^- + 2n$; (b) $e^+ + e^- \to 2p + n + 2\gamma$; (c) $e^+ + e^- \to e^+ + e^- + p + \bar{p} + 2\gamma$; (d) $n + p \to e^- + p + \bar{p}$?

8. (I) How much charge is contained in 1 g of protons?

9. (II) The electric charge of a body is independent of the body's motion. Suppose that this were not true, but that the charge of a particle such as an electron or a proton that moves at speed v has the form $e = e_0[1 + (v^2/c^2)]$, where e_0 is the particle's charge when at rest and $c \simeq 3 \times 10^8$ m/s is the speed of light. What would the net charge on a hydrogen atom be, assuming that the atom consists of a proton at rest and an electron orbiting the proton at average speed $(v/c) \simeq (1/137)$?

22–3 *Coulomb's Law*

10. (I) How far apart must two protons be for the force on each other to be the same as the weight of one proton on the surface of the earth?

11. (I) A proton is believed to consist of two "up" quarks of charge $+\frac{2}{3}e$ and one "down" quark of charge $-\frac{1}{3}e$. Assume that all three quarks are equidistant from each other at the distance of 1.5×10^{-15} m. What are the electrostatic forces between each pair of the three quarks?

12. (I) Two identical charged sodium ions separated by 2.3×10^{-9} m have a force between them of 2.3×10^{-10} N. What is the charge of each ion, and how many electron charges does this represent?

13. (I) Two tiny cork balls, both of mass 0.05 g, each have just one electron charge, $q = -1.6 \times 10^{-19}$ C. They are separated by 10 cm, which is much greater than their sizes. What is the ratio of the magnitudes of the Coulomb force between them to the gravitational force they exert on each other? Why is this result so different from that of Example 22–3?

14. (II) Suppose that we were to measure a charge in some new unit, which we will call the esu, so defined that Coulomb's law reads, in magnitude, $F = q_1 q_2/r^2$, and so that $F = 1$ dyne (10^{-5} N) when $q_1 = q_2 = 1$ esu and $r = 1$ cm. (a) How many esu are there in 1 C? (b) What is the charge of the electron in esu? (The esu is an actual unit, the *electrostatic unit*.)

15. (II) An electron and a proton attract each other with a $1/r^2$ electric force, just like the gravitational force. Suppose that an electron moves in a circular orbit about a proton. (a) If the period of the circular motion is 24 h, what is the radius of the orbit? (b) If the period is 4×10^{-16} s, as it is in a hydrogen atom, what is the radius of the orbit?

16. (II) A charge q is split into two parts, $q = q_1 + q_2$. In order to maximize the repulsive Coulomb force between q_1 and q_2, what fraction of the original charge q should q_1 and q_2 have?

17. (II) An alpha particle nucleus (composed of 2 protons and 2 neutrons) is directed onto a particular uranium nucleus (^{238}U, with 92 protons and 146 neutrons). The alpha particle stops and turns around at a distance of 10^{-13} m from the uranium nucleus. Ignore the effects of electrons, and treat the alpha particle and uranium nucleus as pointlike. What is the Coulomb force on the alpha particle at its closest approach to the nucleus?

18. (II) An electron orbits in uniform circular motion about a much heavier—and therefore nearly stationary—proton at a distance of 5×10^{-11} m. (a) What are the magnitude and direction of the Coulomb force exerted on the electron by the proton? (b) What is the speed of the electron in its circular orbit? (c) the frequency of the circular orbit? (d) Calculate the spring constant of a spring with an electron at its end and the frequency of part (c).

19. (II) Two pointlike particles are placed 8.75 cm apart and are given equal charge. The first particle, of mass 31.3 g, has an initial acceleration of 1.93 m/s² toward the second particle. (a) What is the mass of the second particle if its initial acceleration toward the first is 5.36 m/s²? (b) What charge does each particle have?

20. (II) Two cork balls, each of mass 0.20 g, are hung by insulating threads 20.0 cm long from a common point. The cork balls are given an equal charge by a Teflon rod. The balls repel and deflect as shown in Fig. 22–18. What charge q was given to each cork ball?

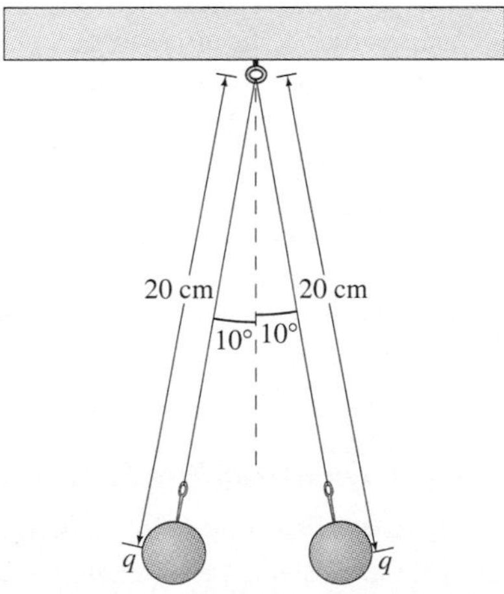

FIGURE 22–18 Problem 20.

21. (II) Astronomical data tell us that the radius of the earth is 6.3×10^6 m, that its mass is 5.98×10^{24} kg, that the moon's mass is 7.36×10^{22} kg, and that the mean earth–moon separation is 3.8×10^8 m. Suppose that, instead of being

electrically neutral, as we believe, the earth and moon each have an excess positive charge of 5.7×10^{13} C. (a) What is the magnitude of the electrical repulsion between the earth and the moon? (b) What is the ratio of this repulsive force to the attractive gravitational force? (c) If the charge on the earth were distributed uniformly throughout the volume of the earth, what would the excess charge density be, in coulombs per cubic meter (C/m^3)? (d) Assume that the excess positive charge is due to excess protons, which have an electric charge of 1.6×10^{-19} C. Calculate the density of protons, in units of protons per cubic meter, that corresponds to the conditions in part (c). (e) The mean density of the earth is 5.52×10^3 kg/m^3, and a proton has mass 1.67×10^{-27} kg. Protons account for about half the mass of the earth. Compute the density of all protons in the earth, and compare this to your answer in part (d).

22. (II) Three unknown charges q_1, q_2, and q_3 exert forces on each other. When q_1 and q_2 are 12.0 cm apart (q_3 is absent), they attract each other with a force of 0.91×10^{-2} N. When q_2 and q_3 are 25 cm apart (q_1 is absent), they attract with a force of 7.2×10^{-3} N. When q_1 and q_3 are 12.0 cm apart (q_2 is absent), they repel each other with a force of 5.6×10^{-3} N. Find the magnitude and sign of each charge.

23. (II) An electron has a mass of 0.9×10^{-30} kg and a charge of -1.6×10^{-19} C. The mass of the earth is 6×10^{24} kg, and its radius is 6.4×10^6 m. Suppose that the earth has a net negative charge Q at its center. (a) How large would Q have to be for the charge repulsion on an electron to cancel the gravitational attraction at the earth's surface? (b) Suppose that this net charge is due to a discrepancy between the positive proton charge and the negative electron charge. Assume that half the mass of the earth is due to protons, each of which has a mass of 1.6×10^{-27} kg (the rest is neutrons, assumed to be neutral; electrons do not contribute much to the mass). What is the size of the charge discrepancy, compared to the electron charge?

24. (III) Use the similarity between Coulomb's law and the law of universal gravitation to calculate the distance of closest approach between a point charge of $+10^{-6}$ C, which starts at infinity with kinetic energy of 1 J, and a fixed point charge of $+10^{-4}$ C. Assume that the moving point charge is aimed straight at the fixed point charge. (*Hint:* The similarity to gravity consists of using the notions of potential energy and energy conservation.)

22–4 *Forces that Involve Multiple or Continuous Charges*

25. (II) A charge of $-2q$ is fixed on a plane at the origin (0, 0) of an xy-coordinate system, and a charge $-q$ is fixed at (-2 cm, $+2$ cm). Where must a charge of $-3q$ be placed at rest for it to be in equilibrium (that is, so that it remains at rest)? Is the equilibrium stable?

26. (II) What is the total force on each of the three quarks in Problem 11 due to the other two quarks?

27. (II) Three positive charges of magnitude 1.2 μC are placed at the corners of an equilateral triangle of sides 6 cm. What is the net force on a charge of -2 μC placed at the midpoint of one of the sides?

28. (II) Four positive charges $+q$ sit in a plane at the corners of a square whose sides have length d, as in Fig. 22–19. A negative charge, $-q$, is placed in the middle of the square. (a) What is the net force on the negative charge? (b) Is the equilibrium point at the center a stable equilibrium for motion of the negative charge in the plane of the square? (c) for motion of the negative charge perpendicular to the plane of the square?

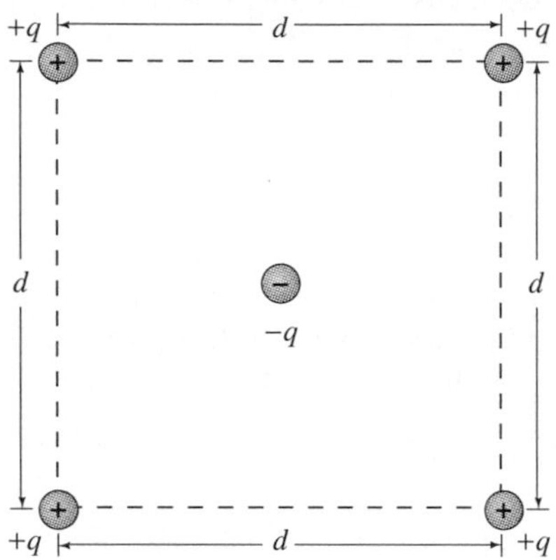

FIGURE 22–19 Problem 28.

29. (II) Charges q, $2q$, $-4q$, and $-2q$ (q is positive) occupy the four corners of a square of sides $2L$, centered at the origin of a coordinate system (Fig. 22–20). (a) What is the net force on charge q due to the other charges? (b) What is the force on a new charge Q placed at the origin?

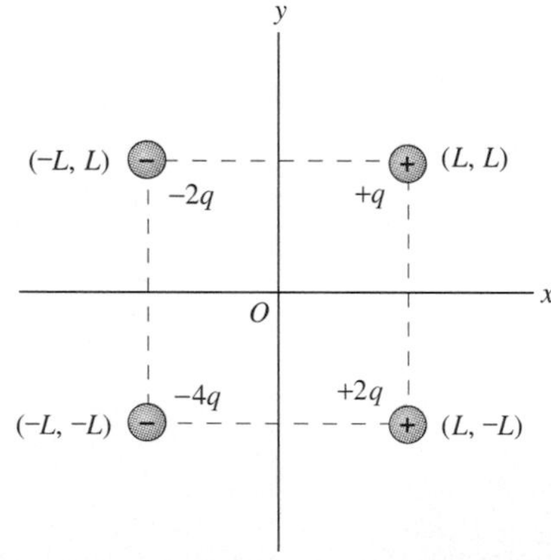

FIGURE 22–20 Problem 29.

30. (II) A charge Q is distributed uniformly along a rod of length $2L$, extending from $y = -L$ to $y = L$ (Fig. 22–21). A charge q is placed on the x-axis at $x = D$. (a) In what direction is the force on q, given that Q and q have the same sign? (b) What is the charge on a segment of the rod of infinitesimal length dy? (c) What is the force vector on charge q due to the small segment dy? (d) Express an integral

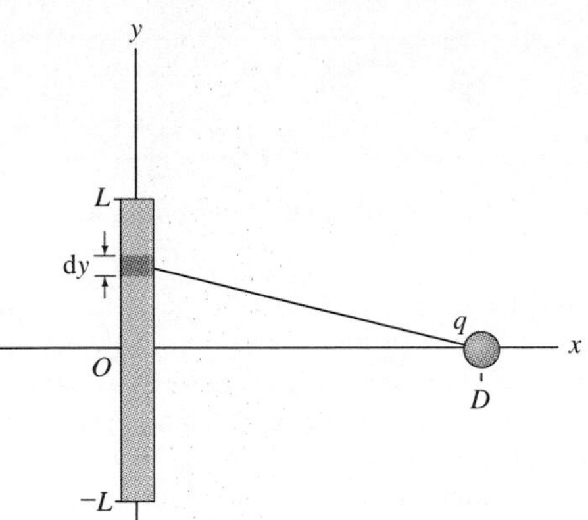

FIGURE 22–21 Problem 30.

that describes the total force in the x-direction. (e) Compute the integral in order to find the total force in the x-direction.

31. (II) A charge is spread uniformly along the y-axis, stretching infinitely far in both the positive and negative directions. The charge density (charge per unit length) on the y-axis is λ. Find the force on a point charge q placed on the x-axis at $x = x_0$.

32. (II) A charge is spread uniformly along the y-axis from $y = 0$ to $y = +\infty$. The charge density on the y-axis is λ. Find the force on a point charge q placed on the x-axis at $x = x_0$.

33. (II) A long, thin rod of length L that contains a uniform distribution of charge Q points away from a point charge q. The nearest part of the rod is a distance d from the point charge. What is the electric force exerted on the charge q by the rod?

34. (II) A charge Q is distributed uniformly over a thin ring of radius R. The ring is oriented in the xy-plane, with its center at the origin. Find the force on a charge q located at the origin, and discuss the stability of its motion in the xy-plane. How does this compare with the case of a point charge placed at the center of a sphere whose surface is uniformly charged?

35. (II) Use the results of Example 22–6 to calculate the force on a positive point charge of magnitude 2.4 μC located 4 cm above the center of a uniformly charged solid plate of radius 6 cm that carries a total positive charge of 10 μC. [*Hint:* Break the disk into concentric rings, use the results of Example 22–6 for each ring, and sum over the forces due to the rings.]

36. (II) Calculate the force exerted on a charge q by an infinite plane sheet with surface charge density (charge per unit area) σ. [*Hint:* You may use the results of Example 22–6.]

37. (II) Consider an infinite vertical sheet that carries a charge of 10^{-4} C/m^2. A cork ball of mass 5 g is suspended by a string 60 cm long at a distance of 20 cm from the charged sheet. What is the string orientation (a) if a charge $q = 5 \times 10^{-9}$ C is placed on the cork ball? (b) if instead a charge $q = -2.4 \times 10^{-9}$ C is placed on the ball?

38. (II) A total charge of 3.1 μC is distributed uniformly over a thin, semicircular wire of radius 10.0 cm. What is the force on a charge of 2.0 μC located at the center of the circle?

39. (II) A succession of $n + 1$ alternating positive and negative charges q are located along the x-axis at the points $x = 0$, $x = d$, $x = 2d, \ldots, x = nd$. An isolated charge Q is placed as shown in Fig. 22–22 at the point $x = D$ a very long distance away from the origin ($D \gg nd$). (a) Write a general expression for the electric force on charge Q. (b) Approximate your result, using the condition $D \gg nd$. Keep only leading and next-to-leading terms. [*Hint:* Use $(1 + x)^{-2} \simeq 1 - 2x$ for $x \ll 1$.]

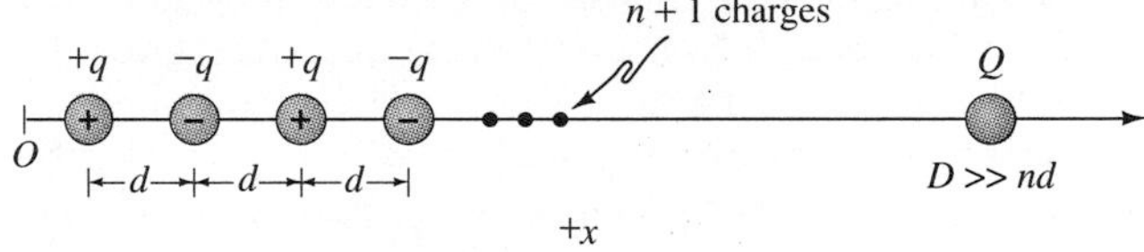

FIGURE 22–22 Problem 39.

40. (III) What is the force per unit area between two infinite, uniformly charged plates with a surface charge density of $+10^{-5}$ C/m^2 and -10^{-5} C/m^2, respectively, when the distance between the plates is 10 cm? What if the distance between the plates is doubled? [*Hint:* You may use the result of Problem 36.]

General Problems

41. (II) How much charge $+Q$ should be distributed uniformly over a square, horizontal plate of dimensions 1 m × 1 m if a ball of mass 1 g and charge 1 μC is to remain suspended 1 mm over the surface of the plate? Take gravity into account in this problem. How would your answer change if the ball were to be suspended 2 mm over the plate? *Qualitatively*, how would your answer change if the ball were to be suspended 1 m over the plate?

42. (II) A single charge $q_1 = +10^{-7}$ C is fixed at the base of a plane that makes an angle θ with the horizontal direction. A small ball of mass $m = 2$ g and charge $+10^{-7}$ C is placed in a smooth, frictionless groove in the plane that extends directly to the fixed charge (Fig. 22–23). It is allowed to move up and down until it finds a stable position $\ell = 10$ cm from the fixed charge. What is θ?

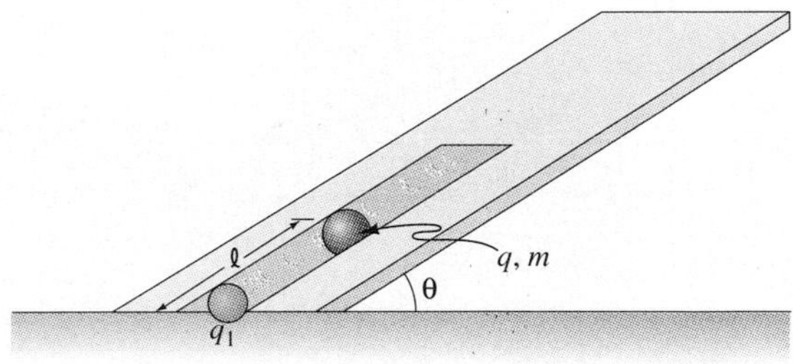

FIGURE 22–23 Problem 42.

43. (II) The nucleus of an iron atom contains 26 protons within a sphere of radius 4×10^{-15} m. What is the Coulomb force between two protons at opposite sides of this nucleus? The answer to this problem illustrates that the force that holds the nucleus together against the Coulomb repulsion of its constituents must be strong indeed.

44. (II) An electron moves in a circular planetary orbit around a proton. (a) If the centripetal force is the attractive Coulomb force, what is the speed of the electron in terms of the charge e and the radius of the circular orbit? (b) What is the angular momentum, L, of the electron in the orbit? (c) Express the speed in terms of e and L. (d) Express the radius of the orbit in terms of e and L. (e) Express in terms of e and L the time it takes for the electron to go around the circle once. (f) Evaluate all these quantities, given that $L = 1.05 \times 10^{-34}$ kg·m²/s. This corresponds to a simplified version of the hydrogen atom.

45. (II) Suppose that the proton charge were slightly larger than the electron charge, so that $q_{\text{proton}} = (1 + \delta)e$ and $q_{\text{electron}} = -e$, where $0 < \delta \ll 1$. (a) Given that there are approximately 1.25×10^{57} protons (and electrons) in the sun, and approximately 1.15×10^{44} protons and electrons in the earth, what is the upper limit on δ set by the fact that the resultant earth–sun electric repulsion cannot be large enough to cancel the attraction due to gravity? The mass of the sun is approximately 2×10^{30} kg, that of the earth is approximately 6×10^{24} kg, and $G = 6.7 \times 10^{-11}$ N·m²/kg². (b) How would your value of δ change the weight of a football player who contains 3×10^{28} protons and electrons?

46. (II) (a) What is the force on a charge Q, located in the xy-plane at the point $(x, 0)$, due to the following distribution of four charges: q at $(0, 3a)$, $-q$ at $(0, a)$, $-q$ at $(0, -a)$, and q at $(0, -3a)$? (b) Show that for $x \gg a$, the force decreases as x^4. [*Hint:* Use $(1 + z)^{-3/2} \simeq 1 - (3z/2)$ for $z \ll 1$.] (c) Suppose that the charges at the locations in part (a) were q, $-q$, q, $-q$, respectively. What would the force be for $x \gg a$, and why is it so different?

47. (II) Two fixed positive charges q are separated by a length ℓ. A third positive charge q of mass m is constrained to run on a line between the two fixed charges (Fig. 22–24). (a) When the third charge is placed a distance x from the left-hand fixed charge, what is the net force on the third charge? Where is this force zero; that is, where is the equilibrium point? (b) What is the net force as a function of the displacement of the third charge from the equilibrium point of part (a)? (c) For *small* values of the displacement from the equilibrium point, the third charge behaves as if a spring were acting on it. What is the value of the oscillation frequency?

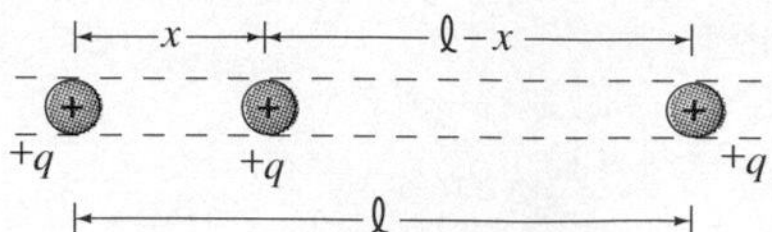

FIGURE 22–24 Problem 47.

48. (II) Show that the force between two spherically symmetric distributions of charge is identical to the force between two point charges that are located at the geometric center of each distribution and have the same total charge. (*Hint:* Use the fact that the force on a point charge due to distribution 1 is the same as if distribution 1 were concentrated at its center, then use similar reasoning for distribution 2, and then use Newton's third law.)

49. (III) Two rods, each of length $2L$, are placed parallel to one another a distance R apart. Each carries a total charge Q, distributed uniformly over the length of the rod. Give an integral for the magnitude of the force between the rods, but do not evaluate it. Without working out any integrals, can you determine the force between the rods for $R \gg L$?

50. (III) Consider an infinite number of identical point charges q located at equally spaced points on the x-axis at the locations $x = na$ (n takes on integer values that range from $-\infty$ to $+\infty$). (a) Write an expression for the force on a charge Q, located at $x = 0$ and $y = R$, due to all the point charges q, and show the direction of the net force. (b) Take the limit of your result when the inter-charge spacing $a \to 0$ and the charge $q \to 0$ such that $q/a = \lambda$ (a fixed charge density). Show that your expression can be written as an integral, and use dimensional analysis to determine the R-dependence of the force on charge Q.

CHAPTER 23

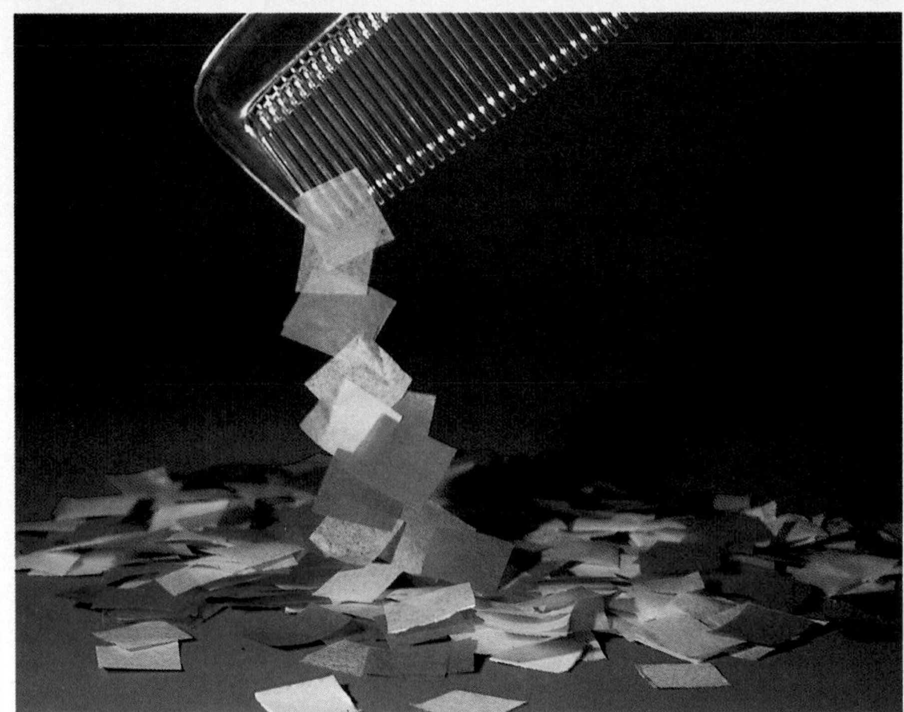

An electric field forms when a comb is rubbed with a cloth. Bits of paper are more likely attracted to the comb because the field has induced a dipole moment in them than because they have any net charge.

ELECTRIC FIELD

Just as the earth is influenced by the sun even though the sun is 93 million miles away, one charge can impose a force on another even if they are separated by a large distance. The concept of action at a distance, in which a force acts across an empty space, has always seemed difficult to accept. Action at a distance suggests that somehow the body responsible for the force on a second body reaches out, measures the distance to the second body, and then acts. Michael Faraday suggested a way around this conceptual difficulty: The first body influences the surrounding space by setting up a *field* around itself that is present whether there is a second body or not. When the second body is located at a given point, the field at this point acts on that body. This important idea can be developed quantitatively and, like any really good idea, leads to further ideas that go far beyond the original concept in utility and insight. In this chapter we shall introduce and develop the concept of an electric field produced by static charges and learn about some of the ways in which it can be useful. We shall continue to use the field concept in future chapters, because it forms a basis for understanding many electrical and magnetic effects.

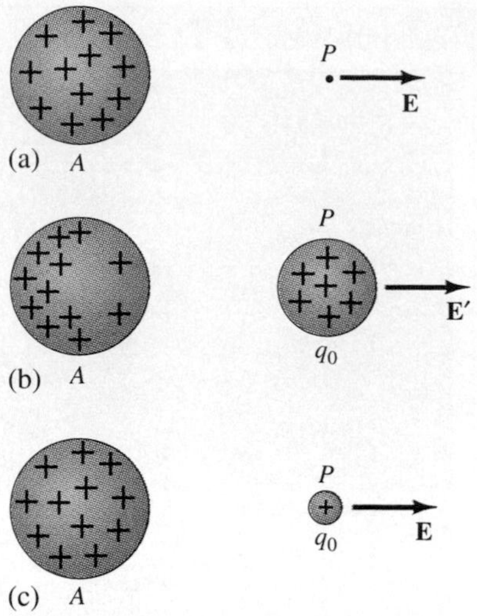

FIGURE 23–1 (a) An electric field exists at a point P due to charges on the sphere A. (b) The test charge q_0 repels the charges on sphere A. A new electric field, $\mathbf{E}'$, is produced at P by sphere A, because the charges on A have been redistributed. (c) The test charge q_0 is now so small that it hardly affects the charges on sphere A. The electric field produced by A at point P is now the same as in part (a). In each case the electric field is due to the charge on sphere A.

23–1 ELECTRIC FIELD

It is useful to think of a distribution of charges, positive or negative, as giving rise to an **electric field**, which acts on any charge placed in the field. The electric field present at any particular point can be detected by placing a small positive **test charge** q_0 at that location and seeing if it experiences a force. A test charge is only a probe: It does not *produce* the electric field that we are trying to measure; the field is due to other charges. The test charge must be at rest, because, as we shall soon see, moving charges experience different forces. The electric field, $\mathbf{E}$, can be defined by measuring the magnitude and direction of the electric force $\mathbf{F}$ acting on the test charge. The definition of the field is

The definition of the electric field

$$\mathbf{E} \equiv \frac{\mathbf{F}}{q_0}. \tag{23–1}$$

The reason we use a *small* test charge is that a large charge could, through its Coulomb interaction, cause the charges responsible for the electric field to move (Fig. 23–1). We might thereby affect the original charge distribution that gives rise to the electric field and hence the field itself. Thus, we use an infinitesimally small test charge q_0 and define the electric field ideally by

$$\mathbf{E} \equiv \lim_{q_0 \to 0} \frac{\mathbf{F}}{q_0}. \tag{23–2}$$

The electric force is a vector, and so is the electric field. We know a vector such as the electric field completely when we know both its magnitude and its direction *at every point in space.*

The SI units of electric field

From the definition in Eq. (23–2), the SI units of electric field are newtons per coulomb (N/C).[†] Table 23–1 gives the magnitudes of the electric fields in various physical situations.

[†] In Chapter 25 we shall learn that the electric field can be expressed alternatively in SI units of volts per meter (V/m), where 1 N/C = 1 V/m.

TABLE 23–1

VALUES OF SOME ELECTRIC FIELDS (N/C)

Interplanetary space	10^{-3}–10^{-2}
Atmosphere at earth's surface in clear weather	100–200
In a thunderstorm	10^3
Electrical breakdown of dry air	3×10^6
In a Van de Graaff particle accelerator†	10^6
In the Fermilab particle accelerator‡	1.2×10^7
In atoms within the radius of an electron orbit	10^9
In the electromagnetic radiation of the most intense laser	10^{12}
At a distance of twice the radius from the center of a uranium nucleus	5×10^{20}

† See Chapter 25.

‡ See Section 23–4.

The Electric Field of a Point Charge

The simplest example of an electric field is the field associated with a point charge q_1. Consider two point charges, q_1 and q_0, located a distance r apart (Fig. 23–2). The Coulomb force on q_0 due to q_1 is

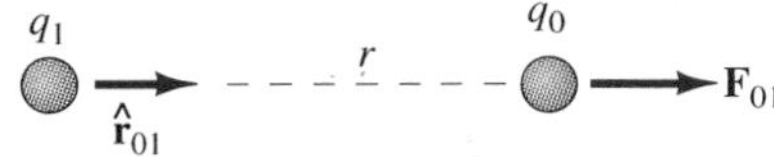

FIGURE 23–2 The force $\mathbf{F}_{01}$ exerted on point charge q_0 by point charge q_1; both charges are positive.

$$\text{for a point charge:} \qquad \mathbf{F}_{01} = \frac{q_1 q_0}{4\pi\epsilon_0 r^2}\,\hat{\mathbf{r}}_{01}, \tag{23–3}$$

after Eq. (22–8). If we take q_0 as our test charge, we can use Eqs. (23–1) and (23–3) to find the electric field due to q_1:

$$\text{for a point charge:} \qquad \mathbf{E}_1 = \frac{\mathbf{F}_{01}}{q_0} = \frac{q_1}{4\pi\epsilon_0 r^2}\,\hat{\mathbf{r}}_{01}. \tag{23–4}$$

The value of the test charge has cancelled, so the limiting process in Eq. (23–2) introduces no complications. Equation (23–4) specifies that $\mathbf{E}_1$ is in the same direction as $\mathbf{F}_{01}$, that of the unit vector $\hat{\mathbf{r}}_{01}$, which points from q_1 to q_0. We drop the subscript on $\hat{\mathbf{r}}_{01}$ in Fig. 23–3, which shows the direction of $\mathbf{E}_1$ as determined by moving our test charge to various points a distance r away from q_1. This field is radial (Fig. 23–3a), and we have used the radial unit vector $\hat{\mathbf{r}}$ (measured from q_1) to specify completely the electric field $\mathbf{E}$ due to a point charge q (we also drop the subscripts on $\mathbf{E}_1$ and q_1):

$$\text{for a point charge:} \qquad \mathbf{E} = \frac{q}{4\pi\epsilon_0 r^2}\,\hat{\mathbf{r}}. \tag{23–5}$$

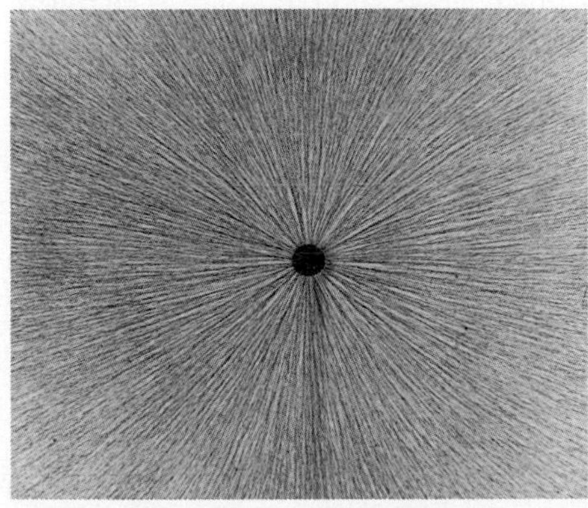

(a)

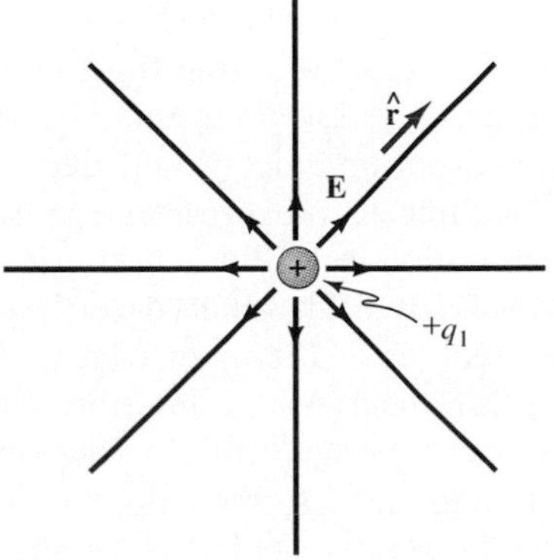

(b)

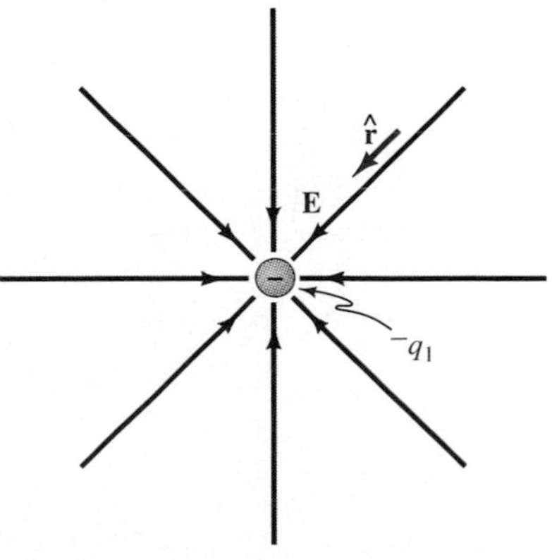

(c)

FIGURE 23–3 (a) Threads floating in oil become aligned with the electric field of this point charge. (b) The direction of the electric field $\mathbf{E}$ due to q_1 is radial. The charge is positive, and the field points away from it. (c) The charge is negative, and the field points toward it.

The electric field due to a point charge q is directed away from a positive charge and toward a negative charge.

The electric field points *away* from a positive charge, as in Fig. 23–3b. When the charge is negative, the electric field has the same magnitude but is opposite in direction. The electric field due to a negative charge points *toward* that charge, as in Fig. 23–3c.

The Usefulness of the Field Concept

Once we know the electric field, **E**, produced by a point charge q, we can find the force on any point charge q' placed in that field by using Eq. (23–1); that is,

$$\mathbf{F} = q'\mathbf{E}. \tag{23–6}$$

More important, *any* distribution of charges, not simply a point charge, produces an electric field throughout space. We use the subscript "ext" (for external) on **E** to emphasize that the external electric field is independent of the charge q' on which the force acts. Once we know $\mathbf{E}_{\text{ext}}$, *the force on any point charge* q' *in the field* is the generalization of Eq. (23–6):

The force on a point charge in an electric field

for a point charge in an external electric field: $$\mathbf{F} = q'\mathbf{E}_{\text{ext}}. \tag{23–7}$$

Equation (23–7) is a general result that is quite useful.

Why do we bother to introduce fields? Why not deal just with forces between charges? We have already mentioned the role the field plays in resolving the conceptual difficulties of action at a distance. There are other reasons why the field concept is useful and even necessary. When a test charge is acted upon by some complicated configuration of other charges, it experiences a force that depends on its location **r**. This force is a complicated function of the vectors that measure the displacement of the test charge from the other charges. It is economical to determine once and for all what the field **E**(**r**) due to the other charges is. Once we know this electric field, it is a simple matter to determine the force on *any* charge placed *anywhere* in the field.

The field concept becomes indispensable when we see in Chapter 25 that the field carries energy. To preserve the important idea of energy conservation, the field is a *necessary* concept. But the real power of the field concept appears when the field arises from *moving charges*. Even if the moving charges are limited to a small region (for example, within the arms of an antenna), the electric field spreads through all of space, and the speed of spreading is the speed of light. The supernova known as 1987A took place approximately 163,000 years ago; electric fields caused by the violent motion of many charges within and around the exploding star reached earth on February 23, 1987. These traveling fields made electrons in terrestrial antennas move, and this was the signal that supernova 1987A had occurred. This description of the process is much easier to grasp than is the idea that an electrical force exists between the charges in the supernova and those in a terrestrial detector, and that this force depends not only on the separation between the charges, but also on a time lag between their respective motions. A force law with a built-in time delay that depends on the distance between two interacting bodies is difficult to express and work with.

The notion of a field is useful in many disciplines. In hydrodynamics we employ a velocity field, which describes the velocity **v** at all points where fluid flow occurs, such as in the pipes of a city water system. In thermal physics we use a temperature field, which describes the temperature at all points in a room. And in acoustics we use a field of variations in air density. In the latter two cases, there is no directionality to the field, so the field is related to a force in a more indirect way than the electric field is related to the electric force. We then have scalar fields instead of vector fields.

EXAMPLE 23–1 Find the electric field due to a point charge $q = +1.4\ \mu\text{C}$ at a distance of 0.10 m from the charge. What is the force on a charge $q' = -1.2\ \mu\text{C}$ placed this far from q?

SOLUTION: The electric field of a point charge is given directly by Eq. (23–5):

$$\mathbf{E} = \left(\frac{q}{4\pi\epsilon_0 r^2}\right)\hat{\mathbf{r}} = \left[\frac{(9.0 \times 10^9\ \text{N}\cdot\text{m}^2/\text{C}^2)(1.4 \times 10^{-6}\ \text{C})}{(0.10\ \text{m})^2}\right]\hat{\mathbf{r}} = (1.3 \times 10^6\ \text{N/C})\hat{\mathbf{r}}.$$

The electric field is directed outward radially from the position of the 1.4-μC charge (Fig. 23–3b). If the charge were negative rather than positive, the field would point radially inward rather than radially outward.

To find the force on q', we treat the electric field determined above as an external field and use Eq. (23–7). The magnitude of the force is the magnitude of the field times the magnitude of q':

$$F = |q'|E = (1.2 \times 10^{-6}\ \text{C})(1.3 \times 10^6\ \text{N/C}) = 1.5\ \text{N}.$$

The sign of q' is negative, so, when it is multiplied by the magnitude of the outward radial field, the resulting force on q' acts radially inward. It is no surprise that opposite charges attract.

Alternatively, we can use Eq. (23–7) directly, including the vector notation. We get

$$\mathbf{F} = (-1.2 \times 10^6\ \text{C})[(1.3 \times 10^6\ \text{N/C})\hat{\mathbf{r}}] = (-1.5\ \text{N})\hat{\mathbf{r}},$$

the same result as before.

If more than one point charge is responsible for producing an electric field, we use the principle of superposition to determine the net electric field. The superposition principle states that the net electric force on a body is the vector sum of the forces due to individual point charges. Therefore the net electric field is the vector sum of the fields of individual charges present. The net force exerted on our test charge q_0 due to all the other charges in the region is

$$\mathbf{F}_{\text{net}} = \mathbf{F}_{01} + \mathbf{F}_{02} + \mathbf{F}_{03} + \cdots = \sum_i \mathbf{F}_{0i}. \qquad (23\text{–}8)$$

Thus

$$\mathbf{E}_{\text{net}} = \frac{\mathbf{F}_{\text{net}}}{q_0} = \frac{\mathbf{F}_{01}}{q_0} + \frac{\mathbf{F}_{02}}{q_0} + \frac{\mathbf{F}_{03}}{q_0} + \cdots \qquad (23\text{–}9)$$

$$= \mathbf{E}_1 + \mathbf{E}_2 + \mathbf{E}_3 + \cdots = \sum_i \mathbf{E}_i. \qquad (23\text{–}10)$$

In Eq. (23–10), $\mathbf{E}_2$, for example, is the electric field due to only the charge q_2 at the point in space where we have placed q_0. By using Eq. (23–5) for each point charge q_i, we find that

for a group of point charges: $$\mathbf{E}_{\text{net}} = \frac{1}{4\pi\epsilon_0}\sum_i \frac{q_i}{r_i^2}\hat{\mathbf{r}}_i. \qquad (23\text{–}11)$$

The net electric field produced by a group of point charges

In this equation, the unit vector $\hat{\mathbf{r}}_i$ is directed from the position of charge q_i to the position where the field is being measured.

EXAMPLE 23–2 Consider three charges placed on a line: $q_1 = +2\ \mu C$ at $x_1 = -2$ cm, $q_2 = +3\ \mu C$ at $x_2 = +4$ cm, and $q_3 = -2\ \mu C$ at $x_3 = +10$ cm (Fig. 23–4). Find the electric field at point A, the origin of the coordinate system.

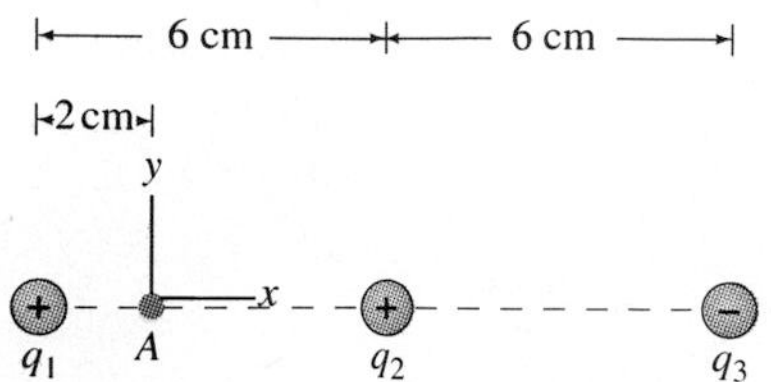

FIGURE 23–4 Example 23–2. The electric field at point A is due to three charges. The distance x is measured from point A.

SOLUTION: The solution is obtained from straightforward application of Eq. (23–11). Although it is generally important to keep in mind that the required sum is vectorial, in this case all the positions lie along a straight line. Place a Cartesian coordinate system at point A. The electric field at point A is

$$\mathbf{E}_A = \mathbf{E}_1 + \mathbf{E}_2 + \mathbf{E}_3,$$

where $\mathbf{E}_j$ ($j = 1$, 2, or 3) is the field due to charge q_j at point A. Application of Eq. (23–11) for the individual electric fields gives

$$\mathbf{E}_A = \frac{1}{4\pi\epsilon_0}\left(\frac{q_1(+\mathbf{i})}{x_1^2} + \frac{q_2(-\mathbf{i})}{x_2^2} + \frac{q_3(-\mathbf{i})}{x_3^2}\right). \tag{23–12}$$

We must pay careful attention to signs. The unit vectors $\pm\mathbf{i}$ in parentheses indicate the direction of the unit vector $\hat{\mathbf{r}}_j$ from the position of charge q_j to point A. The actual direction of the electric field $\mathbf{E}_j$ due to charge q_j is, however, determined by the product $q_j\hat{\mathbf{r}}_j$, and the sign of the charge must be taken into account. For example, the direction of $\mathbf{E}_3$ is $+\mathbf{i}$, because the negative sign of charge q_3 multiplied by $(-\mathbf{i})$ gives a direction $(+\mathbf{i})$. Numerical evaluation of Eq. (23–12) gives

$$\begin{aligned}\mathbf{E}_A &= (9 \times 10^9\ \text{N}\cdot\text{m}^2/\text{C}^2)\\ &\quad\times\left[\frac{(2 \times 10^{-6}\ \text{C})}{(0.02\ \text{m})^2}(\mathbf{i}) + \frac{(3 \times 10^{-6}\ \text{C})}{(0.04\ \text{m})^2}(-\mathbf{i}) + \frac{(-2 \times 10^{-6}\ \text{C})}{(0.10\ \text{m})^2}(-\mathbf{i})\right]\\ &= (3 \times 10^7\ \text{N/C})\mathbf{i}.\end{aligned}$$

The net electric field at point A is in the $+x$-direction, or toward the right.

Electric Dipoles and Their Electric Fields

An electric dipole

An **electric dipole** consists of two charges $+q$ and $-q$, of equal magnitude but opposite sign, separated by a distance L (Fig. 23–5a). The field of one charge decreases as $1/r^2$. If the two opposite charges (q_1 and $-q_2$, say) do not add to zero, their field would have the form $(q_1 - q_2)/r^2$ for $r \gg L$. If equal and opposite charges were to sit precisely on top of one another, the two $1/r^2$ contributions to the field would cancel to a zero electric field. But because the two equal but opposite charges of a dipole are not quite on top of one another, the resulting

(a)

(c)

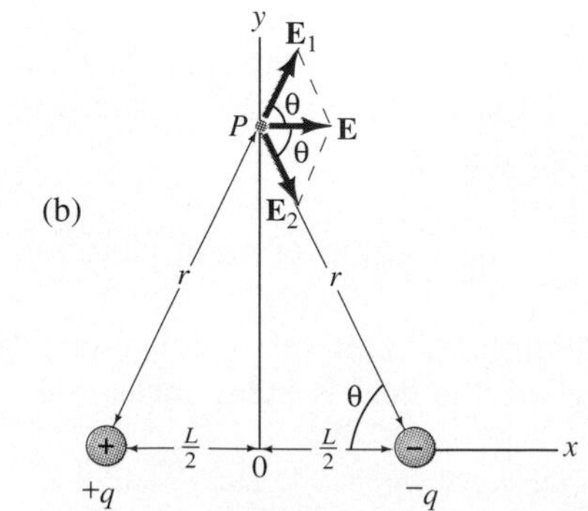

FIGURE 23–5 (a) An electric dipole consists of equal but opposite charges separated by a distance L. (b) The net field at point P, $\mathbf{E}$, acts only along the direction from $+q$ to $-q$. (c) The electric dipole moment $\mathbf{p}$ is directed from the negative charge to the positive charge.

field decreases as $1/r^3$, as we shall see. The electric field depends only on the product qL, which is called the **electric dipole moment** of the neutral pair $(+q, -q)$ and is denoted by the letter p. We make $p = qL$ a vector by defining $\mathbf{L}$ to be directed from $-q$ to $+q$ (Fig. 23–6). Thus the electric dipole moment $\mathbf{p}$ is

The electric dipole moment

$$\mathbf{p} = q\mathbf{L}. \tag{23–13}$$

The vector $\mathbf{p}$ points from the negative charge to the positive charge.

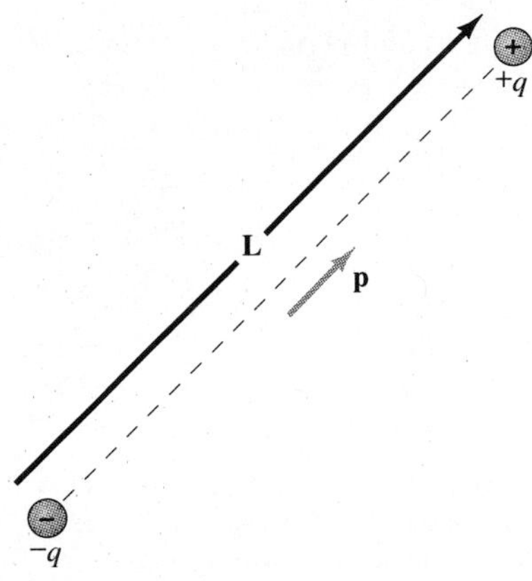

FIGURE 23–6 An electric dipole with dipole moment $\mathbf{p} = q\mathbf{L}$. The direction of $\mathbf{p}$ is from $-q$ to $+q$.

EXAMPLE 23–3 Find the electric field of the electric dipole shown in Fig. 23–5b at a point P that is a large distance $r(r \gg L)$ from each charge. P lies along the perpendicular axis that bisects the line between the two charges.

SOLUTION: The x- and y-axes are shown in Fig. 23–5b, and point P has xy-coordinates $(0, y)$. The net electric field at P is given by $\mathbf{E} = \mathbf{E}_1 + \mathbf{E}_2$, where the field $\mathbf{E}_1$ is due to the charge $+q$ and the field $\mathbf{E}_2$ is due to the charge $-q$. The magnitudes of the two fields are the same, but $\mathbf{E}_1$ points away from $+q$, whereas $\mathbf{E}_2$ points toward $-q$. The y-components of $\mathbf{E}_1$ and $\mathbf{E}_2$ exactly cancel each other, and we are left with only a net x-component toward the right that is twice the x-component of the field due to either charge:

$$\mathbf{E} = E_x\mathbf{i} = (E_{1x} + E_{2x})\mathbf{i} = 2E_{1x}\mathbf{i},$$

where

$$E_{1x} = \frac{q}{4\pi\epsilon_0 r^2}\cos\theta.$$

From Fig. 23–5b we see that $\cos\theta$ is given by

$$\cos\theta = \frac{L/2}{r} = \frac{L}{2r}.$$

Thus the total electric field of the dipole along the perpendicular bisector is

$$\mathbf{E} = \left(\frac{2q}{4\pi\epsilon_0 r^2}\right)\left(\frac{L}{2r}\right)\mathbf{i} = \frac{qL}{4\pi\epsilon_0 r^3}\,\mathbf{i}. \tag{23–14}$$

Equation (23–14) is the correct result along the perpendicular bisector even when the distance from the charge pair is not large. The electric field decreases with r as $1/r^3$. The field at points *along the bisecting axis* is given by

$$\mathbf{E} = -\frac{\mathbf{p}}{4\pi\epsilon_0 r^3}, \tag{23–15}$$

where we have used Eq. (23–13) for the electric dipole moment, $\mathbf{p}$. If $r \gg L$, then $r \simeq y$ and

$$\text{along the bisecting axis:} \quad \mathbf{E} \simeq -\frac{\mathbf{p}}{4\pi\epsilon_0 y^3}. \tag{23–16}$$

The electric field from a dipole is generally *not* antiparallel to the dipole moment, although that is the case here. In Eqs. (23–15) and (23–16), we have given the field only along the bisecting axis.

The Importance of Electric Dipoles

In Example 23–3 we found that the electric dipole field along a bisecting axis depends on neither q nor L alone, but on their product. This is true for the dipole field at any point in space. Only the product $p = qL$ can be determined from the field of an electric dipole, not L or q separately.

Induced and permanent electric dipoles

Electric dipoles are of great interest because they occur so often in nature. An electric field is produced even though the total charge of a dipole is zero. External fields frequently induce charge separations in electrically neutral molecules and materials, leading to an excess of positive (or negative) charge on one side (and the other) and hence to an **induced electric dipole moment** (Fig. 23–7). There are also examples of charge configurations with **permanent electric dipole moments** (dipole moments that are not induced by external fields) in nature. Many molecules that have an extended structure, with negatively charged electrons distributed preferentially in certain regions, have permanent electric dipole moments. The water molecule, H_2O, which has a V shape with the oxygen atom at the apex of the V, is an example (Fig. 23–8). In cases such as common salt (NaCl) and hydrochloric acid (HCl), a molecule is formed by electrons that cluster preferentially around one atom, giving that atom a negative charge. The other atom is left with a positive charge. In such molecules (which are held together by what is called *ionic bonding*), there is always a permanent electric dipole moment. On the molecular level, where the effects of electric dipole fields have great physical importance, permanent dipole moments are always much larger than induced dipole moments. For example, $p \simeq 6 \times 10^{-30}$ C·m for a water molecule, whereas a hydrogen atom in the rather strong field $E = 3 \times 10^6$ N/C acquires an induced dipole moment of $p \simeq 3 \times 10^{-34}$ C·m.

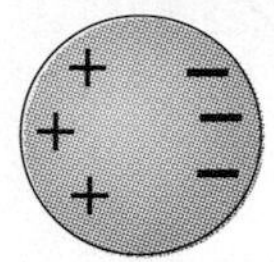
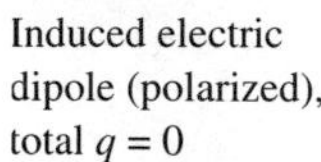

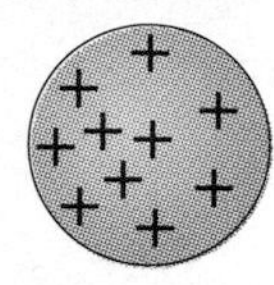

FIGURE 23–7 A nearby charge can induce a polarized charge, and hence an electric dipole, on a neutral body.

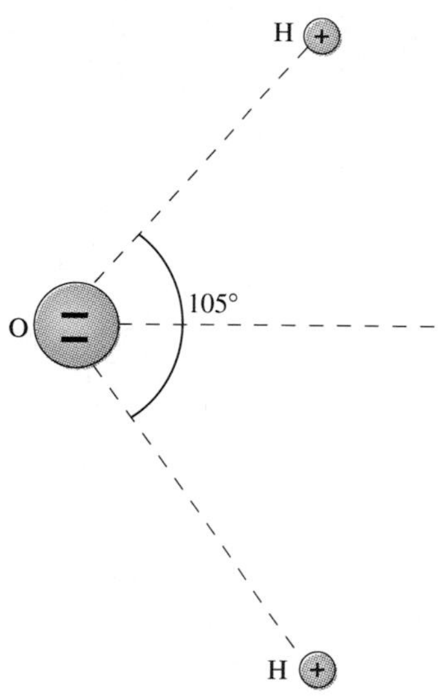

FIGURE 23–8 The water molecule, H_2O, is a permanent electric dipole. Both hydrogen electrons are shared with the oxygen atom, creating a strong electric bond that holds the molecule together (by what is called covalent bonding).

23–2 ELECTRIC FIELD LINES

The electric field due to a charge distribution and the force experienced by charged particles in the field can be visualized in terms of **electric field lines**. Their use was introduced by Michael Faraday around the middle of the nineteenth century, even before the concept of the electric field was clearly understood.[†] Faraday used the phrase "lines of force." Electric field lines are continuous in space, in contrast to the field itself, which is represented by a different vector at each point in space.

We have already seen that we can map out the electric field by moving a test charge around in space. The field is easily expressed in algebraic form and is the best tool for obtaining algebraic or numerical results that concern electric forces. The field is, however, awkward to use in a visual sense. It is not easy to draw a region of space and at each point (or even at nearby points) draw a vector whose varying length and direction represents the field. Electric field lines are an alternative more suited to visual representation.

Electric field lines

Electric field lines are smooth directional lines in space that are determined by the electric field, according to two simple rules:

1. Electric field lines are drawn so that the tangent to the field line at each point specifies the direction of the electric field **E** at that point. This rule relates the *direction* of the electric field lines to the direction of the electric field.
2. The *density* in space of electric field lines around a particular point is proportional to the strength of the electric field at that point.

[†] Later, Faraday, who was a great experimentalist but less strong as a theorist, gave this concept more physical significance than it actually has.

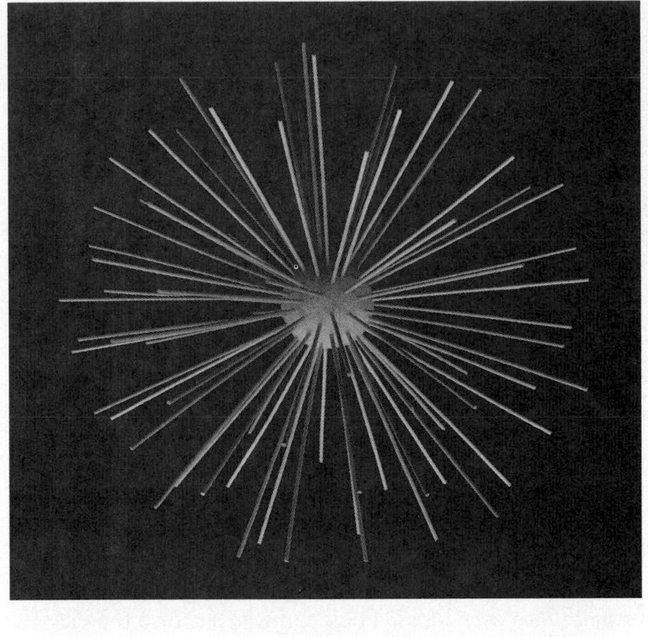

(a)

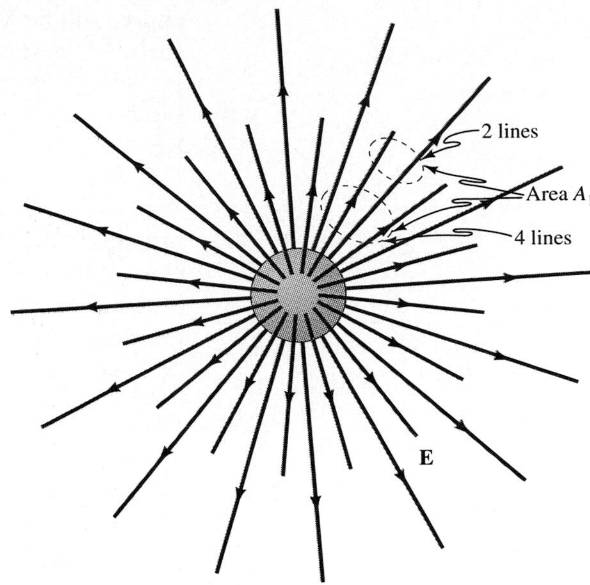

(b)

N total lines,
q total charge

FIGURE 23–9 (a) Representation of the radial electric field lines from a point charge. (b) Fewer field lines pass through the same-size area farther from the charge.

Properties of Electric Field Lines

Let us draw the electric field lines of a positive point charge q. We know that the electric field points radially away from a positive charge at every location in space. The field has the same magnitude all around a sphere centered on the charge, and this magnitude decreases with the distance r from the charge as $1/r^2$. Electric field lines are radial, pointing outward from the charge, and distributed uniformly about the charge (Fig. 23–9a). We can use Fig. 23–9 to illustrate their properties.

Techniques to help you draw electric field lines

Property One. Rule 2 above, which states how electric field lines reflect the strength of the associated electric field, requires some explanation. The electric field does not change its direction abruptly as we move through a region of charge-free space. Thus, in a small region, electric field lines are very nearly parallel to one another. In this small region we can take a small area that is oriented perpendicular to the nearly parallel field lines. The density of electric field lines is then the number of electric field lines that cross this small area, divided by that area. Note that the density is the number of lines *per unit area.*

Property Two. How do we set the number of electric field lines and the density of electric field lines? We can *choose* to draw for convenience N field lines that originate at a given charge q; N is any number. Then the number of lines that leave the other charges is determined. In particular, the number of field lines leaving a positive charge q_i is N_i, where $N_i = (q_i/q)N$. Thus half the number of lines will originate from a positive charge half as big. Now we can use the rule about the density of lines to show that *lines can start or terminate only on charges, never in empty space.* Suppose that N lines originate at the point charge q in Fig. 23–9b, and that lines are neither created nor destroyed. If we consider the number of field lines that pass through an area the size of a dime, we see from the figure that many more field lines would pass through this area when the area is close to the point source than when the same area is far away. *If no new electric field lines are created as we move away from the charge,* then the density of lines at a radius R from the charge will be N divided by the area of the surface perpendicular to the lines. This surface is a sphere of radius R centered on the charge, and the density of lines is $N/4\pi R^2$. We know that the density of lines is proportional to the strength of the field, and that the field falls off as $1/r^2$. Therefore, *the connection between the strength of the field and the density of the electric field lines is automatic if electric field lines are neither created nor destroyed in regions where there is no charge.* We have demonstrated this only for a single point charge, but the superposition principle can be used to show that this is true in general. Note that, for a point charge, the electric field lines go off to infinity. This will be true whenever there is some localized collection of charges. At distances that are large compared with the dimensions of the region that contains the charge, the net charge appears to be localized at a point, and electric field lines will be distributed evenly over a distant sphere that surrounds the net charge.

Property Three. Electric field lines originate on, and run outward from, positive charges. They run toward, and terminate on, negative charges. This reflects the fact that electric charges are the sources of electric fields, which point away from positive charges and toward negative charges.

Property Four. *No two field lines ever cross.* They cannot, because the electric field has a definite magnitude and direction at any point in space. If two or more electric field lines were to cross at some point, then the direction of the electric field at that point would be ambiguous.

Symmetry can be a useful guide to drawing electric field lines. A point charge looks the same when viewed from any direction. It has spherical symmetry, and the field lines follow the only direction that respects this symmetry—namely, they are radial. Similarly, if we are dealing with a long line of charge, there is a symmetry around the line, and the field lines must project radially outward from the line, perpendicular to a cylinder that surrounds the line.

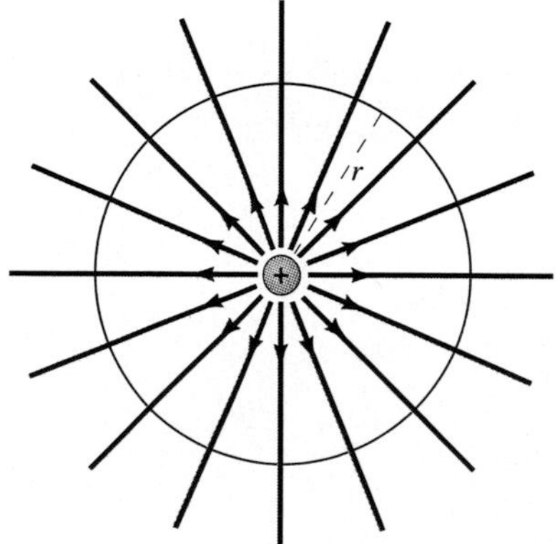

FIGURE 23–10 Electric field lines due to a point charge $+q$. Note the number of field lines that cross the circle (sphere) at radius r.

It is convenient to draw electric field lines that lie in a plane that cuts through a space with a charge or charges. This plane is the plane of the page on which the lines are drawn. For an isolated charge, such a drawing would look like Fig. 23–10. *Such a drawing should not be used carelessly for determining the field strength.* You cannot simply count the field lines that cross a particular line or surface. Figure 23–10 shows a circle of radius r centered on a positive charge. The number of lines that cross this circle is fixed at N, so the density of lines that cross the circle is N divided by the circumference of the circle, or $N/2\pi r$. Yet we know that the field decreases as $1/r^2$, not as $1/r$. The density of lines, which determines the magnitude of the electric field, is a density per unit *area*, not per unit length. It is not always easy to see how the area density of lines varies on a drawing of the lines that form a plane. Nevertheless, these planar drawings of electric field lines are still useful to help us visualize the field and the effect the field would have on a point charge.

Some Examples

The easiest way to demonstrate the usefulness of electric field lines is to look at several examples beyond that of the isolated point charge. Figures 23–11a and 23–11b show the electric field lines on a plane that passes through two positive

(a)

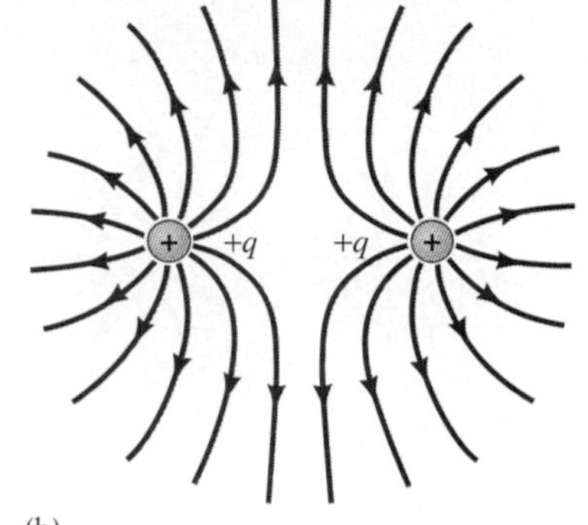

(b)

FIGURE 23–11 (a) The electric field lines due to two point charges $+q$, shown by threads in oil. (b) Schematic diagram of the field lines, which go off to infinity and appear to repel each other.

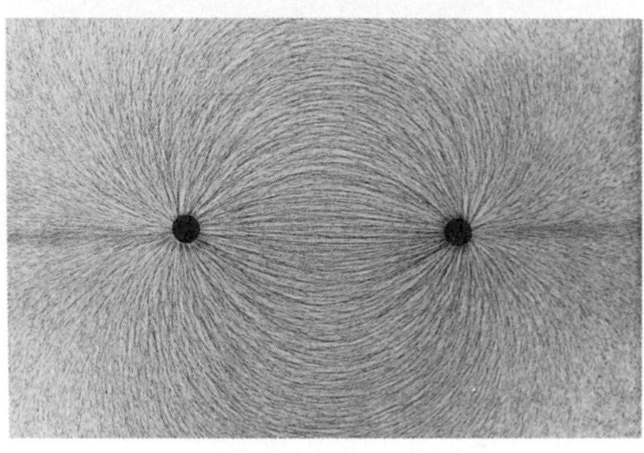

(a)

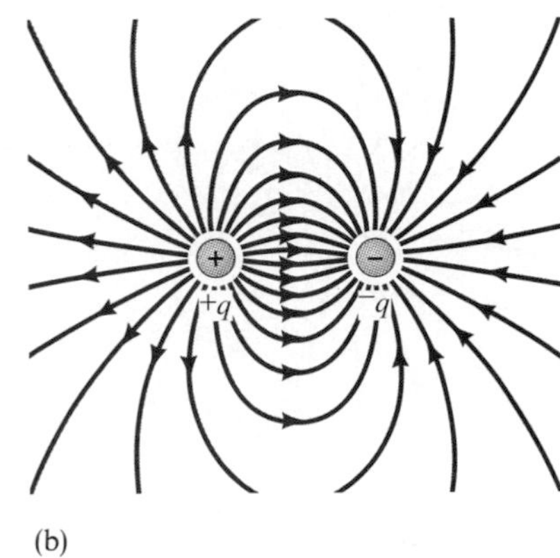

(b)

FIGURE 23–12 (a) The electric field lines due to point charges $+q$ and $-q$, a dipole, shown by threads in oil. (b) Schematic diagram of the field lines, all of which begin on $+q$ and end on $-q$; those that appear to be broken actually continue far from the charge.

charges of equal magnitude. The field lines all extend to infinity, because there are no negative charges on which the lines can terminate. The field lines that approach each other between the two positive charges appear to repel each other, which follows because two field lines cannot cross. If we were to place a positive charge q' in the region shown in Fig. 23–11b, the field lines would show us the direction of the force on the charge (and likewise the acceleration). Once we have the field lines, it is easy to see the direction of force a given charge would have in the electric field. We emphasize that, although the electric field itself has physical meaning, electric field lines are simply an aid to picturing the electric field and how a charge would react when placed in that field.

Figures 23–12a and 23–12b depict the field lines of an electric dipole. The charges have equal magnitude, $\pm q$, so an equal number of field lines are attached to them, and every field line that originates on $+q$ terminates at $-q$. Near each charge the field lines are purely radial, but they must deviate from the radial direction in order to reach the other charge. Notice that the field lines in Fig. 23–12b are consistent with the field **E** determined in Example 23–3 (compare Figs. 23–12b and 23–5).

EXAMPLE 23–4 Draw the electric field lines for a system that consists of two charges, $+2q$ and $-q$.

SOLUTION: The positive charge has twice the charge of the negative one, and we decide arbitrarily to show 12 lines that originate from $+2q$. Then only 6

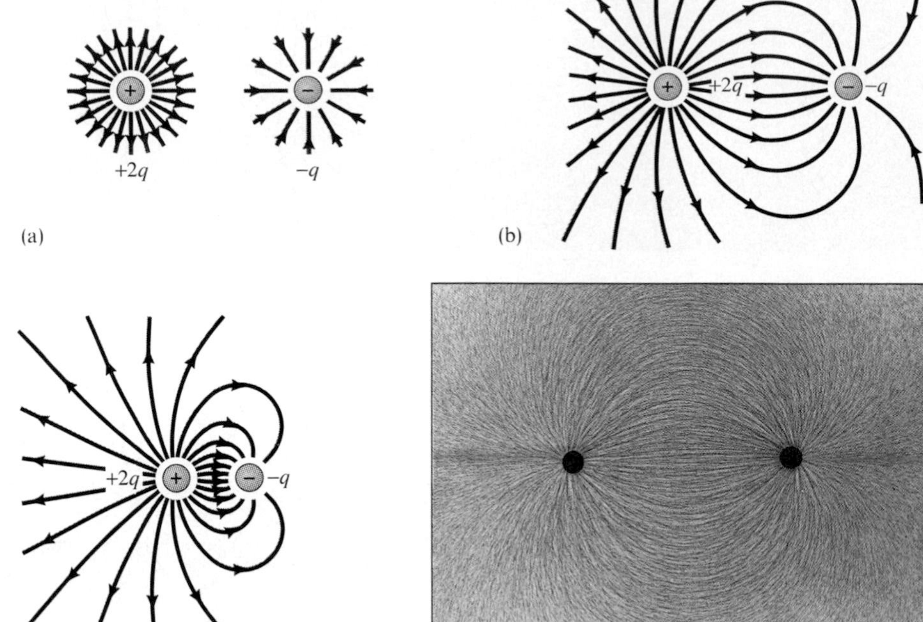

FIGURE 23–13 Example 23–4. (a) The electric field lines close to the $+2q$ and $-q$ point charges are those of a point charge. (b) Half the electric field lines that emerge from $+2q$ end up on $-q$. (c) Far from the point charges, the electric field lines are those of a point charge $+q$. (d) Electric field lines for two charges of opposite sign and different magnitude as shown by threads in oil.

lines will terminate at $-q$ (Fig. 23–13a). We draw these lines in two dimensions. Near the charges we draw the field lines as radial. Six of the lines that originate from $+2q$ must terminate on $-q$, and none of the field lines may cross, so we take the 6 lines of $+2q$ nearest to $-q$ to be the lines that terminate at $-q$ (Fig. 23–13b). What happens to the remaining 6 lines that emerge from $+2q$? Although they will initially curve toward $-q$, we realize that, very far away, they will appear to be coming from a net charge $+q$ $(+2q - q = +q)$ and will therefore be directed radially outward at large distances. The fact that 6 lines remain is consistent with our original choice of 12 lines for the line density (Fig. 23–13c).

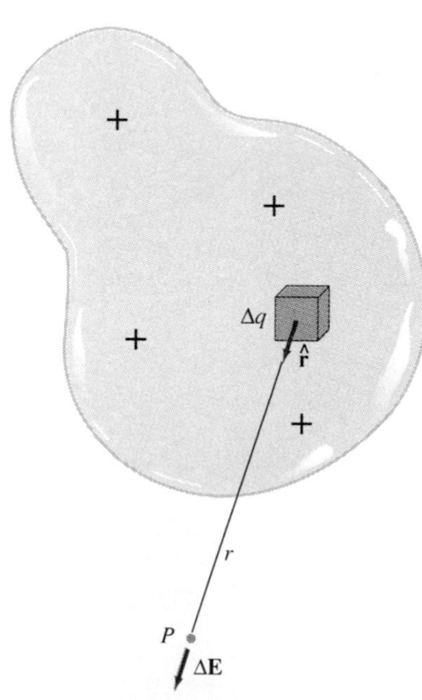

FIGURE 23–14 To find the electric field due to a continuous charge distribution, add all the electric fields $\Delta\mathbf{E}$ due to the charge elements Δq.

23–3 THE ELECTRIC FIELD DUE TO A CONTINUOUS CHARGE DISTRIBUTION

We have thus far concentrated on electric fields due to point charges or collections of them. But *continuous* distributions of charge also produce fields, and such distributions are very important in practice. We will consider charges that are distributed *uniformly* throughout a region in space, whether a line, a surface, or a volume. Distributions where there is symmetry are also emphasized. For charge distributions that are not uniform or not symmetrical, the problem of determining the resulting electric field can be more complex.

It is useful to set up a general framework for calculating electric fields due to line, surface, and volume distributions. Consider the calculation of the electric field at point P due to the charge distribution shown in Fig. 23–14. We divide the charge distribution into tiny elements, each of charge Δq. We first find the electric field $\Delta\mathbf{E}$ at an external point P that is due to a tiny charge element Δq, whose distance from P is r:

$$\Delta\mathbf{E} = \frac{\Delta q}{4\pi\epsilon_0 r^2}\,\hat{\mathbf{r}}. \tag{23–17}$$

Here $\hat{\mathbf{r}}$ is the unit vector pointing away from the charge element. Superposition applies, and the total electric field at P is found by summing the infinitesimal fields $\Delta\mathbf{E}$:

$$\mathbf{E} = \sum \Delta\mathbf{E}. \tag{23–18}$$

In differential notation, Eq. (23–17) becomes

$$\mathrm{d}\mathbf{E} = \frac{\mathrm{d}q}{4\pi\epsilon_0 r^2}\,\hat{\mathbf{r}}. \tag{23–19}$$

The total electric field is found by integrating over the entire charge distribution:

$$\mathbf{E} = \sum_{\substack{\lim\\ \Delta q\to 0}} \Delta\mathbf{E} = \int \mathrm{d}\mathbf{E} = \frac{1}{4\pi\epsilon_0}\int \frac{\mathrm{d}q}{r^2}\,\hat{\mathbf{r}}. \tag{23–20}$$

Some Specific Cases

The expression for the electric field [Eq. (23–20)] is a formal one. The integral is complicated by the presence of the unit vector $\hat{\mathbf{r}}$, which varies in direction as we integrate over the distributed charge. In addition, to be able to perform the integration, we must know just how the charge is distributed in space. This last step is necessary in order to convert the charge element $\mathrm{d}q$ to a volume element. We can make the conversion for charge distributed uniformly over one-, two-, and three-dimensional regions of space as follows.

One Dimension: Line Segment. If charge q is distributed uniformly along a line segment of length L on the x-axis, we denote the linear charge density (charge/length) by λ:

$$\lambda \equiv \frac{q}{L}. \tag{23–21}$$

The infinitesimal charge contained in an infinitesimal length $\mathrm{d}x$ is

$$\text{for a charged line segment:}\qquad \mathrm{d}q = \lambda\,\mathrm{d}x. \tag{23–22}$$

Two Dimensions: Surface. If charge q is distributed uniformly on a surface of total area A, we denote the surface charge density (charge/area) by σ.

$$\sigma \equiv \frac{q}{A}. \tag{23–23}$$

The infinitesimal charge over the differential area $\mathrm{d}S$ is

$$\text{for a charged surface:}\qquad \mathrm{d}q = \sigma\,\mathrm{d}S. \tag{23–24}$$

Three Dimensions: Volume. If charge q is distributed uniformly throughout a volume V, we denote the volume charge density (charge/volume) by ρ:

$$\rho \equiv \frac{q}{V}. \tag{23–25}$$

If dV is differential volume, the infinitesimal charge is

$$\text{for a charged volume:} \qquad dq = \rho\, dV. \tag{23–26}$$

Examples 23–5 and 23–6 illustrate how the integration for the electric field is performed.

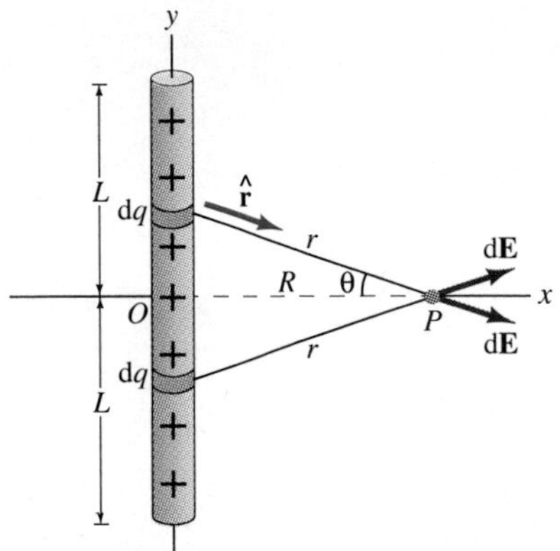

FIGURE 23–15 Example 23–5. The electric field due to a rod of length $2L$ that carries a uniform charge density λ, at a distance R from the rod.

EXAMPLE 23–5 A straight insulating rod of length $2L$ carries a uniform linear charge density λ. Determine the electric field at point P, a distance R from the rod along the perpendicular bisector (Fig. 23–15). First find the field in the limit that the rod is much longer than $R (L \gg R)$. Then find it for a distance very far from the rod $(R \gg L)$.

SOLUTION: Equation (23–22) applies here, because distribution of charge is linear. The origin is the midpoint of the rod, which we place along the y-axis. We use Eqs. (23–20) and (23–22), with $\hat{\mathbf{r}} = \cos\theta\,\mathbf{i} - \sin\theta\,\mathbf{j}$ (Fig. 23–15) to find that

$$\mathbf{E} = \frac{\lambda}{4\pi\epsilon_0}\int_{-L}^{L}\frac{dy}{r^2}(\cos\theta\,\mathbf{i} - \sin\theta\,\mathbf{j}). \tag{23–27}$$

The charge dq at a distance y *below* the x-axis gives rise to a field $d\mathbf{E}$ that is a mirror image of the field $d\mathbf{E}$ due to another charge dq at a distance y *above* the axis. Thus we expect the net y-component of the field to vanish by symmetry. Here we shall demonstrate this formally by performing the integration, although normally we take advantage of the symmetry to reduce the mathematical calculation.

It is often true that the key to performing integrations such as that of Eq. (23–27) is to find the right variables. In this case, the simplest variable to use is the angle θ. Both y and $\mathbf{r}$ depend on θ, and we must change the integration variable from y to θ. We need to find the dependence of both y and r on θ. We have

$$\tan\theta = \frac{y}{R} \tag{23–28}$$

and

$$\cos\theta = \frac{R}{r}. \tag{23–29}$$

From Eq. (23–28) we get

$$dy = R\, d(\tan\theta) = R\sec^2\theta\, d\theta = \frac{R}{\cos^2\theta}\, d\theta.$$

With Eq. (23–29) the combination dy/r^2 that appears in Eq. (23–27) is

$$\frac{dy}{r^2} = \frac{1}{r^2}\frac{R}{\cos^2\theta}\, d\theta = \frac{1}{\cancel{r^2}}\frac{\cancel{R}}{(R/\cancel{r})^2}\, d\theta = \frac{1}{R}\, d\theta.$$

The factor $1/R$ is a constant and comes out of the integral, leaving

$$\mathbf{E} = \frac{\lambda}{4\pi\epsilon_0 R}\int_{-\theta_0}^{\theta_0}(\cos\theta\,\mathbf{i} - \sin\theta\,\mathbf{j})\, d\theta.$$

The limits $-\theta_0$ and θ_0 are the maximum values of θ, corresponding to the two ends of the line of charge. The integrals are elementary. The coefficient of $\mathbf{j}$, which is the y-component of the field, is proportional to $\cos\theta_0 - \cos(-\theta_0) = 0$, and, as we expected, there is no y-component to the field. The coefficient of $\mathbf{i}$

is the x-component,

$$E_x = \frac{\lambda}{4\pi\epsilon_0 R}\int_{-\theta_0}^{\theta_0}\cos\theta\, d\theta = \frac{\lambda}{4\pi\epsilon_0 R}\sin\theta\Big|_{-\theta_0}^{\theta_0}$$

$$= \frac{\lambda}{2\pi\epsilon_0 R}\sin\theta_0. \qquad (23\text{–}30)$$

We can use $\sin\theta_0 = L/\sqrt{L^2 + R^2}$, if desired.

For a rod with length $L \gg R$, $\sin\theta_0 \simeq 1$, and the component E_x from Eq. (23–30) becomes

$$\text{for } L \gg R: \qquad E_x = \frac{\lambda}{2\pi\epsilon_0 R}. \qquad (23\text{–}31)$$

The electric field near a long, uniformly charged, straight rod

Equation (23–31) gives the electric field for an almost infinitely long rod (or for a point very close to a finite rod). The direction of the field is perpendicular to the rod. Notice that Eq. (23–31) states that the electric field varies as $1/R$ for an infinitely long rod, as opposed to the inverse-square dependence ($1/R^2$) for the point charge. The reason is that there is an infinite amount of charge in an infinite rod with a finite charge density. The summation over all the fields from charges in the rod, even the ones that are very distant from the point at which the field is measured, builds up a net field that decreases more slowly than the field of a finite charge distribution.

For the case of $R \gg L$, $\sin\theta_0 \simeq L/R$, and Eq. (23–30) becomes

$$\text{for } R \gg L: \qquad E_x = \frac{\lambda L}{2\pi\epsilon_0 R^2} = \frac{Q}{4\pi\epsilon_0 R^2}, \qquad (23\text{–}32)$$

where $Q = 2\lambda L$ is the total charge on the rod. In this case ($R \gg L$), we have obtained the point-charge result, because a rod of finite length looks like a point when it is viewed from large distances.

EXAMPLE 23–6 Find the electric field at a distance R from an infinite plane sheet with a uniform surface charge density σ.[†]

SOLUTION: Refer to Fig. 23–16a, where we establish the xy-plane in the plane sheet. We want to find the electric field at a point P a distance R above the plane, and we choose the z-axis so that P is on it.

Frequently, a multiple-integration problem can be viewed as an integral over a single-integral result; for example, a double integral over an area may be viewed as an integral over thin strips, each strip being the result of a single integral along its length. Because we have already found **E** due to an infinite wire, we divide the plane into a series of parallel wires, or strips, aligned along the x-axis. Each strip has width dy. Figure 23–16b is the view along the x-axis.

Because σ is the charge per unit area, $\lambda = \sigma\, dy$ is the charge per unit length along each of the parallel strips. From Fig. 23–16a, we see that P is located a closest distance r from each strip, and we can use the results of Example 23–5 for the field at point P due to the pictured strip. Equation (23–31) then gives the field d**E** from the strip. The field has magnitude

$$dE = \frac{\sigma\, dy}{2\pi\epsilon_0 r}.$$

[†] This example is important because it relates to the charged plates of capacitors, elements of electric circuits.

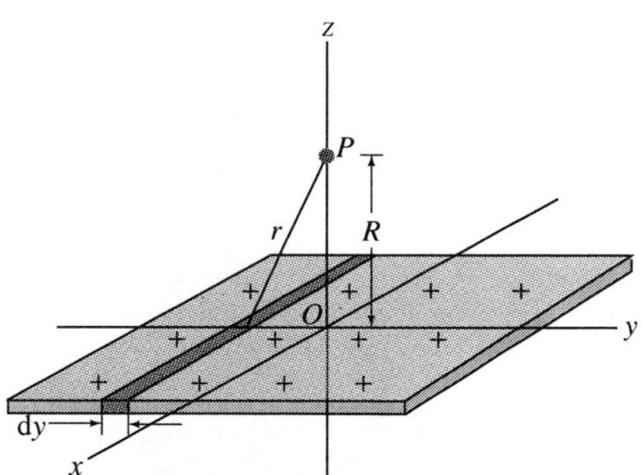

(a)

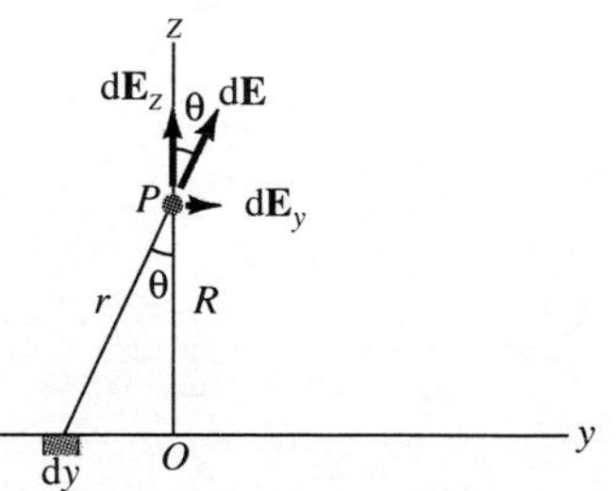

(b)

FIGURE 23–16 Example 23–6. (a) The electric field at a point P a distance R above an infinite charged plane can be found by considering a strip of width dy. (b) The electric field due to a strip of width dy.

The direction of $d\mathbf{E}$ is shown in Fig. 23–16b. We can separate $d\mathbf{E}$ into components

$$dE_y = \frac{(\sin\theta)\sigma\,dy}{2\pi\epsilon_0 r},$$

$$dE_z = \frac{(\cos\theta)\sigma\,dy}{2\pi\epsilon_0 r}.$$

As in Example 23–5, we can see from symmetry that the total component $E_y = 0$, because another strip of width dy on the other side of P will exactly cancel the contribution of the strip we are considering. We need to determine only $\mathbf{E} = E_z\mathbf{k}$:

$$E_z = \int dE_z = \frac{\sigma}{2\pi\epsilon_0}\int_{-\infty}^{\infty}\frac{\cos\theta}{r}\,dy.$$

The integration becomes much simpler if we use θ rather than y as a variable. As in Example 23–5, we have the trigonometric relation $\tan\theta = y/R$, so that again $dy = R\,d(\tan\theta) = R\sec^2\theta\,d\theta = R/\cos^2\theta\,d\theta$. In addition, $r = R/(\cos\theta)$, so the combination that appears in the integral simplifies greatly:

$$\frac{\cos\theta}{r}\,dy = \frac{\cancel{\cos\theta}}{\cancel{R}/(\cancel{\cos\theta})}\,\frac{\cancel{R}}{\cancel{\cos^2\theta}}\,d\theta = d\theta.$$

We need to know the end points of the θ integration, and from the figure we see that when y runs from $-\infty$ to $+\infty$, θ runs from $-\pi/2$ to $+\pi/2$. Thus

$$E_z = \frac{\sigma}{2\pi\epsilon_0}\int_{-\pi/2}^{\pi/2} d\theta = \frac{\sigma}{2\epsilon_0},$$

or

The electric field from a large, uniformly charged plane

$$\text{for a uniformly charged plane:}\qquad \mathbf{E} = \frac{\sigma}{2\epsilon_0}\mathbf{k}. \tag{23–33}$$

The final result has the electric field everywhere perpendicular to the plane and *constant* in both magnitude and direction: The field $\mathbf{E}$ does not even depend on how far the point is from the plane. This is reasonable physically: If the plane is infinite and has a uniform charge distribution, it looks the same from everywhere.

In reality, we cannot have planes of infinite extent. For finite planes, the result above holds for distances much closer to the finite plane than the distance to the edge of the plane.

The Electric Field between Two Uniformly Charged Planes with Opposite Charge

Example 23–6 shows that the electric field for a positively charged plane of uniform surface charge density σ is uniform and directed away from the plane perpendicularly (Fig. 23–17a). If the plane were negatively charged, the field would be similar but would be directed *toward* the plane (Fig. 23–17b). What happens if we place the two planes, oppositely charged but with the same magnitude of charge density σ, parallel to each other? As shown in Fig. 23–17c, the fields outside the parallel planes will exactly cancel each other, but the fields between the planes are additive. The resulting field is shown in Fig. 23–17d. For two parallel, oppositely charged planes, the electric field is zero everywhere except between the

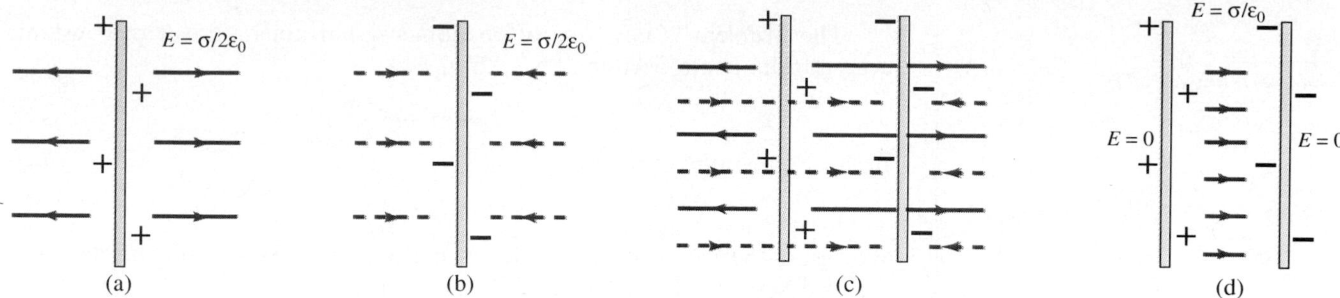

FIGURE 23–17 (a) The electric field due to a positively charged plane is directed away from the plane; (b) that due to a negatively charged plane is directed into the plane. (c) With two parallel planes carrying equal but opposite charge, the electric field cancels to zero outside the planes but is additive inside. (d) The field inside is σ/ϵ_0 and is directed from the positive plane to the negative plane.

planes, where the field has magnitude

$$\text{for parallel planes of opposite uniform charge:} \quad E = \frac{\sigma}{\epsilon_0} \qquad (23\text{–}34)$$

and is directed from the positively to the negatively charged plane (remember that the direction of the electric field is always the direction of the force on our positive test charge q_0).

23–4 MOTION OF A CHARGED PARTICLE IN AN ELECTRIC FIELD

We have been concerned with the construction of the electric field of a given collection of charges. Let us turn now to the force that charged particles will experience in an external electric field. Newton's second law becomes

$$\mathbf{F} = q\mathbf{E}_{\text{ext}} = m\mathbf{a}, \qquad (23\text{–}35)$$

where a particle of mass m and charge q has an acceleration $\mathbf{a}$ due to a given external electric field $\mathbf{E}_{\text{ext}}$. We then solve Newton's second law as usual. Example 23–7 demonstrates the procedure.

EXAMPLE 23–7 Consider two oppositely charged parallel plates (Fig. 23–18). The magnitude of the surface charge density on each plate has a constant value of $\sigma = 1.0 \times 10^{-6}\ \text{C/m}^2$, and the plates are 1.0 cm apart. (a) If a proton is released from rest near the positively charged plate, with what speed will it strike the negatively charged plate? (b) What will the proton's transit time be?

SOLUTION: (a) We first calculate the electric field and the acceleration of the proton; then we can use kinematic relations to determine the speed and transit time. Refer to the coordinate system in Fig. 23–18. The field $\mathbf{E}$ has only an x-component, $E_x = \sigma/\epsilon_0$, according to Eq. (23–34). From Eq. (23–35), the acceleration a_E due to the electric field is

$$a_E = a_x = \frac{qE_x}{m} = \frac{q\sigma}{m\epsilon_0}$$

$$= \frac{(1.6 \times 10^{-19}\ \cancel{\text{C}})(1.0 \times 10^{-6}\ \cancel{\text{C}}/\cancel{\text{m}^2})}{(1.67 \times 10^{-27}\ \text{kg})(8.85 \times 10^{-12}\ \cancel{\text{C}^2}/\text{N}\cdot\cancel{\text{m}^2})} = 1.08 \times 10^{13}\ \text{m/s}^2, \qquad (23\text{–}36)$$

where we use the known charge q and mass m of the proton.

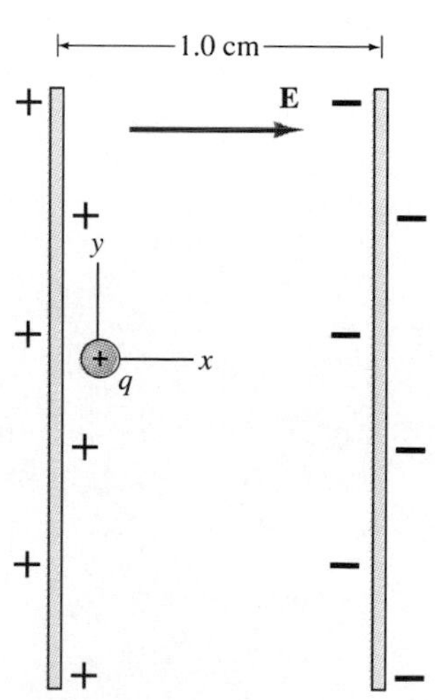

FIGURE 23–18 Example 23–7. A charge $+q$ moving between parallel plates.

The problem is now one of one-dimensional kinematics with constant acceleration. From Section 2–5, with $a_E = a$,

$$v^2 - v_0^2 = 2ax.$$

Because the initial speed v_0 is zero,

$$v^2 = 2ax. \tag{23–37}$$

We insert the value of the acceleration from Eq. (23–36) and the distance traveled between the plates ($x = 1.0$ cm) to find that

$$v^2 = 2(1.08 \times 10^{13}\ \text{m/s}^2)(1.0 \times 10^{-2}\ \text{m}) = 2.2 \times 10^{11}\ \text{m}^2/\text{s}^2;$$

$$v = 4.7 \times 10^5\ \text{m/s}.$$

(b) Because the proton starts from rest, the transit time is found by dividing the final speed by the acceleration:

$$t = \frac{v}{a} = \frac{4.7 \times 10^5\ \text{m/s}}{1.08 \times 10^{13}\ \text{m/s}^2} = 4.3 \times 10^{-8}\ \text{s}.$$

The plates accelerate protons and thus represent a charged-particle accelerator.

Deflection of Moving Charged Particles

Let us consider what happens when we inject a charged particle into a region of uniform **E** between two plates. The particle has initial velocity $\mathbf{v}_0$ perpendicular to **E** (Fig. 23–19). For practical consideration, assume that the particle has negative charge (an electron, for example), so that in Fig. 23–19, it will be deflected up. From Eq. (23–35) the acceleration vector is

$$\mathbf{a} = a_x\mathbf{i} + a_y\mathbf{j} = \frac{qE}{m}\mathbf{j}. \tag{23–38}$$

Note that the x-component of the acceleration is zero. Because the initial velocity is only in the x-direction ($\mathbf{v}_0 = v_0\mathbf{i}$), the velocity vector becomes

$$\mathbf{v} = v_x\mathbf{i} + v_y\mathbf{j} = v_0\mathbf{i} + \frac{qE}{m}t\mathbf{j}. \tag{23–39}$$

The charged particle travels a horizontal length L between the charged plates in

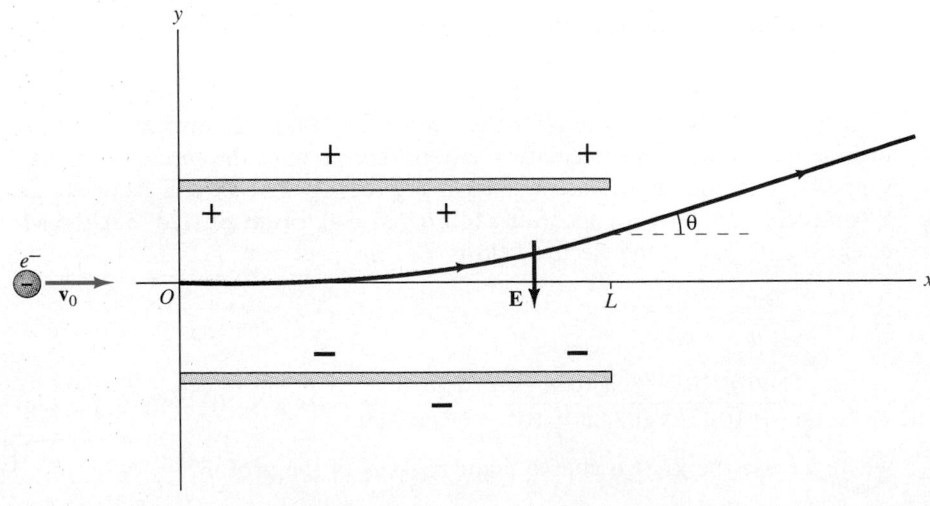

FIGURE 23–19 A particle is deflected when it passes between parallel, oppositely charged plates.

the time T, determined by

$$x = v_0 T = L; \tag{23–40}$$

$$T = \frac{L}{v_0}. \tag{23–41}$$

The particle's deflection in the y-direction is then

$$y = \frac{1}{2} a_y t^2 = \frac{1}{2}\frac{qE}{m} T^2 = \frac{1}{2}\frac{qE}{m}\frac{L^2}{v_0^2}. \tag{23–42}$$

The charged particle emerges from the plates at a position (x, y) given by Eqs. (23–40) and (23–42). The charged particle is then free from the influence of any force (ignoring gravity) and continues past the plates in a straight line at an angle θ from its initial direction:

$$\tan\theta = \frac{v_y}{v_x} = \frac{(qE/m)(L/v_0)}{v_0} = \frac{qEL}{mv_0^2}. \tag{23–43}$$

APPLICATION

The Oscilloscope

The scheme for the deflection of charged particles described in the text is the basis for the functioning of the *oscilloscope* (Fig. B1–1). This device measures the magnitudes and time dependences of electronic signals through an observable deflection of electrons. Figure B1–2 is a schematic diagram of the most important part of an oscilloscope, the cathode-ray tube, or CRT. The tube has two sets of plates to deflect electrons both horizontally and vertically. In an *electron gun*, the electrons leave a heated filament (*cathode*) and are accelerated to an initial speed v_0 (see Example 23–8). They are directed to a fluorescent screen on which their arrival is visible. In normal operation the signal to be studied is converted to an electric field between the vertical-deflection plates, the plates responsible for vertical deflection. Note that these plates are actually horizontal! Another electric field, linearly proportional to time (called a time base), is set up between the horizontal-deflection plates. The change in signal magnitude (vertical) as a function of time (horizontal) can then be observed on the oscilloscope screen.

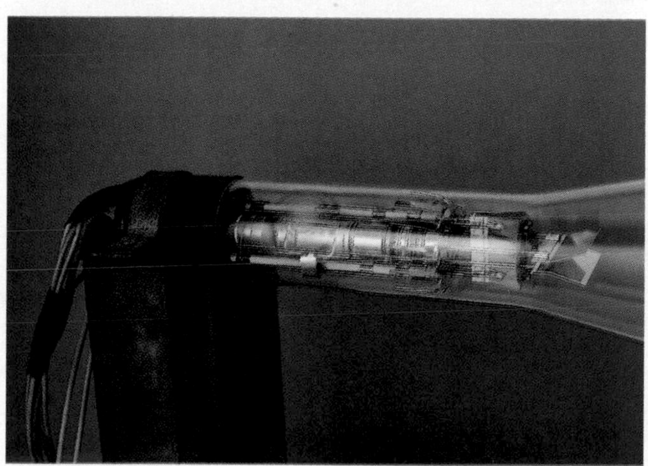

FIGURE B1–1 Cutaway view of an oscilloscope.

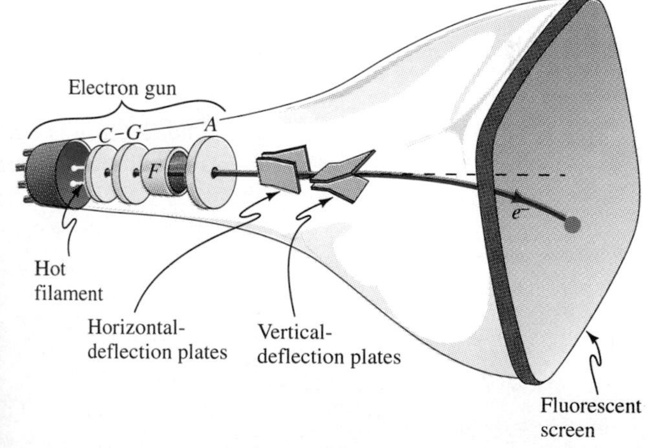

FIGURE B1–2 Schematic diagram of a cathode-ray tube, an element of oscilloscopes, televisions, and computer display terminals. Electrons are emitted from the cathode (point C), a hot filament; controlled by the grid (G); focused by the focusing anode (F); and then accelerated (A) by a high voltage while being formed into beams (collimated) through small apertures. Vertical-deflection plates (which are horizontal) deflect the beam vertically according to the voltage applied between the plates. Voltage is supplied also to the horizontal-deflection plates to sweep the beam regularly across the screen at a rate that can be varied. The electrons land on the screen and cause the spots on which they land to fluoresce.

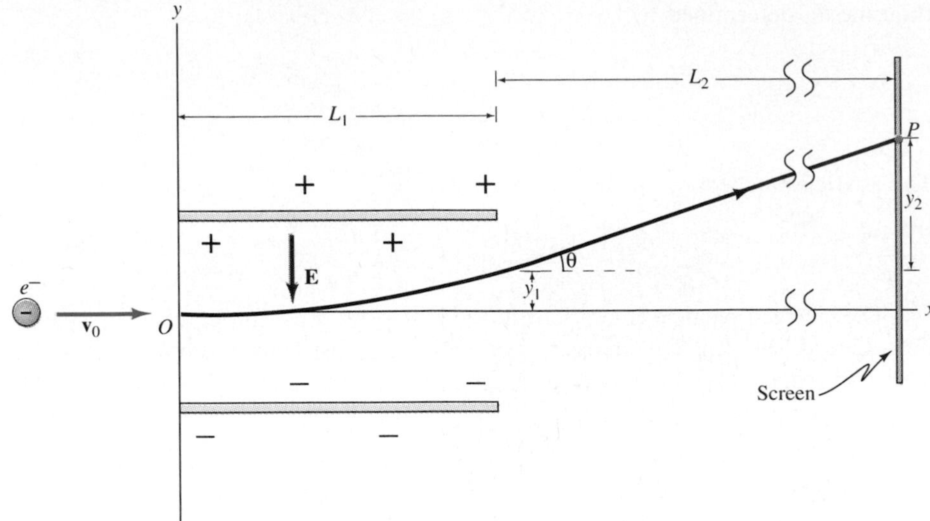

FIGURE 23–20 Example 23–8. An electron passing between vertical-deflection plates.

EXAMPLE 23–8 An electron at a speed $v_0 = 3 \times 10^6$ m/s enters the region between the vertical-deflection plates of a CRT; the plates have length $L_1 = 3$ cm (see "The Oscilloscope" box). A fluorescent screen is located $L_2 = 12$ cm past these plates. Find the electron's total vertical deflection on the screen from its initial direction if the electric field between the plates points downward with a magnitude of $E = 10^3$ N/C. There is no horizontal deflection.

SOLUTION: Figure 23–20 illustrates the situation. The total vertical deflection of the electron is the deflection y_1, acquired in passing between the plates, as well as the additional deflection y_2, a result of the straight path after the electron leaves the plates. We use Eq. (23–42) to find the deflection y_1 of the electron while it is between the plates:

$$y_1 = \frac{1}{2}\frac{qE}{m}\frac{L_1^2}{v_0^2}.$$

The electron is deflected; after it leaves the region between the plates, it travels at an angle to its original direction, given by Eq. (23–43) as $\tan\theta = qEL_1/mv_0^2$. The y-deflection over the distance L_2 to the screen is then

$$y_2 = L_2 \tan\theta = \frac{qEL_1L_2}{mv_0^2}.$$

Finally, the total deflection is

$$y = y_1 + y_2 = \frac{1}{2}\frac{qEL_1^2}{mv_0^2} + \frac{qEL_1L_2}{mv_0^2} = \frac{qEL_1}{mv_0^2}\left(\frac{1}{2}L_1 + L_2\right).$$

Numerical evaluation with a minus sign for E (**E** points downward), gives

$$y = \frac{(-1.6 \times 10^{-19}\ \text{C})(-10^3\ \text{N/C})(3 \times 10^{-2}\ \text{m})}{(9.11 \times 10^{-31}\ \text{kg})(3 \times 10^6\ \text{m/s})^2} \times \left[\frac{1}{2}(3 \times 10^{-2}\ \text{m}) + (12 \times 10^{-2}\ \text{m})\right]$$

$$= 8.0 \times 10^{-2}\ \text{m}.$$

Of this 8.0 cm, the deflection $y_1 = 0.9$ cm, and the deflection $y_2 = 7.1$ cm.

23–5 THE ELECTRIC DIPOLE IN AN EXTERNAL ELECTRIC FIELD

In Section 23–1 we introduced the electric dipole, which has a total charge of zero but a positive and a negative center of charge separated by a distance L. Permanent electric dipoles (for example, polar molecules such as NaCl and H_2O) exist in nature. In addition, many neutral materials, such as the cork balls discussed in Chapter 22, have induced dipole moments when they are in an external electric field. Because of their importance, we want to discuss the reaction of permanent electric dipoles in external electric fields.

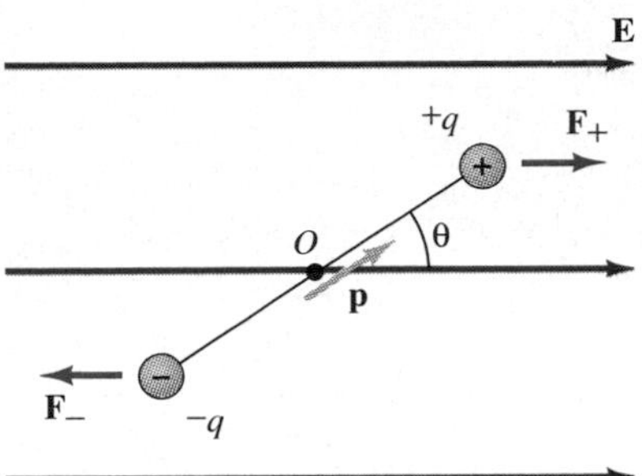

FIGURE 23–21 A dipole placed in a uniform external electric field experiences no net force but can experience a torque.

Consider the permanent electric dipole discussed in Example 23–3. The electric dipole moment of the dipole has a vector character, and we assigned its direction as indicated in Fig. 23–6. The expression for the dipole moment $\mathbf{p}$ is given by Eq. (23–13). We place the electric dipole in a uniform external field (Fig. 23–21). The forces on $+q$ (F_+) and $-q$ (F_-) are

$$\mathbf{F}_+ = q\mathbf{E},$$

$$\mathbf{F}_- = -q\mathbf{E} = -\mathbf{F}_+.$$

We notice that the two forces are equal and opposite and therefore cancel. *There is no net force on the dipole.*

However, there is a torque that tends to rotate the dipole. To calculate the torque and the corresponding rotation, we must choose a reference point, and it is convenient to choose this point to be the midpoint of the dipole, at point O in Fig. 23–21. The actual motion will be independent of the choice of reference point. The torque, $\boldsymbol{\tau}$, about a point due to a force that acts on another point a displacement $\mathbf{r}$ away is given by Eq. (10–6):

Recall from Chapter 10 that the motion of a system on which torque acts is independent of the choice of origin.

$$\boldsymbol{\tau} = \mathbf{r} \times \mathbf{F}, \tag{23–44}$$

where $\mathbf{r}$ is measured from point O. The resulting torque from the force on each charge is then clockwise, with magnitudes

Recall that the right-hand rule determines the direction of a vector product.

$$\tau_+ = \left(\frac{L}{2}\right) qE \sin\theta,$$

$$\tau_- = \left(\frac{L}{2}\right) qE \sin\theta,$$

where the subscripts + and − refer to the charges. Because both τ_+ and τ_- are clockwise rotations, the total torque is also clockwise, with magnitude

$$\tau = \tau_+ + \tau_- = qLE \sin\theta. \tag{23–45}$$

We can represent this expression for the torque on a dipole as the vector product of $\mathbf{p}$ and $\mathbf{E}$:

$$\boldsymbol{\tau} = \mathbf{p} \times \mathbf{E}, \tag{23–46}$$

The torque on a dipole in an electric field

which gives both the magnitude ($pE \sin\theta$) and the direction (into the page in Fig. 23–21) of the torque.

The maximum torque ($\tau = pE$) occurs when $\mathbf{p}$ and $\mathbf{E}$ are perpendicular ($\theta = \pi/2$). The torque is zero when $\mathbf{p}$ and $\mathbf{E}$ are parallel ($\theta = 0$) or antiparallel ($\theta = \pi$). The torque tends to rotate the electric dipole until $\mathbf{p}$ is parallel to $\mathbf{E}$. The position $\theta = 0$ corresponds to a stable equilibrium, but the position $\theta = \pi$ is one of unstable equilibrium, because a small deviation will cause the dipole to rotate toward $\theta = 0$.

TABLE 23–2

AN ELECTRIC DIPOLE ROTATING IN A UNIFORM ELECTRIC FIELD

Electric Field	*Torque, τ*	*Angular Velocity, ω*
	Maximum, into page	Zero
	Decreasing, into page	Increasing, into page
	Zero	Maximum, into page
	Changed direction, out of page, increasing	Decreasing, into page
	Maximum, out of page	Zero
	Decreasing, out of page	Changed direction, increasing, out of page

Without a mechanism to dissipate the dipole's energy, the dipole will oscillate about $\theta = 0$ forever if it starts at some nonzero value of θ. As the dipole rotates toward $\theta = 0$, it is gains kinetic energy and passes through $\theta = 0$ to the other side. The torque, however, then becomes counterclockwise, and the dipole slows down, stops, returns to $\theta = 0$, and passes through it again to the original side. Table 23–2 illustrates a time sequence for a rotating dipole in an electric field.

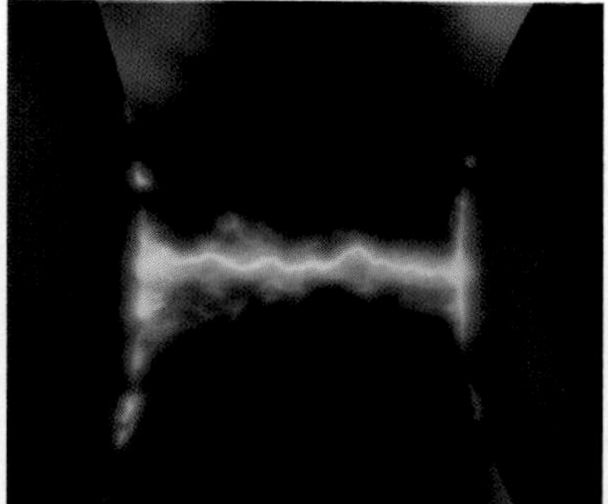

FIGURE 23–22 The electric field between the metal pieces is so large that there is electrical breakdown in the air: a spark.

Breakdown. The forces due to a constant external electric field on the *components* of a dipole, either permanent or induced, tend to pull the dipole apart. Whether this happens or not depends on the strength of the external field compared to the strength of the forces that hold the dipole together. For the molecules of air (mostly nitrogen and oxygen), there is *breakdown* when the external field is roughly 3×10^6 N/C. At this point the components of the induced molecular dipole break into two pieces of opposite charge that are pulled *away* from each other by the external field. As the molecules are broken up, the resulting charged components are accelerated through the air and collide with other molecules, helping to break them up. The result is a cascade of flowing charges that produce a spark (Fig. 23–22). Lightning is a familiar large-scale example of this process.

The Energy of a Dipole in an External Electric Field

Work must be done for an electric dipole to rotate in an external electric field. The electric field, for example, does positive work to rotate the dipole from $\theta = \pi/4$ to $\theta = 0$. An external agent would have to do positive work (and the electric field would do negative work) to rotate the dipole from $\theta = \pi/4$ to $\pi/2$. We learned in Section 10–6 that the work, W, done by the external agent while it exerts a torque τ on the system and moves it from angle θ_0 to θ is given by

$$W = \int dW = \int_{\theta_0}^{\theta} \tau \, d\theta.$$

Thus, for a dipole with dipole moment p,

$$W = \int_{\theta_0}^{\theta} pE \sin\theta \, d\theta = pE(\cos\theta_0 - \cos\theta), \qquad (23\text{–}47)$$

where E is the external electric field. The work performed by the external agent is transformed into potential energy of the electric dipole, so the change in potential energy, $\Delta U = U - U_0$, is

$$U - U_0 = pE(\cos\theta_0 - \cos\theta). \qquad (23\text{–}48)$$

We are free to choose the constant U_0, and we choose it such that $U_0 = 0$ at $\theta_0 = \pi/2$. Thus the potential energy at angle θ is given by

$$U = -pE\cos\theta. \qquad (23\text{–}49)$$

Notice that Eq. (23–49) is consistent with our choice for the zero of the potential energy, because $U_0 = -pE\cos\theta_0$, which is zero when $\theta_0 = \pi/2$.

Equation (23–49) can be written more compactly by using the scalar dot product of **p** and **E**:

$$\text{for a dipole:} \quad U = -\mathbf{p}\cdot\mathbf{E}. \qquad (23\text{–}50)$$

The potential energy of a dipole in an external electric field

Earlier we discussed the stability of the equilibrium of the dipole in an external field. We can see directly from Eq. (23–50) that the orientation in which **p** is aligned with **E** is a point of stable equilibrium, because U has a minimum there. In contrast, U has a maximum when **p** is antiparallel to **E**, and therefore this is a point of unstable equilibrium.

EXAMPLE 23–9 The water molecule, H_2O, has a permanent electric dipole moment of $p \simeq 6 \times 10^{-30}$ C·m (Fig. 23–8). Calculate the maximum torque on a water molecule in a uniform electric field of 10^5 N/C. Also calculate its minimum potential energy, and compare this energy to a typical molecular thermal energy kT when the molecule is part of a gas at a temperature of 400 K. What can you conclude about the alignment of water molecules of the gas at this temperature?

SOLUTION: The calculation of the maximum torque and minimum potential energy is a straightforward application of our results. From Eq. (23–45), the maximum torque occurs for $\theta = \pi/2$ and has the value

$$\tau_{max} = pE = (6 \times 10^{-30}\ \text{C}\cdot\text{m})(10^5\ \text{N/C}) = 6 \times 10^{-25}\ \text{N}\cdot\text{m}.$$

From Eq. (23–49) the minimum potential energy occurs when $\theta = 0$; that is, when **p** and **E** are parallel:

$$U_{min} = -pE = -6 \times 10^{-25}\ \text{N}\cdot\text{m} = -6 \times 10^{-25}\ \text{J},$$

where we have used the previous result for τ_{max}. Because this is a point of stable equilibrium, the dipole moment of the water molecule would tend to align itself with **E** if it were isolated and undisturbed.

Molecular collisions may disturb the molecules significantly. At a temperature T, the average kinetic energy of a molecule is the number of degrees of freedom in its motion times $kT/2$—this is the equipartition theorem (see Chapter 19). We can *estimate* this energy with the expression kT. At $T = 400$ K,

$$kT = (1.38 \times 10^{-23}\ \text{J/K})(400\ \text{K}) \simeq 6 \times 10^{-21}\ \text{J}.$$

The average kinetic energy is some 10^4 times greater than the minimum potential energy due to the interaction of the dipole in the electric field. Thus random, disorienting collisions between molecules with kinetic energies 10^4 times greater than the alignment energies will mask any tendency for the molecules to line up in the field.

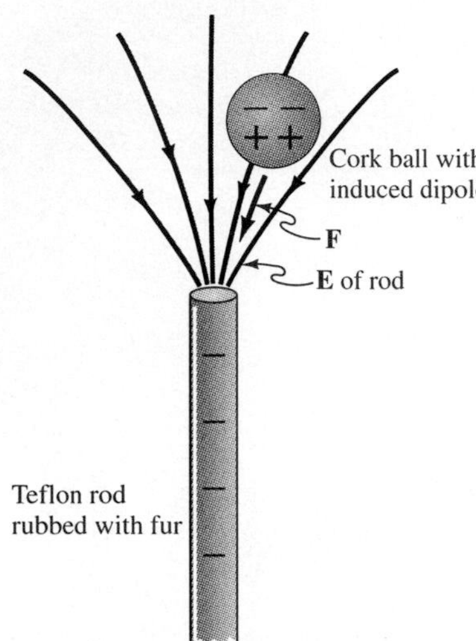

FIGURE 23–23 A dipole placed in a nonuniform external electric field can experience a net force. In this case the external electric field induces a dipole in the cork ball, which then experiences a force due to the electric field. This effect can be understood by using Coulomb's law.

The Electric Dipole in a Nonuniform Electric Field

If a dipole is placed in a nonuniform external electric field, then, in addition to a torque, there may be a net force on the dipole. The resulting motion would be a combination of linear acceleration and rotation. The details of the motion depend critically on the particular electric field configuration.

The effect of a nonuniform electric field on an induced dipole explains the attraction of a neutral cork ball coated with conducting paint to a Teflon rod rubbed against fur, as discussed in Section 22–1. The charged Teflon rod induces an electric dipole on the cork ball, which is in the nonuniform electric field of the rod. The attraction can be explained equally well by the direct use of forces on the dipole as well as by the Coulomb force between like and unlike charges (Fig. 23–23). Another example of the action of a nonuniform electric field on an induced dipole is the attraction between small bits of paper and a comb that has just been charged by passing it through hair. The bits of paper have induced dipole moments and are attracted to the comb in its nonuniform field.

SUMMARY

Charge distributions set up electric fields in the space around them. The electric field vectors can be mapped out by moving a small, positive test charge q_0 around in this field. The field **E** is defined as the force **F** on this test charge, divided by q_0:

$$\mathbf{E} \equiv \lim_{q_0 \to 0} \frac{\mathbf{F}}{q_0}. \qquad (23\text{–}2)$$

The electric field has units of N/C (or V/m). The force on a point charge q' in a given external electric field is

$$\mathbf{F} = q'\mathbf{E}_{\text{ext}}. \qquad (23\text{–}7)$$

Electric field lines aid in the visualization of the direction and magnitude of the electric field produced by various charge configurations. Electric field lines begin and end on positive and negative charges, respectively, but are otherwise continuous. At a given point an electric field line is tangent to the electric field at that point, and the density in area of the electric field lines is proportional to the strength of the field.

From Coulomb's law, the electric field due to a point charge q is

$$\mathbf{E} = \frac{q}{4\pi\epsilon_0 r^2}\,\hat{\mathbf{r}}. \qquad (23\text{–}5)$$

Electric fields obey the principle of superposition:

$$\mathbf{E}_{\text{net}} = \mathbf{E}_1 + \mathbf{E}_2 + \mathbf{E}_3 + \cdots = \sum_i \mathbf{E}_i, \qquad (23\text{–}10)$$

where $\mathbf{E}_i$ labels the field of the components that make up a charge distribution.

In its simplest form, an electric dipole consists of a positive charge q separated by a distance L from a negative charge $-q$. Such a configuration, or a configuration that is electrically neutral but has an imbalance of positive and negative charge from one side to another, occurs often in nature and produces an electric field. This field decreases with distance r as $1/r^3$ and, for the simple dipole, is proportional to the magnitude of the electric dipole moment $\mathbf{p}$, given by

$$\mathbf{p} = q\mathbf{L}. \qquad (23\text{–}13)$$

The direction of $\mathbf{L}$ (and of $\mathbf{p}$) is from the negative charge to the positive charge. This direction determines the angular dependence of the electric dipole field.

The electric field due to a continuous charge distribution is

$$\mathbf{E} = \frac{1}{4\pi\epsilon_0}\int \frac{dq}{r^2}\,\hat{\mathbf{r}}. \qquad (23\text{–}20)$$

Here r is the distance of a charge element dq from the point where the field is measured. To use this result, it is necessary to know how dq varies throughout space.

The electric field due to an infinitely long wire is radial and perpendicular to the wire. A charged plane, infinite in area, with a charge per unit area of σ has an electric field that is uniform and directed perpendicular to the plane, with

$$E = \frac{\sigma}{2\epsilon_0}. \qquad (23\text{–}33)$$

In addition to producing an electric field, an electric dipole experiences a torque in a uniform external electric field:

$$\boldsymbol{\tau} = \mathbf{p} \times \mathbf{E}. \qquad (23\text{–}46)$$

The dipole in the external field has a potential energy of

$$U = -\mathbf{p}\cdot\mathbf{E}. \qquad (23\text{–}50)$$

QUESTIONS

1. Why do gasoline trucks drag metal wires along the road?
2. Why can two electric field lines never cross?
3. We have introduced the concept of an electric field. Why might it be useful to introduce an analogous gravitational field?
4. An inflated rubber balloon is charged by rubbing it with fur. Explain what happens when that balloon is placed against (a) a metal wall; (b) an insulating wall.
5. Electric field lines originate from positive charges and terminate at negative charges, as exemplified by the field lines due to a dipole. Does this statement contradict the depiction of field lines due to a single positive point charge?
6. A pair of equal and opposite charges forms a dipole, and the electric field of a dipole is not zero. But if we were to look at a dipole from very far away, the two charges would appear to be on top of one another and to cancel; that is, we would see no charge, and hence we should see no field. How do you reconcile these statements?
7. Can the electric dipole induced on a spherical conducting ball cause the ball to rotate? How about the electric dipole induced on a long rod?

8. After you comb your hair, the comb can often attract small pieces of paper. The act of combing may induce an electric charge on the comb, but the combing does not itself affect the paper. What accounts for the attraction?
9. Explain how the water molecule, H_2O, acts as an electric dipole (see Fig. 23–8), given that there are *two* spatial regions (around the H atoms) with negative charge.
10. Explain why the electric-field-line technique would not be useful for a point charge if Coulomb's experiments had shown the electric force to decrease as $1/r$ or as $1/r^{2+\delta}$.
11. The internal motion of a liquid can be described by a velocity field, which is the velocity vector of the element of fluid at a given point. In what ways is this field like an electric field, and in what ways is it different?
12. Can you invent an arrangement of charges whose electric field would be radially directed into a point in some region of empty space? The correct answer has implications for the stability of charges placed in static electric fields.
13. Suppose that a small electric dipole ($+q$ and $-q$) is placed somewhere on a line that is perpendicular to, and bisects, a second (fixed) dipole ($+Q$ and $-Q$), as seen in Fig. 23–24. If the small dipole is free to pivot about its center, what will it do?

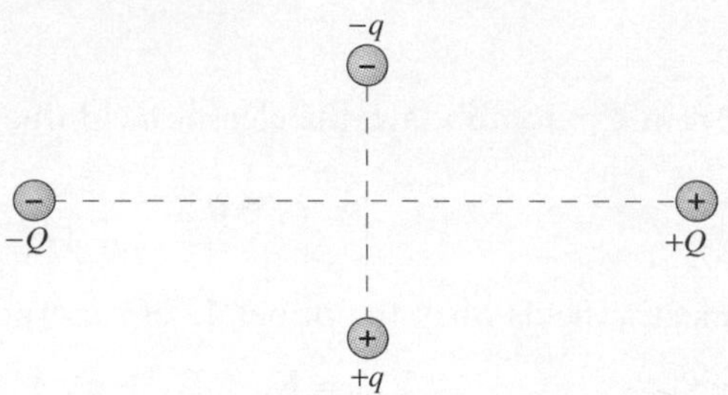

FIGURE 23–24 Question 13.

14. Consider a large number of identical dipoles centered in the xy-plane and pointing in the z-direction, distributed with uniform density. What is the electric field in the limit that the dipoles form a continuous distribution?
15. A large, flat, positively charged plate (of uniform charge density) is placed on the ground. A positively charged pellet, starting from rest, is released from above the plate. Ignore all air resistance. Qualitatively describe the motion of the pellet, according to the height from which it is dropped.
16. Suppose that a positively charged pellet is dropped from above onto the north pole of a large, positively charged sphere (of uniform charge density). Disregarding air resistance, and any small instabilities that would make the pellet move away from the vertical, describe the motion of the pellet.

PROBLEMS

23–1 *Electric Field*

1. (I) A 3-μC charge is located at $(x, y) = (0 \text{ cm}, 3 \text{ cm})$. Determine the electric field at (4 cm, 9 cm).
2. (I) Calculate the electric field at the origin due to the following distribution of charges: $+q$ at $(x, y) = (a, a)$, $+q$ at $(-a, a)$, $-q$ at $(-a, -a)$, and $-q$ at $(a, -a)$.
3. (I) A charge of -12 μC is located at the point $x = 0$ m, and a second charge, $q = 0.5$ nC, at the point $x = 0.1$ m. What are the magnitude and direction of the electric field (a) at $x = 1$ m? (b) at $x = 0.11$ m? (c) at $x = 0.10001$ m? (d) at $x = 0.09999$ m?
4. (II) Charges of $+2$ μC, -4 μC, -6 μC, and $+8$ μC are located at the four corners of a square 4 cm on each side (Fig. 23–25). Calculate the electric field at the center of the square.

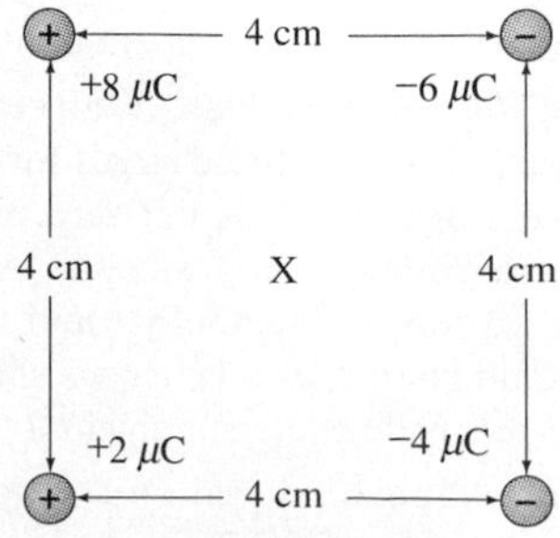

FIGURE 23–25 Problem 4.

5. (II) Calculate the electric field at a point 3.5 cm vertically above the center of the square in Problem 4.
6. (II) A charge $-q$ is located at $y = -\ell/2$, and a second charge $+q$ is located at $y = +\ell/2$ (Fig. 23–26). (a) What is the electric field at the origin? (b) If the charge at $-\ell/2$ were instead $+q$, what would the electric field be at the origin? (c) For part (b), what would the electric field be in the entire xz-plane specified by $y = 0$?

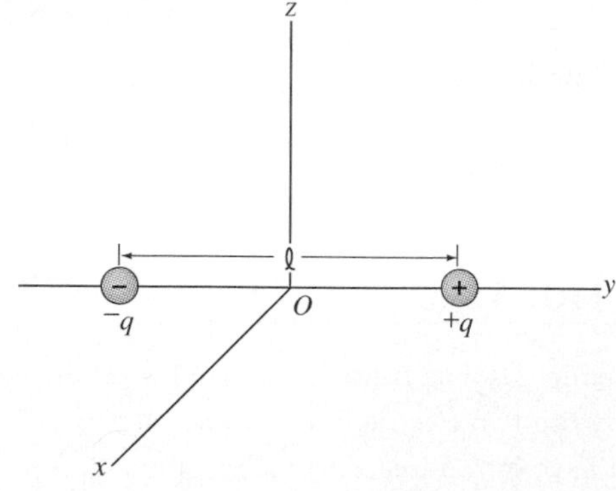

FIGURE 23–26 Problem 6.

7. (II) Identical charges Q are placed at $x = a$ and $x = -a$, respectively. (a) What is the electric field at $x = 0$? (b) Suppose that a positive test charge q_0 is placed at $x = 0$. Will it be in stable or unstable equilibrium? [*Hint:* Assume that the test charge is displaced a distance δ in a direction per-

pendicular to the x-axis. What will the net force on the test charge be at the new location?]

8. (II) Calculate the electric field *along the axis* of a dipole at a distance r from the center of the dipole shown in Fig. 23–5. Work out the field for $r \gg L$.

9. (II) A succession of n alternating positive and negative charges q are placed along the x-axis, each a distance d from its neighbors. The arrangement is symmetrical about the y-axis, with the first $+q$ charge at $x = d/2$, the first $-q$ charge at $x = -d/2$, the second $-q$ charge at $3d/2$, the second charge $+q$ at $-3d/2$, and so forth (Fig. 23–27). What is the field at a distant point $y = Y$ (where $Y \gg nd$) on the y-axis?

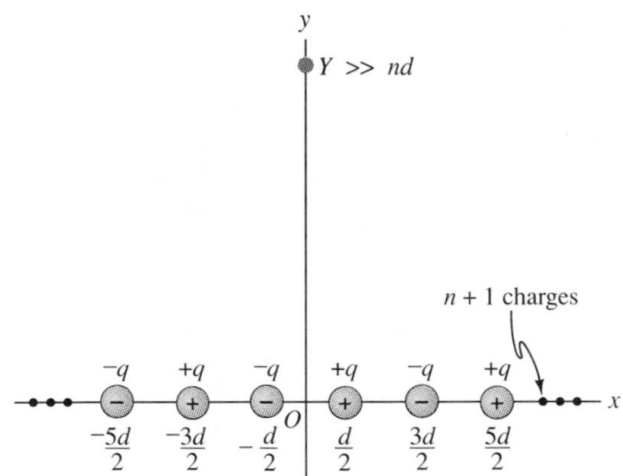

FIGURE 23–27 Problem 9.

10. (III) Suppose that the positive test charge in Problem 7 is constrained to move along the x-axis only. Will $x = 0$ be a stable equilibrium position? If it is, then the test charge should oscillate about $x = 0$ for small enough displacements. If that were the case, what would the frequency of oscillation be for a test charge of mass m? [*Hint:* Assume that the charge is displaced to a point $x = \delta$, where $\delta \ll a$, and calculate the magnitude and the direction of the electric field there. Use the approximation $1/(a + \delta)^2 = (1/a^2) - (2\delta/a^3) + \cdots$, valid for $\delta \ll a$.]

23–2 *Electric Field Lines*

11. (I) Draw the electric field lines due to charges of $+3$ C and $+1$ C located 4 cm apart.

12. (II) Consider charges q placed along the x-axis at $x = na$, with $n = 0, \pm 1, \pm 2, \pm 3, \ldots$. Sketch the electric field lines.

13. (II) The field lines due to an electric dipole **p** are shown in Fig. 23–12b; by definition, the direction of **p** points from $-q$ to $+q$. Sketch the field lines for (a) a dipole $-$**p** adjacent and parallel to the dipole **p**; (b) a dipole **p** adjacent and parallel to the dipole **p**; (c) a dipole $-$**p** on the axis of **p** some distance away past the $-q$ charge; (d) a dipole **p** on the axis of **p** some distance away past the $-q$ charge.

23–3 *The Electric Field Due to a Continuous Charge Distribution*

14. (I) Calculate the electric field due to an infinitely long, thin, uniformly charged rod with a charge density of 4 μC/m at a distance 50 cm from the rod. Assume that the rod is aligned with the x-axis.

15. (I) A thin rod uniformly charged with a total charge of 5 μC and length 10 cm is placed along the z-axis, centered at the origin. Find the electric field at $(x, y, z) = (5$ cm, 0 cm, 0 cm) and (0 cm, 5 cm, 0 cm).

16. (I) Sketch the electric field lines between a point charge Q and a uniformly charged, flat square of area L^2 and total charge $-Q$. The point charge is located a distance L above the center of the plane.

17. (II) A negative charge is distributed uniformly on a long cylindrical shell. Sketch the field lines both inside and outside the shell. Do not include the ends of the cylinder.

18. (II) Consider positive charges distributed uniformly with a charge density λ on a circle of radius R. (a) Use symmetry arguments to deduce the direction of the electric field at a point in the plane of the circle but outside the circle. (b) What is the magnitude of the electric field at a distance L along the axis of the circle for $L \gg R$?

19. (II) A rod with a uniform negative charge is bent into a semicircle. Make a rough sketch of the electric field lines in the plane of the rod.

20. (II) Two infinite plates with a uniform charge density of 3 μC/m^2 are placed along the yz-plane with one plate passing through $x = 2$ cm and the other through $x = -2$ cm. Determine the electric field at $(x, y, z) =$ (a) (0 cm, 0 cm, 0 cm); (b) (5 cm, 0 cm, 0 cm); (c) (5 cm, 2 cm, 3 cm).

21. (II) Two large, flat, vertically oriented plates are parallel to each other, a distance d apart. Both have the same uniform positive charge density σ. What is the electric field in the space around and between them?

22. (II) The axis of a hollow tube of radius R and length L is aligned with the y-axis; the tube's left-hand edge is at $y = 0$, as shown in Fig. 23–28. It carries a total charge q distributed uniformly along its surface. By integrating the result for a field due to a hoop of charge along the axis of the hoop (see Example 22–6), find the electric field along the y-axis due to the tube as a function of y.

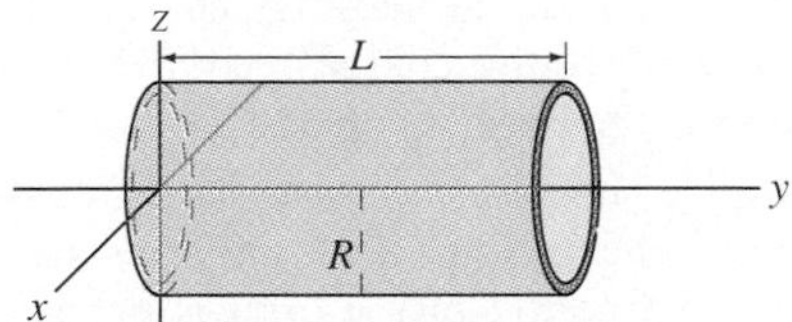

FIGURE 23–28 Problem 22.

23. (II) A total charge Q is distributed uniformly over a rod of length L. The rod is aligned on the x-axis, with one end at the origin and the other at the point $x = L$. Calculate the electric field at a point $(0, D)$, and compare this result with the field at the point $(L/2, D)$.

24. (II) A thin, circular disk of radius R is oriented in the xy-plane with its center at the origin. A charge Q on the disk is distributed uniformly over the surface. (a) Find the electric field due to the disk at the point $z = z_0$ along the z-axis. (b) Find the field in the limit $z_0 \to \infty$. (c) Find the field in the limit that $R \to \infty$. Are the limits of parts (b) and (c) the same?

25. (II) Consider a thin, uniformly charged rod 50 cm long that is bent into a semicircle. The total charge on the rod is 2 μC. What are the magnitude and direction of the electric field at the center of the semicircle?

26. (II) A rod 80 cm long is charged uniformly with a charge density of 40 μC/m. A charge of 20 μC is placed 80 cm from the midpoint of the rod along a line perpendicular to the rod. Calculate the electric field at a point halfway between the point charge and the center of the rod.

27. (III) Consider a point at a height z_0 directly above the midpoint of a square with sides of length $2L$. The (nonconducting) square carries a uniform charge density σ. (a) Use the method of Example 23–6 to write an integral for the electric field at z_0. (b) How does the integral simplify in the limit $L \to \infty$? (c) $z_0 \to 0$?

23–4 *Motion of a Charged Particle in an Electric Field*

28. (I) An infinite plate carries a uniform charge density $\sigma = 6.42 \times 10^{-7}$ C/m^2. A pellet of mass 4.75 g is placed at rest 0.866 m from the plate. The pellet carries a negative charge $q = -3.69 \times 10^{-6}$ C. What is its speed when it reaches the plate? Ignore all forces except the electrostatic attraction.

29. (I) A large, flat plate with unknown, uniform charge density σ is placed on a horizontal tabletop. A cork ball of mass 1.55 g, carrying a charge 4.5×10^{-7} C, is placed at rest above the plate and remains at rest. What is σ?

30. (II) Consider an infinite wire with uniform charge density λ along the z-axis. A negatively charged particle moves in a circle in the xy-plane centered on the wire. Calculate the particle's speed, and show that the speed is independent of the radius of the circle. Ignore all forces except those due to the wire.

31. (II) A negative charge $-q$ is restricted to move in a plane in which there is a continuous line of positive charge and a charge density λ. The negative charge, of mass m, can pass the line of positive charge freely. What is the equation of motion for the negative charge?

32. (II) A positive charge q can travel in a circular orbit about a negatively charged line with uniform charge density λ. Show that the period of the orbit is proportional to the radius of the orbit. Compare this to the dependence of the period of a circular orbit on the radius of the orbit for a point charge that interacts with another point charge.

33. (II) In the Olympic games of the year 2020, a diver of mass 70 kg who carries a positive charge of 1 μC dives with zero initial velocity from 50 m above a uniform sheet of positive charge and charge density σ. Assume the sheet to be infinite. What must σ be so that the diver takes 1 min to fall to the sheet's surface?

34. (II) A cork ball of mass 0.40 g is placed between two large horizontal plates. The bottom plate has a uniform charge density of $+0.80 \times 10^{-6}$ C/m^2, whereas the upper plate has a uniform charge density of -0.50×10^{-6} C/m^2. The cork ball, which carries an unknown charge, is placed between the plates and is observed to float motionlessly. What are the sign and magnitude of the charge on the ball?

35. (II) Consider the cathode-ray tube of Example 23–8. This time an electron enters the region between the vertical-deflection plates with a total speed of $v_0 = 3.0 \times 10^6$ m/s. The direction is such that the velocity has a vertical component $v_{0y} = +3.0 \times 10^5$ m/s. Find the total vertical deflection of the electron when it reaches the screen.

36. (II) A cork ball of mass 5 g, carrying a charge of -2 μC, is suspended from a string 1 m long above a horizontal, uniformly charged plate of charge density 1 μC/m^2. The ball is displaced from the vertical by a small angle and allowed to swing. Show that the ball moves in simple harmonic motion, and calculate the angular frequency of that motion.

37. (III) A proton moves at speed $v = 5 \times 10^5$ m/s in the $+x$-direction and enters a certain region. An electric field in the region also is oriented in the $+x$-direction. The field's strength drops linearly with x: At the beginning of the region, $x = 0$ m, the field strength is 500 N/C; at $x = 3$ m, the field strength is zero. How much time does it take for the proton to traverse this region? [*Hint:* The equation of motion will be more familiar in terms of the variable $x' = x - 3$.]

23–5 *The Electric Dipole in an External Electric Field*

38. (I) An electric dipole consists of two opposite charges of magnitude 2 μC placed 10 cm apart (Fig. 23–29). The dipole

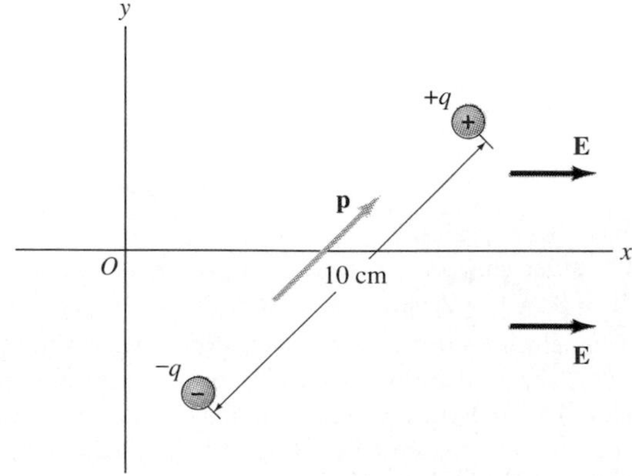

FIGURE 23–29 Problem 38.

is placed in a uniform electric field of 10 N/C along the x-axis, with the direction of **p** at an angle of $+45°$ from the x-axis in the xy-plane. Determine the torque on the dipole.

39. (I) The magnitude of the two opposite charges that form an electric dipole is decreased by a factor of 10 while the separation between the charges is halved. What is the change in magnitude of the torque on the dipole in a uniform electric field?

40. (II) Describe the motion of the dipole in Problem 38. How much work does the electric field do when the dipole moves from its initial position to alignment with the electric field?

41. (II) Assume that the electrons from the hydrogen atoms of H_2O each spend half the time in the spheres of the hydrogen and oxygen atoms. If the electric dipole moment of H_2O is $p = 6 \times 10^{-30}$ C·m and the angle between the hydrogen atoms and the oxygen atom is 105° (see Fig. 23–8), what is the distance between (a) each hydrogen atom and the oxygen atoms; (b) the hydrogen atoms? [*Hint:* Calculate the electric field at a point far away, and obtain an expression for the coefficient of $1/r^3$.]

42. (II) A molecule of lithium fluoride (LiF) has a permanent dipole moment. The molecule is placed in a uniform electric field of strength 10^4 N/C, and the difference between the maximum and minimum potential energies of the molecule in this field is 4.4×10^{-25} J. What is the electric dipole moment of the LiF molecule?

General Problems

43. (II) A point charge $-q$ is fixed at the center of a hollow spherical conductor of charge $+q$. Draw the electric field lines both inside and outside the sphere.

44. (II) A point charge $+q$ is fixed at the center of a hollow spherical conductor also of charge $+q$. Draw the electric field lines both inside and outside the sphere.

45. (II) Draw the electric field lines for a point charge $+q$ near an infinitely long, positively charged wire.

46. (II) A cork ball of radius 0.5 cm and a charge of +5.0 nC is covered with conducting paint. What is the electric field strength just outside the surface? A uranium nucleus, with a radius of 10^{-14} m, has a positive charge of $92e$. What is the electric field strength just outside the surface of the nucleus?

47. (II) A charge of 2 μC is located at the position $(x, y) = (2, 0)$. A charge of -3 μC is located at $(-3, 0)$. All positions are in centimeters. Find (a) the electric field at (0, 4) due to each charge; (b) the force due to each charge on a charge of 1 nC at (0, 4); (c) the total force on the 1-nC charge at (0, 4); (d) the total electric field at (0, 4).

48. (II) Two infinitely long, uniformly charged rods, with charge densities of λ and $-\lambda$, respectively, are lined up parallel to each other and separated by a distance R. What are the magnitude and direction of the electric field due to the two rods at points that lie (a) on a line joining the two rods, and (b) along a perpendicular bisector of that line? Draw a figure to show the configuration, and use symmetry.

49. (II) What is the force per unit length that one of the two rods in Problem 48 exerts on the other?

50. (II) Two uniformly charged infinite plates with charge densities 2 μC/m^2 and -3 μC/m^2 are placed at right angles, the first one along the xz-plane, the second along the yz-plane. A test particle of mass 1 g and charge 2×10^{-7} C is placed a distance of 1 m from both planes; that is, its initial position is $(x, y, z) = (1\text{ m}, 1\text{ m}, 0\text{ m})$. What is the location of the test particle after a time t?

51. (II) Two infinite lines of charge density 3 μC/m are parallel to the z-axis. One line passes through $(x, y) = (2\text{ cm}, 0\text{ cm})$; the other, through $(x, y) = (-2\text{ cm}, 0\text{ cm})$. Find (a) the electric field at the origin; (b) the force on a 1-μC charge at the origin; (c) the force on a 2-μC charge located at $(x, y) = (6\text{ cm}, -4\text{ cm})$.

52. (II) A proton with kinetic energy of 10^6 eV is fired perpendicular to the face of a large metal plate that has a uniform surface charge density of $\sigma = 5.0 \times 10^{-6}$ C/m^2. (a) Calculate the magnitude and direction of the force on the proton. (b) How much work must the electric field do on the proton to bring it to rest? (c) From what distance should the proton be fired so that it stops right at the surface of the plate?

53. (II) The electric charge with the smallest magnitude that can be isolated is the charge on the electron or the proton. In 1909, Robert A. Millikan developed a classic method to measure this charge, known as the *oil drop experiment.*† Millikan was able to place charges on tiny droplets of oil, which would fall at a given terminal velocity under the influence of gravity and air drag. By placing these droplets between parallel, horizontal charged plates, as in Fig. 23–30, the electric field between the plates produces a force on the charged droplet that is directed upward and can partly cancel the gravitational force. If the mass and size of the droplet are known, then, by seeing how fast droplets fall with and without the electric field, the charge can be measured.

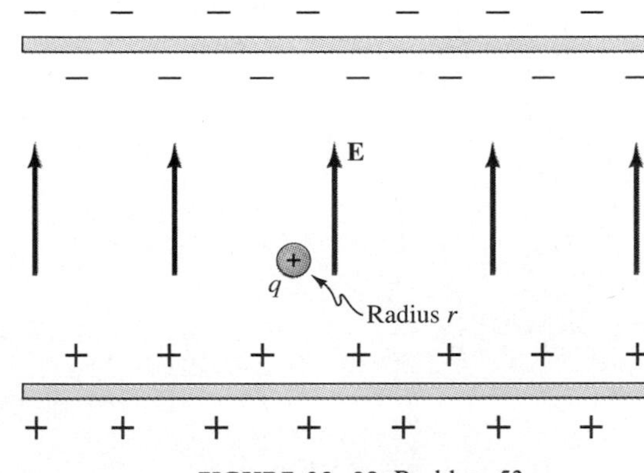

FIGURE 23–30 Problem 53.

† Millikan measured many droplets with different amounts of charge. He found that there was a minimum charge; all the charges he observed were integer multiples of the minimum charge. Millikan interpreted that minimum charge as the charge on the electron (if the charge was negative) or the proton (if positive).

The drag force on a droplet of radius r that falls at a steady speed v through air is also directed upward and is given by *Stokes's law*, $F_{drag} = 6\pi\eta r v$, where η is the viscosity of air. (a) Show from Newton's second law that the terminal velocity v_0 of the *uncharged* drop is $v_0 = \frac{2}{9}r^2\rho g/\eta$, where ρ is the density of the oil and g is the acceleration due to gravity. (b) Suppose that the charge on the drop, q, is positive and that the field is directed vertically upward, as in the figure, so that the electric force points up. Show by using Newton's second law that the charge is given by

$$q = \frac{18\pi(v_0 - v_1)}{E}\sqrt{\frac{v_0\eta^3}{2\rho g}},$$

where v_1 is the terminal velocity when the electric field E is imposed. (c) Take the minimum charge as 1.6×10^{-19} C, the oil's density as 0.85 g/cm^3, and the radius of the droplet as 2.0×10^{-4} cm. The droplet has the minimum charge. Find the value of E that will hold the droplet stationary between the plates.

54. (II) We will learn in Chapter 24 that the electric field near a conductor *must be perpendicular to the conducting surface.* Using this fact, draw the electric field lines for the following configurations: (a) a point charge $+q$ above an infinite, uncharged conducting plane; (b) a point charge $-q$ near an infinitely long, positively charged conducting wire. (c) a point charge $+q$ a distance $L/2$ above a charged conducting plane of area L^2 and charge $+q$.

55. (II) The field due to a line of uniform charge density λ varies with a radial distance r from the line as $1/r$. Suppose that a point charge q of mass m is placed at rest a distance R from the line, and that the force on the point charge due to the field of the line is attractive. Use dimensional analysis to calculate how the time it will take for the charge to drop to the charged line depends on λ, q, m, R, and ϵ_0.

56. (III) Consider the straight, nonuniformly charged rod of length L aligned along the x-axis, with the ends at $x = \pm L/2$, in Example 22–7. We showed there that the force on a charge q located at a point $x = R$ on the x-axis, to the right of the right-hand end of the rod, is

$$\mathbf{F} = \frac{q\lambda_0}{2\pi\epsilon_0 L}\left\{\ln\left[\frac{R-(L/2)}{R+(L/2)}\right] + R\left[\frac{1}{R-(L/2)} - \frac{1}{R+(L/2)}\right]\right\}\mathbf{i}.$$

Show that for $R \gg L$, the force reduces to that of a dipole acting on q, $\mathbf{F} \simeq (q\lambda_0 L^2/12\pi\epsilon_0 R^3)\mathbf{i}$. What is the dipole moment? [*Hint:* Use the approximate forms $(1 - x)^{-1} \simeq 1 + x + x^2 + x^3 + \cdots$ and $\ln(1 + x) \simeq x - (x^2/2) + (x^3/3) - \cdots$, both appropriate for $x \ll 1$.]

57. (III) The field of an electric dipole decreases as $1/r^3$ when the distance of a given point to the dipole, r, is much larger than the separation between the charges. The only way to arrange two charges with a total charge of zero is to form a dipole. There are, however, many ways to arrange four charges with a total charge of zero in a compact pattern. An arrangement with an electric field that behaves at great distances as $1/r^4$ is an *electric quadrupole.* (a) For four charges aligned with alternating sign (such as $+ - - +$ so that the combination acts like dipoles of opposite orientation) along an axis, show that the field on the axis perpendicular to the line of charges decreases as $1/r^4$, where r is much larger than any separation distance within the quadrupole. [*Hint:* Use the approximation

$$(r^2 + \delta^2)^{-3/2} \simeq \frac{1}{r^3} - \frac{3\delta^2}{2r^5} + \cdots$$

(good when $\delta \ll r$) for each of the four charges; do not forget the sign of the charge in determining the field.] (b) For the arrangement shown in Fig. 23–31, sketch the field, using the field-line technique.

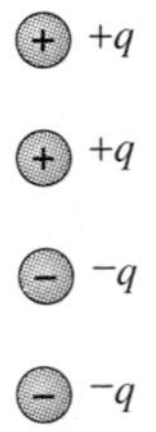

+ $+q$

+ $+q$

− $-q$

− $-q$

FIGURE 23–31 Problem 57.

CHAPTER 24

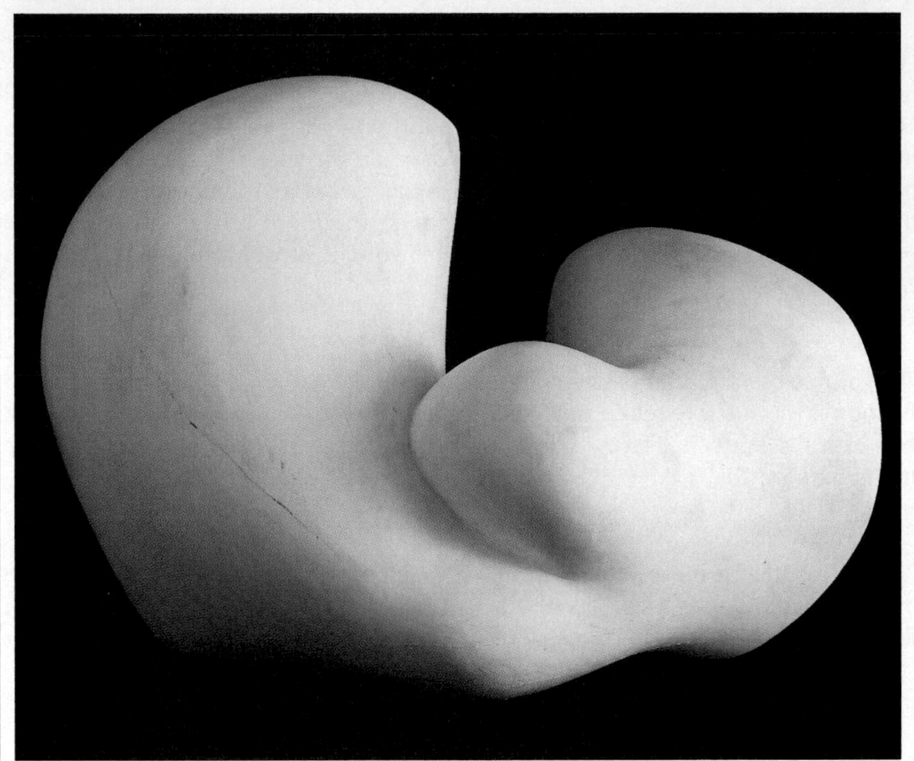

Gauss' law relates the electric charge enclosed by a closed surface of any shape—even the shape of the plaster sculpture Human Concretion by Jean Arp—to an integral of the electric field over the surface.

GAUSS' LAW

Gauss' law is a fundamental re-expression of Coulomb's law.† In Chapter 23 we learned the meaning of the electric field and how to use Coulomb's law to calculate the field due to a stationary distribution of charges. As well as being a fundamental law in electromagnetism, Gauss' law facilitates the calculation of electric fields in many cases. In particular, it greatly simplifies the calculation of electric fields when there is symmetry in the charge distribution. Gauss' law also gives us insight into the behavior of conductors. To use Gauss' law, we need to be able to calculate and use the electric flux, a quantity analogous to the flux we encountered in our study of fluid flow. Also, we shall examine the extent to which Gauss' law and Coulomb's law have been verified experimentally.

† Among the contributions to physics of Karl Friedrich Gauss, a great mathematician of the nineteenth century, are his work on celestial mechanics, electromagnetism, optics, and the theory of errors.

24–1 ELECTRIC FLUX

Flux was first discussed when we studied fluid flow in Chapter 16.

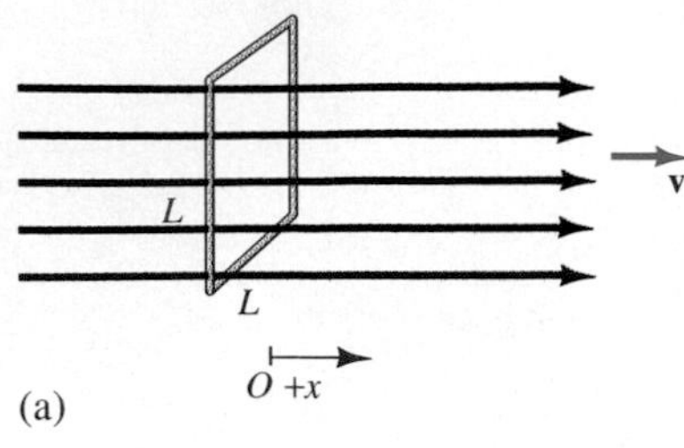

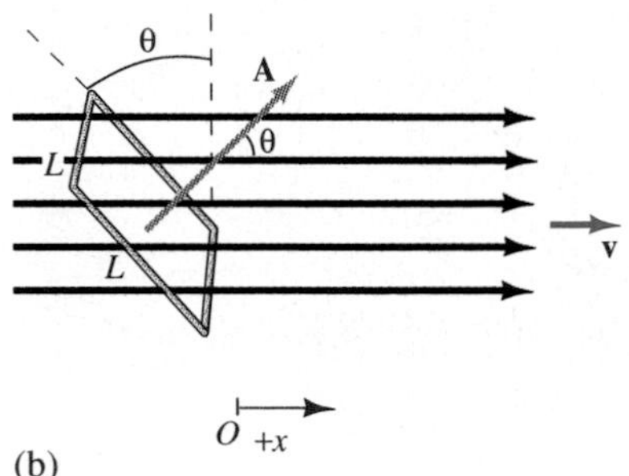

FIGURE 24–1 (a) A jet of water moving in the x-direction passes through a square loop. (b) When the loop is tilted, less water flows through it, by a factor $\cos\theta$. The vector **A** is perpendicular to the loop.

The concept of flux goes beyond its application to the electric field. The electric flux is easiest to understand by analogy with the flow of water. Consider a flowing river. We assume that the water flows smoothly with a uniform velocity in the horizontal, or x-, direction (Fig. 24–1a). We could dip into the water a wire loop in the form of a square of area $A = L^2$. Depending on the orientation of our loop, different volumes of water would pass through the loop in a given time. If the loop is oriented perpendicular to the flow, the volume of water that flows through the loop in 1 s is given by Φ_w, where

$$\Phi_w = vL^2 = vA.$$

The volume of water that passes per unit time is the *flux*. Less water passes through the loop if it is tilted to make an angle θ with the vertical, as in Fig. 24–1b, because the vertical length that faces the stream is reduced from L to $L\cos\theta$. Thus the flux is reduced to

$$\Phi_w = vL^2\cos\theta = vA\cos\theta.$$

We can find a more general form for the flux by assigning a direction to the area of the loop. The area becomes a vector **A** with magnitude A and direction perpendicular to the loop (Fig. 24–1b). We can also remove the restriction that the velocity **v** is in the x-direction. Because $\mathbf{v}\cdot\mathbf{A} = vA\cos\theta$, where θ is now the angle that **A** makes with the water velocity, we can express the flux as

$$\Phi_w = \mathbf{v}\cdot\mathbf{A}.$$

If the surface is not a flat, square one, but an irregular one like a butterfly net (Fig. 24–2), whose orientation is different in different places, then we can find the total flux by summing the fluxes through infinitesimal areas d**A**. Each area forms a tiny plane, and the direction of d**A** is perpendicular to its plane. We sum the infinitesimal fluxes $d\Phi_w = \mathbf{v}\cdot d\mathbf{A}$ by using integration. Thus the flux of water, the volume that passes through some nonplanar surface S per unit time, is

$$\Phi_w = \iint_S d\Phi_w = \iint_S \mathbf{v}\cdot d\mathbf{A}. \tag{24–1}$$

We use the double integral sign to emphasize that we are integrating over a two-dimensional surface. Note that in this expression the velocity **v** may be different at each point over the surface. We can indicate that **v** varies with position by writing it as a function of y and z: $\mathbf{v}(y, z)$. For example, suppose that water flows in the $+x$-direction, but with a speed that depends on y and z. If the surface S through

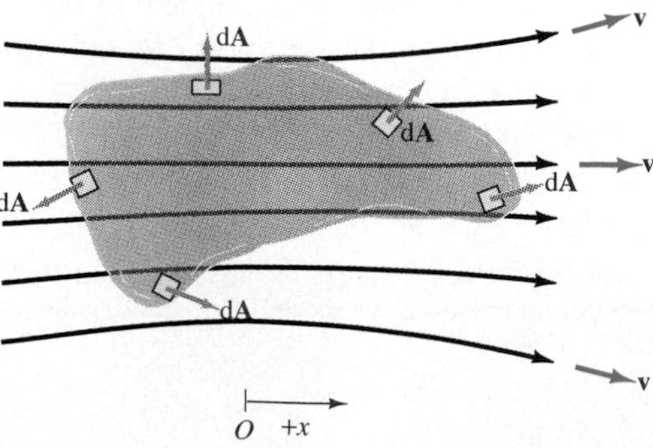

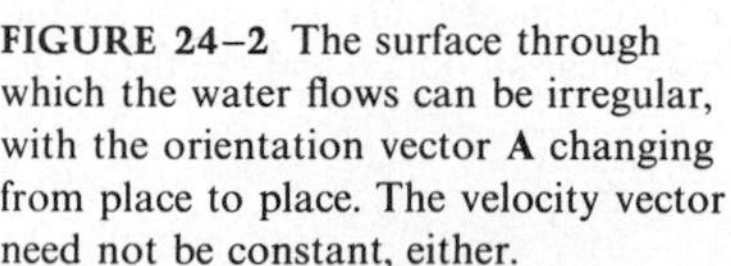
FIGURE 24–2 The surface through which the water flows can be irregular, with the orientation vector **A** changing from place to place. The velocity vector need not be constant, either.

which the flux is calculated is in the yz-plane, the differential cross-sectional area is $dA = dy\,dz$, the direction of $\mathbf{dA}$ is the x-direction (so that $\mathbf{v} \cdot \mathbf{dA} = v\,dA$), and the flux is

$$\Phi_w = \iint_S v(y, z)\,dy\,dz. \tag{24–2}$$

There is a similarity between fluid flow and the electric field. This can be seen in a comparison of streamlines in a fluid (see Fig. 16–19) and electric field lines (Fig. 24–3). We can extend the notion of flux to the electric field. The **electric flux**, Φ (or Φ_E, when we need to distinguish it from some other kind of flux), is defined as

Definition of electric flux

$$\Phi \equiv \iint_S \mathbf{E} \cdot \mathbf{dA}. \tag{24–3}$$

The electric flux proves to be an enormously useful quantity: We can use it to help find electric fields, and it appears in the formulation of the fundamental laws of electricity and magnetism.

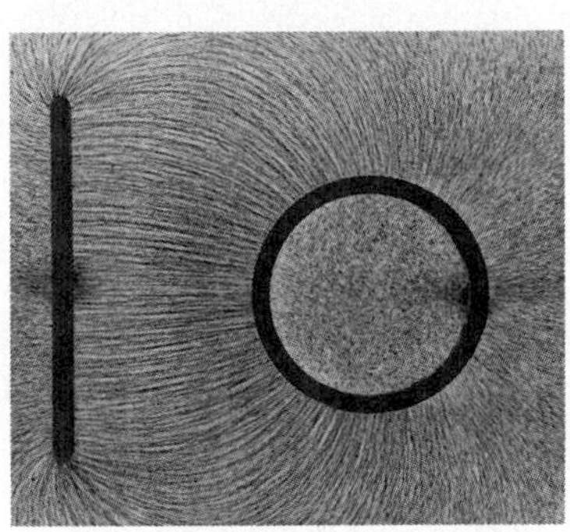

FIGURE 24–3 The electric field lines due to a charged conducting cylinder close to an oppositely charged conducting plate, shown by threads in oil.

The similarity between fluid flow and electric flux is not perfect. Although moving water may actually pass through a surface, electric fields do not represent anything that moves physically. *No physical movement is involved in the electric flux.* Note also that the "surface" that we use to calculate the flux is generally imaginary. No actual body has to form the surface. We shall be imagining many different surfaces for our convenience.

As we discussed in Chapter 23, the magnitude of the electric field is proportional to the density of electric field lines through an area perpendicular to the lines. Let N be the number of electric field lines that pass through a surface S, where the area perpendicular to $\mathbf{E}$ is $A_\perp$. For this simple calculation, we suppose that E is constant over the surface area, so that $E \propto N/A_\perp$, and $N \propto EA_\perp = \Phi$. Because the flux defined in Eq. (24–3) is proportional to the electric field times the area through which it passes, that equation tells us that *the electric flux through a surface is proportional to the number of electric field lines that pass through the surface.*

The electric flux through a surface is proportional to the number of field lines that pass through the surface.

The Gaussian Surface

As we shall see, in order to use Gauss' law, we need to determine the electric flux through a closed surface. Such surfaces, which will generally be imaginary, may have the shape of a sphere, or a cylinder, or any other shape. We refer to such an imaginary surface as a **Gaussian surface**. Figure 24–4 shows a Gaussian surface together with electric field lines that pass into and out of the surface area. The electric flux passing through a closed surface has the same form as the flux through an open surface (such as the surface that spans our wire loop), with only one refinement: We define the direction of an infinitesimal surface element $\mathbf{dA}$ of the Gaussian surface to be perpendicular to the surface and *pointing to the outside of the closed surface.* Thus, from Eq. (24–3), the electric flux through a closed surface is

$$\text{through a closed surface:} \qquad \Phi = \oiint d\Phi = \oiint \mathbf{E} \cdot \mathbf{dA}, \tag{24–4}$$

where the circle on the double integral indicates that we are integrating over a closed surface.

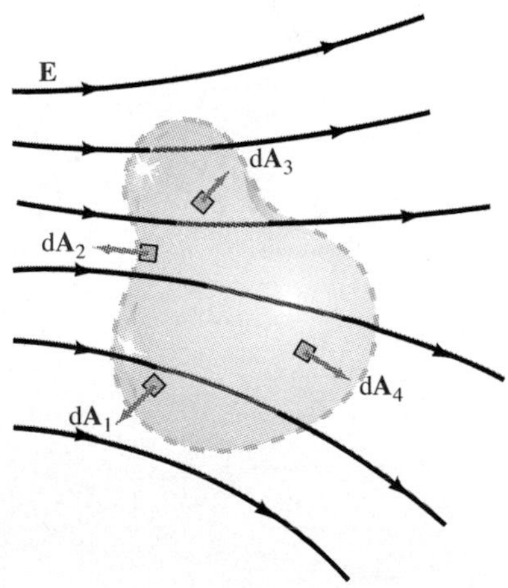

FIGURE 24–4 By convention, the directions of the areas $\mathbf{dA}$ are away from, and perpendicular to, the surface area. The electric field lines pierce the closed surface area. The total electric flux for this surface, which is dashed to remind you that it is imaginary, is zero.

Figure 24–4 shows the direction of area elements $\mathbf{dA}$ at four different spots on the Gaussian surface. Notice that for $\mathbf{dA}_1$ and $\mathbf{dA}_2$, $\mathbf{E}$ and $\mathbf{dA}$ are oriented so that the product of $\mathbf{E} \cdot \mathbf{dA}$ is negative; for $\mathbf{dA}_3$ and $\mathbf{dA}_4$, the dot product is positive. The flux is negative for the part of the surface where the electric field lines enter the closed surface and positive where the electric field lines exit the surface. The total flux for the case shown in Fig. 24–4 is zero, because all the field lines that enter the

FIGURE 24–5 A closed fish net in water is a mechanical analogue to a Gaussian surface in an electric field. Unless there is a source (a faucet) or sink (a drain) within the net, all the water that flows into the net flows out.

surface exit it. We do not need to do the difficult integration indicated by Eq. (24–3) for this case, because physical reasoning is much easier. In Chapter 23 we learned that electric field lines must originate from and end on charges. If we have a Gaussian (closed) surface that surrounds no charges, then no electric field lines can originate from or end inside that closed surface. The same number of field lines that enter must exit, so the integration in Eq. (24–4) must be zero. Thus, *if there is no charge inside a closed surface, the electric flux through the surface is zero.* Here again, the analogy to fluid flow is helpful. Suppose that our closed surface is formed by a fish net, and that net is placed in a river (Fig. 24–5). If there is no source (for example, a faucet) or sink (for example, a drain) inside the net, then all the water that flows into the net must flow out of it. The water flowing in the river in which the net is placed is analogous to the electric field in the region where the Gaussian surface is placed.

Gauss' Law Refers to Net Charge

We can again make use of the fluid analogy to argue that the flux through a Gaussian surface is also zero if no *net* charge is enclosed by the surface. Consider a closed wire basket inside a river. Two hoses lead into the interior of the basket. One hose pumps water in at a certain rate, while the other hose pumps water out at the same rate. The flowing river is analogous to an external electric field. The end of the inflow hose is analogous to a positive electric charge. The end of the outflow hose is analogous to a negative electric charge. Again, all the water that flows into the region enclosed by the basket flows out, and the flux of water through the basket is zero. In just the same way, if a closed surface surrounds equal amounts of positive and negative charge, then the electric flux through that surface is zero. Figure 24–6 shows the electric field due to a dipole, described in Chapter 23. Imagine a series of Gaussian surfaces of any convenient shape placed wherever we

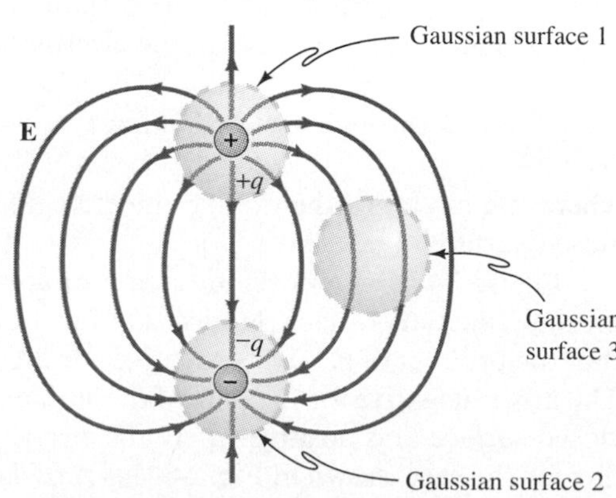

FIGURE 24–6 Three Gaussian surfaces (imaginary and therefore dashed) in the electric field of a dipole. For surface 1, which surrounds the $+q$ charge, the electric flux is positive; for surface 2, which surrounds the $-q$ charge, the flux is negative; and for surface 3, which surrounds no charge, the flux is zero.

choose. For example, if we place an imaginary Gaussian surface (surface 1) around charge $+q$ of a dipole, all the electric field lines exit the Gaussian surface, and the total electric flux is positive. If we place a second Gaussian surface (surface 2) around charge $-q$, all the electric field lines enter the Gaussian surface, and the electric flux is negative. Any Gaussian surface, such as surface 3, that surrounds *neither* charge has no net electric flux through it, because the same number of electric field lines enter and exit such a surface. If the Gaussian surface surrounds *both* charges, again the number of field lines that enter and exit the surface is equal, and the total flux is zero. This observation is an important one. Our result applies to any Gaussian surface that surrounds a configuration of charges, as long as there is no *net* charge. We summarize:

The electric flux through a closed surface that encloses no net charge is zero.

The electric flux through a closed surface is zero if the surface encloses no net charge.

Remember, we are not referring to a *real* closed body that is inserted into the electric field lines. A Gaussian surface is purely an imaginary closed surface that we place wherever we choose. The electric flux through the closed surface depends on whether or not there is a net electric charge inside the Gaussian surface, and, if there is, on its magnitude and sign. This is the basis of Gauss' law.

24–2 GAUSS' LAW

We have seen that the electric flux through a closed surface that encloses no net charge is zero. When the surface does enclose a net charge, the electric flux through the surface is not zero, and *Gauss' law* expresses that flux in terms of the charge enclosed. We begin by looking at the flux through a Gaussian surface that encloses a point charge. Figure 24–7 shows an (imaginary) Gaussian sphere of radius R centered on a static point charge q. The centered sphere is chosen because the electric field has constant magnitude at a fixed distance from a charge, and it will be easy to find the flux through the sphere. We use Eq. (24–3) to find the electric flux that passes through the Gaussian surface. The electric field due to a point charge q was found in Eq. (23–5) to be

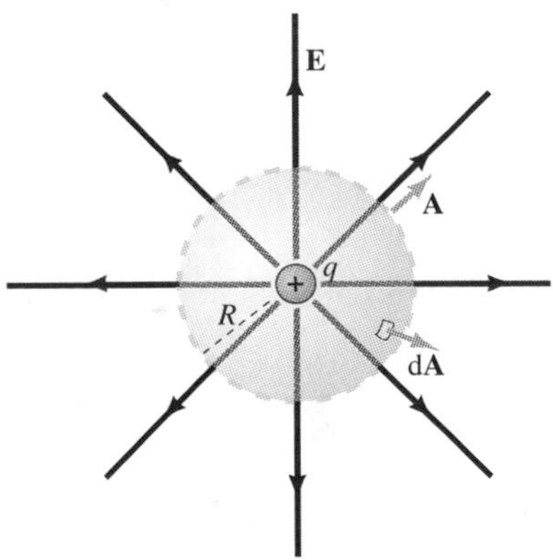

FIGURE 24–7 A simple choice for a Gaussian surface for a point charge q is a sphere of radius R.

$$\mathbf{E} = \left(\frac{q}{4\pi\epsilon_0 r^2}\right)\hat{\mathbf{r}}.$$

The electric field points in the radial direction, outward for positive q. Because the direction of the infinitesimal area $\mathbf{dA}$ for a small area on the sphere also points outward in the radial direction, the product $\mathbf{E}\cdot\mathbf{dA} = E\,dA$. Because the electric field has the constant value $q/4\pi\epsilon_0 R^2$ everywhere on our sphere, the infinitesimal electric flux through the infinitesimal area dA is

$$d\Phi = E\,dA = \frac{q}{4\pi\epsilon_0 R^2}\,dA.$$

We can now pull the (constant) field E outside the integral sign in the integral that represents the total flux:

$$\Phi = \oint\!\!\oint d\Phi = \oint\!\!\oint \mathbf{E}\cdot\mathbf{dA} = \oint\!\!\oint E\,dA = \oint\!\!\oint \frac{q}{4\pi\epsilon_0 R^2}\,dA = \frac{q}{4\pi\epsilon_0 R^2}\oint\!\!\oint dA.$$

The integral over the closed surface of dA is just the area of the closed surface, $A = 4\pi R^2$. Thus

$$\Phi = \frac{q}{4\pi\epsilon_0 R^2}A = \frac{q}{4\pi\epsilon_0 R^2}4\pi R^2 = \frac{q}{\epsilon_0}. \qquad (24\text{–}5)$$

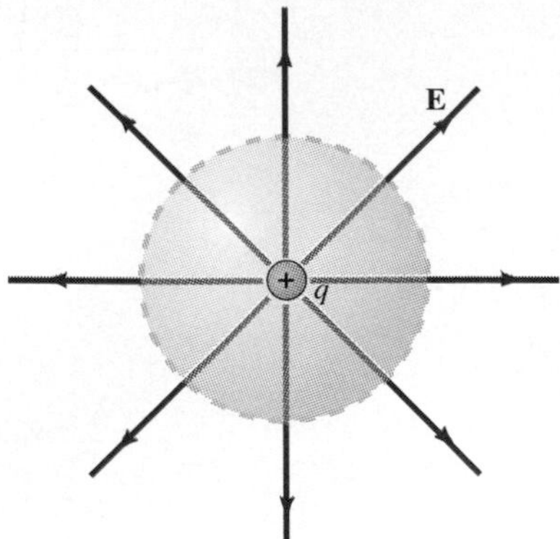

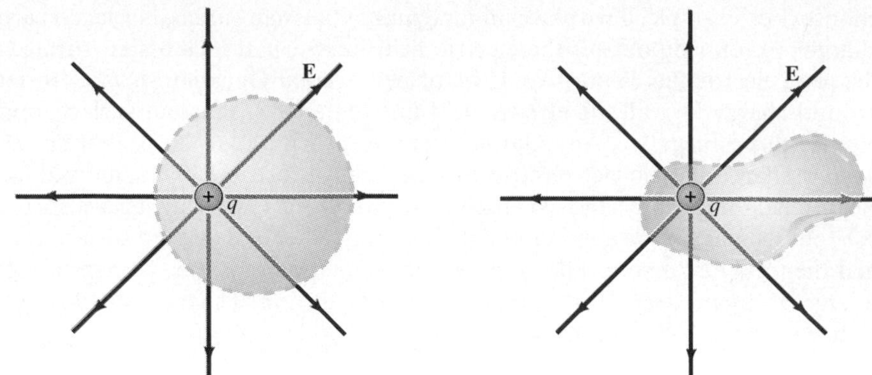

FIGURE 24–8 All the Gaussian surfaces shown give the same result for the electric flux. The same number of electric field lines pierce each surface.

This result is independent of the radius of our Gaussian sphere. The electric flux emanating from a point charge is q/ϵ_0.

We have considered a sphere centered on a charge and already have the surprising result that the electric flux is independent of the sphere's radius. But we can go much further and show that the flux through *any* closed surface that surrounds the charge gives this same result! Such surfaces include Gaussian spheres off center from the point charge, or, indeed, any regular or irregular surface around the charge (Fig. 24–8). To calculate the flux by direct analytic or numerical integration for such surfaces could be a monumental task! In order to establish our result, recall that we have shown in Section 24–1 that the flux through a surface is proportional to the number of electric field lines that pass through that surface. Now, because electric field lines originate from and stop on charges, the number of electric field lines that pass through any surface that surrounds our single charge is the same as the number of electric field lines that pass through a sphere centered on the point charge. So the flux through any of these Gaussian surfaces is exactly the same, as Fig. 24–8 illustrates. Equation (24–5) is thus established for any Gaussian surface *S as long as S surrounds the point charge q*:

The electric flux through any closed surface that encloses a point charge q is proportional to q.

$$\oiint \mathbf{E} \cdot d\mathbf{A} = \frac{q}{\epsilon_0}. \tag{24–6}$$

Gaussian surface 3 in Fig. 24–6 is one case in which the charge is outside the Gaussian surface. Here, as many lines leave as enter, and the charge gives no net flux through the surface.

We need to generalize Eq. (24–6) for the case of multiple point charges and continuous charge distributions. We know that the *total or net charge Q* can be broken down into an assembly of point charges q_i. And from the superposition principle, we know that the total electric field $\mathbf{E}$ is the sum of the fields $\mathbf{E}_i$ due to point charges q_i. The total flux Φ through a Gaussian surface due to the net charge is then just the sum of the fluxes Φ_i due to the charges q_i:

$$\Phi = \sum_i \Phi_i = \oiint \sum_i \mathbf{E}_i \cdot d\mathbf{A} = \frac{1}{\epsilon_0} \sum_i q_i = \frac{Q}{\epsilon_0}.$$

Our result is generally referred to as **Gauss' law**,

Gauss' law

$$\oiint \mathbf{E} \cdot d\mathbf{A} = \frac{Q}{\epsilon_0}. \tag{24–7}$$

The closed surface is *any* Gaussian surface that surrounds the *net* charge Q. The case in which the net charge is zero is included here, either because no charge whatsoever is enclosed by S or because there is an equal amount of positive and negative charge.

We can now more easily understand the example of the dipole discussed in Section 24–1. We now know that the electric flux through the surface that surrounds only the negative charge $-q$ is $-q/\epsilon_0$; we know that the flux through the surface that surrounds only the positive charge $+q$ is $+q/\epsilon_0$. And we also know why the flux through any other closed surface, including surfaces that surround both charges, is zero.

Coulomb's Law and Gauss' Law

Our treatment of Gauss' law has followed from Coulomb's law, because the derivation of Gauss' law used the electric field of the point charge that was determined by using Coulomb's law. This procedure can be reversed, and we can derive Coulomb's law from Gauss' law. To do so, we center a Gaussian sphere on a point charge q (Fig. 24–7). The electric field of the charge, **E**, is assumed to be unknown. Gauss' law tells us only that the electric flux integrated over the surface of the sphere is q/ϵ_0. This alone is insufficient to determine the field, because the flux through any tiny surface element of the sphere depends on the value of the field in that region. However, we can use a symmetry argument. All directions around a point charge should be equivalent. The only configuration of field around a charge that does not favor some particular direction is a radial field. The surface element **dA** of a Gaussian sphere is also radial. If we assume that, at all locations, **E** is along the direction of **dA**, we have at worst made a sign error, and we can repair this later. Thus

$$\mathbf{E} \cdot \mathrm{d}\mathbf{A} = E\,\mathrm{d}A.$$

Moreover, symmetry, or the assumption that there is no preferred direction, also implies that **E** will have the same magnitude everywhere on the centered sphere. We can then remove E from the integral that expresses the total flux through the sphere:

$$\oiint \mathbf{E} \cdot \mathrm{d}\mathbf{A} = \oiint E\,\mathrm{d}A = E \oiint \mathrm{d}A = EA = E(4\pi r^2) = \frac{q}{\epsilon_0},$$

where r is the radius of the Gaussian sphere. The last term in this equality is just Gauss' law. The equation can be solved for the magnitude of the electric field:

$$E = \frac{q}{4\pi\epsilon_0 r^2}.$$

This result is consistent with Eq. (23–5). Because E is positive, we correctly chose the direction of **E** to be radially outward for a positive charge. The symmetry of the situation tells us only that the electric field must be radial, either outward or inward. Gauss' law determines the orientation of **E** to be radially outward. Coulomb's law follows directly from the previous equation if we put another charge, q', in the electric field and use $\mathbf{F} = q'\mathbf{E}$.

The derivation of Coulomb's law from Gauss' law for a point charge is a particularly simple application of the use of Gauss' law and symmetry. We implied in Chapter 23 in our discussion of the usefulness of fields that Coulomb's law becomes difficult to use when the charges are moving, because that law involves distances between charges, but "information" about the distance propagates at a finite speed. In fact, Coulomb's law as such ceases to make sense for rapidly moving

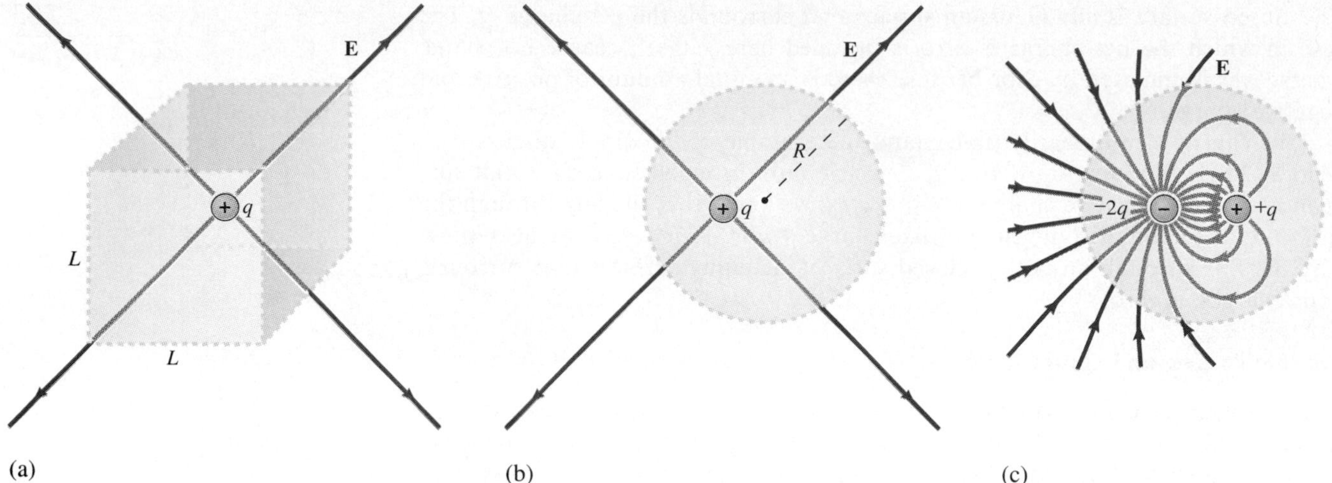

FIGURE 24–9 Example 24–1.

charges, whereas Gauss' law continues to hold. Gauss' law is more general than Coulomb's law.

In Examples 24–1 and 24–2, we use the fact that Gauss' law does not require us to use any particular surface. This is important because the flux through one surface may be much easier to calculate than the flux through another.

EXAMPLE 24–1 Find the electric flux through the Gaussian surfaces in Fig. 24–9: (a) a cube of sides L that surrounds the charge q; (b) a sphere of radius R that surrounds the charge q, which is off center; (c) a sphere of radius b that surrounds the charges $-2q$ and $+q$.

SOLUTION: (a) We do not need to do the direct integration of the electric field over the cube. According to Gauss' law, the total electric flux is simply q/ϵ_0, because q is the net charge enclosed by the cubic Gaussian surface. The shape of the surface is immaterial, as is the position of the charge inside.

(b) It does not matter that the Gaussian sphere is off center. The total electric flux is still q/ϵ_0.

(c) We need not concern ourselves with the positions of the two charges within the cube. The total net charge Q enclosed by the Gaussian surface is $-2q + q = -q$, and the total electric flux through the Gaussian surface is $-q/\epsilon_0$.

EXAMPLE 24–2 Consider a point charge $q = 1$ mC placed at a corner of a cube of sides 10 cm in an electric field **E**. Determine the electric flux through each face of the cube.

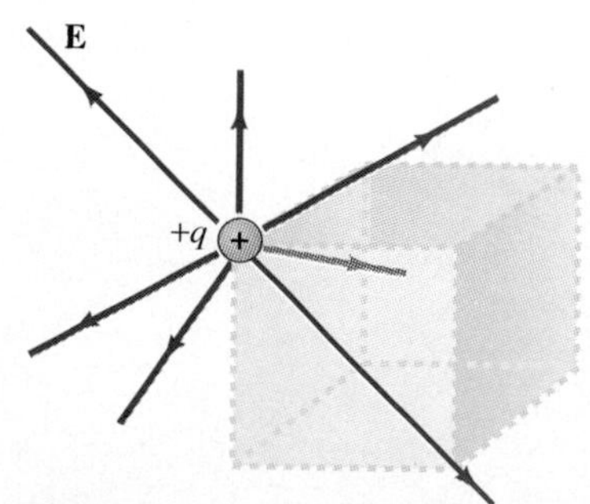

FIGURE 24–10 Example 24–2.

SOLUTION: The situation is sketched in Fig. 24–10. This problem appears difficult, but the use of symmetry makes it easy to solve. First, consider the three faces of the cube that the charge touches. For each of these faces, the product $\mathbf{E} \cdot \mathbf{A} = 0$, because the electric field acts along each of the three faces. The electric flux through each of the remaining three faces of the cube must be equal by symmetry, because nothing distinguishes these faces from one another.

What is the total electric flux through the cube? It would take seven other similarly placed cubes to surround the point charge q completely. Because each of the eight cubes is placed symmetrically about the charge, each of the cubes has an electric flux of $q/8\epsilon_0$. Therefore, each of the three faces of the cube that touch the charge must have an electric flux of $q/24\epsilon_0$. Note that the electric flux through each face, as well as the total electric flux through all faces, is independent of the size of the cube.

Numerical evaluation gives

$$\Phi_{\text{face}} = \frac{q}{24\epsilon_0} = \frac{10^{-3}\ \not{C}}{24(8.85 \times 10^{-12}\ \text{C}^{\not 2}/\text{N}\cdot\text{m}^2)} = 4 \times 10^6\ \text{N}\cdot\text{m}^2/\text{C}.$$

24–3 APPLICATIONS OF GAUSS' LAW

Gauss' law is a fundamental law in its own right. It is also a powerful tool for the determination of electric fields in situations where there is a high degree of symmetry. If there is enough symmetry so that the electric field is constant over a simple surface and can be removed from the integral that expresses the flux, then we can solve the equation that expresses Gauss' law for the field magnitude. Under these circumstances we do not need to perform complex integrations. We illustrate this technique with several examples that involve continuous charge distributions. Consider the line of charge and the charged plane sheet. We discussed these examples in Chapter 23 and used integration over the charge distribution to find the field. We shall also look at the spherical shell and uniformly charged sphere, cases for which we quoted results in Chapter 23 and (for gravitation) in Chapter 12. We shall see that Gauss' law determines the fields briefly and simply. But the real power of solution by Gauss' law is revealed when we discuss conductors in Section 24–4. There we find the fields in situations that are entirely new.

PROBLEM-SOLVING TECHNIQUES

To use Gauss' law to find electric fields given a charge distribution, the following steps are helpful:

1. Make a sketch of the charge distribution. It will help you to recognize any appropriate symmetry.
2. Identify any spatial symmetry of the charge distribution and the electric field it produces. For example, a point charge has spherical symmetry because it looks the same from all around a sphere centered on it. The spherical symmetry of the point charge implies that the field must be radial.
3. Choose a Gaussian surface that is matched to the symmetry. This is the most important step in determining electric fields by Gauss' law. A good choice of surface makes the solution easy. The experience we have gained so far in dealing with and visualizing electric fields is helpful here. It is most useful if the Gaussian surface is chosen so that the field is either parallel to ($d\Phi_E = 0$), or perpendicular to ($d\Phi_E = E\,dA$), the various elements of the surface, and so that the field is constant over the part of the surface to which it is perpendicular. For example, the Gaussian surface best suited to a point charge is a sphere centered on the charge.
4. With surfaces chosen as in step 3, it should be possible to remove the electric field from inside the integral that expresses the flux. Then Gauss' law becomes an algebraic expression for the *magnitude* of the field.

In Examples 24–3 through 24–6, we use these techniques together with Gauss' law, Eq. (24–7), to determine the field.

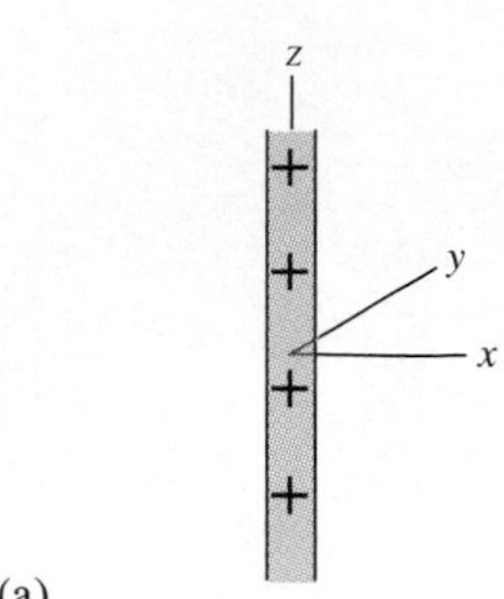

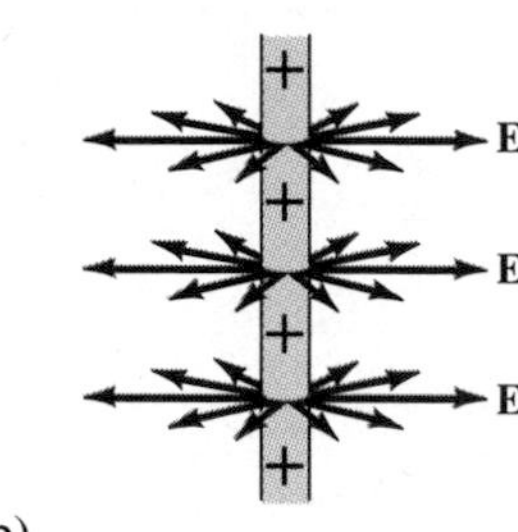

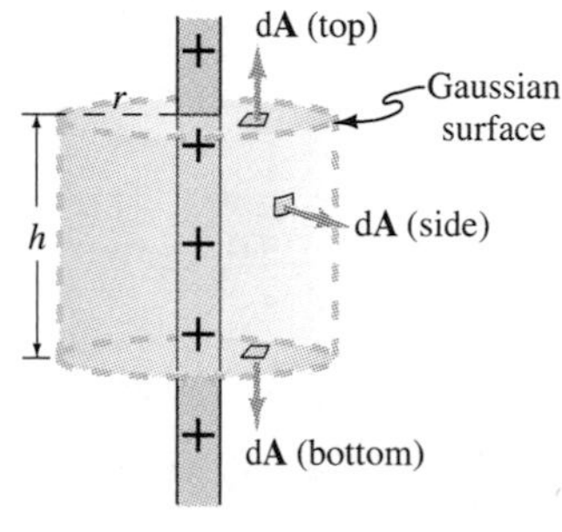

FIGURE 24–11 (a) Example 24–3. A line of charge is oriented along the z-axis. (b) By symmetry, the direction of the electric field **E** is radial in the xy-plane. (c) The best Gaussian surface to use to determine the electric field of a line charge is a cylinder. The directions of the areas **dA** for the various surfaces of the cylinder are shown.

EXAMPLE 24–3 *Line of charge.* Determine the electric field due to an infinitely long, straight charged rod with positive, constant charge density λ.

SOLUTION: Figure 24–11a illustrates the situation. We have oriented the rod along the z-axis. To find the appropriate Gaussian surface, we want to see what symmetry tells us about the direction and magnitude of the electric field lines. These lines must leave the positively charged rod, and, to be symmetric, the electric field lines must extend away from the rod radially in the xy-plane (Fig. 24–11b). The electric field lines cannot have a component along the rod, because there is no way to decide whether the field would be oriented in the $+z$- or $-z$-direction. The direction of the field can easily be identified by visualizing the force on a positive test charge placed outside the rod; the rod will repel it by the force $\mathbf{F} = q\mathbf{E}$. Moreover, again by symmetry, the magnitude of the field must be the same on every point of a circle centered on the rod. Thus the field magnitude can depend only on the radial distance from the rod. The Gaussian surface that takes advantage of the symmetry is a closed cylinder of radius r and height h, centered on the rod (Fig. 24–11c). This surface will enable us to find the field at a distance r from the rod.

We now want to find the flux through the cylinder. We can express this flux as

$$\Phi = \iint_{\text{top}} \mathbf{E} \cdot d\mathbf{A} + \iint_{\text{bottom}} \mathbf{E} \cdot d\mathbf{A} + \iint_{\text{side}} \mathbf{E} \cdot d\mathbf{A}.$$

For the top and bottom, note that **E** is parallel to these surfaces, so the surface element **dA** is perpendicular to **E**. Thus

$$\text{for the top and bottom surfaces:} \qquad \mathbf{E} \cdot d\mathbf{A} = 0.$$

The flux through the top and bottom is zero. For the curved side, the electric field is perpendicular to the surface, so

$$\text{for the side surface:} \qquad \mathbf{E} \cdot d\mathbf{A} = E\,dA.$$

This expression must be integrated over the side to find the flux through the side of the cylinder. But we have chosen the cylindrical surface so that the electric field has constant magnitude over the surface, and the field magnitude at a distance r from the wire can be removed from the integral. Thus

$$\Phi = \iint_{\text{side}} \mathbf{E} \cdot d\mathbf{A} = E \iint_{\text{side}} dA.$$

The remaining integral over the side is just the side area of a right cylinder of height h, namely, $2\pi rh$. We thus have, for the total flux through the cylinder,

$$\Phi = 2\pi rhE.$$

Now that the flux has been calculated, let us apply Gauss' law. The net charge inside the cylinder, q, is the charge on a portion of the rod of length h. This charge is just the charge density, λ, times the length, $q = \lambda h$. Equation (24–7), Gauss' law, now reads

$$2\pi rhE = \frac{q}{\epsilon_0} = \frac{\lambda h}{\epsilon_0}.$$

We can solve for E:

$$E = \frac{\lambda}{2\pi\epsilon_0 r}. \qquad (24\text{–}8)$$

The arbitrary height h has canceled. In SI units, the charge density is in coulombs per meter. Thus $\epsilon_0 E$ has the units of coulombs per square meter, as it must.

We obtained Eq. (24–8) much more easily than we obtained Eq. (23–30), which we found by direct integration. Equation (24–8) also gives the field for charged wires of finite length, as long as the radial distance r from the wire is much less than the distance from an end of the wire. In this case the line is effectively infinite, as we pointed out in Chapter 23.

Why can we not use Gauss' law to find the field of a finite line of charge? Gauss' law continues to hold for *any* distribution of charge, but, for a finite line of charge, the symmetry that allows us to determine the direction of **E** and remove it from the flux integration is not present. If the ends of the line are in view, they provide a guide to tell us where we are along the wire—for example, we can look to see that we are close to one end or the other. The symmetry along the wire is lost. This loss of symmetry has two consequences: First, the electric field will have a component along the wire; and second, the magnitude of the field will vary *along* the line.

EXAMPLE 24–4 *Spherical shell.* Determine the electric field both inside and outside a spherical shell of radius R that has a total charge Q distributed uniformly on its outer surface.

SOLUTION: We show the configuration in Fig. 24–12a. First we find the electric field outside the shell. From symmetry the electric field must be directed radially outward for a positive charge Q and must have a constant magnitude everywhere at the distance r. We take a Gaussian surface that is a sphere of radius r centered on the shell. The product $\mathbf{E} \cdot d\mathbf{A} = E\, dA$, because **E** and $d\mathbf{A}$ are in the same direction. Gauss' law becomes

$$\frac{Q}{\epsilon_0} = \oiint \mathbf{E} \cdot d\mathbf{A} = E \oiint dA = E4\pi r^2. \qquad (24\text{–}9)$$

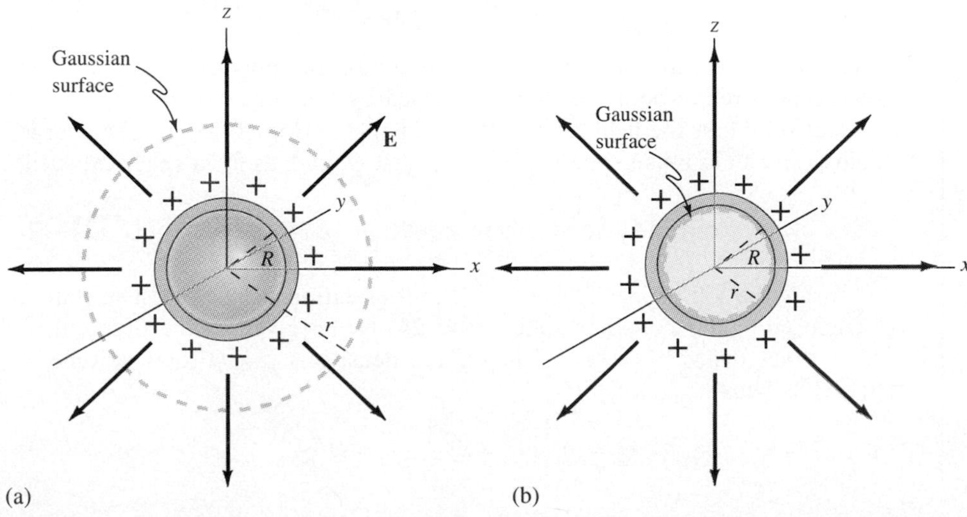

FIGURE 24–12 (a) Example 24–4. The best Gaussian surface to determine the electric field outside a uniformly charged spherical shell. The symmetry is spherical. (b) The best Gaussian surface to determine the electric field inside a uniformly charged spherical shell is a sphere inside the shell.

Here the total area of the Gaussian sphere is $4\pi r^2$, and the total charge enclosed by the Gaussian sphere is the charge Q on the surface of the spherical shell. The electric field outside the shell is, from Eq. (24–9),

$$\text{outside a spherical shell, } r \geq R: \qquad E = \frac{Q}{4\pi\epsilon_0 r^2}. \tag{24–10}$$

Thus the electric field is the same as the field of a point charge of the same total magnitude Q at the center of the spherical shell.

For a point inside the spherical shell, we again have spherical symmetry and so we draw another Gaussian sphere inside the shell (Fig. 24–12b). The integration of the electric flux proceeds as before. However, in this case no charge is enclosed by the Gaussian sphere, so the left side of Eq. (24–9) must be zero. Therefore, *the electric field inside a uniformly charged spherical shell must be zero*:

$$\text{inside a spherical shell, } r < R: \qquad E = 0. \tag{24–11}$$

We noted in Chapter 12 that these same results hold for the force of gravity due to a spherical shell of matter. The mathematical problem is identical, because the gravitational force has the same inverse-square form as the Coulomb force. In Chapter 12 we gave only the results, without derivation, because the direct integration technique is fairly complicated. The Gauss' law derivation provided here is a very simple one. It is interesting to note that Newton delayed the publication of his theory of gravitation by some 20 years because of his lack of a simple proof of these results. If he had known Gauss' law, Newton would have saved a lot of time!

EXAMPLE 24–5 *Solid sphere.* Find the electric field outside and inside a solid, nonconducting sphere of radius R that contains a uniformly distributed total charge Q.

SOLUTION: This charge distribution, shown in Fig. 24–13a, exhibits the same symmetry as does Fig. 24–12a: The electric field must be purely radial and can vary only with the distance r from the center of the sphere. All Gaussian surfaces will then be best taken as spheres centered on the charged sphere, and the flux through any of these spheres will have the form

$$\Phi = \iint_{\text{sphere at } r} \mathbf{E} \cdot \mathrm{d}\mathbf{A} = E \iint_{\text{sphere}} \mathrm{d}A = E 4\pi r^2.$$

Here E is the field at a distance r from the center. In applying Gauss' law, we need to be careful about the charge enclosed by the Gaussian surface.

Consider first the field outside the solid sphere (Fig. 24–13a). The charge enclosed by a Gaussian sphere at $r > R$ is just Q, and, as for a spherical shell,

$$\text{outside a solid sphere, } r > R: \qquad E = \frac{Q}{4\pi\epsilon_0 r^2}. \tag{24–12}$$

Inside the solid sphere, where $r < R$, the situation is different. The charge Q' enclosed by our Gaussian sphere (Fig. 24–13b) is given by the charge density ρ times the volume $\frac{4}{3}\pi r^3$. The charge density is $\rho = Q/\text{total volume} = Q/(\frac{4}{3}\pi R^3)$. Thus

$$Q' = \rho \frac{4}{3}\pi r^3 = Q \frac{\frac{4}{3}\pi r^3}{\frac{4}{3}\pi R^3} = Q \frac{r^3}{R^3}.$$

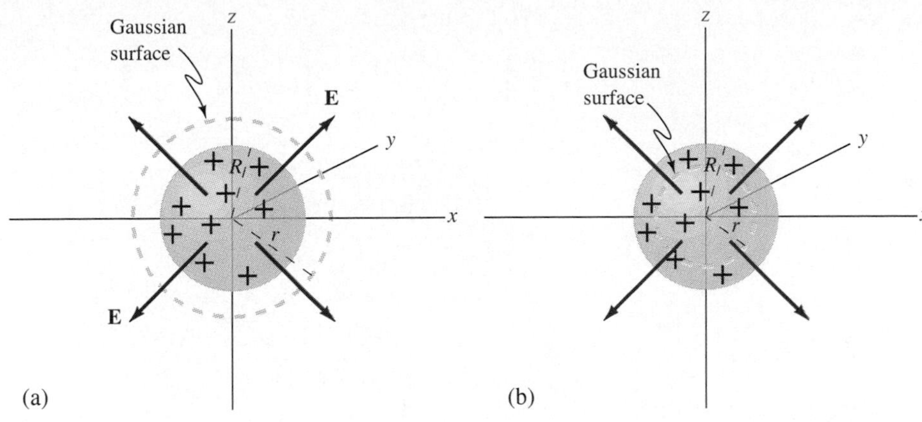

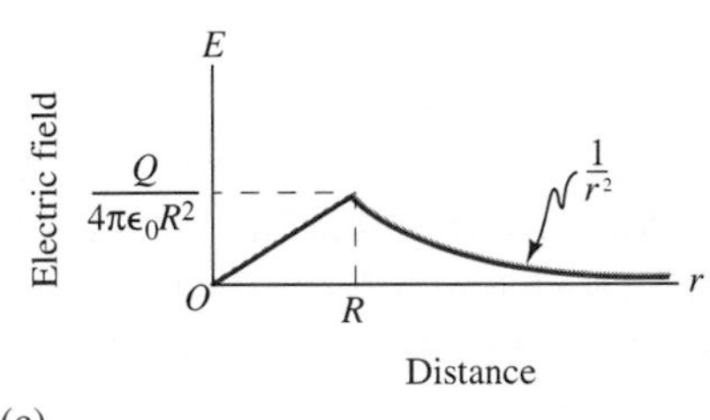

FIGURE 24–13 (a) Example 24–5. The best Gaussian surface to determine the electric field outside a uniformly charged, nonconducting sphere. The symmetry is spherical. (b) The best Gaussian surface to determine the electric field inside a uniformly charged, nonconducting sphere is a Gaussian sphere inside the solid sphere. Only the charge inside the Gaussian sphere contributes to the electric field at r. (c) the electric field due to a uniformly charged, nonconducting sphere as a function of the distance from the center of the sphere.

From Gauss' law the field at a radius $r < R$ is

$$E = \frac{Q'}{4\pi\epsilon_0 r^2} = Q\frac{r^3}{R^3}\frac{1}{4\pi\epsilon_0 r^2}$$

inside a solid sphere, $r < R$:
$$= \frac{Q}{4\pi\epsilon_0}\frac{r}{R^3}. \tag{24–13}$$

Here the charge increases with radius as r^3, whereas the area increases with radius as r^2, so $E \propto Q/\text{area} \propto r$. The charge located outside r gives a net field that adds to zero at r. The electric field due to a solid sphere has the radial dependence displayed in Fig. 24–13c. As symmetry demands, the field is zero at the center of the sphere. The field increases linearly with r up to the radius of the sphere and then decreases inversely as the square of r. The fields in Eqs. (24–12) and (24–13) match at the point $r = R$.

Generally speaking, for a spherically symmetric charge distribution, the field at a radius r is that of a point charge at the center whose magnitude is the total charge within the sphere of radius r. This is, as we have seen, easy to prove by using Gauss' law. It holds not only for thin shells and solid spheres but indeed for *any* distribution of charge whose charge density varies only with the radius.

EXAMPLE **24–6** *Plane of charge.* Find the electric field outside an infinite, nonconducting plane of charge with uniform charge density σ.

SOLUTION: We show a charged plane in Fig. 24–14. If the plane is positively charged, then, from symmetry, we can see that the electric field will be perpen-

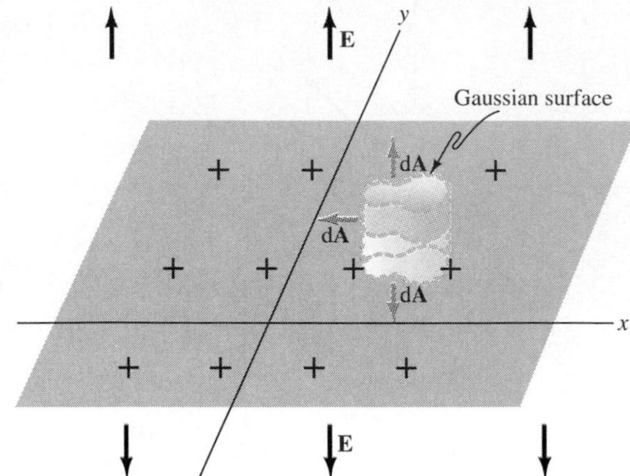

FIGURE 24–14 Example 24–6. A convenient Gaussian surface for a uniformly charged infinite plane can be any shape whose sides are perpendicular to the plane and whose top and bottom are parallel to the plane.

dicular to, and point away from, the plane. We can verify this fact by placing a test charge near the plane. The force on the test charge will be either directly away from or directly toward the plane. Symmetry also dictates that the electric field has a magnitude that depends at most on the perpendicular distance from the plane. Because the electric field is perpendicular to the plane, a good choice for the Gaussian surface is any right solid (such as a cylinder) with its top and bottom (area A) parallel to the charged plane (Fig. 24–14). Every facet of this Gaussian surface is either parallel or perpendicular to the electric field. The differential areas **dA** for the top and bottom of the Gaussian surface also point away from the charged plane, so the product $\mathbf{E} \cdot \mathbf{dA}$ for the three surfaces is

$$\text{for the top:} \qquad \mathbf{E} \cdot \mathbf{dA} = E\,dA;$$

$$\text{for the bottom:} \qquad \mathbf{E} \cdot \mathbf{dA} = E\,dA;$$

$$\text{for the side:} \qquad \mathbf{E} \cdot \mathbf{dA} = 0.$$

The last equation follows because **dA** for the side points everywhere parallel to the plane, but **E** is everywhere perpendicular to the plane.

Equation (24–7) for Gauss' law now becomes

$$\frac{Q}{\epsilon_0} = \oint\!\!\oint \mathbf{E} \cdot \mathbf{dA} = \iint_{\text{top}} \mathbf{E} \cdot \mathbf{dA} + \iint_{\text{bottom}} \mathbf{E} \cdot \mathbf{dA} + \iint_{\text{side}} \mathbf{E} \cdot \mathbf{dA} = 2EA,$$

where we have used the fact that E is constant over the top and bottom area A of the Gaussian surface.

The total charge enclosed by the Gaussian surface is the charge on the plane within the surface. Because the charge density is σ and the area enclosed is A, we must have $Q = \sigma A$. The previous equation becomes

$$\frac{Q}{\epsilon_0} = \frac{\sigma A}{\epsilon_0} = 2EA;$$

$$E = \frac{\sigma}{2\epsilon_0}. \tag{24–14}$$

In SI units, σ is measured in coulombs per square meter. Thus the units of $\epsilon_0 E$ are C/m^2, a correct result. Equation (24–14) is the same result that we found with much more difficulty by direct integration in Chapter 23 [Eq. (23–33)]. Note that E is independent of the distance from the plane.

Equation (24–14) also expresses the electric field at a given point due to a *finite* charged plane, as long as the distance from the point to the ends of the plane is much greater than the perpendicular distance of the point from the plane.

24–4 CONDUCTORS AND ELECTRIC FIELDS

A good conductor, such as silver, copper, or aluminum, has a large number of "free" electrons, which can move within the (electrically neutral) material. Any electric field that may appear inside the metal as the result of the presence of an external electric field will cause the electrons to move. In less than a microsecond they rearrange themselves into a configuration that cancels the electric field inside the material. If any field whatsoever remained inside, it would cause the electrons of the conductor to move until they came to equilibrium. (We refer to electrostatic equilibrium.) *Conductors have no internal static electric field.*

There is no static electric field inside a conductor.

This property of conductors is illustrated in Fig. 24–15. A conductor is placed in a spatially constant and static external field that points to the right (Fig. 24–15a). In this case, some electrons in the metal move to the left side of the conductor, leaving a deficiency of electrons on the right side of the conductor. The arrangement of excess electrons on the left and deficient electrons on the right forms a new, internal electric field that points to the left. This internal field will precisely cancel the external field, so that there is no net field within the conductor (Fig. 24–15b). The movement of charges in response to applied electric fields is called *induction.*

The fact that there are no static electric fields within conductors has implications for the behavior of conductors when charges are put on or near them, or when they are placed in external electric fields. This behavior is determined with the help of Gauss' law. Consider what happens when an excess charge is added to a conductor. We show such a conductor in Fig. 24–16, together with a Gaussian

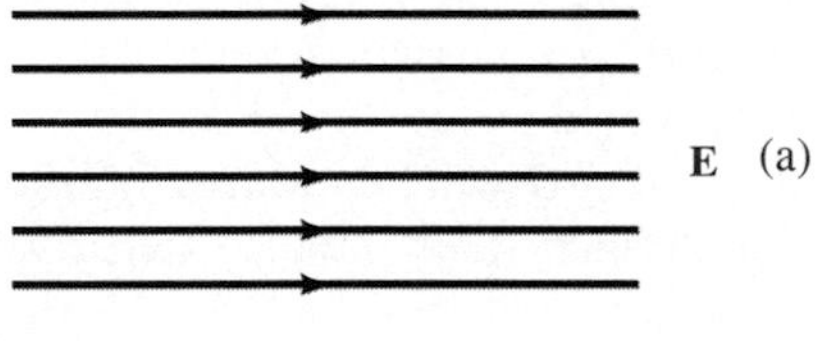

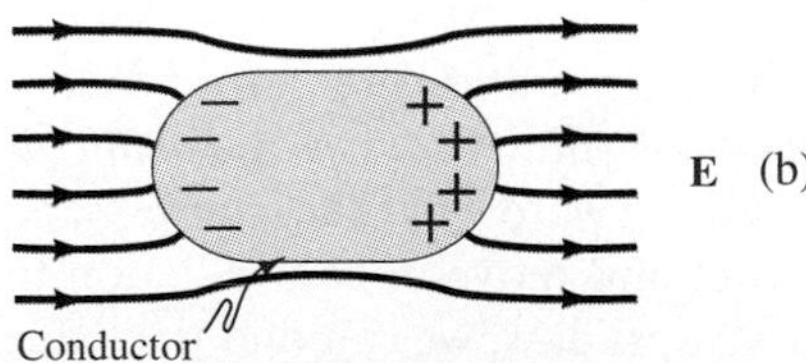

FIGURE 24–15 An uncharged conductor in an external electric field. (a) The electric field before a conductor is introduced. (b) Charges are induced on the surface of the conductor such that the electric field inside the conductor is zero. The induced charges modify the field outside the conductor, so the field no longer has its original form.

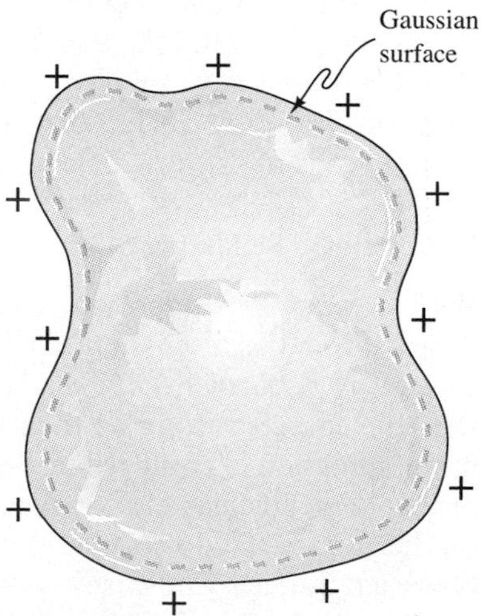

FIGURE 24–16 To find the electric field inside a conductor of arbitrary size and shape, choose a Gaussian surface just inside the surface, so that the closed surface encloses no charge.

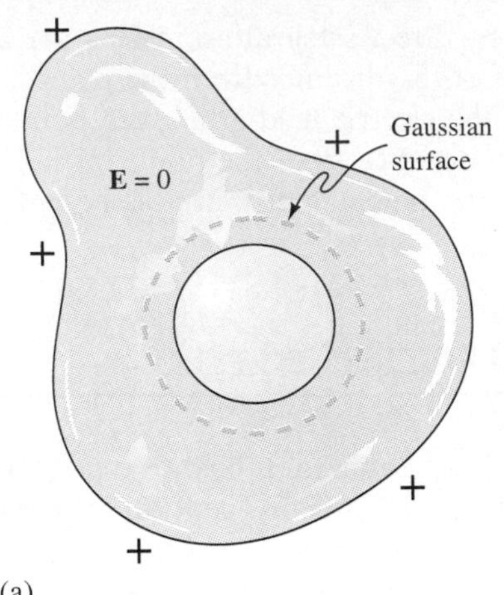

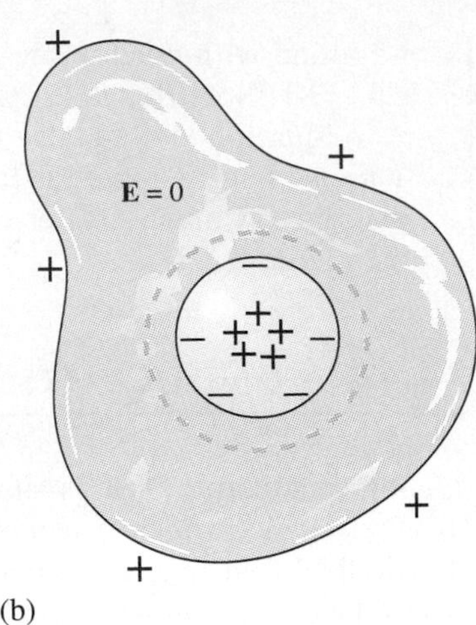

FIGURE 24–17 (a) A nonconducting, hollow space inside a conductor with no charge inside. All the charge must be on the outer surface of the conductor. (b) If we place a charge inside the hollow space, an induced charge will appear on the inside surface of the conductor, making the electric field zero inside the conducting material. A Gaussian surface drawn just outside the hollow space helps to show these results.

surface just inside the metal surface. If we apply Gauss' law to this surface, we find that, because there is no field, there is no flux, and hence there is no net charge inside the metal. Where is the excess charge? *In electrostatic equilibrium, all excess charge is on the outside surface of a conductor.*

Free charges move to the outside surfaces of conductors.

Imagine a bubble that contains a nonconducting medium (such as air) within a conductor (Fig. 24–17a), and suppose that there is no excess charge inside the bubble. It is only on the surface of the bubble that charge might accumulate. A Gaussian surface surrounding the bubble, but drawn within the conductor, has no electric flux through it, because there is no field in the conductor. Thus there is no net charge within that Gaussian surface. We have thereby shown that there can be no net charge on the surface of the bubble. *Any excess charge placed on a conductor, even if the conductor contains nonconducting bubbles, moves to the outside surface of the conductor*, provided there is no charge within the nonconducting bubbles.

We must modify our reasoning when there is charge within nonconducting bubbles in the conductor. Suppose that a nonconducting bubble inside a conductor contains a charge $+Q$ (Fig. 24–17b). Again, draw a Gaussian surface within the metal to surround the bubble. Because there is no field inside the metal, the net charge enclosed must be zero. In this case, a charge of $-Q$ will be induced on the *inner* surface of the metal; that is, on the bubble surface. This induced negative charge keeps the electric field zero *inside the conductor*.

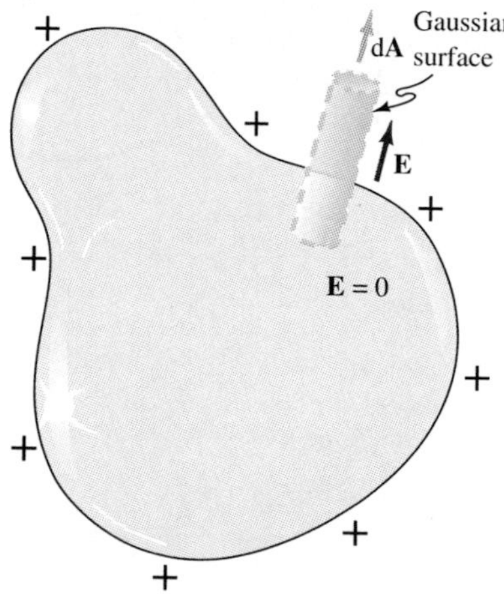

FIGURE 24–18 To find the electric field outside a conductor of arbitrary size, we choose a small right circular cylinder for the Gaussian surface. The only part of the cylinder through which there is a nonzero electric flux is the outside end of the cylinder.

Electrostatic Fields near Conductors

The electric field near a conductor is perpendicular to the surface.

We can draw two important conclusions about electrostatic fields around conductors from this discussion. First, *the electric field immediately outside a conductor must be perpendicular to the conductor's surface.* If there were a parallel component, then the charges resting on the surface would react and move in contradiction to our assumption of equilibrium. Second, by using Gauss' law, we can find the value of this perpendicular electric field near the surface in terms of the charge density on the surface. To do so, consider the conductor shown in Fig. 24–18, with a tiny Gaussian surface whose side is perpendicular and whose top is parallel to the conductor's surface. It is tiny because the surface charge density σ may vary over the conductor, and we shall be referring to σ only at the point where the Gaussian surface is erected. The electric field is zero inside the metal surface, and it is parallel to the side of the Gaussian surface. Thus the only contribution to the flux comes from the top. If the Gaussian surface is small enough, **E**, which is perpendicular to the top surface, can be regarded as constant over it, and

$$\frac{Q}{\epsilon_0} = \oiint \mathbf{E} \cdot d\mathbf{A} = EA,$$

where A is the area of the top of the Gaussian surface. The total charge Q enclosed by the Gaussian surface is σA, so the previous equation becomes

$$\frac{Q}{\epsilon_0} = \frac{\sigma \cancel{A}}{\epsilon_0} = E\cancel{A}.$$

The area cancels. The electric field just outside the surface is proportional to the local charge density σ:

The electric field just outside the surface of a conductor

$$E = \frac{\sigma}{\epsilon_0}. \qquad (24\text{–}15)$$

This result holds only near the conductor's surface. Whether this result is useful or not depends on our knowing the charge density σ, and the magnitude of the field will vary around the surface of the conductor as σ does. The electric field will always be perpendicular to the conductor near its surface (Fig. 24–19). It is useful to check this result by considering a conductor that is a sphere of radius R and total charge Q. In this case symmetry demands that the charge is spread evenly over the surface, and

$$\sigma = \frac{Q}{\text{area}} = \frac{Q}{4\pi R^2}.$$

For the field just outside the sphere, Eq. (24–15) would then give $E = Q/4\pi\epsilon_0 R^2$, which agrees with Eq. (24–12), a result derived earlier.

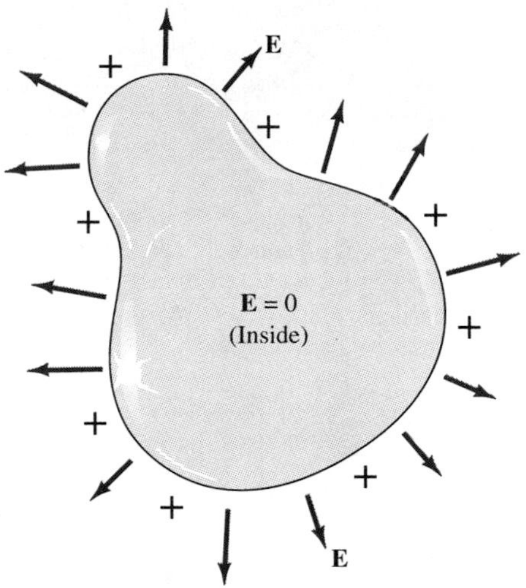

FIGURE 24–19 The electric field in and around a conductor in equilibrium. The electric field inside the conductor is zero, and, just outside the conducting surface, it must be perpendicular to the surface. The magnitude of the electric field varies according to the surface charge density σ, which may not be constant everywhere on the surface.

The field just outside a conductor ($E = \sigma/\epsilon_0$) is twice as large as the field of a nonconducting charged plane with the same charge density ($E = \sigma/2\epsilon_0$). This is simple to understand. The charge on a surface of area dA, $\sigma\, dA$, gives rise to a certain number of field lines. For a nonconducting plane, the field lines divide equally between the two sides of the plane. For a conducting plane, there are no field lines on the conducting side, so all the field lines must emerge on the open side.

We can summarize what we have learned about conductors as follows:

Properties of electric fields in and around conductors

1. The electrostatic field inside a conductor is zero.
2. The electrostatic field immediately outside a conductor is perpendicular to the surface and has the value σ/ϵ_0, where σ is the local surface charge density.
3. A conductor in electrostatic equilibrium, even one that contains nonconducting bubbles, can have charge only on its outer surface, as long as the bubbles contain no net charge.

We can add one more important result. Suppose that we have a bubble in a metal, with no charge in it. We now know that there is no field within the metal and, moreover, no net charge on the inner surface of the metal surrounding the bubble. Even for nonsymmetric situations, it can be shown that, as long as there is no charge within the cavity, *the electric field is zero everywhere within the bubble.*

The fact that there are no electric fields within charge-free cavities in metals has practical applications. Research laboratories often have enclosures formed by copper screens or sheets. These "shielded" rooms are necessary for taking careful electronic measurements unaffected by outside electrical interference (Fig. 24–20). The enclosure is a cavity within the metallic solid formed by the copper screens. There is no electric field within the enclosure due to any external effects, as long as there is zero net charge inside. If there were a net charge inside, charge would be induced on the inside of the copper screens, forcing the electric field inside the copper to be zero, and there would be an electric field inside the enclosure.

The consequences of these properties go beyond the laboratory. The interior of your car is a safe place in the event of nearby lightning, but, for the same reason, your car radio works less well when the car is in the "cage" formed by a metal bridge.

FIGURE 24–20 By Gauss' law, there is no static electric field in an empty cavity in a metal. Researchers take advantage of this fact by working within a metal cage, inside of which fields due to outside sources are minimized.

*24–5 HOW WELL DO WE KNOW GAUSS' AND COULOMB'S LAWS?

Gauss' law is equivalent to Coulomb's law only because Coulomb's law is an inverse-square law. Gauss' law is one of the cornerstones of our understanding of electricity and magnetism, and we must ask just how well it is known and test it as precisely as possible. The errors implicit in any measurements of Coulomb's law set limits on our knowledge of Gauss' law, and these limits have been improved by various means right up to the present. It is one of the characteristics of science to be eternally skeptical of yesterday's experiment. It is not so much that yesterday's experiment is wrong as it is that yesterday's result might be only an approximation, and that, with a more modern apparatus, a more accurate experiment can be done. Here we review the precision with which the equations of electrostatics are known. We look in detail at one particularly sensitive technique for testing Gauss' law.

The earliest tests are associated with the discovery of Coulomb's law itself. Gauss' and Coulomb's law were discovered in a peculiar order. Joseph Priestley knew that there is no gravitational field entirely within a spherically symmetric mass distribution. He made the inspired speculation that a similar behavior of the gravitational and electric force laws would explain why a charged cork ball placed inside a charged metal container is not attracted to the walls of the container. This effect had first been seen, if not understood, in 1755 by Benjamin Franklin, who reported it to Priestley. Priestley thus had experimental evidence of Coulomb's law, and he reported it in 1767, although he made no truly quantitative test.

Franz Aepinus read of Franklin's work and guessed at an inverse-square dependence. Aepinus published his idea in Latin in a Russian journal in 1759. Somehow John Robison heard about this idea and in 1769 did direct experimental tests of the distance dependence of the force between charges. Robison expressed the uncertainties in his result as a *deviation* from Coulomb's law; that is, he supposed that the force is not an inverse-square law but rather has the distance dependence

$$F \propto \frac{1}{r^{2 \pm \delta}}.$$

He then gave limits on the parameter δ, as in Table 24–1. When $\delta = 0$, the inverse-square law is exact. The smaller the limit on δ, the closer the law is known to be an inverse-square law. Unfortunately, Robison made his results known only at an obscure meeting, and it was not until 1801, well after Coulomb's publication, that Robison published his results.

TABLE 24–1

EXPERIMENTAL MEASUREMENTS OF DEVIATION FROM AN INVERSE-SQUARE FORCE LAW†: FORCE $\propto 1/r^{2\pm\delta}$

Investigators	*Date*	*Maximum* δ
Robison	1769	0.06
Cavendish	1773	0.02
Coulomb	1785	0.10
Maxwell	1873	5×10^{-5}
Plimpton and Lawton	1936	2×10^{-9}
Williams, Faller, and Hill	1971	3×10^{-16}

† For more information on this subject, see A. S. Goldhaber and M. M. Nieto, "The Mass of the Photon," *Scientific American*, p. 86, May 1976.

The next discovery of Coulomb's law was made in 1773 by the rather eccentric Henry Cavendish. Through his knowledge of gravitation, he knew that the presence or absence of charge on the inner surface of a closed conductor is a consequence of what we now call Gauss' law, and hence a direct test of a $1/r^2$ force law. He placed one conducting sphere inside another and connected the two by a wire. After placing a charge on the apparatus, he disconnected the wire and looked for any charge that remained on the inner sphere. To the accuracy of his experiment, he found none. He was able to describe his result by saying that δ must be smaller than a certain value. Cavendish was even worse than Robison at publishing his results, which did not appear in print until more than 100 years later! Cavendish's experiment now goes under the general name of the *Faraday "ice-pail" experiment*, after Michael Faraday; it is the basis for many of the modern high-precision tests of Gauss' law.

It was not until 1785 that Charles Coulomb came into the picture, but he published his results promptly and now bears the credit for the force law. He tested the force law directly by using a torsion balance much like that used by Henry Cavendish in 1798 to test the gravitational force (see Chapter 12). Coulomb's limit on δ was in fact worse than either Robison's or Cavendish's, as Table 24–1 shows. Notable improvements in experiments similar to Cavendish's were made by James Clerk Maxwell in 1873, by Samuel J. Plimpton and Willard E. Lawton in 1936, and by Edwin R. Williams, James E. Faller, and Henry A. Hill in 1971. As a glance at Table 24–1 shows, the degree of precision of these experiments, which are direct tests of Gauss' law, are astonishing.

A Null Experiment

Let us describe a simple version of the Faraday ice-pail experiment, which can easily be performed in a lecture demonstration or an undergraduate laboratory.[†] The equipment required is an *electroscope*, a detector of free charge (Fig. 24–21a). When the electroscope receives excess charge, the charge is distributed throughout,

(a)

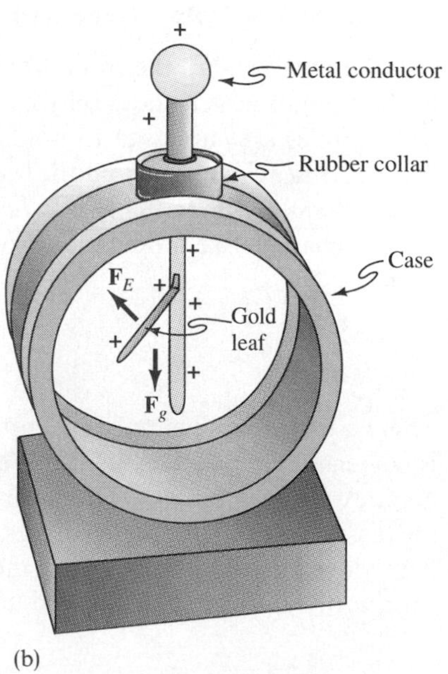

(b)

FIGURE 24–21 (a) An electroscope, a device that detects the presence of charge. (b) Schematic diagram of an electroscope. When the metal conductor is charged, the thin gold leaf also becomes charged and is repelled from the conducting rod.

[†] The "ice pail" was presumably a metal bucket, used for ice, that was handy in Faraday's lab. It served as a metal container to surround charge.

including to the metal rod and the gold leaf inside the case (Fig. 24–21b). The leaf is repelled from the metal rod until the vertical component of the electrostatic repulsion is balanced by the force of gravity on it. Addition of more charge moves the leaf farther away from the rod. A hollow metal container with a hole in the top, as in Fig. 24–22a, is also necessary. Charge can be introduced to the inside of the container with a small metal ball on the end of an insulating rod. The electroscope is attached to the outside of the container and thus indicates charge on the outside.

Next a positive charge, $+Q$, is placed on the metal ball, which is inserted through the small hole into the hollow container without touching it (Fig. 24–22b). Gauss' law states that there is no net charge *inside* the nearly closed metal container, so that a charge of $-Q$ is induced on the inside surface of the container.[†] Because the metal container is neutral, a charge of $+Q$ is induced on the outside, and the electroscope indicates this. Moving the ball around leads to no change whatsoever in the electroscope, consistent with Gauss' law. Then the metal ball is touched to the interior of the hollow container (Fig. 24–22c). If Gauss' law is correct, the charge on the ball neutralizes the $-Q$ charge induced on the inside surface, leaving the $+Q$ charge on the outside surface. The electroscope indicates this result by not changing in the least. When the metal ball is removed from the container, the container's outer surface remains charged (Fig. 24–22d). By touching the metal ball to another electroscope, we can verify that it carries no charge.

The description of this experiment shows why it is potentially so precise: If Gauss' law is correct, there is *no change* in the position of the gold leaf when the inner surface is touched. Equivalently, the Cavendish experiment tests for the absence of charge on the inner of two spheres. Experiments such as Coulomb's, which look for small changes in comparison with larger effects, are inherently less precise than experiments such as Cavendish's, which look for small changes in comparison with no effect. Experiments that test for small change versus no change are called **null experiments**. It is far easier to make a precision test of Gauss' law than of Coulomb's law, because a null experiment can be done.

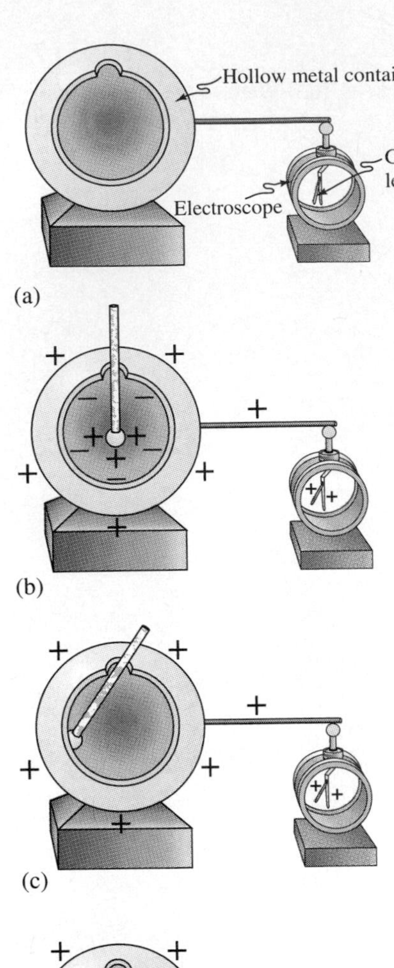

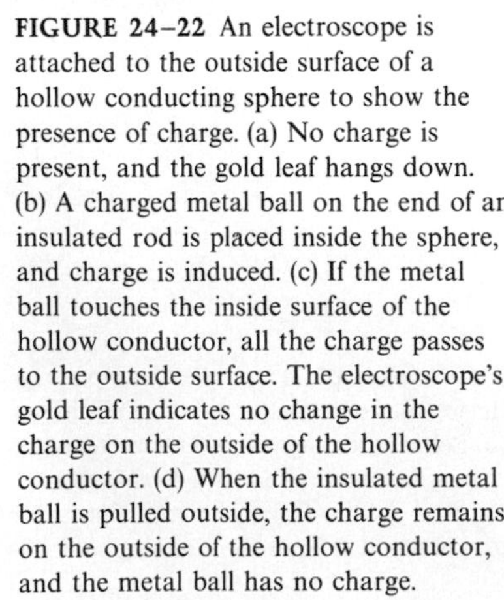

FIGURE 24–22 An electroscope is attached to the outside surface of a hollow conducting sphere to show the presence of charge. (a) No charge is present, and the gold leaf hangs down. (b) A charged metal ball on the end of an insulated rod is placed inside the sphere, and charge is induced. (c) If the metal ball touches the inside surface of the hollow conductor, all the charge passes to the outside surface. The electroscope's gold leaf indicates no change in the charge on the outside of the hollow conductor. (d) When the insulated metal ball is pulled outside, the charge remains on the outside of the hollow conductor, and the metal ball has no charge.

Coulomb's Law Holds over Small and Large Distances

This is not the end of the story, however. First, the experiments that we have listed in Table 24–1 test the laws only over a distance of about 1 m. Yet the laws of electrodynamics are supposed to hold in atomic systems and over galactic distances. Second, other evidence about the framework of the laws of physics suggests strongly that *a deviation from Coulomb's law of the form* $1/r^{2+\delta}$ *is not possible*. Instead, a way to characterize a deviation from Coulomb's law is by using the *approximate* form

$$F \propto \frac{e^{-\mu r}}{r^2},$$

where e is the exponential constant 2.78 . . . and μ is a constant. If Coulomb's law is correct, the parameter $\mu = 0$. We have seen the exponential form earlier; it is a function that decreases with r over a distance that depends on μ. The larger μ is, the faster the exponential decreases, and the larger the violation of Coulomb's law. Any violation is, we now know, more properly expressed by limits on μ. We can determine limits on μ, and hence tests of the accuracy of Coulomb's law, from the

[†] The hole can be made smaller and smaller until its presence does not matter.

previously reported experiments. The experiment of Williams, Faller, and Hill, for example, implies that μ is smaller than $6 \times 10^{-8}\ m^{-1}$. These limits can be extended by observing the space dependence of the earth's magnetic field and also of Jupiter's magnetic field, as measured by the spacecraft *Pioneer 10*. Although we have not yet studied magnetism, we can say that the limits on μ found thereby are indeed those associated with Gauss' law. The planetary measurements, besides being direct, give values of μ that are smaller by an order of magnitude or more than those given by the laboratory experiments, and have the further advantage of testing Gauss' law out to large distances.

Finally, how well do we know Gauss' law at short distances? The colors of light given off by excited hydrogen atoms are very sensitive indicators of the Coulomb force at distances on the atomic scale, about 10^{-10} m. The accuracy with which Gauss' (and therefore Coulomb's) law is known is comparable to the accuracy of the experiments of Plimpton and Lawton (see Table 24–1), that is, to about one part in 1 billion. Even down to nuclear distances, about 10^{-15} m, experiments indicate consistency with the basic theory that leads to Coulomb's law.

SUMMARY

The electric flux due to the electric field **E** that passes through any surface is

$$\Phi = \iint_S \mathbf{E} \cdot d\mathbf{A}. \tag{24–3}$$

Gauss' law relates the electric flux through a closed Gaussian surface—an imaginary closed surface—to the total charge enclosed by the surface, Q:

$$\oint\!\!\oint \mathbf{E} \cdot d\mathbf{A} = \frac{Q}{\epsilon_0}. \tag{24–7}$$

Gauss' law is equivalent to Coulomb's law for static situations, and, unlike Coulomb's law, it holds even when we consider nonstatic fields. It is thus one of the fundamental equations of electromagnetism.

Gauss' law is also a powerful tool for determining electric fields due to charge distributions with a high degree of symmetry. It can be used to derive in simple fashion the electric fields due to a straight-line charge or due to a conducting plane. For a general spherically symmetric charge distribution centered at the origin of a coordinate system, Gauss' law gives a simple derivation of the field at a distance r from the origin. If q is the total charge contained within a Gaussian sphere of radius r, then the electric field at r is the same as that of the field of a point charge q at the origin, $E = q/(4\pi\epsilon_0 r^2)$.

Conductors react in special ways to electric fields and to charges:

1. The electrostatic field inside a conductor is zero.
2. The electrostatic field immediately outside a conductor is perpendicular to the surface and has the value σ/ϵ_0, where σ is the local surface charge density (which is not necessarily constant).
3. If there are no nonconducting holes that contain charge, a conductor in electrostatic equilibrium can have charge only on its outer surface.

Gauss' law (and its equivalent, Coulomb's law) has been subjected to many experimental tests since the mid-eighteenth century. The inverse-square law dependence on distance has been verified to a precision that ranges from one part in 10^9 to one part in 10^{16} over distances between 10^{-10} m and 10^9 m.

QUESTIONS

1. A temperature field is defined when the temperature of every point of a region of space is specified. Is it possible to compute a flux associated with this field?
2. In the text we refer to the Faraday ice-pail experiment and discuss one version of it in detail. The discussion concerns a sphere with a hole cut in it, and we speak of the inside and outside of this open sphere (Fig. 24–22a). Yet an open sphere does not have a clear inside and outside, because, unlike a closed, hollow sphere, it can be deformed continuously to a plane. Why is it possible to talk of the inside and outside of an open sphere, and why does the open sphere behave like a closed, hollow sphere (a bubble) in the experiment?
3. Use Gauss' law to show that electric field lines must be continuous and must originate from and end on charges.
4. Describe the way in which Gauss' law would fail if the field of a point charge were to decrease as $1/r$ rather than as $1/r^2$.
5. If a large, flat plate is positively charged, the field extends in both directions from the plate and has a magnitude of $\sigma/2\epsilon_0$. If a second plate of equal but opposite charge is placed parallel to the first plate, the field around the first plate extends only toward the second plate and has a magnitude of σ/ϵ_0, where σ is exactly the same as before. How do you reconcile this second case with Gauss' law?
6. A first Gaussian surface is a sphere of radius r, and a second Gaussian surface is a concentric sphere of radius $r - \delta$, where δ is very small. The electric flux through the first surface is Q/ϵ_0, and the electric flux through the second surface is zero. Can you conclude from this that the space between the surfaces is filled uniformly with a charge Q?
7. Consider an electric field **E** that is zero at every point on a closed surface S. Does this mean that there are no charges within this surface? Give an example for which there are charges inside a surface while $\mathbf{E} = 0$ on the surface.
8. What would Gauss' law look like for fluid flow? What would it look like for the gravitational field, defined by force/(unit mass)?
9. Charge is distributed uniformly on a circular wire that is surrounded by a torus (doughnut) for which the wire serves as an axis. Does the symmetry allow us to say anything about the electric field due to the charge on the circular wire?
10. A positive point charge and a negative point charge of equal magnitude are fixed on the surface of a conductor of arbitrary shape. What, if anything, can be said about the resulting electric field lines?
11. A region in space has a uniform electric field. What can we say about whether or not any charges are inside the region?
12. To derive the electric field of an infinitely long line of charge, we used a Gaussian surface in the form of a right cylinder centered on the line. Why does the use of such a surface not allow us to find the field of a line of charge of *finite* length?
13. Suppose that the electric field in some region is known to have only an x- and a y-component, and that the components depend only on x and y, not on z. What can you deduce about the charge distribution that gives rise to this field?
14. You have a probe that measures the electric field at any point in space. For a region in which you know independently that the charge density is constant, how can you use the probe to measure that charge density?

PROBLEMS

24–1 Electric Flux

1. (I) An infinitely large, nonconducting, thin plate carries a uniform charge density σ. (a) What is the electric flux through a circle of radius R placed parallel to the plane? (b) What is the flux through that circle if the plane of the circle is tilted at a 30° angle with respect to its original orientation?
2. (I) The electric field due to an infinitely long, straight line of charge with uniform charge density λ points straight away from the line and has magnitude $E = \lambda/2\pi_2 r$, where r is the distance from the wire. Calculate the flux of this electric field through a right cylinder of height h and radius R, concentric with the charged line. Repeat the calculation for a cylinder of radius $2R$.
3. (I) The electric field in a certain region of space points in the z-direction and has magnitude $E = 4xz$, where x and z are measured from some origin. Calculate the flux of that field through a square perpendicular to the z-axis; the corners of the square are at $(x, y, z) = (1, 1, 3), (1, 2, 3), (2, 2, 3)$, and $(2, 1, 3)$. (All fields are measured in N/C, all distances in m.)
4. (I) An electric field has the components $E_x = -2x$, $E_y = -2y$, and $E_z = 3z$. Calculate the electric flux through the sides of a unit cube, whose corners are $(x, y, z) = (0, 0, 0)$,

(1, 0, 0), (1, 1, 0), (0, 1, 0), (0, 0, 1), (1, 0, 1), (1, 1, 1), and (0, 1, 1). (All fields are measured in N/C, all distances in m.)

5. (II) An electric field that is constant in direction is perpendicular to the plane of a circle of radius R. The maximum magnitude of the field in that plane occurs at the axis of the circle. Suppose that the magnitude of the electric field in the plane decreases from an axial value as $1/r$. Find the electric flux through the plane of the circle.

6. (II) By direct calculation (that is, without using Gauss' law), find the flux of a constant electric field $\mathbf{E}$ through a hemispherical surface of radius R whose circular base is perpendicular to the direction of the field. Your result should be the same as the flux through the top surface of a cylinder whose circular base, of radius R, is oriented perpendicular to the field direction. [*Hint:* The area of an infinitesimal strip at a latitude θ and a thickness $Rd\theta$ is $2\pi R^2 \sin\theta \, d\theta$; θ varies from 0 at the north pole to $\pi/2$ at the equator.]

7. (II) A charge q is placed just above the center of a horizontal circle of radius r, and a hemisphere of this radius is erected about the charge. Compute the electric flux through the closed surface that consists of the hemisphere and the planar circle (Fig. 24–23).

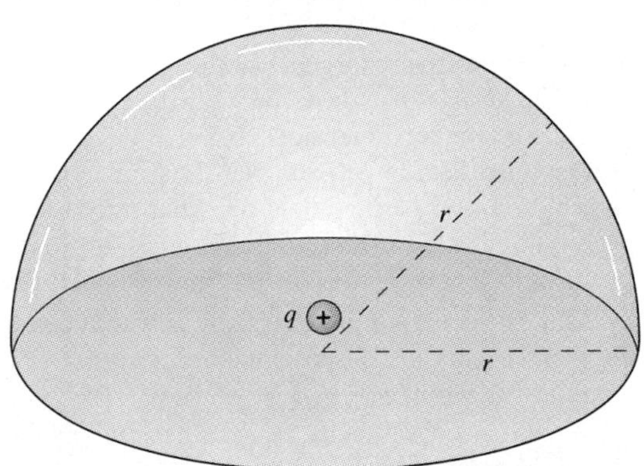

FIGURE 24–23 Problem 7.

8. (III) Consider an infinitesimal parallelepiped located at the point (x, y, z) with sides dx, dy, and dz along the x-, y-, and z-axes (Fig. 24–24). Show that the electric flux of the electric field given by $\mathbf{E} = E_x\mathbf{i} + E_y\mathbf{j} + E_z\mathbf{k}$ through the surface that bounds this volume is given by

$$\Phi = \left(\frac{\partial E_x}{\partial x} + \frac{\partial E_y}{\partial y} + \frac{\partial E_z}{\partial z}\right) dx\, dy\, dz.$$

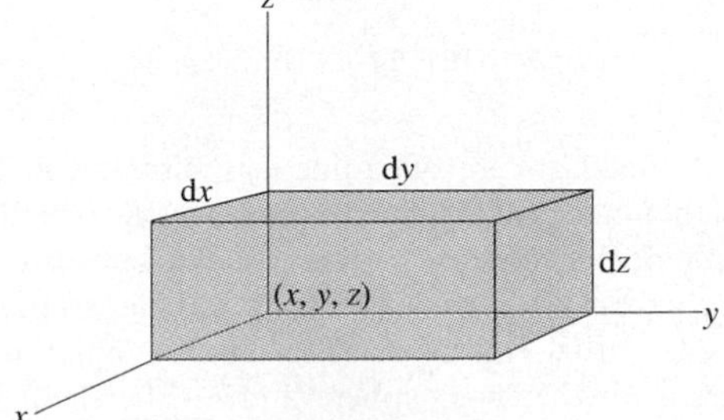

FIGURE 24–24 Problem 8.

The quantity in parentheses (the coefficient of $dx\, dy\, dz$) is called the *divergence* of the vector field $\mathbf{E}$.

24–2 *Gauss' Law*

9. (I) A charge of 10^{-3} C is distributed uniformly on the surface of a sphere of radius 1 cm. Calculate the total electric flux through a concentric sphere (a) just within the charged surface and (b) just outside the charged surface.

10. (I) A 5-μC point charge is placed just inside the center of one face of an imaginary Gaussian cube. What is the flux that passes through the sum of all six faces of the cube?

11. (I) The net electric flux passing through a given closed surface is -4×10^2 N·m²/C. What charge is contained inside the surface if that surface is (a) a sphere of radius 3 cm, (b) a cube of sides 3 cm, (c) a right circular cylinder of height 3 cm and radius 1 cm?

12. (II) A 3-μC charge is placed at the center of a cube of sides 4 cm. Determine the electric flux through each of the sides.

13. (II) A given region has an electric field that is a sum of two contributions: a field due to a charge $q = 3 \times 10^{-10}$ C at the origin, plus a uniform field of strength $E_0 = 50$ N/C in the $-x$-direction. Calculate the flux through each side of a cube with sides of length 10 cm that are parallel to the x-, y-, and z-directions; the cube is centered at the origin.

14. (II) The *gravitational field* $\mathbf{g}$ due to a point mass M may be obtained by analogy with the electric field by writing an expression for the gravitational force on a test mass, and dividing by the magnitude of the test mass, m. Show that Gauss' law for the gravitational field reads $\Phi = \oint \mathbf{g} \cdot d\mathbf{A} = -4\pi GM$, where G is the gravitational constant. Use this result to calculate the gravitational field at a distance r from the center of a sphere of radius R and uniform density for $r < R$ and for $r > R$.

15. (II) A point charge q is located at the center of a tetrahedron with sides of length L (Fig. 24–25). What is the average value of the electric field over one face of the tetrahedron?

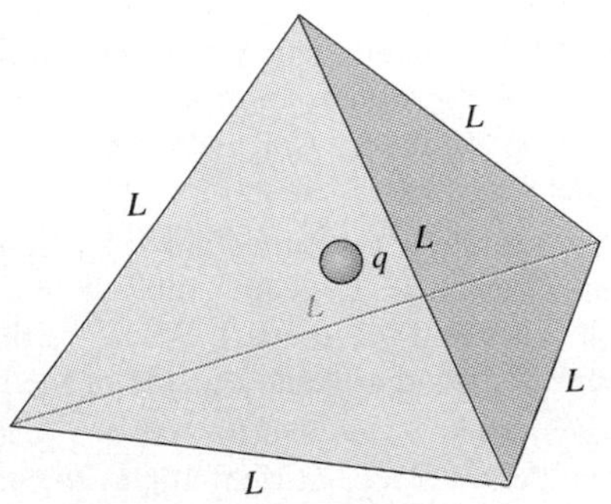

FIGURE 24–25 Problem 15.

24–3 *Applications of Gauss' Law*

16. (I) Calculate the electric field outside a long cylinder of finite radius R with a uniform (volume) charge density ρ spread throughout the volume of the cylinder.

17. (I) Use Gauss' law to show that the electric field outside a large, thin, nonconducting plate with uniform charge density σ is given by $E = \sigma/2\epsilon_0$.

18. (II) An infinitely long cylinder of radius R carries a uniform (volume) charge density ρ. Calculate the field everywhere inside the cylinder.

19. (II) On a clear day in Nebraska, the electric field just above the ground is 90 N/C and points toward the ground. The earth is itself a reasonable conductor and contains no electric field. How much net charge is contained on the surface of a 40-acre corn field (1 acre $\simeq$ 4000 m^2)?

20. (II) Two long, thin cylindrical shells of radii r_1 and r_2, respectively, are oriented coaxially (one cylinder is centered inside the other). The cylinders carry equal and opposite linear charge densities λ. Describe the resulting electric field inside the smaller cylinder, between the cylinders, and outside the larger cylinder.

21. (II) A balloon of radius 30 cm carries a charge of 3×10^{-8} C distributed uniformly over its surface. What is the electric field at a distance of 40 cm from the center of the balloon? Suppose that the balloon shrinks to a radius of 10 cm but loses none of its charge. What is the electric field at a distance of 40 cm from the center?

22. (II) A thin, cylindrical copper shell of diameter 4 cm has along its axis a thin metal wire of diameter 3×10^{-3} cm. The wire and the shell carry equal and opposite charges of 10^{-9} C/cm, distributed uniformly. Calculate the electric field in the region between the wire and the cylinder, and the magnitude of the electric field at the surface of the wire.

23. (II) A long, cylindrical shell of inner radius r_1 and outer radius r_2 carries a uniform volume charge density ρ. Find the electric field due to this distribution of charge everywhere in space.

24. (II) A Teflon rod of radius 4.0 cm and height 20.0 cm is being charged uniformly over its surface. How much charge can the rod hold before the surrounding air breaks down electrically, which happens when the electric field in air is 3.0×10^6 N/C? Ignore the likelihood of breakdown at the sharp edges.

25. (II) A thick, nonconducting spherical shell with a total charge of Q distributed uniformly has an inner radius R_1 and an outer radius R_2. Calculate the resulting electric field everywhere.

26. (II) Two infinite-plane nonconducting, thin sheets of uniform surface charge 12 $\mu C/m^2$ and -8 $\mu C/m^2$ are parallel to each other and 0.1 m apart. What are the electric fields between the two sheets and outside of them?

27. (II) Two infinite-plane sheets that are just like those of Problem 26 are placed at right angles to each other. What are the fields in the four regions into which space is divided by the planes?

28. (II) A slab of nonconducting material forms an infinite plane. The slab has a thickness t and carries a uniform positive charge density ρ. It is oriented parallel to the xy-plane, with its upper surface at $z = t/2$ and its lower surface at $z = -t/2$. Use Gauss' law to find the electric field both above and below the surface, as well as at an arbitrary value of z in the interior of the slab.

29. (II) Consider a sphere of radius 4 cm that carries a negative charge of 40 μC distributed uniformly. The sphere is embedded in another, larger sphere of radius 10 cm. The outer shell has a positive charge of 50 μC distributed uniformly. Calculate the electric field as a function of radius r for $0 < r < 15$ cm.

30. (II) Charge is distributed throughout a sphere with the charge density given by $\rho = \rho_0$ for $r < a$, $\rho = \rho_0(r - R)/(a - R)$ for $a < r < R$, and $\rho = 0$ for $R < r$. Calculate the flux through the spherical surfaces at $r = a$, $r = R$, and $r = 10R$, and calculate the corresponding electric fields at these radii.

24–4 *Conductors and Electric Fields*

31. (I) Two concentric metallic shells, perfect conductors, have radii of R and $2R$, respectively. A charge q is placed on the inner shell, and a charge $-2q$ is placed on the outer shell. What are the electric fields in all of space due to the two shells?

32. (I) Two oppositely charged, parallel metal plates give rise to a field of 3×10^{11} N/C between them. The plates are square and have dimensions 0.5 m $\times$ 0.5 m. How much charge must there be on each plate? Assume that the charge distribution and electric field are uniform, as if the plates were infinite in size. This will be a good approximation if the distance between the places is much smaller than 0.5 m.

33. (I) What is the maximum charge density that can be placed on a large conducting plate in order to avoid electrical breakdown in air ($E_{max} = 3 \times 10^6$ N/C)?

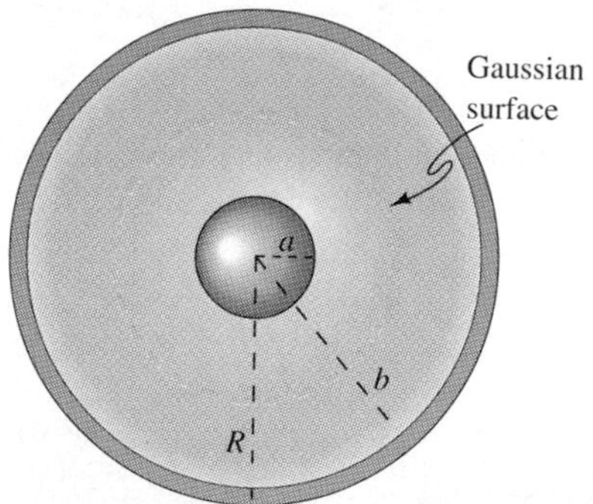

FIGURE 24–26 Problem 34.

34. (II) A metal sphere of radius a is surrounded by a metal shell of inner radius b and outer radius R. The flux through a spherical Gaussian surface located between a and b is Q/ϵ_0, and the flux through a spherical Gaussian surface just outside radius R is $2Q/\epsilon_0$ (Fig. 24–26). What are the total charges on the inner sphere and on the shell? Where are the charges located, and what are the charge densities?

35. (II) The electric field near the earth's surface on a given day is 100 V/m, pointing radially inward. If this were true everywhere on the surface of the earth, what would the sign and magnitude of the total charge on earth be? If the earth is treated as a conductor, where is the charge located? What is the charge density?

36. (II) A point charge q is placed a distance $L/2$ over the center of a conducting square plate of area L^2. (a) Draw the electric field lines on both sides of the plate, which has charge $-q$. (b) Repeat part (a) for a charge on the plate of $2q$.

37. (II) The center of a solid conducting sphere of radius 4 cm and charge 2 μC is placed 10 cm above and away from the center of a flat, horizontal conducting square plate of area 25 cm^2 and charge 1 μC. Draw the electric field lines.

General Problems

38. (II) Consider a cube of sides a located at the origin (Fig. 24–27). Suppose that an electric field is present and given by $\mathbf{E} = bx^2\mathbf{i}$, where b is a constant. Calculate the flux through each side of the cube, and use this to find the charge within the cube.

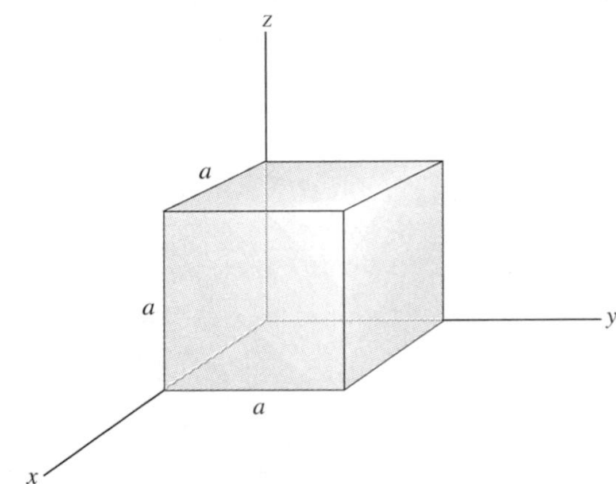

FIGURE 24–27 Problem 38.

39. (II) Repeat the calculation of Problem 38 for $\mathbf{E} = bx^2\mathbf{i} + cxz\mathbf{k}$. The quantities b and c are constants.

40. (II) Repeat the calculation of Problem 38 for $\mathbf{E} = bx^2\mathbf{i} - dxy\mathbf{j} + cxz\mathbf{k}$. The quantities b, c, and d are constants.

41. (II) Consider a solid sphere of radius R with a charge Q distributed uniformly. Suppose that a point charge q of mass m, with a sign opposite that of Q, is free to move within the solid sphere. Charge q is placed at rest on the surface of the solid sphere and released. Describe the subsequent motion. In particular, what is the period of the motion, and what is the total energy of the point charge? [*Hint:* Recall the properties of the motion for which the force varies linearly with the distance from a fixed point and is a restoring force.]

42. (II) A constant electric field is inside a tube of square cross section with sides of length L and is parallel to the sides of the tube. A plane surface cuts the interior of the tube at an angle θ (Fig. 24–28). Show by explicit calculation that the flux through this surface is independent of the angle θ. How would you show this without calculating explicitly the flux through the surface?

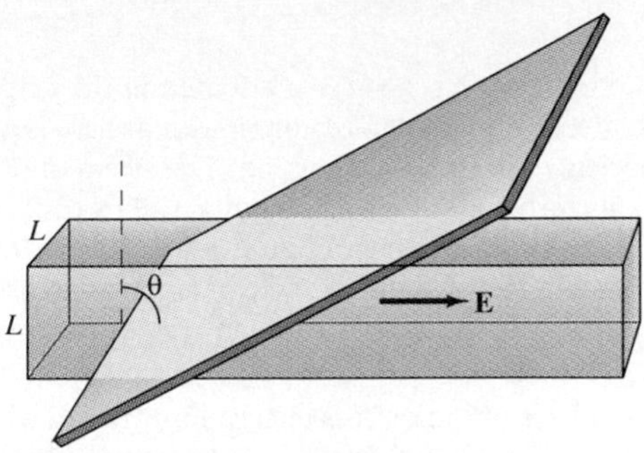

FIGURE 24–28 Problem 42.

43. (II) A conducting sphere of radius 0.25 m is centered at the origin of a coordinate system, as is a surrounding conducting shell of radius 0.75 m. The inner sphere has a charge density of 0.10 mC/m^2 over its surface, and the outer sphere has a uniform charge density twice that large. (a) Find the electric field at a distance 0.30 m from the origin; (b) at a distance 0.50 m from the origin. (c) How would your answers to parts (a) and (b) change if the outer shell were not present? (d) What is the electric field at a distance 1.0 m from the origin?

44. (II) A constant electric field E that points in the $+z$-direction passes through an equilateral tetrahedron whose base is in the xy-plane and whose six edges have length L (Fig. 24–25). Calculate the total flux through the three upper sides of the tetrahedron.

45. (II) How should the charge density of a sphere of radius R vary with the distance from the center of the sphere to give a radial field of constant magnitude within the sphere? What happens at the origin, and why?

46. (II) A right solid conducting cylinder has a charge of 10 mC. Inside the cylinder a -3-mC charge rests at the center of a hollow spherical space (Fig. 24–29). (a) What is the charge on the surface of the hollow spherical space? (b) What is the charge on the outside surface of the cylinder?

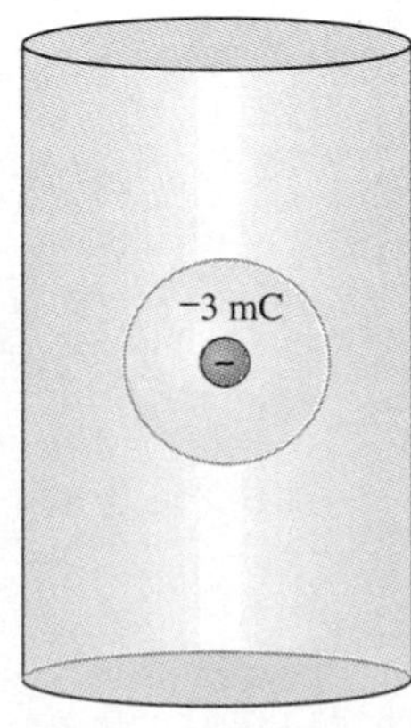

FIGURE 24–29 Problem 46.

47. (II) A conductor has a surface oriented in the yz-plane that marks the boundary of a region in which there is an electric field oriented in the $+x$-direction. The strength of this field drops linearly as x increases from $x = 0$ m to $x = 3$ m. At the beginning of the region, at $x = 0$ m, the field strength is 500 N/C; at $x = 3$ m, the field strength has dropped to zero. Describe the distribution in the x-direction of the charge that produces this field.

48. (III) A sphere of radius R is charged uniformly with charge density ρ. Use Gauss' law to show that the electric field inside the sphere at a point P whose displacement vector from the sphere's center is $\mathbf{r}$ is given by $\mathbf{E} = (\rho/3\epsilon_0)\mathbf{r}$. A small sphere centered at the point whose displacement from the origin is $\mathbf{a}$ is cut out of the sphere (Fig. 24–30). Use the superposition principle to calculate the electric field inside the cavity. [*Hint:* The cavity can be created by inserting in the original sphere a sphere of opposite charge density, $-\rho$, and radius b, centered at $\mathbf{a}$.]

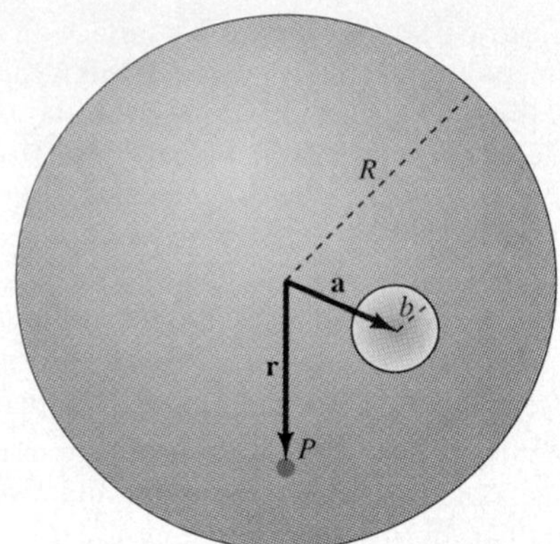

FIGURE 24–30 Problem 48.

49. (III) Use Gauss' law to show that a test charge in the electric field due to any given static charge distribution cannot be in stable equilibrium. [*Hint:* At an equilibrium point, the net electric field must be zero. What must the fields in the vicinity of that point be so that the equilibrium is stable?]

50. (III) In a two-dimensional world, the electric field due to a point charge is radial and has the magnitude $E = Q/2\pi\epsilon_0 r$, just like the electric field due to an infinite wire. What form does Gauss' law take in the two-dimensional world? [*Hint:* Try replacing a closed surface by a closed contour.]

CHAPTER 25

The discharge of lightning bolts provides an impressive demonstration that there is energy in electric fields. A significant electric potential difference exists between the earth and the clouds or between different clouds when lightning forms.

ELECTRIC POTENTIAL

The electric (Coulomb) force, like the gravitational force, arises from fundamental laws of nature. The electric force is conservative, so a collection of charges will have a potential energy. This potential energy can become kinetic energy, just as the potential energy of a rock balanced on the edge of a cliff becomes kinetic energy when the rock falls. In this chapter we describe electric potential energy. The concepts of conservative forces, work, and potential energy have already been developed in Chapters 6 and 7. Moreover, many of the results we develop here are similar to those for gravitation (Chapter 12), because the gravitational force and Coulomb's law have the same inverse-square form.

Electric forces concern the interaction of a charge distribution and a second charge. We found it useful to employ the electric field, which isolates the effect of the charge distribution alone, instead of the force. The force is the product of the second charge and the electric field. Similarly, the electrical potential energy is the energy of the charge distribution together with a second charge. In this chapter we define the electric potential, measured in volts, which is a property of the charge distribution alone. The potential bears the same relation to the electric field as the potential energy bears to the force.

25–1 ELECTRIC POTENTIAL ENERGY

The properties of conservative forces and potential energy were studied in Chapter 7.

We have already learned that the concept of an energy of position, or a potential energy, is extremely useful. For example, we know that a mass m at a height h (much less than the earth's radius) above the surface of the earth has a potential energy that can be written as $U(h) = mgh$. This helps us to determine the object's speed at any height if we know its speed at one height. *Any* force that is a function of position only is a conservative force, which means that an object under the influence of such a force has a potential energy associated with it. This potential energy is a function of position, and it can be converted to kinetic energy in accordance with the conservation of energy: The total energy is $E = K + U$, where K is the kinetic energy. Conservation of energy means that the change in E is zero, so $\Delta E = 0 = \Delta K + \Delta U$, or $\Delta K = -\Delta U$. Thus any change in U will be matched by an equal but opposite change in K.

The electric force on charge q_0 due to charge q, separated by a distance r, is

$$F = \frac{qq_0}{4\pi\epsilon_0}\frac{1}{r^2}\,\hat{\mathbf{r}}, \tag{25–1}$$

where $\hat{\mathbf{r}}$ is the unit vector that points radially outward from the position of q. This force bears a striking resemblance to the gravitational force between a mass m_0 and a mass m separated by a distance r,

$$\mathbf{F} = -Gmm_0\frac{1}{r^2}\,\hat{\mathbf{r}}. \tag{25–2}$$

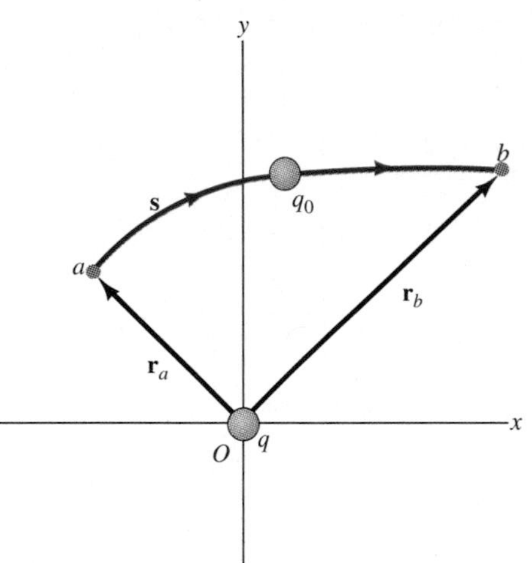

FIGURE 25–1 When a test charge q_0 moves from point a to point b in the presence of a charge q that is fixed in place, the potential energy of the system changes.

(The gravitational force is always attractive, whereas the electric force is attractive or repulsive according to whether qq_0 is negative or positive.) Both forces are conservative, so both have a potential energy U. This potential energy, which is a function of position, takes the same form for both cases.

Only *changes* in potential energy have meaning. From Eq. (7–9), we can express the change in potential energy of our system as the charge q_0 (or, in the case of gravitation, the mass m_0) moves from an initial point a at position $\mathbf{r}_a$ to a final point b at position $\mathbf{r}_b$ through the displacement $\mathbf{s}$ (Fig. 25–1) as

$$\Delta U = U_\mathrm{f} - U_\mathrm{i} = U(\mathbf{r}_b) - U(\mathbf{r}_a) = -\int_{\mathbf{r}_a}^{\mathbf{r}_b} \mathbf{F}\cdot d\mathbf{s}. \tag{25–3}$$

For conservative forces the integral in this expression is a line integral whose value is *independent of the path of integration* between points a and b.

Let us now evaluate the change in electric potential energy for the point charge q at the origin and the point charge q_0 that moves from point a to point b. We start with the simplest situation (Fig. 25–2a), in which point a is on the same radius as point b. Then we take the path from a to b along the dashed line shown in Fig. 25–2a. Because the Coulomb force points outward along the radial direction we have chosen, we have for our path

$$\mathbf{F}\cdot d\mathbf{s} = F\,dr.$$

Then, from Eq. (25–3), the potential energy change when charge q_0 moves from a to b is

$$\Delta U = -\int_{r_a}^{r_b} F\,dr = -\int_{r_a}^{r_b} \frac{qq_0}{4\pi\epsilon_0 r^2}\,dr$$

$$= -\frac{qq_0}{4\pi\epsilon_0}\int_{r_a}^{r_b}\frac{dr}{r^2} = -\frac{qq_0}{4\pi\epsilon_0}\left(\frac{-1}{r}\right)\Bigg|_{r_a}^{r_b} = \frac{qq_0}{4\pi\epsilon_0}\left(\frac{1}{r_b} - \frac{1}{r_a}\right). \tag{25–4}$$

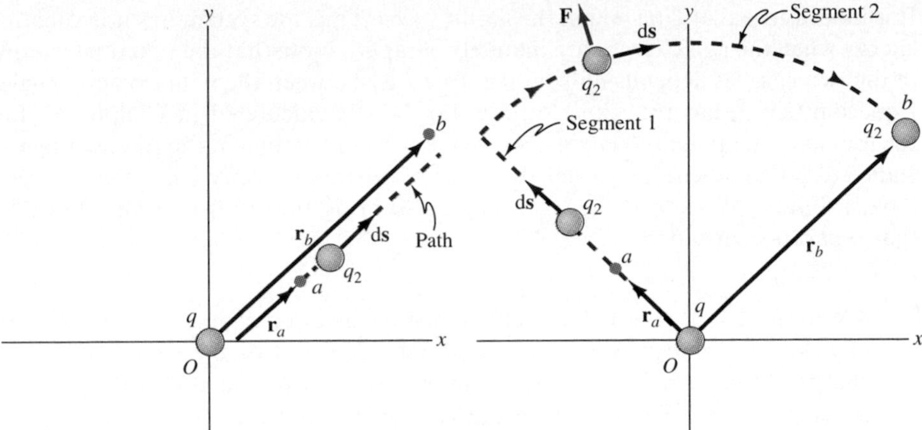

FIGURE 25–2 The change in potential energy of the system of two charges q and q_0 when the charge q_0 moves from point a to point b, in terms of a path-independent line integral. (a) Charge q_0 moves along a radius, and the path is taken directly along the radius. (b) The two points are not along the same radius. The path runs radially outward to the radius of point b, then follows the circumference at that radius.

What if charge q_0 moves between two points that do not lie on the same radius, as in Fig. 25–2b? In that case we follow the dashed path shown. [Remember, the result of the integration in Eq. (25–3) is path independent.] For segment 1, which runs outward radially from a to a distance r_b from the origin, the result is identical to Eq. (25–4). For segment 2, which follows a circumference at a distance r_b from the origin, the integral is zero, because the force is perpendicular to the path segment **ds** everywhere. The result for the change in potential energy is still given by Eq. (25–4).

Equation (25–4) is the change in electrical potential energy when charge q_0 moves from any point a distance r_a from charge q to any other point a distance r_b from charge q.

Let us look at the physical content of Eq. (25–4). Suppose first that the charges move closer together ($r_a > r_b$). If the charges repel (qq_0 is positive), the change in potential energy is positive. This is like moving a mass *up* a mountain. If the charges attract (qq_0 is negative), the system loses potential energy when the charges move closer together. This is like moving a mass *down* the mountain. As with any potential energy, electric potential energy can be converted into kinetic energy. If there are no additional forces acting, then like-sign charges slow down, or lose kinetic energy, when they move closer together. Similarly, charges of opposite sign speed up, or gain kinetic energy, when they move closer together. We can redo our analysis when the charges move farther apart ($r_a < r_b$). Charges that repel lose electric potential energy and, if there are no other forces, gain kinetic energy. Opposite charges (which attract) gain electric potential energy when they move farther apart and lose kinetic energy in the absence of other forces.

Equation (25–3) shows that the change in electric potential energy is given by the difference of two functions, $U(r_b)$ and $U(r_a)$. We know from Chapter 7 that only *changes* in potential energy have physical consequences. We are therefore free to choose zero of the potential energy function to be at whatever value of r we like. It can be convenient and natural to choose zero potential energy to be at infinity. We can do this if we let $r_a \to \infty$ and let r_b take on a general value r in Eq. (25–4):

$$\Delta U = U(r) - U(r_a)\Big|_{r_a \to \infty} \to \frac{qq_0}{4\pi\epsilon_0}\frac{1}{r}.$$

We then say that the potential energy of a charge q_0 a distance r from charge q is the difference in potential energy between that point and infinity. When we reverse the roles of q and q_0, the potential energy of q at a distance r from q_0 is again $qq_0/4\pi\epsilon_0 r$. We may thus say that the **electric potential energy** $U(r)$ for a system of two point charges q and q_0 separated by a distance r is

Electric potential energy between two point charges. Zero potential energy is chosen to be at infinity.

$$U(r) = \frac{qq_0}{4\pi\epsilon_0}\frac{1}{r}. \tag{25–5}$$

It is indeed true that $U(r) = 0$ in the limit $r \to \infty$. Thus the system has no potential energy when the two charges are infinitely far apart. Note that the potential energy of the two charges depends *only* on the distance r between them, not on any angle. Equation (25–5) has the same form as Eq. (12–9), calculated in Chapter 12 for the gravitational potential energy. As we know from Chapter 7, the physical significance of potential energy is that the value of potential energy when two charges have a finite separation is the work needed to bring the charges from infinity to that separation distance.

EXAMPLE 25–1 Nuclear fission is the breakup of nuclei. Fission occurs in heavy nuclei because of the Coulomb repulsion of the many positively charged protons in the nucleus. The more protons are in a nucleus, the stronger is the repulsion. The attractive nuclear force is not quite strong enough to overcome the repulsion between all the protons in a heavy nucleus. In calculating the repulsive forces between the protons, we want to find the potential energy between the protons. Find the electrostatic potential energy between two of the 92 protons in a ^{236}U nucleus in which the protons are as close as they can be inside a uranium nucleus, about 2×10^{-15} m apart. The radius of a uranium nucleus is about 8×10^{-15} m.

SOLUTION: We take zero potential energy to be at infinity and use Eq. (25–5) to calculate the electrostatic potential energy. The proton charge (see Table 22–1) is 1.6×10^{-19} C. Then

$$U = \frac{(+e)^2}{4\pi\epsilon_0 r} = \frac{(9 \times 10^9\ \text{N}\cdot\text{m}^2/\text{C}^2)(1.6 \times 10^{-19}\ \text{C})^2}{(2 \times 10^{-15}\ \text{m})} \simeq 10^{-13}\ \text{J}.$$

This is a typical energy value on the nuclear scale and is about 10^5 times larger than the energy that holds the proton and electron together in a hydrogen atom (see Example 25–3). When fission occurs, this potential energy is converted into kinetic energy of the various fragments. This kinetic energy is in turn the energy source of nuclear reactors.

There is an important difference between gravitational potential energy and electrostatic potential energy. The gravitational force is always attractive. The potential energy is always negative, if its value at infinity is taken to be zero. Electrostatic forces are attractive only when the charges are of opposite sign. If the two charges have like signs, the force is repulsive, and the potential energy is positive if its value at infinity is zero.

25–2 ELECTRIC POTENTIAL

A point charge q is the source of an electric field **E** that exists in the surrounding space. The electric field affects any charge q_0 introduced into that space because there is a force **F** on q_0 given by $\mathbf{F} = q_0\mathbf{E}$. We saw in Section 25–1 that the introduction of a charge q_0 a distance r from q gives rise to the potential energy $U(r)$ of Eq. (25–5). If we write $U(r) = q_0V(r)$, we can make a statement analogous to the statement about the electric field: A charge q is the source of an **electric potential** (or just **potential**) $V(r)$, which affects any charge q_0 a distance r from q by creating potential energy $U(r) = q_0V(r)$. Strictly speaking, we should deal with a small *test charge* q_0, so that its presence does not disturb charge q or indeed any more general charge distribution that gives rise to the electric potential. The definition of electric potential, which we shall see below is a *work per unit charge*, due to a charge distribution is then

$$V(\mathbf{r}) \equiv \frac{U(\mathbf{r})}{q_0}, \qquad (25\text{–}6)$$

Definition of electric potential

where $U(\mathbf{r})$ is the potential energy of the test charge q_0 in the presence of the charge distribution. The potential $V(\mathbf{r})$ is independent of q_0, just as the electric field, defined by $\mathbf{E} \equiv \mathbf{F}/q_0$, is independent of the test charge:

The electric potential is a property only of the charge distribution that produces it.

Electric potential, like the electric field, is a property only of the charge or charges producing it, not of the test charge q_0.

The Electric Potential of a Point Charge

Let us calculate the electric potential of the simplest possible system, namely, one point charge. Consider two point charges q and q_0 separated by a distance r. As Eq. (25–5) shows, the potential energy of the system is $U(r) = q_0/4\pi\epsilon_0 r$. If we think of q_0 as a test charge, then $U/q_0 = q/4\pi\epsilon_0 r$. We have found *the electric potential of a point charge q at a point a distance r from the charge*:

$$V(r) = \frac{q}{4\pi\epsilon_0 r}. \qquad (25\text{–}7)$$

The electric potential of a point charge

In Eq. (25–7) we have assumed that zero potential energy is at infinity, and, as a consequence, *we have taken zero electric potential due to a charge q to be at infinity.* As we claimed, the electric potential depends only on charge q, not on the test charge q_0.

The *electric potential difference* due to the charge q between the points a and b at locations $\mathbf{r}_a$ and $\mathbf{r}_b$ is given by

$$\Delta V = V_b - V_a = \frac{U_b - U_a}{q_0} = \frac{q}{4\pi\epsilon_0}\left(\frac{1}{r_b} - \frac{1}{r_a}\right). \qquad (25\text{–}8)$$

Here we have abbreviated V as a function of r_a, or $V(r_a)$ as V_a, and so forth.

We can obtain another formulation of the electric potential difference by using Eqs. (25–3) and (25–8) and substituting $\mathbf{F} = q_0\mathbf{E}$:

$$\Delta V = \frac{U_b - U_a}{q_0} = -\int_{\mathbf{r}_a}^{\mathbf{r}_b} \mathbf{E} \cdot d\mathbf{s}. \qquad (25\text{–}9)$$

Here the potential energy difference is expressed as a *path-independent* integral over an electric field. There is no reference in Eq. (25–9) to the electric field of the point charge. Equation (25–3) is the potential energy change when a test charge q_0 moves from point a to point b in the field of *any* charge distribution. Thus Eq. (25–9) is a general expression for the electric potential difference between two points. Any charge distribution produces an electric field, and any charge distribution will have an electric potential. Electric potential is a useful concept in part because it is a scalar quantity. It is easier to deal with than the vector quantity that determines it, the electric field.

The electrical potential difference is an integral over the electric field.

We recall that the change in the potential energy of a system is equal to the negative of the work done by the system in moving an object from point a to point b. Equivalently, $U_b - U_a$ is the work done by an external agent to move the object. These relations hold for the changes in electric potential energy when a test charge moves. Therefore we can interpret Eq. (25–9) to mean that

The electrical potential difference $V_b - V_a$ is the work per unit charge that must be done to move a test charge from point a to point b without changing its kinetic energy.

This work is performed by an external agent; for example, we may literally push the charge. If there is no external agent, then a change in potential, which corresponds to a change in potential energy of the test charge, must be accompanied by a corresponding change in the kinetic energy of the test charge.

If we know the electric potential $V(\mathbf{r})$ due to a charge distribution and we know the magnitude of a test charge q_0, then we also know the potential energy $U(\mathbf{r})$ when q_0 is placed at a point a displacement $\mathbf{r}$ away:

$$U(\mathbf{r}) = q_0 V(\mathbf{r}). \tag{25–10}$$

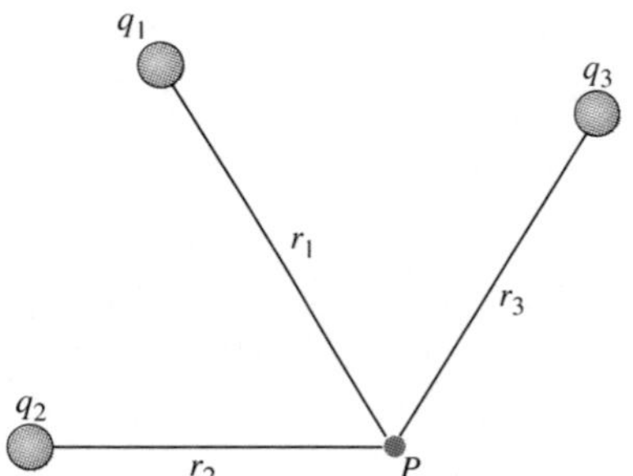

FIGURE 25–3 The superposition principle determines the potential at point P due to multiple charges. We simply add the potential due to each of the charges.

This equation tells us that, in the absence of other forces, a positive test charge q_0 in the presence of an electric potential *will move toward lower values of the potential*, because in that way the potential energy decreases. The charge speeds up as it moves to lower potentials.

The Electric Potential of Different Charge Distributions

The electric field obeys the superposition principle. Therefore the electric potential of a system of charges can also be determined from the superposition principle. The superposition principle states that the electric field of a collection of charges is the sum of the electric fields of each charge. Thus *the electric potential at a point P due to n point charges $q_1, q_2, \ldots, q_n$* (Fig. 25–3 shows three charges) at distances $r_1, r_2, \ldots, r_n$ from point P is just

$$V_P = \frac{q_1}{4\pi\epsilon_0 r_1} + \frac{q_2}{4\pi\epsilon_0 r_2} + \cdots + \frac{q_n}{4\pi\epsilon_0 r_n}$$

or

Electric potential of a collection of point charges

$$V_P = \frac{1}{4\pi\epsilon_0}\sum_{i=1}^{n}\frac{q_i}{r_i}, \tag{25–11}$$

where r_i is the distance from point charge q_i to point P. The electric potential due to a collection of charges is the scalar sum of the potentials due to single charges. This scalar sum is much easier to perform than the vector sum that expresses the electric field due to a collection of point charges.

The calculation of the electric potential due to a continuous charge distribution is also straightforward. We first find the electric potential $\mathrm{d}V$ at a point P due to a small charge Δq (Fig. 25–4). Because electric potential is a scalar quantity, the addition of all the tiny potentials $\mathrm{d}V$ is given by scalar integration. Thus *the potential due to a continuous charge distribution* takes the symbolic form

Electric potential due to a continuous charge distribution

$$V = \int \mathrm{d}V = \frac{1}{4\pi\epsilon_0}\int\frac{\mathrm{d}q}{r}. \tag{25–12}$$

The integration must be done over the entire charge distribution. In Section 25–5 we shall discuss techniques for its calculation for specific cases.

Units of Electric Potential

The dimension of electric potential is energy per charge, so its SI unit is joules per coulomb (J/C). Electric potential is used frequently, so it has a separate name in the SI, the **volt** (V). It is named after Alessandro Volta, who did research at the beginning of the nineteenth century on the nature of electric energy:

Units of electric potential

$$1\ \text{V} \equiv 1\ \text{J/C}. \qquad (25\text{–}13)$$

In Chapter 23 we mentioned that the electric field has the units volts per meter (V/m) as an alternative to newtons per coulomb (N/C). Equation (25–9) shows us why this is so: Electric potential has the dimensions of electric field times length, so the dimensions of electric field must be the dimensions of potential divided by length (units V/m):

$$1\ \text{N/C} = 1\ \text{V/m}. \qquad (25\text{–}14)$$

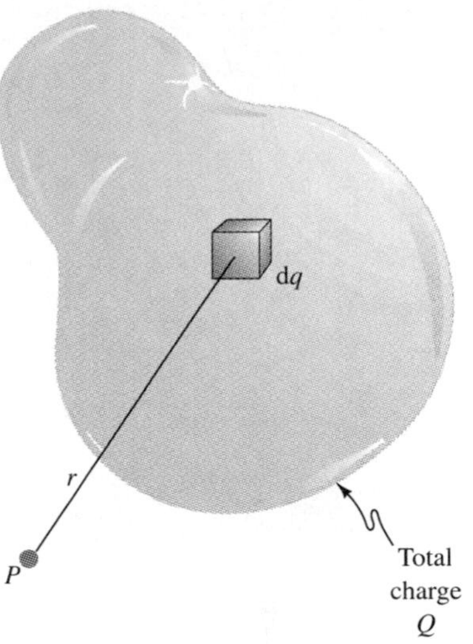

FIGURE 25–4 To find the potential at point P due to a continuous charge, integrate over the differential charges dq as if each dq were a point charge.

The Potential Energy in a System of Charges

Equation (25–10) gives the potential energy $U(r) = q_0V(r)$ of a test charge q_0 placed in the electric potential of a charge distribution. If the charge distribution is a collection of charges, then the electric potential V_P is given by Eq. (25–11), and the potential energy of the test charge is $U(r) = q_0V_P$. But it would be a mistake to call this the potential energy of the entire system of charges $q_0, q_1, q_2, \ldots, q_n$, because the product q_0V_P just represents the work that needs to be done to bring charge q_0 in from infinity. It does *not* take into account the work that must be done to bring the charges $q_1, q_2, \ldots, q_n$ in from infinity. To calculate the potential energy of a collection of three charges, for example, we assemble the three of them one by one. To bring the first charge, q_1, in to the point P_1 requires no work by the external agent if the kinetic energy of the charge is unchanged. To bring the second charge, q_2, in from infinity to the point P_2 does take work, because of the potential due to q_1. For our two charges, the work the external agent must do to bring q_2 in from infinity—the potential energy—is given by

$$U_{12} = q_2V_1 = \frac{q_1q_2}{4\pi\epsilon_0 r_{12}}, \qquad (25\text{–}15)$$

where r_{12} is the distance between charges q_1 and q_2.

What happens if we bring a third charge, q_3, in from infinity? We have to calculate the additional work done by an external force to bring q_3 in. This work is given by the product of q_3 and electric potentials V_1 and V_2 due to q_1 and q_2 in place. Thus the additional contribution to the potential energy of the system is

$$U_{13} + U_{23} = \frac{q_1q_3}{4\pi\epsilon_0 r_{13}} + \frac{q_2q_3}{4\pi\epsilon_0 r_{23}}, \qquad (25\text{–}16)$$

where r_{13} and r_{23} are the distances between q_3 and q_1, q_3 and q_2, respectively. The total potential energy U of the system is the sum of U_{12}, U_{13}, and U_{23}:

$$U = \frac{1}{4\pi\epsilon_0}\left(\frac{q_1q_2}{r_{12}} + \frac{q_1q_3}{r_{13}} + \frac{q_2q_3}{r_{23}}\right). \qquad (25\text{–}17)$$

This can be generalized to any number of charges, and the resulting formula for the *electric potential energy of the system* is a simple generalization of Eq. (25–17):

The potential energy of a system of charges

$$U = \frac{1}{4\pi\epsilon_0}\sum_{i<j}\frac{q_iq_j}{r_{ij}}, \qquad (25\text{–}18)$$

where r_{ij} is the distance between the locations of the charges q_i and q_j. The sum over i and j includes all charge pairs in the system, and the inequality $i < j$ avoids the counting of pairs more than once. We can eliminate that restriction by writing the equivalent expression

$$U = \frac{1}{2} \sum_{\substack{i,j \\ i \neq j}} \frac{q_i q_j}{4\pi\epsilon_0 r_{ij}}.$$

Now the sum is unrestricted, except that we omit the case $i = j$, which is not in the original sum, Eq. (25–18). Thus we can rewrite Eq. (25–18) as

$$\begin{aligned} U &= \frac{1}{2} q_1 \sum \frac{q_j}{4\pi\epsilon_0 r_{1j}} + \frac{1}{2} q_2 \sum \frac{q_j}{4\pi\epsilon_0 r_{2j}} + \frac{1}{2} q_3 \sum \frac{q_j}{4\pi\epsilon_0 r_{3j}} + \cdots \\ &= \frac{1}{2} q_1 V_1 + \frac{1}{2} q_2 V_2 + \frac{1}{2} q_3 V_3 + \cdots, \end{aligned} \qquad (25\text{–}19)$$

where V_1 is the electric potential due to all the other charges at the location of charge q_1, and so on. It should be stressed that the potential energy of q_1 in a given potential V_1 is still $q_1 V_1$; this means that $q_1 V_1$ can be converted into the kinetic energy of the particle that carries charge q_1. This potential energy must be distinguished from the potential energy of the *entire* charge configuration, Eq. (25–18) or (25–19). The potential energy of the entire charge configuration is the energy that would be made available if *all* the charges that appear in the problem were to escape to infinity. It is because the potential energy of any one charge depends on the "environment" created by all the other charges that the total potential energy of the system is not equal to the sum of the individual potential energies of the particles.

In Examples 25–2 and 25–3, we illustrate calculation techniques for the electric potential energy and the electric potential when two or more point charges are involved.

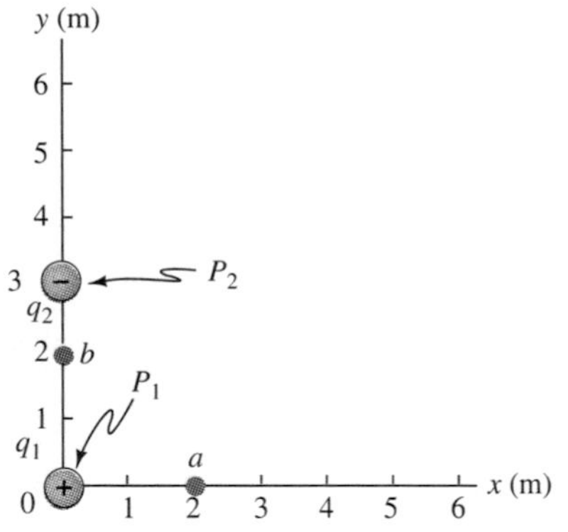

FIGURE 25–5 Example 25–2.

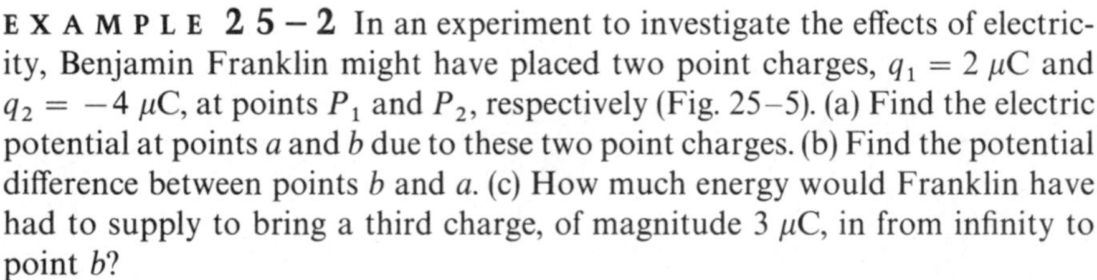

EXAMPLE 25–2 In an experiment to investigate the effects of electricity, Benjamin Franklin might have placed two point charges, $q_1 = 2\ \mu C$ and $q_2 = -4\ \mu C$, at points P_1 and P_2, respectively (Fig. 25–5). (a) Find the electric potential at points a and b due to these two point charges. (b) Find the potential difference between points b and a. (c) How much energy would Franklin have had to supply to bring a third charge, of magnitude $3\ \mu C$, in from infinity to point b?

SOLUTION: (a) We use Eq. (25–11) to determine the electric potential. Let us label the 2-μC charge, located at P_1, as q_1, and the -4-μC charge, located at P_2, as q_2. First, we find the potential at point a. The distance from point a to point P_1 is $r_{1a} = 2$ m, and the distance from point a to point P_2 is $r_{2a} = \sqrt{(2\text{ m})^2 + (3\text{ m})^2} = 4$ m. The electric potential V_a at point a is then

$$\begin{aligned} V_a &= \frac{1}{4\pi\epsilon_0}\left(\frac{q_1}{r_{1a}} + \frac{q_2}{r_{2a}}\right) \\ &= (9 \times 10^9\ \text{N·m}^2/\text{C}^2)\left(\frac{2 \times 10^{-6}\ \text{C}}{2\text{ m}} + \frac{-4 \times 10^{-6}\ \text{C}}{4\text{ m}}\right) = 0\ \text{V}. \end{aligned}$$

The units of potential are volts, a consequence of our use of SI units everywhere. In this case, the unit combination is N·m/C = J/C = V. Such a check is always useful.

Next we find the electric potential at point b. The distance from charge q_1 to b is $r_{1b} = 2$ m; similarly, $r_{2b} = 1$ m. Therefore, the potential V_b is

$$V_b = \frac{1}{4\pi\epsilon_0}\left(\frac{q_1}{r_{1b}} + \frac{q_2}{r_{2b}}\right)$$

$$= (9 \times 10^9\ \text{N}\cdot\text{m}^{\not{2}}/\text{C}^{\not{2}})\left(\frac{2 \times 10^{-6}\ \not{\text{C}}}{2\ \not{\text{m}}} + \frac{-4 \times 10^{-6}\ \not{\text{C}}}{1\ \not{\text{m}}}\right)$$

$$= -2.7 \times 10^4\ \text{V} = -27\ \text{kV}.$$

(b) The potential difference $V_b - V_a = -27\ \text{kV} - 0\ \text{kV} = -27\ \text{kV}$. Thus the electric potential is higher at point a than at point b.

(c) The new charge acts like a test charge, $q_0 = 3\ \mu\text{C}$. We now know the electric potential of the original system of two charges, so we use Eq. (25–10), $U_b = q_0 V_b$, to find the potential energy of the new charge at point b:

$$U_b = q_0 V_b = (3\ \mu C)(-27\ \text{kV}) = (3 \times 10^{-6}\ \text{C})(-27 \times 10^3\ \text{V}) = -0.08\ \text{J}.$$

The answer is in units of joules, because we have used SI units throughout.

The work that Franklin would have done to bring q_3 in from infinity is equal to the change in the potential energy of the system, or -0.08 J. Does this make sense? The electric potential at point b is negative. The new charge is positive and will be attracted to the negative potential. Franklin would not have done positive work to bring the charge in to point b; on the contrary, he would have done negative work, just as we have calculated. He would have done positive work to bring the same charge back out to infinity. It is useful to think of what is happening physically rather than to rely solely on the signs in the equations. It is all too easy to make a sign error.

EXAMPLE 25–3 The hydrogen atom in its normal, unexcited configuration has an electron that revolves around a proton at a distance of 5.3×10^{-11} m (Fig. 25–6). At the position of the electron, what is the electric potential due to the proton? Determine the electrostatic potential energy between the two particles. This energy is relevant to understanding the chemical activity of atoms.

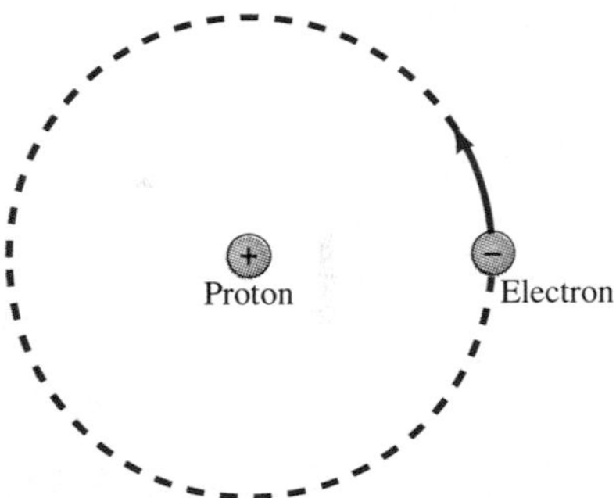

FIGURE 25–6 Example 25–3. A simplistic representation of an electron that orbits a proton in the hydrogen atom.

SOLUTION: The electric potential V_p due to the proton can be found by using Eq. (25–7). We have

$$V_p = \frac{+e}{4\pi\epsilon_0 r} = \frac{(9 \times 10^9\ \text{N}\cdot\text{m}^{\not{2}}/\text{C}^{\not{2}})(1.6 \times 10^{-19}\ \not{\text{C}})}{5.3 \times 10^{-11}\ \not{\text{m}}} = 27\ \text{V}.$$

The electrostatic potential energy is found by using Eq. (25–15), and we simply multiply the potential by the charge of the electron (the moving particle):

$$U = (-e)V_p = (-1.6 \times 10^{-19}\ \text{C})(27\ \text{V}) = -4.3 \times 10^{-18}\ \text{J}. \quad (25\text{–}20)$$

The Electron-Volt

We often determine energy by multiplying charge times voltage, as we have done in every example thus far in this chapter. Because the charge on an electron is needed so frequently, a useful unit of energy is that of the charge of an electron (or proton) times 1 V. We call this unit of energy an **electron-volt** (eV). The electron-volt is not an SI unit. The relation between the electron-volt and the SI unit joule is

The electron-volt

$$1\ \text{eV} = (1.6 \times 10^{-19}\ \text{C})(1\ \text{V}) = 1.6 \times 10^{-19}\ \text{J}.$$

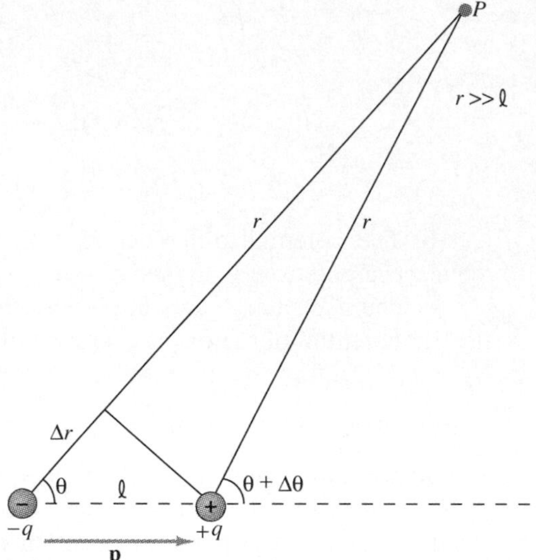

FIGURE 25–7 Example 25–4. Geometry to find the potential at a point P for an electric dipole. The dipole moment $p = q\ell$.

This unit is especially valuable for calculations in atomic, nuclear, and particle physics. In Example 25–1 the electrostatic potential energy between the two close protons becomes 6×10^5 eV, or 0.6 MeV. In Example 25–3 the electrostatic potential energy between the proton and electron of the hydrogen atom becomes -27 eV. In atomic physics we find that energy is typically on the scale of an electron-volt, whereas in nuclear physics it is MeV (10^6 eV), and in particle physics it is GeV (10^9 eV).

EXAMPLE 25–4 Calculate the electric potential due to an electric dipole whose dipole moment has magnitude p at an arbitrary point P (Fig. 25–7).

SOLUTION: A dipole consists of two pointlike charges, so Eq. (25–11) determines the potential, which will be zero at infinity. Let the distance from the $+q$ charge to point P be r, and the distance from the $-q$ charge to P be $r + \Delta r$. The point is also specified in Fig. 25–7 by the angle θ between **p** and the line between the $-q$ charge and P. Equation (25–11) gives

$$V = \frac{+q}{4\pi\epsilon_0 r} + \frac{-q}{4\pi\epsilon_0(r + \Delta r)} = \frac{+q(r + \Delta r) - qr}{4\pi\epsilon_0 r(r + \Delta r)} = \frac{q}{4\pi\epsilon_0} \frac{\Delta r}{r(r + \Delta r)}. \quad (25\text{–}21)$$

If **p** is the dipole moment, the distance between the charges is $\ell = p/q$, and the distance Δr is

$$\Delta r = \ell \cos\theta = \frac{p\cos\theta}{q} \quad (25\text{–}22)$$

When this result is substituted into Eq. (25–21), we have

$$V = \frac{qp\cos\theta}{4\pi\epsilon_0}\left[\frac{1}{r(qr + p\cos\theta)}\right]. \quad (25\text{–}23)$$

As we have mentioned, the dipole charge distribution occurs repeatedly in nature. For example, polar molecules behave as electric dipoles.† The exact electric

† Although such molecules (H_2O, for example) are neutral, they have an excess of positive charge on one end and an excess of negative charge on the other end. A more complete discussion is given in Chapter 26.

dipole potential derived in Example 25–4 takes a simple approximate form far from the dipole, when $r \gg \ell$. The condition $r \gg \ell$ is equivalent to $qr \gg q\ell = p$, and we can drop the second term in the denominator of Eq. (25–23). The result is

$$\text{for } r \gg \ell: \qquad V = \frac{p \cos \theta}{4\pi\epsilon_0 r^2}, \tag{25–24}$$

where we now measure θ from anywhere between the two charges of the dipole. Note that the potential of the dipole for distant points decreases as $1/r^2$, as compared to the $1/r$ dependence for a point charge.

25–3 EQUIPOTENTIALS

Equipotentials defined

Regions where the electric potential of a charge distribution has constant values are called **equipotentials**. They are particularly interesting and worth investigating. Suppose that a system of charges produces a certain potential. The positions in space that have the same electric potential form surfaces in three dimensions and lines in two dimensions. We say that the places where the potential has a constant value form **equipotential surfaces** in three dimensions or **equipotential lines** in two dimensions. As an example, consider the equipotential surfaces formed by a point charge. The electric potential is proportional to $1/r$ and has a constant value at any fixed radial distance from the charge. Therefore, a sphere centered on the charge forms an equipotential surface (Fig. 25–8). Any other sphere centered on the charge forms a different equipotential, because the potential varies according to the radius of the sphere.

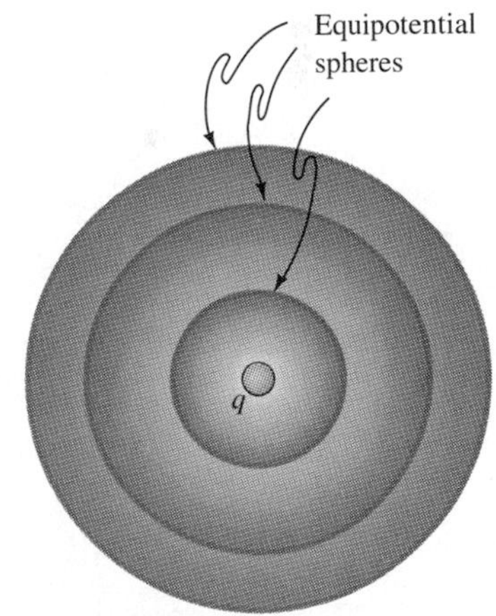

FIGURE 25–8 The equipotential surfaces for a point charge are spheres that surround the point charge.

Equipotentials are analogous to contour lines on a topographic map, lines for which the elevation difference from sea level is constant (Fig. 25–9). Because the gravitational potential energy of a mass depends only on the mass's elevation, the gravitational potential energy does not change when a mass moves along a contour line. Consequently, the force of gravity has no component along contour lines. Gravity acts in a direction perpendicular to a contour line, and a ball that starts on a particular contour line will accelerate in a direction perpendicular to

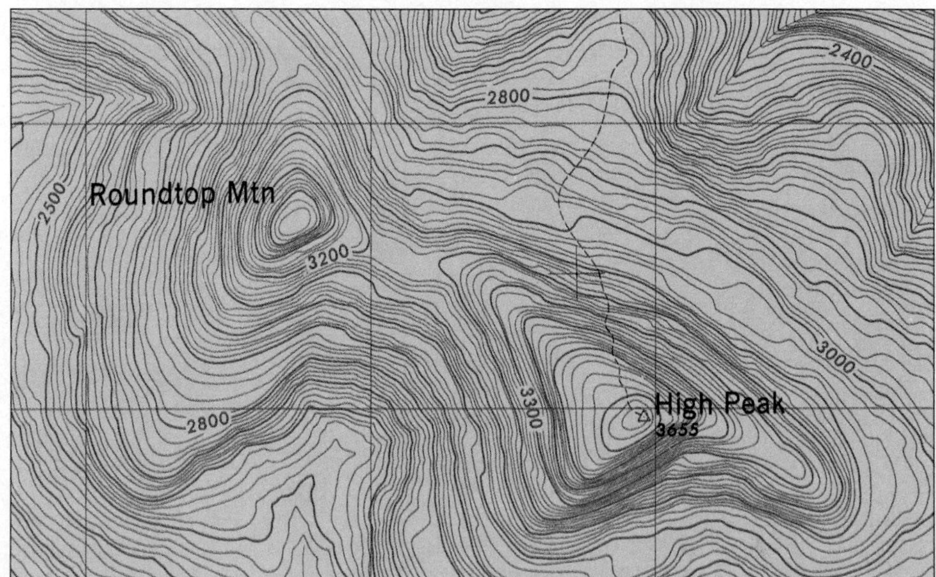

FIGURE 25–9 The contour lines on topographic maps are lines of constant elevation. These are also lines of constant gravitational potential energy. The force of gravity has no component *along* contour lines, only perpendicular to them. This map shows the contours of two peaks in the Catskill Mountains of New York.

the line, or what we would call straight down the hill (a skier would call this the fall line). What holds for contour lines holds for any equipotential surface or line: Any conservative force acts in a direction perpendicular to the equipotential, because it can have no component along the equipotential.

Because the potential has exactly the same value over an equipotential, so does the potential energy of a test charge. No work is done when the test charge moves at constant speed on an equipotential surface or line. For the point charge discussed above, the equipotentials are spheres centered on the charge (Fig. 25–8). A test charge can move freely about any one such surface without work being done by the electric field.

Electric field lines are perpendicular to the equipotential surfaces due to a system of charges.

Because no work is done by the electric force when a test charge moves on an equipotential, we can understand why the electric field cannot have a component along an equipotential surface. If it did, then that component of the electric field would do work to move a charge on the equipotential surface. This is not possible. Thus *the electric field must be everywhere perpendicular to the equipotential surface.* Furthermore, because all the charge on a conductor in equilibrium resides on the surface, a potential difference between two points on the surface would be quickly equalized by a flow of free charge, so *the surface of a conductor must be an equipotential.* In fact, the same reasoning shows that the entire conductor will then be at that same electric potential.

Electric Field Lines from Equipotentials

The fact that the electric field and the equipotentials are everywhere perpendicular to each other is very helpful in finding equipotential surfaces if the field is known, and in finding the electric fields if the equipotentials are known. We can illustrate this for some charge configurations for which the fields are known. Consider the point charge. In Fig. 25–8 we showed spherical equipotential surfaces. In Fig. 25–10, we add the electric field lines, which extend outward radially for a positive charge. The equipotential surfaces are necessarily spheres (perpendicular to the radii/vectors that emerge from the origin).

Consider the electric field lines between two oppositely charged plates (Fig. 23–11). The equipotential surfaces are planes parallel to the charged plates. If the

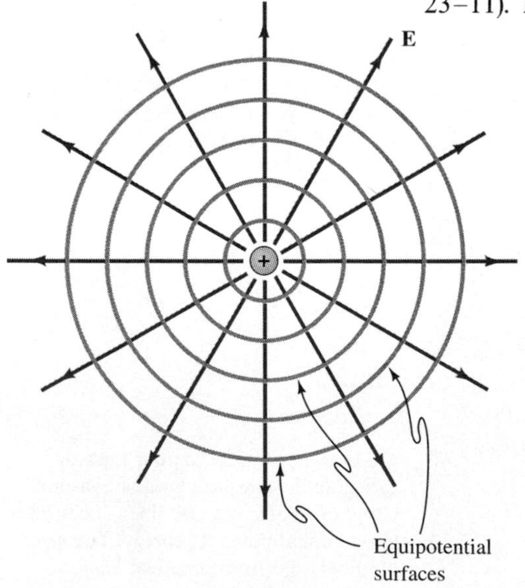

FIGURE 25–10 The electric field lines for a positive (negative) point charge extend outward (inward) radially, perpendicular to the equipotentials.

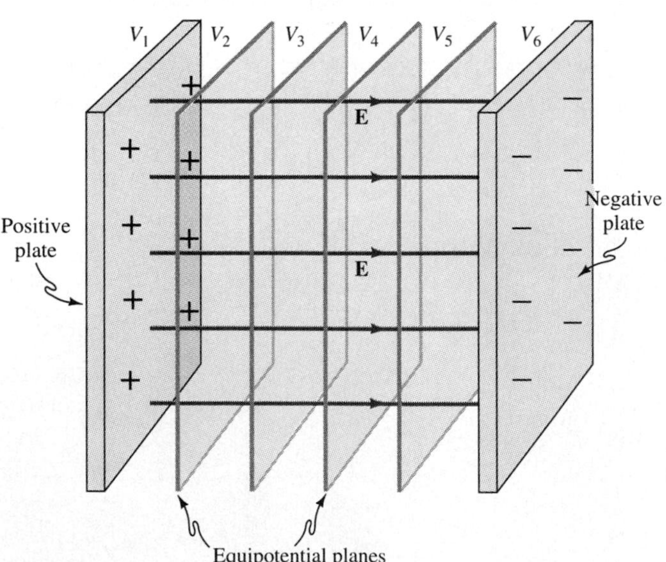

FIGURE 25–11 The electric field lines (burgundy) and the equipotentials (blue) for two oppositely charged parallel plates.

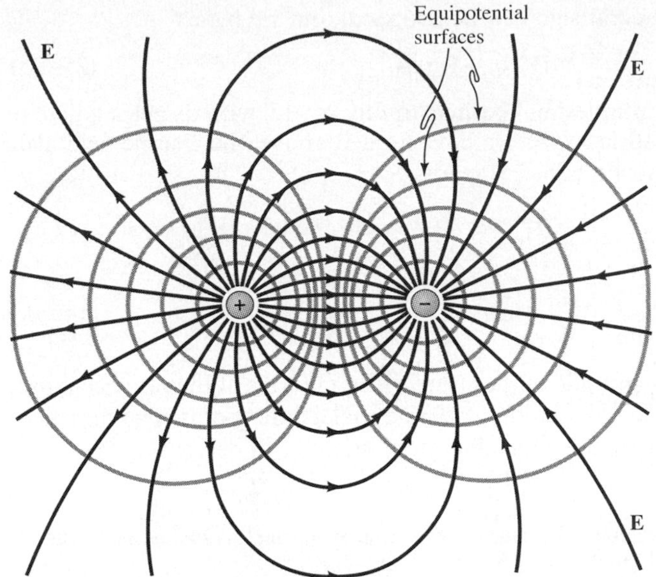

FIGURE 25–12 The electric field lines and equipotentials for an electric dipole.

charged plates are conductors, then they must also be equipotential surfaces. Thus we have a series of n equipotential planes $V_1, V_2, V_3, \ldots, V_n$ between and including the two charged plates.

Let us next examine the equipotential surfaces due to an electric dipole, such as a polar molecule (Fig. 25–12). If we draw our equipotential surfaces everywhere perpendicular to the electric field lines, we arrive at the dark blue lines shown in Fig. 25–12. Even without using the electric dipole potential derived in Example 25–4, we have already determined visually what the equipotential surfaces must look like.

25–4 DETERMINING ELECTRIC FIELDS FROM ELECTRIC POTENTIALS

As we have already seen, if we know the electric field, **E**, Eq. (25–9) determines the potential difference $V_b - V_a$ between any two points a and b:

$$V_b - V_a = \int_{\mathbf{r}_a}^{\mathbf{r}_b} dV = -\int_{\mathbf{r}_a}^{\mathbf{r}_b} \mathbf{E} \cdot d\mathbf{s}.$$

Because electrostatic forces are conservative, the potential difference is independent of the path taken between a and b in the line integral, and we can choose this path for convenience.

We learn in this section that we can reverse this procedure and calculate the electric field if the potential is known. Such a calculation is completely analogous to finding the force between objects if their potential energy is known as a function of position. Consider Fig. 25–13, which shows a set of electric field lines and two closely spaced equipotential surfaces. These equipotential surfaces are, by construction, perpendicular to the field lines. If the spacing between the equipotentials is very small, then so is the potential difference between them, which we write as dV. From Eq. (25–9), if the distance between the initial and final points a and b is infinitesimally small, then we are no longer integrating over a finite path in the integral $\int_{\mathbf{r}_a}^{\mathbf{r}_b} \mathbf{E} \cdot \mathrm{d}\mathbf{s}$. The integral sign can be dropped, and we have

$$\mathrm{d}V = -\mathbf{E} \cdot \mathrm{d}\mathbf{s}. \tag{25–25}$$

It is simplest to take our infinitesimal path as in Fig. 25–13, with ds perpendicular to the two equipotential surfaces. As we have argued above, the electric field also points in that direction, so Eq. (25–25) reads

$$\mathrm{d}V = -E\,\mathrm{d}s.$$

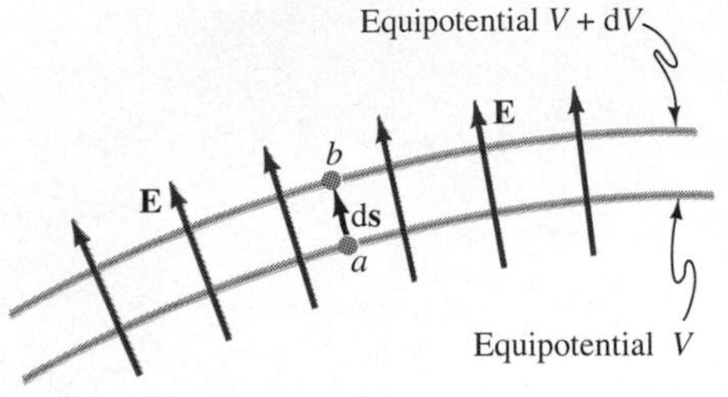

FIGURE 25–13 Two equipotentials differ by d*V*. The displacement **ds** along the direction of **E** between the equipotentials is perpendicular to them.

Equivalently,

$$E = -\frac{\mathrm{d}V}{\mathrm{d}s}. \tag{25–26}$$

This equation gives the magnitude of the electric field in terms of the rate of change of V in a direction perpendicular to the equipotential at that point. The direction of the field is along that perpendicular direction, oriented to decreasing values of the potential. In other words,

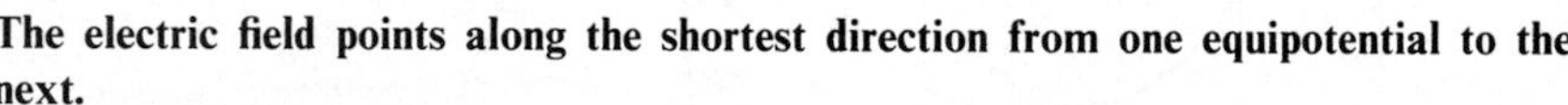

The electric field points along the shortest direction from one equipotential to the next.

When equipotentials form concentric spheres, as they do for a point charge, the perpendicular direction lies along a radius. Therefore the electric field points in the radial direction, with magnitude

$$E = -\frac{\mathrm{d}V}{\mathrm{d}r}.$$

The same expression holds for equipotentials that form concentric cylinders, but in this case the variable r is the distance to the cylindrical axis.

How the Potential Determines the Field in Cartesian Coordinates

We can look at this from a different point of view by supposing that an arbitrary displacement vector **ds** is decomposed into Cartesian coordinates:

$$\mathrm{d}\mathbf{s} = \mathrm{d}x\,\mathbf{i} + \mathrm{d}y\,\mathbf{j} + \mathrm{d}z\,\mathbf{k}.$$

Here **i**, **j**, and **k** are the unit vectors in the x-, y- and z-directions, respectively. Then the scalar product in Eq. (25–25) takes the form

$$\mathrm{d}V = -\mathbf{E} \cdot \mathrm{d}\mathbf{s} = -E_x\,\mathrm{d}x - E_y\,\mathrm{d}y - E_z\,\mathrm{d}z, \tag{25–27}$$

where we have separated the field **E** into Cartesian components. In general, the potential depends on all three space coordinates, $V = V(x, y, z)$. The change in V in going from an initial position $\mathbf{r} = x\mathbf{i} + y\mathbf{j} + z\mathbf{k}$ to a new position $\mathbf{r} + \mathrm{d}\mathbf{s} = (x + \mathrm{d}x)\mathbf{i} + (y + \mathrm{d}y)\mathbf{j} + (z + \mathrm{d}z)\mathbf{k}$ is

$$\mathrm{d}V = \frac{\partial V}{\partial x}\mathrm{d}x + \frac{\partial V}{\partial y}\mathrm{d}y + \frac{\partial V}{\partial z}\mathrm{d}z. \tag{25–28}$$

Note the use of the partial derivatives here, a usage that is necessary because V depends on all three Cartesian coordinates. Recall that partial derivatives are very simple to use: The partial derivative with respect to x, for example, means that y and z are held fixed while the ordinary derivative with respect to x is taken. To illustrate, if $V = xz^2$, then $\partial V/\partial x = z^2$, $\partial V/\partial y = 0$, and $\partial V/\partial z = 2xz$.

We can equate the coefficients of dx, dy, and dz in Eqs. (25–27) and (25–28):

$$E_x = -\frac{\partial V}{\partial x}, \qquad E_y = -\frac{\partial V}{\partial y}, \qquad E_z = -\frac{\partial V}{\partial z}.$$

Equivalently, *the electric field vector is given in terms of derivatives of the electric potential* by

$$\mathbf{E} = -\frac{\partial V}{\partial x}\mathbf{i} - \frac{\partial V}{\partial y}\mathbf{j} - \frac{\partial V}{\partial z}\mathbf{k}. \qquad (25\text{–}29)$$

The electric field in terms of derivatives of the electric potential

Equation (25–29) gives the Cartesian components of the electric field in terms of the potential. We have found a way to express a particular vector, the electric field, in terms of the derivatives of a scalar, the electric potential.† The derivative operation in Eq. (25–29) produces an electric field vector that points in the direction of the greatest decrease in the potential. This direction is perpendicular to the equipotential surfaces.

EXAMPLE 25–5 Use the electric potential of a point charge q to find its electric field.

SOLUTION: We know the electric potential and are asked to find the electric field. In this case the potential is a function only of the radial distance from the charge, $V = q/4\pi\epsilon_0 r$. The equipotential surfaces are therefore spheres at a constant distance from the charge. According to our discussion of Eq. (25–26), the electric field is therefore directed outward radially—perpendicular to the equipotentials, in the direction of decreasing potential—and has magnitude

$$E = -\frac{\mathrm{d}V}{\mathrm{d}r} = -\frac{q}{4\pi\epsilon_0}\frac{\mathrm{d}}{\mathrm{d}r}\left(\frac{1}{r}\right) = \frac{q}{4\pi\epsilon_0 r^2}.$$

We are merely reproducing what we already know here, but the technique is useful in contexts where we do not already know the answer!

EXAMPLE 25–6 A potential distribution in space is described by the function $V = Axy^2 - Byz$, where A and B are constants. Find the electric field.

SOLUTION: Here again we are given the electric potential and must find the electric field. We require a simple application of Eq. (25–29). First we find the partial derivatives:

$$\frac{\partial V}{\partial x} = Ay^2;$$

$$\frac{\partial V}{\partial y} = 2Axy - Bz;$$

$$\frac{\partial V}{\partial z} = -By.$$

The electric field is therefore

$$\mathbf{E} = -Ay^2\mathbf{i} - (2Axy - Bz)\mathbf{j} + By\mathbf{k}.$$

We conclude this section with a comment on the electric dipole. Equation (25–24) gives the dipole potential, which is proportional to the factor $\cos\theta$ in Fig. 25–7. On the bisecting axis, $\theta = 90°$, and, because $\cos 90° = 0$, the electric potential

† A mathematical shorthand for these operations has been developed. The electric field is given by the *gradient* operator ∇ that acts on the potential, $\mathbf{E} = -\nabla V$, where the gradient operator is defined to be $\nabla \equiv \mathbf{i}\frac{\partial}{\partial x} + \mathbf{j}\frac{\partial}{\partial y} + \mathbf{k}\frac{\partial}{\partial z}$.

will be zero there. However, this does not mean that the electric field will be zero on the bisecting axis. The electric field is determined by *derivatives* of the potential at a particular r or θ. What counts in determining the field is how fast the potential is changing, not whether it is zero at some point. We have $\partial V/\partial\theta \propto \sin\theta$. Along the bisecting axis, where $\theta = 90°$, the electric field is a maximum.

25–5 CALCULATING THE POTENTIALS OF EXTENDED CHARGE DISTRIBUTIONS

Only rarely do we deal with the electric field and potential of a single point charge. More often, we have collections of charges spread over regions of space, as when a charge spreads over the surface of a metal, or when the field of a complicated ionic molecule determines its chemical or biological behavior. It is therefore necessary for us to be able to find the potentials of continuous charge distributions. These charge distributions may not be simple ones, and we must develop strategies for calculating the corresponding electric potentials. In this section we first summarize the underlying techniques, then illustrate them with a series of examples.

Techniques for calculating electric potentials

The qualitative shapes of equipotential surfaces due to a charge distribution are most easily found by graphical techniques. For quantitative calculations, we have thus far learned two different ways to determine the electric potential of a charge distribution.

1. If the electric field is known, then Eq. (25–9) can be used to determine the potential:
$$\Delta V = V_b - V_a = -\int_{\mathbf{r}_a}^{\mathbf{r}_b} \mathbf{E} \cdot d\mathbf{s}.$$
2. If the electric field is not known, we may calculate the potential directly by using one of various forms:

for one point charge, Eq. (25–7):
$$V = \frac{q}{4\pi\epsilon_0 r};$$

for many point charges, Eq. (25–11):
$$V = \frac{1}{4\pi\epsilon_0}\sum_i \frac{q_i}{r_i};$$

for a continuous charge distribution, Eq. (25–12):
$$V = \frac{1}{4\pi\epsilon_0}\int \frac{dq}{r}.$$

In a direct calculation of electric potential, a decision must be made about the location of zero potential. In fact, the convention that zero potential is at infinity is already implicit in Eqs. (25–7), (25–11), and (25–12). This is almost always the most convenient choice for a charge distribution that does not extend all the way to infinity. If a potential *difference* is calculated directly, no decision need be made about the zero level.

Parallel Plates

Let us first look at the relation between electric field and potential for two parallel conducting plates (a *parallel-plate capacitor*—we shall learn more about these in Chapter 26), each brought to different potentials (Fig. 25–14a). We suppose that the plates are close enough together or large enough that we can ignore the distortions of the field near the edges. This case is therefore an approximation to the central regions of real parallel-plate capacitors. In the figure the left-hand plate is at a lower potential than the right-hand plate. The electric field between parallel plates is known to be constant, and it runs from regions of higher potential to lower potential, from right to left in this case. In Eq. (25–9), we take the path to be a straight line from the left to the right plate, so that **E** is antiparallel to d**s**. Let the direction from left to right define the x-axis, with the left plate at $x = 0$. Then the potential difference between the plates is given in terms of the field E between the plates and the separation ℓ of the plates by

$$\Delta V = V_{\text{right}} - V_{\text{left}} = -\int_{\text{left}}^{\text{right}} \mathbf{E} \cdot d\mathbf{s} = +\int_0^{\ell} E\, dx = E \int_0^{\ell} dx = E\ell.$$

In turn, the electric field between parallel plates has magnitude

The electric field between parallel plates

$$E = \frac{\Delta V}{\ell}. \tag{25–30}$$

Equation (25–30) is an important practical result. It states that

The constant electric field between parallel conducting plates is the potential difference between the plates divided by the distance between the plates.

This result gives the electric field between two parallel-plane conducting plates whose potential difference is ΔV. We can also use it to find the equipotential surfaces associated with *any* constant field. These surfaces are planes perpendicular to the field, and the potential difference for a plane a distance ℓ from a reference equipotential changes *linearly* with ℓ, $\Delta V = E\ell$. Note the sign: The potential decreases along the direction in which the electric field points.

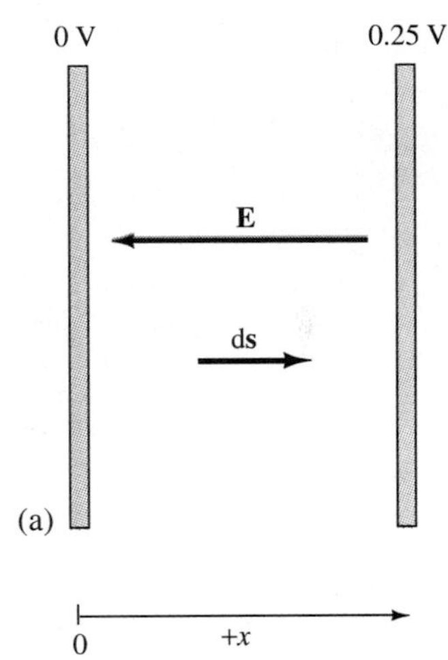

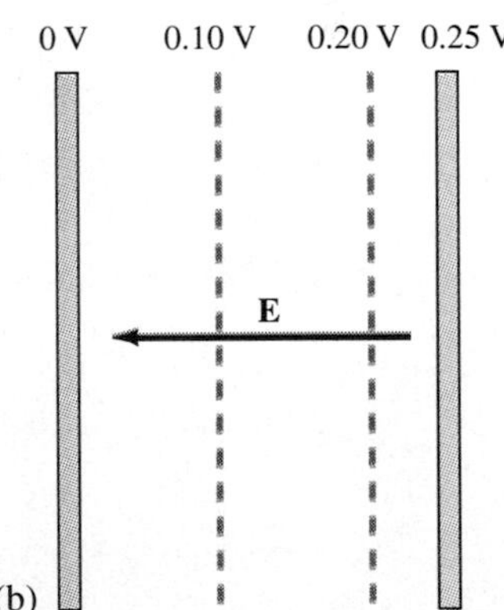

FIGURE 25–14 (a) Two parallel plates, viewed from the side, have a potential difference of 0.25 V. A differential displacement d**s** is indicated. (b) Example 25–7. Equipotentials for $V = 0.1$ V and 0.2 V. The plates are themselves equipotentials. The electric field points from right to left when the potential of the left-hand plate is lower than that of the right-hand plate.

EXAMPLE 25–7 Two parallel metal plates have area $A = 225\ \text{cm}^2$ and are separated by $\ell = 0.50$ cm. They have a potential difference of 0.25 V (Fig. 25–14a). Find the numerical value of the electric field. What are the charge density and total charge on each plate? Draw the equipotential surfaces at 0.10 V and 0.20 V.

SOLUTION: Equation (25–30) applies directly here. We know the potential difference ΔV between the plates, as well as their separation. Thus the magnitude of the electric field is

$$E = \frac{\Delta V}{\ell} = \frac{0.25\ \text{V}}{0.0050\ \text{m}} = 50\ \text{V/m},$$

and it points from right to left (Fig. 25–14b).

We know from Chapter 23 that the electric field between the plates is σ/ϵ_0:

$$E = 50\ \text{V/m} = \frac{\sigma}{\epsilon_0};$$

$$\sigma = 50\epsilon_0\ \text{V/m} = (50\ \text{V/m})(8.85 \times 10^{-12}\ \text{C}^2/\text{N}\cdot\text{m}^2) = 4.4 \times 10^{-10}\ \text{C/m}^2.$$

The answer has to be in coulombs per square meter, because we used SI units consistently. We know the area of the plates, so we can calculate the total charge on each plate to be

$$Q = \sigma A = (4.4 \times 10^{-10}\ \text{C/}\cancel{\text{m}^2})(225\ \cancel{\text{cm}^2})(10^{-4}\ \cancel{\text{m}^2}/\cancel{\text{cm}^2}) = 1.0 \times 10^{-15}\ \text{C}.$$

Because the electric field is constant between the parallel plates, we have from Eq. (25–30) that, at a distance d from the left plate, the potential differs by an amount $\Delta V = Ed$ from its value at the left plate, $d = 0$. The equipotential surface for 0.10 V is then

$$d = \frac{\Delta V}{E} = \frac{0.10 \not{V}}{50 \not{V}/\text{m}} = 0.20 \text{ cm}$$

from the left plate. For 0.20 V, we similarly determine a distance of 0.40 cm (Fig. 25–14b).

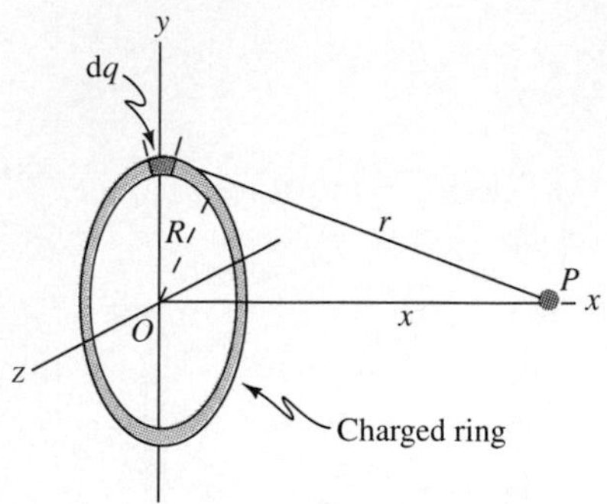

FIGURE 25–15 Example 25–8. Geometry to find the potential at a point P on the axis of a charged ring of radius R by using a differential charge dq.

EXAMPLE 25–8 *The Charged Ring.* Find the electric potential due to a uniformly charged ring of radius R and total charge Q at a point P on the axis of the ring.

SOLUTION: We use the geometry shown in Fig. 25–15 and find the electric potential at point P a distance x along the axis from the center of the ring. This problem is a straightforward application of Eq. (25–12). We set up a differential charge dq along the ring that is a constant distance $r = \sqrt{R^2 + x^2}$ from point P. Because r is constant, we can bring it outside the integral of Eq. (25–12) to obtain

$$V = \frac{1}{4\pi\epsilon_0 r} \int \mathrm{d}q = \frac{Q}{4\pi\epsilon_0 r} = \frac{Q}{4\pi\epsilon_0 \sqrt{R^2 + x^2}}. \qquad (25\text{–}31)$$

Had we wished to find it, the electric field along the axis could be determined by applying the derivative operations of Eq. (25–29) (see Problem 33). This method is easier than the direct integration technique presented in Chapter 23 for the electric field.

EXAMPLE 25–9 *The Charged Disk.* Find the electric field due to a thin, flat, uniformly charged disk of radius R and total charge Q at a point P along its axis by first calculating the electric potential at this point.

SOLUTION: The situation is shown in Fig. 25–16. In order to find the electric field, we first find the electric potential. Because we know the potential due to a ring from Example 25–8, we divide the disk into a series of concentric rings, with the intent of using the superposition principle. In Fig. 25–16 we divide the disk into a series of rings of radius r and width dr. The constant charge density over the disk is $\sigma = Q/\pi R^2$, and the total charge contained in

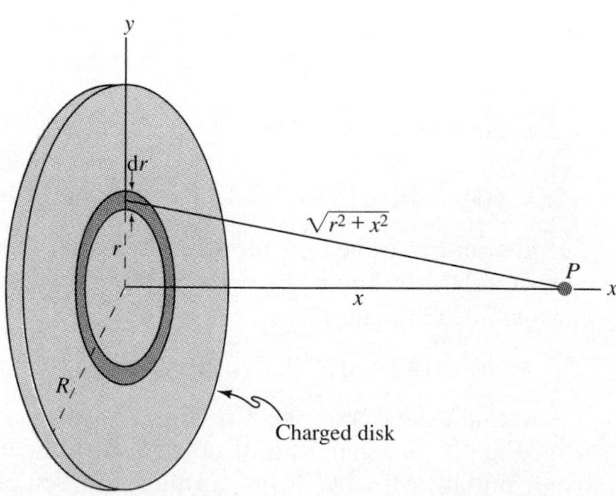

FIGURE 25–16 Example 25–9. Geometry to find the potential at a point P on the axis of a charged disk of radius R. The potential due to the ring of radius r and width dr is first found and then integrated.

a ring of differential area dA is

$$dq = \sigma\, dA = \sigma 2\pi r\, dr.$$

We use Eq. (25–31) for the potential due to the ring, and then integrate over all the rings to determine the total potential of the disk:

$$dV = \frac{dq}{4\pi\epsilon_0\sqrt{r^2+x^2}} = \frac{2\pi\sigma}{4\pi\epsilon_0}\frac{r\,dr}{\sqrt{r^2+x^2}};$$

$$V = \frac{\sigma}{2\epsilon_0}\int_0^R \frac{r\,dr}{\sqrt{r^2+x^2}} = \frac{\sigma}{2\epsilon_0}\sqrt{r^2+x^2}\Big|_0^R = \frac{\sigma}{2\epsilon_0}(\sqrt{R^2+x^2}-x)$$

$$= \frac{Q}{2\pi\epsilon_0 R^2}(\sqrt{R^2+x^2}-x). \tag{25–32}$$

The integration is a more difficult one than that of Example 25–8. Nevertheless, it required no vectorial manipulation.

Because the potential depends only on x, the electric field has only an x-component, $\mathbf{E} = E_x\mathbf{i}$. The component E_x is given by

$$E_x = -\frac{\partial V}{\partial x} = -\frac{Q}{2\pi\epsilon_0 R^2}\left(\frac{x}{\sqrt{R^2+x^2}} - 1\right).$$

That $\mathbf{E}$ has only an x-component should not come as a surprise: By symmetry, only the x-component of the field receives contributions that do not cancel out. As a check, this result can be shown in the limit that $x \gg R$ to reduce correctly to the point charge limit $Q/4\pi\epsilon_0 x^2$.

EXAMPLE 25–10 *The Charged Line.* Find the electric potential as a function of the radial distance R from an infinite charged line of uniform charge density λ.

SOLUTION: We have previously found the electric field for an infinite charged line, and can use Eq. (25–9) to find the potential from the electric field. Equation (23–32) gives us the electric field, which has only a radial component. We integrate along a radial direction, so that $\mathbf{ds} = \mathbf{dr}$ (Fig. 25–17a). Equation (25–9) becomes

$$\Delta V = -\int E_r\, dr = -\frac{\lambda}{2\pi\epsilon_0}\int\frac{dr}{r}.$$

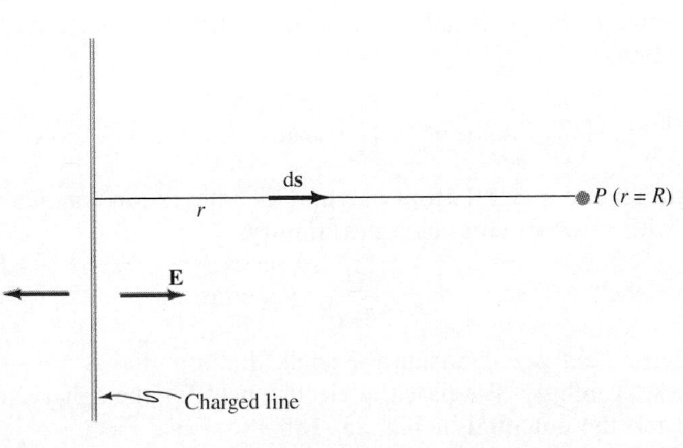

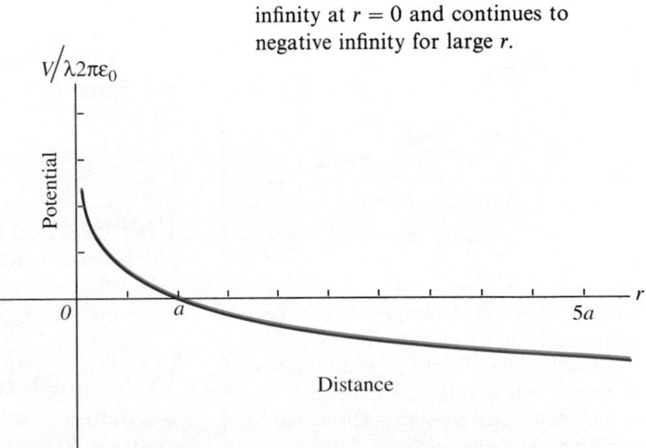

FIGURE 25–17 Example 25–10. (a) An infinite charged line has a radial electric field. To find the potential at the point P, we consider a displacement $\mathbf{ds}$ in the electric field $\mathbf{E}$ and use the known electric field. (b) The resulting potential, defined to be zero at $r = a$, goes to positive infinity at $r = 0$ and continues to negative infinity for large r.

The potential difference depends, of course, on the end points of the integration. Let zero potential be at $r = a$, so that

$$\Delta V = V_R - V_a \equiv V = -\frac{\lambda}{2\pi\epsilon_0}\int_a^R \frac{dr}{r} = -\frac{\lambda}{2\pi\epsilon_0}\ln r\Big|_a^R;$$

$$V = -\frac{\lambda}{2\pi\epsilon_0}\ln\frac{R}{a}. \qquad (25\text{–}33)$$

Note that in this case it is not possible to set zero potential at infinity, because at $a = \infty$, the logarithm is infinite. Physically, this is because the line itself reaches to infinity; we can never get far away from it. We graph the potential of Eq. (25–33) in Fig. 25–17b, where we have assumed that the charge of the line is positive.

EXAMPLE 25–11 *The Charged Sphere.* Find the potential for a uniformly charged spherical shell of total charge Q and radius R at positions both inside and outside the shell. Set zero potential at infinity.

SOLUTION: We already know the electric field of the spherical shell from Example 24–4 and can use Eq. (25–9) to determine the electric potential. As a second fixed point, we choose $r = \infty$ and determine the potential difference, ΔV, at r with the potential at ∞. From Example 24–4 we have Eqs. (24–10) and (24–11):

$$\text{outside a spherical shell, } r > R: \qquad E = \frac{Q}{4\pi\epsilon_0 r^2};$$

$$\text{inside a spherical shell, } r < R: \qquad E = 0.$$

The electric field is purely radial, so in Eq. (25–9), which gives the potential difference, we integrate along a radius from infinity to an arbitrary radial distance r. The differential element $d\mathbf{r}$ points out from the origin, so $\mathbf{E}\cdot d\mathbf{s} = E_r\,dr = E\,dr$. For a point outside the spherical shell,

$$\Delta V = V(r) - V(\infty) = -\int_\infty^r E\,dr = -\frac{Q}{4\pi\epsilon_0}\int_\infty^r \frac{dr}{r^2} = \frac{Q}{4\pi\epsilon_0}\left(\frac{1}{r} - \frac{1}{\infty}\right) = \frac{Q}{4\pi\epsilon_0 r}.$$

If we choose $V(\infty) = 0$, then the potential $V(r)$ equals ΔV:

$$\text{outside the shell, } r > R: \qquad V = \frac{Q}{4\pi\epsilon_0 r}. \qquad (25\text{–}34)$$

The difference between the potential for a position inside the shell and the potential at infinity is

$$\Delta V = -\left(\int_\infty^R E_{\text{outside}}\,dr + \int_R^r E_{\text{inside}}\,dr\right).$$

Because $E_{\text{inside}} = 0$, the second integral drops out, and the integration is similar to the previous one with zero potential again at infinity,

$$\text{inside the shell, } r \le R: \qquad V = \frac{Q}{4\pi\epsilon_0 R} = \text{a constant}. \qquad (25\text{–}35)$$

Even though the electric field is zero inside the shell, the potential is not, if we define it to be zero at infinity. We plot the electric field for the spherical shell in Fig. 25–18a and the potential in Fig. 25–18b.

E
Electric field
1/r²
O
R
r
Distance
(a)

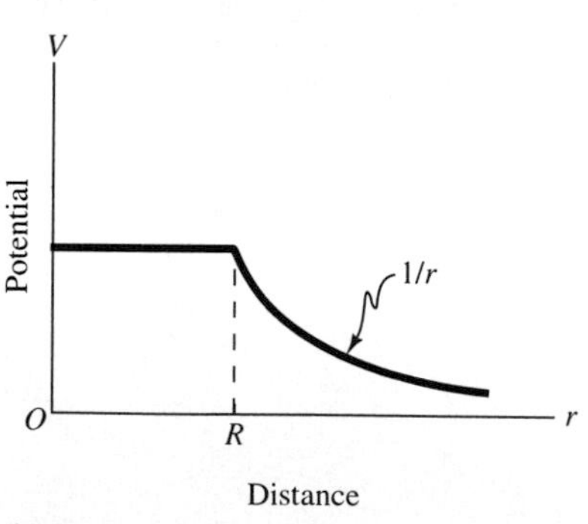

(b)

FIGURE 25–18 (a) Example 25–11. The electric field and (b) electric potential for a spherical shell of radius R. Even though the electric field is zero inside the shell, the potential has a constant value equal to that on the shell's surface.

TABLE 25–1

ELECTRIC FIELDS AND POTENTIALS FOR VARIOUS CHARGE CONFIGURATIONS

Charge Configuration	*Magnitude of Electric Field*	*Electric Potential*	*Location of Zero Potential*
Point charge	$\frac{q}{4\pi\epsilon_0 r^2}$	$\frac{q}{4\pi\epsilon_0 r}$	∞
Infinite line of uniform charge density λ	$\frac{\lambda}{2\pi\epsilon_0 r}$	$-\frac{\lambda}{2\pi\epsilon_0}\ln\frac{r}{a}$	$r = a$
Parallel, oppositely charged plates of uniform charge density σ, separation d	$\frac{\sigma}{\epsilon_0}$	$\Delta V = -Ed = -\frac{\sigma d}{\epsilon_0}$	Anywhere
Charged disk of radius R, along axis at distance x	$\frac{Q}{2\pi\epsilon_0}\left(\frac{\sqrt{R^2+x^2}-x}{\sqrt{R^2+x^2}}\right)$	$\frac{Q}{2\pi\epsilon_0 R^2}(\sqrt{R^2+x^2}-x)$	∞
Charged spherical shell of radius R	$r \geq R$: $\frac{Q}{4\pi\epsilon_0 r^2}$	$r > R$: $\frac{Q}{4\pi\epsilon_0 r}$	∞
	$r < R$: 0	$r \leq R$: $\frac{Q}{4\pi\epsilon_0 R}$	∞
Electric dipole	Along bisecting axis only, far away: $\frac{p}{4\pi\epsilon_0 r^3}$	Everywhere, far away: $\frac{p\cos\theta}{4\pi\epsilon_0 r^2}$	∞
Charged ring of radius R, along axis	$\frac{Qx}{4\pi\epsilon_0(R^2+x^2)^{3/2}}$	$\frac{Q}{4\pi\epsilon_0\sqrt{R^2+x^2}}$	∞
Uniformly charged nonconducting solid sphere of radius R	$r \geq R$: $\frac{Q}{4\pi\epsilon_0 r^2}$	$r \geq R$: $\frac{Q}{4\pi\epsilon_0 r}$	∞
	$r < R$: $\frac{Qr}{4\pi\epsilon_0 R^3}$	$r < R$: $\frac{Q}{8\pi\epsilon_0 R}\left(3-\frac{r^2}{R^2}\right)$	∞

We have now calculated the electric field and potential for several different distributions of charge. We summarize these results in Table 25–1.

25–6 ELECTRIC POTENTIALS AND FIELDS AROUND CONDUCTORS

Probably the most important cases of continuous charge distributions occur on metals; in Section 25–7 we shall give several practical examples. These distributions are rarely uniform, because charges are free to move on and within metals. Nevertheless, from what we already know, we can learn a surprising amount about the electric potentials near metals.

We have already learned that in electrostatics the electric field inside a conducting material must be zero, that the net charge on a conductor must lie on its outside surface, and that the electric field just outside the conductor's surface must be normal to the surface. We have also noted that, because there is no component

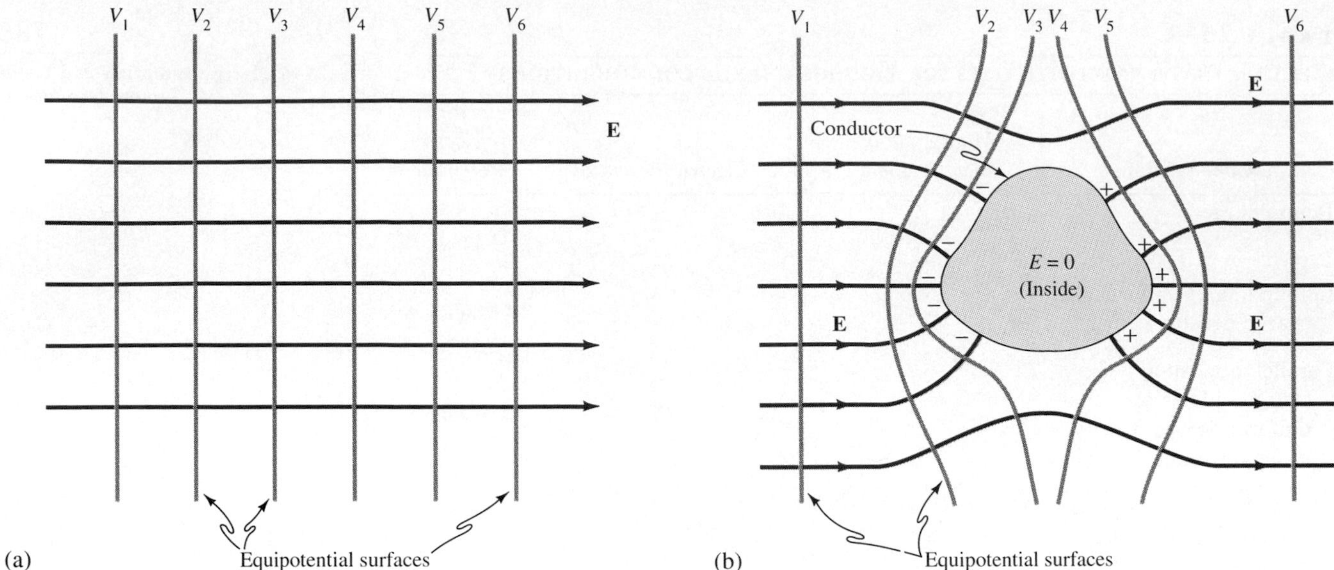

FIGURE 25–19 (a) A uniform electric field before an uncharged conductor is placed in the field. (b) Afterward, the electric field is changed dramatically, with no electric field inside the conductor. Induced charges, which make the electric field inside the conductor zero, appear on the outside surface of the conductor. These charges affect the electric field outside the conductor.

of the electric field along the conducting surface, the conducting surface must itself be an equipotential. Because the electric field everywhere inside a conducting material is zero, the potential inside must have the same value as it has at the surface. We have already seen an effect of this nature in Example 25–11, the charged spherical shell. The potential at the surface and everywhere inside the shell has a constant value (that is, it is an equipotential).

We can also say something about the electric potential outside a conductor, whether the conductor is charged or not. If it is charged, the fact that the electric field is perpendicular to the surface means that the equipotentials near the surface will be parallel to the surface. This will be true even if the conductor is uncharged. Consider, for example, the electric field shown in Fig. 25–19a due to two parallel plates (not shown). If we place an uncharged conductor of arbitrary size in this electric field, the field around the conductor will be greatly modified (Fig. 25–19b). Charge will be induced on the outside of the conductor in equilibrium, thereby forcing the electric field to be normal to the conducting surface, and again *the equipotential surfaces near a conductor of arbitrary shape must be parallel to the conductor's surface.*

Near a conductor, the equipotential surfaces are parallel to the conductor's surface.

In Chapter 24 we found the magnitude of the electric field near the surface of a conductor in terms of the charge density at that point. The charge density, however, can vary for an irregular surface. Here we shall see how the concept of potential allows us to say more about the charge density and hence the fields near irregularly shaped charged conductors.

The Role of Sharp Points on Conducting Surfaces

Consider the irregular conductor shown in Fig. 25–20. The left side has a more pointed configuration than the right side. We can characterize these two sides by inscribing spheres in the ends and measuring their respective radii. The left region is more sharply curved than the right region, and $r_1 < r_2$. We model this conductor with a two-step process. First consider the two spherical conductors shown in Fig. 25–21a. The sizes of these spheres match the two ends of our irregular conductor of Fig. 25–20: The charges q and q' are placed on the two spheres. The electric potentials at the spheres are, respectively,

$$V_1 = \frac{q}{4\pi\epsilon_0 r_1}$$

and

$$V_2 = \frac{q'}{4\pi\epsilon_0 r_2}.$$

Now connect the two spheres by a long conducting wire (Fig. 25–21b). The entire system will come to the same potential after charge flows rapidly between the two spheres. This potential is

$$V = \frac{q_1}{4\pi\epsilon_0 r_1} = \frac{q_2}{4\pi\epsilon_0 r_2},$$

where q_1 and q_2 are the equilibrium charges on the two spheres. The charges and radii are related by

$$\frac{q_1}{r_1} = \frac{q_2}{r_2}.$$

Because we have used only the fact that the entire system is at a single potential, this calculation will apply also to our irregular conductor of Fig. 25–20. The connected spheres form a model for the relative size of the electric fields at the two ends of the conductor. Note that q and q' no longer appear; all that remains is the charge-conservation requirement that $q + q' = q_1 + q_2$.

The surface charge density σ on a sphere is determined by the charge on the sphere and the surface area:

$$\sigma = \frac{q}{\pi r^2}.$$

For our two spheres, therefore, the equation $q_1/r_1 = q_2/r_2$ is

$$\frac{\sigma_1 \not{\pi} r_1^{\not 2}}{\not{r}_1} = \frac{\sigma_2 \not{\pi} r_2^{\not 2}}{\not{r}_2};$$

$$\sigma_1 r_1 = \sigma_2 r_2. \qquad (25\text{–}36)$$

The electric field E_i just outside each conducting sphere is equal to σ_i/ϵ_0, so that we can replace σ_i by $\epsilon_0 E_i$. Thus $\epsilon_0 E_1 r_1 = \epsilon_0 E_2 r_2$, or

$$E_1 r_1 = E_2 r_2;$$

$$\frac{E_1}{E_2} = \frac{r_2}{r_1}. \qquad (25\text{–}37)$$

The electric fields are inversely related to the radii. *For a small radius, the surface charge density and the corresponding electric field are large.*

The effect just described is important in conductors with sharp points

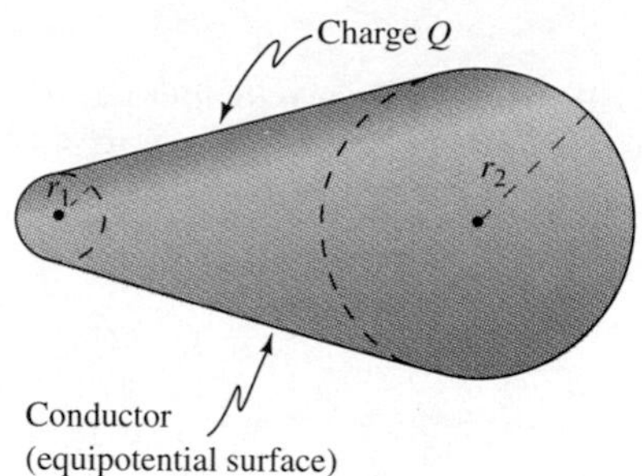

FIGURE 25–20 A conductor of odd shape is approximated by spheres of radii r_1 and r_2 at its ends.

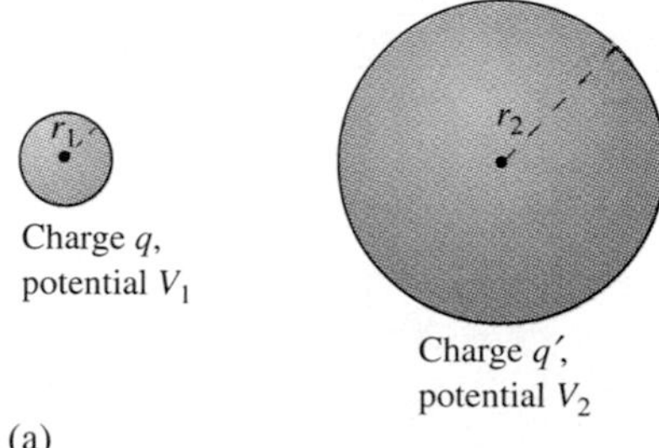

Conducting wire

r_1 r_2

Charge q_1, potential V

Charge q_2, potential V

(b)

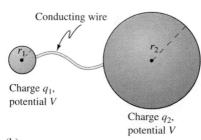

FIGURE 25–21 (a) Two conductors initially are at different potentials dependent on their respective charges. (b) If the conductors are connected by a wire, charge must flow to make the potentials equal everywhere.

The electric field near a conductor is larger near regions of sharp curvature.

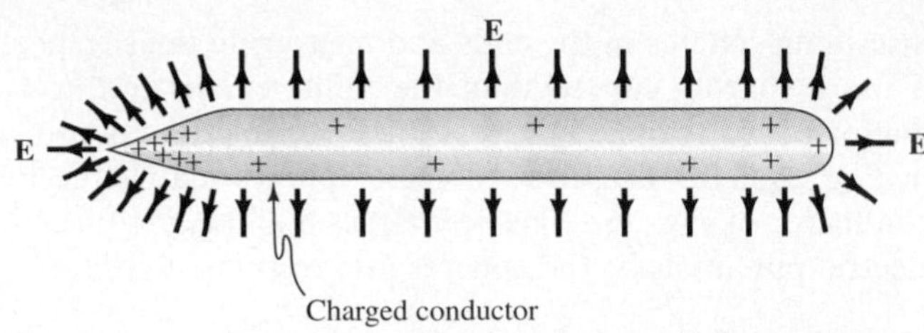

FIGURE 25–22 The electric field near small radii of curvature can be quite large, as seen for the point of a charged needle.

(Fig. 25–22). Even if the conductor is at a low electric potential, particular regions of the surface with small radii of curvature can have large electric fields nearby. When an electric field is strong enough to overcome the attraction between ions and electrons, *corona discharge* occurs. The electrons are ripped loose from the molecules, which are said to be *ionized*. In air, this occurs for fields on the order of 3×10^6 V/m. The ionization of air molecules causes a greenish glow. Sailors long ago saw these glows at the pointed tops of their masts and spars and dubbed the phenomenon *St. Elmo's fire*. The phenomenon is more generally called *dielectric breakdown* (see Fig. 23–22, and the opening photograph in Chapter 26).

The positively charged ionized molecules are accelerated by the large electric fields and move away from the region of breakdown. Thus after dielectric breakdown, the air in effect becomes a conductor that carries away the excess charge. This lowers the electric potential around the original conductor. In thunderstorms there are large potential differences between the ground and clouds because of a buildup of charges in the clouds (see the chapter-opening photograph). Lightning rods cause a local dielectric breakdown and lower the potential difference between the rod and the clouds. The rods do not *attract* the lightning strikes—by lowering the potential difference, they prevent lightning from striking in the vicinity.†

For spherical conductors the potential and electric field on the surface are related by $V = RE$. Dielectric breakdown can thus be produced by putting a relatively low potential on conductors with sharp needle points. The maximum potential that can be put on a metal without air ionization is $V_{max} = R(3 \times 10^6 \text{ V/m})$. For a needle point with a radius of curvature $R = 0.1$ mm, the maximum potential is only 300 V, but for a domed structure of $R = 3$ m, the maximum potential is about 10^7 V.

25–7 ELECTRIC POTENTIALS AND ELECTROSTATIC FIELDS IN TECHNOLOGY

Even though electrostatics represents only a small portion of our study of electromagnetism, it has important applications. We briefly mention a few of them here to indicate how the understanding of basic principles can be put to good practical use.

The Van de Graaff Accelerator

If we place a charge anywhere in a conductor, the charge will move to the outside surface; the field inside the conductor will be zero. Robert Van de Graaff took

† Conversely, a high, isolated object such as a tree or a tall building can provide a path to the ground that requires less energy; these objects are often struck by lightning.

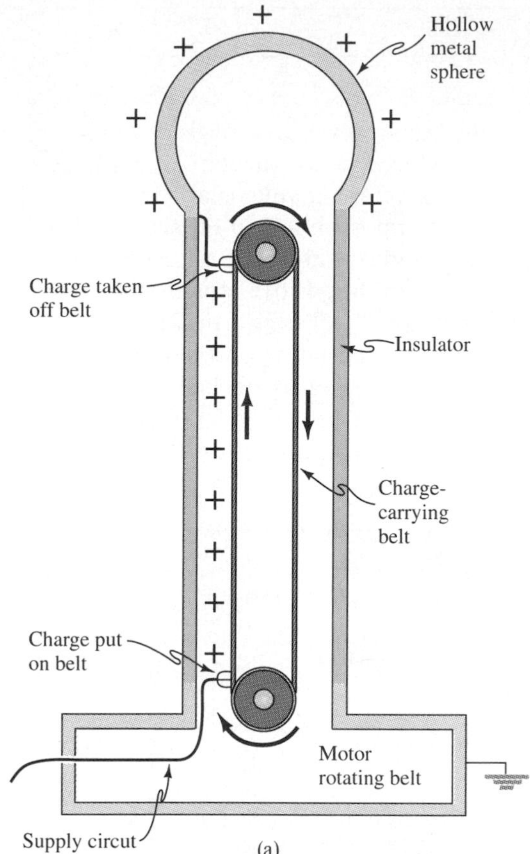

FIGURE 25–23 (a) Schematic diagram of a simple Van de Graaff accelerator. Charge is sprayed on the rotating belt at the bottom and taken off at the top. The charge goes to the outside surface of the conductor, and the potential continues to build up to high values. The symbol in the bottom right indicates that the base of the accelerator has been grounded. (b) The children touching this Van de Graaff generator are brought to a high electric potential. The individual hairs behave like the leaves of an electroscope.

advantage of this concept in 1931 to build an *accelerator*, an apparatus that produces highly energetic charged particles. Such particles are useful for, among other things, microscopic probes of matter and as cancer treatments. Van de Graaff used a device similar in concept to the apparatus shown schematically in Fig. 25–23a. An insulated belt (or chain) continuously brings charge to the inside of a hollow conductor where brushes extract charge from the belt. Because the brush holders are connected by a wire to the hollow conductor, the charge from the belt will proceed from the brushes along the wire to the outside surface of the conductor. The electric potential on the spherical conducting surface increases as charge flows to its surface ($V = q/4\pi\epsilon_0 R$).

An *ion source* located inside the hollow conductor produces charged atoms of the same sign as the potential. These charged atoms are repelled from the region of high potential and are thus accelerated. Such devices are called **Van de Graaff accelerators** or **Van de Graaff generators** (Fig. 25–23b). Similar devices called *tandems* have a high positive potential in the center; negative atoms, attracted to the positive potential, start from a potential of zero and are then stripped of two or more electrons inside the hollow conductor as the atoms pass through a thin target. The positive atoms are then accelerated again as they are repelled by the high potential back to zero, gaining additional energy. By surrounding the entire accelerator with large, high-pressure tanks filled with gases that resist breakdown, potentials as high as 25 million V have been reached.

APPLICATION

The Field-Ion Microscope

The phenomenon that large electric fields occur at sharp points on a conductor is carried to its extreme in **field-ion microscopy**, in which high electric fields allow us to produce images of individual atoms in the crystalline structure of the sharp point, or tip, of a metal. A fine tip of the crystalline material is prepared, commonly by dipping a mechanically formed tip in an electrolyte, a substance that dissolves atoms off the end. These tips are as small as 200 nm across, depending on the particular metal and crystal being prepared. On the 200-nm scale such a tip looks smooth, but at the atomic level it is still very rough. The tip is then introduced into a vacuum, and a large positive potential of several kilovolts is applied (Fig. B1–1). The end of the tip is smoothed off even further as atomically sharp points of atoms of the metal itself are driven off in the form of positive ions by the large fields. The smoothing process leaves a tip like that in Figs. B1–2a and B1–2b, depending on the precise orientation of the crystal. Oranges stacked in layers into a semipyramidal shape provide a familiar example of the possible structure of a tip (Fig. B1–2c).

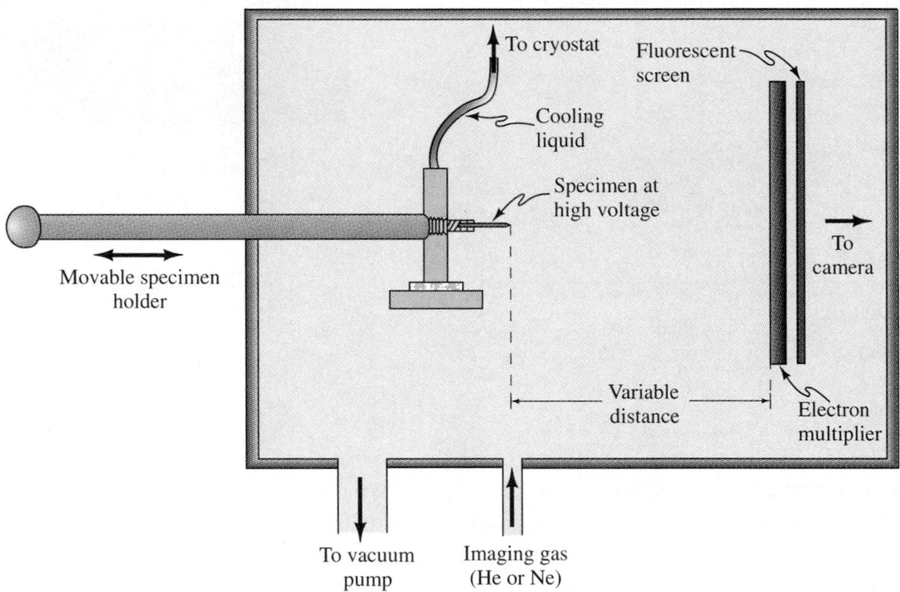

FIGURE B1–1 Schematic diagram of a field-ion microscope. The cryostat maintains a steady, low temperature in the chamber.

Xerography

Photocopying machines take advantage of electrostatics in several steps of **xerography**, or photoreproduction. The process begins with a positively charged plate coated with photoconducting material—something that is a good conductor in light but not in the dark—such as selenium (Fig. 25–24a). Light reflected from the original to be copied passes through a lens onto the charged plate, where the dark areas remain charged, but in areas that receive light, charges flow to the plate underneath (Fig. 25–24b). The resulting image of the dark areas is represented by the remaining charges. Negatively charged toner (a black powder) is added to the positively charged plate, leaving the original dark areas with black toner on the plate (Fig. 25–24c). In the next step, paper that has also been positively charged is placed over the plate and attracts the black, negatively charged toner (Fig. 25–24d). Heat is used to fuse the toner (and thus the image) to the paper. That is why the freshly peeled paper (Fig. 25–24e) feels warm and is sometimes still charged when it comes out of the photocopying machine.

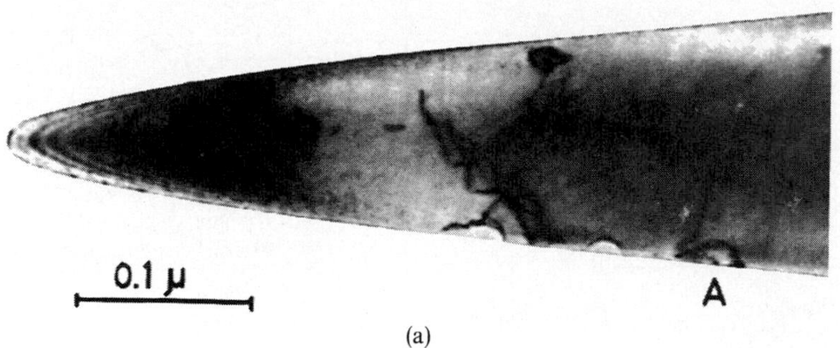

(a)

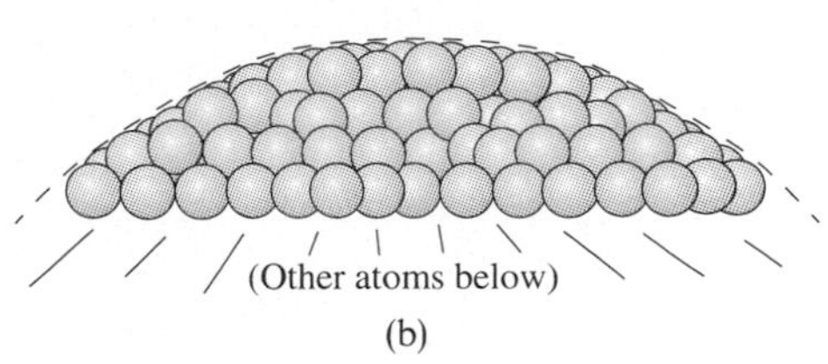

(b)

FIGURE B1–2 (a) Magnified iron tip. (b) The spheres represent individual atoms in a field-ion tip.

In the next stage, a dilute gas such as helium or neon, called the *imaging* gas, is introduced into the chamber that contains the tip. The positive potential on the tip is increased, again to several kilovolts, until the gas atoms just begin to be ionized. This happens only where the field is largest: right above individual tip atoms. The field is strong enough to ionize the gas. The gas ions are then driven away from the tip, following the electric field lines out from the tip atoms to a grounded screen, where the impinging gas ions leave a visible trace. The image formed corresponds to the position of the individual atoms of the tip, which thus become visible in a picture such as Fig. B1–3. Field-ion microscopy is useful for observing crystal structures and the effects of impurities and defects in crystals. Even some noncrystalline materials can be investigated this way.

FIGURE B1–3 Field-ion micrograph of an iridium tip. The features are formed by individual atoms.

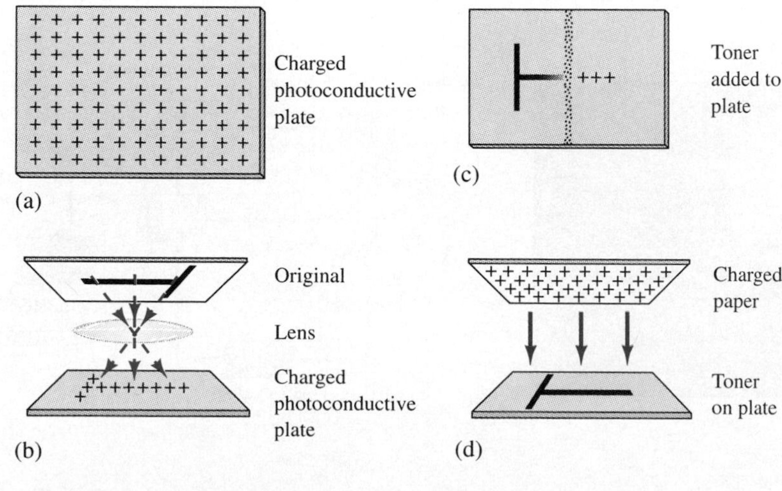

FIGURE 25–24 Schematic diagram of xerography. (a) A positively charged photoconductive plate. (b) Light from the white areas on the original neutralizes the positive charges on the plate. (c) The negatively charged toner is attracted to the positive charge. (d) The positively charged paper picks up the toner. Heat seals the toner on the paper, (e) which is then peeled off the plate.

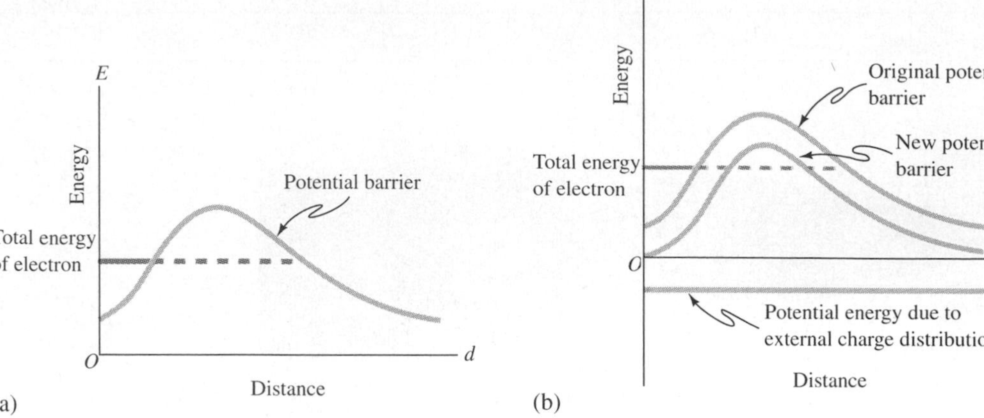

FIGURE 25–25 Energy of electrons versus distance from the outside surface of a metal. (a) Electrons are held within a metal by a potential barrier near the surface. If the barrier is higher than the total energy of the electrons, the electrons can escape only by tunneling, a very unlikely event. (b) The presence of an external object that attracts the electrons reduces the barrier effect and makes the probability of tunneling much higher.

Electric Potentials and Quantum Engineering

Electrons are trapped within a metal by electrostatic attractions to their parent ions. These forces can be represented in classical physics by a potential barrier that the electrons cannot cross. Figure 25–25a is an energy diagram of the potential energy of an electron as well as the electron's (constant) total energy. Classically, the electron cannot enter the region where its total energy is less than its potential energy. However, as we described in Chapter 7, the electron has some very small probability of "tunneling" through the barrier due to quantum effects.

Some positively charged object brought near the metal surface pulls on the electrons. The positively charged object has an electric potential with respect to the metal, and an electron has a potential energy due to the external object (Fig. 25–25b). When the potential energy due to the external object is added to the original potential energy that holds the electron within the metal, the barrier is in effect reduced. Even if the external potential is too weak to lower the maximum potential energy below the electron's total energy and allow the electrons to escape classically, the fact that it lowers the barrier makes it easier for the electrons to tunnel through the barrier. This "potential-assisted tunneling" is utilized in the *scanning tunneling microscope* (Fig. 25–26a). A weak positive potential is placed on an ultrafine tungsten needle. The needle scans the surface of a sample and provides the necessary potential to help electrons escape the sample by tunneling. These

FIGURE 25–26 (a) The scanning head of a scanning tunneling microscope. (b) Schematic diagram of a scanning tunneling microscope. The fine-tipped needle in the scanning head comes to within 1 nm of the sample; this distance is the tunnel gap. The tunneling current across this gap holds the gap distance constant as the tip scans the sample surface and thereby provides a map of that surface. The base voltage leads to the tunneling current, which can be used to form the driving voltage. The driving voltage moves the tip through piezoelectricity, a phenomenon in which a voltage is applied to and thereby moves a crystal.

(a)

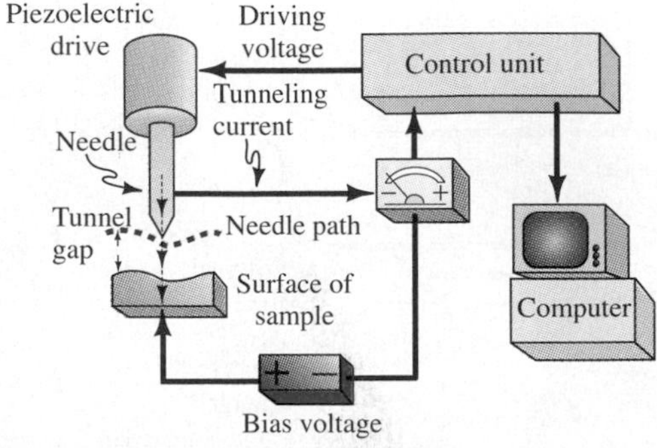

(b)

FIGURE 25–27 False-color scanning tunneling micrograph of the surface of a sample of graphite. The regular pattern of individual atoms is evident.

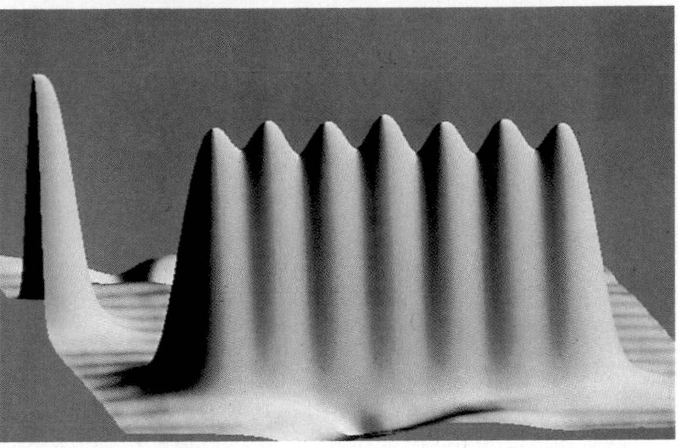

FIGURE 25–28 Individual xenon atoms (whose size is on the order of tenths of a nanometer) have been moved to line up in a row.

electrons are attracted to the needle and form a current through it whose magnitude depends on the distance between the needle and the surface (Fig. 25–26b). This effect is used in two ways:

1. A feedback mechanism that constantly repositions the needle can be set up so that the current is constant. The distance between the needle tip and the sample's surface is therefore constant. The repositioning can be measured, and the topography of the surface is thereby mapped (Fig. 25–27).
2. The potential on the needle can exert a slight pull on whole atoms. Even though atoms are neutral, the negatively charged electrons and the positively charged nuclei form an induced dipole. The same mechanism that allows a comb, charged by being pulled through dry hair on a dry winter day, to attract pieces of electrically neutral paper attracts the atoms out of the sample material. In this way atoms can be pulled along the surface of the sample and moved *one at a time* to new positions (Fig. 25–28). This effect holds the promise of allowing new molecules and ultrasmall logic circuits, switching circuits in computers, to be constructed. The combination of electrostatics and quantum mechanics is rapidly becoming a tool in what is aptly called *quantum engineering.*

SUMMARY

The Coulomb force is conservative, so a potential energy—electric potential energy—is associated with it. If a test charge q_0 moves from point a to point b in the presence of a point charge q at the origin, the change in potential energy is given by

$$\Delta U = \frac{qq_0}{4\pi\epsilon_0}\left(\frac{1}{r_b} - \frac{1}{r_a}\right). \tag{25–4}$$

The electric potential difference due to any charge distribution between points

a and b is defined as the change in potential energy divided by the magnitude of a test charge q_0:

$$\Delta V = V_b - V_a = \frac{U_b - U_a}{q_0} = -\int_{r_a}^{r_b} \mathbf{E} \cdot d\mathbf{s}. \qquad (25\text{–}9)$$

Here $\mathbf{E}$ is the electric field due to the charge distribution. The integral in Eq. (25–9) is independent of the path between the end points. The potential difference $V_b - V_a$ is the work done per unit charge by an external agent in moving a test charge from point a to point b with no change in kinetic energy. The potential is independent of the test charge.

The electric potential can be determined by the following methods, in addition to graphical methods:

1. If the electric field is known, then Eq. (25–9) may be used.
2. If the electric field is not known, it is generally easier to calculate the potential directly by using one of these forms:

for one point charge:
$$V = \frac{q}{4\pi\epsilon_0 r}; \qquad (25\text{–}7)$$

for many point charges:
$$V = \frac{1}{4\pi\epsilon_0} \sum_i \frac{q_i}{r_i}; \qquad (25\text{–}11)$$

for a continuous charge distribution:
$$V = \frac{1}{4\pi\epsilon_0} \int \frac{dq}{r}. \qquad (25\text{–}12)$$

In each of these cases, zero potential is chosen to be at infinity.

The SI unit of electric potential is the volt (V); 1 V = 1 J/C. A useful unit of energy for atomic and subatomic systems is the electron-volt (eV); 1 eV = 1.6×10^{-19} J.

The electric field can be determined if the potential is known in terms of derivatives of the potential:

$$\mathbf{E} = -\frac{\partial V}{\partial x}\mathbf{i} - \frac{\partial V}{\partial y}\mathbf{j} - \frac{\partial V}{\partial z}\mathbf{k}. \qquad (25\text{–}29)$$

The electric field between two parallel plates is constant and is given by the potential difference divided by the distance between the plates:

$$E = \frac{\Delta V}{\ell}. \qquad (25\text{–}30)$$

The electric field and potential for several charge configurations are given in Table 25–1.

Equipotential surfaces are surfaces at a fixed potential. The electric field is perpendicular to equipotentials. The surfaces of conductors form equipotentials, and the potentials inside conductors in equilibrium are everywhere the same as the potential on the surface. Electric fields just outside conductors are inversely proportional to the radius of curvature, so there are high electric fields near sharp points on conductors even if the conductors are at low potentials.

Applications of electrostatics include the Van de Graaff accelerator, the field-ion microscope, xerography, and the control of tunneling phenomena.

QUESTIONS

1. How many joules are in 1 V·C?
2. An infinite plane is uniformly charged with positive charge of density σ. How would you use a known negative test charge to measure σ?
3. How would you create an electric field inside the hollow space of a spherical metal shell?
4. In good weather, the electric field in the lower atmosphere is approximately 100 V/m, pointing downward. What happens when a 3-m metal rod is planted in the ground?
5. When an electric field moves a charge by doing work on it, what is the source of the energy to do the work? Where did this energy come from originally?
6. In describing the potential difference as the work per unit charge to move a test charge, we added the phrase "without changing the kinetic energy" (see the boldface statement on p. 730). Why is this important?
7. Will a conductor always be an equipotential? If not, under what circumstances will that occur?
8. Using Eq. (25–29), explain why changing the location of zero potential does not affect the value of the electric field.
9. A small Van de Graaff generator can be used as a lecture demonstration device. If a person touches the dome, his or her hair stands up (see Fig. 25–23b). Explain why. Why should the person stand on an insulated mat during this demonstration?
10. The earth is typically defined to be at zero potential. Does this mean that the earth can have no net charge? If the earth does have a net charge, can it still be at zero potential?
11. If we know the electric potential at a certain point, do we also know the electric field? What can we know about the electric field if we know the electric potential at two points arbitrarily close to one another?
12. Is the electric potential energy of a system of point charges independent of the order in which the system is assembled?

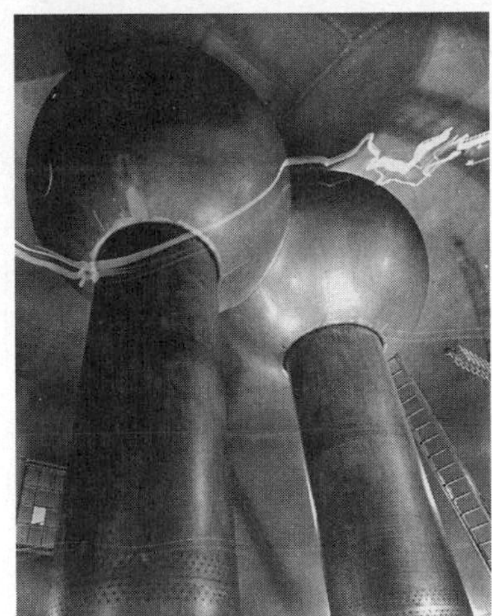

FIGURE 25–29 Question 13.

13. Why are there so many curved surfaces on the Van de Graaff generator in Fig. 25–29?
14. How do we really know that electric forces are conservative?
15. In the potential associated with a point charge, we chose zero potential to be infinitely far from the charge. What would change in our predictions about electric charges if we had chosen the potential to be zero at $r = 10^{-10}$ m from the charge?
16. If we start with point charges, for each of which zero potential is at infinity, is it possible for a superposition of charges to have zero potential other than at infinity?
17. The potential of a configuration of point charges is zero at certain points. Does this mean that the force on a test charge is zero at these points?
18. Is it possible to arrange charges so that the potential is zero over a finite contiguous region?

PROBLEMS

25–1 *Electric Potential Energy*

1. (I) A charge of 2.0×10^{-4} C is fixed at the origin of a coordinate system. A charge of 2.0×10^{-6} C is placed on a raisin of mass 1.0 g. The raisin is then brought from far away to a point 45 cm from the origin. What is the electric potential energy of the system?
2. (I) Suppose that the raisin of Problem 1 is released from rest from its position 45 cm from the origin. If no other forces act on the raisin, where will it move? What will its final kinetic energy be?
3. (I) A 3-μC charge is brought in from infinity and fixed at the origin of a coordinate system. (a) How much work is done? (b) A second charge, of 5 μC, is brought in from infinity and placed 10 cm away from the first charge. How much work does the electric field of the first charge do when the second charge is brought in? (c) How much work does the external agent do to bring the second charge in if that charge moves with unchanging kinetic energy?
4. (I) Charges $q_1 = 6.0 \times 10^{-5}$ C and $q_2 = -4.0 \times 10^{-4}$ C are placed at rest 0.50 m apart. How much work must be done by an outside agent to move these charges slowly and steadily until they are 0.40 m apart?
5. (II) A positive charge of magnitude 5.0×10^{-5} C is placed 1.0 cm above the origin of a coordinate system, and a negative charge of the same magnitude is placed 1.0 cm be-

low the origin, both on the z-axis. What is the potential energy of a positive charge of magnitude 4.0×10^{-6} C placed at the position $(x, y, z) = (10 \text{ cm}, 0 \text{ cm}, 15 \text{ cm})$? at (10 cm, 0 cm, 0 cm)?

6. (II) Repeat the calculation of Problem 5 for the case that (a) both charges on the z-axis are positive and the third charge is negative; (b) the signs and magnitudes of all charges are the same.

7. (II) A charge of 4 μC is placed at the point $x = 2$, $y = 3$, $z = 0$ (all distances given in centimeters). Calculate the work done in bringing a charge of -8 μC from $x = 2$, $y = 15$, $z = -30$ to the point $x = 2$, $y = 12$, $z = 6$, assuming that the charge is moved at a steady speed.

8. (II) Use potential energy arguments to show that charges of the same sign cannot form a system with a closed circular orbit.

25–2 *Electric Potential*

9. (I) Two equal charges of -4 mC are placed along the y-axis at -3 mm and 4 mm, respectively. Where is the electric potential zero?

10. (I) Two charges are placed along the x-axis: 4 μC at 2 cm, and -2 μC at 4 cm. Find those points along the x-axis where the potential is zero.

11. (I) A proton moves from point A to point B under the sole influence of an electric field, losing speed as it does so from $v_A = 3 \times 10^4$ m/s to $v_B = 3 \times 10^3$ m/s. What is the potential difference between the two points?

12. (I) An external force steadily moves a point charge of $+10^{-6}$ C from a negatively to a positively charged plate. The plates are large and parallel, and the negatively charged plate is at a potential of -20 kV, whereas the positively charged plate is at a potential of $+10$ kV. How much work does the external force do?

13. (I) Consider two very long coaxial cylinders that carry opposite charges. The interior cylinder, negatively charged, is at a potential of -20 kV, whereas the exterior cylinder, positively charged, is at a potential of $+10$ kV. An external force steadily moves a point charge of $+10^{-6}$ C from the negatively to the positively charged cylinder. How much work does the external force do?

14. (II) Three charges are at rest on the z-axis—$q_1 = 2$ mC at $z = 0$ m, $q_2 = 0.5$ mC at $z = 1$ m, and $q_3 = -1.5$ mC at $z = -0.5$ m. What is the potential energy of this system?

15. (II) Charges $+q$, $-q$, $+q$, and $-q$ are placed on successive corners of a square in the xy-plane. Plot all the locations in the xy-plane where the potential is zero.

16. (II) Consider two charges of 24×10^{-2} μC and -10×10^{-2} μC, respectively, at opposite ends of a diameter of a circle of radius 25 cm. (a) What is the potential on a point of the circle that is 30 cm from the positive charge? (b) How much work is required to bring a charge of -0.2 μC from infinity to that point on the circle?

17. (II) The origin of a coordinate system is at the intersection point of the perpendicular bisectors of the sides of an equilateral triangle of sides 10 cm. Calculate the potential at the origin due to three identical charges of 0.8 μC placed at the corners of the triangle.

18. (II) Consider a square of sides 20 cm. Charges are placed on the corners of the square as follows: 12×10^{-2} μC at (0 cm, 0 cm); -24×10^{-2} μC at (0 cm, 20 cm); 36×10^{-2} μC at (20 cm, 20 cm); -24×10^{-2} μC at (20 cm, 0 cm). What is the potential at the point (40 cm, 40 cm)?

19. (II) A 2-μC charge is fixed at $(x, y) = (2 \text{ mm}, 3 \text{ mm})$, a -4-μC charge is fixed at (2 mm, 6 mm), and a -5-μC charge is fixed at (5 mm, 3 mm). What is the potential energy of the system? Does the order in which the charges are brought in from infinity matter?

20. (II) Two parallel conducting plates are brought to a potential difference of 3000 V, and a small pellet of mass 2 mg carrying a charge of 10^{-7} C accelerates from rest at one plate. With what speed will it reach the other plate?

21. (II) A charge Q is distributed uniformly over the surface of a spherical shell of radius R. How much work is required to move these charges to a shell with half the radius? The charges are again distributed uniformly.

22. (III) Calculate the potential inside and outside a sphere of radius R and charge Q, in which the charge is distributed uniformly throughout. (*Hint:* The additive constant for the potential inside the charged sphere must be chosen so that the two potentials, inside and outside, agree at $r = R$.)

25–3 *Equipotentials*

23. (I) Draw the equipotential surfaces for (a) a thin disk charged uniformly over its area and (b) a charged ring.

24. (I) Draw four equipotential surfaces for the charges shown in Fig. 25–30.

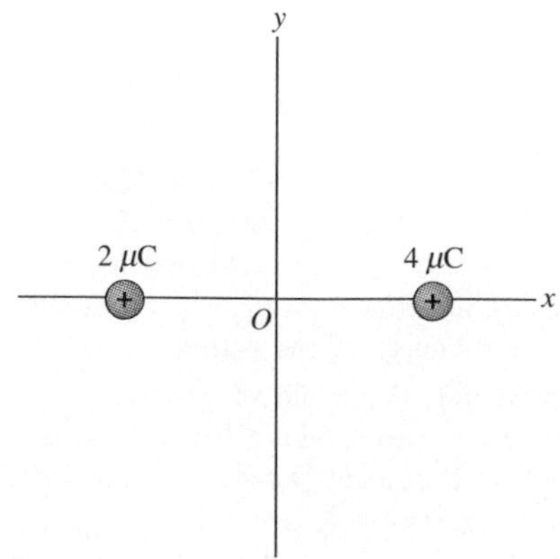

FIGURE 25–30 Problem 24.

25. (I) Sketch the equipotential surfaces for the charges shown in Fig. 25–31. Assume that the rod is an insulator.

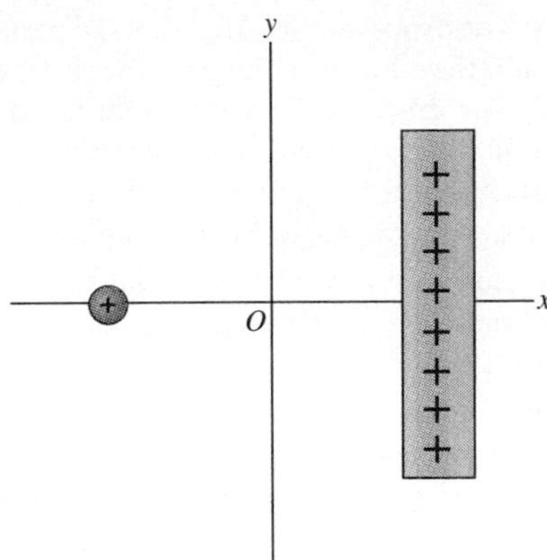

FIGURE 25–31 Problem 25.

26. (I) Sketch the electric fields and the equipotentials for the charge distribution shown in Fig. 25–32. Assume that the rod (of infinite length) is an insulator.

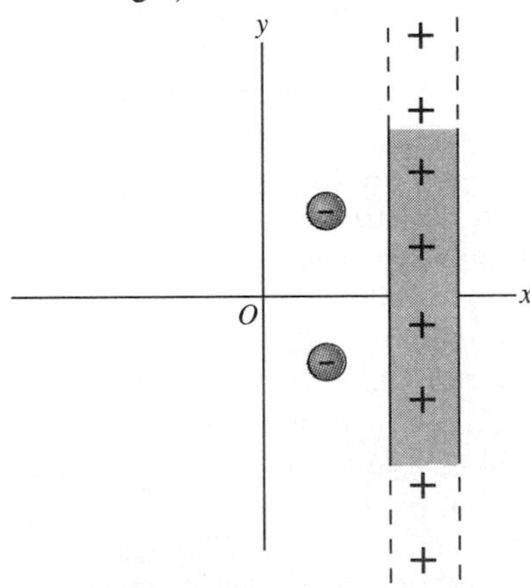

FIGURE 25–32 Problem 26.

27. (II) A uniformly charged metal rod is placed parallel to an infinite, uncharged metal plate. Sketch the equipotentials in a plane perpendicular to the plate and to the rod, and in a plane perpendicular to the plate but parallel to the rod.

28. (II) Sketch the equipotentials in the xy-plane due to an infinite number of identical point charges q that lie on a line and are separated by a distance a, so that the coordinates of the point charges are $x_n = na$ and $y_n = 0$, where $n = 0, \pm 1, \pm 2, \pm 3, \ldots$.

29. (II) Two charges of equal magnitude but opposite sign are separated by a distance L. Sketch the equipotentials. What equipotential surfaces will have a potential of zero when the separate potentials for the two charges are chosen to be zero at infinity?

30. (II) Two infinite plates, each charged uniformly with charge density σ, are placed at right angles to each other and are almost touching. What are the equipotential surfaces? What are the equipotential surfaces if one of the plates has charge density $-\sigma$?

25–4 *Determining Electric Fields from Electric Potentials*

31. (I) The electric potential of a charge distribution within some region of space is $V(x, y, z) = Q/4\pi\epsilon_0 x$. Find the electric field in this region.

32. (I) Find the electric field of a charge distribution if the electric potential of the distribution is $V = Ax^3z + By^2z^2 + C$, where A, B, and C are constants.

33. (II) Starting from the solution in Example 25–8 of the potential due to a uniformly charged ring, use the derivative operations in Eq. (25–29) to find the electric field along the axis of the ring.

34. (II) Find the electric field far away along the bisecting axis of an electric dipole from the potential given in Eq. (25–24).

35. (II) A certain distribution of static charges in space produces an electric potential of the form $V(x, y, z) = a_1 + a_2xz + a_3z^2$, where the coefficients a_i are constants. Find the electric field, **E**, at the origin and at the (x, y, z) point (0 m, 0 m, 1 m).

36. (II) Consider charge distributed in an infinitely long cylinder of radius R whose axis forms the z-axis. The charge distribution depends only on the distance r from the z-axis. The potential is given for $r < R$ by $V(r) = (Q/2\pi\epsilon_0)[A(r/R) + B(r/R)^2 + C]$, where A, B, and C are constants. What is the electric field within the rod? What is the value of C if the potential is defined to be zero on the cylinder's surface?

37. (II) The potential $V(r)$ of a spherically symmetric charge distribution is given by $V(r) = (Q/4\pi\epsilon_0 R)[5 - 4(r/R)^2]$ for $r < R$ and $V(r) = Q/4\pi\epsilon_0 r$ for $r > R$. (a) Calculate the electric field. (b) Where is the charge, and how is it distributed? [*Hint:* To determine the charge, use Gauss' law for a variety of values of r.]

38. (III) The potential in the xy-plane due to a certain charge distribution is given by

$$V(x, y) = \frac{Q}{4\pi\epsilon_0 L} \times$$

$$\left[\arctan\left(\frac{y}{x - a_0}\right) - 2\arctan\left(\frac{y}{x}\right) + \arctan\left(\frac{y}{x + a_0}\right)\right],$$

where L and a_0 are constant lengths. Show that the electric field at distances $x \gg a_0$, $y \gg a_0$ is proportional to a_0^2, and find its dependence on x and y. Express your answer in terms of r, the distance to the origin, and θ, the angle that the line from the origin to the point (x, y) makes with the x-axis.

25–5 *Calculating the Potentials of Extended Charge Distributions*

39. (I) Two large, metal, parallel plates have a potential difference of 50 kV, and the electric field between them has magnitude 10^5 V/m. What is the separation distance between the plates?

40. (I) In fair weather there is a constant electric field near the earth's surface whose magnitude is roughly 100 V/m, directed downward. (a) Find the potential associated with this field. (b) What is the most convenient point to choose for zero potential? (c) How does the potential energy of a test charge near the earth compare in form with the potential energy of gravity? (d) How much negative charge would have to be placed on a person of mass 50 kg to have the electric force balance the force of gravity?

41. (II) Find the potential as a function of the perpendicular distance R from an infinite line of uniform charge density by using Gauss' law and Eq. (25–9).

42. (II) Charges are distributed with uniform charge density λ along a semicircle of radius R, centered at the origin of a coordinate system. What is the potential at the origin?

43. (III) Find an expression for the electric potential at all points due to a rod of length L and uniform charge density λ, using Eq. (25–12). The rod is oriented along the z-axis, with its center at the origin. Show that at distances much greater than L from the rod, the potential reduces to that of a point charge $Q = \lambda L$ at the origin.

44. (III) A charge $3q_0$ is placed on the x-axis at the point $x = x_0$ (where x_0 is positive), and a second charge, $-q_0$, is placed on the x-axis at the point $x = -x_0/2$. (a) What is the potential on the x-axis of this distribution of charges? Assume that zero potential for a point charge is at infinity. (b) Show that your result for part (a) can be approximated for large x by a term proportional to $1/x$, plus a term proportional to $1/x^2$, plus higher powers of $1/x$. (c) Show that the expansion of part (b) is that of a point charge at the origin, plus an electric dipole oriented along the x-axis and centered at the origin, plus other terms. Find the strength of the point charge as well as the dipole moment of the electric dipole. (d) How large must x be so that the approximation of a point charge plus a dipole comes within 1 percent of the exact answer? [*Hint:* Use the approximation $(1 + z)^k = 1 + kz + \frac{1}{2}k(k - 1)z^2 + \cdots$ for $z \leq 1$.]

25–6 *Electric Potentials and Fields around Conductors*

45. (I) A thin disk of radius 23 cm carries a total charge of 1.5×10^{-7} C spread evenly over its surface. What is the minimum work required to bring a charge $q = 2.0 \times 10^{-8}$ C at rest from infinity to a distance of 78 cm from the disk along its axis?

46. (I) A thin ring of radius 42 cm carries a uniformly distributed charge of 4.7×10^{-7} C. A negative charge $q = -4.8 \times 10^{-8}$ C is placed on the axis of the ring 34 cm from the plane of the ring. How much work must an external agent do to move the charge slowly and steadily to a distance 120 cm away, also on the axis?

47. (I) The same charges are placed on two identical drops of mercury. The drops are isolated and take perfectly spherical shapes, and the electric potential at the surface of each drop is 900 V. The drops coalesce into a larger drop with a net charge double that of either smaller charge. What is the potential at the surface of this larger charge?

48. (I) Two conducting spheres of different sizes are connected by a thin conducting wire. The radius of the larger sphere is three times that of the smaller sphere. If a total charge Q is placed on this apparatus, what fraction of Q sits on each sphere?

49. (I) An electric field of 3×10^6 V/m is sufficiently large to cause sparking in air. Find the highest potential to which a conductor of radius 10 cm can be raised before breakdown occurs in the air surrounding it. Assume that zero potential is taken at infinity.

50. (II) Concentric metal shells, perfect conductors, have radii R and $2R$, respectively. A charge q is placed on the inner shell, and a charge $-2q$ is placed on the outer shell. (a) What are the electric fields in all space due to the two shells? (b) What is the potential difference between the two shells? (c) If a thin, perfectly conducting wire now joins the two shells, how does the charge redistribute itself?

51. (II) Two spherical conductors of radii 20 mm and 100 mm are connected by a thin wire and carry charges q_1 and q_2, respectively. If the wire is cut and the centers of the spheres are 250 mm apart, there is a repulsive force of 3.5 N between them. Use this information to calculate (a) q_1 and q_2 and (b) the electric fields at the surfaces of the conductors when they are connected by the wire.

52. (II) A balloon of radius 20 cm is sprayed with a metallic coating so that the surface is conducting. A charge of 4×10^{-8} C is placed on the surface. (a) What is the potential on the balloon's surface? (b) Suppose that some air is let out of the balloon, so that its radius shrinks to 10 cm. What is the new potential on the balloon's surface? (c) What happens to the energy associated with the change in potential energy?

25–7 *Electric Potentials and Electrostatic Fields in Technology*

53. (I) A proton is accelerated from rest in a Van de Graaff accelerator through a potential of 5.5×10^6 V. (a) What energy does the proton have, in electron-volts and joules? (b) What is the proton's final speed?

54. (I) A small Van de Graaff generator is used to demonstrate the effects of high potential. The device has a radius of 11 cm and stands in air. What is its maximum potential, and how much charge does the dome hold?

55. (II) Early Van de Graaff accelerators were built to operate in air without high-pressure gases. (a) How much voltage could an accelerator with a domed surface of radius 1 m produce? (b) How much kinetic energy could the protons produced by such an accelerator have? (c) What is the total charge on the accelerator dome when the maximum field is attained?

General Problems

56. (I) We have a high-voltage power supply capable of producing 5000 V, and we want to ionize the air molecules between parallel plates. What plate separation will give us electrical breakdown?

57. (II) Write an expression for the total energy of two point charges—one positive and of magnitude Q, fixed at the origin; the other negative and of magnitude q and mass m, located at a point a distance r from the origin. What is the energy if charge q moves in a circular orbit of radius r around charge Q?

58. (II) A nonconducting sphere of radius R carries a charge $+Q$ distributed uniformly throughout its volume. What is the potential energy of a point charge $-q$ a distance r $(r < R)$ from the center of the sphere? Show that q oscillates as though it were attached to a spring, and find the spring constant.

59. (II) Three electrons are located along the x-axis at positions -2 mm, 0 mm, and 2 mm, respectively. How much energy was required to move each of the electrons in turn from infinity? Does the order in which they were moved matter?

60. (II) Calculate the potential at the point (x, y) due to the dipole in Fig. 25–33, which consists of a charge $+q$ placed at $(0, a)$ and a charge $-q$ placed at $(0, -a)$, and use this potential to calculate the electric field at that point. What is the magnitude of the field at point P, which is a distance $3a$ from the midpoint of the dipole along a line that makes an angle of 45° with the dipole axis?

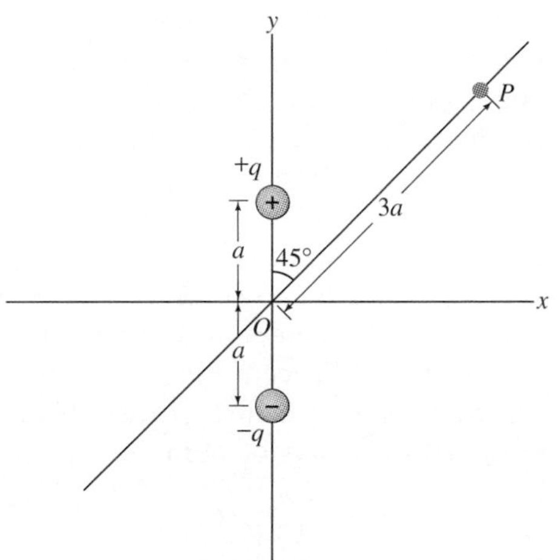

FIGURE 25–33 Problem 60.

61. (II) Find the electric field along the axis of a uniformly charged ring of radius R and total charge Q by taking the appropriate derivatives of the potential found in Example 25–8. Set up the problem by using the direct integration techniques presented in Chapter 23. Compare the difficulties of the two ways of calculating the electric field.

62. (II) A positron, charge $+e$, is released from rest a distance $r_0 = 10^{-10}$ m from a proton of charge $+e$. The positron accelerates away from the proton because of the repulsive Coulomb force. At what distance from the proton will the positron have exactly the same speed an electron, which has the same mass as the positron but opposite charge, would have if it were moving in a circular orbit of radius r_0 around the proton? Assume that the proton is so massive that it is fixed.

63. (II) An electron is moving in the field of a helium nucleus (atomic number $Z = 2$). What is the change in the electron's potential energy when it moves from a circular orbit of radius 3×10^{-10} m to one of radius 2×10^{-10} m? What is the change in kinetic energy? in the total energy of the electron? This energy change is carried off by light emitted by the electron.

64. (II) An electric dipole fixed in space consists of a charge $+q$ at the point $x = -1$ m and a charge $-q$ at the point $x = +1$ m, where $q = 3$ C. A test charge $q_0 = 0.01$ C is steadily moved from the point $x = +10$ m to the point $x = -5$ m by following a semicircular path of radius 7.5 m that takes the test charge through the y-axis. How much work is required to move the test charge?

65. (II) Two identical cork balls of charge 2.0 μC are suspended from the same point by thin threads 0.80 m long. (a) Calculate the mass of the cork balls if the threads each make a 30° angle with the vertical. (b) Calculate the potential energy of the system of two balls due to the presence of charges and to the presence of gravity as a function of the angle θ the threads make with the vertical. Choose zero gravitational potential energy to correspond to $\theta = 0$.

66. (II) A large, square plane with sides of length L, parallel to the yz-plane and located at x_1, has charge density σ_1. A similar plane, located at x_2, has charge density σ_2. How much work must be done to bring the second plane to within a distance a of the first one? Neglect end effects; that is, calculate the fields as though the planes were infinite.

67. (II) An infinitely long cylinder of radius R is filled with uniform charge density ρ. Calculate the potential inside and outside the cylinder.

68. (II) The inner radius of a spherical dielectric shell is 10 cm, and the outer radius is 12 cm. The shell carries a charge of 10^{-8} C, distributed uniformly. Sketch the shape of the potential for all values of r, the distance from the center of the shell, and evaluate it at the center and at the inner and outer radii.

69. (II) A solid sphere of radius R has uniform charge density ρ. Calculate the total potential energy by calculating the energy required to bring a spherical shell of thickness dr and charge density ρ from infinity to a distance r from the sphere's center in the potential due to a uniformly charged sphere of radius r.

CHAPTER 26

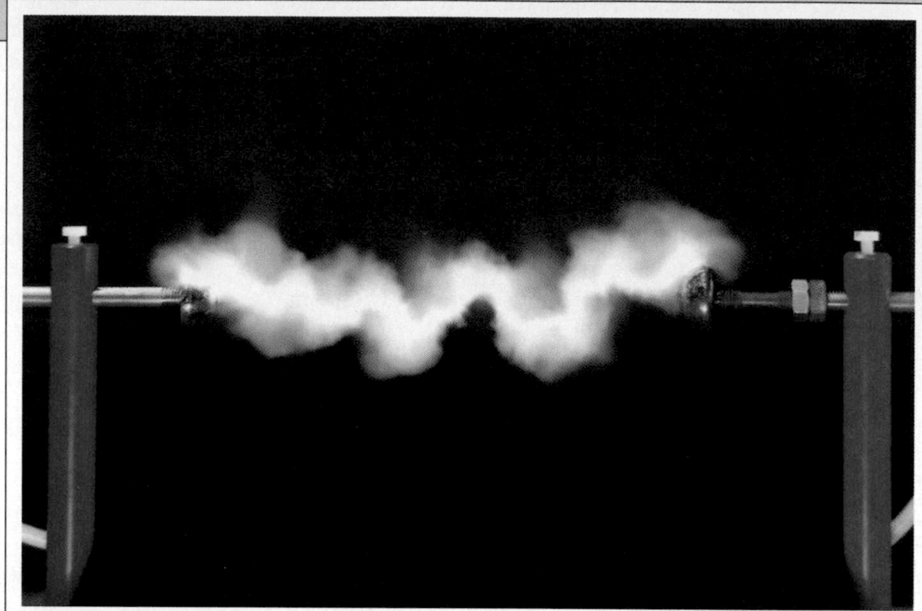

A test of high-voltage insulators used in overhead power transmission lines. The sparks indicate that the air between the charged metals has suffered dielectric breakdown.

CAPACITORS AND DIELECTRICS

Any conducting object is characterized in electrostatics by an electric potential that is constant everywhere on and within the object. The potential difference of two charged conductors can accelerate test charges, and thus the system stores energy. A capacitor is a device of this type; it stores energy because it stores charge. Capacitors are important in electric circuits. The relation between the amount of charge a capacitor stores and the potential difference of its components depends on the geometry of the capacitor. The relation between stored charge and potential difference is affected by the insulating (nonconducting) material, called dielectric material, placed between the charged components of the capacitor. In this chapter we shall study how geometry and dielectric materials affect capacitors. We also want to consider the microscopic structure of dielectrics, and thereby extend our fundamental knowledge of the behavior of matter.

26–1 CAPACITANCE

The definition and properties of a capacitor

A pair of conductors, whether separated by empty space or by a nonconducting material, forms a **capacitor**. Capacitors store charges. In their most common and useful form, capacitors are made of two conductors with equal but opposite

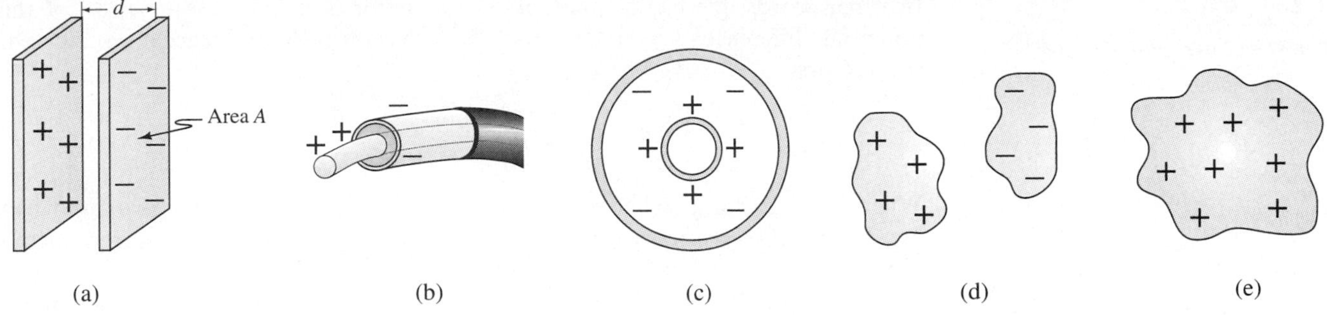

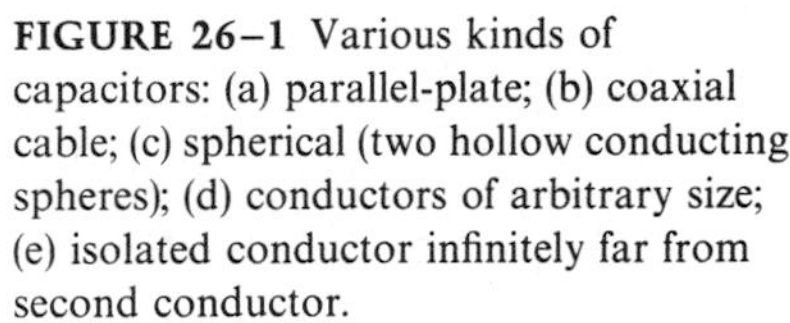

FIGURE 26–1 Various kinds of capacitors: (a) parallel-plate; (b) coaxial cable; (c) spherical (two hollow conducting spheres); (d) conductors of arbitrary size; (e) isolated conductor infinitely far from second conductor.

charges Q. There is a potential difference V between the conductors.† We showed in Chapter 25 that this potential difference is linearly dependent on the charge; that is, $V \propto Q$. Thus, if we double the charge, we double the potential difference between the two conductors. In other words, the ratio of Q to V between two conductors is constant. The constant ratio Q/V depends on the shape and arrangements of the two conductors of a capacitor—that is, on geometry—and on the material between the conductors. Figures 26–1a to 26–1e illustrate different configurations of conductors that can act as capacitors. In some cases a capacitor is formed when one conductor with charge Q induces a corresponding charge $-Q$ on an adjacent conductor. The second conductor may be very far away, even at infinity (Fig. 26–1e).

Capacitors are important for several reasons. Different forms of capacitors can hold different amounts of charge for a given potential difference or can maintain different potential differences for a given amount of charge. With the appropriate capacitor, we can therefore control the storage and delivery of charge. We can similarly use capacitors to control potential differences. Almost any device with an electronic circuit contains capacitors. Because they involve a potential difference, capacitors store energy as well as charge. A lightning strike is the spectacular discharge of a large capacitor formed by the system of a cloud and the earth. Capacitors are particularly useful for short-term storage of charge and energy. A camera photoflash contains a capacitor that stores energy and then discharges it when the flash takes place. Another important use of capacitors is the slow but smooth delivery of energy when capacitors are coupled with other circuit elements. Emergency backup systems for computers depend on this usage of capacitors.

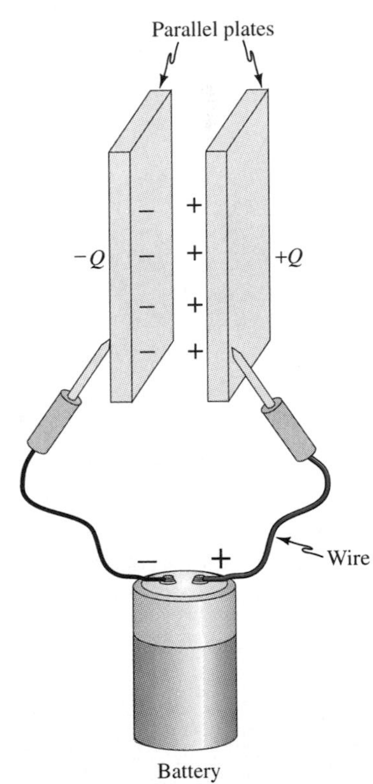

FIGURE 26–2 We can place charge $+Q$ on one conductor and charge $-Q$ on another conductor by using a battery.

Let us place equal but opposite charges, $+Q$ and $-Q$, on two conductors that form a capacitor. This is easily done by touching the two conductors with wire leads from the + and − terminals of a battery (Fig. 26–2). The amount of charge accumulated depends on the shape of the conductors and on their relative positions as well as on how long we leave the wires connected. We call Q the *charge on the capacitor* even though the *net* charge on the oppositely charged pair of elements is zero.

We argued above that the charge on a capacitor is proportional to the potential difference. The proportionality constant is called the **capacitance**, C, determined by the relation

$$Q = CV. \tag{26–1}$$

The charge on a capacitor

† It is conventional (and convenient) to write V, not ΔV, for the potential difference between the plates of a capacitor. When there is no possibility for confusion, we also use in this context (here and in the following chapters) the terms "potential," "voltage drop," or "voltage" rather than the formally correct term "potential difference."

In other words, the capacitance of the capacitor is defined as the ratio of the potential difference, V, that results when charges $\pm Q$ are placed on the two conductors:

The definition of capacitance

$$C \equiv \frac{Q}{V}. \tag{26–2}$$

When there is charge (or potential) on a capacitor, we say it is *charged*, and when a capacitor *discharges*, as when a flashbulb fires, it can deliver its stored energy rapidly.

C is always taken to be positive; that is, Eq. (26–2) should properly contain absolute values. The unit of capacitance is coulombs per volt (C/V), but capacitance occurs so frequently that it has been given its own SI unit, the **farad** (F), in honor of Michael Faraday:

The SI unit of capacitance

$$1\ \text{F} = \frac{1\ \text{C}}{1\ \text{V}}. \tag{26–3}$$

In practice, the farad is inconveniently large, and, for practical use, units of μF, nF, and pF (often called "puffs") are much more common.[†]

Calculating Capacitance

The capacitance of a capacitor can be calculated in simple form if the geometry is simple enough. The simplest capacitor consists of two parallel conducting plates of area A with charges $+Q$ and $-Q$, respectively, distributed uniformly over the plates. If the dimensions of the plates are large compared with the separation between the plates, d, then the electric field between the plates is to a very good approximation constant. Neglecting (small) edge effects, we found previously that $E = \sigma/\epsilon_0$, where σ is the charge density Q/A, ϵ_0 is the permittivity of free space, and the potential difference between the plates is $V = Ed$. We combine these results to find that

$$V = Ed = \frac{\sigma}{\epsilon_0} d = \frac{Q}{A}\frac{d}{\epsilon_0} = Q\frac{d}{\epsilon_0 A}.$$

Thus $Q/V = \epsilon_0 A/d$, and the capacitance $C \equiv Q/V$ of a parallel-plate capacitor is

The capacitance of a parallel-plate capacitor

$$C = \frac{Q}{V} = \frac{\epsilon_0 A}{d}. \tag{26–4}$$

If the parallel plates were smaller or the separation larger, then the electric field might bulge out near the edges. The fields near the edge are called *fringe fields* (Fig. 26–3). The presence of fringe fields will not affect the linear relationship between the charge and the potential, but it will modify the simple form of Eq. (26–4). The effect of fringe fields will be small as long as all the linear dimensions of the plates are much larger than the separation d.

We can use Eq. (26–4) to derive a commonly used unit for the permittivity of free space, ϵ_0, different than the one we had derived previously, $\text{C}^2/\text{N}\cdot\text{m}^2$. Equation (26–4) implies that the units of ϵ_0 are alternatively those of capacitance divided by length, farads per meter (F/m):

$$\epsilon_0 = 8.85 \times 10^{-12}\ \text{C}^2/\text{N}\cdot\text{m}^2 = 8.85 \times 10^{-12}\ \text{F/m} = 8.85\ \text{pF/m}.$$

Either unit is consistent with the SI.

[†] Commercial capacitors are marked with these values, typically in rather arcane notation. The abbreviation for a microfarad can be mF, uF, or μF, and picofarads are abbreviated mmF, uuF, $\mu\mu$F (for micro-microfarads), or pF.

EXAMPLE 26–1 Calculate the capacitance C of parallel plates of area $A = 100\ \text{cm}^2$ separated by a distance $d = 1$ cm.

SOLUTION: We know both A (100 cm^2) and d (1 cm), so we can use Eq. (26–4) to determine the unknown capacitance. We find that

$$C = \frac{\epsilon_0 A}{d} = \frac{(8.85\ \text{pF/m})(10^{-2}\ \text{m}^2)}{10^{-2}\ \text{m}} = 8.85\ \text{pF}.$$

The plate area is rather large, yet the capacitance is only 8.85 pF. Most common practical capacitors have capacitances much smaller than 1 F, illustrating why the farad alone is hardly used as a unit.

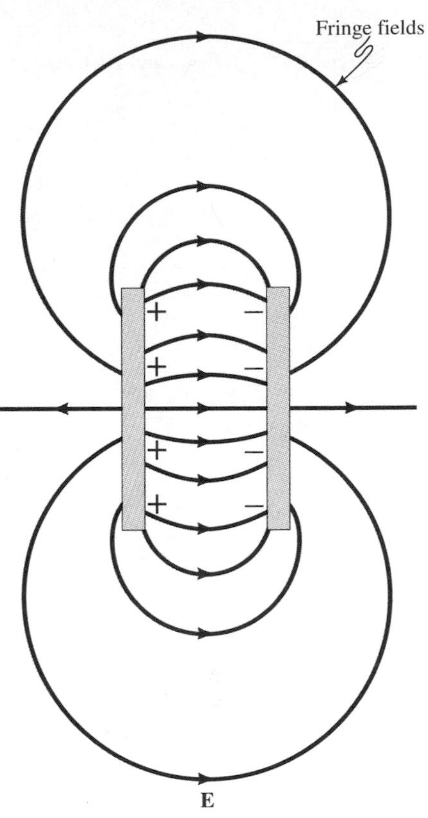

FIGURE 26–3 The electric field, including the fringe fields, between the two conducting plates of a parallel-plate capacitor.

EXAMPLE 26–2 *The Coaxial Cable.* A *coaxial cable* is a type of cable useful for transmitting information, as in cable television systems (Fig. 26–4a). It is made of a solid (or stranded) cylindrical conducting wire of radius a surrounded by a coaxial conducting sheath of radius b (Fig. 26–4b). Find the capacitance per unit length of a coaxial cable, assuming that there is a vacuum between the central wire and the sheath.

SOLUTION: We want to use the geometry shown in Fig. 26–4b and find the potential difference for a given charge. The ratio of these quantities determines the capacitance. However, for an infinitely long coaxial cable we can specify only the charge *per unit length*, so we must similarly calculate the capacitance per unit length.

To find the capacitance per unit length, we set a charge per unit length $+\lambda$ on the outer (and $-\lambda$ on the inner) conductor of the cable and calculate the resulting potential between the conductors. This calculation was already done in Example 25–10. There we saw how Gauss' law gives the electric field and hence the potential outside a wire of finite radius. That example applies directly to the potential in the space *between* the conductors of the coaxial cable. From Example 25–10, we have

$$V_b - V_a = \frac{\lambda}{2\pi\epsilon_0} \ln \frac{b}{a}.$$

Dividing the charge per unit length λ by the potential difference $V = V_b - V_a$ gives the capacitance per unit length:

$$\frac{C}{\text{length}} = \frac{\lambda}{V_b - V_a} = \frac{2\pi\epsilon_0}{\ln(b/a)}. \qquad (26\text{–}5)$$

The capacitance of a coaxial cable

Equivalently, the capacitance of a length L of the cable is L times this value.

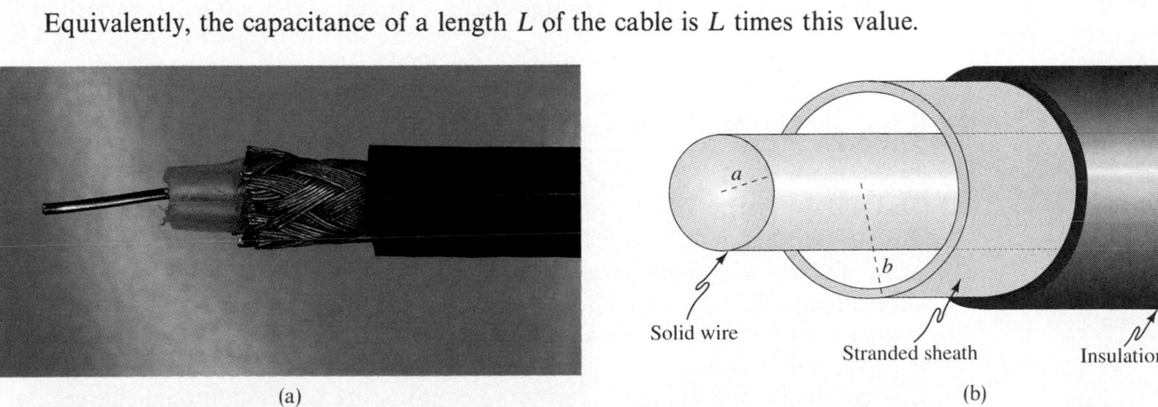

FIGURE 26–4 (a) A coaxial cable. (b) Schematic diagram of a coaxial cable with an inner solid wire of radius a and an outer sheath of radius b. Normally, an insulator is placed between the two conductors, but here we assume that a vacuum (or air) occupies that space.

An isolated conductor can have a capacitance because, when such a conductor is charged, the charge must be brought in from infinity. In effect, the second conductor is at infinity. Example 26–3 illustrates how to calculate the capacitance of an isolated conductor.

EXAMPLE 26–3 *The Isolated Sphere.* What is the capacitance of an isolated conducting sphere of radius R? Calculate the capacitance of the earth.

SOLUTION: We find the potential that results from placing a charge Q on the sphere, and the ratio of these quantities is, by definition, the capacitance. In this case the potential difference involved is just the potential of the sphere if zero potential is at infinity. We found in Example 25–11 that the potential for a conducting sphere is

$$V = \frac{Q}{4\pi\epsilon_0 R}.$$

The capacitance of the isolated sphere is Q divided by V:

The capacitance of an isolated sphere

$$\text{for an isolated sphere:} \qquad C = \frac{Q}{V} = 4\pi\epsilon_0 R. \qquad (26\text{–}6)$$

The dependence of the capacitance on R in Eq. (26–6) could have been obtained from dimensional analysis. The permittivity, ϵ_0, has units F/m and so must be multiplied by a length to give an acceptable capacitance. The only length in this problem is the radius of the sphere.

The earth's capacitance is determined by setting R equal to the radius of the earth, $R_e = 6.38 \times 10^6$ m. Therefore

$$C = 4\pi\epsilon_0 R = 4(3.14)(8.85 \times 10^{-12}\ \text{F}/\cancel{\text{m}})(6.38 \times 10^6\ \cancel{\text{m}}) = 7.10 \times 10^{-4}\ \text{F}.$$

The farad is indeed a large unit.

All these examples illustrate that the capacitance of any particular configuration of conductors depends on geometry but is independent of charge and voltage. However, we shall soon learn that, even if the geometry remains the same, the capacitance can also be changed by changing the materials between the conductors.

A technique for determining capacitance

The technique for determining the capacitance is always the same. We assume that the conductors have a charge $\pm Q$, then we find the potential difference V between the conductors due to this charge. The ratio Q/V gives the capacitance.

26–2 ENERGY IN CAPACITORS

There is an electric field between the two conductors of a charged capacitor, and this field can accelerate a test charge. Thus, a charged capacitor is capable of doing work and must contain energy. We can determine the energy contained in a charged capacitor by seeing how much work is required to charge it initially. We charge a capacitor by taking a positive charge dq from one conductor and moving it to the other conductor. The second conductor then has charge $+dq$, and the first conductor has charge $-dq$. As we continue to move additional charges dq_i, the existing charges on the conductors oppose the transfer of still more charge, and we will have to do work to move the additional charge.

To calculate the work needed to charge a capacitor, assume that the conduc-

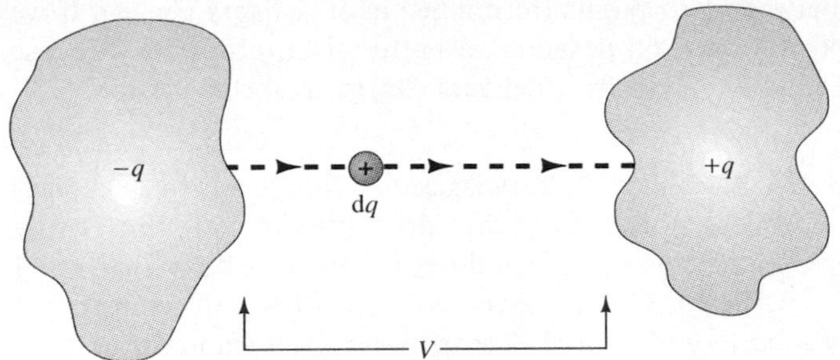

FIGURE 26–5 When an infinitesimal charge dq is moved between the two conductors of a capacitor charged to a potential difference V, work must be done.

tors are already charged to a potential difference V with charge q (Fig. 26–5). The capacitance, C, is related to q and V by $V = q/C$. A charge dq is now moved from the negatively charged conductor to the positively charged one. The potential difference opposes the transfer of charge dq, and the work that must be done is

$$dW = V\,dq = \frac{q}{C}\,dq.$$

The total work done when we start with zero charge and end up with charges $\pm Q$ is obtained by integrating the above expression from $q = 0$ to $q = Q$:

$$W = \int dW = \int_0^Q \frac{q}{C}\,dq = \frac{1}{C}\int_0^Q q\,dq = \frac{Q^2}{2C}. \qquad (26\text{–}7)$$

Note that we have not restricted the derivation to parallel-plate capacitors: This is a general result for all capacitors.

Because charging a capacitor requires work, this work is now stored in the capacitor as potential energy capable of doing work. It is this potential energy that is able to move a test charge placed between the conductors or to cause a flashbulb to flash (Fig. 26–6). The potential energy of a charged capacitor is

Forms for the energy contained in a charged capacitor

$$U = \frac{Q^2}{2C}. \qquad (26\text{–}8)$$

Because $Q = CV$, this result is equivalent to

$$U = \frac{CV^2}{2}. \qquad (26\text{–}9)$$

The first form is used when the charge is known, and the second when the potential is known. Another equivalent form that is useful when the charge and voltage are known is

$$U = \frac{QV}{2}. \qquad (26\text{–}10)$$

As we know from Chapter 25, the electric potential energy associated with the movement of a charge Q through a fixed potential V is $U = QV$. This expression differs by a factor of 2 from the capacitor energy $QV/2$, Eq. (26–10). The reason for this difference is that as a capacitor is charged, the potential is steadily increasing, so in effect the average potential as the charging takes place is $V/2$.

FIGURE 26–6 The energy stored in a capacitor is released in the forms of visible light and heat when the charge on the capacitor passes through the photoflash of a camera. Here we show a typical charging/discharging circuit for such a device.

Batteries versus Capacitors

A *battery* is a chemical device for the storage of energy. Whereas the potential of a capacitor decreases as the capacitor delivers its charge, a battery is able to maintain a fixed potential between two points (terminals) as it delivers charge. If we want to raise a capacitor to a certain potential, a battery is appropriate, because it can hold the desired potential even as it delivers charge to the capacitor.

EXAMPLE 26–4 A 12-V car battery is used to charge a 100-μF capacitor. (a) How much energy is stored in the capacitor? (b) Compare this energy with the energy stored in the battery itself, if the battery is capable of delivering a total charge of $Q = 3.6 \times 10^5$ C at the given voltage. (This is the charge that can be delivered by a battery rated at 100 ampere-hours, a standard unit for charge.)

SOLUTION: (a) When a battery charges a capacitor, the voltage difference between the capacitor plates is the voltage rating of the battery. We therefore want to find the energy stored in the known capacitor when it is at 12 V, given by Eq. (26–9):

$$U = \frac{CV^2}{2} = \frac{(100 \times 10^{-6}\ \text{F})(12\ \text{V})^2}{2} = 7.2 \times 10^{-3}\ \text{J}.$$

The answer is in joules, because we have used SI units throughout.

(b) As we know from Chapter 25, the electric potential energy associated with the movement of a charge Q through a fixed potential V is $U = QV$. Thus the potential energy in the battery is

$$U = QV = (3.6 \times 10^5\ \text{C})(12\ \text{V}) = 4.3 \times 10^6\ \text{J}.$$

The battery contains a factor of 6×10^8 more energy than is stored in the capacitor!

FIGURE 26–7 Large capacitors can release their energy quickly. These capacitors are part of a capacitor bank that supplies energy to Lawrence Livermore National Laboratory's NOVA laser, a device used in the study of nuclear fusion as a possible energy source.

We saw in Example 26–4 that a car battery has a potential energy of around 10^6 J, far more energy than that of a practical capacitor of 100 μF charged to a moderate potential. A potential more like 10^{10} V would be required to store the equivalent amount of energy in this capacitor. The largest available commercial capacitors have capacitance on the order of 1 F, but such capacitors can be taken to a potential of only several volts, whereas it would require a potential of around 1000 V for a 1-F capacitor to contain as much energy as the car battery has. A battery's energy is stored in chemical bonds rather than in the macroscopic separation of charge. Batteries are a practical way to store large amounts of energy for long periods, but not a practical way to deliver the energy quickly. Conversely, as we noted, the possibility of quick energy delivery is one advantage of a capacitor (Fig. 26–7). Capacitors can also be used for a slower energy delivery, in conjunction with other circuit elements. For energy backup for computers, capacitors with a capacitance as large as several farads and volumes of only a few cubic centimeters operate at a potential of only a few volts; they can very steadily deliver a charge at a rate of about 1 μC/s for about $\frac{1}{2}$ h.

26–3 ENERGY IN ELECTRIC FIELDS

The electric field is of fundamental physical significance. In this section we develop the concept that the energy of a capacitor is contained in the electric field itself. Because the two conductors of a capacitor are charged, electric field lines point from the positively charged conductor to the negatively charged one. It is this electric field that causes the acceleration of a test charge placed between capacitor plates.

Let us relate the expression for the energy in a capacitor to the strength of the electric field in the capacitor. The parallel-plate capacitor is convenient for this purpose, because both the capacitance and the field are known (Fig. 26–8). Equation (26–4) gives the capacitance for this case, $C = \epsilon_0 A/d$, where A is the area of the plates and d is their separation. The field has constant strength E, and the potential difference between the plates is $V = Ed$. Thus Eq. (26–9) gives the energy

$$U = \frac{CV^2}{2} = \frac{\epsilon_0 A}{2d}(Ed)^2$$

$$= \frac{\epsilon_0 E^2}{2}(Ad). \tag{26–11}$$

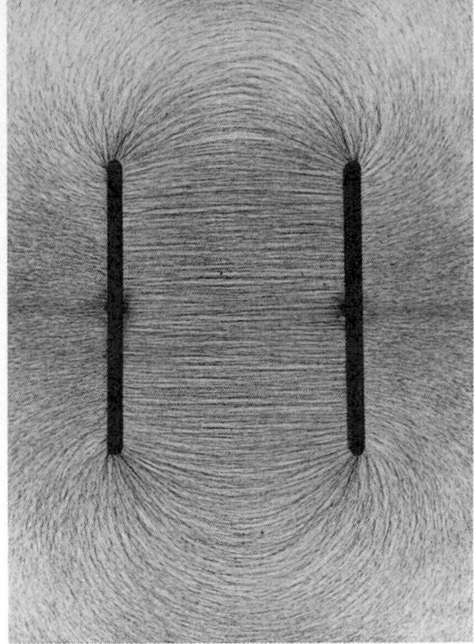

FIGURE 26–8 The electric field lines due to a charged parallel-plate capacitor, shown by threads in oil.

We have written Eq. (26–11) so that the volume of the space between the plates, Ad, stands out. This is the volume that contains the electric field, and, because the field is constant, the coefficient of the volume in Eq. (26–11) is the **energy density**, u, or energy per unit volume:

$$u \equiv \frac{U}{\text{volume}} = \frac{\epsilon_0 E^2}{2}. \tag{26–12}$$

The energy density in an electric field

The energy of a capacitor is thus located where the electric field is in space (Fig. 26–9a). We could now imagine forgetting about the plates, as long as there is a field (Fig. 26–9b). In fact, *Eq. (26–12) is a general expression in electrostatics for the local energy density in free space even for a variable electric field.* Wherever there is an electrostatic field, even one that varies throughout space, the energy density, or energy per unit volume, at a particular location in space is found by squaring the electric field and multiplying by $\epsilon_0/2$. Later we shall see that we must modify the coefficient $\epsilon_0/2$ when dielectrics are present, but the energy density is always proportional to the square of the field.

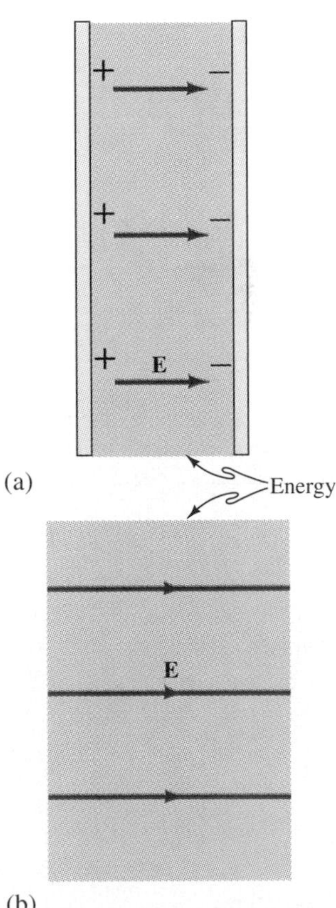

FIGURE 26–9 Electric fields have energy whether they are (a) inside a capacitor or (b) in free space.

EXAMPLE 26–5 (a) Determine the energy density due to an isolated, charged spherical conductor of radius R at each point in space as a function of the distance r from the sphere's center. (b) Use this energy density to compute the system's total energy. (c) Compare this total energy to the expression found by treating the sphere as a capacitor. (d) Compare this total energy to the work done in charging the sphere.

SOLUTION: (a) Equation (26–12) expresses the energy density in terms of the electric field. Equation (23–5) gives the electric field at a radius r outside the charged sphere. This field is radial and has magnitude

$$E = \frac{Q}{4\pi\epsilon_0 r^2}.$$

Inside the conducting sphere, the field is zero. The energy density is then

$$u = \frac{\epsilon_0 E^2}{2} = \frac{Q^2}{32\pi^2\epsilon_0 r^4}$$

outside the sphere and zero inside.

(b) The total energy of the electric field is the integral over space of the energy density. Because $u = 0$ outside the sphere, we need integrate only over radii $r > R$. Because of the spherical symmetry, it is useful to regard the volume as a series of concentric spherical shells. The volume dV of a thin shell of thickness dr is the product of dr and the surface area of the shell, $4\pi r^2$. Then the volume element $dV = 4\pi r^2\, dr$, and

$$U = \int u\, dV = \frac{Q^2}{32\pi^2\epsilon_0} \int_R^\infty \frac{4\pi r^2}{r^4}\, dr = \frac{Q^2}{8\pi\epsilon_0} \int_R^\infty \frac{1}{r^2}\, dr$$

$$= -\frac{Q^2}{8\pi\epsilon_0} \frac{1}{r}\bigg|_R^\infty = \frac{Q^2}{8\pi\epsilon_0 R}.$$

(c) Equation (26–6) gives the capacitance of an isolated sphere, $C = 4\pi\epsilon_0 R$. Using Eq. (26–8) for the energy, we have

$$U = \frac{Q^2}{2C} = \frac{Q^2}{8\pi\epsilon_0 R},$$

a result identical to that of part (b).

(d) To calculate the work required to charge the sphere by bringing charge in from infinity, suppose that the sphere already has charge q and is at a potential of $V = q/4\pi\epsilon_0 R$. The additional work to bring charge dq in is $dW = V\, dq$, and the total work is

$$W = \int dW = \int V\, dq = \int_0^Q \frac{q}{4\pi\epsilon_0 R}\, dq = \frac{1}{4\pi\epsilon_0 R} \int_0^Q q\, dq = \frac{Q^2}{8\pi\epsilon_0 R}.$$

The work done to bring the charge in from infinity is indeed equal to the total energy of the system found in part (b). This is a good check on the validity of our calculations.

FIGURE 26–10 The symbols used to indicate (a) capacitors and (b) batteries in electric circuits.

26–4 CAPACITORS CONNECTED IN PARALLEL AND IN SERIES

Parallel and series combinations of circuit elements

Now we begin our discussion of *electric circuits*. We have already mentioned two *circuit elements*: capacitors and batteries. Universal symbols for batteries and capacitors, shown in Fig. 26–10, have been adopted for *circuit diagrams*, schematic drawings that illustrate the connection of various circuit elements. The lines connecting circuit elements in these diagrams are assumed to be perfectly conducting wires, meaning that they form equipotentials. The different capacitor combinations used in circuits can be drawn with this notation. In a **parallel** combination, capacitors are connected as in Fig. 26–11a; in a **series** combination, capacitors are connected as in Fig. 26–11b. We want to find the capacitance of the single *equivalent capacitor* that can replace these combinations without changing the potential across them. By learning about equivalent capacitance, we can simplify our treatment of circuits.

Parallel Connection

The battery in Fig. 26–12a maintains a potential V across the points a and b. Because the connecting wires are perfect conductors, the line ace is an equipotential, as is the line bdf. The potential difference from points c to d and from e to f

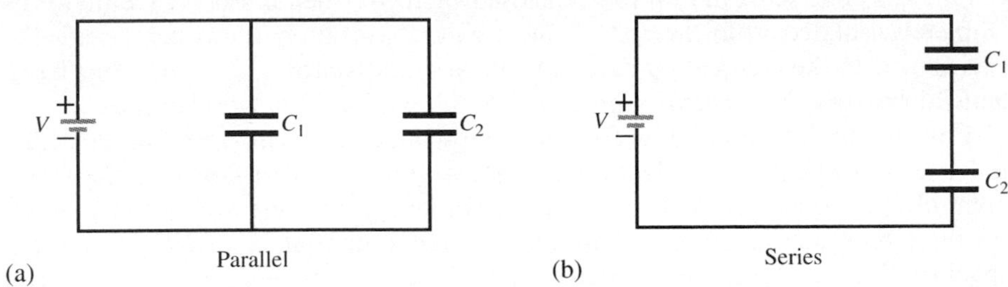

FIGURE 26–11 (a) Circuit with two capacitors connected in parallel. (b) Circuit with two capacitors connected in series.

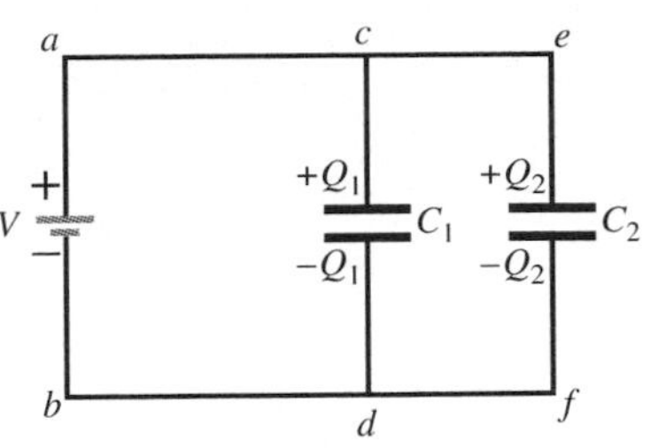

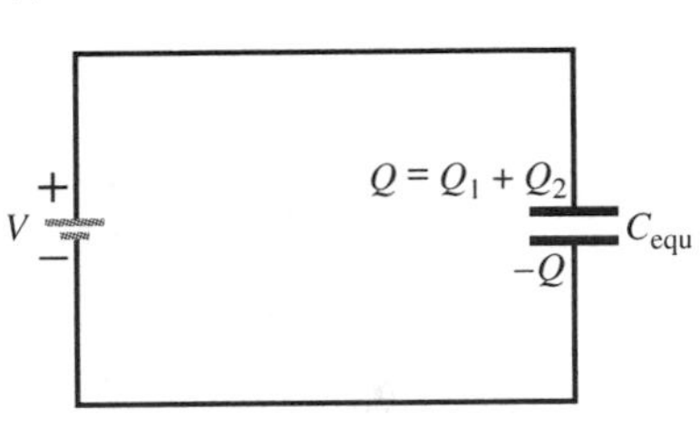

FIGURE 26–12 (a) A battery is used to place the same voltage V across the two capacitors connected in parallel. The charge on each capacitor depends on their individual capacitances. The two capacitors can be replaced (b) by an equivalent capacitor C_{equ}.

must therefore be V also. *Capacitors connected in parallel have the same potential between their conductors.*

What is the equivalent capacitance, C_{equ}, of the single capacitor that replaces the parallel combination of capacitors C_1 and C_2, as defined by Fig. 26–12b? The same potential difference V and total charge Q must be maintained on the equivalent capacitor as are on the parallel combination of C_1 and C_2. The charges on C_1 and C_2 are related to the voltage across each capacitor:

$$Q_1 = C_1 V \quad \text{and} \quad Q_2 = C_2 V.$$

The charge for the equivalent capacitor is $Q = C_{equ} V$, and the total charge produced by the battery for the circuit in Fig. 26–12a is $Q_1 + Q_2$, so

$$Q = Q_1 + Q_2 = C_1 V + C_2 V = (C_1 + C_2)V = C_{equ} V.$$

The last equality shows that

$$C_{equ} = C_1 + C_2. \tag{26–13}$$

With n capacitors connected in parallel, we can similarly show that the equivalent capacitance is

$$C_{equ} = C_1 + C_2 + \cdots + C_n. \tag{26–14}$$

The equivalent capacitance for parallel connection

When capacitors are arranged in parallel, the total capacitance is larger than any of the individual capacitances.

Series Connection

Now let us determine the single capacitor equivalent to capacitors connected in series (Fig. 26–13a). Because the battery maintains a fixed potential V, charge $+Q$ appears at c and charge $-Q$ at f. The positive charge at c induces a charge $-Q$ at d, and similarly the negative charge at f induces a charge $+Q$ at e. If we

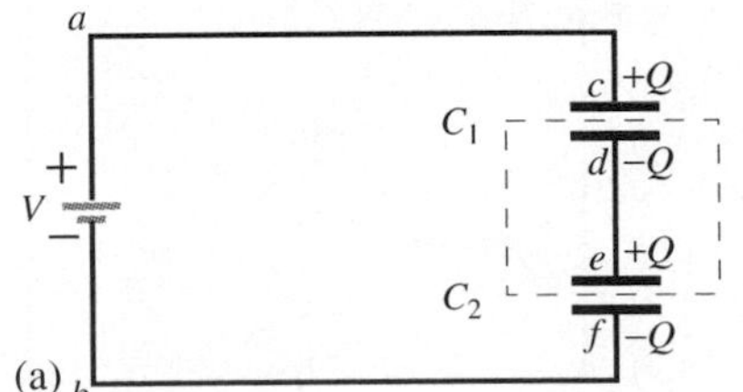

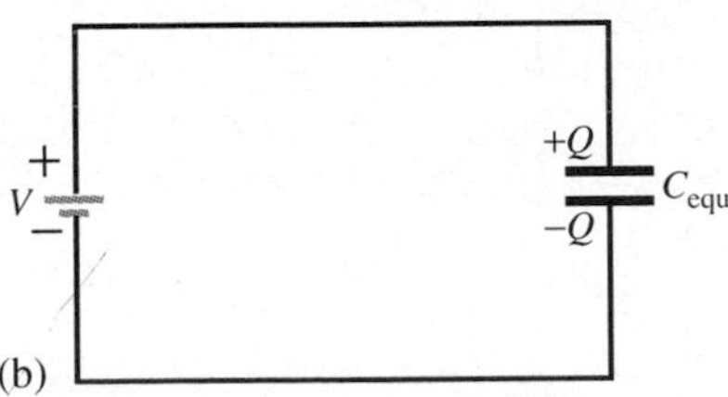

FIGURE 26–13 (a) Two capacitors connected in series must have identical charges $\pm Q$ but can have different voltages. The net charge within the dashed region is zero. The two capacitors can be replaced (b) by an equivalent capacitor C_{equ}.

examine the conductor from d to e (enclosed by the dashed lines), we see that it is completely isolated from the rest of the circuit. Its total charge is zero, perfectly consistent with the charge $-Q$ at d and $+Q$ at e. Capacitors C_1 and C_2 thus have identical charges Q. *Capacitors connected in series have identical charges.*

For the circuit containing the single equivalent capacitor, in Fig. 26–13b, the battery must cause the identical charges $+Q$ and $-Q$ to flow and the same potential V to be across the equivalent capacitor C_{equ} as there are between c and f in the circuit of Fig. 26–13a. Capacitor C_1 has potential $V_1 = Q/C_1$; similarly, capacitor C_2 has potential $V_2 = Q/C_2$. The total potential $V = V_1 + V_2$, so we have

$$V = V_1 + V_2 = \frac{Q}{C_1} + \frac{Q}{C_2} = Q\left(\frac{1}{C_1} + \frac{1}{C_2}\right) = \frac{Q}{C_{\text{equ}}}.$$

The last equality shows that the value of the equivalent capacitance is determined by

$$\frac{1}{C_{\text{equ}}} = \frac{1}{C_1} + \frac{1}{C_2}; \tag{26–15}$$

$$C_{\text{equ}} = \frac{C_1 C_2}{C_1 + C_2}. \tag{26–16}$$

With n capacitors connected in series, the equivalent capacitance is given by

The equivalent capacitance for series connection

$$\frac{1}{C_{\text{equ}}} = \frac{1}{C_1} + \frac{1}{C_2} + \cdots + \frac{1}{C_n}. \tag{26–17}$$

When capacitors are arranged in series, the total capacitance is the reciprocal of the expression in Eq. (26–17), and this capacitance is less than any of the individual capacitances.

EXAMPLE 26–6 Determine the equivalent capacitance for the capacitors in the circuit shown in Fig. 26–14a.

SOLUTION: In this circuit, the capacitors are combined both in series and in parallel, and we want to find the single equivalent capacitance for the whole combination. We proceed in steps. We first combine the two capacitors in parallel into one capacitor of value C'_{equ} (Fig. 26–14b) and then combine the three remaining capacitors in series into one final equivalent capacitor (Fig. 26–14c).

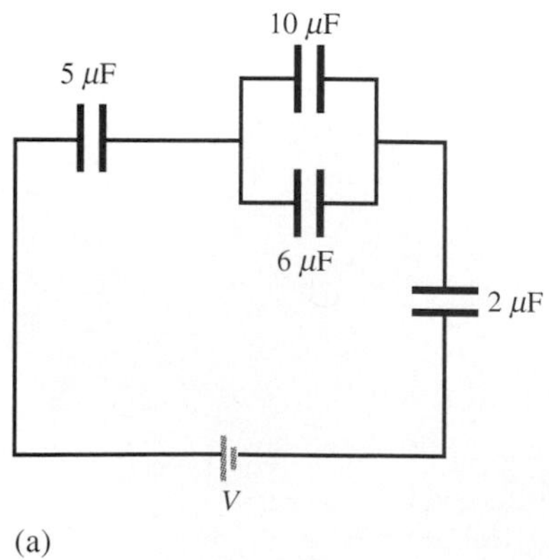

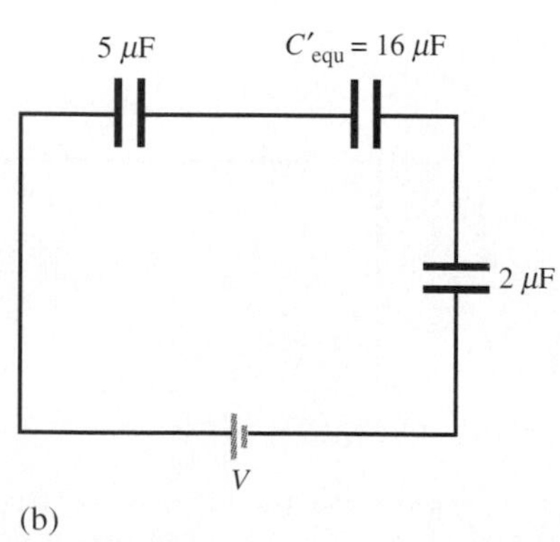

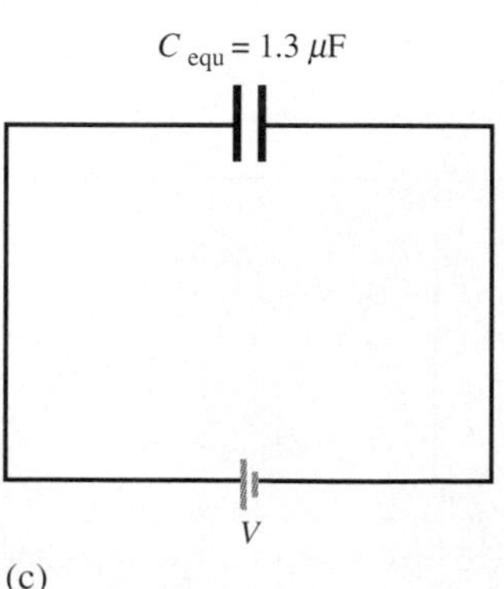

FIGURE 26–14 (a) Example 26–6. Four capacitors of this electric circuit can be combined into one. (b) The two parallel capacitors are combined, giving three capacitors in series, (c) which are then combined into the equivalent capacitor C_{equ}.

We use Eq. (26–14) to combine the parallel capacitors:

$$C'_{\text{equ}} = 10\ \mu\text{F} + 6\ \mu\text{F} = 16\ \mu\text{F}.$$

Having reduced the capacitor arrangement to the one shown in Fig. 26–14b, we combine the series capacitors by using Eq. (26–17), as in Fig. 26–14c:

$$\frac{1}{C_{\text{equ}}} = \frac{1}{5\ \mu\text{F}} + \frac{1}{16\ \mu\text{F}} + \frac{1}{2\ \mu\text{F}} = \frac{61}{80}(\mu\text{F})^{-1};$$

$$C_{\text{equ}} = 1.3\ \mu\text{F}.$$

26–5 DIELECTRICS

Many materials, such as paper, plastics, and glass, do not conduct electricity easily; we referred to them earlier as *insulators*. Nevertheless, they modify the external electric fields in which they are placed. In this context we call these materials **dielectrics**. We shall see that a dielectric placed in a capacitor allows larger charges on the capacitor for a given voltage. A solid dielectric placed between a capacitor's two conductors also lends strength and mechanical stability to the capacitor. Finally, a dielectric can reduce the possibility of sparking across the plates of a capacitor.

Michael Faraday is generally given credit for performing the first experiments that showed that when *insulating materials are placed between the two conductors of a capacitor, the capacitance increases* (Fig. 26–15). If C_0 is the capacitance of a

The presence of a dielectric in a capacitor increases the capacitance.

FIGURE 26–15 Faraday used this capacitor in the mid-1800s during his investigations of the properties of dielectrics.

given capacitor in a vacuum (or in air), the capacitance, C, with a dielectric placed between its conductors is larger than C_0 by a factor called the **dielectric constant**, κ. We have

$$C = \kappa C_0; \tag{26–18}$$

Definition of the dielectric constant

$$\kappa \equiv \frac{C}{C_0}. \tag{26–19}$$

The dielectric constant, which is larger than unity for all materials, depends on the material as well as on external conditions such as temperature. Under certain conditions and for some materials, κ is very close to one, whereas for other conditions and materials, it can be as large as several hundred. For example, κ for air under normal conditions is 1.0005. This dielectric constant is almost indistinguishable from that of a vacuum, for which $\kappa = 1$. Table 26–1 gives a representative set of values of κ, but the temperature dependence of many of these values is so strong that the values must be used with care.

TABLE **26–1**

DIELECTRIC PROPERTIES OF MATERIALS†

Material	*Dielectric Constant,* κ	*Dielectric Strength,* E_{max} (10^6 V/m)
Vacuum	1.0	
Air	1.00054	3
Paraffin	2.0–2.5	10
Teflon	2.1	60
Polystyrene	2.5	24
Lucite	2.8	20
Mylar	3.1	
Plexiglas	3.4	40
Nylon	3.5	14
Paper	3.7	16
Fused quartz	3.75–4.1	
Pyrex	4–6	14
Bakelite	4.9	24
Neoprene rubber	6.7	12
Silicon	12	
Germanium	16	
Water	80	
Strontium titanate	332	8

† Values for some materials depend strongly on temperature and the frequency of oscillating fields.

Experimental Evidence for the Behavior of Dielectrics

To simulate the experiment that led Faraday to his conclusions about dielectrics, we first use a battery to charge a parallel-plate capacitor in air to a potential V_0 and a charge $Q_0 = C_0V_0$ (Fig. 26–16a). Here the subscripts refer to the quantity in air—for example, C_0 is the capacitance when there is air between the plates. We disconnect the battery and use a voltmeter, a device that measures the voltage, or potential difference (Fig. 26–16b). We then slide a dielectric, such as Plexiglas, between the plates (Fig. 26–16c). We observe that *the voltage is then reduced* by a factor we call κ to the new value

$$V = \frac{V_0}{\kappa}.$$

Because Plexiglas is an insulator, the charge Q_0 on the capacitor plates cannot change, yet the voltage does change. The capacitance must therefore change from the original value C_0 to a new value C when the Plexiglas is inserted:

$$C = \frac{Q_0}{V} = \frac{Q_0}{V_0/\kappa} = \kappa\frac{Q_0}{V_0} = \kappa C_0.$$

Our experimental result verifies Eq. (26–18), and the factor κ by which the voltage is reduced (when the charge is held fixed) is the dielectric constant.

We can do one further experiment. This time we leave the battery connected to the capacitor after it is charged in air (Fig. 26–17a). The potential continues to be the battery voltage V_0 after we insert the Plexiglas (Fig. 26–17b). However, we observe that *the charge on the conducting plates increases* by a factor of κ ($Q = \kappa Q_0$). Our experimental result remains in agreement with Eq. (26–18):

$$C = \frac{Q}{V_0} = \frac{\kappa Q_0}{V_0} = \kappa C_0.$$

A reinterpretation of these results in terms of permittivity is useful. If we take a parallel-plate capacitor, for which $C_0 = \epsilon_0 A/d$, we have

$$C = \kappa C_0 = \frac{\kappa\epsilon_0 A}{d}. \tag{26–20}$$

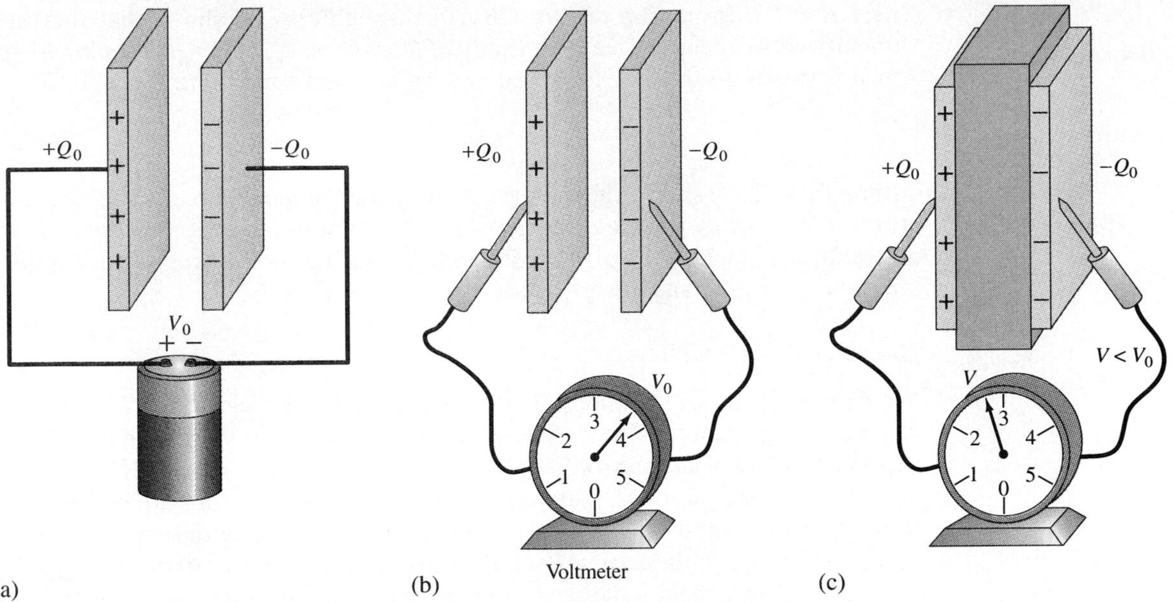

FIGURE 26–16 (a) A battery charges a capacitor to charge Q_0 and potential V_0. (b) If we take the battery away and measure the voltage with a voltmeter, we measure V_0 for the voltage. (c) If a dielectric is inserted into the capacitor, the voltage drops to $V < V_0$.

If the charge is held fixed, the new voltage is

$$V = \frac{V_0}{\kappa} = \frac{Q_0}{\kappa C_0} = \frac{\sigma_0 A}{\kappa C_0} = \frac{\sigma_0 d}{\kappa \epsilon_0}. \tag{26–21}$$

The new electric field between the plates is reduced in magnitude to

When there is a dielectric between the plates of a capacitor, the electric field is decreased.

$$E = \frac{V}{d} = \frac{\sigma_0}{\kappa \epsilon_0} = \frac{E_0}{\kappa}, \tag{26–22}$$

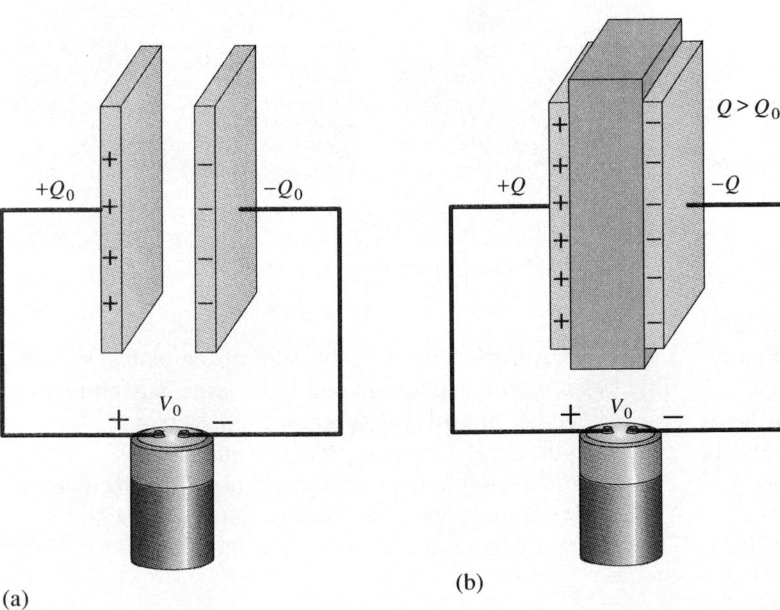

FIGURE 26–17 (a) Again a battery charges the capacitor to charge Q_0 and potential V_0, but (b) this time we leave the battery connected when we insert the dielectric. The potential must remain at V_0, but the new charge is $Q > Q_0$.

where $E < E_0$. Inspection of Eqs. (26–20) through (26–22) shows that they are summarized *by replacing the permittivity of free space,* ϵ_0, *by a new permittivity,* ϵ, which depends on the dielectric used and on external conditions:

The effect of a dielectric is that ϵ_0 is replaced by ϵ in all formulas of electrostatics.

$$\epsilon = \kappa\epsilon_0, \tag{26–23}$$

Although we have shown this simple rule only for the parallel-plate case, the substitution of ϵ for ϵ_0 when a dielectric is involved applies to all geometries and to all equations in which the permittivity appears, such as in the expressions for field strength, potential, and energy density.

Dielectrics in Capacitors

We have spoken of the important possibility of using a dielectric to increase the charge storage in a capacitor. Dielectrics also play a role in *voltage breakdown* across the plates of a capacitor (see the chapter-opening photograph). If the electric field in (or voltage across) a material becomes too great, electrons are pulled away from their atoms and cascade across the material, discharging the capacitor. Such events will damage dielectrics. Each dielectric has a *dielectric strength*, E_{max}, the maximum electric field a dielectric will support without breakdown. Table 26–1 contains some representative values. The dielectric strengths of commercial capacitors are indicated with a maximum allowable voltage.

APPLICATION

Capacitors

Commercial capacitors take various forms (Fig. B1–1).[†] Two metal strips separated by a thin layer of Mylar or paraffin-impregnated paper is an older type. When this "sandwich" is rolled up, it makes a compact capacitor of up to a few microfarads. Today's capacitors are more sophisticated, however. Two main types of capacitors achieve large capacitance in a small space by different strategies. In *multilayer ceramic capacitors*, metal sheets are in effect folded into a compact form. The sheets are separated by ceramic insulators with dielectric constants as high as 20,000. These high dielectric constants account for the large capacitances, on the order of thousands of microfarads, in multilayer ceramic capacitors.

Capacitances of roughly the same size can be achieved in even smaller volumes with *electrolytic capacitors*. This class of capacitors is constructed by depositing the dielectric, a nonconducting metal oxide, in a thin layer on a sheet of metal. The second conductor is a conducting paste or liquid that adheres well to the metal oxide. The metal-oxide layer between the conductors can be made quite thin, as thin as 10^{-8} m. Moreover, by etching the metal before the metal-oxide layer is deposited, a series of sharp valleys are created in the metal, greatly increasing its surface area. If we recall that the capacitance of parallel plates is inversely proportional to the distance between the plates and proportional to the area of the plates, we see that electrolytic capacitors can have large capacitances.

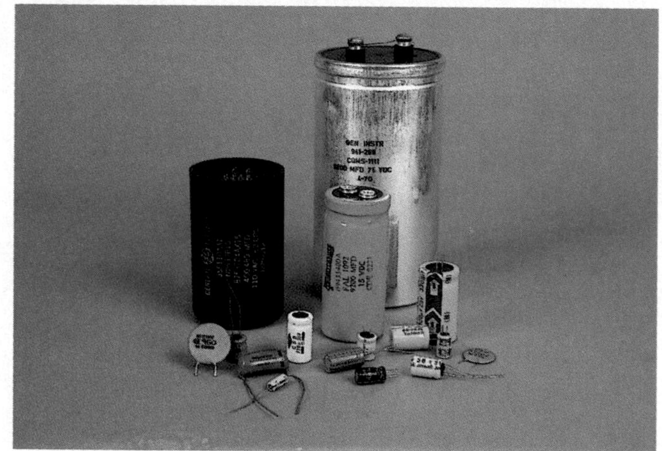

FIGURE B1–1 Some commercial capacitors.

Together, multilayer ceramic capacitors and electrolytic capacitors comprise the bulk of all capacitors available for commercial use. The electronics industry continues to demand new capacitors with even larger capacitances per unit volume.

[†] For further information, see "Capacitors" by Donald M. Trotter, Jr., *Scientific American*, July 1988, p. 86.

EXAMPLE 26–7 A parallel-plate capacitor has area $A = 20.0\ \text{cm}^2$ and a plate separation $d = 4.0$ mm. (a) Find the capacitance in air and the maximum voltage and charge the capacitor can hold. (b) A Teflon sheet is slid between the plates, filling the entire volume. Find the new capacitance and maximum charge. (c) Before the insertion of the Teflon, the plates are set to a voltage of 24 V by a battery that is then disconnected. What are the energies in the capacitor before and after the Teflon is inserted? Was work done in inserting the Teflon?

SOLUTION: (a) Equation (26–4) gives the capacitance in air, which we denote here by C:

$$C = \frac{\epsilon_0 A}{d} = \frac{(8.85 \times 10^{-12}\ \text{F}/\cancel{\text{m}})(2.00 \times 10^{-3}\ \cancel{\text{m}^2})}{4.0 \times 10^{-3}\ \cancel{\text{m}}} = 4.4 \times 10^{-12}\ \text{F} = 4.4\ \text{pF}.$$

The maximum charge depends on the maximum voltage. From Table 26–1, the dielectric strength of air is 3×10^6 V/m, so

$$V_{max} = E_{max}d = (3 \times 10^6\ \text{V}/\cancel{\text{m}})(4.0 \times 10^{-3}\ \cancel{\text{m}}) \simeq 10^4\ \text{V} = 10\ \text{kV}.$$

In turn,

$$Q_{max} = CV_{max} = (4.4 \times 10^{-12}\ \text{F})(10^4\ \text{V}) \simeq 5 \times 10^{-8}\ \text{C}.$$

(b) From Table 26–1, the dielectric constant of Teflon is 2.1. If we denote new values by primes, we have

$$C' = \kappa C = (2.1)(4.4\ \text{pF}) = 9.2\ \text{pF}.$$

Table 26–1 gives the dielectric strength of Teflon to be 6.0×10^7 V/m, so

$$V'_{max} = E'_{max}d = (6.0 \times 10^7\ \text{V}/\cancel{\text{m}})(4.0 \times 10^{-3}\ \cancel{\text{m}}) = 2.4 \times 10^5\ \text{V}$$

and

$$Q'_{max} = C'V'_{max} = (9.2 \times 10^{-12}\ \text{F})(2.4 \times 10^5\ \text{V}) = 2.2\ \mu\text{C}.$$

Both the maximum voltage and maximum charge are greatly increased after the Teflon is inserted.

(c) Equation (26–9) determines the energy in a charged capacitor. Before the Teflon is inserted, the energy is

$$U = \frac{CV^2}{2} = \frac{(4.4 \times 10^{-12}\ \text{F})(24\ \text{V})^2}{2} = 1.3 \times 10^{-9}\ \text{J}.$$

After the Teflon is inserted, C increases by the factor κ, whereas V decreases by the same factor. The product CV^2 thus *decreases* by a factor of κ:

$$U' = \frac{C'V'^2}{2} = \frac{(\kappa C)(V/\kappa)^2}{2} = \frac{U}{\kappa} = \frac{1.3 \times 10^{-9}\ \text{J}}{2.1} = 6 \times 10^{-10}\ \text{J}.$$

Because the potential energy has decreased, the capacitor does positive work as the Teflon is inserted.

When the material in the space between the plates of a capacitor is replaced by a dielectric of higher dielectric constant, the energy decreases (see Example 26–7). The capacitor therefore does positive work as the new dielectric is inserted, and so there must be a force that pulls the dielectric in. With sensitive instruments, the tug on the dielectric can be measured.

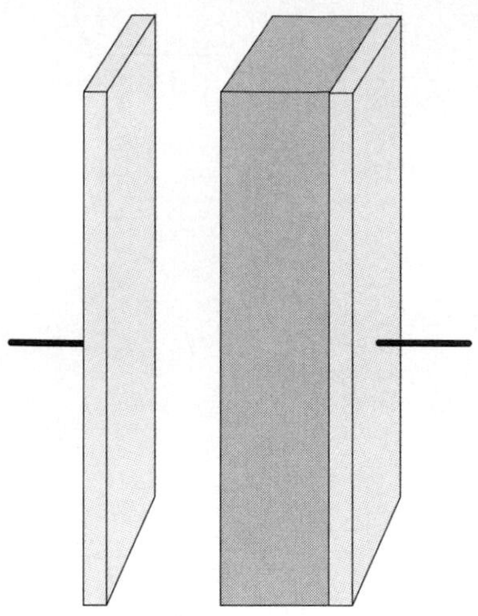

FIGURE 26–18 Example 26–8. A dielectric inserted in a parallel-plate capacitor fills only half the volume.

EXAMPLE 26–8 Suppose that the Teflon sheet inserted between the capacitor plates in Example 26–7 is only 2 mm thick and fills only half the volume (Fig. 26–18). Before the Teflon is inserted, the disconnected capacitor carries a charge of 1 nC. Find the electric field everywhere inside and the new capacitance.

SOLUTION: Although the Teflon sheet is on the right in Fig. 26–18, we shall see that this choice is immaterial. By Gauss' law, the electric field strength in air remains

$$E_0 = \frac{\sigma}{\epsilon_0} = \frac{Q}{\epsilon_0 A} = \frac{10^{-9}\ \text{C}}{(8.85 \times 10^{-12}\ \text{F/m})(2 \times 10^{-3}\ \text{m}^2)} \simeq 6 \times 10^4\ \text{V/m}.$$

The field inside the Teflon must be reduced by the factor κ:

$$E_{\text{Tef}} = \frac{E_0}{\kappa} = \frac{6 \times 10^4\ \text{V/m}}{2.1} \simeq 3 \times 10^4\ \text{V/m}.$$

To find the total voltage drop across the plates (which we need to find the capacitance), we calculate the integral

$$\begin{aligned} V &= \int_0^d E\,dx = \int_0^{2\ \text{mm}} E_0\,dx + \int_{2\ \text{mm}}^{4\ \text{mm}} E_{\text{Tef}}\,dx \\ &= E_0 \int_0^{2\ \text{mm}} dx + E_{\text{Tef}} \int_{2\ \text{mm}}^{4\ \text{mm}} dx = E_0\,(2\ \text{mm}) + E_{\text{Tef}}\,(2\ \text{mm}) \\ &= (5.6 \times 10^4\ \text{V/m})(2 \times 10^{-3}\ \text{m}) + (2.7 \times 10^4\ \text{V/m})(2 \times 10^{-3}\ \text{m}) \simeq 170\ \text{V}. \end{aligned}$$

This calculation is independent of the precise location of the Teflon sheet.

Finally, the capacitance is by definition $C = Q/V$:

$$C = \frac{10^{-9}\ \text{C}}{170\ \text{V}} = 6\ \text{pF}.$$

This value is intermediate to the capacitances of the system empty or filled with Teflon. This capacitor is equivalent to two capacitors in series—one empty and of width $d/2$; the other filled with Teflon and of width $d/2$.

EXAMPLE 26–9 Consider the coaxial cable of Example 26–2. There are equal but opposite charges per unit length λ on the two elements of the cable, which therefore acts as a capacitor. A plug of material of dielectric constant κ is inserted between the wire and the sheath to a depth x from the end. What is the change in potential energy of the charged cable? What electric force acts on the plug as it is inserted?

SOLUTION: In Example 26–2 we identified the potential difference between the conductors in the coaxial cable as

$$V_b - V_a = V = \frac{\lambda}{2\pi\epsilon_0} \ln \frac{b}{a}.$$

The potential energy per unit length is half the product of the voltage difference and the charge per unit length:

$$\frac{U}{\text{unit length}} = \frac{1}{2}\lambda V = \frac{\lambda^2}{4\pi\epsilon_0} \ln \frac{b}{a}$$

[see Eq. (25–19)]. In the region in which the dielectric is located, ϵ_0 is replaced by $\epsilon = \kappa\epsilon_0$. Thus the change in potential energy per unit length when the dielectric plug is inserted is

$$\frac{\Delta U}{\text{unit length}} = \frac{\lambda^2}{4\pi}\left(\frac{1}{\epsilon_0} - \frac{1}{\epsilon}\right)\ln\frac{b}{a}.$$

If the plug penetrates by a depth x, the total change in potential energy is

$$\Delta U = \frac{x\lambda^2}{4\pi}\left(\frac{1}{\epsilon_0} - \frac{1}{\epsilon}\right)\ln\frac{b}{a}.$$

Because $\kappa > 1$, $\epsilon > \epsilon_0$ and $\Delta U < 0$. The energy decreases when the plug is inserted, so we expect the plug to be pulled into the space between the conductors. The force exerted on the dielectric as it moves into the cable is obtained from

$$F = -\frac{dU}{dx} = -\frac{\lambda^2}{4\pi}\left(\frac{1}{\epsilon_0} - \frac{1}{\epsilon}\right)\ln\frac{b}{a}.$$

F is positive because $\epsilon > \epsilon_0$. The plug is indeed pulled into position.

26–6 MICROSCOPIC DESCRIPTION OF DIELECTRICS

Molecules with permanent electric dipole moments, such as H_2O, are called *polar* molecules. In the absence of an external electric field, the directions of the dipole moments of polar molecules in a material are randomly distributed (Fig. 26–19a). However, when the material is placed in an external electric field, as in Fig. 26–19b, the dipoles are subject to a torque due to the electric field and tend to align

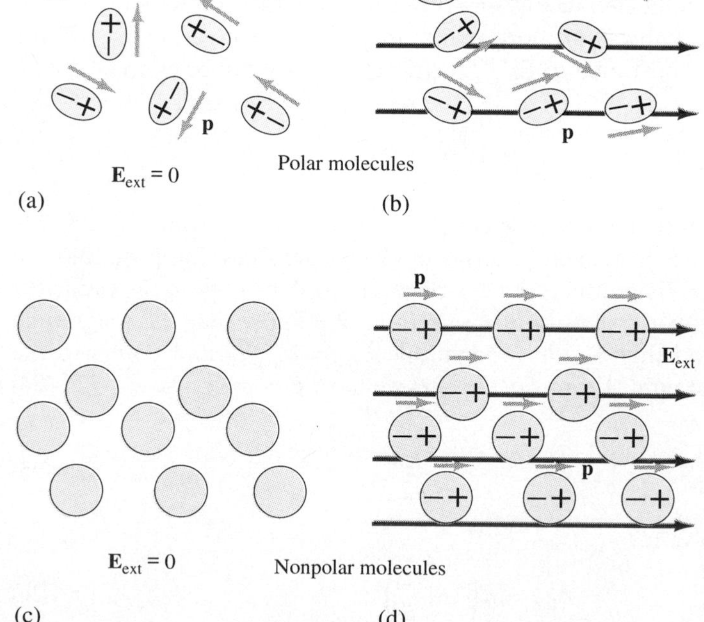

FIGURE 26–19 (a) For polar molecules not in an external electric field, the dipole moments, **p**, are randomly oriented. (b) In an external electric field, the dipole moments align themselves with the field. (c) For nonpolar molecules not in an electric field, there is no indication of a charge distribution. (d) In an external electric field, nonpolar molecules obtain an induced dipole moment aligned with the electric field.

themselves with the field. Because of thermal agitation, this alignment is imperfect and occurs only on average. It will be more pronounced for stronger electric fields and lower temperatures. There is a wide range of values of the external field for which the average degree of alignment grows *linearly* with the external electric field.

Molecules without a permanent dipole moment are called *nonpolar*. In the absence of an external electric field, their charge distributions are symmetric; that is, no particular direction is picked out (Fig. 26–19c). As we discussed in Chapter 23, when these molecules are placed in an external electric field, they acquire an induced dipole moment that is fully aligned with the field (Fig. 26–19d). The magnitude of this dipole moment increases as the external field increases, and again there is an important range of values of the external electric field for which the dipole moment grows *linearly* with the field.

If an insulating slab made of either polar or nonpolar molecules is placed in a charged capacitor, the effect on the capacitor is as shown in Fig. 26–20a. There is no free charge to move large distances in an insulator. However, the dipole moments, either permanent or induced, become aligned with the electric field. The inside of the dielectric remains electrically neutral, but the charge distribution is *polarized* so that *induced charge* appears on the two outside surfaces of the slab. We denote the induced surface charge density as σ_{ind}. The two dielectric surfaces have equal but opposite charge densities. The induced surface charge density is proportional either to the degree to which the permanent dipole moments of polar molecules are aligned, or to the magnitude of the induced dipole moments of nonpolar molecules. There will thus be an important range of values of the external field for which σ_{ind} *is proportional to the external field.*

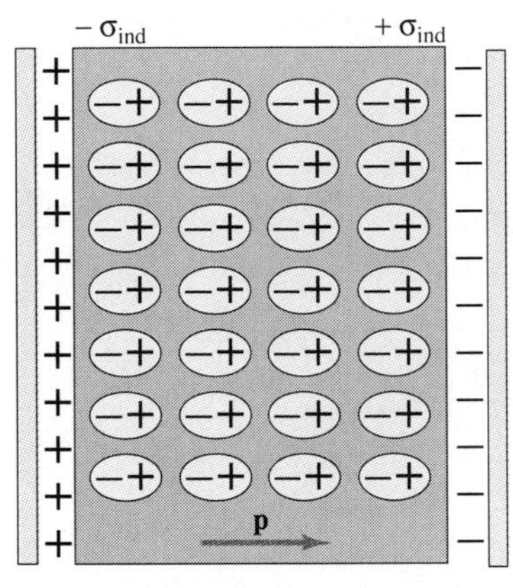

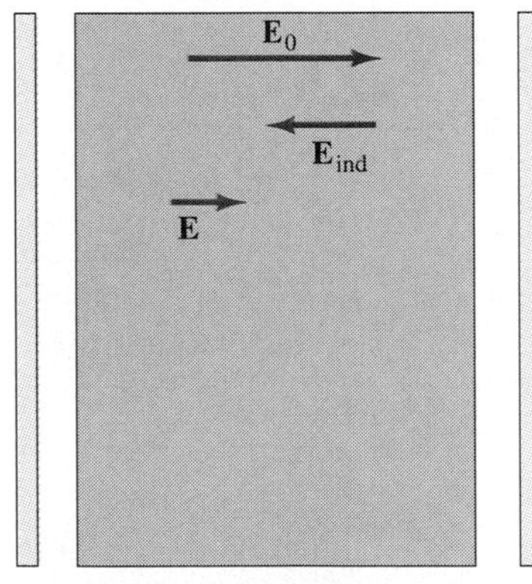

FIGURE 26–20 (a) When a dielectric is inserted in a capacitor, induced charges appear on the surface. (b) The induced charges cause an induced electric field $\mathbf{E}_{\text{ind}}$ opposite to the external electric field $\mathbf{E}_0$ caused by the free charge on the capacitor plates. The net effect is a reduced electric field $\mathbf{E} = \mathbf{E}_0 - \mathbf{E}_{\text{ind}}$ within the dielectric. In the region between the conducting plates where there is no dielectric, the net electric field remains $\mathbf{E}_0$.

We can now sort out the electric fields that appear in dielectrics. We can distinguish three fields. The external field $\mathbf{E}_0$ would be present whether the dielectric were inserted or not. An *induced electric field* $\mathbf{E}_{\text{ind}}$ is produced by the induced surface charge, which is due to the external field. Because of the way the induced charge forms, $\mathbf{E}_{\text{ind}}$ is opposite to $\mathbf{E}_0$. Finally, the net electric field inside the dielectric, $\mathbf{E}$, is

$$\mathbf{E} = \mathbf{E}_0 + \mathbf{E}_{\text{ind}}. \tag{26–24}$$

The direction of $\mathbf{E}$ is indicated in Fig. 26–20b. From Eq. (26–22) we can see that only if $\mathbf{E}_{\text{ind}}$ (which we know is proportional to σ_{ind}) is proportional to $\mathbf{E}_0$ will the resultant field $\mathbf{E}$ be proportional to $\mathbf{E}_0$. This proportionality can be expressed with the constant κ:

$$\mathbf{E} = \frac{\mathbf{E}_0}{\kappa}.$$

The constant κ is in fact the dielectric constant, as we can see from Eq. (26–22).

Let us translate this argument in terms of charge densities. Suppose that the original electric field $\mathbf{E}_0$ is produced by a surface charge density σ on the capacitor plates. We refer to σ as the density of *free charge*. We know that the magnitude of $\mathbf{E}_0$ is $E_0 = \sigma/\epsilon_0$ and that $\mathbf{E}_{\text{ind}}$ has magnitude $E_{\text{ind}} = \sigma_{\text{ind}}/\epsilon_0$ and points to the left if $\mathbf{E}_0$ points to the right. From Eq. (26–22) we have $E = \sigma/\kappa\epsilon_0$, so Eq. (26–24) becomes

$$\frac{\sigma}{\kappa\epsilon_0} = \frac{\sigma}{\epsilon_0} - \frac{\sigma_{\text{ind}}}{\epsilon_0}. \tag{26–25}$$

We cancel ϵ_0 and solve for σ_{ind}:

$$\sigma_{\text{ind}} = \sigma\left(1 - \frac{1}{\kappa}\right). \tag{26–26}$$

Because $\kappa > 1$ for all dielectrics, the induced charge is always less than the free charge. This is evident in our microscopic model; if E_{ind} were to exceed E_0, we would actually reverse the field in the material.

Gauss' Law and Dielectrics

We have mentioned that the presence of a dielectric causes ϵ_0 to be replaced by ϵ. How does this result affect Gauss' law? For the Gaussian surface drawn in Fig. 26–21, Gauss' law in its original form gives

$$\oiint \mathbf{E} \cdot d\mathbf{A} = \frac{Q_{\text{encl}}}{\epsilon_0} = \frac{Q - Q_{\text{ind}}}{\epsilon_0}. \tag{26–27}$$

The enclosed charge is the free charge minus the induced charge. From Eq. (26–25),

$$\frac{Q}{\kappa} = Q - Q_{\text{ind}}.$$

Thus we have an alternative form of Gauss' law when dielectrics are present:

$$\oiint \mathbf{E} \cdot d\mathbf{A} = \frac{Q}{\kappa\epsilon_0}. \tag{26–28}$$

Gauss' law in the presence of a dielectric

The constant ϵ_0 is replaced by $\epsilon = \kappa\epsilon_0$ *provided* Q *is the free charge.* This is a general result, even though Fig. 26–21 refers to a parallel-plate capacitor. If the dielectric is not uniform throughout, κ will not have the same value throughout. In this case, κ should be brought under the integral, and Gauss' law reads

$$\oiint \kappa\mathbf{E} \cdot d\mathbf{A} = \frac{Q}{\epsilon_0}. \tag{26–29}$$

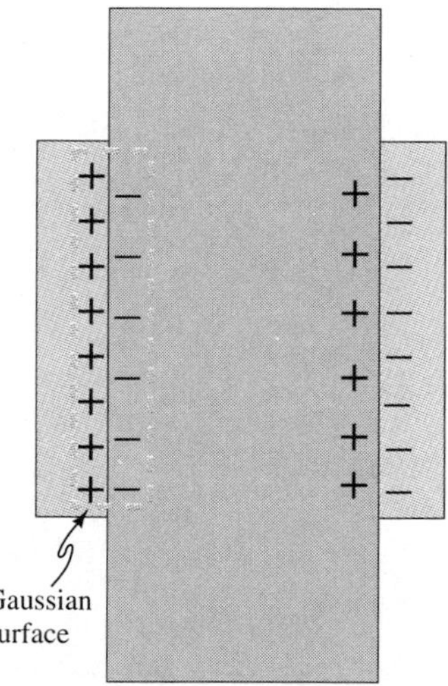

FIGURE 26–21 A dielectric fills the entire volume of a capacitor. A Gaussian surface surrounding the interface region between the dielectric and each plate surrounds both free and induced charges. The total charge enclosed by the Gaussian surface when Gauss' law is applied includes both charges.

Consequences of the Microscopic Model of Dielectrics

Our model of induced charges explains the experimental behavior of the dielectrics we have described to this point. This model is also the basis for a variety of further experimental predictions, all of which can be tested:

1. There are two classes of dielectrics, those made of either nonpolar or polar molecules. The dipole moments of induced dipoles are generally much smaller than those of permanent dipoles, so the value of κ for nonpolar dielectrics should be much closer to one than that for polar dielectrics.
2. Because of disruptive thermal effects, polar dielectrics should line up more easily—have larger values of κ—at lower temperatures. Kinetic theory (see Chapter 19) shows that the dielectric constant takes the more precise form

$$\kappa = 1 + \frac{\text{a constant}}{T}. \tag{26–30}$$

   This temperature dependence holds rather well. Equation (26–30) is called *Curie's law.* Nonpolar dielectrics should not obey such a law.
3. The polarization of solid substances with a permanent dipole moment can change if the planes of their lattice structure are stressed by being twisted or pressed. Under such stress, the internal electric fields change, and the changing fields produce an electrical signal. This phenomenon, known as *piezoelectricity,* is the principle behind the operation of microphones and strain gauges.

SUMMARY

Capacitors are devices for storing electric charge and energy and typically consist of two conductors with equal and opposite charges Q and potential difference V. Capacitance is defined as

$$C \equiv \frac{Q}{V}. \qquad (26\text{–}2)$$

The capacitance of a parallel-plate capacitor in air is given by

$$C = \frac{\epsilon_0 A}{d}, \qquad (26\text{–}4)$$

where A is the plate area and d is the plate separation. The SI unit of capacitance is the farad: 1 F = 1 C/V.

The potential energy of a capacitor can be written as

$$U = \frac{Q^2}{2C} = \frac{CV^2}{2} = \frac{QV}{2}. \qquad (26\text{–}8, 26\text{–}9, 26\text{–}10)$$

The energy density, or energy per unit volume, of an electric field is

$$u = \frac{\epsilon_0 E^2}{2}. \qquad (26\text{–}12)$$

Capacitors connected in parallel can be replaced by an equivalent capacitor with capacitance

$$C_{\text{equ}} = C_1 + C_2 + \cdots + C_n. \qquad (26\text{–}14)$$

Capacitors connected in series can be replaced by an equivalent capacitor according to the relation

$$\frac{1}{C_{\text{equ}}} = \frac{1}{C_1} + \frac{1}{C_2} + \cdots + \frac{1}{C_n}. \qquad (26\text{–}17)$$

Dielectrics are insulators with a characteristic property called the dielectric constant, κ; $\kappa > 1$. When a dielectric fills the space between the two conducting plates of a capacitor, the value of the capacitance is increased:

$$C = \kappa C_0, \qquad (26\text{–}18)$$

where C_0 is the value of the capacitor with a vacuum (or air) between its conductors. Our previous results can be modified for the presence of dielectrics by replacing the permittivity of free space, ϵ_0, by the permittivity ϵ given by

$$\epsilon = \kappa\epsilon_0. \qquad (26\text{–}23)$$

Each insulator also has a characteristic property called the dielectric strength, which gives the approximate maximum electric field that the insulating material can withstand before it breaks down and ionizes.

The behavior of capacitors can be understood by considering the molecular structure of matter. Polar and nonpolar molecules of a dielectric become aligned with the external electric field, reducing the effects of that field. An alternative form of Gauss' law when dielectrics are present in a capacitor is

$$\oint\!\!\oint \mathbf{E} \cdot d\mathbf{A} = \frac{Q}{\kappa\epsilon_0}, \qquad (26\text{–}28)$$

where Q is the free charge.

QUESTIONS

1. There are two common ways to write SI units of permittivity. Does the fact that there is more than one way present a problem?
2. You have two parallel plates, a battery, a voltmeter, and a piece of unknown plastic. Devise a method to determine the dielectric constant of the plastic.
3. What argument can you give to show that the electric field of a parallel-plate capacitor cannot drop abruptly to zero as we pass outside the region between the plates? Recall the fact that the voltage drop around any closed path must be zero.
4. What is the meaning of a capacitor with zero capacitance?
5. If the radius of the outer sheath of the coaxial cable in Example 26–2 is taken to infinity, the capacitance per unit length of the system increases to infinity. In other words, the capacitance per unit length of a single charged wire is infinite. Why? Physically, what prevents this from causing a real problem?
6. It is not possible to break up every combination of capacitors into a sequence of parallel and series capacitors. Find an example of a combination that cannot be decomposed in this way.
7. From our discussion of the physical nature of dielectrics, can you imagine a physical system in which the dielectric constant is less than one?
8. The plates of a charged parallel-plate capacitor are disconnected from the charging battery and are pushed together. What happens to the potential difference, the capacitance, and the stored energy?
9. The plates of a parallel-plate capacitor, still connected to a battery with potential difference V, are pushed together. What happens to the charge on the plates, the capacitance, and the stored energy?
10. You are given a thin, metal sheet of area A. You can make it into a spherical shell, roll it into two concentric cylinders, or cut it to make a parallel-plate capacitor. Which arrangement would give the largest possible capacitance?
11. What happens if you short out (connect with a conductor) the two plates of a large, charged capacitor? Could this be dangerous?
12. For finite parallel plates there is a fringe field (see Figure 26–3). What effect would you expect this phenomenon to have on the capacitance of a parallel-plate capacitor?
13. Is it possible for a pair of nonconductors carrying equal but opposite charges to act as a capacitor? In what ways would such an arrangement differ from, or be similar to, the capacitors treated in this chapter?
14. Why is it a good idea to short out (connect with a conductor) the plates of a large capacitor when the capacitor is not in use?
15. Would you expect the term "dielectric strength" to have meaning for a vacuum?
16. Air, particularly on humid days, can cause charge leakage. Why, then, can capacitors with air between their plates hold charge in a way that is useful for circuits?
17. A combination of capacitors in a circuit may be equivalent to a single capacitor whose capacitance can be determined in terms of the original capacitances. If we were to replace the combination of capacitors with a single equivalent capacitor, would we see any difference in the behavior of the circuit?

PROBLEMS

26–1 Capacitance

1. (I) (a) What is the capacitance of two square metal plates, each 100 cm^2 in area, separated by 0.5 cm? (b) What is the radius of a conducting sphere with the same capacitance?
2. (I) At different times a 4-μF capacitor has a charge of (a) 4 μC, (b) 10 μC, and (c) 1 mC. What is the voltage across the capacitor in each case?
3. (I) How much charge can be stored on the plates of a 1-μF capacitor if the plates are attached to a battery that can give a potential difference of (a) 2 V? (b) 12 V?
4. (I) You must design a capacitor capable of storing 10^{-7} C of charge, but you have only a 100-V power supply and two metal plates of area 0.2 m^2 each. What limits do you put on the separation between the plates?
5. (I) What is the capacitance of a piece of coaxial cable 25 cm long for which the radius of the inner conductor is 0.50 mm and the radius of the outer conducting sheath is 1.5 mm?
6. (II) Calculate the capacitance of two concentric spherical conductors of radii r and R, respectively. Discuss the limits of (a) finite r, $R \to \infty$; (b) $(R - r) \ll r$.
7. (II) Two concentric conducting spheres have radii of 5.0 cm and 23 cm, respectively, and an equal but opposite charge of 6.3×10^{-7} C. What is the potential difference between them? (*Hint:* Use the results of Problem 6.)
8. (II) A parallel-plate capacitor of area 0.040 m^2 carries a charge $q = 4.0 \times 10^{-8}$ C. The potential across the plates increases with time t according to the equation $V = 50.0\text{ mV} + (0.10\text{ mV/s})t$, as a result of a time-dependent increase of the separation between the plates. Find the function of time that describes the separation.
9. (II) A parallel-plate capacitor has square plates 40 cm on a side, separated by 5 mm. The capacitor is charged to 230 V, then disconnected from the charging battery. What is the charge density on the plates? the total charge on each plate?
10. (II) The capacitance of a variable capacitor used in a radio varies from 0.2 μF to 0.01 μF. The capacitor is charged to

a potential difference of 300 V at maximum capacitance and then isolated. At minimum capacitance, what is the voltage?

26–2 *Energy in Capacitors*

11. (I) A thundercloud has a charge of 900 C and a potential of 90 MV with respect to the ground 1 km below. (a) What is the capacitance of the system? (b) How much energy is stored in the thundercloud system?

12. (I) How much energy is stored on a metal sphere of radius 12 cm when a charge of 4.0×10^{-5} C is placed on it?

13. (II) A coaxial cable with an inner wire of diameter 3 mm and an outer sheath wire of diameter 8 mm has a potential of 1 kV between the wires. (a) What is the capacitance of 10 m of the cable? (b) How much energy is stored in the 10-m piece of cable? in a 1 km piece?

14. (II) Two concentric conducting spheres of radii 7.0 cm and 18 cm, respectively, are given equal but opposite charges of 4.2×10^{-8} C. How much energy is stored in the system?

26–3 *Energy in Electric Fields*

15. (I) The electric field in a large thunderstorm is 125,000 V/m. How much energy is contained in 1 m^3? in 1 km^3?

16. (I) A Van de Graaff accelerator with a spherical dome of radius 2.0 m has a potential of 300,000 V in air. Assume that the accelerator is in effect a charged sphere. How much energy is stored in its electric field?

17. (I) Approximately how much energy is stored in a cube of sides 5 cm that is 1.0 m from a point charge of magnitude 5×10^{-4} C?

18. (II) An isolated metal sphere of radius 15 cm is at potential 5000 V. What is the charge on the sphere? What is the energy density of the electric field outside the sphere? Integrate this to obtain the total energy in the electric field.

19. (II) A metal sphere of radius 0.10 m carries a charge of 8.5×10^{-6} C. How much energy is contained in a spherical region of radius 25 cm that is concentric with the sphere?

20. (III) A nonconducting sphere of radius 0.10 m carries a uniformly distributed charge of 8.5×10^{-6} C. How much energy is contained in a spherical region of radius 25 cm that is concentric with the sphere?

21. (III) The plates of a parallel-plate capacitor are 600 cm^2 in area and 0.2 cm apart. The potential difference between the plates is 800 V. (a) What is the field between the plates? (b) the charge on each plate? (c) the force exerted by the field on one of the plates? (d) Suppose that the plates are pulled apart so that the separation increases by 10 percent. What is the change in the stored energy? Is this consistent with the answer to part (c)? [If not, you probably answered (c) incorrectly.]

22. (III) Assume that an electron consists of a sphere of radius R with its charge distributed uniformly on the surface. (a) What is the electric field outside of the radius R? (b) What is the total electrostatic energy stored in the electric field? (c) Assume that all the energy of part (b) is solely responsible for the rest energy of the electron. (Rest energy is the energy associated with an object's mass, according to the theory of special relativity, even if the object is at rest. It takes the form mc^2, where in this case m is the electron's mass, 0.9×10^{-30} kg, and c is the speed of light, 3×10^8 m/s.) What must the radius R of the electron be?

26–4 *Capacitors Connected in Parallel and in Series*

23. (II) Find the capacitance of the parallel-plate system in Fig. 26–22. Can this system be represented by two pairs of parallel plates of half the total area connected in series or in parallel?

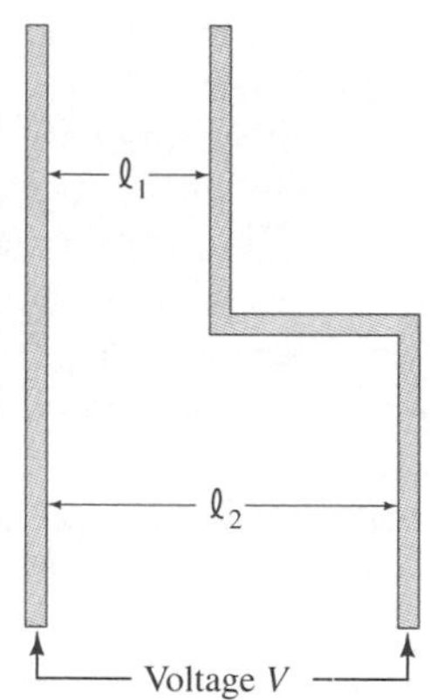

FIGURE 26–22 Problem 23.

24. (II) Find the equivalent capacitance of the circuit shown in Fig. 26–23.

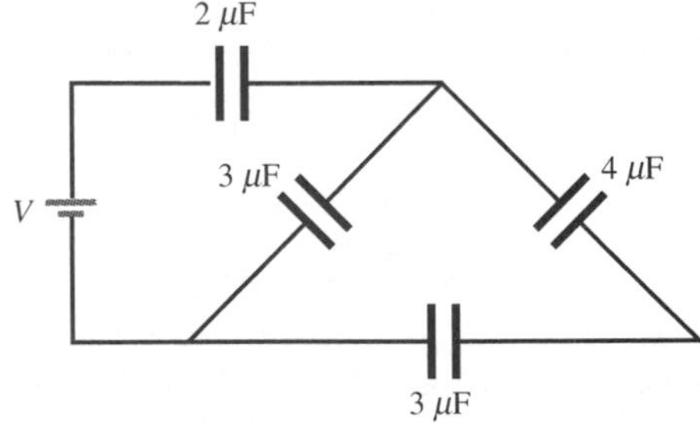

FIGURE 26–23 Problem 24.

25. (II) Two large, thin metal plates of area A and thickness d, carrying charges Q and $-Q$, respectively, are placed a distance D apart. Suppose that an uncharged, thin metal plate of the same area and thickness is placed between them, such that the distance between the uncharged plate and the positively charged plate is x. What is the capacitance of the combined system as a function of x?

26. (II) What is the capacitance of the two concentric spherical conductors of radii 3.0 mm and 12 mm, respectively, con-

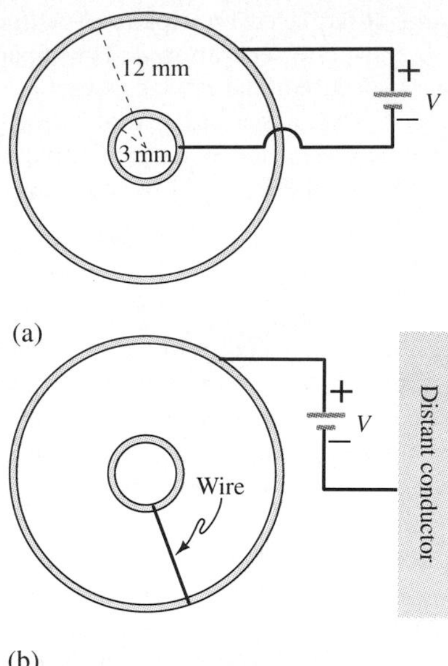

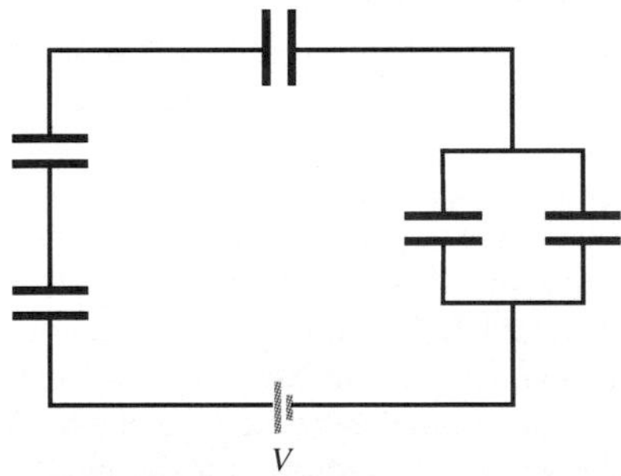

FIGURE 26–24 Problem 26.

nected as shown in Fig. 26–24a? Suppose that the conductors are connected as shown in Fig. 26–24b. What is the capacitance now?

27. (II) Find the equivalent capacitance of the circuit shown in Fig. 26–25. The capacitance of each capacitor is 5 μF.

FIGURE 26–25 Problem 27.

28. (II) Figure 26–26 illustrates a set of five capacitors (their capacitances indicated) connected together across the points a and b. What is the value of a single capacitor that could replace this system and collect the same total charge for a given voltage drop V_{ab}?

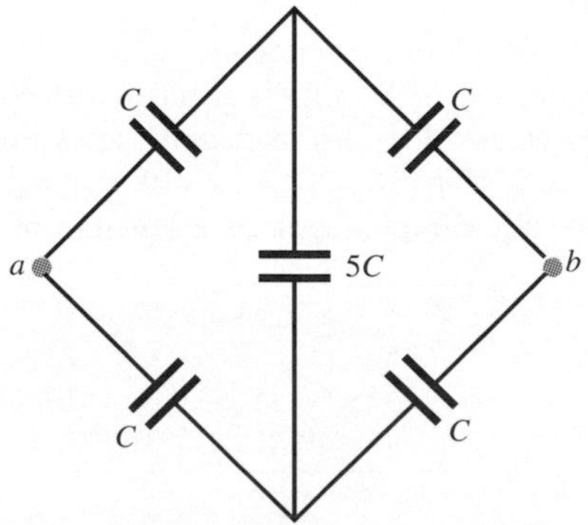

FIGURE 26–26 Problem 28.

29. (II) Capacitor C_1 has a capacitance of 2 μF; capacitor C_2 has a capacitance of 3 μF. A charge of $q = 10$ μC is placed on C_1, whereas C_2 is brought to a potential difference between its plates of 50 V. (a) What is the total energy stored in the two capacitors? (b) The negatively charged plate of C_1 is connected to the positively charged plate of C_2. What will change in the system, if anything? Neglect the fringe fields at the ends of the capacitors.

30. (II) Consider the capacitors of Problem 29, with C_2 modified so that it holds a charge of 10 μC at a potential difference of 50 V between the plates. (a) How is the capacitance of C_2 modified? (b) What is the charge on the capacitor equivalent to the whole system when the negatively charged plate of C_1 is connected to the positively charged plate of C_2?

31. (II) You have four capacitors whose capacitances are 2 μF, 3 μF, 4 μF, and 5 μF, respectively. Describe a circuit with an equivalent capacitance smaller than the 5-μF capacitor by 0.032 μF.

26–5 Dielectrics

32. (I) With two parallel plates, a 12-V battery, and a voltmeter, you charge the plates with the battery and then disconnect it. You have a piece of plastic whose dielectic constant you want to measure. After you slide the plastic into the full volume between the plates, the voltmeter indicates a voltage drop from 12 V to 3.4 V. What is the dielectric constant?

33. (I) A 12-V automobile battery can store 4×10^6 J of energy. Find the area of a parallel-plate capacitor that that can store the same amount of energy, if the separation between the plates is 1 mm and a dielectric with dielectric constant $\kappa = 3$ is between the plates.

34. (II) Repeat Problem 13 for polystyrene placed between the wires of the coaxial cable.

35. (II) Calculate the change in capacitance of an isolated sphere that becomes embedded in a dielectric with dielectric constant κ. If the capacitance change is due to a charge induced on the surface of the dielectric, what is the ratio of the induced charge density to that of the original surface charge density?

36. (II) Two large, parallel metal plates have a potential difference of 120 kV, and the electric field between them has a magnitude of 1.0×10^6 V/m. A material with a dielectric constant of 1.5 is inserted between the plates, with the plate separation adjusted so that the capacitance is unchanged. Calculate the new plate separation.

37. (II) A parallel-plate capacitor carrying charge q_0 is modified by the insertion of a dielectric with $\kappa = 1.8$ between the plates. As a consequence, the energy stored in the capacitor triples. What will the charge be after the dielectric is inserted?

38. (II) A coaxial cable has an inside wire of radius 2.0 mm and an outside metal sheath of radius 3.5 mm. The intermediate region is filled with a material of dielectric constant 2.2. (a) What is the capacitance of such a cable 100 m long? (b) If the potential difference between the inner and outer conductors is 1200 V, what is the charge on the inner conductor, and how much energy is stored in 100 m of cable?

39. (II) A dielectric slab of thickness d and dielectric constant κ is inserted in the middle of a parallel-plate capacitor of plate separation D. What is the new capacitance of the capacitor, given that the area of each plate is A?

40. (II) A parallel-plate capacitor of dimensions 30 cm × 40 cm and separation distance 1 cm contains a dielectric slab of thickness 0.4 cm and dielectric constant 2.4. The potential difference between the plates is 200 V. What are the electric fields in the empty space and inside the dielectric?

41. (II) A parallel-plate capacitor of area 10 cm^2 and plate separation 5 mm holds how much free charge if the voltage between its plates is 300 V, and the following materials are inserted between its plates: air, paper, Neoprene, Bakelite, and strontium titanate? (Use Table 26–1.)

42. (II) Two parallel plates of area 100 cm^2 with Plexiglas inserted between them break down when a voltage of 10 kV is applied to the plates. How much charge will the plates hold when the Plexiglas is removed? (Use Table 26–1.)

43. (II) A capacitor consists of two concentric spherical shells of radii r_1 and r_2, respectively. Calculate the capacitance if the space between the shells is filled with a dielectric of dielectric constant κ. If the capacitor starts out with air between the shells and carries a charge Q, and if the space is then filled with the dielectric, what is the change in energy?

44. (II) A parallel-plate capacitor has area $L \times L$ and separation $D \ll L$. One-half the space between the plates is filled with a dielectric for which $\kappa = \kappa_0$, and the other half with a dielectric for which $\kappa = \kappa_1$ (Fig. 26–27). Find the capacitance of this capacitor.

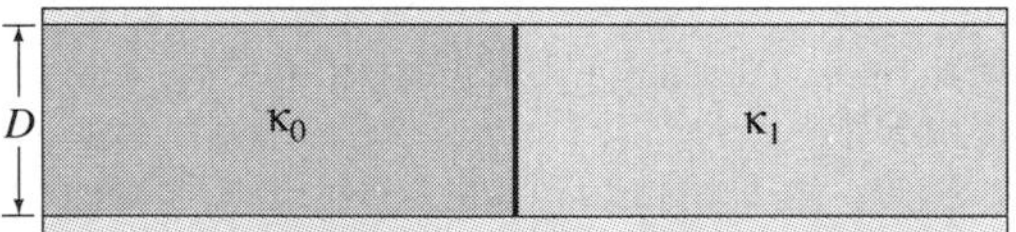

FIGURE 26–27 Problem 44.

45. (II) A capacitor consists of 12 plates attached alternately to a positive and negative terminal. The plates are 8.0 cm × 15 cm in size and are 0.30 mm apart. What is the capacitance? Suppose that the region between the plates is filled with material of dielectric constant 2.5. What will the capacitance be?

26–6 *Microscopic Description of Dielectrics*

46. (II) Use Gauss' law and Eq. (26–22) to show from Fig. 26–20 that $E_{ind} = \sigma_{ind}/\epsilon_0$.

47. (II) A charge Q is placed on a parallel-plate capacitor of area $L \times L$ and plate separation d. The capacitor is then filled with a dielectric of dielectric constant κ. If $L = 0.5$ m, $d = 1$ mm, $Q = 5\ \mu$C, and $\kappa = 2.3$, what is the surface charge induced on the dielectric? What is the magnitude of the electric field in the dielectric? How much energy is stored in this capacitor?

General Problems

48. (II) *Estimate* how much charge you pick up when you walk across a carpet on a dry winter day. (*Hint:* View yourself as a good conductor, spherical in shape, and notice how close your hand has to come to a doorknob before the inevitable spark occurs. Use Table 26–1.)

49. (II) You have 100 cm^2 of aluminum plate (which you can cut into pieces) and a 200-cm^2 sheet of 1-cm thick Bakelite (which you can also cut). Neither material can be sliced into thinner sheets or rolled, and the minimum separation between any aluminum plates you cut is 1 cm. You have a power supply of a single voltage, 500 V. (a) Design a system that will hold the maximum amount of charge. What charge and energy can this system hold? (b) Design a system that has the maximum electric field, and find this field. Is this the same system as part (a)?

50. (II) Calculate the energy of a composite capacitor that consists of N identical capacitors of capacitance C_1 that are connected (a) in series; (b) in parallel. In parts (a) and (b), the total potential difference across the composite capacitor is V. (c) Assume that the total charge is Q, and repeat the calculation.

51. (II) A capacitor consists of two flat metal plates of area 0.25 m^2 and plate separation of $d = 3.0$ cm. A flat metal plate of the same area and of thickness 1.0 cm is inserted midway between the plates of the capacitor, leaving two spaces of thickness 1.0 cm each. (a) Find the new capacitance. (b) If the original capacitor has charge Q, what is the surface charge density induced on the intermediate plate? (c) Suppose that the original charge on the external plates remains the same. How does the energy of the new system compare to the energy of the system without the inserted plate? (d) Compare the capacitor with the metal inserted to the same capacitor with a dielectric of the same dimensions inserted.

52. (II) A parallel-plate capacitor has an area of $L \times L$ and a plate separation of $D \ll L$. It is filled with a nonuniform dielectric whose dielectric constant varies linearly across the capacitor (Fig. 26–28). At $x = 0$, $\kappa = \kappa_0$, and at $x = L$, $\kappa = \kappa_1$. We can express κ as a function of x: $\kappa = \kappa_0 +$

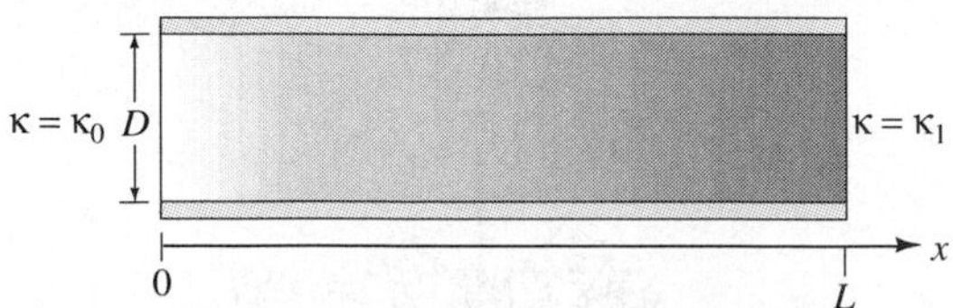

FIGURE 26–28 Problem 52.

$[(\kappa_1 - \kappa_0)x/L]$. Treat the capacitor plates as broken into a set of capacitors connected in parallel with plates that are strips of width dx, and calculate the capacitance.

53. (II) A thunderstorm is a fairly complicated phenomenon in terms of the distribution of charges, but we can say roughly that there is a voltage drop of as much as 10^8 V between the earth and the bottom of a thundercloud, and the charges involved may run into the hundreds of coulombs. Estimate the capacitance of the earth–cloud system and the energy contained in the space between the cloud and the earth.

54. (II) You are given four capacitors of capacitance 2.0 μF each. How many different effective capacitors can you make with them, using one, two, three, or all four in different configurations (in series and/or in parallel), and what are their capacitances?

55. (II) Consider the arrangement of four capacitors shown in Fig. 26–29. Capacitors A, B, C, and D have capacitances 2.4 μF, 3.6 μF, 1.2 μF, and 4.0 μF, respectively. Suppose that a battery applies a potential difference of 600 V across the circuit, which is then disconnected from the battery. What is the potential difference across each capacitor?

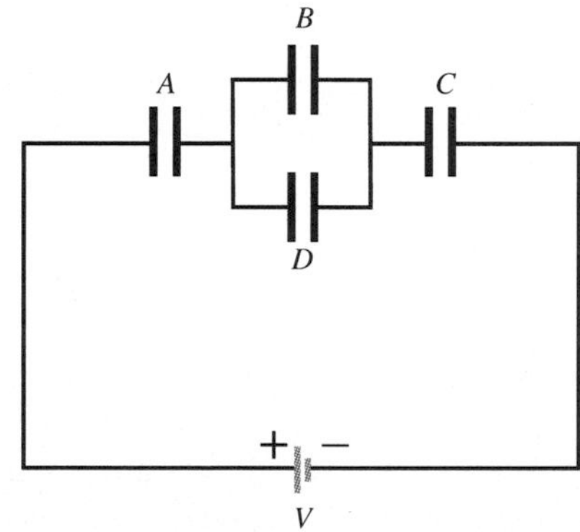

FIGURE 26–29 Problem 55.

56. (II) Consider a parallel-plate capacitor of plate area 1.0 m^2 and plate separation 4.0 mm. (a) Assume that the maximum electric field strength (before breakdown) in air is 3.0×10^6 V/m. What are the capacitance and the charge stored at the maximum voltage? (b) Suppose that the capacitor is immersed in oil of dielectric constant $\kappa = 2.4$, and the maximum charge that can be stored is a factor of 10 larger than that without the oil. What is the maximum field strength the oil can maintain?

57. (II) A parallel-plate capacitor has a capacitance of 4.0 μF. The plates are charged to 600 V. What is the energy stored in the capacitor? How much work is required to insert a dielectric of $\kappa = 2.0$ between the plates? Assume that the capacitor is disconnected from the voltage source before the dielectric is inserted.

58. (II) A dielectric of dielectric constant κ is inserted a distance x into a parallel-plate capacitor with square plates of area A and plate separation d. What is the capacitance as a function of x? Calculate the amount of energy stored in the capacitor for a potential difference V.

59. (II) A parallel-plate capacitor has an area of $L \times L$ and a plate separation of $D \ll L$. It is filled with a nonuniform dielectric whose dielectric constant varies linearly from one plate to another (Fig. 26–30). At the bottom plate, the dielectric constant is κ_0, whereas at the upper plate, it is κ_1. If y is the distance measured up from the bottom plate to the top plate, then $\kappa = \kappa_0 + [(\kappa_1 - \kappa_0)y/D]$. Treat the capacitor as a set of capacitors connected in series, and calculate the capacitance.

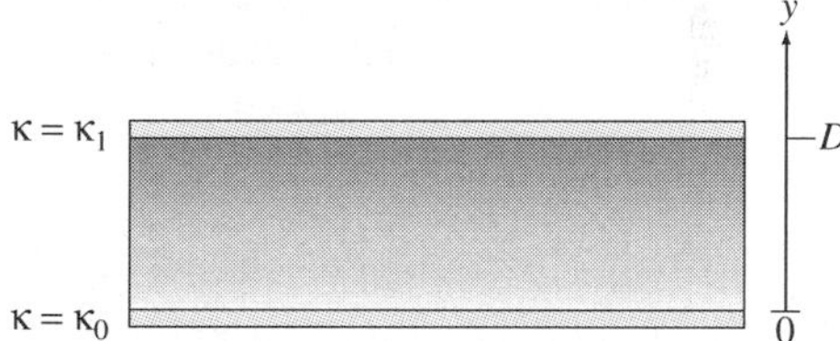

FIGURE 26–30 Problem 59.

60. (II) Two identical capacitors of capacitance C are connected in series across a total potential V. A dielectric slab of dielectric constant κ can fill one of the two capacitors and is slowly inserted into that capacitor. Compute the changes in the total electric energy of the two capacitors, in the charge on each capacitor, and in the potential drop across each capacitor. Account for any energy change by a corresponding change in energy in some other part of the system.

61. (II) Three capacitors of strengths 1 μF, 2 μF, and 4 μF, respectively, can be connected in various ways between two points. What arrangement gives the smallest equivalent capacitance, and what arrangement gives the largest capacitance?

62. (II) Show that when capacitors are arranged in series, the total capacitance is less than any of the individual capacitances.

CHAPTER 27

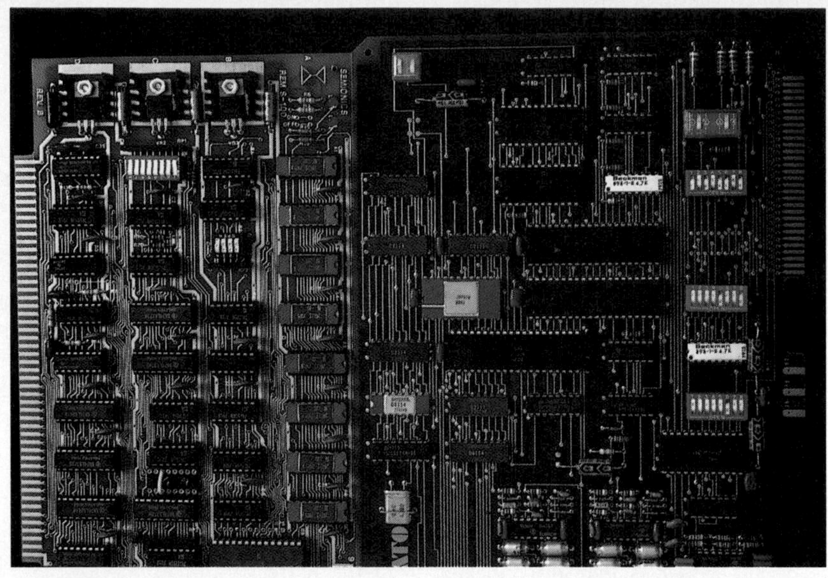

Electric circuits such as this one, which is part of a computer, govern the workings of our technological society.

CURRENTS IN MATERIALS

Charges move under the influence of electric fields, and that movement is described as an electric current. The motion of charged particles in free space—for example, the electron beam of an oscilloscope or a television tube—is familiar to us from Chapter 23. More frequently, however, we use electric currents within the materials that make up circuits. The motion of charges within a material is complicated by the presence of forces in addition to those due to external electric fields. The additional forces are due to collisions within the material and to internal electric fields. The effect of these forces is like the effect of drag: The charges move at constant speed. Because of the draglike forces, we must expend energy to make charges pass through materials, and we produce thermal energy. To describe the flow of currents in materials macroscopically, we introduce resistance, resistivity, and conductivity, characteristics of the materials. On the microscopic level, resistance to current flow can be described qualitatively with a classical model of electron movement in a crystalline lattice of atoms. A fundamental understanding of resistance, however, requires the ideas of quantum physics. Quantum physics also explains the differences among conductors, insulators, semiconductors, and superconductors.

27–1 ELECTRIC CURRENT

Electric current (or just **current**) is the total charge that passes through some cross-sectional area A per unit time. In Fig. 27–1 we have drawn the charge that passes through a wire. Because *charge is conserved* (see Chapter 22), the same amount of

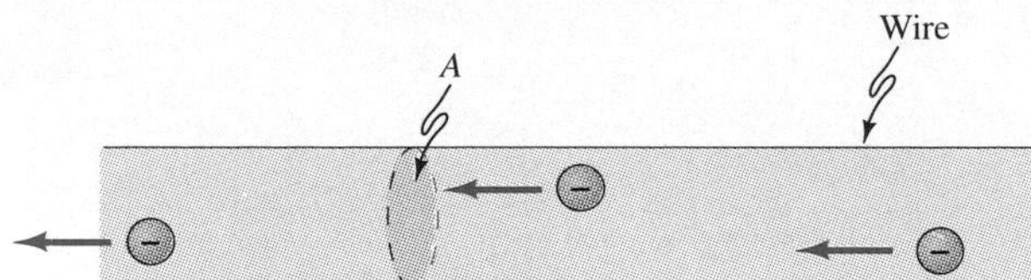

FIGURE 27–1 Charges move in a cross section of wire.

charge flows through a cross section that is normal to the axis of the wire as flows through a tilted cross section. Thus, if all we are interested in is the total charge flow, the shape or orientation of the area need not be specified. Even when charges flow through a region of empty space, the conservation of charge allows us to follow the flow systematically. Here we concentrate on the general notions of current, whether that current describes the motion of charges within free space or within conducting materials.

If ΔQ is the amount of charge passing through an area in a time interval Δt, then the *average current*, I_{av}, is defined as

$$I_{av} \equiv \frac{\Delta Q}{\Delta t}. \tag{27–1}$$

If the current changes with time, we define the *instantaneous* current, I, by taking the limit $\Delta t \to 0$, so that the current is the instantaneous rate at which charge passes through an area:

$$I \equiv \frac{dQ}{dt}. \tag{27–2}$$

Definition of electric current

Units of Current

The unit of current is coulomb per second, but this unit is also called the *ampere* (A), after André Marie Ampère, who performed pioneering work in electricity and magnetism early in the nineteenth century.[†] The ampere will be defined more precisely in Chapter 29, but that definition will be equivalent to the simple relation

$$1\ \text{A} = 1\ \text{C/s}.$$

The definition of the ampere

The term "amp" is often used for the ampere. Because the ampere is a rather large unit, current is expressed also in milliamps (mA), or 10^{-3} A; microamps (μA), or 10^{-6} A; or even nanoamps (nA), or 10^{-9} A.

Currents occur over a wide range of values (Table 27–1). Currents that have

TABLE 27–1

VALUES OF VARIOUS CURRENTS

Situation	*Current (A)*
Advanced computer technology chips	10^{-12} to 10^{-6}
Electron beam of a TV set	10^{-3}
Proton beam of the Fermilab accelerator	3×10^{-3}
Current that is dangerous when it passes through the human body	10^{-2} to 10^{-1}
Flashlight bulb	0.3
Household light bulb	1
Automobile starter	200
Peak current in a lightning strike	10^4
Maximum current carried by a superconducting niobium wire of 1 cm^2 cross section	10^7

[†] The ampere is one of the seven base units of the SI (see Appendix I–1). The significance of a base unit is that derived units, such as the coulomb and the farad, are defined in terms of the base units. Thus it would be more correct to say that 1 C is defined as 1 A·s.

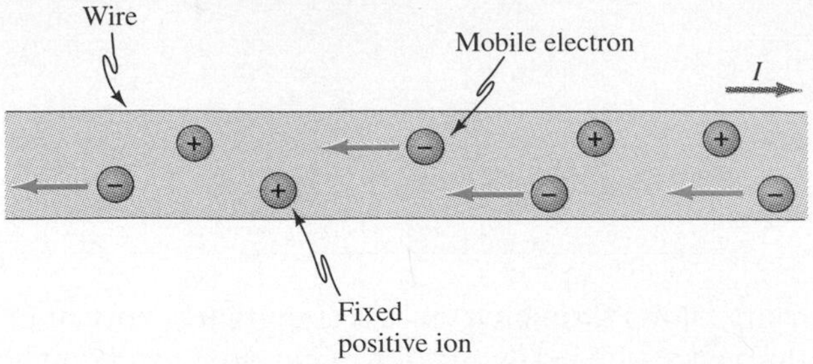

FIGURE 27–2 By convention the current direction is the direction in which positive charge appears to move, even though in conductors it is the negative charge, electrons, that move.

a harmonic time dependence are called alternating currents (AC), a phenomenon we shall study in more detail in Chapter 34. For such currents, the values in Table 27–1 represent the average magnitude of the oscillating current.

Current is a scalar quantity, but it has a sign associated with it. It is useful to indicate the sign of the current by a directional arrow. By convention, we associate the direction of the arrow with the flow of positive charges,[†] even though in metals it is actually the negative charges that move (Fig. 27–2). The positive charges—the atomic ions the electrons leave behind—are fixed in an ordered crystal lattice (see Chapter 21). This arbitrary convention for current direction causes no real problem; a flow of positive charge to the right and a flow of the same amount of negative charge to the left represents the same current. It is *not possible* by measuring the current alone to determine the sign of the charges that move (the *charge carriers*). By convention, then,

Currents flow in the direction that positive charges would flow.

Currents are depicted as if the positive charges are moving, even though in conductors the carriers (the electrons) are negatively charged.

EXAMPLE 27–1 An accelerator used for research on the treatment of tumors ejects protons at the rate of 2.0×10^{13} protons/s. What current does this beam of protons form?

SOLUTION: The current is the charge per second carried by the beam. This current is

$$\begin{aligned} I &= (\text{number of protons/s})(\text{charge per proton}) \\ &= (2.0 \times 10^{13}\ \text{protons/s})(1.6 \times 10^{-19}\ \text{C/proton}) \\ &= 3.2 \times 10^{-6}\ \text{C/s} = 3.2 \times 10^{-6}\ \text{A}. \end{aligned}$$

Current Density

It is often necessary to deal with the *details* of charge motion, not just an overall movement of charge. Then we must work with **current density**, **J**, the rate of charge flow per unit area through an infinitesimal area. Note that the flow rate can vary from one point to another, and to define the current density we must take into account the local magnitude and direction of the charge flow. Unlike current, which is a scalar, current density is a *vector*, with units of amperes per square meter. The direction of **J** is defined to be the direction of the net flow of positive charges at the particular infinitesimal element of area.

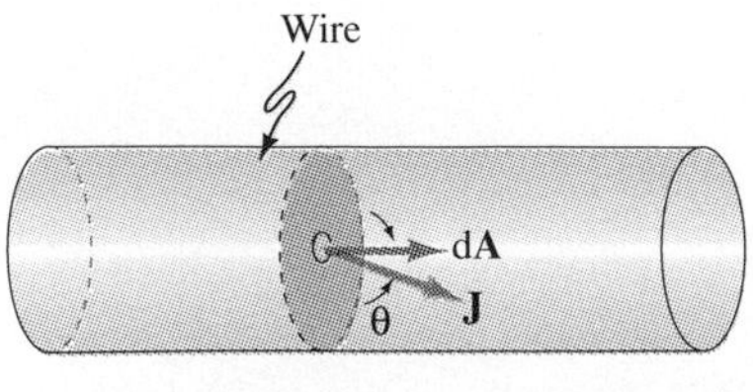

FIGURE 27–3 The area of a finite wire is divided up into differential areas d**A** with the current density, **J**, defined at every point. The direction of d**A** is normal to the differential area.

What is the relation between current density and current? We determine this relation in a wire by dividing the finite area A through which the current flows into infinitesimal areas d**A** (Fig. 27–3). This procedure is analogous to one we followed in treating fluid flow (in Chapter 16) or electric flux (in Chapter 24). The

[†] We have Benjamin Franklin to blame for this slight inconvenience.

differential current dI flowing through $\mathbf{dA}$ is

$$dI = \mathbf{J} \cdot \mathbf{dA} = J\, dA \cos\theta, \tag{27–3}$$

where θ is the angle between $\mathbf{J}$ and the area element $\mathbf{dA}$. From Eq. (27–3), we see that dI is a maximum when $\mathbf{J}$ and $\mathbf{dA}$ are parallel and is zero when $\mathbf{J}$ is perpendicular to $\mathbf{dA}$. The total current passing through the area A is a sum over the differential currents dI:

$$I = \iint_A \mathbf{J} \cdot \mathbf{dA}. \tag{27–4}$$

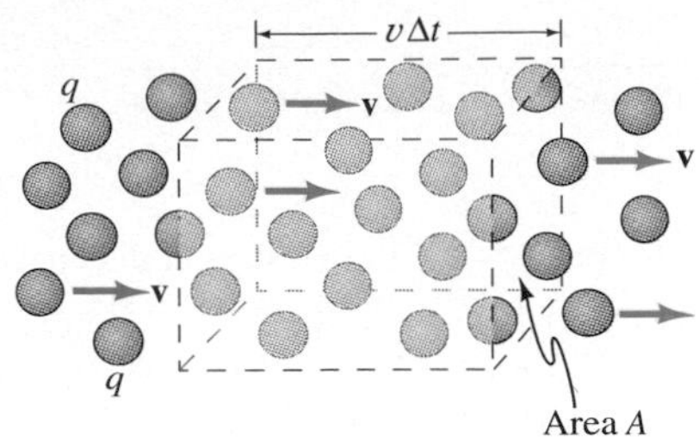

FIGURE 27–4 A collection of particles (each with charge q) with number density n_q all move to the right with velocity $\mathbf{v}$. The total charge passing through an area A in time Δt is $\Delta Q = n_q qvA\,\Delta t$.

Current Density of Moving Charges

Let us find the current density of a group of moving charges. Suppose that we have a collection of particles with charge q. In some small region the number of these charged particles per unit volume, the *number density*, is n_q. Suppose also that these particles all move with velocity $\mathbf{v}$. Then, in a time interval Δt, the amount of charge passing through a given area A perpendicular to $\mathbf{v}$, as in Fig. 27–4, is ΔQ, the charge contained in the volume $A(v\,\Delta t)$ swept out by the moving charges:

$$\Delta Q = \left(\frac{\text{charge}}{\text{volume}}\right)(\text{volume}) = (n_q q)(Av\,\Delta t) = n_q qvA\,\Delta t. \tag{27–5}$$

Here we have used the fact that the charge per unit volume is the number density of the charge carriers times the charge per particle. Thus the current is given by

$$I = \frac{\Delta Q}{\Delta t} = n_q qvA. \tag{27–6}$$

Finally, the current density is I divided by A in the limit of *small* A, $J = I/A$. The direction of $\mathbf{J}$ is specified by the direction of $\mathbf{v}$:

$$\mathbf{J} = n_q q\mathbf{v}. \tag{27–7}$$

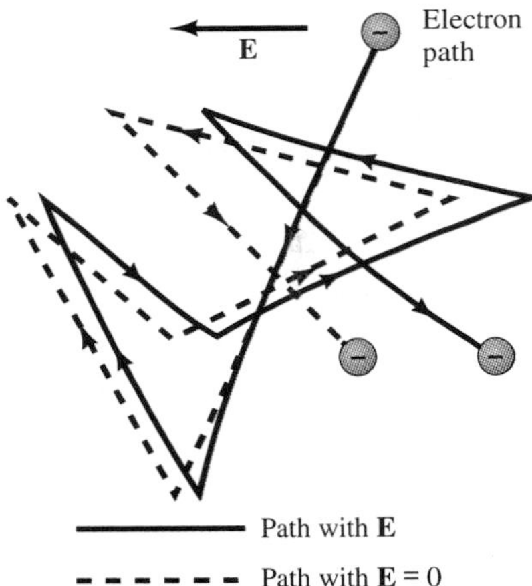

FIGURE 27–5 An electron collides frequently with the ions and impurities in a metal and scatters randomly. In an electric field, the electron picks up a small component of velocity opposite the field. The differences in the paths are exaggerated. The electron's path in an electric field is slightly parabolic.

27–2 CURRENTS IN MATERIALS

Earlier we defined *conductors* as materials through which charge moves easily, *insulators* as materials through which charge does not move easily, *semiconductors* as materials intermediate to conductors and insulators, and *superconductors* as materials that under certain circumstances—in particular, at low temperatures—carry charge with no inhibition whatsoever. How materials carry charge is of obvious importance. A simple model (to be developed further in Section 27–4) explains that metals are good conductors because they contain electrons that behave as though they were free. A free electron in a metal experiences a force $\mathbf{F} = -e\mathbf{E}$ and thus an acceleration in a direction opposite to an electric field $\mathbf{E}$.† Such electrons undergo frequent collisions with the positive ions that form the crystal lattice of the metal whether the field is present or not (Fig. 27–5). When there is no field, the electrons do not, *on average*, move in any particular direction. Their motion is random, like the motion of air molecules. When the electric field is present, there is a *net* movement of electrons in the direction of the electric force they experience. The collisions in effect give rise to a drag force on the flow of electrons. As with the fall of a parachute, drag acts to settle the motion to a steady flow in

The classification of materials according to how well they carry charge was made in Chapter 22.

† Although it remains true that, in electrostatics, metals and other conductors contain no electric fields, here we are not in an electrostatic situation: Charges are moving continuously.

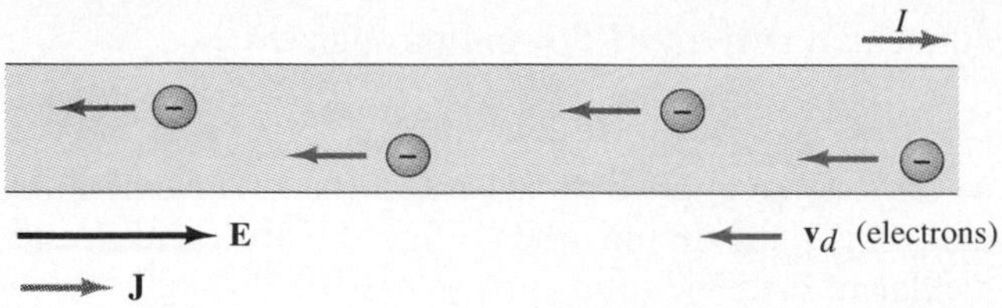

FIGURE 27–6 Electrons drift in the direction opposite that of the current, I, current density, $\mathbf{J}$, and electric field, $\mathbf{E}$.

the direction of the force. The electrons move with a constant terminal velocity, here called the **drift velocity**, $\mathbf{v}_d$. Equation (27–6) gives the relation between drift speed and current. In the special case that n_q equals n_e, the density of free electrons in the metal, Eq. (27–6) gives

$$I = n_e q v_d A, \tag{27–8}$$

where A is the cross-sectional area of a metal wire.

We can solve Eq. (27–8) to find the drift speed in terms of the current:

$$v_d = \frac{I}{n_e q A}. \tag{27–9}$$

Equations (27–8) and (27–9) are the desired relations between current and drift speed. Remember that the direction of the electron's drift velocity is opposite to the defined direction of the current density because of the positive charge-carrier convention.

We show in Fig. 27–6 the relationships among the external electric field, $\mathbf{E}$, the current, I, the current density, $\mathbf{J}$, the electron's drift velocity, $\mathbf{v}_d$, and the movement of electrons. For the case of the wire, $J = I/A$; from Eq. (27–9) we have

$$\mathbf{J} = n_e q \mathbf{v}_d, \tag{27–10}$$

where $\mathbf{J}$ is indeed opposite to the direction of $\mathbf{v}_d$ because of the negative sign of charge $q\,(-e)$.

EXAMPLE 27–2 Estimate the drift speed, v_d, for electrons in a copper wire of diameter $d = 1$ mm that carries a current of 100 mA. Copper has about one free electron per atom available to carry charge and has a mass density of 8.92 g/cm^3 and a molecular weight of 63.5 g/mol.

SOLUTION: The situation is similar to that shown in Fig. 27–6. We use Eq. (27–9) to calculate the drift speed. We are given the current, I, and can find $A = \pi r^2$, where $r = d/2$ is the wire's radius. However, we must find n_e from the given information about copper. Because copper has about one free electron per atom, the density of free electrons, n_e, is identical to the density of copper atoms, n. The atomic density is derived from the mass density of copper, $\rho_{\text{Cu}} = 8.92$ g/cm^3; the number of atoms per mole, N_A; and the molar weight of copper, $M = 63.5$ g/mol:

$$n_e = n = \frac{N_A \rho_{\text{Cu}}}{M} = \frac{(6.02 \times 10^{23}\ \text{atoms/mol})(8.92\ \text{g/cm}^3)}{63.5\ \text{g/mol}} \frac{(1\ \text{electron})}{\text{atom}}$$

$$= 8.5 \times 10^{22}\ \text{electrons/cm}^3.$$

If we assume that the current and drift speed are constant across the wire, Eq. (27–9) gives

$$v_d = \frac{I}{n_e q A} = \frac{100 \times 10^{-3}\ \text{A}}{(8.5 \times 10^{22}\ \text{electrons/cm}^3)(1.6 \times 10^{-19}\ \text{C/electron})\pi(0.05\ \text{cm})^2}$$

$$= 9.4 \times 10^{-4}\ \text{cm/s} = 9.4 \times 10^{-6}\ \text{m/s}.$$

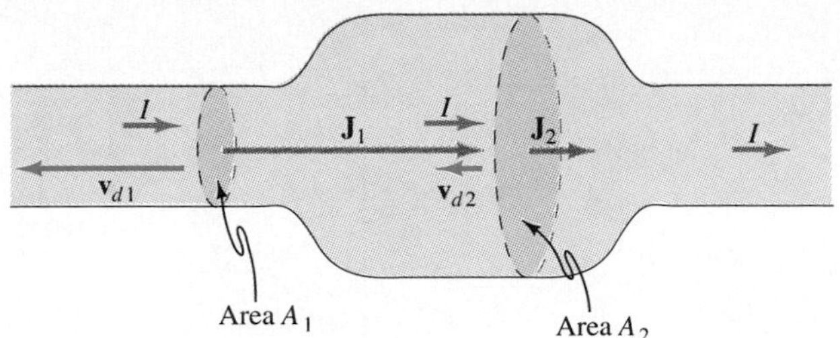

FIGURE 27–7 When a current is steady, it is the same in all parts of a wire, even if the area of the wire varies. The current density and drift velocity are lower, however, in parts of the wire with larger cross section.

The drift velocity is so slow, only about 0.001 cm/s, that you might wonder how a measurable current can even flow. What happens when we switch on a household circuit? Certainly we do not have to wait for hours for the electrons to drift several feet. When the switch is thrown, the electric field that influences the electrons to move in the wire is set up throughout the wire at speeds approaching the speed of light. The free electrons are spread throughout the wire, and they all start moving at once in response to this field—those nearest the switch as well as those nearest the electrical appliance. A similar effect occurs in fluid flow. If you want to move a sprinkler while you water the lawn, you turn the water off, move the sprinkler, and then turn the water back on. Because the hose is already full of water, the sprinkler starts immediately: The force of the water at the faucet end is quickly transmitted all along the hose, and the water at the sprinkler end of the hose flows from the sprinkler almost the moment the faucet is opened.

Current and the Conservation of Charge

How does the conservation of charge affect currents in materials? The conservation of mass implies that, in a steady state, the rate at which fluid enters a system of pipes is the rate at which it leaves the system. Similarly, the conservation of charge leads to the principle of the *conservation of current*. For steady flow, in which currents do not change with time, the total current that enters some section of wire is the total current that leaves that section. Thus the current is the same everywhere along a wire, even if the wire changes in area (Fig. 27–7). Because the current is the same everywhere along a wire of varying area, the current density is inversely proportional to the area. In other words, if at two points along a wire that wire has cross-sectional areas A_1 and A_2, respectively, then the conservation of current implies that $J_1A_1 = J_2A_2$, where J_1 and J_2 are the current densities at these respective regions, assumed to be constant across the wire's cross section. The density of charge carriers and their charge are fixed in a given metal. Thus, by Eq. (27–10), if the current density is inversely proportional to the area of the wire, so is the drift speed. A familiar analogue to this situation occurs when you are driving and approach a region of the road where three lanes narrow into one. Motion in the three-lane section is painfully slow, but once you are in the one-lane section, your speed can increase.

Current, like charge, is conserved.

27–3 RESISTANCE

We have seen that when an electric field is applied to a conductor, a current flows. We can consider the potential difference V due to the electric field as the source of the motion. The amount of current that flows through a material for a given potential difference across that material depends on the material's properties and geometry.

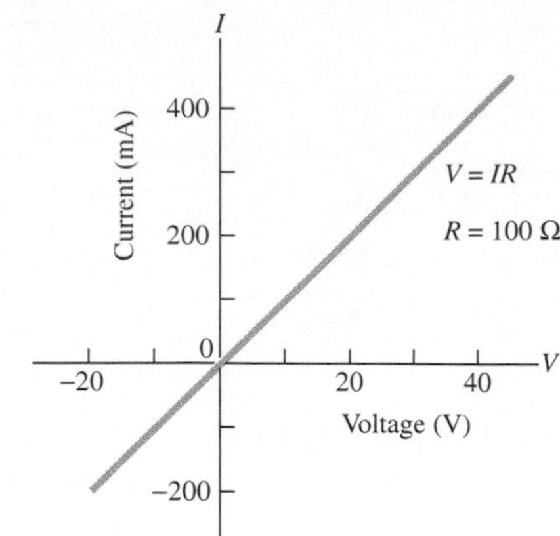

FIGURE 27–8 For ohmic materials, Ohm's law states that the ratio V/I is a constant.

The **electrical resistance**, R, of a piece of material is a measure of how easily charge flows within that material. The electrical resistance is defined to be the ratio of the voltage (potential difference) across the material to the current that flows through it:

The definition of resistance

$$R \equiv \frac{V}{I}. \tag{27–11}$$

The units of resistance are volts per ampere, but a separate SI unit called the **ohm** (Ω) has been defined as the resistance through which a current of 1 A flows when a potential difference of 1 V is applied:

Units of resistance

$$1\ \Omega = 1\ \frac{\text{V}}{\text{A}}.$$

Ohm's law applies whenever the resistance of a material is a constant.

Georg Simon Ohm was the first to study the resistance of different materials systematically. In 1826 he published his experimental result that for many materials, including most metals, *the resistance is constant over a wide range of potential differences.* This statement is called *Ohm's law.* It is really not a law at all but an empirical statement about how materials behave. When the resistance of a material is constant over a range of potential differences, we say that the material is *ohmic.* We shall continue the traditional practice of referring to this linear relation between voltage and current for these materials as a "law" and writing it as

$$V = IR, \tag{27–12}$$

where R is independent of V. *The resistance, R, is understood to be independent of V here.* Figure 27–8 illustrates the consequence of a constant R.

FIGURE 27–9 Resistors are color coded to indicate the value of their resistance.

Resistors

A piece of ohmic material of significant resistance, a **resistor**, is the most mundane of the elements that make up an electric circuit (Fig. 27–9). ("Significant" depends on the application, but in practice a resistance of a few ohms is small.) A resistor of given resistance R with a given potential V between its terminals allows the flow of a current $I = V/R$. Resistors are represented in circuit diagrams by zigzag lines, —ww—, and they are connected to each other and to other elements, such as capacitors, by conducting wires generally assumed to have negligible resistance. Many resistors in electric circuits are given in kilohms (kΩ or $10^3\ \Omega$) or megohms (MΩ or $10^6\ \Omega$).

There are many *nonohmic materials,* materials for which the voltage and cur-

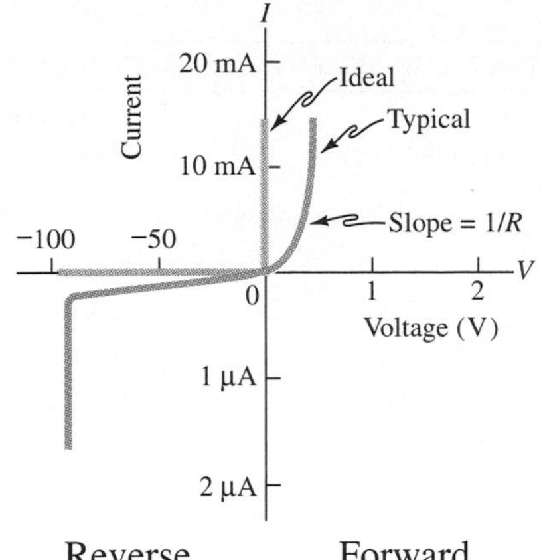

FIGURE 27–10 Many materials are nonohmic (they do not follow the idealized I–V curve), as the I–V curve for a typical diode shows. The resistance is still defined by V/I, but Ohm's law does not apply for such materials, and the resistance varies. Note the scale changes. "Forward" and "reverse" refer to current directions. Data for "Typical" curve are from Paul Horowitz and Winfield Hill, *The Art of Electronics*, Cambridge: Cambridge University Press, 1989.

rent do not obey the linear relation of Ohm's law. Figure 27–10 shows current versus voltage curves (ideal and typical) for a **diode**, a device that transmits current easily when the voltage is positive, but prevents current flow (that is, it has a very high resistance) when the voltage is negative. Diodes are used in many electric devices (Fig. 27–11). For example, they may be used to allow a battery to be charged but to prevent it from discharging, without the need for a switch.

Resistivity and Conductivity

The resistance of a conducting wire of a given material can vary with the wire's shape. Let us again consider a uniform wire. We may think of the resistance to a flow of charge in a conductor as the result of collisions of the moving charge carriers (electrons) with lattice atoms. When the length of the wire is doubled, the number of collisions doubles, just as the distance a pedestrian travels is, on average, proportional to the number of times he or she collides with another pedestrian in a crowd. Thus the *resistance of a wire is proportional to its length L.* Conversely, if the cross-sectional area of a wire is doubled, then twice as much current will flow through it, just as twice as much water will flow out of a bathtub with two identical drains as will flow out of a tub with one drain. As long as the potential remains constant, a doubling of the current implies a halving of the resistance. Therefore, the *resistance of a wire of a given material is inversely proportional to its cross-sectional area A.*

We combine these two results to define the **resistivity** of a material, ρ, by the relation†

$$\rho \equiv R\frac{A}{L}. \tag{27–13}$$

The definition of resistivity

FIGURE 27–11 Diodes on a heat sink. Diodes are devices that allow current to pass in only one direction.

† It is customary to use the same symbol, ρ, for resistivity, mass density, and charge density. There should be no confusion given the context in which these quantities appear.

TABLE 27–2

RESISTIVITIES, CONDUCTIVITIES, AND TEMPERATURE COEFFICIENTS (AT 20°C)

Material	*Resistivity, ρ* $(\Omega\cdot m)$	*Conductivity, σ* $(\Omega\cdot m)^{-1}$	*Temperature Coefficient, α* $(°C)^{-1}$
Conductors			
Elements			
Aluminum	2.82×10^{-8}	3.55×10^{7}	0.0039
Silver	1.59×10^{-8}	6.29×10^{7}	0.0038
Copper	1.72×10^{-8}	5.81×10^{7}	0.0039
Iron	10.0×10^{-8}	1.0×10^{7}	0.0050
Tungsten	5.6×10^{-8}	1.8×10^{7}	0.0045
Platinum	10.6×10^{-8}	1.0×10^{7}	0.0039
Alloys			
Nichrome	100×10^{-8}	0.1×10^{7}	0.0004
Manganin	44×10^{-8}	0.23×10^{7}	0.00001
Brass	7×10^{-8}	1.4×10^{7}	0.002
Semiconductors			
Carbon (graphite)	3.5×10^{-5}	2.9×10^{4}	−0.0005
Germanium (pure)	0.46	2.2	−0.048
Silicon (pure)	640	1.6×10^{-3}	−0.075
Insulators			
Glass	10^{10} to 10^{14}	10^{-14} to 10^{-10}	
Neoprene rubber	10^{9}	10^{-9}	
Teflon	10^{14}	10^{-14}	

Resistivity is a property only of the type of material, whereas resistance depends on both the type of material and its shape.

With this definition, and with the dependence of R on L and A that we have already established, ρ does not depend on the dimensions of the conductor, but only on the type of material. The units of resistivity are ohm-meters ($\Omega\cdot$m); characteristic values for a variety of materials are given in Table 27–2. Equation (27–13) is typically rewritten as

$$R = \rho\frac{L}{A}. \tag{27–14}$$

The reciprocal of the resistivity is the **conductivity**, σ:[†]

The definition of conductivity

$$\sigma \equiv \frac{1}{\rho}. \tag{27–15}$$

Typical values of conductivity are in Table 27–2.

We can write Eq. (27–12) in terms of resistivity and conductivity:

$$V = IR = \rho\frac{L}{A}I = \rho L\frac{I}{A};$$

$$\frac{V}{L} = \rho\frac{I}{A}.$$

We recall that V/L is just the magnitude E of the electric field applied to the material, and I/A is the magnitude of current density, J. Because charges move in the direction of the electric field, we therefore find that

The relation between electric field and current density

$$\mathbf{E} = \rho\mathbf{J}. \tag{27–16}$$

[†] Not to be confused with σ for a surface charge density.

Equivalently, from the definition of conductivity, Eq. (27–15), we have

$$\mathbf{J} = \sigma\mathbf{E}. \tag{27–17}$$

Equations (27–16) and (27–17) are general results, not limited to ohmic materials, for which ρ and σ do not vary with V or $\mathbf{E}$.

The resistivities and conductivities of the materials shown in Table 27–2 vary over many orders of magnitude. The conductivity of metals is a factor of 10^{21} higher than that of a good insulator, such as Teflon. Copper and silver have very high conductivities, but the cost of silver precludes its use in conducting wires (except during national emergencies, such as World War II). Aluminum is used for large conducting wires but is no longer used in household circuits. This is because the oxides of aluminum inhibit the formation of good electrical contacts; there is a large current density in regions where the poor contact reduces the current flow to channels of limited size. Overheating and fire danger are the result.

EXAMPLE 27–3 Determine the current density, resistance, and electric field for the copper wire of Example 27–2 if the wire is 10 m long.

SOLUTION: According to Example 27–2 we know that the wire carries a 100-mA current. The wire has a diameter $d = 1$ mm, so its cross section is $A = \pi(\frac{1}{2}d)^2$. Thus we can compute the unknown current density from the definition of J:

$$J = \frac{I}{A} = \frac{100 \times 10^{-3}\ \text{A}}{\pi(0.5 \times 10^{-3}\ \text{m})^2} = 1.3 \times 10^5\ \text{A/m}^2.$$

If we take the resistivity of copper from Table 27–2, we can use Eq. (27–14) to determine the resistance:

$$R = \rho\frac{L}{A} = \frac{(1.72 \times 10^{-8}\ \Omega\cdot\cancel{\text{m}})(10\ \cancel{\text{m}})}{\pi(0.5 \times 10^{-3}\ \cancel{\text{m}})^2} \simeq 0.2\ \Omega.$$

Finally, given the resistivity, we can determine the electric field from Eq. (27–16):

$$E = \rho J = (1.72 \times 10^{-8}\ \Omega\cdot\cancel{\text{m}})(1.3 \times 10^5\ \text{A/m}^{\cancel{2}}) = 2.2 \times 10^{-3}\ \text{V/m}.$$

Note that both the current density and electric field are independent of the length of the wire. The voltage required to produce both, however, is dependent on the wire's length.

The Temperature Dependence of Resistivity

Resistivities of some materials have a strong temperature dependence; that of copper is shown in Fig. 27–12. We can represent the temperature dependence by the following linear approximation, which is sufficiently accurate for most purposes:

$$\rho \simeq \rho_0[1 + \alpha(T - T_0)]. \tag{27–18}$$

The parameter α is the *temperature coefficient of resistivity*, and ρ_0 is the resistivity at the reference temperature T_0, normally 20°C. Values of ρ, σ, and α are given in Table 27–2 for $T_0 = 20$°C. Resistivities for most metals increase with temperature, as in Fig. 27–12, because, roughly speaking, the higher temperature causes increased vibrations of the lattice atoms, which in turn impede the motion of drifting electrons more effectively.

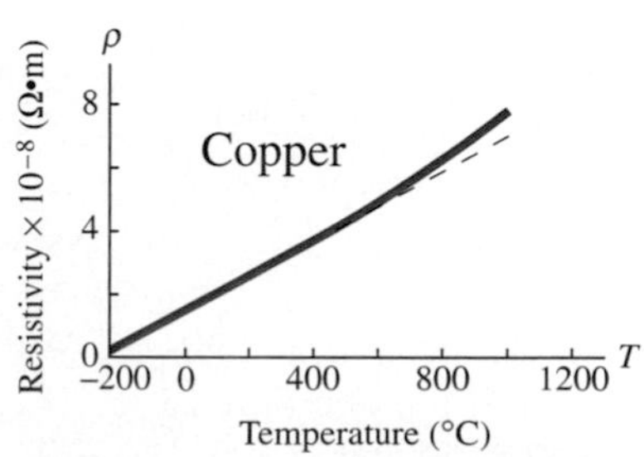

FIGURE 27–12 The resistivity of copper as a function of temperature.

Thermometers can be made by using the temperature dependence of the resistivity of certain stable materials (Fig. 27–13). For example, the resistivity of platinum, which has a high melting point and does not oxidize in air, is the basis of a secondary thermometer standard at high temperatures. For temperatures much greater than 1000°C, the resistivity of platinum is no longer accurately represented by Eq. (27–18), and a modified formula with terms that include the square and cube of $(T - T_0)$ is used along with the appropriate coefficients.

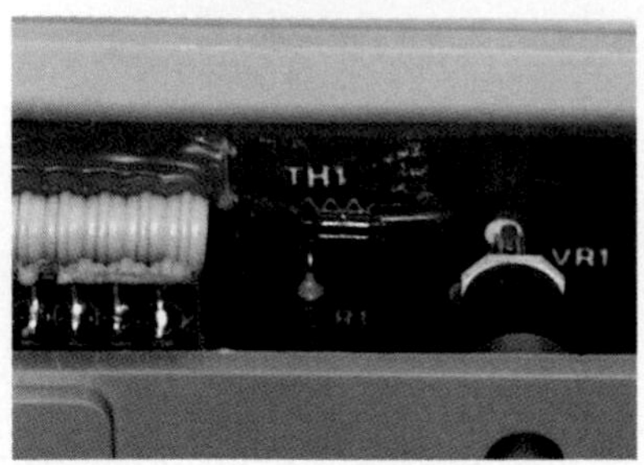

FIGURE 27–13 The resistivity of some materials varies so strongly with temperature that they can be used not only to measure temperature but to provide electrical controls, as in this thermostat.

EXAMPLE 27–4 Calculate the resistance of a coil of platinum wire of diameter 0.5 mm and length 20 m at 20°C. Also determine the resistance at 1000°C.

SOLUTION: We are asked for the resistance, not the resistivity, but the relation between these two quantities is given by Eq. (27–14). We can thus find the resistance at 20°C from the resistivity at 20°C in Table 27–2:

$$R_{20^\circ\mathrm{C}} = \rho\,\frac{L}{A} = (10.6 \times 10^{-8}\ \Omega\cdot\cancel{\mathrm{m}})\,\frac{20\ \cancel{\mathrm{m}}}{\pi[\frac{1}{2}(0.5 \times 10^{-3}\ \cancel{\mathrm{m}})]^2} = 11\ \Omega.$$

To find the resistance at 1000°C, we combine Eq. (27–14), $R = \rho L/A$, with Eq. (27–18) to produce an equation that gives resistance as a function of temperature for conducting wires:

$$R = R_0[1 + \alpha(T - T_0)]. \qquad (27\text{–}19)$$

From Table 27–2 we get the temperature coefficient for platinum and find that

$$R_{1000^\circ\mathrm{C}} = (11\ \Omega)[1 + (0.0039^\circ\cancel{\mathrm{C}^{-1}})(1000^\circ\cancel{\mathrm{C}} - 20^\circ\cancel{\mathrm{C}})] = 52\ \Omega,$$

where we have used $T_0 = 20$°C. The resistance is a factor of nearly 5 greater at the higher temperature.

*27–4 FREE-ELECTRON MODEL OF RESISTIVITY

A simple classical model first proposed in 1900 by Paul Drude and known as the **free-electron model**, or the **Drude model**, can help us understand Ohm's law. The model describes classically the connection between the macroscopic concept of resistivity in metals and microscopic parameters such as the drift velocity of charge carriers. The model was proposed some 25 years before quantum mechanics was sufficiently well established to allow a more correct explanation. Because it does not include quantum behavior, the free-electron model can at best describe only certain qualitative features of resistivity and has fundamental deficiencies. Nevertheless, it is worthwhile for us to look at this model for two reasons: First, the model allows us to focus on the concept of resistivity. Second, the model provides us with an example of how model-building in the physical sciences proceeds, and how we can say whether a model succeeds or fails.

We start with the idea that solids contain "free" electrons, which can move within the material and carry charge. The density of free electrons, n_e, depends on the material, and, as we shall see, is responsible for the differences among conductors, insulators, and semiconductors (which we shall discuss in Section 27–5). In metals, the number of loosely attached electrons per atom (these are the electrons that behave as though they were free) lies in the range 1.0 to 1.3, but it can be as large as 3.5, as in aluminum.

The model postulates that free electrons form a gas of independent particles at temperature T. When a current is produced, the electrons are accelerated by an applied electric field, but collisions with the atoms or ions that form the crystal lattice of the solid slow them down. In an average sense drag forces act on the electrons (Fig. 27–14). The simplest drag force is proportional to the electrons' speed, so Newton's second law for the electron motion is, for the component of motion parallel to the applied field,

Crystal lattices of solids are discussed in Chapter 21.

$$ma = -eE - (\text{a constant})v,$$

where m is the mass of an electron. The constant must have dimensions of mass/time, and we write it as m/τ, where τ is a quantity with dimensions of time. It is reasonable to equate τ with the average time between collisions, because it is the collisions that impede the electrons' motion.† The acceleration drops to zero when the speed reaches the drift speed, v_d. When the acceleration is zero, $ma = -eE - (m/\tau)v_d = 0$, or

$$v_d = -\frac{eE\tau}{m}. \tag{27–20}$$

The minus sign indicates that the direction of the drift velocity is opposite to that of the electric field, as is appropriate for negative charge carriers. When this expression is inserted into Eq. (27–10) for the current density, we obtain

$$J = n_e e v_d = \frac{-(n_e e)eE\tau}{m} = \frac{-n_e e^2\tau}{m}E. \tag{27–21}$$

Comparison with Eq. (27–17) yields

$$\sigma = \frac{n_e e^2\tau}{m} \tag{27–22}$$

for the conductivity. [The directional information carried by the minus sign in Eq. (27–21) is not relevant for the calculation of either conductivity or resistivity, which are positive quantities.] Equivalently, the resistivity, $\rho = 1/\sigma$, is predicted to be

$$\rho = \frac{m}{n_e e^2\tau}. \tag{27–23}$$

The quantities e and m are independent of the type of material. The average time between collisions may be expressed in terms of the *mean free path* λ and the average speed v_{av} of the electrons in the free-electron "gas" by using Eq. (19–51):

$$\tau = \frac{\lambda}{v_{av}}.$$

We have used the free-electron model to predict the resistivity of a solid in terms of microscopic parameters of the solid. For normal electric fields, none of the quantities in Eq. (27–23) depend on E, and thus the resistivity (or conductivity) is constant. It was with this argument that the original atomic foundation of Ohm's law was laid down by Drude (and independently by Hendrik Lorentz) in 1900.

FIGURE 27–14 The water in the stream is analogous to the electrons in the free-electron model of electric current. The water flows downhill, but it has a terminal velocity—or drift velocity—because the rocks of this riprap impede its motion.

† A rigorous analysis confirms this result. See D. E. Tilley, *American Journal of Physics*, **44**, 1976, p. 597.

EXAMPLE 27–5 What is the free-electron model's prediction for the collision time of current-carrying electrons in copper, given that the resistivity of copper is $1.7 \times 10^{-8}\ \Omega\cdot\text{m}$? You may use the parameters of Example 27–2.

SOLUTION: The connection between microscopic parameters such as resistivity and collision time in the free-electron model is given in Eq. (27–23). In Example 27–2 we estimated that the number density of current-carrying electrons is $n_e = 8.5 \times 10^{22}$ electrons/cm^3 $= 8.5 \times 10^{28}$ electrons/m^3. From Eq. (27–23),

$$\tau = \frac{m}{n_e e^2 \rho} = \frac{0.91 \times 10^{-30}\ \text{kg}}{(8.5 \times 10^{28}\ \text{electrons/m}^3)(1.6 \times 10^{-19}\ \text{C})^2(1.7 \times 10^{-8}\ \Omega\cdot\text{m})}$$
$$= 2.4 \times 10^{-14}\ \text{s}.$$

Temperature Dependence in the Free-Electron Model

According to the kinetic theory of gases, discussed in Chapter 19, the average of the velocity squared (the root-mean-square, or rms, velocity) is given by Eq. (19–33) as

$$v_{\text{av}} = v_{\text{rms}} = \sqrt{\frac{3kT}{m}}, \tag{27–24}$$

where k is Boltzmann's constant.† We can use Eq. (27–24) for the average velocity to find that $\tau = \lambda/v_{\text{av}}$, and then use this result in Eq. (27–23) for the resistivity:

$$\rho = \frac{m}{n_e e^2}\frac{1}{\tau} = \frac{m}{n_e e^2}\frac{v_{\text{av}}}{\lambda} = \frac{m}{n_e e^2}\frac{1}{\lambda}\sqrt{\frac{3kT}{m}}. \tag{27–25}$$

Because the mean free path is, according to the kinetic theory, independent of temperature, the free-electron model predicts that *the resistivity should be proportional to* $\sqrt{T}$. We can see why resistivity should increase with temperature by considering an analogy in which a person moves through a crowd. If the people that make up the crowd move around more briskly, it is more difficult to pass through the crowd, because collisions are more frequent. Resistivity increases with temperature in the free-electron model because collisions are more frequent.

Equation (27–24) gives for the rms speed of an electron in a metal at room temperature (293 K)

$$v_{\text{rms}} = \sqrt{\frac{3kT}{m}} = \sqrt{\frac{3(1.38 \times 10^{-23}\ \text{J/K})(293\ \text{K})}{9.11 \times 10^{-31}\ \text{kg}}} = 1.2 \times 10^5\ \text{m/s}.$$

The small mass of electrons has a marked effect here. As we know from our study of the kinetic theory (Chapter 19), the average velocity of an electron is itself *zero*, because all directions of motion are equally likely, despite the fact that the electron covers 120 km in 1 s! The drift speed estimated in Example 27–2 is *10^{10} times smaller* than the thermal speed we have just found. The electron attains only a very small drift-velocity component opposite to the direction of the electric field between each collision.

† Strictly speaking, the rms speed is not the average speed. This degree of refinement is not warranted here.

EXAMPLE 27–6 In Example 27–5 we found the mean time between collisions for electrons in copper to be 2.4×10^{-14} s, according to the free-electron model. Find an approximate value for the mean free path at room temperature.

SOLUTION: The mean free path is given approximately by $\lambda = v_{\text{rms}}\tau$. Previously we calculated v_{rms} to be 1.2×10^5 m/s at room temperature. Thus

$$\lambda = v_{\text{rms}}\tau = (1.2 \times 10^5 \text{ m/}\cancel{\text{s}})(2.4 \times 10^{-14} \cancel{\text{s}}) = 2.9 \times 10^{-9} \text{ m}.$$

Note that this value is equivalent to only a few atomic spacings (see Problem 38).

The Failure of the Free-Electron Model

The free-electron model helps us understand charge conduction in materials. However, if we take a closer look at the model's predictions by comparing them with experimental results, we find significant discrepancies. For example, the random thermal speed of electrons in metals is *more than a factor of 10* higher than the model predicts for copper at room temperature. Experimental values for the mean speed are essentially independent of temperature, rather than having the $\sqrt{T}$ dependence the model predicts. In addition, the actual mean free path is much larger than expected and has a T^{-1} dependence.

Although the free-electron model is *qualitatively* correct in many aspects, it cannot be taken too literally. A correct model of electrical conduction requires the use of quantum mechanics. Conduction electrons do not act as a classical gas of noninteracting electrons. They obey a velocity-distribution law based on quantum physics, and the movement of electrons depends on these quantum ideas. Quantum physics requires us to treat electrons as though they were waves scattering from the lattice structure of the material. Quantum physics predicts that *in a perfectly ordered crystal with no impurities, at a temperature of zero, there would be no resistance to electron flow, and conductivity would be infinite.* Finite conductivities occur because of the effects of impurities and of the thermal vibrations of lattice atoms at finite temperatures. At high temperatures, resistivity to electron flow is caused primarily by thermal vibrations. At low temperatures, resistivity is due to electrons being scattered by impurities.

There is ample evidence that the quantum physics ideas are correct. Quantum physics predicts the mean free path to be about two orders of magnitude greater than the classical prediction and the mean collision time to be about one order of magnitude longer. Because the mean speed $v_{\text{av}} = \lambda/\tau$, we then expect v_{av} to be about an order of magnitude greater than that predicted by the simple classical theory. All these quantum predictions, as well as the prediction of the resistance itself, are borne out by experimental measurements.

*27–5 INSULATORS, CONDUCTORS, AND SEMICONDUCTORS

Materials differ over an enormous range in their ability to conduct electricity. A good conductor might have a resistivity of $10^{-8}\ \Omega\cdot\text{m}$; a good insulator, about $10^{14}\ \Omega\cdot\text{m}$. The resistivity of semiconductors ranges from 10^3 to $10^{-5}\ \Omega\cdot\text{m}$ and

FIGURE 27–15 Energy diagram that shows the possible energy levels of an electron in a solid; it takes no account of the crystalline structure. Classical physics predicts a continuum of possible energies, but quantum mechanics shows that the possible levels are actually discrete but so closely spaced that they are hard to distinguish.

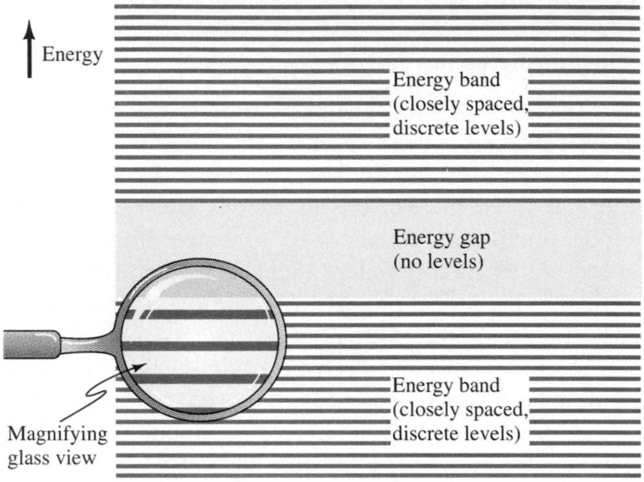

FIGURE 27–16 Energy diagram that shows the possible energy levels of an electron within a material made of a regular lattice of atoms. In contrast to the possibilities of Fig. 27–15, the electron energies are restricted to lie within allowed bands, and there is a large energy gap where no electrons are allowed. Even within the allowed bands, the possible electron energies are closely spaced discrete levels, as the magnified view shows. In the pink regions, the electron energy levels are filled; in the blue regions, electron levels are present but are unfilled.

depends sensitively on temperature. A proper quantitative explanation of the resistivity of all materials requires quantum physics. In this section we use a minimal amount of quantum mechanics to describe the critical properties that distinguish conductors, insulators, and semiconductors.†

In classical physics the energy of an electron in a metal can take on any value; we say that the energy values form a *continuum*. In contrast, a quantum description of electrons in metals shows that the possible energy values of electrons confined to a metal are *quantized*; that is, the possible energies have discrete values. In other words, an electron cannot have *any* energy value, much as the frequencies of standing waves on a string cannot have any value, just a set of discrete values. In a sample of material whose size is large compared with atomic sizes (10^{-10} m), these energy values are so close together that they appear to be continuous, just as the separate dots in a newspaper photograph are not distinguishable from a large distance. Figure 27–15, an energy diagram, illustrates the allowed energy levels. It is important to keep in mind that this diagram illustrates only the *possible* energy levels. We do not necessarily have electrons in each energy level.

When a set of atoms forms a regular background lattice, the possible energy values of the electrons are modified still further. The allowed energies of an electron are still discrete, but instead of a tiny separation between neighboring levels, there are *energy gaps*, which are large regions of energy forbidden to the electron. The regions where the energy levels are close together are called *allowed bands* of energy levels (Fig. 27–16). The gaps are quite sizable on the scale of atomic physics, of the magnitude of electron-volts. Again, specifying the bands as we have done

† More details can be found in Chapter 41, where we discuss quantum physics.

does not by itself specify which levels have electrons (whether the energy levels are *occupied* or not). The bands specify only the possible values of electron energies.

According to quantum physics, *there are at most two electrons in any one energy level.* This property, proposed by Wolfgang Pauli in 1925 and called the *Pauli exclusion principle,* has no counterpart in classical physics, and it plays a crucial role in determining the properties of materials. Consider a solid. There are many "free" electrons, and in an equilibrium state of that material, they fill the lowest energy levels available in the allowed bands, up to two in each level. When all the electrons are placed in the lowest possible energy state, we are left with two possible situations. In the first, the highest level to be filled is somewhere in the middle of a band; in the second, the electrons just fill one or more bands completely. This description assumes that the material is at a low enough temperature so that the electrons cannot "jump" to higher energy levels due to thermal effects.

Suppose that we now add some energy to the free electrons—by imposing an electric field, for example. The electrons in the lower energy levels cannot accept that energy, because they cannot move into an already-filled higher energy level. The only electrons that can accept energy are those that lie in the top levels, and then only if there are nearby empty levels into which they can move. Materials with a partly filled band are *conductors.* When the top layer of their electrons moves freely in the empty energy levels immediately above, there is a current. The electrons that jump from a lower level to a higher level are said to be *excited.* The energy-band structure for conductors is shown in Fig. 27–17a. *Conductors are characterized by electrons in only partly filled bands.*

If the highest-energy electrons of a material fill a band completely, then a small electric field will not give these electrons enough energy to jump the large energy gap to the bottom of the next (empty) band. We then have an *insulator* (Fig. 27–17b). An example of a good insulator is diamond (a form of carbon), whose energy gap is 6 eV.

In semiconductors, the highest-energy electrons fill a band (the valence band) at $T = 0$, as in insulators. Unlike insulators, semiconductors have a small energy

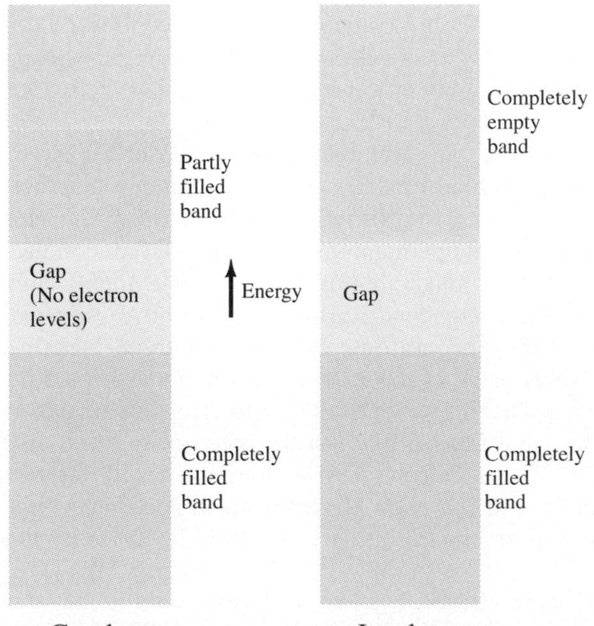

FIGURE 27–17 (a) Conductors have electrons in partly filled bands, whereas (b) insulators have an energy gap between a completely filled band and the next completely empty band. The pink and blue regions indicate where the allowed energy levels are filled and unfilled, respectively. Within each of the allowed bands, the possible energy levels form a set of closely spaced discrete levels.

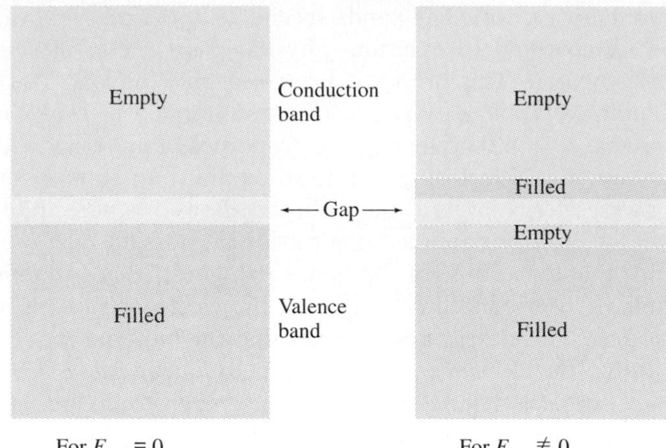

FIGURE 27–18 (a) For zero temperature and no external electric field, semiconductors have only a small energy gap between a completely filled band and the next highest, completely empty band. (b) A modest electric field $\mathbf{E}_{ext}$ or finite temperatures are enough to give some of the electrons sufficient energy to jump the energy gap, leaving holes in the valence band and conduction electrons in the previously empty conduction band.

gap between that band and the next, the conduction band (Fig. 27–18a). Because the energy gap is so small, a modest electric field (or finite temperature) will allow some electrons to jump the gap and thereby conduct electricity (Fig. 27–18b). Thus there is a minimum electric field under the influence of which a material changes from insulator to conductor. Silicon and germanium have energy gaps of 1.1 eV and 0.7 eV, respectively, and are semiconductors. For semiconductors, an increase in temperature will give a fraction of the electrons enough thermal energy to jump the gap. For an ordinary conductor a rise in temperature *increases* the resistivity, because the atoms, which are obstacles to electron flow, vibrate more vigorously. A temperature increase in a semiconductor allows more electrons into the empty band and thus *lowers* the resistivity.

When an electron in the valence band of a semiconductor crosses the energy gap and conducts electricity, it leaves behind what is known as a **hole**. Other electrons in the valence band near the top of the stack of energy levels can move into this hole, leaving behind their own holes, into which still other electrons can move, and so forth. The hole behaves like a positive charge that conducts electricity on its own as a positive charge carrier. An electron excited from the valence band to the conduction band is thus doubly effective at conducting electricity in semiconductors.

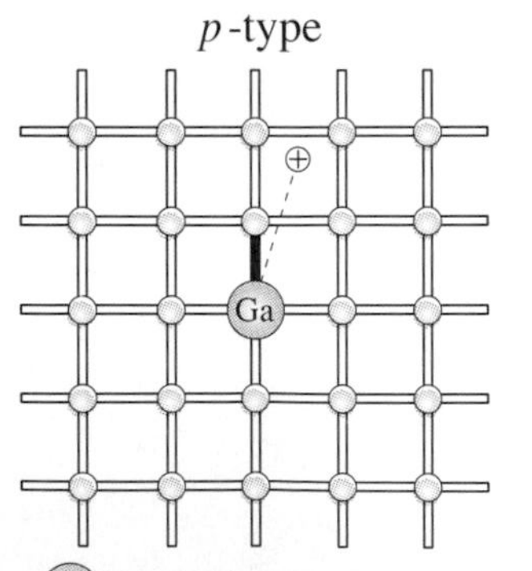

FIGURE 27–19 An (a) *n*-type and a (b) *p*-type semiconductor are created by doping the original lattice with atoms that have, respectively, more and less valence electrons than the atoms of the original lattice have.

One of the major advances in materials technology has been our ability to produce new semiconductors. Semiconductor materials that are compounds, such as gallium arsenide, are called *hybrid* semiconductors, as opposed to *intrinsic* elemental semiconductors, such as silicon and germanium. Other special semiconductors are made by introducing impurities, small amounts of different elements, into the lattice. For example, an atom in the chemical group of phosphorus, arsenic, and antimony can replace one of the silicon atoms in a lattice without affecting the lattice itself too much. However, each of these impurity atoms has one more electron in its valence level than a silicon atom has; this extra electron, for which there is no room in the valence band, takes a place in the conduction band and can conduct electricity (Fig. 27–19a). A semiconductor with impurities of this sort is called an *n*-type semiconductor, and the extra electrons are called *donor* electrons. The semiconducting material, silicon in this case, is said to be *doped* by the impurity atoms.

Atoms of elements in the same chemical group as boron, aluminum, and gallium have one less valence electron than silicon has. If, as in Fig. 27–19b, such an atom is added to a lattice of silicon as an impurity, there is one less electron than is needed to form a bond that holds the lattice together. This electron must be provided by the electrons of the valence band of the lattice material, and holes are created in this band. These holes act as positive charge carriers. The impurity atoms are called *acceptors*, and a semiconductor with impurities of this sort is called a *p*-type semiconductor.

Many electronic devices, such as the diode mentioned earlier, depend heavily on the properties of semiconductors. Probably the best known of these devices are transistors, which can amplify electronic signals.

*27–6 SUPERCONDUCTORS

In 1911 H. Kammerlingh Onnes, who in 1908 had been the first to produce temperatures low enough to liquefy helium, found that mercury loses *all* its resistance abruptly at a *critical temperature* T_c of 4.1 K (Fig. 27–20). This state of affairs persists at temperatures below T_c. When a material attains zero resistance at some critical temperature, it is called a *superconductor*. Detailed measurements on a superconducting ring in which a current had been induced showed that there was no observable decrease in the current after a year. From the measurements it was possible to deduce that, if there were any resistive decrease of the current, it had to occur over a period of at least 10^9 years!

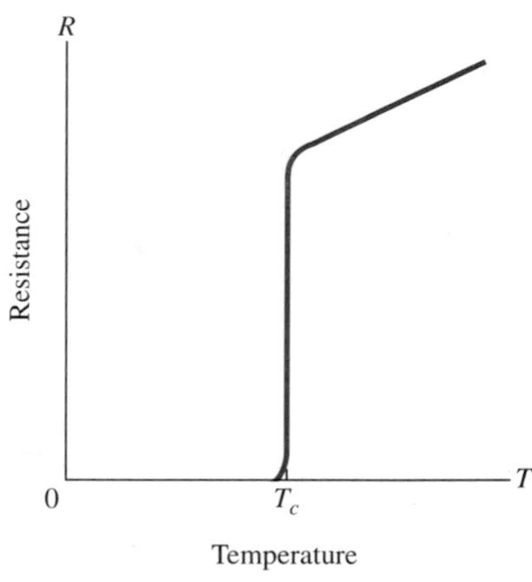

FIGURE 27–20 For superconductors the resistance drops to zero at the critical temperature T_c.

The prospect of having an *electric current that lasts forever* is an enticing one. It implies, among other things, the cheap transmission of electricity. The phenomenon of superconductivity cannot be understood as an extension of ordinary conductivity. The abruptness with which resistance disappears completely suggests that at T_c, an ordinary conductor makes a transition to a totally different state of matter, much as liquid water turns into a crystal (ice) at 273 K.

In 1957 John Bardeen, Leon Cooper, and Robert Schrieffer satisfactorily explained the *superconducting phase* with quantum physics, in what is now known as the BCS theory. Their explanation postulates that pairs of electrons move in a coordinated way through a superconductor, forming what are known as *Cooper pairs*. These pairs are well correlated even if each electron of the pair is rather far apart. Two free electrons will repel each other due to the Coulomb force. However, in a lattice, a detailed application of quantum principles shows that there is a subtle *attraction* between electrons that allows them to form pairs. What is crucial about superconductors is not, however, that pairs are possible, but that in the superconducting phase, *all* the electrons are involved together—just as when water freezes, all the water molecules take part in forming the ice crystal.

Why does this phase transition result in a loss of resistance? Normally, the resistance of electrons to the acceleration of an electric field is due to the collision of individual electrons with the atoms of the lattice. If the superconductor consisted of individual Cooper pairs, the same could happen to individual pairs of electrons, but because the state is a collective one, *all* the Cooper pairs would have to be slowed down or broken up *at the same time*. This would require a large amount of energy, so the uninterrupted flow of current, with no resistance, is favored.

The BCS theory can explain why some materials become superconducting but others do not, although it does not predict the precise values of T_c very well. In particular, until 1986 the materials with the highest known values of T_c became superconducting at 23 K. Helium is liquid at such temperatures and is thus used

FIGURE 27–21 Superconducting magnets are used in practical tools, such as magnetic resonance imagers (MRI) and particle accelerators. The operating cost is associated more with refrigeration than with direct power loss in the magnet wires. These superconducting magnets are designed for use in the Superconducting Supercollider accelerator.

for cooling superconductors. However, liquid helium is so expensive that superconducting devices have been limited to fairly specialized applications, such as magnets for particle accelerators (see Chapter 30) or nuclear magnetic resonance imaging machines in hospitals (see Chapter 32) (Fig. 27–21). In 1986 K. Alex Muller and J. George Bednorz discovered a new class of materials for which T_c is much higher; superconductors have been discovered with a T_c above 120 K (−153°C).† (Interestingly, it appears that these new materials cannot be explained by straightforward application of the BCS theory.) This discovery has great technological implications, because such materials can be cooled relatively cheaply with nitrogen (which is liquid at 77 K). It seems likely that these materials will be used extensively in small devices, such as switches in supercomputers. The outlook for the use of the new high-temperature superconductors is undoubtedly a bright one, and research and development in the field of superconductors will be extensive.

27–7 ELECTRIC POWER

Electric energy is sent to our homes and workplaces and supplies much of the energy used in our society. Efficient delivery of this energy is of paramount importance. In this section we shall look at the ways in which resistance affects the delivery of electric energy.

We have likened electrical resistance to mechanical drag. When there is drag in mechanical motion, mechanical energy is converted to thermal energy. The second law of thermodynamics (Chapter 20) shows that some of this thermal energy is irretrievably lost, in the sense that it cannot all be converted to mechanical work. Similarly, some electric energy is lost because of resistance. Just as mechanical friction generates heat, the passage of a current through a resistor generates heat. Sometimes it is the thermal energy that we want to use, as in the heating element of an electric stove. But it cannot all be converted to useful mechanical work.

Power is discussed in Chapter 6.

To calculate the energy lost per unit time (the power lost) when a current flows in a material, consider a small charge dq that moves through a potential

† See, for example, "Superconductors beyond 1–2–3," Robert J. Cava, *Scientific American*, August 1990, p. 42.

difference V. The change in the potential energy of the charge (dU), which is equal to the work done (dW) by the electric force due to the potential difference, is given by $dU = V\,dq$. It follows that the power, the rate at which energy is expended by the force that pushes the charge, is

$$P = \frac{dW}{dt} = V\frac{dq}{dt}. \qquad (27\text{–}26)$$

Because the current $I = dq/dt$, the electric power lost, which is the power that must be delivered to move I through the potential V, is

The power lost in resistance

$$P = VI. \qquad (27\text{–}27)$$

This result is a general one, independent of the type of material—in particular, whether the material is ohmic or nonohmic—and of the nature of the charge movement. Power has SI units of watts (W), with 1 W = 1 J/s. By using Eq. (27–27) we have another unit for power:

$$1\ \text{W} = 1\ \text{V}\cdot\text{A}. \qquad (27\text{–}28)$$

For ohmic materials $V = IR$, where R is a constant. Thus the power expenditure for ohmic materials is

$$P = VI = V\left(\frac{V}{R}\right) = \frac{V^2}{R}. \qquad (27\text{–}29)$$

Equivalently, we can use $V = IR$ in Eq. (27–29) to find that

Alternative forms for the power lost in resistance

$$P = VI = (IR)I = I^2R. \qquad (27\text{–}30)$$

Whether Eq. (27–29) or Eq. (27–30) is used depends on what is known in a particular application. The power lost (rate of energy loss) in a resistor appears in the form of thermal energy and is variously called *ohmic heating*, *Joule heating*, and *I^2R loss*.

EXAMPLE 27–7 Nichrome is an alloy of nickel, chromium, and iron often used as a heating element in electrical devices. A nichrome wire 1.0 m in length is crisscrossed along the bottom of a toaster oven that can carry a maximum current of 16 A when there is a 120-V potential difference from one end of this wire to the other.† If the resistivity of nichrome is $1.0 \times 10^{-6}\ \Omega\cdot\text{m}$, what is the radius of the wire? What power does the toaster use?

SOLUTION: In this example, a wire of an unknown resistance carries a known current when a known voltage difference is applied. We can therefore solve for the resistance, R. We then find the area of the wire from the known value of the resistance, the length of the wire, and the resistivity of the material. The resistance of the nichrome wire is determined by setting $I = 16$ A when $V =$ 120 V. From the definition of resistance,

$$R = \frac{V}{I} = \frac{120\ \text{V}}{16\ \text{A}} = 7.5\ \Omega.$$

† Real household electricity involves an oscillating voltage difference and an oscillating (or alternating) current. Ignore these effects here and in Example 27–8.

We now use Eq. (27–13), which relates the dimensions of the wire and the resistivity to the resistance. This equation is solved for the cross-sectional area A of the wire:

$$A = \frac{\rho L}{R} = \frac{(1.0 \times 10^{-6}\ \Omega\cdot\text{m})(1.0\ \text{m})}{7.5\ \Omega} = 1.3 \times 10^{-7}\ \text{m}^2.$$

The radius r of the wire is determined from $A = \pi r^2$ to be 0.20 mm.

The power consumed by the toaster is determined by Eq. (27–27) from V and I:

$$P = VI = (120\ \text{V})(16\ \text{A}) = 1900\ \text{W}.$$

Electric power is used in a context other than the one on which we have concentrated here. When electric energy is delivered to a home, the energy delivered per unit time is also called the electric power. This power is not always the power lost in resistance. We pay the electric company by the amount of energy we purchase from them. The energy unit in the electric power industry is the kilowatt-hour (kWh):

$$1\ \text{kWh} = (1\ \text{kW})\left(\frac{1000\ \text{W}}{1\ \text{kW}}\right)\left(\frac{1\ \text{J/s}}{\text{W}}\right)(1\ \text{h})\left(\frac{3600\ \text{s}}{\text{h}}\right) = 3.6 \times 10^6\ \text{J}. \quad (27\text{–}31)$$

The kWh is not an SI unit.

EXAMPLE 27–8 A 100-W bulb is left on in an outdoor storage room to keep paint from freezing. The 100-W rating refers to the power dissipated in the bulb's filament, which is a simple resistor (Fig. 27–22). If electricity costs 8 cents/kWh, about how much does it cost to burn the light bulb for 3 months during winter?

SOLUTION: We are given the power used by the bulb and the length of time over which the power is dissipated. From this information we can find the energy used, and because we are given the price rate of electric energy, we can find the total cost. The total number of hours during the 3-month time $\simeq$ (3 months)(30 d/month)(24 h/d) = 2160 h. The amount of energy used is then the power multiplied by the length of time over which it is dissipated:

$$\text{energy} = (\text{power})(\text{time}) = (100\ \text{W})(2160\ \text{h}) \simeq 220\ \text{kWh}.$$

Thus

$$\text{cost} = (220\ \text{kWh})\left(\frac{\$0.08}{1\ \text{kWh}}\right) = \$17.60.$$

Although the primary purpose of a light bulb is to produce light, most of the electric energy it dissipates is converted into heat, not light.

FIGURE 27–22 Example 27–8. The filament of a light bulb acts as a resistor when current passes through it. The light emitted is part of the energy dissipated by the filament.

Resistors used in circuits are characterized not only by their resistance, but also by a power rating. This power rating states the maximum power that the resistor can dissipate without being damaged due to overheating. The power rating is measured in watts. According to Eq. (27–30), which states that the power dissipated in a resistor is $P = I^2R$, we can deduce the maximum allowed current from the power rating. One class of relatively inexpensive resistors, composition and carbon film resistors, is limited to about 2 W, whereas a second, more expensive type known as wire-wound resistors have a power rating up to 50 W.

Electric current is the rate at which charge passes. The instantaneous current is given by

$$I \equiv \frac{dQ}{dt}. \qquad (27\text{–}2)$$

The unit of current is the ampere (A), 1 C/s. Currents are depicted as though the positive charges are moving, but it is actually the (negative) electrons that are mobile.

The current density, **J**, is a vector quantity that gives the current that passes through an area per unit time. The current is related to the current density by

$$I = \iint_A \mathbf{J} \cdot d\mathbf{A}. \qquad (27\text{–}4)$$

The free-electron model of conduction is useful as a qualitative description of current flow in a solid. The average, or drift, speed of the electrons that pass through the material is

$$v_d = \frac{I}{n_e q A}, \qquad (27\text{–}9)$$

where n_e is the density of free electrons and q is the charge of an electron.

Electrical resistance, R, is the ratio of voltage to current

$$R \equiv \frac{V}{I}. \qquad (27\text{–}11)$$

Many conducting metals show a linear relationship between voltage and current. The resistance is then constant over a wide range of voltages. This relation is called Ohm's law, $V = IR$.

Resistivity, ρ, is the quantity that distinguishes the part of the resistance that is intrinsic to each particular type of material. For wires of length L and area A, we have

$$R = \rho \frac{L}{A}. \qquad (27\text{–}14)$$

The inverse of the resistivity is the conductivity, σ, which expresses how well a type of material conducts current:

$$\sigma \equiv \frac{1}{\rho}. \qquad (27\text{–}15)$$

Both ρ and σ depend on temperature.

The electric field and current density are related by

$$\mathbf{E} = \rho \mathbf{J} \qquad (27\text{–}16)$$

and by

$$\mathbf{J} = \sigma \mathbf{E}. \qquad (27\text{–}17)$$

The correct explanation of electrical conduction in metals depends on how electrons fill allowed energy bands and on the energy gap between these bands. Materials that conduct current easily have electrons in partly filled bands. Semiconducting materials, such as silicon and germanium, can be doped by impurity atoms to increase the density of charge carriers. The explanation of superconductivity requires both quantum mechanics and the presence of a new phase of matter in which electrons collectively transport electric current.

When a current moves through a potential difference, electric power, P, is dissipated (or produced), given by

$$P = VI. \tag{27–27}$$

For resistive materials the power is also given by

$$P = \frac{V^2}{R} = I^2R. \tag{27–29, 27–30}$$

QUESTIONS

1. Consider the electron beam in a cathode-ray tube. The velocity of the electrons in the beam changes as the electrons are accelerated. Is the current the same everywhere in the beam?
2. How does the free-electron model for electrical resistance account for power dissipation? Does our microscopic picture agree with the voltage/current result?
3. The same current passes through two similar wires of unequal areas. Which wire will get hotter, and why?
4. The same current passes through two wires of the same area. One of the wires is made of aluminum, whereas the other is made of brass. Which wire will get hotter, and why?
5. What factors determine the differences in drift velocity of electrons in wires if the dimensions and current are the same?
6. If the movement of charges in a wire is similar to the flow of water in a hose, why, when a new hose is hooked up to a faucet, do we have to wait for a while until the water comes out, but when we hook a new wire up to a circuit, we do not have to wait for charge to come out the other end when the switch is turned on?
7. According to Eq. (27–25), the resistivity in the free-electron model should vary with the square root of the temperature and thus should be zero at $T = 0$. Is this reasonable? How would you interpret this result?
8. We know that the resistivity of a metal is temperature dependent, and so therefore is the resistance of a wire. In Chapter 21 we saw that the dimensions of a piece of metal such as a wire change when the wire is heated. Does this provide an additional reason to change the resistance of a wire as it undergoes Joule heating? Would you expect the effect to be large?
9. When you throw a switch and current flows in a household wire, does the wire become charged?
10. Suppose that we orient a wire between the plates of a charged capacitor so that there is an electric field along the cross section of the wire. Will the resistance of the wire change because all the charge-carrying electrons crowd to one side of the wire, thus effectively reducing the wire's cross section?
11. Gauss' law states that free charge in a conductor moves to the surface of the conductor. Does this mean that the current flowing through a wire is actually on the wire's surface?
12. The resistivity of most metals is on the order of $10^{-8}\ \Omega\cdot\text{m}$. Discuss why this might be so in terms of the result given by Eq. (27–23).
13. What is likely to happen when a current is so large that the power dissipation in a resistor through which the current flows exceeds the resistor's power rating? What mechanism is responsible for such a disaster scenario?

PROBLEMS

27–1 *Electric Current*

1. (I) A wire of diameter 2.2 mm carries a current of 0.46 A. What is the average current density? How much charge crosses a fixed point in the wire per second?
2. (I) Three straight wires of area 2 mm², 3 mm², and 4 mm², respectively, are aligned along the x-axis. They carry current densities of magnitude 14 A/m², 7 A/m², and 8 A/m², respectively, also along the x-axis. Find the current in each wire.
3. (I) A wire of radius 1.6 mm carries a current of 0.092 A. How many electrons cross a fixed point in the wire in 1 s?
4. (I) Charge carriers in a semiconductor have a number density $n_q = 2.3 \times 10^{24}$ carriers/m³. Each carrier has a charge whose magnitude is that of an electron's charge. If the current density is 1.2×10^4 A/m², what is the speed of the charge carriers?
5. (I) The density of charge-carrying electrons in copper is 8.5×10^{28} electrons/m³. If a current of 1.2 A flows in a

wire 1.8 mm in radius, what is the speed of the electrons? How does that speed change in a second wire, of diameter 2.4 mm, connected end-to-end with the first wire?

6. (I) An electron accelerator in which electrons travel at a speed of 0.90×10^8 m/s produces a beam of electrons that carries a current of 1.0 mA. The effective area occupied by the beam is 3.0 cm^2. What is the density of electrons in the beam? Ignore all relativistic effects.

7. (II) In the National Synchrotron Light Source X-ray device at Brookhaven National Laboratory, there is an electron beam with an average current of 200 mA. The electrons have a kinetic energy of 2.5 GeV and a speed extremely close to the speed of light. How many electrons pass a given point in the accelerator per hour? How many electrons are contained in a 1-m length of the beam? Ignore all relativistic effects (a poor approximation, in this case).

8. (II) A cube of material is placed with one corner at the origin of a coordinate system, its sides, 1 cm long, parallel to the three axes. The current density is $A\mathbf{i} + B\mathbf{j} + C\mathbf{k}$ throughout the cube. The units of A, B, and C are mA/cm^2. What are the currents along the x-axis, y-axis, and z-axis?

9. (II) In a plasma containing equal densities n of electrons and (positive) ions, the ions move to the right. Their speed is a factor of 10^{-3} smaller than the speed with which the electrons move to the left. What is the (net) current density? Give its direction and magnitude.

27–2 *Currents in Materials*

10. (I) Calculate the drift speed of electrons in the conduction cables of an automobile starter cable, which is made of copper and has a diameter of 4 mm, supposing that the cable carries 100 A. How would this speed change if the diameter of the wire were doubled? (*Hint:* Useful data are contained in Example 27–2.)

11. (II) An aluminum wire of area 50 mm^2 placed along the x-axis passes 10,000 C in 1 h. Assume that there is one free electron for each aluminum atom. Determine the current, current density, and drift speed. The mass density of aluminum is 2.7 g/cm^3.

12. (II) Silver has one electron per atom available to carry charge. Given that the mass density of silver is 10.5×10^3 kg/m^3 and that its molecular weight is 108 g/mol, calculate the drift speed of the electrons in a silver wire that carries 1 A and has a circular cross section 0.6 mm in radius.

13. (II) Two parallel metal wires of diameter 0.5 cm and a charge-carrier density $n_e = 10^{23}$ electrons/cm^3 carry a current of 5 A each. The wires join and then split into five identical but separate wires with a radius one-tenth that of the original wire. All the wires are made of the same material. What are the drift speeds in both the larger and smaller wires? Can you explain the difference in speeds in terms of the speeds of water flow in pipes?

14. (II) The charge carriers in a certain wire of circular cross section and radius R have a drift speed down the wire that is not constant across the wire. Instead, the drift speed rises linearly from zero at the circumference ($r = R$) to v_0 at the center ($r = 0$) (Fig. 27–23). Compare the total current carried by this wire with the current carried by a wire of the same radius, same density of charge carriers, and a constant drift speed of $v_0/2$.

FIGURE 27–23 Problem 14.

15. (III) Charges q move longitudinally down a rod of circular cross section and radius R. The density of the charge carriers, n, decreases as a function of the radial distance r from the center of the rod according to $n = n_0 - n'r$. The speed v of the charge carriers varies with r according to $v = v_0 - v'r^2$, where n_0, n', v_0, and v' are constants. Calculate the current that passes through the rod.

27–3 *Resistance*

16. (I) An underground wire made of aluminum is 91.4 m long and has an area of 0.30 cm^2. (a) What is its resistance? (b) What is the radius of a copper wire of the same length and resistance?

17. (I) An old house is wired with AWG #18 copper wire, which has a diameter of 0.0403 in. (a) What is the wire's resistance per 1000 ft? (b) One circuit consists of only one wire behind walls and has a resistance of 2.3 Ω. How long is this wire?

18. (I) The resistivity of copper is 1.72×10^{-8} Ω·m. What is the resistance of a section of gauge #10 wire (diameter 0.2588 cm) that is 10 m long?

19. (I) A carbon rod used in a welding machine is 2.0 mm in diameter and 10.0 cm in length. What is its resistance, and how much current will pass through it, if the welding machine puts a voltage of 220 V across it?

20. (II) An electrician tests for a short circuit by putting a potential difference of 6.0 V across two neighboring parallel wires that, if there were no short, would be independent of each other. A current of 0.26 A then flows in the wires. The wires consist of material with a resistivity of 1.7×10^{-8} Ω·m and have a diameter of 1.5 mm. Given that the short effectively makes the wires act like a single wire, how far away is the short?

21. (II) A nichrome wire of diameter 0.5 mm and length 50 cm is connected to a 50-V battery. What current passes through the wire at room temperature (25°C) and after the wire heats up to 400°C?

22. (II) An aluminum wire of length L and a copper wire of length $2L$ have precisely the same resistance. Given that the resistivity of aluminum and copper are $2.8 \times 10^{-8}\ \Omega\cdot\text{m}$ and $1.7 \times 10^{-8}\ \Omega\cdot\text{m}$, respectively, what is the ratio of the radii of the two wires?

23. (II) A copper wire is stretched by 1 percent. Assuming that its volume is unchanged, what is the percentage change in the resistance?

24. (II) You have a 100-m-long wire of area 0.5 mm^2 with a thin coating of insulation; you cannot identify the type of material that makes up the wire. You have a 12.0-V battery and a device to measure current. When the battery is placed across the two ends of the wire, you measure a current of 1.07 A. What is the wire material? (Use Table 27–2.)

25. (II) A coil used to produce a magnetic field is made of copper wire of area 0.7 mm^2 wound many times around a spool of diameter 40 cm. The resistance of the wire is 3.7 Ω. As we shall learn in Chapter 30, to know the magnetic field, we must know the number of turns of wire. How many turns of wire are there on the spool?

26. (II) You wish to double a current that flows through a wire of fixed length, but you can increase the voltage that drives the current by only a factor of 1.5. You have other wires made of the same material but of different radii. What is the smallest factor by which the radius of a replacement wire should differ from the radius of the original wire?

27. (II) Aluminum has a density of $2.7 \times 10^3\ \text{kg/m}^3$. What is the resistance of an aluminum wire 2 cm in diameter and 250 m long? What is the mass of the wire? What is the mass of a copper wire, of density $8.9 \times 10^3\ \text{kg/m}^3$, with the same length and same total resistance?

28. (II) What are the length and the radius of a copper wire (of circular cross section) whose resistance is 2 Ω and whose mass is 1.5 kg?

29. (II) How much silver (of density $10.5 \times 10^3\ \text{kg/m}^3$) would be needed to make a wire 1 km long, with a resistance of 5 Ω?

30. (II) A copper pipe has an inside diameter of 5 cm and an outside diameter of 6 cm. What length of copper pipe will have a resistance of $10^{-2}\ \Omega$?

31. (II) A copper resistor has the shape of a cylindrical shell. What is the resistance of this resistor if its length is 1 m, its inner radius is 0.1 cm, and its outer radius is 0.2 cm? What is the radius of a solid wire of circular cross section with the same length and the same resistance? Compare the masses of the two resistors.

32. (II) A *zener diode*, named for Clarence Zener, has the I–V curve shown in Fig. 27–24. Sketch the resistance of the diode versus both current and voltage. What is special about the critical voltage V_c?

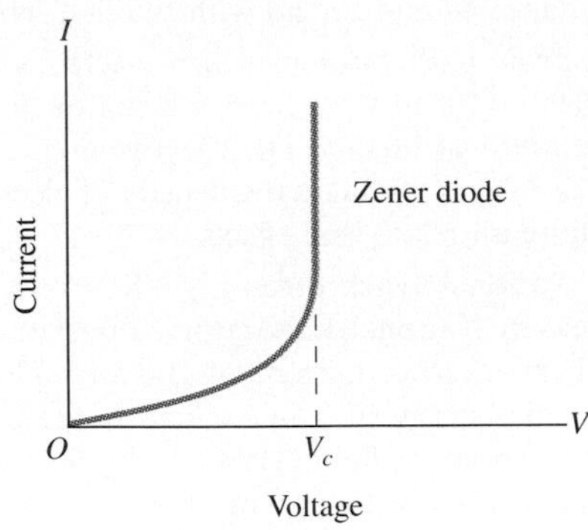

FIGURE 27–24 Problem 32.

33. (III) The resistivity of copper increases by 0.4 percent per °C as its temperature is raised from 20°C. Its coefficient of linear expansion is $17 \times 10^{-6}\ (°\text{C})^{-1}$. What is the fractional change in the resistance of a copper wire when it is heated from 20°C to 200°C? How much of that change is due to the thermal expansion? Ignore any change in area due to thermal expansion.

34. (III) A sphere of radius r and resistivity ρ is connected to external wires of radius r_0 ($r_0 \ll r$) at the sphere's north and south poles. Calculate the resistance of the sphere to currents that run across these two points. (*Hint:* Break the sphere into slices connected in series, and integrate to sum over the slices.)

*27–4 *Free-Electron Model of Resistivity*

35. (I) Using the average time between collisions as calculated in Example 27–5, determine the drift speed of charge carriers for a material in which the electric field is 2.0×10^{-3} V/m.

36. (I) Recall Equation (19–51), which relates the collision cross section to the mean free path of a particle. Assuming the mean free path for an electron in copper that was calculated in Example 27–6, estimate the collision cross section for an electron that collides with an ion.

37. (II) In Problem 14 we described a wire of radius R within which the drift speed of charge carriers varies with the distance from the center of the wire as $v_d = v_0[1 - (r/R)]$. Supposing that this wire is made of ohmic material, describe how the resistivity must vary with r to produce this drift-speed profile.

38. (II) In Example 27–6 we calculated the mean free path between collisions of a free electron in copper to be 2.9×10^{-9} m. Assume that copper, whose mass density is 8.9 g/cm^3, forms a cubic lattice. Calculate the distance be-

tween copper atoms. How many atomic spacings is the mean free path?

39. (III) Consider a ball bearing of diameter 0.80 cm rolling down a smooth plane inclined at an angle $\phi = 4.0°$ to the horizontal. On the plane, steel pins are distributed evenly with a density $\sigma = 0.075$ per cm^2. When the ball hits a pin, it rebounds and slows down. Estimate the mean speed of the ball as it rolls down the plane by supposing that every time it hits a pin, it comes to rest and then resumes its motion. Assume for simplicity that the velocity is one-dimensional.

*27–5 *Insulators, Conductors, and Semiconductors*

40. (II) Treating electrons as a gas of independent particles, at what temperatures would an average electron have sufficient energy to cross the energy gap for silicon (1.1 eV), germanium (0.7 ev), and carbon (6 eV)?

27–7 *Electric Power*†

41. (I) What is the resistance of a 75-W bulb used on a 120-V line?

42. (I) What is the maximum voltage that can be applied to a 1000-Ω resistor rated at 1.5 W?

43. (I) A graduate student in engineering has a collection of 100-Ω resistors with different power ratings of 1/8, 1/4, 1/2, 1, and 2 W. What is the maximum current that the student should use in each resistor?

44. (I) What is the maximum allowable current for (a) a 5-kΩ, 2-W resistor? (b) a 10-kΩ, 1/2-W resistor?

45. (I) Consider a resistor of resistance R. If the maximum allowed power dissipation is P, what is the maximum allowed operating voltage?

46. (II) A heater uses nichrome wiring ($\rho = 10^{-6}\ \Omega\cdot$m) and generates 250 W when connected across a 110-V line. How long must the wire be if its cross-sectional area is 10^{-6} m^2?

47. (II) Consider the terminals of a 12 V-battery connected by a copper wire. How long must the wire be if its cross-sectional area is 3×10^{-5} m^2 and if the power dissipated is 1.2 kW?

48. (II) Automobiles have circuit breakers, devices that switch the current off when it exceeds a critical value, to protect the electrical system from damage. One circuit for a car's lights has a 4-A breaker. (a) What is the maximum power that can be delivered by the 12-V battery to this circuit? (b) How many light bulbs, each requiring 5 W, can this circuit handle?

49. (II) A 12-V battery is connected to two metal wires dipped in a pot of water. A current of 100 mA flows for 240 h. How much energy is taken out of the battery during that time?

50. (II) A 500-W electric heater is designed to operate on a line of 115 V. As the result of a brownout, a partial interruption of electrical power, the line voltage drops to 105 V. Assuming that the heating unit has a fixed resistance, what is the power of the heater now?

† When we refer to household electricity applications here, assume in each case that their currents and voltage differences are of the simple, constant type discussed in this chapter.

General Problems†

51. (II) An electric hot plate is used to boil water. The current drawn by the hot plate is 10 A. Make a rough estimate of the voltage and resistance, based on how long it takes to make a pot of tea.

52. (II) One month's electricity bill for an apartment is $25.33, and the cost of electricity is 8 cents/kWh. All appliances used in this apartment work at 120 V. How many electrons passed through the apartment's electrical meter that month?

53. (II) A wire of resistance r is drawn—pulled like taffy—to double its length. Assuming a constant voltage and a fixed volume, by how much does the power dissipation change?

54. (II) A Van de Graaff accelerator delivers 6-MeV protons at a current of 3 μA through a target onto a piece of tungsten that serves to stop the proton beam. (a) How many protons stop in the tungsten in 24 h? (b) How much energy is delivered to the tungsten in 1 d? (c) What is the power of the proton beam?

55. (II) A piece of brass is machined into a long, tapering cylinder. Its radius is expressed by $r = r_0 + \alpha x$, where α is a constant and x is measured from the narrow end of the tapering cylinder and runs from 0 to L (Fig. 27–25). Find an expression for the resistance of this piece, and compare it to the resistance of a cylinder of equal length and of a radius equal to the radius of the tapering rod at its midpoint.

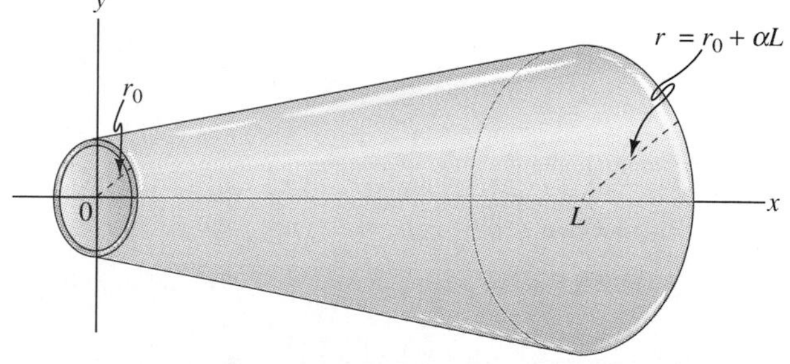

FIGURE 27–25 Problem 55.

56. (II) A bus bar (a conducting bar that supplies current to several circuits) made of copper, of resistivity 1.72×10^{-8} Ω·m, is meant to carry 100 A over a distance of 0.25 m at a temperature of 300°C. What is the minimum cross section of the bus bar if no more than 0.2 W of power is to be dissipated?

57. (II) A generator delivers 60 A at a voltage of 110 V. What power does the generator deliver? How long would it take to raise the temperature of 10^{-3} m^3 of water by 80°C? How long would it take to boil away 1 L of water, starting at 20°C?

58. (II) A given wire has a resistance of 7.2 Ω per 1000 ft. A 100-ft extension cord made of this wire is used simultaneously with an electric drill that draws 3 A and a table saw that draws 10 A. (An extension cord has two wires in it.) If the voltage source is a constant 120 V, how much voltage is available at the tools? How much power is lost in the extension cord? Would the extension cord feel warm?

59. (II) A potential is set up from one end of a copper wire to the other; as a result, current flows. The copper is thermally isolated to some extent. As the current flows, the wire heats up, causing the resistivity to increase. Suppose that, during a short time period, the temperature of the wire as a function of time t is given by $T = T_0 + kt^2$. (a) Describe the current in the wire during this period. (b) What is the power dissipated by the wire as a function of time? (c) From the change with time of the dissipated power, will the wire continue to heat up and, perhaps, melt?

60. (II) The density of charge-carrying electrons in copper is 8.5×10^{28} electrons/m^3, its resistivity is 1.7×10^{-8} Ω·m, and the drift speed in a copper wire is 1.2×10^{-5} m/s. The wire has a diameter of 1 mm and a length of 3 m. At what rate must thermal energy be carried off by a cooling medium if the wire is to maintain its temperature?

61. (II) A single layer of 100 turns of closely spaced wire of radius $r_1 = 0.2$ mm is wound in a coil of diameter $D_1 =$ 2 cm. A second coil of the same length but of diameter $D_2 =$ 4 cm is composed of a single layer of closely spaced wire of radius $r_2 = 0.5$ mm. The wires are made of the same material. Find the ratio of the resistances of the two coils.

62. (III) A thin wire of length L and cross-sectional area A oriented in the x-direction is made of an ohmic material whose resistivity varies along the wire according to the empirical law $\rho = \rho_0 e^{-x/L}$. (a) Describe how the field within the wire varies with position if the end at $x = 0$ is at a potential V_0 greater than the end at $x = L$. (b) How does the potential vary as you move along the wire? (c) What is the total resistance of the wire?

63. (III) If all the energy lost from Joule heating stays in a wire, and the temperature increases as a result, the resistivity will increase according to Eq. (27–18). The current will therefore change as a function of time, the Joule heating will change, and so forth. If the wire material has a constant heat capacity, the rate of energy loss in the wire will be proportional to the rate of change of temperature. Assuming that the potential stays constant, set up a differential equation that describes the rate of temperature change. If this equation is solved, how can the current be found as a function of time?

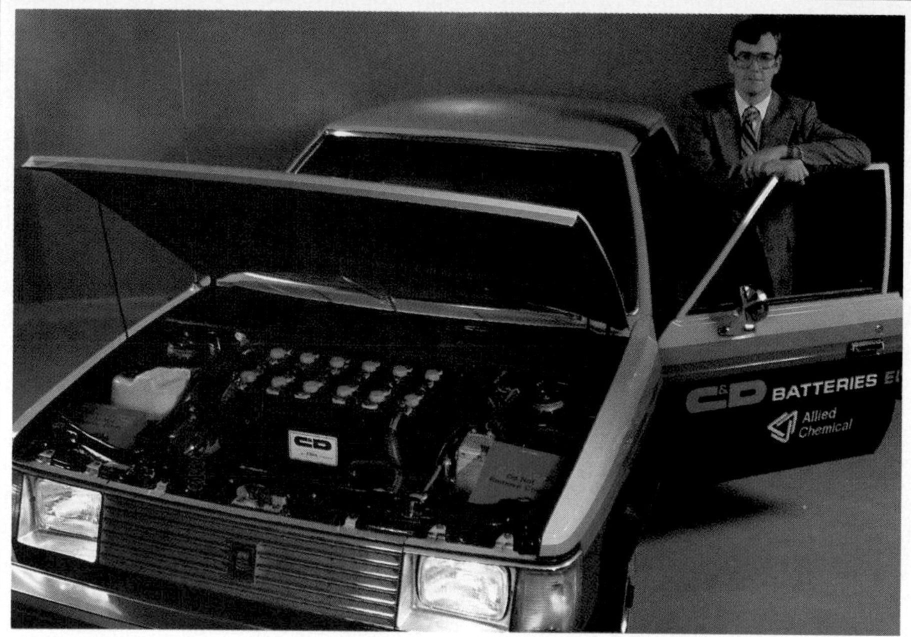

Batteries store energy in chemical form and release that energy in electrical form. That they can maintain a given potential difference over long time periods enables us to produce battery-operated electric cars.

DIRECT-CURRENT CIRCUITS

We have seen how charges move under the influence of potential differences, and how resistors and capacitors influence the flow of current and the movement of charge. When resistors, capacitors, and batteries or other sources of electric energy are connected together by conducting wires, they form electric circuits. We can understand the flow of currents in circuits by applying just two simple physical principles, the conservation of current and of energy. In this chapter we learn to apply these principles systematically to the analysis of circuits. We also discuss the common instruments that measure and monitor the current, voltage, and resistance of electric circuits. The flow of energy to and from circuit elements provides an important theme for this chapter and leads us to the concept of time-varying currents and voltages.

28–1 EMF

The sources of electric energy that cause charges to move in electric circuits have historically been called sources of **electromotive force**. They are actually sources of energy, not of force, and to avoid the word force, we employ the term **emf** instead. When we think of sources of emf, we usually have in mind batteries, but

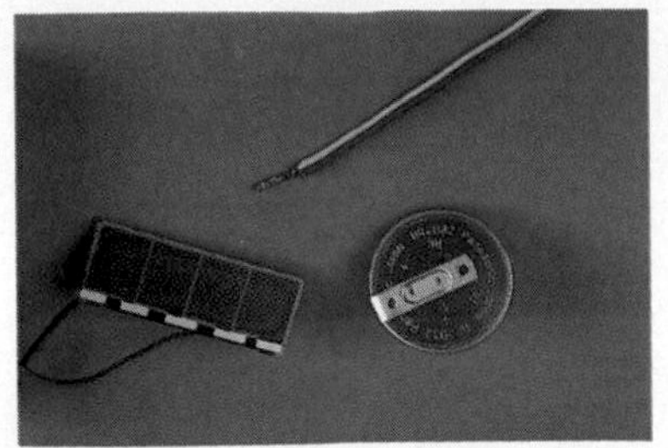

FIGURE 28–1 A collection of sources of emf, including a chemical battery, a thermocouple wire, and a solar cell.

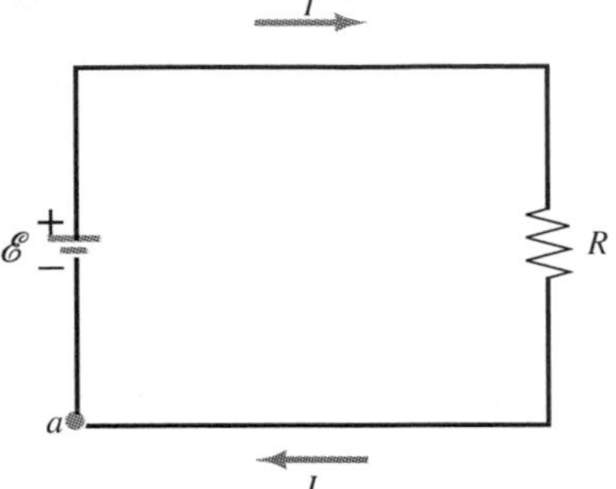

FIGURE 28–2 A simple circuit with a source of emf, $\mathscr{E}$, and a resistor, R.

FIGURE 28–3 Alessandro Volta, who introduced the phrase "electromotive force," invented the first electric battery, the voltaic pile, in 1800. Here he is showing his invention to the emperor Napoleon.

there is a wide variety of sources of electric energy made by humans. A battery converts chemical energy to an emf; a solar cell converts the energy in sunlight to an emf; a thermocouple produces an emf as a result of a difference in temperature; a large commercial electric power plant may burn coal, gas, or nuclear fuel to drive a generator that produces an emf, or it may use the kinetic energy from falling water for the same purpose (Fig. 28–1).

In this chapter, we use the term "battery" for any source of emf. We shall restrict ourselves to batteries for which the emf is *constant* with time. At least until Section 28–5, we shall also restrict our attention to phenomena such as current flows or potential differences that are similarly constant in time. Such constant behavior is called *equilibrium* or *steady-state* behavior. When all the currents, fields, potentials, and so forth in a circuit are constant in time, we are speaking of **direct-current**, or **DC**, behavior.

Circuits

When batteries, resistors, capacitors, or other circuit elements (to be introduced later) are all connected by idealized resistanceless wires, they form a **circuit**. For example, when a switch is closed and a battery sends current through the filament of a light bulb, a circuit has been formed. Figure 28–2 illustrates a simple circuit by using the conventions for resistors, ideal wires (wires with no resistance), and batteries; the light bulb is a simple resistor. At this stage we deal with steady currents, and we consider only circuits without capacitors.

Problems involving circuits typically involve relating the currents and potential differences in them. We may want to know, for example, the potential drop across a capacitor or the current that passes through a resistor when there is a particular emf in the circuit.

The Role of Batteries

When a battery is part of a simple circuit such as the one in Fig. 28–2, a current flows from the terminal of the battery at the higher potential, the one marked positive (Fig. 28–3).† How much current flows depends on the rest of the circuit. In Fig. 28–2 the remainder of the circuit consists of a single resistor; we refer to its resistance, R, as the **load** resistance. There is current flow here because negative charge carriers (electrons) are attracted to the positive terminal. Because the current is defined as moving in a direction opposite to that of the electrons, it may be helpful to imagine the positive charges flowing to the negative terminal (or the terminal at the lower potential). The battery has a potential difference across its terminals called the **terminal voltage**.

Inside a chemical battery, a chemical process carries the positive charges back to the positive terminal (Fig. 28–4). The battery can be thought of as a device that expends energy to pump charges, just as a water pump expends energy to pump water uphill to a tank with a higher gravitational potential energy. It is the internal pumping action of the battery that gives a precise definition of the emf. Suppose that it takes work $\mathrm{d}W$ to move a charge $\mathrm{d}q$ from the negative to the positive terminal. Then the emf of the battery is defined to be

The definition of emf

$$\mathscr{E} \equiv \frac{\mathrm{d}W}{\mathrm{d}q}. \tag{28–1}$$

The SI unit of emf is the volt, or joules per coulomb. The word *voltage* is some-

† The two "terminals" of a battery are simply points across which the battery maintains a potential difference. In an ordinary D-cell, the terminals are at opposite ends.

times used loosely to describe the emf, $\mathscr{E}$, but voltage more properly refers to the potential difference or terminal voltage across the emf terminals, which may be different from $\mathscr{E}$, as we shall see below.

When a battery sets charges into motion, driving a current from the positive (higher-potential) terminal around the circuit to the negative (lower-potential) terminal, we say that the battery *discharges*. In discharging, the battery is expending its chemical energy. If charge is instead driven from the negative to the positive terminal, a process that can be accomplished in conjunction with a larger battery, the smaller battery is said to be *charging*.

FIGURE 28–4 Old car batteries present an environmental hazard when they are not disposed of properly. Acid is part of a battery's chemical process.

Suppose that the potential difference across the battery terminals in Fig. 28–2 is $\mathscr{E}$. According to our reasoning, a current will flow around the circuit. To find this current, we can use the fact that the electric potential is associated with a conservative force. Therefore the net work done by this force in sending a charge around a closed loop is zero. In turn, the total potential drop involved in any round trip that starts from any point on a closed loop must be zero. Let us make such a round trip that starts at point a and follow the current around the circuit. There is no change in potential as we pass through the ideal (resistanceless) wire. In crossing the battery from the negative to the positive terminal, the potential *increases* by $\mathscr{E}$. When we cross the ohmic resistance, the potential *decreases* by an amount IR [see Eq. (27–12)]. The potential drop implies a decrease in the potential energy of the charges. This potential energy is converted into thermal energy in the resistor. The net potential change as we go once around the circuit is zero, so

$$\mathscr{E} - IR = 0. \tag{28–2}$$

This equation determines the current, I:

$$I = \frac{\mathscr{E}}{R}. \tag{28–3}$$

Internal Resistance

There is some energy loss in the operation of any actual, as opposed to ideal, source of emf. For example, a car battery heats up to some extent when it discharges, a result of resistive heating. Thus a real battery contains an *internal resistance* r in addition to maintaining an emf. This resistance is sometimes shown separately from the emf (Fig. 28–5). If we calculate the net potential change around the circuit as before, we find that Eq. (28–2) becomes

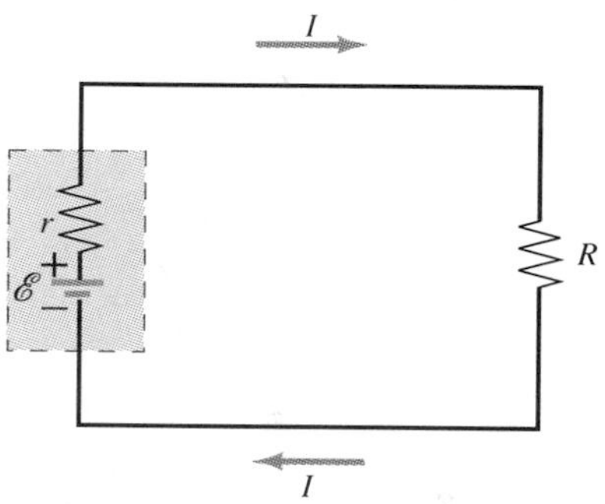

FIGURE 28–5 A source of emf also contains an internal resistance r. The shaded region includes both.

$$\mathscr{E} - Ir - IR = 0. \tag{28–4}$$

Two modifications result from the internal resistance. One, *the potential difference across the battery terminals is no longer just* $\mathscr{E}$; it is given instead by

$$\text{with internal resistance:} \qquad V = \mathscr{E} - Ir. \tag{28–5}$$

Due to internal resistance, the potential across the terminals of a battery is different from its emf.

This potential difference is a function of the current. Depending on the direction of current flow, the voltage across the terminals of a battery can be greater or less than the battery's emf.[†] Two, the current depends on the internal resistance. From Eq. (28–4), the current in our circuit is

$$I = \frac{\mathscr{E}}{r + R}. \tag{28–6}$$

Compare this to Eq. (28–3).

[†] If there is a second source of emf in a circuit, current can flow into the positive terminal of a battery, as we shall see in Section 28–2.

We can see from Eq. (28–5) that for small internal resistance, the potential across the terminals is approximately the same as the emf. This is equally true if there is no current at all, so a reading of the potential across the terminals is also a reading of the emf when the external, or load, resistance in the circuit is very large. It is desirable that the internal resistance be small, but it is useful to remember that when we say "small," we mean small in comparison with external resistances, which vary depending on the application. The typical internal resistance of a car battery is less than 0.01 Ω, but it may be as large as 0.1 Ω for a flashlight battery. In many situations the internal resistance is so small that it can be ignored in electric circuit analyses. Curiously, ordinary batteries run down with age not because their emf decreases, but because their internal resistance increases, so that the current they can supply decreases.

EXAMPLE 28–1 One of two different resistors with respective resistances $R_1 = 5.00\ \Omega$ and $R_2 = 10.00\ \Omega$ can be placed into the circuit shown in Fig. 28–5. The emf, $\mathscr{E}$, and internal resistance, r, of the battery are unknown. When only R_1 is inserted, the current is $I_1 = 0.291$ A, and when only R_2 is inserted, the current is $I_2 = 0.147$ A. Find $\mathscr{E}$ and r.

SOLUTION: There are two unknowns here, $\mathscr{E}$ and r, and two different expressions for the current by which to determine them. Equation (28–4) used twice for the two different currents and resistances gives

$$\mathscr{E} - I_1 r - I_1 R_1 = 0,$$

$$\mathscr{E} - I_2 r - I_2 R_2 = 0.$$

Multiplying the first equation by I_2 and the second by I_1 and then subtracting to solve for $\mathscr{E}$, we find that

$$\mathscr{E} = \frac{I_1 I_2}{I_1 - I_2}(R_2 - R_1).$$

If we insert this into either of the first two equations, we can in turn solve for r:

$$r = \frac{I_2 R_2 - I_1 R_1}{I_1 - I_2}.$$

Numerical evaluation gives

$$\mathscr{E} = \frac{(0.291\ \text{A})(0.147\ \not{\text{A}})}{0.291\ \not{\text{A}} - 0.147\ \not{\text{A}}}(10.00\ \Omega - 5.00\ \Omega) = 1.48\ \text{V}$$

and

$$r = \frac{(0.147\ \not{\text{A}})(10.00\ \Omega) - (0.291\ \not{\text{A}})(5.00\ \Omega)}{0.291\ \not{\text{A}} - 0.147\ \not{\text{A}}} = 0.10\ \Omega.$$

Electric Power and Batteries

A source of emf (or electric energy) is also a source of *electric power*. The power is the rate at which the source delivers energy. From Eq. (27–30), the power of the source is the potential drop across the source times the current that passes through. For a source of emf, $\mathscr{E}$, we have

$$P = \mathscr{E} I. \qquad (28\text{–}7)$$

Let us see how this works for the circuit of Fig. 28–5. Equation (28–6) tells us that $\mathscr{E}$ is given by $I(r + R)$, where I is the current in the circuit and r and R are the internal and load resistances, respectively. If we use this relation for $\mathscr{E}$ in Eq. (28–7), we find that

$$P = I^2R + I^2r. \tag{28–8}$$

Energy conservation dictates this result. The electric power of the source of emf is balanced by the sum of the power dissipated in both the internal and load resistances.

28–2 SINGLE-LOOP CIRCUITS AND KIRCHHOFF'S LOOP RULE

A **single-loop circuit** is a circuit that consists of a single path for the current. The simple circuit discussed in Section 28–1 (Fig. 28–5) is an example of a single-loop circuit. It is useful to repeat the exercise of going around the loop and examining the potential change at every step, and to do so we redraw the circuit in Fig. 28–6a with various points indicated. The potential is graphed in Fig. 28–6b, where we follow the circuit along the current direction, with zero potential chosen arbitrarily at point a. The region inside the battery is normally not accessible. Because we are not interested in its details, in Fig. 28–6a we draw the internal resistance, r, as though it were separate from the emf, and in Fig. 28–6b we draw the rise of the emf as gradual. There is no potential change in an ideal conducting wire. In a real wire there is some very small resistance, but it is generally ignored unless very long wires are used.

That the potential change in traversing the complete circuit is zero is just a reflection of the conservation of energy. If we had followed the circuit in the opposite direction, against the current, the changes would all be of the opposite sign, but the end result would still be that the potential change in a complete circuit is

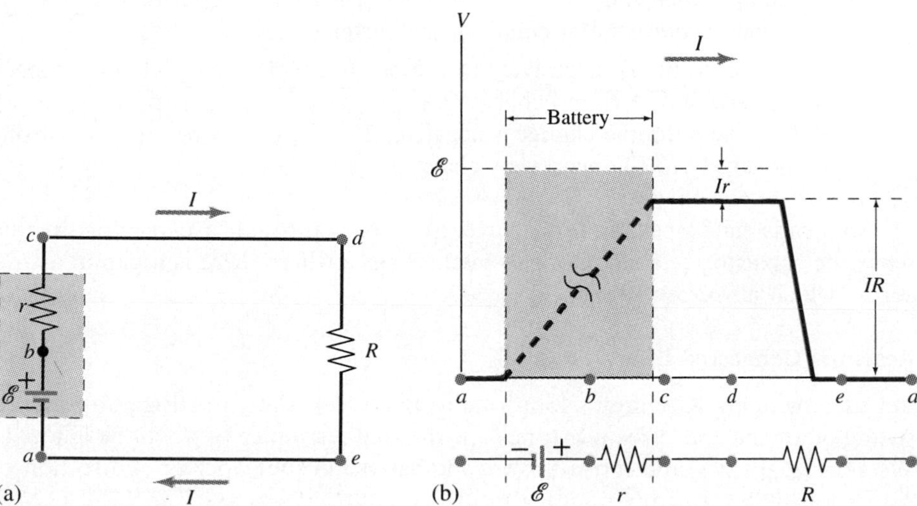

FIGURE 28–6 (a) A single-loop circuit showing the emf ($\mathscr{E}$), internal resistance (r), and load resistor (R). (b) The potential differences for the points labeled in part (a).

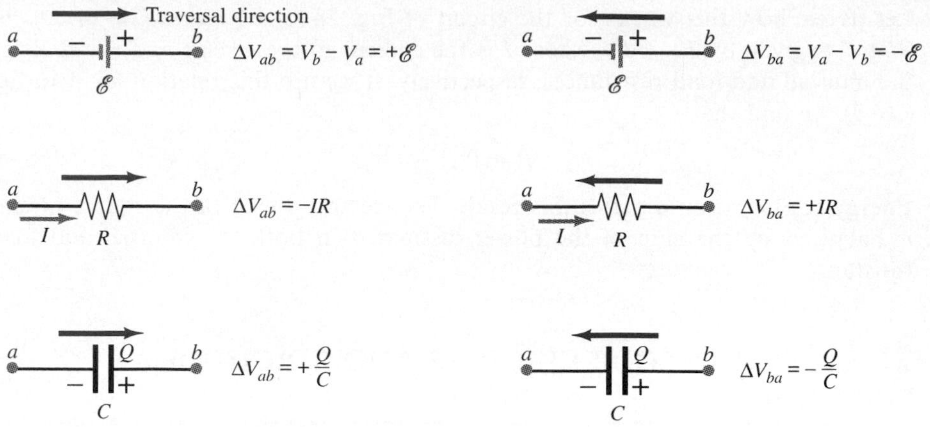

Traversal direction

$\Delta V_{ab} = V_b - V_a = +\mathscr{E}$ $\Delta V_{ba} = V_a - V_b = -\mathscr{E}$

$\Delta V_{ab} = -IR$ $\Delta V_{ba} = +IR$

$\Delta V_{ab} = +\frac{Q}{C}$ $\Delta V_{ba} = -\frac{Q}{C}$

FIGURE 28–7 Rules for potential differences across circuit elements.

zero. In the context of circuits, this simple law is given a special name, **Kirchhoff's loop rule**:†

Kirchhoff's loop rule

The sum of the potential changes around a closed path is zero, or

$$\sum_{\text{closed path}} \Delta V = 0. \tag{28–9}$$

The loop rule is applicable to any closed path in any electric circuit. When a circuit is laid out in a diagram, many closed paths may be possible, and the loop rule applied to these many loops is an important tool for finding the desired circuit parameters. In Fig. 28–6a there is only one closed loop.

In applying the loop rule, we need to know the potential differences across various parts of a circuit. Therefore it is useful to summarize what we have learned here and in previous chapters about the potential changes ΔV across individual circuit elements (Fig. 28–7):

Rules for potential differences across circuit elements

1. In going from the negative to the positive terminal of a battery with emf $\mathscr{E}$, the potential change is positive, $\Delta V = +\mathscr{E}$. In going from the positive to the negative terminal, the potential change is negative, $\Delta V = -\mathscr{E}$.
2. In moving across a resistance R *along* the direction of the current, I, the potential change is negative, $\Delta V = -IR$. The sign is opposite, $\Delta V = +IR$, in moving *against* the direction of the current.
3. In moving from the negatively to the positively charged plate of a capacitor of capacitance C and charge Q, the potential change is positive, $\Delta V = +Q/C$. The potential change is negative, $\Delta V = -Q/C$, when we move from the positively to the negatively charged plate.

In steady-state operation, no current can pass through a capacitor. In this sense the capacitor acts as an open switch, a place where there is a gap in a wire across which no current flows.

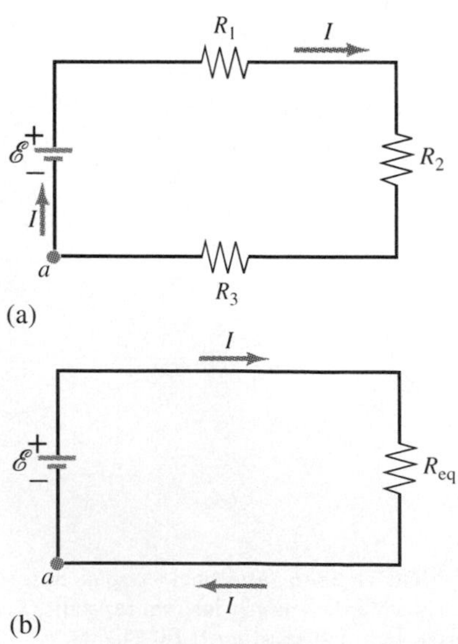

FIGURE 28–8 (a) The circuit with three resistors connected in series is equivalent to (b) the circuit with one resistor of value R_{eq}.

Resistors Connected in Series

Let us now apply Kirchhoff's loop rule to the circuit shown in Fig. 28–8a, consisting of an emf and three resistors. (The internal resistance of $\mathscr{E}$ will be ignored.) We start at point a and move toward the battery in the (clockwise) direction of

† This rule, named after the nineteenth-century physicist Gustav Kirchhoff, is also called Kirchhoff's second law.

the assumed current. (We could just as well follow the circuit in the opposite direction.) Kirchhoff's loop rule, Eq. (28–9), gives

$$\sum \Delta V = \mathscr{E} - IR_1 - IR_2 - IR_3 = 0.$$

The solution for the current is

$$I = \frac{\mathscr{E}}{R_1 + R_2 + R_3}. \tag{28–10}$$

We compare the circuit of Fig. 28–8a with a second circuit, shown in Fig. 28–8b. The three resistors in Fig. 28–8a are said to be connected *in series.* We want to find a single equivalent resistor R_{eq} that, when it replaces the series combination, allows the same current to flow in the circuit. (The operation should remind you of the combination of capacitors connected in series and replaced by a single equivalent capacitor, from Chapter 26). The current in Fig. 28–8b is simply

$$I = \frac{\mathscr{E}}{R_{eq}}.$$

Comparison of the two preceding equations shows that the resistance equivalent to a set of n individual resistors connected *in series* is found by addition:

The resistance equivalent to resistors connected in series

for n resistors connected in series: $$R_{eq} = R_1 + R_2 + R_3 + \cdots R_n. \tag{28–11}$$

The analysis of single-loop circuits can be very simple. For example, consider again the circuit of Fig. 28–6a, which includes a resistor R, an emf $\mathscr{E}$, and an internal resistance r. The two resistors are connected in series, and the equivalent series resistance is $r + R$. The current is the emf divided by the total resistance, $I = \mathscr{E}/(r + R)$, which is the result of Eq. (28–6).

EXAMPLE 28–2 Consider the circuit shown in Fig. 28–8a, where the internal resistance of the emf is small enough to ignore. Find the current in the loop if $\mathscr{E} = 12$ V, $R_1 = 2.0\ \Omega$, and $R_2 = R_3 = 6.0\ \Omega$.

SOLUTION: Starting at point a, we use Kirchhoff's loop rule and traverse the closed loop in a clockwise manner. We have

$$\sum \Delta V = +\mathscr{E} - IR_1 - IR_2 - IR_3 = 0,$$

where I is the (unknown) current. Solving this equation for the current gives

$$I = \frac{\mathscr{E}}{R_1 + R_2 + R_3} = \frac{12\text{ V}}{2.0\ \Omega + 6.0\ \Omega + 6.0\ \Omega} = \frac{12\text{ V}}{14\ \Omega} = 0.86\text{ A}.$$

We could equally well realize that the three resistors are connected in series, and, using Eq. (28–11), add their resistances to obtain the equivalent resistance, $2.0\ \Omega + 6.0\ \Omega + 6.0\ \Omega = 14\ \Omega$. The result for the current is then $(12\text{ V})/(14\ \Omega) = 0.86$ A as before.

EXAMPLE 28–3 Find the current for the two-battery circuit shown in Fig. 28–9. The values of the emfs and resistances are $\mathscr{E}_1 = 6$ V, $r_1 = 0.4\ \Omega$, $R_3 = 3\ \Omega$, $r_2 = 0.1\ \Omega$, $\mathscr{E}_2 = 12$ V, and $R_4 = 10\ \Omega$.

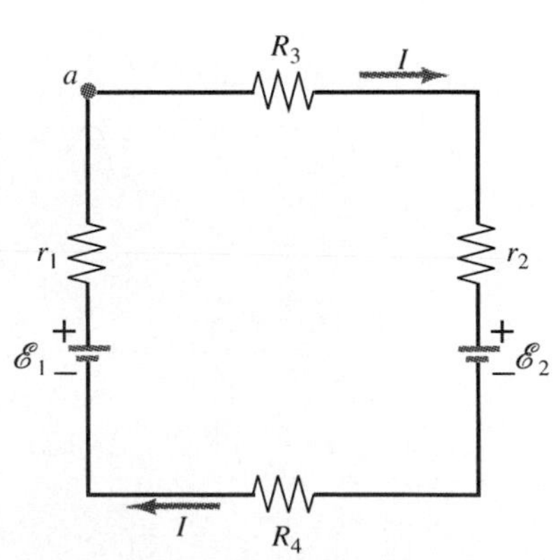

FIGURE 28–9 Example 28–3.

SOLUTION: We assume that the current flows in the direction shown, and we proceed counterclockwise around the circuit from point a (just to show that the results that follow from Kirchhoff's loop rule depend neither on the direction followed nor on the assumed direction of the current).[†] The loop rule gives

$$\sum \Delta V = +Ir_1 - \mathscr{E}_1 + IR_4 + \mathscr{E}_2 + Ir_2 + IR_3 = 0;$$

$$I = \frac{\mathscr{E}_1 - \mathscr{E}_2}{r_1 + R_4 + r_2 + R_3} = \frac{6\ \text{V} - 12\ \text{V}}{0.4\ \Omega + 10\ \Omega + 0.1\ \Omega + 3\ \Omega} = \frac{-6\ V}{13.5\ \Omega} \simeq -0.4\ \text{A}.$$

The minus sign indicates that we assumed the wrong direction for the current. The actual direction taken by the current is opposite to that in Fig. 28–9. Because the emf $\mathscr{E}_2$ is larger than the emf $\mathscr{E}_1$, we could have realized from physical reasoning that the current will run counterclockwise. It is not normal for two emfs to oppose each other in single-loop circuits, except when one battery is charging another.

28–3 MULTI-LOOP CIRCUITS AND KIRCHHOFF'S JUNCTION RULE

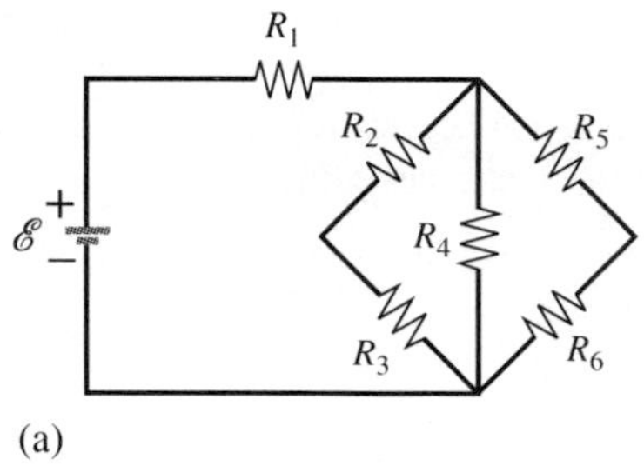

(a)

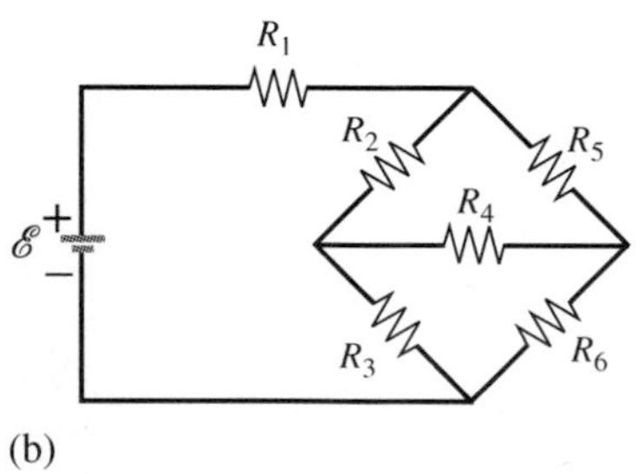

(b)

FIGURE 28–10 Two multi-loop circuits.

Not all circuits are as simple as those we have discussed so far. Consider the two circuits shown in Fig. 28–10. These are examples of **multi-loop circuits**, circuits that have **junctions**, or *nodes*, which are places in the circuit where at least three lines (or wires) meet. Currents entering a junction divide their flow into the various branches of the junction. The circuit in Fig. 28–10a has two four-line junctions, and the circuit in Fig. 28–10b has four three-line junctions.

Although Kirchhoff's loop rule alone is enough to tell us all we need to solve a single-loop circuit, it is insufficient for multi-loop circuits. In order to formulate the equations that will enable us to solve for the behavior of multi-loop circuits, we need to count the number of loops. How can this be done? The circuits in Fig. 28–10 are both three-loop circuits. For planar circuits (circuits that can be drawn in a plane with no two wires crossing each other), the number of loops is just the number of enclosed areas through which you could poke a pencil.

In steady-state operation, the current moving along a wire in an electric circuit is constant. If it were not, charge would build up at some point and change the electric field—in disagreement with our assumption of a steady state. This *conservation of current* also holds at a junction in a circuit, where three or more wires come together. We know that charge is conserved, so that at any given time the rate at which charge enters a junction is equal to the rate at which charge leaves the junction. **Kirchhoff's junction rule** (also known as Kirchhoff's first law) states that

Kirchhoff's junction rule

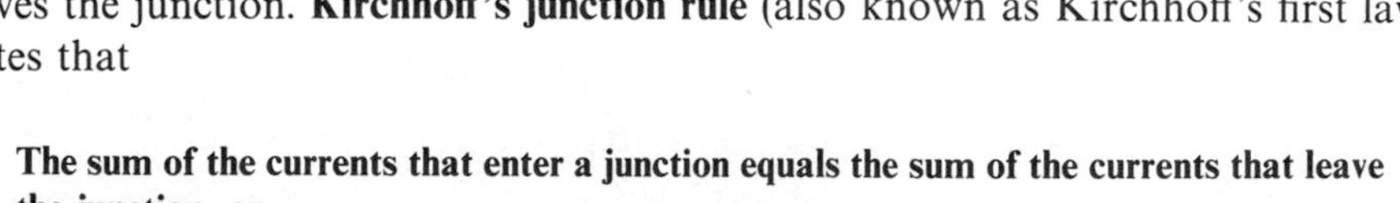

The sum of the currents that enter a junction equals the sum of the currents that leave the junction, or

$$\sum I_{\text{in}} = \sum I_{\text{out}}. \tag{28–12}$$

Equation (28–12) is true for each junction in a multi-loop circuit. Several times we have used water flow as a useful mechanical analogue to current flow. The conservation of current at junctions is analogous to the conservation of water flux

[†] When we assume a direction for the current, we are doing no more than enabling ourselves to set up the equations that will ultimately determine that current. The solutions of those equations determine the sign of the current—or in other words, its direction. What happens is no different than finding that the solution of an algebraic equation for an unknown quantity x gives a negative rather than positive value for x.

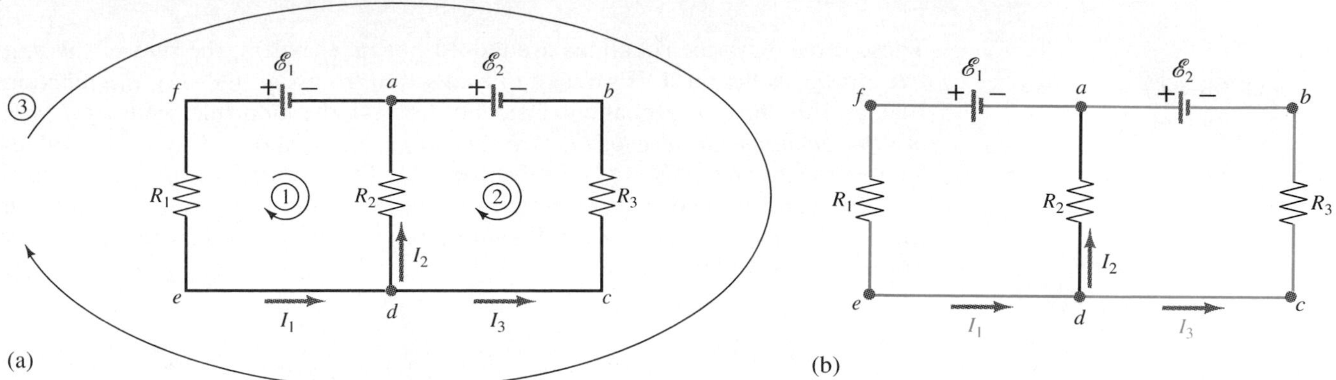

FIGURE 28–11 (a) A multi-loop circuit with two loops and two junctions. (b) The three legs of the circuit are shown in different colors.

at the junction of two rivers: The total water flow into the meeting point must equal the total water flow out.

Solving for the Behavior of Multi-Loop Circuits

Let us look at how the junction rule and loop rule help us solve for the unknown values in multi-loop circuits. Consider the circuit in Fig. 28–11a. Current flows between every junction. Thus we have currents I_1, I_2, and I_3 in the three separate *legs* (or *branches*) *afed*, *da*, and *dcba*, respectively (Fig. 28–11b). We can postulate the direction of the three currents arbitrarily, but, as we shall see, the algebraic equations for the currents will determine them for us. If these equations give a negative value for any of the currents, that current actually travels in the opposite direction than the one we postulated. The junction rule can be applied at junctions *a* and *d* to obtain

$$\text{for junction } a: \quad I_2 + I_3 = I_1; \tag{28–13a}$$

$$\text{for junction } d: \quad I_1 = I_2 + I_3. \tag{28–13b}$$

These two equations are not independent of each other; in fact, they are identical! This sometimes happens when the junction rule is applied, but it is best to go ahead and write an equation for each junction, then check whether or not they are independent.

We now need to apply the loop rule to each of the loops in the circuit in Fig. 28–11a. According to the idea that the number of independent loops is the number of different holes in the circuit through which a pencil could pass, this circuit has two *independent* loops. But we can draw three possible loops, those indicated by the circled integers 1, 2, and 3, where 3 is around the perimeter of the circuit. Only two of these three loops can be independent. As our independent loops we can choose *any* two of the three possible loops. Before we choose, let us apply the loop rule for all three loops and verify that only two are independent. Point *a* is part of each of the three loops, so for convenience we begin at point *a* and apply the loop rule of Section 28–2 (remembering the conventions for algebraic signs for the potential differences). The internal resistances for the two emfs are neglected. We traverse each loop in the direction of the loop arrow:

$$\text{for loop 1:} \quad +I_2R_2 + I_1R_1 - \mathscr{E}_1 = 0; \tag{28–14a}$$

$$\text{for loop 2:} \quad -\mathscr{E}_2 + I_3R_3 - I_2R_2 = 0; \tag{28–14b}$$

$$\text{for loop 3:} \quad -\mathscr{E}_2 + I_3R_3 + I_1R_1 - \mathscr{E}_1 = 0. \tag{28–14c}$$

These three loop-rule equations are indeed not independent; the sum of the first two produces the third. If we take one equation from Eqs. (28–13), the junction rule, and the first two equations from Eq. (28–14), the loop rule, *we have a total of three equations to solve for the three unknowns* I_1, I_2, and I_3. If we insert values for resistances and emfs, we can solve the three linear equations for the currents.

The example involving the circuit of Fig. 28–11 illustrates features common to many circuit problems. We can summarize the important features with a set of problem-solving techniques.

PROBLEM-SOLVING TECHNIQUES

In problems associated with multi-loop circuits, we must find unknown circuit parameters (such as resistance or current) when other parameters are given. To solve these problems, the following procedure may be helpful.

1. Draw a diagram with sources of emf, resistors, capacitors, and so forth clearly labeled. List which parameters are known and which are unknown.
2. Assign a separate current for each leg of the circuit, and indicate that current on the diagram. A leg is any line of the circuit that starts and finishes on a junction but does not itself include a junction. (For example, the three possible lines that connect *a* and *d* in Fig. 28–11b are all legs.) It may not be immediately obvious which way the current flows. Any direction can be assumed for the current, and the final algebraic solution will determine the correct direction. If the solution for a current turns out to be negative, our initial guess for the current direction was wrong.
3. Apply the junction rule for the currents at each junction. If the circuit has N' junctions, then $N' - 1$ of the equations relating currents at the junctions will be independent.
4. Identify the number of loops N by counting the number of different ways that a pencil can poke through the circuit, a simple procedure for planar circuits. Indicate N loops on the diagram (for example, the loops labeled 1 and 2 on Fig. 28–11a).
5. Apply the loop rule to each of these loops.
6. Check to see that the number of linear equations from steps 3 and 5 matches the number of unknowns.
7. Solve these equations for the unknowns, whether they are currents or other parameters of the circuit. It is usually best to solve these equations *algebraically* and only later substitute numerical values. Any special cases or simple limits can easily be checked this way.

Resistors Connected in Parallel

Just as we were able to connect capacitors in parallel, we say that the three resistors in the circuit shown in Fig. 28–12a are connected *in parallel.* We want to learn whether or not a single equivalent resistor could replace these without affecting the current I supplied by the battery, as in Fig. 28–12b. From the single loop of Fig. 28–12b, we see that $R_{eq} = \mathscr{E}I$, so we can determine R_{eq} if we can find I. To do so we turn to the analysis of the three-loop, four-junction circuit of Fig. 28–12a. This figure is labeled with the currents that we will need to find the equivalent resistance. No current is labeled in the legs from point *a* to point *b* or from point *d* to point *c*. We do not need to label these currents, because application of the junc-

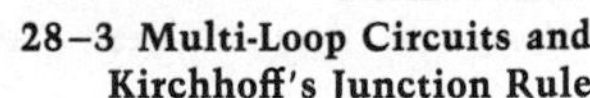

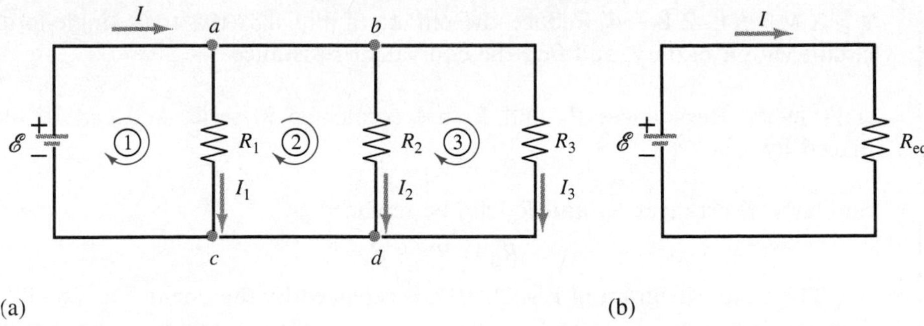

FIGURE 28–12 (a) The circuit with three resistors connected in parallel is equivalent to (b) the circuit with one resistor of value R_{eq}.

tion rule at junctions b and d indicates that the current from a to b is $I_{ab} = I_2 + I_3$, and the current from d to c is $I_{dc} = I_2 + I_3$. Therefore, $I_{ab} = I_{dc} = I_2 + I_3$. Application of the junction rule for a gives

$$I = I_1 + I_{ab} = I_1 + I_2 + I_3. \tag{28–15}$$

Exactly the same result would follow for application of the junction rule at junction c, so the junctions at a and c are not independent of each other. We have thus applied all the junction rules (step three of the problem-solving techniques).

Application of the loop rule for the three loops indicated in Fig. 28–12a gives the following:

for loop 1: $\mathscr{E} - I_1R_1 = 0;$

for loop 2: $-I_2R_2 + I_1R_1 = 0;$

for loop 3: $-I_3R_3 + I_2R_2 = 0.$

These three equations can be written as

$$\mathscr{E} = I_1R_1 = I_2R_2 = I_3R_3.$$

Each of the three resistors has the same voltage, $V = \mathscr{E}$, across them, so the currents are

$$I_1 = \frac{\mathscr{E}}{R_1}, \qquad I_2 = \frac{\mathscr{E}}{R_2}, \qquad \text{and} \quad I_3 = \frac{\mathscr{E}}{R_3}.$$

The current from Eq. (28–15) becomes

$$I = \frac{\mathscr{E}}{R_1} + \frac{\mathscr{E}}{R_2} + \frac{\mathscr{E}}{R_3} = \mathscr{E}\left(\frac{1}{R_1} + \frac{1}{R_2} + \frac{1}{R_3}\right).$$

The circuit in Fig. 28–12a can be replaced by an equivalent circuit of the same emf, $\mathscr{E}$, and current, I, and with a single resistor with value R_{eq} (Fig. 28–12b). The current for this circuit has the value

$$I = \frac{\mathscr{E}}{R_{eq}}.$$

Comparison of the two previous equations shows that the equivalent resistance of n resistors connected in parallel is

for n resistors connected in parallel: $$\frac{1}{R_{eq}} = \frac{1}{R_1} + \frac{1}{R_2} + \frac{1}{R_3} + \cdots + \frac{1}{R_n}. \tag{28–16}$$

The resistance equivalent to resistors connected in parallel

Note that the resistance of the equivalent resistor is less than that of *any* of the individual resistors: $R_{eq} < R_i$.

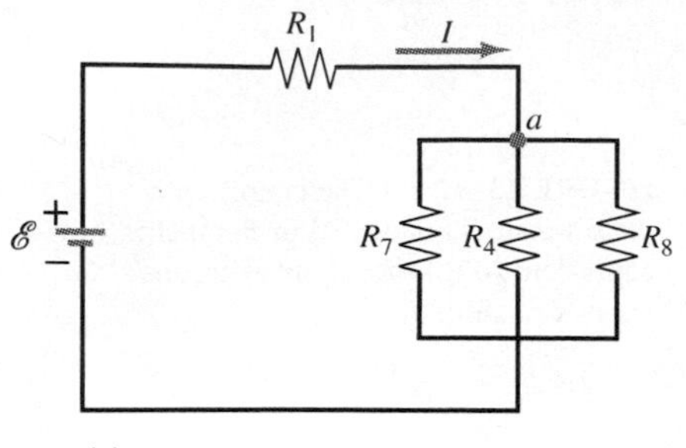

(a)

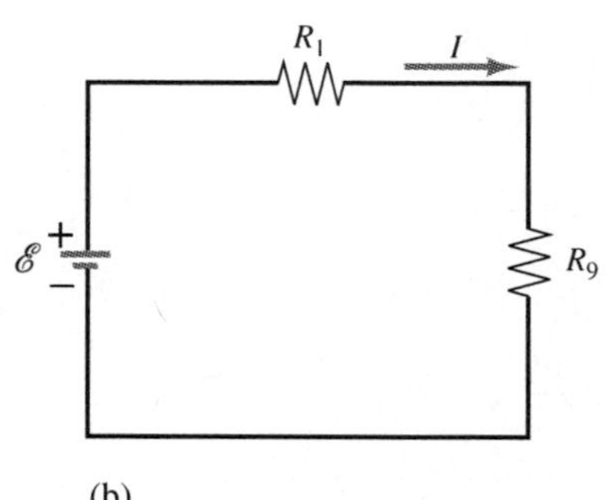

(b)

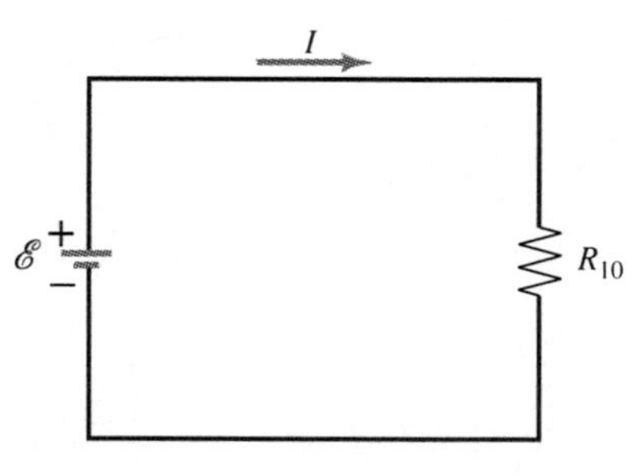

(c)

FIGURE 28–13 Example 28–4. A succession of circuits equivalent to the circuit of Fig. 28–10a.

EXAMPLE 28–4 Reduce the circuit of Fig. 28–10a to a single-loop circuit with a battery, and find the equivalent resistance.

SOLUTION: Resistances R_2 and R_3 are connected in series and can be replaced by

$$R_7 = R_2 + R_3.$$

Similarly, resistances R_5 and R_6 can be replaced by

$$R_8 = R_5 + R_6.$$

The circuit diagram in Fig. 28–10a is replaced by the one in Fig. 28–13a. The resistances R_7, R_4, and R_8 are connected in parallel and can be replaced by R_9, given as in Fig. 28–13b by

$$\frac{1}{R_9} = \frac{1}{R_7} + \frac{1}{R_4} + \frac{1}{R_8}$$

Finally, as in Fig. 28–13c, the series resistances R_1 and R_9 can be combined to give R_{10},

$$R_{10} = R_1 + R_9.$$

EXAMPLE 28–5 Find the currents for the circuit of Fig. 28–11, given that $\mathscr{E}_1 = 6.00$ V, $\mathscr{E}_2 = 12.0$ V, $R_1 = 100.0\ \Omega$, $R_2 = 10.0\ \Omega$, and $R_3 = 80.0\ \Omega$.

SOLUTION: Let us follow the problem-solving techniques. The circuit diagram is already given, and all parameters are labeled on Fig. 28–11. In this case the unknowns are I_1, I_2, and I_3. Moreover, we have already written the junction and loop equations for this circuit, Eqs. (28–13) and (28–14), respectively. According to step 3, there is a single independent junction, and we choose Eq. (28–13b). According to step 4, there are two independent loops, and we choose Eqs. (28–14a) and (28–14b) for loops 1 and 2. We have three equations to solve for the three unknowns.

To solve, we substitute $I_1 = I_2 + I_3$ from the junction equation into the loop equations:

$$I_2R_2 + (I_2 + I_3)R_1 - \mathscr{E}_1 = 0; \tag{28–17a}$$

$$-\mathscr{E}_2 + I_3R_3 - I_2R_2 = 0. \tag{28–17b}$$

Solve Eq. (28–17b) for I_3:

$$I_3 = \frac{I_2R_2 + \mathscr{E}_2}{R_3}. \tag{28–18}$$

When this result is inserted into Eq. (28–17a), we find an equation for I_2:

$$I_2(R_1 + R_2) + (I_2R_2 + \mathscr{E}_2)\frac{R_1}{R_3} - \mathscr{E}_1 = 0.$$

Solving for I_2, we get

$$I_2 = \frac{R_3\mathscr{E}_1 - R_1\mathscr{E}_2}{R_1R_2 + R_1R_3 + R_2R_3}. \tag{28–19}$$

If we substitute this result into Eq. (28–18), we get a final expression for I_3:

$$I_3 = \frac{R_2(\mathscr{E}_1 + \mathscr{E}_2) + R_1\mathscr{E}_2}{R_1R_2 + R_1R_3 + R_2R_3}. \tag{28–20}$$

Finally, I_1 is the sum of I_2 and I_3:

$$I_1 = \frac{R_2(\mathscr{E}_1 + \mathscr{E}_2) + R_3\mathscr{E}_1}{R_1R_2 + R_1R_3 + R_2R_3}. \tag{28–21}$$

Equations (28–19), (28–20), and (28–21) constitute the desired algebraic solution. Is there a limit in which we can check them? If we take R_2 to be large, we would expect that there is so much resistance in this leg that current I_2 should drop to zero. Indeed, in this limit $I_1 \to (\mathscr{E}_1 + \mathscr{E}_2)/(R_1 + R_3)$, I_3 tends to the same result, and $I_2 \to 0$. This is precisely what we would expect for the circuit if the segment containing R_2 were eliminated.

Straightforward numerical substitution of the known circuit parameters into Eqs. (28–19), (28–20), and (28–21) gives

$$I_1 = 68 \text{ mA}, I_3 = 141 \text{ mA, and } I_2 = -73 \text{ mA}.$$

Note that I_2 is negative, meaning that it flows in a direction opposite to the one we assumed in Fig. 28–11.

EXAMPLE 28–6 Assume that the emfs and resistors are known for the circuit in Fig. 28–10b. Express the linear equations that can be solved to find all currents.

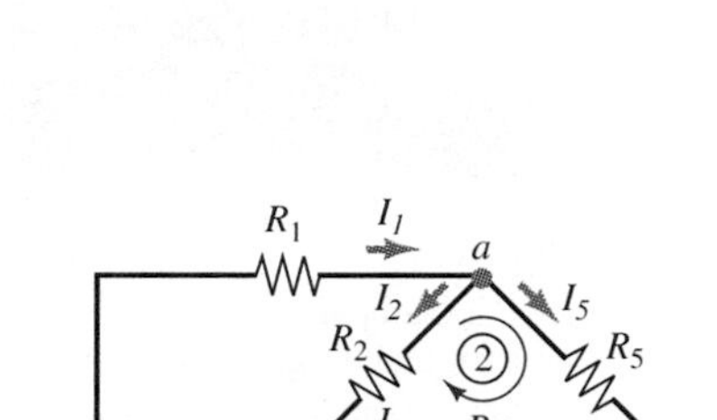

FIGURE 28–14 Example 28–6. The circuit of Fig. 28–10b is labeled with loops and junctions to be solved by circuit analysis.

SOLUTION: Because the currents are not indicated in the circuit diagram, we redraw it with the currents in each separate segment indicated as well as with the junctions labeled (Fig. 28–14). We have indicated three independent loops in the circuit. There are six currents, so we need six linear equations. There are four junctions, but, when the junction rule is applied, one will give a relation that is not independent of the others. Thus we may choose any three junctions to apply the junction rules, giving three equations. The three loop equations provide the necessary remaining three equations.

The six equations are

for junction a: $I_1 = I_2 + I_5$;

for junction b: $I_2 = I_3 + I_4$;

for junction c: $I_4 + I_5 = I_6$;

for loop 1: $\mathscr{E} - I_1R_1 - I_2R_2 - I_3R_3 = 0$;

for loop 2: $-I_5R_5 + I_4R_4 + I_2R_2 = 0$;

for loop 3: $-I_6R_6 + I_3R_3 - I_4R_4 = 0$.

With sufficient patience, we can solve these six equations for the six unknown currents.

EXAMPLE 28–7 Find the steady-state currents I_1 and I_2 in the circuit drawn in Fig. 28–15, as well as the resistance of resistor R_3 that will give a steady-state current $I_3 = 50$ mA. Find in addition the potential drop across the capacitor. The values of the known circuit elements are $\mathscr{E} = 6$ V, $R_1 = 100\ \Omega$, $R_2 = 80\ \Omega$, and $C = 2\ \mu F$.

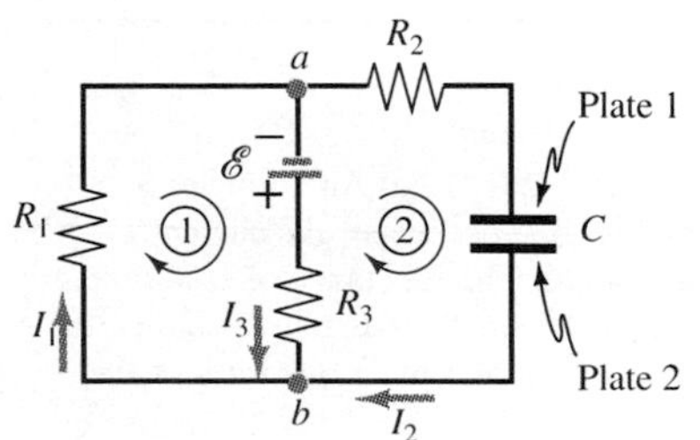

FIGURE 28–15 Example 28–7. Notice that in steady-state operation, the capacitor acts as an open switch, despite the fact that it has voltage across its plates.

SOLUTION: This problem is unusual in two respects. There is a capacitor, and we are asking for the values of two currents, I_1 and I_2, and a resistor R_3, rather than three currents. There are two loops in the circuit and two junctions, but only one junction is independent. There will thus be three equations, enough to find a solution. As we have already remarked, no steady current can pass through a capacitor, which therefore acts as an open switch. Current

FIGURE 28–16 Despite having been replaced in many applications by digital instruments, analog meters such as this multimeter remain useful.

I_2 will be zero, so in effect this is a simple single-loop circuit. We no longer need to worry about the equation for loop 2.

The junction equation is particularly simple. Either junction gives $I_1 = I_2 + I_3$, and because $I_2 = 0$, $I_1 = I_3 = 50$ mA. The resistance R_3 is found from the loop equation

$$\mathscr{E} - I_3R_3 - I_1R_1 = 0;$$

$$R_3 = \frac{\mathscr{E} - I_1R_1}{I_3} = \frac{\mathscr{E} - I_3R_1}{I_3} = \frac{6\text{ V} - (50\text{ mA})(100\ \Omega)}{50\text{ mA}} = 20\ \Omega.$$

To find the potential drop V_C across the capacitor, note that the loop equation for loop 2 is valid even if there is no current in the loop. Because this equation expresses potential changes around this loop, it will give us V_C. Starting at point a and following the loop as indicated in Fig. 28–15,

$$-I_2R_2 + V_C + I_3R_3 - \mathscr{E} = 0.$$

The first term does not contribute, because $I_2 = 0$, so

$$V_C = \mathscr{E} - I_3R_3 = 6\text{ V} - (50\text{ mA})(20\ \Omega) = 5\text{ V}.$$

Note the sign of the potential difference between the two plates. The potential increases when we go from plate 1 to plate 2, meaning that plate 2 is at a higher potential than plate 1. Positive charges have accumulated on plate 2.

28–4 MEASURING INSTRUMENTS

We have referred freely to the currents and voltages in various circuit elements, but thus far we have not said how these quantities are measured. A wide variety of instruments exists for this purpose, but we shall concentrate on just a few. Whether quantities such as current, voltage, or resistance are measured with analog (the continuous movement of the hands of a watch is an example) or digital (numeric) displays, and whatever the detailed mechanism of these measurements, certain general principles must be respected so that the measuring devices do not distort the operation of the circuit being measured. We shall emphasize these principles rather than the details of the instruments themselves.

Ammeters measure current, **voltmeters** measure voltage, and **ohmmeters** measure resistance. These devices are often combined into one instrument called a **multimeter** or **VOM** (*v*olt-*o*hm-*m*illiammeter) (Fig. 28–16). Analog versions of these three devices typically utilize a *galvanometer* (to be studied in Chapter 30), which relies on magnetic effects. A galvanometer, indicated by a circled "G," consists of a coil of wire that rotates in a magnetic field produced whenever a current passes through the wire. A needle is deflected by an amount proportional to the current that passes through the coil.

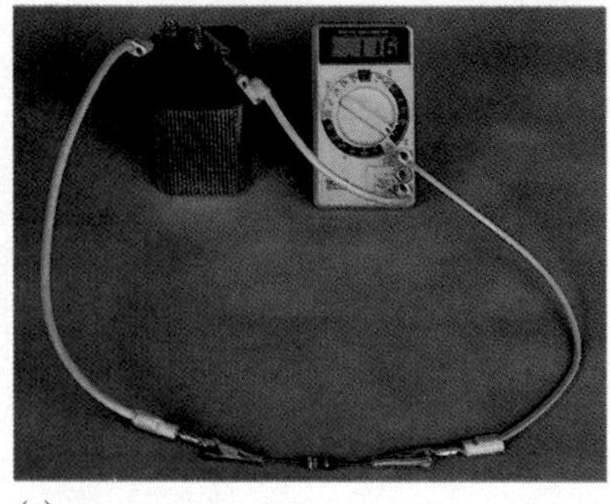

(a)

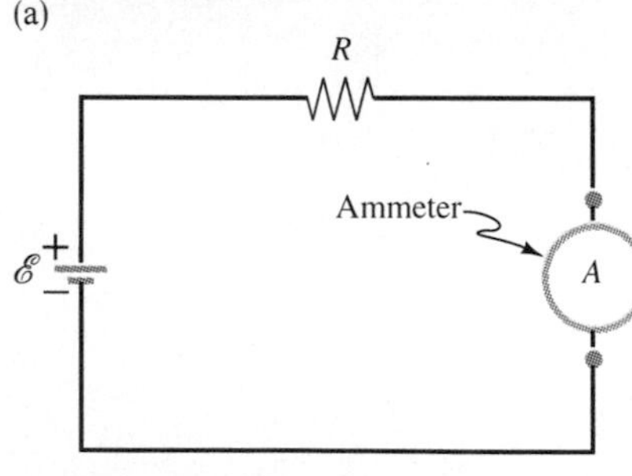

(b)

FIGURE 28–17 (a) An ammeter is placed in series where the current is to be measured. The resistance of an ammeter should be small so as not to change the current. (b) Schematic diagram of the circuit.

Ammeters

An ammeter should have a small resistance.

Ammeters measure currents in circuit wires. That a galvanometer reacts to current does not mean that it is itself a good ammeter. *A good ammeter should have a resistance that is small compared to other resistances in the circuit.* This is necessary because the ammeter is placed in series in the circuit segment through which the current to be measured passes. Only if the ammeter has a small resistance will it not affect the current. Figure 28–17a shows a circuit being probed by an ammeter,

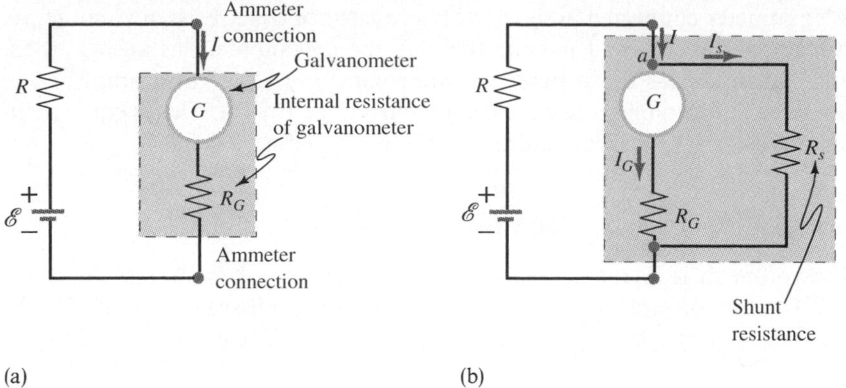

FIGURE 28–18 (a) The galvanometer of an ammeter, contained in the shaded region, may be rotated to its limit if too large a current passes through it. (b) By adding a shunt resistor, the ammeter becomes usable, because most of the current passes through the shunt resistor. That resistor also serves to decrease the total internal resistance of the ammeter within the shaded region.

indicated by a circled "A" in Fig. 28–17b. Suppose that the resistance of the ammeter is R_A and that the series resistance is R. Before the ammeter is inserted, the current is

$$I = \frac{\mathscr{E}}{R}.$$

After the ammeter is inserted, the current is

$$I = \frac{\mathscr{E}}{R + R_A}. \tag{28–22}$$

Only if $R_A \ll R$ will the current be the same with or without the ammeter attached.

In some cases the current flowing through the galvanometer of an ammeter is so large that the needle that measures the rotation of the galvanometer is pinned at a maximum deflection (Fig. 28–18a). This can be controlled with the aid of another resistor, a *shunt resistor* of resistance R_s, placed in parallel with the ammeter (Fig. 28–18b). Some of the current flowing through the ammeter is diverted through the shunt resistor. The smaller the value of R_s, the more current is diverted through it. Knowing R_s allows us to use a measurement of the reduced current flowing through the galvanometer, I_G, to find the current I (see Problem 38).

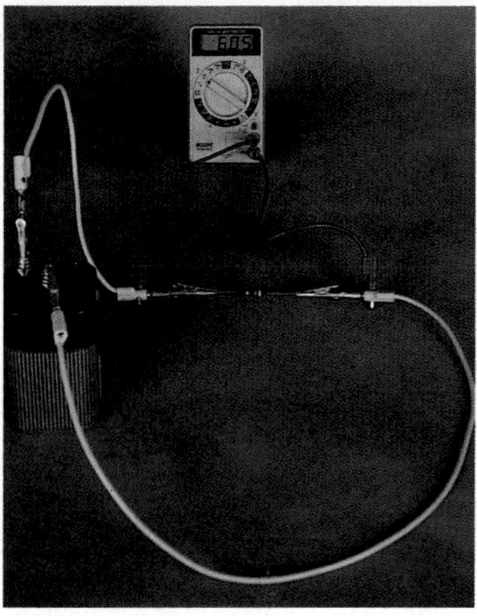

(a)

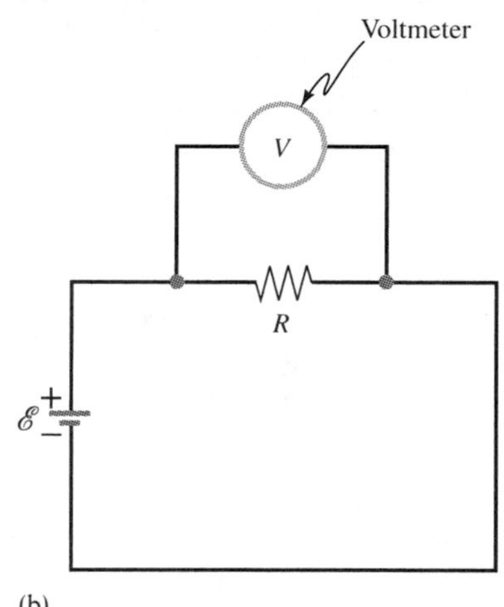

(b)

FIGURE 28–19 (a) A voltmeter is placed in parallel across the circuit element whose potential drop is to be measured. The resistance of a voltmeter should be large so as not to change the circuit. (b) Schematic diagram of the circuit.

Voltmeters

Voltmeters measure potential differences across circuit elements with which they are placed in parallel. Figure 28–19a shows a voltmeter, indicated by the circled "V" in Fig 28–19b, used to measure the potential across a resistor. The figure makes it evident why the voltmeter, which measures the potential difference *across* a circuit element, is placed in parallel with that element. *A good voltmeter should have a large resistance*, so that it does not affect the circuit. If the internal resistance of the voltmeter is R_V, the combination of voltmeter and resistance in Fig. 28–19b forms a parallel resistance circuit with equivalent resistance

A voltmeter should have a large resistance.

$$\frac{1}{R_{\text{eq}}} = \frac{1}{R_V} + \frac{1}{R}.$$

When $R_V \gg R$, then $R_{\text{eq}} \simeq R$, and none of the parameters of the original circuit are affected.

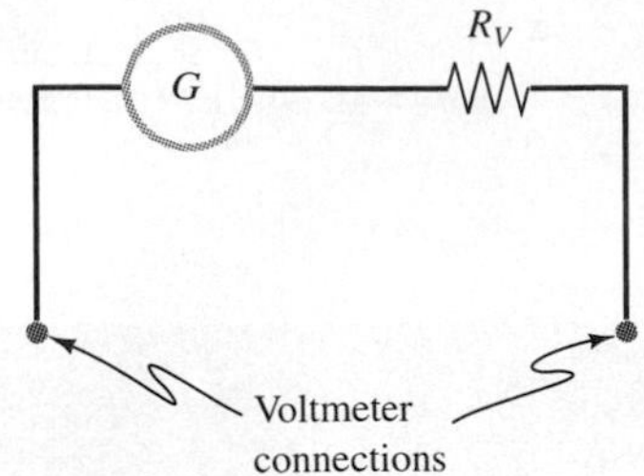

FIGURE 28–20 A galvanometer with a large series resistance can serve as a voltmeter.

A galvanometer connected in series with a resistor of large resistance can serve as a voltmeter. If the current I passing through the galvanometer is known, then the potential drop across the voltmeter is approximately IR_V. For example, take a galvanometer that can measure a maximum current of 100 μA. This is equivalent to a measurement of 10 V if the internal resistance is set at

$$R_V = \frac{10\ \text{V}}{100 \times 10^{-6}\ \text{A}} = 10^5\ \Omega.$$

An analog voltmeter is then a galvanometer with a series resistor of resistance R_V (Fig. 28–20). In our example we used a 10^5-Ω resistor for a full-scale reading of 10 V. If we had wanted to measure a full-scale voltage of 1000 V, we would have taken $R_V = 10^7\ \Omega$.

EXAMPLE 28–8 A voltmeter with an internal resistance of $10^5\ \Omega$ is used to measure the voltage across resistor R_1 in the circuit of Fig. 28–21. Compare the potential drop with and without the voltmeter for $\mathscr{E} = 6$ V, $R_1 = 10$ kΩ, and $R_2 = 5$ kΩ. This describes the error in measurement caused by the voltmeter itself.

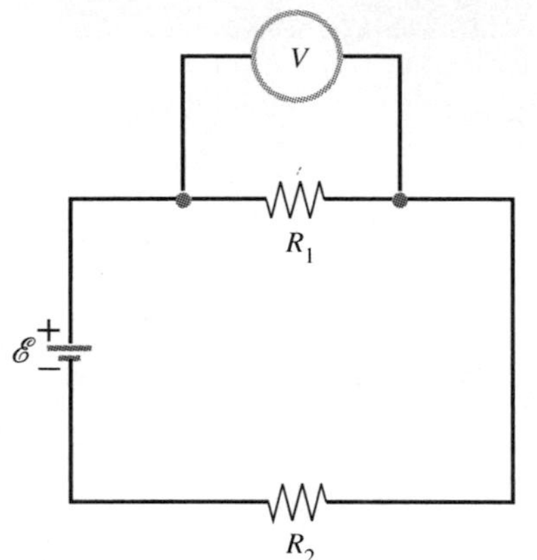

FIGURE 28–21 Example 28–8.

SOLUTION: Without the voltmeter, the current flowing through the circuit is $\mathscr{E}/(R_1 + R_2)$, or (6 V)(10 kΩ + 5 kΩ) = (6 V)/(15,000 Ω) = 0.4 mA. The voltage V_1 across R_1 is then

$$V_1 = IR_1 = (4 \times 10^{-4}\ \text{A})(10^4\ \Omega) = 4\ \text{V}.$$

When the voltmeter is connected, resistance R_1 is replaced by the equivalent resistance R_1', given by

$$\frac{1}{R_1'} = \frac{1}{R_1} + \frac{1}{R_V} = \frac{1}{10^4\ \Omega} + \frac{1}{10^5\ \Omega} = 1.1 \times 10^{-4}\ \Omega^{-1};$$

$$R_1' \simeq 9 \times 10^3\ \Omega.$$

The change in R_1 affects the current in the circuit and hence the voltage drop across it. This drop is just the voltage drop of the equivalent resistance R_1':

$$V_1' = IR_1' = \frac{\mathscr{E}R_1'}{R_1' + R_2} = \frac{(6\ \text{V})(9 \times 10^3\ \cancel{\Omega})}{(9 \times 10^3\ \cancel{\Omega}) + (5 \times 10^3\ \cancel{\Omega})} \simeq 3.9\ \text{V}.$$

The difference between 3.9 V and 4 V is 0.1 V, equivalent to a 3-percent error due to the voltmeter. The larger the voltmeter's resistance, the smaller the error so introduced.

Ohmmeters

The ohmmeter, which measures resistance, also contains a galvanometer and an internal reference resistance R_{ref} but includes in addition a battery of known emf (Fig. 28–22a). The ohmmeter is first partly calibrated by touching the ohmmeter terminals together (we *short out* the terminals) so that the ohmmeter measures *zero resistance.* The current that flows is $I = \mathscr{E}/R_{\text{ref}}$. This deflection is marked "0" on the scale of an ohmmeter. The resistor to be measured, R, is isolated and placed across the ohmmeter terminals, as in Fig. 28–22b. The current is now reduced to $\mathscr{E}/(R_{\text{ref}} + R)$. By calibrating the remainder of the scale with known resistances, the current reading is converted to a reading of R, and we have an ohmmeter. Note

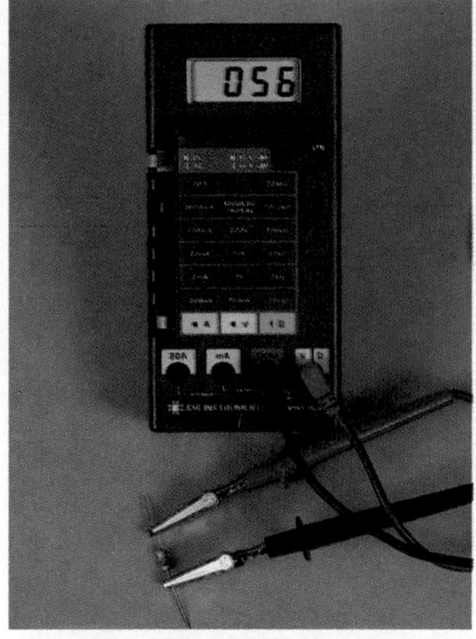

(a)

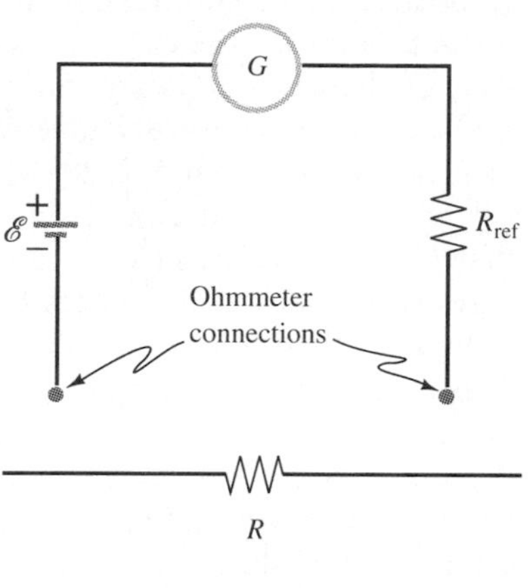

(b)

FIGURE 28–22 (a) A source of emf, a galvanometer, and a reference resistor serve as a simple ohmmeter. (b) Schematic diagram of the ohmmeter.

that if R is much greater than R_{ref}, there is little current, and the galvanometer needle is deflected very little, so that readings of large resistances are concentrated near the beginning of the deflection scale—the ohmmeter scale is not linear (Fig. 28–23). If R is much less than R_{ref}, the needle is close to its position when $R = 0\ \Omega$. This instrument is therefore not very accurate when it comes to measuring resistances either much greater or much less than R_{ref}. However, we are always free to change the internal resistance R_{ref} so that the instrument can be used to measure a wider range of R. The ohmmeter described here is not the only type possible. Several other possibilities are discussed in the end-of-chapter problems.

For most applications analog devices are being supplanted by digital devices, which are on the whole cheaper and more accurate. Digital multimeters with 0.1-percent accuracy are not uncommon. Such devices use transistorized electronics and have typical internal resistances of 100 MΩ in their voltmeter mode. However, analog devices (dials)—the gasoline gauge of your automobile, for example—remain superior for the visual recognition of trends in measured parameters.

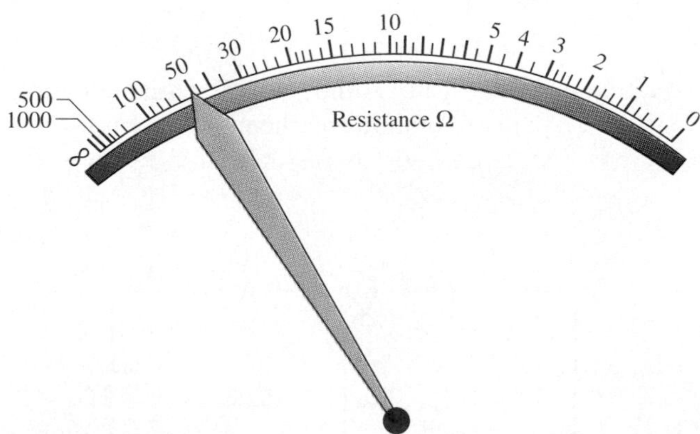

FIGURE 28–23 The response of this ohmmeter is nonlinear; it is not useful to measure resistances much smaller or much larger than that of the reference resistor.

28–5 *RC* CIRCUITS

RC circuits are circuits that contain both resistors and capacitors. They are interesting because their currents and potentials exhibit time-varying behavior. For steady-state currents, the capacitor acts as an open switch, as we have said earlier. Even for circuits containing time-*independent* sources of emf, we introduce time dependence in a circuit every time we open or close a switch. Circuits with capacitors, such as the circuit in Fig. 28–24, have interesting time-dependent effects. Such effects are useful for the control of motors, machinery, or computers.

We first observed the effect of a capacitor in an electric circuit in Example 28–7, where we concentrated on steady-state operation, with a fully charged capacitor. Now we want to examine the more interesting, transient behavior that occurs when the capacitor is being charged and discharged. Consider the circuit shown in Fig. 28–24, with an initially uncharged capacitor. When the switch is closed (to position a) at $t = 0$, current begins to flow from the positive terminal of the battery, and positive charge begins to collect on plate 1 of the capacitor while an equal amount of negative charge collects on plate 2. Current flows everywhere in the circuit, *except* through the plates of the capacitor. Immediately after the switch is closed, the current has its maximum value, but the charge that builds up on the capacitor plates opposes further charge flow, and the current decreases. When the potential across the capacitor plates equals the emf and equilibrium is reached, the current falls to zero. This occurs when the charge on the capacitor plates, Q_0, is such that $\mathscr{E} = Q_0/C$.

After equilibrium has been reached and the current has become zero, we change the switch to position b and take the battery out of the circuit. The circuit now consists only of the charged capacitor and the resistor. Current flows through the circuit from plate 1 of the capacitor to plate 2. The rate of flow is limited by the resistor. At first the current is high, but it decreases as the capacitor *discharges* through the resistor. Eventually the capacitor discharges completely, and the current again falls to zero at equilibrium.

We first apply Kirchhoff's loop rule to the circuit of Fig. 28–24 for the switch at position a, when the capacitor is being charged. The loop rule gives

$$\mathscr{E} - IR - \frac{Q}{C} = 0. \tag{28–23}$$

In this equation neither the current nor the charge on the capacitor is constant while the capacitor charges. Because $I = \mathrm{d}Q/\mathrm{d}t$, we can rewrite Eq. (28–23) as

$$\mathscr{E} - R\frac{\mathrm{d}Q}{\mathrm{d}t} - \frac{Q}{C} = 0. \tag{28–24}$$

The single variable in this equation is the charge Q. Although differential equation (28–24) is straightforward to solve, we prefer to omit the mathematical complexities and present its solution:

$$Q = C\mathscr{E}(1 - e^{-t/RC}). \tag{28–25}$$

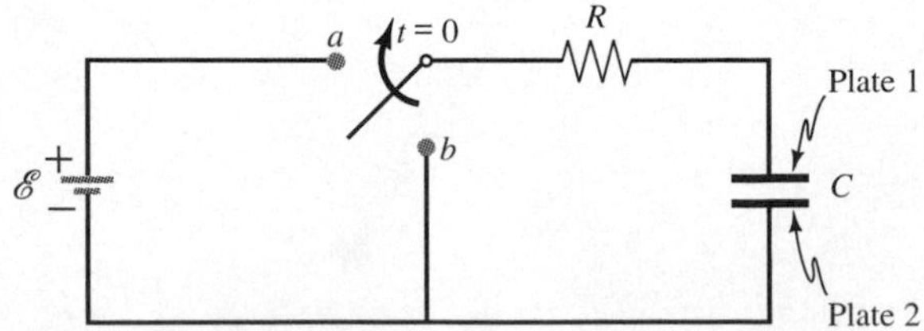

FIGURE 28–24 A circuit used to charge and discharge a capacitor through a resistor. When the switch is closed at a, the capacitor is charged by the source of emf, whereas the capacitor discharges through R when the switch is thrown to b.

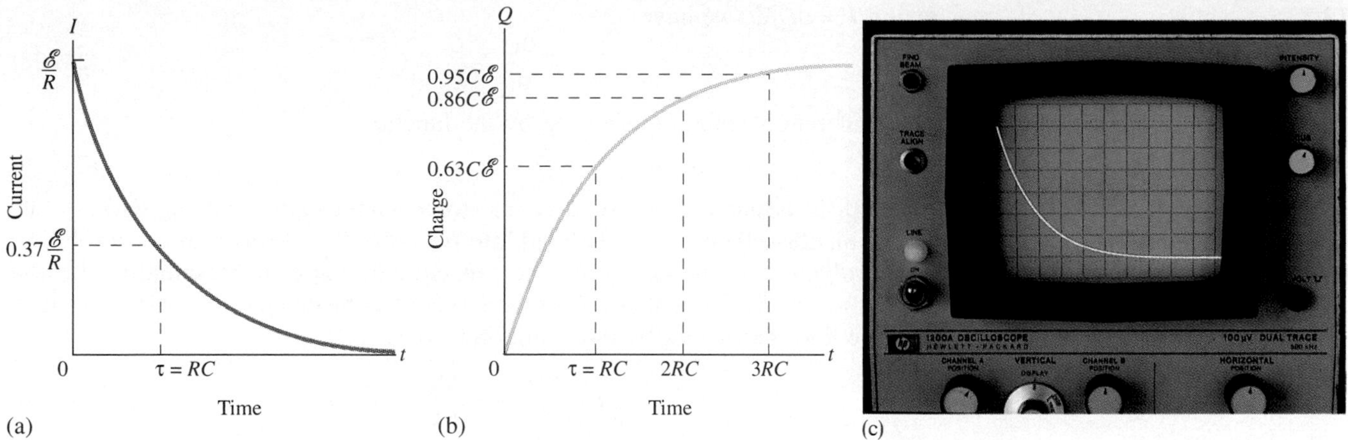

FIGURE 28–25 The time response of (a) the current I and (b) the charge Q across a capacitor as the capacitor is charged. The characteristic time response of the exponential behavior is RC. The value 0.37 in the graph of current is the factor e^{-1}; the value 0.63 in the graph of charge is the factor $(1 - e^{-1})$. (c) This oscilloscope screen shows the exponential current drop on a charging capacitor.

By differentiating Eq. (28–25) with respect to time and substituting into Eq. (28–24), we can see that it is a solution (see Problem 46). More importantly, does it agree physically with what we expect? According to Eq. (28–25) the charge on the capacitor is zero at $t = 0$ and builds smoothly to $C\mathscr{E}$ at $t = \infty$, in agreement with our earlier discussion.

We can find the current in the circuit by differentiating Eq. (28–25) with respect to time:

$$I = \frac{dQ}{dt} = \cancel{C}\mathscr{E}\left(\frac{1}{R\cancel{C}}\, e^{-t/RC}\right) = \frac{\mathscr{E}}{R}\, e^{-t/RC}. \tag{28–26}$$

The sign of the current is positive, so we chose the correct current direction (clockwise). The maximum value of the current is $\mathscr{E}/R$ at $t = 0$, and the current is zero at $t = \infty$, also in agreement with our earlier discussion. Just after the switch is closed, the emf of the battery is $\mathscr{E} = IR$, and no potential drops across the capacitor, because it is uncharged. As the capacitor charges, the current drops *exponentially* to zero.

Equations (28–25) and (28–26) show that the time dependence of both charge and current is determined by the product RC, which is called the **time constant**. It has the units of time; with R and C in SI units, RC will be in seconds. The time constant determines how fast a capacitor charges and discharges. The smaller the value of RC, the more quickly the exponentials in the equations for Q and I fall; similarly, the larger the value of RC, the more slowly the exponentials change. Figure 28–25a shows the current in the circuit, and Fig. 28–25b shows the charge on the capacitor as a function of time while the capacitor is being charged (Fig. 28–25c). After a time RC, the current has dropped to $e^{-1} \simeq 0.37$ times its original value. After this same amount of time, the capacitor is $(1 - e^{-1}) \simeq 63\%$ fully charged. It is 86% charged at time $2RC$ and 95% charged at time $3RC$.

The time scale of the time dependence in RC circuits is given by the product RC.

Return once again to the circuit of Fig. 28–24. Suppose that the switch has been in position *a* for a long time, the capacitor is fully charged, and there is no current. At time $t = 0$, we throw the switch to position *b*. Only the discharging capacitor and the resistor are now in the circuit (Fig. 28–26). The positive charge is on plate 1, and we assume as before that the current is clockwise. The loop rule now gives

$$IR + \frac{Q}{C} = 0. \tag{28–27}$$

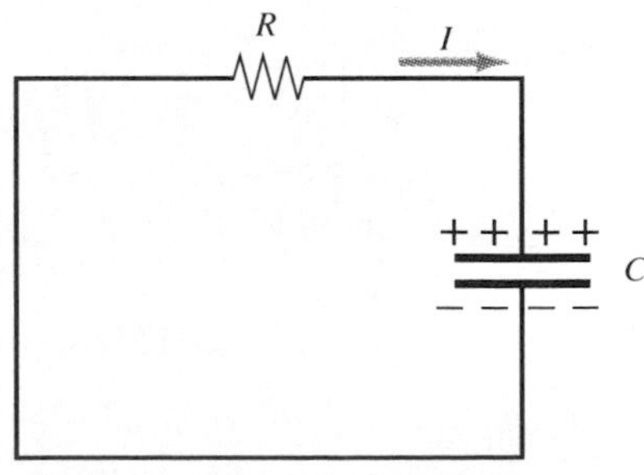

FIGURE 28–26 The circuit of Fig. 28–24 after the switch has been thrown to position *b*.

Using $I = \mathrm{d}Q/\mathrm{d}t$, we have

$$R\frac{\mathrm{d}Q}{\mathrm{d}t} + \frac{Q}{C} = 0. \tag{28–28}$$

This differential equation is solved by the function

$$Q = Q_0 e^{-t/RC}, \tag{28–29}$$

where Q_0 is the initial charge on the capacitor when the switch is changed, $Q_0 = C\mathscr{E}$. Equation (28–29) may be substituted into Eq. (28–28) to verify that it is a solution (see Problem 47). The charge on the capacitor decreases exponentially with the time constant RC, and, after a long time, there will be no charge on the capacitor.

We find the current by differentiating Eq. (28–29):

$$I = \frac{\mathrm{d}Q}{\mathrm{d}t} = -\frac{Q_0}{RC}e^{-t/RC}. \tag{28–30}$$

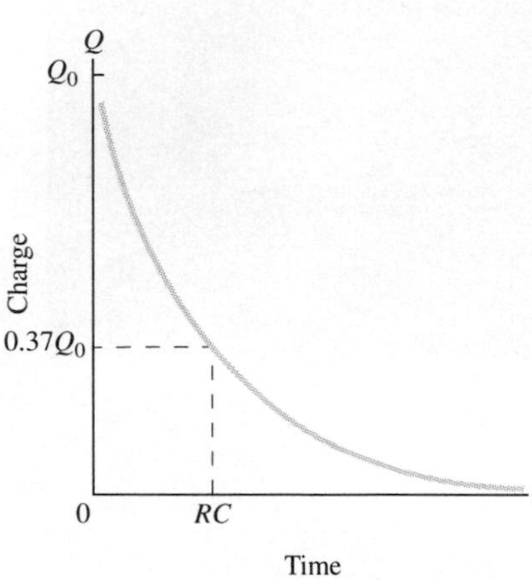

FIGURE 28–27 The capacitor charge as the capacitor of Fig. 28–26 discharges through the resistor as a function of time. The characteristic time response of the exponential behavior is again RC.

The current in this case is negative, indicating that the actual current is counterclockwise, opposite in direction to the current we assumed when we drew the diagram. It is again a maximum at $t = 0$, when the magnitude of the current is $Q_0/RC = \mathscr{E}/R$. After a long time, the current is again zero.

The behavior of the charge and current for the capacitor that discharges through a resistor is qualitatively what we expected from our earlier discussion. The magnitude of the current for this case is just as shown in Fig. 28–25a for the charging capacitor. The charge on the capacitor is plotted as a function of time in Fig. 28–27. Again, the factor 0.37 is e^{-1}.

Energy in RC Circuits

It is interesting to examine the role of energy in the case of a charging capacitor. From the definition of potential, the amount of work done by the battery emf during the charging process is $\mathscr{E}$ times the total charge processed by the battery. This charge is the final charge $C\mathscr{E}$ on the capacitor plates after a long period of time. Thus, the work done by the battery, W_{bat}, is

$$W_{\text{bat}} = \mathscr{E}(C\mathscr{E}) = C\mathscr{E}^2. \tag{28–31}$$

Where is the energy that matches this work? In part, the energy is stored in the capacitor. We know from Eq. (26–9) that the total energy stored by a capacitor is $CV^2/2$. The voltage V in this case is $\mathscr{E}$, so the energy stored by the capacitor, E_{cap}, is

$$E_{\text{cap}} = \frac{C\mathscr{E}^2}{2}. \tag{28–32}$$

Where has the other half of the work done by the battery gone? The only other circuit element is the resistor, and the other half of the work has gone into Joule heating of that resistor. From Eq. (27–30) we know that the power loss in the resistor is $P = I^2R$. We can integrate the power over time to find the energy loss in the resistor, E_{res}:

$$E_{\text{res}} = \int_0^\infty I^2R\,\mathrm{d}t = \frac{\mathscr{E}^2}{R}\int_0^\infty e^{-2t/RC}\,\mathrm{d}t$$
$$= \frac{C\mathscr{E}^2}{2}\left(-e^{-2t/RC}\right)\Big|_0^\infty = \frac{C\mathscr{E}^2}{2}. \tag{28–33}$$

The thermal energy loss in the resistor indeed accounts for the other half of the work done by the battery. This 50-percent split of energy between the resistor and the capacitor is *independent* of $\mathscr{E}$, R, and C. For the case of the discharging capacitor, all the energy stored in the capacitor dissipates as heat in the resistor.

EXAMPLE 28–9 The charging circuit shown in Fig. 28–24 (with a switch thrown to position a at $t = 0$) has the circuit elements $\mathscr{E} = 12$ V, $R = 100.0\ \Omega$, and $C = 10.0\ \mu$F. (a) Find the time constant, the final charge on the capacitor, and the work done by the battery. (b) How long does it take for the capacitor to be charged to 99.9 percent of its final charge?

SOLUTION: (a) The time constant is RC:

$$RC = (100.0\ \Omega)(10.0 \times 10^{-6}\ \text{F}) = 1.00 \times 10^{-3}\ \text{s} = 1.00\ \text{ms}.$$

The final charge on the capacitor, Q_f, according to Eq. (28–25) in the limit of large t, is $C\mathscr{E}$:

$$Q_f = C\mathscr{E} = (10.0 \times 10^{-6}\ \text{F})(12\ \text{V}) = 1.2 \times 10^{-4}\ \text{C}.$$

According to Eq. (28–31), the work done by the battery, W_{bat}, is

$$W_{\text{bat}} = C\mathscr{E}^2 = (10 \times 10^{-6}\ \text{F})(12\ \text{V})^2 = 1.4 \times 10^{-3}\ \text{J}.$$

(b) We use Eq. (28–25) to find how long it takes the capacitor to become 99.9 percent charged:

$$Q = 0.999Q_f = Q_f(1 - e^{-t/RC}).$$

We eliminate the maximum charge, Q_f, from this equation and rearrange the equation to obtain

$$e^{-t/RC} = 1 - 0.999 = 0.001.$$

Taking the natural logarithm of both sides gives

$$-\frac{t}{RC} = -6.91.$$

The time to reach 99.9 percent of its final charge is

$$t = 6.91RC = (6.91)(1.00\ \text{ms}) = 6.91\ \text{ms}.$$

SUMMARY

Sources of emf (electromotive force), $\mathscr{E}$, such as chemical batteries, are sources of electric energy. The emf is defined by the amount of work it can do to move charge:

$$\mathscr{E} \equiv \frac{dW}{dq}. \qquad (28\text{–}1)$$

Batteries cause charges to move in circuits. The simplest circuits to analyze are direct-current circuits, in which no circuit parameters change with time.

Analysis of single- or multi-loop circuits is accomplished by the use of Kirchhoff's two rules. Kirchhoff's loop rule is "circuit language" for the conservative nature of electric force. It states that the sum of the potential changes around a closed path of a circuit is zero:

$$\sum_{\text{closed path}} \Delta V = 0. \qquad (28\text{–}9)$$

We can specify the potential change across batteries, resistors, and capacitors. Any source of emf has an internal resistance r that may be large enough to require consideration.

Kirchhoff's junction rule is "circuit language" for the conservation of electric current. It states that the sum of currents that enter a junction equals the sum of the currents that leave the junction:

$$\sum I_{in} = \sum I_{out}. \tag{28–12}$$

A total of n resistors connected in series can be combined into an equivalent resistance by the relation

$$R_{eq} = R_1 + R_2 + R_3 + \cdots + R_n. \tag{28–11}$$

A total of n resistors connected in parallel can also be combined into an equivalent resistance by the relation

$$\frac{1}{R_{eq}} = \frac{1}{R_1} + \frac{1}{R_2} + \frac{1}{R_3} + \cdots + \frac{1}{R_n}. \tag{28–16}$$

Ammeters, voltmeters, and ohmmeters measure current, voltage, and resistance, respectively. Ammeters must have a small internal resistance so that they do not affect the circuit leg in which the current is measured. Conversely, voltmeters need a large internal resistance, because they are used in parallel with the circuit element being measured, and they too should not affect the circuit being measured.

A circuit exhibits time-varying behavior when a capacitor is being charged with a source of emf or when the capacitor is being discharged. For a simple circuit with an emf $\mathscr{E}$, a resistor R, and a capacitor C, an initially uncharged capacitor has charge

$$Q = C\mathscr{E}(1 - e^{-t/RC}) \tag{28–25}$$

and current

$$I = \frac{\mathscr{E}}{R} e^{-t/RC}. \tag{28–26}$$

When the source of emf is disconnected from the circuit and the capacitor is allowed to discharge through the resistor, the charge decreases exponentially:

$$Q = Q_0 e^{-t/RC}. \tag{28–29}$$

The time constant RC determines the time dependence of exponential increase or decrease of the charge and the current during the charging and discharging of a capacitor.

QUESTIONS

1. Why is it dangerous to be in a full bathtub when an electrical appliance is standing on the edge of the tub?

2. An inexpensive voltmeter measures the voltage of a flashlight battery as 0.9 V, whereas a high-quality digital voltmeter measures 1.5 V. What might cause this difference?

3. Show that it is irrelevant whether, in drawing a circuit diagram, the internal resistance of a source of emf is placed before or after the emf itself.

4. Why does the potential drop linearly along the length of a resistor that is a uniform cylinder?

5. In Chapter 27 we showed that the resistance of a piece of material is proportional to its length and inversely proportional to its cross-sectional area. Explain how this result is equivalent to the rules for series and parallel resistances.

6. What sense does it make to draw a circuit diagram with resistanceless wires when real wires always have some resistance?

7. By taking a special combination of batteries of constant emf, resistors, and capacitors, is it possible to construct a circuit in which the emf around a closed loop is not zero?

8. Is it possible to break up any combination of resistors into a sequence of parallel and series resistances? If not, give an example of a combination that cannot be so decomposed.

9. A flash unit on a camera discharges by means of an RC circuit. Find out (say, from a local camera store) what the R and C values are, and obtain the RC time.

10. It would appear superficially that, in Section 28–5, if we had chosen the current in the wrong direction, we would have ended up with the equation $IR - (Q/C) = 0$ rather than Eq. (28–7) to describe the current in an RC circuit.

This would be disastrous, because the solution of our new equation would be a rising exponential, $e^{+t/RC}$, rather than a falling exponential, and this expression would grow without limit. What is wrong with this reasoning?

11. When you reverse the polarities of all batteries in a circuit, the magnitudes of all currents stay the same. Why?

12. Is an RC circuit that charges a capacitor used by a flashbulb likely to be the same as the circuit through which the bulb fires?

13. Suppose that you connect the terminals of two batteries of different emfs + to + and − to −. What do you expect to happen?

14. It might appear that the only effect of the internal resistance of a battery in a circuit is to change the battery's emf from $\mathcal{E}$ to $\mathcal{E}' = \mathcal{E} - Ir$, where $\mathcal{E}'$ acts as though there were no internal resistance. Could this be true? If not, why not?

15. In Example 28–4 we showed that the circuit of Fig. 28–10a can be reduced to a single-loop circuit by using an equivalent resistance. Is this true of the circuit of Fig. 28–10b?

PROBLEMS

28–1 EMF

1. (I) A 12-V car battery is rated at 110 A, meaning that it will send 110 A through a wire connected to its terminals. What is its internal resistance?

2. (I) The Magellan spacecraft that studied Venus in 1990 used two solar panels capable of producing 1200 W. If the solar array was capable of producing a total of 40 A, what was the emf of the device?

3. (I) Nickel–cadmium batteries used in space flight are rated at 30 A·h (they can put out 30 A for 1 hr) and 30 V. How much energy do the batteries contain?

4. (II) A flashlight battery with an internal resistance of 0.10 Ω produces a 100-mA current through a 15-Ω resistor. What is the emf of the battery? What is the terminal voltage of the battery in this usage?

5. (II) A certain automobile battery has an emf of 12 V. When it produces a current of 100 A, the terminal voltage reads 9.0 V. Calculate the internal resistance of the battery. What is the power dissipated in the battery when it produces this current?

6. (II) The internal resistance of a battery whose emf is 3.0 V varies with the current according to the equation $r = (\alpha + \beta I)$, where $\alpha = 0.10\ \Omega$ and $\beta = 0.031\ \Omega/\text{A}$. Find the terminal voltages and the power dissipated in the battery when $I = 2.0$ A and $I = 5.0$ A.

28–2 Single-Loop Circuits and Kirchhoff's Loop Rule

7. (I) For what value of R_2 in the circuit of Fig. 28–28 is the voltage across the points a and b zero? For what value is the current in the circuit zero?

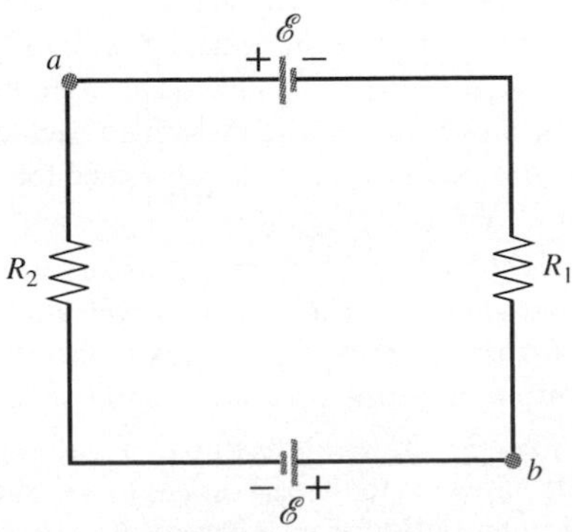

FIGURE 28–28 Problem 7.

8. (I) Six resistors of 150 Ω each are connected in series. If the potential difference between the ends of this set of resistors is 600 V, what current flows through the resistors? What is the power expended in the circuit?

9. (I) Two 60-Ω resistors are placed in series across two terminals whose potential difference is 120 V. What is the total power dissipated?

10. (II) A portion of a larger circuit is shown in Fig. 28–29. The potential drop between points b and a, known as V_{ba}, is $V_{ba} = 2$ V; $V_{cb} = 3.5$ V, $V_{cd} = 2$ V, and $V_{df} = -0.5$ V. Find the potential differences V_{gf}, V_{ag}, and V_{ca}.

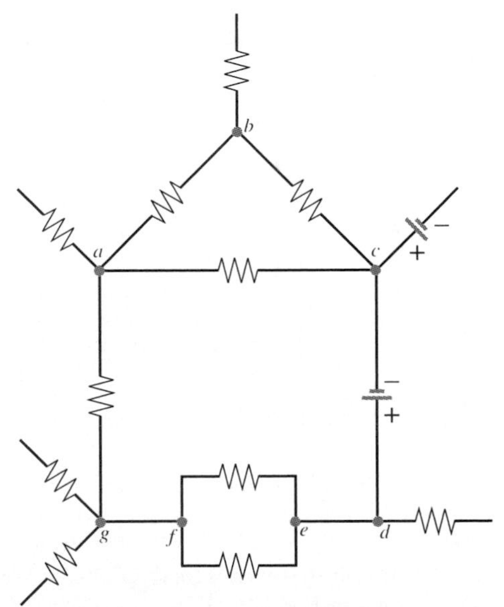

FIGURE 28–29 Problem 10.

11. (II) A generator (a "battery" that uses mechanical rather than chemical energy) of emf 220 V and internal resistance 1.2 Ω is used to charge a series of 20 batteries, each with emf 6 V and internal resistance 0.05 Ω. (a) What is the terminal voltage of the generator? (b) the terminal voltage of the bank of batteries? (c) What series resistance must be included to allow a charging current of 40 A? (d) What is the power dissipated in all the resistors?

12. (II) A flashlight consists of two 1.5-V batteries connected in series to a bulb with resistance 10 Ω. (a) What is the power delivered to the bulb? (b) Batteries run down when they in effect contribute an additional (internal) resistance to the circuit. How large is the additional resistance if the power delivered to the bulb has decreased by one-third of its initial value?

28–3 *Multi-Loop Circuits and Kirchhoff's Junction Rule*

13. (I) In Fig. 28–30 the currents $I_1 = 2$ A, $I_2 = 0.5$ A, $I_3 = -3$ A, $I_4 = -0.5I_6$, and $I_5 = -I_6$. Find the unknown currents I_4, I_5, and I_6.

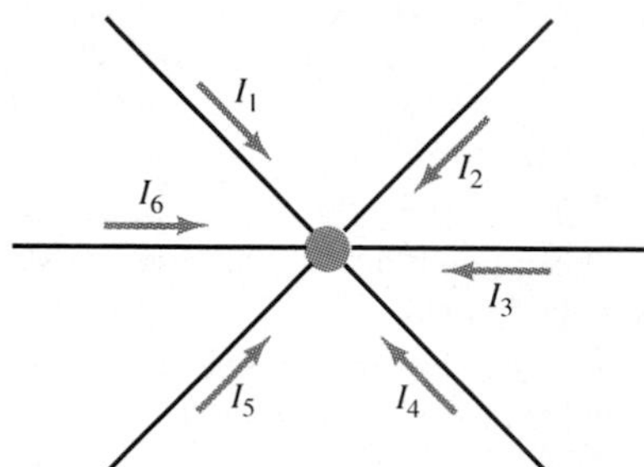

FIGURE 28–30 Problem 13.

14. (I) Find the equivalent resistance of the circuit shown in Fig. 28–31.

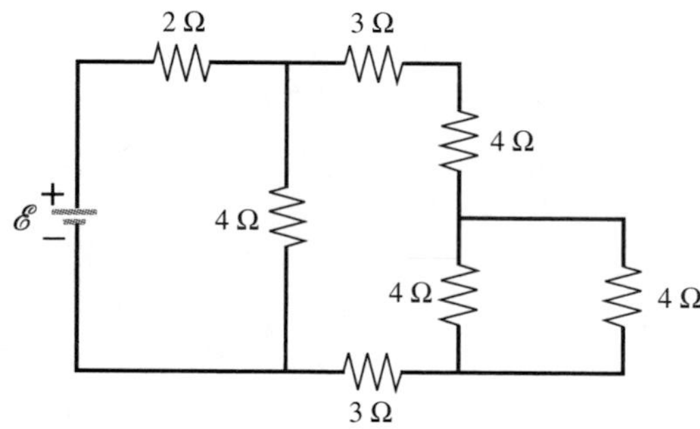

FIGURE 28–31 Problem 14.

15. (II) Find the current that passes through the 4-Ω resistor in the circuit shown in Fig. 28–32.

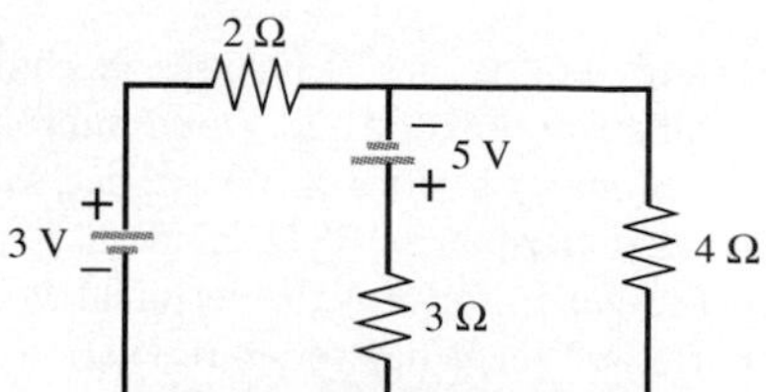

FIGURE 28–32 Problem 15.

16. (II) Find the current that passes through each of the resistors in the circuit shown in Fig. 28–33.

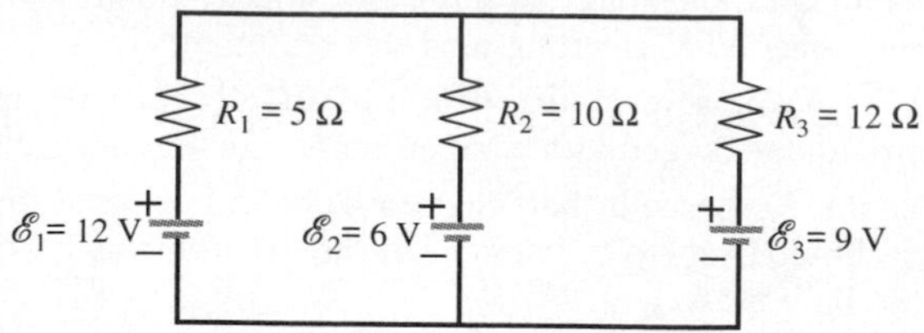

FIGURE 28–33 Problem 16.

17. (II) Can the resistors of the circuit in Fig. 28–34 be reduced to a single equivalent circuit by application of the rules for circuits with connections in parallel and in series? Solve for the currents through the three resistors.

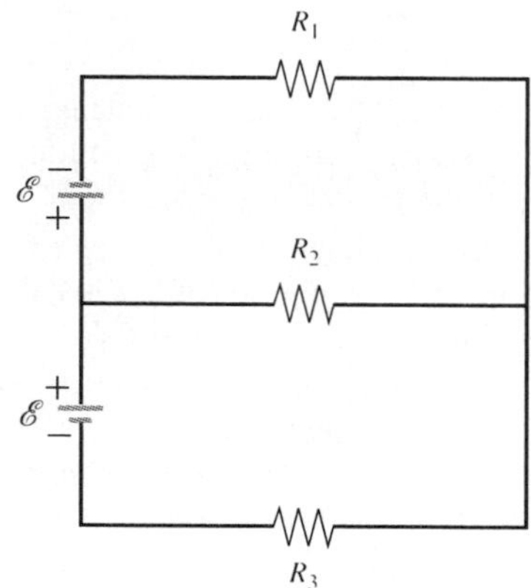

FIGURE 28–34 Problem 17.

18. (II) Three resistors connected in parallel have resistances of 20 Ω, 35 Ω, and 67 Ω, respectively. The total current passing through the set is 10 A. What is the potential difference across the set, and what are the currents in each of the resistors?

19. (II) The voltage at an electrical outlet is a constant 240 V. You have ten identical light bulbs whose maximum power consumption is 60 W. (a) What is the resistance in each bulb if the power consumption is 600 W when the bulbs are connected in series? (b) If the ten bulbs are connected in parallel, an additional resistor is needed so that the bulbs do not burn out. The resistor may be connected either in parallel or in series with the total set of light bulbs. What would the value of the resistance have to be in each case? What is the power loss in the resistor in each case? (These bulbs are unusual in that they are designed for a maximum voltage of 24 V.)

20. (II) N identical batteries, with emf $\mathscr{E}$ and internal resistance r, are connected in parallel across a resistance R. Obtain the value for the current, and compare its value with that obtained if the batteries are connected in series.

21. (II) Figure 28–35 shows an example of a *voltage divider*, a device that allows a reduced voltage to be obtained. Calculate the potential difference across the line CD in terms of the potential difference across the line AB.

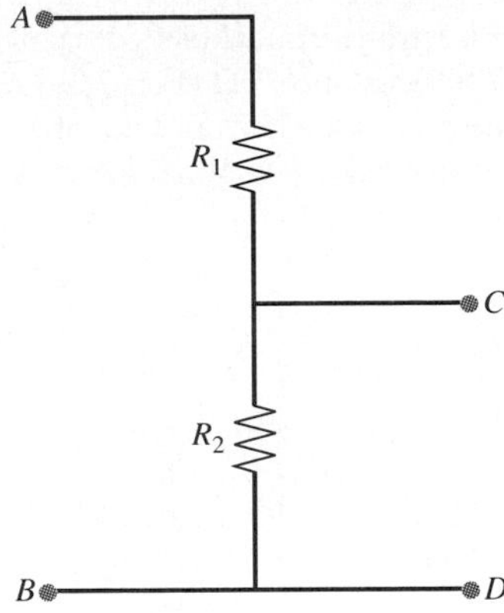

FIGURE 28–35 Problem 21.

22. (II) The circuit shown in Fig. 28–36 is an example of a loaded voltage-divider circuit (see Problem 21). By varying the values of R_1 and R_2, different values of V_L can be obtained; R_L represents the load. Let $R_1 = R_2 = 10\ \text{k}\Omega$ and $\mathscr{E} = 6$ V. For load resistances of 50 kΩ, 500 kΩ, and 5 MΩ, how much different is V_L than 3 V?

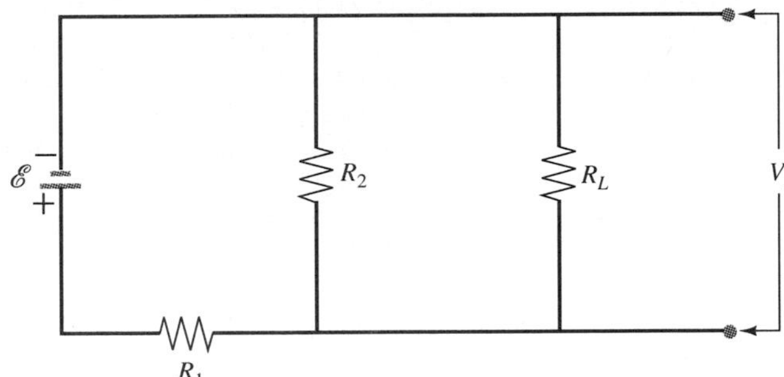

FIGURE 28–36 Problem 22.

23. (II) Consider the circuit shown in Fig. 28–12a. If $\mathscr{E} = 1.5$ V, and if all three resistors have identical resistances, find the resistance that will ensure that the current I_3 will be 10 mA.

24. (II) How many independent junctions are there in the circuit shown in Fig. 28–37? To verify your answer, solve for all currents.

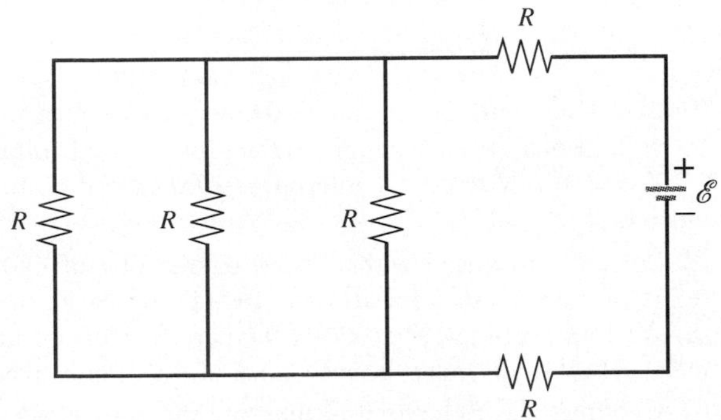

FIGURE 28–37 Problem 24.

25. (II) Replace the network of resistors in Fig. 28–38 by a single equivalent resistor. Is the combination of resistors one that can be reduced to a single resistor by successive application of the rules for parallel and series resistors?

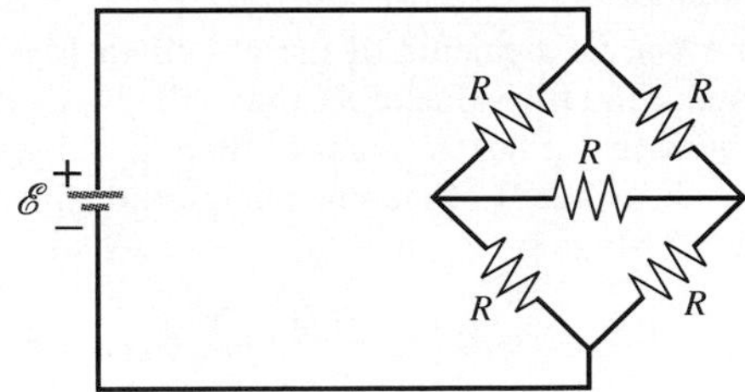

FIGURE 28–38 Problem 25.

26. (II) Two batteries with emf $\mathscr{E}_1 = 12$ V and $\mathscr{E}_2 = 24$ V, respectively, are connected to resistors with the resistances as marked in Fig. 28–39. (a) Calculate the power dissipated in the 6-Ω resistor. (b) Assume that the terminals on the 12-V battery are reversed, and repeat your calculation.

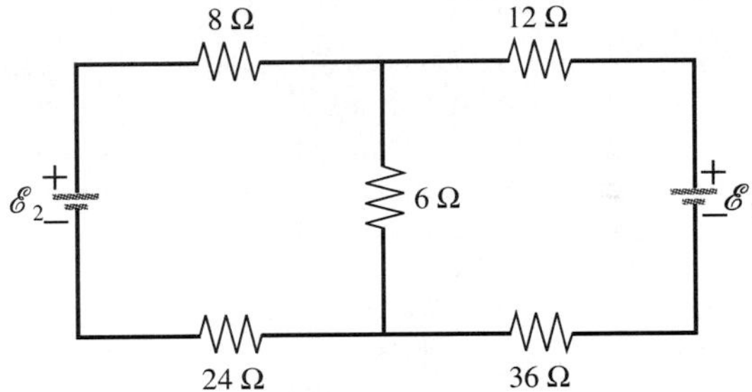

FIGURE 28–39 Problem 26.

27. (II) Consider the circuit shown in Fig. 28–40. Calculate the current and the power dissipated in the 4-Ω resistor as a function of the unknown resistance R_x.

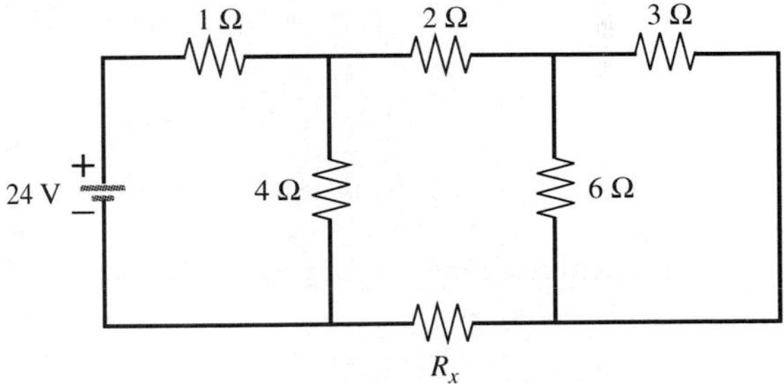

FIGURE 28–40 Problem 27.

28. (II) Consider the circuit shown in Fig. 28–41. Calculate the

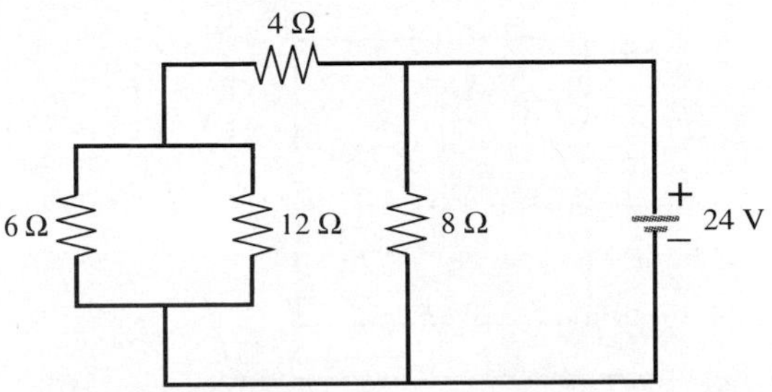

FIGURE 28–41 Problem 28.

current through the 6-Ω resistor (a) by calculating the equivalent resistance for the circuit, and (b) by using Kirchhoff's rules.

29. (II) The known elements of the circuit in Fig. 28–42 are indicated. Find the value of R_3 that will give a current I_3 of 0.1 A with the indicated sign. Is there a value of R_3 that will give a current I_3 of the same magnitude but of opposite sign? If so, what is it?

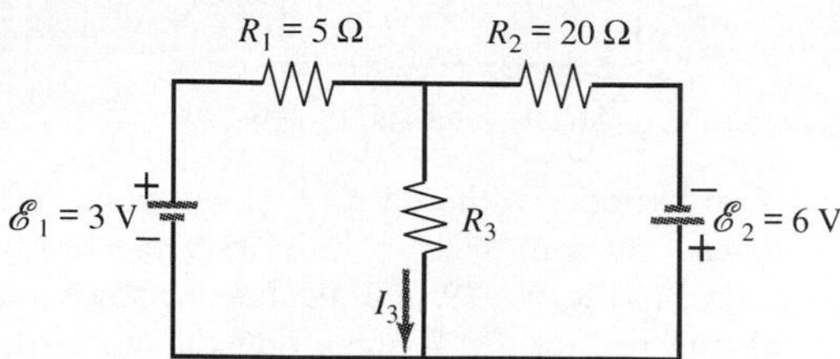

FIGURE 28–42 Problem 29.

30. (II) Points a and b are connected by the system of resistors shown in Fig. 28–43. A battery of 25 V and negligible internal resistance is connected across points a and b. (a) What is the equivalent resistance between points a and b? (b) the potential difference across the 1.8-Ω resistor? (c) the current flowing through the 5.1-Ω resistor?

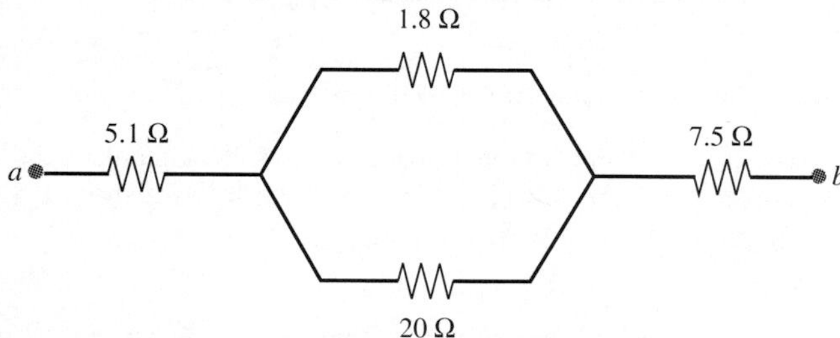

FIGURE 28–43 Problem 30.

31. (III) A cube consisting of identical wires, each of resistance R, is put across a line with voltage V (Fig. 28–44). What is the equivalent resistance of the cube? What is the current in each of the wires?

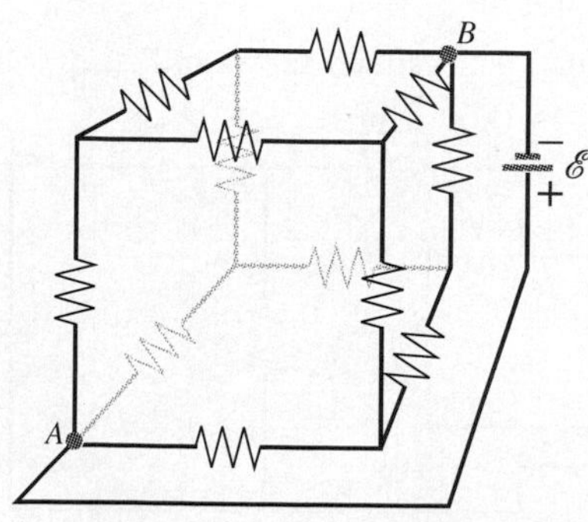

FIGURE 28–44 Problem 31.

32. (III) Consider a tetrahedron whose sides consist of identical wires, each with resistance 1 Ω (Fig. 28–45). Suppose that this arrangement is attached at two of its corners to a generator with potential 4 V. What is the power generated in each of the wires?

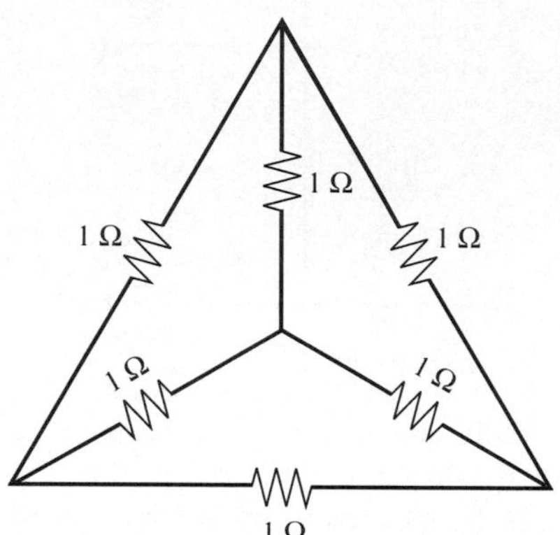

FIGURE 28–45 Problem 32.

33. (III) Figure 28–46 shows a ladder of resistors with n rungs. (a) Find the equivalent resistance between points P_1 and P_2 for $n = 1$; (b) for $n = 2$; (c) for $n = 3$; (d) for the limit $n \to \infty$. [*Hint:* Write an expression for R_n (the equivalent resistance of a ladder of n rungs) in terms of R_{n-1} (the equivalent resistance of a ladder of $n - 1$ rungs) and R, and use that equation in the limit $n \to \infty$.]

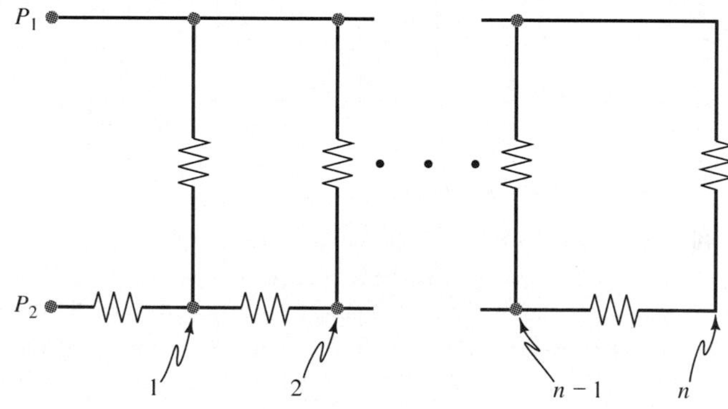

FIGURE 28–46 Problem 33.

28–4 Measuring Instruments

34. (I) A voltmeter with an internal resistance of 5 kΩ measures the voltage of a D-cell flashlight battery (of nominal voltage 1.5 V) as 1.1 V. What is the internal resistance of the worn-out battery?

35. (I) Currents produced with a 12-V source of emf and a range of resistances from 10 Ω to 1000 Ω are to be measured to an accuracy of at worst 0.1 percent with an ammeter. How small must the resistance of the ammeter be?

36. (I) A voltmeter is to be used to measure the voltage across a range of resistances from 100 Ω to 10,000 Ω. What is the minimum value of the internal resistance of the voltmeter such that a measurement can be carried out to 0.1 percent accuracy?

37. (II) An ammeter that can measure a maximum current of 50 μA has an internal resistance of 10^{-4} Ω. What series resistance will convert it to a 0-to-100-V voltmeter?

38. (II) Suppose that the current to be measured by an ammeter is so large that a galvanometer deflected by the current would be pinned at its maximum reading. This problem can be resolved by the use of a shunt resistor. Refer to Fig. 28–18, and show that with the shunt resistor (resistance R_s) present, the current I is given in terms of a reduced current I_G flowing through the galvanometer by the formula $I = I_G[1 + (R_G/R_s)]$, where R_G is the resistance of the galvanometer. Thus a reading of the reduced current I_G allows us to determine the current I.

39. (II) The output of the voltage-divider network shown in Fig. 28–47 is to be measured with two voltmeters of internal resistances 5 kΩ and 100 MΩ, respectively. What voltage will each indicate?

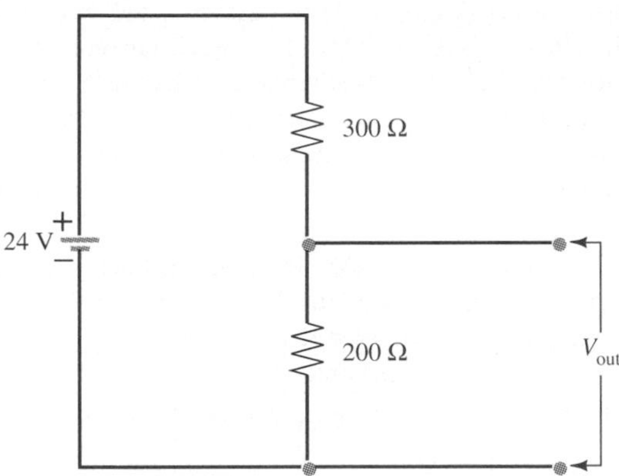

FIGURE 28–47 Problem 39.

40. (II) A *Wheatstone bridge* is a device that measures resistances. In the circuit shown in Fig. 28–48, R is an unknown resistance. The resistances R_1, R_2, and R_3 are variable. A galvanometer, G, can be used to determine when the potential difference between B and C is zero, given that the battery is connected between A and D. The variable resistances are varied until there is no current in the galvanometer when the circuit is closed at the switch, S. Obtain an expression for R in terms of R_1, R_2, and R_3.

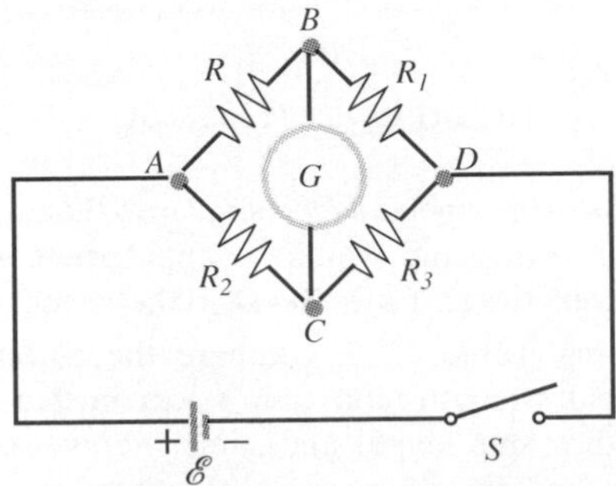

FIGURE 28–48 Problem 40.

41. (II) The circuit shown in Fig. 28–49 is used to measure the resistance R_x. Draw the circuit including internal resistances. V and I are the voltage and current measured, respectively. Find an exact expression for R_x in terms of the internal resistances of the voltmeter and ammeter. Under what conditions is $R_x = V/I$?

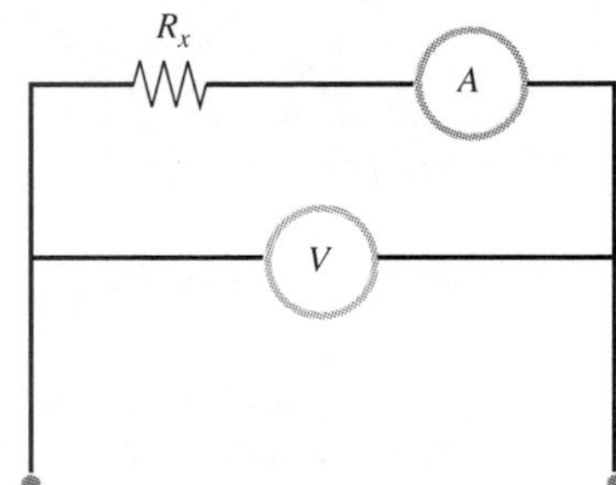

FIGURE 28–49 Problem 41.

42. (II) Repeat Problem 41 for the circuit shown in Fig. 28–50.

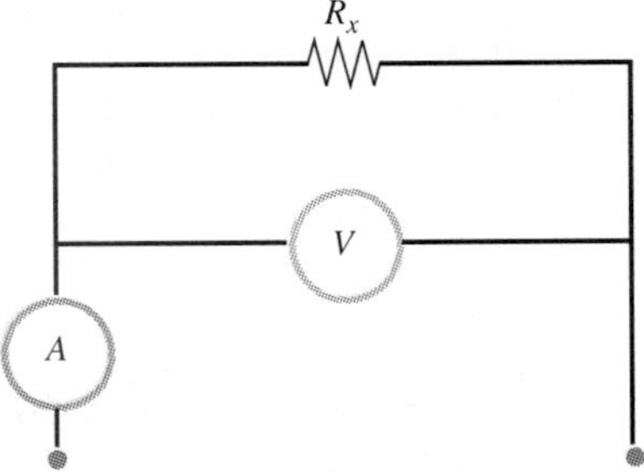

FIGURE 28–50 Problem 42.

28–5 RC Circuits

43. (I) A flashbulb in an RC circuit discharges with a time constant of 10^{-3} s. If the capacitor has a capacitance of 10 μf, what is the resistance in the RC circuit?

44. (I) A flashbulb mechanism operating through an RC circuit has a capacitor charged with a time constant of 2.0 s. If the resistance in the RC circuit is 10^5 Ω, what is the capacitance of the charging mechanism?

45. (I) Show that the product RC has units of seconds. Find the time constants for the following values of R and C: 10 MΩ, 2 μF; 3 Ω, 2 μF; 1000 Ω, 3 pF.

46. (II) Show by direct substitution that Eq. (28–25) is a solution for the differential equation (28–24).

47. (II) Show by direct substitution that Eq. (28–29) is a solution for the differential equation (28–28).

48. (II) A resistor of resistance 4×10^4 Ω and a capacitor of capacitance 80 μF are connected in series to a 6-V battery.

Calculate (a) the time constant and (b) the current at a time when the charge on the capacitor has acquired 70 percent of its maximum value.

49. (II) Calculate the current in the battery as a function of time for the circuit shown in Fig. 28–51 if the switch S is closed at time $t = 0$.

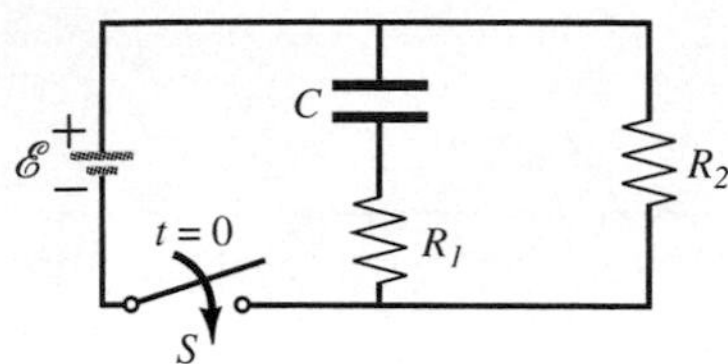

FIGURE 28–51 Problem 49.

50. (II) The circuit of Fig. 28–24 has $\mathscr{E} = 12$ V, $R = 20$ kΩ, and $C = 100\ \mu$F. What are the voltage across the resistor and the charge on the capacitor 0.5 s after the switch is closed to position a?

51. (II) You have two capacitors of capacitance 20 μF and three resistors, one of resistance 4 Ω and the remaining two of resistance 2 Ω. Find the connection between these elements that will make a circuit whose time constant is 5×10^{-5} s.

52. (II) Show that the time constant of a parallel-plate capacitor filled with a dielectric with a finite resistivity is independent of the area and separation of the plates.

53. (II) Polycarbonate, a so-called polar polymer, is a material with a dielectric constant $\kappa = 3.2$. It has a resistivity $\rho = 2 \times 10^{14}\ \Omega \cdot$m. Suppose that it is used to fill the space in a parallel-plate capacitor of area 0.25 m^2 and plate separation 0.25 mm. A charge of $Q = 5\ \mu$C is placed on the plates of the isolated capacitor. How long does it take for 90 percent of the charge to leak away?

General Problems

54. (II) Consider the circuit shown in Fig. 28–52, in which a 12-V battery is used to charge a 6-V battery. The resistance in the circuit is 20 Ω. Calculate (a) the current in the circuit; (b) the rate at which the energy of the smaller battery increases; (c) the total rate of energy dissipation in the resistor.

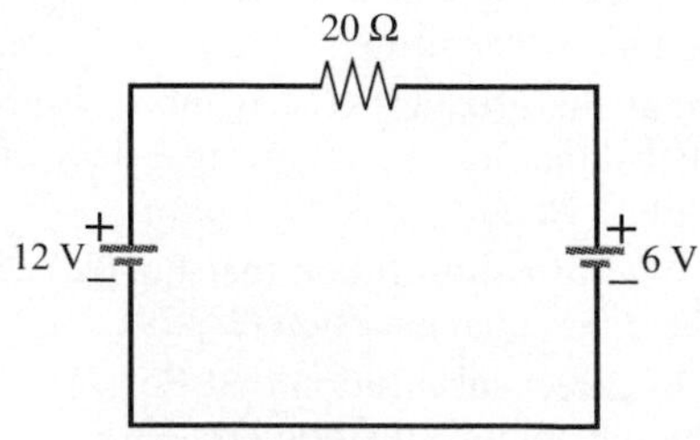

FIGURE 28–52 Problem 54.

55. (II) A student has a wide range of resistors, all rated for $\frac{1}{2}$ W. How can a student combine identical resistors to obtain an effective resistance of 20 Ω rated for 2 W?

56. (II) Show that if a battery of fixed emf and internal resistance r is connected to an external resistor of resistance R, then the maximum power delivered to the external resistor occurs when $R = r$.

57. (II) When separately connected across a line with voltage V, two resistors generate power P_1 and P_2, respectively. What is the power generated when the two resistors are connected in series? in parallel?

58. (II) A resistor R forms a single loop with an arrangement of two batteries of emf $\mathscr{E}$ and internal resistance r. The batteries are arranged (a) in series and (b) in parallel. Find the current through the resistance in both cases. Which arrangement gives the larger current for large R? for small R?

59. (II) A 1200-W kitchen mixer, a 480-W vacuum cleaner, and a lamp with a 150-W bulb are all plugged into the same outlet in a 120-V circuit. A *fuse* acts as a switch that opens if the current exceeds 15 A. How much current does each device draw? If the bulb is replaced, what is the minimum wattage of the new bulb that will blow the circuit? Do not worry about the oscillations of the current and voltage in real household circuits, but assume that all currents and voltages are DC.

60. (II) Consider three resistors of 4 Ω each. Each resistor can dissipate at most 20 W. Consider the four possible ways of of arranging the three resistors, and calculate the maximum power that can be dissipated in each of the ways.

61. (II) Find the currents in each leg of the circuit shown in Fig. 28–53.

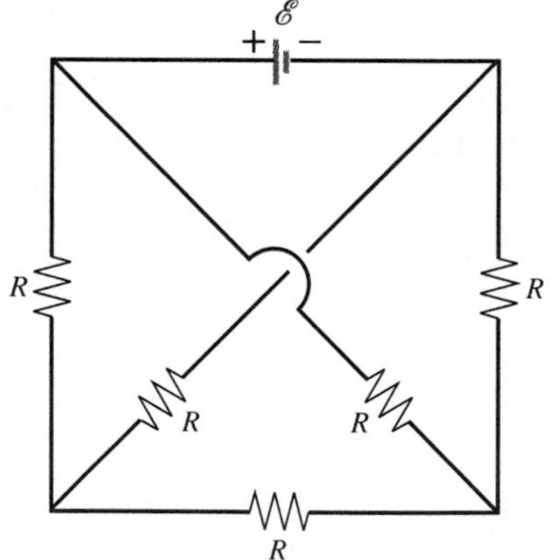

FIGURE 28–53 Problem 61.

62. (II) Design an ohmmeter circuit similar to that in Fig. 28–22 that has a switch and would be appropriate to measure resistances of 100 Ω, 1 kΩ, 10 kΩ, 100 kΩ, and 1 MΩ.

63. (II) By using Table 27–2, compare the current density, electric field strength, and power loss in two cylindrical wires of the same length and same radius, one made of aluminum and the other made of copper. The wires are connected (a) in series and (b) in parallel. (c) If wires of a given length and radius were constructed from all the materials listed in Table 27–2, and if the same current were passed through each, which wire would have the largest current density, the weakest electric field, and the least power loss? Assume throughout that the temperature dependence is unimportant.

64. (II) A simple *potentiometer* circuit used to measure unknown voltages accurately is shown in Fig. 28–54. Here V_S is the known source voltage, V_x is the unknown voltage, and the resistor is a variable one from which the values R_1 and R_2 can be read from the position of the pointer. These resistances are varied until the current in the ammeter is zero. Show that the unknown voltage then has the value $V_x = V_S R_2/(R_1 + R_2)$.

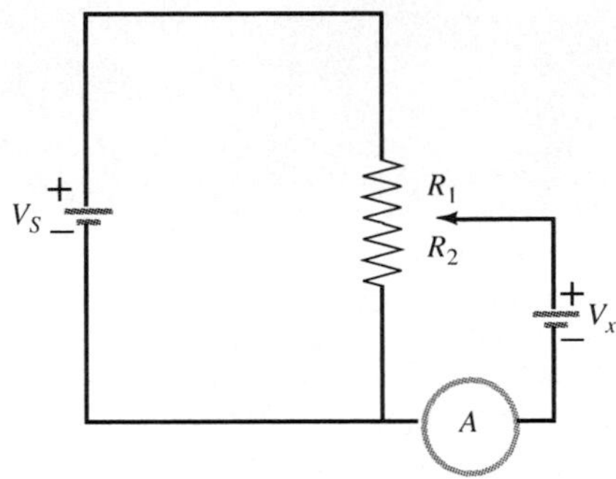

FIGURE 28–54 Problem 64.

65. (II) A single-loop circuit contains a battery of emf V_0 and negligible internal resistance and, connected in series with the battery, a capacitor filled with a material of conductivity σ. The capacitor consists of two circular plates of radius r separated by a distance $d \ll r$. (a) What is the electric field between the capacitor plates? (b) What is the current flowing in the circuit?

66. (III) Consider the infinite network of resistors shown in Fig. 28–55a. Calculate the resistance, R^*, of the network by noting that with an infinite set of resistors, adding one more rung to the ladder does not change the resistance. Thus the network may be broken up as shown in Fig. 28–55b.

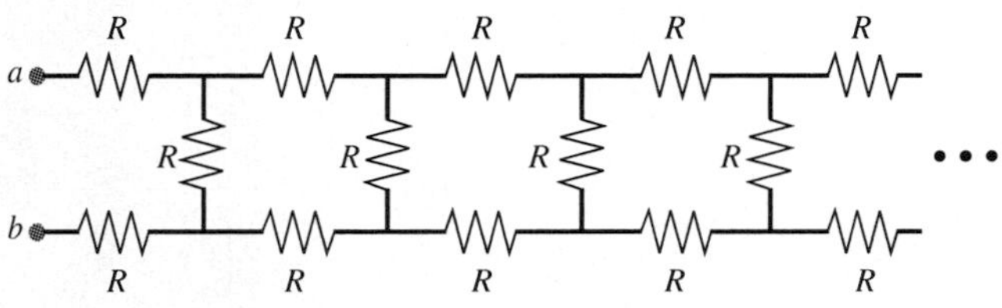

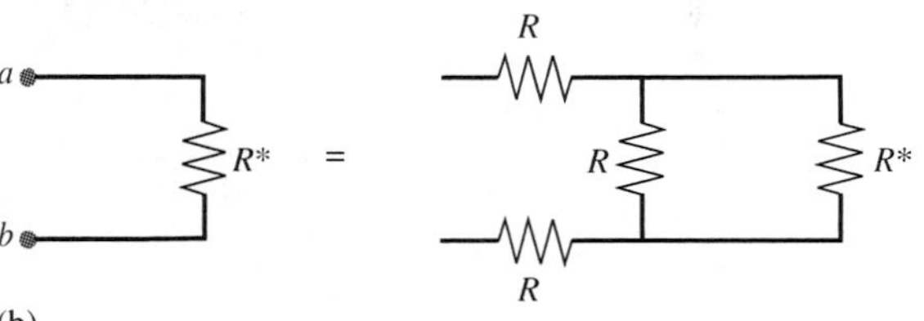

FIGURE 28–55 Problem 66.

CHAPTER 29

A spectacular manifestation of the earth's magnetic field is the aurora, a result of the movement of charged particles high in the atmosphere, under the influence of that field. This photograph of the Aurora Australis was taken from the space shuttle Discovery in 1991.

THE EFFECTS OF MAGNETIC FIELDS

It has been known for thousands of years that lumps of the natural mineral magnetite, called lodestones, exert on each other what we now know to be magnetic forces. Lodestones can impart some of their properties to bits of iron that are brought near. Sailors have used the navigational compass, made from lodestone or treated iron, for at least 800 years, and possibly much longer. In 1600 William Gilbert, an English physician who made a systematic study of electrical and magnetic phenomena, suggested that the compass behaves as it does because the earth is itself a giant lodestone, or magnet. Indeed, our use of the words "magnetic north and south poles" in connection with bar magnets comes from Gilbert's association of magnetism with the earth's geographic North and South Poles.

Magnetism was not associated with electricity until 1820, when André Ampère used experiments of his own and those of Hans Christian Oersted to show that magnetic effects arise when electric charges move. In fact, electrical and magnetic phenomena are both aspects of the interactions of electrically charged objects. In the 1820s Michael Faraday did crucial work to uncover the connection between electricity and magnetism, but it was James Clerk Maxwell who, in the late 1860s, made the ultimate synthesis of electricity and magnetism. This synthesis is described entirely by Maxwell's equations, which, together with Newton's work, the ideas of thermodynamics, and Einstein's special theory of relativity, summarize virtually all of classical physics. Our understanding of light and other electromagnetic waves rests on Maxwell's great achievement. In Chapters 29 to 35 we shall study magnetic phenomena, their connection to electrical phenomena, their practical applications, and other remarkable consequences of Maxwell's equations.

We shall discover the laws of magnetic forces by describing experiments with magnets and with electric currents. We shall learn how magnetic forces are associated with magnetic fields, just as electric forces are associated with electric fields. Magnetic fields are generated by magnets and by moving charges. In this chapter we concentrate on the effects of magnetic fields on test objects, leaving for Chapter 30 the question of how magnetic fields are generated.

29–1 MAGNETS AND MAGNETIC FIELDS

When two bar **magnets** are brought close to each other, the forces between them, **magnetic forces**, become evident. (As we shall see, these forces are of a type we have not yet encountered.) Bar magnets have an orientation, or an axis. In some positions they attract each other, in others they repel, and in still others they exert torques on each other. We arbitrarily label the end of a magnet that is attracted to a point very near the earth's geographic South Pole as the *south pole*, S, and the other end as the *north pole*, N. If we experiment with two bar magnets labeled in this way, we find that the N end of one attracts the S end of the other, whereas the two N ends repel each other, as do the two S ends (Fig. 29–1). (Note that this means that the earth's geographic North Pole behaves like a magnetic south pole, whereas the geographic South Pole acts as a magnetic north pole.) Based on our experience with charges, we might be tempted to conclude that a bar magnet contains "magnetic charges" (or *magnetic monopoles*) at each end, and that we could somehow extract them. Experiment shows that this is not possible; if you were to saw a bar magnet in two, you would end up with two bar magnets, not two separate magnetic charges.

Iron and a few other materials have a peculiar property: If we place such a material near a lodestone (a "natural" magnet), that material becomes a new magnet. Something about the very presence of a nearby magnet causes iron to become magnetic. Whatever the underlying cause, we can turn tiny shavings of iron (iron filings) into tiny magnets that can be used as test probes of magnetic forces.

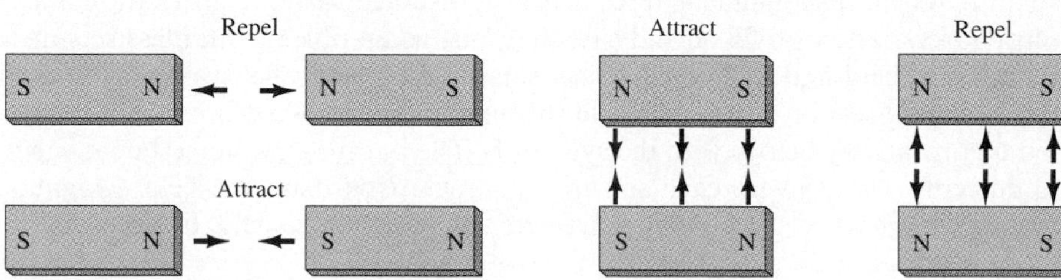

FIGURE 29–1 Bar magnets exert magnetic forces on each other.

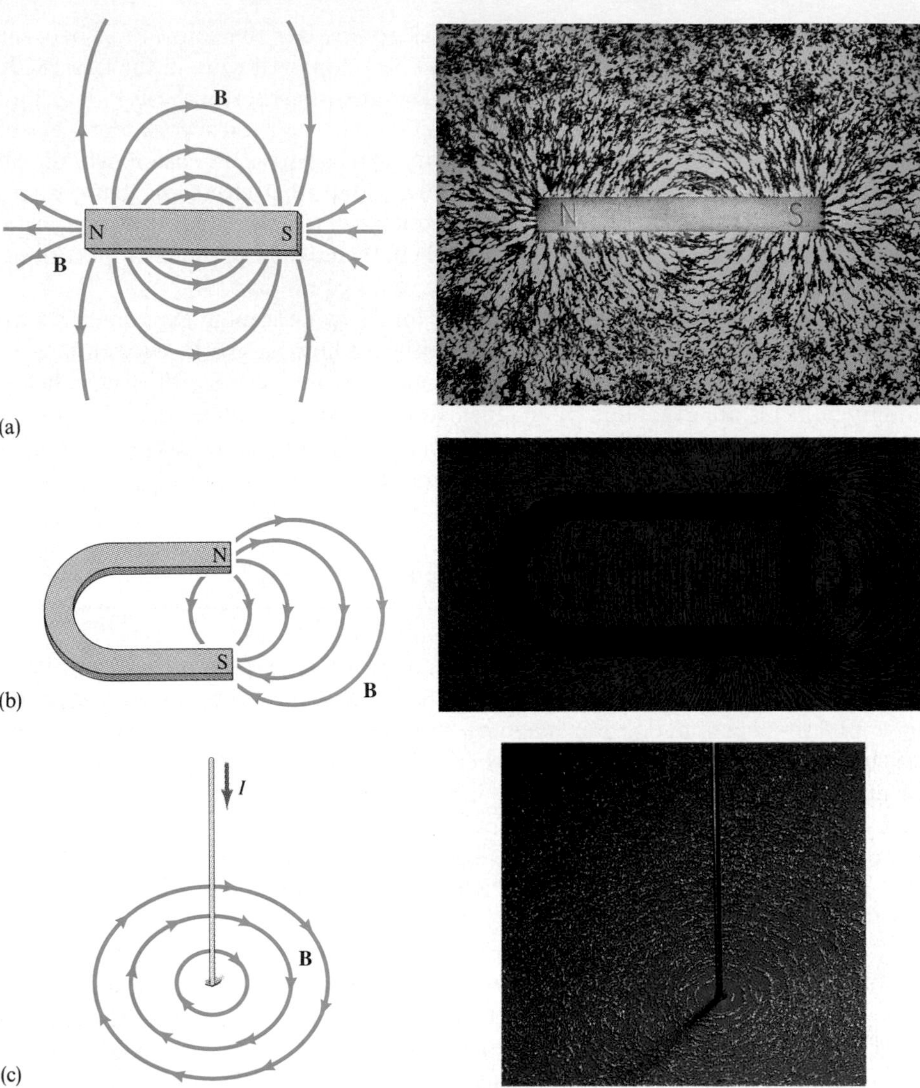

FIGURE 29–2 Iron filings map the magnetic field for (a) a straight bar magnet, (b) a horseshoe magnet, and (c) a current-carrying wire.

Magnetic Fields

If, as in Fig. 29–2, we scatter iron filings on a sheet of plastic above a bar magnet, the filings become aligned in certain directions—the magnetic forces line them up—and they clump more densely in certain regions, such as near the poles, than in others. For different magnets, the scattered filings have different densities and alignments. Figure 29–2 shows the distribution of iron filings around a straight bar magnet, a horseshoe magnet, and—surprise!—a wire that carries current. The last case is a clue that magnetic forces are associated with moving charges as well as with magnets. The magnetic force acts at a distance, just like gravitational and electric forces. The iron filings indicate that, just as an object with mass sets up a gravitational field and a charged object sets up an electric field, a magnet or an electric current sets up a **magnetic field** throughout space. We denote this field, defined quantitatively below, with the symbol **B**. The bar magnet, horseshoe magnet, and current-carrying wire each set up a characteristic magnetic field. Magnetic forces have a marked directional character, so the magnetic field, like the electric field, is a vector.

Magnets and electric currents set up magnetic fields.

Just as the force on a small electric test charge can be used to map out an electric field, the alignment and density of iron filings are a measure of the direction and strength of any magnetic field that may be present. The connection between the iron filings and the magnetic field is straightforward. **B** is oriented along the alignment direction of the filings, and its magnitude is proportional to the density of the filings. Just as the electric field can be visualized with electric field lines, the magnetic field can be visualized with magnetic field lines, continuous lines that run parallel to the direction of the field at every point and whose density (the number of lines per unit area) is proportional to the strength of the field. The iron filings thus map out the magnetic field lines. We see in Fig. 29–2 that the iron filings align themselves between the poles of magnets, and so therefore do magnetic field lines. We take the direction of the magnetic field to run *from the N pole to the S pole*, just as we assign the electric field to run from positive electric charges to negative charges. Note that if the N end of a small test magnet is placed near the N pole of a large magnet, the test magnet turns away, consistent with an intuitive scenario in which the magnetic field of the large magnet points away from its N pole. However, notice that the magnetic field around a current-carrying wire has no magnetic pole. Once we have mapped out a magnetic field in this way, we can further investigate its effects and find the force laws associated with magnetism.

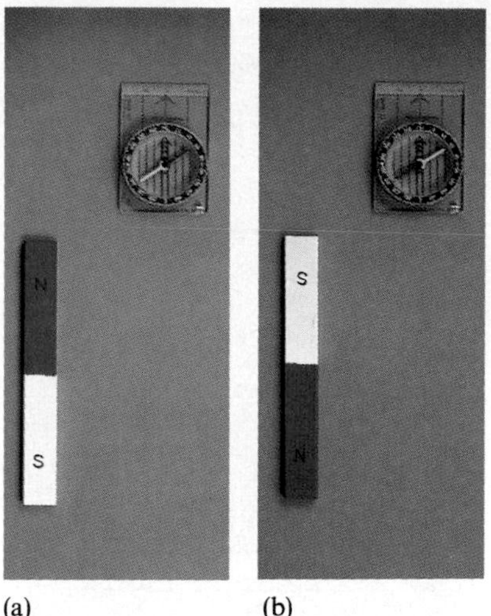

FIGURE 29–3 A compass needle can be used to determine the direction of a magnetic field, such as that of a bar magnet. (a) The N pole of the bar magnet is not attracted to the N end of the needle, but (b) the S pole of the bar magnet is.

29–2 MAGNETIC FORCE ON AN ELECTRIC CHARGE

Experiments show that electric charges as well as bar magnets experience forces in the presence of magnetic fields; that is, they are accelerated by those fields. This phenomenon is easy to demonstrate in the classroom or the laboratory (Fig. 29–3). We can use a bar magnet to deflect the electron beam of an oscilloscope.[†] When a bar magnet is placed in different orientations near the beam, the beam deflects in various ways. The deflection allows us to measure the magnetic forces on the beam.

Consider a bar magnet with its N pole oriented so that the magnetic field is in the $+y$-direction (Fig. 29–4). The magnitude of the magnetic field at the posi-

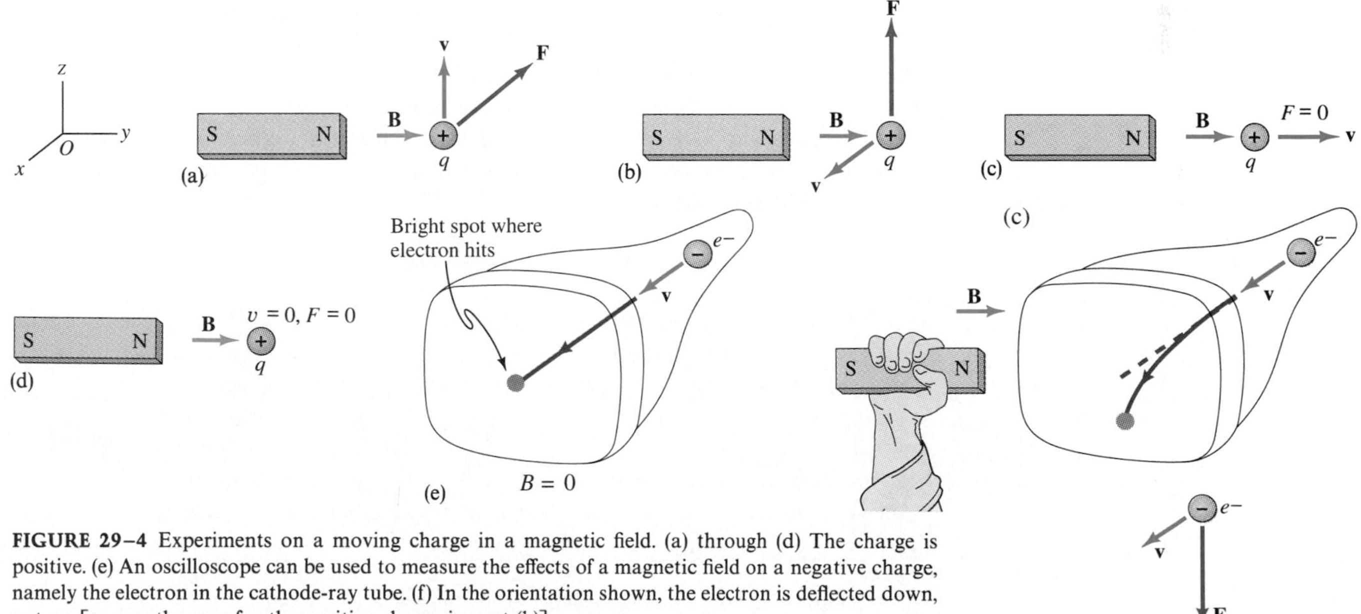

FIGURE 29–4 Experiments on a moving charge in a magnetic field. (a) through (d) The charge is positive. (e) An oscilloscope can be used to measure the effects of a magnetic field on a negative charge, namely the electron in the cathode-ray tube. (f) In the orientation shown, the electron is deflected down, not up [as was the case for the positive charge in part (b)].

[†] Do not use your TV set for this purpose: The picture may remain distorted!

tion of a moving *positive* electric test charge q can be varied by varying the distance of the magnet from the charge. We observe the following phenomena (but remember that when these experiments are done with a bar magnet and an oscilloscope, the electron charge is negative):

1. If q moves at speed v in the $+z$-direction, then q is deflected in the $-x$-direction (Fig. 29–4a). Furthermore, the larger v is, the stronger the force **F** is. Detailed measurements show that the magnitude of **F** due to the magnetic field is proportional to v.
2. If q moves in the $+x$-direction, **F** is in the $+z$-direction, again proportional to v (Fig. 29–4b).
3. If q moves in the y-direction (+ or −), there is no change in the charge's direction or speed; that is, there is no force (Fig. 29–4c).
4. If q moves at speed v in an arbitrary direction, **F** is proportional to the velocity component perpendicular to the magnetic field, $v_\perp$, and perpendicular to the directions of both $\mathbf{v}_\perp$ and **B**. This result summarizes parts (a) through (c). In particular, if the charge is at rest, so that $v = 0$, there is no force (Fig. 29–4d).
5. **F** is proportional to the magnitude of **B**.
6. **F** is proportional to the sign and magnitude of q. We can use an oscilloscope to study the effect of a negative charge, namely, that of an electron (Figs. 29–4e and 29–4f).

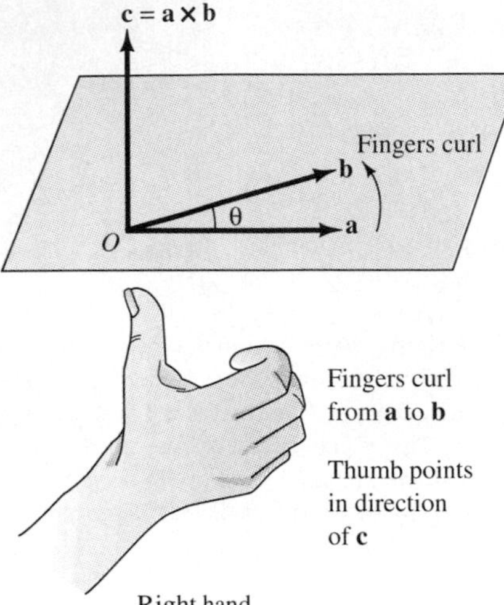

FIGURE 29–5 The vector product $\mathbf{c} = \mathbf{a} \times \mathbf{b}$ and the right-hand rule that determines the direction of the vector product. See Chapter 10, Fig. B1–1.

The crucial feature of this seemingly complicated collection of results is the dependence on **v**. Result 4 contains within it the three results 1 to 3. In results 1 and 2, the initial velocity is purely in the z- or x-directions and is therefore perpendicular to **B**. In result 3, there is no $\mathbf{v}_\perp$ and also no force. That the force is perpendicular to both **v** and **B** is indicated on Fig. 29–4.

The vector product is summarized in Chapter 10.

To summarize, the magnetic force **F** is proportional in magnitude to q, $v_\perp$, and B, and the direction of **F** is perpendicular to both **B** and **v** and depends on the sign of q. A direction perpendicular to both **v** and **B** can be represented by the *vector product*, which was discussed extensively in connection with torques and rotational motion. Recall that a vector $\mathbf{c} = \mathbf{a} \times \mathbf{b}$ (the vector product of **a** and **b**) has magnitude $ab \sin \theta$, where θ is the angle between vectors **a** and **b** and is always taken to be less than 180° (Fig. 29–5). Vector **c** is perpendicular to both **a** and **b**, in a direction determined by the right-hand rule. Thus our experiments have determined that the magnetic force on a test charge q moving with velocity **v** in a magnetic field **B** is given by

The magnetic force law

$$\mathbf{F}_B = q\,\mathbf{v} \times \mathbf{B}. \tag{29–1}$$

This important result is the **magnetic force law.**[†] We henceforth drop the subscript B unless we need to distinguish the magnetic force from some other force. If θ is the angle between vectors **v** and **B**, the magnitude of **F** is given by

$$F = qvB \sin \theta = qv_\perp B. \tag{29–2}$$

Figure 29–6 shows how, according to a right-hand rule, **F** is perpendicular to the plane formed by **v** and **B**. Comparison with the observations in Fig. 29–4 shows that Eq. (29–1) is completely consistent with our experimental results. Recall that

[†] Strictly speaking, a precise *definition* of the magnetic field produced by a source such as a magnet or a current-carrying wire is given by measurement of the force on a moving test charge. Thus Eq. (29–1) is used as a definition of **B**. The treatment of the magnetic field in terms of the orientation and density of iron filings we gave earlier turns out to be compatible with this definition.

the vector product of two *parallel* vectors is zero; that is why there is no magnetic force on a charge that moves along the axis of a bar magnet and no magnetic force associated with the component of **v** parallel to **B**.

Equation (29–1) shows that the dimensions of **B** are quite different from those of the electric field, **E**. The SI unit of magnetic field is called the *tesla* (T), in honor of Nikola Tesla, who made important contributions to the technology of electrical energy generation. In terms of previously defined SI units,

The unit of magnetic field

$$1\ \mathrm{T} = 1\,\frac{\mathrm{kg}}{\mathrm{C\cdot s}}. \qquad (29\text{–}3)$$

Another (non-SI) unit in common use is the *gauss* (G); 10^4 G = 1 T.

Table 29–1 contains some representative values of magnetic fields.

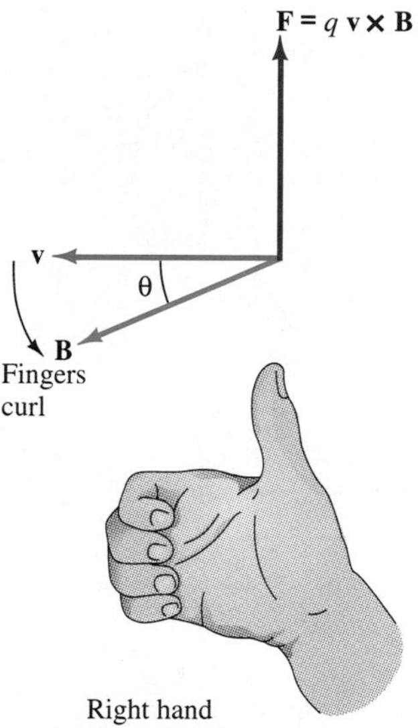

FIGURE 29–6 The right-hand rule for the magnetic force law $\mathbf{F} = q\,\mathbf{v} \times \mathbf{B}$.

EXAMPLE 29–1 The undisturbed electron beam of an oscilloscope moves along the x-direction (Fig. 29–7). The south pole of a bar magnet approaches the cathode-ray tube from above and deflects the beam. The magnitude of the magnetic field from the magnet is 0.05 T in the vicinity of the beam, and the speed of the electrons in the beam is 2×10^5 m/s. What is the magnitude of the magnetic force on the electrons? What is the direction of this force; that is, which way is the beam deflected?

SOLUTION: This exercise is a straightforward application of the magnetic force law, Eq. (29–1). We want to find the unknown force, given the charge q, its velocity **v**, and the magnitude of the magnetic field. The magnetic field is directed toward the south pole, so, as Fig. 29–7 indicates, the field of the bar magnet, **B**, is in the $+y$-direction. The velocity **v** is perpendicular to **B**, and the magnetic force is

$$\mathbf{F} = q\,\mathbf{v} \times \mathbf{B} = q(v\mathbf{i} \times B\mathbf{j}) = qvB(\mathbf{i} \times \mathbf{j}) = qvB\mathbf{k}.$$

Here **i**, **j**, and **k** are the unit vectors in the x-, y-, and z-directions, respectively. With numerical values,

$$\mathbf{F} = (-1.6 \times 10^{-19}\ \mathrm{C})(2 \times 10^5\ \mathrm{m/s})(5 \times 10^{-2}\ \mathrm{T})\mathbf{k} = (-1.6 \times 10^{-15}\ \mathrm{N})\mathbf{k}.$$

Using the right-hand rule, we can verify that the vector product $\mathbf{v} \times \mathbf{B}$ is in the $+z$-direction. The electron charge, however, is negative, so the force on the electron is in the $-z$-direction, and the beam is deflected in this direction, as shown in Fig. 29–7.

TABLE 29–1

SOME MAGNETIC FIELDS

Location or Source	*Magnitude (T)*
Interstellar space	10^{-10}
Near surface of earth	5×10^{-5}
Refrigerator magnet that holds notes	10^{-2}
Bar magnet near poles	10^{-2}–10^{-1}
Near surface of sun	10^{-2}
Large scientific magnets	2–4
Largest steady-state magnet	30
Largest pulsed field in laboratory	500–1000
Near surface of pulsar	10^8
Near surface of atomic nucleus	10^{12}

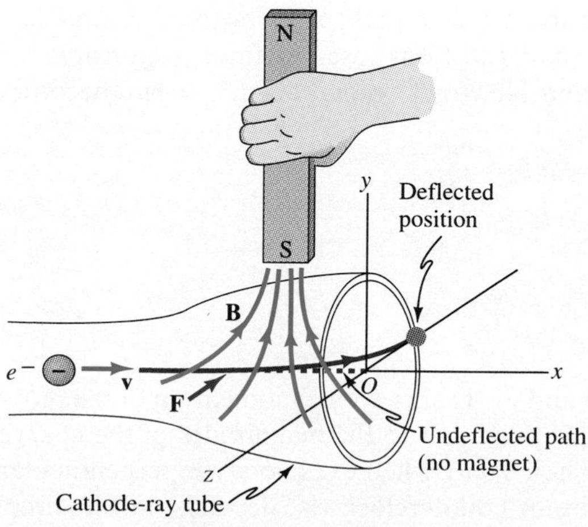

FIGURE 29–7 Example 29–1. The magnetic field points toward the south pole of the bar magnet and is therefore oriented in the $+y$-direction. The spot at which the beam reaches the screen is deflected in the $-z$-direction.

The Lorentz Force

Further experimentation shows that charges react independently to electric and magnetic fields. Thus if an electric field is present in addition to the magnetic field, it produces an additional force $\mathbf{F} = q\mathbf{E}$ on the charge. The net force is

The Lorentz force law

$$\mathbf{F} = q[\mathbf{E} + (\mathbf{v} \times \mathbf{B})]. \tag{29-4}$$

This equation, whose implications we shall study in Section 29–3, is known as the **Lorentz force law**, named after the late-nineteenth-century physicist Hendrik A. Lorentz, who influenced the development of many areas of classical physics.

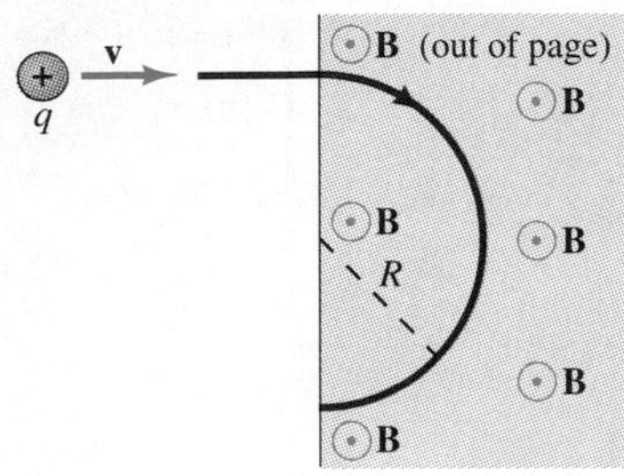

(a)

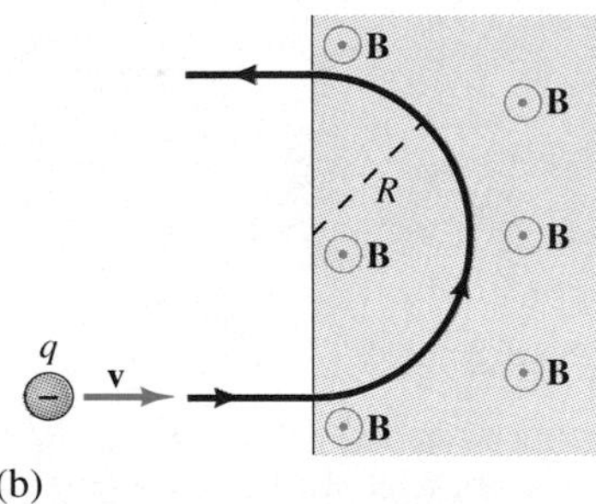

(b)

FIGURE 29–8 (a) A charged particle moves perpendicularly to a constant magnetic field, **B**, shown in a view from above. The charged particle traces a circular path in the plane perpendicular to **B**, which is directed out of the plane of the page. (b) The direction of the curvature is opposite for opposite charges.

29–3 MAGNETIC FORCE ON A CHARGE: APPLICATIONS

Magnetic forces on charged particles have important implications that range from the functioning of electronic devices to phenomena in astrophysics and plasma physics. In this chapter we assume that the magnetic fields are independent of time: Thus we are dealing with **magnetostatics**.

Circular Motion in a Constant Magnetic Field

The magnetic force law, Eq. (29–1), states that only the velocity in the plane perpendicular to **B** contributes to the expression for the force. *The component of the velocity of a charged particle parallel to the magnetic field is not affected by the field,* and it is therefore unchanging in the absence of any other forces (such as electric forces). In addition, Eq. (29–1) states that *the force on the charge, and hence the charge's acceleration, is perpendicular to* **B** *and thus acts only in the plane perpendicular to* **B**.

To explore the consequences of these observations more closely, consider a magnetic field **B** that is uniform in some region of space and a test charge q that enters this region with a velocity **v** perpendicular to the field (Fig. 29–8). What is the consequent motion of the charge? According to Eq. (29–1) the force will be perpendicular to **v** and have magnitude $F_B = qvB$. We saw in Chapter 3 that, when the acceleration (and hence the force) is constant in magnitude and perpendicular to the velocity, there is *circular motion at constant speed.* A charged particle moving perpendicularly to a constant, spatially uniform magnetic field will move in a circle (Fig. 29–8a). The magnitude of the acceleration for circular motion is $a = v^2/R$, where R is the radius of the particle's circular path; the direction of the acceleration is toward the center of the circular path. By Newton's second law the force must have magnitude $ma = mv^2/R$. In our case, the force responsible for the acceleration has magnitude F_B, and Newton's second law ($F_B = ma$) becomes

The kinematics associated with acceleration perpendicular to the velocity was covered in Chapter 3.

$$F_B = qvB = \frac{mv^2}{R}.$$

We solve for R;

The radius of the circle traced out by a charged particle moving perpendicular to a constant magnetic field.

$$R = \frac{mv}{qB}. \tag{29-5}$$

R is proportional to the product of m and v—that is, to the momentum of the moving particle, $p = mv$—and inversely proportional to the magnitudes of the charge q and of the field, **B**. Equation (29–5) holds *only* when the velocity is perpendicular to **B**. Whether the motion is clockwise or counterclockwise depends on the direc-

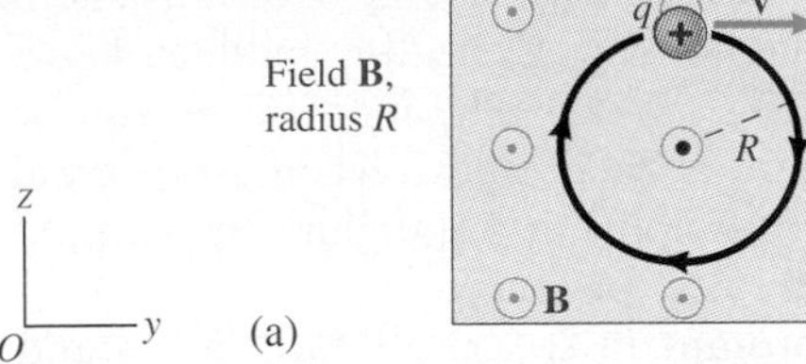

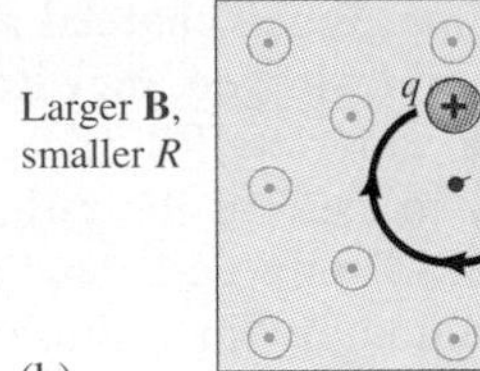

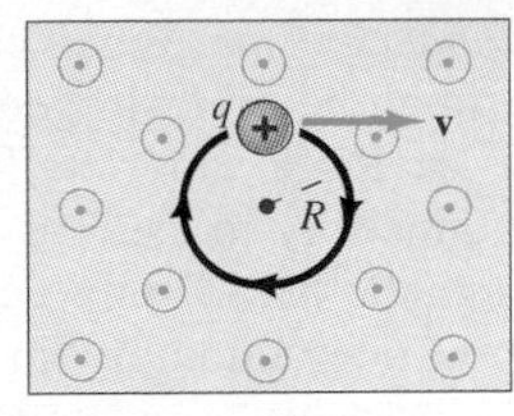

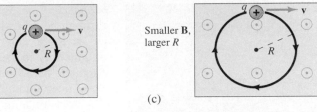

FIGURE 29–9 The radius of a circle traced out by a charged particle that moves perpendicularly to a constant magnetic field **B** is inversely proportional to *B*.

tion of **v** and on the sign of the charge, according to the right-hand rule. Figure 29–8b depicts the motion for a test charge of opposite sign to the test charge in Fig. 29–8a. As Fig. 29–9 illustrates, the larger *B* is, the larger the magnetic force is, and the "tighter" the curved path is, corresponding to a smaller radius of curvature (the radius of the fragment of a circle along which the charge moves at a given moment). The smaller the magnetic field, the smaller the force, and the larger *R* is.

The circular motion has a period $T = 2\pi R/v$, or, from Eq. (29–5),

$$T = \frac{2\pi}{\not{v}}\frac{m\not{v}}{qB} = \frac{2\pi m}{qB}. \tag{29–6}$$

Equivalently, the frequency $f = 1/T$ is

$$f = \frac{qB}{2\pi m}. \tag{29–7}$$

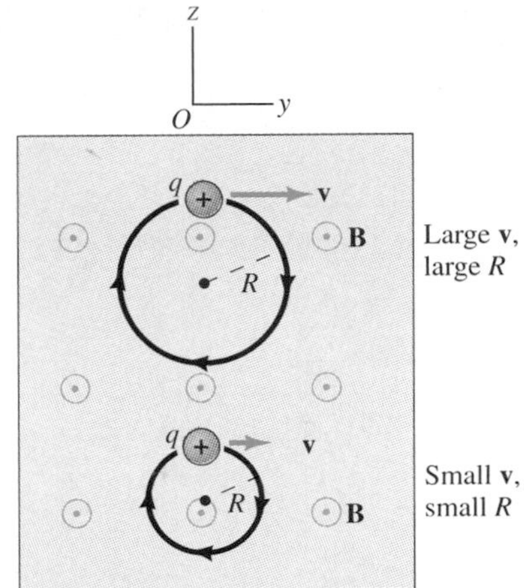

FIGURE 29–10 For a given charge and a constant magnetic field perpendicular to the direction of motion, the time for a particle to make one revolution is independent of the particle's speed.

This frequency is called the **cyclotron frequency**. Notice that the period and frequency are *independent* of the speed. A slow particle traces out a tight circle in the same time that a fast particle traces out a large circle (Fig. 29–10). The fact that the cyclotron frequency is constant is a guiding principle of the *cyclotron*, a device in which charged particles are accelerated by an electric field while they stay in a near-circular orbit because they move between the poles of a large magnet (Fig. 29–11; see Problem 24).

Equation (29–5), which specifies the radius of the circular path of a charged particle, finds application in many particle-detection devices—for example, in the

FIGURE 29–11 A cyclotron at the Lawrence Berkeley Laboratory. This photo was taken shortly before the cyclotron first operated in 1939.

FIGURE 29–12 The tracks left by charged particles moving in a magnetic field.

bubble chamber. When charged particles produced in high-energy collisions speed through liquid hydrogen, they leave tracks that consist of very tiny bubbles, like a jet leaving a vapor trail in the atmosphere (Fig. 29–12). The momentum can be obtained by measuring the radius of curvature of their tracks when an external magnetic field is imposed. This information is helpful in untangling the complicated interaction in which those particles were first produced.

Even for a magnetic field that is not constant in space, we can understand the motion of a charged particle qualitatively by assuming that the field is nearly uniform (this is always a good approximation if we look at a sufficiently small region of space). Let us consider a magnetic field whose magnitude varies in space but whose direction does not vary. Only the velocity in the plane perpendicular to the field direction matters, so we assume that the initial velocity lies in that plane. If the field in a region is weak, the particle will move in a large circular path. When this motion brings it into a region where the field is stronger, the circular path tightens. The radius of curvature changes from place to place (Fig. 29–13).

Finally, when there is a component $v_{\|}$ of velocity $\mathbf{v}$ that does not lie in the plane perpendicular to $\mathbf{B}$, that component of the velocity does not change. The particle advances along $\mathbf{v}_{\|}$ while it moves in a circle in the plane formed by $\mathbf{v}_{\perp}$ (Fig. 29–14a).

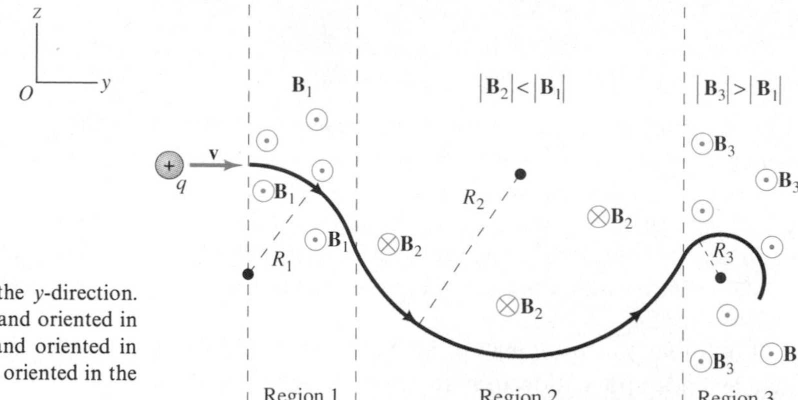

FIGURE 29–13 A particle of charge q moves with $\mathbf{v}$ in the y-direction. In region 1 the magnetic field $\mathbf{B}_1$ is of medium magnitude and oriented in the $+x$-direction. In region 2 $\mathbf{B}_2$ is small in magnitude and oriented in the $-x$-direction. In region 3 $\mathbf{B}_3$ is of large magnitude and oriented in the $+x$-direction.

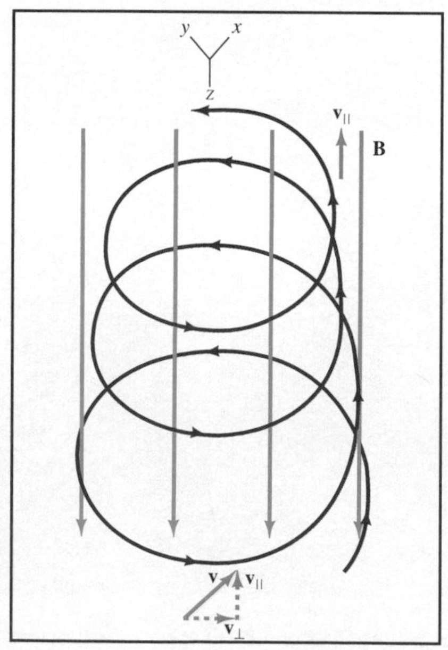

(a)

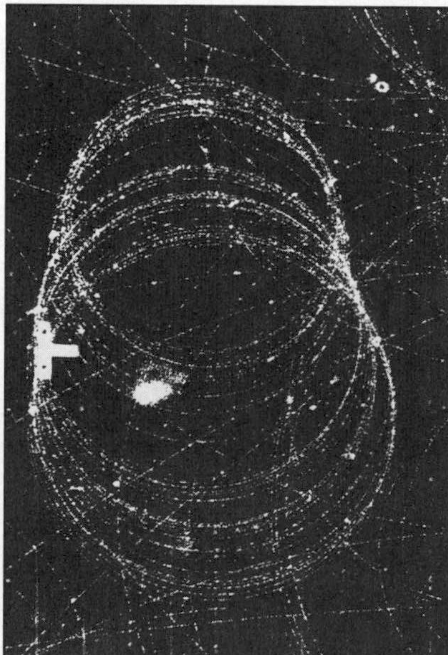

(b)

FIGURE 29–14 (a) A charged particle in a region where the magnetic field is constant follows a helical path. (b) An electron in a cloud chamber produced this 10-m-long spiral track. The electron's path begins at the bottom. The helix tightens about halfway up because the electron has radiated a photon and has thereby lost energy.

The resulting trajectory forms a *spiral* (or *helix*) with its axis along **B** (Fig. 29–14b). The circular motion in the plane perpendicular to **B** has a radius given by

$$R = \frac{mv_\perp}{qB}. \tag{29–8}$$

EXAMPLE 29–2 A particle of unknown charge q and unknown mass m moves at speed $v = 4.8 \times 10^6$ m/s in the $+x$-direction into a region of constant magnetic field (Fig. 29–15). The field has magnitude $B = 0.5$ T and is oriented in the $+z$-direction. The particle is deflected in the $-y$-direction and traces out a fragment of a circle of radius $R = 0.1$ m. What is the sign of the particle's charge, and what is the ratio q/m?

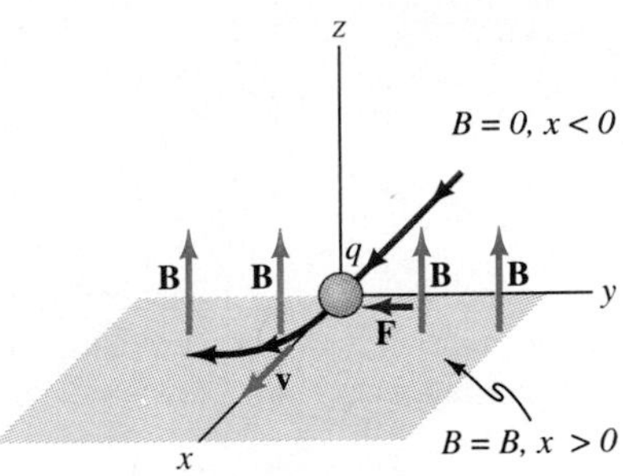

FIGURE 29–15 Example 29–2. If the speed of a particle of unknown charge and mass that moves in a region of constant magnetic field is known, then a measurement of the radius of curvature of the particle's path gives the charge-to-mass ratio of the particle.

SOLUTION: We are given the velocity **v** and the magnetic field, **B**, and can use the magnetic force on the particle to find the unknown charge and mass. To determine the sign of the charge, we need only relate the sign of the force, which acts in the $-y$-direction, to the sign of the charge. The force is given by Eq. (29–1), and the vector product $\mathbf{v} \times \mathbf{B}$ points, by the right-hand rule, in the $-y$-direction. In order for the force $q\,\mathbf{v} \times \mathbf{B}$ to point that way, the charge q must be positive.

The magnitude of the ratio q/m is determined by Eq. (29–5), $R = mv/qB$:

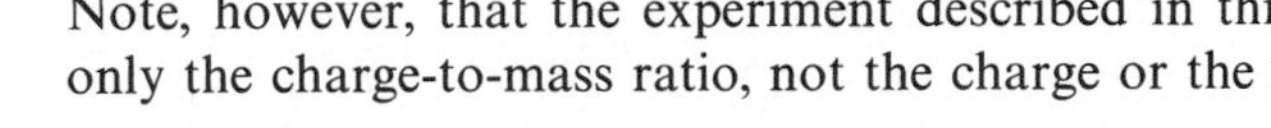

$$\left|\frac{q}{m}\right| = \frac{v}{BR} = \frac{4.8 \times 10^6 \text{ m/s}}{(0.5 \text{ T})(0.1 \text{ m})} = 9.6 \times 10^7 \text{ C/kg}.$$

This particle is a proton. To see why, let us assume that the unknown charge is that of an electron, $|q| = 1.6 \times 10^{-19}$ C. Then $m = |q|/(9.6 \times 10^7 \text{ C/kg}) = (1.6 \times 10^{-19} \text{ C})/(9.6 \times 10^7 \text{ C/kg}) = 1.7 \times 10^{-27}$ kg, just the mass of a proton. Note, however, that the experiment described in this example can measure only the charge-to-mass ratio, not the charge or the mass alone.

Energy of a Charged Particle in a Static Magnetic Field

In the situations we have described, the speed of the charged particle never changes—the velocity component parallel to **B** is unaffected, and the perpendicular velocity component undergoes uniform circular motion in which its direction but not its magnitude changes. Because the kinetic energy is $K = \frac{1}{2}mv^2$, it is generally true that *the kinetic energy of a charged particle in a static magnetic field is constant.* The work–energy theorem states that the work done by a force equals the change in kinetic energy; equivalently, *a static magnetic field does no work on a charge.*

A charged particle moving in a static magnetic field has a constant kinetic energy.

Velocity Selectors

A particular arrangement of electric and magnetic fields makes a **velocity selector**. Consider a region with uniform, mutually perpendicular **E** and **B** fields (Fig. 29–16). (We say that such fields are *crossed.*) A particle of mass m, (positive) charge q, and velocity **v** is directed perpendicularly to both **E** and **B** when it enters this region. We shall show that there is a certain value of v for which the particle traverses the region undeflected. At a speed other than v, the same particle *is* deflected, so in a beam of particles with a variety of speeds, only those particles with a certain speed pass through undeflected.

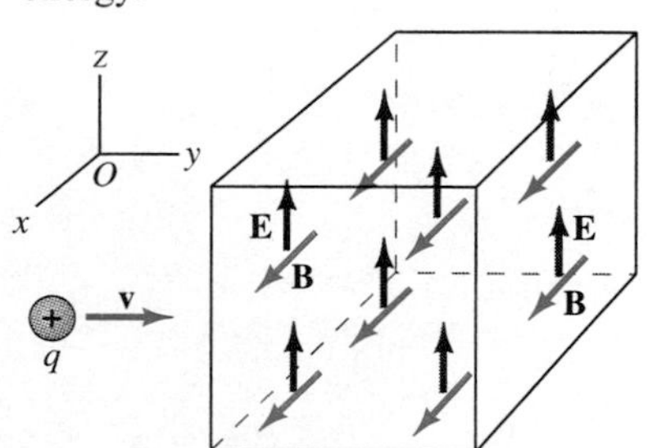

FIGURE 29–16 A charged particle enters a region of crossed electric and magnetic fields. If the speed of the particle is $v = E/B$, the particle will cross the region undeflected.

Both **E** and **B** fields are present, so we need to use the Lorentz force law [Eq. (29–4)] and compute the contributions to the force from both fields. Referring to the coordinates of Fig. 29–16, the electric force contribution $q\mathbf{E}$ is

$$\mathbf{F}_E = qE\mathbf{k}.$$

The magnetic force $q\,\mathbf{v} \times \mathbf{B}$ is, by the right-hand rule,

$$\mathbf{F}_B = -qvB\mathbf{k}.$$

Because the electric and magnetic forces point in opposite directions, they will cancel if their magnitudes are equal, in which case the particle will travel undeflected. This cancellation occurs for $qvB = qE$, so the speed of a charged particle that passes through the crossed fields undeflected is

The speed of a charged particle that moves undeflected, perpendicular to crossed magnetic and electric fields.

$$v = \frac{E}{B}. \qquad (29\text{–}9)$$

If we reverse the sign of charge q, the electric force points in the $-z$-direction while the magnetic force points in the $+z$-direction; the forces still cancel when v is given by Eq. (29–9).

The Charge-to-Mass Ratio of the Electron. Sir Joseph John Thomson, who can correctly be called the discoverer of the electron, performed a series of wide-ranging experiments whose results were crucial to our understanding of the electron and of the electrical nature of matter. He used a velocity selector in 1897 as an important component of his experiment to measure the charge-to-mass ratio of the electron (Fig. 29–17a). He first accelerated electrons in an electric field—not the electric field of the velocity selector—by passing them through an electric potential V. The work thereby done on the electrons is qV. The electrons gained a speed v determined by $mv^2/2 = qV$, so

$$v = \sqrt{\frac{2qV}{m}}. \qquad (29\text{–}10)$$

(a)

FIGURE 29–17 (a) J.J. Thomson at work in his laboratory. (b) Schematic diagram of Thomson's apparatus for measuring the charge-to-mass ratio of the electron. The magnetic field is directed into the plane of the page.

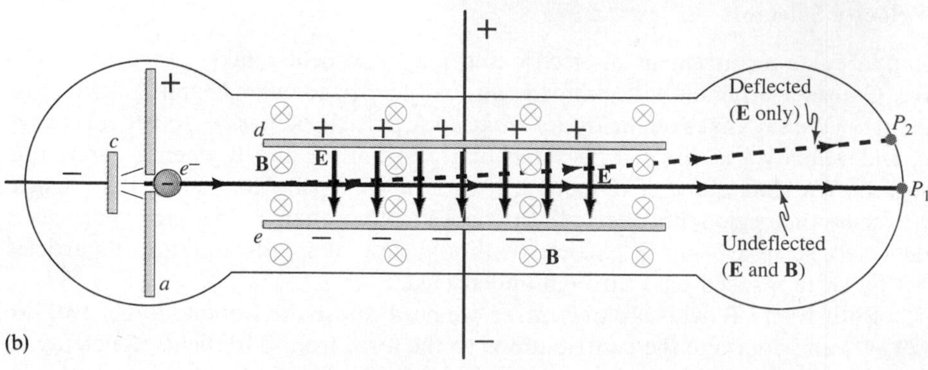

(b)

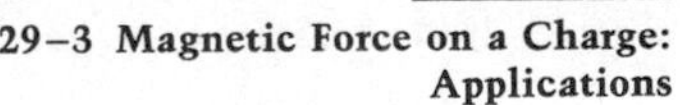

The electrons accelerated in this way continued into a region of crossed electric and magnetic fields. Thomson adjusted the magnitudes of these fields until the electrons passed through the apparatus undeflected. When we combine Eqs. (29–9) and (29–10), we find that

$$v = \frac{E}{B} = \sqrt{\frac{2qV}{m}}.$$

When both sides of this equation are squared, we can solve for q/m:

$$\frac{q}{m} = \frac{E^2}{2VB^2}. \qquad (29\text{–}11)$$

Figure 29–17b represents Thomson's apparatus. The electrons were accelerated between plates c and a and moved on to the region between plates d and e. Thomson did not measure the velocity of the electrons by measuring the voltage between plates a and c. He instead turned off the magnetic field between plates d and e and measured the velocity of the electrons by observing their deflection as they passed through the vertical electric field between plates d and e, as indicated by their arrival at point P_2. He then turned the magnetic field on and adjusted it until the electrons were undeflected; that is, until they arrived at point P_1, which lies on the central axis of the apparatus. His first result for q/m for the electron was 0.77×10^{11} C/kg. Even though Thomson's result was different from the value now accepted, 1.759×10^{11} C/kg, his measurement was a tremendous achievement.

Magnetic Fields in Outer Space

Magnetic fields exist in outer space. Throughout our galaxy, for example, the magnetic field strength is in the range of 10^{-10} T. Charged particles (*cosmic rays*) are generated and accelerated by various stellar processes. If their momentum is less than a certain critical value p_c, they drift in gigantic circles within the galaxy due to the magnetic forces on them.† Cosmic rays with a momentum greater than p_c move on a circle with a radius of curvature greater than the galaxy's radius, and they therefore escape the galaxy. To estimate p_c for a cosmic ray whose charge is the electron charge e, we use the observation that the radius of the galaxy is about 5×10^{21} m. From Eq. (29–5) the critical momentum has magnitude

$$p_c = eBR = (1.6 \times 10^{-19}\ \text{C})(10^{-10}\ \text{T})(5 \times 10^{21}\ \text{m}) = 8 \times 10^{-8}\ \text{kg·m/s}.$$

For a particle such as an electron or a proton, this momentum is enormously large. For comparison, an electron in the beam of a television picture tube typically has a momentum of 10^{-22} kg·m/s, whereas protons in the world's largest proton accelerator (at Fermilab) attain momenta of 5×10^{-16} kg·m/s. Because cosmic rays with a momentum greater than p_c leave the galaxy, we should expect to detect more cosmic rays that strike the earth with momenta lower than p_c than with momenta greater than p_c. Experimental observations of particles arriving from outer space help us to estimate the value of the interstellar magnetic field.

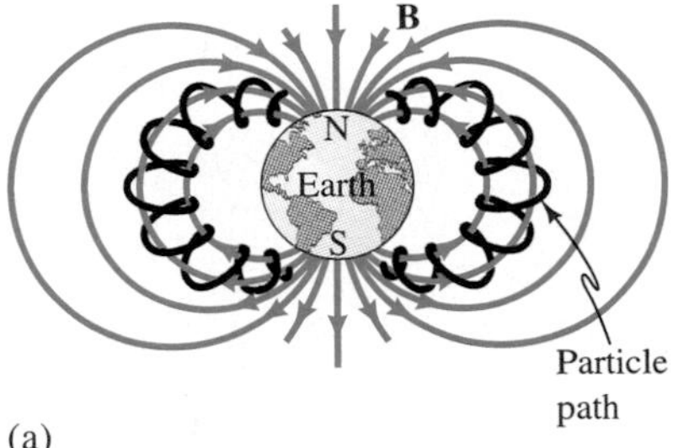

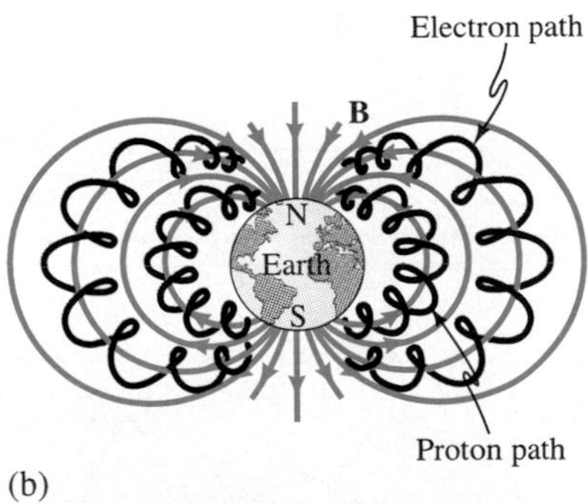

FIGURE 29–18 (a) Charged particles spiraling around the earth's magnetic field. The pitch, or inclination, of the helix decreases to zero as the particles near the poles, where the field lines are denser, and the particles reverse the direction of their helical path. The particles are trapped, bouncing back and forth between the polar regions. (b) They form the two Van Allen belts, one for electrons and one for protons.

Van Allen Belts. Earth has a magnetic field like that of a huge bar magnet, directed from the geographic South Pole to the geographic North Pole. Charged particles far enough from the earth's surface that the atmosphere does not interfere with their motion travel in spiral paths around the earth's magnetic field lines. Figure 29–18a illustrates how the lines become more dense, or "pinch," as they

† We use momentum rather than speed because a calculation of the speed here gives a critical speed greater than the speed of light. This indicates that special relativity is necessary, and special relativity suggests that momentum should be used here.

come nearer to the poles. Detailed analysis of the force on a moving charge shows that when the lines "pinch," the helical path of the particles becomes flatter and eventually turns back on itself. A particle spiraling along the field lines from the South Pole toward the North Pole turns around near the North Pole and spirals back southward. There are, in effect, **magnetic mirrors** near the poles. Because this process repeats itself in reverse, charged particles bounce back and forth between the poles. These particles are thereby trapped and accumulate around the earth's magnetic field lines into regions called **Van Allen belts**. James Van Allen discovered these belts in 1958 by using data from the Explorer I satellite.

There are two Van Allen belts, one containing protons at a mean height of 3000 km above the earth's surface, and the other containing electrons at about 15,000 km (Fig. 29–18b). The belts have densities up to 10^5 particles/cm^3. The particles in the belt originate mostly from the *solar wind*, which consists of particles streaming outward from the sun. The protons and electrons are separated into two regions because they have different speeds and momenta. One consequence of the belts is the *aurora*, a phenomenon that occurs wherever the charged particles enter the atmosphere and excite the atoms in the air (see the chapter-opening photograph). The auroras are more noticeable near the poles, where the magnetic field lines dip toward the earth, carrying the Van Allen belts with them.

EXAMPLE 29–3 Assume that a proton of speed 1.5×10^7 m/s approaches the earth at an angle of 40° to the earth's magnetic field lines and is captured in the lower Van Allen belt (at a mean altitude of 3000 km) without a change in speed. If the mean strength of the field at this altitude is 10^{-5} T, find the cyclotron frequency and the radius of curvature for the proton's motion.

SOLUTION: The cyclotron frequency, from Eq. (29–7), is

$$f = \frac{qB}{2\pi m} = \frac{(1.6 \times 10^{-19}\ \text{C})(10^{-5}\ \text{T})}{2\pi(1.67 \times 10^{-27}\ \text{kg})} = 150\ \text{Hz}.$$

The proton moves in a spiral whose radius of curvature depends on the component of velocity perpendicular to the magnetic field. This component has magnitude $v_\perp = v \sin 40° = (1.5 \times 10^7\ \text{m/s})(0.64) \simeq 10^7$ m/s. Using Eq. (29–8) we then find that

$$R = \frac{mv_\perp}{qB} = \frac{(1.67 \times 10^{-27}\ \text{kg})(10^7\ \text{m/s})}{(1.6 \times 10^{-19}\ \text{C})(10^{-5}\ \text{T})} = 10^4\ \text{m} = 10\ \text{km}.$$

The radius of curvature is much less than the altitude of the Van Allen belt.

Magnetic mirrors analogous to those of the Van Allen belts but made by humans may be an important ingredient of any future thermonuclear fusion power generation, because they can trap the charged particles used in thermonuclear fusion reactions. Indeed, much of the current research effort in controlled thermonuclear fusion relies heavily on the effects of magnetic fields on charged particles.

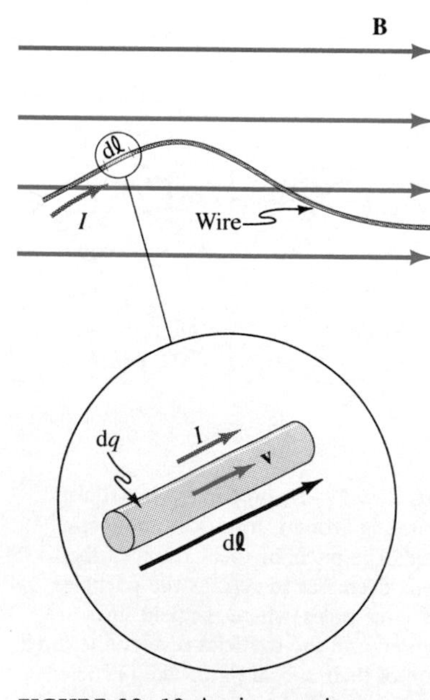

FIGURE 29–19 A wire carrying a current in a magnetic field. We isolate a segment $d\ell$ of the wire that contains the moving charge dq.

29–4 MAGNETIC FORCES ON CURRENTS

We have learned that there may be a force on moving charges in a magnetic field. Because electric currents in wires consist of moving charges, we expect that a magnetic field will exert a force on the charges in a current-carrying wire, and thus on the wire itself (Fig. 29–19). Experiment bears out this expectation.

A wire has moving charges throughout, and the magnetic field may vary significantly along its length. The total force on a current-carrying wire is the vector sum of the magnetic forces on all of the moving charges within it. To find the total force, we first determine the force on a small segment of a current-carrying wire. We subsequently sum (integrate) the infinitesimal force on each segment.

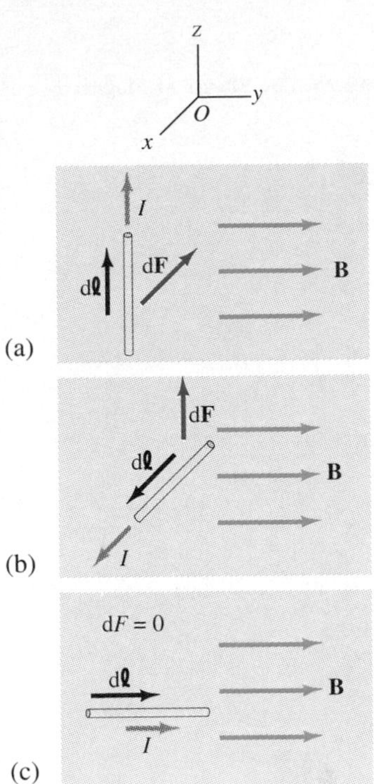

FIGURE 29–20 A wire segment, aligned with the (a) z-axis, (b) x-axis, and (c) y-axis, in a magnetic field, with the infinitesimal forces $\mathbf{dF}$ that act on the segment.

Magnetic Forces on Infinitesimal Wires with Currents

Let us denote the small segment of a thin current-carrying wire by $\mathbf{d}\boldsymbol{\ell}$: It has both an infinitesimal magnitude $d\ell$ and a direction along the instantaneous current carried by the wire at segment $\mathbf{d}\boldsymbol{\ell}$. If the moving charge dq contained in a segment of wire $\mathbf{d}\boldsymbol{\ell}$ has velocity $\mathbf{v}$ along the wire (Fig. 29–19), its displacement $\mathbf{d}\boldsymbol{\ell}$ in time dt is $\mathbf{d}\boldsymbol{\ell} = \mathbf{v}\,dt$, so

$$\mathbf{v} = \frac{\mathbf{d}\boldsymbol{\ell}}{dt}. \tag{29–12}$$

Because the current, I, is by definition dq/dt, the amount of charge within the segment is

$$dq = I\,dt. \tag{29–13}$$

Note that the magnetic field will be uniform over the length of the segment if the segment is small enough. With Eqs. (29–12) and (29–13) we can calculate the magnetic force $\mathbf{dF}$ that acts on our charge element dq and hence on the wire element:

$$\mathbf{dF} = dq\,\mathbf{v} \times \mathbf{B} = (I\,\cancel{dt})\left(\frac{\mathbf{d}\boldsymbol{\ell}}{\cancel{dt}} \times \mathbf{B}\right).$$

We cancel the factor dt to find the infinitesimal force on a wire element $\mathbf{d}\boldsymbol{\ell}$ carrying current I in a magnetic field $\mathbf{B}$:

$$\mathbf{dF} = I\,\mathbf{d}\boldsymbol{\ell} \times \mathbf{B}. \tag{29–14}$$

Note that the current is the same everywhere along the wire, because current is conserved.[†] The magnitude of the magnetic force $\mathbf{dF}$ is given by

The magnetic force on a segment of current-carrying wire

$$dF = I\,d\ell\,B\sin\theta, \tag{29–15}$$

where θ is the angle between the direction of the wire segment (the current's direction) and the direction of the magnetic field. As for the direction of the force, Fig. 29–20 shows three different placements of a segment of current-carrying wire in a uniform magnetic field that points in the $+y$-direction. In each case the direction of the force on the wire segment is given by the right-hand rule and Eq. (29–14). In Fig. 29–20a, $\mathbf{d}\boldsymbol{\ell}$ points in the $+z$-direction, so the force points in the $-x$-direction. In Fig. 29–20b, $\mathbf{d}\boldsymbol{\ell}$ points in the $+x$-direction, so the force points in the $+z$-direction. In Fig. 29–20c, $\mathbf{d}\boldsymbol{\ell}$ points in the $+y$-direction, parallel to $\mathbf{B}$, so there is no force.

Magnetic Forces on Finite Wires with Currents

The *net* force $\mathbf{F}$ on a *finite* section of wire is the vectorial sum of the forces on the various infinitesimal segments that make up the wire. We find the net force by integrating $\mathbf{dF}$ [Eq. (29–14)] over the total length of the wire. Because I is the

[†] We have not considered thick wires, for which we might need to worry about variations in the current density inside the wire.

same everywhere along the wire, the summation of Eq. (29–14) has the form

$$\mathbf{F}_B = I \int (\mathrm{d}\boldsymbol{\ell} \times \mathbf{B}). \tag{29–16}$$

Whether the integral is easy to perform or not depends on the particular situation, as we shall see in Example 29–4.

EXAMPLE 29–4 A long, straight wire carrying a current I passes through a region of constant magnetic field perpendicular to the wire (Fig. 29–21a). The length of the wire in the region of the magnetic field is L. Find the net force on the wire. Repeat the problem for a wire oriented at an angle θ to the field (Fig. 29–21b).

SOLUTION: All the elements of the expression for the total force, Eq. (29–16), are known here; we need only perform the integration. Each segment $\mathrm{d}\boldsymbol{\ell}$ of the wire points in the same direction, $\mathrm{d}\boldsymbol{\ell} = \mathrm{d}\ell\,\mathbf{k}$, and $\mathbf{B}$ is constant at each segment, $\mathbf{B} = B\mathbf{j}$. The infinitesimal force on each segment is therefore identical:

$$\mathrm{d}\mathbf{F} = I[(\mathrm{d}\ell\,\mathbf{k}) \times (B\mathbf{j})] = I\,\mathrm{d}\ell\,B(\mathbf{k} \times \mathbf{j}) = I\,\mathrm{d}\ell\,B(-\mathbf{i}),$$

a vector that points in the $-x$-direction. The direction is easily checked with the right-hand rule. The net force $\mathbf{F}$ is an integral over $\mathrm{d}\mathbf{F}$:

$$\mathbf{F} = \int \mathrm{d}\mathbf{F} = \int I\,\mathrm{d}\ell\,B(-\mathbf{i}) = IB(-\mathbf{i}) \int_0^L \mathrm{d}\ell = -IBL\mathbf{i}. \tag{29–17}$$

The net force points in the $-x$-direction, and its magnitude is IBL.

The only change when the wire makes an angle θ with $\mathbf{B}$ is in the expression for the magnitude of the infinitesimal force element: There is an additional factor $\sin\theta$ in $\mathrm{d}F$ [Eq. (29–15)]. The direction of the force remains, by the right-hand rule, in the $-x$-direction. Because θ is constant, it does not enter into the integration, which thus remains as above. We have

$$F = IBL\sin\theta, \text{ in the } -x\text{-direction.} \tag{29–18}$$

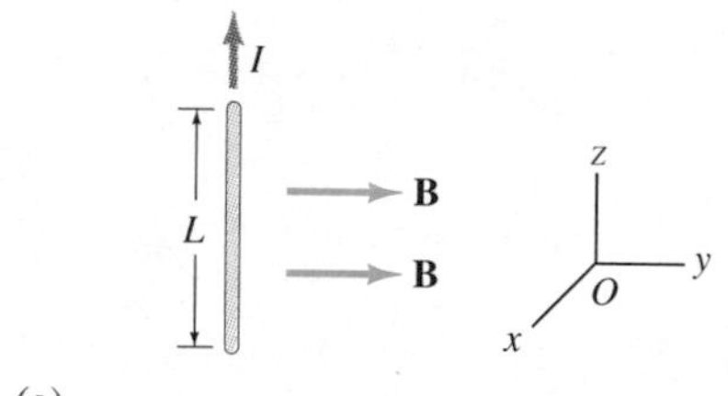

(a)

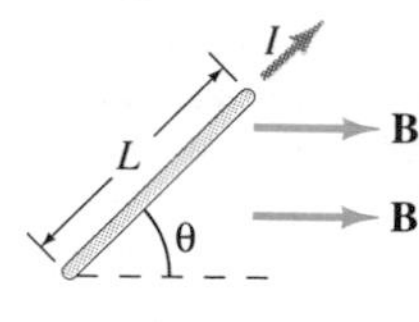

(b)

FIGURE 29–21 (a) Example 29–4. (b) The same wire segment as in part (a), but at an angle.

Equations (29–17) and (29–18) are useful and important results. They can be combined into one vectorial equation that gives the force on a thin, straight wire of length L in a magnetic field, namely,

$$\mathbf{F} = I\,\mathbf{L} \times \mathbf{B}, \tag{29–19}$$

where the vector $\mathbf{L}$ is oriented along the wire in the direction of the current.

The expression for the magnetic force on an isolated moving charge has led us directly to the expression for the magnetic force on a current-carrying wire. Historically, the order of discovery was the reverse: The first experiments on magnetic forces, those of Oersted, François Arago, and Ampère, were performed on thin wires that carried electric currents in the presence of magnetic fields. The results of these experiments then led to an understanding of magnetic forces on moving charges.

29–5 MAGNETIC FORCE ON CURRENT LOOPS

Magnetic fields exert forces on all kinds of current-carrying wires, including those of closed loops. As we shall see, a uniform magnetic field actually exerts only a torque on a current loop. This phenomenon provides the torque that runs direct-

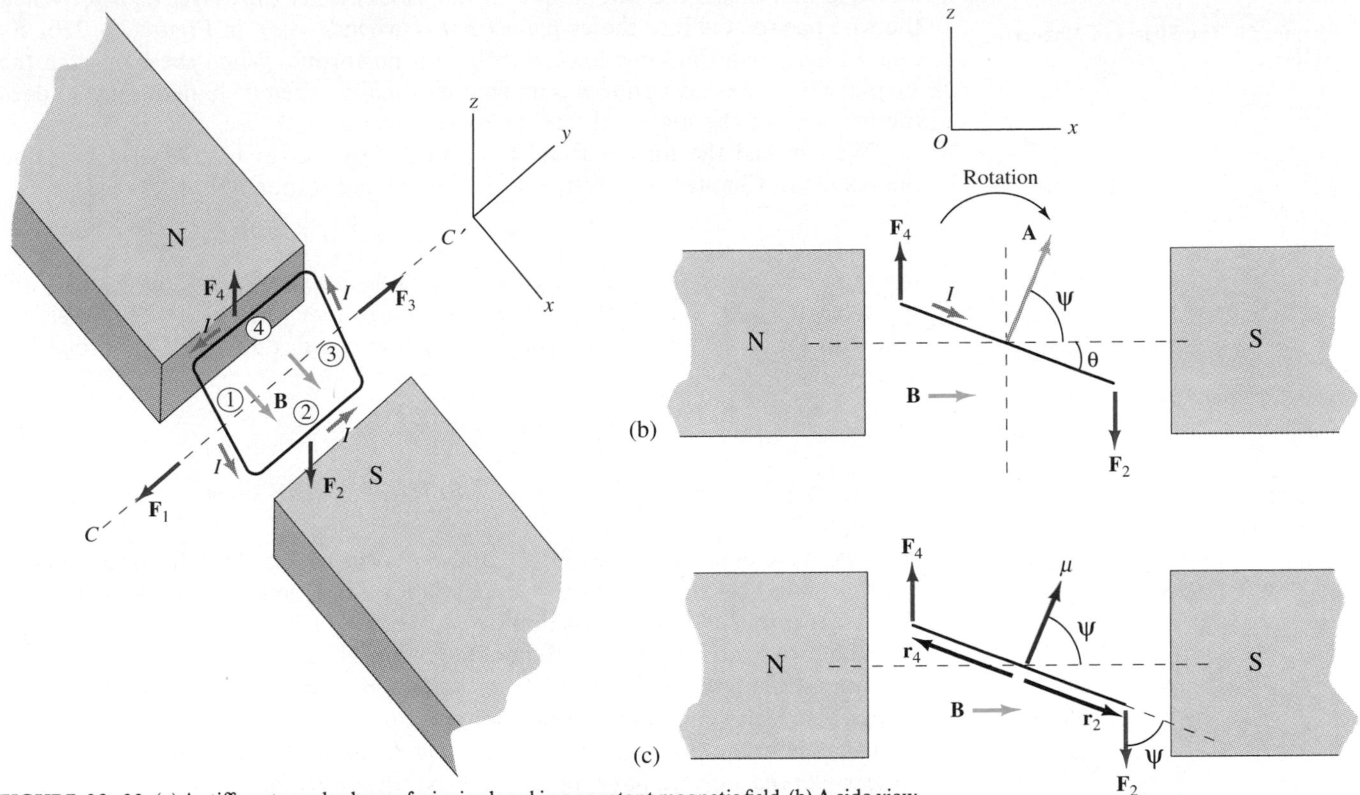

FIGURE 29–22 (a) A stiff, rectangular loop of wire is placed in a constant magnetic field. (b) A side view, looking along the y-axis, of the loop. (c) Geometry of the loop that allows us to calculate the torque on the loop. The torque tends to align the vector $\boldsymbol{\mu}$ with the magnetic field, **B**.

current electric motors and the galvanometer, the device cited in Chapter 28 for use in ammeters and voltmeters. Consider a stiff rectangular loop of wire carrying current I (Fig. 29–22a). The magnetic field is oriented in the $+x$-direction. The sides of the wire loop are denoted 1, 2, 3, and 4; sides 1 and 3 have length a, and sides 2 and 4 have length b. Figure 29–22b is a side view of the apparatus along the $+y$-direction. The direction perpendicular to the plane of the loop (the direction of the thumb when the fingers of the right hand follow the current direction) makes an angle ψ with the magnetic field.

We can calculate the force on each leg of the loop by using Eq. (29–19). The angle θ between the direction of the magnetic field and the direction of the current is shown in Fig. 29–22b for leg 1; legs 2 and 4 are perpendicular to the magnetic field, into and out of the page, respectively. Therefore the force on each leg is

$$F_1 = IaB \sin\theta, \text{ in the } -y\text{-direction}; \tag{29–20a}$$

$$F_2 = IbB, \text{ in the } -z\text{-direction}; \tag{29–20b}$$

$$F_3 = IaB \sin\theta, \text{ in the } +y\text{-direction}; \tag{29–20c}$$

$$F_4 = IbB, \text{ in the } +z\text{-direction}. \tag{29–20d}$$

These forces are indicated in Figs. 29–22a and 29–22b. Forces $\mathbf{F}_1$ and $\mathbf{F}_3$ are equal and opposite, as are forces $\mathbf{F}_2$ and $\mathbf{F}_4$, so there is no net force. However, there is an important difference between these two sets of forces: $\mathbf{F}_1$ and $\mathbf{F}_3$ act along the same axis (CC' in Fig. 29–22a) and exert no torque on the loop. $\mathbf{F}_2$ and $\mathbf{F}_4$ act along different axes, as emphasized in Fig. 29–22b, and therefore produce a torque that causes the wire loop to rotate clockwise in the magnetic field. When the wire has rotated into the yz-plane (that is, when $\theta = 90°$ in Figure 29–22b), $\mathbf{F}_2$ and $\mathbf{F}_4$ act along the same axis, and there is no torque. When the loop is in the xy-plane ($\theta = 0°$), the torque is a maximum. Finally, when θ changes sign, so does the torque, and the loop will tend to rotate counterclockwise.

We can find the torque about the central axis CC' in Fig. 29–22a by using the results of Chapter 10. From Eq. (10–6), the net torque about this axis, $\boldsymbol{\tau}$, is

$$\boldsymbol{\tau} = (\mathbf{r}_2 \times \mathbf{F}_2) + (\mathbf{r}_4 \times \mathbf{F}_4),$$

where $\mathbf{r}_2$ and $\mathbf{r}_4$ are the perpendicular vectors from axis CC' to legs 2 and 4, respectively (Fig. 29–22c). Both $\mathbf{r}_2$ and $\mathbf{r}_4$ have magnitude $a/2$. Figure 29–22c shows that ψ is the angle between $\mathbf{r}_2$ and $\mathbf{F}_2$ and between $\mathbf{r}_4$ and $\mathbf{F}_4$. The torque has magnitude

$$\tau = r_2F_2 \sin\psi + r_4F_4 \sin\psi$$

$$= \frac{a}{2}(IbB)\sin\psi + \frac{a}{2}(IbB)\sin\psi = IabB\sin\psi. \quad (29\text{–}21)$$

Here we have used the values for F_2 and F_4 given by Eq. (29–20). According to the right-hand rule, both terms in the equation for $\boldsymbol{\tau}$ point in the $+y$-direction, so the net torque is in this direction.

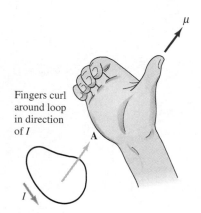

Fingers curl around loop in direction of I

FIGURE 29–23 A right-hand rule indicates the direction of the magnetic dipole moment, $\boldsymbol{\mu}$, of a current loop.

The torque on a current loop in a magnetic field as given by Eq. (29–21) can be summarized and generalized. First, the factor ab is the area A of the current loop. This result generalizes to *any* planar loop of area A, whatever its shape, by the calculus technique of decomposing a planar loop of any shape into tiny rectangles like the loop we have discussed. Next, the vectorial nature of the torque is handled neatly by defining a vector $\boldsymbol{\mu}$ perpendicular to the plane of the loop.[†] Two vectors are perpendicular to any plane; which of them do we choose for $\boldsymbol{\mu}$? We choose to define $\boldsymbol{\mu}$ with a right-hand rule: Curl the fingers of the right hand in the direction of the current around the loop, and the right thumb gives the direction of $\boldsymbol{\mu}$ (Fig. 29–23). Try it for Fig. 29–22c, where we have indicated $\boldsymbol{\mu}$. The angle between $\boldsymbol{\mu}$ and $\mathbf{B}$ is ψ. We have thus shown that the direction and magnitude of the *torque on a current loop* are given by

The torque on a current loop in a constant magnetic field

$$\boldsymbol{\tau} = \boldsymbol{\mu} \times \mathbf{B}, \quad (29\text{–}22)$$

provided that the magnitude of $\boldsymbol{\mu}$ is taken to be

$$\mu = IA. \quad (29\text{–}23)$$

From these equations, the magnitude of the torque is

$$\tau = \mu B \sin\psi,$$

exactly as in Eq. (29–21), and the direction of the torque is the $+y$-direction. A torque in this direction acts to align $\boldsymbol{\mu}$ and $\mathbf{B}$. It is generally true that *the torque tends to rotate a current loop or coil in such a way that* $\boldsymbol{\mu}$ *and* $\mathbf{B}$ *become aligned.*

The torque on a current loop in a uniform magnetic field is completely analogous to the torque on an electric dipole (a pair of equal and opposite electric charges) in a uniform electric field, discussed in Chapter 23. The electric dipole was described by the *electric dipole moment* $\mathbf{p}$, a vector aligned with the two charges

[†] Do not confuse μ with the permeability of free space, μ_0.

and equal in magnitude to the charge times the distance of charge separation. In terms of the response to an external magnetic field, *the current loop is in all measurable respects a* **magnetic dipole**.[†] We therefore call $\boldsymbol{\mu}$, which plays a role analogous to **p**, the **magnetic dipole moment** of the loop.

One more generalization is possible. Suppose that, instead of one turn of wire, our loop consists of N turns (each surrounding the same plane area). We might now refer to a *coil* rather than a loop. *Each* turn of the coil experiences the forces we have described, and the torque is multiplied by N. This factor is included with the other factors intrinsic to the coil, so the magnetic dipole moment will now take into account N:

$$\mu = INA. \tag{29–24}$$

The magnetic dipole moment of a coil of current

We have seen another system that aligns itself with external magnetic fields: the iron filings we used to make our preliminary definition of magnetic fields. These iron filings, we argued, behave like little bar magnets that are also rotated by magnetic fields. Bar magnets react to fields just like current loops do. As we shall see in Chapter 30 when we find the magnetic fields *produced* by magnetic dipoles, bar magnets are themselves magnetic dipoles. This is so because, at a microscopic level, metals contain circulating currents, formed by the electrons that orbit atomic nuclei. The magnetic behavior of materials is a very active area of research. We shall study this behavior in more detail in Chapter 32.

Magnetic Charges

Our procedure in studying magnetic fields has been to probe the effects of these fields with test objects such as bar magnets or tiny magnetic dipoles. This is similar to the procedure we followed in probing the effects of electric fields, but there is one very fundamental difference. For the case of electric fields, we are able to use test objects with single electric charges, or electric monopoles. With single charges, it is quite direct to learn the force laws for electric fields. If we had used electric dipoles as probes, the corresponding force laws would have appeared to be much more complicated, and our job would have been much more difficult. Yet this is just what we have done for magnetism! Why did we not use magnetic *charges*, or magnetic monopoles, as probes? The answer is that, although many experiments have been performed to find magnetic monopoles, no one has ever observed a magnetic monopole unambiguously, and it may be that such objects simply do not exist. If indeed they do not exist, there is a fundamental asymmetry between electricity and magnetism. It is remarkable that electricity and magnetism are nevertheless linked so intimately.

Motors and Galvanometers

With the apparatus shown in Fig. 29–22a, as soon as the wire loop rotates past the position in which it is aligned with the yz-plane, the torque on it changes sign and becomes counterclockwise. In fact, the torque changes direction when ψ goes through 0° or 180°. However, if we can make the *current* switch directions every time the loop passes $\psi = 0°$ or 180°, then the torque will continuously produce a

[†] One of the measurable properties—the defining property—of a dipole, magnetic or electric, is the characteristic field that the dipole itself *produces*. We deal with the calculation of the magnetic dipole field in Chapter 30.

(a)

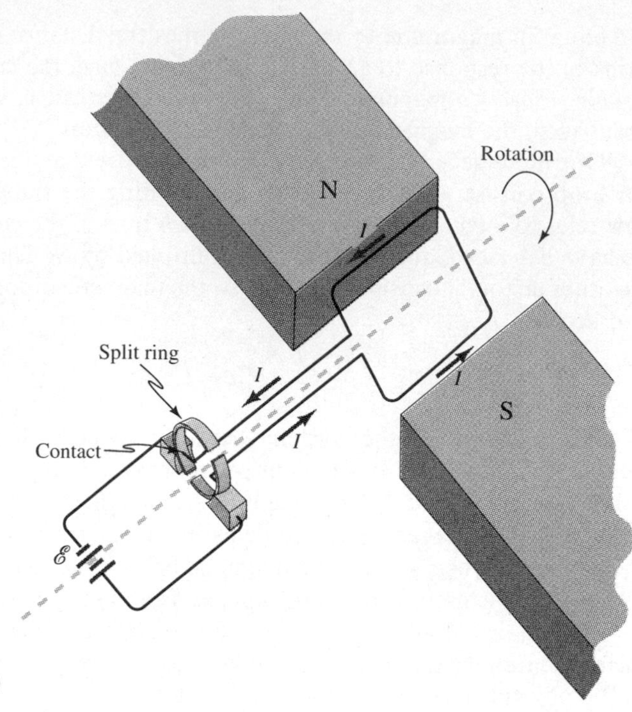

(b)

FIGURE 29–24 (a) A current-carrying loop aligned in a magnetic field is fitted with a split-ring commutator. (b) Schematic diagram of a split-ring commutator. The torque on the loop serves to turn the loop and makes a motor.

clockwise rotation. Such a device is known as a *split-ring commutator* (Fig. 29–24). The loop continues to rotate under the influence of a torque whose sign does not change, and we will have created a type of *electric motor*.

The fact that a magnetic field exerts a torque on a current loop suggests also that such a loop can be made into a device that measures currents. In fact, we have already used such a device in Chapter 28, a *galvanometer*. For example, we can attach a spring to a loop to balance the torque due to a known magnetic field (Fig. 29–25a). The amount the spring stretches is a measure of the torque of the loop and hence of the current that passes through the loop. The galvanometer, shown schematically in Fig. 29–25b, is the basis for the electrical instruments we studied in Chapter 28.

Energy and the Torque on Loops

When a magnetic field rotates a current loop, the field does work. For a constant field, the only variable in the work is the angle of rotation, ψ. We know from Chapter 7 that when the force (or the torque) depends only on position, the concept of potential energy is useful because we can apply the principle of the conservation of energy. Thus we can associate a potential energy $U(\psi)$ with an oriented loop in a magnetic field, where ψ is the angle between $\boldsymbol{\mu}$ and $\mathbf{B}$. As always, only *changes* in the potential energy have physical consequences. The change in potential energy in rotating the coil from some initial angle ψ to a final angle of 90° is given by the negative of the work done by the magnetic field in moving the coil through these angles:

$$U(\psi) - U(90°) = -\int_{\psi}^{90°} \tau \, d\psi' = -\int_{\psi}^{90°} (\mu B \sin \psi') \, d\psi'$$

$$= -\mu B \int_{\psi}^{90°} \sin \psi' \, d\psi' = \mu B \cos 90° - \mu B \cos \psi.$$

The cosine of 90° is zero, so

$$U(\psi) - U(90°) = -\mu B \cos\psi. \tag{29–25}$$

Zero U can be chosen for convenience. It is customary to choose U to be zero at $\psi = 90°$; that is, when $\boldsymbol{\mu}$ is perpendicular to **B**. Thus we set $U(90°) = 0$ in Eq. (29–25), giving us *the potential energy of a current loop with a given magnetic dipole moment* $\boldsymbol{\mu}$ *in a constant magnetic field* **B**:

$$U(\psi) = -\boldsymbol{\mu} \cdot \mathbf{B}. \tag{29–26}$$

The energy of a current loop in a constant magnetic field. The lowest value of potential energy occurs when the magnetic dipole moment and the magnetic field are aligned.

The potential energy has a *minimum* when $\boldsymbol{\mu}$ is aligned along **B** (that is, when $\psi = 0°$). Thus the orientation in which $\boldsymbol{\mu}$ is aligned with **B** is a stable equilibrium point. This agrees with our earlier result that the torque tends to rotate the loop to line up $\boldsymbol{\mu}$ and **B**.

EXAMPLE 29–5 A current loop in a constant magnetic field is initially aligned so that $\boldsymbol{\mu}$ points in a direction slightly different from that of **B**, and the loop is then released. There is no mechanism such as friction for energy loss; that is, no damping. Find the subsequent motion of the loop.

SOLUTION: We start with the loop oriented slightly away from the stable equilibrium point, that is, with small but nonzero ψ, where ψ is the angle between $\boldsymbol{\mu}$ and **B**. To describe the behavior of the loop, we express $U(\psi)$ for small values of ψ. Using the small-angle approximation $\cos\psi \simeq 1 - (\psi^2/2)$ from Appendix IV–9, we have

$$U(\psi) = -\mu B \cos\psi \simeq -\mu B + \frac{\mu B \psi^2}{2}. \tag{29–27}$$

The term $-\mu B$ is a constant and therefore has no bearing on the loop's motion. The term $\mu B\psi^2/2$ is characteristic of a familiar physical system, the harmonic oscillator. This result is not surprising, because we know that small motions about almost *any* stable equilibrium are harmonic. The harmonic oscillator of Chapter 13 has potential energy $kx^2/2$, a form associated with harmonic motion in the variable x about the equilibrium point $x = 0$. Here, ψ plays the

(a)

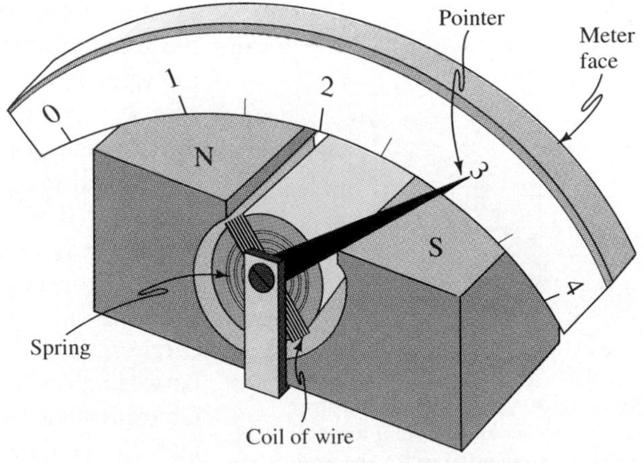

(b)

FIGURE 29–25 (a) A galvanometer measures the current in a loop. A spring balances the torque from the magnetic forces on the loop. (b) Schematic diagram of the galvanometer. The amount by which the spring distorts, as measured on the meter face, indicates the current.

role of x, whereas μB substitutes for the spring constant k. The loop in the magnetic field acts like a physical pendulum, so we should also substitute for the mass m in the harmonic oscillator the rotational inertia of the loop, I_M, about the rotation axis. (We use I_M to avoid any possible confusion with the current.) The motion will then be

$$\psi = \psi_m \sin(\omega t + \phi)$$

[see Eq. (13–1)]. Here ψ_m and ϕ are the (given) amplitude and phase, respectively, and the angular frequency is

$$\omega = \sqrt{\frac{\mu B}{I_M}}.$$

We have harmonic motion in ψ about the stable equilibrium point $\psi = 0$, where $\boldsymbol{\mu}$ is aligned with **B**. There is an interchange between the potential energy, Eq. (29–27), and the kinetic energy term associated with the motion of the loop: When the potential energy is large, at the maximum value of ψ, the kinetic energy is zero; when the potential energy is zero, at $\psi = 0$, the kinetic energy is a maximum (see Problem 47).

An undamped compass needle behaves much like the loop in this example, because the needle is a bar magnet and therefore acts as a current loop.

29–6 THE HALL EFFECT

The direction of a current does *not* itself determine the sign of the charge carriers in that current, because a current to the right can be produced by the movement either of positive charges to the right or of negative charges to the left. The *Hall effect* will allow us to find this sign. Consider a strip of metal of length L on which an electric potential is applied from one end to the other, so that a current flows in the strip. The strip is placed in a uniform magnetic field perpendicular to the strip (Fig. 29–26). Let us describe qualitatively the potential difference between points a and b.

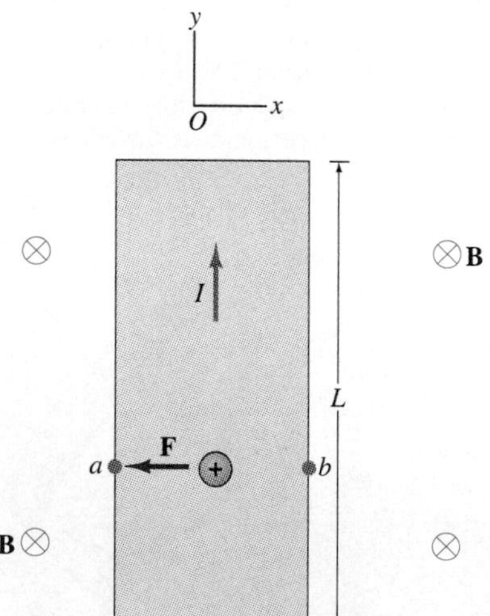

FIGURE 29–26 A conducting strip perpendicular to a constant magnetic field develops a potential called the Hall voltage between points a and b when the strip carries a current I.

Equation (29–19) gives the total force on the strip, $\mathbf{F} = I\,\mathbf{L} \times \mathbf{B}$. This force, by the right-hand rule, is directed in the $-x$-direction—it acts to the *left* in Fig. 29–26. By using the equivalent force law $\mathbf{F} = q\,\mathbf{v} \times \mathbf{B}$, we can show that the force on the charge carriers acts to the left, whatever the sign of the charge carriers. If the moving charges are positive and the current flows in the $+y$-direction, then the velocity of the charges is also in the $+y$-direction. According to the right-hand rule, $\mathbf{F} = q\,\mathbf{v} \times \mathbf{B}$ is then directed toward point a in Fig. 29–26. However, if the moving charges are negative (electrons), then the velocity of the charges is in the $-y$-direction. The vector product $\mathbf{v} \times \mathbf{B}$ is directed to the right, but q is negative, and $q\,\mathbf{v} \times \mathbf{B}$ again points to the left, moving these negative charges toward a. Either way, there is a buildup of the charge carriers at the left side of the strip. This buildup cannot continue indefinitely: Once enough charge carriers have moved to the left, they will supply a repulsive Coulomb force against the movement of other charge carriers there. An equilibrium is established once an electric potential is set up between points a and b that prevents further leftward drift of charge carriers. Charges then move up the strip as they would if there were no magnetic field. In fact, the charge separation leads to an electric field between points a and b. We therefore have crossed **E** and **B** fields, and the charges travel undeflected up the wire, just as in the velocity selector discussed in Section 29–3.

The sign of the potential difference between points a and b determines the sign of the charge carriers. This phenomenon is known as the **Hall effect**. If the charge carriers are negative, negative charges build up on the left side of the metal strip, and point a is at a lower potential than point b. Conversely, if the carriers are positive, positive charges build up on the left side of the strip, and point a is at a higher potential than point b. The first measurement of the sign of this *Hall potential* was performed by the physicist Edwin H. Hall in 1879. His measurement proved that the carriers of current in metals are negatively charged.

The Hall effect allows us to determine the sign of the charge carriers in a metal.

EXAMPLE 29–6 The apparatus of Fig. 29–26 that demonstrates the Hall effect sits in a magnetic field of 2.0 T. The width of the strip is 1.0 cm, and a voltage of magnitude 7.2 μV is measured across the strip. What is the speed v of the charge carriers in the strip?

SOLUTION: We know the magnitude of the magnetic field, and, from the voltage, we can find the magnitude of the crossed electric field. As we have already seen, the electric field set up across the strip is just such that the velocity-selector condition [Eq. (29–9)] holds, so the unknown speed is given by

$$v = \frac{E}{B}.$$

The electric field that is set up across the strip is determined from the electric potential, V, by

$$E = \frac{V}{d}$$

[recall Eq. (25–30)], where d is the width of the strip. Thus the charge-carrier's speed is

$$v = \frac{V}{Bd} = \frac{7.2 \times 10^{-6}\ \text{V}}{(2.0\ \text{T})(1.0 \times 10^{-2}\ \text{m})} = 3.6 \times 10^{-4}\ \text{m/s}.$$

This is a measurement of an electron's drift speed.

SUMMARY

Magnets, moving electric charges, and electric currents all experience magnetic forces. These forces can be described in terms of a magnetic field, $\mathbf{B}$, whose spatial dependence can be mapped out with iron filings or by observing its effect on a moving electric test charge or a test current element. In terms of this field, the magnetic force on an electric charge q depends on the charge's velocity according to the magnetic force law,

$$\mathbf{F}_B = q\,\mathbf{v} \times \mathbf{B}. \qquad (29\text{–}1)$$

The SI unit of magnetic field is the tesla, T: 1 T = 1 kg/(C·s). When both magnetic and electric fields are present, the Lorentz force law holds:

$$\mathbf{F} = q[\mathbf{E} + (\mathbf{v} \times \mathbf{B})]. \qquad (29\text{–}4)$$

In a static magnetic field, the component of a charged particle's velocity parallel to the field is unaffected by that field. The magnitude of the force on the particle due to the field is proportional to the component of the velocity perpendicular to the field, and the direction of the force is perpendicular to this component of the velocity and to the field itself. It follows that the kinetic energy of a charged particle in a magnetic field is unchanging. When the field is constant, a charged particle traveling perpendicular to the field moves in a circle of radius

$$R = \frac{mv}{qB}. \tag{29–5}$$

The frequency of the particle's circular motion is the cyclotron frequency,

$$f = \frac{qB}{2\pi m}, \tag{29–7}$$

which is independent of the particle's velocity. The general path followed by a moving charge is a spiral around the magnetic field lines. Van Allen belts are regions near the earth where charged particles accumulate in spiraling paths under the influence of the earth's magnetic field.

When a charged particle has a particular velocity perpendicular to constant crossed electric and magnetic fields, the electric and magnetic forces cancel, and the particle passes through the fields undeflected. The magnitude of this special velocity is

$$v = \frac{E}{B}. \tag{29–9}$$

With the help of a velocity selector, an apparatus based in part on this phenomenon, the charge-to-mass ratio of the electron can be measured.

The infinitesimal magnetic force on an infinitesimal length of thin wire $d\boldsymbol{\ell}$ that carries a current I in the presence of a constant magnetic field is

$$d\mathbf{F} = I\, d\boldsymbol{\ell} \times \mathbf{B}. \tag{29–14}$$

To find the net force on a wire of finite length in a magnetic field, Eq. (29–14) is integrated. For example, the force on a straight wire of length L in a uniform magnetic field is given by Eq. (29–19), $\mathbf{F} = I\,\mathbf{L} \times \mathbf{B}$. A second example concerns a wire that carries a current and is formed into a loop (or coil) of N turns; the area of the face of the loop is A. When it is placed in a constant magnetic field, such a loop experiences a torque

$$\boldsymbol{\tau} = \boldsymbol{\mu} \times \mathbf{B}. \tag{29–22}$$

The loop reacts as a magnetic dipole with magnetic dipole moment μ. For a coil of N turns, μ has magnitude

$$\mu = INA \tag{29–24}$$

and direction perpendicular to the face of the coil, oriented by a right-hand rule on the current. The torque tends to rotate the loop so that $\boldsymbol{\mu}$ and $\mathbf{B}$ become aligned. The potential energy of the loop in a constant magnetic field can be expressed as

$$U(\psi) = -\boldsymbol{\mu} \cdot \mathbf{B}, \tag{29–26}$$

where ψ is the angle between $\boldsymbol{\mu}$ and $\mathbf{B}$.

The Hall effect exploits the equivalence between the force on a moving charge and the force on a current-carrying wire. This effect proves that the current carriers in metals are negatively charged.

QUESTIONS

1. A wire carrying a current is electrically neutral, yet a magnetic field acts on it. Why?
2. Explain how you might define and measure a magnetic field if magnetic monopoles existed.
3. An electron beam in an oscilloscope is deflected to the right on the screen. Could this be caused by an electric field *or* by a magnetic field? Explain how you could distinguish these possibilities.
4. An electron beam makes a spot in the center of the screen of a cathode-ray tube. A bar magnet is brought in from the left side (as seen from the front of the tube), with the S pole nearest to the beam. Which way will the spot move? Suppose that the N end of the bar magnet is brought near the beam from above. Which way will the spot move?
5. Much of the description of magnetic forces depends on the use of a right-hand rule. Does the magnetic force depend fundamentally on the fact that we have chosen the right rather than the left hand?
6. If you have just used a velocity selector for electrons and you wish to use it to choose positrons with the same speed, do you have to change any settings on the selector? Positrons are like electrons, but positively charged.
7. Induced charges give rise to electric forces even between electrically neutral objects. How do we know that the forces between bar magnets are not induced electric forces?
8. Imagine that an electrically neutral wire carrying a current moves in the presence of an external magnetic field. Do you expect that there will be an additional force on the wire due to the movement?
9. You have a fixed length of wire and want to use it to make a magnetic dipole with the largest possible magnetic dipole moment. Into what shape should you wind it? Are you better off making a single loop or N loops?
10. A small bar magnet forms a magnetic dipole; a current-carrying wire in the shape of a small loop also forms a magnetic dipole. If that is the case, the current loop should give rise to a magnetic field. Use this analogy to sketch the magnetic field lines that would be generated by such a current loop.
11. The earth's magnetic south pole is near the geographic North Pole. Why would the geographic North Pole have been called a magnetic south pole?
12. Suppose that the coil of a direct-current electric motor consists of many turns rather than one turn of wire that carries a current I. Does the coil rotate faster than a single loop would? Does the split-ring commutator still work?
13. You have a large pail of water, a bar magnet with its N and S ends unmarked, a straight pin, and a cork. How could you make a compass? One of the things you need to know to construct this compass is how to distinguish north from south; you are allowed to watch the sun to help with this part of the question.
14. Do magnetic north poles repel positive electric charges?
15. A classmate tells you that 1 T is 1 N/A·m. Is your classmate correct?

PROBLEMS

29–1 *Magnets and Magnetic Fields*

1. (II) Sketch the magnetic fields for the arrangements of bar magnets shown in Fig. 29–27.

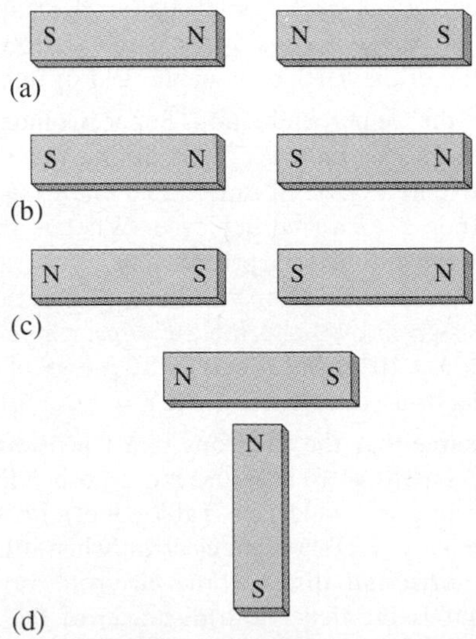

FIGURE 29–27 Problem 1.

29–2 *Magnetic Force on an Electric Charge*

2. (I) A proton with velocity $\mathbf{v} = (2 \times 10^3)\mathbf{i} + 10^2\mathbf{j} + (5 \times 10^2)\mathbf{k}$ m/s moves through a magnetic field $\mathbf{B} = 0.3\mathbf{i} + 0.4\mathbf{j} + 0.5\mathbf{k}$ T. Calculate the force on the proton.
3. (I) A proton of energy 100 keV enters a region of uniform magnetic field. The proton moves uniformly in the $+x$-direction perpendicular to the field. It then experiences an acceleration of 3×10^{10} m/s^2 in the $+y$-direction. What are the magnitude and direction of the magnetic field?
4. (II) A cork ball carrying charge q has a mass of 2.0 g and is set in straight-line motion perpendicular to a uniform magnetic field of 0.10 T. What is the value of q if its direction of motion changes by 1° in 1.0 s?
5. (II) (a) A rapidly moving charged particle of charge e, mass m, and speed v passes through a region of magnetic field $\mathbf{B}$, which points in a direction perpendicular to the motion (Fig. 29–28). The particle spends a time interval Δt in the region. Estimate the angle θ through which it will be deflected during Δt, assuming that θ is small. (b) The particle is a proton, with $m = 1.7 \times 10^{-27}$ kg, and $e = 1.6 \times 10^{-19}$ C, and the speed is 1.4×10^7 m/s. The size of the

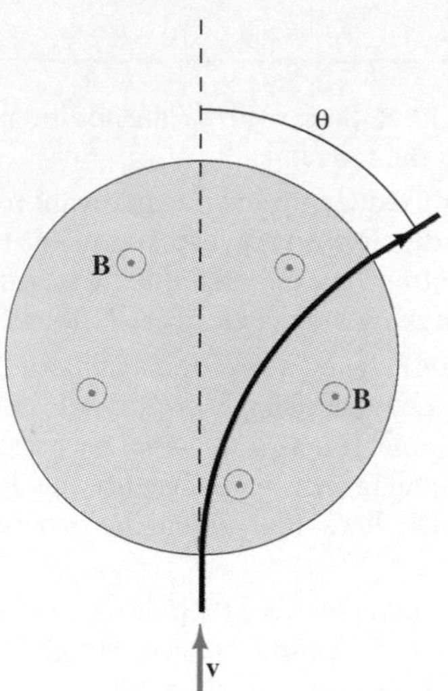

FIGURE 29–28 Problem 5.

magnetized region is 0.1 m across. How large must B be to give rise to a deflection of 0.1 rad?

6. (II) In an oversimplified model of the earth's magnetic field, the field is parallel to the rotation axis, has a constant magnitude of 10^{-4} T up to a height of 100 km, and then quickly drops to zero (Fig. 29–29). A cosmic-ray particle with charge $+e$ and mass 9.5×10^{-26} kg moves at a speed of 10^8 m/s directly toward the equatorial region from above. (a) In what direction is the particle deflected? (b) Estimate how much it will be deflected from the point of impact it would have if it were uncharged. (In fact, this is not a realistic example. Cosmic-ray particles as massive as this have greater charge.)

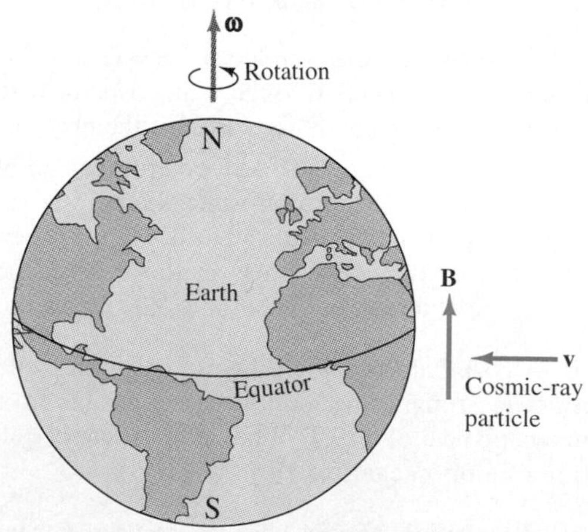

FIGURE 29–29 Problem 6.

7. (II) Undisturbed electrons in a television tube would travel at a speed of 7×10^7 m/s along the $+x$-direction. The television is located at a spot on the earth's surface where the magnetic field has a vertical component of 13 μT and a horizontal component of 22.5 μT. The horizontal component is aligned along the $+x$-direction and we define the vertical direction as the z-direction. Given that the tube is 0.3 m long, calculate the deflection—direction and magnitude—of the electron beam due to the earth's field.

29–3 *Magnetic Force on a Charge: Applications*

8. (I) A proton is sent into a region of constant magnetic field, oriented perpendicular to the proton's path. There the proton travels at a speed of 3×10^6 m/s in a circular path of radius 20 cm. What is the magnitude of the magnetic field?

9. (I) (a) Suppose that electrons from an electron gun with a voltage of 1600 V are injected into a region of constant magnetic field perpendicular to the electrons' velocity. What magnetic field will give the electrons a radius of curvature of 6 cm? (b) A magnetic field of what magnitude is necessary to give an alpha particle (charge $q = 2e$ and mass $m_\alpha = 7360$ times the mass of an electron) with a kinetic energy of 1200 eV a path with radius of curvature of 20 cm?

10. (I) Show that the radius of curvature of a proton moving at a velocity of 25 km/s in a magnetic field of 10^{-10} T is small compared to interplanetary distances. The protons therefore spiral around interplanetary magnetic field lines; we say that the protons are "tied to the magnetic field lines" in cosmic magnetic fields.

11. (I) The magnetic field at the surface of a neutron star has magnitude 10^8 T. What is the radius of the circular orbit of an electron that moves there at 1 percent of the speed of light? What is the magnitude of the magnetic force on the electron?

12. (I) With what frequency will deuterons, which have the same charge as protons but twice the mass, circulate in a cyclotron with a magnetic field of 1.2 T?

13. (I) In a certain region the average radius of curvature of the trajectory of electrons trapped in the Van Allen belt is 300 m and the average electron energy is 100 keV. What is the value of the earth's magnetic field in this region?

14. (II) In the Superconducting Super Collider accelerator scheduled to be built in Texas during the 1990s, protons will move in a circle of radius 13.8 km. Magnetic fields of magnitude 5 T can be achieved. What is the magnitude of the momentum of a proton that moves at a radius of 13.8 km in this field? For protons with this momentum, the energy is given by the formula $E = pc$, where c is the speed of light, 3×10^8 m/s. Calculate the energy of the proton in megaelectron-volts, where 1 MeV $= 1.6 \times 10^{-13}$ J.

15. (II) Assume that the electrons in a television picture tube have an energy of 10 keV and move perpendicularly to the earth's magnetic field (see Table 29–1). (a) Calculate the final velocity (vector) of an electron when it hits the screen if the horizontal distance the electron travels is 40 cm. (b) What is the deflection (distance) of the electron perpendicular to its original direction?

16. (II) The earth acts as a giant magnet whose field lines are like those of a bar magnet, running from the magnetic north pole to the magnetic south pole. Thus the magnetic field at the equator is approximately constant, of magnitude 5×10^{-5} T, and runs from the geographic South Pole to the geographic North Pole. If we ignore air resistance and the gravitational force, a charged body could orbit the earth at the equator as a result of the magnetic force if it has just the right velocity. Suppose that such a body has a charge of 0.1 C and a mass of 0.1 g. (a) What would its velocity have to be for it to travel in such an orbit? (b) Suppose that the gravitational force acts on this body as well. What is the ratio of the gravitational force to the magnetic force?

17. (II) In Section 29–3 we calculated the critical momentum for an electron to stay within the galaxy. (a) Given that the energy of a high-energy particle is related to its momentum by $E = pc$ (see Problem 14), what is the energy of an electron with the critical momentum? What are the critical momentum and energy for (b) an alpha particle (charge $2e$ and mass 4 times the mass of a proton? (c) an ion of uranium, with charge e and mass 240 times the proton mass?

18. (II) A proton moves horizontally perpendicular to a constant magnetic field oriented so as to deflect the proton upward. The magnitude of the field is 0.010 T. What is the speed of the proton so that the magnetic force just cancels the gravitational force on the proton, leaving it in horizontal flight? This problem illustrates how very weak the gravitational force is compared to electromagnetic forces.

19. (II) An electron moves at a speed $v = 10^6$ m/s in a region of constant magnetic field of magnitude 0.3 T. The direction of the electron when it enters this region is at 20° to the field, and the electron follows a helical path. When you look along the direction of the magnetic field, the path is a projected circle. How far has the electron traveled along the direction of **B** when one projected circle has been completed?

20. (II) An electron enters a bubble chamber that contains a constant magnetic field of strength 0.10 T and follows a helical path. The spacing between the turns of the path is 3.0 mm, as is the radius of the circular part of the path. Find the components of the velocity parallel and perpendicular to the field.

21. (II) A proton and an alpha particle, which has twice the charge and four times the mass of the proton, are each accelerated through the same potential difference and enter a region of constant magnetic field perpendicular to their paths. (a) What is the ratio of the radii of their orbits? (b) What is the ratio of the frequencies of their orbits?

22. (II) An electron is injected at $t = 0$ s with velocity $\mathbf{v}_0 = (10^5 \text{ m/s})\mathbf{i}$ into a region with parallel electric and magnetic fields $\mathbf{E} = (200 \text{ V/m})\mathbf{j}$ and $\mathbf{B} = (0.05 \text{ T})\mathbf{j}$, respectively. Calculate the subsequent motion.

23. (II) You want a velocity selector to be tunable, with the capacity to select electrons that have been accelerated from rest by a potential that runs from 1200 V to 12,000 V. If the magnetic field, B, is fixed at 0.1 T, what range of electric field strengths must be available? If the electric field strength were fixed at 100 V/cm, what range of magnetic field strengths must be available?

24. (II) Figure 29–30 is a schematic diagram of a cyclotron. A charged particle starts out at the central point and, for a given magnetic field perpendicular to the plane of motion, follows a circular path. The cyclotron takes advantage of the fact that the time for the particle to execute a half-circle is independent of the particle's velocity. An alternating voltage is applied across the gap between the two "dees" (the semicircular regions), so that, when the particle crosses the gap, the voltage acts to accelerate it. When the particle gets to the gap again after having completed a half-circle, the voltage has changed sign, and the particle is once again accelerated. The frequency of the oscillating voltage must match the cyclotron frequency. In this way, the particle is always accelerated, completing ever bigger circles in the same time, until the beam is extracted at the maximum radius. (a) If the magnetic field has strength 1 T and the circulating particle is a proton, $q = +e$ and $m = 1.7 \times 10^{-27}$ kg, what is the cyclotron frequency? (b) What is the maximum velocity of the proton for a maximum radius of 50 cm? (c) the corresponding maximum kinetic energy? (d) If the maximum voltage across the gap is 50 kV, how many full circles does the proton make before it reaches its maximum energy? (e) How much time does the proton spend in the accelerator?

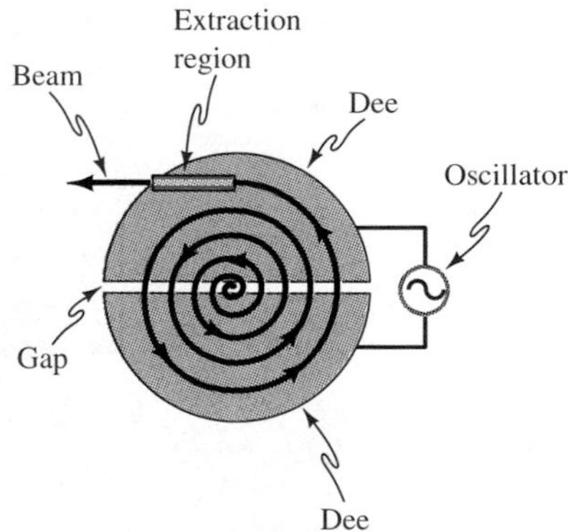

FIGURE 29–30 Problem 24.

25. (II) A cyclotron used for accelerating protons has a magnetic field of magnitude 2 T. The circular region in which the magnetic field exists has a radius of 70 cm. (a) What is the cyclotron frequency? (b) What is the largest kinetic energy that a proton accelerated in this machine can have? (c) Repeat parts (a) and (b) for a doubly ionized helium nucleus, $^4He^{2+}$, with four times the mass of a proton and twice the charge.

26. (II) The particle accelerator at Fermilab, the Fermi National Accelerator Laboratory in Batavia, Illinois, can accelerate protons to relativistic speeds. The accelerator is circular

and holds the protons in circular paths by increasing the strength of a magnetic field perpendicular to this path as the protons' momentum increases. (The momentum increases because the protons pass repeatedly through regions of electric potential.) The radius of the main Fermilab accelerator is 6.2 km, and the magnets are capable of maintaining magnetic field strengths between 1 T and 4.5 T. Given that the magnitude of a proton's electric charge is 1.6×10^{-19} C, what range of momenta can be accommodated in this accelerator? Because protons of such momenta are highly relativistic, their energies are given by the approximate relativistic formula $E = pc$. What range of energies can be reached at Fermilab? What would the speed of a baseball, mass 0.5 kg, be if it had the energy of the most energetic protons at Fermilab? (For the baseball, use the normal nonrelativistic formulas that relate energy and speed.)

27. (II) A proton, with charge q_p and mass m_p, is accelerated through an electric potential V. The proton then enters a region of constant magnetic field **B** oriented perpendicular to its path. In this region, the proton's path is circular with radius of curvature R_p. Another particle with the same charge as the proton but with mass m_x follows under the same conditions. Its radius of curvature in the magnetic field, R_x, is 1.4 times as large as R_p. What is the ratio of m_x to m_p?

The device we have described is a type of *mass spectrometer*, which can be used to identify a material by the masses of that material's constituent molecules (Fig. 29–31). Sometimes, instead of a simple electrostatic potential as in our example, a velocity selector of crossed **E** and **B** fields is used to select particles of a given speed.

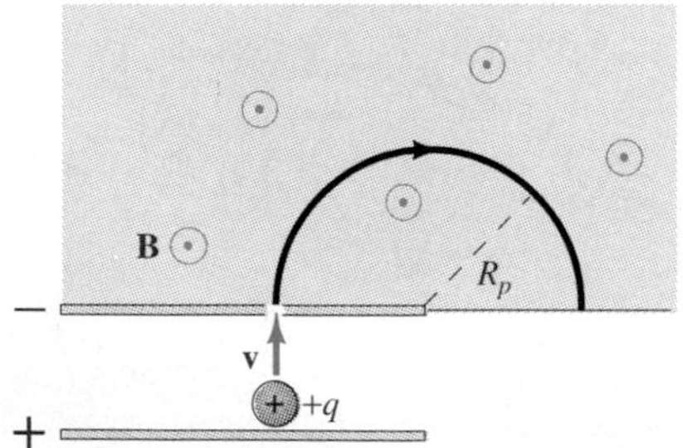

FIGURE 29–31 Problem 27.

28. (II) A helium leak detector consists of a mass spectrometer (see Problem 27) and a vacuum system that can be connected to the vacuum system of an apparatus suspected to have the type of leak in which gases can enter from the outside and degrade its vacuum. Helium gas is sprayed from outside around the region suspected of leaking. If helium gas is sucked inside, this helium can make its way to the leak detector, where it ionizes into $^4He^+$, accelerates, and is then analyzed by passing it through a magnetic field. If the acceleration is performed across a potential difference of 100V and the radius of curvature characteristic of singly ionized helium is 20 cm, what is the value of the magnetic field?

29. (III) You have an apparatus that can form an electric field of 1000 N/C and a magnetic field of 0.5 T. You want to build a velocity selector to select electrons of speed 4×10^3 m/s. (a) Draw the orientation of your apparatus, showing **E**, **B**, and **v**. (b) What are the minimum and maximum values of v that you can select? [*Hint:* Set the apparatus up so that **v** and **B** are not perpendicular to each other.]

29–4 *Magnetic Forces on Currents*

30. (I) A long wire carries a current of 15A. A bar magnet is brought near the wire so that the charge carriers, of speed 10^{-3} cm/s, experience a magnetic field of 5×10^{-2} T perpendicular to their direction of motion. Calculate the force (a) on each moving charge carrier (electron) and (b) on a 1-m length of the wire.

31. (I) A thin, straight wire carries a current of 10 mA and makes an angle of 60° with a constant magnetic field of magnitude 10^{-6} T. The portion of the wire in this field has a length of 10 cm. Calculate the force, both direction and magnitude, on this segment of the wire.

32. (I) A straight wire is placed in a uniform magnetic field of magnitude 0.010 T. The direction of the field makes an angle of 30° with that of the wire, which carries a current of 10 A. What is the force on a 1.0-m segment of the wire?

33. (II) A current I flows through a circular wire loop of radius R that lies in the xy-plane (Fig. 29–32). Consider a constant magnetic field of magnitude B that points in the x-direction. Calculate the force on an element of the loop formed by an angle $d\theta$, located at an angle θ from the $+x$-axis.

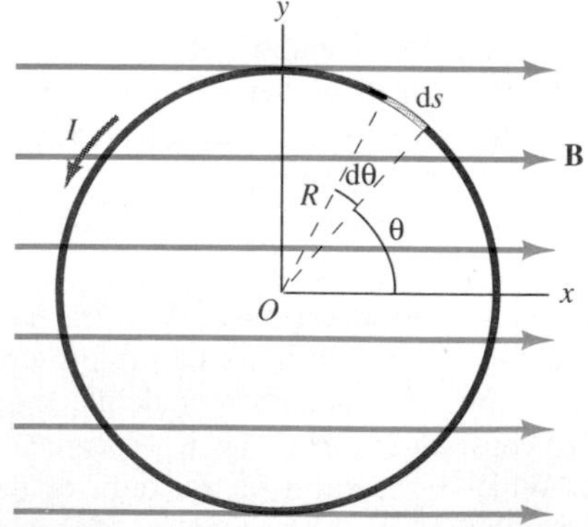

FIGURE 29–32 Problem 33.

34. (II) A wire of length L is suspended from two springs of spring constant k attached to a current source (Fig. 29–33). A magnetic field B, in a horizontal direction perpendicular to the wire (out of the page), is turned on; a current I then flows in the wire, which moves to a new equilibrium position. Which way will the wire move, and by how much?

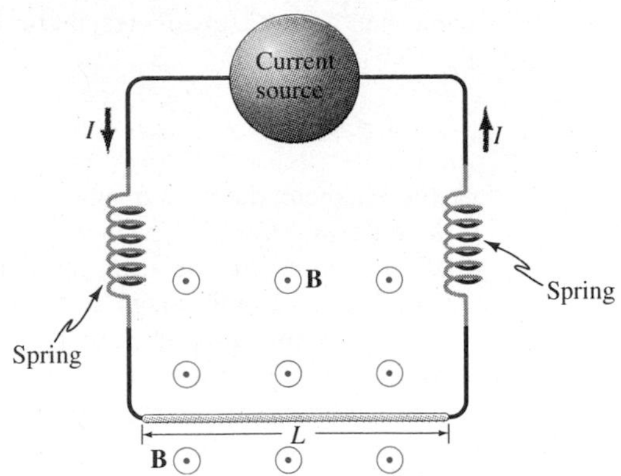

FIGURE 29–33 Problem 34.

35. (II) Figure 29–34 shows a possible device for measuring magnetic fields. A loop carrying a current I is dipped into a region of magnetic field. The loop is suspended from a spring of spring constant k that stretches if the magnetic field points in a certain direction. Here the loop has width $\ell = 1$ cm, $I = 1$ mA, the spring stretches 0.5 cm, $k = 4 \times 10^{-4}$ N/m, and the magnetic field is uniform. What is the magnitude of the field? How could such a device be used, or modified, to measure fields that are not uniform?

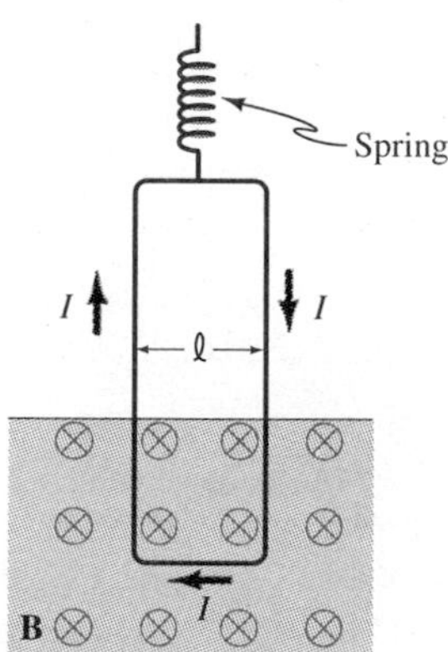

FIGURE 29–34 Problem 35.

29–5 *Magnetic Force on Current Loops*

36. (I) A wire coil of area 10 cm^2 with 220 turns experiences a maximum torque of 10^{-3} N·m when placed in a magnetic field of 0.01 T. What is the current through the coil?

37. (I) A rectangular wire loop of height 30 cm and width 20 cm consists of 240 turns and carries a current of 0.40 A. What are the magnitude and direction of the magnetic dipole moment? If a uniform magnetic field of 0.060 T is applied to the loop, and the field's direction makes an angle of 57° with the normal to the current loop, what is the torque (magnitude and direction) that acts on the loop?

38. (I) A circular coil of diameter 3.0 cm, consisting of 250 turns of wire, carries a current of 240 mA. How much work must be done to flip the coil through 180° when it is placed in a uniform magnetic field of 0.40 T? The field makes an initial angle of 30° with the direction of the coil's dipole moment.

39. (I) A wire forms a circular coil of N turns and radius R and carries a current I. The coil's magnetic dipole moment is initially aligned with a fixed external magnetic field, **B**. How much work must be done by an external force to rotate the coil through an angle θ?

40. (I) An atom can have a magnetic dipole moment of 10^{-23} J/T. Such an atom is placed in a magnetic field of 10 T. What is the range of potential energies involved?

41. (II) A wire carrying a current I splits into two channels of resistance R_1 and R_2, respectively, forming a circuit. The wire enters the space between the two poles of a magnet with a uniform magnetic field that runs from one pole piece to the other (Fig. 29–35). The circuit forms a loop; the field lies in the plane of the loop. What is the torque on the circuit about the wire axis, given that the wires are a distance d apart and that the length of the split is L?

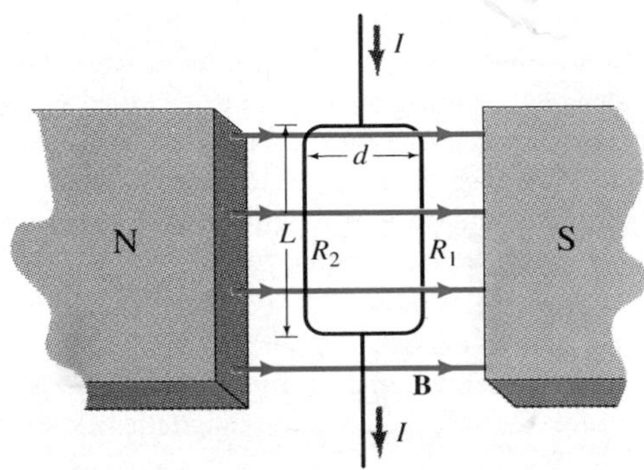

FIGURE 29–35 Problem 41.

42. (II) A circular wire coil of area 20 cm^2 has 20 turns. When the coil is placed in a magnetic field of 1 T, the maximum torque is 10^{-6} N·m. (a) What is the current in the coil? (b) What work is required to rotate the coil 180° in the magnetic field? Does the work depend on the initial angle?

43. (II) An electric motor consists of a current-carrying wire loop in a constant magnetic field **B** (Fig. 29–36). The field produces a torque that tends to rotate the loop so that the loop's magnetic dipole moment, $\boldsymbol{\mu}$, and **B** become aligned. When that happens, a split-ring commutator reverses the current direction, so that $\boldsymbol{\mu}$ changes its orientation by 180°, and the torque acts to continue the rotation. Suppose that $\boldsymbol{\mu}$ and **B** start out almost antiparallel. Plot the magnitude of the torque as a function of the angle between $\boldsymbol{\mu}$ and **B**, as this angle runs from $-180°$ to 0°. At 0° the commutator reverses the current. Plot the torque through another half turn. What is the average value of the torque through a full turn if the current in the motor is 2.2 A, the magnitude of **B** is 0.10 T, and the area of the loop is 80 cm^2?

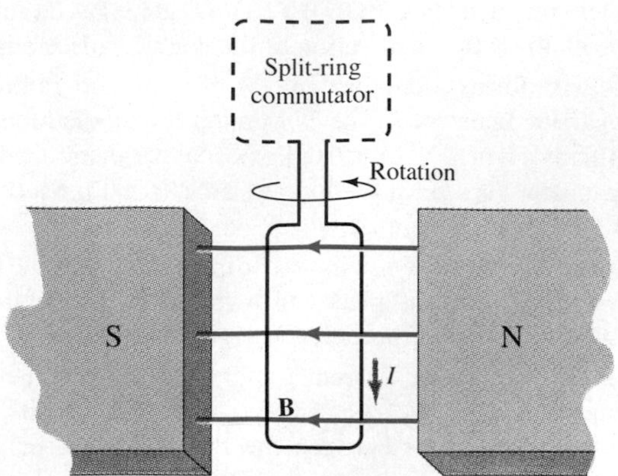

FIGURE 29–36 Problem 43.

44. (II) An electron, of charge $q = -1.6 \times 10^{-19}$ C, has a "size" of about 3×10^{-15} m, called its classical radius. The magnetic dipole moment of the electron is roughly 10^{-23} A·m^2. (a) Suppose that this magnetic moment were due to the entire charge q orbiting at the classical radius. What would the speed of the charge be to generate this magnetic moment? (b) Suppose that the electron's magnetic moment were perpendicular to a magnetic field of magnitude 1 T. What is the torque on the electron?

45. (II) The current loop shown in Fig. 29–37 lies in the xy-plane and consists of a straight segment α of length $2R$ in the x-direction and a semicircular segment β, which has a radius of curvature R. There is a constant magnetic field of strength B into the page. (a) Compute the magnetic force on segment α. (b) Find the magnetic force on segment β. You may wish to use symmetry arguments to simplify your task. (c) Add the results of parts (a) and (b) to find the net force on the loop. (d) How could you generalize your results to a loop of any shape in the xy-plane?

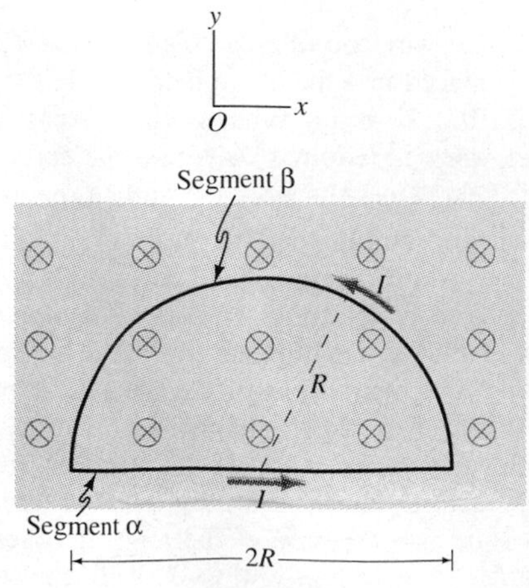

FIGURE 29–37 Problem 45.

46. (II) (a) Calculate the magnetic dipole moment of a single atom, based on the following model: One electron travels at speed 2.2×10^6 m/s in a circular orbit of diameter 10^{-10} m. (b) The individual atomic magnetic dipoles of magnetic materials (such as iron) are preferentially lined up to point in the same direction. If a fraction f of the dipoles are so aligned along the long axis (with the rest oriented randomly so that their magnetic dipole moments add vectorially to zero), what is the net magnetic dipole moment of a piece of such material 1 cm^2 in area and 10 cm long (Fig. 29–38)? (The material may be viewed as an array of cubes, each of which contains one atom and is 10^{-10} m on a side.) (c) What is the torque experienced by the piece of material in part (b) in a field of 10^{-3} T when the magnetic field is directed at right angles to the long axis of the material?

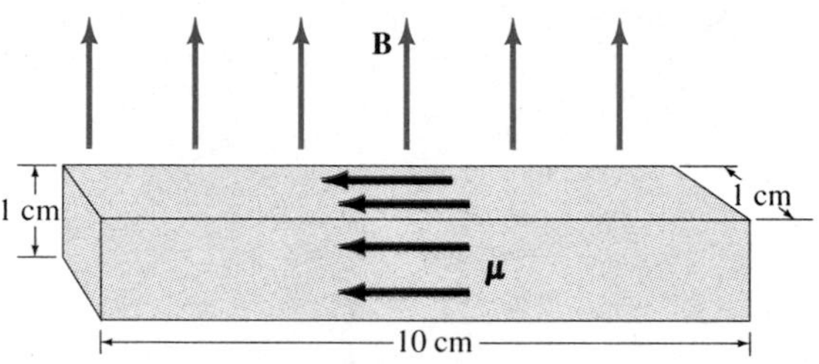

FIGURE 29–38 Problem 46.

47. (II) We showed in Example 29–5 that, when a current loop with magnetic dipole moment $\boldsymbol{\mu}$ is displaced slightly from perfect alignment of $\boldsymbol{\mu}$ and a magnetic field **B**, the rotational motion of the current loop due to the torque of the field is harmonic, with angular frequency $\omega = \sqrt{\mu B / I_M}$. I_M is the moment of inertia about the rotation axis. Calculate the kinetic energy, K, associated with this motion, and show that the sum of potential energy and kinetic energy is a constant.

48. (II) A coil carrying current $I = 1.0$ mA has moment of inertia $I_M = 1.2 \times 10^{-8}$ kg·m^2 about a rotational axis and an area of 4.0×10^{-4} m^2. The coil is placed in a magnetic field of magnitude 0.10 T, displaced 10° from alignment between its magnetic dipole moment, $\boldsymbol{\mu}$, and the field, and released from rest. Describe the subsequent motion. What is the maximum angular speed of the coil in that motion?

29–6 *The Hall Effect*

49. (II) Suppose that the strip of metal used in the apparatus that demonstrates the Hall effect has a cross section of width w and depth d_0. (The width is the space across which the Hall voltage ΔV is measured.) Show that the density n of charge carriers with charge e is independent of the width and is given by $n = IB/(d_0 e\,\Delta V)$. Knowing the density of carriers, find an expression for the drift speed as measured by a Hall apparatus.

50. (II) The probe that demonstrates the Hall effect is used to measure the density of charge carriers in an unknown sample of metal. A sample of the material 0.10 mm thick is placed in a magnetic field of 0.50 T. When a current of 100 mA passes through the material, a Hall voltage of 1.1 μV is measured. What is the density of charge carriers?

51. (II) Consider a Hall-effect experiment done with an aluminum strip, whose density is 2.7 g/cm^3. The strip is 1.0 cm wide (the distance between the points where the Hall voltage is measured) and 0.50 mm thick. When the magnetic field is 0.050 T, the current is 96 A, and the Hall voltage is 1.0 μV. How many electrons per atom are free to carry current in aluminum?

52. (II) A Hall-effect probe can be used to measure the magnitude of a magnetic field. A researcher has lost her instruction booklet and forgotten the calibration procedure. However, when she places the Hall probe inside a known magnetic field of 2000 G, she measures a Hall voltage of 234 mV. What is the field of a magnet with a Hall voltage of 498 mV?

General Problems

53. (II) The wire coil of a galvanometer has an area of 1 cm^2 and 200 turns. The coil is placed in a magnetic field of magnitude 0.1 T and oriented so that its plane is initially parallel to the field. The restoring torque of the galvanometer spring is proportional to the angular deflection, with a proportionality constant of 10^{-7} N·m/° (see Example 29–5). What current corresponds to a deflection of 50°?

54. (II) The masses of atomic ions of known charge can be precisely measured by finding the time an atom takes to complete a circular trajectory in a known magnetic field. With a magnetic field of magnitude 1 T and an apparatus capable of measuring times to an accuracy of 10^{-9} s, how accurately can the mass of an ion with charge $+e$ be measured in 1 rev? If the mass is to be measured to an accuracy of 10^{-30} kg, how many revolutions must be measured?

55. (II) When an electron orbits a proton, the smallest circular orbit is one with a radius of about 0.5×10^{-10} m, the Bohr radius. The proton's electric field must have what magnitude to make the electron follow this orbit? Compare the magnitude of the magnetic field that would be required to make an electron move in a circle of the same radius at the speed it would have if it were orbiting a single proton.

56. (II) A massive charge Q is fixed at the origin of a coordinate system. A magnetic field **B** points in the $+z$-direction. A light particle of charge q and mass m moves in a circular orbit of radius r about the origin. For what value of **B** (as a function of r) is such motion possible, if Q and q have the same sign and if the angular momentum of the motion is a fixed constant L?

57. (II) Consider a parallel-plate capacitor with charge density $\pm 6.0 \times 10^{-5}$ C/m^2 on the two plates and an electric field that points in the $+z$-direction. What magnetic field is necessary to provide a velocity selector for 40-keV deuterons that move in the $+y$-direction? A deuteron has a mass of 3.2×10^{-27} kg and a charge of 1.6×10^{-19} C; 1 keV = 1.6×10^{-16} J.

58. (II) A narrow beam of particles of mass m and charge q travels in free space at speeds between v_1 and v_2. It enters a region of length L with a constant magnetic field that points perpendicular to the beam direction and parallel to the boundary between the field-free region and the region with the field (Fig. 29–39). There it follows a circular path with radius of curvature R until it exits that region. Show that the beam widens when it emerges from the region with the field, and calculate the spread in terms of a range of angles.

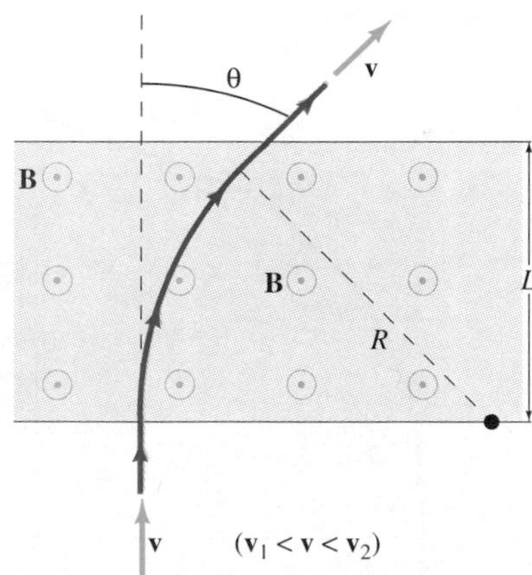

FIGURE 29–39 Problem 58.

59. (II) An electron moving in the xy-plane is subject to forces due to a constant magnetic field **B** that points in the $+z$-direction. Assuming that the electron loses 10 percent of its energy after 20 turns, as a consequence of frictional forces, what will the fractional change in the radius of the orbit be after 20 turns?

60. (II) For the motion described in Problem 59, (a) what will the fractional angular-momentum change be during the 20 turns? (b) What is the torque exerted by the frictional forces in terms of the initial kinetic energy?

61. (II) Electrons are injected into a region with a constant

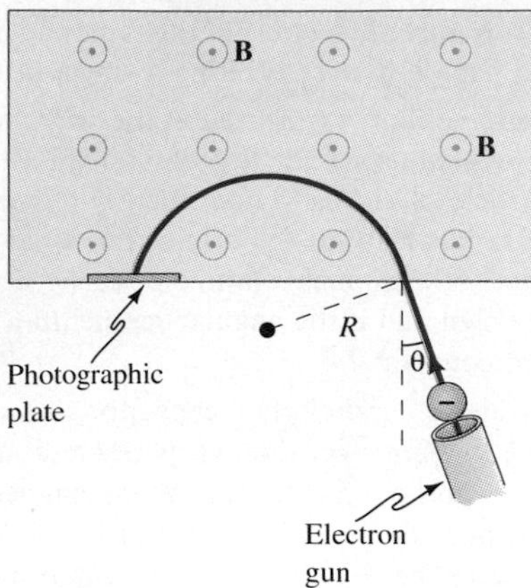

FIGURE 29–40 Problem 61.

magnetic field **B** by an electron gun with known voltage V (Fig. 29–40). The electrons move in a plane perpendicular to **B** and follow an arc of radius R. Determine the charge-to-mass ratio e/m for the electrons in terms of the given parameters.

62. (II) Particles with mass $M_A = A(1.6 \times 10^{-27}$ kg) and charge $q = 1.6 \times 10^{-19}$ C are accelerated by a potential difference of 2×10^4 V and directed perpendicularly into a region of uniform magnetic field of strength 0.2 T (Fig. 29–41). The region with the field is 30 cm deep. Calculate the angular deflection of the particles, θ, as a function of A.

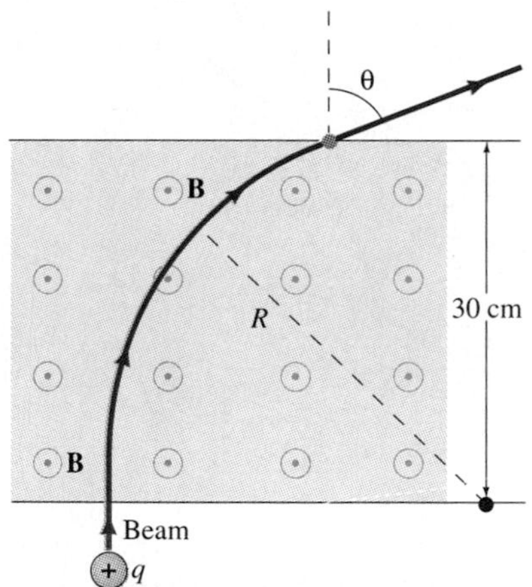

FIGURE 29–41 Problem 62.

63. (II) N electrons move at speed v in a circular orbit of radius R. (a) What is the angular momentum of the system of electrons? (b) the magnetic dipole moment associated with the current loop? (c) the ratio of the quantities in parts (a) and (b)?

64. (II) A rectangular wire loop of width a and height b is connected to a current source that, when turned on, gives rise to a current I in the wire. The loop is suspended in a uniform magnetic field **B** that points in a vertical direction (Fig. 29–42), and it would hang vertically if there were no current. We assume that the wire is massless, but two masses m are suspended at the lower corners. What is the angle θ at which the loop is in equilibrium? Calculate this in two ways: by using torques, and by expressing the potential energy as a function of θ and minimizing it. What happens if the direction of the current is reversed?

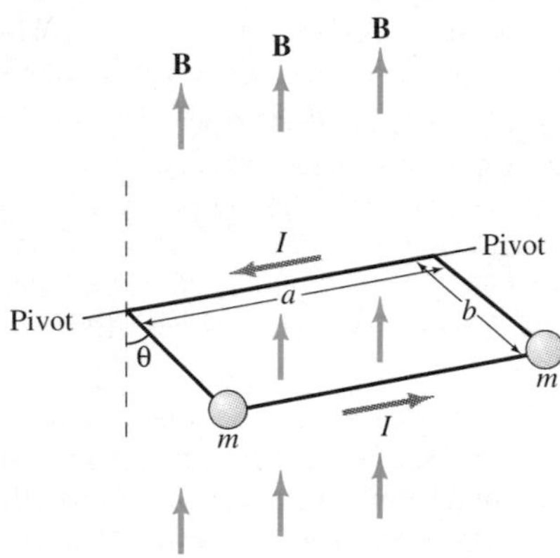

FIGURE 29–42 Problem 64.

65. (III) Suppose that an experimental apparatus can have both electric and magnetic fields constant in magnitude and direction. In this apparatus a proton moving at a speed of 1.0×10^4 cm/s in the $+z$-direction does not accelerate, whereas a proton moving at a speed of 2.0×10^4 cm/s with no x-component at an angle of 35° with respect to the z-axis experiences an initial acceleration of magnitude 7.2×10^8 m/s² in the $-x$-direction. A proton moving in the xy-plane has a circular orbit. Find the values of **E** and **B** in the apparatus.

CHAPTER 30

In the tokamak, an experimental device for the study of fusion-generated power, magnetic fields are used to contain a gas of positive ions. These fields, and the windings of wire that produce them, are topologically complex.

THE PRODUCTION AND PROPERTIES OF MAGNETIC FIELDS

We learned in Chapter 29 that moving charges and current-carrying wires are subject to forces when they are in the presence of a magnetic field. But how is the magnetic field produced? We have already learned that electric charges exert forces on each other, and that electric charges produce electric fields. In this chapter we shall see that moving charges and currents exert magnetic forces on each other, and that magnetic fields are themselves produced by moving charges or, equivalently, by currents. We describe and explore the ways in which these magnetic fields are produced. We shall learn about Ampère's law (including Maxwell's generalization) and the Biot–Savart law, which describe the magnetic fields produced by moving charges or currents. We shall also begin an exploration of the intimate relation between electric and magnetic fields that will eventually lead to Maxwell's equations and to an understanding of the phenomenon of light.

(a)

(b)

FIGURE 30–1 (a) The compass needle continues to point north when there is no current in the wire, but (b) when the switch is closed, the needle reacts to the magnetic field produced by a current in the wire.

30–1 AMPÈRE'S LAW

Hans Christian Oersted discovered during the winter of 1819 to 1820 that electric currents can influence compass needles (Fig. 30–1). Before this discovery, there was only a suspicion of a connection between electricity and magnetism. Oersted, as well as André-Marie Ampère, soon showed that *current-carrying wires exert forces on each other*. Because such wires are electrically neutral, these forces are not electric. We recall from Chapter 29 that a current-carrying wire aligns iron filings on a plane perpendicular to the wire in a circular pattern (Fig. 29–2c). This suggests that a current-carrying wire is a source of a magnetic field. The force acting on one current-carrying wire in the presence of another is actually due to the magnetic field generated by the second wire. In this section we shall find an expression for this magnetic field.

The Magnetic Field of a Straight Wire

In order to find the magnetic field produced by a single straight wire, we find the force between two parallel current-carrying wires by assuming that one is the source of a magnetic field and that the force on the other one is due to this field. Consider two wires with parallel straight segments of length L labeled 1 and 2, respectively. The forces between the wires weaken so rapidly as the separation between the wires, d, increases that we need worry only about forces between segments closest to one another. When the currents are parallel, the wires attract each other (Figs. 30–2a, b); when the currents are antiparallel, they repel (Figs. 30–2c, d).

Suppose that now the wire segments are long, $L \gg d$. In Fig. 30–3a, the force on wire 2, which is directed to the left, is due to the magnetic field of wire 1. Ac-

(a)

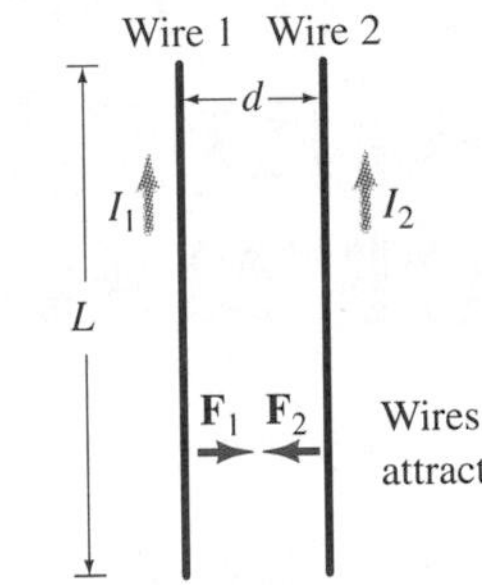

(b)

(c)

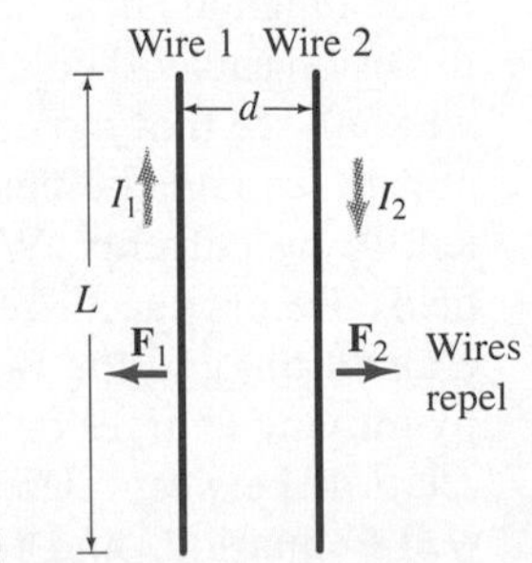

(d)

FIGURE 30–2 Two parallel wires that carry currents exert forces on one another; these forces are larger when the wires are closer. (a), (b) The currents are parallel, and the forces are attractive. (c), (d) The currents are antiparallel, and the forces are repulsive.

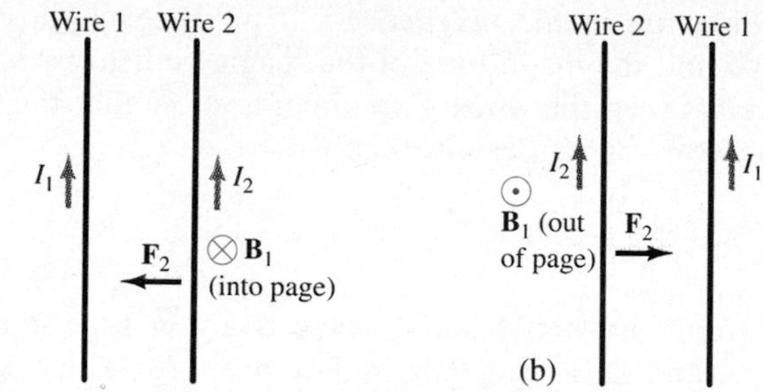

FIGURE 30–3 Determining the direction of the magnetic field due to wire 1. Currents I_1 and I_2 are parallel to each other. (a) According to the right-hand rule, $\mathbf{B}_1$ due to wire 1 is directed into the page when wire 2 is to the right of wire 1. (b) $\mathbf{B}_1$ due to wire 1 is directed out of the page when wire 2 is to the left of wire 1.

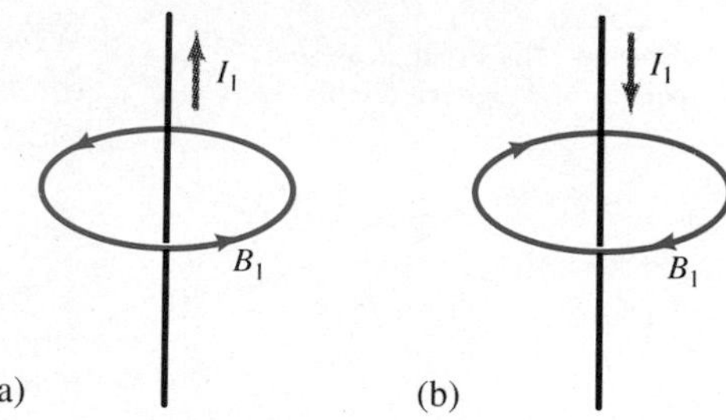

FIGURE 30–4 (a) The magnetic field due to wire 1, $\mathbf{B}_1$, traces out a circle around the wire in the direction shown. (b) If the current in wire 1 were reversed, the orientation of $\mathbf{B}_1$ would change as determined by a right-hand rule.

cording to Eq. (29–19), which describes the force on a segment of current-carrying wire, the force on wire 2 is of the form

$$\mathbf{F}_2 = I_2\,\mathbf{L}_2 \times \mathbf{B}_1, \tag{30–1}$$

provided only that the field $\mathbf{B}_1$ due to wire 1 is the same all along wire 2. This assumption is justified for long wires. Here the vector $\mathbf{L}_2$, of magnitude L, is oriented along the direction of I_2. We find that as we move wire 2 around wire 1, always holding them parallel and at the same separation, the force is always attractive and always has the same magnitude. Equation (30–1) and this observation about the force on wire 2 imply that *the magnetic field $\mathbf{B}_1$ due to wire 1 traces out a circle around wire 1.* Equation (30–1) shows that force $\mathbf{F}_2$ is insensitive to any component of $\mathbf{B}_1$ that is *parallel* to the wires, because the vector product of two parallel vectors is zero. As for the component of $\mathbf{B}_1$ perpendicular to wire 2, application of a right-hand rule in Eq. (30–1) shows that $\mathbf{B}_1$ must be directed into the page when wire 2 is in its original position to the right of wire 1 (Fig. 30–3a). However, if wire 2 is moved to the left of wire 1, field $\mathbf{B}_1$ at wire 2 will be directed out of the page, because the two wires continue to attract each other (Fig. 30–3b). By using this argument for other positions, we find that, as we suspected, the magnetic field $\mathbf{B}_1$ due to wire 1 traces out a circle around wire 1 (Fig. 30–4a; see Fig. 29–2c).

We can demonstrate with one further experiment that there is no component of $\mathbf{B}_1$ oriented parallel to wire 1. If we wrap wire 2 in a circle around wire 1, as in Fig. 30–5, a component of $\mathbf{B}_1$ parallel to wire 1 would cause a force to be exerted on wire 2. As Eq. (30–1) indicates, such forces would tend to expand or contract the circle traced by wire 2, and this effect would be measurable. The result is that, in this configuration, wire 2 experiences no forces. The magnetic field $\mathbf{B}_1$ about wire 1 indeed traces out circles around the wire.

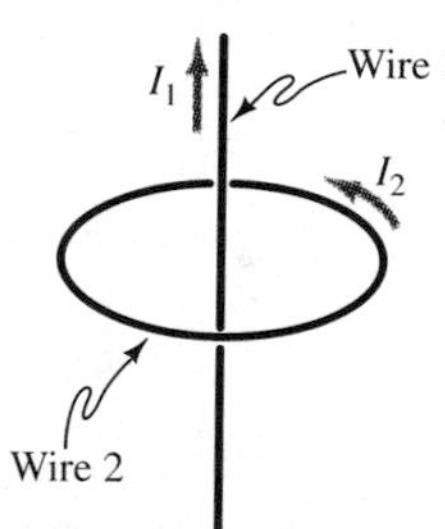

FIGURE 30–5 If wire 2 traced a circle around wire 1, it would react to any components of $\mathbf{B}_1$ due to wire 1 that are parallel to wire 1. No such forces are found.

Let us return to the parallel configuration of the wires. If we reverse the direction of the current in wire 1, then the force on wire 2 also reverses. A repeat of the exercise by which we determined the direction of the magnetic field about wire 1 would show that the magnetic field lines again trace circles around the wire, but in the opposite direction (Fig. 30–4b). We can summarize by saying that the direction of $\mathbf{B}$ is determined by a right-hand rule (Fig. 30–6):

If the thumb of the right hand is oriented along the direction of current flow in a wire, the fingers curl in the direction of the magnetic field.

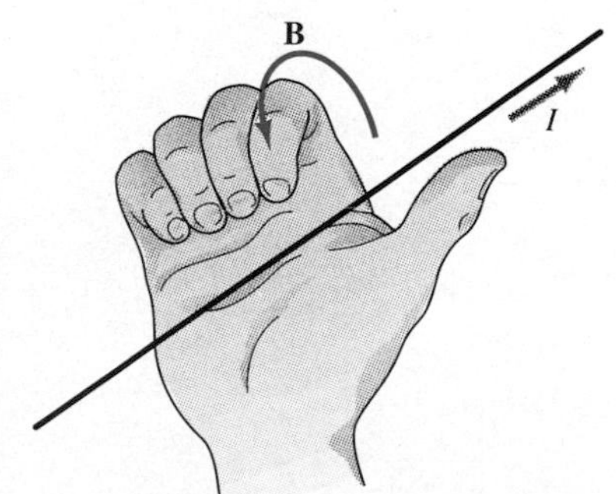

FIGURE 30–6 A right-hand rule determines the direction of the magnetic field around a current-carrying wire.

We have found the direction of the magnetic field produced by a current. What is its magnitude? We find the magnitude of the magnetic field by studying the magnitude of the force between the wires. Experiments show that the magnitude of the force between two parallel, straight segments of wire is

$$F = \frac{CI_1I_2L}{d},$$

where I_1 and I_2 are the currents in wires 1 and 2, respectively, d is the separation between the wire segments, and L is their length. The proportionality constant C depends on how the units of current are defined. Conversely, if we use a *defined* proportionality constant, the force between two current-carrying wires determines the units of current. In the SI, this latter course is the one followed. C is defined according to

$$F = \frac{\mu_0 I_1 I_2 L}{2\pi d}, \tag{30–2}$$

where the constant μ_0, called the **permeability of free space**, is in turn defined to be

$$\mu_0 \equiv 4\pi \times 10^{-7}\ \text{T}\cdot\text{m/A}. \tag{30–3}$$

The definition of the ampere

With this definition of μ_0, 1 A is defined as the current that travels in two long, parallel wires of length L that are 1 m apart, such that the attractive force between them is $(2 \times 10^{-7}$ N/m$)L$. Is this result consistent with other definitions already given? In Chapter 22 we defined the coulomb as the charge on two pointlike objects such that there is a certain force between them. In Chapter 27 we provisionally defined 1 A as 1 C/s. The definition of the coulomb in terms of a force between charges depends on another defined constant, ϵ_0, in exactly the same way that the definition of the ampere depends on μ_0. Thus, for our relations to be consistent, *the definitions of μ_0 and ϵ_0 must be consistent.* If both μ_0 and ϵ_0 are defined, the same must be true for their product, which is given by

$$\mu_0\epsilon_0 \equiv c^{-2} = (2.99792458 \times 10^8\ \text{m/s})^{-2}. \tag{30–4}$$

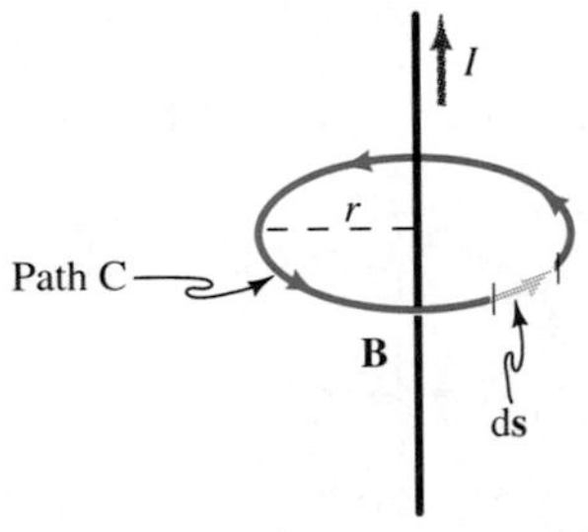

FIGURE 30–7 Path C circles a current-carrying wire at a constant distance r from the wire and follows the direction of the magnetic field, **B**, around the wire.

The constant c is precisely the speed of light![†] We shall see in Chapter 35 why this is so.

Comparison between Eqs. (30–1) and (30–2) shows that a wire that carries a current I gives rise to a magnetic field whose magnitude is

The magnetic field produced by a straight wire

$$B = \frac{\mu_0 I}{2\pi r} \tag{30–5}$$

at a distance r from the wire. The direction of this field is given by the right-hand rule as described above.

Ampère's Law

We can find a more universal form for the magnetic field produced by a current by expressing Eq. (30–5) differently. Imagine a line integral over the magnetic field, **B**, that follows a circular path of radius r around a wire. The integration path follows the direction of **B**. This path is labeled C and shown in Fig. 30–7. The

[†] In the SI, the speed of light and the second are *defined* mathematically, and the meter is measured in terms of them. Specifically, 1 m is the distance light travels in a vacuum in 1/299,792,458 of 1 s. See *Physics Today*, August 1989, p. 23.

sign $\oint$ indicates that the path of the line integral goes all the way around the circle, or is closed. For the path chosen, **B** and the infinitesimal distance element **ds** of the integral are parallel, so $\mathbf{B} \cdot \mathbf{ds} = B\,ds$. Because B is a constant when the distance r from the wire is constant,

$$\oint \mathbf{B} \cdot \mathbf{ds} = B \oint ds = B(2\pi r). \tag{30–6}$$

The factor $2\pi r$ is the length of the path, which is the circumference of the circle of radius r. If we use Eq. (30–5), we find that

$$\oint \mathbf{B} \cdot \mathbf{ds} = \frac{\mu_0 I}{2\pi r} 2\pi r = \mu_0 I. \tag{30–7}$$

Equation (30–7) includes a right-hand-rule convention in which path C must be taken in the direction of the fingers of the right hand when the thumb is oriented along I.

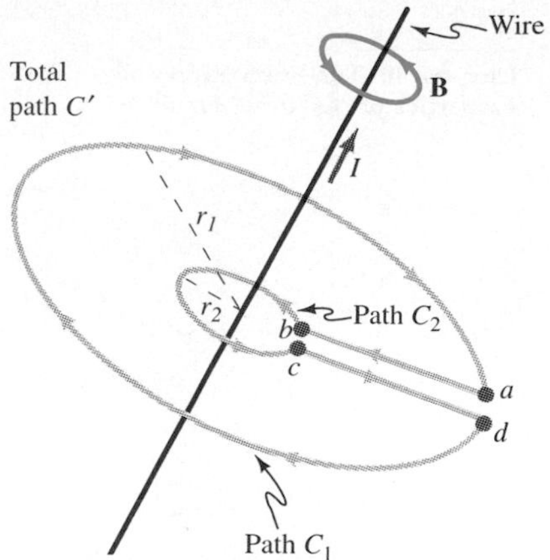

FIGURE 30–8 Path C' consists of a clockwise circle C_1 of radius r_1, a leg a that moves in to a distance r_2 from the wire, a counterclockwise circle C_2 of radius r_2, and a leg b that moves out to r_1. The magnetic field follows a counterclockwise circle.

A current passes through the path described above. Let us now consider a path through which *no* current passes. In particular, consider path C' shown in Fig. 30–8. Path C' is broken into the segment from a to b, the nearly full circle C_2, the segment from c to d, and the nearly full circle C_1. We wish to compute $\oint_{C'} \mathbf{B} \cdot \mathbf{ds}$. The total contribution of the two paths from a to b and c to d is zero, because **B** is perpendicular to the path there. Thus

$$\oint_{C'} \mathbf{B} \cdot \mathbf{ds} = \int_{C_1} \mathbf{B} \cdot \mathbf{ds} + \int_{C_2} \mathbf{B} \cdot \mathbf{ds}$$
$$= -B_1(2\pi r_1) + B_2(2\pi r_2), \tag{30–8}$$

where B_1 is the magnitude of the magnetic field at a distance r_1 from the wire and B_2 is the magnitude of the field at a distance r_2. The first term is negative because **B** is oriented opposite to the path direction on portion C_1. From Eq. (30–5) we see that the two contributions to Eq. (30–8) cancel:

$$\oint_{C'} \mathbf{B} \cdot \mathbf{ds} = -\frac{\mu_0 I}{2\pi r_1}(2\pi r_1) + \frac{\mu_0 I}{2\pi r_2}(2\pi r_2)$$
$$= -\mu_0 I + \mu_0 I = 0. \tag{30–9}$$

FIGURE 30–9 A late-nineteenth-century romanticization of André-Marie Ampère, standing, and his friend and collaborator François Arago, experimenting with the magnetic effects of electric currents.

What is the difference between Eq. (30–7) and Eq. (30–9)? Earlier, the path C enclosed current I, whereas we see from Fig. 30–8 that path C' encloses no current. We have taken the first steps toward the generalization that follows (which more detailed consideration justifies). Let the quantity I_{enclosed} be the total current enclosed by *any closed path.* Then

$$\oint \mathbf{B} \cdot \mathbf{ds} = \mu_0 I_{\text{enclosed}}, \tag{30–10}$$

Ampère's law. The magnetic field produced by a current obeys this law.

where the integral is taken around that closed path. Equation (30–10), which was formulated by Ampère in the 1820s during his extensive work on magnetism, is known as **Ampère's law** (Fig. 30–9). The direction of the loop integral must be specified: If the fingers of the right hand curl in the same sense as the integral path, the thumb points in the direction a positive current takes in passing through the loop. The total current may thus include both positive and negative contributions. Remember that *the path does not have to be circular, just closed.* The generalization we have made includes an experimental result that is worth singling out: *The magnetic fields produced by different currents superpose,* just as the electric fields of different charges add according to the superposition principle.

Using Ampère's Law

If there is some symmetry that suggests that the integral over a particular path is simple, then Ampère's law [Eq. (30–10)] can be used to find the magnetic field, in analogy with the way we use Gauss' law to find electric fields. In the case of Gauss' law, the integral is taken over a closed surface, and **E** is related to the electric charge enclosed. In the case of Ampère's law, the integral taken is along a closed path, and **B** is related to the electric current enclosed by the path.

EXAMPLE 30–1 The current, I, within a wire whose cross section has radius R is known to be distributed uniformly over that cross section. What is the magnetic field as a function of the distance r from the wire's axis outside the wire and within the wire?

SOLUTION: The wire looks the same as we move around it (it has cylindrical symmetry). Therefore we expect any magnetic field not to vary with the angle around the wire, but to be a function only of the radial distance r from the central axis. If we apply Ampère's law, Eq. (30–10), for a circular path of radius r centered on the middle of the wire, **B** will be the same all along this path, and we can use information about the current enclosed by the path to determine the field as a function of r. The amount of current enclosed depends on whether the path lies outside or inside the wire.

Figure 30–10a shows the path *outside* the wire that will determine the field outside. By the right-hand rule, **B** is oriented in the direction of the path, so $\mathbf{B} \cdot d\mathbf{s} = B\,ds$. The magnetic field magnitude is constant over the chosen path and thus comes out of the integral, leaving just the circumference of the path. The current enclosed is the *total* current carried by the wire. Thus Ampère's law, Eq. (30–10), becomes

$$\oint \mathbf{B} \cdot d\mathbf{s} = \oint B\,ds = B \oint ds = B(2\pi r) = \mu_0 I.$$

We can solve for B to find that

$$B = \frac{\mu_0 I}{2\pi r}.$$

Ampère's law shows that the magnetic field outside the wire is independent of the size of the wire, just as Gauss' law shows that the electric field outside a spherically symmetric charge distribution is independent of the size of the distribution.

To find the field inside the wire, we continue to use symmetry, but this time we take our circular path *inside* the wire (Fig. 30–10b). The current enclosed by the path is I times the ratio of the area of the circle of radius r to the area of the wire:

$$I_{\text{enclosed}} = I\frac{\pi r^2}{\pi R^2}.$$

Then, as before, Ampère's law gives

$$\oint \mathbf{B} \cdot d\mathbf{s} = B(2\pi r) = \mu_0 I\left(\frac{\pi r^2}{\pi R^2}\right).$$

If we solve for B, we find that

$$B = \frac{\mu_0 I}{2\pi R^2}r.$$

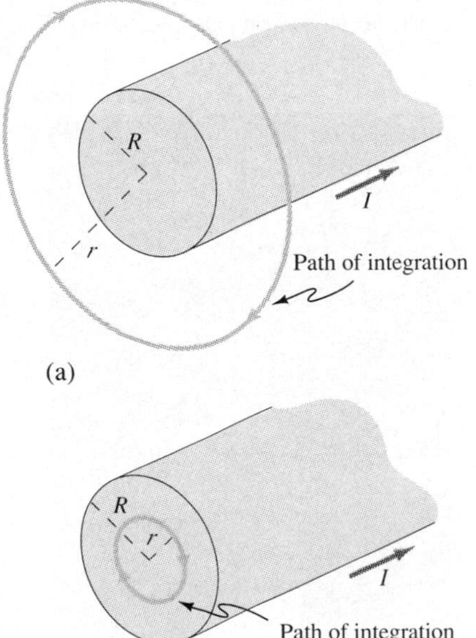

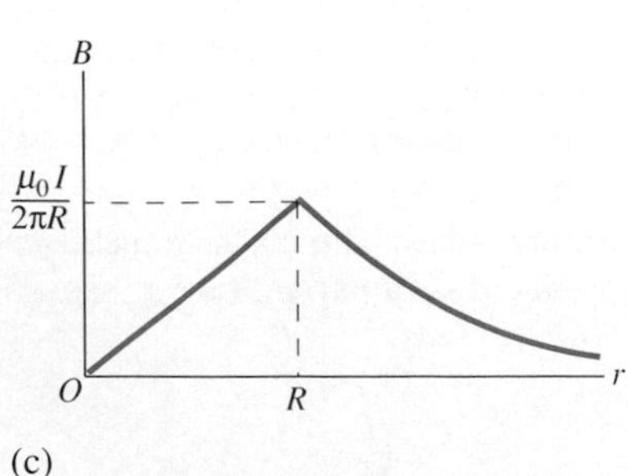

FIGURE 30–10 (a) Example 30–1. A circular path of radius r is used to determine the magnetic field outside a wire that carries a current I. (b) A similar path inside the wire. (c) The magnitude of the magnetic field versus r.

By analogy with Gauss' law for electricity, any current outside a circle of radius r makes no contribution to the magnetic field at radius r. Inside the wire, the magnetic field decreases linearly to zero as r approaches zero. Also, the results for outside and inside the wire agree at $r = R$. Figure 30–10c is a graph of the magnitude of the magnetic field.

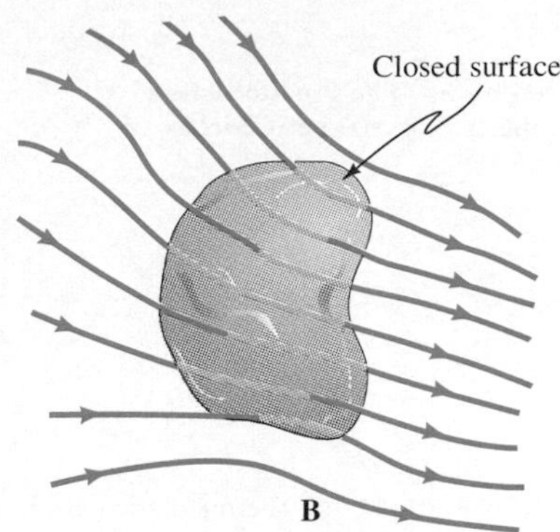

FIGURE 30–11 Magnetic field lines are everywhere parallel to the magnetic field; their density measures the field's strength. There are no magnetic charges, so magnetic field lines do not end, and the magnetic flux through a closed surface is zero. This is Gauss' law for magnetism.

30–2 GAUSS' LAW FOR MAGNETISM

Gauss' law for electricity expresses an important relation obeyed by the flux of electric field, that is, by an integral over an area of the electric field through that area. Gauss' law relates the electric charge contained within a closed surface to the flux through that closed surface. In this section we shall look at an analogous expression for magnetic fields. Static electric fields begin and end on electric charges. *Unlike electric field lines, magnetic field lines form closed curves.* If magnetic charges existed, analogous to electric charges, then magnetic field lines would originate and terminate on magnetic charges, just as electric field lines originate and terminate on electric charges. However, *despite much experimental effort, magnetic charges have never been discovered.* Any magnetic charges would represent isolated magnetic north or south poles. As we noted in Chapter 29, when a bar magnet is cut in two, we find two new bar magnets rather than isolated magnetic charges. Because magnetic charges apparently do not exist, there is nothing analogous to electric charge on which magnetic field lines can begin or end.

Magnetic charges have never been discovered. Consequently, magnetic field lines form closed curves.

With no magnetic charges, a relation like Gauss' law for electricity holds for magnetism, but with the electric charge replaced by zero. Let us define the **magnetic flux**, Φ_B, for a magnetic field **B** over a surface S, open or closed, by

$$\Phi_B(S) \equiv \iint_S \mathbf{B} \cdot d\mathbf{A}. \qquad (30\text{–}11)$$

The magnetic flux

Then **Gauss' law for magnetism** is

$$\text{for a closed surface:} \qquad \Phi_B = \oiint \mathbf{B} \cdot d\mathbf{A} = 0. \qquad (30\text{–}12)$$

Gauss' law for magnetism

As for the electric flux, infinitesimal surface elements d**A** are perpendicular to the surface and, for a closed surface, are oriented outward. This relation can be interpreted to mean that the number of magnetic field lines that enter a closed surface, minus the number that leave the surface, is zero (Fig. 30–11). Any magnetic field line entering a closed surface must leave it somewhere, because there are no magnetic charges on which magnetic field lines can begin or end.

The SI unit for magnetic flux is the unit of magnetic field times area; that is, tesla–square meters ($\text{T}\cdot\text{m}^2$). This unit occurs often enough to be given its own name in SI, the **weber** (Wb), after Wilhelm Eduard Weber:

$$1 \text{ Wb} \equiv 1 \text{ T}\cdot\text{m}^2. \qquad (30\text{–}13)$$

The SI unit of magnetic flux

Because $1 \text{ T} = 1 \text{ N}\cdot\text{s/C}\cdot\text{m}$, we can write equivalently $1 \text{ Wb} = 1 \text{ kg}\cdot\text{m}^2/\text{C}\cdot\text{s}$.

The Field Lines of a Bar Magnet

In Chapter 29 when we drew the magnetic field lines for a bar magnet, it seemed natural to think of the field lines as starting on the north pole and ending on the south pole. In light of what we have just learned, we know that magnetic field lines never start or stop—they are *continuous.* Therefore we need to reconsider

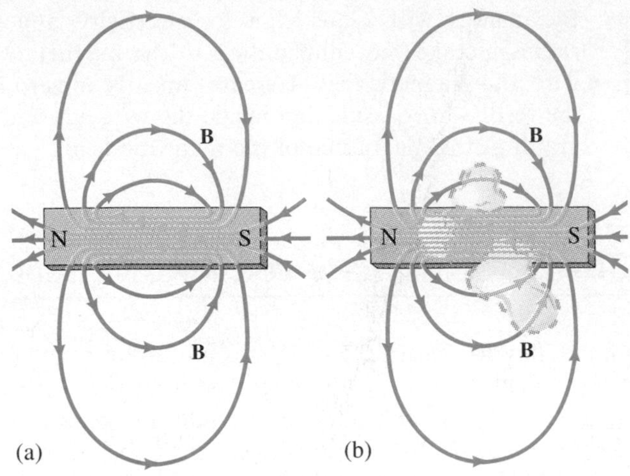

FIGURE 30–12 (a) The magnetic field lines of a bar magnet are continuous, forming closed loops. They continue within the magnet, running from the south pole to the north pole. (b) The continuity of the magnetic field lines follows from Gauss' law for magnetism: The same number of magnetic field lines that enter any closed surface will leave it.

our view of the magnetic field lines for a bar magnet. In fact, the field lines do not start or stop at the poles *but pass through the bar magnet*. The field lines that run from the north pole to the south pole outside the magnet return within the magnet to form closed loops (Fig. 30–12a). This view is consistent with Gauss' law for magnetism, Eq. (30–12), because for any closed surface that can be drawn in and around a bar magnet, the same number of field lines enter the surface as leave it. Three surfaces are shown in Fig. 30–12b. We stated previously that the form of the magnetic field lines outside a bar magnet is just like the electric field lines outside an electric dipole, which consists of equal and opposite electric charges. Now we see that if we look *between* the charges of the electric dipole and compare the electric field there to the magnetic field within a bar magnet, there is a crucial difference. As Fig. 30–13 shows, the fields are in different directions, and hence the fluxes are of different signs in the two cases.

Using Gauss' Law

Gauss' law for magnetism is useful for limiting the forms a magnetic field may take. As an example, we shall use Gauss' law for magnetism to show that the magnetic field around a straight current-carrying wire can have no radial component. We need a suitable closed (imaginary) surface to construct about the wire in order

FIGURE 30–13 Although electric field lines outside an electric dipole resemble magnetic field lines outside a bar magnet, the fields within are quite different. The electric field lines begin and end on the electric charges, whereas the magnetic field lines are continuous.

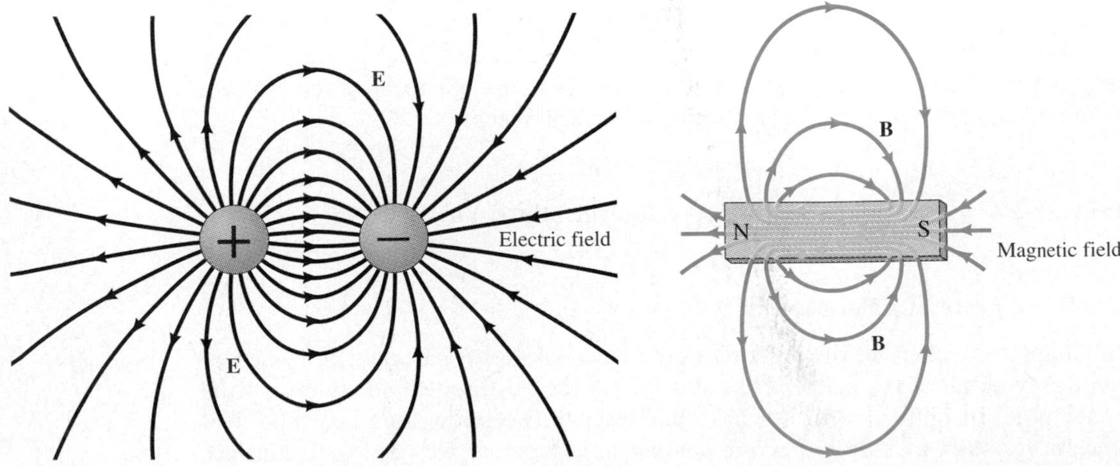

to exploit any symmetries. For a straight wire the appropriate symmetry is cylindrical symmetry. Our closed surface will therefore be a right cylinder of radius R and length L whose central axis lies on the wire (Fig. 30–14). The wire looks the same from any point on the surface of the cylinder, so the magnetic field cannot depend on the angle around the axis of the cylinder. Thus, if there were a radial component B_r at some fixed radial distance, it would have to be the same all around the wire.

To find the net magnetic flux through the closed cylinder, Φ_B, we must consider contributions from its ends and sides. Only a component of **B** along the wire, a longitudinal component, would contribute to the flux at the ends. But the contribution from one end must cancel the contribution from the other end—if the longitudinal component enters the surface at one end, it must leave the surface at the other end. The flux through the ends is therefore zero.

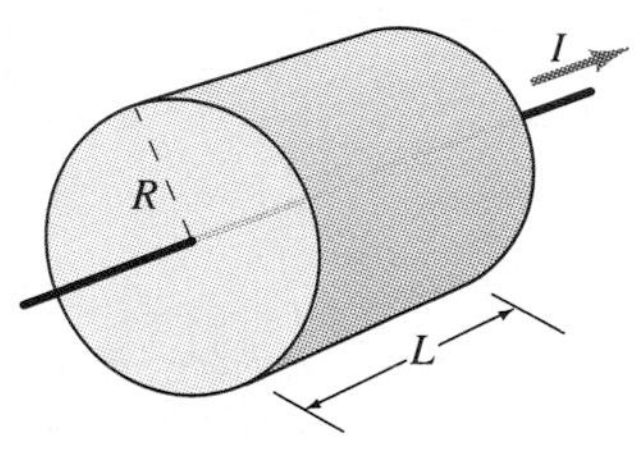

FIGURE 30–14 A Gaussian surface that exploits the symmetry of a long, straight wire.

The contribution to the magnetic flux from the sides is due to the radial component of the field, B_r. We have for the net flux

$$\Phi_B = \iint_{\text{ends}} \mathbf{B} \cdot d\mathbf{A} + \iint_{\text{sides}} \mathbf{B} \cdot d\mathbf{A} = \iint_{\text{sides}} \mathbf{B} \cdot d\mathbf{A} = B_r(2\pi RL),$$

where $2\pi RL$ is the area of the cylinder's sides. By Gauss' law this must equal zero, and because the area of the cylinder sides is not zero, B_r must be zero. We have shown by Gauss' law that there can be no radial component of the magnetic field.

This discussion shows one way in which Gauss' law for magnetism can be used. We shall see in later chapters that the magnetic flux plays a central role in other fundamental laws of electromagnetism. It is therefore important to know how to calculate the magnetic flux for both closed and open surfaces. Example 30–2 is an exercise of this type. Because the calculation of magnetic flux is similar mathematically to that of electric flux, this exercise is not really a new one.

EXAMPLE 30–2 The region between the poles of a tabletop electromagnet (Fig. 30–15a) contains a constant magnetic field, $B = 0.0030$ T, oriented in the $+x$-direction. A square wire loop of sides $L = 1.0$ cm is oriented at a 30° angle to the field (Fig. 30–15b). Find the magnetic flux through the loop.

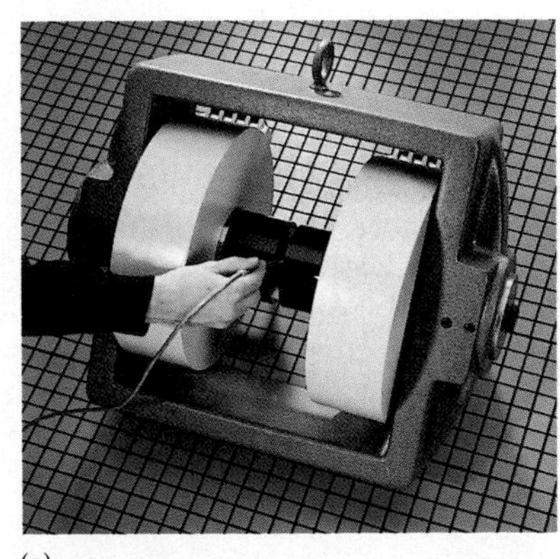

(a)

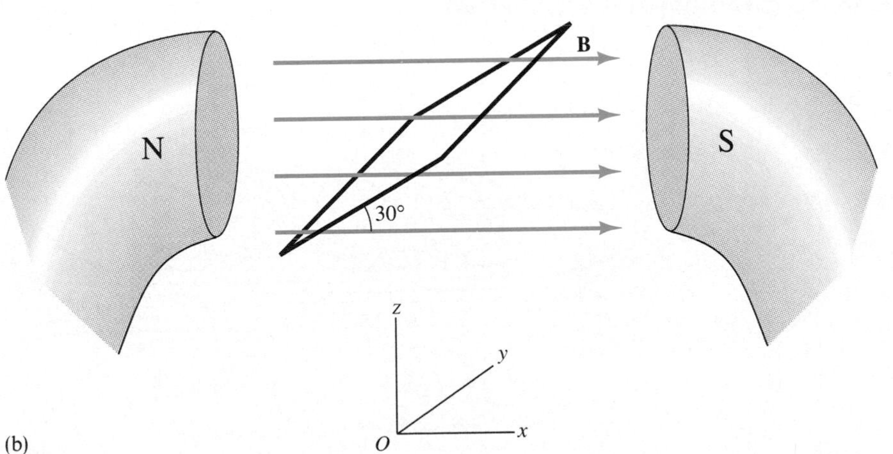

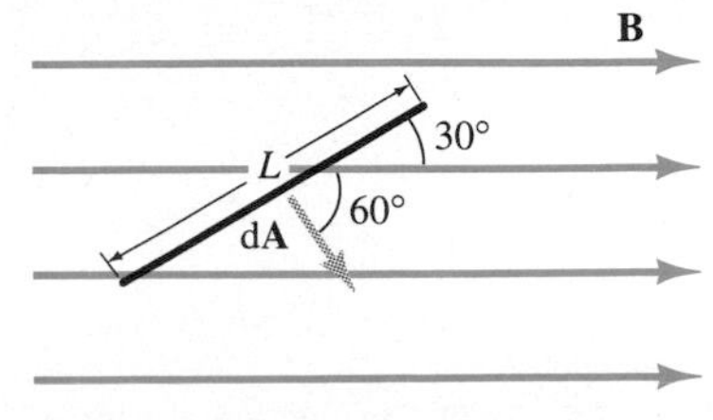

(c) Side view

FIGURE 30–15 (a) Example 30–2. A tabletop electromagnet (b) for which **B** is oriented in the $+x$-direction. (c) The surface element $d\mathbf{A}$ is oriented perpendicular to the surface.

SOLUTION: To find the flux through a given surface, we need to specify an area element $\mathbf{dA}$ of the surface, find its scalar product with the magnetic field, and integrate over the entire surface. Figure 30–15c shows that the surface element $\mathbf{dA}$, which is perpendicular to the wire loop, makes an angle $\theta = 60°$ with $\mathbf{B}$. Thus $\mathbf{B} \cdot \mathbf{dA} = B \cos\theta\, dA$. Because B and θ are constants, they can be removed from the area integral in Eq. (30–11) for the magnetic flux:

$$\Phi_B(S) = \iint_S \mathbf{B} \cdot \mathbf{dA} = \iint_S B \cos\theta\, dA = B\cos\theta \iint_S dA = B \cos\theta\, L^2.$$

The numerical value is

$$\Phi_B(S) = (0.0030\ \text{T})(\cos 60°)(1.0 \times 10^{-2}\ \text{m})^2 = 1.5 \times 10^{-7}\ \text{Wb}.$$

30–3 SOLENOIDS

A **solenoid** is an electrical element that generates magnetic fields similar to the electric fields generated by a parallel-plate capacitor (Fig. 30–16a). Just as a parallel-plate capacitor sets up a constant electric field in space, so a solenoid generates a constant magnetic field in a region of space. An ideal solenoid consists of a coil of wire wound uniformly into an infinitely long cylinder, as in Fig. 30–16b. In this figure we have exaggerated the spacing between the wires, which normally are closely wound. The diameter of the cylinder is d, the current carried is I, and the wires are wound so that there are n turns per unit length, where the length is measured along the axis of the solenoid.

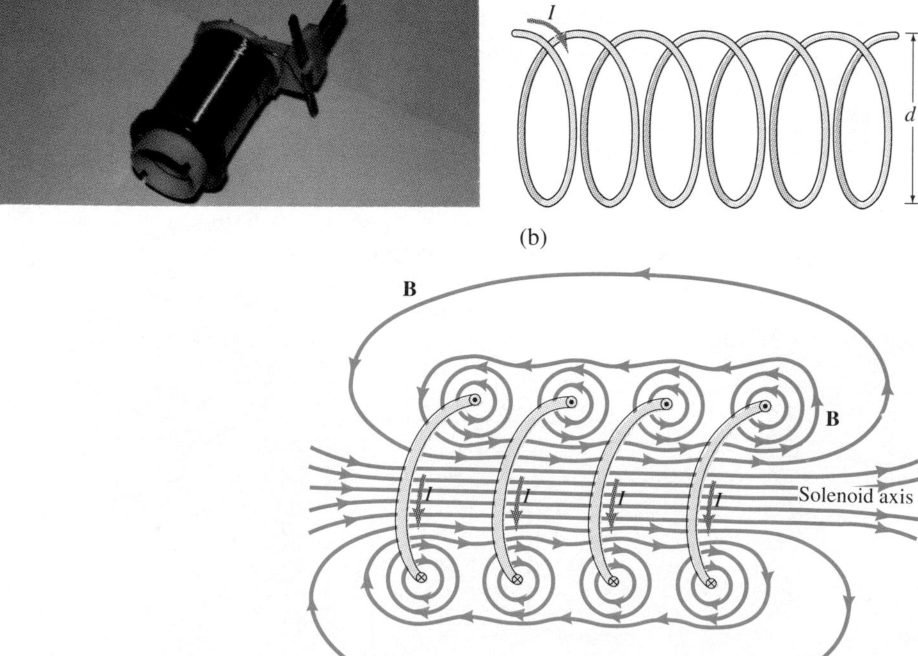

FIGURE 30–16 (a) An ideal solenoid is an infinitely long cylinder made from (b) a uniformly wound coil carrying current I (view shown is exaggerated).

FIGURE 30–17 One section of a three-turn solenoid, showing the superposed magnetic fields (view shown is exaggerated).

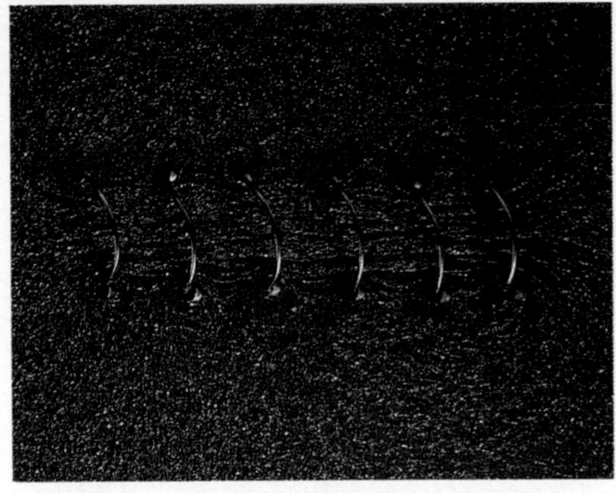

(a)

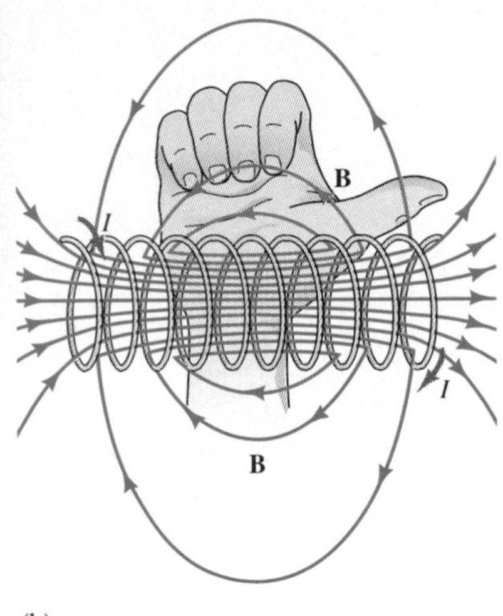

(b)

FIGURE 30–18 (a) Magnetic field lines of a solenoid, as shown by iron filings that align with the field. (b) A right-hand rule gives the direction of the magnetic field within a solenoid.

Let us sketch what the magnetic field of a solenoid might look like. Figure 30–17 is a cross-sectional view of several loops of a solenoid, again spaced more widely than they normally would be. Very near the wires, the magnetic fields form circles around the wires, because the field approximates that of a single straight wire. Figure 30–17 shows that between adjacent turns of the wire, the fields from adjacent wires tend to *cancel* each other. Inside the solenoid, these fields *add together* to form a large component that points to the right along the axis of the solenoid. Every loop of wire contributes constructively to make this interior field along the axis a strong one. Outside the cylinder the scenario is somewhat different. A region at the top gets constructive contributions from the circular fields due to the wires at the top, making a field that points to the left at the top of Fig. 30–17 and to the right at the bottom of that figure. The circular fields from the wires at the bottom contribute also and make a field outside the cylinder that points to the right at the top and to the left at the bottom. This field tends to cancel the contribution from the wires at the top.

The emerging qualitative model of the field lines is that the fields from the different loops of the coil reinforce inside the cylinder and make a net field that is parallel to the cylinder axis and whose direction is determined by a right-hand rule: If the fingers curl in the direction of the current, the thumb shows the direction of the magnetic field (Fig. 30–18). Outside the cylinder, the field points primarily in the opposite direction and is much weaker. Another way to see that the field is weaker outside the cylinder is to recall that magnetic field lines close. Because these lines cannot cross each other (why not?), the lines are squeezed together in the limited space inside the cylinder but spread throughout space on the outside (Fig. 30–19). *Even though the magnetic field is not exactly zero outside a real solenoid, it is a good approximation to take the field there to be insignificant.* In an ideal, infinitely long solenoid, the exterior magnetic field is indeed zero.

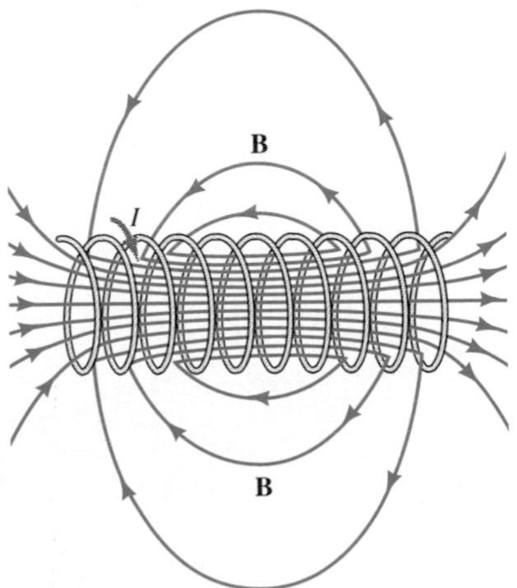

FIGURE 30–19 The magnetic field lines are densely spaced within a solenoid but widely spaced outside it. If the solenoid were infinitely long, the field outside would be zero.

Using Ampère's Law to Find the Magnetic Field in a Solenoid

Now that we understand qualitatively that a long solenoid has a large magnetic field inside—parallel to the solenoid axis—and a weak field outside, we can apply Ampère's law to calculate quantitatively the magnetic field inside the solenoid.

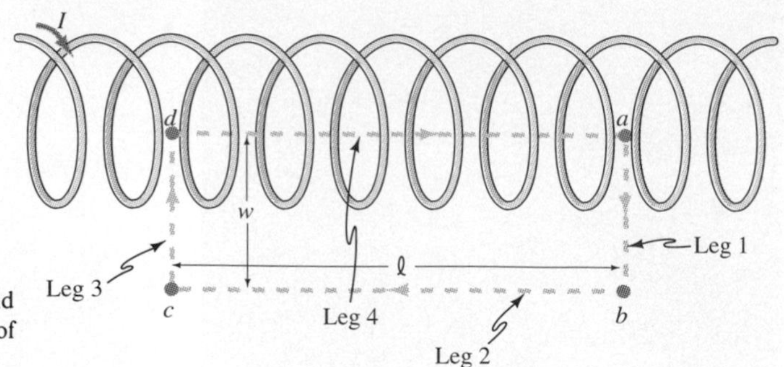

FIGURE 30–20 An imaginary rectangular loop is drawn half inside and half outside a solenoid. This loop provides a path for the application of Ampère's law.

Figure 30–20 shows a solenoid that carries a current I, and a closed *imaginary* loop consisting of four legs in a rectangle of length ℓ and height w on which to apply Ampère's law. The wire of the solenoid passes N times from above through the imaginary loop. The path about the imaginary loop is taken to be clockwise, so by the right-hand rule the net current into the imaginary loop, NI, is positive. We now calculate the line integral on the left-hand side of Eq. (30–10). There is only a very small contribution from leg 2 (point b to point c), because the field outside is insignificant. There is no contribution from leg 1 (point a to point b) or from leg 3 (point c to point d) for two reasons. First, the field outside is insignificant, and the field inside is parallel to the cylinder axis and hence perpendicular to the path. Second, any contributions from these two legs would cancel each other, because they are in opposite directions. From point d to point a (leg 4), the field is parallel to the path. Along this portion of the path, the field has a constant unknown value B. The contribution to the integral is $B\ell$, and thus Ampère's law gives

$$\oint \mathbf{B} \cdot d\mathbf{s} = B\ell = \mu_0 I_{\text{enclosed}} = \mu_0 NI. \tag{30–14}$$

We can eliminate the explicit dependence on ℓ by noting that the number of times the solenoid wire passes through the imaginary loop, N, is the length ℓ times the number of solenoid turns per unit length, n: $N = n\ell$. The quantity n is the *turn density* of the solenoid. We have

$$B\not{\ell} = \mu_0 n\not{\ell} I,$$

and the *interior magnetic field of a long solenoid* has magnitude

The magnetic field within a solenoid

$$B = \mu_0 nI. \tag{30–15}$$

Note that Eq. (30–15) contains no reference to the distance from the axis on the inside of the loop. Our derivation is completely independent of how close the path in Fig. 30–20 comes to the solenoid axis, and any choice of this distance would give the same field. *The magnetic field inside a long solenoid is uniform throughout the interior.* The magnetic field depends linearly on the current.

EXAMPLE 30–3 A solenoid consists of wires each of diameter $d = 0.6$ mm that can carry a maximum current of $I = 0.03$ A; the wires are tightly wound in a single layer. (a) What is the maximum magnitude of the field inside the solenoid? (b) Assume that the maximum current a wire can carry is proportional to the area of the wire, and that the wire's diameter is the only variable under consideration. What should the wire diameter be in order to double the magnetic field inside?

SOLUTION: (a) From Eq. (30–15) we can find the magnitude of the unknown magnetic field inside the solenoid, given the current, I, and the turn density, n. We know I, but we must calculate n. If the wire has diameter d, then we have one turn every length d, and $n = 1/d$. Thus

$$B = \mu_0 nI = \frac{\mu_0 I}{d} = \frac{(4\pi \times 10^{-7}\ \text{T}\cdot\not{\text{m}}/\not{\text{A}})(0.03\ \not{\text{A}})}{0.6 \times 10^{-3}\ \not{\text{m}}} = 0.6\ \text{T}.$$

(b) Because the solenoid is tightly wound, the turn density n depends on the diameter d as $n = 1/d$. Thus, if the diameter increases by a factor f, n changes (decreases) by a factor $1/f$. We also need the dependence of the maximum current on d. If we increase d by a factor f, then the cross-sectional area of the wire increases by a factor f^2, and so does the current. The field depends on the product nI, so if d increases by a factor f, the magnetic field increases by a factor $(1/f)(f^2) = f$. If the field is to be doubled, the diameter of the wire used in the solenoid should also be doubled.

The technique of using Ampère's law to calculate the magnetic field inside a long, cylindrical solenoid is directly applicable to noncylindrical geometry, with exactly the same results. Equation (30–15) holds *even if the cross section of the winding is not circular*. We require only that the solenoid is long and that the cross-sectional area is constant.

The results we have found for the ideal solenoid hold rather well even for a solenoid of finite length. Figure 30–21 shows the magnetic field lines of a solenoid in a plane that cuts through the center of the solenoid. The solenoid's length is four times its diameter.

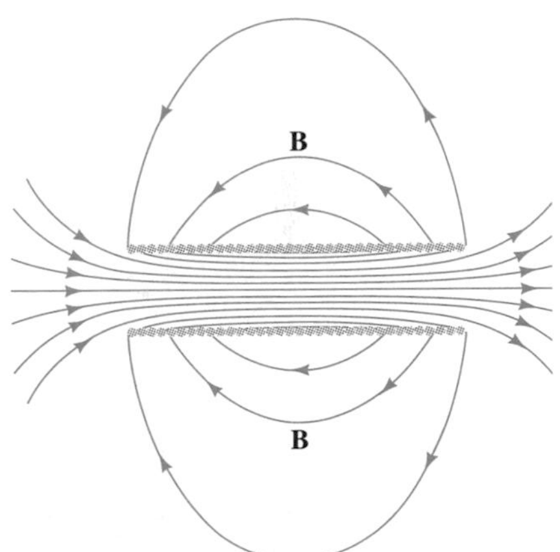

FIGURE 30–21 The magnetic field of a solenoid of finite length. (After E. M. Purcell, *Berkeley Physics Course: Electricity and Magnetism*, McGraw-Hill, 1990, p. 229.)

The *exterior* field of the solenoid of finite length illustrated in Fig. 30–21 looks just like the magnetic field of a bar magnet, Fig. 29–2a. Does this mean that the field of a bar magnet has the same physical origin as that of a solenoid? The answer is yes. A bar magnet is made of aligned, atom-sized current loops. Ampère suggested that permanent magnets are associated with internal currents. In fact, the interior field of a bar magnet is also the same as the uniform interior field of a solenoid. The origin of magnetism in matter is discussed further in Chapter 32.

A Toroidal Solenoid

A real solenoid has finite length, and therefore its magnetic field has some end effects. These end effects can be eliminated by making the solenoid into a doughnut shape, a *torus* (Fig. 30–22). Giving it this shape does introduce some variation

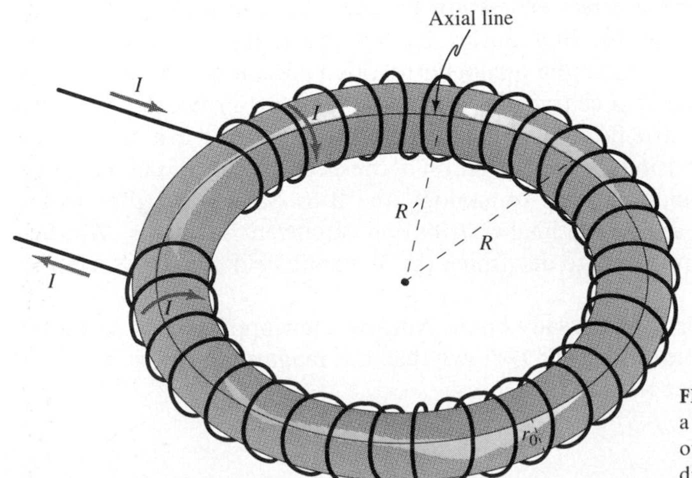

FIGURE 30–22 A torus wrapped with a wire that carries a current I has a magnetic field inside, which we can calculate by using Ampère's law. The overall radius of the torus is R, whereas the radius of the coil is r_0. The distance R' is not equal to R.

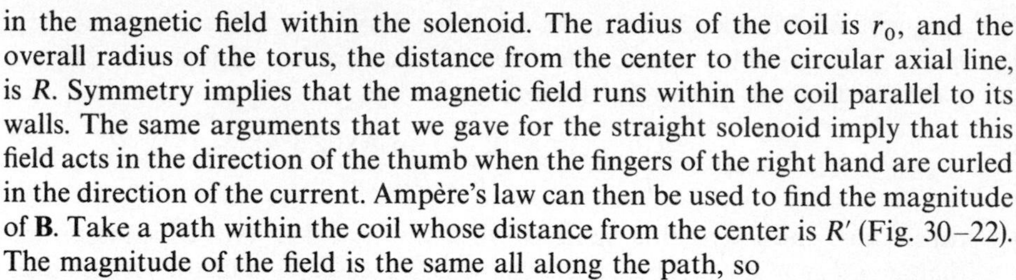

in the magnetic field within the solenoid. The radius of the coil is r_0, and the overall radius of the torus, the distance from the center to the circular axial line, is R. Symmetry implies that the magnetic field runs within the coil parallel to its walls. The same arguments that we gave for the straight solenoid imply that this field acts in the direction of the thumb when the fingers of the right hand are curled in the direction of the current. Ampère's law can then be used to find the magnitude of **B**. Take a path within the coil whose distance from the center is R' (Fig. 30–22). The magnitude of the field is the same all along the path, so

$$\oint \mathbf{B} \cdot d\mathbf{s} = B(2\pi R') = \mu_0 NI.$$

Here N is the total number of loops that form the coil. Thus the magnitude of the field at a distance R' is

$$B = \frac{\mu_0 NI}{2\pi R'}. \tag{30–16}$$

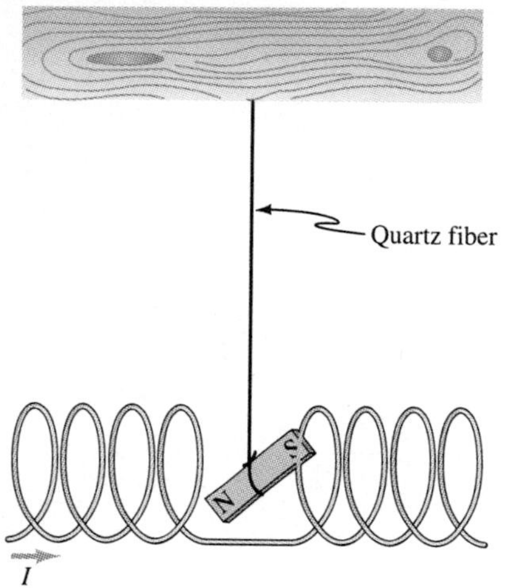

FIGURE 30–23 A torsion balance whose pointer is a permanent bar magnet can be placed in the magnetic field of a solenoid. The amount of deflection of the magnet measures the current in the coil and makes a galvanometer.

Although there are no end effects, the field is not constant across the cross-sectional area of the torus. The field at one value of R' is different from the field at a different value of R'. If the coil's radius r_0 is much less than the overall radius R of the torus, the possible values of R', from $R - r_0$ to $R + r_0$, do not vary much. The magnetic field within the torus will not vary very much either.

The magnetic field produced by a solenoid provides the means for making a *galvanometer*, an instrument we first discussed in Chapter 28. In Chapter 29 we saw how a permanent bar magnet aligns itself in a magnetic field, because there is a torque on the magnet. In Fig. 30–23 we show a magnet suspended from a quartz fiber. The fiber resists being twisted. This arrangement makes a torsion balance; when it is placed in the constant magnetic field in the central region of a solenoid, the torque due to the magnetic field balances the restoring torque due to the quartz fiber. The deflection can then be calibrated to the magnetic field. Because the magnetic field of the solenoid is directly proportional to the current that flows through it, the deflection of the magnet is a direct measure of the current.

30–4 THE BIOT–SAVART LAW

Ampère's law is a general one. However, its usefulness as a tool for calculating magnetic fields depends on symmetry in the system of currents that create the magnetic field. Here we find a direct expression for the magnetic field produced by a current. This expression, the *Biot–Savart law*, can be applied even when there is no symmetry. There is a simple analogue to this procedure in electrostatics. When there is symmetry in a charge distribution, Gauss' law provides a powerful tool for finding the electric field. When there is no symmetry, we can always find the net electric field by using the superposition of the electric fields of point charges (as determined by Coulomb's law). In analogy, the Biot–Savart law gives us the magnetic field due to an infinitesimal distribution of current segments. We then use the superposition principle to determine the magnetic field of a finite arrangement of currents.

We start with a result we already know. Ampère's law applied outside a long, straight wire that carries a current I shows that the magnitude of the magnetic field at a radial distance r from the wire is

$$B = \frac{\mu_0 I}{2\pi r},$$

Eq. (30–5). The field lines form circles around the wire, with the direction given by the right-hand rule. We expect this field to be the sum of the contributions of all the infinitesimal current elements $I\,d\ell$ that make up the wire. To find the form of the individual contributions, we make the observations that follow.

The $1/r$ dependence of the magnetic field resembles the $1/r$ dependence of the electric field due to a long, charged rod of constant linear charge density λ, as given in Eq. (23–31):

$$E = \frac{1}{2\pi\epsilon_0}\frac{\lambda}{r}.$$

This result was obtained by integrating the component of the electric field perpendicular to the wire due to the charge in an infinitesimal length $d\ell$ of the charged rod, from Eq. (23–27):

$$dE_\perp = \frac{1}{4\pi\epsilon_0}\frac{\lambda\,d\ell}{r^2}\cos\phi.$$

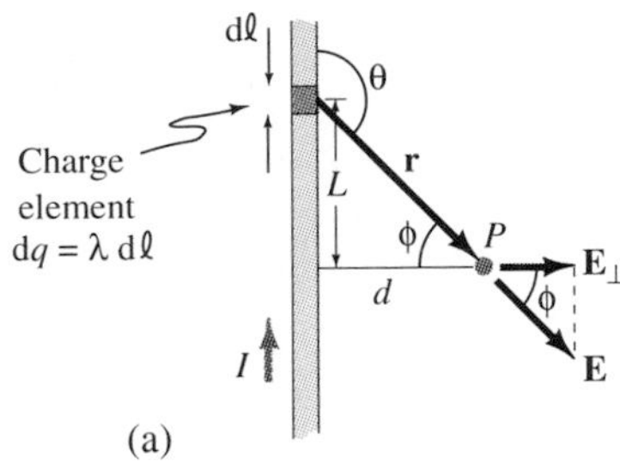

Here the particular element of the charged rod is a distance $r = \sqrt{L^2 + d^2}$ from the point where the field is measured, and ϕ is as shown in Fig. 30–24a. The factor $(1/4\pi\epsilon_0)\,(\lambda\,d\ell)/r^2$ is just the electric field strength for a point charge $dq = \lambda\,d\ell$. The second factor, $\cos\phi$, is present only because we are looking at the perpendicular component. The resemblance between Eqs. (23–31) and (30–5), in which $1/\epsilon_0$ is replaced by μ_0 and λ is replaced by I, suggests that we try a similar procedure for finding the magnetic field that results from a length $d\ell$ of a wire that carries a current I. The result has the same form as that for the electric field of the charged wire, Eq. (23–27), with the replacements specified above:

$$dB = \frac{\mu_0}{4\pi}\frac{I\,d\ell}{r^2}\cos\phi. \qquad (30\text{–}17)$$

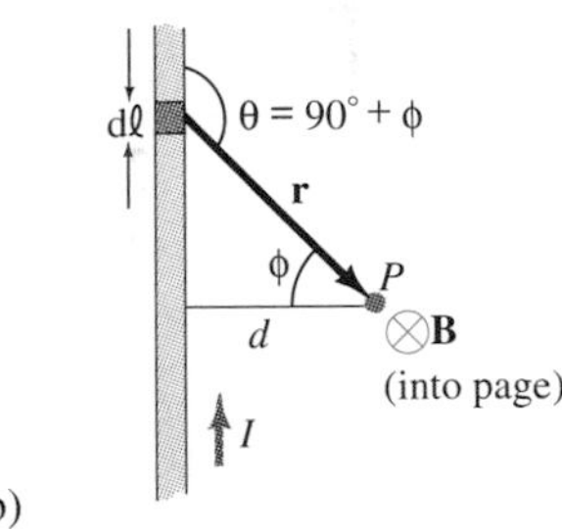

FIGURE 30–24 (a) A thread that carries a net charge density λ can be broken up into segments that contribute an electric field at any point. Integration over the contributions from the segments gives the net electric field. (b) A wire carrying current I similarly has a magnetic field that can be calculated by integrating the contributions of segments of the thread. For a segment of length $d\ell$, the vector $\mathbf{d\ell}$ is oriented along the segment in the direction of I.

The electric field is directed radially away from the charge element. The magnetic field still forms circles, and at point P in Fig. 30–24b, $\mathbf{B}$ is directed into the page, as required by the right-hand rule. The direction of the current is indicated by making the infinitesimal length $d\ell$ a vector $\mathbf{d\ell}$ whose direction is along I. Note that $\mathbf{B}$ is perpendicular to both $\mathbf{d\ell}$ and $\mathbf{r}$. When we speak of a vector perpendicular to two other vectors, we are reminded of the vector product. A short exercise in trigonometry shows that if $\theta = \phi + 90°$ (as in Fig. 30–24b), then $\cos\phi = \cos(\theta - 90°) = \cos\theta\cos 90° + \sin\theta\sin 90° = \sin\theta$. Equation (30–17) becomes

$$dB = \frac{\mu_0}{4\pi}\frac{I\,d\ell\sin\theta}{r^2} = \frac{\mu_0}{4\pi}I\frac{d\ell\,r\sin\theta}{r^3} \qquad (30\text{–}18)$$

in the direction perpendicular to $\mathbf{d\ell}$ and $\mathbf{r}$. The presence of a $\sin\theta$ factor confirms our guess that a vector product is involved. We have found *the magnetic field* $\mathbf{dB}$ *produced by a segment of wire* $\mathbf{d\ell}$ *that carries a current I at a displacement* $\mathbf{r}$ *from the segment:*

$$\mathbf{dB} = \frac{\mu_0}{4\pi}\frac{I\,\mathbf{d\ell}\times\mathbf{r}}{r^3}. \qquad (30\text{–}19)$$

The Biot–Savart law

The vector product $\mathbf{d\ell}\times\mathbf{r}$ has a magnitude $d\ell\,r\sin\theta$ and the proper direction for this situation, into the page. Notice that there is no ambiguity as to the vector $\mathbf{d\ell}$. If the segment is short enough, it may be treated as a straight line. Equation (30–19) is the **Biot–Savart law**, named after the two physicists who first formulated it, Jean-Baptiste Biot and Félix Savart. This law is analogous to Coulomb's law in electricity. It even has the same overall distance dependence of $1/r^2$—note that the magnitude of r appears in the numerator of Eq. (30–19). The angular factors are quite different, however.

Using the Biot–Savart Law

The Biot–Savart law can be used to find the magnetic field due to nonsymmetric current distributions. It is thus used in the same way Coulomb's law is used in electricity. The net magnetic field is found by integrating over **dB**:

$$\mathbf{B} = \int d\mathbf{B} = \frac{\mu_0}{4\pi}\int \frac{I\, d\boldsymbol{\ell} \times \mathbf{r}}{r^3}. \qquad (30\text{–}20)$$

Equation (30–20) is often too complicated for practical use. In practice, magnetic fields due to nonsymmetric current distributions are measured experimentally, although sophisticated computer programs are now available to design the current windings needed to produce a desired field. Such programs superpose infinitesimal field contributions **dB**, each given by the Biot–Savart law.

EXAMPLE 30–4 A straight segment of wire of length L carries a current I. Use the Biot–Savart law to find the magnetic field in the plane perpendicular to the wire and passing through the midpoint of the wire segment.

SOLUTION: Orient the wire along the x-axis, as in Fig. 30–25a, with its center positioned at the origin. We then wish to find the field in the yz-plane. At a given distance D from the wire, the wire looks the same from anywhere in this plane. Thus we can calculate the field anywhere around the wire and choose the point $y = D$, $z = 0$. The right-hand rule shows that the quantity $d\boldsymbol{\ell} \times \mathbf{r}$ points out of the page, where $d\boldsymbol{\ell}$ and $\mathbf{r}$ are as shown in the figure. The field forms circles around the wire, as before. Let us now concentrate on finding the magnitude at point D.

From Eq. (30–18), the magnetic field from the segment $d\ell$ has magnitude

$$dB = \frac{\mu_0 I}{4\pi}\, dx\, \frac{\sin(\pi - \theta)}{r^2} = \frac{\mu_0 I}{4\pi}\, dx\, \frac{\sin\theta}{r^2} = \frac{\mu_0 I}{4\pi}\, dx\, \frac{\cos\phi}{r^2},$$

where we have replaced $d\ell$ by dx and used the angle ϕ defined in Fig. 30–25a. To find the net magnetic field, we sum over the contributions of segments from $x = -L/2$ to $x = L/2$:

$$B = \frac{\mu_0 I}{4\pi}\int_{-L/2}^{L/2} \frac{\cos\phi}{r^2}\, dx.$$

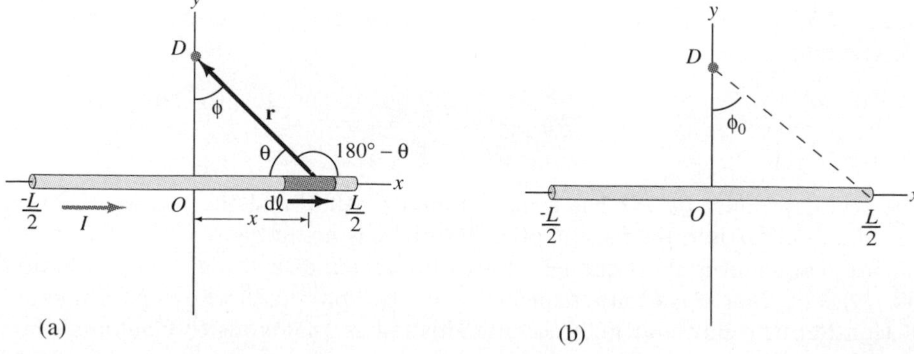

FIGURE 30–25 (a) Example 30–4. A straight segment of wire of length L, carrying a current I, is oriented on the x-axis and centered at the origin. (b) The limits of the integration are $-\phi_0$ and $+\phi_0$; here we show the limit $+\phi_0$.

Both ϕ and r depend on x. The integral is computed most simply if we use trigonometric variables, and we therefore change variables from x to ϕ. We have

$$\frac{x}{D} = \tan\phi;$$

$$dx = D\,d(\tan\phi) = D\sec^2\phi\,d\phi = \frac{D}{\cos^2\phi}\,d\phi.$$

In addition, $r = D/\cos\phi$, so the combination that appears in the integral above becomes

$$\frac{\cos\phi}{r^2}\,dx = \cos\phi\,\frac{1}{(D/\cos\phi)^2}\,\frac{D}{\cos^2\phi}\,d\phi = \frac{1}{D}\cos\phi\,d\phi.$$

Thus

$$B = \frac{\mu_0 I}{4\pi D}\int_{-\phi_0}^{+\phi_0}\cos\phi\,d\phi,$$

where the limit of integration $+\phi_0$ is shown in Fig. 30–25b. The integral of the cosine is the sine, and

$$B = \frac{\mu_0 I}{4\pi D}[\sin\phi_0 - \sin(-\phi_0)] = \frac{\mu_0 I}{2\pi D}\sin\phi_0.$$

We can reexpress this result in terms of L and D by using

$$\sin\phi_0 = \frac{L/2}{\sqrt{(L/2)^2 + D^2}}$$

to find that

$$B = \frac{\mu_0 I}{4\pi}\,\frac{L}{D\sqrt{(L/2)^2 + D^2}}. \qquad (30\text{–}21)$$

The form given in Eq. (30–21) shows that the field depends not only on the distance D from the wire but also on the relative magnitudes of D and L. We can check this result by taking the limit that L is large; we should then recover Eq. (30–5). In the limit $L \gg D$, the square root in Eq. (30–21) becomes $L/2$, and

$$B \to \frac{\mu_0 I}{4\pi}\,\frac{\not{L}}{D(\not{L}/2)} = \frac{\mu_0 I}{2\pi D},$$

the correct result as given by Ampère's law.

Perhaps you wonder why, in Example 30–4, we could not have used Ampère's law directly. The reason is that there are contributions from the current-carrying wires that bring the current into our finite segment of length L and take it out (not shown in Fig. 30–25a). These wires destroy the cylindrical symmetry. It is only such symmetry that makes Ampère's law useful. Ampère's law is still valid, of course, but it is not useful here. The Biot–Savart law can *always* be used. When the segment becomes very long, so that the contributions from other segments of wire are small, then Ampère's law is again useful. As we have seen, the Biot–Savart law and Ampère's law then give the same result.

PROBLEM-SOLVING TECHNIQUES

In the examples of this chapter and Chapter 29, we have studied two aspects of problems on static magnetic fields. We may want to find the magnetic fields produced by a given time-independent set of currents, or we may want to find the magnetic forces on currents or on moving charges. Two sets of key laws contain all that is generally necessary to approach such problems, and you should understand the symbols in these formulas and what the laws mean. First, we have the laws that determine the magnetic field due to moving charges or currents, which can be written in the two forms

$$\text{Ampère's law:} \qquad \oint \mathbf{B} \cdot d\mathbf{s} = \mu_0 I_{\text{enclosed}}, \tag{30–10}$$

$$\text{Biot–Savart law:} \qquad d\mathbf{B} = \frac{\mu_0}{4\pi} \frac{I \, d\boldsymbol{\ell} \times \mathbf{r}}{r^3}. \tag{30–19}$$

Second, we have the laws that express the force on a moving charge or a current due to a given magnetic field, namely,

$$\mathbf{F}_B = q \, \mathbf{v} \times \mathbf{B}, \tag{29–1}$$

$$d\mathbf{F} = I \, d\boldsymbol{\ell} \times \mathbf{B}. \tag{29–14}$$

Each set of laws involves a right-hand rule, and you should know how to use it. For Ampère's law, if the thumb of the right hand follows the current, the fingers curl in the direction of the integration path. For the force laws and the Biot–Savart law, the right-hand rule for a vector product applies.

Based on these laws, we can suggest a list of habits to develop. Many of these are the very same habits that are useful for solving *any* problem in physics.

1. Draw a figure that indicates the physical situation with the quantities known, including directions if appropriate.
2. Write down what is known and what is to be determined. Are you dealing with moving charges or with currents?
3. What physical principles connect the unknown quantities to the known ones?
4. If the problem concerns a force, see if there is enough information to determine the force directly from the force laws. Otherwise, you may need to compute a magnetic field or integrate an infinitesimal force.
5. In a situation with enough symmetry (for example, for long, straight wires), we can use Ampère's law to calculate a magnetic field. When applicable, Ampère's law will usually give the answer more easily than the Biot–Savart law.
6. If the system is not sufficiently symmetric, the Biot–Savart law is always available. In using it, be sure that the infinitesimal element $d\ell$ and the position vector $\mathbf{r}$ are identified properly. A partial symmetry may rule out one or more directions for the magnetic field. If only one direction is indicated by symmetry, then the other components will cancel in the calculation of the integral over $d\ell$, and you need only integrate for the component desired.
7. Checks of dimensions and units are always appropriate.
8. Substitute numbers only at the last stage; any checks you can find of limits or special cases are always helpful.

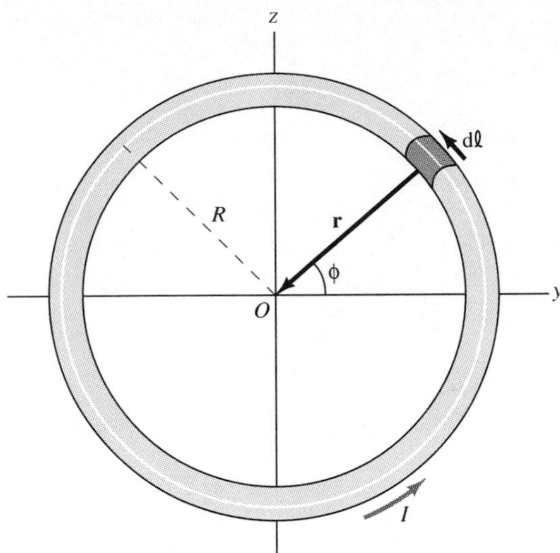

FIGURE 30–26 Example 30–5. An integration is required to find the net magnetic field at the center of the loop.

EXAMPLE 30–5 A wire forms a circular loop of radius $R = 12$ cm. A current $I = 8.0$ A flows counterclockwise in the wire (Fig. 30–26). Find the magnetic field at the center.

SOLUTION: There is not enough symmetry here to allow us to draw a path along which the magnetic field is constant, so we cannot use Ampère's law. We must use the Biot–Savart law and integrate over the contributions $\mathbf{dB}$ of the different elements of the wire to find the unknown total field $\mathbf{B}$. Vector $\mathbf{r}$ runs from the current element $d\ell$ to the center of the circle, the point where we want to find $\mathbf{B}$. The quantity $d\ell \times \mathbf{r}$ for the current element shown in Fig. 30–26 is out of the page, and this will be true for all the current elements that make up the loop. The net magnetic field at the center is thus out of the page.

The magnitude of the field due to the element shown is given according to Eq. (30–18) by

$$dB = \frac{\mu_0}{4\pi}\frac{I\,d\ell}{R^2}.$$

There is no sine factor here because $d\ell$ is perpendicular to $\mathbf{r}$. The integral of $d\ell$ around the circle is the circumference $2\pi R$, so the net field at the center has magnitude

$$B = \int dB = \int \frac{\mu_0}{4\pi}\frac{I\,d\ell}{R^2} = \frac{\mu_0 I}{4\pi}\frac{1}{R^2}\int d\ell = \frac{\mu_0 I}{4\pi}\frac{2\pi R}{R^2}$$

$$= \frac{\mu_0 I}{2R}. \tag{30–22}$$

The numerical value of this field is

$$B = \frac{(4\pi \times 10^{-7}\ \mathrm{T\cdot m/A})(8.0\ \mathrm{A})}{2(0.12\ \mathrm{m})} = 4.2 \times 10^{-5}\ \mathrm{T}.$$

This value is about that of the earth's magnetic field at the earth's surface.

The orientation of the magnetic field we have calculated is consistent with a right-hand rule in which the fingers of the right hand curl with the current and the thumb indicates the direction of the field. This orientation rule is the same one that we found for the magnetic field of a solenoid, and indeed the loop can be viewed as a highly compressed solenoid.

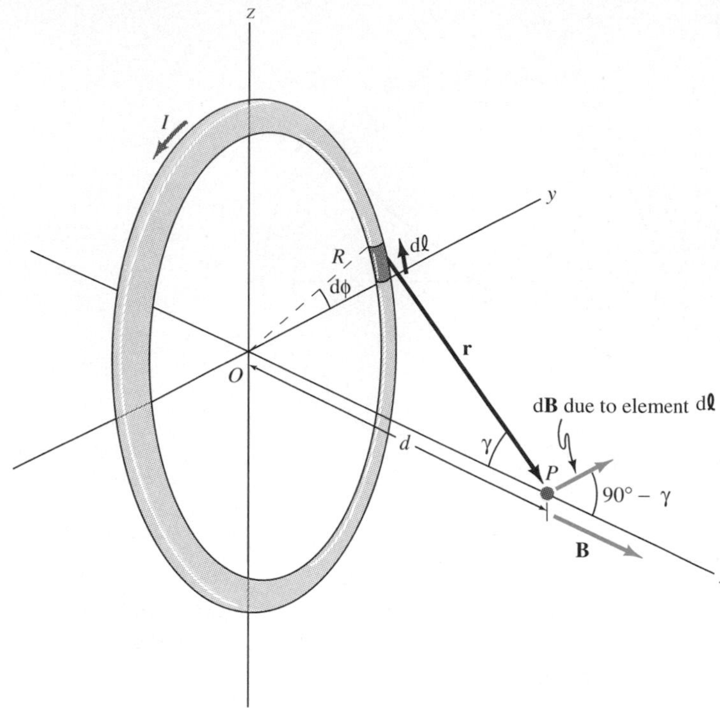

FIGURE 30–27 Example 30–6. The net magnetic field along the loop axis due to a current loop carrying a current I is oriented along the axis, according to a right-hand rule. The contribution from an infinitesimal element $d\ell$ has components along the axis and in other directions as well.

EXAMPLE 30–6 Consider again the circular wire loop of Example 30–5, drawn in Fig. 30–26. Find the magnitude of the magnetic field all along the axis of the loop, oriented as in Fig. 30–27. What is the limit of your result (a) at the center and (b) at large distances along the axis?

SOLUTION: In this case we want to find the magnetic field all along the axis of the loop, not just at the center, as in Example 30–5. As we stated there, Ampère's law is not useful here. We must first use the Biot–Savart law to find the magnetic field on the axis from a particular current element, and then sum over the contributions of the elements. The loop is now aligned so that its axis is along the x-axis and its center is at the origin (Fig. 30–27). Consider a point P a distance d from the center. We have chosen an element $d\ell$ where the loop passes through the $+y$-axis. Its contribution $\mathbf{dB}$ is perpendicular to both $d\ell$ and $\mathbf{r}$, so it has *both* a component along the loop axis and a $+y$-component. If we had chosen an element on the opposite side of the loop, where the loop cuts the $-y$-axis, we would have found a $\mathbf{dB}$ with a component along the loop axis in the same direction as the contribution from the first loop element, and also a component in the $-y$-direction. The y-component of $\mathbf{dB}$ from the first element cancels the y-component from the second element. This will be true for all pairs of elements around the loop, so we must calculate only the component of $\mathbf{dB}$ along the loop axis, which is the x-component. This understanding of the role played by symmetry is important, and a moment or two spent looking for such symmetries is time well spent.

From the geometry shown, we have

$$dB_x = \frac{\mu_0}{4\pi}\frac{I\,d\ell}{r^2}\cos(90^\circ - \gamma).$$

The infinitesimal length $d\ell = R\, d\phi$, the magnitude of $\mathbf{r}$ is $r = \sqrt{R^2 + d^2}$, and

$$\cos(90^\circ - \gamma) = \cos 90^\circ \cos \gamma + \sin 90^\circ \sin \gamma = \sin \gamma = \frac{R}{\sqrt{R^2 + d^2}}.$$

Collecting terms, we have

$$dB_x = \frac{\mu_0}{4\pi} \frac{IR^2\, d\phi}{(R^2 + d^2)^{3/2}}.$$

The net field is

$$B_x = \int dB_x = \frac{\mu_0 I R^2}{4\pi (R^2 + d^2)^{3/2}} \int_0^{2\pi} d\phi = \frac{\mu_0 I}{2} \frac{R^2}{(R^2 + d^2)^{3/2}}. \qquad (30\text{–}23)$$

As our symmetry argument states, the magnetic field points along the $+x$-axis.

(a) For the center of the loop, Eq. (30–23) correctly reduces to Eq. (30–22) when d is zero.

(b) At large distances, $d \gg R$, the axial magnetic field in Eq. (30–23) reduces to

$$B = \frac{\mu_0 I}{2} \frac{R^2}{d^3} = \frac{\mu_0}{2\pi} \frac{I\pi R^2}{d^3}. \qquad (30\text{–}24)$$

Magnetic Dipoles

Equation (30–24) is similar to our expression for the electric field along the axis of an electric dipole, Eq. (23–14). Indeed, careful calculation of the magnetic field over all space due to the current-carrying loop of Examples 30–5 and 30–6 shows that the magnetic field lines, as in Fig. 30–28, are just like the electric field lines of the electric dipole, Fig. 23–12. The current loop forms a *magnetic dipole*, and the strength of the magnetic dipole is characterized by the *magnetic dipole moment*, μ. (See Chapter 29, where we studied the *response* of such a loop to an external field.) In place of the electric dipole moment, $p = dL$, for the electric field of an

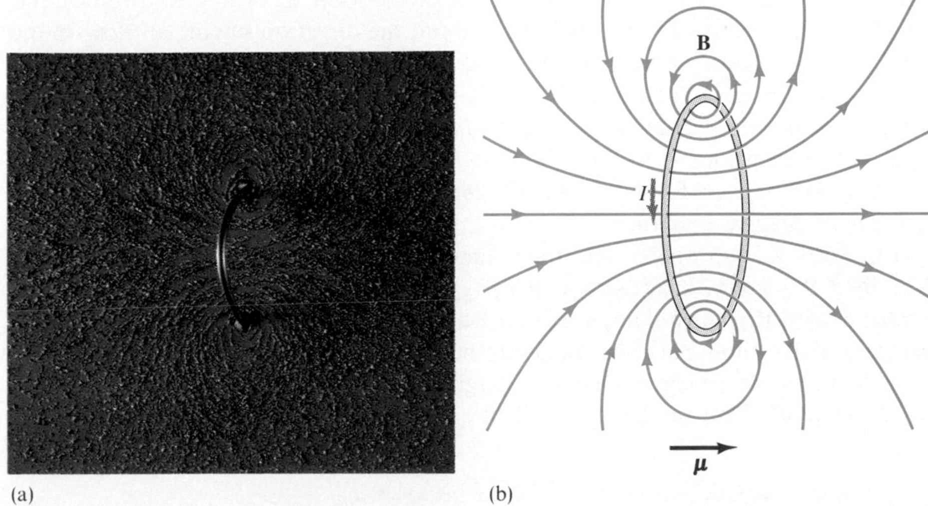

FIGURE 30–28 (a) Magnetic field lines for a circular loop of current, as shown by iron filings. (b) The field for such a loop is a magnetic dipole field.

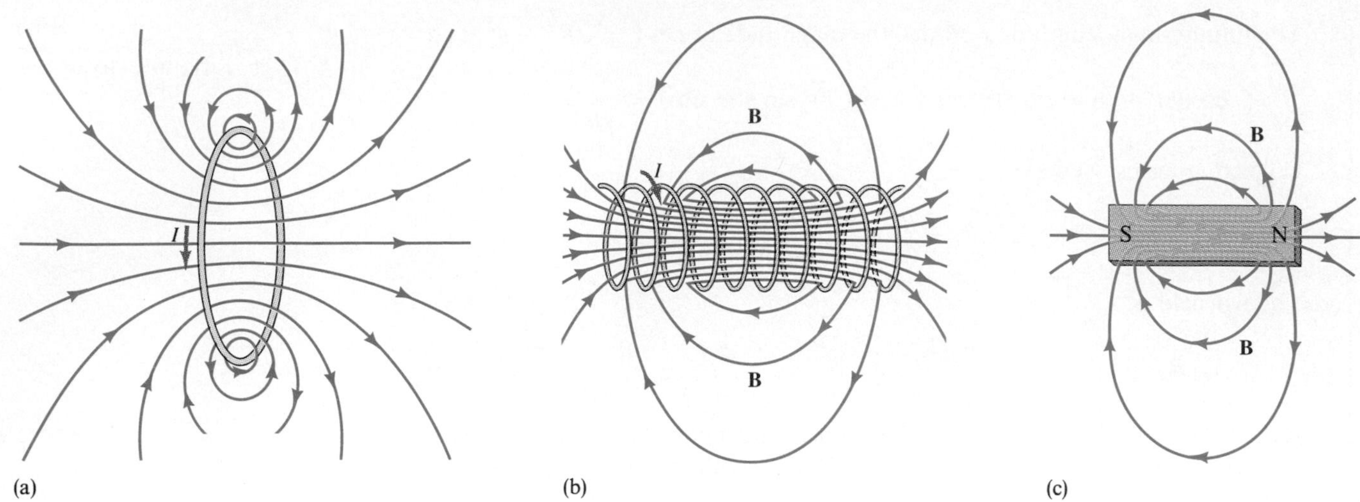

FIGURE 30–29 (a) The magnetic field lines of a circular loop of current have the same form as the magnetic field lines of (b) a solenoid and of (c) a bar magnet.

electric dipole, we find the quantity μ,† which is defined by

$$\text{for a circular current loop:} \qquad \mu \equiv I\,\pi R^2. \tag{30–25}$$

The magnetic field is shown in Fig. 30–28. At a distance d far from the loop, the magnitude of the magnetic field along the axis is

$$\text{along the axis:} \qquad B = \frac{\mu_0}{2\pi}\frac{\mu}{d^3}. \tag{30–26}$$

In fact, for *any* closed loop of area A that carries a current I, there is an analogous result: The magnetic field decreases as $1/d^3$ far from the loop, and the strength of the field is characterized by the more general dipole moment given by the product of the current in the loop and the area of the loop:

The magnetic dipole moment. Magnetic dipoles are formed from current loops.

$$\mu = IA. \tag{30–27}$$

The direction of the field is determined if we form a vector $\boldsymbol{\mu}$ from μ. For any plane loop, $\boldsymbol{\mu}$ is perpendicular to the plane of the loop according to a right-hand rule: Curl the fingers of the right hand around the direction of current flow in the loop, and the thumb points in the direction of $\boldsymbol{\mu}$. Note that because magnetic charges do not exist, the magnetic dipole is not formed from a pair of equal but opposite magnetic charges, but rather from a closed loop of current. The magnetic field of a current loop (Fig. 30–29a) has the same form as the fields of a solenoid and of a bar magnet (Figs. 30–29b, 30–29c). Indeed, both solenoids and bar magnets act as magnetic dipoles.

In Section 29–5 we saw that the magnetic dipole responds to an external field: The field $\mathbf{B}_{ext}$ exerts a torque $\boldsymbol{\tau} = \boldsymbol{\mu} \times \mathbf{B}_{ext}$ on the loop that tends to line up the vector $\boldsymbol{\mu}$ with $\mathbf{B}_{ext}$. As a comparison of Eqs. (29–22) and (30–27) shows, the same vector $\boldsymbol{\mu}$ determines both the magnetic field of the loop and the reaction of the loop to an external magnetic field. A magnetic dipole, like an electric dipole, both produces a field and responds to a field.

† Not to be confused with the permeability of free space, μ_0.

30–5 THE MAXWELL DISPLACEMENT CURRENT

In this section we shall learn that there is a logical flaw in Ampère's law when there is time dependence in the current, and we present the modification James Clerk Maxwell proposed in 1865 that removes the flaw. Maxwell's modification was crucial to the completely unified theory of electricity and magnetism that will be discussed in Chapter 35. In circuits with capacitors, we may have currents that carry charges to capacitor plates and terminate there. In Chapter 28, we saw that the currents in RC circuits may be time dependent. As currents flow into a capacitor, the charge builds up, and eventually the current that brings the charge must decrease. This is consistent with charge conservation, a fundamental law of nature. If there is no charge buildup anywhere, then currents must be steady, because charge must flow out of a region (no matter how large or small) at the same rate it flows into that region.

Ampère's law is applied with an integration over some closed path. The right-hand side of Ampère's law, Eq. (30–10), contains what we called the current enclosed by a path. By "the current enclosed by a path" we mean the rate of charge flow through a surface whose boundary is the closed path. Such a surface can be chosen in many different ways (Fig. 30–30), but *when the current is continuous, the current that crosses any one of these surfaces must be the same as the current that crosses any other*. The freedom in how the surface is chosen therefore presents no problem. There is a situation, however, in which the freedom to choose the surface presents a difficulty, and that situation arises when the current deposits charge on the plates of a capacitor. Figure 30–31 shows two surfaces with the same loop as their boundary. A current I crosses surface 1 in the positive sense, while no current crosses surface 2. Yet there is no unambiguous way to distinguish the two surfaces in the application of Ampère's law.

Maxwell noted that even if no current passes through surface 2, there is nevertheless a distinguishing feature for this surface, namely, *there is a changing electric flux through it*. As the charge builds up on the plates of the capacitor, so does the resulting electric field, and hence so does the electric flux in the space between the plates.

Suppose that we have a capacitor with large, planar plates of area A (Fig.

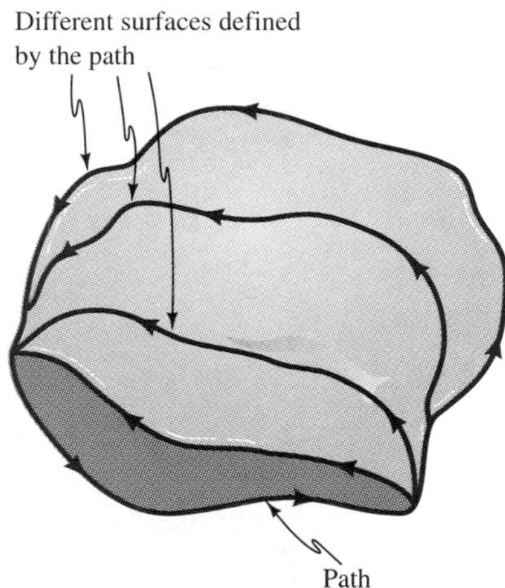

FIGURE 30–30 A closed path defines an infinite number of surfaces.

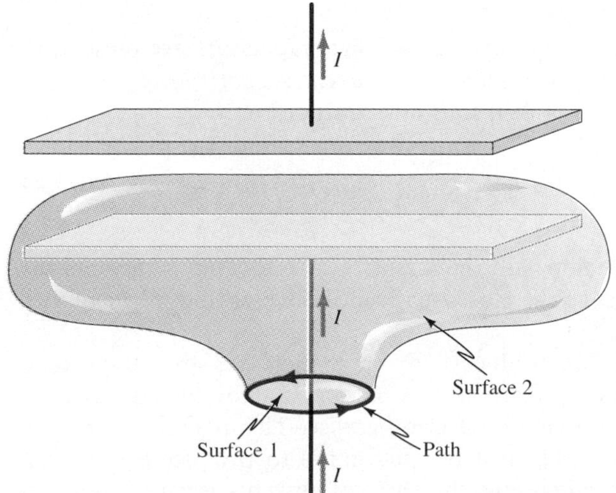

FIGURE 30–31 Two surfaces bounded by the same closed path. A current passes through surface 1 but not through surface 2.

30–32). As the current I accumulates on one of the plates, it deposits a charge $+q$. We can then use the results of Chapters 23 and 26, that the field **E** of such a capacitor is uniform between the plates and small outside the region between the plates. The field points from the plate with positive charge (where the charge accumulates) to the plate with negative charge. The electric flux associated with a surface that passes between the plates, Φ_E, is then just EA. The magnitude of the electric field is given by $E = (1/\epsilon_0)(q/A)$. Because $\Phi_E = EA$ in this case, this result is equivalent to

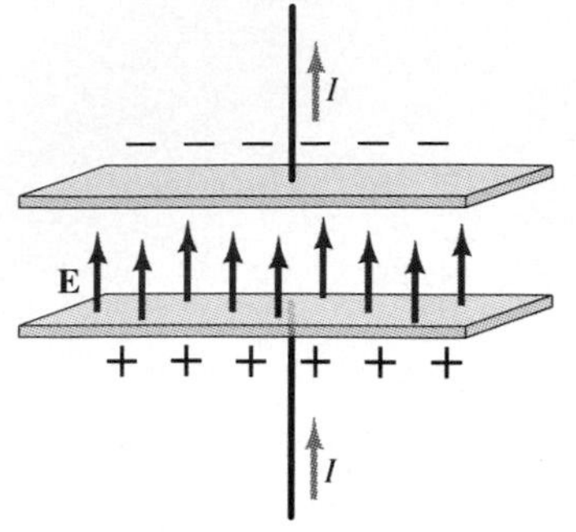

FIGURE 30–32 The electric field between two parallel plates that carry charges $+q$ and $-q$, respectively, is uniform and small outside the region between the plates.

$$\epsilon_0 \Phi_E = q. \qquad (30\text{–}28)$$

If we take a time derivative, we find the relation

$$\epsilon_0 \frac{d\Phi_E}{dt} = \frac{dq}{dt} = I. \qquad (30\text{–}29)$$

Equation (30–29) implies that whatever the value of the current that passes through the wire that leads to the capacitor, that current equals the quantity $\epsilon_0\, d\Phi_E/dt$ between the plates. Therefore if we replace I in Ampère's law by the *sum* of the two terms in Eq. (30–29),

$$I + \epsilon_0 \frac{d\Phi_E}{dt},$$

Ampère's law would be satisfied for *any* surface we could draw for the path of Fig. 30–31. For surface 1, only the term I in this sum applies, whereas for surface 2, only the changing flux term applies. The second term, $\epsilon_0\, d\Phi_E/dt$, is written in a way that does not refer explicitly to the plane geometry. Indeed, Maxwell was able to show that if the sum of these two terms is used, any surface gives the same answer in Ampère's law. Maxwell called the changing flux term the **displacement current,** I_d:

The displacement current

$$I_d \equiv \epsilon_0 \frac{d\Phi_E}{dt}. \qquad (30\text{–}30)$$

Note that the displacement current is present only when there are changing currents; a truly steady current cannot involve changing flux. But any current that enters a capacitor *must* be changing. For example, the current decreases as the capacitor plates are charged, until the plates are charged to saturation, at which point the current stops.

The sum of the ordinary current and the displacement current through any closed loop is unchanging.

Even though the current is not continuous when capacitors are present, *the sum of the ordinary current and the displacement current is continuous.*

Maxwell's generalized form of Ampère's law is accordingly

The generalized Ampère's law. It applies even when the currents change with time.

$$\oint \mathbf{B} \cdot d\mathbf{s} = \mu_0(I + I_d) = \mu_0 I + \mu_0\epsilon_0 \frac{d\Phi_E}{dt}. \qquad (30\text{–}31)$$

Here the surface through which passes the quantity whose value is the sum of I and I_d is *any* surface that spans the closed path along which the magnetic field is integrated.

That a changing electric flux produces a magnetic field has great importance for electromagnetic waves, as we shall see in Chapter 35. However, as Example 30–7 shows, it has very little practical effect otherwise. This is because the time rate of change of the electric field must be quite large to produce a significant magnetic field. In laboratory conditions the effect of currents is much more important than the effect of changing electric fields.

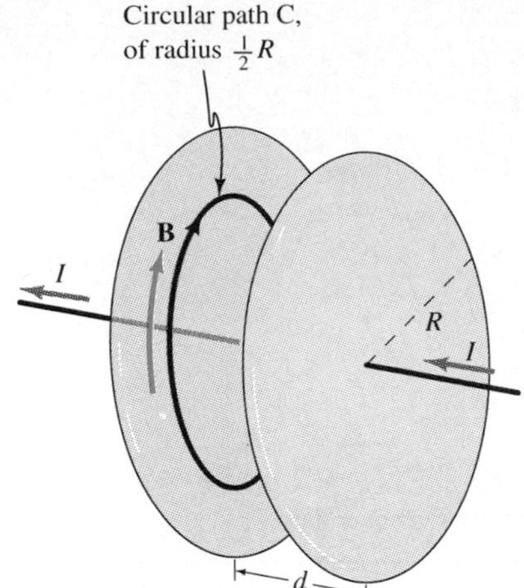

FIGURE 30–33 Example 30–7. Symmetry requires that the value of the magnetic field is the same everywhere on path C.

EXAMPLE 30–7 The planar circular plates of a capacitor are being charged. At a given moment the charge is being built up at the rate of 1 C/s. The plates have a radius $R = 0.1$ m and a separation $d = 1$ cm. Calculate the magnetic field due to the displacement current midway between the plates at a radius equal to half the plate radius.

SOLUTION: Because the plates are circular, symmetry requires that the value of the magnetic field is the same everywhere on path C, a circular path centered on the plates' axis and of radius $R/2$ (Fig. 30–33). The line integral in Ampère's law is taken in the sense drawn. Because **B** has constant magnitude on this path and points along the path, we may remove it from the integral. The remaining integral is the circumference of the path, $2\pi(R/2)$:

$$\oint \mathbf{B}\cdot d\mathbf{s} = B\oint ds = B\left(2\pi\frac{R}{2}\right).$$

In order to calculate the displacement current, we note that, in terms of the charge q on the plates, the electric field is constant and has magnitude given by

$$E = \frac{1}{\epsilon_0}\frac{q}{\pi R^2}.$$

We must now calculate the electric flux through the area bounded by path C (and not the *total* electric flux in the capacitor). The flux through path C is E times the area $\pi(R/2)^2$:

$$\Phi_E = \frac{1}{\epsilon_0}\frac{q}{\cancel{\pi}\cancel{R^2}}\cancel{\pi}\left(\frac{\cancel{R}}{2}\right)^2 = \frac{1}{4\epsilon_0}q.$$

Thus the displacement current is

$$I_d = \epsilon_0\frac{d\Phi_E}{dt} = \cancel{\epsilon_0}\left(\frac{1}{4\cancel{\epsilon_0}}\frac{dq}{dt}\right) = \frac{1}{4}\frac{dq}{dt}.$$

We can find the magnitude B by using Eq. (30–31):

$$B\cancel{2}\pi\frac{R}{\cancel{2}} = \mu_0 I_d = \frac{\mu_0}{4}\frac{dq}{dt};$$

then

$$B = \frac{\mu_0}{4\pi}\frac{1}{R}\frac{dq}{dt}.$$

Numerically, $dq/dt = 1\ \text{C/s} = 1\ \text{A}$ and $R = 0.1\ \text{m}$, so

$$B = \frac{4\pi \times 10^{-7}\ \text{N/A}^2}{4\pi}\frac{1}{0.1\ \text{m}}(1\ \text{A}) = 10^{-6}\ \text{T}.$$

This is indeed a small field.

*30–6 CONSISTENCY DIFFICULTIES: THE FRAME DEPENDENCE OF FORCES AND NEWTON'S THIRD LAW

We now know that a moving charge produces a magnetic field, and that a magnetic field produces a force on a moving charge. If we combine these two concepts to compute the magnetic forces that two moving charges exert on each other, we arrive at some surprising conclusions. These conclusions turn out to be cornerstones of the special theory of relativity. To study them, we first want to express the magnetic field produced by a single moving charge.

The Magnetic Field of a Moving Charge

We can start with the Biot–Savart law and arrive at the magnetic field produced by a single charge q. A wire element $d\ell$ carrying a current I contains moving charges. Because $I = dq/dt$, the term $I\,d\ell$ in the Biot–Savart law is of the form $I\,d\ell = (dq/dt)\,d\ell = dq\,(d\ell/dt) = dq\,\mathbf{v}$. In this case the infinitesimal charge dq is the single charge q. When we substitute $I\,d\ell = q\mathbf{v}$ into the Biot–Savart law, Eq. (30–19), we find *the magnetic field,* $\mathbf{B}$, *at a point P located a displacement* $\mathbf{r}$ *from a charge q that moves with velocity* $\mathbf{v}$:

$$\mathbf{B} = \frac{\mu_0}{4\pi}\frac{q\,\mathbf{v} \times \mathbf{r}}{r^3}. \tag{30–32}$$

The magnetic field forms circles around $\mathbf{v}$ in the planes perpendicular to $\mathbf{v}$ (Fig. 30–34).

The Magnetic Forces between Two Moving Charges

Magnetic forces are velocity dependent; their direction and magnitude both depend on the velocity of the charge being acted upon. They are different from frictional

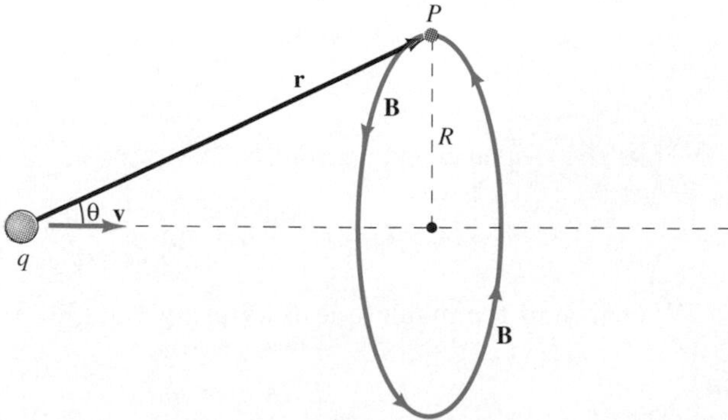

FIGURE 30–34 A charge q moves with velocity $\mathbf{v}$. The magnetic field forms circles about the axis along which $\mathbf{v}$ extends, in the plane perpendicular to $\mathbf{v}$.

forces, such as the drag on a parachute, which may depend on the velocity in some approximate way but which in fact represent a gross average over a variety of forces that act on a moving body, slowing it down. The difference is in the fundamental nature of the magnetic force law, Eq. (29–1). This velocity dependence is what led Einstein to develop the special theory of relativity.

Let us use the magnetic field of a moving charge [Eq. (30–32)], together with the magnetic force law, to find the magnetic force between two moving charges q_1 and q_2. Suppose that the charges move with velocities $\mathbf{v}_1$ and $\mathbf{v}_2$, respectively. The displacement vector from charge 2 to charge 1 is $\mathbf{r}_{21}$, and the displacement vector from charge 1 to charge 2 is $\mathbf{r}_{12}$; $\mathbf{r}_{12} = -\mathbf{r}_{21}$, and each has magnitude d.

To find the magnetic force $\mathbf{F}_1 = q_1 \mathbf{v}_1 \times \mathbf{B}$ on charge q_1, we first compute the magnetic field $\mathbf{B}_2$ *due to* q_2 at the position of q_1. From Eq. (30–32),

$$\mathbf{B}_2 = \frac{\mu_0}{4\pi} \frac{q_2 \mathbf{v}_2 \times \mathbf{r}_{21}}{d^3},$$

so

$$\mathbf{F}_1 = \frac{\mu_0}{4\pi} q_1 q_2 \frac{\mathbf{v}_1 \times (\mathbf{v}_2 \times \mathbf{r}_{21})}{d^3}. \qquad (30\text{–}33a)$$

Similarly, the force on q_2 is

$$\mathbf{F}_2 = \frac{\mu_0}{4\pi} q_2 q_1 \frac{\mathbf{v}_2 \times (\mathbf{v}_1 \times \mathbf{r}_{12})}{d^3}. \qquad (30\text{–}33b)$$

Application and comparison of these equations in some simple special cases will lead us to our surprising results.

Magnetic Forces Appear to Depend on the Reference Frame

Let us first apply Eqs. (30–33) to the case of two charges of the same sign. They both move at speed v along the $+x$-direction (Fig. 30–35a). The magnetic field $\mathbf{B}_2$ due to q_2 at the position of q_1 is directed out of the page, whereas the magnetic field $\mathbf{B}_1$ due to q_1 at the position of q_2 is directed into the page. The magnetic forces are computed from the Lorentz force law or directly from Eqs. (30–33) (Fig. 30–35b). They are equal but opposite and attract each other (or repel each other, if the charges have opposite signs) with magnitude

$$F_B = \frac{\mu_0}{4\pi} q_1 q_2 \frac{v^2}{d^2}. \qquad (30\text{–}34)$$

There is also an electric force between these charges, with magnitude

$$F_E = \frac{1}{4\pi\epsilon_0} \frac{q_1 q_2}{d^2}. \qquad (30\text{–}35)$$

The constants ϵ_0 and μ_0 are related according to Eq. (30–4) by the speed of light, c, so

$$\frac{F_B}{F_E} = \frac{(\mu_0/4\pi) q_1 q_2 (v^2/d^2)}{(1/4\pi\epsilon_0)(q_1 q_2/d^2)} = \mu_0 \epsilon_0 v^2 = \frac{v^2}{c^2}. \qquad (30\text{–}36)$$

Unless v is comparable to c, the magnetic force is insignificant compared to the electric force. As we know, the speed of light is large ($c \simeq 3 \times 10^8$ m/s).

In mechanics we established the principle that no experiment can determine an

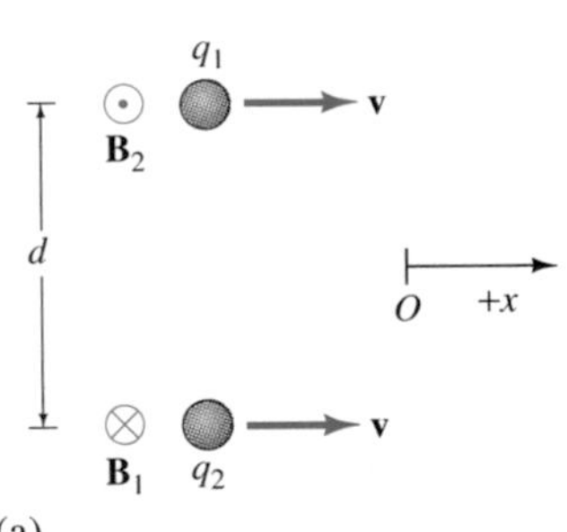

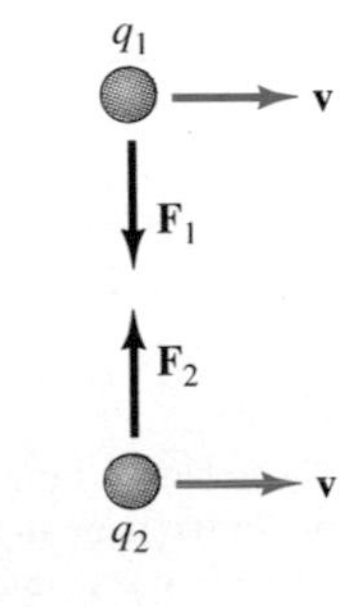

FIGURE 30–35 Two charges, each with speed v and moving in the $+x$-direction, separated by a distance d. (a) The magnetic field at each charge due to the other moving charge. (b) The force on each charge due to the magnetic field of the other charge.

absolute reference frame at rest. Equivalently, no experiment can enable us to distinguish one uniformly moving frame from another. This principle is the *relativity principle.* As an example, by observing the trajectory of a thrown ball, you cannot tell whether the platform from which you throw the ball is stationary or in motion with constant velocity.

Application of the relativity principle to the magnetic forces calculated above leads us into difficulty. We could imagine changing inertial reference frames and, by moving along with the two charges, make our observations from the frame in which the two charges are at rest. In this frame there will be no magnetic fields, and hence no forces due to magnetic fields, even though the electric forces would appear to be the same. If there were no further frame-dependent effect, *the relativity principle would be violated*, because the total force would depend on whether we are moving or not.

If we wish to save the relativity principle (this was Einstein's starting point and is confirmed by experiment), some modification of the laws we have developed to this point is necessary. We must demand that *when we change inertial reference frames, the electric and magnetic fields* "*mix*" in such a way that the (observable) acceleration of charges is unaffected. The term "mix" (we say that the fields *transform among themselves*) means that a combination of electric and magnetic fields in one frame becomes a different combination in another frame. Thus a pure electric field in the frame in which the charges are at rest must become a mixture of magnetic and electric fields in the frame in which the charges are moving. As our calculation of the relative magnitude of the electric and magnetic forces shows, it is only when the speed with which the charges are moving is comparable to the speed of light that the mixing effects have important physical consequences. In this circumstance, considerable modification of the electric and magnetic fields is necessary. Many modern applications in materials research and medicine involve fast-moving charged particles, so these modifications have technological as well as scientific importance.

The Failure of Newton's Third Law

A second application of Eqs. (30–33) leads to equally interesting conclusions. Suppose that instead of moving parallel to each other, the two charges are moving at some angle, as in Fig. 30–36. They are still in the plane of the page and separated by a distance d at the time we consider them. The magnetic field $\mathbf{B}_1$ due to q_1 at the position of q_2 is directed into the page, but *there is no field* $\mathbf{B}_2$ *due to* q_2 *at the position of* q_1. In other words, there is a force $\mathbf{F}_2$ on q_2 due to q_1, but *there is no magnetic force* $\mathbf{F}_1$ *on* q_1 *due to* q_2.

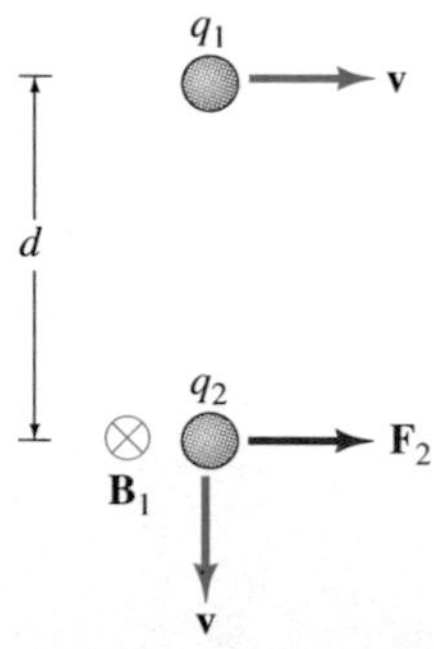

FIGURE 30–36 Two charges move at speed v in the plane of the page in the directions shown. They are separated instantaneously by a distance d. There is no magnetic field due to q_2 at the position of q_1, but there is a magnetic field due to q_1 at the position of q_2. Thus there is a magnetic force $\mathbf{F}_2$ on q_2, but none on q_1.

In this situation, *Newton's third law fails.* We saw in Chapter 8 that Newton's third law in mechanics is equivalent to the conservation of momentum. We shall see in Chapter 35 that when *both* electric and magnetic fields are present, the combination of those fields carries momentum. As our earlier discussion of the relativity principle underlines, the only frame of reference in which both fields are not present is the frame in which all charges are at rest, and in this frame there is no problem with Newton's third law. Thus in all frames in which Newton's third law is in danger, the electromagnetic fields carry momentum. Although we are not now equipped to show it in detail, *conservation of momentum continues to hold when the momentum carried by the fields is taken into account.* As for the relativity principle, these considerations are also intimately connected with special relativity. Conservation of momentum, not Newton's third law, is a bedrock of physical law.

Magnetic fields are produced by electric currents and, equivalently, by moving charges. The magnetic field lines about a long, straight wire that carries a constant current form circles around the wire in the plane perpendicular to the wire. The direction of the field lines in these circles is determined when the thumb of the right hand points along the direction of the current flow: The fingers then curl in the direction of the magnetic field. The magnitude of the field at a radial distance r from the wire is

$$B = \frac{\mu_0 I}{2\pi r}. \tag{30–5}$$

The defined constant $\mu_0 = 4\pi \times 10^{-7}\ \text{T}\cdot\text{m/A}$ is the permeability of free space.

The magnetic fields produced by unchanging currents obey Ampère's law:

$$\oint \mathbf{B} \cdot \mathrm{d}\mathbf{s} = \mu_0 I_{\text{enclosed}}. \tag{30–10}$$

Here the line integral follows any closed path through which the current I_{enclosed} passes. A second law obeyed by the magnetic field springs from the absence of magnetic equivalents to the electric charge. Because there are no magnetic charges on which magnetic field lines begin or end, magnetic field lines must close on themselves. This fact is expressed by Gauss' law for magnetism:

$$\text{for a closed surface:} \qquad \Phi_B = \oiint \mathbf{B} \cdot \mathrm{d}\mathbf{A} = 0. \tag{30–12}$$

This law states that the magnetic flux, Φ_B, through any closed surface is zero, or equivalently that the number of magnetic field lines that enter a closed surface is the same as the number of lines that leave the surface.

Ampère's law is an important practical tool for determining magnetic fields when there is enough symmetry to allow a path choice in which the integral simplifies, as in the determination of the interior field of a long solenoid. A solenoid is a wire wound uniformly into a coil to form a tube. When current flows, a magnetic field is produced within the tube, has constant magnitude, and is aligned with the tube axis. The magnitude of the interior field is

$$B = \mu_0 n I, \tag{30–15}$$

where n is the number of windings of wire per unit length of the solenoid. Because its interior magnetic field is constant, a solenoid is to magnetism what a capacitor is to electricity.

When there is not enough symmetry to allow Ampère's law to be used to determine the magnetic field produced by a given configuration of currents, the Biot–Savart law can be used instead. According to this law, the magnetic field $\mathrm{d}\mathbf{B}$ produced by a segment of wire $\mathrm{d}\ell$ that carries a current I at a displacement $\mathbf{r}$ from the segment is given by

$$\mathrm{d}\mathbf{B} = \frac{\mu_0}{4\pi}\frac{I\,\mathrm{d}\boldsymbol{\ell} \times \mathbf{r}}{r^3}. \tag{30–19}$$

The magnetic field from an infinitesimal segment can be integrated to find the net magnetic field due to a finite segment of wire. A closely related result is the magnetic field of a single moving charge,

$$\mathbf{B} = \frac{\mu_0}{4\pi}\frac{q\,\mathbf{v} \times \mathbf{r}}{r^3}. \tag{30–32}$$

Application of the Biot–Savart law shows that the magnetic field due to a ring of current, or the exterior field of a solenoid of finite length, is a magnetic dipole field, whose form is the same as that of a bar magnet or equivalently has the same form as the exterior electric field produced by an electric dipole. The current loop forms a magnetic dipole, characterized by a magnetic dipole moment, $\boldsymbol{\mu}$, which is aligned perpendicular to the surface of the loop according to a right-hand rule. Its magnitude is

$$\mu = IA, \tag{30–27}$$

where A is the loop area.

If currents are not constant, as when wires are interrupted by the presence of charging capacitor plates, then one surface that spans a closed path might not cross the wire that another surface might cross, and the concept of the current enclosed by a path becomes ambiguous. This ambiguity is remedied by Maxwell's modification of Ampére's law to

$$\oint \mathbf{B} \cdot d\mathbf{s} = \mu_0(I + I_d) = \mu_0 I + \mu_0\epsilon_0 \frac{d\Phi_E}{dt}. \tag{30–31}$$

The quantity I_d, proportional to the rate of change of electric flux, is known as the Maxwell displacement current. The surface through which the sum of I and I_d passes is any surface that spans the closed integration path.

Electric and magnetic fields transform among themselves when viewed from different reference frames, a result that is a cornerstone of the special theory of relativity. Newton's third law fails when magnetic forces are concerned, even though the conservation of momentum continues to be valid when the momentum contained in the electromagnetic fields is included.

QUESTIONS

1. A compass needle is moved near a straight wire that carries a current. What is the orientation taken by the compass needle in various locations about the wire?

2. A magnetic field about a current is oriented according to a right-hand rule. Does this mean that right-handedness has some intrinsic meaning in physics—that nature prefers one hand over the other?

3. In the definition of the ampere, does it matter how long the two parallel wires are?

4. A wire connected to a battery is placed in the yoke of a tabletop electromagnet when a switch is open (see Fig. 30–15a). When the switch is closed, the wire may take a big jump upward or it may not, according to which side of the battery terminals the wires are attached. Why?

5. The magnetic field outside a straight solenoid is small, but not zero, whereas the field outside a toroidal (doughnut-shaped) solenoid is exactly zero. Why?

6. In the definition of the ampere, do we have to worry about the Coulomb forces between the charges in the two wires?

7. Why is the Biot–Savart law written in differential form? Explain why it cannot be written as in Eq. (30–19) but without the differential signs.

8. Why is it preferable to define current in terms of the force between two long, parallel wires rather than in terms of the rate at which charge passes a point?

9. Is it possible to arrange a set of electric currents and produce a magnetic field that, at large distances from the apparatus responsible, is everywhere directed radially away from the apparatus? Feel free to choose your apparatus, and give either a proof that it is impossible or a description of the apparatus.

10. Suppose that the space between the plates of the capacitor discussed in Section 30–6 is not empty but is filled with a dielectric. How would the treatment in that section, and the determination of the displacement current, change?

11. Suppose that magnetic charges were discovered. What would some practical consequences be?

12. What are the SI units of the ratio E/B, where E is an electric field and B is a magnetic field?

13. When two bar magnets are placed side by side, they will (a) attract or (b) repel if the adjacent poles are (a) opposite or (b) the same. If you draw magnetic field lines for the combination of two magnets in both cases, the net magnetic field between the magnets will tend to (a) cancel or (b) be doubled. What conclusions can you draw?

PROBLEMS

30–1 *Ampère's Law*

1. (I) What is the force per meter between two long, parallel wires, each carrying 10.0 A but in opposite directions, if the two wires are 2.0 cm apart?
2. (I) Find the dimensions of μ_0 and ϵ_0, and use your expressions to show that the product $\mu_0\epsilon_0$ has dimensions of $(1/\text{speed})^2$. Find the value of that speed in SI units.
3. (I) A coaxial cable consists of a central wire that carries current I to the right and a tube centered on the central wire that carries the same current to the left. Find the magnetic field outside the cable.
4. (II) (a) In a thick, straight wire carrying a current that is uniform through its cross section, where is the magnetic field the greatest? (b) If the radius of the wire is R and the current is I, what is the value of the maximum magnetic field? (c) What is the minimum magnetic field, and where does this occur? Consider regions both inside and outside the wire.
5. (II) Plot the curves of constant magnetic field in the xy-plane for values B_0, $2B_0$, $3B_0$, and $4B_0$ of the field about a straight wire that carries current along the z-axis. B_0 is some field value that you can choose. These curves are the intersections with the xy-plane of the surfaces of constant field.
6. (II) An electron beam contains electrons that move along the $+x$-axis at $0.01c$. The beam enters a region of length 10 cm and runs parallel to a wire that carries a 0.50-A current in the $+x$-direction. The beam is 1 cm from the wire. (a) Specify the direction in which the beam is deflected, if at all. (b) Find the deflection of the beam as it passes through the 10-cm region by calculating the impulse it receives during its brief passage through that region. (c) After it has passed the wire, does the beam have the same energy it had when it entered the region that contains the wire?
7. (II) A very thin, infinitely long metal sheet lies in the xy-plane, between $x = -w$ and $x = w$. A current of density h A/m flows in the x-direction. What are the magnitude and direction of the magnetic field at a distance $z \ll w$ above and below the sheet? Neglect end effects.
8. (II) Consider two parallel metal sheets, such as the sheet of Problem 7, with currents flowing in opposite directions. What are the magnetic fields between and outside the sheets? What is the situation when the currents are parallel rather than antiparallel?
9. (II) Consider a wire that passes through the origin along the z-axis and carries a current I. (a) Calculate the x- and y-components of the magnetic field at a point whose coordinates are $(x, y, 0)$. (b) Use this result to obtain the magnetic field due to two wires that are parallel to the z-axis, cross the xy-plane at $(a, 0)$ and $(-a, 0)$, and carry current I in the $+z$-direction. (c) What are the fields when the currents are in opposite directions? Sketch the magnetic field lines for the two cases.
10. (II) A uniform current with current density h A/m flows parallel to the z-axis on a cylindrical metal sheath, where the radius of the cylinder is R. What is the magnetic field outside the sheath? inside the sheath?
11. (II) Current flows down the inner cylinder of a coaxial cable and returns on the outside cylinder (Fig. 30–37). The radius of the inner cylinder is 0.5 cm, and the radius of the thin outer cylindrical shell is 0.8 cm. Calculate the magnetic field on the cylindrical surface midway between the inner and outer surfaces, given that the current is 5 A. Ignore end effects.

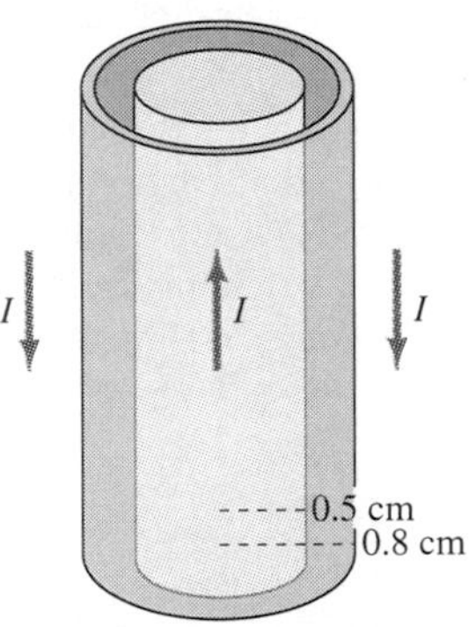

FIGURE 30–37 Problem 11.

12. (II) Two long, parallel wires carrying a current I in the same direction each have a mass density λ. The wires are initially a distance D apart and are then released. Write an equation that describes the relative motion of the wires. Ignore all forces other than the magnetic force.

30–2 *Gauss' Law for Magnetism*

13. (I) Figure (30–38) shows the magnetic field lines that emerge from one pole of a bar magnet; these lines resemble the electric field lines that emerge from one end of an electric dipole. Sketch the magnetic field lines in the region of the pole inside the magnet. For comparison, also sketch the electric field lines in the central region of an electric dipole. Comment on the differences between the sketches.

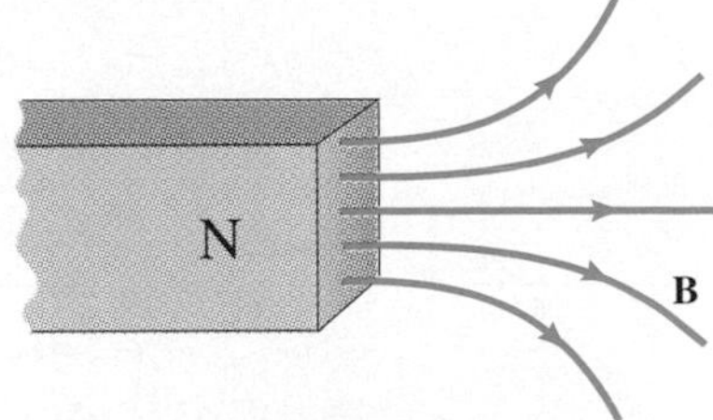

FIGURE 30–38 Problem 13.

14. (II) A long current-carrying wire is oriented vertically; next to it is drawn a square whose area lies in the same plane

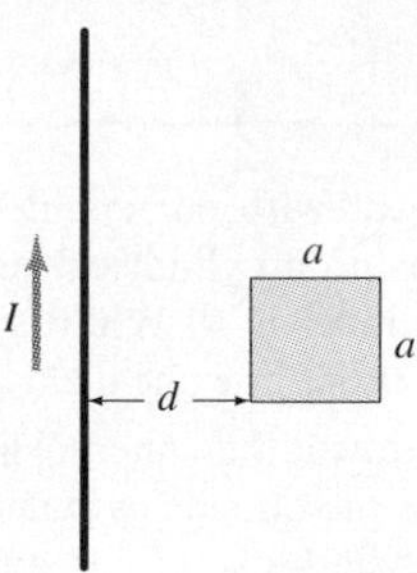

FIGURE 30–39 Problem 14.

as the wire (Fig. 30–39). Using the distances indicated, find the magnetic flux through the square.

15. (II) Show by using Gauss' law for magnetism that a magnetic field with only an x-component must be constant as x varies.

16. (II) Show that Gauss' law is satisfied for the magnetic field due to a straight wire that carries a current I in the $+z$-direction for a volume that represents a portion of a cylindrical shell of height h, extending from a radius r to a radius R and formed by an angle θ (Fig. 30–40).

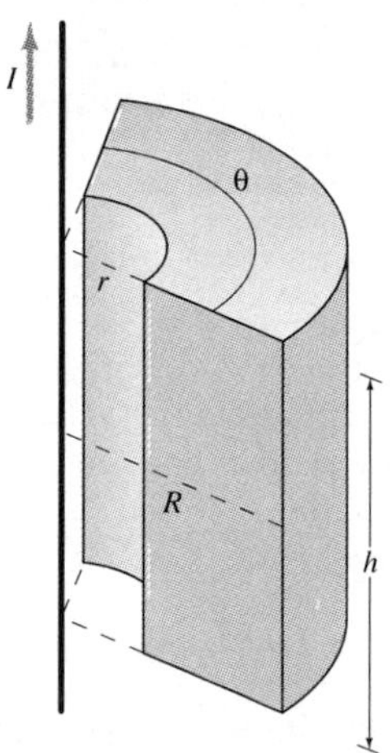

FIGURE 30–40 Problem 16.

17. (III) Apply Gauss' law for magnetism to a parallelepiped of dimensions a, b, and c, one of whose corners is located at point (x, y, z), as in Fig. (30–41). Assume that the dimensions in the x-, y-, and z-directions (a, b, and c) are small enough so that $B(x + a, y, z) = B(x, y, z) + a\,\partial B/\partial x$, and so on. Show that Gauss' law leads to the condition $(\partial B_x/\partial x) + (\partial B_y/\partial y) + (\partial B_z/\partial z) = 0$ in this limit.

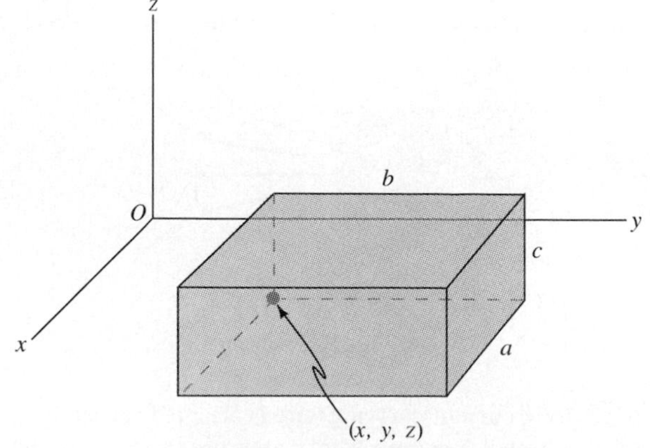

FIGURE 30–41 Problem 17.

30–3 *Solenoids*

18. (I) A solenoid of diameter 1.5 cm has a length of 60 cm and 450 turns of wire. What is the magnetic field at the center of the solenoid when the current in the coil is 5 A?

19. (I) A long, superconducting solenoid is wound with fine niobium-tin wire so that there are 3×10^4 turns/m. If a power supply produces 50 A, what is the magnetic field inside the solenoid?

20. (II) A magnetic field is needed over a cylindrical volume 2 cm in radius by 10 cm in length. You have 300 m of insulated wire 1 mm in diameter and a 1-A current source. Design a system to produce the maximum field along the axis of the cylindrical volume. What is the magnetic field? In doing the calculation, use the average radius of the resulting coil.

21. (II) The magnetic field inside a cylindrical solenoid of area 10 cm^2 is 2.5 T along the axis of the solenoid. What is the magnetic flux through a disk of radius 6 cm placed perpendicular to the solenoid axis (Fig. 30–42)?

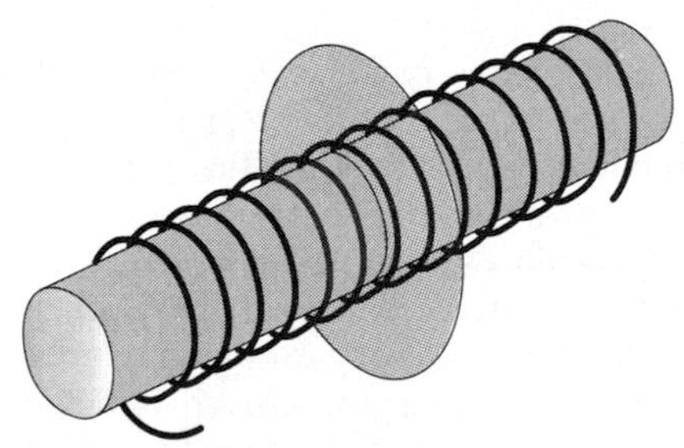

FIGURE 30–42 Problem 21.

22. (II) Show that the magnetic flux through an ideal cylindrical solenoid of radius R is given by the formula $\Phi_B = \mu_0 n I \pi R^2$, where n is the turn density.

23. (II) Consider a toroidal solenoid with a square cross section, each side of which has length L. The inner wall of the torus forms a cylinder of radius R. The torus is wound evenly with N loops of wire, and a current I flows through the wire. What is the total magnetic flux through the torus?

30–4 *The Biot–Savart Law*

24. (I) Two long wires are placed along the y- and z-axes, respectively. They carry the same current I in the positive directions. Calculate the magnetic field along the x-axis.

25. (II) A segment of wire forms a straight line of length L and carries a current I. Find the magnetic field due to the wire segment in the plane perpendicular to it and passing through one end.

26. (II) An infinitely long L-shaped wire is placed so that a current I flows in along the y-axis toward the origin, then out from the origin along the x-axis. What is the magnetic field at a point on the z-axis at a height H above the origin?

27. (II) Consider a straight segment of wire of length L that carries a current I. Use the Biot–Savart law to find the magnetic field along the axis of the wire, beyond the wire itself, due to this segment.

28. (II) A differential length $d\mathbf{L}$ of wire carrying a current of 10 A is positioned at the origin of a coordinate system and points in the $+x$-direction. Find the magnetic field due to this wire segment at the following (x, y, z) positions, given in centimeters: (a) (0, 0, 5), (b) (0, 5, 0), (c) (5, 0, 0), and (d) (5, 0, 5). Give both the magnitude and direction of the magnetic field.

29. (II) Consider a thin dielectric ring 2 cm in diameter that rotates around a stem perpendicular to the plane of the ring and through its center at the rate of 250 rev/s. Assume that the ring is charged uniformly and carries a total charge of 10^{-5} C. What is the magnetic field produced at the center of the ring by the rotating charge?

30. (II) Repeat the calculation of Problem 29 for a solid disk 4 cm in diameter, with the same total charge.

31. (II) A current loop consists of a square with sides of length L. A current I circulates counterclockwise around the loop. Find the direction and magnitude of the magnetic field at the center of the square. Compare this to the field at the center of a circular loop of diameter L that carries the same current.

32. (II) Consider the wire shown in Fig. 30–43. Calculate the magnetic field at point P, the center of the half-circle of radius R around which the wire turns, as a function of R and the current I carried by the wire.

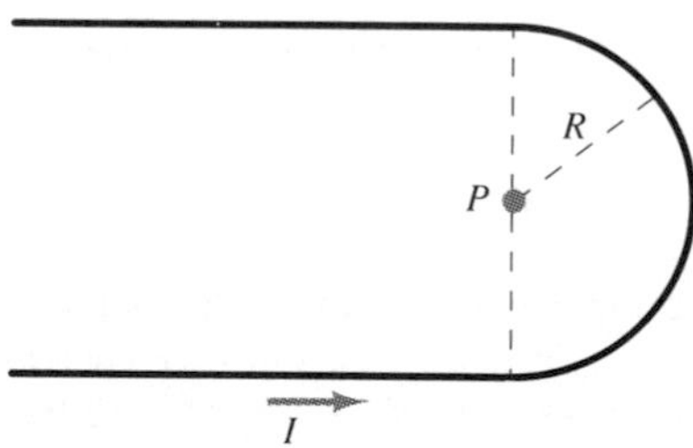

FIGURE 30–43 Problem 32.

33. (II) A very long wire is aligned along the $+x$- and $+y$-axes, making a right angle at the origin (Fig. 30–44). A current I travels in the $-y$-direction and continues in the $+x$-direction. What is the magnetic field at the point (x, y), where both x and y are positive?

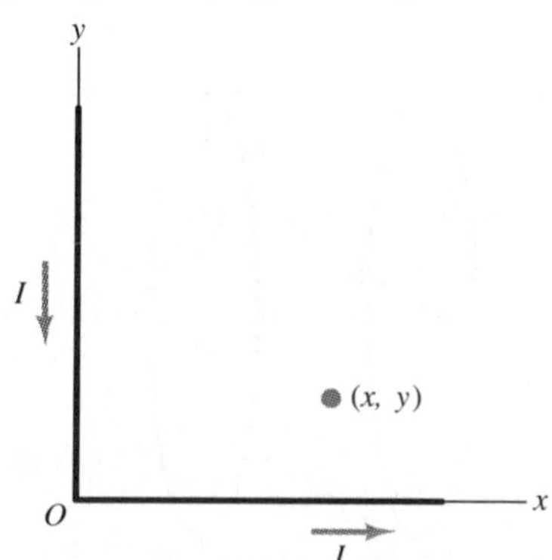

FIGURE 30–44 Problem 33.

34. (II) Consider the wire shown in Fig. 30–45 with the inner and outer radii of the semicircle given as 40 cm and 60 cm, respectively. Given that the current in the wire is 4 A, what is the magnetic field at point P, the center of the semicircles?

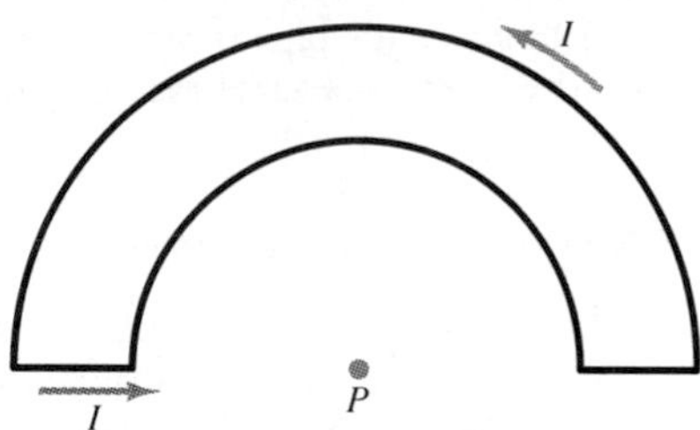

FIGURE 30–45 Problem 34.

35. (II) Calculate the magnetic field at the center of a wire square that consists of 20 loops and has sides of length 25 cm and carries a current of 4 A.

36. (II) A charge q moves at instantaneous speed v when it crosses the axis of a ring of current with a magnetic dipole moment μ. At that instant q is located a distance d from the center of the ring in the direction of the dipole moment vector and is moving perpendicular to the axis (Fig. 30–46). What is the resulting instantaneous motion of the charge? Find the instantaneous radius of curvature of its motion.

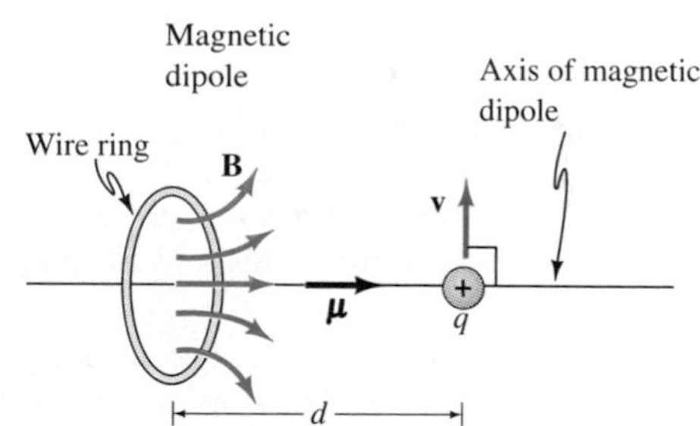

FIGURE 30–46 Problem 36.

37. (II) A circular current loop of radius R produces a magnetic field. At what distance along the axis of the loop does the field have magnitude 1/2 the magnitude at the center of the loop? At what distance is the magnitude of the field reduced to 1/100 the value at the center? Give your answer in units of R.

38. (II) Find the magnetic field at point P in Fig. 30–47 if a

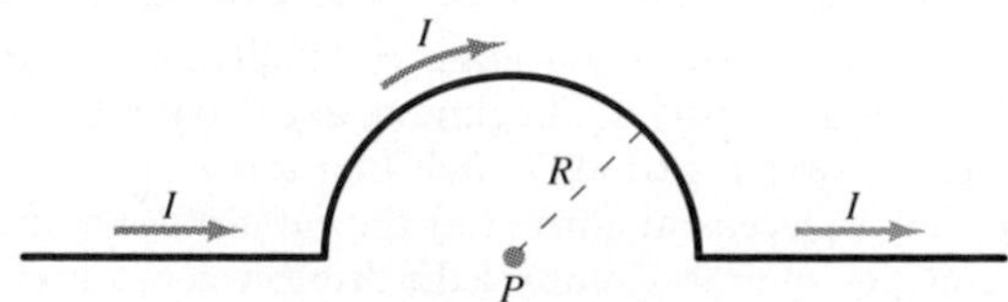

FIGURE 30–47 Problem 38.

current of 5 A flows in the infinitely long wire; the radius R of the semicircle is 5 cm.

39. (III) By integration, find the magnetic dipole moment of a spherical shell of radius R that carries a total charge Q, distributed uniformly, if the shell rotates with angular velocity ω oriented along the z-axis.

40. (III) Consider a long, thin-walled metal pipe that carries a total current I distributed evenly along the walls of the pipe. A simple application of Ampère's law indicates that the magnetic field inside the pipe is zero. Show by a simple geometric argument that the same result follows from the Biot–Savart law.

30–5 *The Maxwell Displacement Current*

41. (I) Consider the RC circuit shown in Fig. 30–48. Switch S is closed at time $t = 0$. Calculate the displacement current in the capacitor. After a time that is long compared with the product RC, the switch is opened again. What is the displacement current then?

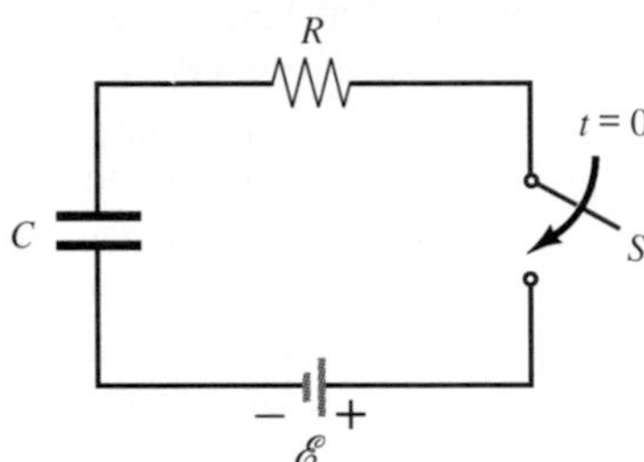

FIGURE 30–48 Problem 41.

42. (I) A parallel-plate capacitor is being charged at a rate of $I = 1.5$ A. The plates have an area of 0.60 m^2 and are separated by 3.0 cm. What is the value of $\int \mathbf{B} \cdot d\boldsymbol{\ell}$ for a closed path midway between the plates and covering an area of 1.0×10^{-2} m^2?

43. (II) A 1-μA current starts flowing in a circuit with a 5×10^{-11} F capacitor of area 4 cm^2 at $t = 0$ s. (a) How fast is the voltage across the capacitor plates changing at $t = 0$ s? (b) Use the result of (a) to calculate *explicitly* $d\Phi_E/dt$ and the displacement current at $t = 0$ s.

44. (II) An alternating voltage of the form $V = V_0 \cos(\omega t)$ is connected across a capacitor C. What is the displacement current in the capacitor?

45. (II) A voltage of the form $V = V_0 \cos(\omega t)$, with $\omega = 250$ rad/s and $V_0 = 120$ V, is applied across the plates of a 3-μF capacitor; the plates are 0.8 cm apart. (a) What is the maximum rate of change in electric field between the plates? (b) the maximum value of current leading to the capacitor?

46. (II) A conducting sphere of radius R initially has a uniform surface-charge density σ_0. Beginning at $t = 0$ this charge is drained off over a period t_0 such that $\sigma = \sigma_0 [1 - (t/t_0)]$. Find the displacement current at the surface of the sphere as a function of time. Compare the displacement current to the current carried off by the wire.

General Problems

47. (II) Calculate the force per unit area between two metal sheets that carry identical currents in the same direction. The sheets carry a current of linear density h A/m as in Problem 7.

48. (II) Equal but opposite currents I travel in the inner and outer wires of a coaxial cable. As a function of the distance from the central axis, find the magnetic field (a) inside the inner wire; (b) in the region between the wires; (c) in the outer (tubular) wire; (d) outside the outer wire.

49. (II) A hydrogen atom may be described as consisting of an electron that moves in a circular orbit around a proton. The force that gives rise to the motion is the Coulomb attraction between the proton and the electron, which have charges $\pm e$, respectively, where $e = 1.6 \times 10^{-19}$ C. The motion is further constrained by the requirement that the angular momentum has the value $nh/2\pi$, where n is an integer and $h = 6.63 \times 10^{-34}$ J·s, Planck's constant. Calculate the magnitude and direction of the magnetic field at the location of the proton. What is the magnetic moment of the current loop?

50. (II) Find the force between the long, straight wire and the rectangular wire loop shown in Fig. 30–49 for currents $I_1 = 5$ A and $I_2 = 2$ A.

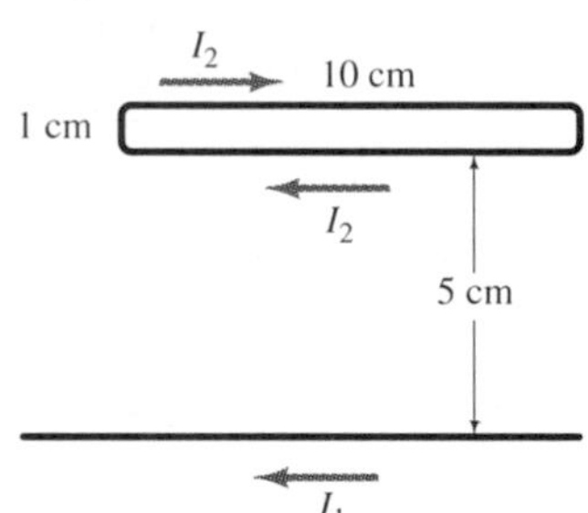

FIGURE 30–49 Problem 50.

51. (II) In Example 30–6 we found the magnetic field due to a circular wire of radius R, carrying a current I, at a point a distance d away from the center of the ring but along the axis to be

$$B = \frac{\mu_0 I}{2} \frac{R^2}{(R^2 + d^2)^{3/2}}.$$

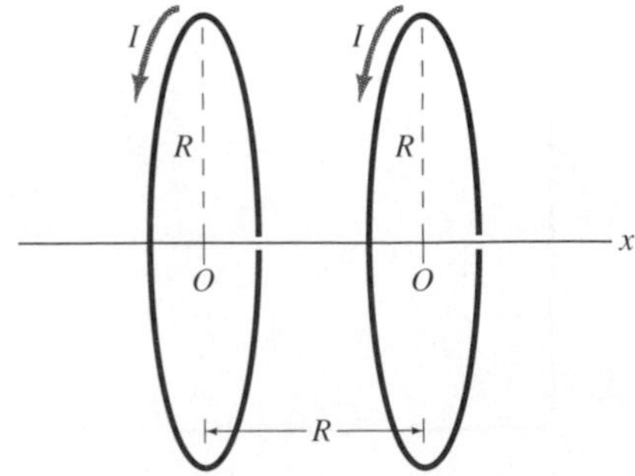

FIGURE 30–50 Problem 51.

A pair of such coils placed coaxially a distance R apart makes up a *Helmholtz coil*, for which the magnetic field everywhere inside is fairly constant (Fig. 30–50). (a) Determine the magnetic field on the axis as a function of x, and evaluate the field at $x = 0$, $x = R/4$, and $x = R/2$. (b) Show that $dB_x/dx = 0$ and $d^2B_x/dx^2 = 0$ at $x = R/2$.

52. (II) Consider two parallel wires spaced a distance $d = 1$ cm apart and each carrying a current $I = 1$ A. (a) Compare the magnetic force between these wires to the electric force they would exert on each other if the current carriers (electrons) were not neutralized by a background of positive charges. Use 10^{21} per cm as the linear density of charge carriers in the wire. (b) What excess of electrons per unit length over the positive background would make the electric force equal the magnetic force between the wires? (c) What fraction of the total number of charge carriers is the excess calculated in part (b)?

53. (II) A long wire carries a current I_1. A segment of a second wire, which carries a current I_2, is oriented radially away from the first wire. The segment has length L, and its closest end is a distance d from the first wire (Fig. 30–51). Calculate the torque, direction and magnitude, on the wire segment about the axis defined by the long wire.

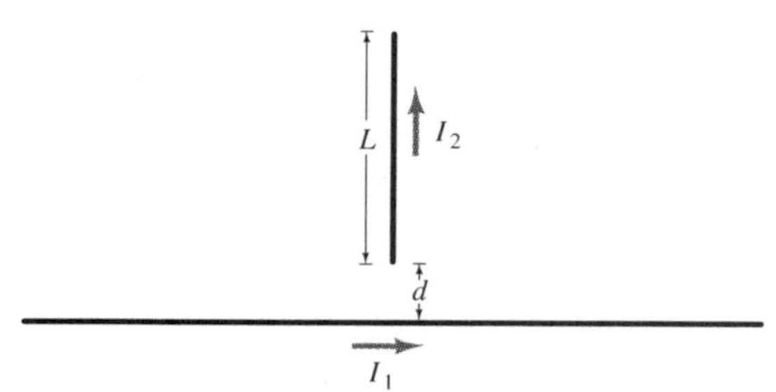

FIGURE 30–51 Problem 53.

54. (III) A certain electric current distribution produces a magnetic field of the form $\mathbf{B} = \beta(y\mathbf{i} - x\mathbf{j})$ near the origin of a coordinate system. Find the current distribution responsible.

55. (III) You have a power supply capable of producing 10 W, but it is limited to 0.2-A current. You have to build a cylindrical solenoid 1 cm in radius and 12 cm long. You are able to obtain only 3 kg of copper but luckily find a wire-extruding machine that allows you to make any size wire you desire. Explain how to build a solenoid that maximizes the magnetic field. Calculate the maximum magnetic field, and give the parameters of the solenoid winding. In calculating the field, use the average radius of the windings around the solenoid.

56. (III) A long wire carrying a current I consists of a straight section, a semicircle of radius R, and another straight section (Fig. 30–52). Find the magnetic field at a point P a distance x from O along the axis that passes through the center of curvature of the semicircle (point O) and is perpendicular to the wire.

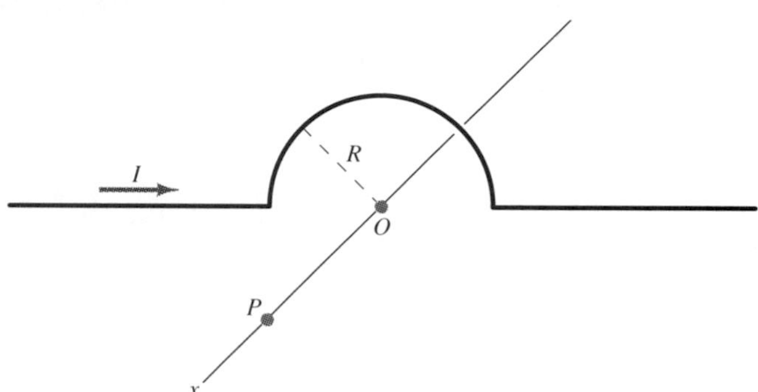

FIGURE 30–52 Problem 56.

CHAPTER 31

Michael Faraday delivering one of his famous public lectures in 1856. These lectures earned him great popular success.

FARADAY'S LAW

In the previous chapters we dealt with the concept of the magnetic field, its sources, and its effects on moving charges and currents. We began to get hints of the intimate connection between electricity and magnetism. In this chapter we introduce an entirely new physical law, Faraday's law. This law states that a changing magnetic field, or, more precisely, a changing magnetic flux, produces an electric field (or, equivalently, an emf) around a path. Faraday's law connects electric and magnetic fields and predicts the existence of electric fields not associated with conservative forces. Faraday's law is a crucial ingredient in our eventual understanding of electromagnetic waves and how they are generated. In addition, Faraday's law has far-reaching technological applications. It lies behind our entire system of electrical power generation and plays a role in most of the electronic devices we use.

31–1 MICHAEL FARADAY AND MAGNETIC INDUCTION

A great experimentalist, as opposed to a merely competent one, recognizes when an odd or unexpected measurement is truly significant. He or she is open to the idea that an unexpected phenomenon does not always represent grounds to send the apparatus back to the manufacturer for repair, and understands that a small

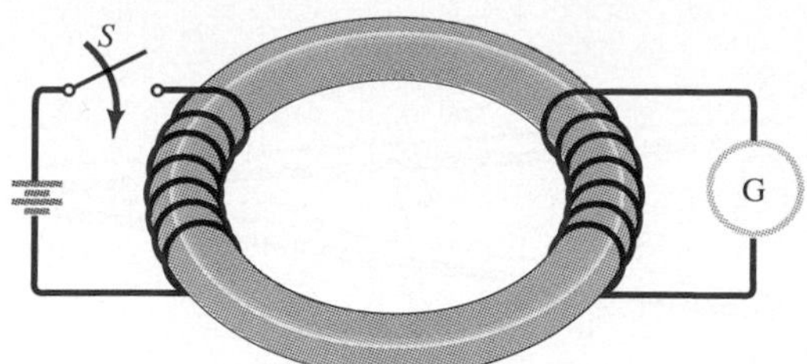

FIGURE 31–1 Faraday's ring. A changing magnetic flux in the iron ring induces a current in the galvanometer coil on the right; the changing flux is due to the opening or closing of a switch connected to the battery of the coil on the left.

effect is not always experimental error. He or she has the persistence to pursue the effect systematically, check its reality, and pursue its ramifications from as many points of view as possible.

Michael Faraday exhibited all these qualities in the discovery of what we now know as Faraday's law. What he accomplished is a classic example of the process of scientific discovery. In grand outline, Faraday's law states that *changing magnetic fields generate electric fields.* More precisely, the *change* of the magnetic flux through any surface bounded by a closed line causes an emf around that line. This emf can induce a current in a wire. Faraday called the current produced by a changing magnetic flux an **induced current**, and he called the general phenomenon **magnetic induction**. Within several days of his first observations, he completed a series of experiments that revealed essentially all the aspects of magnetic induction. His discovery is all the more remarkable because the effect he observed is such a small one, certainly not as obvious as the effects of Coulomb's law.

Faraday's discovery of 1831 was not an accident.† At the time, it was known that an electric current produces a magnetic field. Ampère's law demonstrated this; for example, a current passing through the wires of a solenoid produces a magnetic field inside the solenoid. In a period of enthusiastic experimentation on electricity and magnetism, it was natural to ask if magnetic fields could themselves make currents flow. Faraday had pursued this sort of question for several years—an entry dated 1822 in his notebook sets the goal "convert magnetism into electricity"—when in 1831 he made the experiment shown schematically in Fig. 31–1. All the elements of this experiment are familiar to us except one. The battery sends a current through the coil on the left side of an iron ring, which acts as a solenoid. The galvanometer is used to indicate any current in the coil on the right side of the ring. The only unfamiliar element is the iron ring. The iron does two things: It carries the magnetic field set up by the left-hand coil within the torus, and hence through the right-hand coil (recall that magnetic field lines are closed), and it *magnifies* the size of the field set up in the left-hand coil. (We shall see why in Chapter 32, where we discuss the magnetic properties of materials.) We say that the iron ring *links* the two coils. To Faraday's disappointment, he observed no effect on the galvanometer when a *steady* current passed through the left-hand coil. Faraday's intuition served him well, however, when he noticed a very small twitch of the galvanometer *when the switch that controlled the flow of current in the left-hand coil was opened or closed.* He replaced the galvanometer he was using with one more sensitive to momentary changes, verified that his observation was not a case of mechanical vibration upon opening or closing the switch, and launched into a tenacious and systematic study of the effect he had observed.

In this chapter we shall begin to discuss Faraday's findings. One of the first things he did was to eliminate the possibility that the battery was itself important

†Interestingly, Joseph Henry discovered the law of induction at about the same time, although he was late in making his results known. Henry made other discoveries, and his name will come up again in our discussion of electricity and magnetism.

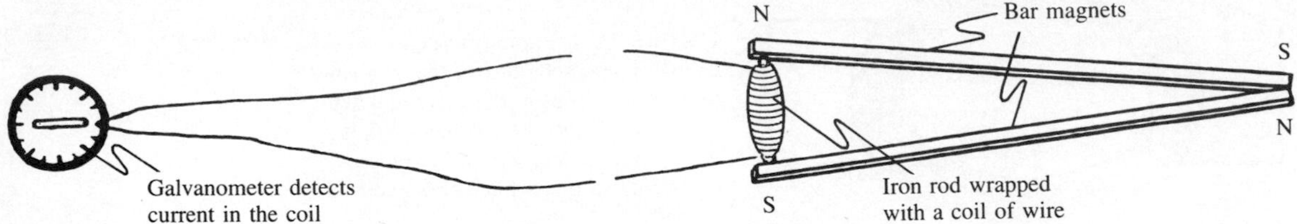

FIGURE 31–2 A figure from one of Faraday's lectures, indicating that a changing magnetic flux induces a current even when no battery is present. The figure was published in Faraday's book *Experimental Researches in Electricity* in 1839; the labeling is our own.

to the effect. In a second experiment, illustrated in Fig. 31–2, a drawing from a publication by Faraday (with our labeling added), two bar magnets make a vee shape. An iron rod links the magnetic field of the bar magnets when it touches the ends of the two magnets as shown, and a large magnetic flux passes through the rod. This rod is surrounded by a coil attached to a galvanometer. When the rod is brought into contact with the ends of the two bar magnets at the open end of the vee, or when the rod is pulled away, the galvanometer deflects, indicating a current in the coil. The crucial phenomenon leading to the current is that the magnetic flux through the coil *changes*.

The announcement of Faraday's discovery was greeted with great excitement, because in addition to the discovery of a new physical law, the possibility of converting mechanical energy to electric energy thereby became a reality.† Indeed, electricity generation worldwide is based on Faraday's results, so we might hazard that Faraday's discovery has had a greater effect on the material welfare of humans than has any other discovery before or since.

31–2 FARADAY'S LAW OF INDUCTION

With our modern capacity to control and measure currents, it is easy to illustrate some of the phenomena in Faraday's law of induction. Figure 31–3a shows a changing distance between a bar magnet and a wire loop; the current in the loop is measured by a galvanometer. There is an induced current in the wire loop only when there is movement, or, more precisely, when the magnetic flux through the loop changes. A faster movement—a more rapidly changing flux—results in a larger current than a slower movement does. Figure 31–3b shows a switch closed in the circuit of one loop; a current momentarily appears in the second loop. Figure 31–3c shows a changing distance between a loop through which a current passes and a second loop; a current appears in the second loop. Figure 31–3d shows a changing orientation between a current-carrying loop and a second loop; a current appears in the second loop. Note that it is the change not just of the magnetic field but of the magnetic flux that induces a current. For example, if the bottom coil of Fig. 31–3c is squeezed, *simply changing its area*, the magnetic flux through it changes even though the magnetic field due to the upper coil does not change, and an emf is induced in the bottom coil.

Faraday's experiments describe these observations qualitatively and quantitatively. Faraday's law, which summarizes his findings, states that the negative of the *time rate of change* of the magnetic flux through a surface, Φ_B, equals an emf around the closed loop that bounds the surface. We know from Chapter 28 that the emf, $\mathcal{E}$, is an electric potential change; that is, a line integral of an electric field.

The magnetic flux and its properties were first discussed in Chapter 30.

† The early-nineteenth-century public was fascinated by demonstrations of electrical phenomena. When a woman asked Faraday about the usefulness of his discoveries, he replied, "Madam, of what use is a baby?"

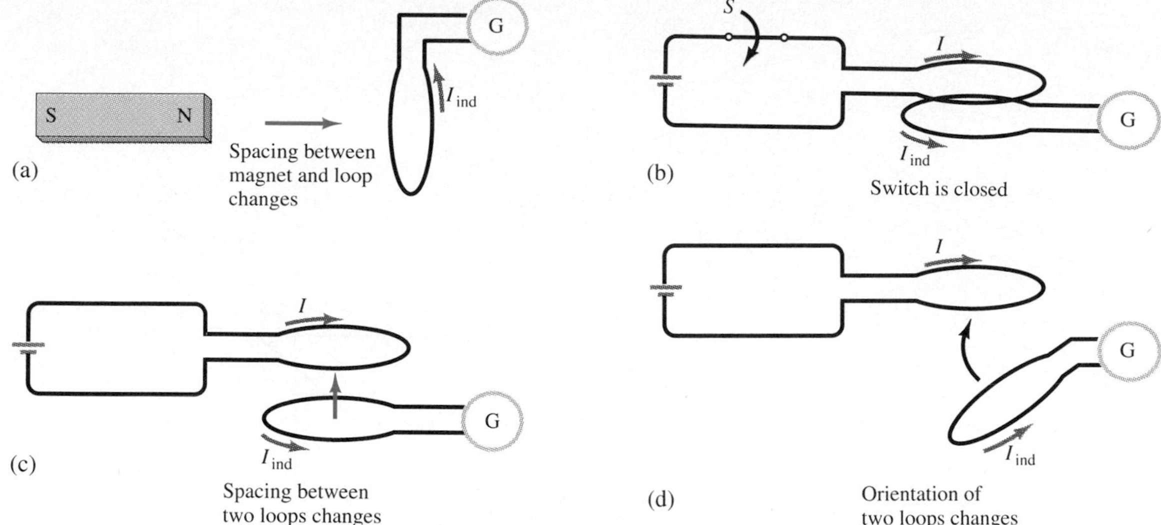

FIGURE 31–3 Ways to make a magnetic flux through a loop change and thereby induce a current I_{ind}. (a) The distance between a wire loop and a bar magnet changes, and a current is induced in the loop. (b) A switch is closed to start a current in one loop, and a current is induced in a second, nearby loop. (c) The distance between a current-carrying loop and a second loop changes, and a current is induced in the second loop. (d) A second loop rotates toward the current-carrying loop, and a current is induced in the second loop.

In this case we are interested in the line integral around a closed loop:

$$\mathcal{E} = \oint \mathbf{E} \cdot d\mathbf{s}. \tag{31–1}$$

This line integral—indeed, any line integral—must include a specification of a direction in which we follow the line (Fig. 31–4). The precise statement of **Faraday's law of induction** is

$$\mathcal{E} = \oint \mathbf{E} \cdot d\mathbf{s} = -\frac{d\Phi_B}{dt}. \tag{31–2}$$

Faraday's law of induction

Here Φ_B is the magnetic flux through the surface S that spans the loop:

$$\Phi_B = \iint_S \mathbf{B} \cdot d\mathbf{A}.$$

The loop around which the emf is defined, Eq. (31–1), must bound the surface through which the flux is calculated, and the orientation of that surface is determined by the direction of the loop integral and a right-hand rule. This right-hand rule works as follows: If the fingers of the right hand curl in the direction of the loop, the thumb indicates the direction of the surface for calculating the flux—that is, the direction the surface element **dA** takes (Fig. 31–4). The minus sign in Eq. (31–2) is critical, and we shall return to it.

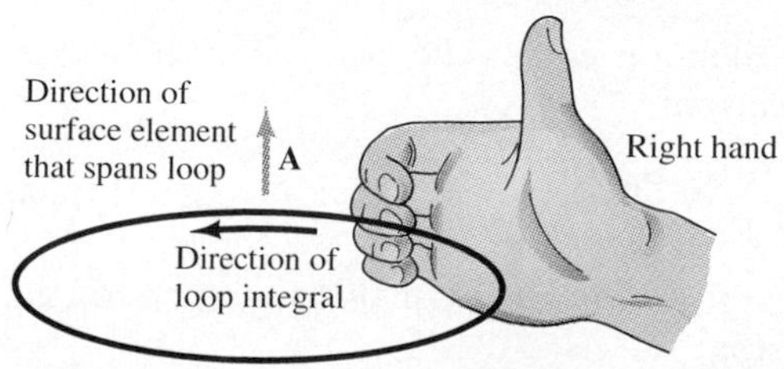

FIGURE 31–4 When the direction around a loop is given, the orientation of the surface that spans the loop is specified by a right-hand rule. Here we show a single vector **A** for the entire surface, which is flat. For a curved surface, the directions of infinitesimal areas **dA** vary from point to point.

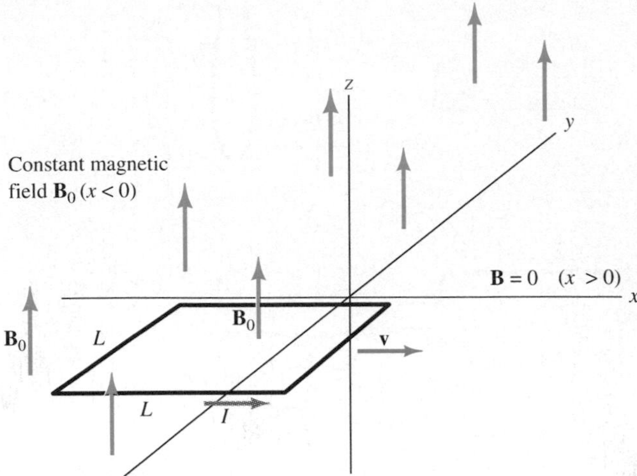

FIGURE 31–5 Example 31–1.

EXAMPLE 31–1 A constant magnetic field has only a z-component B_0 in the region $x < 0$, and is zero for $x > 0$ (Fig. 31–5). A square metal loop with sides of length L is oriented in the xy-plane and pulled through the field with steady velocity $\mathbf{v} = v\mathbf{i}$. The total resistance of the loop is R. Find the induced current in the wire as a function of time, assuming that the front edge of the square crosses the line $x = 0$ at $t = 0$. Evaluate your result for $B_0 = 1.0$ T, $L = 0.10$ m, $R = 0.065\ \Omega$, and $v = 10.0$ cm/s.

SOLUTION: As the wire loop passes out of the region $x < 0$, the magnetic flux upward through the loop decreases, and an emf is induced in the loop, given by Faraday's law, Eq. (31–2). It is easiest to calculate the flux through the planar surface of the loop. The loop integral as seen from above is taken to be counterclockwise. With this orientation for the surface element, the infinitesimal surface elements $d\mathbf{A}$ that make up the integral are oriented upward.

For $t < 0$, the flux through the loop takes on the constant value

$$\Phi_B = \iint_S \mathbf{B} \cdot d\mathbf{A} = B_0 \iint_S dA = B_0 L^2,$$

because the same magnetic field passes through the entire loop. This is unchanging with time for $t < 0$, so there is no induced emf and no current.

In the time period $t = 0$ to $t = L/v$, the loop is in the process of leaving the region of the magnetic field. Thus the flux changes. Only that part of the loop for which $x = L - vt$ remains in the region of the field. In other words, only an area $(L - vt)L$ remains in the field, and

$$\Phi_B = B_0(L - vt)L.$$

This flux is not constant, and for this time period

$$\frac{d\Phi_B}{dt} = -B_0 vL.$$

The emf counterclockwise around the loop equals the *negative* of this value: $\mathcal{E} = +B_0 vL$. A counterclockwise current

$$I = \frac{\mathcal{E}}{R} = \frac{B_0 vL}{R} \tag{31–3}$$

is induced in the loop between $t = 0$ and $t = L/v$. In this time period, the numerical value of the induced current is

$$I = \frac{(1.0\ \text{T})(10.0 \times 10^{-2}\ \text{m/s})(0.10\ \text{m})}{0.065\ \Omega} = 0.15\ \text{A}.$$

For $t > L/v$, the loop has moved out of the region of constant field, so the flux takes on a constant value (zero), and there is neither an induced emf nor a current.

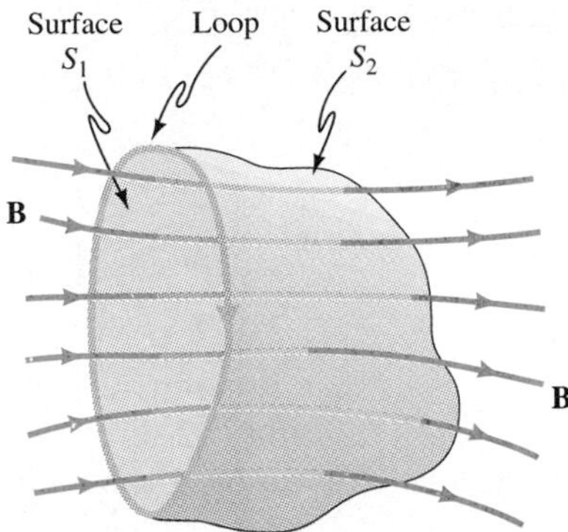

FIGURE 31–6 The same net number of magnetic field lines pass through any two surfaces S_1 and S_2 that bound a closed loop.

The Surface Used in Faraday's Law

In our statement of Faraday's law, we did not specify the surface formed by the loop in question. We shall now show that *it does not matter which surface is used, as long as that surface is bounded by the loop in question. The flux is the same through any such surface.* To see how this comes about, let us recall some of the features of magnetic flux. The magnetic flux through a surface counts the magnetic field lines that pass through that surface. Because, as we know by experiment, no magnetic charges exist, *all magnetic field lines are continuous; they neither begin nor end on charges.* Moreover, *the magnetic flux through a surface is proportional to the net number of field lines that pass through a surface.* (Recall that the number of field lines that pass through a surface can have a sign: The sign of the lines that pass through a surface in one direction is opposite that of the lines that pass through with the opposite orientation.)

The magnetic flux takes on the same value for any surface bounded by a given loop.

Let us now examine the consequences of these properties of magnetic field lines. Consider magnetic field lines and a loop bounded by two surfaces, S_1 and S_2 (Fig. 31–6). Because the lines are continuous, the number of lines that pass through the two surfaces must be the same for both surfaces, and hence the flux through the two surfaces must be the same. Our argument holds for any pair of surfaces, not just the two in Fig. 31–7. Some lines may not pass through S_1, such as line 1, but if a line passes into some third surface, S_3, and not into surface S_1, then the line must also pass out of surface S_3 and therefore does not contribute to the net flux through S_3. Note that for our argument to hold, all surfaces must have the same orientation with respect to the loop. This orientation is specified by the right-hand rule (Fig. 31–4).

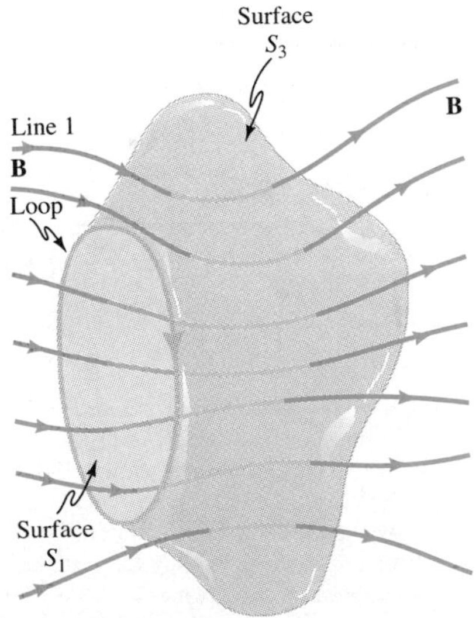

FIGURE 31–7 Surfaces S_1 and S_3 both bound a loop. If magnetic field line 1 passes through surface S_3 but not through surface S_1, it must pass through surface S_3 again and hence does not contribute to the net magnetic flux through surface S_3.

We have come to the important conclusion that *the magnetic flux through one surface bounded by a closed loop is the same as the magnetic flux through any other surface bounded by the same loop. Both surfaces must be oriented by the right-hand rule.* This is a very helpful result because it shows that we can choose for convenience the surface through which to calculate the magnetic flux.

The best surface over which to calculate flux

A good problem-solving technique is to find a surface over which the flux is easily calculated when it is necessary to compute the flux, as it is in Faraday's law. Example 31–2 illustrates this technique.

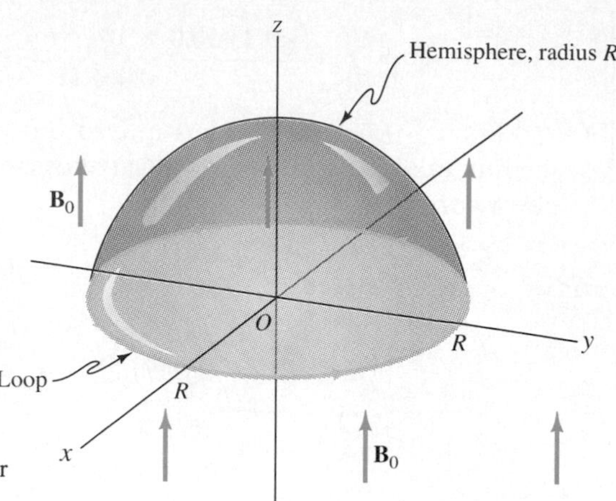

FIGURE 31–8 Example 31–2. The vector $\mathbf{B}_0$ is constant and vertical.

EXAMPLE 31–2 Suppose that a certain region has a magnetic field with the constant value $\mathbf{B} = B_0\mathbf{k}$, where $\mathbf{k}$ is the unit vector in the z-direction. Find the magnetic flux upward through the hemisphere of radius R shown in Fig. 31–8.

SOLUTION: We can find the flux through the hemisphere by finding the flux through any other surface bounded by the same loop. The boundary of the hemisphere is a circle of radius R in the xy-plane, and the simplest surface with which to calculate the flux is that planar circle in the xy-plane. For the circle, $\mathbf{B}$ is parallel to $d\mathbf{A}$. Thus the sign of the flux is positive if B_0 is positive. Mathematically, the equality of the flux through the hemisphere and through the planar circle is

$$\Phi_B\Big|_{\text{hs}} = \iint_{\text{pc}} \mathbf{B} \cdot d\mathbf{A} = B_0 \iint_{\text{pc}} d\mathbf{A} = B_0 \pi R^2.$$

Although the direct calculation of the flux through the dome is difficult, when the flat surface is used, the calculation becomes trivial.

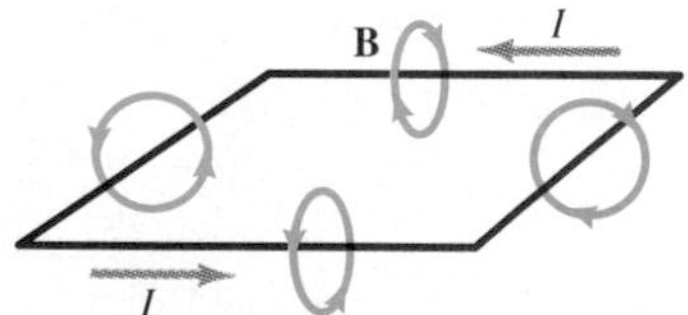

FIGURE 31–9 The current induced in the wire loop of Example 31–1 produces a magnetic field oriented as shown. This field increases the total flux upward through the loop and thus counteracts the decrease in magnetic flux that induces the current in the first place.

Lenz's Law and the Direction of Induced Current

It is instructive to look more closely at the induced current of Example 31–1. We know that currents produce magnetic fields, and the induced current is no exception. By using the right-hand rule, we can see that the magnetic field produced by the induced current is directed up through the loop (Fig. 31–9). This field tends to increase the magnetic flux through the loop, which in Example 31–1 was decreasing. In effect, the induced current acts to oppose the decreased flux that caused it. Analysis of Faraday's law shows that it is always true that the induced current tends to keep the flux from changing. This way of thinking about Faraday's law, due to Heinrich Emil Lenz, is a general physical principle called **Lenz's law:**

Lenz's law

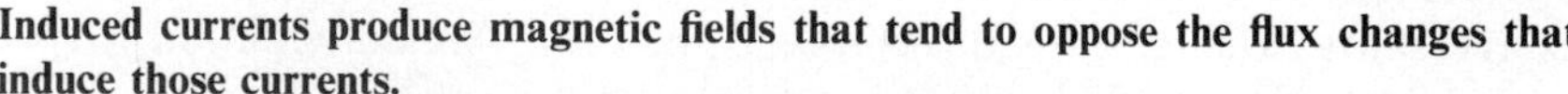

Induced currents produce magnetic fields that tend to oppose the flux changes that induce those currents.

Lenz's law is very useful in determining the *direction* of an induced current. Although this can be done with Faraday's law, it is all too easy to make a sign error. Lenz's law helps to avoid such errors.

EXAMPLE 31–3 The north pole of a bar magnet is thrust at the face of a fixed metal ring (Fig. 31–10). Use Lenz's law to determine the direction of any induced current in the ring.

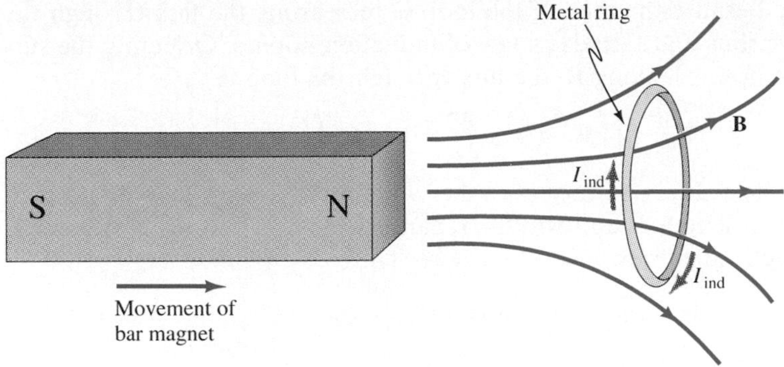

FIGURE 31–10 Example 31–3. When the bar magnet approaches the ring, a current is induced in the ring.

SOLUTION: As the north pole of the magnet approaches the ring, the magnetic field lines near the ring become more dense. The magnetic flux through the surface of the ring, which is perpendicular to the magnet, *increases*. Lenz's law states that the current induced will *oppose* the change of magnetic flux that passes through the ring. The induced current must therefore produce a field that serves to decrease the magnetic flux (and magnetic field). The induced magnetic field will be directed to the left. If we use the right-hand rule, it is easy to see that the current in the ring that will produce this field is oriented counterclockwise as seen from the north pole of the magnet.

You should be able to derive this result by applying Faraday's law.

EXAMPLE 31–4 A closed loop is constructed of a fixed U-shaped wire and a conducting crossbar free to move in the x-direction, all in the xy-plane (Fig. 31–11a). The closed end of the loop is at $x = 0$. A magnetic field oriented in the z-direction varies with x according to $\mathbf{B} = Cx\mathbf{k}$; it is zero at $x = 0$. Suppose that the movable crossbar is pulled at a constant speed v to the right from $x = 0$ when $t = 0$. Its position at any time is $x = vt$. If the resistance of the loop varies with the total length L according to $R = \alpha L$, where α is some constant coefficient, what is the current in the loop as a function of time?

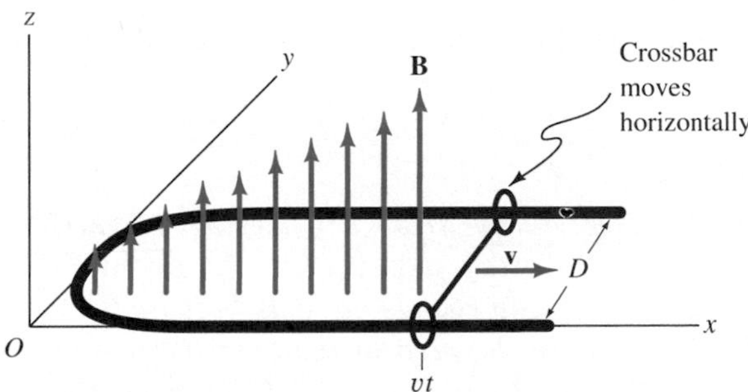

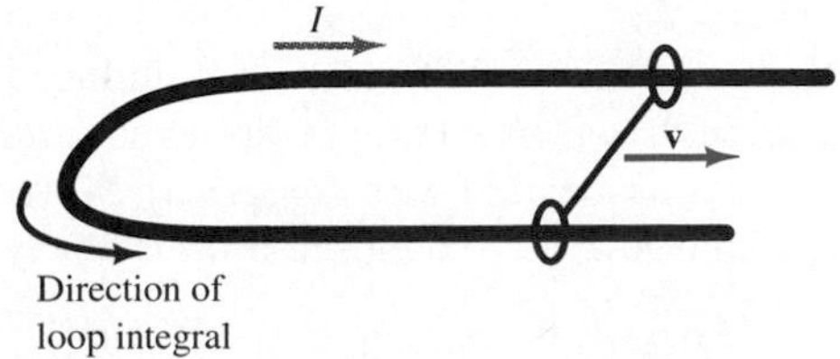

FIGURE 31–11 (a) Example 31–4. (b) Lenz's law shows that the current induced in the circuit that contains the moving crossbar will be in the direction shown.

SOLUTION: Because the area of the loop is increasing, the flux through the loop is increasing, and Faraday's law of induction applies. Orienting the surface element upward, along **B**, the flux through the loop is

$$\Phi_B = \iint \mathbf{B} \cdot d\mathbf{A} = \iint B\, dA = \iint Cx\, dA.$$

Because the field does not vary over the y-direction, we can take dA to consist of strips of length D and width dx parallel to the y-direction. At time t, the loop length runs from $x = 0$ to $x = vt$. Thus

$$\Phi_B = \int_0^{vt} CxD\, dx = \tfrac{1}{2}CDv^2t^2;$$

$$\frac{d\Phi_B}{dt} = CDv^2t.$$

The magnetic flux increases with time.

Lenz's law states that a current will be induced in the clockwise direction, because the magnetic field due to the induced current must be directed down in order to decrease the flux (Fig. 31–11b). This is in accord with Faraday's law, for which the loop integral $\oint \mathbf{E} \cdot d\mathbf{s}$, which by convention is counterclockwise, is negative, corresponding to an induced emf in the direction of the current. Thus we have

$$\oint \mathbf{E} \cdot d\mathbf{s} = \mathscr{E} = -\frac{d\Phi_B}{dt} = -CDv^2t.$$

To relate the emf to the induced current, we must find the total resistance of the loop, R. This resistance is

$$R = \alpha L = \alpha(2D + 2vt) = 2\alpha(D + vt).$$

The induced current is

$$I = I_{\text{ind}} = \frac{\mathscr{E}}{R} = -\frac{CDv^2t}{2\alpha(D + vt)}.$$

You should be able to verify that this current has the correct dimensions.

Some Comments

A Changing Flux Does Not Necessarily Mean a Changing Magnetic Field. Example 31–4 illustrates several noteworthy physical features about Faraday's law, among them that *the magnetic flux can change not only because the magnetic field changes with time, but also because the area of the loop through which the flux is calculated may change with time.* In Example 31–4 the magnetic field is constant, and it is not reasonable to say that this constant magnetic field has produced an electric field in space. The only place an electric field can be said to appear is in the moving loop.

Induced Electric Fields Are Nonconservative. We note that induced fields differ fundamentally from the electric fields we have previously encountered. In our earlier work, electric fields were always associated with *conservative* forces. The work done by those fields in moving a charge around a closed loop is always zero:

$$\text{conservative:} \quad \mathscr{E} = \oint \mathbf{E} \cdot d\mathbf{s} = 0.$$

This is precisely what is *not* true for the fields that result from Faraday's law; the emf here about a closed loop is specified by the changing flux:

$$\text{nonconservative:} \qquad \mathscr{E} = \oint \mathbf{E} \cdot d\mathbf{s} = -\frac{d\Phi_B}{dt}.$$

The induced electric field is not conservative, and it cannot be described by a potential that is a function of space. There is an apparent failure here of the conservation of energy. We noted at the end of Chapter 30 that for the law of momentum conservation to hold true, electric and magnetic fields must carry momentum. Similarly, the apparent failure of energy conservation is rectified because the magnetic field, as well as the electric field, contributes to the total energy.

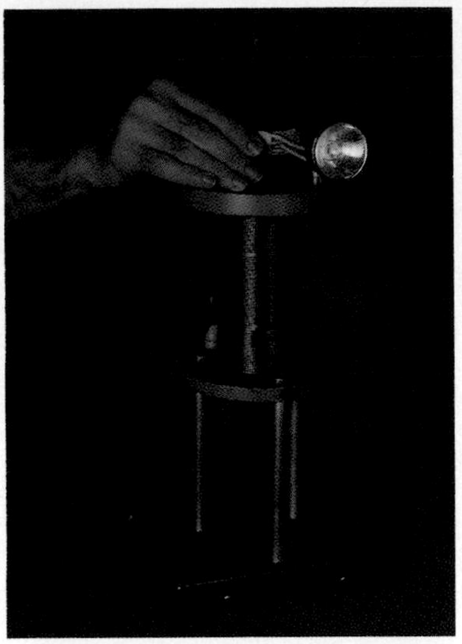

FIGURE 31–12 Although there is no direct connection between the wire coil and the current-carrying iron core, the bulb lights. The changing current in the core induces a large enough current in the coil connected to the bulb so that the bulb lights.

A Symmetry between Ampère's Law and Faraday's Law. If we now recall the form of Ampère's law that includes the Maxwell displacement current (see Section 30–5), we see a degree of symmetry in the laws of electricity and magnetism (Fig. 31–12). The displacement current is produced by a changing electric flux, and Ampère's law describes a relation between the integral of the magnetic field around a loop (in terms of a current that passes through the loop) and a changing electric flux through any surface bounded by the loop, Eq. (30–32). If there is no ordinary electric current, then we can drop the term proportional to I in Eq. (30–32) and write Ampère's law as

$$\frac{d\Phi_E}{dt} = \frac{1}{\epsilon_0\mu_0}\oint \mathbf{B} \cdot d\mathbf{s}.$$

Except for the factor $1/\epsilon_0\mu_0$ and the sign, this law is analogous to the Faraday law of induction.[†]

31–3 MOTIONAL EMF

In this section we shall study further the fact that when a conductor moves in a magnetic field, an emf is induced in the conductor. We call this emf a **motional emf.** Consider a conducting rod that moves with constant speed in a magnetic field (Fig. 31–13a). We shall show that Faraday's law leads to an accumulation of posi-

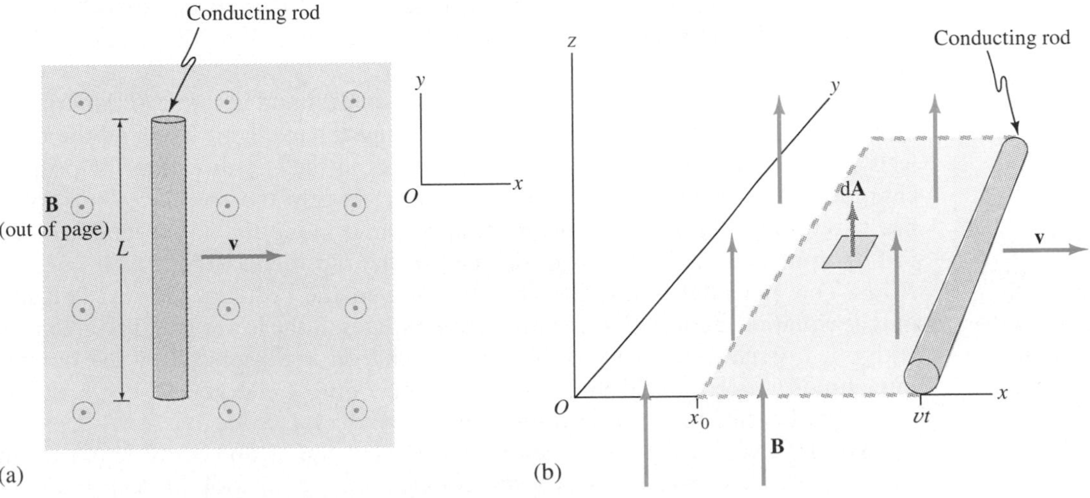

FIGURE 31–13 (a) A conducting rod of length L moves with constant speed v in the $+x$-direction through a constant magnetic field directed in the z-direction. The rod is oriented in the y-direction. (b) A motional emf is induced along the moving rod; this emf can be computed by Faraday's law. The dashed lines represent the imaginary closure of a loop, to which we apply Faraday's law.

[†] The symmetry between Faraday's law and Ampère's law allows the equations of electromagnetism to have wavelike solutions. These so-called electromagnetic waves will be fully explored in Chapter 35.

tive charges at the bottom of the rod and negative charges at the top, leading to an emf along the rod. We shall then show that the same effect may be viewed as a consequence of the Lorentz force law. In Fig. 31–13a the rod has length L and moves at speed v through the field. We can choose our axes so that the position of the rod along the x-axis is $x = vt$. How can this be a situation in which Faraday's law applies? The answer is that Faraday's law applies to *any* loop, whether it be a real conducting loop, an imaginary loop in space, or a loop that is part real conductor and part imaginary, the type we shall now consider. Let us imagine attaching the moving rod to a fixed imaginary line, indicated in Fig. 31–13b by a dashed line. The rod and line form a closed loop. The (fixed) portion of the dashed line parallel to the rod is at $x = x_0$. The loop formed by the imaginary line and the rod is situated in the xy-plane with its area elements $d\mathbf{A}$ oriented in the $+z$-direction and with its area equal to $L(vt - x_0)$. The flux through the closed loop is therefore given by

$$\Phi_B = B_0 L(vt - x_0),$$

and the rate of change of this flux is

$$\frac{d\Phi_B}{dt} = B_0 Lv. \qquad (31\text{–}4)$$

This rate of change is independent of x_0 and of any other part of the imaginary closure of the loop, indicating that our choice for the position of the dashed part of the loop is unimportant. The emf taken in the counterclockwise direction, the direction that must be taken if we orient the flux in the $+z$-direction, is then

$$\mathscr{E} = -B_0 Lv. \qquad (31\text{-}5)$$

If the dashed line were conducting wire, this emf would drive a current clockwise (as seen from above) around the circuit. This would be consistent with the application of Lenz's law. Because there is no actual closed circuit, current flows only for a short while from the top to the bottom of the rod. Positive charges accumulate toward the bottom end of the rod until a canceling emf is set up by the accumulating charges and equilibrium is achieved. Because the rod is the only "real" part of our circuit, the entire emf acts along its length. This is a reflection of our earlier remarks: If the magnetic field is itself unchanging, then there is no reason to have an induced electric field in empty space. The effect is due to the motion of the rod.

Motional emf can be viewed as the result of the magnetic force on moving charges.

The resulting *motional emf* is also simple to understand in terms of the Lorentz force law. Each charge carrier in the rod is moving in a magnetic field and therefore feels a force that equals $q\,\mathbf{v} \times \mathbf{B}$. This force acts in the $-y$-direction for positive charges. The force per unit charge, or the electric field that produces this force, has magnitude vB_0. Because this force is constant along the entire length of the rod, the potential difference from one end of the rod to the other is $\Delta V = EL = B_0 vL$. This potential is just the emf that acts in the rod, Eq. (31–5). The sign in that equation dictates that the higher potential is at the lower end, in agreement with our Lorentz force analysis. In this case, charges are displaced by the Lorentz force until an electric field that produces a force on the charge carriers, just canceling the Lorentz force, is set up within the conductor.

We have now seen that we can view motional emf as due equivalently to the Lorentz force law or to Faraday's law of induction. Let us look back at Example 31–1, in which a square loop moves out of a region of constant magnetic field into a region with no field. Figure 31–14 shows a different orientation of the arrangement in Fig. 31–5, where we have labeled the sides of the square a through

d. We stated that there was no current in the loop for $t < 0$, when the loop was entirely within the constant field. Yet the Lorentz force law still holds. These statements are reconciled by observing that the forces on the charge carriers in the different segments cancel when the loop is entirely within the field region. The force on positive charges in leg a is directed downward, as is the force on the positive charges in leg c, and no charges can move around the loop. Similarly, the forces on the positive charges in legs b and d are also directed downward, so there is no force at all that tends to circulate the charge. Once leg a passes out of the field region, however, there is a net force due to the charges in leg c, and this tends to push the charges in the counterclockwise direction, as we deduced in Example 31–1. The emf will be B_0vl, just as we found in the example. Finally, there are no magnetic forces once the loop has left the field region.

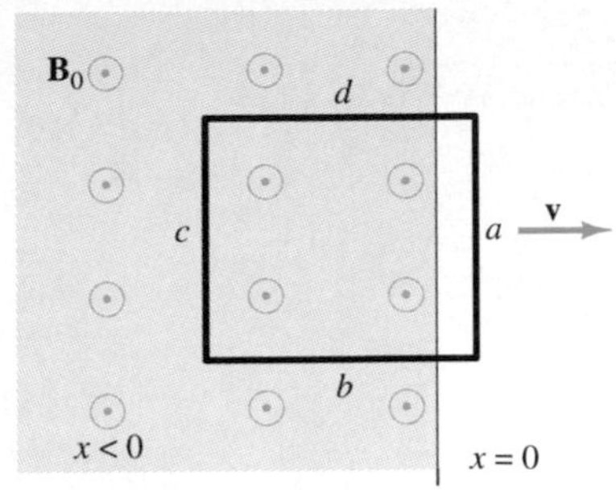

FIGURE 31–14 An analysis of the moving loop of Example 31–1 in terms of the Lorentz force law.

EXAMPLE 31–5 A rod of length L lying in the xy-plane pivots with constant angular velocity ω counterclockwise about the origin (Fig. 31–15). A constant magnetic field of magnitude B_0 is oriented in the z-direction. Find the motional emf in the rod by applying Faraday's law of induction.

SOLUTION: We complete a loop by drawing the dashed lines shown in Fig. 31–15. The area element will be oriented upward, so the loop integral representing the emf must be taken to be counterclockwise. Here $\theta = \omega t$, and the area of the loop is the area of the segment swept out by the rod through the angle θ:

$$\text{area} = \tfrac{1}{2}\theta L^2 = \tfrac{1}{2}\omega t L^2.$$

The magnetic flux through the loop is B_0 times the area, and the rate of change of the flux is

$$\frac{d\Phi_B}{dt} = \frac{d}{dt}\left(\tfrac{1}{2}B_0\omega L^2 t\right) = \tfrac{1}{2}B_0\omega L^2.$$

The motional emf is then

$$\mathscr{E} = -\tfrac{1}{2}B_0\omega L^2.$$

The sign indicates that the emf is directed radially out from the origin. We can verify this direction by use of the right-hand rule to indicate the direction of the Lorentz force vector $\mathbf{v} \times \mathbf{B}$.

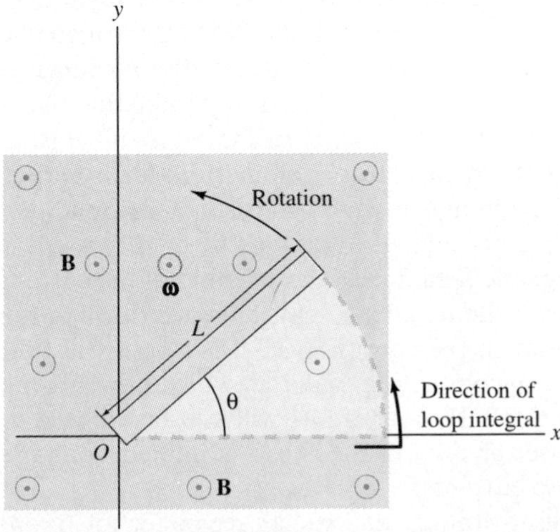

FIGURE 31–15 Example 31–5. The dashed lines represent the imaginary closure of a loop, to which we apply Faraday's law.

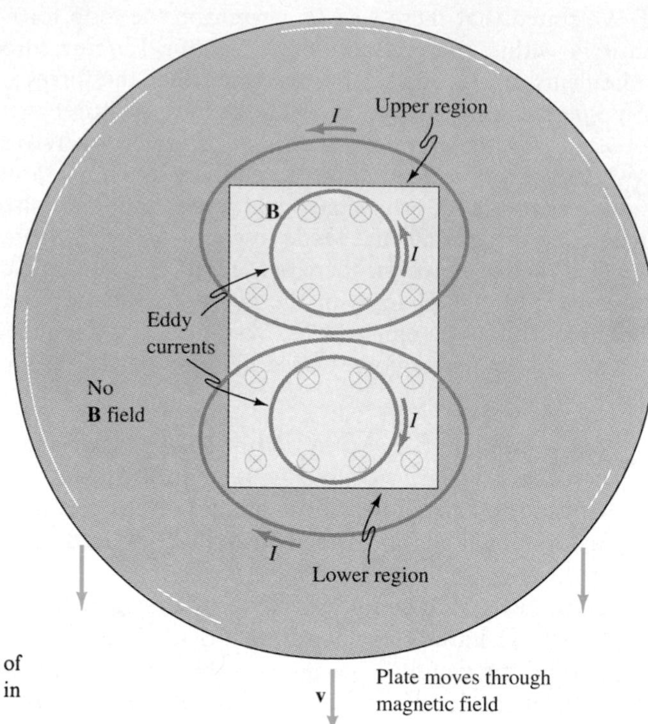

FIGURE 31–16 As the circular metal plate drops through a small region of constant magnetic field directed into the plate, eddy currents are induced in the plate. The direction of these currents is given by Lenz's law.

Eddy Currents

We have talked thus far about induced emfs in the motion of wires and rods. But we can also move large pieces of metal in a magnetic field. What is the effect in this case? The physical origin of motional emf suggests that here, too, currents will be induced but distributed throughout the conductor. These currents are known as **eddy currents**. Figure 31–16 shows the eddy currents set up in a flat, vertically oriented plate that moves through a limited region of magnetic field directed into the page. Two regions of magnetic field are indicated in Fig. 31–16, the upper region and the lower region. As the plate moves through the space where there is magnetic field (between the poles of a magnet, for example), the magnetic flux increases in the upper region and decreases in the lower region. Currents are induced in the metal plate that oppose the change in magnetic flux; these currents are counterclockwise in the upper region, clockwise in the lower region.

The eddy currents thus induced are dissipated in Joule heating through the resistivity of the metal plate. This heating can be a significant disadvantage in certain applications; when it is undesirable, it can be reduced by eliminating paths for the current flow. This is done either by cutting slots in the metal plate (Fig. 31–17) or by laminating the metal with an insulator. Lamination, for example, may be used in the iron core of an electromagnet, which consists of a wound coil of wire—a solenoid—surrounding an iron core. For reasons to be discussed in Chapter 32, the core increases the magnetic field due to the solenoid. If the current in the wire varies with time, then so will the magnetic field that runs through the iron core. Induced eddy currents appear in the core (Fig. 31–18a). To avoid this, the iron can be laminated with some nonconducting material, which prevents the induced currents from developing (Fig. 31–18b). Currents will still be induced in the iron sheets of the core, but because the area of the sheets is limited, so is the flux change, and therefore so are these currents.

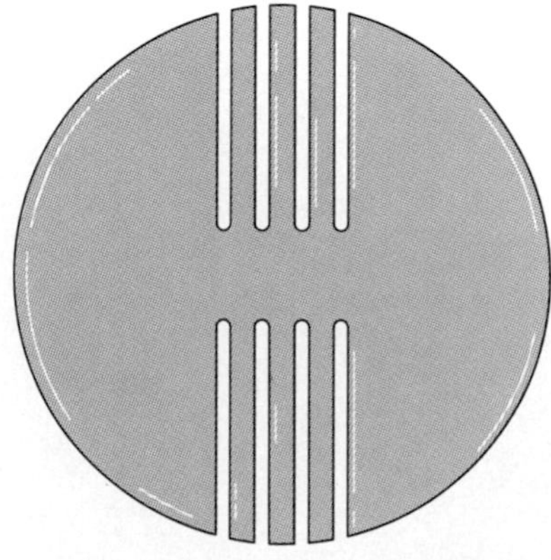

FIGURE 31–17 To inhibit the development of eddy currents in the moving metal plate, slots can be cut in the plate.

Eddy currents are not always a disadvantage, as we shall see in Section 31–4.

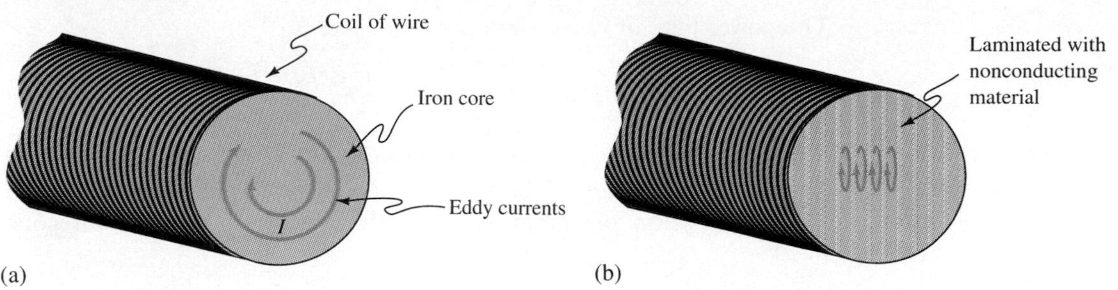

FIGURE 31–18 (a) Eddy currents are set up in the iron core of an electromagnet if the current in the surrounding solenoid is not steady. (b) To inhibit these currents, the core can be constructed of iron sheets laminated with sheets of nonconducting material.

31–4 FORCES, ENERGY, AND POWER IN MOTIONAL EMF

The presence of induced currents because of changing magnetic fluxes has one further implication: We already know that currents experience forces in the presence of magnetic fields, and this must be true for the induced currents. Thus wires in which currents are induced will experience forces. Moreover, *the magnetic force on the induced current always inhibits the motion that produces the motional emf*. This is a consequence of the sign implied by Lenz's law.

Magnetic forces on induced currents inhibit motion.

In the moving loop of Example 31–1, an induced current, I, appears as the loop leaves the region of magnetic field. The force on the wire is given by Eq. (29–16),

$$\mathbf{F}_B = I\int(\mathrm{d}\boldsymbol{\ell} \times \mathbf{B}), \tag{31–6}$$

where $\mathrm{d}\boldsymbol{\ell}$ describes a length element of the wire and **B** describes the magnetic field at that wire element. Let us once again draw the loop, this time including the forces on each leg (Fig. 31–19). There is no force on those legs or portions of legs that are out of the magnetic field. Using the right-hand rule, the force on the portion of leg b that is located in the field is directed down, and it is canceled by the force on the portion of leg d located in the field. The only contribution to the net force comes from leg c. Here application of Eq. (31–6) gives a force

$$F_c = ILB_0 \quad \text{to the left.} \tag{31–7}$$

Recall from Example 31–1, Eq. (31–3), that the magnitude of the current is $I = B_0 vL/R$, where R is the total resistance of the loop and N is the speed with which the loop moves. The force on the loop is thus

$$F_c = \frac{vL^2B_0^2}{R}. \tag{31–8}$$

This force acts to slow down the loop. Energy must be supplied at a certain rate to the loop to keep it moving with the constant velocity **v**. In other words, power must be expended by an external force to keep the loop moving. The external force will have the magnitude F_c, and it will be directed to the right, so the power expended is

$$P = \mathbf{F}\cdot\mathbf{v} = F_c v = \frac{v^2L^2B_0^2}{R}. \tag{31–9}$$

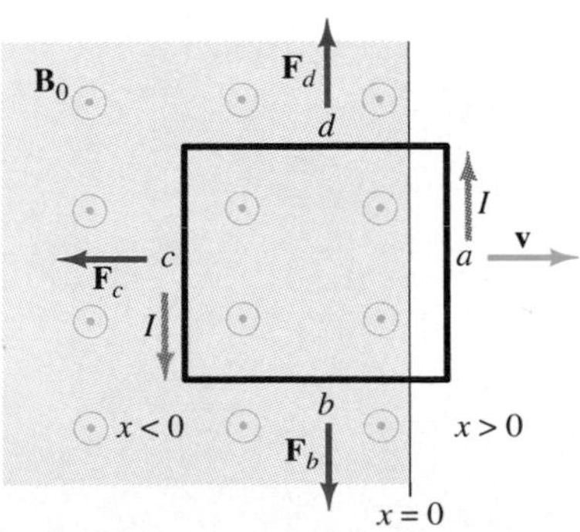

FIGURE 31–19 The magnetic forces on the loop of Example 31–1 act to slow down the loop.

Let us compare the power expended by the external force to the power lost in the resistor. The power loss, or energy loss per unit time, is

$$P = I^2R = \left(\frac{B_0vL}{R}\right)^2 R = \frac{v^2L^2B_0^2}{R}. \qquad (31\text{–}10)$$

Equations (31–9) and (31–10) are the same. *The power loss due to the current flow through the resistor is matched by the power required to keep the loop moving.* The principle of conservation of energy suggests that this result was a foregone conclusion.

Magnetic Drag

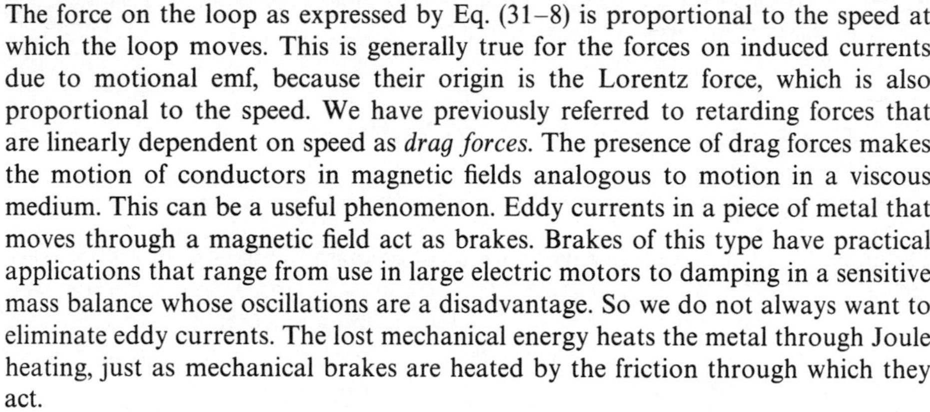

The force on the loop as expressed by Eq. (31–8) is proportional to the speed at which the loop moves. This is generally true for the forces on induced currents due to motional emf, because their origin is the Lorentz force, which is also proportional to the speed. We have previously referred to retarding forces that are linearly dependent on speed as *drag forces.* The presence of drag forces makes the motion of conductors in magnetic fields analogous to motion in a viscous medium. This can be a useful phenomenon. Eddy currents in a piece of metal that moves through a magnetic field act as brakes. Brakes of this type have practical applications that range from use in large electric motors to damping in a sensitive mass balance whose oscillations are a disadvantage. So we do not always want to eliminate eddy currents. The lost mechanical energy heats the metal through Joule heating, just as mechanical brakes are heated by the friction through which they act.

FIGURE 31–20 A levitated magnet. A bar magnet moves toward the superconducting material, inducing persistent currents in the superconductor. The magnetic forces between the superconductor and the magnet are repulsive and sufficiently strong to support the magnet's weight.

The metal plate falling through a magnetic field (see Section 31–3) and pictured in Fig. 31–16 illustrates magnetic drag. The drag force works against gravity and leads to a terminal velocity. Instead of accelerating through the magnet's poles, the plate eventually falls at a steady rate.

We can carry this to the extreme by supposing that the metal plate is a superconductor, a material with no resistance and with the property that the magnetic field does not penetrate its interior. If we drop such a piece of material into the field region, the piece is more than slowed down by its entry into the field; it is repelled. It will bounce back up from the field region. This bounce will be perfectly elastic with no energy loss, because there is no mechanism for energy loss in a material with no resistance. The repulsion of such materials from regions where there is a magnetic field is the principle that underlies magnetic levitation (Fig. 31–20), used in magnetically levitated trains (see Fig. 2–28).

EXAMPLE 31–6 A square loop of wire with sides of length 5.0 cm falls at speed v under the influence of gravity past a region with a constant magnetic field of magnitude 15 T—only large scientific magnets can attain such a high field—into a region with no magnetic field (Fig. 31–21). The field is oriented perpendicular to the loop, into the page. The loop is constrained to remain vertically oriented. The total resistance of the loop is 1.0 Ω, and its mass is 150 g. (a) Find the terminal velocity of the loop as it passes the boundary between the two fields. (b) Calculate the total energy lost to Joule heating in the loop during this period. Assume as an approximation that the loop moves at its terminal velocity when it enters the magnetic field region and that this velocity remains constant during the loop's passage.

SOLUTION: (a) Whether the loop enters or leaves the region with the magnetic field, the drag force is oriented upward and is given by Eq. (31–8), $F = vL^2B_0^2/R$.

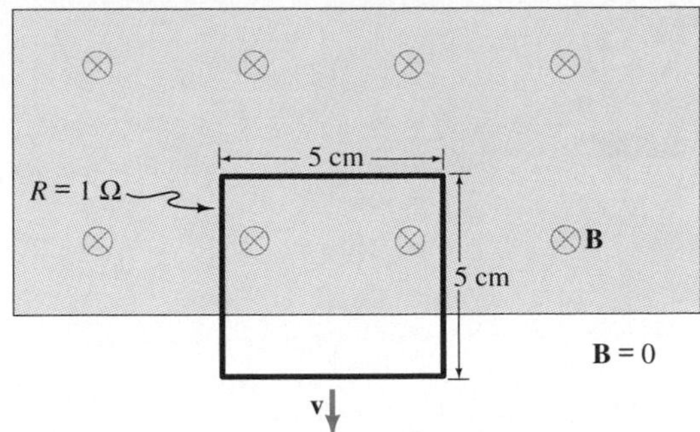

FIGURE 31–21 Example 31–6.

The force of gravity is directed downward with magnitude mg. The terminal velocity v_t is reached when the drag force equals the force of gravity:

$$\frac{v_t L^2 B_0^2}{R} = mg;$$

$$v_t = \frac{mgR}{L^2 B_0^2}. \tag{31–11}$$

Numerically,

$$v_t = \frac{(0.15\ \text{kg})(9.8\ \text{m/s}^2)(1.0\ \Omega)}{(5.0 \times 10^{-2}\ \text{m})^2(15\ \text{T})^2} = 2.6\ \text{m/s}.$$

(b) As long as the loop is moving at its terminal velocity, the total energy lost to Joule heating will be a constant given by the product of power and the time the loop spends in the transition region. The power dissipated in Joule heating is, according to Eq. (31–10),

$$P = I^2 R = \frac{v_t^2 L^2 B_0^2}{R}.$$

The time spent in transition is $t = L/v_t$, so the energy loss is

$$\Delta E = Pt = \frac{v_t^2 L^2 B_0^2}{R}\left(\frac{L}{v_t}\right) = \frac{v_t L^3 B_0^2}{R}.$$

Substituting for v_t from Eq. (31–11), we find that

$$\Delta E = \frac{mgR}{L^2 B_0^2}\,\frac{L^3 B_0^2}{R} = mgL.$$

This is just the change in gravitational potential energy. Because the loop moves through this change without a corresponding change in kinetic energy, all this energy goes into Joule heating. We have

$$\Delta E = (0.15\ \text{kg})(9.8\ \text{m/s}^2)(0.050\ \text{m}) = 0.074\ \text{J}.$$

Even with a very high value for the magnetic field, the terminal velocity in Example 31–6 is relatively large. An effective brake of this type requires a large magnetic field. Note that the terminal velocity is proportional to the resistance of the conducting loop. Even though a higher resistance means more energy is dissipated for a given current, the induced current decreases as the resistance increases, and v_t ends up growing as R grows. It is easier to construct large brakes with solid pieces of metal, which contain large distributed eddy currents and hence large resistive losses.

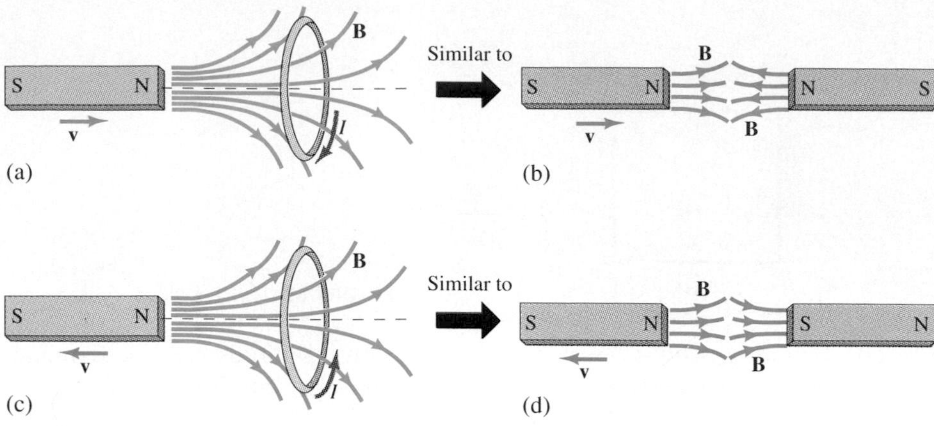

FIGURE 31–22 (a) The magnetic flux through a conducting ring is increasing because the north pole of a bar magnet moves toward it. The ring is repelled. (b) The direction of the induced current in the ring gives the ring the field of a bar magnet with its north pole to the left, and two north poles repel. (c) The bar magnet is pulled away. In this case, a current is induced in the ring in the opposite direction. (d) The magnetic field of the ring is then that of a bar magnet with its south pole to the left, and it is attracted to the receding bar magnet.

Forces and Lenz's Law

Lenz's law gives us a second way to think about the forces on induced currents. The magnetic field of a current loop or solenoid is the same as that of a bar magnet. We have already seen that both magnets and current loops produce magnetic dipole fields. Suppose that the north pole of a bar magnet moves toward a conducting ring that initially has no current. The magnetic flux through the ring increases, and an emf is induced in the ring, causing a current to flow (Fig. 31–22a). The induced current in turn produces a magnetic field whose flux tends to cancel the increase of flux due to the moving bar magnet. The direction of this induced magnetic field is thus opposed to the field of the bar magnet. We can think of it as the magnetic field of a bar magnet, with a field as shown in Fig. 31–22b. The situation is one of two north poles meeting, and we know that two north poles repel each other.

If we now pull the bar magnet away from the ring, the induced current in the ring points in the opposite direction (Fig. 31–22c). The induced magnetic field changes direction, and a bar magnet equivalent to the ring would be turned around with its south pole adjacent to the moving bar magnet (Fig. 31–22d). The situation is one of adjacent north and south poles, and we know that opposite poles attract each other. The magnet pulls the ring along with it.

31–5 EFFECTS OF TIME-VARYING MAGNETIC FIELDS

The magnetic flux through a loop can change in a variety of ways. (i) The loop can move or rotate in the presence of a magnetic field that is not changing with time. (ii) The source of the magnetic field can move, moving the field along with it, such as when a bar magnet moves. (iii) The source of the magnetic field and hence the field itself can have explicit time dependence, such as when the current through a solenoid is made to change. In cases (ii) and (iii), it is not enough to invoke the Lorentz force, as is possible in case (i). Yet there should be no way to tell whether the loop moved or the field changed. Experiment confirms this: Faraday's law of induction holds even when the changing flux is due to time dependence in the fields. Because only motional emf can be interpreted in terms of the Lorentz force law, the fact that an induced emf is also produced by a time-varying magnetic field is a truly new aspect of Faraday's law.

Faraday's law shows that if a magnetic field changes with time, then an electric field is induced in space, such that Eq. (31–2) is satisfied. *If* there is sufficient symmetry in a situation, then it is possible to calculate the induced electric field similarly to the way in which we used Ampère's law to determine a magnetic field.

EXAMPLE 31–7 The two circular pole faces of an electromagnet, both of radius $R = 0.5$ m, are oriented horizontally with the north pole underneath (Fig. 31–23a). The electromagnet produces a field that is uniform throughout the volume between the faces. The field is increased linearly from 0.1 T to 1.1 T over a period of 10 s. Describe the electric field that results in the region between the poles.

SOLUTION: The unknown electric field is induced by the changing magnetic flux. There is cylindrical symmetry between the pole faces, so the induced electric field can vary only with the distance r from the central axis of the pole faces, not with the angle around this axis. We must find the magnetic flux through a horizontal circle of radius r centered on the axis of the pole faces. The magnetic field strength, **B**, is constant over the area of the circle and runs upward from the north pole, so the flux upward through the circle is

$$\Phi_B = \pi r^2 B.$$

As described in the problem statement, the magnetic field has a linear time dependence of the form

$$B = B_0 + \alpha t.$$

The rate of change of flux is therefore

$$\frac{d\Phi_B}{dt} = \pi r^2 \frac{dB}{dt} = \pi r^2 \frac{d}{dt}(B_0 + \alpha t) = \pi r^2 \alpha.$$

Faraday's law, Eq. (31–2), gives the integral around the circle of the induced electric field:

$$\oint \mathbf{E} \cdot d\mathbf{s} = -\frac{d\Phi_B}{dt} = -\pi r^2 \alpha. \qquad (31\text{–}12)$$

Because the flux is oriented upward, the direction of the loop integration is counterclockwise as we look down the north pole (Fig. 31–23b). With the minus sign in Eq. (31–12), the induced electric field points clockwise. Symmetry demands that the electric field have the same magnitude all around the circle, so the electric field is given by

$$\oint \mathbf{E} \cdot d\mathbf{s} = E 2\pi r = -\pi r^2 \alpha.$$

Thus the magnitude of **E** is

$$E = \frac{\not{\pi} r^{\not{2}} \alpha}{2 \not{\pi} \not{r}} = \frac{1}{2}\alpha r.$$

The induced electric field increases as we go out from the center.

Numerically, the coefficient α is found by knowing that B increases from 0.1 T to 1.1 T in 10 s, so the rate of increase is $\alpha = (1.1\ \text{T} - 0.1\ \text{T})/10\ \text{s} = 0.1$ T/s. The induced electric field is then

$$E = \tfrac{1}{2}(0.1\ \text{T/s})r = (0.05\ \text{T/s})r,$$

which increases from 0 N/C at $r = 0$ m to a maximum of $E_{max} = (0.05\ \text{T/s})(0.5\ \text{m}) = 2.5 \times 10^{-2}$ N/C at $r = 0.5$ m.

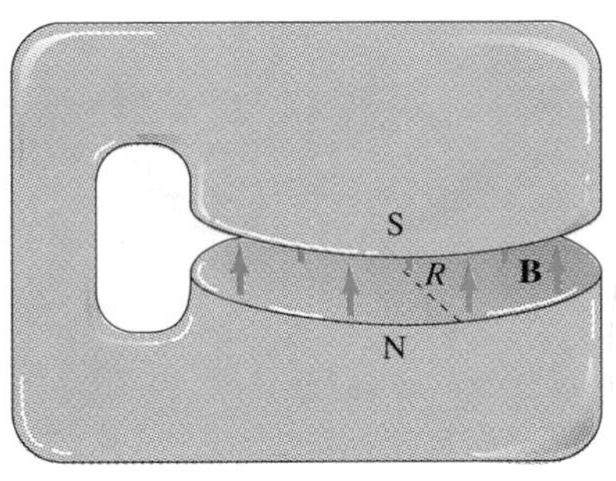

(a)

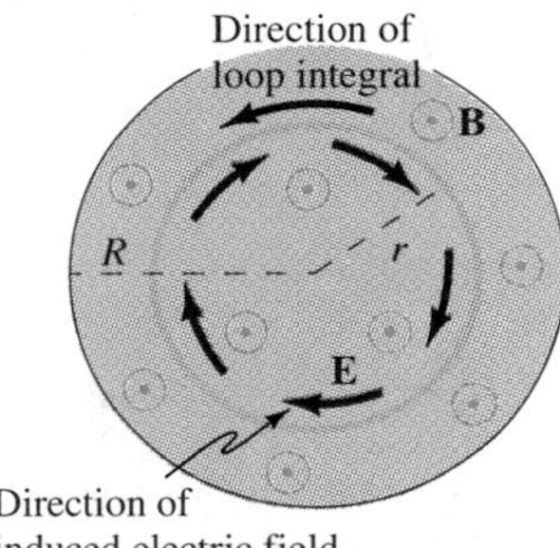

(b)

FIGURE 31–23 (a) Example 31–7. The current windings that produce the field are not shown. (b) A view straight down at the north pole. A changing magnetic field between the pole faces of an electromagnet induces an electric field. Symmetry allows us to specify the electric field, not just its integral around an arbitrary loop.

FIGURE 31–24 An AC generator.

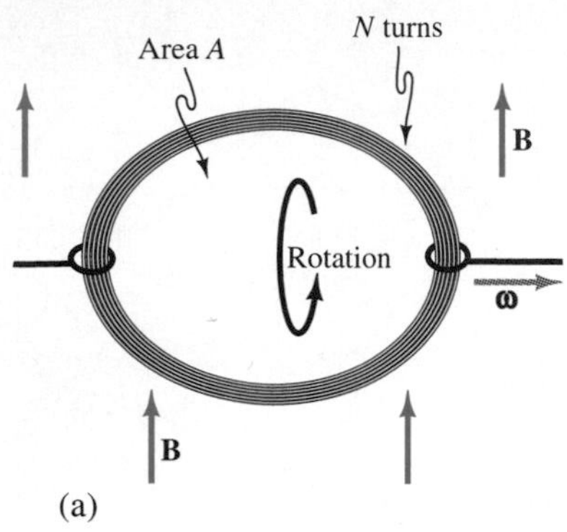

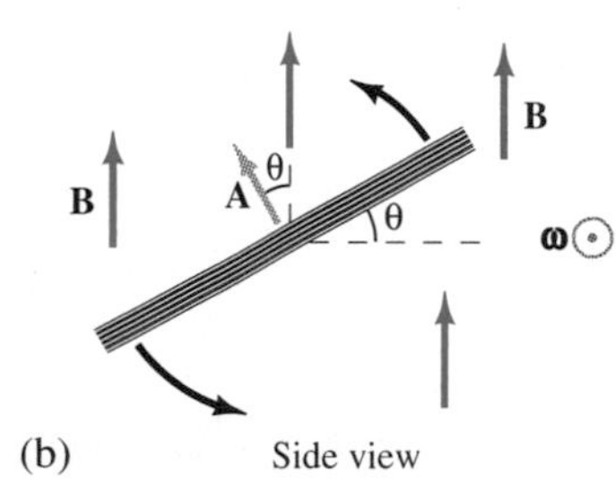

FIGURE 31–25 (a) An external force rotates a coil with angular velocity $\boldsymbol{\omega}$ in a magnetic field. An emf with sinusoidal time dependence of angular frequency ω is induced in the coil. (b) Side view of the process, looking at $\boldsymbol{\omega}$.

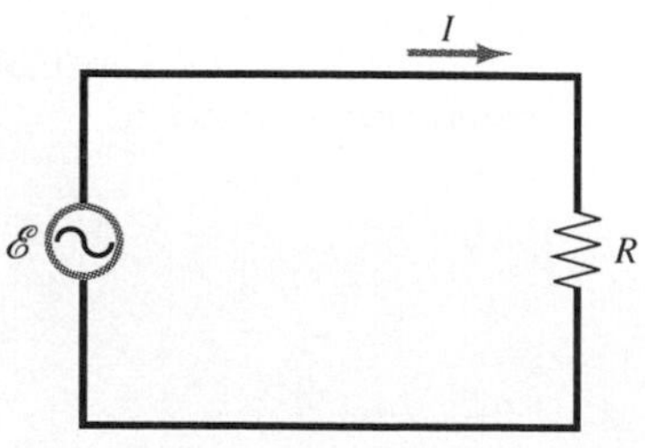

FIGURE 31–26 When the sinusoidally varying emf that results from rotating a coil in a constant magnetic field (as in Fig. 31–25a) is part of a circuit, a sinusoidal current with the same angular frequency ω results.

Is a Magnetic Field Needed Where a Current Is Induced?

One aspect of Faraday's law is not very obvious. Suppose that charged particles such as protons are moving but are so far away from the region of the pole faces of a magnet that the magnetic field is very small where the protons are. Would they still accelerate when the magnetic field is changed? Is an electric field induced even in fringe regions, where the magnetic field is small, or in regions with no magnetic field whatsoever, as in the space outside a toroidal solenoid? According to Faraday's law, all that counts is the change of the flux through the loop in question, no matter where the loop is. Experiment bears out that the answers to both questions are affirmative. To understand this observation, it is useful to recall that magnetic field lines must be closed, and that any change in the field in a small region may be associated with a corresponding change in the total number of field lines outside that region. A *changing* magnetic flux may induce electric fields in regions far from where the magnetic field is large.

31–6 GENERATORS AND MOTORS

The generation of electric energy in our society is based largely on the Faraday induction law. The conversion of mechanical energy (for example, the rotating blades of a steam turbine) to electric current is accomplished with the **alternating-current (AC) generator**, whose principle we now outline (Fig. 31–24).

Imagine a coil of N turns that makes a circle of area A. The coil is placed in a constant magnetic field, $\mathbf{B}$, and rotated at angular speed ω around an axis perpendicular to the field (Fig. 31–25a). The ends of the wire that makes up the coil are brought to the exterior through some sort of sliding contact with a fixed wire. As the coil rotates in the magnetic field, the magnetic flux through it changes, and an emf is induced. According to Fig. 31–25b, the magnetic flux through the loop is $\Phi_B = \mathbf{A} \cdot \mathbf{B} = AB \cos \theta$. We imagine starting the rotation at $t = 0$, so $\theta = \omega t$. Then the time derivative of the magnetic flux is

$$\frac{d\Phi_B}{dt} = AB \frac{d}{dt} \cos(\omega t) = -AB\omega \sin(\omega t). \tag{31–13}$$

There are N turns of wire, so the total emf induced across the two ends of the coil is

$$\mathcal{E} = -N \frac{d\Phi_B}{dt} = NAB\omega \sin(\omega t). \tag{31–14}$$

This arrangement makes up our generator, which is denoted by a circle enclosing a sine wave (Fig. 31–26).

If the wire of the generator coil is connected as a series element of a circuit with a resistance R, then a current

$$I = \frac{\mathcal{E}}{R} = \frac{NAB\omega}{R} \sin(\omega t) \tag{31–15}$$

is generated in the circuit. This *alternating current* oscillates in sign and has a maximum magnitude of $NAB\omega/R$.

The power delivered to this circuit, P, is given by the product of the emf and the current:

$$P = \mathcal{E}I = INAB\omega \sin(\omega t). \tag{31–16}$$

The mechanical force that rotates the loop must be the source of this power. We know that a loop that carries a current forms a magnetic dipole, and that a magnetic dipole experiences a torque that tends to align it with the direction of the magnetic field. Thus the force that rotates the coil must do work against this torque. Let us compute the rate at which this work is done. The torque on a dipole of magnetic dipole moment $\boldsymbol{\mu}$ in a field $\mathbf{B}$ has magnitude

$$\tau = |\boldsymbol{\mu} \times \mathbf{B}| = \mu B \sin\theta,$$

where we refer to Fig. 31–25b for θ and recall that the magnetic dipole moment is perpendicular to the current loop. The mechanical power P_{mech}, or work per unit time, that must be expended by the force that rotates the loop against this torque is

$$P_{\text{mech}} = \tau\omega = \mu B\omega \sin\theta.$$

The magnetic dipole moment of a current loop with N turns is INA [Eq. (29–24)]. Thus

$$P_{\text{mech}} = INAB\omega \sin\theta. \qquad (31\text{–}17)$$

This result is the same as that given by Eq. (31–16). As expected, the electric power is accounted for entirely by the mechanical power expended.

The explicit time dependence of the power is found by taking the product of the current, Eq. (31–15), and the potential, Eq. (31–14), or more simply by evaluating the product V^2/R. We then have

$$P = \frac{(NAB\omega)^2}{R}\sin^2(\omega t). \qquad (31\text{–}18)$$

This quantity is always positive, as opposed to the emf or the current, both of which alternate in sign. The distinction is illustrated in Fig. 31–27, a plot of the current and the power in the circuit of Fig. 31–26 as a function of time.

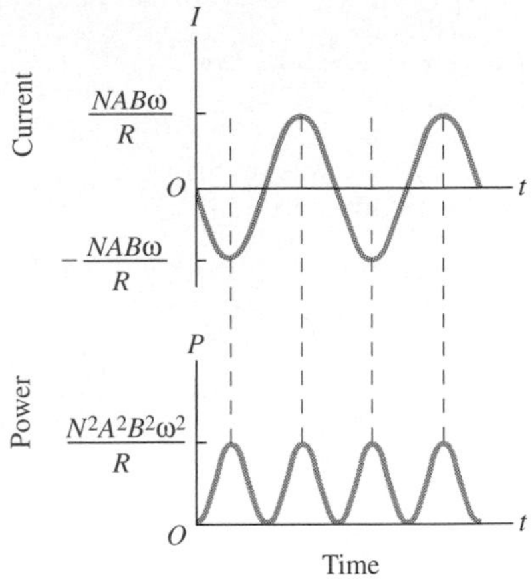

FIGURE 31–27 The power dissipated in the circuit of Fig. 31–26 is always positive, unlike the current, which alternates in sign.

We may want to consider the *time average* of the power. To do so, note that the sine-squared function oscillates between 0 and 1 and averages over one period to 1/2. The average power dissipated in this circuit is then

$$P_{\text{av}} = \frac{1}{2}\frac{(NAB\omega)^2}{R} = \frac{1}{2}\frac{V_{\text{max}}^2}{R}. \qquad (31\text{–}19)$$

We have described only the *principle* by which electric generators operate. The actual practice, in which highly efficient generators use the mechanical power of hot steam or falling water to turn gigantic turbines in huge magnetic fields, involves great engineering skills (Fig. 31–28).

Electric power is transmitted in AC form. Once it has been transmitted, it is typically used to run motors based on Faraday's law of induction that reconvert electric energy back to a mechanical form. The generator described above uses mechanical energy to produce electric current. If we run a generator in reverse, we can convert electric energy in the form of a current into mechanical energy: We have a *motor*. We already saw an example of one such device in Chapter 29 when we discussed how, if a direct current runs through a loop with a split-ring commutator, there is a torque on the loop and the loop turns; the commutator reverses the current after a half turn and ensures that the torque is always in the same direction. If we have a motor with an alternating current, we can dispense with the commutator, because the current automatically reverses direction after one-half cycle. The rate at which such motors turn is tied to the frequency of the alternating current. The design of efficient motors is of great technological importance, but a detailed discussion of motors would take us too far afield.

FIGURE 31–28 A hydroelectric generator, which produces electric power by using the mechanical power of steam to turn turbines in a magnetic field.

*31–7 THE CONNECTION BETWEEN ELECTRIC AND MAGNETIC FIELDS IN MOVING REFERENCE FRAMES

This problem was discussed at the end of Chapter 30.

We are now in a position to resolve the problem of the consistency of the laws of electricity and magnetism with the principle of Galilean invariance—the relativity principle. According to this principle, there should be no way to tell which of two inertial frames is moving and which is not, if they are moving with respect to one another. The laws of motion should be the same in all inertial frames, so they do not provide a way to tell which frame is moving. Thus an observer O and an observer O′, who is moving with constant velocity $\mathbf{u}$ relative to observer O, should express the laws of motion in exactly the same way. In particular, Newton's second law,

$$\frac{d\mathbf{p}}{dt} = \mathbf{F},$$

with $\mathbf{p} = m\mathbf{v}$, must look the same to observer O, who sees a particle move with velocity $\mathbf{v}$, and to observer O′, who sees the same particle move with velocity $\mathbf{v}' = \mathbf{v} - \mathbf{u}$. This will be the case if $\mathbf{u}$ is constant (as it must be for inertial frames) and if the force on the particle is the same to both observers.

If the particle is charged and subject to a Lorentz force, there is an apparent difficulty, because the Lorentz force depends on the velocity of the particle. The problem would be only an apparent one if both observers saw the same Lorentz force. Let us determine what is necessary for the Lorentz force to be the same for both observers. Observer O sees the Lorentz force [Eq. (29–4)]

$$\mathbf{F} = q[\mathbf{E} + (\mathbf{v} \times \mathbf{B})]. \qquad (31\text{–}20)$$

Observer O′ sees the force

$$\mathbf{F}' = q\{\mathbf{E}' + [(\mathbf{v} - \mathbf{u}) \times \mathbf{B}']\}. \qquad (31\text{–}21)$$

Here we have allowed for the possibility that the fields seen by observer O (**E** and **B**) are different from those seen by observer O′ (**E**′ and **B**′). This possibility is what makes it feasible to reconcile the forces. The forces seen by the two observers will be the same *provided that* the fields they see are related by

$$\mathbf{E}' = \mathbf{E} + (\mathbf{u} \times \mathbf{B}) \qquad (31\text{–}22)$$

and

$$\mathbf{B}' = \mathbf{B}. \qquad (31\text{–}23)$$

Although Eqs. (31–22) and (31–23) do not represent a unique solution to the problem of reconciling the forces, they do turn out to be the correct relations when speeds are not large compared to the speed of light. We conclude that under the transformation from one inertial frame to another, electric and magnetic fields get mixed up (transform among themselves) in a very special way.

Electric and magnetic fields mix to an observer who is moving between inertial frames.

The discussion above shows how stationary and moving observers view the source and effect of motional emf. Suppose that an observer O is in a reference frame where there is a magnetic field but no electric field. If a conducting rod moves through this field, there is a motional emf. Its source, according to observer O, is the magnetic force on the conducting electrons in the rod (Section 31–3). Observer O′ moving with the rod sees the rod at rest and moreover sees the same constant magnetic field seen by observer O, from Eq. (31–23). Therefore observer O′ sees no magnetic force. However, observer O′ also sees an electric field **E**′ [Eq. (31–22)], and this electric field has a magnitude that makes the conducting electrons in the rod accelerate in just the way that observer O sees them accelerate. Each observer attributes the observed effects to different combinations of fields.

SUMMARY

Faraday's law describes how a changing magnetic flux through a loop, Φ_B, causes an emf, $\mathcal{E}$, to be induced around the loop:

$$\mathcal{E} = \oint \mathbf{E} \cdot d\mathbf{s} = -\frac{d\Phi_B}{dt}. \tag{31–2}$$

The minus sign in Faraday's law is restated in more physical terms as Lenz's law: Induced currents produce magnetic fields that tend to cancel the flux changes that induce them.

The flux change to which Faraday's law refers can occur either because the magnetic field changes with time or because the area or orientation of the surface through which the flux is calculated changes with time. In the latter case, the induced emf is called a motional emf, and it can be derived directly from application of the Lorentz force law. Application of Lenz's law to motional emfs shows that the induced current must lead to magnetic forces that inhibit the motion of the object in which the emf is induced. The power loss due to resistive flow of induced currents is matched by the power required to keep the conductor moving. When an induced emf occurs because magnetic fields change with time, Faraday's law is a new physical principle.

When a changing magnetic flux passes through a conducting solid, Faraday's law manifests itself by the induction of eddy currents in the material.

The AC generator, the foundation of electrical power generation, is an application of Faraday's law. When a coil rotates in a magnetic field, an emf is induced in the coil. The mechanical energy of the rotation is thus transformed into electric energy in the form of a current in circuits connected to the coil.

QUESTIONS

1. A spherical surface is placed in a changing magnetic field. Will there be an induced electric field along the equator?
2. Must there be a real conducting loop in a region with a changing magnetic flux in order for an electric field to be induced?
3. Electric leads (the wires that run from one part of the apparatus used for experiment to another) for sensitive experiments are almost never separated but are close together or even twisted around one another. Explain why this might be done.
4. Can the magnetic field change over some region without a change in the magnetic flux through a surface in the region? If so, give as many examples as you can.
5. Each part of Fig. 31–29 shows a current being induced in a conducting loop by a changing magnetic flux through the loop. In each case, is the direction of the induced current correct as shown? (a) A magnet approaches a loop; (b) a current-carrying conducting loop approaches a loop at rest; (c) a switch is closed in the first loop, causing a current to flow; (d) a switch is closed in the straight wire, causing a current to flow. Here the wire and loop are both on a flat table.

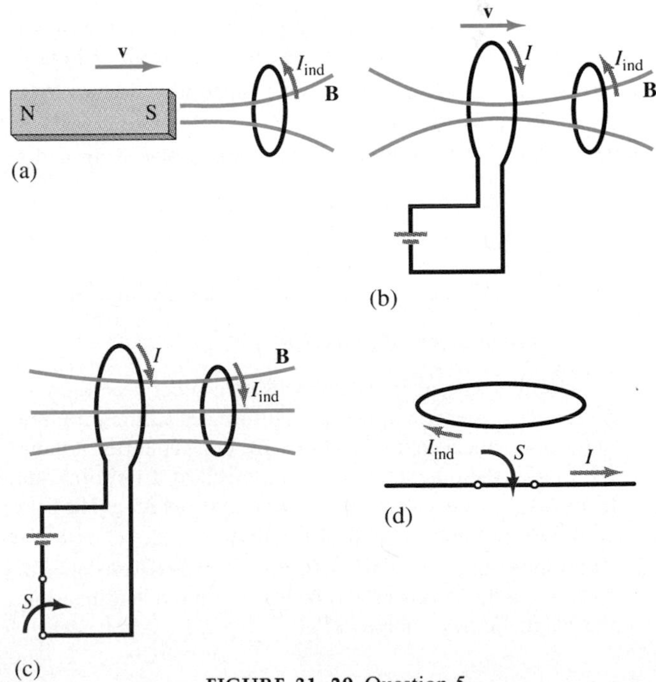

FIGURE 31–29 Question 5.

6. A conducting hoop is rolled in a straight line at a constant speed in an east–west direction in the Northern Hemisphere through the earth's magnetic field. Will a current flow in the hoop? If so, in what direction will it circulate?
7. A rectangular loop is moving across a uniform magnetic field such that the induced emf is zero. What can you say about the angle between the normal to the surface of the loop and the direction of the magnetic field during the motion?
8. A sheet of metal is placed between the pole pieces of a permanent magnet, perpendicular to the direction of the field lines. Does it take positive work to pull the sheet of metal out? If so, why?
9. When a bar magnet is moved toward a current loop, a current is induced in the loop. How, if at all, will physically measurable quantities change if the loop is moved toward the magnet rather than vice versa?
10. If a flat plate hung by a cord and oriented parallel to the pole faces of a magnet—the faces are in a vertical plane, parallel to each other—moves as a pendulum bob through the pole faces, it slows down. If the magnet is sufficiently strong, the plate comes to rest. Why? How could this phenomenon be prevented?
11. Can you describe a situation in which a motional emf is induced in a loop even though there is no change in the magnetic flux through the loop?
12. When a gas is heated sufficiently, its atoms ionize into electrons and positive ions. The material forms a plasma. If the plasma is forced to flow in a channel perpendicular to a magnetic field, an electric potential builds up across the channel. The device based on this phenomenon is the *magnetohydrodynamic (MHD) generator*. Given the magnetic field strength and the potential, we calculate the velocity of the plasma. The charge and density of the charge carriers in the plasma do not enter into the result. Why not?
13. What happens when a bar magnet is dropped down a long, vertical copper tube?
14. In a demonstration an aluminum ring is placed around a projection of an iron core wound with a wire and connected to a battery (Fig. 31–30). The ring jumps when the circuit is closed. Why? What happens if a gap is cut in the ring?

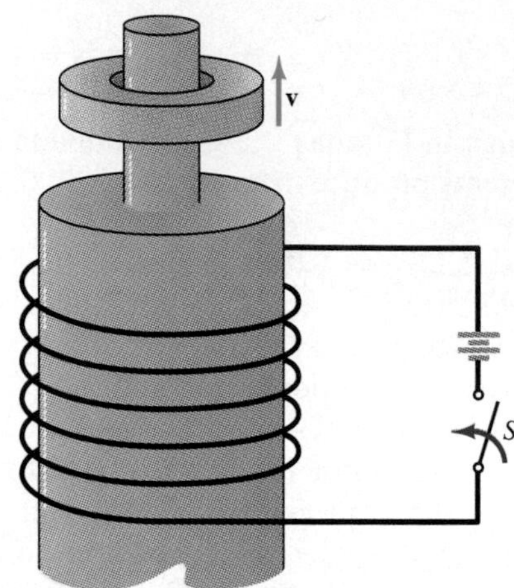

FIGURE 31–30 Question 14.

15. A cylindrical piece of iron is inserted inside a solenoid to increase the magnetic field. A voltage varying harmonically with time is placed across the solenoid leads. A copper ring is slipped down over the solenoid, so that the solenoid passes through the ring, and is held there. (a) Explain why the copper ring becomes hot even though nothing touches it. (b) What is the source of the thermal energy? (c) Explain how energy is conserved in this case.
16. In Question 15, the solenoid axis is vertical. It is possible to find a particular ring of copper that, when slipped over the solenoid and placed horizontally, remains suspended in space around the solenoid. (a) Why does this work? (b) What are the criteria for selecting the particular piece of copper?
17. Describe all the ways you can think of to produce magnetic flux changes when there is a constant current.

PROBLEMS

31–2 Faraday's Law of Induction

1. (I) A loop of wire of area 0.35 m^2 is placed between the pole pieces of an electromagnet, at right angles to the direction of the magnetic field lines. What is the emf generated around the loop if the magnetic field is changed at a uniform rate from 2.0 T to 4.0 T in 9.3 s? Assume that the magnetic field is uniform across the area of the loop.
2. (I) Suppose that the wire in Problem 1 has a resistance of 23 Ω. How much power will be lost to ohmic heating while the magnetic field increases?
3. (I) A magnetic field that changes with time but is uniform in space is directed along the x-axis. A conducting ring of diameter 7 cm and resistance 1.5×10^{-3} Ω is placed in the yz-plane. If the current in the ring is 2 A, how fast is the magnetic field changing?
4. (II) A square wire loop of dimensions $L \times L$ oriented in the xy-plane enters a region where the magnetic field is first oriented in the $+z$- and then in the $-z$-direction (Fig. 31-31). The width of each region is L. The loop moves at speed v in the $+x$-direction. Find the emf, sign and

magnitude, induced in the loop as it enters and passes through the regions with the magnetic field.

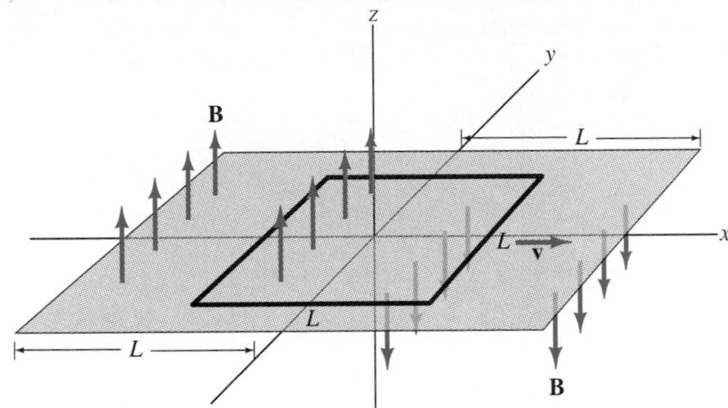

FIGURE 31–31 Problem 4.

5. (II) A long, straight wire oriented in the y-direction carries a current of 0.50 A. A square loop with sides of length 1.0 cm is in the xy-plane with its nearest edge 30 cm from the wire (Fig. 31–32). In a time of 0.10 s the square loop moves uniformly 10 cm closer to the wire. What is the emf induced in the loop while it is moving? Ignore the variation in the wire's magnetic field *across* the loop.

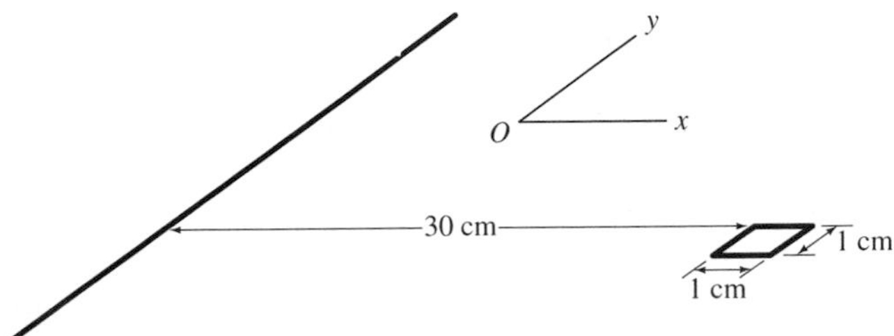

FIGURE 31–32 Problem 5.

6. (II) What is the peak emf produced by a 60-turn square coil 25 cm on each side, rotating on a diagonal axis with a frequency of 10 Hz in a magnetic field of 0.50 T perpendicular to the axis?

7. (II) A coil with 125 turns, a radius of 2.0 cm, and a resistance of 3.0 Ω is rotating about a diameter in a uniform magnetic field of 0.50 T. How fast must it rotate to produce a maximum current of 6.0 A in the coil?

8. (II) There is a constant magnetic field $\mathbf{B} = B_0(\mathbf{i} + \mathbf{j} + \mathbf{k})$ in the region $x > 0$, $y > 0$, $z > 0$. A square loop of dimensions $L \times L$ whose sides are parallel to the x- and y-axes moves with constant velocity $\mathbf{v} = v_0(\mathbf{i} + \mathbf{j})$ in the xy-plane such that its center moves along the line $x = y$. Calculate the emf induced in the loop, given that its leading corner passes the origin at the time $t = 0$.

9. (II) A vertical loop rotates with angular velocity ω as shown in Fig. 31–33. At time $t = 0$ it is aligned perpendicular to a constant magnetic field oriented in the x-direction. Use Lenz's law to find the direction of the emf induced in the loop at $t = 0$, $t = T/4$, $t = T/2$, and $t = 3T/4$, where T is the rotation period of the loop.

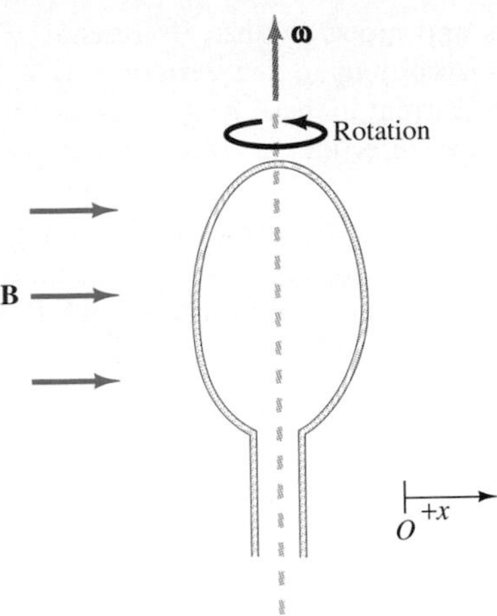

FIGURE 31–33 Problem 9.

10. (II) A closed loop is constructed of a fixed U-shaped wire and a crossbar free to move in the x-direction, all in the xy-plane. The end of the loop is at $x = 0$. A constant magnetic field is oriented in the z-direction, $\mathbf{B} = B_0\mathbf{k}$. The situation and the relevant dimensions are as in Fig. 31–11a, except that here the magnetic field is constant. Suppose that the movable crossbar is pulled at a constant speed v to the right, starting at $x = 0$ when $t = 0$. Its position at any time is $x = vt$. If the resistance of the loop varies with the total length L according to $R = \alpha L$, what is the current in the loop as a function of time? Compare your answer with Example 31–4, and explain any differences.

11. (II) Let the magnetic field in Example 31–4 be a constant field oriented in the y-direction, $\mathbf{B} = B_0\mathbf{j}$. Find the induced current as a function of time.

12. (II) Let the magnetic field in Example 31–4 vary linearly with y and be oriented in the z-direction, $\mathbf{B} = Cy\mathbf{k}$. Find the induced current as a function of time.

13. (II) A metal ring is constructed so as to expand or contract freely. In a region with a constant magnetic field $\mathbf{B}_0$ oriented perpendicular to it, the ring expands, with its radius growing linearly with time as $r = r_0(1 + \alpha t)$ (Fig. 31–34). As the ring

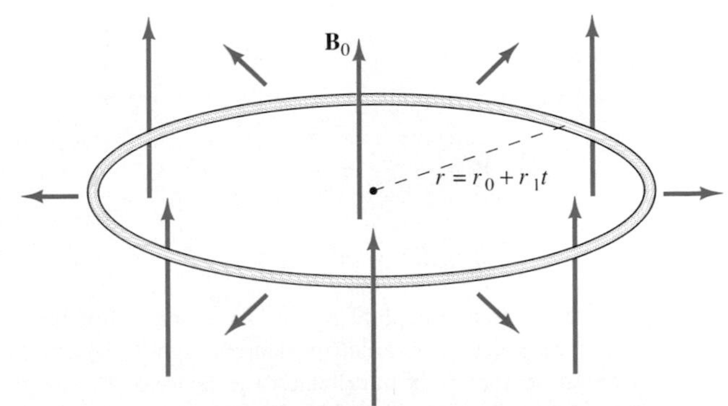

FIGURE 31–34 Problem 13.

expands and grows thinner, its resistance *per unit length* changes according to the empirical rule $R = R_0(1 + \beta t)$. Find the current induced in the ring as a function of time. Specify the direction as well as the magnitude of the current.

14. (II) A circular loop of area A rotates with angular frequency ω about its vertical diameter. The rotating loop is placed in a horizontal constant magnetic field, B. What is the emf induced in the loop?

15. (II) Work Example 31–2 by direct computation of the magnetic flux through the hemispherical surface.

31–3 *Motional EMF*

16. (I) The spacecraft *Voyager I* is moving through interstellar space, where the magnetic field is 10^{-10} T. Assume that *Voyager I* has an antenna 3 m long. If the spacecraft moves so that the antenna rod is perpendicular to the magnetic field when *Voyager I* has a speed of 10^4 m/s, what is the emf induced across the antenna?

17. (I) A 747 is flying due north at 900 km/h in a location where the earth's magnetic field consists of an upward vertical component of 2×10^{-5} T and a southward component of 3×10^{-5} T. If the wingtip-to-wingtip length of a 747 is 35 m, find the emf induced across the wings. If the airplane were flying due east instead of due north, how would your answer change?

18. (II) A metal disk 30 cm in diameter rotates about its axis of symmetry at an angular speed of 450 rad/s. The disk is situated in a uniform magnetic field of 4×10^{-2} T perpendicular to the plane of the disk. What is the induced voltage between the axis and the rim of the disk?

19. (II) A metal bar of length 0.3 m is moved to the right at a speed of 0.1 m/s. The bar makes an angle of 45° with respect to its direction of motion. It is passing through a region of uniform magnetic field of magnitude 10^{-2} T oriented perpendicular to the plane swept out by the bar (out of the page in Fig. 31–35). What is the potential difference between the two ends of the bar as it moves through the magnetic field?

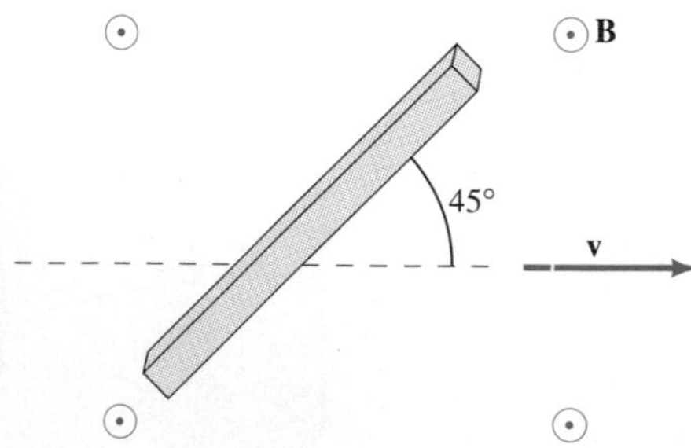

FIGURE 31–35 Problem 19.

20. (II) A circular metal plate moves as a pendulum bob between the poles of a tabletop electromagnet. The plate is oriented so that it is parallel to the faces of the magnet. Describe qualitatively the eddy currents induced in the plate as it moves.

21. (II) A rod of length L moves at constant speed v into the region between the poles of a horseshoe magnet, where there is a constant magnetic field perpendicular to the rod in a circular region (Fig. 31–36). $L = 2R$, the radius of the circular region. What is the emf induced in the rod as a function of time?

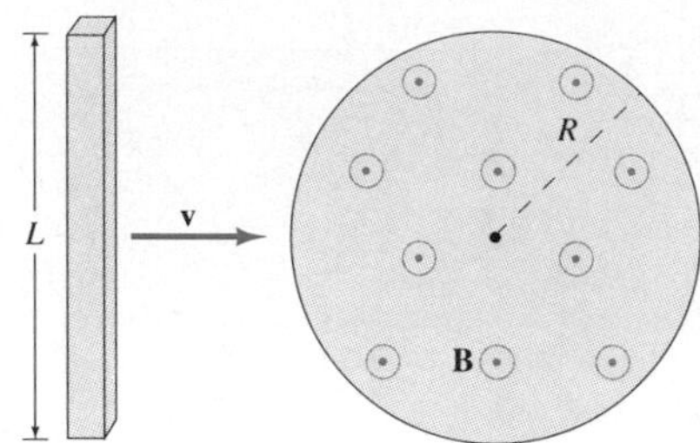

FIGURE 31–36 Problem 21.

22. (II) A *rotating coil* is a common device for measuring magnetic fields. Consider a coil of area A and N turns that is rotated at angular frequency ω in a magnetic field. The position of the coil is adjusted so as to produce a maximum induced current I_{max}, which can be measured by using an appropriate ammeter. R is the total resistance of the coil circuit. Find the relationship between the unknown magnetic field and I_{max}.

23. (II) If the rotating coil of Problem 22 is used with the split-ring commutator described in Chapter 29, DC current can be measured with a sensitive galvanometer. Find the relationship between the unknown magnetic field and the average measured DC current, I.

24. (II) A long, straight wire carries a current of 20 A. A thin metal rod 10 cm long is oriented perpendicular to the wire and moves with a speed of 1.5 m/s in a direction parallel to the wire. What are the size and direction of the emf induced in the rod if the nearest point of the rod is 1 cm away from the wire, and if the rod moves in a direction opposite to the current?

31–4 *Forces, Energy, and Power in Motional EMF*

25. (II) A square wire loop of dimensions $L \times L$ lies in a plane perpendicular to a constant magnetic field. The field exists only in a certain region, with a sharp boundary (Fig. 31–37). The sides of the loop make a 45° angle with this boundary,

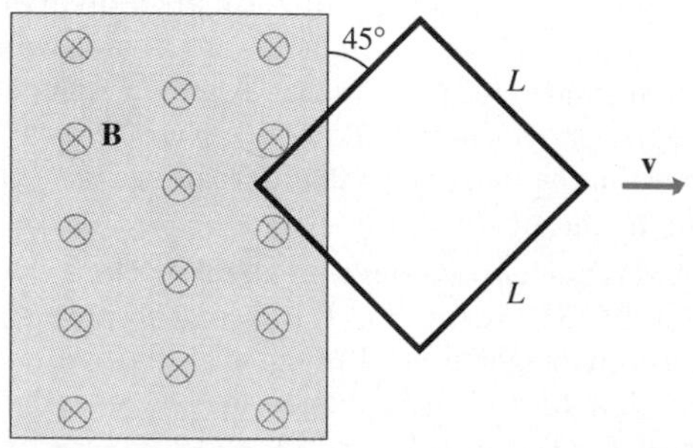

FIGURE 31–37 Problem 25.

and an external force moves the loop at a speed v out of the region of constant field. How much power must be supplied by the external force as a function of time?

26. (II) A conducting bar slides frictionlessly on two parallel horizontal rails 75 cm apart. The bar and rails form a closed circuit with a resistor of resistance 0.1 Ω, assumed to be constant throughout the motion. The circuit is placed in a uniform vertical magnetic field of 0.15 T perpendicular to the circuit's plane. The bar is pulled at a constant speed of 40 cm/s along the rails. (a) What is the magnitude of the force required to pull the bar? (b) What is the rate of Joule heating in the resistor?

27. (II) A long, straight wire carries a constant current I_0. A square loop with sides of length L and two sides parallel to the wire is pulled away at uniform speed v in a direction perpendicular to the wire. The nearest side of the loop is initially a distance D from the wire; the resistance of the loop is R. (a) Calculate the force necessary to pull the loop. (b) At what rate is work being done by the force? (c) How does your answer to part (b) compare with the Joule heating in the loop?

28. (II) In Example 31–6, what happens if the initial speed of the loop is (a) less than v_t, and (b) greater than v_t?

29. (II) Household electricity meters are based on the principle that the drag force due to eddy currents induced when a moving piece of metal passes through a magnetic field is proportional to the velocity of the metal. In this case a thin metal disk rotates, driven by a motor whose torque is proportional to the power consumed in the house. The disk passes between the poles of a magnet. Show that in general the equilibrium velocity of the disk is a measure of the power consumed. The total number of turns of the disk is thus equivalently a measure of the energy consumed.

31–5 *Effects of Time-Varying Magnetic Fields*

30. (I) A long, straight wire carries a current $I = I_0 \cos(120\pi t)$, where t is time. Two sides of a fixed rectangular loop are 50 cm long and are parallel to the wire; the other sides are 2 cm long. The nearest long side is 10 cm from the wire. What is I_0 if the maximum emf induced in the loop is 5 μV? (Ignore the small variation of the magnetic field across the loop.)

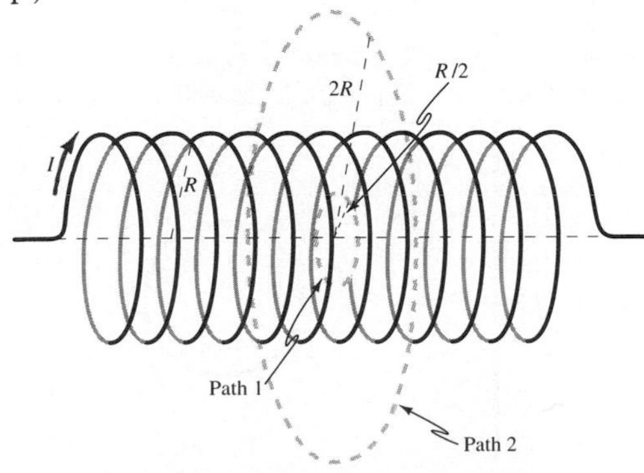

FIGURE 31–38 Problem 31.

31. (II) A long solenoid of radius R and n turns per unit length carries an alternating current $I = I_0 \sin(\omega t)$ (Fig. 31–38). What are the electric fields induced within the solenoid at a distance $R/2$ and outside the solenoid at a distance $2R$? [*Hint:* Apply Faraday's law to the two paths shown, and use symmetry.]

32. (II) A very long cylindrical solenoid of radius r made from n turns of wire per unit length carries a current with the time dependence $I = I_0 e^{-t/t_0}$. Coaxial with and surrounding the solenoid are two turns of wire that make a circular loop slightly larger than the circular cross section of the solenoid (Fig. 31–39). The loop with two turns is far from the ends of the solenoid and has a resistance R. Find the current in the loop with two turns, I', as a function of time.

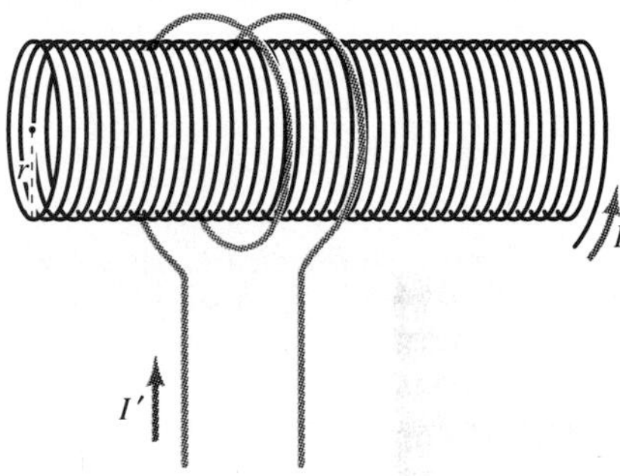

FIGURE 31–39 Problem 32.

33. (II) The uniform magnetic field of the electromagnet of Example 31–7, with circular pole faces of radius $R = 0.5$ m, decreases linearly from 2.0 T to 1.2 T in 1.5 ms. What is the emf induced around the path drawn in Fig. 31–40 that consists of quarter arcs at radial distances $R/4$ and $R/2$, connected by radial lines? The path is clockwise.

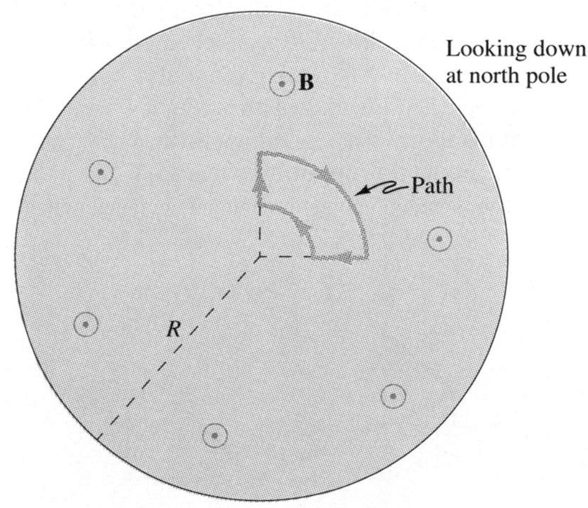

FIGURE 31–40 Problem 33.

34. (II) A solenoid of radius r wound with n turns per unit length carries a current given by $I = I_0 \cos(\omega t)$, where t is the time. What are the magnitude and direction of the induced electric field just outside the solenoid?

31–6 *Generators and Motors*

35. (I) A coil of area 5.0 cm^2 with 50 turns of wire is connected to a resistor of resistance 50 Ω. It is rotated by hand at a frequency of 1.0 rev/s in a magnetic field of 0.50 T. (a) What is the maximum amount of current produced? (b) the average power produced?

36. (II) You have 25 m of wire, a constant magnetic field of 0.15 T, and a device that can rotate a coil at a fixed frequency of 90 Hz. What size circular coil will produce an AC emf of maximum voltage 120 V?

37. (II) A bicycle wheel of radius $R = 35$ cm rotates at angular speed 65 rad/s in a plane perpendicular to a constant magnetic field of magnitude 0.20 T. What is the emf generated between the center of the wheel and its rim? When one end of a wire is attached to the center and the other end to a circular track in contact with the rim, a direct current is generated in the wire. Such a device is called a *homopolar generator.*

*31–7 *The Connection between Electric and Magnetic Fields in Moving Reference Frames*

38. (II) Suppose that observer O sees an electric field $\mathbf{E} = E\mathbf{i}$ and a magnetic field $\mathbf{B} = B\mathbf{k}$. In what direction and at what (constant) speed u should a second observer move so as to see no electric field whatsoever? Use the nonrelativistic relation Eq. (31–22). If $E = 10^3$ V/m, for what range of values of B is the nonrelativistic approximation appropriate?

General Problems

39. (II) A 50-cm-long wire of square cross section with a mass of 6 g and a resistance of 1.0 Ω slides without friction down parallel conducting rails of negligible resistance (Fig. 31–41). The rails are connected to each other at the bottom by a resistanceless rail parallel to the wire, so that the wire and rail form a closed rectangular conducting loop. The plane of the rails makes an angle of 45° with the horizontal, and a uniform vertical magnetic field of 0.50 T, pointing upward, exists throughout the region. What is the steady speed of the wire?

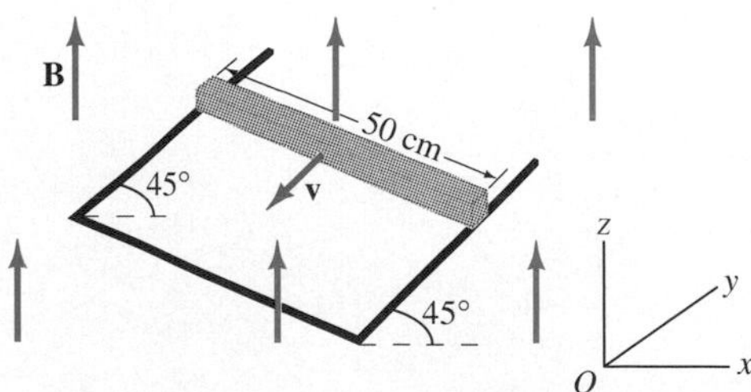

FIGURE 31–41 Problem 39.

40. (II) A straight wire carries a current $I = 10$ A near a rod that moves across two conducting wires (Fig. 31–42). The resistor has $R = 5.0$ Ω, and the rod moves at speed 3.0 cm/s. (a) What is the emf induced in the rod? (b) What is the current in the circuit? (c) How much work is done to move the rod 10 cm to the right? What force does this work?

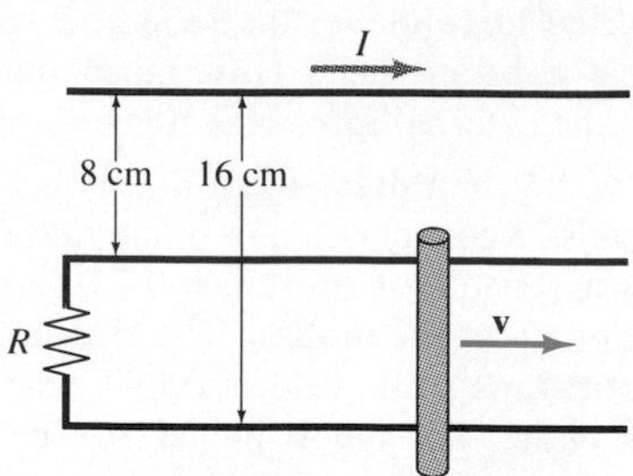

FIGURE 31–42 Problem 40.

41. (II) A large, circular coil of N turns and radius R carries a steady current I and is rotated at a constant angular speed ω about a horizontal diameter. At the center of this coil is a small, fixed, horizontal circular ring of radius r. (a) What is the emf induced in the small ring? (b) What is the angle between the plane of the coil and that of the ring when this emf is a maximum?

42. (II) Consider a 6.00-V battery attached to two conducting, frictionless rails 0.100 m apart. There is a magnetic field **B** of magnitude 0.300 T perpendicular to the rails, and a conducting bar can slide over the rails perpendicular to them as well as to the field (Fig. 31–43). The bar is placed on the rails, starts from rest, and accelerates. (a) What is the direction of its motion? (b) the direction of the emf induced? (c) Given that the total resistance of the closed circuit is 0.200 Ω, calculate the current in the bar when its speed is 1.00 m/s.

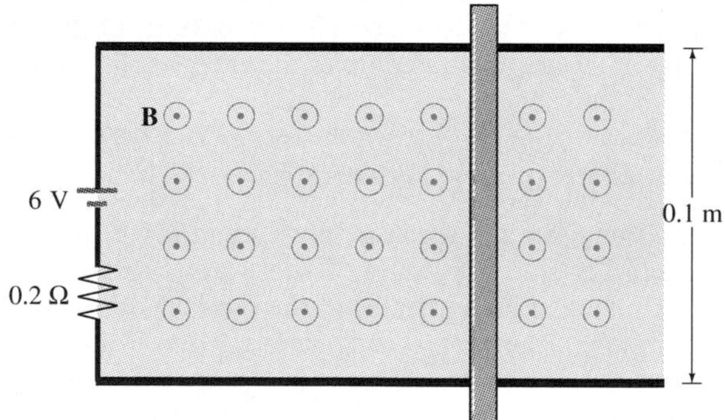

FIGURE 31–43 Problem 42.

43. (II) A wire carrying a current I is oriented in a vertical direction. To its side is a wire loop oriented so that it and

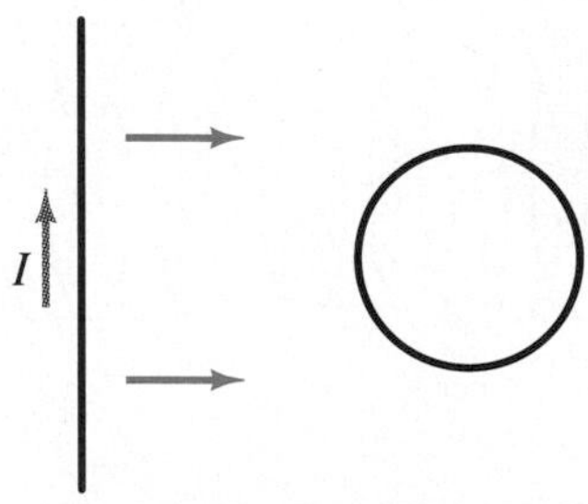

FIGURE 31–44 Problem 43.

the straight wire lie in the same vertical plane (Fig. 31–44). The straight wire is moved toward the loop. If a current is induced in the loop, what is its direction, and what is the direction of the force on the loop?

44. (II) If the plasma in a magnetohydrodynamic generator (see Question 12) is forced to flow in a channel perpendicular to a magnetic field, an electric potential builds up between points a and b, which are 1 m apart (Fig. 31–45). If the magnetic field has a strength 2.5 T, what must the speed of the plasma be in order that the potential be 1000 V?

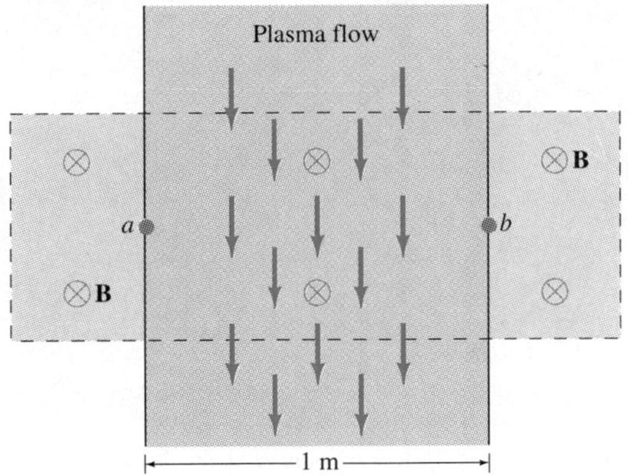

FIGURE 31–45 Problem 44.

45. (II) A coil with 300 turns, a diameter of 10 cm, and a resistance of 20 Ω is placed perpendicular to a uniform magnetic field of 1.0 T. The magnetic field suddenly reverses direction. What is the total charge that passes through the coil?

46. (II) A constant magnetic field of 0.50 T is directed along the x-axis. A wire coil of 30 turns and area 10 cm^2 is placed in the yz-plane. The coil of wire, called a *flip coil*, is then turned over (in other words, rotated by 180°). (a) If the total charge that passes through the coil when it is flipped is 0.030 C, what is the resistance of the coil circuit? (b) The same flip coil is used to measure an unknown magnetic field. The coil is flipped in several directions until it attains its maximum charge of 0.045 C, when the coil is flipped with its face in the xy-plane. What is the magnitude of the magnetic field? (c) What is the direction of the magnetic field in part (b)?

47. (II) A current $I = I_0 \cos(\omega t)$ passes through a solenoid of area 10 cm^2 and 10^5 turns/m. The frequency is 60 Hz, and $I_0 = 10$ A. A small coil—a sense coil—is used to sense the changing flux. This sense coil has an area of 20 cm^2 with 10 turns and is placed across the solenoid so that the face of the coil is perpendicular to the solenoid axis; the two coils are concentric (Fig. 31–46). (a) What is the emf induced in the sense coil? (b) If the resistance of the sense coil circuit is 5 Ω, what is the current?

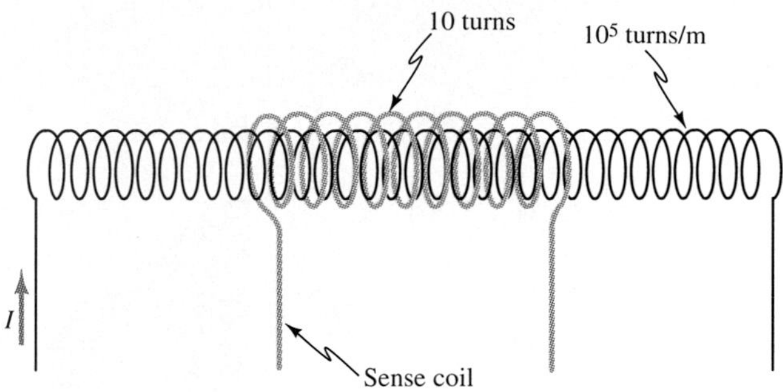

FIGURE 31–46 Problem 47.

48. (II) A wire that is bent into a semicircle is rotated with angular velocity ω about the diameter shown in Fig. 31–47. The bent wire and its supports are placed in a uniform magnetic field perpendicular to the plane of the supports. What is the emf induced in the circuit shown? If the resistance of the closed loop is R, what is the average power dissipated?

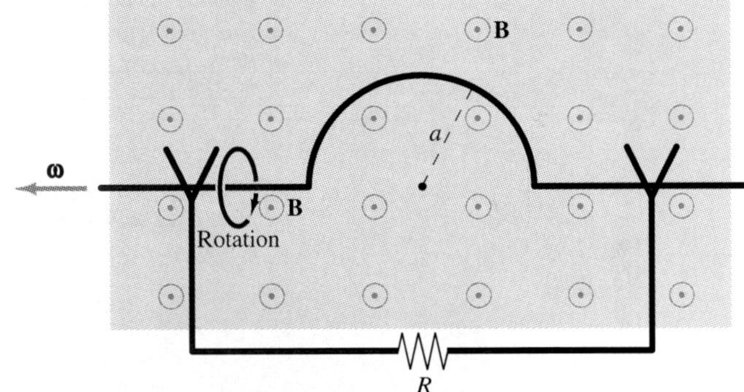

FIGURE 31–47 Problem 48.

49. (III) An electron follows a circular path of radius $R = 1$ m while traveling in a plane perpendicular to a spatially constant magnetic field of magnitude 10^{-6} T. As viewed along the magnetic field lines, the electron follows a counterclockwise path. (a) What is the speed of the electron? (b) Assuming that the motion of the electron is nonrelativistic, what is the energy, E, of the electron? (c) The magnitude of the magnetic field is reduced smoothly by a certain percentage during an interval Δt. Show that the fractional energy change of the electron, $\Delta E/E$, is independent of the radius of the electron's orbit as well as of the electron's speed. (This effect is the basis for low-energy operation of a particle accelerator called the *betatron*.) (d) If the magnetic field is reduced in time $\Delta t = 5$ s by 10 percent, estimate $\Delta E/E$.

CHAPTER 32

Large magnets have a ferromagnetic core. Here such a magnet is used to lift scrap metal in a junkyard.

MAGNETISM AND MATTER

Why can a slab of soft iron be turned into a bar magnet, whereas a piece of aluminum cannot? Why do lodestones (made of magnetite) have magnetic fields? Why will magnets pick up needles but not pieces of paper or plastic? The answers to these questions and many more of great technological import lie in understanding the magnetic properties of matter. This understanding is necessary for the construction of computer memories, electric motors, generators, transformers, particle accelerators, and medical scanners. Just as the dielectric properties of materials depend on the polarizability of atoms and molecules, the magnetic properties of materials depend on the magnetic properties of atoms and molecules. In this chapter we shall relate the different types of magnetic behavior exhibited by bulk materials to the atoms that make up these materials. We shall describe the mechanism responsible for ferromagnetism, a characteristic of permanent magnets. Ferromagnetic materials represent the most important of three major classes of magnetic materials. We shall explore the magnetic properties of superconductors and describe the atomic origin of nuclear magnetic resonance, an important tool in materials science and in medicine.

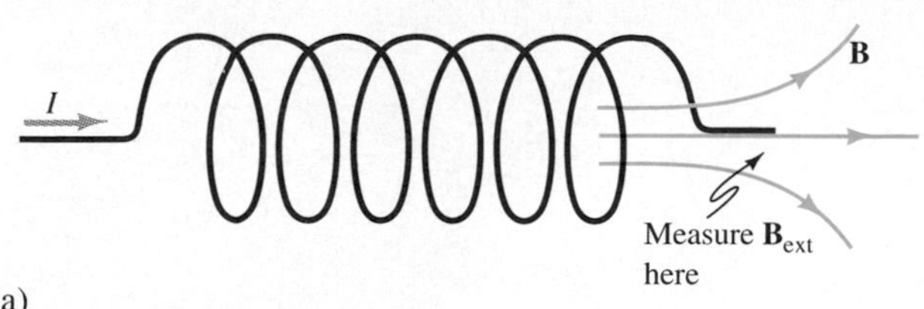

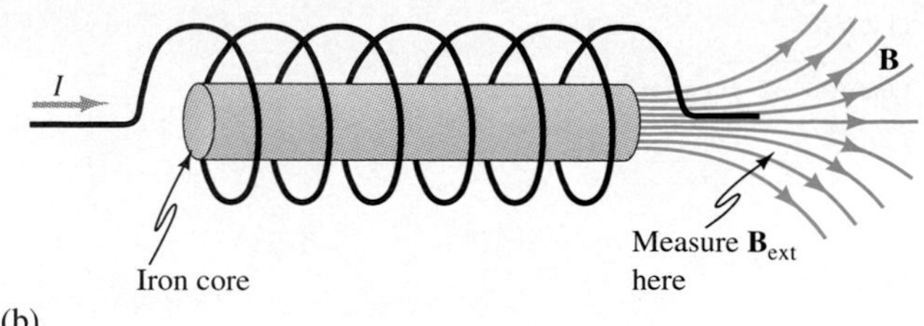

FIGURE 32–1 (a) A solenoid that carries a current has a magnetic field. (b) When the cylindrical volume of the solenoid is filled with an iron core, the magnetic field may be greatly amplified.

32–1 THE MAGNETIC PROPERTIES OF BULK MATTER

Suppose that we measure the magnetic field $\mathbf{B}_{ext}$ produced by, and near one end of, a current-carrying solenoid (Fig. 32–1a). Next we insert a wooden or copper core into the solenoid and repeat the measurement. Only a very precise measurement would show a change of one part in 10^6 in the measured field. Now we replace that core by a soft iron core (Fig. 32–1b). The measured field is increased by a factor of many thousands from the original field. Furthermore, if the iron core is removed, the iron will then act like a bar magnet, even if it showed no such properties before it was inserted. The last observation shows that the solenoid somehow magnetizes the soft iron core, but it is not able to do so for the wooden or copper cores. Materials that can form permanent magnets, such as iron, are called **ferromagnetic**, whereas materials that display magnetic properties only in the presence of external magnetic fields are *nonferromagnetic*.

Ferromagnetism

The bulk magnetic behavior of materials—gases and liquids as well as solids—is characterized by their **magnetization**, **M**, which we define as the *net magnetic dipole moment per unit volume*. The field outside the material is that of a magnetic dipole, the same as that of a bar magnet. Because a magnetic moment has dimensions of current times area, magnetization has dimensions of current times area divided by volume, or current per length. The units of **M** are amperes per meter (A/m). Although the magnetization can vary from point to point within a material, we limit ourselves here to uniform magnetization.

The net magnetic field of a solenoid S_1 with an iron core inside is the vector sum of contributions from the external magnetic field of the solenoid, $\mathbf{B}_{ext}$, and the contribution of the magnetization of the core. The core may itself be viewed as a second solenoid, S_2, because any magnetic dipole is equivalent to a solenoid. But we can express the magnetic field of S_2 in terms of its magnetic moment per unit volume. Suppose that S_2 can be treated as a solenoid of area A, length L, and N loops that carry current I. From Eq. (30–15), the magnetic field of S_2 is

The magnetic field of a solenoid was given in Chapter 30.

$$B_2 = \frac{\mu_0 NI}{L} = \frac{\mu_0 NIA}{LA} = \mu_0 \frac{m_2}{V},$$

where m_2 is the magnetic moment of S_2, $m_2 = NIA$,† and μ_0 is the permeability of free space. The factor m_2/V is the magnetic moment per unit volume—the magnetization, **M**. Thus *the contribution of the core to the total magnetic field is* $\mu_0\mathbf{M}$, and the total field of the solenoid with the core is

$$\mathbf{B} = \mathbf{B}_{ext} + \mu_0\mathbf{M} \tag{32–1}$$

The net magnetic field in the presence of a material with magnetization **M**

† To avoid confusion with the magnetic permeability, μ, to be introduced below, we use the notation m for the magnetic dipole moment throughout this chapter. Do not confuse it with a mass!

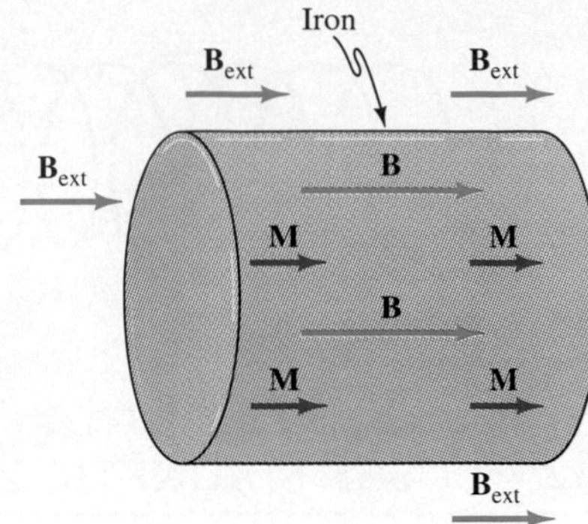

FIGURE 32–2 The net magnetic field, **B**, inside a material is different from the external field, $\mathbf{B}_{\text{ext}}$. The magnetization, **M**, which represents the effect of the magnetic properties of the material's constituents, contributes to the net field inside.

(Fig. 32–2). Of course, a solenoid is not required: Eq. (32–1) applies whenever $\mathbf{B}_{\text{ext}}$ is any external field in which ferromagnetic material is placed. This situation is similar to that of dielectrics, where an interior electric field has contributions from an external electric field due to free charges and from an internal, induced distribution of charge. For that case, it is convenient to separate the effects of internal, induced charges from free charges. We saw in Chapter 26 that this is done through the introduction of a dielectric constant and a generalized permittivity. Here we want to separate the effects of the internal magnetization from the external magnetic fields due to ordinary currents. (Here the ordinary, or external, current is that of solenoid S_1.) We refer to such currents as *free*, or *real*, currents. The effects of free currents are isolated by defining the **magnetic intensity**, **H**, a quantity that depends on the *difference* between a term that involves the net internal field and the magnetization of the material:

The definition of magnetic intensity

$$\mathbf{H} \equiv \frac{\mathbf{B}}{\mu_0} - \mathbf{M}. \tag{32–2}$$

The dimensions of **H** are *not* those of **B** but of **M**. By replacing **B** in Eq. (32–2) with $\mathbf{B}_{\text{ext}} + \mu_0\mathbf{M}$ according to Eq. (32–1), we see that the external magnetic field is related to the magnetic intensity in a material by

$$\mathbf{B}_{\text{ext}} = \mu_0\mathbf{H}. \tag{32–3}$$

The magnetic intensity isolates the effects of free currents.

Equation (32–3) shows that no matter what the material, whether ferromagnetic, nonferromagnetic, or even empty space, *the magnetic intensity measures the magnetic field due to free currents.* Another form for the relation among **B**, **H**, and **M** is found by combining Eq. (32–1) and (32–3):

$$\mathbf{B} = \mu_0\mathbf{H} + \mu_0\mathbf{M}. \tag{32–4}$$

Nonferromagnetic materials have no magnetization unless there is an external magnetic field $\mathbf{B}_{\text{ext}}$ that induces it. We know by experiment that, over a large range of conditions, *the magnitude of the magnetization of nonferromagnetic materials varies linearly with the external magnetic field.* The direction of the magnetization is more complicated: The original field $\mathbf{B}_{\text{ext}}$ and the field $\mu_0\mathbf{M}$ due to the magnetization are parallel for one class of materials but antiparallel for a second class. For nonferromagnetic materials **M** depends linearly on $\mathbf{B}_{\text{ext}}$. Therefore **M** also has a linear dependence on **H**. The **magnetic susceptibility**, χ_m, is defined as the coefficient of the linear relation between the magnetization and the magnetic intensity:

The definition of magnetic susceptibility

$$\mathbf{M} \equiv \chi_m\mathbf{H}. \tag{32–5}$$

Thus a material with a large magnetic susceptibility has a large magnetization in the presence of an external field, and one with a small magnetic susceptibility has a small magnetization. If the susceptibility of a material is positive, its magnetization is aligned along the external field; if the susceptibility is negative, the magnetization is aligned opposite to the external field. The definition of the susceptibility holds for any material, ferromagnetic or otherwise, but for nonferromagnetic materials, χ_m is to a good approximation constant over a wide range of external magnetic fields. Because **M** and **H** have the same dimensions, χ_m is dimensionless. Table 32–1 gives a range of susceptibilities found in nature.

With our definition of χ_m, we can express the relation between the magnetic field in a material and the magnetic intensity. From Eq. (32–1),

$$\mathbf{B} = \mathbf{B}_{ext} + \mu_0\mathbf{M} = \mu_0\mathbf{H} + \mu_0\chi_m\mathbf{H}$$
$$= \mu_0(1 + \chi_m)\mathbf{H}. \tag{32–6}$$

We define the coefficient of **H** in this equation as the **permeability**, μ, of the material:

$$\mu \equiv \mu_0(1 + \chi_m). \tag{32–7}$$

The relation between the total magnetic field in a material and the magnetic intensity, which is a measure of the effect of free currents, is then

$$\mathbf{B} = \mu\mathbf{H}. \tag{32–8}$$

Just as the electric permittivity, ϵ, replaces the permittivity of free space, ϵ_0, in expressions for electric fields in materials if the charge is free charge, so μ replaces μ_0 when the current is the free current in materials. From Eq. (32–3) we see that when there is a magnetic field in a vacuum, that field is related to the intensity by a relation like that of Eq. (32–8), but with μ_0 appearing in the place of μ; thus μ_0 is the permeability of the vacuum. As we can deduce from Table 32–1, μ is very close to μ_0 for nonferromagnetic materials.

The various quantities we defined above are all useful in characterizing the bulk magnetic behavior of materials. In Table 32–2 we summarize these quantities, their physical significance, and their most useful relations.

TABLE **32–1**

SOME MAGNETIC SUSCEPTIBILITIES (at 20°C unless indicated otherwise)

Material	*Susceptibility,* χ_m
Diamagnetic	
Water	-9.1×10^{-6}
Copper	-9.6×10^{-6}
Silver	-2.4×10^{-5}
Carbon (diamond form)	-2.2×10^{-5}
Bismuth	-1.7×10^{-4}
Paramagnetic	
Sodium	7.2×10^{-6}
Cupric oxide	2.6×10^{-4}
Aluminum	2.2×10^{-5}
Liquid oxygen (90 K)	3.5×10^{-3}
Ferromagnetic	
Iron (annealed)	5.5×10^{3}
Permalloy (55% Fe, 45% Ni)	2.5×10^{4}
Mu-metal (77% Ni, 16% Fe, 5% Cu, 2% Cr)	1×10^{5}

TABLE **32–2**

MAGNETIC BULK PROPERTIES AND THEIR RELATIONS

Symbol	*Property*
$\mathbf{B}_{ext}$	External magnetic field, produced independently of type of material by a nearby magnet or currents
H	Magnetic intensity, proportional to the external magnetic field
M	Magnetization, the magnetic dipole moment per unit volume of a material
B	Net magnetic field, the sum of the external magnetic field and a term proportional to the magnetization
μ_0	Permeability of free space
χ_m	Susceptibility of a material
μ	Permeability of a material, $\mu = \mu_0(1 + \chi_m)$

Some Relations

	$\mathbf{B}_{ext}$	**H**	**M**	**B**
$\mathbf{B}_{ext} =$	—	$\mu_0\mathbf{H}$	Not used	Not used
$\mathbf{H} =$	$\mathbf{B}_{ext}/\mu_0$	—	$\mathbf{M}/\chi_m$	$\mathbf{B}/\mu$
$\mathbf{M} =$	$\chi_m\mathbf{B}_{ext}/\mu_0$	$\chi_m\mathbf{H}$	—	Not used
$\mathbf{B} =$	$(1 + \chi_m)\mathbf{B}_{ext}$	$\mu_0(1 + \chi_m)\mathbf{H}$	$\frac{\mu_0(1 + \chi_m)}{\chi_m}\mathbf{M}$	—

EXAMPLE 32–1 A straight solenoid of diameter 5 cm and length 25 cm is wrapped with 200 turns of wire that carries a current of 5 A. The solenoid is filled with a material of magnetic susceptibility $\chi_m = 10^{-5}$. Find (a) the magnetic intensity within the solenoid and (b) the magnetic field within the solenoid. (c) By what factor is the magnetic field changed due to the presence of the material?

SOLUTION: (a) The magnetic intensity is associated with the free currents of the solenoid. It is found from Eq. (32–3) to have magnitude

$$H = \frac{B_0}{\mu_0} = \frac{\mu_0 n I}{\mu_0} = nI,$$

where we have used Eq. (30–15) for the interior field of a solenoid. Here n is the number of turns per unit length:

$$n = \frac{200 \text{ turns}}{0.25 \text{ m}} = 800 \text{ turns/m}.$$

Thus

$$H = (800 \text{ turns/m})(5 \text{ A}) = 4000 \text{ A/m}.$$

(b) The total magnetic field, **B**, includes the effect of the field with that of the material that fills the solenoid. B can be found from, for example, Eqs. (32–8) and (32–7):

$$\begin{aligned} B &= \mu H = \mu_0(1 + \chi_m)H \\ &= (4\pi \times 10^{-7}\ \text{T}\cdot\text{m/A})(1 + 10^{-5})(4000\ \text{A/m}) = 5 \times 10^{-3}\ \text{T}. \end{aligned}$$

(c) The factor by which the field changes is

$$\frac{\Delta B}{B} = \frac{\mu_0(1 + \chi_m)H - \mu_0 H}{\mu_0 H} = \chi_m.$$

The susceptibility is small, so the change in the field is also small. The insertion of a material such as copper or silver inside a solenoid hardly changes the magnetic field, because these materials have small magnetic susceptibilities (see Table 32–1).

The Magnetic Properties of Materials

Table 32–1 reveals that materials break down into three broad classes. The first class is comprised of ferromagnetic materials, which have large positive susceptibilities. As we have already described, these substances can form permanent magnets.

Diamagnetism

Substances with very small negative susceptibilities are called **diamagnetic** materials. In such materials the magnetization direction is *opposite* to the direction of the inducing field. The magnetic field inside such materials is *reduced* from its value outside the material. If a diamagnetic material is placed near the north pole of a magnet, the magnetization produces a field that points toward the pole (Fig. 32–3). The diamagnetic material acts as though it has a north pole adjacent to the external north pole: The diamagnetic material is *repelled* by the magnet. The behavior of diamagnetic substances is similar to that of dielectrics, for which polarization effects tend to cancel the electric field associated with free charges.

Paramagnetism

Substances with small positive susceptibilities are called **paramagnetic** materials. For them, the external magnetic field aligns the atomic magnetic dipole moments parallel to itself. The magnetization points in the same direction as the

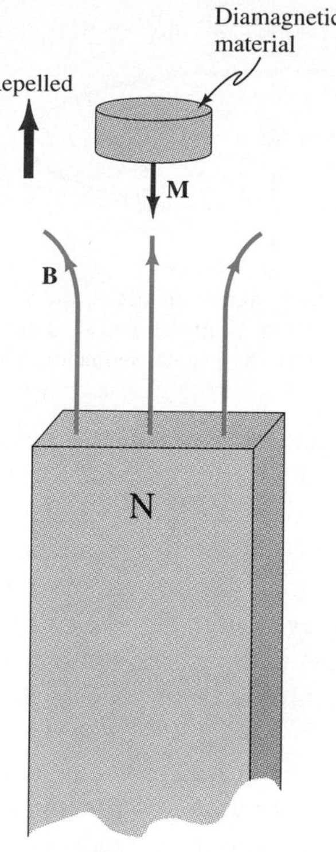

FIGURE 32–3 Diamagnetic substances are repelled by one pole of a nearby bar magnet.

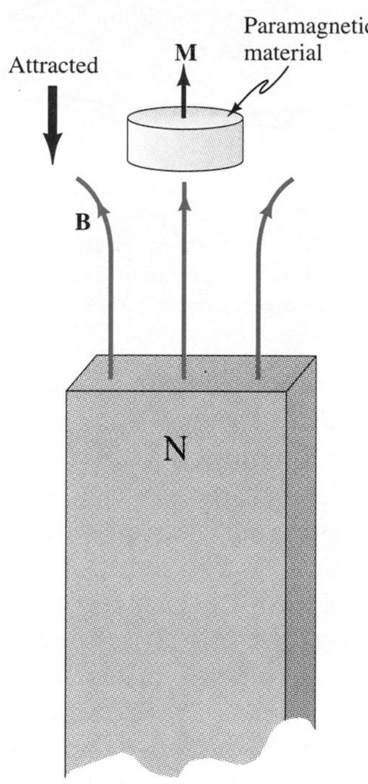

FIGURE 32–4 Paramagnetic substances are attracted to one pole of a nearby bar magnet.

field of an external magnet (Fig. 32–4), and it is as if the paramagnetic substance has a south pole oriented toward the magnet's north pole: the piece of paramagnetic material is *attracted* to the magnet. Ferromagnetism, diamagnetism, and paramagnetism will be discussed further in Sections 32–3 to 32–5, respectively.

32–2 ATOMS AS MAGNETS

The magnetic field of a ring of current is the same as that of a bar magnet. How can two such apparently dissimilar systems have the same magnetic field? The answer lies in the atomic structure of matter and in the magnetic properties of atoms. A complete understanding of the magnetic properties of atoms requires quantum physics, the subject of Chapter 41. It is, however, possible to understand the magnetic properties of atoms qualitatively by starting with a classical planetary model and by adding to this model some features of the quantum mechanical description of atoms. The orbiting electrons in atoms form currents, and the atoms then act as magnetic dipoles. Thus we expect atoms to have magnetic dipole moments. The magnetic moments in a large assembly of atoms point in random directions, and thus the net magnetic moment of a macroscopic material is generally zero. The exception occurs in ferromagnetic materials, in which the atomic magnetic moments tend to line up with their neighbors. Ferromagnetic materials can be used to make permanent magnets. It is possible to destroy "permanent" magnetic effects by heating a magnetic material or by pounding it with a hammer. Such action tends to randomize the collective atomic alignments.

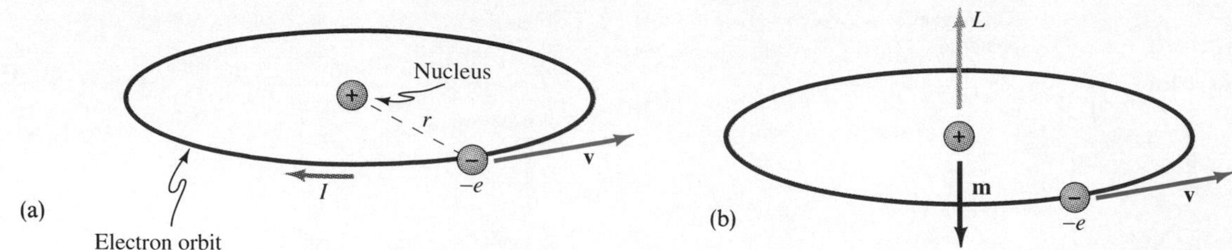

FIGURE 32–5 (a) An electron in a circular orbit around a nucleus. Over numerous orbits, the effect is the same as a continuous current ring about the nucleus. (b) The circulating electron forms a magnetic dipole with magnetic moment **m** ($\mathbf{m}_{\text{orbital}}$) oriented downward, opposite to the angular momentum, **L**.

The magnetic properties of individual atoms are affected by a tendency of their electrons to pair off such that the magnetic moments due to their orbits point in opposite directions. The net result is that a pair of electrons has no magnetic moment. In effect, for each electron with a clockwise orbit around the nucleus, there is an electron with the same orbit in a counterclockwise sense. We then expect that, in terms of orbital motion, atoms with an even number of electrons will have no magnetic dipole moment, whereas in atoms with an odd number of electrons, only a last unpaired electron matters. This simplification persists even when quantum mechanics is included.

An electron acts as a tiny magnet.

As we saw above, electrons contribute to the magnetic dipole moment of atoms through their orbital motion. But electrons also carry an *internal* magnetic moment that is not due to any current structure. *The electron itself behaves as a tiny magnet!* The internal magnetic moments of the paired-off electrons also tend to point in opposite directions, so in most cases the total magnetic moment is still due to the odd, unpaired electron.†

The Magnetic Dipole Moment of Atoms

Suppose, as in Fig. 32–5a, that a single electron of charge $-e$ orbits at speed v in a circular orbit of radius r around a heavy nucleus. The period of the motion is $T = 2\pi r/v$. Because electric current is charge per unit time, the current around the nearly stationary nucleus is

$$I = \frac{-e}{T} = -\frac{ev}{2\pi r}. \tag{32–9}$$

The minus sign indicates that the current is in a direction opposite to that of the motion of the electron. The current loop has an area of πr^2; thus, following Eq. (30–28), the magnitude of the *orbital magnetic dipole moment* is given by

$$m_{\text{orbital}} = I\pi r^2 = \frac{ev}{2\pi r}\pi r^2 = \frac{1}{2}evr. \tag{32–10}$$

The direction of the magnetic moment vector **m** is determined by a right-hand rule (Fig. 32–5b).

EXAMPLE 32–2 Estimate an atomic orbital magnetic moment by taking the radius of the orbit to be of roughly atomic size, 10^{-10} m, and the kinetic energy to be a typical atomic energy of 1 eV. Compare this to the magnetic moment of a macroscopic loop of area 1 cm^2 that carries a current of 1 mA.

† The pairing of electrons does not always take place, and iron, which has an even number of electrons, is strongly magnetic. Whether the pairing occurs or not depends on a subtle interplay of the Coulomb forces between electrons.

SOLUTION: The unknown magnetic moment is determined from Eq. (32–10). We are given the radius r of the orbit and can deduce the speed v of the electron in its orbit from the given energy. It is convenient to convert to SI units, in which case our energy estimate of 1 eV is 1.6×10^{-19} J. From this energy, v is estimated by

$$\tfrac{1}{2}m_e v^2 = 1.6 \times 10^{-19}\ \text{J}.$$

To the same accuracy to which we have given the orbit radius, the electron mass is $m_e \simeq 10^{-30}$ kg, so

$$v = \sqrt{\frac{2(1.6 \times 10^{-19}\ \text{J})}{m_e}} \simeq \sqrt{\frac{3.2 \times 10^{-19}\ \text{J}}{10^{-30}\ \text{kg}}} = 5 \times 10^5\ \text{m/s}.$$

Equation (32–10) then gives the magnetic moment

$$m_{\text{orbital}} = \tfrac{1}{2}(1.6 \times 10^{-19}\ \text{C})(5 \times 10^5\ \text{m/s})(10^{-10}\ \text{m}) \simeq 5 \times 10^{-24}\ \text{A}\cdot\text{m}^2.$$

For comparison, the magnetic moment of the macroscopic loop is

$$m = (10^{-3}\ \text{A})(10^{-4}\ \text{m}^2) = 10^{-7}\ \text{A}\cdot\text{m}^2,$$

which is 2×10^{16} times larger than the magnetic moment of the single atom.

It is generally true that a magnetic dipole moment is due to a circulating charge, and the magnetic moment is proportional to the angular momentum carried by the circulating charge. It is useful to express magnetic moments in terms of angular momentum, because angular momentum is a fundamental physical quantity. For an electron in an atom, the relation goes as follows: Eq. (32–10) can be written in the form

The magnetic moment of a single-electron atom

$$m_{\text{orbital}} = \frac{1}{2}evr = \frac{e}{2m_e}m_e vr = \frac{e}{2m_e}L, \tag{32–11}$$

where m_e is the electron's mass and $L = m_e vr$ is the angular momentum of the electron in its circular orbit. If we include the vectorial properties of both the angular momentum and the magnetic moment, then Eq. (32–11) becomes

$$\mathbf{m}_{\text{orbital}} \equiv g_L \mathbf{L}. \tag{32–12}$$

The coefficient g_L connecting the magnetic moment and the angular momentum is known as the **gyromagnetic ratio**. For the orbital motion we have just seen that

$$g_L = -\frac{e}{2m_e}. \tag{32–13}$$

The minus sign is present because $\mathbf{m}_{\text{orbital}}$ and $\mathbf{L}$ point in opposite directions (Fig. 32–5b).

According to the quantum mechanical quantization rules for circular orbits (Section 10–5), the magnitude of L is $\ell\hbar$, where $\hbar \equiv h/2\pi$, h is Planck's constant, and ℓ is an integer. Thus the vector $\mathbf{m}_{\text{orbital}}$ has magnitude

The magnetic moment of single-electron atoms is quantized.

$$m_{\text{orbital}} = \left(\frac{e}{2m_e}\hbar\right)\ell \equiv m_B \ell. \tag{32–14}$$

The factor m_B is called the **Bohr magneton**, in honor of Niels Bohr, one of the founders of quantum mechanics. Its value is

$$m_B = \frac{e}{2m_e}\hbar = 9.27 \times 10^{-24}\ \text{A}\cdot\text{m}^2. \tag{32–15}$$

To this orbital magnetic moment we must add the contribution of the electron's *intrinsic magnetic moment*, $m_{\text{intrinsic}}$. The value of $m_{\text{intrinsic}}$ turns out to be just m_B. Protons and neutrons also have magnetic moments, with magnitudes given by a formula like Eq. (32–15), but with their own masses rather than the electron mass. Because their masses are approximately 2000 times larger than that of an electron, their magnetic moments are so small that they play no role in the magnetism of bulk matter. The nuclear magnetic moment is important in certain effects, including nuclear magnetic resonance, which we shall discuss later. The magnetism of matter is generally due only to the orbital and intrinsic magnetic moments of electrons.

Bulk Effects Are Due to the Alignment of Atomic Magnetic Dipoles

The magnetic fields produced by individual atoms are small compared to the magnetic fields we see in a bar magnet. How, then, can magnetism be significant in bulk material? If all the atomic magnetic moments were perfectly aligned in a material, we would obviously have a large effect. As Example 32–3 shows, the alignment of the atomic magnetic moments needs be only very slight to produce noticeable bulk effects.

EXAMPLE 32–3 Consider 1 mol of atoms with individual magnetic moments $m_0 = 10^{-23}\ \text{A}\cdot\text{m}^2$. Assume that the magnetic moments can point only in the $+z$- and $-z$-directions with a fraction f pointing "up" and $1 - f$ pointing "down." What value of f gives the same magnetic moment as a 1-cm² wire loop that carries a current of 10 mA?

SOLUTION: The magnetic moment of the specified loop is given by

$$IA = (10^{-2}\ \text{A})(10^{-4}\ \text{m}^2) = 10^{-6}\ \text{A}\cdot\text{m}^2.$$

Now we find the net magnetic moment of the sample for a given f. Where the magnetic moment of an atom points "up," the magnetic moment is $+m_0$; where it points "down," the magnetic moment is $-m_0$. The fraction up is f, and the fraction down is $1 - f$. Thus the net magnetic moment of the atoms is

$$m = N_A m_0[f - (1 - f)] = N_A m_0(2f - 1),$$

where N_A is Avogadro's number. We equate this net magnetic moment to that of the current loop to find that

$$N_A m_0(2f - 1) = IA.$$

We solve for f to find that

$$f = \frac{1}{2}\left(1 + \frac{IA}{2N_A m_0}\right) = \frac{1}{2} + \frac{m}{2N_A m_0}$$

$$= \frac{1}{2} + \frac{10^{-6}\ \cancel{\text{A}\cdot\text{m}^2}}{2(6 \times 10^{23}\ \cancel{\text{atoms}})(10^{-23}\ \cancel{\text{A}\cdot\text{m}^2/\text{atom}})} = \frac{1}{2} + (8 \times 10^{-8}).$$

A random distribution would have $f = \frac{1}{2}$, and *a departure from complete randomness of one part in 10 million gives rise to macroscopic effects.*

Example 32–3 shows that even a small deviation from complete randomness can lead to significant magnetic dipole moments for bulk matter. Just before the example, we wondered how it was possible to have *any* bulk effect. Now we can wonder about the other extreme. If only a small deviation from randomness leads to significant bulk effects, why are most materials not magnetic? The reason is that in a sample of 10^{24} atoms, statistical fluctuations away from an average magnetic moment of zero are expected to lead, on average, to an excess of only 10^{12} atoms that point in a particular direction. (In statistics, $\sqrt{N}$ is a typical fluctuation from the mean when N objects or events are involved.) Thus the typical value for the fraction f of atoms that point in a particular direction is $10^{12}/10^{24} = 10^{-12}$, and this leads to an infinitesimally small net magnetic moment. Our argument must be reexamined when atoms form a solid or a liquid and are thus closely spaced. In that case, forces between the atoms may cause neighboring atoms to line up with each other and lead to significant magnetic moments in large regions of the material. When this happens, permanent magnets form.

The Connection between Microscopic and Macroscopic Quantities

A piece of material will have significant magnetic properties (that is, be a magnet) if the directions of the magnetic dipole of its many component atoms or molecules are not completely random. In that case the vector sum of the atomic magnetic moments will not be zero. If we divide the vector sum of the magnetic moments by the number of atoms, we get a *net* magnetic moment $\mathbf{m}_0$ per constituent (atom or molecule). The magnetization is then

$$\mathbf{M} = n\mathbf{m}_0, \tag{32–16}$$

where n is the number of constituents per unit volume. Once we have determined $\mathbf{M}$, we can determine the other bulk magnetic properties from the discussion in Section 32–1.

32–3 FERROMAGNETISM

Ferromagnetic materials, which include the elements iron, cobalt, nickel, gadolinium, and dysprosium, together with alloys of these and other elements, can have a large permanent magnetization. The direction and magnitude of the magnetization can be set by an external magnetic field.

Ferromagnetism is due to the large-scale alignment of the magnetic moments of electrons.

In ferromagnetic materials, the intrinsic magnetic dipole moments of the electrons in atoms align themselves in large numbers and lead to large magnetic effects. There is no classical mechanism that can align the intrinsic magnetic moments sufficiently strongly, and the explanation is purely quantum mechanical. Werner Heisenberg, one of the creators of the quantum theory, suggested in 1928 that, as a consequence of the exclusion principle (introduced in Chapter 27), electrons with parallel intrinsic magnetic moments arrange themselves in orbits that tend to maximize the distance between them. This reduces the potential energy of Coulomb repulsion between them and thus makes a state with parallel magnetic moments a state of lower energy. As a consequence, under the right circumstances there is a preference for the intrinsic magnetic moments of electrons to line up parallel with one another.

The intrinsic magnetic moments of the unpaired electrons of different atoms—each iron atom, for example, has two such electrons—do not ordinarily become aligned throughout a piece of ferromagnetic material. Rather, the alignment takes place between adjacent atoms in regions called *magnetic domains*, which may contain 10^{17} to 10^{21} atoms and occupy a volume on the order of 10^{-12} to 10^{-8} m^3 (in other words, volumes from 0.1 mm to 1 mm on a side). The magnetic field within these domains is quite large, but the material may be made up of thousands

FIGURE 32–6 Photomicrograph of magnetic domains in a sample of iron with 3% silicon. A strong net magnetic field is associated with each domain. Domains with different orientations appear in different colors.

of such domains, each with a magnetization aligned differently, so the magnetization of the entire material will average to zero unless some special mechanism is present. A sample of magnetic material may look something like Fig. 32–6, which clearly shows the boundaries between domains, called *domain walls*. Figure 32–7a is a schematic diagram of the domains with their individual magnetic moments.

The special mechanism that can align the magnetizations of different domains is provided by an external magnetic field. If a field $\mathbf{B}_{ext}$ is applied to a piece of ferromagnetic material, two things can happen to transform the material into a permanent magnet. First, the size of domains with their magnetic moments already aligned with $\mathbf{B}_{ext}$ may enlarge at the expense of neighboring domains. Second, the magnetic moments of some of the domains may rotate to the direction of $\mathbf{B}_{ext}$ through an overall realignment of their constituents (Fig. 32–7b). (Remember that the state with a magnetic moment aligned along $\mathbf{B}_{ext}$ is a state of lower energy.)

The process we have described can be understood by a simple analogy. Imagine a large marching band whose members face in random directions (Fig. 32–8a). The band leader orders them by loudspeaker to face the same direction but fails to say *which* direction. Perhaps influenced by the random choices of a few band members (*A*, *B*, or *C* in Fig. 32–8b), the immediate neighbors take their cue from these band members and align themselves. The result in this case is three separate regions of alignment. An analogous situation applies in ferromagnets, in which case the arrows in Fig. 32–8 represent the intrinsic dipoles. Before an external field is applied, the dipoles are aligned over differently oriented domains because of the quantum effects mentioned above. How can we get alignment over larger distances? It is only when the precise direction is given over the loudspeaker that the band members all align themselves in that direction. In a ferromagnetic material, the instructions for a precise direction are provided by the external field. Just as the band members will remain aligned even when the loudspeaker is turned off, the atomic magnetic dipole moments remain aligned even when the external field is removed—the magnetization remains. An aligning external field has been applied sometime in the past to any ferromagnet that acts as a permanent magnet.

When a ferromagnet is heated, the increased movement of the atoms leads to a randomization of their orientation and thus to a decrease in the alignment. At

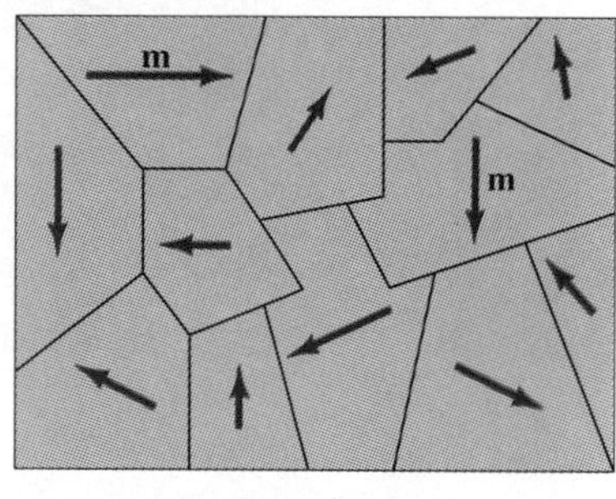

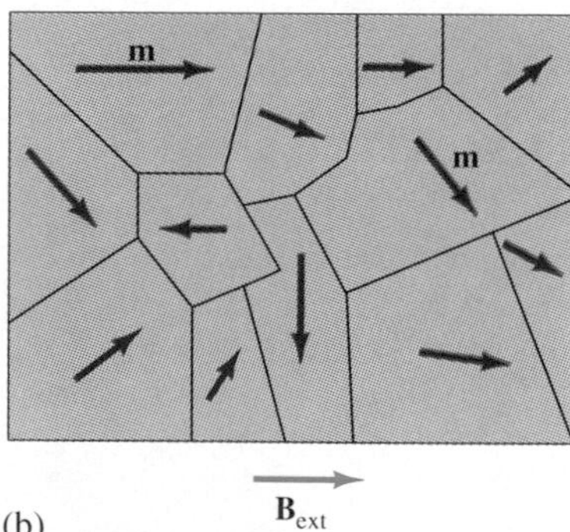

FIGURE 32–7 (a) Domain formation in ferromagnetic materials in the absence of an external magnetic field. The arrows indicate the magnetic moments of individual domains. (b) The presence of an external magnetic field influences the domains, making some larger and realigning others.

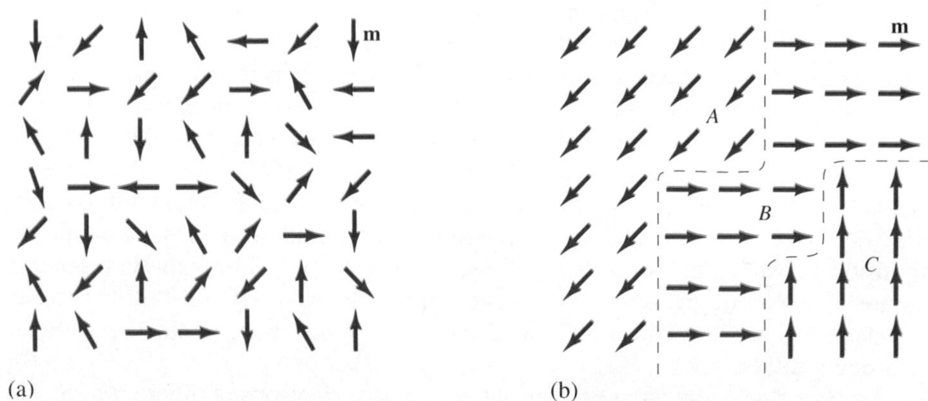

FIGURE 32–8 (a) The members of a marching band (or atomic magnetic moments in a ferromagnet) are oriented randomly. (b) The band members (or the atomic magnetic moments) influence one another in aligning themselves. Unless there is some external guide, the alignment will occur in small regions called domains.

the *Curie temperature*, T_c (named after Pierre Curie), the randomization is complete, and the material is no longer a ferromagnet. The value of T_c varies from material to material; in iron $T_c = 1043$ K, in gadolinium $T_c = 292$ K. Below T_c ferromagnetism appears, just as below 273 K, water forms the ordered lattice we know as ice. When a ferromagnet cools below T_c, it does not automatically become a permanent magnet for the same reason that when a lake freezes, it does not form one huge ice crystal. The transition to ferromagnetic behavior, like freezing, takes place in domains, as described by our marching-band analogy.

EXAMPLE 32–4 Estimate the maximum possible magnetization in a single domain of iron.

SOLUTION: In a single domain with maximum magnetization, the atomic magnetic moments are aligned perfectly. We first find the atomic magnetic moment of iron. The value of the intrinsic magnetic moment of an electron is given by Eq. (32–15), $m_{\text{intrinsic}} = m_B = 9.3 \times 10^{-24}\ \text{A}\cdot\text{m}^2$. The maximum possible magnetization comes when $m_{\text{intrinsic}}$ of the two unpaired electrons in an atom of iron are aligned with each other and with all the atoms' unpaired electrons. The number density n of unpaired electrons in iron is

$$n = \left(\frac{2\ \text{unpaired electrons}}{1\ \cancel{\text{atom}}}\right)\left(\frac{6.02 \times 10^{23}\ \cancel{\text{atoms}}}{1\ \cancel{\text{mol}}}\right)\left(\frac{1\ \cancel{\text{mol}}}{56\ \cancel{\text{g}}}\right)\left(\frac{7.8\ \cancel{\text{g}}}{1\ \cancel{\text{cm}^3}}\right)\left(\frac{10^6\ \cancel{\text{cm}^3}}{1\ \text{m}^3}\right)$$

$$= 1.7 \times 10^{29}\ \text{unpaired electrons/m}^3.$$

We multiply by m_B for each unpaired electron to find a total magnetization of

$$M_{\max} = nm_B = (1.7 \times 10^{29}\ \text{unpaired electrons/m}^{\cancel{3}})(9.3 \times 10^{-24}\ \text{A}\cdot\cancel{\text{m}^2})$$

$$\simeq 1.6 \times 10^6\ \text{A/m}.$$

This result can be compared to the experimental $M_{\max}$ of annealed (tempered) iron, 1.7×10^4 A/m. Because this experimental result involves many domains, the difference of a factor of 100 means that the domains are never perfectly aligned.

Hysteresis

The relation between the magnetic field, B, and the magnetic intensity, H, is more complicated in ferromagnets than in other materials. In order to measure the relation between B and H in a ferromagnetic material, that material is demagnetized by heating. It is cooled, shaped into a ring, and wound with a wire that carries a current I. This experimental arrangement is called a *Rowland ring*. Without the ferromagnetic material, the magnetic field inside the ring, or toroidal solenoid, has the nearly constant value

$$B_0 = \mu_0 H = \mu_0 nI, \tag{32–17}$$

provided that the torus is "thin." Here n is the number of windings per unit length. As we know, when the ferromagnetic material is inserted into the torus, the magnetic field increases tremendously to a new value, B. We measure B by using a sense coil outside the torus (Fig. 32–9). The sense coil measures an induced emf proportional to the time rate of change of the magnetic field. As we raise the current in the toroidal coil at a given rate, we know the magnetic intensity $H = nI$, and the sense

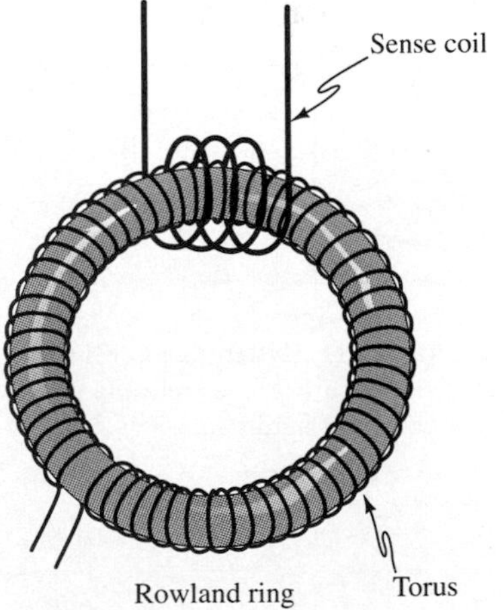

FIGURE 32–9 A Rowland ring is a wound core (solenoid) of material that may be used to measure the relation between **B** and **H** of that material. A sense coil measures changes in **B**.

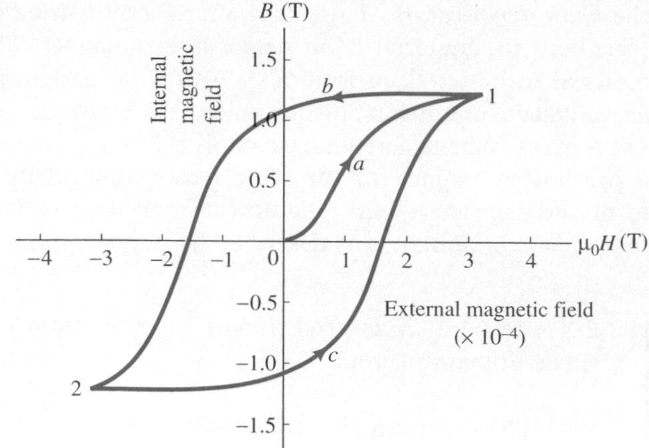

FIGURE 32–10 A magnetization curve illustrates the phenomenon of hysteresis in ferromagnetic materials. The material starts at the origin with zero magnetization. When a magnetic intensity H is applied, the material responds by becoming magnetic and is magnetic even when H is again zero.

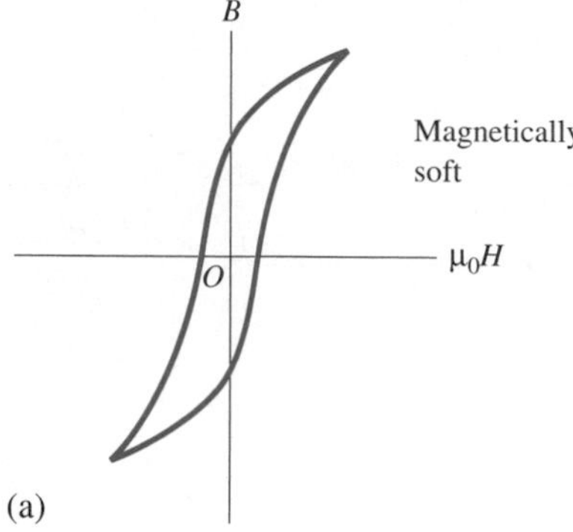

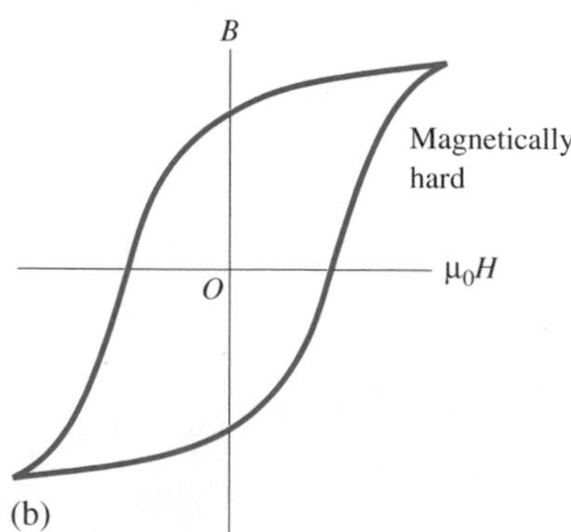

FIGURE 32–11 Hysteresis loops for materials that are (a) magnetically soft and (b) magnetically hard.

coil measures B. Figure 32–10 shows one example of a measured relation between H and B; a plot of H versus B is called a *magnetization curve.* Knowing H and B, we can determine the magnetization from the relation $B = \mu_0 H + \mu_0 M$.

What is observed for ferromagnetic materials is the curve in Fig. 32–10. A magnetization curve of this shape, known as a **hysteresis loop**, indicates the phenomenon of **hysteresis**. The presence of hysteresis shows that there is an irreversibility to the magnetization process. When the current I in the solenoid is changed by a little and then changed back again, the original magnetization is generally not attained. For example, if in Fig. 32–10 we start on curve c at a value of 10^{-4} T for $\mu_0 H$, B in the ferromagnetic material is negative. If $\mu_0 H$ is increased to 3×10^{-4} T and then brought back to 10^{-4} T, B is now positive, following curve b. Hysteresis expresses the fact that the magnetic domains do not return to their original zero-external-field status when the current decreases. They "remember" the rise in field and do not automatically revert to their original alignments.

Some materials have narrow hysteresis loops, meaning that the alignment of the domains follows the external field rather closely (Fig. 32–11a). Materials for which this type of curve holds, such as iron, are considered to be *magnetically soft.* These materials are often used in transformer cores. (As we shall see in Chapter 34, transformers are devices that transform AC currents—or voltages—from one value to another.) Other materials have broad hysteresis loops, meaning that their domains respond only to large external fields (Fig. 32–11b). Such materials, including carbon and tungsten, are said to be *magnetically hard.* They are difficult to magnetize, but they make good permanent magnets once magnetized, because they are equally difficult to demagnetize. Magnetically hard materials are especially important for making computer memories, magnetic tapes, or floppy disks, because such materials are stable against changes due to nearby magnetic fields (Fig. 32–12).

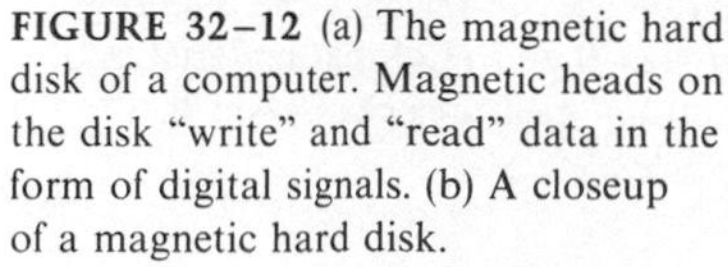

(a)

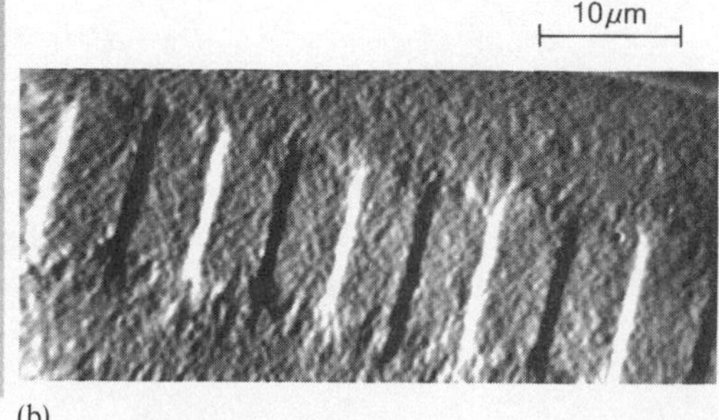

(b)

FIGURE 32–12 (a) The magnetic hard disk of a computer. Magnetic heads on the disk "write" and "read" data in the form of digital signals. (b) A closeup of a magnetic hard disk.

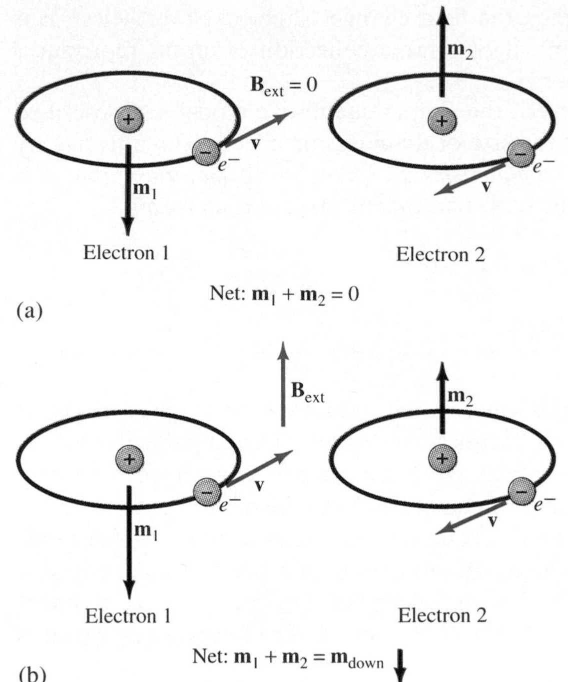

FIGURE 32–13 (a) A two-electron atom with a net orbital magnetic moment of zero in the absence of an external magnetic field. (b) In the presence of an external magnetic field, the orbital magnetic moment associated with each electron is changed, and there is a net magnetic moment that points down, opposite to the external field.

*32–4 DIAMAGNETISM

In diamagnetic materials, the induced magnetic field caused by $\mathbf{B}_{ext}$, the external field, points in a direction *opposite* to the external field, so the magnetic field inside such materials is *less* than $\mathbf{B}_{ext}$. Diamagnetism, as distinguished from ferromagnetism and paramagnetism, occurs in materials whose atoms have no permanent magnetic dipole moments, either orbital or intrinsic. We can understand the phenomenon qualitatively in terms of a classical model in which electrons orbit the nucleus of atoms. Consider two electrons that orbit the same nucleus, with their intrinsic magnetic moments aligned in opposite directions. We suppose that their orbits are identical, except that in one the motion is counterclockwise and in the other clockwise (Fig. 32–13a). With no external magnetic field, the orbital magnetic moments of the two electrons cancel ($\mathbf{m}_1 + \mathbf{m}_2 = 0$), and there is no magnetization.

Now suppose that we turn on an external magnetic field, $\mathbf{B}_{ext}$, perpendicular to the orbits of the electrons (Fig. 32–13b). For the electron on the left in the figure, as the external field increases, the flux through its orbit increases; by Lenz's law, the electron responds so as to counter the increasing flux. The electron (negatively charged) speeds up accordingly, increasing its angular momentum as well as the magnitude of its orbital magnetic moment $\mathbf{m}_1$, which points downward. When the external field levels off, angular momentum conservation ensures that the change in $\mathbf{m}_1$ will persist. The electron on the right must slow down to oppose the increase in flux through its orbit, so the magnitude of its magnetic moment $\mathbf{m}_2$ is reduced. The result is that $\mathbf{m}_1 + \mathbf{m}_2$ now has a net value that points downward, and a magnetic field is produced that opposes the increasing external field. This is the origin of the negative magnetic susceptibility. The classical model needs to be revised for a proper quantum mechanical treatment of the atom, because in quantum mechanics, the angular momentum is *quantized*, as are its possible changes.

It cannot change only slightly when the field changes slightly. Nevertheless, it is reassuring that a statistical treatment of a large collection of atoms reproduces the effect of the classical discussion.

When it is applied quantitatively, the simple qualitative model just described leads to reasonable estimates for the size of diamagnetic effects. The universality of the discussion implies that *diamagnetism is present in all materials*, but it is masked for materials whose atoms have permanent magnetic moments.

*32–5 PARAMAGNETISM

Paramagnetism occurs in materials whose molecules have permanent magnetic dipole moments due to the intrinsic magnetic moments of unpaired electrons. In the absence of an external magnetic field, these dipoles will be oriented randomly because of thermal motion, and the net magnetization of the materials will be zero. An external magnetic field **B** exerts a torque on the atomic dipoles, which tends to align their magnetic moments along **B** and produces a positive magnetic susceptibility. Recall from Chapter 29 that the energy of a magnetic dipole moment **m** in a magnetic field **B** is, by Eq. (29–24), $U = -\mathbf{m} \cdot \mathbf{B}$. The lowest energy occurs when **m** and **B** are *parallel.*

There are two effects at work in determining the extent to which the permanent magnetic dipoles become aligned. The first is the external field, which encourages alignment, and the second is the thermal motion, which randomizes the alignment. The relative importance of these two factors is measured by the relative size of the magnetic energy factor mB and the thermal energy factor kT, where T is temperature. If T is so large that $kT \gg mB$, the average alignment over a large number of electrons will be weak. Conversely, if T is so low that $kT \ll mB$, the average alignment will be strong. For intermediate temperatures the average alignment is proportional to the ratio of these energies, $m_{av} = (\text{a constant})(mB)/(kT)$. At room temperature the intrinsic magnetic moments of most paramagnetic materials are only very slightly aligned, but as we saw in Example 32–3, large bulk effects come from very small alignments. In 1895 Pierre Curie observed the linear relation between magnetization and the ratio of the magnetic field to temperature, now called *Curie's law:*

$$\mathbf{M} = C\frac{\mathbf{B}}{T}, \tag{32–18}$$

where C is *Curie's constant.* This law is often expressed in terms of magnetic susceptibility, defined according to Eq. (32–5) as $\mathbf{M} = \chi_m \mathbf{H}$. If we anticipate that the susceptibility will be small, as it is for paramagnetic materials, then we can replace **B** in Eq. (32–18) by $\mu_0 \mathbf{H}$:

$$\mathbf{M} = C\frac{\mu_0 \mathbf{H}}{T},$$

or

$$\chi_m = \frac{\mu_0 C}{T}. \tag{32–19}$$

The susceptibility is positive, characteristic of paramagnetism.

The temperature dependence in Eq. (32–18) is the same as that of the very similar phenomenon for dielectrics [see Eq. (26–30)]. There, too, this dependence is called *Curie's law*. We expect that the linear relation between magnetization and applied field must fail at sufficiently low temperatures and/or large field. If the intrinsic magnetic moments are aligned perfectly with the field, they cannot be still further aligned to produce still higher magnetization. This *saturation* phenomenon is well known. A more quantitative version of the arguments we have presented can be used to predict the size of the paramagnetic susceptibility. These predictions are well satisfied.

Unlike diamagnetism, paramagnetism is not a universal phenomenon, because relatively few materials have molecules with unpaired electrons (Fig. 32–14). When it is present, paramagnetism is normally a larger effect than diamagnetism. However, diamagnetism dominates at high enough temperatures.

FIGURE 32–14 Oxygen is paramagnetic and is therefore attracted by the poles of a magnet. Here liquid oxygen poured between two poles is held in place by the forces between it and the permanent magnet.

*32–6 MAGNETISM AND SUPERCONDUCTIVITY

Superconductors have magnetic properties that are just as extraordinary as their electric properties are. Remarkably, the same collective quantum physical mechanism that makes any electric field inside a superconductor exactly zero *also makes the magnetic field inside zero.* A superconductor acts as a perfect diamagnet in the sense that currents will always be induced that precisely cancel any magnetic field inside. Alternatively, we say that the magnetic field lines are *expelled* from the superconductor, a phenomenon known as the *Meissner effect.* In Type I superconductors, the field is expelled entirely (Fig. 32–15). In Type II superconductors, the field is isolated in nonsuperconducting filamentary structures within the material (Fig. 32–16). When such filaments are present, the resistivity of the material is no longer exactly zero. Currents circulate on the surfaces of these filaments, shielding the rest of the material from the magnetic field. The amount of magnetic flux within

The magnetic field within a superconducting material is zero.

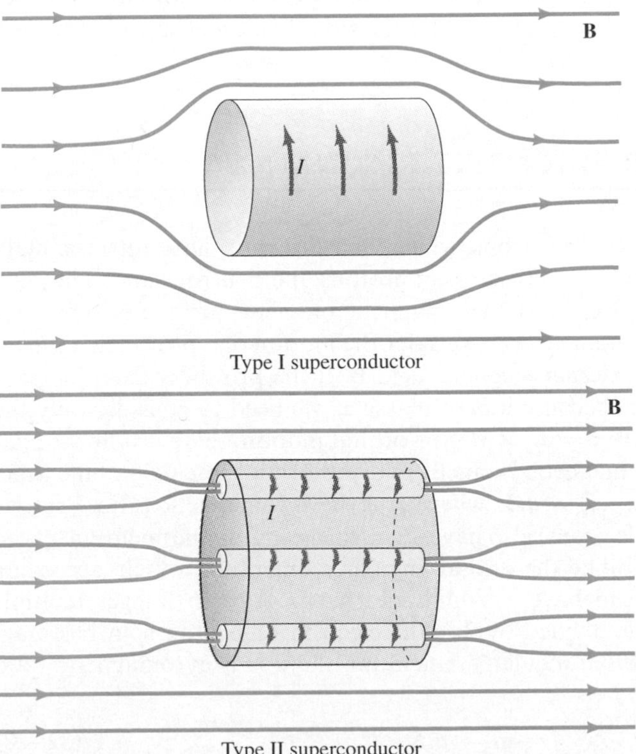

FIGURE 32–15 A Type I superconductor expels magnetic field from its interior by acting as a perfect diamagnet: Surface currents that just cancel the applied field inside are established.

FIGURE 32–16 In Type II superconductors, the magnetic field is confined to filamentary structures. Inside the filaments, the material is not in its superconducting phase.

each filament is a minimum flux permitted by quantum physics, proportional to Planck's constant.

We can translate the statement that a superconductor expels the magnetic field from its interior into a statement about the magnetic susceptibility of the material in its superconducting phase. Equation (32–6) expresses the internal field in terms of the magnetic intensity, **H**, and this internal field must be zero:

$$\mathbf{B} = \mu_0(1 + \chi_m)\mathbf{H} = 0.$$

The magnetic intensity is due to the free currents that produce the external magnetic field, and is not zero. Thus we must have $1 + \chi_m = 0$, and the magnetic susceptibility is

$$\chi_m = -1.$$

The fact that there is no magnetic field inside a superconductor shows that there can be no currents inside. Imagine, as in Fig. 32–16, that some internal region of a superconductor carries current. Then we can draw a path around this region and apply Ampère's law, Eq. (30–10), which expresses the integral of the magnetic field around a loop in terms of the current that passes through the loop. If there were a current, there would be a magnetic field, and this is not possible. We conclude that *all current carried by superconductors must be carried on their surfaces.* A surface is any boundary between superconducting and nonsuperconducting phases of the material; for example, current can be carried on the walls of the filaments in Fig. 32–16.

In the presence of a large enough magnetic field (a *critical field*), a material in a superconducting phase jumps back to the normal (nonsuperconducting) phase, even if the temperature is held fixed. This is a serious problem, because one of the major uses of superconductors is in the construction of electromagnets that do not undergo Joule heating. The magnetic field may itself destroy the superconductivity. Happily, type II superconductors, which channel the magnetic field into filaments, provide a way to make superconducting materials with much higher critical fields. These are the materials used for the construction of superconducting magnets.

*32–7 NUCLEAR MAGNETIC RESONANCE

Nuclei consist of protons and neutrons, which, like electrons, have intrinsic magnetic dipole moments. Because the masses of protons and neutrons are some 2000 times larger than those of electrons, their magnetic moments are some 2000 times smaller than the electron moment m_B. Consider the motion of a proton, with magnetic moment $\mathbf{m}_p$, in an external magnetic field **B**. If the proton is fixed in space (not a bad approximation, because it is so massive), we need to consider only the torque exerted on $\mathbf{m}_p$ by **B**, $\boldsymbol{\tau} = \mathbf{m}_p \times \mathbf{B}$. For orbital motion, there is a linear relation between the orbital magnetic moment and the orbital angular momentum, and that is Eq. (32–12), $\mathbf{m}_{\text{orbital}} = g_L\mathbf{L}$, where g_L is the gyromagnetic ratio. Protons (as well as neutrons and electrons) also have an intrinsic angular momentum, called the *spin* and labeled **S**. Unlike the orbital angular momentum, which, according to quantum mechanics, can have a magnitude that is only an integer multiple of $\hbar$ (Planck's constant, h, divided by 2π), the electron or proton spin has magnitude $\frac{1}{2}\hbar$. As for the orbital angular momentum, there is a gyromagnetic ratio for the spin, g_S, defined by

$$\mathbf{m}_p \equiv g_S\mathbf{S}. \tag{32–20}$$

The torque on the proton is the rate of change of the proton's internal angular momentum, $d\mathbf{S}/dt$, so with $\mathbf{S} = \mathbf{m}_p/g_S$ we have

$$\frac{1}{g_S}\frac{d\mathbf{m}_p}{dt} = \mathbf{m}_p \times \mathbf{B}. \tag{32–21}$$

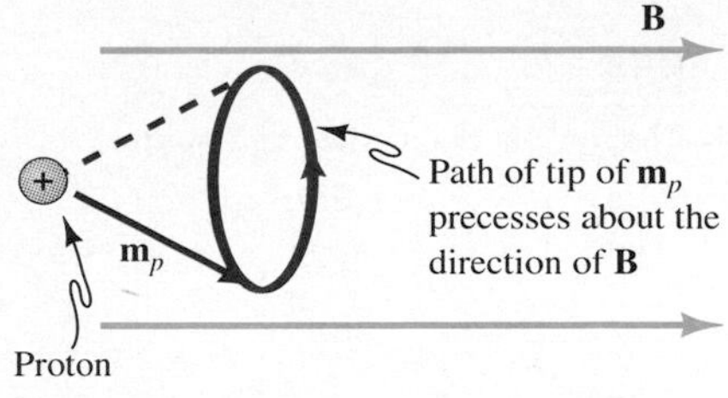

FIGURE 32–17 The magnetic moment of a proton, $\mathbf{m}_p$, precesses about the direction of an external magnetic field.

This equation gives the rate of change of the magnetic moment of a proton in a magnetic field. The magnitude of $\mathbf{m}_p$ cannot change, but its direction can, and Eq. (32–21) describes the *precessional* motion of $\mathbf{m}_p$ about the direction of **B** (Fig. 32–17).† If $\mathbf{m}_p$ is not aligned along the direction of **B**, then the torque is in a direction perpendicular to both $\mathbf{m}_p$ and **B**, and this will have the effect of moving $\mathbf{m}_p$ in such a way that its motion traces a cone around the direction of **B**, as in Fig. 32–17. The magnetic moment may move clockwise or counterclockwise about the axis defined by **B**, depending on the sign of the gyromagnetic ratio. It is straightforward to show that the angular speed of precession is given by

$$\omega_0 = g_S B \tag{32–22}$$

(see Problem 36). The potential energy of a magnetic moment **m** in an external field is given by $U = -\mathbf{m} \cdot \mathbf{B}$, and the motion with the lowest energy is motion in which the magnetic moment tends to point along rather than against the direction of **B**. If **B** is directed along the z-axis, there is conservation of the z-component of angular momentum, because the torque has no component in the z-direction. Thus the magnetic moment cannot flip from an "up" cone to a "down" cone.

The situation changes when an additional, oscillating magnetic field that points in the x-direction, say, is imposed. Such a field can be introduced by the presence of electromagnetic waves, which we shall see in Chapter 35 consist of oscillating electric and magnetic fields. Here only the oscillating magnetic field, of the form $B_1 \cos(\omega t)\,\mathbf{i}$, is important. There is a torque due to the oscillating field given by $\boldsymbol{\tau}_1 = (\mathbf{m}_p \times \mathbf{i})B_1 \cos(\omega t)$. This torque has a z-component. Because of the $\cos(\omega t)$ factor, sometimes the z-component of the torque acts "upward," sometimes it acts "downward," and on average it will not have much of an effect. However, in the special case that the angular frequency ω of the oscillating field *exactly matches* the angular frequency ω_0 of the precession (a condition known as *resonance* and described in Chapter 13), the torque has a steady, long-term effect on the magnetic moment and can flip its direction. The oscillating magnetic field supplies in this case *just the precise amount of energy*, $2\mathbf{m}_p \cdot \mathbf{B}$, necessary to flip the spin of the proton from up to down, or absorbs this amount of energy, the condition required to flip the spin from down to up.

This effect is called **nuclear magnetic resonance** (NMR). When the spin flips due to a transfer of energy between the oscillating field and the proton, there is a detectable signal. (For example, it is possible to measure an energy change in the oscillating field.) Thus, by tuning the frequency of the oscillating magnetic field, the frequency $\omega_0 = g_S B$ can be measured with very high precision. Classically, any cone of motion is feasible for the magnetic moment. However, the quantum nature of intrinsic spins dictates that, for $S = \hbar/2$, only two well-defined cones are possible, and thus we are justified in describing the spin flip as the only possible change in the system.

If the external magnetic field is known, the NMR method may be used to measure the gyromagnetic ratio, g_S. It is in this way that the gyromagnetic ratios for protons and neutrons are known to contain the coefficients 2.79 and -1.91,

† This precession, called *Larmor precession*, is analogous to the precession of a spinning top or tilted gyroscope (see Chapter 10). In that case gravity is the force responsible for the torque.

respectively, in addition to classical factors. These coefficients suggest that protons and neutrons have structures more elaborate than electrons do. Such measurements help us understand the internal workings of the proton and the neutron. NMR measurements can also be extended to nuclei, where they are used to study nuclear structure and the forces that give rise to it. For example, the deuteron nucleus, which may be described as a bound state of a proton and neutron structured in such a way that the intrinsic spins, and therefore the intrinsic magnetic moments, are parallel to each other, is expected to have a magnetic moment that is the sum of the moments of the proton and the neutron. NMR measurements give a slightly smaller result than the sum, indicating that the deuteron magnetic moment must have a small contribution from the relative motion of the proton and the neutron. As a consequence, the deuteron, which had been expected to be spherical in shape, is now known to be slightly cigar-shaped. The nuclear forces that can give rise to such a shape must include some spin-dependent forces, of a magnitude that can be determined from NMR measurements.

APPLICATION

Nuclear Magnetic Resonance

NMR has important applications in the study of materials and in medical diagnostics, where the procedure is called *magnetic resonance imaging* (MRI). (The word "nuclear" was dropped because of patients' fears that nuclear radiation was being used. In fact, MRI is regarded as an especially safe procedure.) In the study of materials, including animal tissue, the gyromagnetic ratios of nuclei are well known, but the magnetic field, B, contains a contribution from the electrons as well as the nuclei of the material. Thus a measurement of B via NMR gives us information about the atoms and molecules to which the nuclei belong (Fig. B1-1). This can be translated into information about crystal lattices and about organic materials. In the case of medical diagnostics, MRI works primarily on hydrogen atoms, whose nuclei contain single protons. MRI locates concentrations of hydrogen atoms in patients. Fat, which has a high concentration of hydrogen, can be distinguished from muscle, which has a much lower hydrogen concentration. Tumors can be distinguished from nerve tissue, and bones, which have little hydrogen, are hardly seen at all (Fig. B1-2). MRI gives results that are complementary to other diagnostic tools, including X-rays. MRI has proven effective in diagnosing neural diseases such as multiple sclerosis.

The development of NMR—from an esoteric application of quantum mechanics, to a method to measure angular momenta and magnetic dipole moments, to an increasingly common technological tool—shows the growing importance of quantum physics in engineering and applied science.

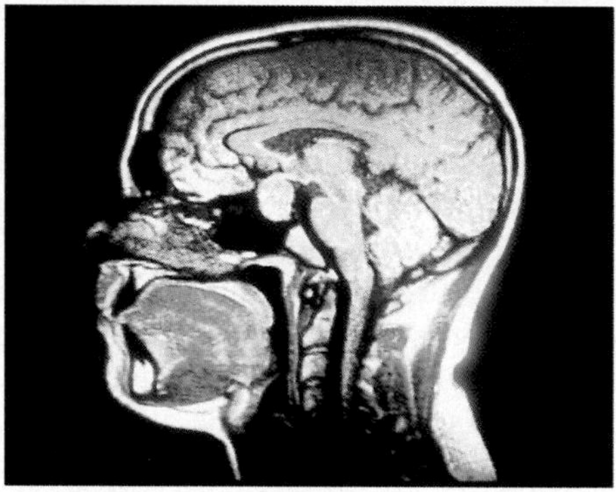

FIGURE B1–2 A magnetic resonance image of the head that indicates a pituitary tumor.

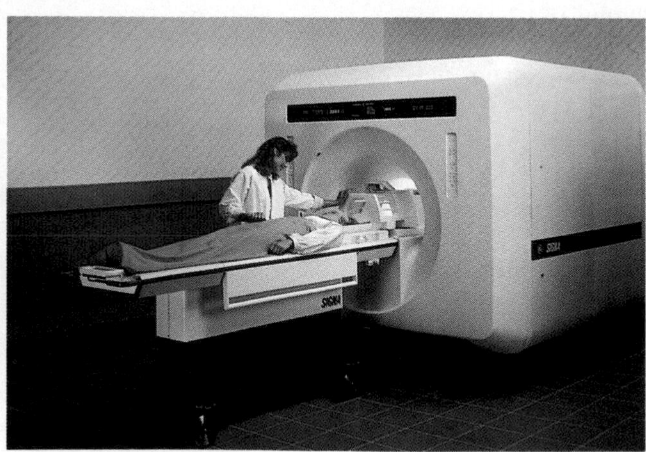

FIGURE B1–1 This machine is used to produce cross-sectional images of patients by magnetic resonance imaging. At its "heart" is a superconducting magnet.

The magnetic properties of bulk matter are summarized in the magnetization, **M**, the magnetic dipole moment per unit volume. In the presence of an external magnetic field $\mathbf{B}_{ext}$, there is a field in a material given by

$$\mathbf{B} = \mathbf{B}_{ext} + \mu_0 \mathbf{M}. \qquad (32\text{–}1)$$

The effect of free (real) currents (as opposed to the induced atomic effects) is contained in the magnetic intensity, $\mathbf{H} = \mathbf{B}_{ext}/\mu_0$:

$$\mathbf{H} \equiv \frac{\mathbf{B}}{\mu_0} - \mathbf{M}. \qquad (32\text{–}2)$$

The magnetic susceptibility, χ_m, describes the response of a material to a magnetic field of external origin:

$$\mathbf{M} \equiv \chi_m \mathbf{H}. \qquad (32\text{–}5)$$

In terms of χ_m, the net magnetic field is given by

$$\mathbf{B} = \mu_0(1 + \chi_m)\mathbf{H} = \mu\mathbf{H}. \qquad (32\text{–}6), (32\text{–}8)$$

Here μ is the permeability of the material:

$$\mu \equiv \mu_0(1 + \chi_m). \qquad (32\text{–}7)$$

Magnetism in matter is due ultimately to the magnetism of its atomic constituents, and particularly to the unpaired electrons of atoms. An orbiting electron produces an atomic orbital magnetic moment

$$\mathbf{m}_{orbital} \equiv g_L \mathbf{L}, \qquad (32\text{–}12)$$

where g_L is the gyromagnetic ratio. Quantum mechanics implies that these magnetic moments take the value

$$m_{orbital} = \left(\frac{e}{2m_e}\hbar\right)\ell \equiv m_B \ell, \qquad (32\text{–}14)$$

where the factor m_B is the Bohr magneton and ℓ is an integer. In addition, electrons have intrinsic magnetic moments equal in magnitude to m_B. Even a very slight alignment of atomic magnetic moments leads to large magnetic effects in bulk matter.

Ferromagnetic materials have large permeabilities. The atomic dipole moments are lined up in small regions called domains due to forces of quantum mechanical origin. The imposition of an external field leads to the dipole moments of the domains lining up together and produces permanent magnets. The fact that a ferromagnetic material "remembers" the orientation of the external field that magnetizes it leads to the phenomenon of hysteresis, in which the magnetization curve depends on how the magnetization was produced.

Diamagnetic materials have small negative susceptibilities, due ultimately to Faraday's law. Diamagnetism is always present but may be masked by other effects. Paramagnetic materials have small positive susceptibilities, due to the intrinsic magnetic moments of unpaired electrons, which find it energetically favorable to line up with an external field. Paramagnetism is strongly temperature dependent. Superconductors expel magnetic field from their interiors.

In nuclear magnetic resonance (NMR), an important tool for materials science and medical applications, the intrinsic magnetic moments of nuclei and nuclear constituents precess about an applied magnetic field. This precession is detected by imposition of an electromagnetic wave of just the right frequency, a frequency ultimately characteristic of the material involved.

QUESTIONS

1. Why, in our calculation of the magnetic dipole moment associated with orbital motion, was it reasonable to think of the nucleus as stationary while the electron circulates around it?
2. When an electron circulates around the nucleus in a planetary model, the system forms an electric dipole. Why does this electric dipole not produce a measurable electric dipole field around the atom?
3. Under what circumstances will Gauss' law for the magnetic field also hold for the magnetic intensity?
4. In a Rowland ring measurement of the magnetic field inside a piece of magnetic material, is it helpful to wrap the sense coil around the material many times?
5. Does iron exhibit diamagnetic properties? How could you determine them?
6. Aluminum is separated in junk yards by using large magnets. How is this possible?
7. Should the magnetic latch on a refrigerator door be made from magnetically hard or soft material?
8. Why should computer floppy disks not be made from magnetically soft material?
9. At sufficiently high temperatures diamagnetism dominates over paramagnetism. Why?
10. Explain how a permanent bar magnet attracts an unmagnetized iron needle.
11. You are given two identical iron rods—one magnetized, the other not. How can you determine which is the magnet, without using a third magnet (for example, the earth)?
12. Suppose that an electron in a circular orbit around a nucleus is placed in an external magnetic field. Will the angular momentum of the electron change if the field is aligned perpendicular to the plane of motion? parallel to the plane of motion?
13. Is is possible to arrange for a classical current loop to have a magnetic moment but no angular momentum? Assume first that you have both positive and negative charge carriers to work with, and then that you have only negative ones.
14. What is the value of **H** in an isolated permanent magnet?
15. It takes an external field to establish a macroscopic magnetization inside a permanent magnet cooled below its Curie temperature. What could have done this for lodestones, permanent magnets found in nature?
16. In a uniform magnetic field, a magnetic dipole experiences no net force, only a torque. How do two bar magnets repel or attract each other?

PROBLEMS

32–1 The Magnetic Properties of Bulk Matter

1. (I) A cylindrical rod of palladium, with magnetic susceptibility $\chi = 8 \times 10^{-4}$, of radius 1 cm and length 5 cm is placed in and aligned with a uniform magnetic field of 1.0 T. What is the magnetic dipole moment of the rod?
2. (I) A piece of iron is placed between the poles of a magnet. The magnetic field was 0.0010 T before the iron was inserted but 1.0 T afterward. What is the magnetic intensity inside the iron?
3. (I) A thin, toroidal coil of total length 55 cm is wound with 1100 turns of wire. A current of 1.7 A flows through the wire. What is the magnitude of **H** inside the torus if the core consists of a ferromagnetic material of magnetic susceptibility $\chi_m = 1.2 \times 10^3$?
4. (II) In a vacuum, a solenoid with a current I has a magnetic field B_0. (a) If copper is placed inside the solenoid, what is the change in the magnetic field? (b) What happens if aluminum is placed inside the solenoid?
5. (II) A 1.0-cm^3 cube of copper is placed between the poles of a magnet with a magnetic field of 6.0 T. What is the induced magnetization in the copper?
6. (II) A long solenoid filled with ferromagnetic material of permeability $\mu = 850\mu_0$ is wound with wire so that there are 12 turns per cm. What current must flow through the wire to produce a magnetic field of 2.4 T within the solenoid?

32–2 Atoms as Magnets

7. (I) Suppose that 1 mol of atoms in a material have individual magnetic moments of 10^{-23} A·m^2. In the absence of any alignment, the magnetic moments form an *average* angle of 90° with some external axis. By how much does the average angle differ from 90° if the material has the same magnetic moment as a 1-cm^2 loop of wire that carries a current of 10 mA? (The magnetic moment of the loop is aligned with the external axis.) Assume that the components of the atomic magnetic moments add algebraically.
8. (I) The atomic number of aluminum is 13, and its density is 2.7 g/cm^3. (a) How many electrons are there in 10 cm^3 of aluminum? (b) If each electron has the magnetic moment estimated in Example 32–2, 5×10^{-24} A·m^2, and all the magnetic moments are aligned, what would the magnetization of the aluminum be?
9. (II) Consider an electron in a circular orbit around a single proton (a hydrogen nucleus) whose total energy is −13.5 eV. Find the value of the orbital magnetic moment.
10. (II) Suppose that an electron in a circular orbit around a single proton (a hydrogen nucleus) has an energy of −13.5 eV, where zero potential energy is at infinity. (a) What is the radius of the orbit? (b) What is the speed of the motion? Is it justifiable to say that the motion is nonrelativistic? (c) What is the period of the orbit? Is it reasonable that observation of the magnetic effects of atoms

takes place over times that are long compared to the periods of electrons?

11. (II) According to quantum mechanics, in the ground state of a hydrogen atom, an electron has no orbital angular momentum. (a) What is the orbital magnetic moment of this electron? (b) In an excited state of the hydrogen atom, the orbital angular momentum is given by $L = 2\hbar$. What, in units of $\hbar$, is the orbital magnetic moment of an electron in this excited state?

12. (II) The intrinsic magnetic moment of a proton has magnitude $m_{\text{intrinsic}} = 5.56(e/2M)(\hbar/2)$, where M is the proton mass. One way to explain the factor 5.56 is to say that for a fraction of time f, the proton is a point charge e at rest, with $m_{\text{intrinsic}} = 2(e/2M)(\hbar/2)$, and that for a fraction of time $1 - f$, the proton consists of a particle of charge e and mass $0.14M$ circling a neutral core such that the orbital angular momentum is $\hbar$. Calculate f.

13. (II) The electron has a *classical radius* given by $r_0 \equiv e^2/4\pi\epsilon_0 m_e c^2 = 2.8 \times 10^{-15}$ m. This quantity is suggested by dimensional analysis: The particular combination of classical quantities is the only one that can be formed with dimensions of length. Use Eq. (32–11), with m_{orbital} equal to the Bohr magneton, m_B, to show that any charge at the distance of the classical radius will be moving faster than the speed of light. (Assume that all the charge is concentrated at a belt of radius r_0.) Treatment of the magnetic moment of an electron as a classical quantity leads to trouble!

14. (II) The current I in a circular loop of radius R is due to the flow of free electrons (one per atom) in the loop. Show that the gyromagnetic ratio of the loop is independent of I, R, and the density of atoms.

32–3 *Ferromagnetism*

15. (I) A torus is wound with 2000 turns/m of wire. A current of 1 A runs through the wire. If the core of the torus is iron, the internal magnetic field in the core is 1.4 T. What is the magnetization? What is the value of μ/μ_0 for the iron core?

16. (II) A current of 2.0 A flows through a solenoid with 500 turns/m. An iron bar, with $\mu/\mu_0 = 3.5 \times 10^3$, is placed along the solenoid axis. (a) What is the magnetic field inside the iron bar? (b) outside the iron bar, but still within the solenoid?

17. (II) A disk-shaped permanent magnet has a thickness of 1.0 cm and a diameter of 5.0 cm. The magnetic field on the axis near its north pole has the value 0.020 T. What is the current carried by a 50-turn coil of the same dimensions that gives this same value of the magnetic field on the axis?

18. (II) A dipole magnetic field has the value 0.005 T in air. A piece of iron with a susceptibility of 1100 is inserted between the dipole faces. Determine B, H, and M inside the iron.

19. (II) A torus with a central radius of 16 cm and a tube radius of 1 cm is filled with iron of permeability $2200\mu_0$ (Fig. 32–18). There are 1000 turns around the torus. How much current must flow in the winding coil to produce a magnetic field of 1 T inside the torus? Treat the torus as having a constant magnetic field equal to the field at the central radius of the torus.

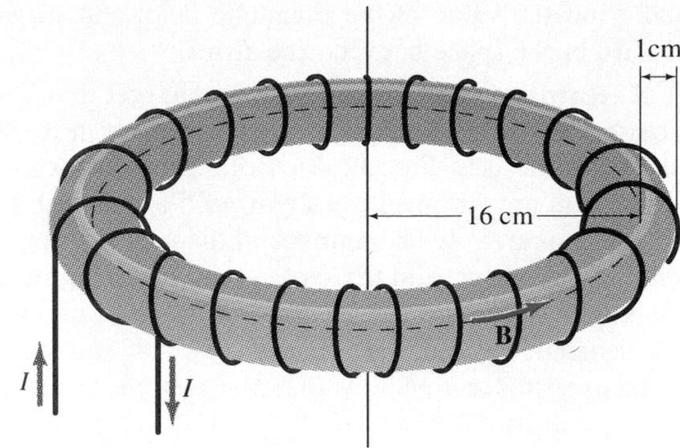

FIGURE 32–18 Problem 19.

20. (II) Figure 32–19 shows a common hysteresis curve for an alloy of ferromagnetic material often used in building magnets. (a) What is the maximum external magnetic field needed to reach the maximum magnetic field inside the alloy? (b) the maximum magnetic field attainable inside the alloy? (c) the maximum magnetic field attainable with no external magnetic field?

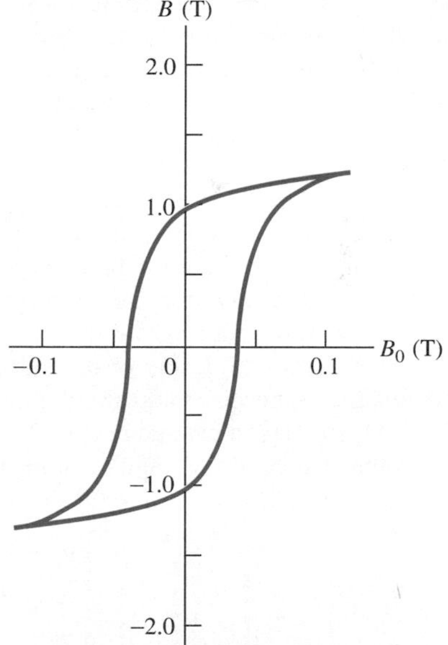

FIGURE 32–19 Problem 20.

21. (II) The maximum value (*saturation value*) of the magnetic field of the material illustrated in Fig. 32–10 is 1.25 T. Find the permeability μ when $\mu_0 H = 3 \times 10^{-4}$ T, a point at which the magnet has reached saturation.

22. (II) Two parallel conducting strips are each 6.0×10^{-4} m thick and 10 cm wide and are separated by a distance of 5.0×10^{-3} m. The space between the strips is filled with a ferromagnetic material whose permeability is $500\mu_0$. Each

strip carries a uniform current of 2.0 A, in opposite directions. Find the value of the magnetic field, and magnetic intensity in the space between the strips.

23. (II) A Rowland ring measures the charge that passes through a sense coil by integrating the current in the sense coil over time (see Fig. 32–9). Both the sense coil and primary coil are wrapped tightly around a material. These coils have an area A. The number of turns in the sense coil is N. The emf induced in the sense coil by a changing magnetic flux in the material (due to a switch that, on closing, passes current through the primary coil) is $\mathscr{E}$, and the sense coil has resistance R. Show that the change in magnetic field is given by

$$\Delta B = -\frac{1}{NA}\int \mathscr{E}\,dt = -\frac{R}{NA}\int I\,dt = -\frac{QR}{NA}.$$

24. (II) A sense coil with a resistance of 0.2 Ω is wrapped tightly in 10 turns around a magnetic material of area 0.12 m^2. When a switch is closed in the primary coil, a charge of 0.6 mC is measured in the sense coil. If the magnetic field was initially zero, what is the new magnetic field in the material? (See Problem 23.)

*32–4 *Diamagnetism*

25. (III) An electron under the influence of some central force moves at speed v_i in a counterclockwise circular orbit of radius R. A uniform magnetic field **B** perpendicular to the plane of the orbit is turned on (Fig. 32–20). Suppose that the magnitude of the field changes at a given rate dB/dt. (a) What are the magnitude and direction of the electric field induced at the radius of the electron orbit? (b) The tangential force on the electron due to the induced electric field increases the electron's speed. Find the value of dv/dt. (c) Assuming that the initial orbital speed was v_i, find the final speed v_f as the magnitude of the magnetic field steadily increases from zero to a final value B_f by integrating dv/dt with respect to time. (d) Using your result for the change in speed, find the change in orbital angular momentum. (e) Use Eq. (32–11) to relate a change in the orbital magnetic moment to the change in the angular momentum.

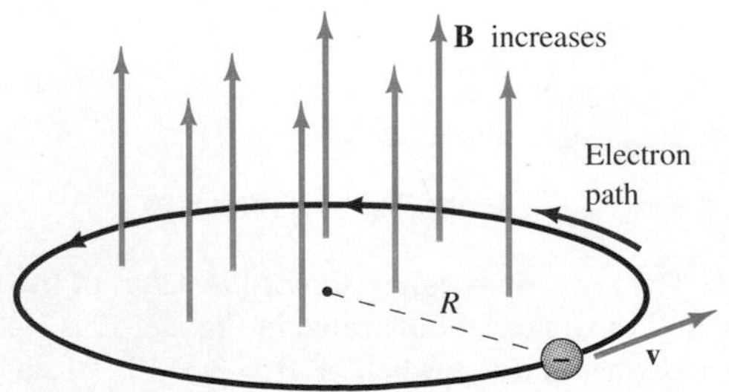

FIGURE 32–20 Problem 25.

26. (III) Refer to Fig. 32–20, but this time assume that the electron circulates clockwise rather than counterclockwise at speed v_i. By applying the same sequence of steps, show that the change in the magnetic moment of the electron's orbit is opposite the direction of change in the external field, just as in the case in which the electron circulates counterclockwise.

27. (III) Refer to Problems 25 and 26. Suppose that there are now two electrons moving at speed v_i in circular orbits of radius R, one clockwise and one counterclockwise. (a) What is the net orbital magnetic moment when the external field is zero? (b) After the external field has reached $\mathbf{B}_f$? (c) Show that the magnetic susceptibility for this system is $\chi_m = -(\mu_0 e^2 R^2/4m_e)\rho_e$, where ρ_e is the electron density.

28. (III) Using the techniques of Problems 25 through 27, estimate the magnetic susceptibility of copper, which has 29 electrons per atom. Assume that all the electrons move in orbits of the same radius, and that 14 move clockwise while 15 move counterclockwise. You will need to calculate the number density of electrons in copper.

*32–5 *Paramagnetism*

29. (II) The temperature of an aluminum sample inside a magnetic field is held constant as the field is increased. Sketch the induced magnetic moment as the magnetic field is increased.

30. (II) A long, straight conducting wire is embedded within an insulating paramagnetic material of magnetic susceptibility 2.6×10^{-4} at 300 K and carries a current of 10 mA. For a temperature of 300 K, find the value of the magnetic intensity as a function of the distance from the wire, as well as the magnetic field.

31. (II) The constant C in Curie's law, Eq. (32–18), can be found by using the techniques of statistical physics to be $C = nm_B^2/k$, where k is Boltzmann's constant and n is the number density of electrons that enter into the paramagnetic effect. Estimate the paramagnetic susceptibility for molybdenum at room temperature by supposing that the density of unpaired electrons equals the number density of atoms. Find the latter by using the atomic weight of molybdenum, 96, and the mass density, 9 g/cm^3.

*32–7 *Nuclear Magnetic Resonance*

32. (I) If the neutron's gyromagnetic ratio is $-1.91e/m_n$, calculate the magnetic moment of the neutron.

33. (II) Assume that it is possible to align perfectly the magnetic moments of protons in 1 mol of hydrogen gas at standard temperature and pressure. What are the magnetization and magnetic field inside the gas?

34. (II) Assume that it is possible to align the magnetic moments of all the neutrons and protons in the nuclei of oxygen. What would the magnetization of 1 mol of the isotopes ^{16}O and ^{18}O be? Assume that oxygen forms a gas of O_2 at standard temperature and pressure.

35. (II) Express the equation $d\mathbf{m}/dt = g_s\,\mathbf{m} \times \mathbf{B}$ relevant to NMR in component form for the case that $\mathbf{B} = B\mathbf{k}$ and

$\mathbf{m} = m_x\mathbf{i} + m_y\mathbf{j} + m_z\mathbf{k}$. (a) Show that m_z is a constant; (b) that $m_x^2 + m_y^2 + m_z^2$ is a constant; and (c) that $m_x = m_1 \cos(\omega t)$ and $m_y = -m_1 \sin(\omega t)$, where ω is *the angular frequency of precession*, satisfy the equation of motion.

36. (II) Calculate the frequency of precession f (see Problem 35) for a proton's magnetic moment in a field of 10^{-1} T. This frequency is in the so-called rf (radio-frequency) range.

General Problems

37. (II) Estimate the diamagnetic susceptibility of the diamond form of carbon, using the formula derived in Problem 27. Take the density of diamond to be 3.5 g/cm^3; the atomic weight, 12; the atomic number, 6; and the atomic radius, 0.75×10^{-10} m. Assume that all the electrons circulate at this radius. Your estimate should be rather good. Compare this result to that found in Problem 28.

38. (II) Large magnets typically consist of wound toruses of ferromagnetic material. There is a gap in the torus, forming a space between pole faces. Show that the magnetic field across the pole faces is the same as the magnetic field inside the ferromagnet by applying Gauss' law for magnetism to a closed surface partly in and partly out of one of the pole faces.

39. (II) The earth's magnetic field is that of a dipole, and the strength of the field at the magnetic north pole is about 0.6×10^{-4} T. Calculate the magnetic moment of the earth. Supposing that this magnetic moment is due to a magnetized iron core whose radius is half the earth's radius, what is the magnetization of the core? If the magnetic moment were due to a circulating belt of current at the radius of the core, what would the magnitude of this current be?

40. (II) A torus of central radius 20 cm and tube radius 1 cm is filled with tungsten. It is wound with 1000 turns of wire and has a 1-A current. The magnetic susceptibility of tungsten is 7.8×10^{-5}. Determine (a) the external magnetic field, $\mathbf{B}_{\text{ext}}$; (b) the magnetic intensity, $\mathbf{H}$; (c) the magnetic field, $\mathbf{B}$; (d) the magnetization, $\mathbf{M}$. (e) Repeat parts (a) through (d) for a torus filled with iron instead of tungsten. The susceptibility of iron is 5.5×10^3.

41. (II) In one of two simple classical models of the electron spin, the charge circulates at the classical electron radius r_0 (see Problem 13). In the other, the total electron charge is spread uniformly over a disk whose radius is the classical radius. Calculate the ratio of magnetic moments for the case in which the overall charge occurs entirely at the classical radius versus the case in which the charge is spread over the disk.

42. (II) The neutron has an internal spin $\mathbf{S}$ of magnitude $\hbar/2$ and a magnetic moment related to the spin by the gyromagnetic ratio, g_S, as in Eq. (32–20). The gyromagnetic ratio is $g_S = -1.91(e/m_n)$. Suppose that a neutron consists of a heavy, positively charged particle of mass M and magnetic moment $e\hbar/2M$, with a lighter, negatively charged particle of mass m orbiting the heavier particle with orbital angular momentum $\hbar$. What would mass m have to be to explain the observed magnetic moment of the neutron? (For simplicity, ignore the motion of the heavier particle about the center of mass.)

43. (III) In our discussion of kinetic theory from Chapter 19, we noted that, according to Boltzmann, the number of systems with a given energy E in a collection of systems in equilibrium at temperature T is given by $Ce^{-E/kT}$. Here C is a constant determined by the requirement that, when all the systems are summed, we find the same total number of systems that we started with. For a collection of N magnetic dipoles at rest in an external magnetic field, we have $N(T) = Ce^{-(-\mathbf{m}\cdot\mathbf{B})/kT} = Ce^{(mB\cos\theta)/kT}$, where θ is the angle between the direction of the dipole and that of the external magnetic field. C is determined by the requirement that the total number of systems N is $N = C\int_0^\pi 2\pi \sin\theta \, d\theta \, e^{(mB\cos\theta)/kT}$. (a) Calculate C. (b) Calculate the average value of $\cos\theta$. (c) Plot $\langle -\cos\theta \rangle$ as a function of mB/kT.

CHAPTER 33

The electrical arc shown here does not form when a switch is closed: It forms only when the switch is opened, because Faraday's law acts to maintain an existing current. The arc is a manifestation of the existing current jumping an air gap.

INDUCTANCE AND CIRCUIT OSCILLATIONS

In electric circuits, resistors are the cause of energy loss, and capacitors are the means by which energy is stored in an electric field. Energy can also be stored in a magnetic field, and inductors are the circuit elements used for this purpose. Their operation is based on Faraday's law, which describes the effects of changing magnetic fields. Inductors are active only when currents change. For this reason, inductors allow a crucial degree of control over circuits with time-varying currents. Electric circuits containing inductors, capacitors, and resistors are analogous to damped harmonic oscillators, and we can understand the behavior of such circuits in terms of the mechanical behavior of the damped harmonic oscillator. The time dependences of currents and voltages are at the heart of the technological applications of circuits. In this chapter we shall concentrate on circuits in which the sources of emf are constant.

33–1 INDUCTANCE AND INDUCTORS

When a circuit contains a changing electric current, that current produces a changing magnetic field. According to Faraday's law, an additional emf may be induced in the circuit. The effects on electric circuits are common ones, because so many

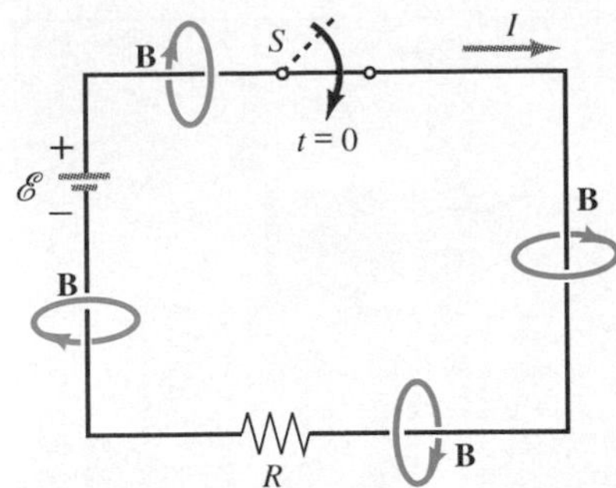

FIGURE 33–1 When the current in a circuit changes, the flux through the circuit changes, leading to an induced emf, according to Faraday's law.

applications of circuits (from computers to televisions to the electrical systems of automobiles) involve time dependence.

Any closed electric circuit in which the current has some time dependence serves to illustrate the general principle. Figure 33–1 shows a circuit with a switch that closes at $t = 0$. When the current increases from zero, the magnetic field around the wire also increases. As the magnetic field grows, the magnetic flux (directed into the page) through the area enclosed by the loop increases. According to Faraday's law, as formulated by Lenz, an emf is induced in the loop and opposes this increase in flux. The induced emf therefore *opposes* the emf of the battery and slows down the flow of current. The principle illustrated here is a simple one: Lenz's formulation of Faraday's law tells us that induced currents always oppose any change in magnetic flux. The result is that *changing currents in circuits lead to induction effects that act to reduce the rate of change of those currents.*

In addition to the effect on the single-loop circuit illustrated above, there is another possible effect, associated with the changing flux through surfaces other than those defined by the current loop. If a second circuit loop is in the general vicinity of the first, the changing magnetic field due to the first circuit can change the magnetic flux through the second circuit, and a current will be induced in the second circuit. This effect in turn produces a changing flux that can affect the first circuit, and so forth. In this case, the two circuits are said to be **linked**. When the first loop induces an emf in itself, we say that there is a **self-inductance**, or **inductance** for short. When the first loop induces a current or emf in a second loop, we say that there is a **mutual inductance** between the two loops. Note that the physical principle behind both self-inductance and mutual inductance is Faraday's law.

Self-Inductance. Because the magnetic field set up around a wire that carries a current I is proportional to I, the magnetic flux through a loop is also proportional to I. The proportionality constant is defined to be the inductance, which we write as L.[†] L depends on the particular surface involved; that is, on the geometry of the loop around which the emf is induced. For a single loop through which a current I flows, the inductance is defined by expressing the magnetic flux, Φ_B, that passes through a surface bounded by the loop as

$$\Phi_B \equiv LI. \qquad (33\text{–}1)$$

The definition of inductance

According to Faraday's law, the emf induced in this loop, $\mathscr{E}$, is the rate of change of the flux through the loop:

$$\mathscr{E} = -\frac{d\Phi_B}{dt} = -L\frac{dI}{dt}. \qquad (33\text{–}2)$$

The emf associated with an inductance

[†] Our use of the symbol L for inductance is another instance of a notation established by long use. L should not be confused with either length or angular momentum.

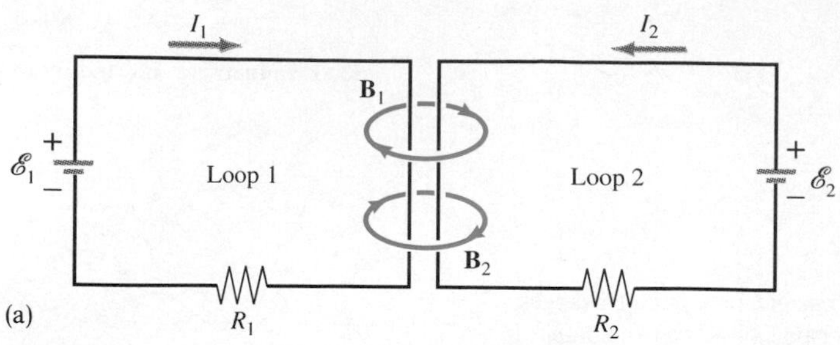

FIGURE 33–2 (a) The flux through a circuit or circuit element may be due to its own current or to the current carried by an adjacent circuit or circuit element. (b) The two coils mounted on the iron core demonstrate mutual induction.

As we shall see, the addition of an induced emf term proportional to the rate of change of a current will have a marked effect on how charges flow through the circuit.

Mutual Inductance. Let us now consider the two adjacent circuits shown in Fig. 33–2. If a current I_1 flows in loop 1 and a current I_2 flows in loop 2, there is a magnetic flux $\Phi_B(1)$ through the area of loop 1 given by

The definition of mutual inductance

$$\Phi_B(1) = L_1 I_1 + M_{12} I_2. \tag{33–3}$$

The first term is due to the current flowing in loop 1, and the constant of proportionality L_1 is the self-inductance of loop 1. The second term is due to the current flowing in loop 2, and the constant M_{12} is the mutual inductance of loop 1 due to loop 2. Equation (33–3) defines the mutual inductance. Both L_1 and M_{12} depend *only* on geometry and, as we shall see, on the medium in which the circuit is embedded. The inductances do not depend on the currents themselves.

The term *mutual* implies a degree of symmetry between the two loops. The magnetic flux through loop 2 has a term proportional to its own current (Fig. 33–3a) and also a term proportional to the current in loop 1 (Fig. 33–3b):

$$\Phi_B(2) = L_2 I_2 + M_{21} I_1. \tag{33–4}$$

The second term involves what might appear to be a new constant, M_{21}. Although it is not obvious, it is nevertheless true that *the mutual inductances are equal—*

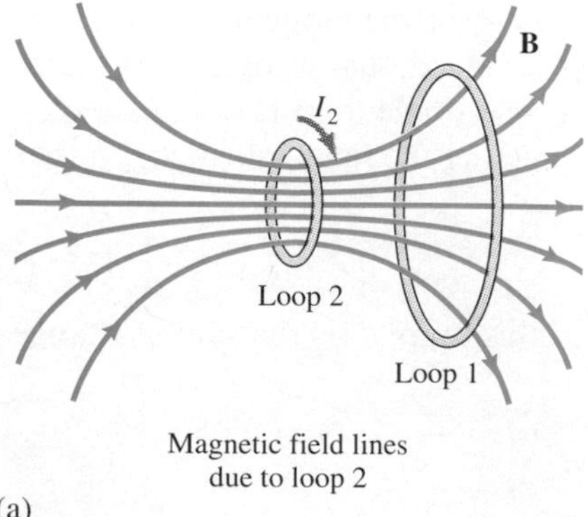

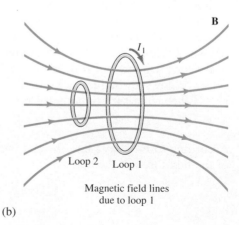

FIGURE 33–3 Mutual inductance. The magnetic flux through loop 2 is due to (a) the magnetic field from its own current, I_2, and (b) the magnetic field from the current I_1 in loop 1.

FIGURE 33–4 Joseph Henry, as depicted in a stained-glass window in the First Presbyterian Church of Albany, New York, the site of Henry's baptism.

$M_{12} = M_{21}$. We will not provide the proof here. It is customary to drop the subscripts and write $M = M_{12} = M_{21}$, the mutual inductance of two loops.

Faraday's law gives the emf induced in loop 2 due to the change in current of loop 1:

$$\mathscr{E}_{21} = -M \frac{dI_1}{dt}. \tag{33–5}$$

The emf associated with a mutual inductance

A similar expression gives the emf induced in loop 1 due to the current in loop 2.

The inductances L and M have SI units of magnetic flux divided by current, or webers per ampere (Wb/A). Like resistance and capacitance, inductance occurs so frequently that it is given its own unit in the SI, the **henry** (H):

$$1 \text{ H} = 1 \text{ Wb/A} = 1 \text{ T}\cdot\text{m}^2/\text{A}. \tag{33–6}$$

The SI unit of inductance

This unit is named after the nineteenth-century investigator Joseph Henry, who contemporaneously with Faraday investigated many effects of induction (Fig. 33–4). An inductance of 1 H is large but not unrealizable. For example, a cylindrical solenoid of area 10 cm^2, length 20 cm, and a winding density of 10 turns/cm has an inductance of 0.25 mH. (We shall soon compute the inductance of a solenoid.)

Elements within circuits with a significant self-inductance provide another source of emf to be taken into account when the loop rule is used for potential changes around a circuit. Such elements (usually in the form of solenoids) are as useful as capacitors and resistors and are typically introduced into circuits deliberately. Whether they are introduced or are intrinsically contained in the circuit, as in Fig. 33–1, these elements are called **inductors** (Fig. 33–5). They are represented in circuit diagrams by the symbol ∞∞∞. The potential change across them, given by Eq. (33–2), is such as to oppose any increase or decrease in the current; this expression of Lenz's law is taken care of by the minus sign in Eq. (33–2). Mutual inductance is generally so small that it is not a factor in the loop rule. The role of mutual inductance in linked circuits is a special one that is very important in *transformers*, devices used to change the magnitude of time-varying voltages. Linked circuits will be discussed further in Chapter 34.

FIGURE 33–5 A collection of inductors.

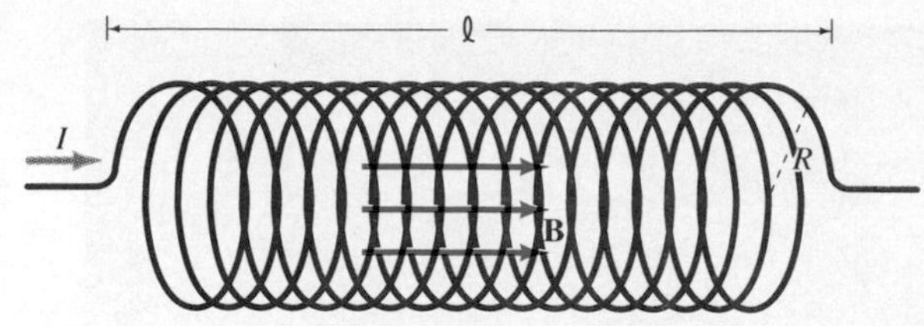

FIGURE 33–6 A solenoid of length ℓ, radius R, and turn density n carries a current I. Only the magnetic field **B** inside is shown; the outside field is zero in the limit where the solenoid is infinitely long.

Finding the Inductance

In order to use the loop rule with inductors present, we must be able to calculate or measure the self-inductance or mutual inductance. Like capacitance, inductance can easily be calculated for only a few simple, but important, geometries. The most important of these is the solenoid. Consider the ideal solenoid of Fig. 33–6, which has length ℓ and radius R. For $\ell \gg R$, the magnetic field within the solenoid is longitudinal and constant and is given by Eq. (30–15):

$$B = \mu_0 n I,$$

where μ_0 is the permeability of free space, n is the number of turns per unit length of wire, and I is the current the solenoid carries. The magnetic flux through one turn of the solenoid is the field B times the cross-sectional area A, $\Phi_B = BA = \mu_0 A n I$. The total magnetic flux is this value times the *total* number of turns $N = n\ell$:

$$\Phi_B = \mu_0 A n^2 \ell I. \tag{33–7}$$

By comparison with Eq. (33–1), the self-inductance is the coefficient of the current:

The inductance of an ideal solenoid

$$\text{for an ideal solenoid:} \qquad L = \mu_0 A \ell n^2. \tag{33–8}$$

EXAMPLE 33–1 During a short time period, the current in a cylindrical coil of length 10 cm, radius 0.5 cm, and 1000 turns of wire is increased at the steady rate of 10^3 A/s. Find the emf induced during this period.

SOLUTION: We are given dI/dt; the induced emf is, according to Eq. (33–2),

$$\mathscr{E} = -L\frac{dI}{dt}.$$

We must next determine the inductance of the solenoid, given by Eq. (33–8). The density of turns is $n = (1000 \text{ turns})/(0.1 \text{ m}) = 10^4$ turns/m. The area of the solenoid is given by $A = \pi r^2$, so

$$L = \mu_0 A \ell n^2 = (4\pi \times 10^{-7}\ \text{T·m/A})[\pi(0.005\ \cancel{\text{m}})^2](0.1\ \text{m})(10^4\ \cancel{\text{m}^{-1}})^2 = 10^{-3}\ \text{H}.$$

The rate of change of current is given as 10^3 A/s, and

$$\mathscr{E} = -(10^{-3}\ \text{H})(10^3\ \text{A/s}) = -1\ \text{V}.$$

As we go around the circuit in the direction of the current, the induced emf will be negative, -1 V. In many electric circuits, 1 V is a significant emf.

As an example of a calculable mutual inductance, consider a solenoid (of length ℓ, radius R_1, winding density n_1, and current I_1) that contains within it a single loop of radius R_2 whose area is oriented perpendicular to the axis of the solenoid (Fig. 33–7). The magnetic field of the solenoid is given by Eq. (30–15), $B = \mu_0 n_1 I_1$. The magnetic flux that passes through the single loop is

$$\Phi_B = BA_2 = B\pi R_2^2 = \mu_0 \pi R_2^2 n_1 I_1.$$

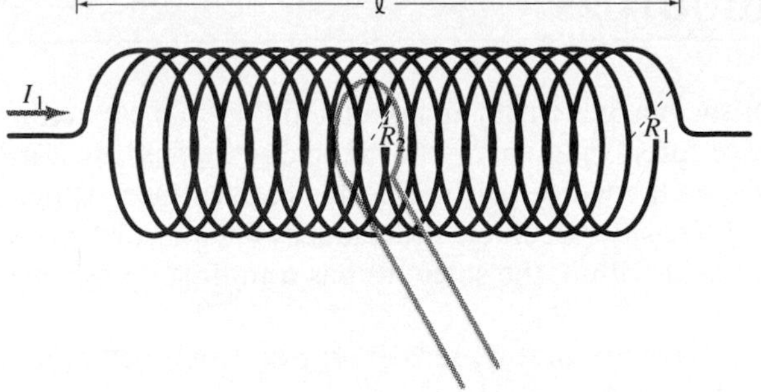

FIGURE 33–7 The area of the single, small loop is oriented perpendicular to the axis of the solenoid.

By definition the mutual inductance, M, is the coefficient of I_1:

$$M = \mu_0 \pi R_2^2 n_1. \tag{33–9}$$

If the single loop is instead a second solenoid with a total number of turns N_2, then the total flux that links that second solenoid contains a factor N_2, and M must be increased by this same factor:

$$M = \mu_0 \pi R_2^2 n_1 N_2. \tag{33–10}$$

EXAMPLE 33–2 Consider the single loop and the solenoid shown in Fig. 33–7. Suppose that the loop carries a current I_2 that is a function of time. Find the emf in the solenoid induced by current I_2.

SOLUTION: In order to apply Faraday's law, we need to find the magnetic flux, Φ_B, due to the loop that links the solenoid. This flux is the mutual inductance of the loop and solenoid times current I_2. The mutuality of the inductance allows us to use the result of Eq. (33–9) for the inductance:

$$\Phi_B = MI_2 = \mu_0 \pi R_2^2 n_1 I_2.$$

The emf in the solenoid is the negative time derivative of this flux:

$$\mathscr{E} = -\frac{d\Phi_B}{dt} = -\mu_0 \pi R_2^2 n_1 \frac{dI_2}{dt}.$$

This example illustrates the usefulness of knowing that the mutual inductance $M_{12} = M_{21}$. To calculate the magnetic flux through the solenoid directly from the dipole field of the single loop would be a formidable task indeed.

The Effects of Magnetic Materials on Inductance

In Chapter 32 we considered modifications due to materials with magnetic properties. We showed how we can write expressions that include the free (or real) current only if we replace the permeability of free space, μ_0, by the permeability of the material, μ. The permeability is given by Eq. (32–7),

$$\mu = \mu_0(1 + \chi_m).$$

Here χ_m is the magnetic susceptibility of the material—negative and small for diamagnets, positive and small for paramagnets, and positive and large for ferromagnets. If a solenoid were filled with a magnetic material, its self-inductance would change by the replacement of μ_0 by μ in Eq. (33–8). For diamagnetic and paramagnetic materials, the susceptibility varies over such a small range that μ remains practically the same as μ_0. For ferromagnetic materials, the magnetic fields (and inductances) may be increased by factors of thousands.

33–2 ENERGY IN INDUCTORS

An inductor plays a role for the magnetic field analogous to that of a capacitor for the electric field: It is a device for storing energy in the magnetic field. Just as parallel plates make a simple capacitor with easily calculable capacitance, a solenoid makes a simple inductor with easily calculable inductance. The parallel-plate capacitor has a uniform electric field within; the solenoid has a uniform magnetic field within.

The work–energy theorem establishes that there is energy in an inductor. Because any emf induced in the inductor opposes the change in current, work must be done by an external source, such as a battery, to cause a current to pass through an inductor. Just how much work is done is a measure of the energy stored in the inductor. To calculate this energy, we proceed in a manner analogous to that for the capacitor in Chapter 26: Calculate the work performed by some external emf, the work required to pass a current through the inductor.

The general expression for the rate dW/dt at which an external emf, $\mathscr{E}_{ext}$, does work when a current I flows (that is, the power, P) is derived from Eqs. (27–26) and (27–27):

$$\frac{dW}{dt} = I\mathscr{E}_{ext}.$$

If the external emf and an inductor are the only two circuit elements, the external emf must be equal but opposite to the induced emf in the inductor, given by Eq. (33–2).† This last equation therefore becomes

$$\frac{dW}{dt} = +LI\frac{dI}{dt}. \tag{33–11}$$

If the current is increasing, the power is positive, meaning that the external source must do work in supplying positive energy to the inductor; the internal energy U_L in the inductor is increasing. If the current is decreasing, the power is negative, meaning that the external source takes energy from the inductor; the inductor's internal energy is decreasing. The net change ΔU_L in the total magnetic energy of the inductor as the current changes from a value I_1 to a value I_2 between the times t_1 and t_2 can be found by integrating the work done by the external source as the current changes. We integrate Eq. (33–11) for dW/dt from an initial time t_1 to a later time t_2:

$$\Delta U_L = \int_{t_1}^{t_2} \frac{dW}{dt}\,dt = \int_{t_1}^{t_2} LI\frac{dI}{dt}\,dt = L\int_{I_1}^{I_2} I\,dI$$
$$= \frac{1}{2}LI_2^2 - \frac{1}{2}LI_1^2. \tag{33–12}$$

In particular, if the inductor carries a current I, then the increase in energy as the current increases from zero, which we refer to simply as the energy of the inductor, is

The energy contained in an inductor

$$U_L = \frac{1}{2}LI^2. \tag{33–13}$$

Equation (33–13) should be compared to the expression for the energy U_C contained in a capacitor of capacitance C that carries charge Q, from Eq. (26–8):

$$U_C = \frac{1}{2}\frac{Q^2}{C}.$$

† In practice there will always be a small resistance in the circuit, at least if superconductors are not involved.

Note that the equation for the energy of a capacitor contains the factor $1/C$, whereas the expression for the energy of an inductor contains the factor L. Recall, however, that the absence of capacitance corresponds to $C = \infty$, whereas the absence of inductance corresponds to $L = 0$. In the limits $C \to \infty$ and $L \to 0$, the respective energies will be zero.

EXAMPLE 33–3 A solenoid is designed to store $U_L = 0.10$ J of energy when it carries a current I of 450 mA. The solenoid has a cross-sectional area A of 5.0 cm^2 and a length ℓ of 0.20 m. How many turns of wire must the solenoid have?

SOLUTION: We want to find the number of turns N, given the current, the length and area of the solenoid, and the energy it stores. To find N, we need to express the energy in terms of it, and for that we need the inductance. The self-inductance of a solenoid is given by Eq. (33–8). This inductance can be written in terms of the total number of turns N rather than the turn density n by using $N = n\ell$, where ℓ is the length of the solenoid. Thus

$$L = \frac{\mu_0 A N^2}{\ell}.$$

The expression for the energy, Eq. (33–13), then reads

$$U_L = \frac{1}{2}\frac{\mu_0 A N^2}{\ell} I^2.$$

We can solve for N:

$$N = \frac{1}{I}\sqrt{\frac{2U_L\ell}{\mu_0 A}} = \frac{1}{4.5 \times 10^{-1}\ \not{A}}\sqrt{\frac{2(0.10\ \text{J})(0.20\ \not{m})}{(4\pi \times 10^{-7}\ \text{N}/\not{A}^2)(5.0 \times 10^{-4}\ \not{m}^2)}}$$
$$= 1.8 \times 10^4 \text{ turns.}$$

The fact that so many turns are needed in Example 33–3 indicates that inductors are less practical for storing energy than capacitors are. The primary role of inductors is not in energy storage but rather in controlling the time dependence of currents in circuits (Fig. 33–8).

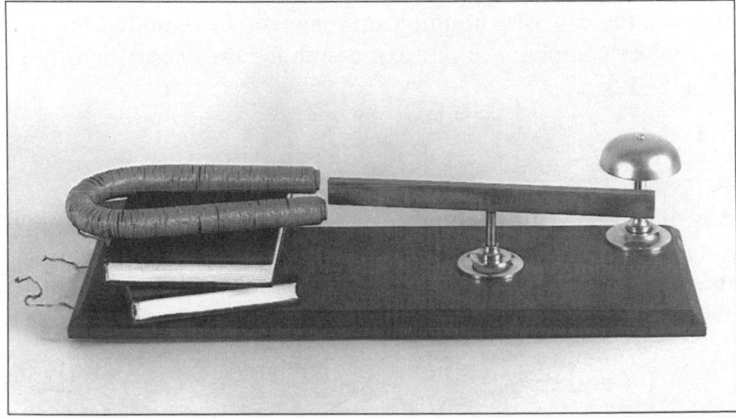

(a)

(b)

FIGURE 33–8 Applications of inductors. (a) This rudimentary apparatus built by Joseph Henry exhibits the principle by which a simple doorbell operates. When the magnet is activated, a current is induced in the metal rod, which swings on a pivot and taps the bell. (b) In a seismograph, a permanent magnet is fixed to the case, while a coil of wire hangs by a spring (or vice versa). When the earth moves as a result of an earthquake or explosion, the fixed magnet vibrates, and a current is induced in the coil. The current activates the recording mechanism of the seismograph.

Note that an inductor has an energy given by Eq. (33–13) even if the current is steady. We have argued that the origin of the effects of inductance are those of Faraday's law, which involves changes in current. How are these two statements reconciled? The energy of an inductor carrying steady current originates in the original buildup of current, even if it occurred in the distant past. Work had to be done to establish the current in the inductor, and if the source of emf is removed suddenly, the inductor then does work to induce a current in the original direction. It is indeed *changes* in current that are at the origin of changes in magnetic energy.

33–3 ENERGY IN MAGNETIC FIELDS

Any current-carrying element in a circuit has a magnetic field and hence a self-inductance. This circuit element has an energy associated with its magnetic field. In Chapter 26 we argued that the electric energy associated with a capacitor is located in the electric field within the capacitor. Similarly, the energy of an inductor is located in its magnetic field. The ideal solenoid presents us with a tool to find the energy density in a magnetic field, because the magnetic field within a solenoid is uniform.

The inductance of an ideal solenoid of area A and length ℓ is given by Eq. (33–8), so from the expression for the total energy of an inductor, Eq. (33–13), we have

$$U_L = \frac{1}{2} LI^2 = \frac{1}{2} \mu_0 A \ell n^2 I^2. \tag{33–14}$$

We also know that the magnetic field in the solenoid is proportional to the current. Equation (30–15) gives the precise connection, $B = \mu_0 nI$. If we substitute for I in terms of B in Eq. (33–14), we get

$$U_L = \frac{1}{2} \frac{B^2}{\mu_0} A\ell. \tag{33–15}$$

The volume within the solenoid is $A\ell$. Because the magnetic field is constant within the solenoid, we may identify the **energy density**, u_B, the energy per unit volume of the magnetic field, as

The energy density of a magnetic field

$$u_B = \frac{1}{2} \frac{B^2}{\mu_0}. \tag{33–16}$$

This result generalizes to the case of a nonuniform magnetic field, no matter how it is produced. It should be compared to our expression for the energy density of an electric field, Eq. (26–12):

$$u_E = \frac{1}{2} \epsilon_0 E^2,$$

a result derived in a similar way. It is important to realize that *energy is located within the electric and magnetic fields themselves.*

When both magnetic and electric fields are present, the energy density is the sum of both magnetic and electric energy densities:

$$u = u_B + u_E = \frac{1}{2}\left(\frac{B^2}{\mu_0} + \epsilon_0 E^2\right). \tag{33–17}$$

EXAMPLE 33–4 A large electromagnet produces a magnetic field of 1 T. Compare the energy density associated with this field to that of the largest electric field in air, about 10^6 V/m.

SOLUTION: Equations (33–16) and (26–12) express the energy density in magnetic and electric fields, respectively. From them, the ratio of the magnetic and electric energy densities is

$$\frac{u_B}{u_E} = \frac{\frac{1}{2}\frac{B^2}{\mu_0}}{\frac{1}{2}\epsilon_0 E^2} = \frac{1}{\mu_0\epsilon_0}\frac{B^2}{E^2}.$$

Note that as in Eq. (30–4), the factor $1/\mu_0\epsilon_0$ has dimensions of speed squared, and this speed is the speed of light, c. In our case, the magnitudes of both the magnetic and electric fields are given, and

$$\frac{u_B}{u_E} = \frac{1}{(4\pi \times 10^{-7}\ \text{N/A})(8.85 \times 10^{-12}\ \text{F/m})}\frac{(1\ \text{T})^2}{(10^6\ \text{V/m})^2} = 9 \times 10^4.$$

This high ratio of magnetic to electric energy densities is not relevant to the energies in a circuit, because the magnetic fields associated with changing currents are normally very small.

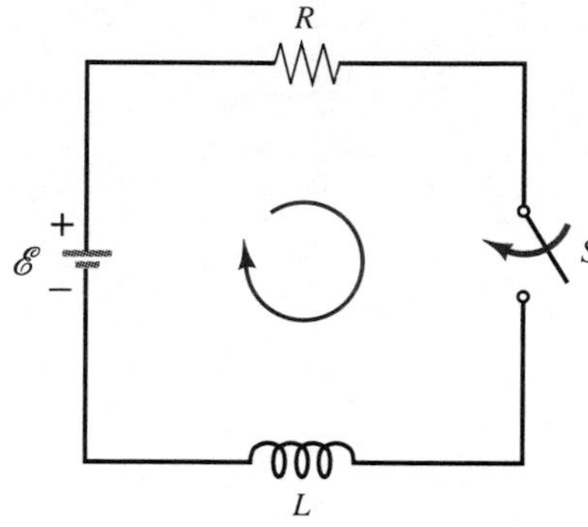

(a)

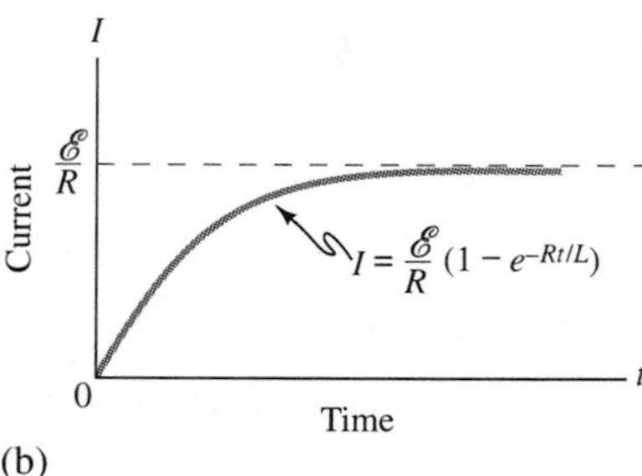

(b)

FIGURE 33–9 (a) An RL circuit, including a source of constant emf. (b) If the switch is closed at $t = 0$, the current rises from zero to a steady-state value only after a period of time determined by the ratio L/R.

33–4 OSCILLATIONS IN CIRCUITS

Because the potential drop across an inductor depends on how rapidly the current passing through it changes, we expect the presence of inductors in circuits to lead to new time-dependent phenomena. A simple case to study is a circuit with a source of emf $\mathcal{E}$, a resistor of resistance R, and an inductor of inductance L (Fig. 33–9). We call such a circuit an **RL circuit**. A switch is included to allow us to control the initial conditions. When we close the switch, the inductor senses the changing current, and, by Lenz's law, opposes it; the result is that the inductor forces the current to build up over time rather than to make a sudden jump. To see this more quantitatively, we apply the loop rule to the circuit in the direction of the pink arrow in Fig. 33–9a. The sum of potential changes around the circuit is

$$\mathcal{E} - IR - L\frac{dI}{dt} = 0. \tag{33–18}$$

The solution to this differential equation for I will verify our physical reasoning about the behavior of currents in the presence of inductors. To find the solution, compare Eq. (33–18) with Eq. (28–24), which comes from applying the loop rule to the RC circuit shown in Fig. 33–10:

$$\mathcal{E} - \frac{Q}{C} - R\frac{dQ}{dt} = 0.$$

Here Q is the charge on the capacitor. The first differential equation, which determines the current in RL circuits, has exactly the same *form* as the equation for the charge in RC circuits. What characterizes the RC circuit is the transient behavior of the charge with time. The similarities between the RC and RL circuits are

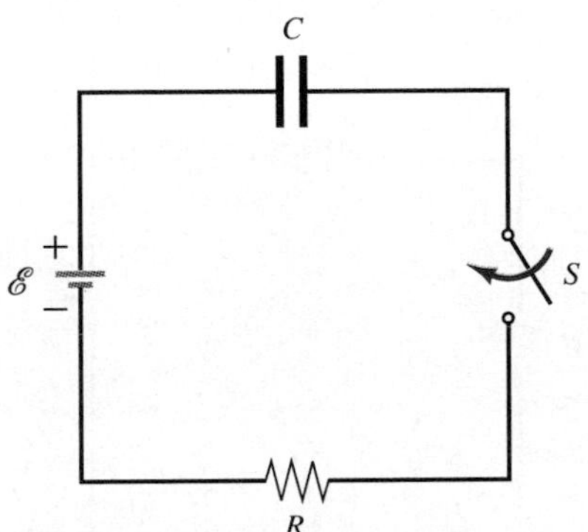

FIGURE 33–10 An RC circuit is analogous to the RL circuit of Fig. 33–9a.

TABLE 33–1

ANALOGY BETWEEN *RC* AND *RL* CIRCUITS

	RC Circuit Parameter	*RL Circuit Parameter*
Variable	Q	I
Coefficient of variable	$1/C$	R
Coefficient of $\frac{d}{dt}$ (variable)	R	L
Time constant	RC	L/R

summarized in Table 33–1, and the solution of the differential equation for current, Eq. (33–18), is the same as the solution of Eq. (28–24) for charge if we make the substitutions indicated in the table. In particular, if we replace RC with L/R, the current in the RL circuit will have time dependence $e^{-t/(L/R)}$. The RL circuit is said to have a *time constant* L/R. Large values of the time constant (large L and/or small R) mean that the transient time dependence is slow to disappear, whereas small values of the time constant (small L and/or large R) mean that the transient behavior quickly disappears. The origin of the transient behavior lies in the role of the inductor. An emf is induced across the inductor as long as the current changes, but not otherwise.

For complete solutions of Eq. (33–18) for the current, including the initial conditions, we can apply directly the results of Chapter 28 for the solution of the differential equation for the RC circuit. As in Chapter 28, it is important to understand *physically* how different initial conditions will affect the solution. As a guiding principle, keep in mind that

The current in an inductor cannot change instantaneously. When current is constant, an inductor plays no role in a circuit.

The current in an inductor never changes instantaneously, and when the current settles down to a constant value, the inductor plays no role in the circuit.

To take an example, suppose that the switch in Fig. 33–9a is closed at $t = 0$. Before $t = 0$, there is no current. After a long time the current settles down to the constant value $I = \mathscr{E}/R$, because when the current is constant, the inductor plays no role in the circuit. The function with exponential time dependence that satisfies these limits is

$$I = \frac{\mathscr{E}}{R}\left[1 - e^{-t/(L/R)}\right] = \frac{\mathscr{E}}{R}(1 - e^{-Rt/L}). \tag{33–19}$$

Figure 33–9b is a graph of current versus time for this case. At $t = 0$, the exponential term is unity and $I = 0$, a result consistent with the principle that the current cannot change instantaneously. As $t \to \infty$, the exponential term drops out and I approaches $\mathscr{E}/R$, as though the inductor were not present at all. Substitute Eq. (33–19) into Eq. (33–18) to verify that it is indeed a solution (see Problem 37).

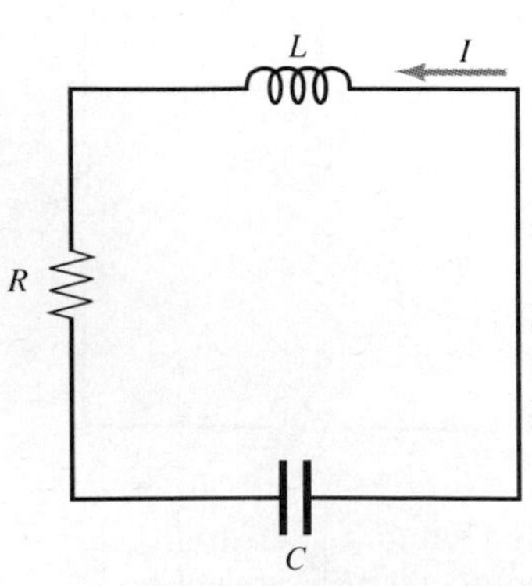

FIGURE 33–11 A prototype *RLC* circuit.

When resistance, inductance, and capacitance are all present in a circuit, as in Fig. 33–11, we have what is generally called an **RLC circuit**. Typically, there will also be a switch and a battery, which provides the current. We assume that a current is already present in the circuit. Following this circuit in the direction of the arrow, the loop rule gives

$$-L\frac{dI}{dt} - IR - \frac{Q}{C} = 0. \tag{33–20}$$

Because

$$I = \frac{dQ}{dt},$$

Eq. (33–20) can also be written as

$$-L\frac{d^2Q}{dt^2} - R\frac{dQ}{dt} - \frac{Q}{C} = 0. \qquad (33\text{–}21)$$

This equation is a differential equation that, with the appropriate initial conditions, determines the charge on the capacitor and hence the current in the circuit.

Equation (33–21), which contains the charge together with its first and second derivatives, has an analogue in mechanics problems that involve masses on springs in the presence of drag. The spring force $-kx$ is proportional to the position x of the mass, measured from the equilibrium position; the drag force $-bv$ is proportional to the velocity or first derivative of x, and the acceleration term ma in Newton's second law is proportional to the second derivative of x. When we put the acceleration term on the same side of the equation as the forces, Newton's second law reads

Damped harmonic oscillators are discussed in Section 13–7.

$$-m\frac{d^2x}{dt^2} - b\frac{dx}{dt} - kx = 0 \qquad (33\text{–}22)$$

for the damped harmonic oscillator. This equation describes the motion of a mass at the end of a spring immersed in a fluid that gives rise to a drag force. This is a physical system about which we have some intuition. The mathematical equivalence of Eqs. (33–20) and (33–22), described in detail in Table 33–2, is immensely helpful in understanding the effects of inductors.

An *RLC* circuit is analogous to a damped harmonic oscillator.

The remarkable feature of the motion of a mass on a spring is, of course, that it is harmonic. Damping modulates the harmonic behavior by changing the period slightly and imposing an envelope on the motion, in the form of a falling exponential. We thus expect, at least if the damping term $-R\,dQ/dt$ is not too large, that the charge on the capacitor of an *RLC* circuit will also have harmonic time dependence within an envelope that falls exponentially with time. Current is the time derivative of charge, and because the derivatives of sines, cosines, and exponentials are cosines, sines, and exponentials, the current will also be harmonic in time within an envelope that falls exponentially with time.

No Resistance. Let us first suppose that there is no damping: We set $R = 0$. In this case we expect the system of inductor and capacitor to behave like a mechanical system of a mass and spring with no damping, which is the case of simple harmonic oscillation. The angular frequency of the analogous mechanical system is $\omega = \sqrt{k/m}$, so, using the equivalences of Table 33–2, the angular frequency of the **LC circuit** is

$$\omega = \frac{1}{\sqrt{LC}}. \qquad (33\text{–}23)$$

TABLE **33–2**

ANALOGY BETWEEN *RLC* CIRCUITS AND DAMPED HARMONIC MOTION

	Circuit	*Harmonic Motion*
Variable	Q	x
Coefficient of variable	$1/C$	k
Coefficient of $\frac{d}{dt}$ (variable)	R	b
Coefficient of $\frac{d^2}{dt^2}$ (variable)	L	m

The general form of the charge on the capacitor is then

$$Q = Q_1 \sin(\omega t) + Q_2 \cos(\omega t), \tag{33–24}$$

or, equivalently,

$$Q = Q_0 \cos(\omega t + \phi). \tag{33–25}$$

The constants, either Q_1 and Q_2 or Q_0 and ϕ, are determined by the initial conditions. For example, if we know that the capacitor has already been charged to a given total charge at time $t = 0$ and that the current at $t = 0$ is also zero (because a switch allowing current to flow in the *RLC* circuit is closed at that point), then these two conditions determine the two unknown constants. Exercises of this type are similar to those we treated in Chapter 13.

EXAMPLE 33–5 The capacitor charge in a circuit containing only a capacitance C and inductance L in series has the form $Q = Q_0 \cos(\omega t)$. (a) Find the voltage across the inductor. (b) If $L = 12\ \mu\text{H}$ and $C = 0.80\ \mu\text{F}$, what is the period for the voltage in part (a)?

SOLUTION: (a) We are given the capacitance and inductance and can compute any changes in the current from the specified changes in the charge on the capacitor plates. In terms of these known quantities, the unknown voltage across the inductor is

$$V_L = -L\frac{dI}{dt} = -L\frac{d^2Q}{dt^2}.$$

We have

$$\frac{d^2Q}{dt^2} = Q_0\frac{d}{dt}\left[\frac{d}{dt}\cos(\omega t)\right] = Q_0\frac{d}{dt}[-\omega \sin(\omega t)] = -\omega Q_0[\omega \cos(\omega t)]$$
$$= -\omega^2 Q_0 \cos(\omega t) = -\omega^2 Q.$$

Thus

$$V_L = \omega^2 LQ = \omega^2 LQ_0 \cos(\omega t).$$

Just as in simple harmonic motion, where the acceleration of the mass on the end of a spring is proportional to the mass's displacement, in *LC* circuits the voltage across the inductor is proportional to the charge on the capacitor.
(b) The unknown period of the voltage across the inductor is identical to the period for the charge on the capacitor:

$$T = \frac{2\pi}{\omega}.$$

With $\omega = 1/\sqrt{LC}$, we have

$$T = 2\pi\sqrt{LC} = 2\pi\sqrt{(12 \times 10^{-6}\ \text{H})(0.80 \times 10^{-6}\ \text{F})} = 1.9 \times 10^{-5}\ \text{s}.$$

Resistance Is Introduced. Let us turn now to the case in which L, C, and R are all present, as in the circuit of Fig. 33–11. We again follow the techniques of Chapter 13, as applied to the damped harmonic oscillator, in finding a solution to Eq. (33–20). In fact, we can apply our solution to the damped oscillator, Eq. (13–54), if we use the equivalences in Table 33–2. The solution to Eq. (33–20) takes the general form

$$Q = Q_0 e^{-\alpha t} \cos(\omega' t + \phi). \tag{33–26}$$

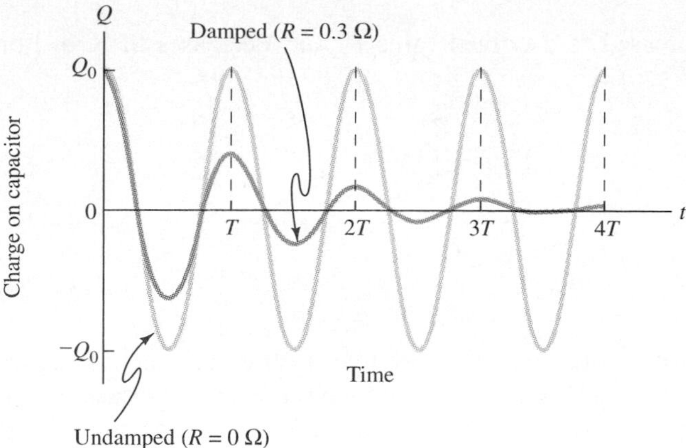

FIGURE 33–12 Comparison of an *RLC* circuit with and without damping: the charge on the capacitor versus time. The damped case has a very slightly larger period. *T* is the period of the undamped oscillator.

The constants α and ω' are determined by substitution back into the original differential equation, Eq. (33–21), to be

$$\alpha = \frac{R}{2L} \tag{33–27}$$

and

$$\omega'^2 = \frac{1}{LC} - \frac{R^2}{4L^2} = \frac{1}{LC} - \alpha^2. \tag{33–28}$$

The constants Q_0 and ϕ are determined from the initial conditions.

The fact that the exponential damping constant α depends only on L and R is consistent with the result we found in studying *RL* circuits, which have transient behavior. The damping factor α for the full *RLC* circuit is *half* the size of the time constant in *RL* circuits. The resistive element R is the crucial element in damping, because energy is dissipated within it. The exponentially falling damping factor $e^{-\alpha t}$ forms a decreasing envelope for the harmonic behavior. In addition, the modified angular frequency ω' of the harmonic behavior is shifted from the angular frequency ω of the undamped circuit (an *LC* circuit) by the α-dependence of ω'. If the damping constant α is *small* compared to the undamped angular frequency $\omega = 1/\sqrt{LC}$, then ω' is only slightly less than ω. In Fig. 33–12 we plot the behavior of the capacitor charge for the case $L = 1$ H, $C = 1$ F, and $R = 0.3\ \Omega$, comparing it to the previous case where we had set $R = 0\ \Omega$. The period in the damped case is only very slightly larger than the period for the undamped case.

Figure 33–13 shows what happens to the charge if R is increased to $1.5\ \Omega$, to $2\ \Omega$, and to $4\ \Omega$. What is the explanation for this behavior? Equation (33–28)

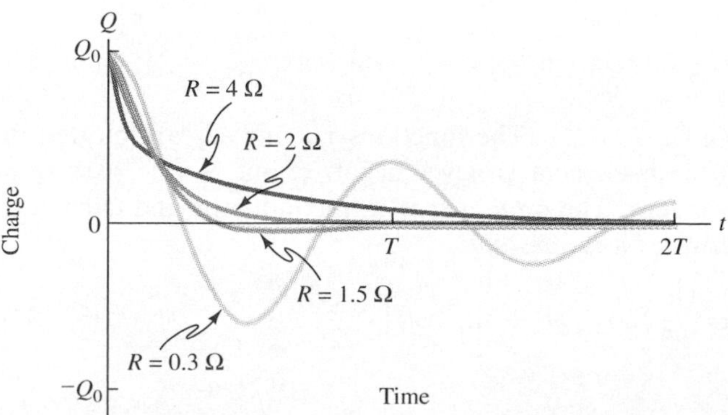

FIGURE 33–13 The *RLC* circuit of Fig. 33–11 for various values of *R*. The critical value R_c is $R = 2\ \Omega$. There is no oscillatory behavior, only damping, when $R > R_c$.

shows that when R is increased to a critical value R_c, ω'^2 decreases to zero. For $R = R_c$, we have

$$0 = \frac{1}{LC} - \frac{R_c^2}{4L^2},$$

which has the solution

$$R_c = 2\sqrt{\frac{L}{C}}. \tag{33–29}$$

When ω'^2 is zero, there is no more oscillation; we refer to this case, as in Chapter 13, as *critical damping*. The value $R = 2\ \Omega$ in Fig. 33–13 represents this case: When $L = 1$ H and $C = 1$ F, we have $R_c = 2\ \Omega$. For values of R larger than the critical value, there is *overdamping*. This kind of motion has its mechanical analogue in the motion of a mass at the end of a spring when the mass is moving in a jar of molasses. The actual value of the exponential decay factor is easily obtained by attempting a solution of the form $Q = Q_0 e^{-\alpha t}$ and solving for α. An example of this behavior is the case $R = 4\ \Omega$ (a value greater than R_c) in Fig. 33–13.

33–5 ENERGY IN *RLC* CIRCUITS

Energy is an important consideration in the harmonic oscillator, and we expect it to be similarly important in *RLC* circuits. We again consider cases in which there is no damping and in which damping is introduced.

No Resistance. We set $R = 0$, meaning that there is no damping. Suppose that the initial conditions are such that the charge on the capacitor is given by

$$Q = Q_0 \cos(\omega t). \tag{33–30}$$

The current flowing through the circuit is then

$$I = \frac{dQ}{dt} = -\omega Q_0 \sin(\omega t). \tag{33–31}$$

One full period of both of these functions is plotted in Fig. 33–14. The energy contained in a capacitor is given by Eq. (26–8), $U_C = Q^2/2C$, or

$$U_C = \frac{Q_0^2}{2C}\cos^2(\omega t). \tag{33–32}$$

Equation (33–14) gives the magnetic energy in an inductor, $U_L = \frac{1}{2}LI^2$:

$$U_L = \frac{1}{2}L\omega^2 Q_0^2 \sin^2(\omega t) = \frac{Q_0^2}{2C}\sin^2(\omega t). \tag{33–33}$$

We have used $\omega = 1/\sqrt{LC}$, Eq. (33–23). The functions U_C and U_L are plotted in Fig. 33–15. The energy in each element is never negative, but as one rises to a maximum, the other falls to zero. The *total* energy in the inductor and capacitor is the sum of these two terms and is *constant*:

$$U = \frac{Q_0^2}{2C}[\cos^2(\omega t) + \sin^2(\omega t)] = \frac{Q_0^2}{2C}. \tag{33–34}$$

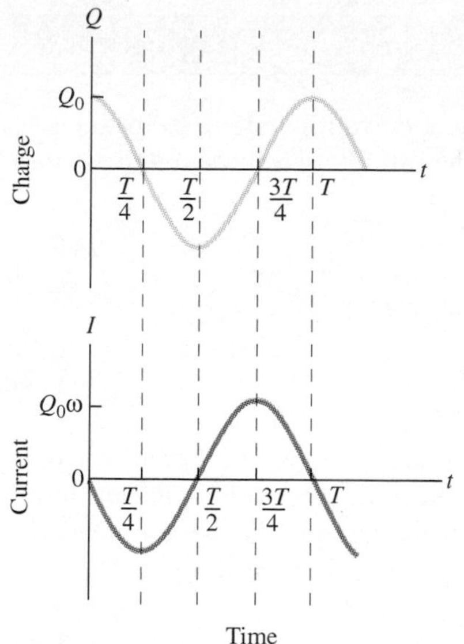

FIGURE 33–14 The capacitor charge and the current in a circuit that contains only inductance and capacitance are harmonic functions of time.

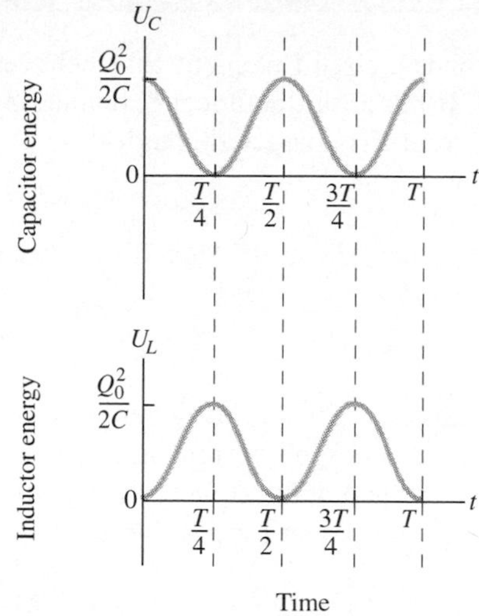

FIGURE 33–15 The energies of capacitor and inductor for charge and current of Fig. 33–11, but with $R = 0$. These oscillate out of phase.

The two circuit elements swap the energy back and forth harmonically, just as in the mechanical oscillator the total energy, a harmonic sum of the potential energy of the spring and the kinetic energy of the attached mass, is constant. In Fig. 33–16a the capacitor starts out fully charged at $t = 0$, and the current flow is zero. The electric field in the capacitor is a maximum, and all the energy is stored there. Current starts to flow, reaching a maximum at $t = T/4$ (Fig. 33–16b). At this point the electric field in the capacitor is zero, and all the energy is stored in the magnetic field of the inductor. At $t = T/2$, the capacitor is again fully charged but with the positive charge on the opposite plate; the current is zero, and all the energy is back in the capacitor (Fig. 33–16c). The process now starts again in reverse, and all the energy is back in the inductor at $t = 3T/4$, when the current is a maximum (Fig. 33–16d). At $t = T$ the entire process starts again (Fig. 33–16e). Without a resistor to dissipate energy, the oscillation will continue forever.

Resistance Is Introduced. The role of resistance is easily understood in terms of energy. Because power, P, is voltage times current, the *rate* at which energy is expended in a resistor is the product of the current and the potential change $V_R = IR$ across it:

$$P_R = IV_R = I^2R. \tag{33–35}$$

This power is proportional to the current squared and is always positive. Energy is *always* lost to Joule heating in a resistor, regardless of the sign of the current. This is the origin of the exponential damping in *RLC* circuits. The power loss in resistors should be contrasted to the equivalent expressions for inductors ($P_L = IV_L$) or capacitors ($P_C = IV_C$). The rate of energy expenditure can be positive or negative in both the inductor and the capacitor, according to the situation [see Eq. (33–11)]. Unlike the resistor, these elements sometimes take energy from the other circuit elements, and sometimes they give it back.

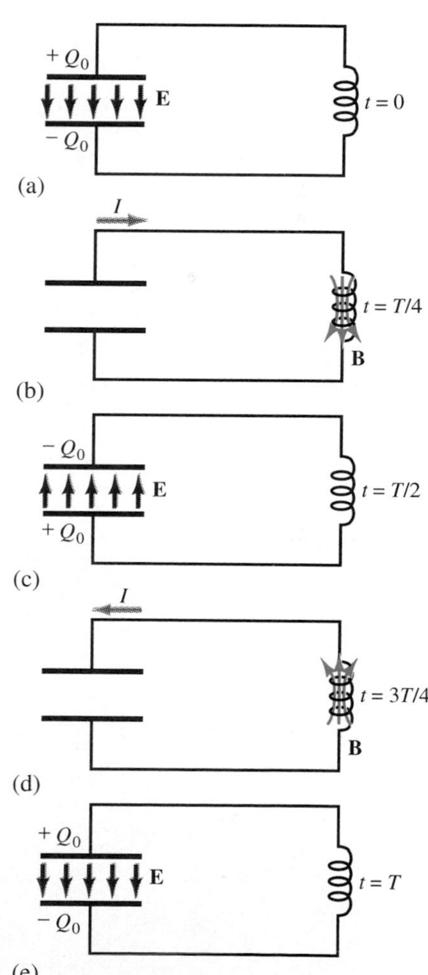

FIGURE 33–16 (a)–(e) Schematic diagrams of how the electric and magnetic fields, and therefore the energies, of Fig. 33–15 are realized within the capacitor and inductor: a time sequence.

An inductor is a circuit element that behaves as a current-carrying loop, such as a solenoid. It has an inductance, L, defined by the ratio between the magnetic flux and the current that passes through it:

$$\Phi_B \equiv LI. \tag{33–1}$$

By Faraday's law, the emf induced in this circuit element is

$$\mathscr{E} = -\frac{d\Phi_B}{dt} = -L\frac{dI}{dt}. \tag{33–2}$$

Faraday's law also shows that when there are adjacent loops in a circuit (or pair of circuits), the changing current in one loop induces an emf in the adjacent loop. In this case the geometrical factor is the mutual inductance, M, which measures both the emf induced in loop 2 due to the change in current in loop 1,

$$\mathscr{E}_{21} = -M\frac{dI_1}{dt}, \tag{33–5}$$

and the emf in loop 1 due to the current in loop 2. Inductance is measured in henries (H) in the SI. The emf in an inductor is one more term to add to the loop rule.

A simple, calculable inductance is that of an ideal solenoid:

$$\text{for an ideal solenoid:} \qquad L = \mu_0 A \ell n^2. \tag{33–8}$$

Here A is the area, ℓ is the length of the solenoid, and n is the number of turns per unit length.

The energy carried in an inductor is given by

$$U_L = \frac{1}{2}LI^2. \tag{33–13}$$

Just as the energy of a capacitor is carried by the electric field in the capacitor, the energy of an inductor is in the magnetic field. The energy density, or energy per unit volume, carried by a magnetic field is found by comparing the known field within a solenoid with the energy carried by the solenoid, and is given by

$$u_B = \frac{1}{2}\frac{B^2}{\mu_0}. \tag{33–16}$$

The combination of inductance, capacitance, and resistance in circuits with and without batteries leads to interesting time dependence for currents and charges. The current in the inductor never changes instantaneously. When an inductor is placed in a circuit with a battery and a resistor, the loop rule produces an equation for the current characterized by transient exponential behavior, with time constant L/R. Such a circuit is called an RL circuit. When a capacitor is added to the circuit, the loop rule produces an equation for the charge on the capacitor that has the same form as the equation for the position of a mass on the end of a spring, with damping proportional to the speed of the mass. The electric circuit, called an RLC circuit, has damped oscillations. In particular, if the resistor is eliminated, the resulting circuit supports harmonic oscillations with a continual exchange of energy between the capacitor and the inductor. The angular frequency of these free oscillations is

$$\omega = \frac{1}{\sqrt{LC}}. \tag{33–23}$$

QUESTIONS

1. Two electric circuits are placed near one another. Each circuit has self-inductance. Must there be a mutual inductance?
2. Does it take more work to cause current to flow through a coil of wire than through the same wire when it is straight?
3. Consider two circular coils that are in a variety of configurations (Fig. 33–17). Assuming that the separation between the coils is roughly the same for the various configurations, can you order the mutual inductances from largest to smallest?

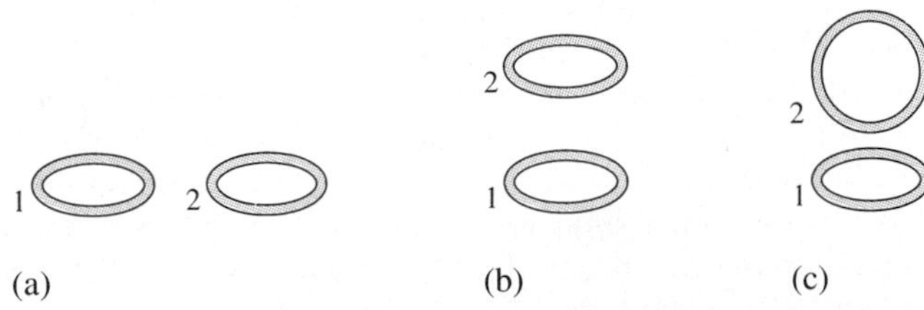

FIGURE 33–17 Question 3.

4. Why might Faraday's law cause the lights in a house to dim when an electrical appliance that uses a lot of energy, such as an electric clothes dryer, is turned on?
5. Is it possible to calculate the mutual inductance between a straight wire and a wire loop?
6. Why do you sometimes see a spark at a light switch when the switch is turned off? Is there a spark when the switch is turned on? Why or why not?
7. Describe how you could measure inductance with a battery of known emf, a known resistor, a voltmeter, and a timer.
8. The energy of a capacitor is in the electric field nearby, and the energy of an inductor is in the magnetic field around it. Can you think of any experiments that would show unequivocally that this energy is in the field itself, distributed locally in space, according to Eq. (33–13)?
9. A light bulb is placed in series with a resistor and in parallel with a coil of large inductance and negligible resistance. When a switch that connects a battery to this circuit is closed, the light bulb flashes before glowing dimly. When the switch is opened, the bulb flashes again before going out. Explain.
10. In the oscillations of an LC circuit, the energy is transferred from the electric field around the capacitor to the magnetic field around the inductor. How does the energy get from one place to the other?
11. Given your knowledge of the largest and smallest practical sizes of capacitors and inductors, what would you estimate is the electronic oscillator with the smallest frequency possible? the largest?
12. A solenoid has magnetic flux outside as well as inside, because magnetic field lines must close on themselves. Does this mean that there is magnetic energy outside the solenoid as well as inside it?
13. In Section 33–3 the magnetic energy calculation for a solenoid used the inductance of a portion of the solenoid, and that was translated into the energy density expression $B^2/2\mu_0$. The magnetic field must come around the outside of the solenoid, because all magnetic field lines are closed. Why does the above calculation give the correct answer for the energy density?
14. We have made an analogy between a damped harmonic oscillator and an RLC circuit. What mechanical quantities are analogous to the energies $LI^2/2$ and $Q^2/2C$ of the RLC circuit?
15. How would you go about finding a generalization of Eq. (33–13) when two circuits with different currents are placed in such close proximity that their mutual inductance plays a role?
16. The magnetic energy density $B^2/2\mu_0$ has the dimensions of pressure and may be viewed as a magnetic pressure. Use this interpretation to justify the attraction/repulsion of two parallel wires that carry currents in the same/opposite directions.

PROBLEMS

33–1 *Inductance and Inductors*

1. (I) A wire loop has an inductance of 2 mH when a current of 30 mA passes through the circuit. What is the value of the magnetic flux that passes through the loop?
2. (I) What is the self-inductance per unit volume of a solenoid?
3. (I) Calculate the mutual inductance of a solenoid 30 cm long of radius 1.5 cm with 500 turns, and a single loop of radius 2.0 cm centered on the solenoid, with its area perpendicular to the axis of the solenoid.
4. (I) An investigator passes through a solenoid a current that changes at the rate of 100 A/s and measures an induced emf of 0.02 V. The length and diameter of the solenoid are 11 cm and 1.3 cm, respectively. What is the number of turns?
5. (I) The emf induced in an isolated circuit when the current in the circuit is changing by 10 A/s is 0.3 V. What is the self-inductance?
6. (II) An electrical engineer needs an inductor capable of producing an emf of 4.0 mV. The current source available produces current of the form $I = I_0 \cos(\omega t)$, with $I_0 = 2.0$ A and $\omega = 3.8 \times 10^2$ rad/s (of frequency 60 Hz). What size inductor should be used?
7. (II) A current of 1 A flows through a circuit placed in isolation. A magnetic flux of 0.010 T·m² passes through the circuit area. When this circuit is placed near another circuit

with a current flow of 2 A, the magnetic flux through the first circuit increases to 0.012 $T \cdot m^2$. (a) What is the mutual inductance of the two circuits? (b) How much magnetic flux passes through the second circuit, whose self-inductance is 1 mH?

8. (II) What is the self-inductance of the single inductor equivalent to two inductors of values L_1 and L_2, respectively, placed in series? Neglect the mutual inductance.

9. (II) Consider two inductors with inductances L_1 and L_2, respectively, connected in parallel. What is the value of the single equivalent inductance that could replace the two inductances, assuming that the mutual inductance can be neglected?

10. (II) Equation (33–8) was derived for an ideal cylindrical solenoid. Show that this result holds also for a solenoid of any shape cross section, provided that the length is large compared to any cross-sectional measure.

11. (II) Consider the cylindrical solenoid and ring illustrated in Fig. 33–18. The solenoid, of diameter $d_1 = 3.0$ cm, has length $\ell = 15$ cm and 20 turns of wire. The ring of wire inside, with diameter $d_2 = 1.0$ cm and area perpendicular to the solenoid's axis, is connected by two wires to a single resistor of resistance $R = 200\ \Omega$. The current I_1 is in the form of a pulse that starts to rise linearly at $t = 0$ s. The current reaches a maximum of 10 A at $t = 0.10$ s, then starts to descend linearly; when the current reaches 0 A at $t = 0.20$ s, it ceases to flow: $I_1 = 0$ for $t < 0$ s; $I_1 = (100\ \text{A/s})t$ for $0 < t < 0.10$ s; $I_1 = (10\ \text{A}) - (100\ \text{A/s})t$ for $0.10\ \text{s} < t < 0.20$ s; $I_1 = 0$ for $t > 0.20$ s. Find the current I_2 induced in the ring as a function of time.

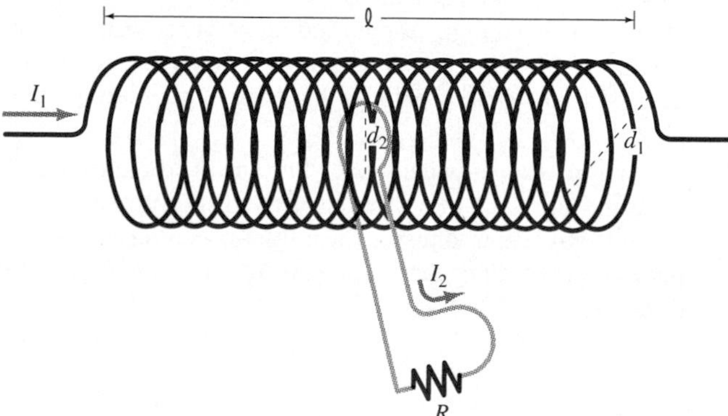

FIGURE 33–18 Problem 11.

12. (II) The ring contained within the cylindrical solenoid of Example 33–2 is replaced by a second cylindrical solenoid, of length ℓ_2, radius R_2, and turn density n_2. Calculate the mutual inductance of this system.

13. (II) The current in an inductor has the periodic triangular form plotted in Fig. 33-19, with an amplitude of 0.50 A and a period of $T = 0.45$ s. What is the voltage across the inductor as a function of time if $L = 2.3 \times 10^{-3}$ H? Either express your answer algebraically or plot it.

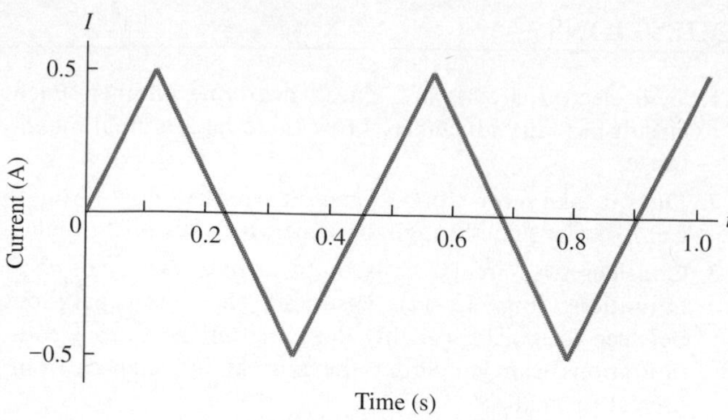

FIGURE 33–19 Problem 13.

14. (II) A coaxial cable has a central conducting wire of radius r_0 surrounded by a conducting tube of radius r_1. The space in between is filled with a material of magnetic permeability μ. Show that if the wire has length ℓ, the self-inductance is $L = (\mu\ell/2\pi)\ln(r_1/r_0)$. [*Hint:* You must calculate the flux in the region between the cylinders.]

15. (II) Consider two identical solenoids placed end to end, with the windings going in the same direction. Prove that the total self-inductance of the combined system is $2(L + M)$, where L is the self-inductance of either solenoid and M is the mutual inductance.

16. (II) A torus of rectangular cross section with width w, height h, and inner radius R is wound with N turns of wire (Fig. 33–20). What is the self-inductance of the torus? Use the approximation $\ln(1 + x) = x$, valid for $x \ll 1$, to discuss the case where $R \gg w$ and its relation to the self-inductance of a solenoid.

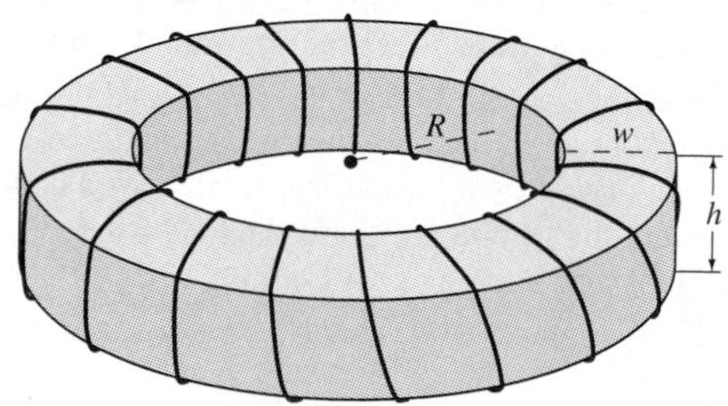

FIGURE 33–20 Problem 16.

17. (II) Consider a torus of square cross section. The radius of the torus (distance from the symmetry axis to the center of the square) is 20 cm; the sides of the square are 3.0 cm. The torus is wound with 1000 turns of wire. (a) What is the self-inductance of the torus? (b) What is the self-inductance if the core of the torus is made of soft iron, with $\mu = 2000\ \mu_0$?

18. (II) Consider a toroidal coil wound around an empty core whose self-inductance is 6.0 mH. The current in the coil changes uniformly by 2.0 A in 0.10 s. (a) What is the induced emf? (b) If the hollow center of the torus is filled with an iron core, with $\mu = 2500\ \mu_0$, what is the induced emf?

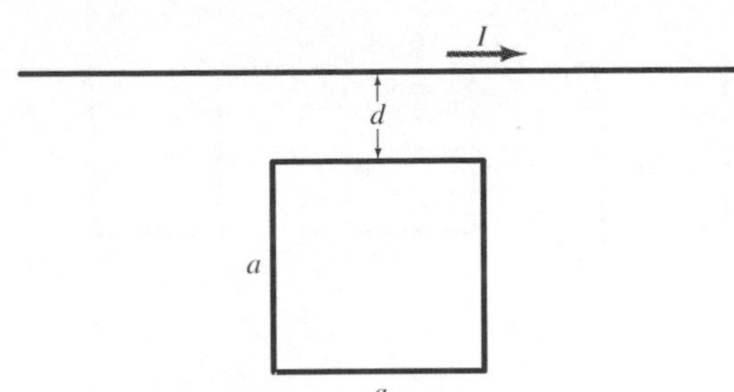

FIGURE 33–21 Problem 19.

19. (II) Figure 33–21 shows a straight wire that carries a current I and a square loop of wire with one side oriented parallel to the straight wire a distance d away. The square has sides of length a. Calculate the mutual inductance of this system. [*Hint:* The magnetic field due to the straight wire through a slice of the square of width dx parallel to the wire is constant, so the flux through this slice is easily calculable. Integrate to find the total flux through the square.]

33–2 Energy in Inductors

20. (I) Consider an inductor with $L = 1$ H and an internal resistance of 0.5 Ω. We wish to use this inductor to store 1 MJ of energy. What is the rate at which energy is lost to Joule heating in this system? It is not practical to store large amounts of energy in large inductors.

21. (I) A capacitor with $C = 0.02\ \mu$F has a charge of 15 μC. What is the equivalent steady current that should be carried by an inductor of $L = 20\ \mu$H if the inductor is to store the same amount of energy?

22. (II) A current with time dependence $I = I_0 e^{-\alpha t}$ passes through an inductor with $L = 100\ \mu$H; $I_0 = 0.1$ A and $\alpha = 0.05\ \text{s}^{-1}$. Compute the power expended in the inductor as a function of time.

23. (II) The voltage across an inductor with $L = 5.0$ mH is fixed at 12 V. The current is increased (a) from 0.0 A to 0.10 A, (b) from 0.10 A to 0.20 A, and (c) from 0.20 A to 0.30 A. What average power must be supplied from an external source in each step?

24. (II) Consider an inductor with $L = 1$ H and a capacitor with $C = 1$ F. (a) Compare the energy contained in the inductor when a current of 10 A flows through it with the energy in the capacitor if the charge is the amount of charge contained in the 10-A current, flowing for 1 s. (b) Repeat part (a) for a current of 1 mA.

25. (II) An electrical engineer constructs a cylindrical solenoid of area 5 cm^2 and length 10 cm from 1000 m of thin wire. The wire will handle a maximum current of 100 mA. (a) What is the inductance of the solenoid? (b) How much energy can the inductor store?

33–3 Energy in Magnetic Fields

26. (I) The earth's magnetic field is about 0.5×10^{-4} T near the surface of the earth (but decreases with altitude above the surface). Approximate the earth's magnetic field above the earth by a constant, and calculate the magnetic energy in a layer 100 km thick above the surface.

27. (I) The two circular pole pieces of a magnet are 63 cm in diameter and 21 cm apart. The magnetic field between them is 0.10 T. What is the magnetic energy stored in the field?

28. (II) A straight wire carries a current $I = 0.25$ A. Find the energy density in the surrounding magnetic field as a function of the distance r from the wire. At what distance from the wire does the energy density equal that of a parallel-plate capacitor with a charge of 10^{-8} C and a capacitance of 4.4×10^{-9} F, if the separation between the plates is 1.0 mm?

29. (II) (a) What is the energy density of the magnetic field outside a straight wire of radius a that carries a current I? (b) What is the total energy per unit length, due to that magnetic field, that is contained in a cylinder of radius R ($R > a$) centered about the wire?

30. (II) Consider a torus of radius R, wound with n turns per unit length of a wire that carries a current I. The cross section of the torus forms a square with sides of length b; $b \ll R$. We know that the magnetic field inside the torus has the nearly constant value $B = \mu_0 n I$. Use this result and the two expressions related to the magnetic energy [Eqs. (33–13) and (33–15)] to show that, for this torus, $L = 2\pi\mu_0 n^2 R b^2$.

31. (II) (a) What is the magnetic field energy density inside a straight wire of radius a that carries current I uniformly over its area? (b) What is the total magnetic field energy per unit length inside the wire?

32. (II) A coaxial cable consists of a wire 1 cm in diameter with a return path for the current in the shape of a very thin cylindrical conductor of diameter 2 cm. A current of 12 A flows through the cable. Calculate the magnetic energy per unit length of cable within the inner wire.

33–4 Oscillations in Circuits

33. (I) An RLC circuit has $R = 10\ \Omega$, $L = 3$ mH, and $C = 10\ \mu$F. (a) Find the damping factor and angular frequency. (b) If the resistance is variable, what value of R will give critical damping?

34. (I) Show that the time constant L/R that characterizes RL circuits has dimensions of time.

35. (I) An electric oscillator consists of a parallel-plate capacitor and a long, cylindrical solenoid. If the resonant frequency of the oscillator is ω_0, what is the frequency of a similar oscillator in which both the capacitance and inductance are reduced by a factor of 2?

36. (II) Consider the RL circuit of Fig. 33–9; the switch is closed at time $t = 0$. For the circuit elements, $\mathcal{E} = 1$ V, $R = 6\ \Omega$, and $L = 1$ mH. Using Eq. (33–19), find how much charge flows in the circuit during the first (a) 1 ms, (b) 1 s, (c) 1 h.

37. (II) Show by direct substitution that Eq. (33–19) is a solution of Eq. (33–18).

38. (II) Show that Eqs. (33–26) through (33–28) solve Eq. (33–20). Suppose that the values of R, L, and C are such that $\omega'^2 < 0$. Given that $Q(0) = Q_0$, what restrictions does Eq. (33–20) impose on the constants Q_1, Q_2, α_1, and α_2 in the trial solution $Q = Q_1 e^{-\alpha_1 t} + Q_2 e^{-\alpha_2 t}$?

39. (II) An open circuit consists of a capacitor C and an inductor L connected in series. A charge q is placed on the capacitor, and the circuit is closed at time $t = 0$ by means of a switch. Find the maximum value of the current, as well as the times for which this maximum value occurs.

40. (II) Consider the basic RLC circuit. By making an appropriate approximation of Eq. (33–28), show that when α is small compared to $\omega = 1/\sqrt{LC}$, the modified angular frequency ω' of the damped RLC circuit is $\omega' \simeq \omega - R^2\sqrt{C/L}/8L$. Find a similar relation for the periods of the undamped and slightly damped cases.

33–5 *Energy in RLC Circuits*

41. (I) Calculate the energy in an LC circuit, assuming that the initial conditions are such that the charge on the capacitor is $Q = Q_0 \cos(\omega t + \delta)$. Show that the energy is constant.

42. (II) A circuit consists of a capacitor of capacitance $C = 500$ pF connected in series with an inductor of inductance $L = 10^{-4}$ H. If a charge of 0.5 μC is put on the capacitor, there is an oscillation in the circuit. (a) What is the maximum current that moves through this circuit? (b) Find the maximum energy within the inductor. (c) What is the ratio of the maximum energy in the inductor to the maximum energy in the capacitor?

43. (II) An LC circuit consists of a 4-mH inductor and a 200-μF capacitor. If the maximum energy stored in the circuit is 10^{-4} J, what are the maximum charge on the capacitor and the maximum current in the circuit? What are the minimum values?

44. (III) The 50-mF capacitor of an LC circuit is initially charged to 2 mC. The 4-mH inductor has a very small resistance. At a particular instant, after 1000 oscillations, the current through the inductor is zero while the capacitor is still charged to 1.0 mC. (a) What is the resistance of the circuit? (b) What are the energies of the circuit before and after the 1000 oscillations? (c) Why are the two values of the energy in part (b) different? Where has the energy gone?

General Problems

45. (II) Show by considering the definition of inductance that if the voltage V across an inductor changes with time, the total current passing through the inductor in that time is given by $I = (1/L) \int V\,dt$.

46. (II) Suppose that a square wave of voltage, as plotted in Fig. 33–22, is set across an inductor with $L = 0.01$ H. Use the result of Problem 45 to plot the current as a function of time.

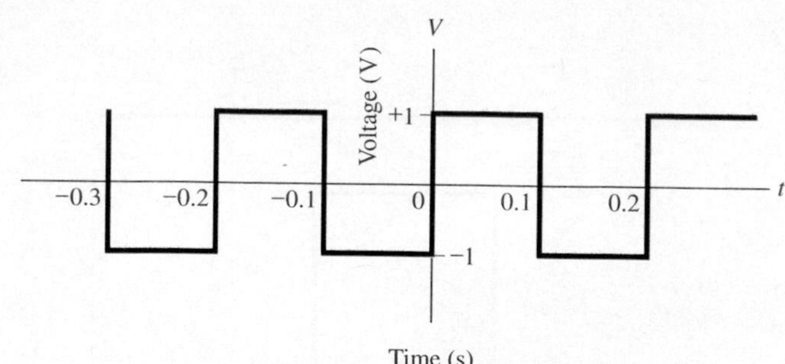

FIGURE 33–22 Problem 46.

47. (II) The switch in the circuit shown in Fig. 33–23 has been closed for a long time. (a) What is the current in each leg of the circuit? (b) When the switch is opened, the current in the inductor drops by a factor of 3 in 5 ms. What is the value of the inductance? (c) What is the current passing in each leg at 10 ms?

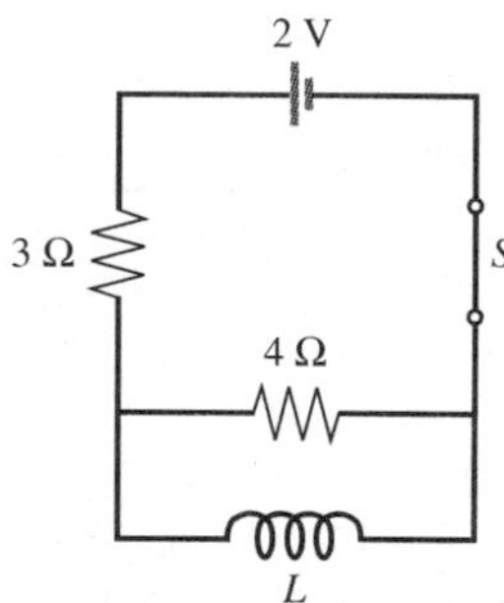

FIGURE 33–23 Problem 47.

48. (II) A coaxial cable has an inner, solid wire of radius r_1 and an outer, hollow wire of radius r_2. A current I flows through the inner wire and returns through the outer wire. Assuming that the cable is infinitely long, find the magnetic field energy per unit length. Include any field energy inside the inner wire and outside the outer wire.

49. (II) Molybdenum is paramagnetic, with a magnetic susceptibility of 1.2×10^{-4} at 300 K, about one-half its value at 20 K. Suppose that the self-inductance of a solenoid filled with molybdenum is $L = 0.1$ mH at 300 K. What is the fractional change in self-inductance between 300 K and 20 K?

50. (II) An electromagnetic light wave consists of a configuration of oscillating magnetic and electric fields. The frequency of these oscillations is the frequency of the light wave, and the amplitude of the magnetic field, B_0, is precisely $1/c$ times the amplitude of the electric field, E_0. Verify that this relation is dimensionally correct. Show that the energy density in the magnetic field is the same as the energy density in the electric field in a light wave.

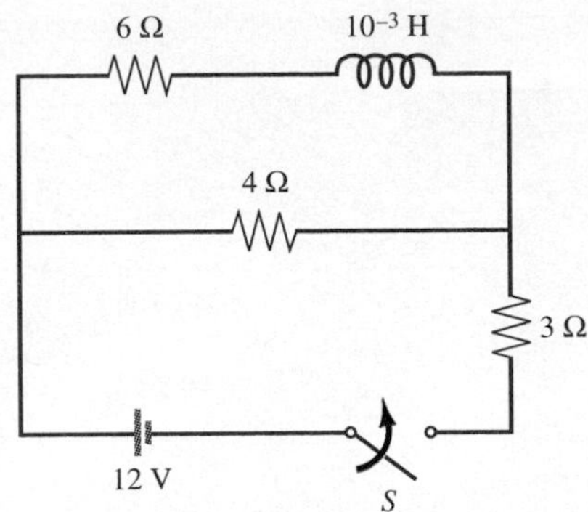

FIGURE 33–24 Problem 51.

51. (II) What are the currents in the three resistors of Fig. 33–24 immediately after the switch is closed? after a long time?

52. (II) Two solenoids are wound on a common soft iron core (Fig. 33–25). Solenoid S_1 is connected in series to a battery and a variable resistor. Starting with the resistor set at A (low resistance), the sliding contact is moved to B (large resistance) and back to A again. Sketch the voltage V across solenoid S_2 while this is happening.

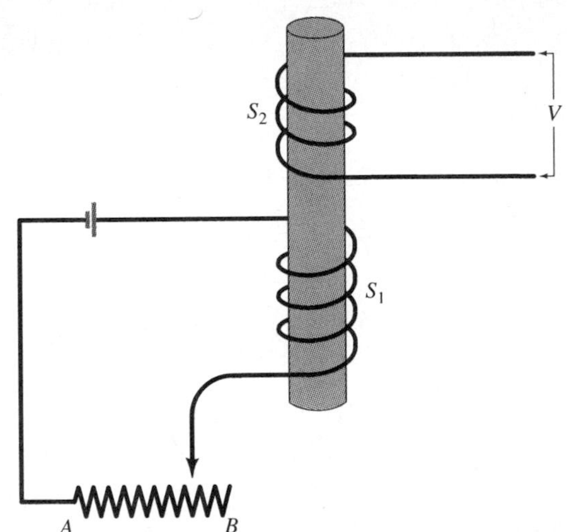

FIGURE 33–25 Problem 52.

53. (II) A ferromagnetic torus is part of a device to be used in a region where the magnetic permeability has the constant value $\mu = 1000\mu_0$. The torus has a circular cross section of 10 cm^2. Over its total length of approximately 80 cm, the torus is wrapped with 640 turns of wire. Immediately surrounding this winding is a secondary winding of 50 turns of (insulated) wire. What is the mutual inductance of the two windings? What is the role of the iron core, if any, in determining this mutual inductance?

54. (II) A torus of inner radius r_i and outer radius r_o has a square cross section (Fig. 33–26). It is wound with N turns of wire that carries a current I. (a) Use Ampère's law to find the magnetic field inside the torus. (b) Calculate the magnetic energy density within the torus. (c) Integrate the magnetic energy density to find the total magnetic energy within the torus. (d) Use the formula $U_L = \frac{1}{2}LI^2$, Eq. (33–13), to compute the self-inductance of this torus.

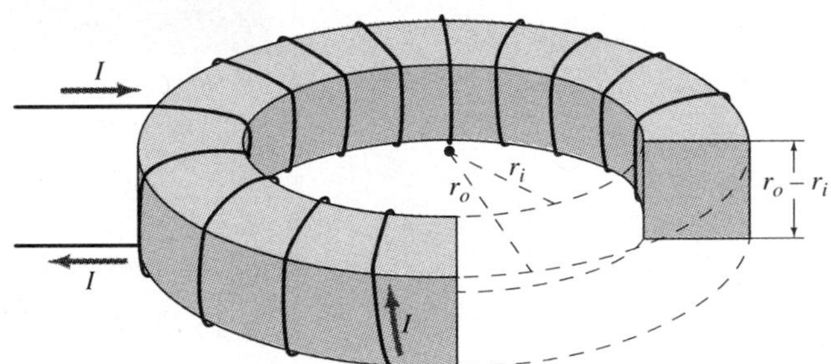

FIGURE 33–26 Problem 54.

CHAPTER 34

The beams of charged particles that circulate within a particle accelerator are guided by magnets whose fields are remarkably stable and noise-free. The AC currents that create those fields are controlled by elaborate circuits.

ALTERNATING CURRENTS

As we saw in Chapter 31, Faraday's law specifies how a changing magnetic flux induces an emf. A coil rotating in the presence of a magnet induces an emf that varies sinusoidally with time. The induced emf produces an alternating current (AC), which is a source of AC power. AC generators convert into electric currents the mechanical energy of falling water or turbines that spin under the pressure of hot steam and are the starting point for the delivery of electric power. We shall see in this chapter how we can vary the maximum voltage of a harmonically oscillating emf, an important element in the delivery of electric power. By including resistors, inductors, and capacitors in circuits with AC sources of emf, currents and voltages with new types of time-dependent behavior become possible. Such circuits are completely analogous to a mechanical system studied earlier, the harmonically driven mass on the end of a spring. Systems of this type exhibit the phenomenon of resonance.

34–1 TRANSFORMERS

In Chapter 31 we described an alternating-current generator based on Faraday's law. Recall that an alternating current is characterized by harmonic (sine and

(a)

(b)

FIGURE 34–1 (a) A commercial transformer, called a Variac, used to obtain output voltages up to 120 VAC. (b) The first transformer for commercial use in the U.S. electric power industry; constructed by George Westinghouse in 1885–1886.

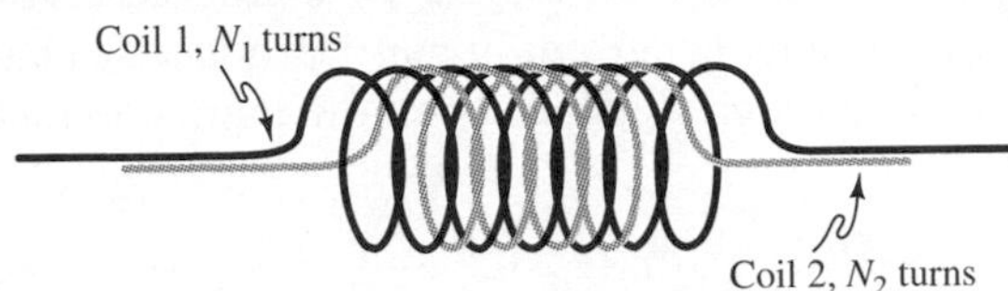

(a)

(b)

FIGURE 34–2 Two methods for creating fully linked coils: (a) one coil tightly wound over the other; (b) two coils wrapped around a common core of ferromagnetic material, which has the property of keeping the magnetic field lines within it.

cosine) time dependence, as are the other variables of the circuit, such as voltage.[†] We would like to be able to vary the maximum AC voltage, because high or low voltages are useful in differing circumstances. For example, it is more economical to transport electric energy at high voltage, but high voltages are dangerous and inefficient in small appliances. Suppose that an AC generator produces an emf of the general form

$$\mathscr{E} = V_0 \sin(\omega t) \tag{34–1}$$

[see Eq. (31–14)]. The factor V_0 is the *voltage amplitude* of the source of emf, and this is the quantity we want to vary. In this section we shall describe a device that can take an AC emf such as that of Eq. (34–1) as an input and produce an AC emf with a *different* voltage amplitude. This device is called a **transformer**, and it can be constructed by using the principle of mutual inductance (Fig. 34–1).

Mutual inductance is discussed in Chapter 33.

Consider two fully linked ideal solenoids. This may be arranged, for example, in either of the two ways shown in Fig. 34–2. The solenoids have equal cross-

[†] We use the abbreviation "AC" to indicate any kind of current or voltage that varies harmonically in time. "AC" is used as an adjective, as in AC current and AC voltage, despite the redundancy.

sectional areas A but a different total number of turns N_1 and N_2. Across the first coil (the *primary coil*) is an AC emf $\mathscr{E}_1$ with an amplitude V_1, as in Eq. (34–1):

$$\mathscr{E}_1 = V_1 \sin(\omega t). \tag{34–2}$$

We shall find the emf $\mathscr{E}_2$ across the second coil (the *secondary coil*) and show that $\mathscr{E}_2$ does indeed have an amplitude different from that of $\mathscr{E}_1$. Figure 34–3 is the standard representation of this situation. Because $\mathscr{E}_1$ is time dependent, the current through coil 1 changes, and there is a changing magnetic flux through it. Faraday's law then implies that the total current I_1 in coil 1 [see Eq. (33–2)] is determined by the equation

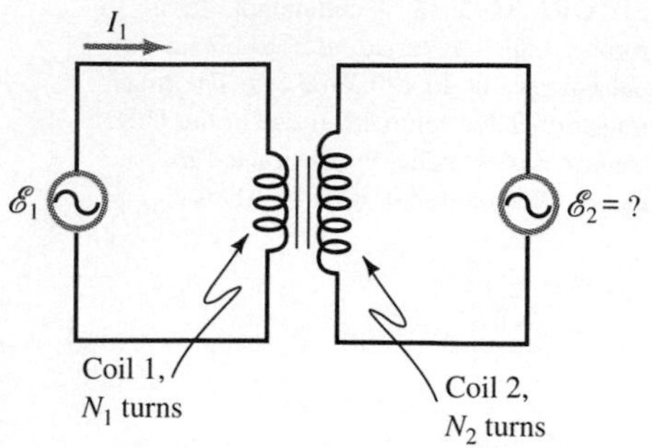

FIGURE 34–3 Schematic circuit-diagram symbol for two fully linked coils. When one coil is in the same circuit with an AC source of emf, this circuit acts as a transformer.

$$\mathscr{E}_1 = -L\frac{dI_1}{dt}, \tag{34–3}$$

where L is the self-inductance of coil 1. We are not interested so much in the value of I_1 as we are in the fact that, at the same time, an emf $\mathscr{E}_2$ is induced across coil 2. This emf is induced because the changing current in coil 1 produces a changing magnetic flux through coil 2. By definition, $\mathscr{E}_2$ depends on the mutual inductance, M:

$$\mathscr{E}_2 = -M\frac{dI_1}{dt}. \tag{34–4}$$

If we substitute for dI_1/dt from Eq. (34–3), we get

$$\mathscr{E}_2 = M\frac{\mathscr{E}_1}{L};$$

$$\frac{\mathscr{E}_2}{\mathscr{E}_1} = \frac{M}{L}. \tag{34–5}$$

Equation (34–5) contains important information: The ratio M/L is a constant, and therefore *$\mathscr{E}_2$ has the same harmonic time dependence as $\mathscr{E}_1$*. If the angular frequency of the current in the primary coil is ω, as in Eq. (34–1), so is that of the current induced in the secondary coil.

Let us now use the fact that the two coils are fully linked ideal solenoids. In this case we know both M and L. The mutual inductance of the two coils is a special case of the mutual inductance of a solenoid and a ring, Eq. (33–9); from that equation we have

$$M = \mu_0 A\frac{N_1}{\ell_1}N_2. \tag{34–6}$$

The self-inductance of coil 1, from Eq. (33–8), is

$$L = \mu_0 A\frac{N_1^2}{\ell_1}. \tag{34–7}$$

Substituting these results into Eq. (34–5), we have

$$\frac{\mathscr{E}_2}{\mathscr{E}_1} = \frac{\cancel{\mu_0 A}\cancel{N_1}N_2}{\cancel{\ell_1}}\frac{\cancel{\ell_1}}{\cancel{\mu_0 A}N_1^{\cancel{2}}};$$

The ratio of the emfs in the two coils of a transformer is equal to the ratio of the number of turns in the coils.

$$\boxed{\frac{\mathscr{E}_2}{\mathscr{E}_1} = \frac{N_2}{N_1}.} \tag{34–8}$$

Because the time dependence of the AC is identical in $\mathscr{E}_1$ and $\mathscr{E}_2$, Eq. (34–8) relates the voltage *amplitudes* V_1 and V_2 (the coefficients of the sinusoidal time dependence of the emfs) in the two coils:

$$\frac{V_2}{V_1} = \frac{N_2}{N_1}. \tag{34–9}$$

The transformer is a tool for manipulating these voltage amplitudes. When $N_2 > N_1$, the transformer is a *step-up transformer*, and the voltage amplitude in the secondary coil is greater than that in the primary coil. When $N_2 < N_1$, the transformer is a *step-down transformer*, and the voltage amplitude in the secondary coil is smaller than that in the primary coil. Note that the terms primary and secondary do not imply any fundamental distinction between the two coils.

How are the currents in the two coils related? Energy must be conserved when the transformer operates. If the transformer is constructed efficiently, meaning that any resistance is reduced to a minimum and there is no loss to Joule heating, then the rate of energy flow, given by the product $I\mathscr{E}$, must be equal in the two transformer coils. We thus have

$$I_1\mathscr{E}_1 = I_2\mathscr{E}_2.$$

We substitute in Eq. (34–8) to find that

$$\frac{I_1}{I_2} = \frac{N_2}{N_1}. \tag{34–10}$$

The ratio of the currents in the two coils of a transformer is equal to the inverse of the ratio of the number of turns in the coils.

In other words, if the voltage amplitude in the secondary coil of a step-up transformer is increased relative to the voltage amplitude in the primary coil, the current carried by the secondary coil decreases by the same factor; if the voltage amplitude in the secondary coil of a step-down transformer is decreased relative to that in the primary coil, the current carried by the secondary coil increases equivalently.

EXAMPLE 34–1 A step-down transformer has 5000 turns in the primary coil, which handles an AC current with voltage amplitude $V_1 = 20{,}000$ V, and 220 turns in the secondary coil. If the current amplitude desired in the secondary coil is 100 A, what is the maximum power that must be delivered by the primary coil to the secondary coil?

SOLUTION: The desired unknown is the power P_2 delivered to the secondary coil. Power is related to the unknown voltage amplitude V_2 and to the known current I_2 by the relation $P_2 = I_2V_2$. We can find V_2 from V_1 and the number of coil turns. V_2 is determined by Eq. (34–9):

$$V_2 = \frac{N_2}{N_1}V_1 = \frac{220}{5000}(20{,}000\ \text{V}) = 880\ \text{V}.$$

If the maximum current carried by the secondary coil is 100 A, the maximum power it carries is $P_2 = (100\ \text{A})(880\ \text{V}) = 88$ kW. If the transformer is constructed efficiently, this is equal to the maximum power that can be carried in the primary coil.

Power Transmission

Low voltages are more practical for local use because they are more easily insulated against breakdown than high voltages are. Conversely, as we show below, it is far more efficient to transmit electric energy at *high* voltages from a generating plant

(a)

(b)

FIGURE 34–4 (a) Generators at power stations produce electric energy in the form of AC current. There transformers such as these convert the electricity from low voltages to high voltages. (b) Power transmission lines carry AC at such high voltages that great care must be taken to keep them isolated from their surroundings electrically. If you stand under one of these lines, you can hear the electrical breakdown in the air that surrounds the wires.

(Fig. 34–4a) to the places where it is to be used locally (Fig. 34–4b). Transformers allow us to reconcile the different voltage requirements of long-distance transmission and local use. The fact that transformers require AC to operate has dictated the role of AC in our use of electricity.

Let us demonstrate briefly that it is more efficient to transmit electric power at high voltages than at low voltages, whether AC or DC. A power transmission line delivers energy for local use at a certain rate, P, and at a certain voltage, V. This means that a current passes through the transmission line, given by $P = IV$. The transmission line will itself have a certain resistance, R, and the power dissipated in the line will be

$$P_{\text{lost}} = I^2R = \frac{P^2R}{V^2}. \tag{34–11}$$

A measure of the efficiency of transmission is the ratio of the power delivered to the power lost in Joule heating. According to Eq. (34–11) this ratio is

$$\frac{P}{P_{\text{lost}}} = \frac{V^2}{PR}. \tag{34–12}$$

This ratio increases rapidly as V increases. To take a realistic example, a transmission line delivering power $P = 1$ MW might have a total resistance of 10 Ω. If the power were delivered at 110 V, the ratio in Eq. (34–12) would be intolerably low: $(110\ \text{V})^2/(1\ \text{MW})(10\ \Omega) = 1.2 \times 10^{-3}$. In this case 99.9 percent of your electricity bill would be for power lost in the wires! If, however, the power were delivered at 500,000 V, the ratio would be $(500{,}000\ \text{V})^2/(1\ \text{MW})(10\ \Omega) = 2.5 \times 10^4$, and most of the electric energy produced would be delivered.

34–2 SINGLE ELEMENTS IN AC CIRCUITS

In this section we shall examine the effects of placing an AC source of emf, for which $\mathscr{E} = V_0 \sin(\omega t)$, in circuits that contain only single elements of resistance, capacitance, and inductance. In this way we can better understand the effects of each element. We shall look at the current that passes through each of the circuit elements and compare that with the voltage.

Resistive Circuit

We begin with the resistive circuit shown in Fig. 34–5a. The loop rule for the potential change around the circuit is

$$V_0 \sin(\omega t) - IR = 0. \quad (34\text{–}13)$$

The voltage across the resistor is then $V_R = IR = V_0 \sin(\omega t)$, and the current I through the resistor is

$$I = \frac{V_R}{R} = \frac{V_0 \sin(\omega t)}{R}. \quad (34\text{–}14)$$

The current through, and voltage across, the resistor have the same sinusoidal time dependence (Fig. 34–5b). We say that they are *in phase*. The amplitude of I is $I_{max} = V_0/R$.

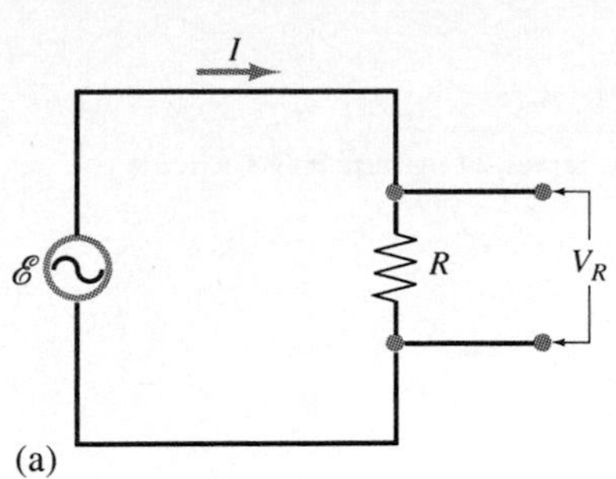

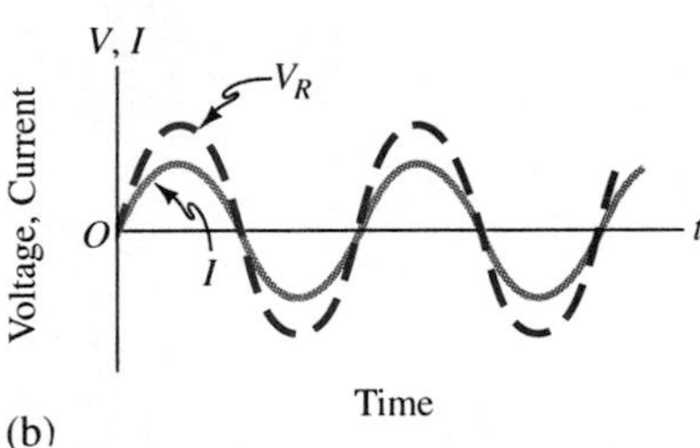

FIGURE 34–5 (a) A resistor connected in series with an AC source of emf. (b) The voltage across the resistor and the current through it are in phase.

Capacitive Circuit

We now place the AC emf across a pure capacitance in the circuit (Fig. 34–6a). We apply the loop rule and calculate the current in the circuit (remember that current does not pass through the capacitor itself) and voltage across the capacitor. The loop rule gives

$$V_0 \sin(\omega t) - \frac{Q}{C} = 0. \quad (34\text{–}15)$$

The voltage across the capacitor is simply the driving emf $V_C = V_0 \sin(\omega t)$. In order to find the current, we first find the charge Q from Eq. (34–15):

$$Q = CV_0 \sin(\omega t). \quad (34\text{–}16)$$

The current I is then the time derivative of the charge:

$$I = \frac{dQ}{dt} = \omega C V_0 \cos(\omega t). \quad (34\text{–}17)$$

With the trigonometric identity $\sin[\theta + (\pi/2)] = \cos\theta$ (from Appendix IV–4), we rewrite Eq. (34–17) in the form

$$I = \omega C V_0 \sin\left(\omega t + \frac{\pi}{2}\right). \quad (34\text{–}18)$$

The maximum value of the current in the circuit is $\omega C V_0$. If we relate the maximum voltage across the capacitor to the maximum current in the circuit, we find that

$$I_{max} = \omega C V_0. \quad (34\text{–}19)$$

This equation can be compared to a similar equation for the resistive circuit, for which the corresponding relation was $I_{max} = V_0/R$. The *effective resistance* for a capacitive circuit is called the **capacitive reactance**, X_C, defined by

$$X_C \equiv \frac{1}{\omega C}. \quad (34\text{–}20)$$

Equation (34–19) now takes the form

$$I_{max} = \frac{V_0}{X_C}. \quad (34\text{–}21)$$

The capacitive reactance has units of ohms.

In Fig. 34–6b we plot the current I and the voltage V_C versus time. Note that at time $t = 0$, the voltage across the capacitor is zero, but the current in the circuit

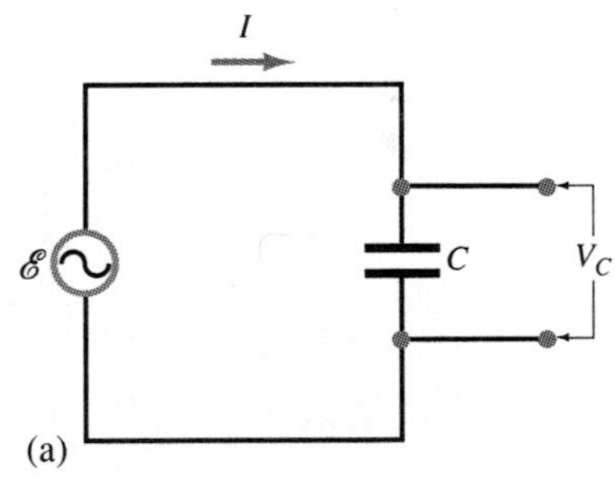

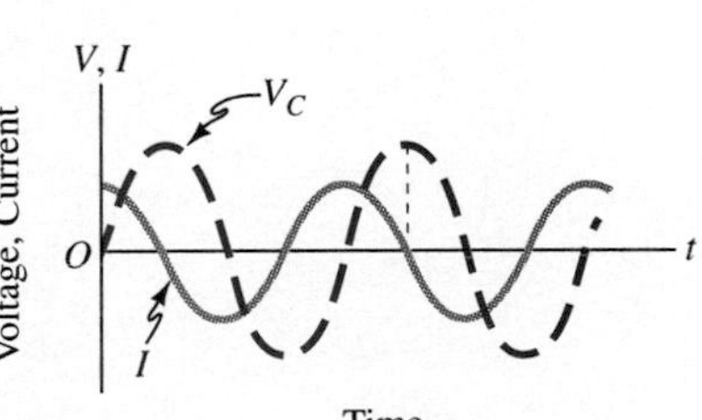

FIGURE 34–6 (a) A capacitor connected in series with an AC source of emf. (b) The current in the circuit leads the voltage across the capacitor by 90°.

is at a maximum. The phase of the current differs by $\pi/2$ rad (90°) from that of the voltage. We say that the current *leads* the voltage by a phase $\pi/2$.

Does the phase difference described above make physical sense? Let us assume that a switch completing the circuit is closed at $t = 0$. At this time there is no charge on the capacitor plates, and the voltage, proportional to $\sin(\omega t)$, starts up from zero. The charge flows readily (at its maximum rate) toward the capacitor plates. As the voltage increases, however, significant charge has already collected on the capacitor plates, and further charge is repelled. *The current then decreases to zero* as the voltage reaches its peak. Throughout the cycle, the sinusoidal curves describing the current and the voltage continue to differ as we have described.

EXAMPLE 34–2 The circuit shown in Fig. 34–6a has an emf given by $\mathscr{E} = V_0 \sin(\omega t)$, with $V_0 = 6.0$ V, and a capacitance $C = 1.0\ \mu$F. (a) What are the peak currents for frequencies of exactly 60 Hz and 6 MHz? (b) What are the currents I and voltages V_C at time 2.0 ms for the 60-Hz frequency?

SOLUTION: (a) The unknown peak current is found from Eq. (34–19) and depends on the angular frequency. The angular frequencies ω for the two cases are found from $\omega = 2\pi f$ and are given by $2\pi(60\text{ Hz}) = 2\pi(60\text{ s}^{-1}) = 377$ rad/s and $2\pi(6 \times 10^6\text{ s}^{-1}) = 3.77 \times 10^7$ rad/s, respectively. Thus, from Eq. (34–19),

$$\text{for 60 Hz:} \qquad I_{max} = (377\text{ rad/s})(1.0 \times 10^{-6}\text{ F})(6\text{ V}) = 2.3\text{ mA};$$

$$\text{for 6 MHz:} \qquad I_{max} = (3.77 \times 10^7\text{ rad/s})(1.0 \times 10^{-6}\text{ F})(6\text{ V}) = 230\text{ A}.$$

The higher frequency makes a significant difference in the maximum current. For the higher frequency, 6 MHz, the capacitive reactive is so small ($X_C = 1/\omega C = 0.027\ \Omega$) that the circuit has almost no resistance to current flow. (b) We want to specify the time in the full expressions for current and voltage, Eqs. (34–18) and $V_0 \sin(\omega t)$, respectively. We can substitute Eq. (34–19) into Eq. (34–18). For $f = 60$ Hz we have

$$I = I_0 \sin(\omega t + \phi)$$

and

$$V = V_0 \sin(\omega t),$$

where $I_0 = 2.3$ mA, $\phi = \pi/2$ rad, $V_0 = 6.0$ V, and $\omega = 377$ rad/s. For $t = 2.0$ ms the current is

$$I = (2.3\text{ mA}) \sin\left[(377\text{ rad/}\cancel{\text{s}})(0.0020\ \cancel{\text{s}}) + \frac{\pi}{2}\text{ rad}\right] = 1.7\text{ mA},$$

and the voltage is

$$V_C = (6.0\text{ V}) \sin[(377\text{ rad/}\cancel{\text{s}})(0.0020\ \cancel{\text{s}})] = 4.1\text{ V}.$$

By $t = 2.0$ ms the current is coming down from its peak toward zero, while the voltage is rising toward its peak. At $t = 2.0$ ms they are both about 70 percent of their peak values.

Inductive Circuit

Now we replace the capacitor with an inductor in the circuit that has an AC emf (Fig. 34–7a). Repeating the previous procedure, we first apply the loop rule to the potentials around the circuit:

$$V_0 \sin(\omega t) - L\frac{dI}{dt} = 0. \quad (34\text{–}22)$$

The voltage drop across the inductor must be the emf $V_L = V_0 \sin(\omega t)$. In order to find the current through the inductor, we need to solve Eq. (34–22) for the current I. We rewrite the equation as

$$dI = \frac{V_0}{L}\sin(\omega t)\, dt.$$

We find the current from the integral of this equation:

$$I = -\frac{V_0}{\omega L}\cos(\omega t).$$

We use the trigonometric identity $\cos\theta = -\sin[\theta - (\pi/2)]$ (Appendix IV–4) to rewrite this equation as

$$I = \frac{V_0}{\omega L}\sin\left[\omega t - \left(\frac{\pi}{2}\right)\right]. \quad (34\text{–}23)$$

The maximum current through the inductor is

$$I_{max} = \frac{V_0}{\omega L}. \quad (34\text{–}24)$$

Comparing this equation with the similar one from the purely resistive circuit, $I_{max} = V_0/R$, we see that the effective resistance for an inductive circuit is ωL. We call this the **inductive reactance**, defined by

$$X_L \equiv \omega L. \quad (34\text{–}25)$$

The inductive reactance has units of ohms.

For an inductive circuit the effective resistance to current flow *increases* at higher frequencies. Physically, this is reasonable, because inductors react to *oppose* any change in the current flow through them. A higher frequency means that the voltage, and therefore the current, is changing more rapidly. A current is induced to oppose the change.

We plot the current and voltage of the inductor versus time in Fig. 34–7b. As for the capacitive circuit, one sinusoidal curve is displaced from the other by a quarter cycle, although the role of the current and voltage curves is reversed in the two cases. This time the current *lags* the voltage. Again let us try to understand what happens when a switch in the circuit is closed at $t = 0$. As the voltage increases from zero, the inductor resists any current flow, and it will induce an opposing current. Thus, as the voltage rises from zero, the current through the inductor will be negative; the phase of the current will be such that the current is "behind" the voltage. When the voltage across the inductor is at its peak, and the voltage just starts to decrease across the inductor, the inductor resists the voltage decrease, so it induces a positive current to help keep the voltage up. When the voltage just starts to decrease from its peak, the current changes from negative to positive. The current stays 90° out of phase with the voltage.

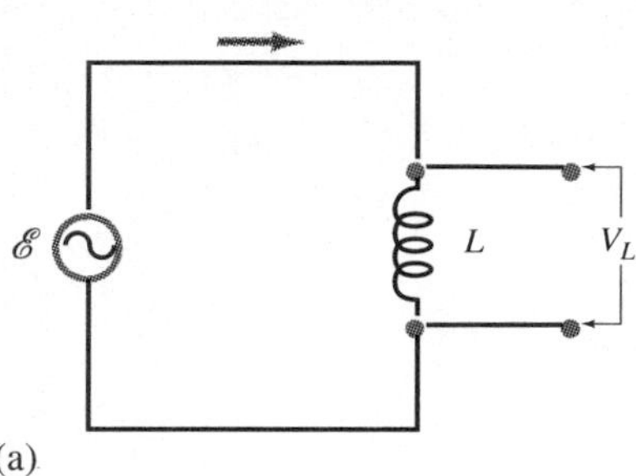

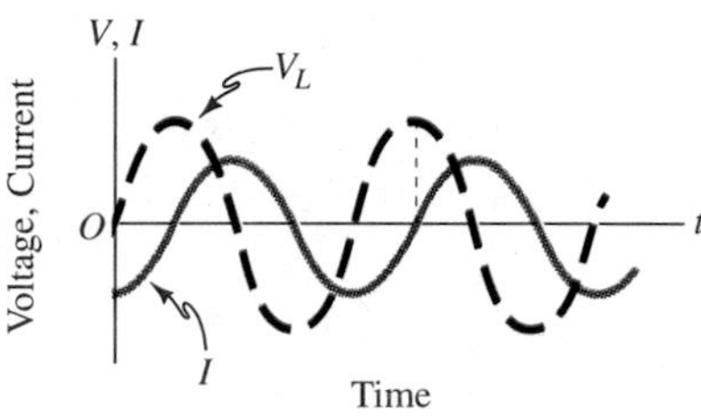

FIGURE 34–7 (a) An inductor connected in series with an AC source of emf. (b) The current in the circuit lags the voltage across the inductor by 90°.

EXAMPLE 34–3 Use the same parameters of Example 34–2, but replace the capacitor with an inductor of inductance $L = 1.00$ mH. Calculate the inductive reactances.

SOLUTION: We found in Example 34–2 that the angular frequencies are 377 rad/s and 3.77×10^7 rad/s, respectively. The inductive reactances are determined from Eq. (34–25):

for 60 Hz: $X_L = \omega L = (377 \text{ rad/s})(1.00 \times 10^{-3} \text{ H}) = 0.377\ \Omega$

for 6 MHz: $X_L = \omega L = (3.77 \times 10^7 \text{ rad/s})(1.00 \times 10^{-3} \text{ H})$
$= 3.77 \times 10^4\ \Omega.$

As expected, the resistance to current flow increases dramatically for the higher frequency, the opposite behavior from that of the capacitive circuit.

34–3 AC IN SERIES *RLC* CIRCUITS

All the results of Section 34–2 can be understood in terms of a mechanical analogue. This analogue gives us new physical insight into the behavior of AC circuits and simplifies the task of combining circuit elements into the more general circuits that make up electronic devices.

The driven harmonic oscillator is discussed in Chapter 13.

In mechanics, a *driven harmonic oscillator* is a device in which a harmonic external force acts on (for example) a mass fixed to a spring. If the driving force has been acting for some time, the mass has no choice but to move with the angular frequency ω of the force, even though the mass is attached to the spring and undergoes some damping. This system illustrates the important physical phenomenon of *resonance*, characterized by a large amplitude when the driving frequency ω is near the natural frequency ω_0.†

In Chapter 33 we noted the similarities between the mass–spring system and series *RLC* circuits without a driving term. If we add an AC source of emf to a series *RLC* circuit, the analogy between the mechanical system and the circuit continues to hold. In particular, *such a circuit exhibits resonant behavior*. The (single-loop) series *RLC* circuit with an AC source of emf is the simplest case of an important class of circuits that involve AC voltages and the circuit elements *R*, *L*, and *C*.

Figure 34–8 illustrates our driven circuit, and application of the loop rule for the potential changes around the circuit gives

$$V_0 \sin(\omega t) - L\frac{dI}{dt} - \frac{Q}{C} - IR = 0. \tag{34–26}$$

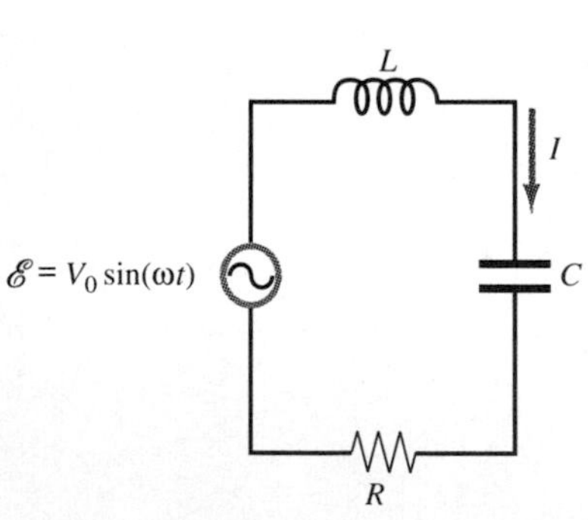

FIGURE 34–8 An *RLC* circuit is driven by an AC emf.

Because $I = dQ/dt$, we can reexpress this result in terms of the single variable Q, the charge on the capacitor:

$$V_0 \sin(\omega t) - L\frac{d^2Q}{dt^2} - \frac{Q}{C} - R\frac{dQ}{dt} = 0. \tag{34–27}$$

The unknown quantity in this differential equation is Q. Once we solve for Q, differentiation with respect to time will give the current. We can then find the voltage drops across the various circuit elements.

Equation (34–27) can be compared to Eq. (13–58), which expresses Newton's second law for a driven harmonic oscillator with damping. It is rewritten here in the form

$$F_0 \sin(\omega t) - m\frac{d^2x}{dt^2} - b\frac{dx}{dt} - kx = 0, \tag{34–28}$$

† When no confusion is possible, we use the term "frequency" rather than "angular frequency" for ω.

where the first term is the driving force, with amplitude F_0. We showed in Chapter 13 that, after the driving force has gone on for a long time, this differential equation for the position x as a function of time has a solution in which the position of the mass oscillates with the angular frequency of the driving force. This solution is given by Eq. (13–59), rewritten here as

$$x = A \sin(\omega t - \phi). \tag{34–29}$$

As we described in Chapter 13, the amplitude A and phase ϕ are determined by direct substitution into the differential equation, with the results

$$A = \frac{F_0}{\sqrt{m^2(\omega^2 - \omega_0^2)^2 + b^2\omega^2}} \tag{34–30}$$

and

$$\frac{1}{\tan\phi} = \frac{1}{b}\left(\frac{k}{\omega} - \omega m\right). \tag{34–31}$$

Here ω_0 is the natural frequency of the oscillator, given by $\omega_0 = \sqrt{k/m}$.

The force and the position are both harmonic with the same frequency, but they are out of phase. For example, the function $\sin(\omega t)$—proportional to the force—rises from zero at $t = 0$, whereas the function $\sin(\omega t - \phi)$—proportional to the position—rises from zero when its argument $\omega t - \phi$ is zero, which occurs when $t = \phi/\omega$.

In order to solve Eq. (34–27) for the charge on the capacitor, we need only make the formal substitutions of Table 34–1, relating the parameters of the harmonic oscillator to the circuit parameters. The solution for the charge is thus

$$Q = Q_{max} \sin(\omega t - \phi), \tag{34–32}$$

where

$$Q_{max} = \frac{V_0}{\sqrt{L^2(\omega^2 - \omega_0^2)^2 + R^2\omega^2}} \tag{34–33}$$

and

$$\frac{1}{\tan\phi} = \frac{1}{R}\left(\frac{1}{\omega C} - \omega L\right). \tag{34–34}$$

TABLE **34–1**

ANALOGY BETWEEN DRIVEN *RLC* CIRCUITS AND SPRING MOTION

	Circuit	*Mass–spring*
Variable	Charge Q	Position x
Coefficient of variable	$\frac{1}{C}$	k
Coefficient of $\frac{\mathrm{d(variable)}}{\mathrm{d}t}$	R	b
Coefficient of $\frac{\mathrm{d^2(variable)}}{\mathrm{d}t^2}$	L	m
Driving term	$V_0 \sin(\omega t)$	$F_0 \sin(\omega t)$
Natural frequency	$\frac{1}{\sqrt{LC}}$	$\sqrt{\frac{k}{m}}$

In this case, the natural frequency ω_0 of the circuit is the frequency of the pure LC circuit (no damping term, $R = 0$), from Eq. (33–23):

$$\omega_0 = \frac{1}{\sqrt{LC}}. \tag{34–35}$$

All the results given here reduce to the cases we treated in Section 34–2, in which only one circuit element is present at a time. We need only replace the values of L, R, or $1/C$ by zero, as appropriate.

Impedance

We have already defined the reactances X_C and X_L in Eqs. (34–20) and (34–25); these quantities play the role of an effective resistance. As we shall see, the effective resistance of our more general circuit is the **impedance**, Z, defined by

Impedance

$$Z \equiv \sqrt{\left(\frac{1}{\omega C} - \omega L\right)^2 + R^2} = \sqrt{(X_C - X_L)^2 + R^2}. \tag{34–36}$$

The impedance has units of ohms. Note that a cable with impedance generally has negligible resistance, so, unlike resistance, impedance is *independent of length.* It is a matter of a little algebra (see Problem 26) to show that, in terms of these quantities, Eq. (34–33) becomes

$$Q_{\text{max}} = \frac{V_0}{\omega Z}. \tag{34–37}$$

In addition,

$$\frac{1}{\tan\phi} = \frac{1}{R}(X_C - X_L). \tag{34–38}$$

From Eq. (34–32), we can find the current in the circuit:

$$I = \frac{dQ}{dt} = \omega Q_{\text{max}} \cos(\omega t - \phi).$$

Thus $I_{\text{max}} = \omega Q_{\text{max}}$. When we insert Q_{max} from Eq. (34–37), we get

$$I = I_{\text{max}} \cos(\omega t - \phi) = \frac{V_0 \cos(\omega t - \phi)}{Z}. \tag{34–39}$$

The current takes the form of an AC emf with voltage amplitude V_0 divided by the impedance; in other words, the current amplitude is the voltage amplitude divided by the impedance. This equation is analogous to the DC equation $I = V/R$. *Impedance thus plays the role of resistance in an AC circuit.*

In contrast to the resistance, the impedance depends on the frequency. The frequency dependence of impedance can be understood on physical grounds. Inductance opposes a change in current, and larger values of angular frequency mean more rapid changes in the current. However, inductance has no effect when static potentials, corresponding to $\omega \to 0$, are involved. These properties are reflected in the frequency dependence of $X_L = \omega L$. A capacitor has just the opposite properties: No current can pass a capacitor in the limit that the current is constant, but the capacitor has little effect when the current changes so rapidly that little charge can accumulate. These properties are reflected in the frequency dependence of $X_C = 1/\omega C$.

Recall from Section 28–5 that a capacitor does not allow a steady current to pass.

EXAMPLE 34–4 The series *RLC* circuit in Fig. 34–8 is driven with an AC source of emf of the form $\mathcal{E} = V_0 \sin(\omega t)$, where V_0 is exactly 110 V, the frequency f is exactly 60 Hz, and $\omega = 2\pi f$. If $R = 20.0\ \Omega$, $L = 5.00 \times 10^{-2}$ H, and $C = 50.0\ \mu$F, find the potential drops across the inductor at $t = 0$ and at $t = t_1$, the first time following $t = 0$ when $\mathcal{E}$ reaches a maximum.

SOLUTION: The quantity we want is the potential drop across the inductor, and it is given by $V_L = -L\,\mathrm{d}I/\mathrm{d}t$. The current I in this circuit is given by Eq. (34–39), so

$$V_L = -L\frac{\mathrm{d}}{\mathrm{d}t}\left[\frac{V_0}{Z}\cos(\omega t - \phi)\right] = \frac{LV_0\omega}{Z}\sin(\omega t - \phi).$$

At the moment $t = 0$, V_L takes the form

$$V_L = \frac{LV_0\omega}{Z}\sin(-\phi) = -\frac{LV_0\omega}{Z}\sin\phi.$$

The emf reaches a maximum at time t_1 when $\omega t_1 = \pi/2$, or

$$t_1 = \frac{\pi}{2\omega} = \frac{\pi}{4\pi f} = \frac{1}{4f}.$$

At $t = t_1$ we have

$$V_L = \frac{LV_0\omega}{Z}\sin\left(\frac{\omega}{4f} - \phi\right) = \frac{LV_0\omega}{Z}\sin\left(\frac{\pi}{2} - \phi\right) = \frac{LV_0\omega}{Z}\cos\phi.$$

To evaluate these results, we have $\omega = 2\pi(60\text{ Hz}) = 377$ rad/s. Then

$$X_C = \frac{1}{\omega C} = \frac{1}{(377\text{ rad/s})(5.00 \times 10^{-5}\text{ F})} = 53.0\ \Omega,$$

$$X_L = \omega L = (377\text{ rad/s})(5.00 \times 10^{-2}\text{ H}) = 18.9\ \Omega.$$

In turn, we have

$$Z = \sqrt{(X_C - X_L)^2 + R^2} = \sqrt{(53.0\ \Omega - 18.9\ \Omega)^2 + (20.0\ \Omega)^2} = 39.6\ \Omega$$

and

$$\cot\phi = \frac{1}{R}(X_C - X_L) = \frac{1}{20.0\ \Omega}(53.0\ \Omega - 18.9\ \Omega) = 1.71,\text{ or }\phi = 30.3^\circ.$$

At $t = 0$,

$$V_L = -\frac{LV_0\omega}{Z}\sin\phi = -\frac{(5.00 \times 10^{-2}\text{ H})(110\text{ V})(377\text{ rad/s})}{39.6\ \Omega}\sin(30.3^\circ)$$
$$= 26.5\text{ V}.$$

At $t = t_1$,

$$V_L = \frac{LV_0\omega}{Z}\cos\phi = \frac{(5.00 \times 10^{-2}\text{ H})(110\text{ V})(377\text{ rad/s})}{39.6\ \Omega}\cos(30.3^\circ) = 45.2\text{ V}.$$

Resonance in Driven RLC Circuits

The current amplitude, like the amplitude of the charge on a capacitor, and like the voltage amplitudes across any of the elements of an *RLC* circuit, exhibits the phenomenon of resonance. The amplitudes are inversely proportional to the impedance ($\propto 1/Z$). When Z is a minimum, these amplitudes are *peaked*, or *resonate*.

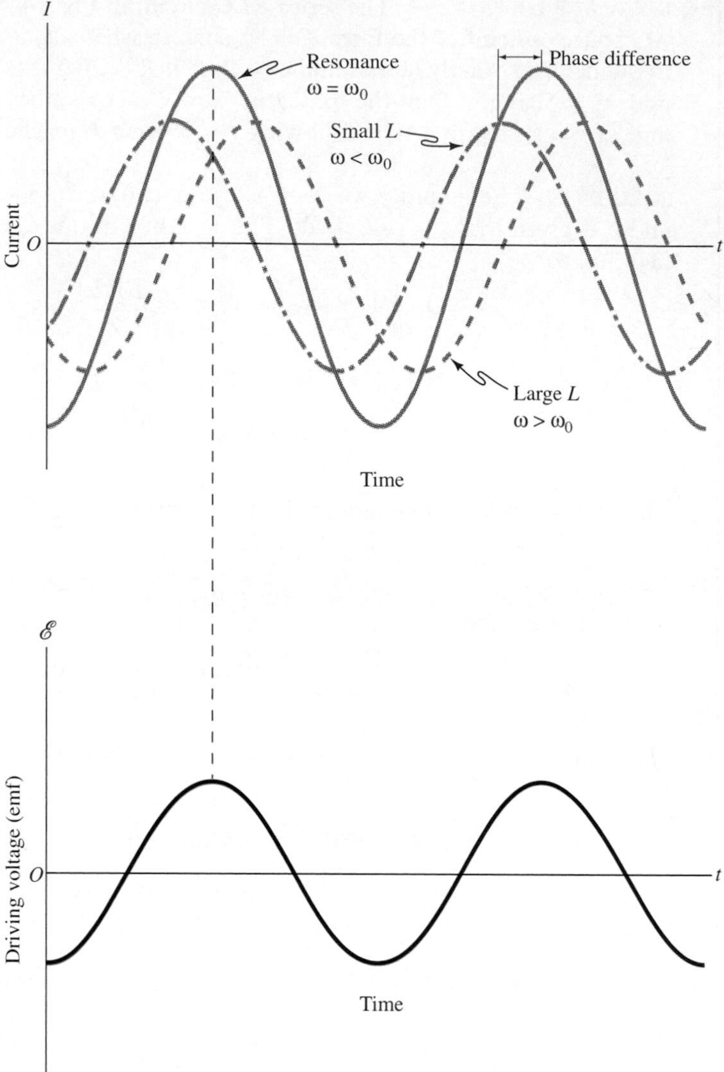

FIGURE 34–9 Given the driving voltage shown, the current through an RLC circuit varies harmonically with the frequency of the driving voltage but differs from it in amplitude and phase in a way that depends on whether or not the driving frequency, ω, equals the undamped natural frequency, ω_0. At resonance ($\omega = \omega_0$), the current amplitude is a maximum and the phase is that of the driving voltage. In the plot of current versus time, we have controlled the natural frequency by varying L.

This occurs when the driving frequency ω is near the undamped natural frequency ω_0 (Fig. 34–9). As for the mechanical oscillator, the amount and sharpness of the peaking depends on the damping factor, in this case the resistance, R. Without R, the amplitudes would be infinite when the driving frequency ω equals ω_0. However, there is always some resistance in real circuits. Figure 13–21 illustrates the resonance peak for several values of the damping parameter b of a mechanical oscillator. Similar plots apply to RLC circuits. The smaller the damping (that is, the smaller the resistance), the larger and sharper the resonance peak is. The maximum in this peak occurs at a driving frequency for which the terms X_C and X_L in Z precisely cancel each other; in other words, for $X_L = X_C$,

$$\omega L = \frac{1}{\omega C}.$$

This condition implies that

$$\omega_{\text{max}}^2 = \omega_0^2 = \frac{1}{LC}. \tag{34–40}$$

At this value of the angular frequency, $Z = R$. The current amplitude at this value of the driving frequency is given simply by

$$\text{at resonance:} \qquad I_{\text{max}} = \frac{V_0}{R}. \tag{34–41}$$

Both the value of I_{max} at resonance *and the sharpness of the resonance peak* increase as R decreases. We shall look at the question of the sharpness more closely in Section 34–4 in our discussion of power.

Resonance phenomena have many applications in circuits. The most familiar concern radio and television reception (Fig. 34–10). One way to tune a receiver is to change the capacitance of a variable capacitor, thus changing the resonant frequency. The receiver is then sensitive to a particular frequency and picks up broadcast signals at that frequency.

FIGURE 34–10 A variable capacitor. Radios can be tuned to pick up specific frequencies by changing the value of C and by using the resonance phenomenon of LC circuits.

34–4 POWER IN AC CIRCUITS

In the description of the voltages across, or currents through, various single circuit elements (Section 34–2), the phase, ϕ, did not seem to play a very important role; it merely described a constant time lag between these quantities and that of the input, or driving, emf. The phase is more important when the power dissipated in the RLC circuit is considered. As we recall, energy is lost only in the resistance of such a circuit, whereas energy held by either the capacitor or the inductor is temporarily stored for some parts of the harmonic cycle but is not lost. Thus the power dissipated in the circuit is

$$P = I^2 R = \frac{[V_0 \cos(\omega t - \phi)]^2}{Z^2} R. \tag{34–42}$$

This power is always positive, but it oscillates between zero and a maximum of $V_0^2 R/Z^2$. For engineering purposes it is more important to know the *average* power dissipated over time. Let us indicate time-averaged quantities with angle brackets, $\langle \ \rangle$. The average of a cosine squared over one cycle is one-half:

$$\langle[\cos(\omega t - \phi)]^2\rangle = \tfrac{1}{2},$$

so

$$\langle P \rangle = \frac{V_0^2 R}{Z^2} \langle[\cos(\omega t - \phi)]^2\rangle = \frac{1}{2}\frac{V_0^2 R}{Z^2}. \tag{34–43}$$

If we substitute Eq. (34–36) for the impedance, we find an explicit form for the average power:

$$\langle P \rangle = \frac{1}{2}\frac{V_0^2 R}{[(1/\omega C) - \omega L]^2 + R^2} = \frac{1}{2}\frac{V_0^2 R \omega^2}{L^2(\omega^2 - \omega_0^2)^2 + \omega^2 R^2}. \tag{34–44}$$

This equation displays the resonant behavior of AC circuits. As the driving frequency ω increases through ω_0, the power dissipated has the typical peaked behavior of resonance. The dissipated power is a *maximum* at resonance, when the driving angular frequency ω equals the angular frequency $\omega_0 = \sqrt{1/LC}$. At resonance, where $\omega = \omega_0$, the power is

$$\langle P \rangle_{\text{res}} = \frac{1}{2}\frac{V_0^2}{R}. \tag{34–45}$$

The concept of the root mean square was first introduced in Chapter 19.

The current displays the same resonant behavior as the power. Equation (34–39) shows that the current oscillates with time. It is useful to characterize the current (and other harmonically varying quantities in AC) with an *rms* (*root mean square*) value. The rms value, x_{rms}, of any quantity x is defined as the square root of the time average of the square of that quantity:

$$x_{rms} \equiv \sqrt{\langle x^2 \rangle}.$$

In particular, if x varies harmonically, that is, if $x = x_0 \cos(\omega t - \phi)$, we can use the fact that the time average of the cosine squared is one-half to show that

$$x_{rms} = \frac{x_0}{\sqrt{2}}. \tag{34–46}$$

When we apply this concept to the AC current, we see from Eq. (34–39) that

$$I_{rms} = \frac{V_0}{\sqrt{2}Z} = \sqrt{\frac{V_0^2\omega^2}{2[L^2(\omega^2 - \omega_0^2)^2 + \omega^2 R^2]}}. \tag{34–47}$$

Note from Eq. (34–43) that I_{rms} and the average power, $\langle P \rangle$, obey the static power–current relation

$$\langle P \rangle = I_{rms}^2 R. \tag{34–48}$$

Figure 34–11 plots I_{rms}^2 as a function of the driving angular frequency ω for three values of resistance. The sharpness in the peak of the average power (or of I_{rms}^2) versus ω is characterized by the *width* of the peak, or, more precisely, the *total width at half-maximum* $\Delta\omega$, commonly called the **bandwidth** in the context of AC. To calculate the bandwidth, we find the angular frequencies at which the power drops to half the peak value and take the difference between these angular fre-

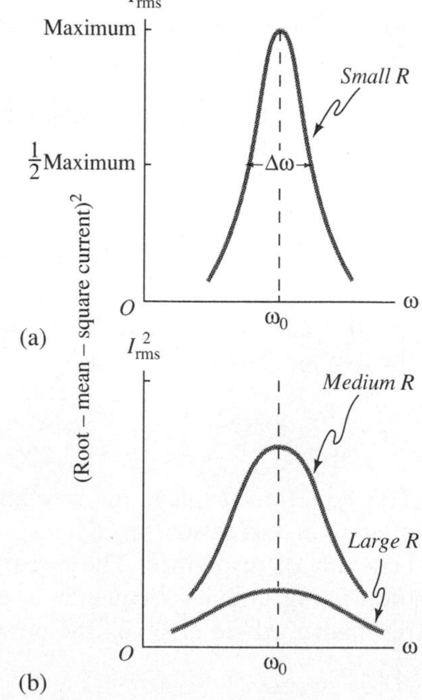

FIGURE 34–11 The rms current squared in a series RLC circuit with an AC source of emf such as the circuit of Fig. 34–8. (a) There is a resonance phenomenon when the driving frequency ω matches the natural frequency ω_0 of the circuit. The bandwidth $\Delta\omega$ measures the width of the rms current squared. (b) As R increases, the peak in the current as a function of ω broadens.

quencies. This calculation shows that, for small values of resistance, the bandwidth is given by

$$\Delta\omega = \frac{R}{L}. \tag{34–49}$$

The smaller the resistance and the larger the inductance, the smaller the bandwidth. The importance of a small bandwidth can be understood by thinking about a radio or television receiver whose tuning circuit depends on the resonance phenomenon. If the resonance is sharp, the receiver will more effectively pick out only the resonant frequency (Fig. 34–11a). This is because the current through the circuit (or the power absorbed from the AC signal) is greatly reduced for frequencies only a little different from the resonant frequency. Conversely, if the resonance is broad, the circuit will respond to frequencies in the AC signal that are far from the desired frequency (Fig. 34–11b).

EXAMPLE 34–5 Two FM radio stations broadcast at the same strength from the same nearby distance, one at a frequency of 91.3 MHz and the other at 91.1 MHz. You like the former very much but do not care for the latter. You want to construct a simple series *RLC* circuit to act as a receiver unique to your favored station, given an inductor with an inductance L of exactly 1 μH and adjustable resistance and capacitance. To limit the power received from the unwanted station to 1 percent of the power received from the desired station, what values would you choose for R and C?

SOLUTION: There are two problems here: One, to make a circuit with resonance at $\omega_0 = 2\pi f = 2\pi(91.3\text{ MHz}) = 5.74 \times 10^8$ Hz; two, to make the resonant peak sharp enough to limit the power from the station broadcasting at $\omega_1 = 2\pi(91.1\text{ MHz}) = 5.72 \times 10^8$ Hz. The resonant frequency is determined from L and C alone. With a known value of L, C is determined: $\omega_0^2 = 1/LC$, or

$$C = \frac{1}{\omega_0^2 L} = \frac{1}{(5.74 \times 10^8\text{ Hz})^2(1\ \mu\text{H})} = 3.04 \times 10^{-12}\text{ F}.$$

The sharpness requirement determines R. The power delivered by the signal at resonance is given by Eq. (34–45), whereas that off resonance is given by Eq. (34–44). The two stations have the same strength, so it is appropriate to use the same value of V_0. Thus

$$\frac{\langle P\rangle_{\omega_1}}{\langle P\rangle_{\text{res}}} = 0.01 = \left[\frac{1}{2}\left(\frac{V_0^2}{R}\right)\right]^{-1}\frac{1}{2}\frac{V_0^2 R\omega_1^2}{L^2(\omega_1^2-\omega_0^2)^2+\omega_1^2R^2} = \frac{R^2\omega_1^2}{L^2(\omega_1^2-\omega_0^2)^2+\omega_1^2R^2}.$$

Thus

$$[L^2(\omega_1^2-\omega_0^2)^2+\omega_1^2R^2](0.01) = R^2\omega_1^2.$$

To a good approximation we can ignore the $\omega_1^2R^2$ term on the left. We can then take the square root of both sides:

$$L(\omega_0^2-\omega_1^2)(0.01) = L(\omega_0-\omega_1)(\omega_0+\omega_1)(0.01) = R\omega_1.$$

The factor $(\omega_0+\omega_1)$ is, to a good approximation, equal to $2\omega_1$, so we have

$$2\omega_1 L(\omega_0-\omega_1)(0.01) \simeq R\omega_1.$$

Thus

$$\begin{aligned}R &= 2L(\omega_0-\omega_1)(0.01) = 2L(2\pi)(f_0-f_1)(0.01)\\ &= 2(10^{-6}\text{ H})(6.28)[(5.74\times10^8\text{ Hz})-(5.72\times10^8\text{ Hz})](0.01) = 2.51\ \Omega.\end{aligned}$$

The Power Factor

The power in AC circuits is commonly given in a form other than that given in Eq. (34–44). We find this form with the aid of the trigonometric identity

$$\sin^2\phi = \frac{1}{(\cot^2\phi) + 1}.$$

If we now use Eq. (34–38) for $\cot\phi$, we find that

$$\sin^2\phi = \frac{1}{[(1/R)(X_C - X_L)]^2 + 1} = \frac{R^2}{(X_C - X_L)^2 + R^2} = \frac{R^2}{Z^2};$$

$$\sin\phi = \frac{R}{Z}. \tag{34–50}$$

Then, using Eqs. (34–47) and (34–49), we see that Eq. (34–48) becomes

$$\langle P\rangle = I_{\text{rms}}^2 R = I_{\text{rms}}^2 Z\sin\phi. \tag{34–51}$$

The term $\sin\phi$ in Eq. (34–51) is called the *power factor*. For a circuit without resistance, it is zero, whereas for a pure resistance it is a maximum at one.

34–5 SOME APPLICATIONS

Most electronic circuits in use today involve elements beyond those we have studied here. These elements may perform amplifying functions, as in transistors, or have resistance that depends on the direction of current flow, as it does in diodes. Modern circuits are typically constructed in integrated form, with many thousands of elements included together from the start, and perform rather general functions. Nevertheless, several principles of such devices beyond those we have already discussed can be understood with a small addition to the elements we have in place.

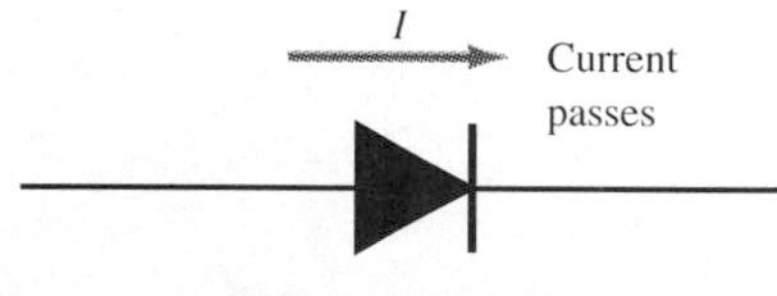

FIGURE 34–12 The symbol for a diode in a circuit diagram. Current can flow only in the direction shown.

Diodes and Rectifiers

Many sources of electric power produce AC voltage. However, many applications of power use require DC voltage. For example, the alternator of an automobile produces AC, but the car battery requires DC to be charged. We need a simple way to change from AC to DC voltage. The process by which this is accomplished is called *rectification*, and the tool used is the *diode*. A **diode** is a circuit element with a high resistance to current that flows in one direction, but a low resistance to current that flows in the other direction, the direction of the arrow in the diode symbol (Fig. 34–12). In effect, the diode allows current flow only in the direction of the arrow. The first diodes were vacuum tubes. Today, diodes are made from semiconductors.

The diode can be used to construct a **rectifier**, a circuit element that changes AC into DC (Fig. 34–13). Consider the circuit shown in Fig. 34–14a. The voltage across the load resistor can be negative or positive. However, when a diode is placed in the circuit, the negative voltages are blocked, leaving only positive voltages across the load resistor (Fig. 34–14b). Such a circuit is called a *half-wave rectifier*. This circuit can suffice as an approximation to a source of DC voltage, although the voltage between points a and b, V_{ab}, is certainly not smooth and constant.

FIGURE 34–13 This DC power supply turns AC current into DC current by means of a rectifier.

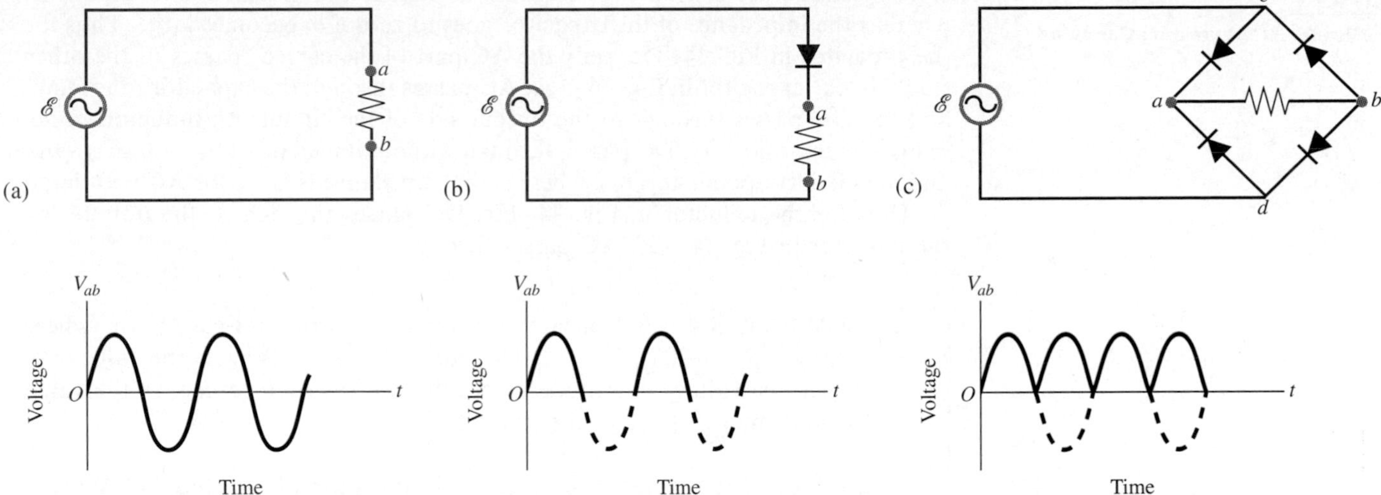

FIGURE 34–14 (a) The AC voltage across a resistor. (b) A half-wave rectifier is applied. (c) A full-wave rectifier is applied.

The foregoing situation can be improved considerably by the circuit shown in Fig. 34–14c, called a *full-wave rectifier*. Although this circuit appears to be more complex, in reality it is quite simple. When the emf produces positive voltage, positive current flows clockwise and passes through the path *cabd*, in the direction of the rectifier arrows. The voltage V_{ab} is positive. When the emf produces negative voltage, *positive current flows counterclockwise*, and the path of the current is *dabc*. In this case also, the voltage V_{ab} is *positive*. Note that now the voltage V_{ab} is positive for all half-cycles, and the rms voltage is much higher than it is for the half-wave rectifier. The use of *filters* (to be discussed next) allows the voltage peaks to be smoothed, producing a more nearly constant voltage.

Filters

A **filter** is a device that takes an input signal from one part of a circuit that may be a mixture of AC and DC and passes only the AC or only the DC signal to a different part of the circuit. Our discussion of the capacitive and inductive reactances shows that either a capacitor or an inductor can act as such a filter. Consider Fig. 34–15, in which the current I from the input is a mixture of DC and AC: for example,

$$I = I_0 + I_1 \sin(\omega t).$$

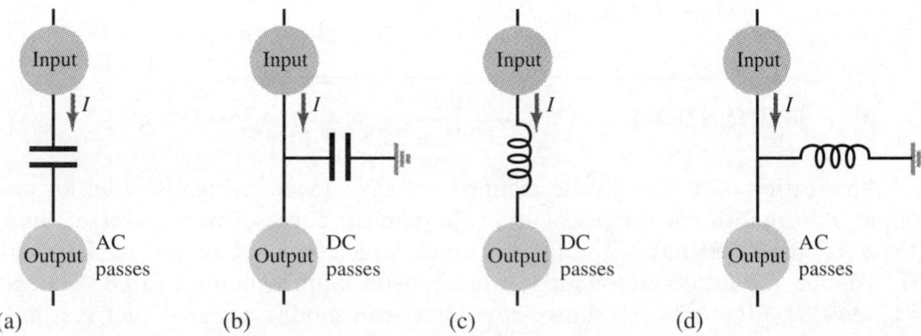

FIGURE 34–15 AC and DC filters formed from capacitors or inductors. With a capacitor: (a) AC current passes through to the output side. (b) DC current passes to the output side. With an inductor: (c) DC passes through to the output side. (d) AC passes through to the output side, but DC does not.

Here I_0 and I_1 are constants. A constant current cannot pass across a capacitor, whereas the impedance of the capacitor goes to zero if ω becomes large. Thus for the capacitor in Fig. 34–15a, only the AC part of the current passes to the other side. For the capacitor in Fig. 34–15b, AC passes through the capacitor to ground, and the DC passes through to the output side of the circuit. An inductor works in just the opposite way: DC passes through without impedance ($X_L \to 0$ as $\omega \to 0$, and $\omega = 0$ corresponds to DC), whereas the impedance is large for AC with large ω. Thus for the inductor in Fig. 34–15c, DC passes through to the output; for the inductor in Fig. 34–15d, AC passes through.

EXAMPLE 34–6 Consider the circuit shown in Fig. 34–16, where $C = 1.0\ \mu\text{F}$, $R = 0.20\ \Omega$, $V_0 = 0.10$ V, and $V_1 = 0.25$ V. What is the value of ω for which the voltage amplitude across the resistor is 50 percent of the value of the maximum voltage of the generator?

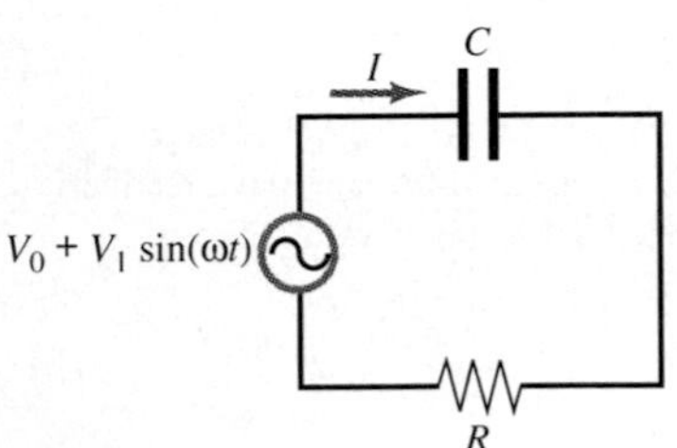

FIGURE 34–16 Example 34–6.

SOLUTION: The generator produces a combination of DC and AC. We can apply the superposition principle by calculating the result of application of the loop rule that corresponds to the DC and AC terms separately, and then add the voltage drops. For the AC term, the current in the circuit is, from Eq. (34–39),

$$I_{AC} = \frac{V_1 \cos(\omega t - \phi)}{Z} = \frac{V_1}{\sqrt{(1/\omega C)^2 + R^2}} \cos(\omega t - \phi).$$

The voltage drop across the resistor from the AC is $I_{AC}R$ and therefore has the amplitude

$$\frac{V_1 R}{\sqrt{(1/\omega C)^2 + R^2}}.$$

The capacitor acts as a perfect filter for the constant term in the input voltage, because no constant current can pass. There is thus no voltage drop across the resistor associated with the V_0 term.

The maximum value of the input voltage is $V_0 + V_1$, and the ratio of the voltage amplitude across the resistor to the maximum input voltage is

$$\frac{V_1 R/\sqrt{(1/\omega C)^2 + R^2}}{V_0 + V_1}.$$

We want this factor to equal 50 percent. We set it to 0.50 and solve for ω:

$$\omega = \frac{1}{RC\sqrt{\dfrac{V_1^2}{(V_0 + V_1)^2(0.50)^2} - 1}}$$

$$= \frac{1}{(0.20\ \Omega)(1.0 \times 10^{-6}\ \text{F})\sqrt{\dfrac{(0.25\ \cancel{V})^2}{(0.10\ \cancel{V} + 0.25\ \cancel{V})^2}\dfrac{1}{(0.50)^2} - 1}} = 0.049\ \text{MHz}.$$

Application of a filter to the rectified voltage shown in Fig. 34–14b, for example, will smooth out the peaks and valleys in the curve of voltage versus time. If the RC time constant of the filter is much larger than the period of the rectified voltage, the resulting voltage is a much better approximation to DC voltage (Fig. 34–17). Figure 34–18 shows how filters can modify a signal that is a mix of many different frequencies.

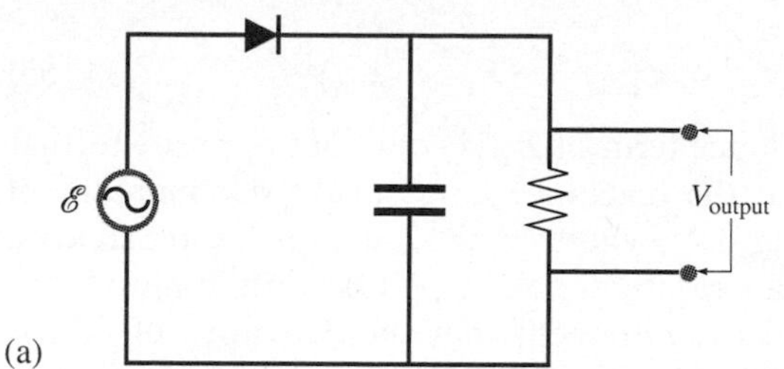

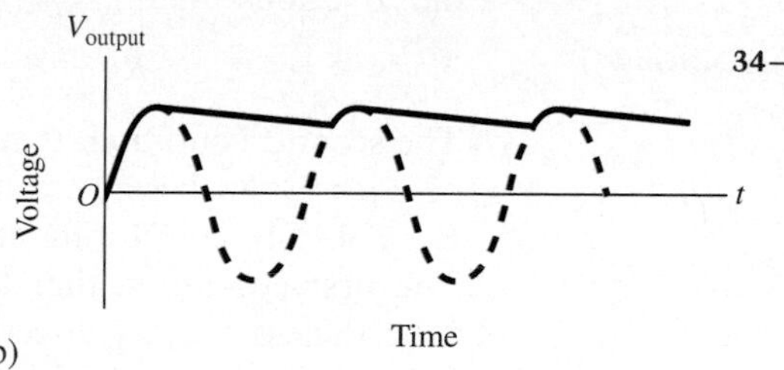

FIGURE 34–17 (a) An *RC* circuit acts as a filter for rectified AC voltage. Such a filter can produce a voltage that is nearly DC. (b) The slow decrease of the nearly constant-voltage segments is governed by the time constant of the *RC* circuit.

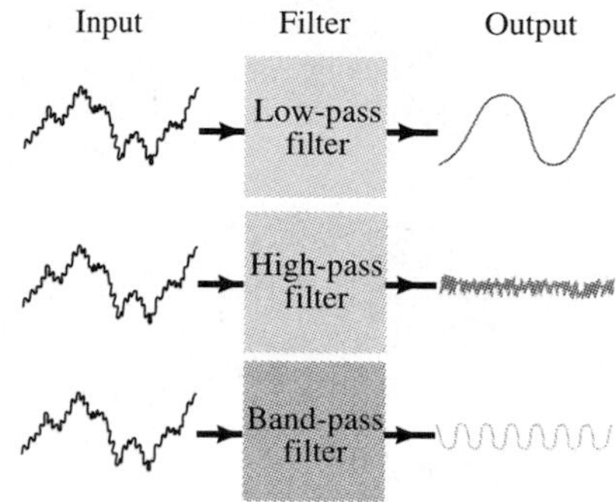

FIGURE 34–18 (a) A low-pass filter allows the low frequencies contained in an input signal to pass. (b) A high-pass filter allows the high frequencies in the signal to pass. (c) A band-pass filter allows a range of frequencies to pass.

Impedance Matching

Another aspect of AC of great practical importance concerns **impedance matching**, which refers, as in our discussion of filters, to the *connection* between different parts of a circuit. Figure 34–19a shows such a situation, in which some combination of circuit elements makes up circuit 1, connected at points a and b to circuit 2. The two circuits have impedances Z_1 and Z_2, respectively. We are not concerned here with the origin of currents in these circuits as much as we are with our ability to deliver power from circuit 1 to circuit 2. We therefore assume that the origin of these currents is within circuit 1 and break that circuit down as in Fig. 34–19b. The primary question is, if Z_1 is fixed, what are the requirements for Z_2 so that the power delivered to circuit 2 is a maximum? If, for example, a stereo amplifier is connected to a loudspeaker, what should the loudspeaker's impedance be in order that maximum power is delivered to it?

The answer is found by computing the average power $\langle P \rangle$ to circuit 2, which, from Eq. (34–48), is $I_{rms}^2 R_2$. The current in the loop of Fig. 34–19b is given by

$$I_{rms} = \frac{\mathcal{E}_{rms}}{Z_{total}}. \tag{34–52}$$

Here $\mathcal{E}_{rms}$ is the rms value of the generator, whose maximum voltage, or amplitude, is V_0. If the generator produces a sinusoidal emf of the form of Eq. (34–1), then Eq. (34–46) shows that $\mathcal{E}_{rms} = V_0/\sqrt{2}$. The total impedance Z_{total} is found by adding separately the capacitive reactances, inductive reactances, and resistances, a result that follows from our knowledge of how series combinations of C, L, and R add (see Problem 32):

$$Z_{total} = \sqrt{[X_{C_1} + X_{C_2} - (X_{L_1} + X_{L_2})]^2 + (R_1 + R_2)^2}. \tag{34–53}$$

Thus the average power delivered to circuit 2 is

$$\langle P \rangle = \frac{\mathcal{E}_{rms}^2 R_2}{Z_{total}^2} = \frac{\mathcal{E}_{rms}^2 R_2}{[X_{C_1} + X_{C_2} - (X_{L_1} + X_{L_2})]^2 + (R_1 + R_2)^2}. \tag{34–54}$$

That there is a value of the parameters of Z_2 that maximizes this power is clear: If Z_2 is too small, the factor R_2 will also be small and $\langle P \rangle$ will be small; if Z_2 is too large, it will dominate the denominator of Eq. (34–54), and $\langle P \rangle$ will again be small. An intermediate value of the parameters of Z_2 will give a maximum value of $\langle P \rangle$. Two independent parameters are involved here: the resistance R_2 and the total reactance of circuit 2, $X_{C_2} - X_{L_2}$. Formally, we find the value of the parameters that maximize $\langle P \rangle$ by taking the derivative of $\langle P \rangle$ with respect to these quantities and setting it equal to zero. From this exercise the power is maxi-

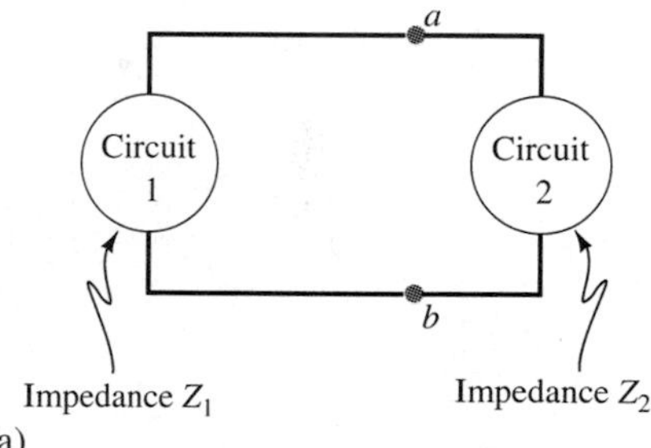

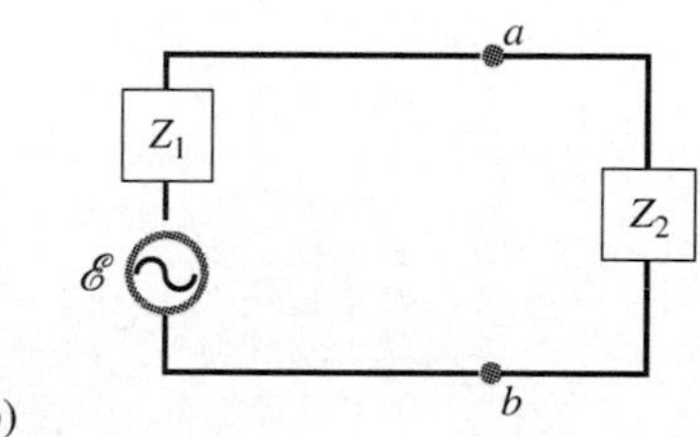

FIGURE 34–19 (a) Circuit diagram to illustrate impedance matching. (b) Circuit 1 of part (a) is broken down into a source of emf $\mathcal{E}$ and an impedance Z_1. Circuit 2 is assumed to include only an impedance Z_2.

mized when

$$R_2 = R_1 \quad \text{and} \quad X_{C_2} - X_{L_2} = -(X_{C_1} - X_{L_1}). \tag{34–55}$$

The second condition, that the reactance term of Z_2 is equal but opposite to that of Z_1, follows because it means that the reactance terms in the denominator of Eq. (34–54) cancel, thus maximizing $\langle P \rangle$ whatever the value of the resistances. The first condition, that the resistances are equal, is perhaps less intuitive but nevertheless follows directly from the requirement that the derivative of $\langle P \rangle$ is zero (see Problem 54). When the conditions of Eq. (34–55) are met, the impedances are said to be matched.

The impedances of two parts of a circuit should be matched when you want to deliver maximum power from one part of the circuit to the other.

Impedance matching is desirable when you wish to deliver maximum power to one part of a circuit. It is worth noting that we do not always wish to deliver maximum power. A voltmeter, for example, should have an impedance *mis*match, because it is desirable that it draw as little current as possible.

The subject of circuit analysis is highly developed. We have been able to do no more than describe its principles, and this chapter will not have taught you to fix, much less design, TVs or computers. But the principles we have described here apply to all electric circuits.

SUMMARY

The presence of AC sources of emf in circuits with resistors, inductors, and capacitors introduces a variety of new possibilities. Transformers allow us to vary the voltage amplitude of AC emfs. The relation between the emfs and the numbers of turns of the primary and secondary coils of a transformer is

$$\frac{\mathcal{E}_2}{\mathcal{E}_1} = \frac{N_2}{N_1}. \tag{34–8}$$

Conservation of energy implies that the currents carried by the respective coils are related inversely:

$$\frac{I_1}{I_2} = \frac{N_2}{N_1}. \tag{34–10}$$

A series *RLC* circuit with an AC source of emf of frequency ω behaves like a damped harmonic oscillator driven by a harmonically varying force. Solutions for currents, voltages, and charges in such circuits can be found by using the solutions already developed for the driven harmonic oscillator. For such circuits the impedance, Z, is a quantity that plays the role of a resistance. The impedance is frequency dependent:

$$Z \equiv \sqrt{\left(\frac{1}{\omega C} - \omega L\right)^2 + R^2} = \sqrt{(X_C - X_L)^2 + R^2}, \tag{34–36}$$

where X_C is the capacitive reactance and X_L is the inductive reactance. The current in the driven circuit is then

$$I = I_{max} \cos(\omega t - \phi) = \frac{V_0 \cos(\omega t - \phi)}{Z}, \tag{34–39}$$

where V_0 is the amplitude of the driving emf and ω is its frequency. The phase ϕ is given by

$$\frac{1}{\tan \phi} = \frac{1}{R}\left(\frac{1}{\omega C} - \omega L\right). \tag{34–34}$$

Such circuits exhibit resonant behavior when the driving frequency is near the natural frequency $\omega_0 = \sqrt{1/LC}$. This type of behavior is most clearly seen in the power dissipated in driven RLC circuits. Averaged over time, the power lost is

$$\langle P \rangle = I_{\rm rms}^2 R = I_{\rm rms}^2 Z \sin\phi, \qquad (34\text{–}51)$$

where $I_{\rm rms} = V_0/\sqrt{2}Z$ is the rms current. Near resonance, the power dissipated is a maximum, and the width of the peak of average power versus driving frequency has a width at half-maximum of

$$\Delta\omega = \frac{R}{L}, \qquad (34\text{–}49)$$

a result that holds as long as the resistance is not too large.

If we add diodes—devices that allow current to pass in one direction only—to our arsenal of circuit elements, we can construct a variety of electronic devices, including rectifiers, which produce a positive (or negative) emf from an AC source, and filters, which take a mixed AC and DC signal and pass predominantly either the constant (DC) part or the variable (AC) part of a given frequency. Impedance matching refers to constraints that describe how different parts of a circuit can be connected together with minimal power loss.

QUESTIONS

1. Why is the material used to make the core of transformers so important?
2. What are some applications for step-up and step-down transformers?
3. Without R, the current amplitude in a series RLC circuit would be infinite when the driving frequency ω equaled ω_0, but this possibility could never happen, because there is always some resistance in real circuits. How do you reconcile this statement with the existence of superconductors?
4. To find the rms current [Eq. (34–47)], we square the current, then take the time average, and then take the square root. Why do we not simply take the time average of the current?
5. In Example 34–6, we took the input emf to be a mixture of DC and AC, then treated the effect of the AC and DC parts separately and added the two parts. How is this procedure justified?
6. A capacitor and a lamp are connected in series with an AC generator of constant voltage but variable frequency (Fig. 34–20). Which of the following three statements is true? The lamp will (a) not light, because the capacitor is connected in series with the lamp; (b) burn brightest when the frequency is high; (c) burn with the same brightness for all frequencies.

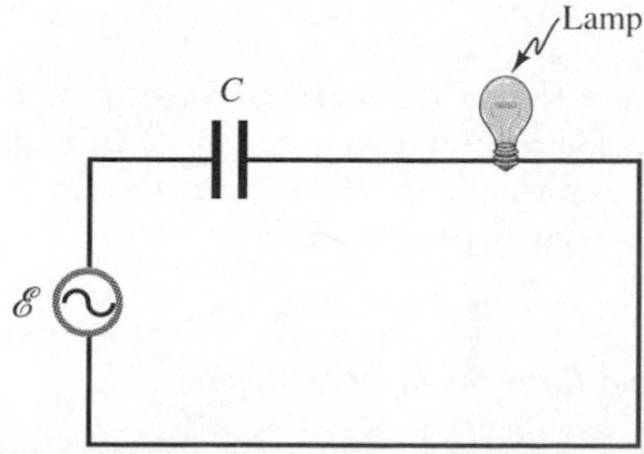

FIGURE 34–20 Question 6.

7. A capacitor, lamp, and resistor are connected to an AC generator of constant voltage but variable frequency (Fig. 34–21). Which of the following statements is true? The lamp will (a) not burn, because the capacitor shorts out the lamp; (b) burn brightest when the frequency is low; (c) burn brightest when the frequency is high; (d) burn with the same brightness for all frequencies.

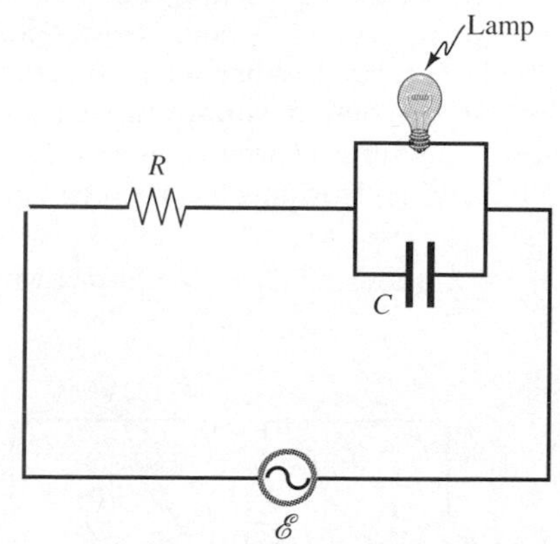

FIGURE 34–21 Question 7.

8. A particular appliance or household circuit is rated for a maximum current. Why is that, and why must currents not exceed that maximum?

9. The primary purpose of an electric heater is to produce heat. Why would such a heater require a 220-V socket rather than a 120-V socket?
10. Two circuits have impedances Z_1 and Z_2, respectively. When these circuits are placed in series with one another, why is the total impedance not given by $Z = Z_1 + Z_2$?
11. The reactance X_C is infinite when the input voltage is DC. Does that mean that the impedance is not defined for this situation?
12. If a capacitor has a large impedance for DC and an inductor has a large impedance for AC, how can a series LC circuit pass any current?
13. Television antenna wires normally have negligible resistance and an impedance of 75 Ω. Why is it important to use antenna wires with the same impedance throughout?
14. Some appliances that operate off a 120-V line yet draw in excess of 20 A of current have different plugs than regular 120-V household devices have. Why are these plugs different? What might happen if they were not different?
15. Do all television sets have transformers? How might transformers be used in a television set?
16. If electricity is transported in power transmission lines at 200 kV to 500 kV, does the power have to be generated at these voltages? Why or why not?

PROBLEMS

34–1 *Transformers*

1. (I) Suppose that electric power costs 15 cents/kWH. Consider a transmission line that delivers 1 MW of power and has a resistance of 10 Ω. Calculate the dollars lost annually due to the transmission line if the power is delivered at (a) 500,000 V and (b) 440 V.
2. (I) Many electrical devices, such as doorbells or buzzers, operate on 12 V AC. A small transformer used to produce this voltage has a primary coil of 1200 turns and takes an input of 120 V AC. How many turns must the secondary coil have?
3. (I) The primary coil of a step-down transformer is connected to house current, 110 V at 60 Hz. If the secondary coil of the transformer delivers a current with an amplitude of 5.0 A at 12 V, what is the current drawn by the primary coil? Ignore losses in the transformer.
4. (I) A transformer whose output voltage can be varied is used to obtain AC power from a 120-V, 10-A supply. The secondary coil consists of 1200 turns of wire. The variable transformer works by connecting different numbers of turns of wire on the secondary coil. The secondary voltage can thereby be regulated. When all 1200 turns act as the secondary coil, the output voltage has an amplitude of 120 V. How many turns of wire should be used to obtain (a) 45 V? (b) 12 V? (c) How much current will flow for each voltage?
5. (II) Figure 34–22 shows an ideal transformer with 110 V on the primary coil supplying power to a resistor of resistance R. If the resistor dissipates 88 W, what is the current in the primary coil?

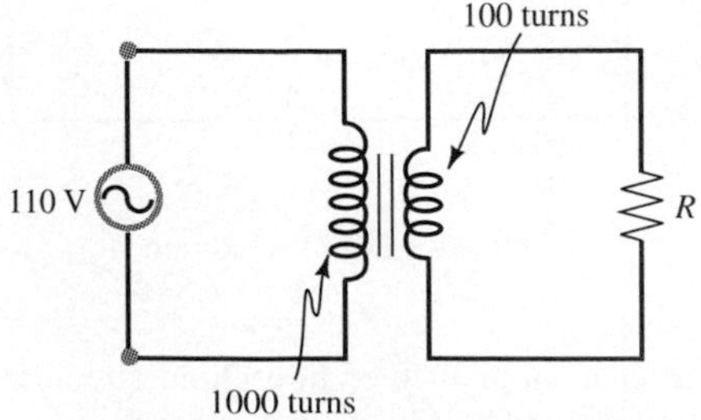

FIGURE 34–22 Problem 5.

6. (II) A step-down transformer has a turn ratio (N_1/N_2) of 5:1. (a) If the primary coil is connected across a 220-V oscillating-voltage generator, what voltage appears across the secondary coil? (b) Assuming that there are no power losses in the transformer, what current would have to flow through the primary coil so that a 40-Ω resistor placed across the secondary coil draws all the power of the circuit? (c) What resistance connected across the 220-V voltage generator would draw the same total power?
7. (II) The transformer shown in Fig. 34–23 has two secondary windings; one supplies 220 V, the other, 11 V. The input voltage at the primary coil is 110 V. If the 220-V secondary coil has 1000 turns, how many turns does the 11-V secondary coil have?

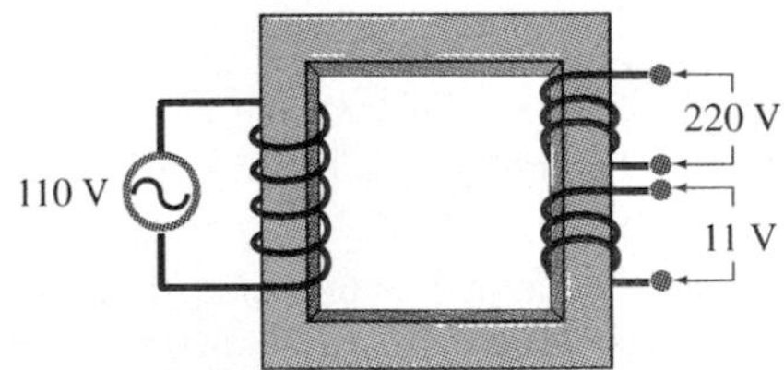

FIGURE 34–23 Problem 7.

8. (II) Suppose that a transformer consists of two separate windings of wire on the same core. The core material has a magnetic permeability μ. How does the ratio of the emfs in the two coils depend on μ?

34–2 *Single Elements in AC Circuits*

9. (I) A 120-Ω resistor is connected across a power supply that produces a voltage of the form $V_0 \sin(\omega t)$, where $f =$

$\omega/2\pi = 60$ Hz and $V_0 = 163$ V. What is the current passing through the resistor?

10. (I) An alternating current of maximum value 10 A in a solenoid of self-inductance $L = 250$ mH induces an emf of maximum value 10 V. What is the angular frequency of the alternating current?

11. (I) An AC power supply with frequency 60 Hz is connected to a capacitor of capacitance $C = 40\ \mu$F. The maximum instantaneous current that passes through the circuit is 2.26 A. What is the maximum voltage?

12. (I) A current flowing through a circuit that contains only a capacitor and an AC power supply has the form $I_0 \cos[2\pi ft - (\pi/6)]$, where $I_0 = 2.45$ A and $f = 180$ Hz. If the maximum voltage supplied by the generator is 95 V, what is the capacitance?

13. (I) An AC power supply operating at a frequency of 50 Hz is connected across an inductor. The maximum voltage of the source is 40 V, and the maximum current is 8.0 A. What is the inductive reactance? What is the inductance of the circuit?

14. (II) A current $I = I_0 \sin[(\omega t + (\pi/4)]$ flows in a circuit for which $I_0 = 4.8$ A and $\omega = 2\pi(30$ Hz). (a) At what times does the peak current flow? (b) If the current flows through a 20-H inductance, what is the peak voltage on the inductor? At what times does this peak voltage occur?

15. (II) The average of the square of the voltage in an inductive circuit (a circuit with no capacitors and no resistors) driven by an AC emf is $(30\text{ V})^2$, and the average of the square of the current is $(2\text{ A})^2$. What is the inductive reactance? If the inductance is 25 mH, what is the frequency of the alternating current?

34–3 *AC in Series RLC Circuits*

16. (I) Consider a radio circuit with a fixed inductance of 50 μH. What is the value of the tunable capacitance for the reception of a 90-m radio wave?

17. (I) What is the range needed for a variable capacitor to be combined with a 0.15-mH coil so that a tuned circuit could be formed to cover the range of broadcast-band frequencies from 540 kHz to 1600 kHz?

18. (I) An AC generator with a voltage amplitude of 10 V and a frequency of 4000 Hz is built to drive a circuit meant to be resonant. The resistance of the circuit is 0.2 Ω, and the inductance is 1 mH. What must the value of the capacitance be?

19. (I) A series *RLC* circuit of frequency 60 Hz has a maximum current of 100 mA. What is the maximum charge on the capacitor? If the impedance is 40 Ω, what is the emf?

20. (II) A series *RLC* circuit has parameters $R = 10.0\ \Omega$, $L = 10.0$ mH, and $C = 20.0\ \mu$F. Find the capacitive reactance, inductive reactance, and impedance for the frequencies (a) 60 Hz, (b) 300 Hz, and (c) 10,000 Hz.

21. (II) A series *RLC* circuit consists of a 60-Hz AC emf with $V_0 = 120$ V; $R = 200\ \Omega$, $L = 150$ mH, and $C = 2$ mF. Find X_C, X_L, Z, Q_{max}, φ, and I_{max}.

22. (II) Find the voltages across the capacitor and inductor in the AC circuit of Problem 21 at $t = 3$ s if the emf is switched on at $t = 0$ s. All circuit elements initially have no charge or current flow.

23. (II) Given that the maximum voltage in the circuit shown in Fig. 34–24 is 110 V and the frequency of oscillation is 60 Hz, calculate the maximum current and the maximum potential drops across the resistor, capacitor, and inductor.

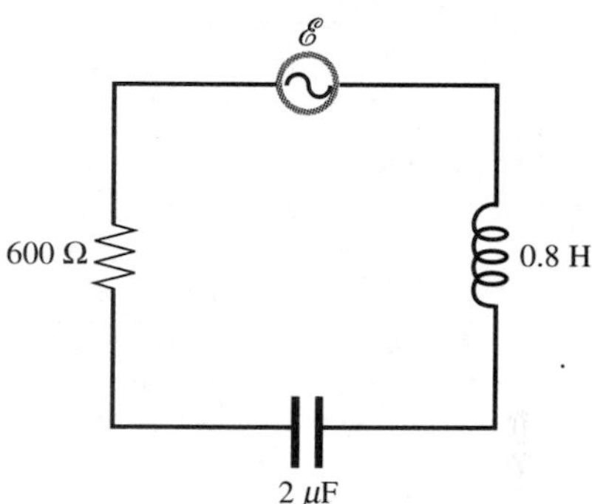

FIGURE 34–24 Problem 23.

24. (II) What is the resonant angular frequency ω_0 of the circuit in Problem 23? Suppose that the voltage generator has a variable angular frequency ω. For what values of ω will the current have half the value it has at resonance?

25. (II) An AC circuit consists of a parallel-plate capacitor and a long, cylindrical solenoid. Suppose that all the dimensions of the apparatus, including the wire sizes, are scaled down by a factor of 2. (Note that the turn density doubles.) How would the resonant frequency of the circuit change? Assume that there are changes in resistance.

26. (II) Show that Eq. (34–32) satisfies Eq. (34–27) by direct substitution. Determine the maximum charge Q_{max} on the capacitor in terms of the impedance.

27. (II) Sketch the current and voltage for the following AC series circuits: (a) a pure capacitive circuit, (b) a pure inductive circuit, (c) an *RL* circuit, (d) an *RC* circuit, and (e) an *LC* circuit.

28. (II) A resistor draws 8 A when connected to a 220-V, 60-Hz line. A capacitor of what capacitance, when connected in series with the resistor, will drop the current to 6 A? What are the voltage drops across the capacitor and the resistor?

29. (II) A 16-μF capacitor is connected in series with a coil whose resistance is 30 Ω and whose inductance can be varied. The circuit is connected across a 12-V, 60-Hz generator. What is the potential difference across the capacitor and across the inductor–resistor combination when the frequency is the resonant frequency?

30. (II) Suppose that the maximum voltages across the resistor, capacitor, and inductor of a series *RLC* circuit driven by an AC generator of frequency f are identical. If the resistor has a resistance R, find the values of C and L in terms of R and f.

31. (II) A series RLC circuit contains a 0.10-μF capacitor and a 75-Ω resistor. If the circuit is resonant at a frequency of 1200 Hz, what is the inductance? If the resistance is increased by 10 percent, by what percentage must the inductance change in order to keep the resonant frequency at the same value?

32. (II) Consider an RLC circuit in which two resistors, R_1 and R_2, are connected in series, as are two capacitors, C_1 and C_2, and two inductors, L_1 and L_2. Show that the resulting total impedance is of the form

$$Z_{\text{total}} = \sqrt{[X_{C_1} + X_{C_2} - (X_{L_1} + X_{L_2})]^2 + (R_1 + R_2)^2}.$$

33. (II) A *phasor* is a vector associated with a harmonic function $f(t) = C\sin(\omega t + \phi)$; any harmonic function can be written in this form. The phasor is in the xy-plane, originates at the origin, and rotates; it is defined to have length C and a y-component $C\sin(\omega t + \phi)$, as in Fig. 34–25. (a) Draw the phasor for the function $D\cos(\omega t + \phi)$ on the graph that contains the phasor for the function $C\sin(\omega t + \phi)$. Which phasor is more advanced in phase—that is, points in a direction corresponding to a larger angle, as measured from the $+x$-direction? (b) What is the phase difference between the two phasors on the plot you drew for part (a)? Draw the phasor for the function $f(t) = A\cos(\omega t) + B\sin(\omega t)$.

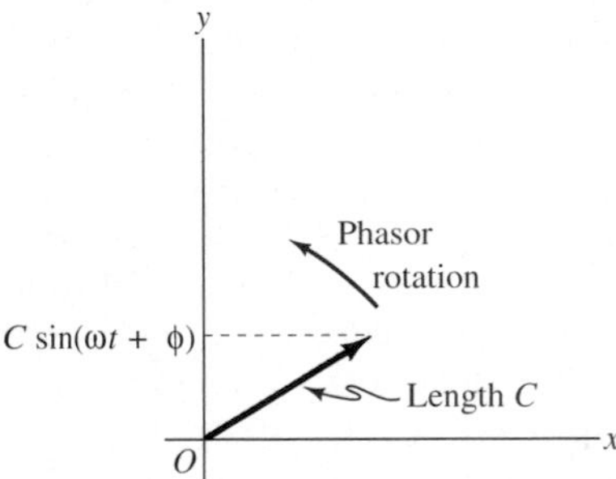

FIGURE 34–25 Problem 33.

34. (III) An alternating current has the form $I(t) = I_0\cos(\omega t - \phi)$. The voltage in a single-loop circuit containing this current has the general form $V(t) = V_0\sin(\omega t)$, with V_0 and ϕ as described in the text. Make a phasor diagram (see Problem 33) for $V(t)$ and $I(t)$ for a circuit that contains (a) only a resistor; (b) only a capacitor; (c) only an inductor.

34–4 *Power in AC Circuits*

35. (I) What is the average power dissipated in the resistor for the circuit in Problem 9?

36. (I) An AC power supply with a frequency of 240 Hz dissipates energy at a rate of 80 W in a 40-Ω resistor. If the current at time 0 s is 2.0 A, what is the current at time 0.05 s?

37. (I) Consider an AC voltage of the form $V_0\sin(\omega t)$ connected to a capacitor of capacitance C. Calculate the instantaneous power VI delivered by the source of emf, and find the average power dissipated in the circuit.

38. (I) A portable electric heater operating on AC voltage of amplitude 115 V is rated at a power of 2 kW. (a) What is the resistance of the heater? (b) Find the rms current. (c) Find the maximum current.

39. (I) What are the power factors for (a) pure capacitive circuits, (b) pure inductive circuits, and (c) pure resistive circuits?

40. (II) Show that, on average, no power is dissipated in a purely inductive circuit (a circuit with neither capacitors nor resistors).

41. (II) What are the power factors for (a) RL circuits, (b) RC circuits, and (c) LC circuits?

42. (II) An AC source of emf operating at a frequency of 60 Hz produces an rms voltage of 115 V. Find the voltage amplitude. The source of emf is connected in series with an impedance of $Z = 25\ \Omega$. Find the rms current and the current amplitude.

43. (II) When a coil draws 200 W from a $V_{\text{rms}} = 110$-V, 60-Hz line, the power factor is 0.6. If the same coil with a capacitor added in series is to draw the same power from a $V_{\text{rms}} = 220$-V, 60-Hz line, what must the capacitance be? If the aim were to maintain the same power factor rather than the same rms power, how would your answer change?

44. (II) An electric motor that consumes 5 kW of power at 220 V with a power factor of 0.80 is to be run at the end of a power transmission line with a total resistance of 2.5 Ω. What voltage and power must be supplied at the input end of the transmission line?

45. (II) An AC transmission line transfers energy to a device with a power factor of 0.8 at the rate of $\langle P \rangle = 12$ kW and a voltage of 480 V. If the transmission line has a resistance of 2.0 Ω, how much energy is lost to Joule heating in the transmission line?

46. (II) A 220-V generator has a current-carrying capacity of 80 A. What is the maximum rate at which energy can be taken from this generator by an impedance with a power factor of 0.55? for a power factor of 0.95?

47. (II) An 80-Ω resistor and a 40-μF capacitor are connected in series to a 220-V, 60-Hz power supply. Calculate the current, power, and power factor. How will these numbers change if an inductance of 0.40 H is connected in series with this circuit?

48. (II) House current, which has an rms voltage of 110 V and a frequency of 60 Hz, drives a resistor of a variable resistance set at $R = 50\ \Omega$, a capacitor of fixed capacitance $C = 20\ \mu$F, and an inductor of variable inductance, connected in series. (a) What is the power absorbed by the circuit if $L = 10$ mH? (b) What would the power drawn be if the resistance were halved without changing the setting of the inductance? (c) By what factor would the inductance need to be changed in order to draw the same power with the new value of R? (d) What is the maximum power drawn in parts (b) and (c)?

49. (II) For a driven series RLC circuit, show that

$$\frac{R}{Z} = \frac{1}{\sqrt{1 + Q^2\left(\dfrac{\omega}{\omega_0} - \dfrac{\omega_0}{\omega}\right)^2}}.$$

The *quality factor*, or *Q-factor*, Q, in the square root is $Q \equiv \omega_0 L/R$. (Here Q is not the electric charge!) This factor is often used by electrical engineers to represent the sharpness of a resonant circuit. This equation shows the resonant characteristic of the power loss when $\omega = \omega_0$. For large values of Q, the resonance is very sharp. For small values of Q, the resonance is broad.

50. (II) For Problem 49 plot R/Z for values of ω/ω_0 from 0.4 to 2.5 and values of Q of 1, 10, and 100. Use a computer program and graphics output, if available.

51. (II) For a driven series RLC circuit, show that Q (see Problem 49) is related to $\Delta\omega$ by the relation

$$\frac{\Delta\omega}{\omega} = \frac{1}{Q}\frac{\omega_0}{\omega} \simeq \frac{1}{Q}.$$

52. (III) Suppose that, instead of following Ohm's law, a resistance is effectively proportional to the frequency—not an unrealistic assumption when radio frequencies are involved. Thus take $R = (\omega/\omega_0)R_0$. How does this change the expression for $\Delta\omega$? Is the peak still centered at $\omega = \omega_0$?

34–5 *Some Applications*

53. (II) Consider the circuit treated in Example 34–5 and drawn in Fig. 34–8. Take $C = 1\ \mu$F and $R = 0.2\ \Omega$, but assume now that the input emf has the purely sinusoidal form $V_1 \sin(\omega t)$, where $V_1 = 0.25$ V. Calculate the potential across the capacitor for (a) $f = 10$ Hz, (b) $f = 10^3$ Hz, and (c) $f = 1$ MHz. What sort of a filter do you conclude this represents?

54. (II) The first condition for impedance matching is that the resistances are equal [Eq. (34–55)]. Show that this is true by starting with Eq. (34–54) in the case that the reactance terms are equal and opposite. Take a derivative of the resulting average power with respect to R_2, set it equal to zero, and show that this gives the equal resistance condition.

55. (II) Two coils are connected in parallel across an AC generator with maximum voltage $\mathscr{E}_0$ and frequency f. The resistance and inductance of the first coil are R_1 and L_1; those of the second coil are R_2 and L_2. What is the current drawn by this circuit? What is the power factor?

56. (II) Given that the driving voltage of the RLC circuit shown in Fig. 34–26 is $V = V_0 \cos(\omega t)$, calculate the currents in the three elements. Is there a resonant frequency? [*Hint:* Write down the circuit equations, and substitute the trial solution $I = I_0 \cos(\omega t + \phi)$].

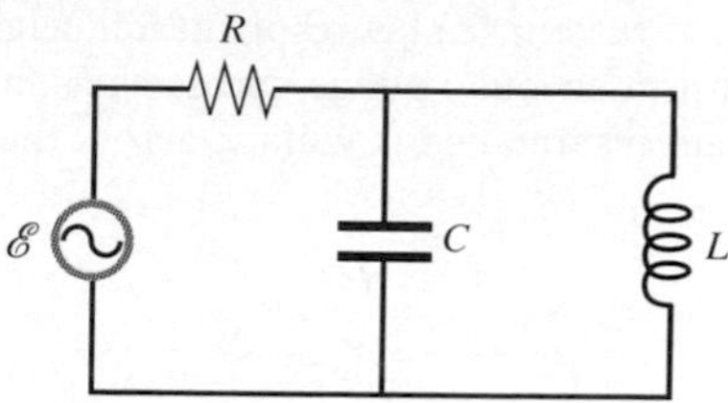

FIGURE 34–26 Problem 56.

57. (II) A diode, through which current can flow only when the emf is positive, acts as a filter for an AC generator of angular frequency ω. The current has maximum magnitude I_0. Find its average and rms values.

58. (III) Design a circuit with a transformer and a full-wave rectifier that will take a 20,000-V AC (60 Hz) power source and produce a good approximation to 400 V DC. Draw a schematic diagram of the circuit, and give the parameters of the transformer.

59. (III) An RC filter circuit like that shown in Fig. 34–16 is called a *high-pass* filter circuit when the voltage output is taken across the resistor. Plot the ratio V_{out}/V_{in} as a function of frequency. Why does such a circuit block signals of low frequency but allow high-frequency signals to pass?

60. (III) An RC filter circuit like that shown in Fig. 34–16 is called a *low-pass* filter circuit when the voltage output is taken across the capacitor. Plot the ratio V_{out}/V_{in} as a function of frequency. Why does such a circuit block high-frequency signals but allow low-frequency signals to pass?

61. (III) Television sets require a high-voltage DC of low current. Replace the inductor in Fig. 34–27 by a large resistor. Show that this "RC" circuit with a rectified AC emf is effective in reducing the AC "ripple" but not the DC component.

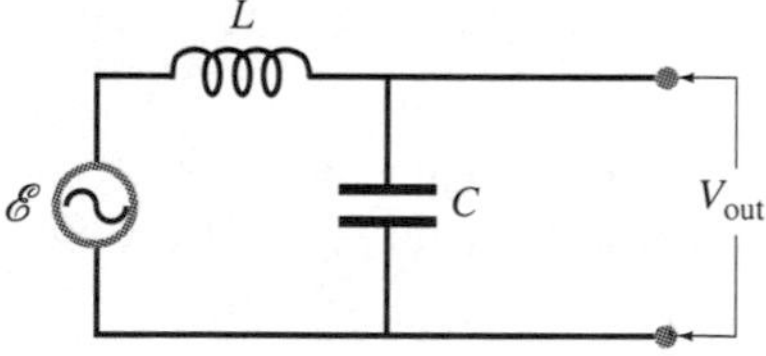

FIGURE 34–27 Problem 61.

62. (III) Consider the LC filter of Fig. 34–27 with the emf $V_0 \sin(\omega t)$. Assume that $X_L \gg X_C$ (or $\omega \gg \omega_0$). (a) Show that $V_{out} = (X_C/X_L)V_0$. (b) Show that the circuit of Fig. 34–27 is generally effective in reducing the AC components, but not the DC components, of emf.

General Problems

63. (II) Consider the circuit shown in Fig. 34–28. The emf has an amplitude of $V_0 = 12$ V and a frequency of 1000 Hz; $L = 20$ mH, $C_1 = 25\ \mu$F, and $C_2 = 45\ \mu$F. Find (a) the maximum current; (b) the resonant frequency; (c) the maximum instantaneous voltage across each capacitor; (d) the maximum instantaneous voltage across the inductor.

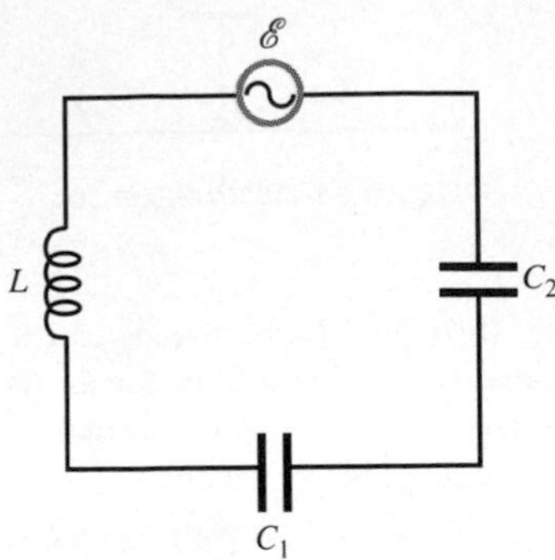

FIGURE 34–28 Problem 63.

64. (II) An amplifier with an equivalent impedance of 15,000 Ω is to be connected to an 8-Ω speaker through a transformer. What should the turn ratio of the transformer be?

65. (II) The impedance Z_1 in Fig. 34–29 can be regarded as a pure resistance $R_1 = 2\ \Omega$, whereas the impedance Z_2 is associated with a series resistance $R_2 = 4\ \Omega$ and a capacitance $C = 5 \times 10^{-8}$ F. If $f = 10^5$ Hz and $V_0 = 50$ V, what is the power dissipated in Z_2?

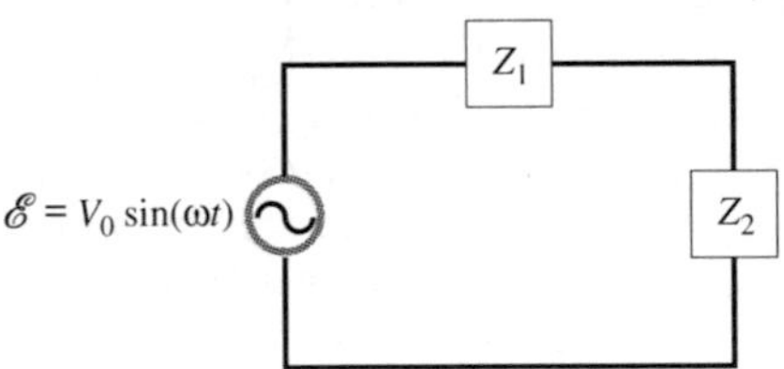

FIGURE 34–29 Problem 65.

66. (II) Consider the circuit shown in Fig. 34–30. The emf has an amplitude of $V_0 = 12$ V and a frequency of 400 Hz; $L =$ 10 mH, $C_1 = 20\ \mu$F, and $C_2 = 30\ \mu$F. Find (a) the maximum current in each leg; (b) the resonant frequency; (c) the maximum instantaneous voltage across each capacitor; (d) the maximum instantaneous voltage across the inductor.

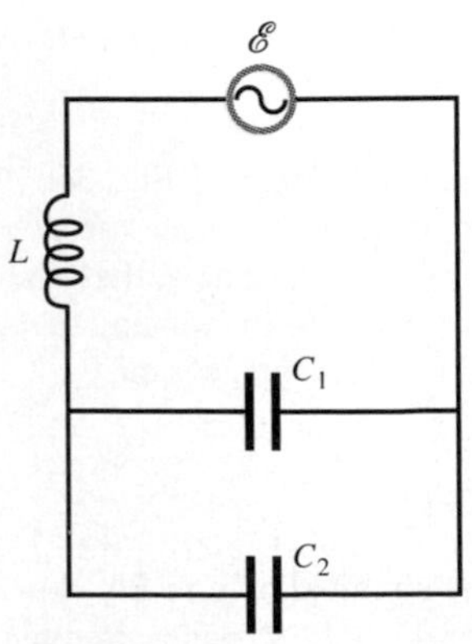

FIGURE 34–30 Problem 66.

67. (II)Write down the two equations that specify the currents I_1 and I_2 in the two loops of the circuit shown in Fig. 34–31.

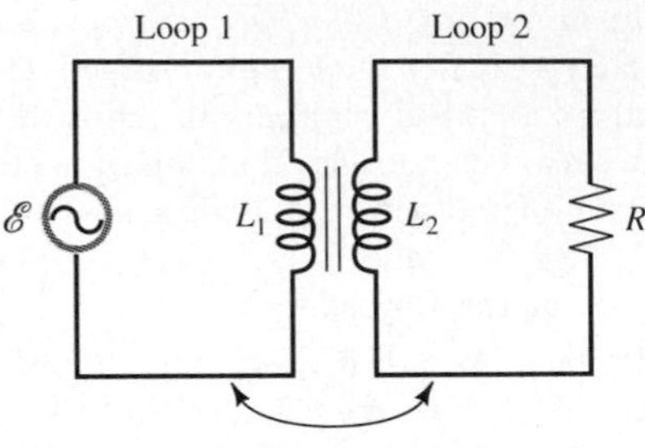

FIGURE 34–31 Problem 67.

68. (II) A series *RLC* circuit is to be designed to have a resonant frequency of 2.2 MHz, and the curve of power versus frequency f is to have a full width of 1100 Hz. If the only capacitor available has a capacitance of 50 pF, what must R and L be?

69. (II) A resistor with $R = 2\ \Omega$ draws a current from a wall plug; a capacitor is connected in parallel with this resistor. The current source has an amplitude of 110 V and a frequency of 60 Hz, and the reactance of the capacitor is 8 Ω at this frequency. What is the current drawn by the parallel combination?

70. (II) A 15-μF capacitor connected in series with a resistor of variable resistance R is connected to a $V_{\mathrm{rms}} = 110$-V, 60-Hz AC supply. Plot the variation of the rms current with R, and calculate the value of R for which the power delivered is maximum.

71. (II) An AC circuit supplies $V_{\mathrm{rms}} = 110$ V at 60 Hz to a 5-Ω resistor, a 40-μF capacitor, and an inductor of variable self-inductance in the 5 mH to 200 mH range, all in series. The capacitor is rated to stand a maximum voltage of 800 V. (a) What is the largest current possible that does no damage to the capacitor? (b) To what value can the self-inductance be increased safely?

CHAPTER 35

These dishes, part of the Very Large Array of radio telescopes in New Mexico, collect radio waves from distant galaxies. Light, radio waves, and other forms of radiation are consequences of the fundamental equations of electromagnetism, Maxwell's equations.

MAXWELL'S EQUATIONS AND ELECTROMAGNETIC WAVES

Faraday's law shows that electricity and magnetism are fundamentally connected. James Clerk Maxwell's introduction of the displacement current enhances this connection and leads to a complete, consistent set of fundamental laws for electricity and magnetism. These laws are known as Maxwell's equations. The individual experiments that led to their establishment never hinted at the wealth of their implications. The most dramatic prediction of Maxwell's equations is the existence of electromagnetic waves that propagate through space at a predictable speed, the speed of light. The realization that light is a form of electromagnetic radiation has led to a complete understanding of all the properties of light. In this chapter we shall also discuss the orientation and relationship of the electric and magnetic fields that propagate in space, forming electromagnetic waves; the energy and momentum carried by these waves; and polarization, a phenomenon that does not appear in any of the waves we have encountered up to now.

35–1 MAXWELL'S EQUATIONS

Let us list and then comment on **Maxwell's equations**, which summarize the content of electricity and magnetism. These equations describe fully electric and magnetic fields in the presence of electric charges and currents.

Maxwell's equations

I. Gauss' law for electric fields

$$\oiint \mathbf{E} \cdot \mathrm{d}\mathbf{A} = \frac{Q}{\epsilon_0}. \tag{35–1}$$

II. Gauss' law for magnetic fields

$$\oiint \mathbf{B} \cdot \mathrm{d}\mathbf{A} = 0. \tag{35–2}$$

III. Generalized Ampère's law

$$\oint \mathbf{B} \cdot \mathrm{d}\mathbf{s} = \mu_0 I + \mu_0 \epsilon_0 \frac{\mathrm{d}}{\mathrm{d}t} \iint_S \mathbf{E} \cdot \mathrm{d}\mathbf{A}. \tag{35–3}$$

IV. Faraday's law

$$\oint \mathbf{E} \cdot \mathrm{d}\mathbf{s} = -\frac{\mathrm{d}}{\mathrm{d}t} \iint_S \mathbf{B} \cdot \mathrm{d}\mathbf{A}. \tag{35–4}$$

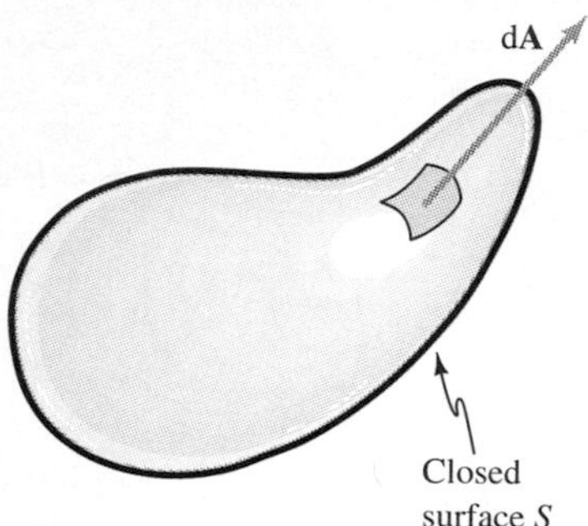

FIGURE 35–1 An infinitesimal surface element d**A** on the closed surface S.

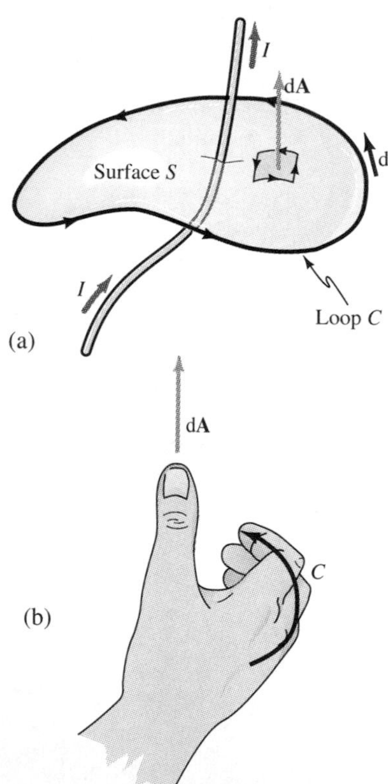

FIGURE 35–2 (a) A surface S bounded by the closed loop C. A current I passes through the surface, but a changing electric flux would also contribute to Ampère's law. If the integration along C proceeds counterclockwise, (b) then the direction of the surface elements d**A** that make up S is given by a right-hand rule.

I. Gauss' law, which, in static situations, is equivalent to Coulomb's law, relates the electric flux through a closed surface (the surface can be imaginary) to the charge enclosed [see Eq. (24–7)]. The surface element d**A** is normal to the surface S and is directed outward with magnitude dA (Fig. 35–1). The charge Q is the total charge contained within the closed surface. The factor ϵ_0 (the permittivity of free space) arises because of our choice of units. This law is actually a generalization of Coulomb's law. Whereas Coulomb's law is correct only for static charges, Gauss' law holds even if the charges are not stationary; that is, even if the electric field varies with time.

II. Magnetic monopoles, which would be the magnetic analogues of electric charge, have never been discovered. Their presumed nonexistence leads to Gauss' law for magnetic fields [see Eq. (30–12)]. This equation holds even for time-dependent magnetic fields.

III. Ampère's law describes the relation between a magnetic field and the current that gives rise to that field. The Maxwell generalization was described in Chapter 30 [see Eq. (30–31)]. The left-hand side of this equation is the expression for the integral of the magnetic field's tangential component along an arbitrary closed loop C (Fig. 35–2). The right-hand side has two contributions: One is the total current flowing through any surface S bounded by the closed loop C; the other is the rate of change of the electric field flux through such a surface, the *displacement current* contribution. Maxwell was responsible for introducing the displacement current. The presence of the parameter μ_0 (the permeability of free space), like that of ϵ_0, is a consequence of our choice of SI units.

IV. Faraday's law describes the induced electric field generated by a changing magnetic flux [see Eq. (31–2)]. The left-hand side is the integral of the tangential component of the induced electric field around an arbitrary closed loop C. The right-hand side measures the rate of change of the magnetic flux through any surface S bounded by C, just as in Fig. 35–2. Equation (35–4), as well as Eq. (35–3), implies a sign convention given by a right-hand rule. The minus sign is very important: It represents the fact that the induced electric field, were it to act on charges, would give rise to an induced current that opposes the change in the magnetic flux (Lenz's law).

Maxwell's equations display a degree of symmetry between electric and magnetic fields. This symmetry is not perfect, because magnetic monopoles apparently do not exist. Faraday's law contains no term like the $\mu_0 I$ term in Ampère's law, because there is no free magnetic charge to form a magnetic current. In a vacuum, where there are no electric charges, the symmetry is perfect.

In the presence of matter, electromagnetic phenomena can be described by a modified form of Maxwell's equations. For most types of materials, we can simply replace ϵ_0 by $\epsilon = \kappa\epsilon_0$, where κ is the dielectric constant. Except for ferromagnetic materials, the additional rule that the permeability of the vacuum (μ_0) is to be replaced by the material's permeability (μ) does not affect matters much, because μ is very close to μ_0.

35–2 THE PROPAGATION OF ELECTROMAGNETIC FIELDS

A glance at Eqs. (35–3) and (35–4) shows that when the electric and magnetic fields are time dependent, they influence each other; they are said to be *coupled*. We will show that as a consequence of this coupling, the electric and magnetic fields can transport energy (and momentum) over very large distances. In fact, these coupled configurations of field can transport energy over much larger distances than might be suggested by the $1/r^2$ falloff of the electric field in Coulomb's law or by the $1/r^3$ falloff of the magnetic field in the Biot–Savart law. The coupled fields produce traveling waves called **electromagnetic waves**. These waves are all around us: Radio and television, microwaves, visible light, and X-rays are examples.

The simple mechanical construction shown in Fig. 35–3 illustrates what we mean by the coupling of electric fields and magnetic fields. Imagine a smooth table with a slot in it and a rope, the vertical rope (in red), along which waves can move in a plane perpendicular to the table and go through the slot. Also imagine a second rope, the horizontal rope (in blue), that lies on the table and is attached to the

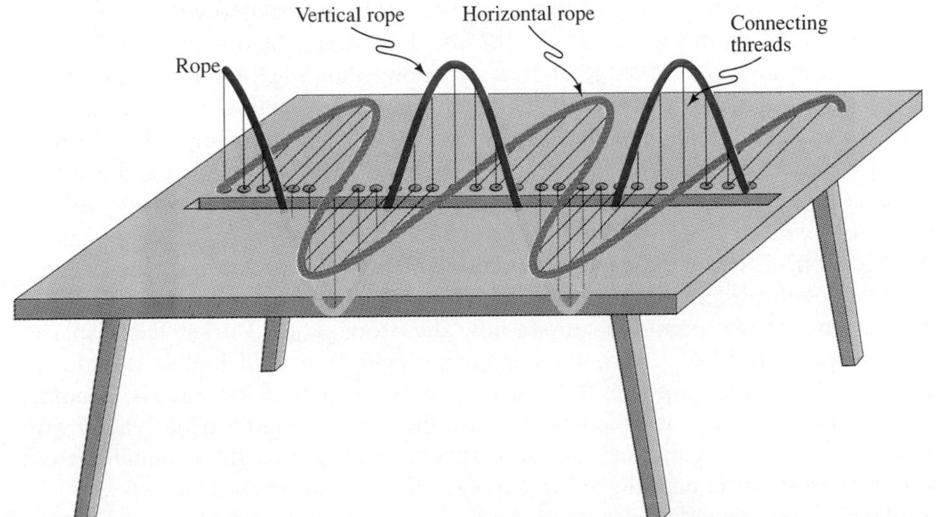

FIGURE 35–3 A mechanical system illustrating coupled waves. When the ropes are tied together with threads, one rope's wave motion is transverse in the vertical direction and produces a wave motion of the other rope, which is transverse in the horizontal direction, and vice versa.

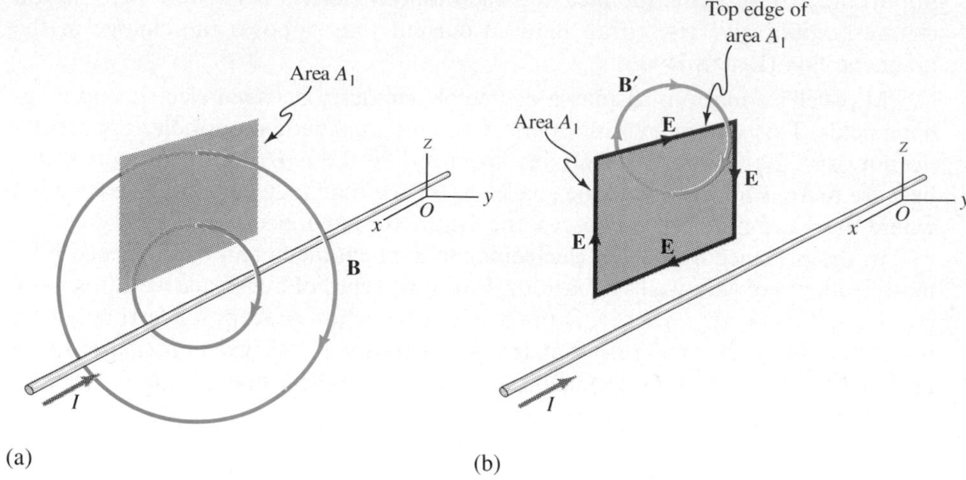

FIGURE 35–4 (a) As we know from Ampère's law, a current-carrying wire aligned in the x-direction has a magnetic field that forms circles in the yz-plane. (b) If the current in the wire changes with time, the magnetic field it produces changes with time, inducing a changing electric field, which in turn induces a changing secondary magnetic field, and so forth.

vertical rope by a set of threads of equal length. These threads pass through tiny hooks in such a way that when the vertical rope is at its highest (or lowest point), the horizontal rope is just above the slot, and when the vertical rope is at table-top level, the horizontal rope is farthest from the slot. Thus, because of the presence of the threads, a wave motion in the vertical rope causes a wave motion in the horizontal rope. The threads couple the two ropes. In the case of electric and magnetic fields, there are no ropes and no threads, but there is still the effect of one field on the other. As a result, oscillations in the electric field cause oscillations in the magnetic field, and vice versa.

We want to get a qualitative feeling for how charges in motion can give rise to electromagnetic waves. We shall make an argument that, although it must eventually be quantified, nevertheless conveys the physical mechanism by which electromagnetic waves propagate. Consider a straight wire that is aligned with the x-axis and carries a current I. With the current as shown in Fig. 35–4a, a magnetic field forms in rings around the wire; as long as the current is constant, this magnetic field is constant.

Suppose that the current is changing. To be definite, we take it to be increasing. The magnetic field increases, as does the magnetic flux through an area A_1 in the xz-plane. According to Faraday's law, Eq. (35–4), a changing magnetic flux induces an emf around the boundary of this area. This emf is associated with the induced electric field shown in Fig. 35–4b. Lenz's law determines the direction of the field.

Consider now the top edge of area A_1. Along that edge, the electric field has been induced in the $-x$-direction. This induced electric field changes, because it is due to a changing magnetic field; in our example, it is increasing. But according to the generalized Ampère's law, Eq. (35–3), we do not need flowing charges to induce a magnetic field. *A changing electric field also produces a magnetic field by giving rise to a displacement current.* The displacement current in this case is along the direction of the changing electric field, which is the direction of the original current. The displacement current is at higher values of z, however, than the original current. At this point we can see how the propagation works: The displacement current produces a secondary magnetic field $\mathbf{B}'$ at still higher values of z (Fig. 35–4b). In the xz-plane, $\mathbf{B}'$ is perpendicular to that plane; that is, it points in either the $+y$- or $-y$-direction. Because the displacement current varies with time, $\mathbf{B}'$ is also changing. Just as the magnetic field due to the original current produces an induced emf aligned in the x-direction if we restrict ourselves to the xz-plane, so the secondary magnetic field produces an induced emf aligned in the

x-direction at still higher values of z, and the process repeats itself to higher and higher z values.

Let us look at some of the qualitative conclusions we can draw from the above discussion.

1. We chose the imaginary loop A_1 to lie in the xz-plane, and this led to a picture in which fields propagate in the z-direction. The electric field was aligned along the x-axis, parallel to the current, and the magnetic field was aligned along the y-axis, perpendicular to both the electric field and to the direction of propagation. With the changing current *restricted to a line* along the x-axis, the fields propagate in a cylindrically symmetric way outward from the current line. The electric field remains parallel to the current direction, and the magnetic field remains perpendicular both to the electric field and to the direction of propagation. *This is a general feature of electromagnetic waves.*
2. The current that is the original source of the fields must change with time. A steady current would simply produce a static magnetic field. Equivalently, *charges that produce propagating electric and magnetic fields must be accelerating.* It is reasonable that if the motion of the charges is harmonic in time, then the electric and magnetic fields will also have a harmonic time dependence. In Section 35–3 we shall verify this expectation.

Only accelerating charges can produce propagating electromagnetic fields.

35–3 ELECTROMAGNETIC WAVES

In this section we shall show how the qualitative discussion at the end of Section 35–2 can be made quantitative. Using Maxwell's equations, we show that the electric and magnetic fields obey wave equations. We shall predict the speeds at which the waves travel, a result that our qualitative argument cannot supply. The electric fields and magnetic fields in these waves are in phase, and their magnitudes are closely related. The electric and magnetic fields that propagate together form an electromagnetic wave. The speed of this wave is the same for all frequencies.

We work from a slightly different starting point than we did in Section 35–2. We replace the current-carrying wire by a *sheet of current*, which can be formed by a set of wires placed side by side and oriented in the xy-plane. The current is aligned with the x-axis (Fig. 35–5). With this configuration of currents, we shall see that the wave propagates only in the z-direction. We expect, according to the

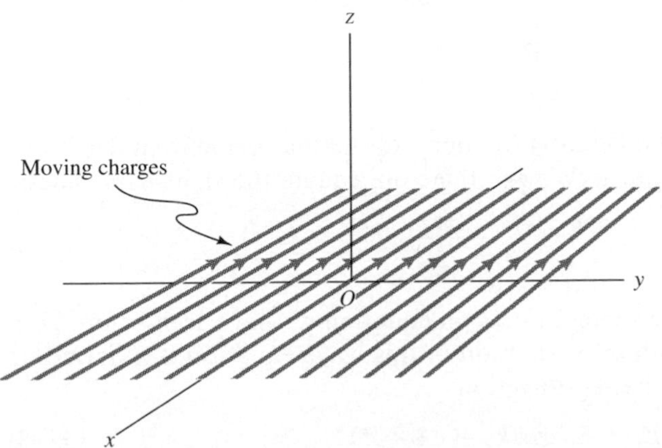

FIGURE 35–5 Current flows in a sheet along the x-direction. It can be approximated by aligning wires side by side in the x-direction. If the current is oscillatory, charges move first in the $+x$-direction, then in the $-x$-direction.

mechanism discussed in Section 35–2, that the electric fields will be oriented in the same direction as the current (parallel to the x-axis), while the magnetic fields will be oriented in the y-direction. For an infinite sheet of current, symmetry dictates that the fields must be the same everywhere on the plane parallel to the current sheet. Thus the fields we find will not depend on x and y, only on z. (Of course, the fields will also depend on time in a way that mirrors the time dependence of the current.) The electromagnetic fields form *plane waves*, which we recall from Chapters 14 and 15 refers to waves that advance along planar wave fronts—in this case, planes parallel to the xy-plane.

We discussed wave equations in Chapters 14 and 15.

To understand quantitatively how we can get time-dependent electric and magnetic fields that behave as described in Section 35–2, we use two of Maxwell's equations, Faraday's law and the generalized Ampère's law, to derive an alternate set of equations for the fields. These alternate equations are wave equations for the electric and magnetic fields. The mathematical procedure for deriving and using the alternate equations is worked out in the box "Maxwell's Equations as Differential Equations Leading to Waves" on pp. 1020–21. Here we summarize the important results derived there.

From Maxwell's equations we can derive

$$-\frac{\partial B_y}{\partial z} = \mu_0\epsilon_0 \frac{\partial E_x}{\partial t} \tag{35–5}$$

and

$$-\frac{\partial B_y}{\partial t} = \frac{\partial E_x}{\partial z} \tag{35–6}$$

for the field components B_y and E_x. Both components depend on the value of z and on the time t. (Recall that partial derivatives appear whenever quantities such as fields depend on two or more variables. In taking a partial derivative with respect to one variable, the other variables are held fixed.)

Equations (35–5) and (35–6) represent two partial differential equations that the x-component of the electric field, E_x, and y-component of the magnetic field, B_y, must obey. These equations couple the two fields, just as threads couple the ropes in our mechanical analogue. The two equations look complicated, but it is possible to combine and simplify them by a straightforward procedure. The partial derivative of Eq. (35–5) with respect to time gives

$$-\frac{\partial^2 B_y}{\partial t\,\partial z} = \mu_0\epsilon_0 \frac{\partial^2 E_x}{\partial t^2}.$$

Similarly, the partial derivative of Eq. (35–6) with respect to z gives

$$-\frac{\partial^2 B_y}{\partial z\,\partial t} = \frac{\partial^2 E_x}{\partial z^2}.$$

Because the order of partial differentiation does not matter, the left-hand sides of these two equations are identical. We can therefore equate the right-hand sides:

$$\frac{\partial^2 E_x}{\partial z^2} = \mu_0\epsilon_0 \frac{\partial^2 E_x}{\partial t^2}. \tag{35–7}$$

This equation for E_x has the same form as an equation we have seen before [Eq. (14–25)]: *It is the wave equation!* A solution of this wave equation is a harmonic plane wave propagating in the $+z$-direction:

$$E_x = E_0 \cos(kz - \omega t + \phi), \tag{35–8}$$

where E_0 is an amplitude, k is a wave number, and ω is an angular frequency. Direct substitution of this expression into the wave equation, Eq. (35–7), will verify that it is indeed a solution. The phase angle ϕ is included because we shall want to look at how the phase of the magnetic field, which also has an oscillating solution, is related to that of the electric field.

Some Properties of the Solution to the Wave Equation

We recall from our discussion in Chapter 14 of wave motion that Eq. (35–8) represents a wave of wavelength $\lambda = 2\pi/k$ and frequency $f = \omega/2\pi$. The speed of the wave's propagation is $v = \lambda f = \omega/k$. This speed is found immediately from the wave equation itself, as comparison with the original form of the wave equation, Eq. (14–25), shows. We have

$$v^2 = \frac{1}{\mu_0\epsilon_0}. \tag{35–9}$$

When we use the numerical values for μ_0 and ϵ_0, we get

$$\begin{aligned} v^2 &= \frac{1}{(1.257 \times 10^{-6}\ \mathrm{T\cdot m/A})(8.854 \times 10^{-12}\ \mathrm{C^2/N\cdot m^2})} \\ &= 8.999 \times 10^{16}\ \mathrm{m^2/s^2} = (3.00 \times 10^8\ \mathrm{m/s})^2 = c^2, \end{aligned}$$

where c is the speed of light. Thus

$$c = \frac{1}{\sqrt{\mu_0\epsilon_0}}. \tag{35–10}$$

The speed of electromagnetic waves

The Relation between E and B in an Electromagnetic Wave

To see how E and B for an electromagnetic wave are related, we can start with Eqs. (35–5) and (35–6) and show that B_y also obeys a wave equation similar to that of E_x, namely,

$$\frac{\partial^2 B_y}{\partial z^2} = \mu_0\epsilon_0 \frac{\partial^2 B_y}{\partial t^2}.$$

Like the x-component of the electric field, the y-component of the magnetic field forms a wave that propagates at speed c in the z-direction. However, because Eqs. (35–5) and (35–6) couple the fields, the waves of B_y do not propagate independently from those of E_x. If we have a wave solution for E_x, Eq. (35–8), then, from Eq. (35–5),

$$\begin{aligned} \frac{\partial B_y}{\partial z} &= -\mu_0\epsilon_0 \frac{\partial E_x}{\partial t} = -\mu_0\epsilon_0 \frac{\partial}{\partial t}[E_0 \cos(kz - \omega t + \phi)] \\ &= -\mu_0\epsilon_0\omega E_0 \sin(kz - \omega t + \phi). \end{aligned} \tag{35–11}$$

Equation (35–6) becomes

$$\frac{\partial B_y}{\partial t} = -\frac{\partial E_x}{\partial z} = -\frac{\partial}{\partial z}[E_0 \cos(kz - \omega t + \phi)] = kE_0 \sin(kz - \omega t + \phi). \tag{35–12}$$

From these two expressions for the derivatives of B_y, it is easy to check that

$$B_y = B_0 \cos(kz - \omega t + \phi) \tag{35–13}$$

has the correct spatial and time dependence.

Relations between Amplitudes. The amplitude B_0 of the magnetic field wave is not independent of the amplitude E_0 of the electric field wave, as Example 35–1 shows.

EXAMPLE 35–1 Consider the electromagnetic traveling wave for which the electric and magnetic fields are given by Eqs. (35–8) and (35–13). Use the derivative relations we have found to show that the amplitudes are related by $E_0 = cB_0$.

SOLUTION: Here we are proving a relation between the two coupled quantities E and B, so we must use the equations that couple them. Equation (35–11) is a coupling equation that relates a derivative of B_y to a derivative of E_x. This equation states that if E_x is given by Eq. (35–8), then

$$\frac{\partial B_y}{\partial z} = -\mu_0\epsilon_0\omega E_0 \sin(kz - \omega t + \phi).$$

With B_y given by Eq. (35–13), we can compute the partial derivative

$$\frac{\partial B_y}{\partial z} = \frac{\partial}{\partial z}[B_0 \cos(kz - \omega t + \phi)] = -kB_0 \sin(kz - \omega t + \phi).$$

We equate these two results:

$$-kB_0 \sin(kz - \omega t + \phi) = -\mu_0\epsilon_0\omega E_0 \sin(kz - \omega t + \phi).$$

The sine factor cancels, and we are left with

$$B_0 = \frac{\mu_0\epsilon_0\omega}{k} E_0.$$

The factor $\omega/k = c$, whereas $\mu_0\epsilon_0 = 1/c^2$, so we are left with $B_0 = E_0/c$, the relation we needed to show.

The relation between the electric and magnetic field amplitudes in an electromagnetic wave is independent of the currents that set up the original wave and that determine the direction of propagation and of the fields. We have in general

The relation between electric and magnetic field amplitudes in an electromagnetic wave

$$E = cB, \tag{35–14}$$

where E and B are the amplitudes of the electric and magnetic fields, respectively, in an electromagnetic wave.

The Transversality of Electromagnetic Waves. The fields described by Eqs. (35–8) and (35–13) form traveling waves that propagate in the z-direction. Even though the fields are oriented in the x- and y-directions, these fields do not depend on x or y. Waves of this type have the same fields everywhere on a plane parallel to the xy-plane and are said to describe *plane waves*. The fact that they are plane waves has to do with the currents we used to set up the waves in the first place, and other forms are possible. In particular, there is nothing special about the x- and y-directions. If we had set up our currents to run in the y- rather than the x-direction, we would have found another set of plane-wave solutions, with **E** in the y-direction and **B** in the x-direction. The wave propagation would still have been in the z-direction. It is generally true that all such solutions display the property we have described, namely, that *the electric field and magnetic field are perpendicular to each other*, or

In an electromagnetic wave, the electric and magnetic fields and the direction of propagation are mutually perpendicular.

$$\mathbf{E} \cdot \mathbf{B} = 0. \tag{35–15}$$

Moreover, an electromagnetic wave is *transverse*, because the direction of the fields involved is perpendicular to the direction of wave propagation. *Neither the electric field nor the magnetic field has a component in the direction of propagation of the wave.*

The Electric Field and Magnetic Field Are in Phase. Note that the phases that appear in the harmonic expressions for B_y and E_x in Eqs. (35–13) and (35–8), respectively, are exactly the same. That the fields are in phase means that when the electric field is a maximum, the magnetic field is also a maximum; when one is zero, the other is zero, and so forth. The fields oscillate together as shown in Fig. 35–6. (For this aspect of electromagnetic waves, the analogy of coupled ropes fails, because as Fig. 35–3 shows, the waves on the two ropes have opposite phases, not the same phase.)

The electric field and magnetic field in an electromagnetic wave are in phase.

Figure 35–7 shows a second view of the wave, in which some important features of electromagnetic waves are emphasized: The fields are in phase, transverse (perpendicular to the direction of propagation), and perpendicular to each other.

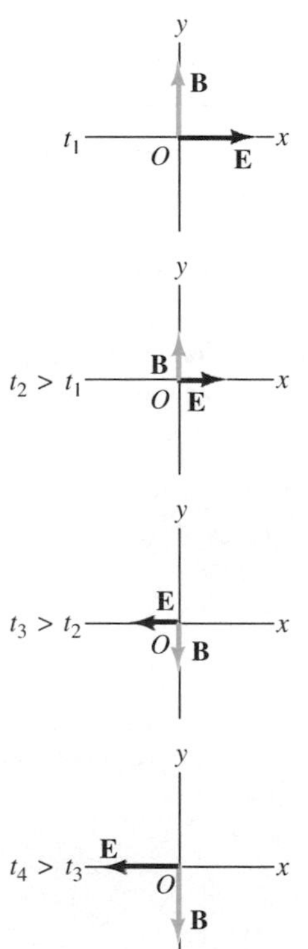

FIGURE 35–6 A view downward from the $+z$-direction of the electric and magnetic fields of an electromagnetic wave in an xy-plane over time.

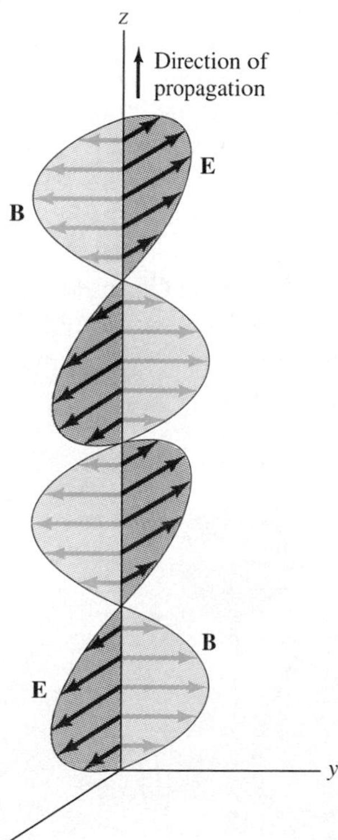

FIGURE 35–7 A view at one particular time of the transverse electric and magnetic fields that propagate along the z-axis.

Electromagnetic Waves Are Real

Maxwell recognized from the numerical evaluation of v, Eq. (35–9), that the wave's speed is *the speed of light*. He immediately drew the conclusion that light (the subject of Chapter 36) is an electromagnetic wave. What is special to us about light is that through evolutionary adaptation our eyes have become particularly good detectors of the range of wavelengths in which the sun emits radiation most strongly and that pass easily through the atmosphere. We call this range of wavelengths the *visible spectrum.* Our eyes can easily distinguish different frequencies within the visible spectrum. We interpret these different frequencies as colors. The shortest wavelengths of the visible spectrum are violet; the longest wavelengths are red.

Maxwell's treatment of the displacement current and his prediction of electromagnetic waves were published in 1864. A number of the leading physicists of his time found the notion of the displacement current difficult to accept, and it was more than 20 years before all resistance to the theory collapsed. It was not possible to confirm the existence of electromagnetic waves at the time Maxwell proposed them. Although the waves propagate without the presence of charges, a changing current is required to get them started. In the mid-1800s there was no technology to create AC currents of a high enough frequency to provide detectable radiation. In 1887 Heinrich Hertz devised the first direct test of Maxwell's waves. Hertz found that when there is dielectric breakdown in air due to a (high) potential difference between two points, it is accompanied by the formation of sparks between the points. The sparks from such a "spark gap" (Fig. 35–8a) appear to have a rhythm that suggests a back-and-forth motion of charge in the gap. To confirm that this oscillatory motion of charges produces radiation, he took a wire bent into a circle with a (second) gap and placed it near the original spark gap (Fig. 35–8b). He found that the electromagnetic wave that propagated in the space between the spark gap and the circular wire loop induced an electric field in the wire and gave rise to sparks in the secondary gap, which therefore acted as a detecting antenna. Hertz also reflected waves from metallic surfaces, focused them by a concave metallic mirror, and found that they generally shared many of the properties of light that we shall study in Chapters 36 and 37. The frequencies of the electromagnetic waves studied by Hertz are quite different from the frequencies of the waves that form visible light. The wave equation for electromagnetic waves admits solutions for

(a)

(b)

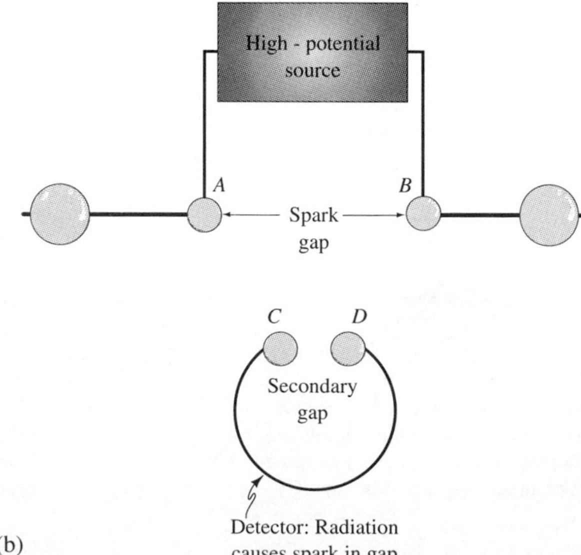

FIGURE 35–8 (a) Hertz's apparatus for the detection of electromagnetic radiation. (b) Schematic diagram of Hertz's apparatus. The radiation propagates from the region between the oscillating spark AB to the gap CD, which detects the radiation produced at gap AB by forming its own sparks.

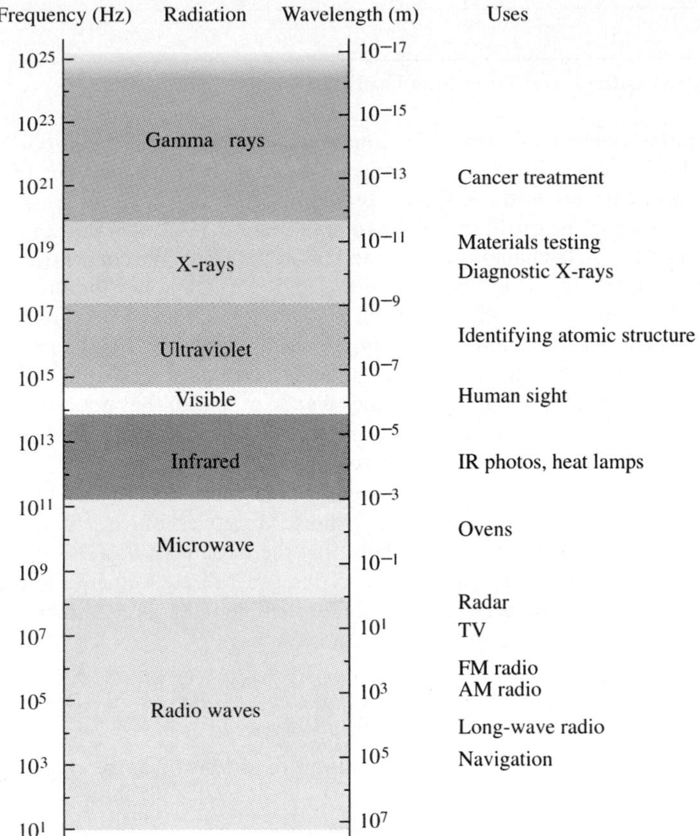

FIGURE 35–9 The spectrum of electromagnetic radiation.

any frequency, and the collection of electromagnetic waves of all frequencies is known as the **electromagnetic spectrum**. In the century since Hertz's work, the electromagnetic spectrum has been explored across an enormous range of frequencies (Fig. 35–9). All evidence points to the fact that Maxwell's equations accurately describe the whole spectrum of radiation. The term *electromagnetic radiation* refers to the entire spectrum of electromagnetic waves, including visible light, ultraviolet radiation, infrared radiation, microwaves, radio waves, X-rays, and gamma rays. Chapter 36 contains a more detailed discussion of the spectrum.

The discussion to this point has involved the formation of waves in a vacuum. Electromagnetic waves also propagate in matter. In (transparent) nonmetallic media, Maxwell's equations are modified only slightly, in that the speed of electromagnetic waves is reduced by a factor n according to

$$v = \sqrt{\frac{1}{\mu\epsilon}} \equiv \frac{c}{n}. \tag{35–16}$$

The quantity n is the **index of refraction** of a given medium. In all except ferromagnetic materials (introduced in Chapter 32), μ is very close to μ_0 and $\epsilon = \kappa\epsilon_0$, where κ is the dielectric constant of the medium. This implies that in most media,

The definition of the index of refraction

$$v = \sqrt{\frac{1}{\mu_0\epsilon_0\kappa}} = \frac{c}{\sqrt{\kappa}}; \tag{35–17}$$

in other words, the index of refraction $n = \sqrt{\kappa}$. It should be noted that the dielectric constant normally depends on the frequency of the electromagnetic wave. When the speed of the wave depends on the frequency, the medium is said to be *dispersive*.

***DERIVATION**

Maxwell's Equations as Differential Equations Leading to Waves

From a set of accelerating charges and Maxwell's equations, we shall derive the important equations that lead us directly to electromagnetic waves. The particular set of charges that we use form currents in the xy-plane, oscillating back and forth in the x-direction, as in Fig. 35–5. As in the qualitative discussion of Section 35–2, we know that the moving charges will give rise to changing electric and magnetic fields. We concentrate on time-dependent fields that vary with z but not with x and y. This implies that for a given z, the fields are the same out to infinity in the x- and y-directions. This cannot strictly be true in a physical situation, and we keep in the back of our minds that somewhere, for large enough values of x and y, the fields actually taper off to zero.

Let us draw an imaginary loop C in the yz-plane (that is, at $x = 0$) that goes from $y = b$ to $y = -b$ at some value of z and returns from $y = -b$ to $y = b$ at $z + \mathrm{d}z$ (Fig. B1–1). We are going to apply the generalized Ampère's law to the loop. Sides $y = \pm b$, going from z to $z + \mathrm{d}z$, are very short. We shall ignore the contribution from the short sides, because we can make these sides infinitesimally short. Moreover, our qualitative argument in Section 35–2 gives us no reason to believe that there is a field B_z. (This can be verified with the help of Gauss' law.) Application of the generalized Ampère's law, Eq. (35–3), now becomes easy. All we need to calculate for the line integral in Ampère's law are the contributions from the long sides of the loop. We have

$$B_y(z + \mathrm{d}z, t)(2b) - B_y(z, t)(2b) = \mu_0\epsilon_0 \frac{\mathrm{d}}{\mathrm{d}t} \iint_{\text{loop area}} \mathbf{E} \cdot \mathrm{d}\mathbf{A}. \tag{B1–1}$$

The difference $B_y(z + \mathrm{d}z, t) - B_y(z, t)$ is, from the definition of a derivative, the rate of change of B_y with respect to z times $\mathrm{d}z$, so

$$2b[B_y(z + \mathrm{d}z, t) - B_y(z, t)] = 2b\left(\frac{\partial B_y}{\partial z}\,\mathrm{d}z\right). \tag{B1–2a}$$

The partial derivative appears because we keep t constant in $B_y(z, t)$.

Now consider the right-hand side of Eq. (B1–1). In using Ampère's law, a right-hand rule dictates that for loop C in the direction shown in Fig. B1–1, the surface element $\mathrm{d}\mathbf{A}$ is oriented in the $-x$-direction, so $\mathbf{E} \cdot \mathrm{d}\mathbf{A} = -E_x\,\mathrm{d}A$. In addition, the area $A = 2b\,\mathrm{d}z$ is infinitesimally small, so we can assume that E_x does not vary over the surface, and we can remove it from the integral. Finally, the time derivative on the right-hand side of Eq. (B1–1) acts only on E_x, because the surface is itself fixed. Thus

$$\mu_0\epsilon_0 \frac{\mathrm{d}}{\mathrm{d}t} \iint_{\text{loop area}} \mathbf{E} \cdot \mathrm{d}\mathbf{A} = -\mu_0\epsilon_0 \frac{\partial}{\partial t} E_x \iint_{\text{loop area}} \mathrm{d}A = -\mu_0\epsilon_0 \frac{\partial E_x}{\partial t} A$$

$$= -\mu_0\epsilon_0 \frac{\partial E_x}{\partial t}\, 2b\,\mathrm{d}z. \tag{B1–2b}$$

We have used a partial derivative because z is a second variable that is held fixed. We now equate the two right-hand sides of Eqs. (B1–2a) and (B1–2b):

$$\cancel{2b}\left(\frac{\partial B_y}{\partial z}\,\cancel{\mathrm{d}z}\right) = -\mu_0\epsilon_0 \frac{\partial E_x}{\partial t}\, \cancel{2b}\,\cancel{\mathrm{d}z};$$

that is,

$$\frac{\partial B_y}{\partial z} = -\mu_0\epsilon_0 \frac{\partial E_x}{\partial t},$$

which is Eq. (35–5).

We next make use of Faraday's law, Eq. (35–4), the fourth of Maxwell's equations. We apply it to a loop C' that goes from $x = a$ to $x = -a$ at some value of z and returns from $x = -a$ to $x = a$ at $z + dz$ (Fig. B1–2). Then a nearly identical derivation to the one that led us to Eq. (35–5) leads us to Eq. (35–6),

$$\frac{\partial B_y}{\partial t} = -\frac{\partial E_x}{\partial z}.$$

Equations (35–5) and (35–6) are the ones we used in Section 35–3 to find electromagnetic waves.

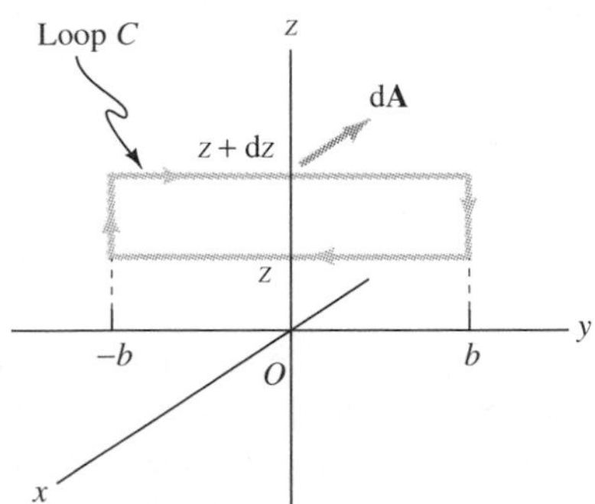

FIGURE B1–1 A loop used to derive a relationship between $\partial B_y/\partial z$ and $\partial E_x/\partial t$ from Ampère's law.

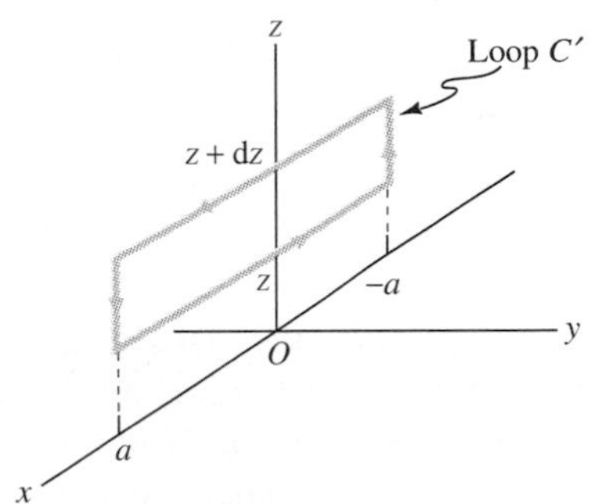

FIGURE B1–2 A loop used to derive a relationship between $\partial B_y/\partial t$ and $\partial E_x/\partial z$ from Faraday's law.

35–4 ENERGY DENSITY, ENERGY FLOW, AND MOMENTUM FLOW

The Energy of Electromagnetic Waves

We can get sunburned because electromagnetic waves carry energy (Fig. 35–10). To find the energy content of electromagnetic waves, recall our earlier result that

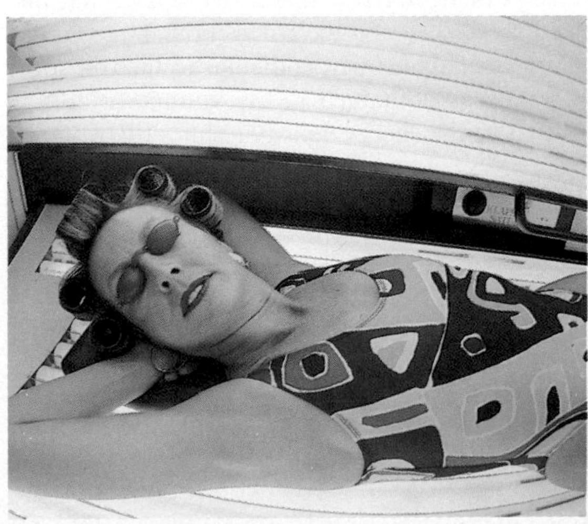

(a)

(b)

FIGURE 35–10 (a) Artificial tanning machines supply energy in the form of intense (and potentially dangerous) ultraviolet radiation. (b) Radiant energy is measured by a device called a radiometer.

electric and magnetic fields store energy. The energy density, given in Eq. (33–17), is

$$u = \frac{1}{2}\left(\frac{B^2}{\mu_0} + \epsilon_0 E^2\right) = \frac{\epsilon_0}{2}\left(\frac{B^2}{\mu_0\epsilon_0} + E^2\right) = \frac{\epsilon_0}{2}(c^2B^2 + E^2). \quad (35\text{–}18)$$

This result is general and therefore must hold for the electromagnetic fields in electromagnetic waves.

Let us apply the energy density result to a wave traveling in the z-direction with the electric field and the magnetic field given by Eqs. (35–8) and (35–13), respectively:

$$E_y = E_0 \cos(kz - \omega t + \phi) \quad \text{and} \quad B_x = -B_0 \cos(kz - \omega t + \phi),$$

where $E_0 = cB_0$. For this wave, the energy density is

$$u = \frac{\epsilon_0}{2}(c^2B_0^2 + E_0^2)\cos^2(kz - \omega t + \phi). \quad (35\text{–}19)$$

The energy contained in the electric field of an electromagnetic wave is equal to the energy contained in the magnetic field.

In this expression, the first and second terms in the coefficient of the cosine-squared factor are the contributions of the magnetic and electric parts of the wave, respectively. Because $E_0 = cB_0$, *the energy contained in an electromagnetic wave is shared equally between the magnetic field and the electric field*. Equivalently, we could take the contribution of either the electric or the magnetic terms to the energy density and multiply by 2 to find the total energy density in an electromagnetic wave:

$$u = \epsilon_0 E^2 = \frac{1}{\mu_0} B^2. \quad (35\text{–}20)$$

The oscillations in electromagnetic waves are so rapid that, for practical purposes, it is enough to consider the average of the energy density over the period of one oscillation, which we write as $\langle u \rangle$. Because the average of the cosine-squared factor in Eq. (35–19) over one period is one-half, we obtain

$$\langle u \rangle = \frac{\epsilon_0}{2} E_0^2 = \frac{1}{2\mu_0} B_0^2. \quad (35\text{–}21)$$

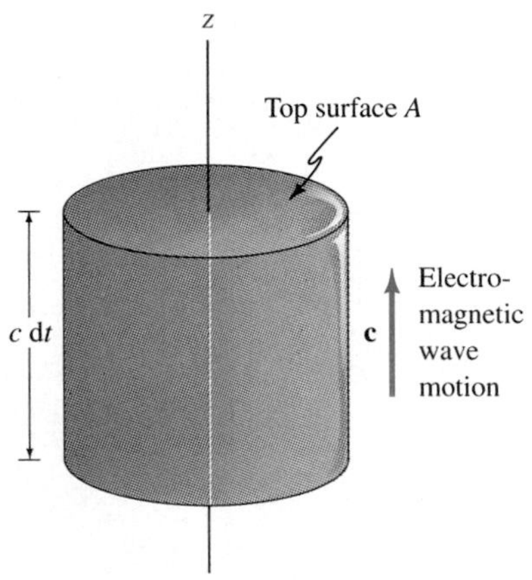

FIGURE 35–11 Electromagnetic energy contained in a volume $Ac\,dt$ is delivered in time dt to the area A.

The Transport of Energy

The $\cos^2(kx - \omega t + \phi)$ time dependence and space dependence of the energy density in Eq. (35–19) shows that the energy in an electromagnetic wave is itself *transported* as a wave that travels at speed $v = \omega/k = c$ in the z-direction. The amount of energy $d\mathscr{E}_t$ transported across a surface of area A perpendicular to the transport direction in a time interval dt is the energy contained in the volume of area A times the distance $c\,dt$ (Fig. 35–11); that is, the energy density u times this volume,

$$d\mathscr{E}_t = u(Ac\,dt).$$

Thus the rate of energy transport, or, equivalently, the power delivered by the electromagnetic wave, is

$$\frac{d\mathscr{E}_t}{dt} = cuA.$$

Finally, the power delivered per unit area to a surface perpendicular to the direction of propagation—that is, the *energy flux*—is given by

The energy delivered by an electromagnetic wave per unit time and area.

$$S = \frac{1}{A}\frac{d\mathscr{E}_t}{dt} = cu. \quad (35\text{–}22)$$

This flux has a direction associated with it and is a vector. The vector **S** that describes the energy flux is known as the **Poynting vector**, after John Henry Poynting. It is given by

$$\mathbf{S} \equiv \frac{1}{\mu_0} \mathbf{E} \times \mathbf{B}. \tag{35–23}$$

Definition of the Poynting vector, which gives the energy flux of electromagnetic fields

Let us check that the magnitude and direction of the vector **S** are indeed correct. We notice that, because the fields **E** and **B** are at right angles to each other, the magnitude of **S** is just EB/μ_0. This can be rewritten in terms of the electromagnetic energy density, $u = \epsilon_0 E^2$, as

$$S = \frac{1}{\mu_0} EB = \frac{\epsilon_0}{\epsilon_0 \mu_0} E\left(\frac{E}{c}\right) = \epsilon_0 c^{\cancel{2}} \frac{E^2}{\cancel{c}} = c\epsilon_0 E^2 = cu. \tag{35–24}$$

This is the same magnitude we found in Eq. (35–22). As for the direction, recall that the vector product of two vectors is perpendicular to both of them. From Fig. 35–7, we see that the direction of the vector product $\mathbf{E} \times \mathbf{B}$ is the $+z$-direction, the direction of wave propagation. More generally, the transversality of the electromagnetic wave will lead to a Poynting vector that lies along the direction of propagation.

See "The Vector Product" box in Chapter 10.

The energy density, and hence the magnitude of the Poynting vector, varies with time. The *average* value of the magnitude of **S** over one cycle of the electromagnetic wave is called the **intensity**, I, of the radiation. Equation (35–24) allows us to relate the intensity to the amplitude E_0 of the electric field in the wave:

The definition of intensity

$$I = \langle S \rangle = c\epsilon_0 \langle E^2 \rangle = \tfrac{1}{2} c\epsilon_0 E_0^2. \tag{35–25}$$

Note from Eq. (35–21) that the intensity is also related to the average energy density in the wave, $I = c\langle u \rangle$.

Momentum in Electromagnetic Waves

An electromagnetic wave carries momentum as well as energy. To see this qualitatively, consider again a plane wave that travels in the z-direction. We simplify matters by employing a wave with electric and magnetic fields along the x- and y-directions, respectively. When such a wave impinges on a particle with charge $+q$, the fields exert forces on the particle. Suppose that at a given time the oscillating electric field of the wave points in the $+x$-direction, so there is a force qE in the $+x$-direction. The charge accelerates and moves with some velocity $\mathbf{v}$ in the $+x$-direction. If **E** points in the $+x$-direction, then **B** (which oscillates in phase with **E**), must point in the $+y$-direction. The magnetic force $q\,\mathbf{v} \times \mathbf{B}$ on the charge acts in the $+z$-direction and pushes the charge in that direction. When the electric field later reverses sign, the electric force on the charge acts in the $-x$-direction, and the velocity picks up a component in the $-x$-direction. The magnetic field has also reversed sign, and the magnetic force *continues to act in the* $+z$*-direction.* All the forces in the x- and y-directions average to zero, but the force in the z-direction is always positive, and there is a net force in the $+z$-direction. The charge has an increased momentum in the $+z$-direction; by momentum conservation, this momentum had to have been brought in by the electromagnetic wave.

In this way the **momentum density** of an electromagnetic wave, the amount of momentum carried by the wave per unit volume, can be shown to be $\mathbf{S}/c^2$. The magnitude of the momentum density is given by

$$\frac{S}{c^2} = \frac{u}{c}, \tag{35–26}$$

The momentum density in electromagnetic waves

and the direction is that of **S**.

In Chapter 19 we used the same method to find the pressure of a gas due to atomic collisions with walls.

Radiation Pressure. One consequence of the fact that electromagnetic waves have momentum is that these waves, when absorbed or reflected, transfer momentum to the material on which they impinge. The rate at which momentum is transferred per unit area is a force exerted per unit area, that is, a pressure: Electromagnetic waves exert **radiation pressure**. When an electromagnetic wave is absorbed, which happens when light falls on a black surface, for example, all the momentum carried by the wave is transferred. The amount of radiation-produced momentum that falls in a perpendicular direction on a surface A in a time interval dt is given by the momentum density multiplied by the volume $A(c\, dt)$. Thus the momentum dp transferred is

$$dp = \left(\frac{S}{c^{\cancel{2}}}\right)(A\cancel{c}\; dt) = \frac{S}{c} A\, dt.$$

The force per unit area (radiation pressure) is given by

$$\frac{F}{A} = \frac{1}{A}\frac{dp}{dt} = \frac{1}{\cancel{A}}\frac{S}{c}\cancel{A} = \frac{S}{c} = u, \qquad (35\text{–}27)$$

where we have used Eq. (35–22). This expresses the radiation pressure when radiation is totally absorbed. When the electromagnetic wave is reflected, which happens when it falls on a shiny, metallic surface, for example, then the momentum of the wave is reversed upon reflection. Thus the momentum density transferred to the metallic surface is $2u/c$ and the radiation pressure is $2u$.

EXAMPLE 35–2 Consider a 10^4-W searchlight that projects a cylindrical beam 0.6 m in diameter. What is the radiation pressure on a metallic mirror placed at right angles to the beam? Ignore the spreading of the beam.

SOLUTION: The power delivered by the electromagnetic wave to a surface of area A at right angles to the beam is given by

$$P = (\text{energy flux})(\text{area}) = SA = cuA,$$

where u is the energy density in the beam at the surface. The area of the beam is $A = \pi r^2 = \pi(0.3\text{ m})^2 = 0.28\text{ m}^2$. Thus

$$u = \frac{P}{Ac} = \frac{10^4\text{ J}/\cancel{\text{s}}}{(0.28\text{ m}^2)(3 \times 10^8\text{ m}/\cancel{\text{s}})} = 1.2 \times 10^{-4}\text{ J/m}^3.$$

In turn the radiation pressure is

$$\frac{F}{A} = 2u = 2.4 \times 10^{-4}\text{ N/m}^2.$$

The pressure is therefore on the order of 10^{-9} atm, a very tiny number.

EXAMPLE 35–3 The intensity (average energy flux) of solar radiation that falls on the earth is $1.4 \times 10^3\text{ W/m}^2$. Compare the force exerted by solar radiation on a totally absorbing dust particle of diameter 10^{-6} m and mass density $3 \times 10^3\text{ kg/m}^3$ with that due to the gravity of the sun. The particle is taken to be at the earth-sun distance. The mass of the sun is $M_{\text{sun}} = 2 \times 10^{30}$ kg, and the distance between the sun and the earth is $R = 1.5 \times 10^{11}$ m.

SOLUTION: The intensity $I = uc$ (we assume time averages throughout) leads to a radiation pressure

$$\frac{F}{A} = u = \frac{I}{c} = \frac{1.4 \times 10^3\text{ W/m}^2}{3 \times 10^8\text{ m/s}} = 0.5 \times 10^{-5}\text{ N/m}^2.$$

The area presented by the dust particle is $A = \pi r^2 = \pi(0.5 \times 10^{-6}\ \text{m})^2 = 0.8 \times 10^{-12}\ \text{m}^2$, so the force is

$$F = uA = (0.5 \times 10^{-5}\ \text{N/m}^2)(0.8 \times 10^{-12}\ \text{m}^2) = 0.4 \times 10^{-17}\ \text{N}.$$

The mass of the dust particle is

$$m = \rho V = \rho \frac{4}{3}\pi r^3 = (3 \times 10^3\ \text{kg/m}^3)\frac{4\pi}{3}(0.5 \times 10^{-6}\ \text{m})^3 = 1.6 \times 10^{-15}\ \text{kg}.$$

The force of gravity has magnitude

$$F_g = \frac{GmM_{\text{sun}}}{R^2} = \frac{(6.67 \times 10^{-11}\ \text{N}\cdot\text{m}^2/\text{kg}^2)(1.57 \times 10^{-15}\ \text{kg})(2 \times 10^{30}\ \text{kg})}{(1.5 \times 10^{11}\ \text{m})^2}$$

$$= 0.9 \times 10^{-17}\ \text{N}.$$

Thus the two forces are comparable, and the radiation pressure may keep the dust particles from falling into the sun. This kind of dust grain is typical of those found in interplanetary space.

FIGURE 35–12 The use of electromagnetic waves of different frequencies has become a normal part of our technology. Three antennas are visible here.

*35–5 DIPOLE RADIATION

We have seen that accelerating charges are necessary to produce electromagnetic waves. We refer to systems in which accelerating charges initiate electromagnetic waves as *broadcasting antennas* and to systems in which we detect the response of charges to the oscillating fields of an electromagnetic wave as *receiving antennas* (Fig. 35–12). In this section we shall describe briefly one of the simplest systems that can act as an antenna, the *dipole antenna*. Radiation emitted with the characteristic pattern of this antenna is called **dipole radiation**.

A dipole antenna is formed by a charge that moves back and forth in harmonic motion along a line. The antenna is itself usually neutral, so the oscillating charge is taken as one of the pair of charges of a dipole. The second charge either is at rest or is oscillating as well (Fig. 35–13). Such an antenna is easy to construct with the help of an AC generator. When the dimensions of the antenna are small compared with the wavelength of the radiation, the current throughout the antenna is in phase. The resulting electric and magnetic fields are oriented as shown in Fig. 35–13. They have the configurations needed to form an outgoing electromagnetic wave whose frequency is that of the oscillating charges.

Dipole antenna

E

I

B (into page)

(a)

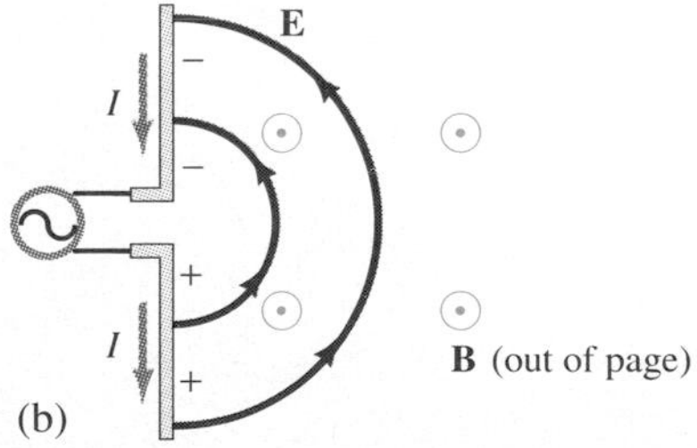

FIGURE 35–13 Pairs of equal but opposite charges move in simple harmonic motion along a line (vertical, here). These pairs of charges form a dipole antenna. In (a) and (b), the two sides of the antenna are oppositely charged, and the electric field directions are reversed.

How the Intensity of Radiation from an Antenna Decreases with Distance

One of the most important characteristics of the electromagnetic waves radiated by an antenna is the rate at which the intensity decreases with increasing distance from the antenna. To understand this feature, it is not important that the antenna be a dipole antenna. Any antenna will do, including an antenna that radiates electromagnetic waves symmetrically in all directions, such as the sun. From a distance, the electromagnetic waves emitted by the sun appear to come from a point source, and we can use this fact to study the magnitudes of the electric and magnetic field strengths. As we learned in Section 35–4, the energy flux (the rate of flow of energy per unit area) is given by $S = cu = c\epsilon_0 E^2$. The total energy flow per unit time (the power) across any surface A is

$$P = \iint_A \mathbf{S} \cdot d\mathbf{A}.$$

If the magnitude of the electric field is independent of direction, as we would expect for a point source, then the *total* rate of energy flow across a sphere of radius R centered on the source is

$$P = c\epsilon_0 E^2(4\pi R^2). \tag{35–28}$$

But all the radiation emitted must eventually pass through any sphere that surrounds the source and has any radius, so P does not depend on R. From our expression for P, we see that this is possible only if the electric field decreases as $1/R$. The magnetic field must similarly fall off as $1/R$, because the magnetic and electric fields are proportional in an electromagnetic wave. Contrast this result with the typical $1/R^2$ behavior of static electric fields (discussed in Chapter 23).

The electric and magnetic fields in the electromagnetic wave emitted by a point source decrease as $1/R$.

We can express the result of Eq. (35–28) in terms of intensity. The quantity $c\epsilon_0 E^2$ is the magnitude of the Poynting vector, and its average value, which is defined as the intensity, I, is one-half this value [Eq. (35–25)]. Thus

$$P = 2I(4\pi R^2). \tag{35–29}$$

Because P is independent of R, the intensity of the electromagnetic wave from a point source decreases as $1/R^2$. Example 35–4 illustrates this important property.

EXAMPLE 35–4 A 100-W light bulb emits electromagnetic radiation equally in all directions. Assume that 10 percent of the 100 W is converted into radiation in the visible spectrum. What is the intensity of the visible radiation 1.5 m from the bulb?

SOLUTION: We must find the intensity of visible radiation at a distance R from a source of radiation, given the total energy per unit time (power) radiated by the source into the visible spectrum. This total power is $P_0 = 10\%$ of $100\ \text{W} = 10\ \text{W}$; it is spread uniformly across the surface of any sphere centered on the source. If R is the radius of such a sphere, from Eq. (35–29) $P_0 = 2I(4\pi R^2)$, where I is the intensity at radius R. Thus, with $R = 1.5$ m,

$$I = \frac{P_0}{8\pi R^2} = \frac{10\ \text{W}}{8\pi(1.5\ \text{m})^2} = 0.2\ \text{W/m}^2.$$

Compare this value to the 1400 W/m^2 in sunlight incident at the top of the earth's atmosphere, or to the 1000 W/m^2 of solar energy that reaches the earth's surface. About half this solar energy is in light in the visible part of the spectrum, whereas the light bulb emits most of its energy in the infrared region of the spectrum.

In our discussion of static electric fields due to a point source, we used symmetry considerations to argue that the *electric field vector points in a radial direction.* This cannot be the case for electromagnetic waves emitted from a point source, because we showed that, for such waves, the fields are transverse to the direction in which they travel. When the waves come from a point source, they travel outward radially. The resolution to what appears to be a paradox is simple: *There are no truly pointlike sources of electromagnetic radiation.* The sun is not a point, and it radiates because charges within it move and accelerate. The sun is really a large collection of dipole antennas whose orientation is random.

The Angular Pattern of Dipole Radiation

The variation of the intensity of electromagnetic radiation with the angle of observation is an important property of radiation from an antenna. In our simple dipole antenna (Fig. 35–13), charges execute simple harmonic motion along the antenna direction (we shall call this the z-axis). The motion of the charge determines a preferred direction, along the z-axis. An observer looking along the z-axis would see no motion. An observer looking along a line perpendicular to the z-axis would see the full range of motion of the charges. An observer at an angle θ to the z-axis would see the charges move harmonically with an amplitude reduced from the full amplitude by a factor $\sin\theta$. The electric field that observer would see is thus proportional to $\sin\theta$. Because the intensity is proportional to the square of the electric field in the wave, the intensity of the radiation emitted by a dipole antenna along the direction of θ is proportional to $\sin^2\theta$:

$$S \propto \frac{\sin^2\theta}{R^2}. \tag{35–30}$$

Here we have also included the $1/R^2$ factor that describes how the intensity varies with the distance R from the antenna. This intensity pattern describes the **angular distribution** of the power emitted by charges oscillating along a line (Fig. 35–14). Figure 35–15 shows the pattern of the electric field lines (plus magnetic field lines) for the electric field radiated by such an antenna for a sequence of five times that each corresponds to one-eighth the oscillation period; one-half cycle is traced out.

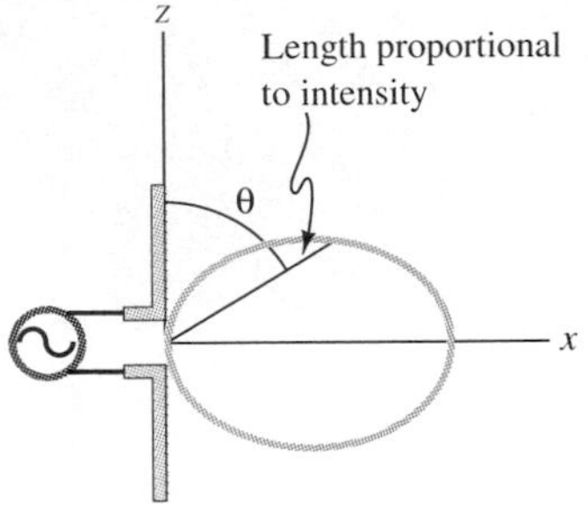

FIGURE 35–14 The intensity distribution S for a radiating dipole antenna. Power is emitted at an angle θ.

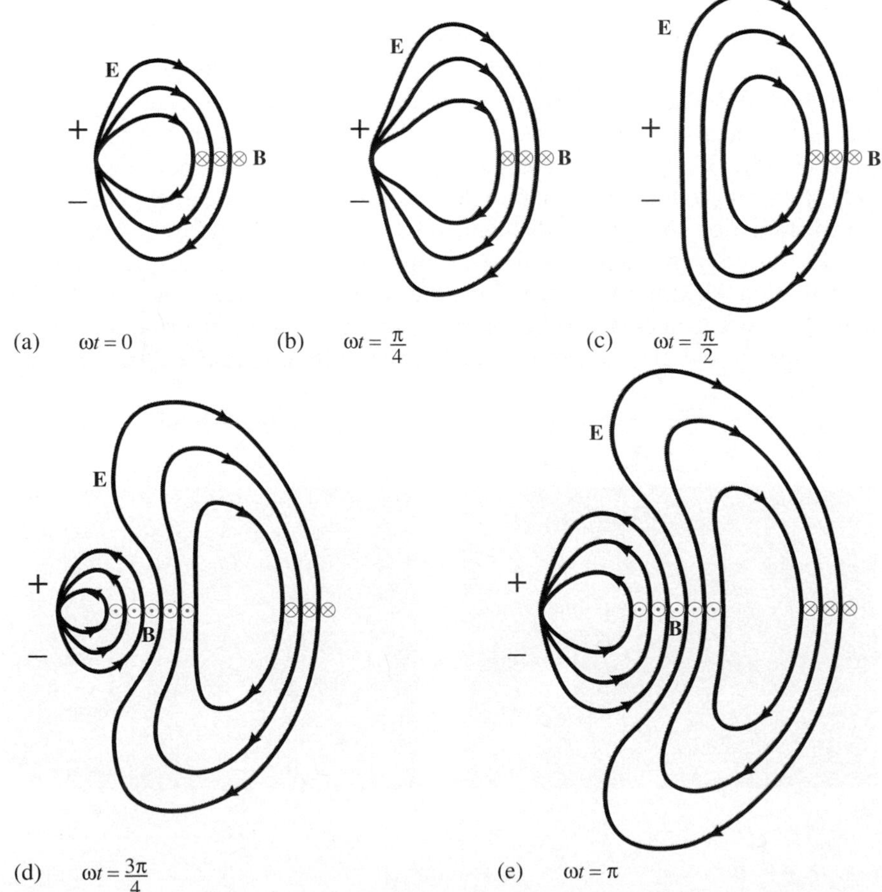

FIGURE 35–15 Electric and magnetic field lines produced by a radiating dipole antenna for a sequence of five times. (After P. Lorrain et al., *Electromagnetic Fields and Waves*, New York: W. H. Freeman, 1988.)

35–6 POLARIZATION

FIGURE 35–16 How well polarizing materials, which have a kind of internal axis, pass light or other electromagnetic waves depends on how that axis is oriented relative to the orientation of the electric field vector of the light. Little light passes through both pairs of glasses here, because the glasses' axes are crossed—light passing through one pair cannot pass through the second pair.

A little experimentation with *polarizing* sunglasses at the seashore shows that a change in the orientation of the glasses' axis results in a change of the intensity of the light transmitted (Fig. 35–16). This occurs because the sunglasses are made of a material that is sensitive to the direction of the vector electric field. Light reflected from water or sand is **polarized**, meaning that its electric field is oriented in a particular way; the glasses "detect" the polarization of the electromagnetic wave (light). In this section we shall study what polarization means and how it is produced and detected.

Let us consider again a charge that oscillates along the z-axis, as in Fig. 35–14. We found that, if we look along the x-direction, we would detect an electromagnetic plane wave that propagates along the x-direction, with an electric field aligned along the z-direction: $\mathbf{E} = E_z\mathbf{k}$, with $E_z = E_1 \cos(kx - \omega t)$. Here we have set the phase $\phi = 0$. We say that the light is **linearly polarized** along the direction of the electric field vector when that vector has a definite orientation. Let us adjust the frequency of the electron motion in our dipole antenna so that the wavelength of its radiation is in the centimeter range. The **polarization** can be detected as follows: A current is induced in a receiving antenna, and the rms current detected can be measured (Fig. 35–17). Place a metal grid (such as an oven rack) between the transmitter and the receiver. The diameter of the wires in the grid should be much less than 1 cm, and the grid spacing should be on the order of 1 cm or less. Then, as we describe below, the intensity of the radiation at the receiver depends on the orientation of the metal grid, and we say that the grid acts as an **analyzer** for the polarization.

The mechanism by which the grid acts as an analyzer is as follows: The electrons in the grid wires are accelerated by the electric field of the wave along the field direction and absorb the energy of the wave. When the wires in the grid are parallel to the electric field, the electrons in the grid wires can move in response to the field (Fig. 35–18a). Because they are set into motion, they *absorb large amounts of energy from the field.* This energy is lost in ohmic heating. The electric field of the radiation that passes through is reduced in magnitude, because energy has been removed from the incident wave. In effect, the grid is opaque to the polarized radiation when it is oriented along the electric field vector. When the wires in the grid are perpendicular to the z-direction (Fig. 35–18b), the electrons in the metal are accelerated across the diameter of the wire. But because the diameter

FIGURE 35–17 (a) A receiver and detector for determining the polarization of microwave radiation. (b) A grid is placed between them, oriented so that the radiation passes. (c) The grid is now oriented so that the radiation cannot pass. The grid's orientation reveals the polarization of the radiation.

(a)

(b)

(c)

is small, the electrons in the grid wires cannot respond fully and cannot absorb large amounts of energy from the incident wave. The energy remains in the transmitted wave, which is therefore strong. The grid acts as if it were transparent when it is oriented perpendicular to the polarization direction of the wave.

Certain materials, such as Polaroid, are analyzers for visible light. They are made of long molecules aligned parallel to each other. Electrons can easily move along the molecules but not across them, and because the molecular spacings are appropriate to the wavelengths of visible light, these materials behave like the microwave grid does.

A microwave grid or a piece of Polaroid is not simply an analyzer; it is also a **polarizer**: The microwave radiation that passes through the grid becomes polarized perpendicular to the grid wires. This is easily understood. Suppose that unpolarized microwave radiation approaches the grid. *Unpolarized radiation* is radiation that consists of a mixture of waves whose electric field vectors are oriented randomly with respect to one another. The electric field is as likely to point in any one direction as in another, as long as it is perpendicular to the direction of wave propagation. Most sources of radiation, such as light bulbs, produce unpolarized radiation, because they radiate from many randomly oriented dipoles. As we have seen, only those waves with the electric field oriented perpendicular to the grid can pass through, whereas the waves with the electric field parallel to the grid are absorbed. Thus the radiation that passes through the grid has become polarized perpendicular to the direction of the grid.

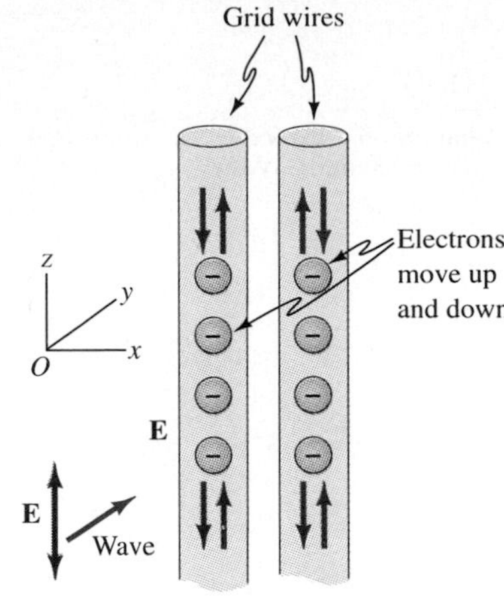

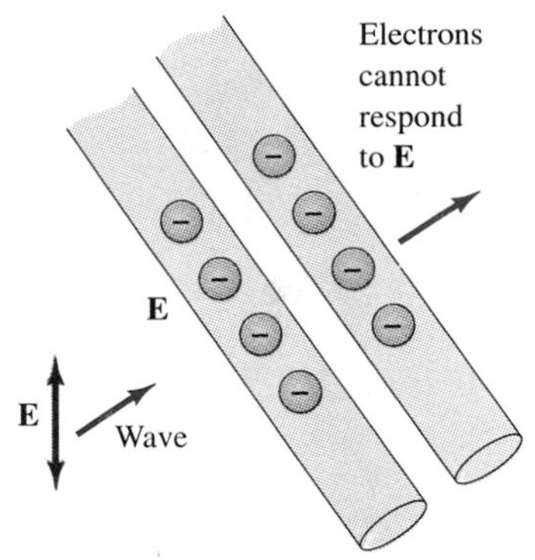

FIGURE 35–18 (a) When the grid wires are oriented in the direction of the electric field of an incoming wave, the electrons of the grid wires can respond and absorb energy from the wave. The transmitted wave is reduced in amplitude. (b) When the wires are perpendicular to the electric field of the wave, the electrons of the grid wires are constrained and cannot respond. Little energy is absorbed, and the wave passes through with little attenuation.

Malus's Law

When unpolarized radiation moving in the z-direction falls on a polarizer whose polarizing axis (the axis perpendicular to the "grid wires" within the polarizing material) makes an angle θ with the x-axis, for example, then only the component of any electric field along the polarizing axis will get through. What emerges is radiation linearly polarized along a line that makes an angle θ with the x-axis. We take the magnitude of the electric field that has passed through the polarizer to be E_0. The corresponding intensity is, from Eq. (35–25),

$$I_0 = \langle S \rangle = (\text{a constant})E_0^2. \qquad (35\text{–}31)$$

Let us now place a second polarizer so that its axis lies along the x-axis (Fig. 35–19). The amplitude for the electric field in the electromagnetic wave incident

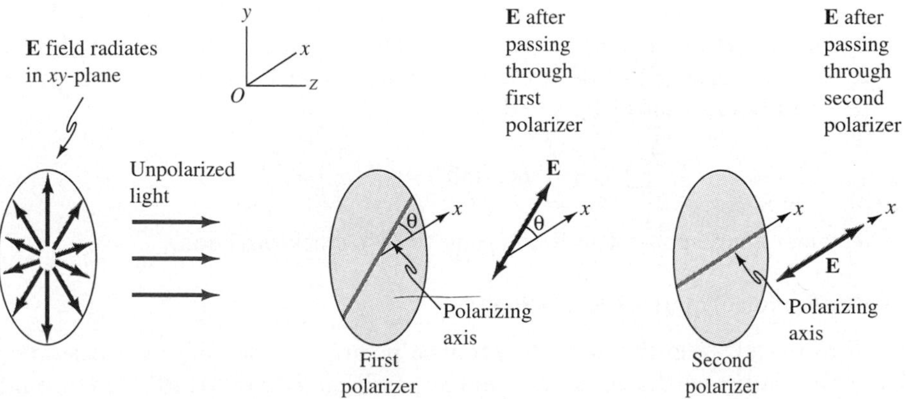

FIGURE 35–19 An unpolarized beam passes first through a polarizer whose axis makes an angle θ with the x-axis, and the beam is then polarized linearly in this direction. A second polarizer aligned with the x-axis allows only the component of the electric field aligned along the x-axis to pass.

on the polarizer is

$$\mathbf{E}_0 = E_0 \cos\theta\, \mathbf{i} + E_0 \sin\theta\, \mathbf{j}. \tag{35–32}$$

Only the component that is parallel to the axis of the second polarizer—that is, the x-axis—passes through. Thus the field behind the second polarizer (which acts here as an analyzer) is given by $E_0 \cos\theta\, \mathbf{i}$. The intensity of the transmitted light is therefore

$$I = (\text{a constant})(E_0 \cos\theta)^2, \tag{35–33}$$

and the intensity of the light is reduced:

$$I = I_0 \cos^2\theta. \tag{35–34}$$

Malus's law

Equation (35–34) is known as **Malus's law**, after Etienne Louis Malus. In particular, when $\theta = \pi/2$ (that is, when the axes of the polarizer and analyzer are perpendicular to each other), radiation is not transmitted.

One of the important consequences of Malus's law is that when unpolarized light passes through a plane polarizer, it has *half* its original intensity. This is shown in Example 35–5, which illustrates the use of Malus's law.

EXAMPLE 35–5 Light passes through the glass plate of a transparency projector and emerges unpolarized with intensity I_0. (a) A Polaroid sheet is placed on the glass plate with its polarizing axis aligned with the 12 o'clock position. What are the polarization and intensity of the emerging light? (b) A second Polaroid sheet, with its polarizing axis along the 2 o'clock position, is placed over the first. Again find the polarization and intensity of the emerging light.

SOLUTION: (a) We can set up the solution by supposing that the light wave propagates in the z-direction and that the 12 o'clock position is aligned along the $+x$-axis. The polarizer then passes light with its polarization in the x-direction, so the emerging light is polarized in the x-direction. To find the intensity, it is necessary to understand that unpolarized light consists of a series of wave bursts with an electric field aligned in different directions, always perpendicular to the propagation direction. If the projection of the incoming electric field on the x-axis is, for some particular burst, $E \cos\theta$, the intensity passed for that burst is $I = I_0 \cos^2\theta$. We must now *average* this intensity over all θ. Because the average of the cosine squared is one-half, the average intensity passed is $I_0/2$.

(b) The 2 o'clock direction is at an angle $\theta = \frac{2}{12}(360^\circ) = 60^\circ$ to the 12 o'clock direction. The intensity of the light incident on the second sheet is $I_0/2$, so, according to Malus's law, the intensity of the radiation that passes through both sheets of Polaroid is

$$I_f = \frac{I_0}{2}\cos^2(60^\circ) = \frac{I_0}{2}\left(\frac{1}{4}\right) = \frac{I_0}{8}.$$

The emerging light is polarized along the 2 o'clock direction.

How to Produce Polarized Radiation

We have already seen two ways to produce polarized radiation: by accelerating charges in an oriented dipole antenna and by passing unpolarized radiation through

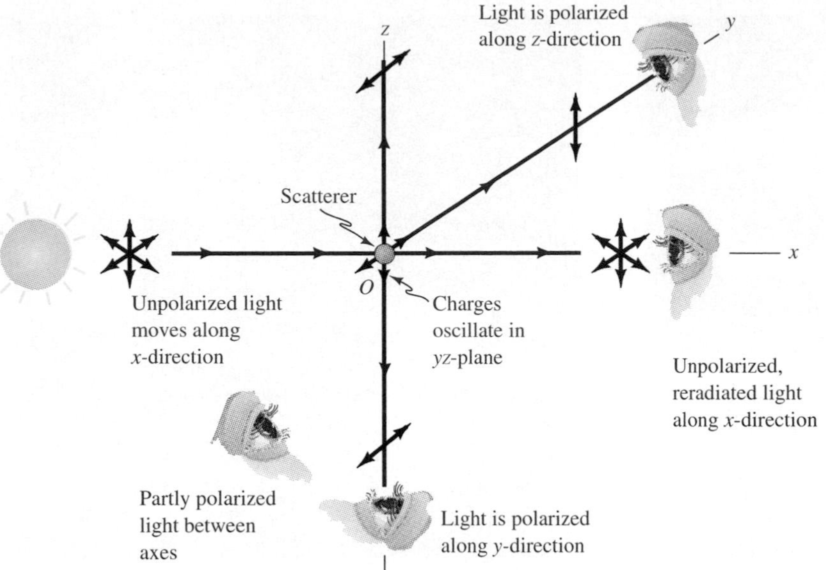

FIGURE 35–20 The polarization of radiation by scattering. An electromagnetic wave can propagate through a material because the wave's electric and magnetic fields cause electrons in the material to oscillate at the radiation frequency; these electrons in turn radiate new waves of the same frequency. The electric field of these new waves is aligned with the electrons' motion. Here unpolarized radiation is perpendicularly incident on the zy-plane in a gas; the electric fields of the radiation lie in that plane but are otherwise unrestricted. An observer along the x-axis sees the full range of motion of the electrons in that plane and hence sees unpolarized light. An observer at 90° to the original wave direction can see a side view of the plane from which the light is radiated and hence sees light fully polarized; the polarization direction is parallel to the plane's edge. At intermediate angles, the reradiated light is partly polarized.

a polarizer. There are other ways to polarize radiation, and we look at two of them now.

Polarization by Scattering. We are aware of the scattering of light when we observe the beam of an automobile headlight in a rainstorm or a snowstorm, because the scattering is quite pronounced there. The headlight beam is much less visible from the side on a dry night (if there is not much dust in the air). Nevertheless, even in the absence of droplets or dust particles, light and other forms of electromagnetic radiation are scattered by air molecules. The scattering mechanism is the following: The oscillating electric field **E** of the incoming radiation sets in motion the electrons in the air molecules. The electrons act like oscillators subject to an external harmonic force and oscillate with the frequency of the incoming field. The electrons move in a plane perpendicular to the incident radiation, and, if the incident wave is unpolarized, then there is no preferred direction to the electron motion as long as it occurs in the plane. An observer looking at an electron in a direction close to that of the incident radiation will see a radiated field that is unpolarized, because there is no preferred direction. An observer looking at the electron in a direction perpendicular to the direction of the incident radiation will see radiation whose electric field is along the direction of the electron motion. This observer will see the electron moving in just one direction (and will not see the component of the motion toward or away from him or her) and thus sees 100 percent linearly polarized light (Fig. 35–20). For angles between these directions, the polarization is partial. A simple way to observe this, if you live where the atmosphere is clear, is to hold a piece of Polaroid and look toward (*but not at*) the sun. The light intensity will change when the Polaroid is rotated, showing that the light scattered by the air molecules is polarized.

Polarization by Reflection. When unpolarized radiation is reflected from a surface such as water, then the reflected light is partly polarized (Fig. 35–21a).

(a)

(b)

FIGURE 35–21 Radiation polarized by reflection. (a) Here we see a shop window with confusing reflections. (b) The same scene, but with the camera lens fitted with a polarizing filter. The reflected light passing through the filter is greatly reduced.

When the angle of incidence is just right, the reflected light is fully polarized (Fig. 35–21b). The reason for this is not very different from the reason for polarization by scattering (Fig. 35–22). Unpolarized light incident at an angle θ_i (the *angle of incidence*) impinges on a surface. In general we may decompose the electric field of the incident wave into two components, each perpendicular to the direction of propagation. One of these directions, the z-direction, is perpendicular to the surface of the page and parallel to the reflecting surface; we label the other the a-direction. When the wave arrives at the surface, its electric field accelerates electrons. These accelerated charges radiate and give rise to both the transmitted and the reflected wave.

Let us first discuss the radiation caused by the component of **E** in the z-direction, perpendicular to the plane of the paper. The electrons accelerated by that component of the incoming electric field move at right angles to the direction of the reflected wave. An observer looking back along the line of the reflected wave sees the full motion of the electrons. Thus there is strong reflection of this

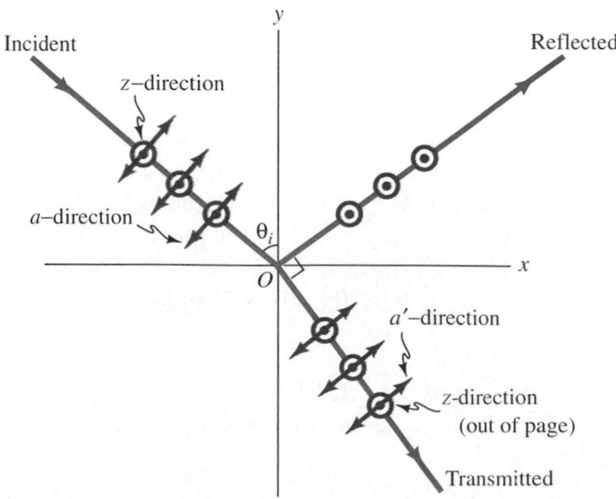

FIGURE 35–22 There is an angle of incidence θ_i for which the reflected wave is fully polarized. The electric field components are perpendicular to the rays, both in the plane of the page (marked by arrows) and out of the plane of the page (marked by circled dots).

part of the incident wave. Next consider the radiation induced by the component of the electric field of the incident wave in the a-direction. The electrons that absorb this incident radiation move parallel to the a-direction. An observer who looks along the line of the reflected wave sees a foreshortened motion of the electrons and thus only a limited amount of reflected radiation. Most of the radiation is transmitted. Thus there is a preferential polarization direction for the reflected light. In the special case that the direction of the reflected wave is at right angles to the a'-direction (perpendicular to the direction of the transmitted radiation), there is no reflected radiation polarized along the a'-direction, because, for this right angle, the motion of the absorbing and reradiating electrons along the a'-direction cannot be seen. *The reflected radiation is plane-polarized with an electric field in the z-direction, parallel to the plane of the reflecting surface.* For this special angle, the reflected and transmitted waves must be at 90° to one another.

The angle of incidence for which the reflected and transmitted (or refracted) rays are perpendicular to one another can be calculated when the rules for these rays (Snell's law) are developed, which we will do in Chapter 36. We give the result here: When the *angle of incidence* is the angle θ_B, known as **Brewster's angle** (named for David Brewster), the reflected ray is plane-polarized. This angle, which is the incident angle in Fig. 35–22, is given by

$$\tan \theta_B = \sqrt{\frac{\epsilon}{\epsilon_0}} = n, \tag{35–35}$$

where we have used the fact that in Eq. (35–16), the definition of the index of refraction, n, $\mu \simeq \mu_0$ (an approximation valid in all transparent materials). As we noted above, the effect is present but less dramatic for other angles. An analyzer whose polarizing axis is oriented in a direction perpendicular to the z-direction (the direction of polarization of the reflected wave) will absorb most of the reflected radiation. Thus Polaroid sunglasses, worn to cut down glare, need to have their polarizing axis aligned in a vertical direction.

*General Forms of Polarization

Up to now, we have concentrated on waves that are linearly polarized; that is, waves whose electric field vectors are aligned in a fixed direction. However, linear polarization is only one special case of a rich set of possibilities. Suppose that we line up two dipoles perpendicular to the z-direction, one along the x-axis and one along the y-axis. Waves of the same frequency are generated from each. A wave will propagate along the z-direction with an electric field given by

$$\mathbf{E} = E_1 \cos(kz - \omega t)\,\mathbf{i} + E_2 \cos(kz - \omega t + \phi)\,\mathbf{j}. \tag{35–36}$$

There is no phase in the first term, because we can set the origin to eliminate this phase, as before. However, we generally cannot eliminate ϕ, the phase of the second term, simultaneously. *The relative phase of the two terms cannot be set to zero by a simple change of origin.*

The electric field of Eq. (35–36) is more complex than that of the simple linear polarization case. Let us follow the tip of the electric field vector, the way we might see the motion of a light bulb at the tip of a moving baton of variable length. What we would see depends on E_1, E_2, and ϕ. The motion of the tip is what determines the polarization of the field. Let us look at the field $\mathbf{E}$ at $z = 0$ for simplicity and consider some special cases:

1. $E_2 = 0$. The electric field vector has only an x-component. This is the simple *linear polarization* case already described (Fig. 35–23a).

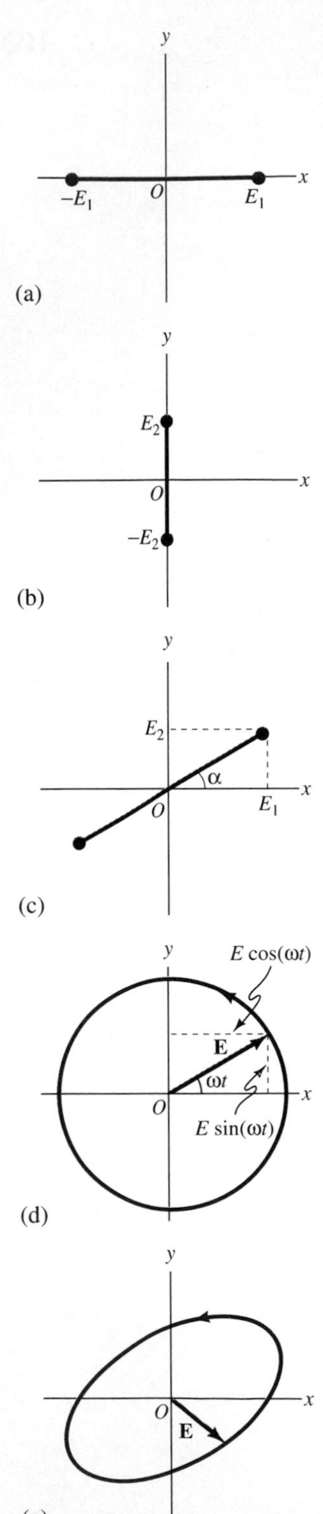

FIGURE 35–23 Various polarizations: (a) linear polarization along the x-axis; (b) linear polarization along the y-axis; (c) linear polarization along a general axis in xy-plane; (d) circular polarization; (e) elliptical polarization.

2. $E_1 = 0$. The tip of the electric field vector oscillates along the y-axis, and the wave is linearly polarized along the y-axis (Fig. 35–23b).
3. The phase $\phi = 0$. The electric field is then given by

$$\mathbf{E} = (E_1\mathbf{i} + E_2\mathbf{j})\cos(\omega t).$$

We would see that the tip of the electric field vector oscillates in a straight line that lies between the x- and y-axes, making an angle α with the horizontal, such that $\tan\alpha = E_2/E_1$. We again have linear polarization, but at an angle (Fig. 35–23c).

4. The situation changes when both E_1 and E_2 are present and $\phi \neq 0$. Consider first $E_1 = E_2$ and the special case $\phi = \pi/2$. In that case, because $\cos[-\omega t + (\pi/2)] = \sin(\omega t)$, we have

$$\mathbf{E} = E_1[\cos(\omega t)\,\mathbf{i} + \sin(\omega t)\,\mathbf{j}].$$

The tip of the electric field traces a counterclockwise circle when we look at the wave from the $+z$-direction (out of the page) with the wave traveling toward us (Fig. 35–23d). We describe this radiation as *circularly polarized*, and by convention we call it *left-circular polarization.* If the relative phase is $\phi = -\pi/2$, the radiation is *right-circularly polarized* and the tip of the electric field vector rotates clockwise circularly as seen by an observer looking at the coming wave.

5. The most general case has $E_1 \neq E_2$ and $\phi \neq 0$, and the tip will execute elliptical motion. In that case we have *elliptically polarized* radiation (Fig. 35–23e).

Note that *all the cases discussed here can be viewed as linear superpositions of radiation polarized along the x- and y-axes, respectively.* The electric field is the vectorial sum of the electric fields from two separate linearly polarized waves that propagate in the same direction with the same frequency. In this sense, circular polarization is based on the linear polarization discussed earlier.

*35–7 ELECTROMAGNETIC RADIATION AS PARTICLES

One of the most astonishing discoveries of the early part of the twentieth century is the discovery that *electromagnetic radiation consists of particles.* The research of Max Planck, of Albert Einstein, and of Arthur Compton established that what we call an electromagnetic wave consists of a large number of individual particles called **photons**. These particles are indivisible: It is not possible to have 0.3 photons, for example. For radiation characterized by a frequency $f = \omega/2\pi$, the energy carried by a single photon is

$$E = hf, \tag{35–37}$$

where $h = 6.63 \times 10^{-34}$ J·s is Planck's constant. A photon also carries momentum, given by

$$p = \frac{E}{c} = \frac{hf}{c} = \frac{h}{\lambda}. \tag{35–38}$$

The particle nature of electromagnetic radiation was established through Compton's experiments on the scattering of radiation by free electrons in carbon. Photons were scattered preferentially through an angle that can be calculated by treating each photon as a relativistic billiard ball that collides elastically with each electron at rest. The momentum of the outgoing photon depends on the collision angle. Equation (35–38) then implies that the frequency of the scattered radiation also depends on the collision angle in a way that can easily be calculated.

That h is small explains why we think of light as a continuous phenomenon rather than a series of individual photons. Someone standing under Niagara Falls does not feel as if he or she is being bombarded by droplets of water!

EXAMPLE 35–6 At what rate does a 60-W light bulb emit photons? Assume for simplicity that the light is emitted with a single wavelength of 590 nm.

SOLUTION: In 1 s the light bulb emits a total energy of 60 J. If there are N photons of frequency $f = c/\lambda = (3 \times 10^8\ \text{m/s})/(5.9 \times 10^{-7}\ \text{m}) = 0.51 \times 10^{15}$ Hz, then

$$N = \frac{E}{hf} = \frac{60\ \text{J}}{(0.51 \times 10^{15}\ \text{s}^{-1})(6.63 \times 10^{-34}\ \text{J}\cdot\text{s})} = 1.8 \times 10^{20}.$$

Thus on the order of 10^{20} photons are emitted every second.

For extremely low light intensities, individual photons can be detected. Similarly, in reactions involving elementary particles, situations in which single photons are emitted occur frequently. Special detectors count them individually. The Had the history of evolution of the human eye been a little different so that the eye could respond to a single photon, the notion of radiation as consisting of particles would have been obvious to everyone.

SUMMARY

Maxwell's equations, comprised of Gauss' laws for electric and magnetic fields, the generalized Ampère's law, and Faraday's law [Eqs. (35–1) to (35–4)], imply that even in the absence of currents and charges, it is possible to have propagating electric and magnetic fields. In the absence of free charges, the electric and magnetic fields obey the wave equation, which has the generic form

$$\frac{\partial^2 E_x}{\partial z^2} = \mu_0 \epsilon_0 \frac{\partial^2 E_x}{\partial t^2}. \tag{35–7}$$

In the case that $E_z = 0$ and $B_z = 0$, the waves propagate along the z-direction. Whatever the direction, the speed of propagation is given by

$$v^2 = \frac{1}{\mu_0 \epsilon_0}. \tag{35–9}$$

This speed is the speed of light, $v = c \simeq 3 \times 10^8$ m/s. In material media characterized by the dielectric constant κ, the speed of propagation is $c/\sqrt{\kappa} = c/n$, where $n = \sqrt{\kappa}$ is the index of refraction. There are solutions of the wave equation (electromagnetic waves) in which the fields have the harmonic form

$$E_x = E_0 \cos(kz - \omega t + \phi) \tag{35–8}$$

and

$$B_y = B_0 \cos(kz - \omega t + \phi). \tag{35–13}$$

These waves propagate in the z-direction. More generally, electric and magnetic fields of waves that propagate in a given direction are transverse to that direction. The electric and magnetic field amplitudes are related by

$$E = cB, \tag{35–14}$$

and the fields are perpendicular to each other:

$$\mathbf{E} \cdot \mathbf{B} = 0. \tag{35–15}$$

Electromagnetic waves carry energy with energy density

$$u = \frac{1}{2}\left(\frac{B^2}{\mu_0} + \epsilon_0 E^2\right). \tag{35–18}$$

This energy is carried in equal amounts by the electric and magnetic fields. Electromagnetic waves also carry momentum, with momentum density $\mathbf{S}/c^2$, where $\mathbf{S}$ is the Poynting vector, given by

$$\mathbf{S} = \frac{1}{\mu_0} \mathbf{E} \times \mathbf{B}. \tag{35–23}$$

Thus radiation can transfer momentum, and when a material absorbs radiation, there is a radiation pressure on the material, given by

$$\frac{S}{c} = u. \tag{35–27}$$

Charged particles radiate when they are accelerated. For a charge q undergoing an acceleration along the z-direction, the energy flux is proportional to

$$S \propto \frac{\sin^2 \theta}{r^2}, \tag{35–30}$$

where θ is the angle with the z-axis and r is the distance from the charge. Radiation with a $\sin^2 \theta$ angular dependence is called dipole radiation.

The polarization of an electromagnetic wave is the direction of the transverse electric field vector. It can be measured because certain materials, called polarizers, transmit electromagnetic waves only along a particular polarization axis. Polarizers may be used to detect as well as to polarize electromagnetic waves. If a second polarizer is placed with its axis making an angle θ with the first one, then the electric field E of the transmitted wave is reduced in magnitude from the electric field E_0 of the incident wave according to $E = E_0 \cos \theta$. Thus the intensity I (the average of the energy flux) of the transmitted light is reduced from the incident intensity I_0 according to Malus's law:

$$I = I_0 \cos^2 \theta. \tag{35–34}$$

Waves can be polarized by reflection. If light falls on a medium of dielectric constant κ at an angle θ_B (Brewster's angle), for which

$$\tan \theta_B = n \tag{35–35}$$

(n is the material-dependent index of refraction), then the reflected light is polarized in a direction perpendicular to both the incoming direction and the reflected direction of the wave. Light can also be polarized by scattering.

QUESTIONS

1. Stable charged particles that move uniformly produce no electromagnetic waves. How does the conservation of energy suggest that this must be true?

2. Short-wave radio signals have wavelengths of several tens of meters. Such waves are particularly well reflected by the earth's ionosphere (an upper layer of the atmosphere with many free charges). Explain why short-wave radio waves can be received over such large distances.

3. The production and detection of polarization depends on the electric field vector. Can there be, in principle, a polarization associated with the magnetic field vector?
4. Can there be standing electromagnetic waves as well as traveling ones? Recall that mechanical standing waves on a string are possible when certain boundary conditions are satisfied, such as the ends of the string being fixed. How can we control the values of electric or magnetic fields on fixed boundaries?
5. If magnetic monopoles were discovered tomorrow, Gauss' law for magnetism, Eq. (35–2), and Faraday's law, Eq. (35–4), would both need to be modified. What would that modification look like?
6. Would the situation of Question 5 change the nature of electromagnetic waves in free space?
7. A *solar sail* is a large surface on which the radiation pressure of the sun's radiation can act and thereby push the sail along (Fig. 35–24). Solar sails have been proposed for space ships to travel throughout our Solar System. What properties would such a sail have to have, and what difficulties do you see in the proposal?

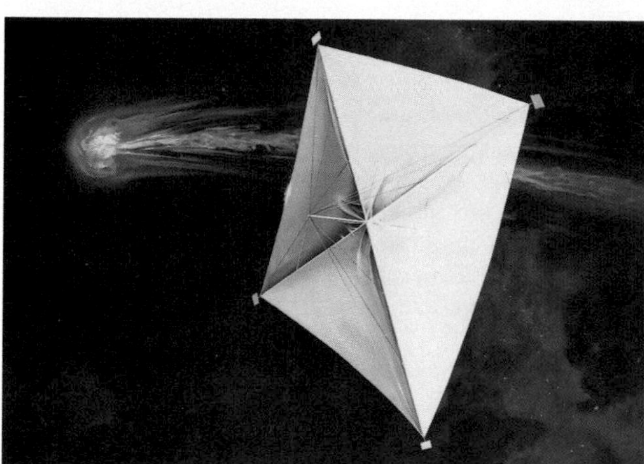

FIGURE 35–24 Question 7.

8. Rockets are propelled forward when mass is ejected backward from them. Could a source of light (or other electromagnetic radiation) be used in place of the mass?
9. How can you tell whether light is plane polarized or not?
10. Can a sound wave in air be polarized?
11. Incident light is linearly polarized along the x-axis. We would like to rotate the direction of polarization so that it lies along the y-axis. Can this be done with one polarizer? Can it be done with two? What is the minimum reduction in intensity when two polarizers are used? Can there be even less intensity reduction with three polarizers?
12. In Hertz's test of the existence of electromagnetic waves, sparks appear in a secondary gap as the result of an AC current in a primary circuit. Hertz interpreted these sparks as due to the effect of electromagnetic waves and not to the effects of Faraday induction. What sort of checks did Hertz need to make in order to rule out Faraday induction?
13. Can you use the example and the arguments given in Section 35–2 to show that electromagnetic waves are transverse? Are there any pitfalls?
14. We showed by thinking about its effect on a free charge that electromagnetic radiation carries momentum. Consider its effect on an electric dipole to see whether it might carry angular momentum as well. Start by orienting the dipole with its axis along the direction of the electric field vector of the electromagnetic wave.
15. Consider the solar sail described in Question 7. Is it better to make a solar sail reflective (shiny) or absorbing (black)?
16. Is the Poynting vector defined only for electromagnetic radiation?

PROBLEMS

35–1 Maxwell's Equations

1. (I) Verify the consistency of the dimensions of both sides of each of the four Maxwell's equations.
2. (II) Gauss' laws for electric fields and for magnetic fields differ due to the lack of magnetic charges. Assume that magnetic monopoles (magnetic charges) exist; denote them by the symbol M. Rewrite Gauss' law for magnetic fields, and give the SI units of M.
3. (II) Ampère's and Faraday's laws differ due to the lack of a currentlike term in Faraday's law. Assume that magnetic monopoles exist (call them M), and rewrite Faraday's law. Discuss the physical significance of any new terms added.
4. (II) Equations (35–1) through (35–4) apply in a vacuum. Write Maxwell's equations for matter instead by using the dielectric constant, κ, and the relative permeability, $\kappa_m = 1 + \chi_m$.

35–3 Electromagnetic Waves

5. (I) If the electric field for a plane wave is given by $E_x = 0$, $E_y = E_0 \cos(kz + \omega t)$, and $E_z = 0$, what are $\mathbf{B}$ and the direction of propagation of the wave?
6. (I) Use dimensional analysis to show that $1/\sqrt{\mu_0 \varepsilon_0}$ has the dimensions of speed, $[\mathrm{LT}^{-1}]$.
7. (II) Find the approximate wavelength, wave number, frequency, and angular frequency for electromagnetic waves associated with (a) your favorite AM station; (b) your favorite FM station; (c) a microwave oven; (d) yellow light; (e) X-rays.
8. (II) Starting from Eqs. (35–5) and (35–6), derive a wave equation for the x-component of the magnetic field. What is the speed of the resulting wave?
9. (II) Write the counterparts of Eqs. (35–5) and (35–6) for

electromagnetic fields B_y and E_z that lie in the yz-plane and propagate in the x-direction. [*Hint:* Start with Figs. B1–1 and B1–2. Then relabel the axes according to $x \to y \to z \to x$.]

10. (II) A plane harmonic wave of electromagnetic radiation with wavelength λ is propagating in the $-x$-direction. The z-component of the electric field has magnitude E_0, and there is no y-component. (a) Write an expression for the electric field. (b) Use this expression and the result of Problem 9 to calculate the magnetic field. What vector components will this field have?

11. (II) A plane wave propagates along the direction in the xy-plane that makes an angle θ with the x-axis. Show that the electric field is given by $\mathbf{E}_0 \cos(kx \cos\theta + ky \sin\theta - \omega t + \phi)$. What directions can $\mathbf{E}_0$ have?

12. (II) A plane wave of wavelength 1.2 m propagates in the z-direction. The electric field points in the y-direction and has an amplitude of 3.0 V/m. Write an expression for the magnetic field, including its amplitude in SI units. Assume that the electric field is at its maximum at $z = 0$, $t = 0$.

13. (II) An electromagnetic traveling wave is generated at the left-hand end of a tube oriented in the z-direction; the wave travels in the $+z$-direction. At the ends of the tube, $z = 0$ and $z = L$, are highly reflective mirrors. The electric field of the incident wave is $\mathbf{E} = E_1 \cos(kz - \omega t)\,\mathbf{i}$, and the electric field of the wave reflected at $z = L$ is $\mathbf{E} = E_1 \cos(kz + \omega t + \phi)\,\mathbf{i}$. Show that the net electric field forms a standing wave, and, by computing B_y, that the associated magnetic field B_y also has the form of a standing wave.

14. (III) A pulse of electromagnetic radiation travels in the $-z$-direction. The electric field is oriented in the x-direction and is given by $\mathbf{E} = E_0 e^{-(z+ct)^2/a^2}\mathbf{i}$. What is the orientation of the magnetic field? Make a guess of the space–time dependence of the magnetic pulse, and use Eqs. (35–5) and (35–6) to find a form for $\mathbf{B}$ that satisfies Maxwell's equations.

35–4 Energy Density, Energy Flow, and Momentum Flow

15. (I) A radio station emits a signal with a power of 50 kW. What are the values of the electric field and magnetic field at distances of 10 km and 1000 km? Assume that the signal far from the antenna is transmitted with equal intensity in all directions. (Real radio stations cannot afford to transmit their energy in this way, and their antennas distribute energy with a high degree of directionality.)

16. (I) The electric field for a given electromagnetic wave has a peak value of 50 mV/m. What is the intensity of the wave?

17. (I) A harmonic plane wave of wavelength 12 μm and an electric field amplitude of 10 V/m impinges on a totally reflecting surface of area 10 cm^2. What is the radiation pressure exerted by the wave?

18. (I) A plane electromagnetic wave with maximum electric field amplitude of 250 V/m is incident on a perfectly absorbing surface perpendicular to the direction of propagation. What is the rate of energy absorption per unit area of the surface?

19. (II) The magnetic field for a given electromagnetic wave has an rms value of 10^{-8} T. What is the intensity of the wave? How much energy is transported per minute through a 0.5-m^2 area?

20. (II) A typical lecture-demonstration laser of power 0.25 mW has a beam of diameter 1.5 mm. (a) What are the peak values of the electric and magnetic fields? (b) Suppose, as is in fact possible, that the beam is focused to an area of one square wavelength. What is the peak value of the electric field, given that $\lambda = 625$ nm?

21. (II) Assume that a 100-W-light bulb emits light equally in all directions. What are the peak and rms values of the electric and magnetic fields at a distance of 1 m?

22. (II) A 150-W lightbulb radiates uniformly in all directions, and 40 percent of this energy is emitted as electromagnetic radiation in the visible light range. What is the electromagnetic energy density of visible light at a distance of 50 cm from the bulb? What are the rms values of the corresponding electric and magnetic fields there?

23. (II) Assume that a 50,000-W radio station emits its signal equally in all directions. (a) What is the intensity of the signal at a distance of 100 km? (b) What is the rms value of the electric field at a distance of 100 km?

24. (II) The total electromagnetic power emitted by the sun is 3.8×10^{26} W. What is the radiation pressure exerted on a totally reflecting surface a distance $r = 10^9$ m from the sun?

25. (II) What are the dimensions and SI units for the Poynting vector? Reduce your answer to the dimensions and units of mass, length, and time, then reexpress it in terms of watts and meters.

26. (II) Solar energy delivered to a horizontal surface in Washington, D.C., averaged over a full year is 160 W/m^2. Assuming that this radiation is fully absorbed on a particular square meter of ground, what is the approximate total momentum delivered to this area in 1 y? Compare this number to an estimate of the momentum absorbed by a baseball catcher in catching a single pitch.

27. (II) Suppose that you want to use the radiation pressure from a beam of light to suspend in a horizontal position a piece of paper of area 100 cm^2 and mass 1 g. Assume that there is no problem with balance, that the paper is dark and absorbs the beam fully, and that the entire beam can be used to hold the paper against the pull of gravity. How many watts must the light produce? Given your answer, what do you suppose would happen to the paper?

28. (II) A light beam with a given Poynting vector falls on a flat, fully reflecting surface at an angle of incidence θ (with respect to the vertical). What is the momentum transferred to the surface per unit area?

29. (II) A laser delivers 10^3 J of energy in a pulse that lasts 10^{-9} s. What are the peak electric and magnetic fields for a laser beam of diameter 10^{-3} m?

30. (III) The short side of a thin, stiff rectangle 2.0 cm $\times$ 0.50 cm is attached to a vertical axis. Half of each side is painted black and is fully light absorbent; the other half is a shiny, reflecting metal (Fig. 35–25). The back of each half is dif-

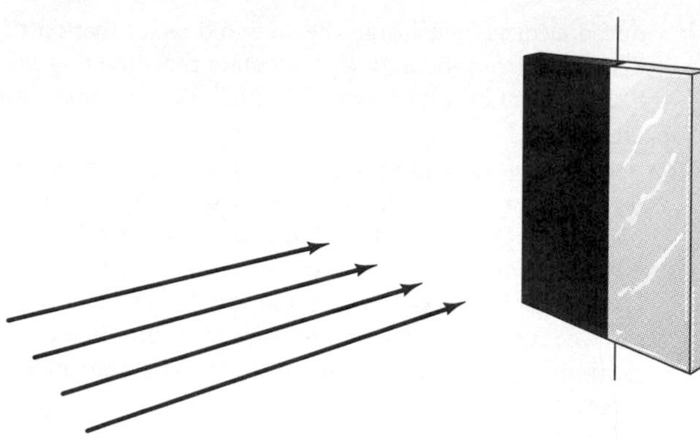

FIGURE 35–25 Problem 30.

ferent from the front. There is no friction at the axis. The apparatus is bathed in a well-collimated (nonspreading) beam of light whose Poynting vector has magnitude 1.0×10^{-3} kg/s^3 and that travels perpendicular to the vertical axis. Is there a net torque on the rectangle's surface? If so, what is its average value due to the light over a full, uniform rotation of the rectangle about the axis?

*35–5 *Dipole Radiation*

31. (II) A broadcasting dipole antenna is oriented along the y-axis. For the geometry shown in Fig. 35–26, give the following information for a point P far away along the z-axis: (a) the direction of the electric field; (b) the direction of the magnetic field; (c) the direction of the Poynting vector. (d) Repeat parts (a) through (c) for the electromagnetic wave a half-cycle later.

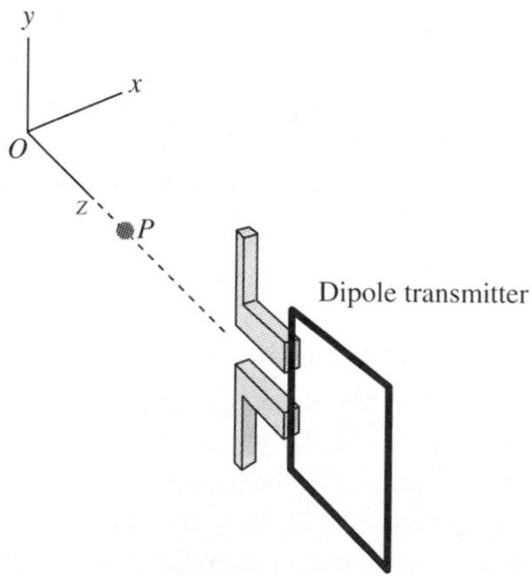

FIGURE 35–26 Problem 31.

32. (II) Suppose that a vertical tower 120 m tall acts as a dipole antenna, with currents running back and forth along the tower to generate electromagnetic waves in a dipole pattern. If the wavelength of each electromagnetic wave is the height of the tower, what is the period of the current oscillation in the tower?

33. (III) A cross-shaped antenna lies in the xy-plane, centered at the origin (Fig. 35–27). The charges oscillate with the same frequency within each arm of the cross. Find the Poynting vector along the z-axis as a function of z if charges moving in the $+x$-direction in the x-arm pass the origin at the same moment that (a) charges moving in the $+y$-direction in the y-arm pass the origin; (b) charges moving in the $-y$-direction in the y-arm pass the origin; (c) charges in the y-arm are at their maximum $+y$-position, ready to turn around.

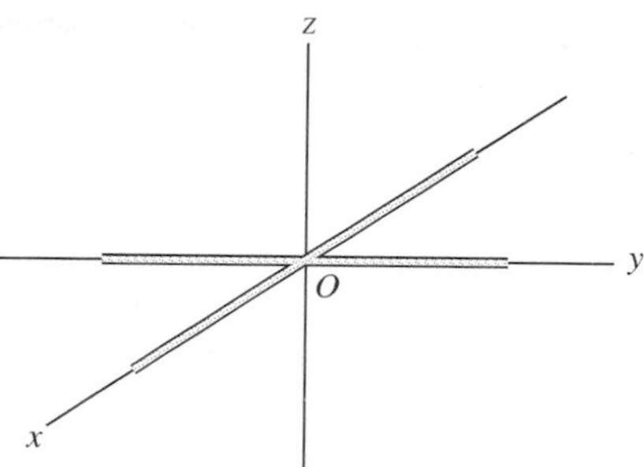

FIGURE 35–27 Problem 33.

35–6 *Polarization*

34. (I) At what angle should the axes of two ideal Polaroid sheets be placed to reduce the intensity of a given source of unpolarized light to (a) 1/2; (b) 1/4; (c) 1/10; (d) 1/100?
35. (I) The axes of four ideal Polaroid sheets are stacked, each at 25° with respect to the previous one. What fraction of initially unpolarized light passes through all four sheets?
36. (I) The moon reflects light off a still pond at night. At what angle above the horizon is the polarization a maximum? The index of refraction of water is 1.33.
37. (II) A beam of light propagating in the z-direction is polarized in the x-direction. Two superposed Polaroid sheets are placed perpendicular to the beam. The polarization axis of one makes a 47° angle with respect to the x-direction, and the axis of the other makes a 63° angle with respect to the axis of the first sheet. What is the intensity of the transmitted beam?
38. (II) What fraction of initially unpolarized light passes through two Polaroid sheets placed at right angles to each other? What happens if a third sheet is placed between the two sheets, with its axis at an angle of 45° to the two?
39. (II) Left-circularly polarized light propagating in the $+z$-direction falls on a piece of Polaroid that is parallel to the xy-plane and whose polarization axis makes a 30° angle with the x-axis. What is the polarization of the transmitted wave?
40. (II) A circularly polarized wave of intensity I_0 passes through an analyzer that passes an electric field in just one direction, so the light comes out polarized linearly. In terms of I_0 calculate the intensity of the light that comes out of the analyzer.
41. (II) If light of intensity I_0 moving in the z-direction is

polarized linearly in the x-direction, it will not pass through a piece of Polaroid that passes light polarized in the y-direction. Figure 35–28 shows a way in which this light can pass the y-direction analyzer if a second analyzer is used. If the lower analyzer makes an angle of θ with respect to the x-direction, what is the intensity of the light that passes the upper analyzer?

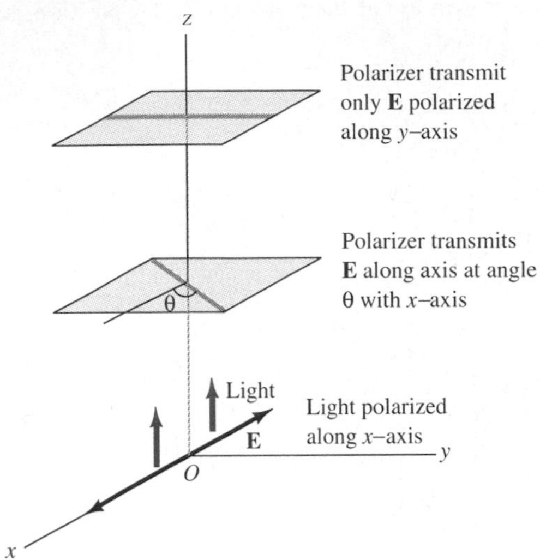

FIGURE 35–28 Problem 41.

42. (II) Unpolarized light of intensity I_0 passes through two pieces of Polaroid successively. (a) What is the intensity of the light after it passes through the first piece of polaroid? (b) The second piece is rotated so that the intensity of the transmitted light goes to zero. What angle does the polarizing axis of the second piece make with that of the first piece? (c) A third piece of Polaroid is inserted between the two pieces in place. Calculate the intensity as a function of the angle θ that the axis of the third piece makes with the axis of the first piece. (d) Show that the intensity of the transmitted light is no longer zero unless the axis of the third piece is parallel to that of either of the other pieces.

*35–7 *Electromagnetic Radiation as Particles*

43. (II) Calculate the number of photons emitted by an AM radio station that broadcasts at a frequency of 10^6 Hz that are required to equal the energy contained in one photon of visible light at a wavelength of 500 nm.

44. (II) The power generated by the sun is 4×10^{26} W. Assuming that it is all emitted at an average wavelength of 500 nm, calculate the number of photons emitted per second.

General Problems

45. (II) Consider a solenoid of n turns/m with radius R. A current $I = I_0 \cos(\omega t)$ goes through the solenoid. (a) Calculate the magnetic field inside the solenoid. (b) Calculate the induced electric field inside the solenoid as a function of the distance r from the axis. (c) Calculate the Poynting vector, **S**. In particular, find its direction at different times during one cycle.

46. (II) The electric and magnetic fields of an electromagnetic wave act on a charge q. With what speed must the charge move so that the magnetic force on the charge is, at most, 10 percent of the electric force? If the electromagnetic wave is traveling in the z-direction and the electric field has only an x-component, what is the direction (or directions) of motion of q so that the magnitude of the magnetic force is greatest?

47. (II) Consider the solar sail described in Question 7. A solar sail can be aligned with its area perpendicular to a radial line from the sun so that the sail is pushed straight outward. Show that in this configuration the force on the sail always has the same sign and is proportional to $1/r^2$, where r is the distance from the sail to the sun. (Assume that only the radiation pressure and the gravitational force due to the sun act on the sail.) This economical method of propulsion has been proposed for travel to the far reaches of the Solar System when transit time is not an important factor.

48. (II) A solar sail (see Question 7) is to be designed such that, when it is aligned perpendicular to the sun's rays and is $1.5 \times 10''$ m from the sun the radiation pressure on it, P, just cancels the gravitational attraction of the sun. The density of the material of the sail, which forms a sheet of constant thickness, is ρ. (a) Find P, given that the energy flux from the sun is 1.4 kW/m^2 at the radius of the earth's orbit. (b) Express the sail's thickness in terms of ρ, P, the mass of the sun, and the gravitational constant. If ρ is 2.0×10^3 kg/m^3, what is the thickness of the sail material? Your result is independent of the sail's area.

49. (II) Find an expression for the electric field of a plane electromagnetic wave with the following properties: (a) the frequency is 10^{12} Hz; (b) the wave travels in a medium of index of refraction 1.5; (c) the wave propagates along a line that lies in the xz-plane and makes a 45°-angle with the z-axis; (d) the wave is polarized along the y-axis; (e) the average value of the Poynting vector is 10^3 W/m^2.

50. (II) A swimming pool has underwater lights. What is Brewster's angle for reflection off the upper surface of water? The index of refraction of water is 1.33.

51. (II) The amount of solar energy reaching your body when you sunbathe on an ocean beach in summer is about 1000 W/m^2. Assume that your body absorbs 50 percent of this incident radiation and that your exposed body area is 0.8 m^2. How much solar energy do you absorb in 2 h?

52. (II) A high-powered, pulsed laser used to confine plasma for nuclear fusion studies is rated at 10 MW. The laser beam is focused on an area of 1 mm^2. Calculate the intensity, peak electric and magnetic fields, and average energy density in this beam. Compare your results to Tables 23–1 and 29–1, which list some values for electric fields and magnetic fields, respectively, in other contexts.

53. (II) What is the number of photons/m^3 contained in a beam of electromagnetic radiation in a plane wave with a wavelength of 2 cm and an electric field amplitude 10 V/m?

54. (II) The solar energy flux at a distance $R_0 = 1.5 \times 10^{11}$ m from the sun (the radius of the earth's orbit) is 1400 W/m^2. (a) What is the total energy flow from the sun in watts? (b) Assume that sunlight is entirely emitted at a wavelength of 600 nm, and calculate the total number of photons/s emitted. (c) Using the result of part (b), find the number of photons/s that strike a 1 mm $\times$ 1 mm surface at a distance R_0. The surface is oriented perpendicular to the sun.

55. (II) A laser emits N photons of frequency f. The beam strikes a mirror that is moving with speed v in the direction of propagation of the laser beam. Assuming that the kinetic energy of the mirror is much larger than that of the beam, use energy conservation and momentum conservation to find the frequency of the reflected beam. Treat the photon as a particle of energy hf and momentum hf/c.

56. (III) Consider a current I that flows through a cylindrical wire of length L, radius b, and resistance R. The current flows uniformly across the cross section of the wire. Calculate the electric fields inside and on the surface of the wire. The current in the wire gives rise to a magnetic field, which you can calculate. Use these fields to find the direction and magnitude of the Poynting vector on the surface of the wire. Show that the rate of energy flow into the wire through its surface is I^2R, the power dissipated in ohmic heating.

57. (III) Consider a capacitor that consists of two circular metal plates of radius R a distance d apart. R is so much larger than d that all fringe fields can be neglected. If the charge on the plates, Q, changes with time, then according to Ampère's law a magnetic field will be induced in the region between the plates. (a) What is the induced magnetic field? (b) Using the induced magnetic field and a calculation of the electric field between the plates, find the Poynting vector. (c) Show that with this Poynting vector, the net energy flow into the capacitor is the rate of change of the capacitor energy $Q^2/2C$.

58. (III) Consider a plane electromagnetic wave of frequency f that propagates in the z-direction in a cubic box whose sides are length L, with L much larger than the wavelength. The electric field of the radiation has the form $\mathbf{E} = E_0 \sin(kz - \omega t)\,\mathbf{i}$. Alternatively, we can say that the radiation consists of N photons, each propagating in the z-direction with energy hf, where h is Planck's constant. Use two alternative expressions for the energy of the radiation to express E_0 in terms of h, f, N, and L.

CHAPTER 36

The impressionist painters made a special effort to capture in their paintings the magical effects of light, as in this scene of boating on the Seine River in France, Regatta at Argentevil, by Claude Monet.

LIGHT

One of the most striking features of light is that it appears to travel in straight lines. This fact manifests itself in the sharpness of shadows and in the rays that appear when light penetrates a dark and dusty room through a tiny opening. The primitive belief that objects are seen because something is emitted from the eye persisted long enough to make a mark on the language (we all "cast a glance"), but even ancient Greek philosophers associated the emission of light with sources such as the sun and identified the eye as the detector. The rectilinear motion of light strongly suggests that light is composed of particles emitted by a source, and Isaac Newton, whose earliest work was on optics, finally supported that view (Fig. 36–1a). The particle model provided an explanation of the observations of Newton's time. It is, in fact, surprising that in Newton's day the idea that light consists of waves could have taken root, but it did, based on Robert Hooke's idea that light is some type of oscillatory activity in an as yet unknown medium. This idea led Christian Huygens in 1687 to propose a wave theory of light. We shall show in this chapter that the wave theory can explain everything that the particle theory of light can. In Chapters 38 and 39 we will present phenomena that can be explained only by the wave theory.

By the early nineteenth century, it had become apparent that certain observations cannot be explained by the particle theory and demand that light behave like a wave. For example, when we look very closely, under controlled conditions, light does not cast *sharp* shadows; to some extent, light bends around corners.

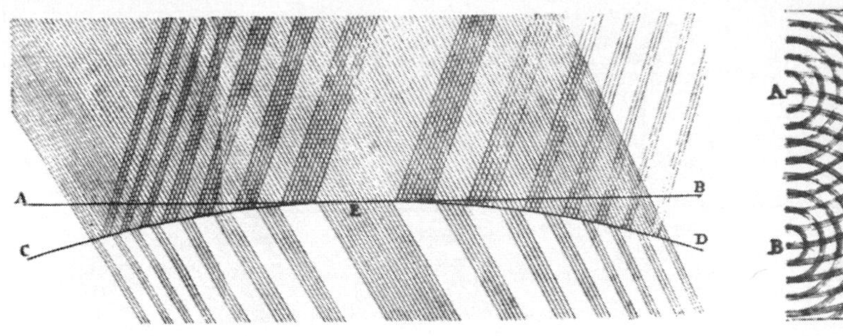

(a)

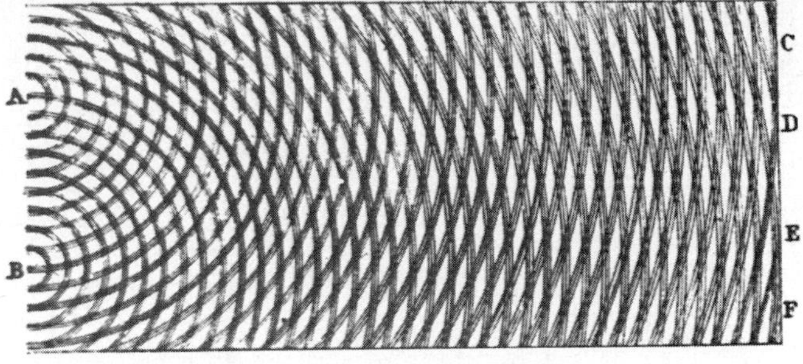

(b)

FIGURE 36–1 (a) Newton's view of the particle nature of light. In this sketch from his *Optics* (1704), the lines AB and CD represent surfaces of glass. (b) Young's view of the wave nature of light. In this sketch published in 1807 from his lectures, points A and B represent pinholes; points C and E represent spots where the waves reinforce each other, and points D and F, spots where they do not.

(Newton did not have the equipment to make this observation, and, in fact, he argued against the wave theory on the basis that light does *not* appear to bend around corners!) Or, again under controlled conditions, we can see that beams of light interfere with each other in just the same way that the waves discussed in Chapter 15 interfere with each other. Definitive experiments by Thomas Young in 1801 on the wave aspects of light eventually established the preeminence of the wave theory (Fig. 36–1b). The subsequent prediction from Maxwell's equations of light and other electromagnetic waves would seem to have settled the question once and for all. In the twentieth century, however, we have had to revise our view once more, as new experimental evidence suggested that some aspects of light can be explained only if light sometimes behaves as particles. Now we are not forced to choose between a particle theory and a wave theory of light. A quantum mechanical explanation encompasses them both.

In this chapter and the next we shall concentrate on those properties of light that were known in Newton's and Huygens' time. They describe the properties of light with which, for the most part, we have some everyday experience: the reflection of light from mirrors, refraction as light passses through glass lenses or water, and the dispersion of light, which leads to rainbows and the prismatic separation of the colors.

36–1 THE SPEED OF LIGHT

To show that light travels at finite speed requires some ingenuity. Although Galileo had admitted the possibility that the speed of light is finite, his early experimental attempts to observe a finite propagation time for the passage of light from one mountaintop to another were failures. Olaus Roemer made observations in 1675 of the eclipses of the moons of Jupiter. He found that the timing of these phenomena as observed at different times of the year could be explained only if a large, but finite value were assumed for the speed of light. Roemer used Solar System data available to him at the time and obtained 2×10^8 m/s for the speed of light, certainly a result of the correct order of magnitude. The first terrestrial measurements were made in 1849 by Hippolyte Fizeau, who used the ingenious device shown in Fig. 36–2a. A light source is placed behind a toothed wheel that can be rotated at high speeds. The light passes through an inclined glass plate and then between two teeth of the rotating wheel (Fig. 36–2b). It then travels to a mirror and is reflected straight back. If the speed of light were infinite, light would come straight back to the gap between cog 1 and cog 2, the same gap through which it entered, independently of the rotational speed of the cogged wheel (Fig. 36–2c). Light traveling at a finite speed would also pass through the same gap if the rotational speed

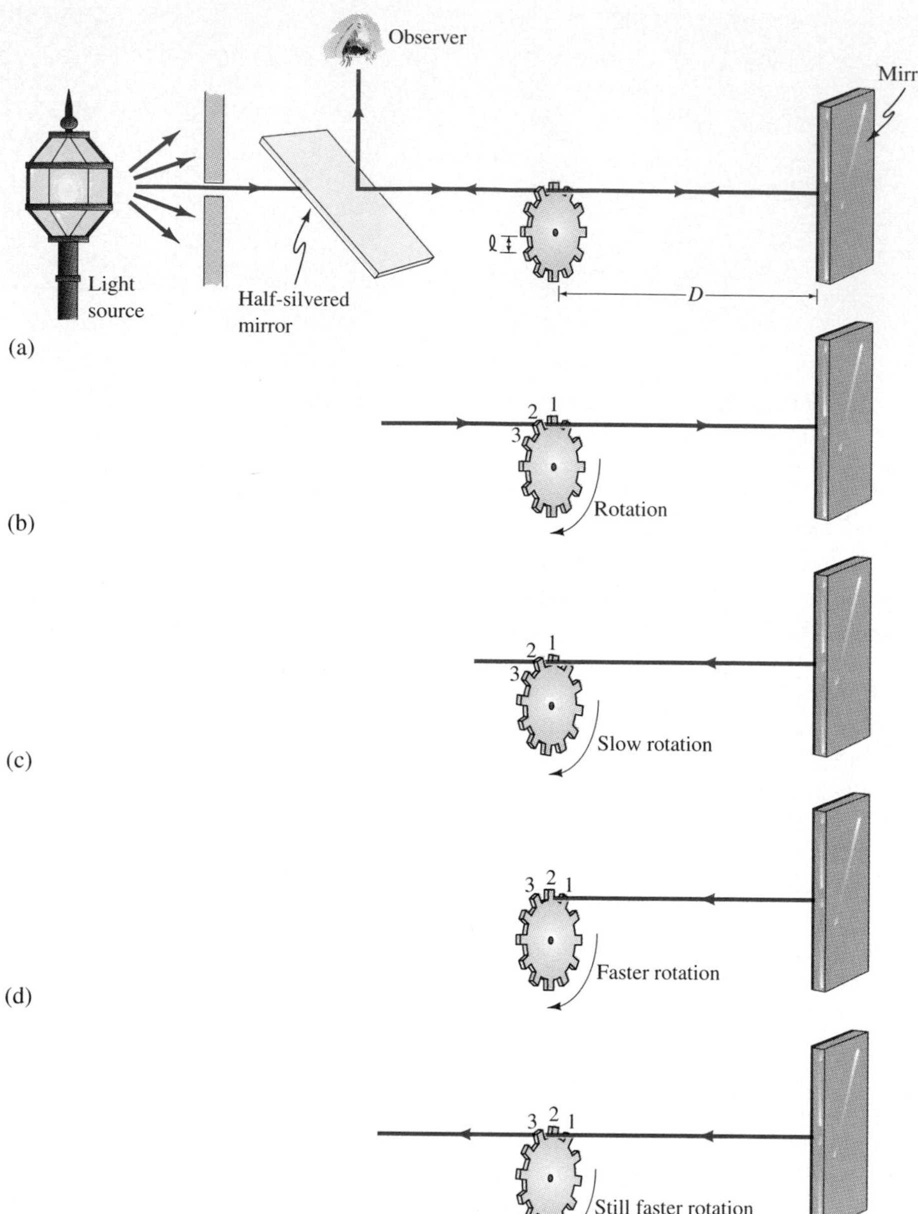

FIGURE 36–2 Fizeau's method for measuring the speed of light: (a) a sketch of the apparatus he used. (b) Incident and (c) reflected light pass through the same gap. (d) When the wheel rotates faster, the reflected light does not manage to pass through the original or the next gap. (e) When the rotation speed is still higher, the reflected light passes through the next gap.

of the wheel were very slow. But if we speed up the wheel and the speed of light has a finite value c, then at a certain value of angular velocity, the wheel moves enough during the time the light travels to the mirror and back that the light will strike cog 2 (Fig. 36–2d). No light reaches the detector. As the wheel rotates even faster, light once again reaches the detector, but this time it passes through the *next* gap in the wheel, the one between cogs 2 and 3 (Fig. 36–2e). If the wheel is moving at speed v when Fig. 36–2e first applies, then we can equate the time $2D/c$ for light to make a round trip from the wheel to the mirror and back with the time ℓ/v for the wheel to move a distance of one gap:

$$\frac{\ell}{v} = \frac{2D}{c}.$$

Here ℓ is the cog spacing and D is the distance from the wheel to the mirror. If D is much larger than ℓ, then a value of v much smaller than c would suffice to measure c accurately.

The best measurement would be limited by our ability to measure time (currently we can measure time to one part in 10^{13}) and distance (currently to four parts in 10^9). Therefore, today we *define* the speed of light in a vacuum to be $c =$ 299,792,458 m/s and use time along with the definition of c to measure distances. In this new system the meter is not defined but measured. One meter is 1/299,792,458 the distance traveled by light in 1 s. For practical purposes it is sufficient to use $c = 3.00 \times 10^8$ m/s.

TABLE 36–1

INDICES OF REFRACTION FOR VARIOUS SUBSTANCES (AT $\lambda = 600$ nm)

Material	*Index of Refraction, n*
Air (1 atm, 0°C)	1.00029
Carbon dioxide (1 atm, 0°C)	1.00045
Ice	1.31
Water (20°C)	1.33
Ethyl alcohol	1.36
Castor oil	1.48
Benzene	1.50
Fused quartz	1.46
Glass (crown)	1.52
Glass (flint)	1.66
Diamond	2.42

The Index of Refraction

Fizeau also found that the speed of light in transparent materials, such as water or glass, is *less* than the speed of light in empty space. We reserve the letter c for the speed of light in empty space, and express the speed of light in a material as

$$v_m = \frac{c}{n}, \tag{36–1}$$

where n is the *index of refraction* of the material, a quantity introduced in Chapter 35. Table 36–1 lists indices of refraction for a variety of materials.

One other aspect of the speed of light in materials deserves special mention: *The index of refraction is a function of wavelength (and frequency).* In glass, violet light, which has a shorter wavelength than red light has, travels more slowly than red light. The dependence of the wave speed on wavelength explains the separation of white light into the colors of a rainbow or a prism, as we shall see in Section 36–5.

The index of refraction depends on the wavelength.

We saw in Chapter 35 [Eqs. (35–16) and (35–17)] that the index of refraction for a material with a dielectric constant κ is

$$n = \sqrt{\kappa}. \tag{36–2}$$

(Here we have assumed that the relative magnetic permeability $\kappa_m \simeq 1$, a good approximation for substances that are transparent to light.) The variation of n with frequency occurs because the dielectric constant can also vary with frequency. We must therefore use κ at the appropriate frequency and not use its static value in Eq. (36–2). Frequency, f, and wavelength, λ, are related by $f\lambda = v$, so in a medium of index of refraction n, we find from Eq. (36–1) that

$$f\lambda = \frac{c}{n}. \tag{36–3}$$

Equation (36–3) shows that it is the product of f and λ that is inversely proportional to n. Note that as radiation passes from one medium to another, *the frequency does not change.* This is easy to understand: Consider two observers on either side of an air–glass interface. Each wave front that passes one observer must pass the other; otherwise there would be a pileup of wave fronts or a disappearance of wave fronts, neither of which can happen. As a consequence, *it is the wavelength of light that changes with the index of refraction,* in such a way that $c/f = n\lambda$ is constant; that is, the same for both media. Thus when light passes between media 1 and 2,

The frequency of a wave remains the same as the wave passes from one medium to another.

$$n_1\lambda_1 = n_2\lambda_2. \tag{36–4}$$

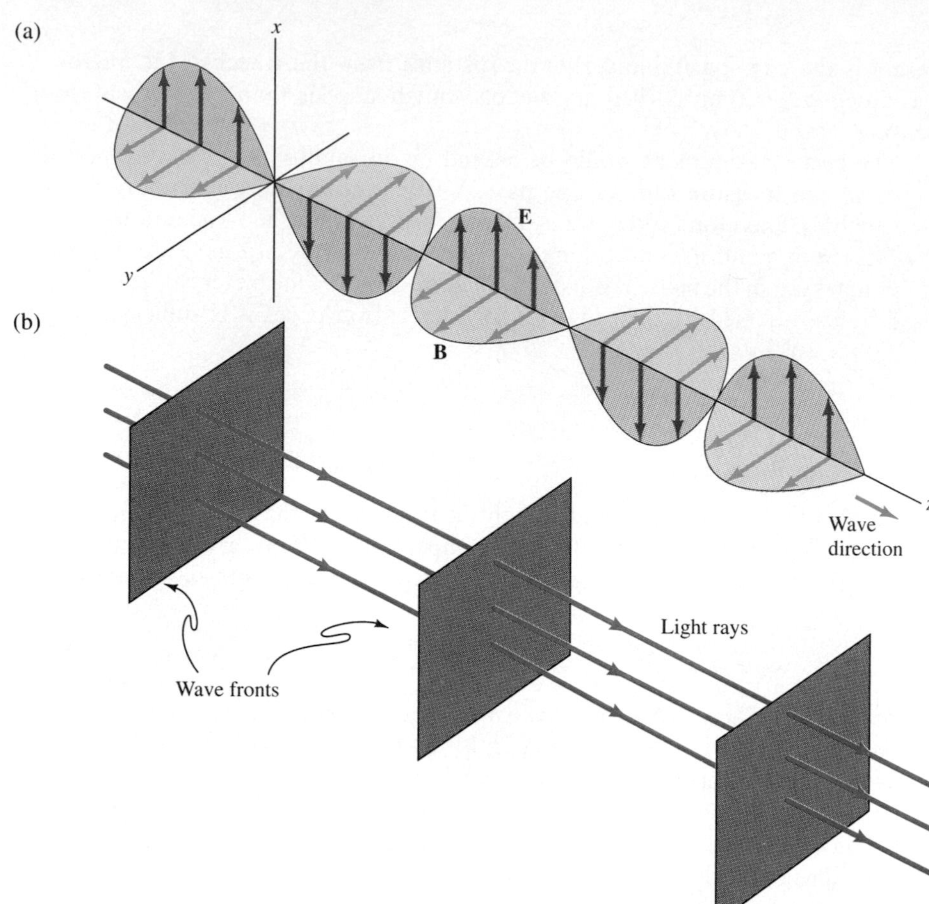

FIGURE 36–3 (a) The electric and magnetic fields of a propagating electromagnetic wave. (b) Wave fronts are chosen arbitrarily at points where the electric and magnetic fields are maximal; the fronts could be chosen at the fields' zero points instead.

36–2 DOES LIGHT PROPAGATE IN STRAIGHT LINES?

One goal for this chapter is to describe a wave model of light. We can rather easily see that a particle model accounts for what might appear to be the most elementary features of the propagation of light, namely, the fact that light seems to travel in straight lines and cast sharp shadows. How can a wave model explain these features, and do these features persist when we look closely?

In Chapter 35 we discussed electromagnetic waves that propagate along the $+z$-axis. The space dependence and time dependence of the electric or magnetic fields are described by a function such as $\cos(kz - \omega t)$. This function has a series of crests and troughs, and one peak occurs at $kz - \omega t = 0$, or

$$z = \frac{\omega t}{k} = ct.$$

Thus the peak propagates at speed c. We called this a plane wave, because all points in the xy-plane defined by a fixed value of z have the same fields, whatever the x- or y-value (Fig. 36–3a). We may thus describe the planes for which the argument $kz - \omega t$ is constant as representing *wave fronts.* Figure 36–3b shows a sequence of wave fronts transverse to the direction of the electromagnetic (light) wave. It is useful to think of these wave fronts as representing a particular set of field values along the wave. For example, the wave fronts could represent the points where the electric field is a maximum or the points where the field is zero

(Fig. 36–3b arbitrarily sets the wave fronts at the points where both fields are maximal). As long as there are no obstacles, the sequence of wave fronts moves at speed c along the original direction of propagation. In other words, the wave moves in a straight line, just as we observe light to move.

Huygens' Principle

We can understand these results from another point of view. Consider a wave front that propagates in the $+z$-direction (Fig. 36–4a). The location of the wave front after a time interval Δt is obtained by viewing every point on the original wave front as a source of light emitting a spherical pulse (or *wavelet*) of radiation (Fig. 36–4b). The radius of the sphere in empty space is $c\,\Delta t$, the distance the light travels in time Δt. In a medium in which the speed of light is c/n, the radius of the sphere is reduced by a factor of n to $(c/n)\,\Delta t$. In the limit that the separation between all the emission points is small, the envelope of all these tiny spheres, *taken in the direction of propagation of the initial wave front*, is the new wave front. In empty space, the wave fronts generated in this way remain planes parallel to the xy-plane. Thus the straight-line propagation of wave fronts is assured. Christian Huygens developed this treatment of waves, which is called **Huygens' principle**.

Huygens' principle can be justified from a detailed study of the behavior of waves in Maxwell's equations, but we shall not do so here. The physical principle is simple if we think of a wave as a propagating disturbance in a medium. One point of a disturbance disturbs the medium next to it, which disturbs the medium next to *it*, and so forth. The disturbance proceeds at a definite speed. Thus the effect of a disturbance at a point is a disturbance that appears later on a sphere that surrounds the point. The only feature of electromagnetic waves that is different is that, as Maxwell's equations show, *a medium is not necessary*.

Let us now see what Huygens' principle gives when a wave front approaches a slot in a wall (Fig. 36–5). When the wave front arrives at the wall, only the part of the wave at the slot can continue to propagate. This part of the wave front generates waves that travel through the slot, with the additional feature that the spherical wavelets emitted near the edges of the slot have no neighbors, and *a wave that spreads away from the slot edges is generated past the slot*. This spreading of the wave around the edges of the slot is known as **diffraction** (to be discussed more thoroughly in Chapter 39). Huygens' principle suggests that the spreading is significant (in terms of the fraction of energy in the bent waves) only if the wavelength is about the same as, or larger than, the size of the slot. If the slot width is much larger than the wavelength, only a small fraction of the energy goes into

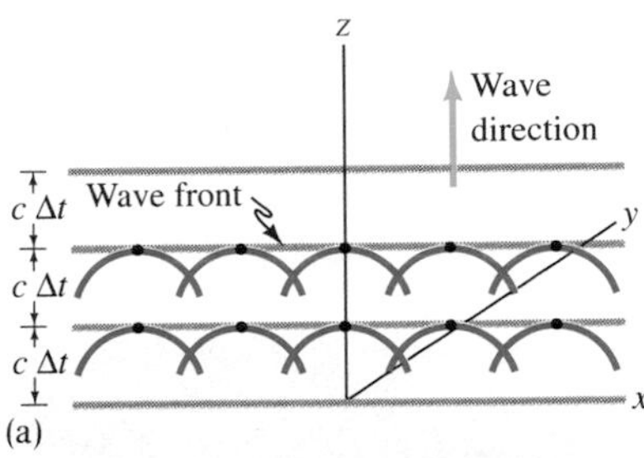

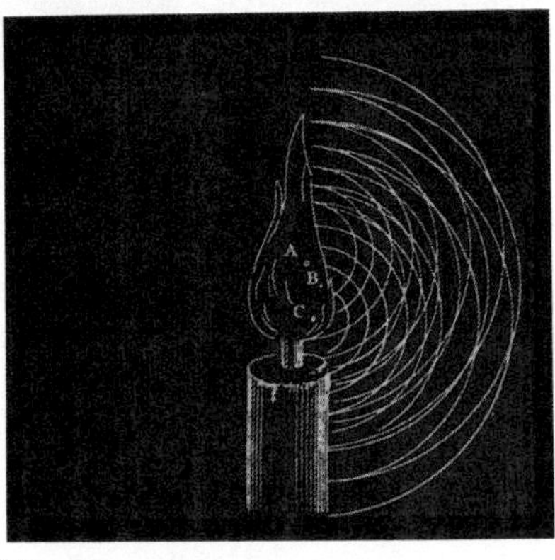

(b)

FIGURE 36–4 (a) Huygens' construction of wave fronts. Wavelets emitted at each point along the wave front add to a new wave front and produce plane waves. (b) Huygens' illustration of wavelets, from his book *Traité de la Lumière* (1678).

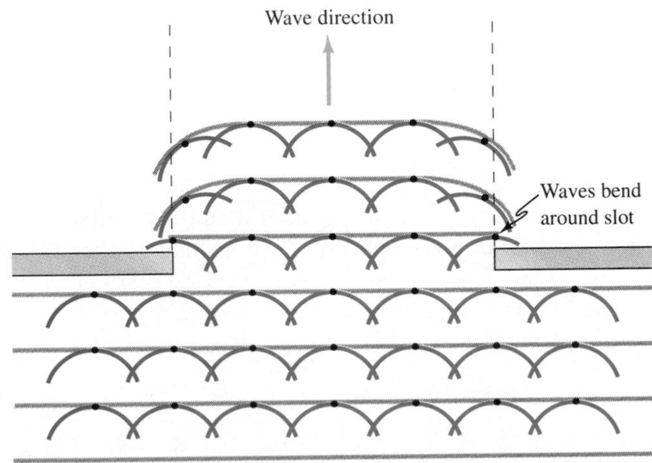

FIGURE 36–5 Huygens' construction of wave fronts that approach and pass through an open slot in a wall. Past the slot, the wave fronts bend around the slot edges.

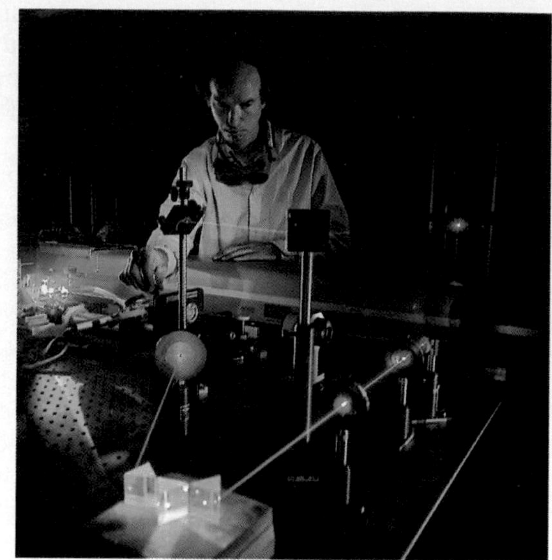

FIGURE 36–6 Rays are lines perpendicular to wave fronts; hence they point along the direction of propagation of light. Here a laser beam is being used to measure impurities in the semiconductor gallium arsenide.

(a)

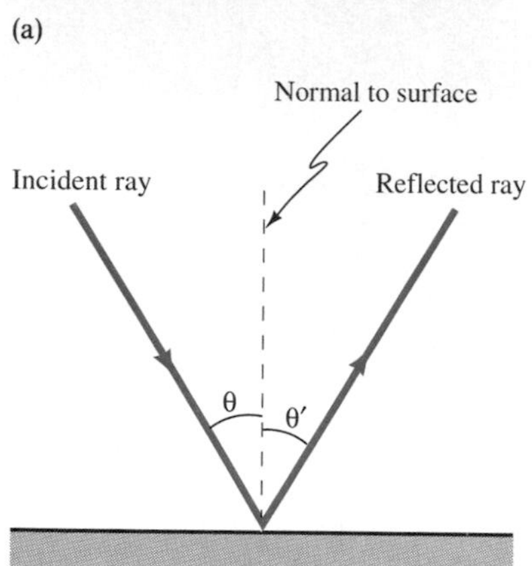

(b)

FIGURE 36–7 (a) A laser beam reflecting off a plane surface. (There are two reflected rays—one reflects off the front surface of the glass plate, the other off the back surface.) (b) The angle of incidence θ equals the angle of reflection θ'.

the bent waves, and it is adequate to view the entire slot as a source of a plane wave front. Because light has wavelengths around 5×10^{-7} m, the slot must not be too much larger than this size for the effect to be significant. This is why the diffraction of light was not observed in Newton's day. A striking illustration of the behavior described above is supplied by the opening photograph in Chapter 15, which shows water waves that pass an island.

Rays

Right now we are not interested in the whole wave front. We may describe the behavior of light purely in terms of the direction of propagation and of any changes in that direction. We can follow this direction by means of **rays**, lines perpendicular to the wave fronts. Light entering a darkened room through a pinhole gives a vivid picture of the propagation of light in the form of rays, as does the light emitted by a laser (Fig. 36–6). We normally do not *see* rays, but a ray can be made visible when light is scattered by, for example, dust particles. We do not see starlight that travels in all directions from a bright star, because space around the star is empty, with no matter to scatter the light. The description of light based on the straight-line propagation of rays is called **geometric optics**.

36–3 REFLECTION AND REFRACTION

Reflection

A light ray **reflects**—it is "thrown back"—when it strikes a smooth surface such as a mirror. The *incident ray* makes an angle θ with a line normal to the surface at the point of reflection. The *reflected ray* lies in the plane formed by the incident ray and the normal (Fig. 36–7a). The angle θ' that the reflected ray makes with the normal obeys the equation known as the **law of reflection**:

The law of reflection

$$\theta' = \theta \tag{36–5}$$

(Fig. 36–7b). The consequences of this law are shown in Fig. 36–8a for the reflec-

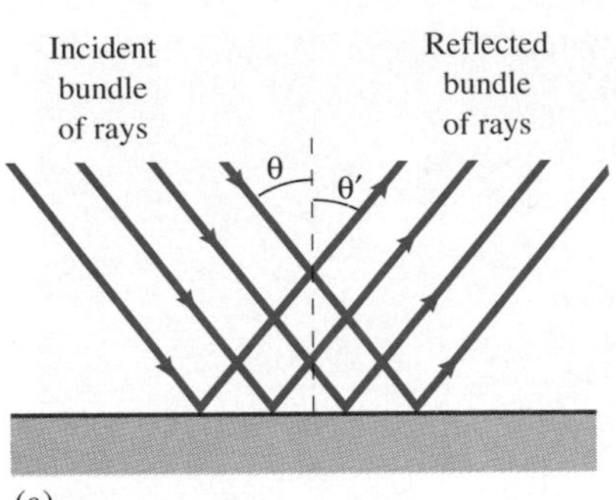

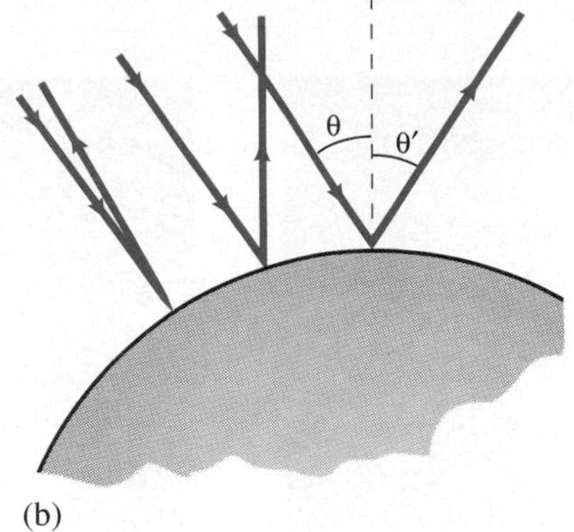

FIGURE 36–8 Reflection of rays from (a) flat and (b) curved surfaces.

tion of a set of parallel incident rays (a *bundle* of rays) from a flat mirror and in Fig. 36–8b for that from a curved surface. For the curved surface, the angles of incidence and reflection are indeed equal, but the direction of the normal to the surface varies from point to point, and the reflected rays radiate in various directions.

EXAMPLE 36–1 A light ray is incident on two mirrors set at 90° with respect to each other. The incident ray is in the plane perpendicular to the two mirrors (Fig. 36–9). Show that after two successive reflections, the outgoing ray will travel along the direction of incidence, no matter what angle the incident ray has with respect to the mirrors.

SOLUTION: We see from Fig. 36–9 that angle ACB, the angle between two normals to surfaces perpendicular to each other, is 90°. Consequently angle BCD is also 90°. Triangles ABC and DBC are congruent, and hence angle CDB, the angle marked ϕ, is equal to θ. Thus the incoming and outgoing rays make the same angle with respect to line AD, and they are parallel.

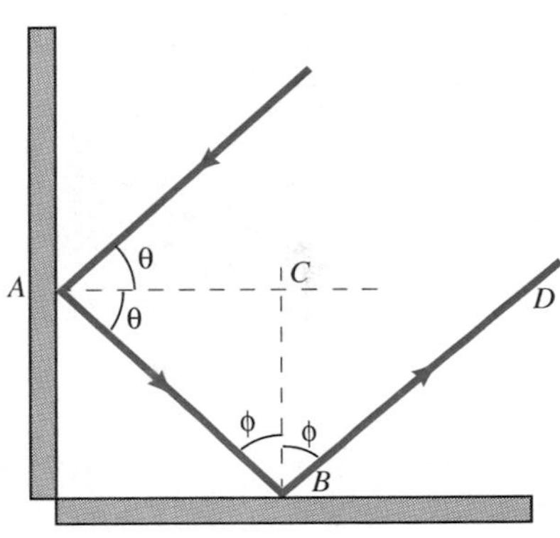

FIGURE 36–9 Example 36–1. The incoming ray that strikes point A eventually emerges along BD.

APPLICATION

Corner Reflectors

Example 36–1 illustrates the principle behind the construction of *corner reflectors*. In a corner reflector, three mirrors are placed together at mutual 90° angles, like the interior corner of a room, or a set of such corners is placed adjacent to one another. Figure B1–1 is a photograph shot into a corner reflector. Geometry of the kind used in Example 36–1 shows that a ray incident at any angle reflects out from a corner reflector along a ray exactly parallel to the incident ray. Corner reflectors are now used routinely as highway markers for the edges of curves and on trucks and trailers, because the light from another vehicle's headlight is automatically reflected back to that vehicle. A corner reflector was placed on the moon by astronauts during the Apollo 11 mission in 1969, and pulses of laser light emitted on earth have been bounced off it and observed (Fig. B1–2). It is possible to measure with great accuracy the time interval between the departure of a pulse and its return. Accordingly it is possible to measure the distance between the laser and the reflector on the moon to an accuracy of 15 cm! One practical application of measurements of such accuracy is in surveying. Laser

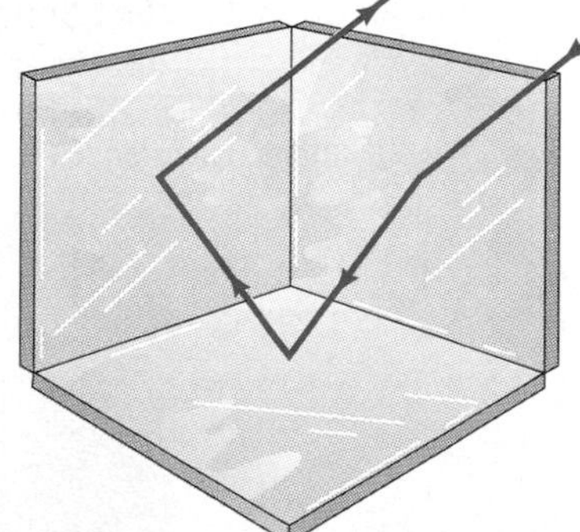

FIGURE B1–1 A corner reflector.

(a)

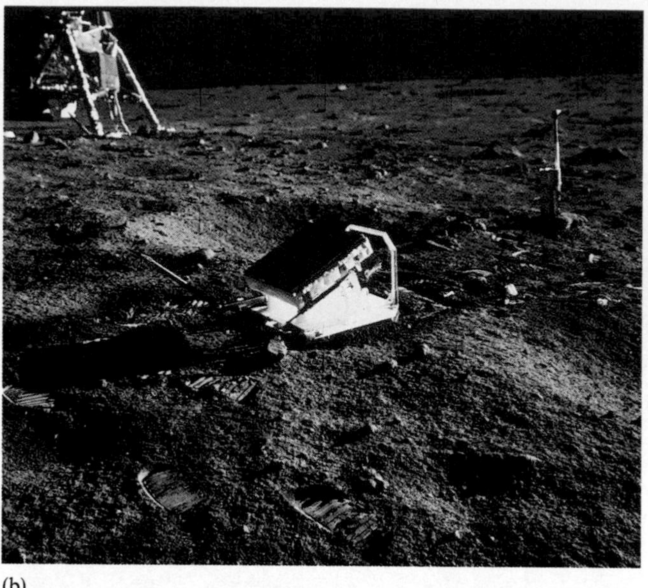

(b)

FIGURE B1–2 (a) A laser beam shot at a corner reflector on the moon from an earth-based observatory. (b) A corner reflector on the moon, used to make precise measurements of the earth–moon distance.

ranging has been used in the study of the relative motion of the two sides of the San Andreas fault in California. No ordinary survey can measure the small changes in the relative positions of points on either side of the fault (on average, up to a few centimeters per year). Corner reflector measurements also show with an accuracy of one part in 10^{11} that the moon and the earth are accelerating toward the sun identically. This is an extremely good test of the equivalence principle (see Chapter 12).

Refraction

When the light forming a ray moves from air to water, or, more generally, from one medium to another, the incident ray is deflected from its original direction at the boundary between the media; the ray is said to undergo **refraction** (Fig. 36–10a). Let the index of refraction of the medium with the incident ray be n_1 and that of the medium with the *refracted ray* be n_2. The angles that the incident and refracted

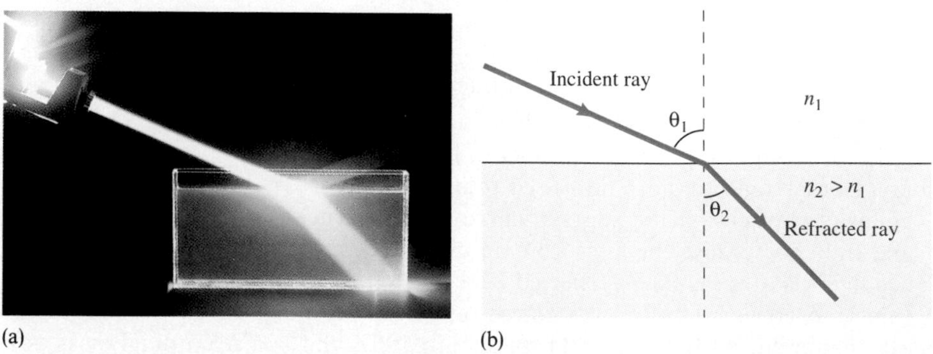

FIGURE 36–10 (a) A beam of light is refracted as it enters a tank of water. (b) Refraction from a medium with index of refraction n_1 into a medium with index of refraction n_2. In this case $n_2 > n_1$, and the refracted ray is bent toward the normal to the boundary surface. If n_2 had been less than n_1, the refracted ray would have bent away from the normal.

rays make with the line normal to the boundary between the media are θ_1 and θ_2, respectively (Fig. 36–10b). Then

Snell's law of refraction

$$n_1 \sin \theta_1 = n_2 \sin \theta_2. \tag{36–6}$$

This result, found by Willebrord Snell in 1621, is known as **Snell's law**. The index of refraction of air is very close to unity, so the angle of the refracted ray θ_2 at the interface for light that passes from air into a medium with index of refraction n is given by

$$\sin \theta_1 = n \sin \theta_2. \tag{36–7}$$

Because n is generally larger than one, it follows that $\theta_2 < \theta_1$; that is, *the light is bent toward the normal to the boundary surface.* Equation (36–6) also shows that when light enters a medium with a lower index of refraction, such as when a ray of light travels from water to air, the ray is bent farther away from the line normal to the boundary (Fig. 36–11).

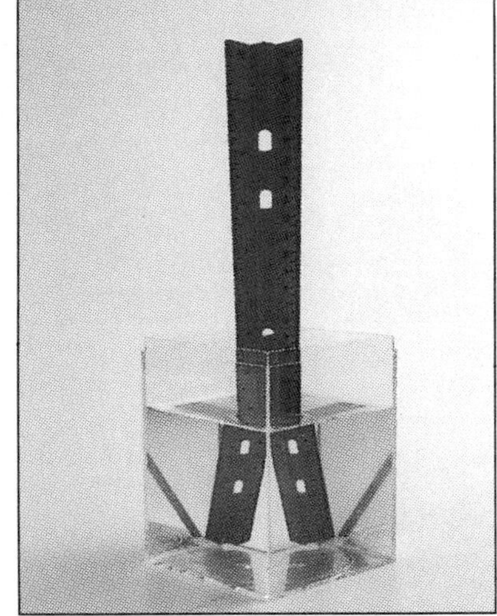

FIGURE 36–11 Water has a higher index of refraction than air does, so the immersed part of this ruler seems to bend away from the normal to the water surface.

EXAMPLE 36–2 Consider light that is refracted by a prism shaped like an equilateral triangle (Fig. 36–12). The incident ray is parallel with the prism's base. What is the total deflection of the ray, given that the index of refraction of the prism material is 1.50?

SOLUTION: The light passes through two surfaces, and we must apply Snell's law at each interface. Figure 36–12 shows that the angle the incident ray makes with the line normal to the first surface is $\theta = 30°$. The angle ϕ that the refracted ray makes with the line normal to the first surface is given by Eq. (36–7) as

$$\sin \theta = n \sin \phi.$$

When $n = 1.50$ and $\theta = 30°$, this expression gives $\phi = 19.5°$.

Figure 36–12 also shows that the refracted ray acts as an incident ray at the second surface. If the angle that the refracted ray makes with the line normal to the second surface is denoted by ψ, then $\phi + \psi + 120° = 180°$, or $\psi = 60° - \phi = 60° - 19.5° = 40.5°$. The angle θ' that the second refracted ray makes with the line normal to the second surface is given by

$$\sin \theta' = n \sin \psi = 1.50 \sin(40.5°) = 0.97.$$

Thus $\theta' = \sin^{-1}(0.97) = 77°$, and from Fig. 36–12 we see that the angle that the outgoing light ray makes with the base of the prism is $\theta' - \theta = 77° - 30° = 47°$.

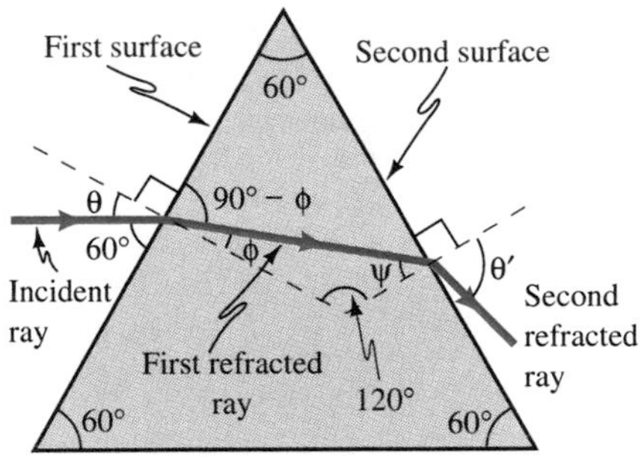

FIGURE 36–12 Example 36–2. Geometric construction of the path followed by a ray incident on a prism.

Refraction is responsible for some curious optical effects. For example, foreshortening of an object in water is a familiar phenomenon (Fig. 36–13a). If you stand by a swimming pool, a person standing in the pool will seem to you to have short legs (Fig. 36–13b). An observer assigns the source of the light ray from the feet of the person in the pool to a point that lies on a straight line along the direction at which the ray enters the observer's eye. Thus the observer assigns the point P to be the source of the ray from the feet.

Energy in Reflection and Refraction

Refraction is generally accompanied by reflection. The incident ray carries electromagnetic energy proportional to the square of the incident electric field $\mathbf{E}_{in}$. At the boundary between media, this energy is divided into energy that is refracted and energy that is reflected, such that the total energy is conserved. It follows from electromagnetic theory that when light is perpendicularly incident on a surface that

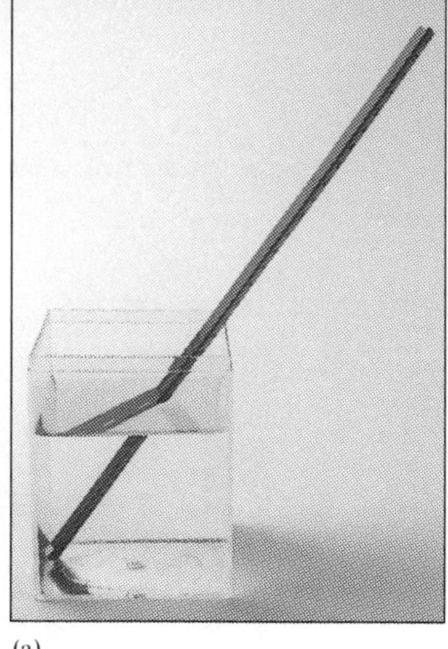

(a)

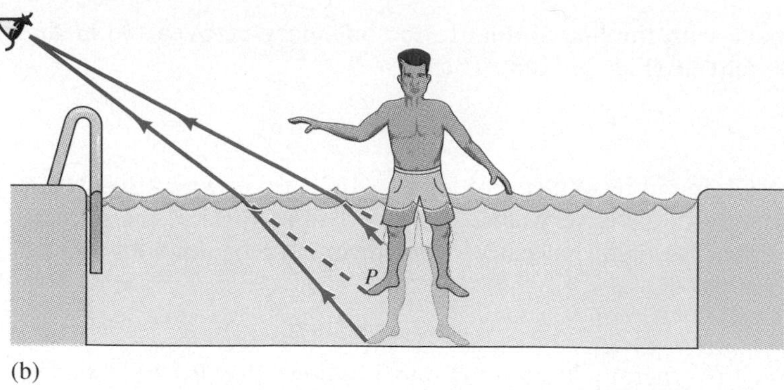

(b)

FIGURE 36–13 (a) From this angle you can see the true depth of the immersed ruler as well as the foreshortened depth. (b) A person standing in a swimming pool appears to have foreshortened legs, as seen by an observer outside the pool.

separates a medium of index of refraction n_1 from a medium of index of refraction n_2, the intensity of the reflected light, I_r, is related to the incident intensity, I_0, by

$$\frac{I_r}{I_0} = \frac{(n_2 - n_1)^2}{(n_2 + n_1)^2}. \tag{36–8}$$

Thus for light perpendicularly incident from air ($n = 1.0$) into glass ($n = 1.5$), only 4 percent of the incident light is reflected. The intensity of the reflected light varies with the angle of incidence.

Total Internal Reflection

For some incident angles, all the incident energy is contained in the reflected ray. This situation, known as **total internal reflection**, can occur only when light travels from a medium with a larger index of refraction toward a medium with a smaller index of refraction, such as when light passes from water toward air. We can understand this phenomenon by simple geometry.

Consider a light ray incident from a medium with an index of refraction n_1 to a medium with an index of refraction n_2, this time with $n_1 > n_2$. Snell's law, Eq. (36–6), may be written in the form $\sin\theta_2 = (n_1/n_2)\sin\theta_1$. Because the factor n_1/n_2 is larger than unity, θ_2 reaches 90° before θ_1 does as θ_1 increases. Figure 36–14 shows what happens for various values of θ_1. When $\theta_2 = 90°$, the ray in medium

FIGURE 36–14 (a) Various rays traveling from a medium with a larger index of refraction (water) to a medium with a smaller index of refraction (air). When the incident angle is θ_c, there is total internal reflection. (b) Refraction and total internal reflection off the air-water interface in a water tank.

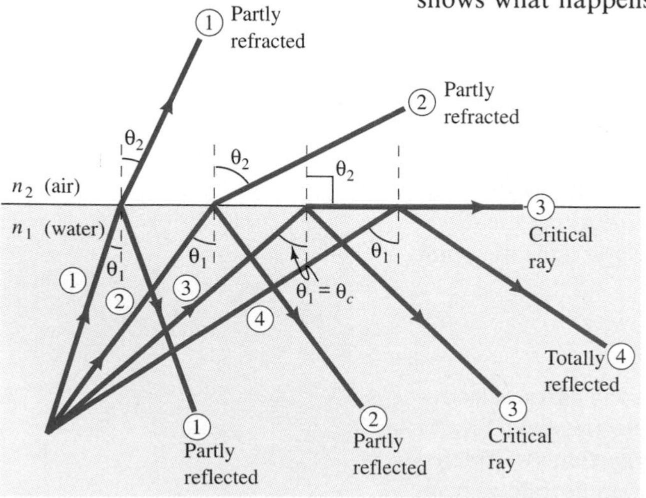

(a)

(b)

2 skims along the interface of the two media. This occurs when θ_1 reaches a critical angle θ_c such that $(n_1/n_2) \sin \theta_c = \sin 90^\circ = 1$, or

$$\sin \theta_c = \frac{n_2}{n_1}. \tag{36-9}$$

The critical angle for total internal reflection

When θ_1 exceeds θ_c, there is no angle θ_2 that can satisfy Snell's law. The electromagnetic energy carried by the incident ray must go somewhere, and the ray is reflected. There is no diminution of the intensity of the reflected ray; the reflection is total.

EXAMPLE 36–3 A swimmer is in a deep pool with her eyes a horizontal distance $R = 1.5$ m from the edge (Fig. 36–15). How far below the surface are her eyes if she is just able to see the full height of a lifeguard who is standing on the pool's edge? The index of refraction of water is 1.33.

SOLUTION: We are interested in a ray from the lifeguard's feet to the swimmer's eyes. (If the swimmer can see the lifeguard's feet, the rest of the lifeguard will be visible.) When the swimmer's eyes are at the deepest point they can be in the pool and she can still see the feet, the reversed ray—that is, the ray from the swimmer's eye to the base of the feet—must be at the critical angle θ_c (Fig. 36–15). We have from Eq. (36–9)

$$\sin \theta_c = \frac{n_{\text{air}}}{n_{\text{water}}} = \frac{1}{n_{\text{water}}} = \frac{R}{\sqrt{R^2 + D^2}},$$

where D is the greatest possible depth of the swimmer's eyes. We can solve this equation for D:

$$D^2 = R^2(n_{\text{water}}^2 - 1) = (1.5 \text{ m})^2(1.33^2 - 1) = 1.7 \text{ m}^2;$$

$$D = 1.3 \text{ m}.$$

Fiber optics represents one of the most important technological applications of total internal reflection. The principle behind this technique of conducting light from one place to another is straightforward: A transparent plastic fiber will serve as a conductor of light if any ray inside the fiber undergoes total internal reflection upon striking the side of the fiber (Fig. 36–16a). (We limit our discussion to fibers whose diameter is large compared with the wavelength conducted—that is, larger

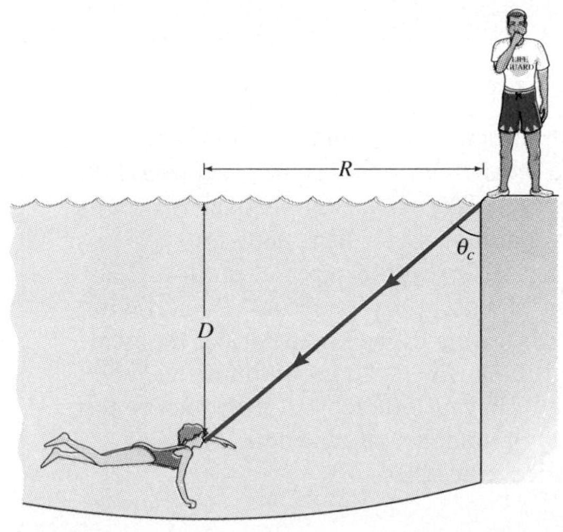

FIGURE 36–15 Example 36–3. When the swimmer's eyes are the deepest they can be in the pool and she can still see the lifeguard's feet, the ray between her eyes and his feet is at the critical angle.

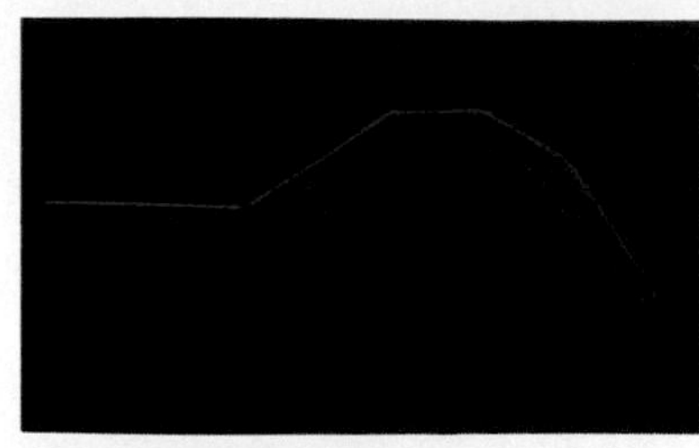

(a)

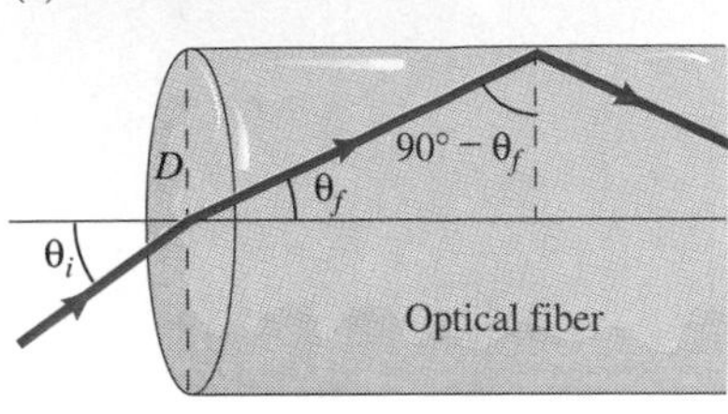

(b)

FIGURE 36–16 (a) Total internal reflection in an optical fiber. (b) Detailed construction of ray angles.

than about 600 nm—so we need not be concerned with complications due to the wave character of light.) Fibers with diameters in the range of 50 μm (roughly the diameter of a human hair) are frequently used. Figure 36–16b shows a ray in air ($n = 1$) entering a cylinder of diameter D at an angle θ_i with the axis of the cylinder. If n_f is the index of refraction of the fiber, then the angle that the ray makes with the axis inside the fiber is θ_f, where $\sin\theta_f = \sin\theta_i/n_f$. This ray will strike the wall of the cylinder at an angle $(90^\circ - \theta_f)$ with the normal to the wall. There will be total internal reflection if $n_f \sin(90^\circ - \theta_f) > 1$; that is, if $n_f \cos\theta_f > 1$. We have

$$n_f \cos\theta_f = n_f\sqrt{1 - \sin^2\theta_f} = n_f\sqrt{1 - \frac{\sin^2\theta_i}{n_f^2}} = \sqrt{n_f^2 - \sin^2\theta_i} > 1.$$

Because $\sin^2\theta_i \leq 1$, we have

$$\sqrt{n_f^2 - \sin^2\theta_i} \geq \sqrt{n_f^2 - 1}.$$

Thus we automatically satisfy the condition for total internal reflection, $n_f \cos\theta_f > 1$, if

$$\sqrt{n_f^2 - 1} > 1. \tag{36–10}$$

Because the largest value of $\sin\theta_i$ is 1 (the light first enters the cylinder from the end), Eq. (36–10) is a condition for internal reflection for *all* of the light that enters the fiber. Equation (36–10) is satisfied for any material with $n_f > \sqrt{2}$. A typical fiber has an index of refraction of 1.62, larger than the critical value. Note that once a ray is in the fiber, it remains inside *even if the fiber curves.*

APPLICATION

Fiber Optics

Optical fibers now play an important role in technology (Fig. B2–1). The ideal situation outlined in the text is modified in a real fiber. The total internal reflection is somewhat less than total if there are impurities such as moisture, dust, or oil on the surface, because electromagnetic energy can leak across the thin "barrier" formed by the air layer between the fiber and the impurity. This problem is controlled largely by coating each fiber with a transparent covering whose index of refraction is lower than that of the fiber. This procedure, known as *cladding*, has allowed for the practical implementation of fiber optics. Light may be reflected thousands of times per meter, and it is therefore important to have no leakage of light. In general the light intensity is attenuated (lessened) as the ray propagates in a medium, and it is important to reduce that attenuation as much as possible. This has been achieved by making the fiber from fused quartz, a highly transparent material, and purifying it to remove all traces of water. For devices such as *fiberscopes*, which are used for internal examinations of the human body, the distance covered by the light does not exceed several meters, and the refinements discussed above are not so critical (Fig. B2–2). For communication by light pulses in optical fibers, much longer distances come into play. For the trans-Atlantic cable TAT-8, which can carry 40,000 conversations over two pairs of glass fibers simultaneously, it is necessary to boost the signal every 50 km with a repeater station. This is still much less expensive than systems of metal wire, which require boosting every kilometer. In the near future, much thinner fibers, 10 μm in diameter, will carry laser light. Because there is so little energy loss in these fibers, the number of repeaters (boosting elements) needed is reduced.

FIGURE B2–1 Optical fibers.

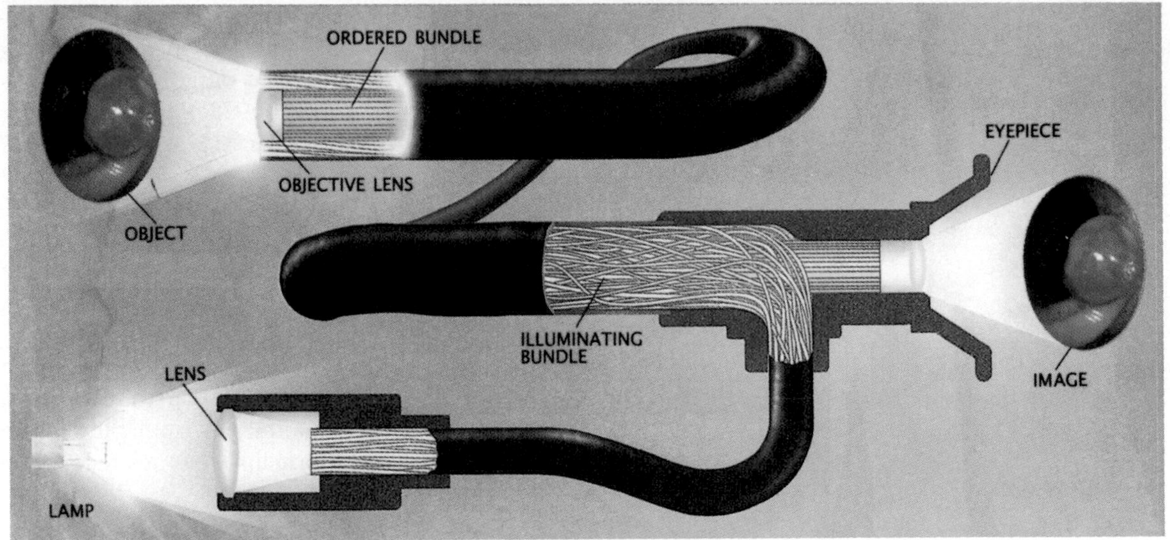

FIGURE B2–2 A fiberscope contains two bundles of optical fibers: one set to carry light from a source to the area to be viewed, and another to carry light from the viewing area to the eye. Here a fiberscope allows us to see a polyp in the stomach.

*36–4 REFLECTION AND REFRACTION FROM FERMAT'S PRINCIPLE

As an example of Huygens' principle in operation, consider the law of reflection [Eq. (36–5)]. Figure 36–17a shows a sequence of wave fronts that approach a mirror. In Fig. 36–17b, point C_2 is the center of a reflected spherical wave, one of

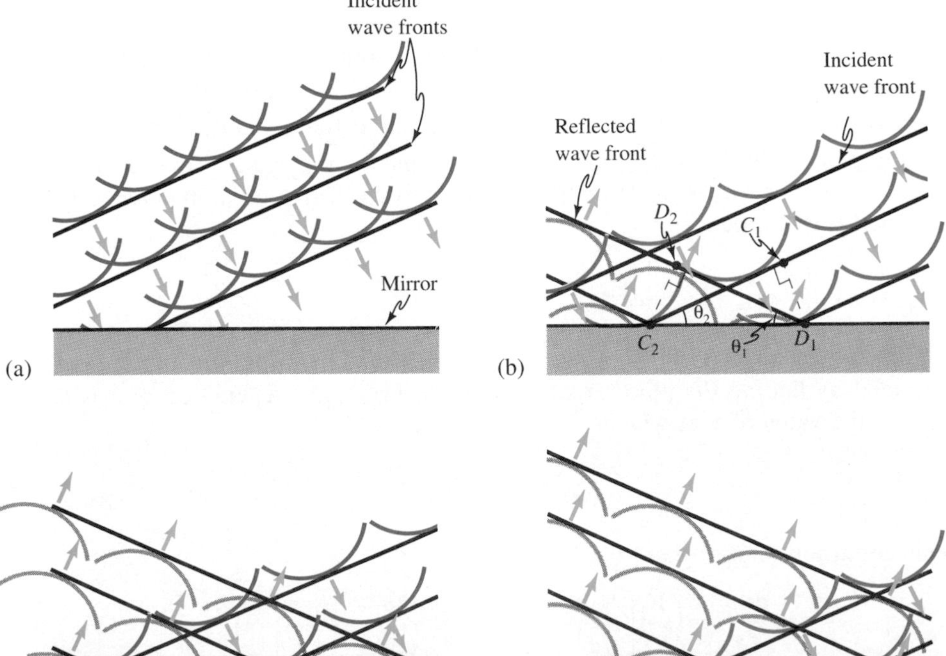

FIGURE 36–17 (a) Incident wave fronts approaching a plane mirror. (b) Later in the sequence, wave fronts reflect. (Note that fewer wavelets are shown, for clarity.) The fronts are generated by Huygens' construction. The relation $C_2D_2 = C_1D_1$ leads by geometrical reasoning to equal angles of incidence and reflection. (c) Later still. (d) Even later, when most of the wave fronts are reflected.

many along the mirror. An outgoing (reflected) wave front—here, the line tangential to point D_2 of the semicircle centered on point C_2—forms. The distance the wave travels in time Δt is the same for incoming and outgoing waves, so a simple geometrical argument yields the result described by Eq. (36–5), namely, that the angle of reflection equals the angle of incidence. Figures 36–17c and 36–17d show later parts of the sequence.

Although Snell's law may also be obtained by an application of Huygens' principle, the geometry is somewhat more complicated. The bending of the wave front is associated with the slowing down of the light waves in the medium. The bending can be visualized by analogy with the direction change of a wide column of soldiers who march at an angle toward a sidewalk, with orders given that as soon as a soldier steps on the sidewalk, he or she must walk slower without changing the distance between soldiers in each row. Rather than go through the geometrical exercise of showing that Snell's law follows from the Huygens construction, we shall demonstrate it from the principle enunciated by Pierre de Fermat in 1657. **Fermat's principle** states that

Fermat's principle

The path of a ray of light between two points is the path that minimizes the travel time.

A similar principle for reflection was enunciated by Hero of Alexandria sometime between 150 B.C. and A.D. 200, and Fermat used the notion of time of traversal of light even before it had been established that light travels at finite speed. Both the straight-line propagation of light in a single medium and the law of reflection can also be derived from Fermat's principle.

To derive Snell's law from Fermat's principle, let us consider a point A in medium 1 with index of refraction n_1, and a point B in medium 2 with index of refraction n_2 (Fig. 36–18a). We want to find the path between points A and B that takes a ray of light the least amount of time to travel. We choose A to be a distance d above the boundary and B a distance d below the boundary, and we choose the horizontal distance between A and B to be $2b$. The straight line connecting A to B crosses the boundary at a distance b from the normal dropped from A onto the boundary, but because the indices of refraction are different, the ray's path crosses the boundary at a point P. Figure 36–18a shows that the distance from A to the intersection point P is $\sqrt{d^2 + x^2}$, and the distance from the intersection point P to B is $\sqrt{d^2 + (2b - x)^2}$. The time for the ray to travel a distance D in a medium of index of refraction n is given by $t = D/v = D/(c/n) = nD/c$. Thus the total travel time is

$$t_{AB} = t_{AP} + t_{PB} = \frac{n_1\sqrt{d^2 + x^2} + n_2\sqrt{d^2 + (2b - x)^2}}{c}. \qquad (36\text{–}11)$$

Figure 36–18b is a graph of t_{AB} as a function of x. The minimum travel time is obtained by finding the place at which the slope of t_{AB} as a function of x is flat; that is, the value of x at which

$$\frac{dt_{AB}}{dx} = 0.$$

This condition implies that

$$\frac{dt_{AB}}{dx} = \left(\frac{1}{c}\right)\left[\frac{n_1 x}{\sqrt{d^2 + x^2}} - \frac{n_2(2b - x)}{\sqrt{d^2 + (2b - x)^2}}\right] = 0. \qquad (36\text{–}12)$$

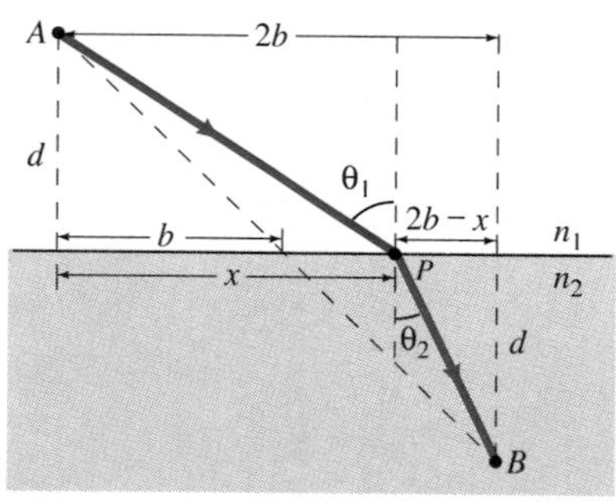

(a)

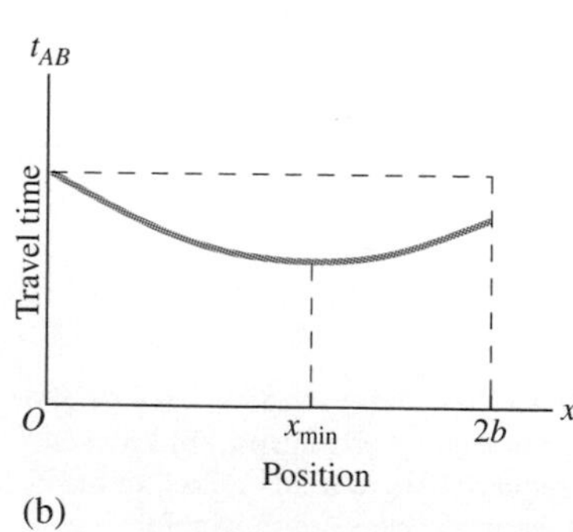

(b)

FIGURE 36–18 (a) Geometry for proving Snell's law by Fermat's principle. (b) The travel time t_{AB} for the ray as a function of x.

Now observe from Fig. 36–18a that

$$\frac{x}{\sqrt{d^2 + x^2}} = \sin\theta_1 \tag{36–13a}$$

and that

$$\frac{2b - x}{\sqrt{d^2 + (2b - x)^2}} = \sin\theta_2, \tag{36–13b}$$

where θ_1 and θ_2 are the angles the two rays make with respect to the normals in their respective media. Thus Eq. (36–12) may be rewritten as

$$n_1 \sin\theta_1 = n_2 \sin\theta_2, \tag{36–14}$$

which is just Snell's law.

It would take us too far afield to derive Fermat's principle directly from Maxwell's equations, but the principle can indeed be shown to follow from them. Fermat's principle is a very general statement about the paths that light rays take, and as Example 36–4 shows, principles such as Fermat's principle, more generally termed minimum principles, can apply in surprising circumstances. Minimum principles form an important part of today's science.

EXAMPLE 36–4 A boy located at point A in Fig. 36–19a spots a ball at point B. Point A is in tall grass, in which the boy can run at 1.1 m/s, and point B is in short grass, in which the boy can run at 2.2 m/s. The whole area is flat. At what point x should he cross the boundary between the grasses so that he retrieves the ball as quickly as possible?

SOLUTION: We let x be arbitrary for the moment and choose it so that the travel time for the child is minimized. Let v_1 be the speed in tall grass and v_2 be the speed in short grass. We have $v_2 = 2v_1$. For arbitrary x, the distance to be covered in tall grass is $L_1 = \sqrt{d^2 + x^2}$, and the time spent there is $t_1 = L_1/v_1$. The distance covered in the short grass is $L_2 = \sqrt{d^2 + (d - x)^2}$, and the time spent there is $t_2 = L_2/v_2 = L_2/2v_1$. Thus the total time spent is

$$t_{\text{total}} = t_1 + t_2 = \frac{L_1}{v_1} + \frac{L_2}{2v_1} = \frac{\sqrt{d^2 + x^2}}{v_1} + \frac{\sqrt{d^2 + (d - x)^2}}{2v_1}.$$

We now want to minimize this time by varying x. This can be done by setting to zero the derivative of t_{total} with respect to x, but the algebra is fairly complicated. Alternatively, we can plot t_{total} as a function of x; in doing so we must use the numerical values of v_1, v_2, and d (Fig. 36–19b). There is indeed a minimum, at about $x = 1.5$ m. Compare this to the 2.5-*m* value x would take if the child were to run in a straight line. Even a small child would intuitively choose a point that might involve a longer distance in the short grass but that nevertheless cuts down the time spent running.

The path followed looks like the path a light ray would take in traveling from a medium with a smaller value for the speed of light to a medium with a larger value for the speed of light. The two situations are in fact analogous!

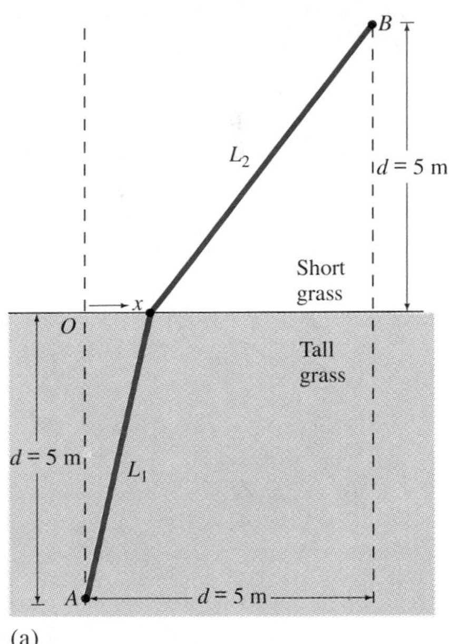

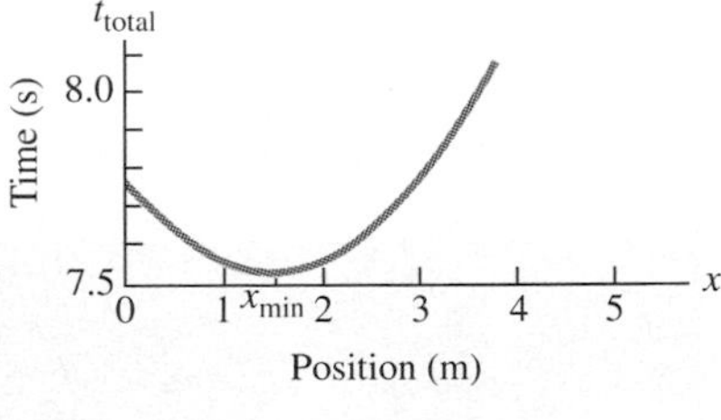

FIGURE 36–19 (a) Example 36–4. (b) The travel time as a function of where the child crosses the border between the tall and short grass.

TABLE 36–2

INDEX OF REFRACTION OF GLASS AS A FUNCTION OF WAVELENGTH

Wavelength in Air (nm)	$\omega^2 = (2\pi c/\lambda)^2$ (rad^2/s^2 $\times$ 10^{31})	n	*Color*
361	2.72	1.539	Near ultraviolet
434	1.89	1.528	Blue
486	1.50	1.523	Blue–green
589	1.02	1.517	Yellow
656	0.82	1.514	Orange
768	0.60	1.511	Red
1200	0.25	1.505	Infrared

FIGURE 36–20 Newton experimenting on the transmission of light through a prism.

36–5 DISPERSION

We shall now explore another property of the index of refraction, a property with some truly spectacular consequences: *The index of refraction depends in general on the wavelength (or color) of the light being transmitted.* Table 36–2 shows how n varies with wavelength for glass, near and including wavelengths of visible light. Different wavelengths are accordingly refracted to different degrees. In this way white light, which consists of a mixture of different wavelengths, can be separated into its constituent colors of the rainbow. Newton was the first to make careful measurements of the transmission of light through a prism (Fig. 36–20). He found that when a pencil-thin beam of sunlight (produced by a small opening in a window shutter) is incident on a prism, the light is split into the colors red, orange, yellow, green, blue, and violet, which make up a rainbow (Fig. 36–21). When he projected each of the colored beams on a second prism, they refracted, but there was no further change in color. Newton also found that when the separated beam passed through a second prism inverted with respect to the first one, the light emerged as white light. The dependence of refraction on the wavelength of light is called **dispersion**.

Dispersion

Rainbows and the Blue Sky

The colors of a rainbow result from dispersion in the scattering of light from individual water droplets in the air. When sunlight falls on a raindrop, light is reflected once before it leaves the drop. Many paths are possible; two are shown in Fig. 36–22a. The geometry is such that no ray can emerge after one reflection at an angle *steeper* than about 42°. Thus when the sun is low to and behind an observer on the ground, no raindrop high in the sky sends light back to the observer.

(a)

(b)

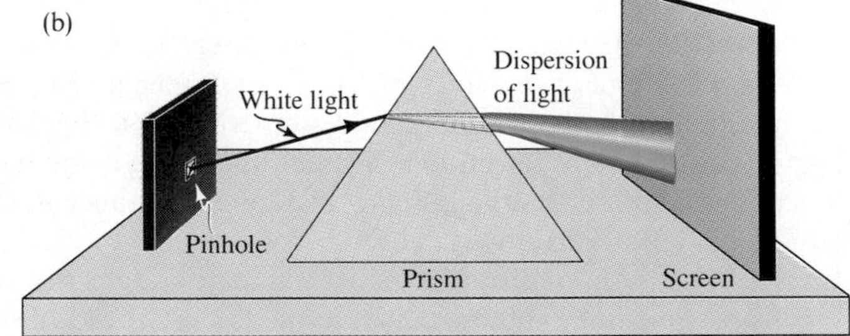

FIGURE 36–21 (a) The dispersion of light by a prism. (b) White light enters the prism, and light of different wavelengths follows different paths. The result is a beam separated by color.

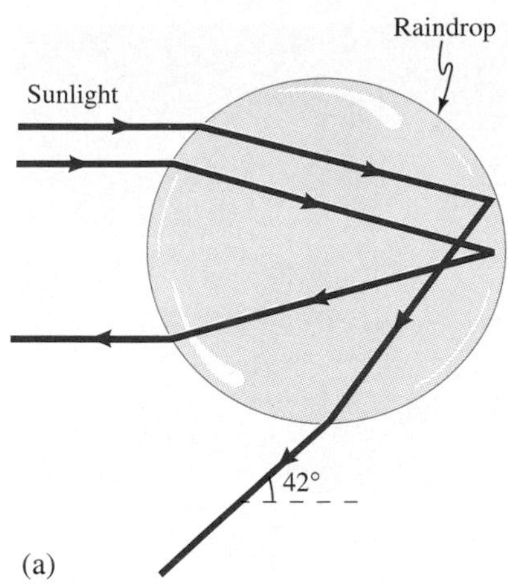

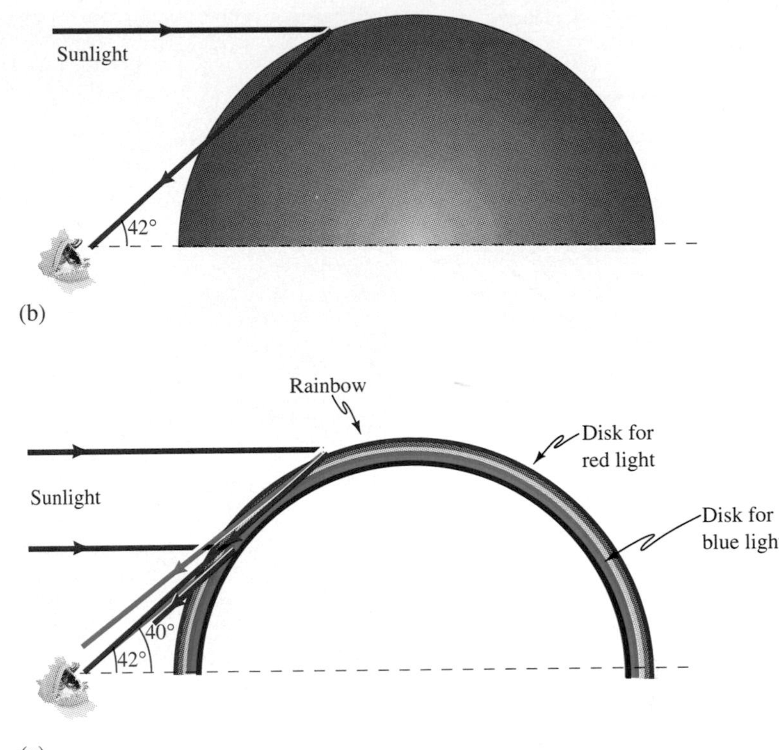

FIGURE 36–22 (a) When sunlight enters a raindrop from the horizontal and exits after one reflection, no light exits at an angle steeper than about 42°. (b) As a result, light comes back to an observer from all the raindrops that lie within a cone with an angle of about 42°, as seen by the observer. (c) Sunlight is a variety of colors. Due to dispersion, the disks that fit into the cones for different colors are of slightly different sizes, with red forming the largest cone and violet, the smallest.

Equivalently, when the sun is behind the viewer, only drops that lie within a cone with an opening angle of about 42° reflect sunlight back to the observer's eye (Fig. 36–22b); moreover, *all* the drops in this cone reflect light to the observer. (We shall refer to a disk that fits into the cone, because the depth of the cone is irrelevant.) One other feature of the disk is that light is reflected most strongly from raindrops at the edge, around 42°. In our discussion thus far, dispersion has played no role. The effect of dispersion is to make the angle of the outer radius of the disk slightly different for different colors. As Fig. 36–22c indicates, the disk for red light is larger than the disk for blue light. Because the intensity of the light in the disk is strongest at the edges, what we see is a red ring outside of a blue ring (with other colors placed accordingly). Inside the rainbow, all the disks overlap, giving white light. A *secondary rainbow* can be produced when there are two internal reflections within the raindrops (Fig. 36–23). The order of the colored disks produced by the raindrops will now be reversed, with red light at the bottom and blue light at the top of the secondary rainbow. Figure 36–24a illustrates how an observer sees rainbows and how the pattern of dispersion leads to the color inversion of a secondary rainbow, compared to a primary rainbow. The light is brightest below the primary rainbow and above the secondary rainbow because the disks overlap, and it is relatively darker between the two rainbows (Fig. 36–24b).

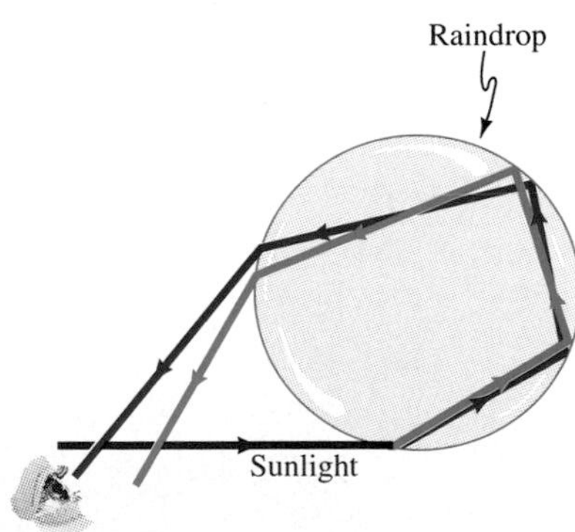

FIGURE 36–23 The light that reaches the eye from a secondary rainbow has undergone two internal reflections in a set of raindrops. Light of shorter wavelengths (blue light) emerges at a steeper angle than light of longer wavelengths (red light) does, in contrast to the light that undergoes only one internal reflection, which forms the primary rainbow.

Dispersion is a widespread phenomenon. Like refraction, the scattering of light by matter has a frequency (or wavelength) dependence. It was shown by Lord Rayleigh in 1872 that the fraction of incident light scattered by air molecules varies as f^4 for light in the visible range. This has immediate observable consequences. As the sun sets, the rays of light pass through more and more atmosphere, and more and more of the high-frequency (low-wavelength) components are scattered. The violet components, then blue, then green, and so forth, of the white light are successively scattered. The sun's color changes from white to yellow to orange and finally to red as the higher frequencies are scattered away from the observer.

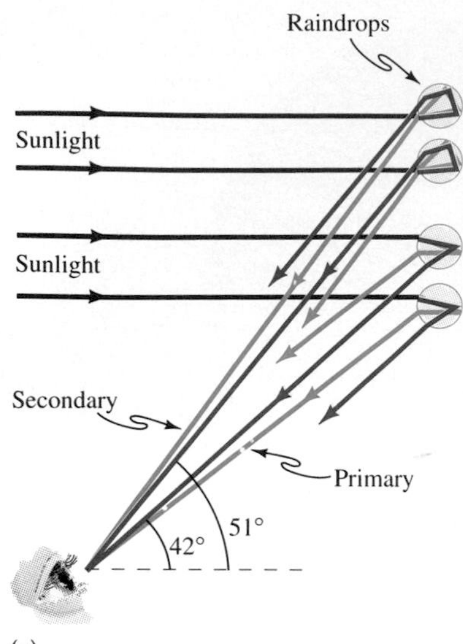

(a)

(b)

FIGURE 36–24 (a) When the eye sees a region of sky that contains raindrops illuminated by sunlight, the light reflected from individual drops forms a primary rainbow (one internal reflection within each drop) and a secondary rainbow (two internal reflections). (b) A double rainbow over the Very Large Array (VLA), a radio telescope observatory in Socorro, New Mexico: The brighter, primary rainbow is on the bottom. The order of colors is reversed in the two rainbows due to the extra reflection, which produces the fainter secondary rainbow. The disks overlap, so it is relatively brighter below the primary rainbow and above the secondary rainbow, but darker between them.

The frequency dependence of the amount of light scattered is also responsible for the fact that the sky looks blue. Because blue light has a higher frequency than red, the blue component of sunlight is scattered into our eyes by the atmosphere. Above the atmosphere, where there are no molecules to scatter the light, astronauts see a black sky (except, of course, for the sun and the stars).

If the sky is blue, why are clouds white? The f^4 law applies to the scattering of light by objects much smaller than the wavelength of the light. Thus it applies to the scattering of light by air molecules. Dust particles and water droplets that make up a cloud are comparable in size to or larger than the wavelength of the light impinging on them, and they act like mirrors. For surfaces larger than the wavelength, reflection rather than scattering occurs, and the geometrical laws of reflection hold for scattering off the droplets. But because the many droplets in a cloud form an irregular surface, we do not get images of the kind we are used to seeing in mirrors. Instead, we get *diffuse* reflection, and the clouds look white or gray (which is just a less intense white).

The Atomic Theory of Dispersion

Dispersion occurs because of the atomic structure of dielectric media. Although the atoms in such media are electrically neutral, they nevertheless have a charge distribution. The positive charges are localized at the center of the atoms, and the negative charges, in the form of electrons, are distributed on the outside, over a region with linear dimensions of 0.1 nm. For our purposes, we shall think of each atom as a single electron (of mass m) that oscillates about a positive ion as though the electron were bound to the ion by a spring. If the spring constant is k, then the angular frequency of oscillation (the natural frequency) is ω_0, given by $\omega_0^2 = k/m$. If no other forces are acting, the motion of the electron along the z-axis would then be of the form

$$z = A \cos(\omega_0 t), \qquad (36\text{–}15)$$

where A is the amplitude of motion.

Now suppose that a plane electromagnetic wave oscillating with angular frequency ω is incident on an atom, with the electric field oriented in the z-direction. The electric force on an electron in the atom is then oscillatory with frequency ω. The situation is that of a driven harmonic oscillator. The motion of the electron is oscillatory with the driving frequency ω. The amplitude exhibits resonance, meaning that it becomes large when $\omega_0 \simeq \omega$. The motion is thus of the form

The driven harmonic oscillator and the associated resonance phenomenon are treated in Section 13–8.

$$z \propto \frac{1}{\omega_0^2 - \omega^2} \cos(\omega t). \tag{36–16}$$

For materials such as water and glass, ω_0 is on the order of 5 to 6 times larger than the characteristic angular frequencies of visible light.

An accelerating charge (the electron) *radiates* electromagnetic energy, and the intensity of the radiation is proportional to the average of the acceleration squared. From Eq. (36–16) the average acceleration is proportional to

$$\frac{d^2z}{dt^2} \propto \frac{\omega^2}{\omega_0^2 - \omega^2} \langle \cos^2(\omega t) \rangle \propto \frac{\omega^2}{\omega_0^2}.$$

In the last step we have used the fact that, for visible light in these materials, $\omega_0^2 \gg \omega^2$. The intensity, I, of the radiation is proportional to the acceleration squared. Thus I varies as ω^4, or, equivalently, as f^4. The wavelength, λ, is related to ω by $\lambda = c/f = 2\pi c/\omega$, so *the intensity of the radiation emitted by a charge set in oscillation by an external electric field is proportional to* $1/\lambda^4$, *where* λ *is the wavelength of the oscillating field.* This is the result first obtained by Lord Rayleigh. We used this result to explain why the sky is blue.

Let us now turn to how the atomic structure leads to a dielectric constant that varies with frequency. When an electron is separated from an equal but oppositely charged ion by a distance z, as given by Eq. (36–16), then this charge separation is responsible for an electric field. For small displacement this electric field is proportional to the displacement z, which is itself proportional to the *external* electric field that leads to the separation. In this case the external electric field is that of the incident electromagnetic wave, with amplitude E_0. As indicated in Section 26–6, the sum of the magnitudes of the incident field and the induced field gives a total electric field reduced from E_0 to E_0/κ. By following the discussion above, we can show that the dielectric constant, κ, is dependent on the angular frequency $\omega = 2\pi f$ as

$$\frac{1}{\kappa} = 1 - \frac{C}{\omega_0^2 - \omega^2},$$

where C is a constant. According to Eq. (36–2), $\kappa = n^2$, where n is the index of refraction. Thus

$$\frac{1}{n^2} = 1 - \frac{C}{\omega_0^2 - \omega^2}. \tag{36–17}$$

This equation describes how the index of refraction varies with frequency. Because $\omega_0 \gg \omega$ for visible light, *the index of refraction increases as the frequency of the light increases.* In fact, atoms and molecules have many resonant frequencies, so a more accurate version of Eq. (36–17) must contain several terms of the form $C_k/(\omega_{0k}^2 - \omega^2)$ added together, which is how the curve in Fig. 36–25 was obtained.

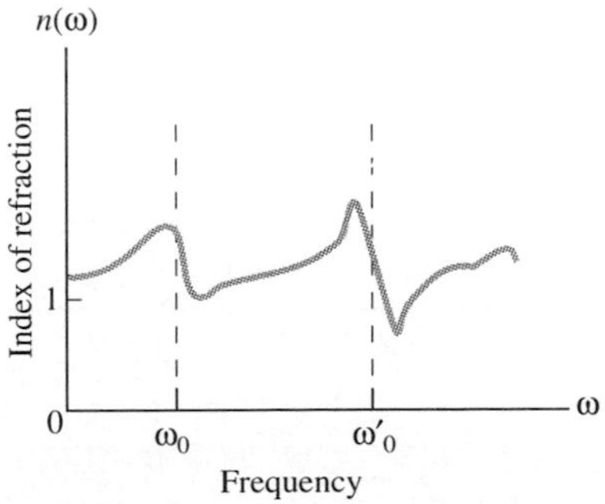

FIGURE 36–25 The index of refraction as a function of light frequency in real atomic materials. Regions where the light frequency ω is near the natural atomic frequencies of the material (here, ω_0 and ω_0') need special treatment, because resonance occurs at those frequencies.

EXAMPLE 36–5 Use the data points in Table 36–2 for the index of refraction in glass to find values for ω_0^2 and for the constant C in Eq. (36–17), and check to what extent that equation fits the rest of the data.

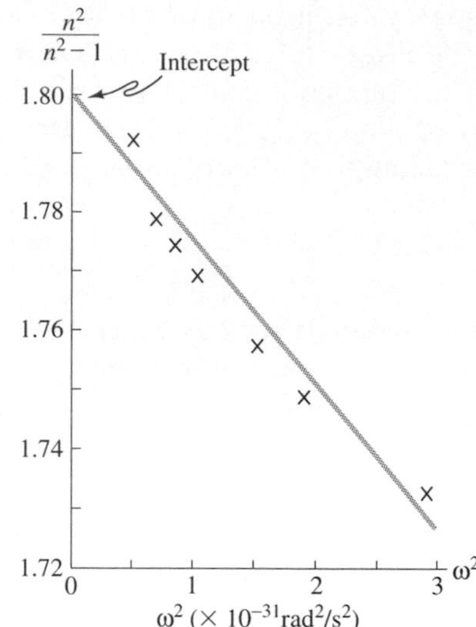

FIGURE 36–26 Example 36–5. The plot of $n^2/(n^2-1)$ versus ω^2 allows us to test the theory we have developed and to extract ω_0^2 and C.

EXAMPLE 36–5

ω^2 (rad²/s² × 10^{31})	n	$n^2/(n^2-1)$
2.72	1.539	1.731
1.89	1.528	1.749
1.50	1.523	1.758
1.02	1.517	1.769
0.82	1.514	1.774
0.60	1.511	1.779
0.25	1.505	1.791

SOLUTION: This problem is actually rather typical of many real-life problems of science and engineering. We are checking here how well a particular theory, the one we have developed for the atomic theory of dispersion, fits real data. We shall use a method of practical value, namely, graphing a function.

We rearrange Eq. (36–17) in order to find a function that is simple to plot:

$$1-\frac{1}{n^2}=\frac{n^2-1}{n^2}=\frac{C}{\omega_0^2-\omega^2}.$$

If we take the inverse of this equation, we get

$$\frac{n^2}{n^2-1}=\frac{\omega_0^2-\omega^2}{C}=\frac{\omega_0^2}{C}-\frac{\omega^2}{C}. \tag{36–18}$$

Next we plot $n^2/(n^2-1)$ versus ω^2 to see if the function is linear, a check of how well the data fit the theory. Assuming that the fit is good, we can find ω_0 and C by looking for the intercept ω_0^2/C and the slope $-1/C$.

Table 36–2 includes the values of ω^2 for each wavelength. Given the values of n, we can calculate the values of $n^2/(n^2-1)$ for each wavelength. In the table at the left, we list ω^2 and the corresponding $n^2/(n^2-1)$ values.

In Fig. 36–26 we have plotted $n^2/(n^2-1)$ versus $\omega^2/10^{31}$ for these data points. To a very good approximation, the points lie along a straight line, and we have drawn a line from which we can extract the slope and intercept. The slope $-1/C=-0.0243\times10^{-31}\,(\text{rad}^2/\text{s}^2)^{-1}$, or $C=4.1\times10^{32}$ rad²/s². From the intercept ω_0^2/C, we find that $\omega_0^2=7.4\times10^{32}$ rad²/s². The value of ω_0 is indeed several times greater than the ω values for visible light.

SUMMARY

The properties of light are well understood in terms of the wave theory of light. The speed of propagation of light waves is $c=3.00\times10^8$ m/s in a vacuum. In transparent media the speed of propagation is c/n, where n is the index of refraction of the medium. In general, the index of refraction depends on the wavelength of the light.

The propagation of light can be described either in terms of wave fronts, which form an envelope of spherical wavelets built upon earlier wavelets (Huygens' principle), or in terms of rays, lines perpendicular to the wave fronts. Light rays travel in straight lines unless they meet boundaries. Upon reflection from a surface, the angle θ that the incident ray makes with the normal to the surface is equal to the angle θ' that the reflected ray makes with the surface (the law of reflection):

$$\theta' = \theta. \qquad (36\text{–}5)$$

In the passage from a medium of index of refraction n_1 to a medium of index of refraction n_2, the incident angle θ_1 and the refracted angle θ_2 are related by Snell's law of refraction:

$$n_1 \sin\theta_1 = n_2 \sin\theta_2. \qquad (36\text{–}6)$$

These results can be established by using the geometry of wave fronts. They can also be derived with the help of Fermat's principle, which states that the path taken by a light ray between two points is the path that takes the shortest time. One consequence of Snell's law is that total internal reflection occurs when light moving in a medium with index of refraction n_1 strikes a boundary of a medium with index of refraction n_2, where $n_1 > n_2$, provided that the angle of incidence is larger than a critical angle θ_c, given by

$$\sin\theta_c = \frac{n_2}{n_1}. \qquad (36\text{–}9)$$

The dependence of the index of refraction on wavelength is called dispersion. Dispersion causes the different frequencies in a beam of white light to refract through different angles. The colors of a prism, the rainbow, and the blue sky are all naturally occurring dispersion phenomena. Dispersion can be understood in terms of the atomic theory of matter.

QUESTIONS

1. If light travels only in straight lines, how does a light burning in one room give light in another room?

2. How difficult would it be to reflect light back to earth from the moon by using two perpendicular plane mirrors? Why does it help if there are three mutually perpendicular mirrors?

3. If fish could think, they might realize that the relative indices of refraction of water and air allow them to outwit fishermen. Why?

4. A person swimming under water sees a lifeguard who is standing in the shallow part of the pool; the water comes up to the lifeguard's waist. In what way does the swimmer see the lifeguard's upper body distorted?

5. A fisherman standing up to his waist in a lake appears to have shorter than normal legs, to a person outside the lake. How will a fish in a horizontal position near the bottom of the lake appear to the observer?

6. A plane wave of radiation has an electric field of the form $\mathbf{E}_0 \cos(kz - \omega t)$ when it propagates in empty space. How do k and ω change when the plane wave enters a medium with index of refraction n?

7. As the sun sets, its color changes from white to yellow to orange and finally to red. As the lowest part of the sun sinks below the horizon, the sun appears squashed, more egg-shaped than circular. Why?

8. The permeability, μ_0, is a defined constant. If c is also a defined constant, is ϵ_0 in turn defined? What implication does this have for the definition of charge (see Chapter 22)?

9. A coin lies at the bottom of a pool of water. Starting from a point immediately above the coin, you observe the coin from the level of the surface. You then move your head horizontally away from the coin. Is there a horizontal distance at which the coin is no longer visible?

10. Light from the sky refracts near the surface of hot sand, giving the impression that there is a bright surface that could be interpreted as water: a mirage (Fig. 36–27). The air near the surface of hot sand is hotter than the surround-

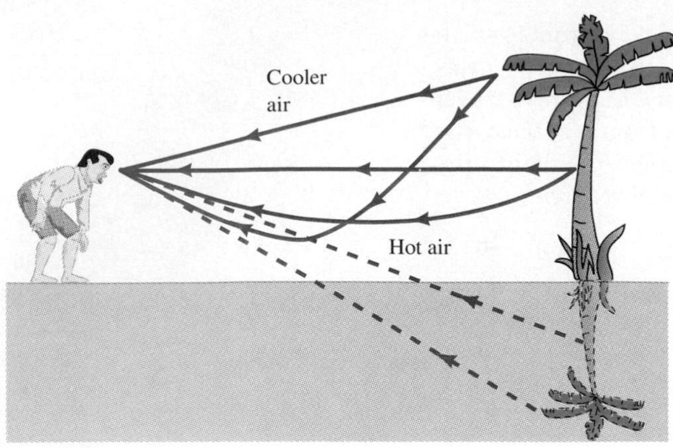

FIGURE 36–27 Question 10.

ing air. Does light travel faster or slower in hot air than in cold air?

11. Mirages can occur when a layer of cold air lies closer to the surface. How would such an air layer affect the appearance of distant houses?

12. Why does the sky look black rather than blue, as it does from earth, to astronauts in orbit?

13. What is the index of refraction of a vacuum?

14. For a moment you are lying in the middle of a circular swimming pool, at the bottom of the pool, which is filled to a depth of 1 m with water and is surrounded by trees. A 2-m tall lifeguard is standing in the water about 3 m from you. What do things look like as you scan in all directions?

15. Laser light directed into the end of a glass rod comes out the other end with almost the original intensity. If another glass rod touches the side of the first rod, making a 30° angle with the lengthwise direction of the first rod, nothing happens. But if the point of contact is lubricated with glycerin, some of the original light beam is "stolen" by the second rod. Explain what happens.

16. White light is incident onto a pane of glass. Is there a dispersion of colors in the reflected light?

17. Stick a pin into the underside of a cylinder of cork, then float the cork in water. Even if you do not stick the pin in very far, you may not be able to see it from outside the water. Why not?

PROBLEMS

36–1 *The Speed of Light*

1. (I) The nearest star to our Solar System (aside from the sun) is Alpha Centauri, some 4.2 ly from earth. How far is this in meters?

2. (I) A light wave of blue light ($\lambda = 460$ nm) passes from air into water, where the index of refraction is 1.33. What are the wavelength and frequency of the light in water?

3. (I) Yellow light, whose frequency is 5×10^{14} Hz, impinges on glass, $n = 1.5$. What are the wavelengths of this light in a vacuum and in glass? What is the index of refraction of a material within which the wavelength of yellow light is one-half its value in a vacuum?

4. (II) Suppose that you have a version of Fizeau's apparatus in which the round-trip distance for the light beam is $2D = 300$ m. The width of the opening between the teeth on the cogged wheel is 1.0 mm, and the center-to-center distance between these gaps is 2.0 mm. The wheel has a radius of 10.0 cm. What would the minimum rotational speed be, in revolutions per minute, so that light entering through the center of one gap would come out through the center of the next gap? Is such an apparatus realizable?

5. (II) Figure 36–28 shows an exaggerated view of the eclipsing of Io, the innermost moon of Jupiter, as seen from two different points on earth's orbit around the sun. If the earth were stationary at a point nearest Jupiter, N, a particular eclipse would begin at a precise time. When the earth is at point F, the eclipse starts somewhat later than expected, because the light has to travel the additional distance of a diameter of the earth–sun orbit. The mean distance from the earth to the sun is 1.50×10^{11} m. How much later will the eclipse be seen at point F compared with point N?

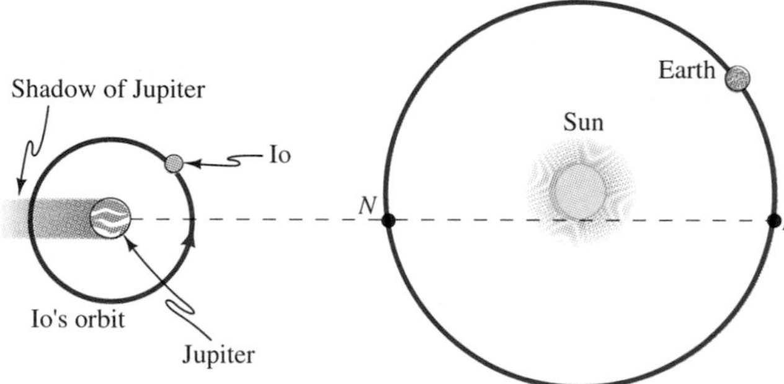

FIGURE 36–28 Problem 5.

6. (II) The speed of light in a vacuum is defined to be 299,792,458 m/s. A lunar ranging experiment measures the time for a light pulse to reach the moon and reflect back to earth. Such experiments allow us to determine the distance between the moon and the earth, which is approximately 3.84×10^8 m, to an accuracy of 15 cm. What is the smallest time interval that can be measured by the clock used to determine the time it takes for light to go to the lunar reflector and back?

7. (II) Galileo attempted to measure the speed of light with the help of lights and a clock on two adjacent mountains. In essence, a shutter over a light was opened on the first mountain, an observer on the second mountain saw that signal and returned a second signal, and the experimenter on the first mountain looked for a delay between the time

the shutter was opened and the time the signal was returned. Use your knowledge of human reaction time to estimate the time measured by the first experimenter for the total round trip. How long would it actually take for light to travel back and forth between two mountaintops separated by 4 km? Your answers explain why Galileo's attempt did not work.

8. (II) In an experiment similar to Fizeau's, we have a cogged wheel of diameter 30 cm. A laser beam shines through one opening, travels 2000 m, and is reflected back. Given that the fastest rotation rate of the wheel is 7.0×10^4 rev/min, what should be the separation between adjacent cogs on the rim of the wheel?

36–3 Reflection and Refraction

9. (I) A fixed projector emits a narrow beam of light onto a plane mirror. At what angle with respect to the beam should you place the mirror in order to turn the beam by 90°?

10. (I) The critical angle for a particular material (used in air) is observed to be 43°. What is the material's index of refraction?

11. (I) A horizontal beam of light is reflected from a plane mirror that revolves about a vertical axis at a rate of 30 rev/min. The reflected beam sweeps across a screen that, at the point nearest the mirror, is 20 m away. With what speed does the spot of light move across the screen at the point nearest the mirror?

12. (II) What is the critical angle for total internal reflection in crown glass (used in air), for which $n = 1.52$? Show that it is possible to use a 45°-45°-90° triangular prism of crown glass to make a perfect reflector of light (Fig. 36–29).

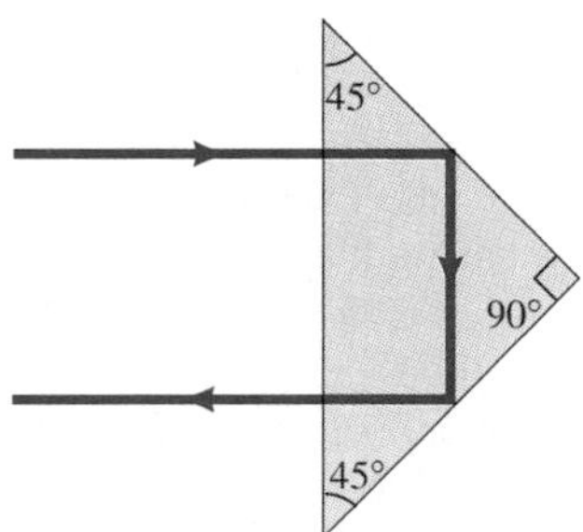

FIGURE 36–29 Problem 12.

13. (II) A thick glass plate ($n = 1.62$) lies on the bottom of a tank of water ($n = 1.33$). A light ray enters the water from air, making an angle of 60° with the normal to the surface. What angle does the ray make with the normal when the ray is in the water? What angle does it make with the normal when it is in the glass?

14. (II) Light in air enters a stack of three parallel plates with indices of refraction 1.50, 1.55, and 1.60, respectively. The incident beam makes a 60° angle with the normal to the plate surface. At what angle does the beam emerge into the air after passing through the stack?

15. (II) A beam of light enters a glass sphere ($n = 1.5$) at a latitude of 30°, parallel to the equatorial plane. Make a careful drawing to determine the angle at which the beam will strike the back of the sphere. Will there be total internal reflection?

16. (II) White light is refracted by the triangular prism shown in Fig. 36–30. A beam of light enters the prism along a path parallel to the prism base. The light is observed on a screen that is located 10 m from the prism and is perpendicular to the emerging rays. How far apart on the screen are the spots of blue light ($n = 1.528$) and red light ($n = 1.514$)?

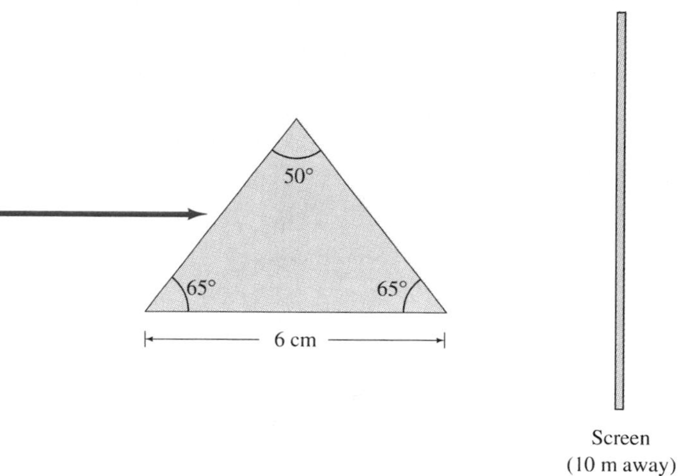

FIGURE 36–30 Problem 16.

17. (II) A lifeguard 1.8 m tall stands in water 60 cm deep. From a vantage point at the bottom of the pool, a swimmer sees the lifeguard's head to be along a line at a 45° angle to the vertical. How far is the swimmer's eye from the lifeguard's feet? (For water, $n = 1.33$.)

18. (II) A narrow beam of light is incident at a 30° angle from the normal onto a glass pane 6 mm thick. Describe the position of the exit beam of light. What is its direction? Is it displaced from the incident beam? If so, by how much? (For the glass, $n = 1.60$.)

19. (II) Light is incident on an equilateral triangular prism ($n = 1.55$) at a 35° angle from the normal to one of the faces. What is the exit angle?

20. (II) Consider a solid glass rod of length 30 cm and diameter 2 cm, with index of refraction 1.53. The ends of the rod are perpendicular to the lengthwise direction. (a) Light enters the center of the end of the rod from air. What is the maximum angle of incidence for which the light is totally reflected inside the rod? (b) Repeat part (a) for a similar rod totally immersed in water ($n = 1.33$).

21. (II) A ray of light impinges at a 45° angle of incidence on a glass pane of thickness 3 mm and index of refraction 1.45. The light is reflected by a mirror that touches the back of the pane. By how much is the beam displaced, compared with the return path it would have if the pane were absent?

22. (II) At noon a 2.0-m-long vertical stick casts a shadow 1.0 m long. If the same stick is placed at noon in a flat-bottomed pool of water half the height of the stick, how long is the shadow on the floor of the pool? (For water, $n = 1.33$.)

23. (II) You have three transparent liquids labeled 1, 2, and 3 that do not mix. When light is sent from liquid i to liquid j, there is an angle of incidence θ_i and an angle of refraction θ_j. Two separate experiments show the following: $1 \rightarrow 2$, $\theta_i = 22°$ and $\theta_j = 32°$; $2 \rightarrow 3$, $\theta_i = 35°$ and $\theta_j = 51°$. Find the ratios of the indices of refraction for each pair of liquids.

24. (II) Consider light that is perpendicularly incident on a triangular prism of the kind shown in Fig. 36–31. The index of refraction of the prism material is $n_1 = 1.645$. Suppose that the two reflecting sides are coated with a thin, uniform layer of a dielectric with index of refraction $n_2 = 1.42$. Will the prism still be totally reflecting? How large can n_2 be so that the prism is still totally reflecting?

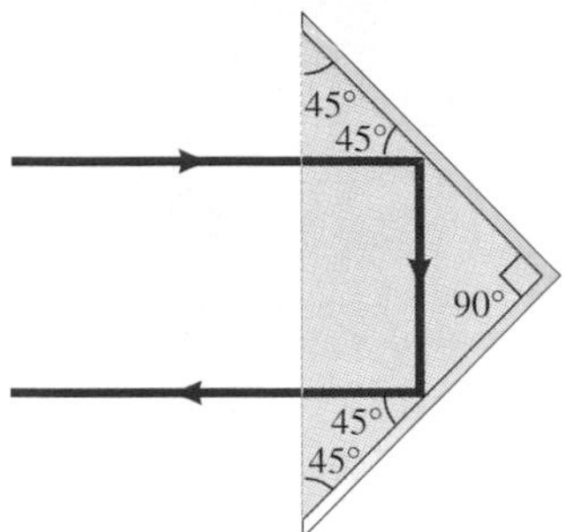

FIGURE 36–31 Problem 24.

25. (II) A prism has a cross section in the shape of an isosceles triangle with a base-to-height ratio of 1/10. A beam of light is incident upon the left side, parallel to the base. At what angle relative to the base will the beam leave the right side of the prism, which is made of glass with $n = 1.52$?

26. (III) Use Huygens' construction to prove Snell's law by working out the geometrical details in Fig. 36–32.

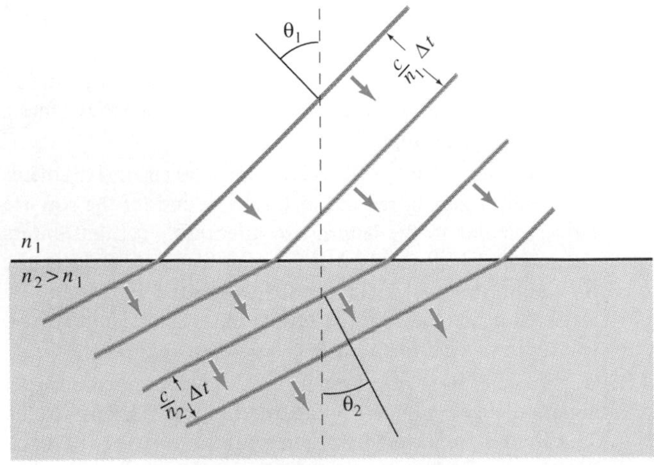

FIGURE 36–32 Problem 26.

*36–4 *Reflection and Refraction from Fermat's Principle*

27. (II) Use Fermat's principle to show that the critical angle for total internal reflection is given by $\sin \theta_c = 1/n$, where n is the index of refraction of the medium in which the light ray originates. The outside medium is air.

28. (II) Show that the law of reflection follows from Fermat's principle.

29. (II) By using Fermat's principle, show that if two media have exactly the same index of refraction, then a beam of light travels in a straight line when it crosses the boundary between them.

30. (II) Use Fermat's principle to show that a beam of light that enters a plate of glass of uniform thickness emerges parallel to its initial direction.

31. (II) Calculate the parallel displacement of a beam of light that strikes a vertical slab of glass, with index of refraction n and thickness D, at an angle ϕ with the horizontal (use Fermat's principle). (In Problem 30 we showed that a ray of light that passes through a slab of glass emerges parallel to its initial direction. The ray is, however, displaced from its original line, and that displacement is what we want here.)

36–5 *Dispersion*

32. (I) By what percent does the speed of red light in a type of glass ($\lambda = 700$ nm, $n = 1.514$) exceed that of blue light in the same glass ($\lambda = 450$ nm, $n = 1.528$)?

33. (II) We wish to select a glass to construct a prism that can separate the yellow ($\lambda = 590$ nm) component of light from the blue–green ($\lambda = 490$ nm) component. The prism is to be a bar with the cross section of an equilateral triangle. If a ray of white light arrives parallel to the base of the prism, it must leave the prism with the two colors separated by at least 2°. What must the difference in indices of refraction be for the two colors? [*Hint:* Because the difference of angles is small, so is the difference of indices of refraction. Keep only leading terms in differences of angle and of index of refraction.]

34. (II) A beam of white light, whose frequencies are mixed with equal intensity, passes within a piece of glass and impinges on a boundary to the air at an angle of incidence θ. The index of refraction of the glass increases with increasing angular frequency according to the formula $n^2 = 1 + [C/(\omega_0^2 - \omega^2 - C)]$, where $C = 443 \times 10^{30}\ \text{rad}^2/\text{s}^2$ and $\omega_0^2 = 795 \times 10^{30}\ \text{rad}^2/\text{s}^2$. (a) What is the largest angular frequency that passes through the glass into the air? (b) At what angle of incidence should the light approach the boundary if we wish to allow only frequencies of $\omega = 2.9 \times 10^{15}$ rad/s (red light) and below to pass through to the air?

35. (II) Use the data in Problem 34 to calculate the critical angles for total internal reflection for five values of wavelengths in the range 430 nm to 770 nm. Plot your results.

36. (II) Atomic theory applied to a gas leads to an expression for C in Example 36–5, in which $n - 1$ is very small. It is

$C = Ne^2/2\epsilon_0 m_e$, where N is the number of atoms per unit volume, m_e is the mass of an electron, and e is the electron charge. In air $N \simeq 3 \times 10^{25}\ \text{m}^{-3}$ and $n = 1.0003$. Use this to calculate ω_0. Compare your value with that calculated in Example 36–5. Would you expect these values to be close?

General Problems

37. (I) Light of wavelength 660 nm enters a piece of crown glass, whose index of refraction is 1.52. What are the wavelength and speed of that light in the glass?

38. (II) Show that if an incident ray of white light that is parallel to the base of a prism in the shape of an isosceles triangle (apex angle 2ϕ) is separated into two components that exit the prism with an angular separation $\Delta\theta \ll 1$, then the difference in the indices of refraction for the two colors is proportional to $\Delta\theta$. Find the equation that expresses the relation between the differences in the indices of refraction and in $\Delta\theta$. [*Hint:* Consider the angle of emergence for a given n, and then find Δn as a function of $\Delta\theta$.]

39. (II) A pin is partly inserted perpendicularly into the flat surface of a cork with a 2.0-cm radius. The cork, with the pin on the underside, is set afloat in a pool. A length of 1.5 cm of cork is under the water surface. Because of the effects of refraction, much of the pin is hidden from view from above the surface. What length of pin can be hidden in this way?

40. (II) A beam of light is incident at an angle of 30° to the vertical on a horizontal glass plate of thickness 2.0 cm. The index of refraction of the glass is $n = 1.52$. The beam emerges on the other side. What is the perpendicular distance between the straight-line extrapolation of the incident ray and the ray refracted by the glass plate?

41. (III) Sound can refract like light does. Suppose that a submarine lies flat 180 m below the water surface, and that there are three thermal layers of water (each 60 m deep) of different temperatures. The speed of sound in water depends on temperature. In the bottom layer, the speed is 1.16 times that in the top layer; in the middle layer, the speed is 1.08 times that in the top layer. A detection device at surface level determines that sound from the submarine arrives at the surface at a 40° angle with the horizontal. What is the horizontal distance between the detector and the submarine?

42. (III) A ray of light is incident at an angle of incidence θ_i on one surface of a prism whose cross section is an isosceles triangle (apex angle 2ϕ). The light exits the prism at a total deflection angle Θ (Fig. 36–33). The prism has index of refraction n and is in a vacuum, which has an index of refraction of exactly 1. For what angle θ_i is the angle of deflection Θ a minimum?

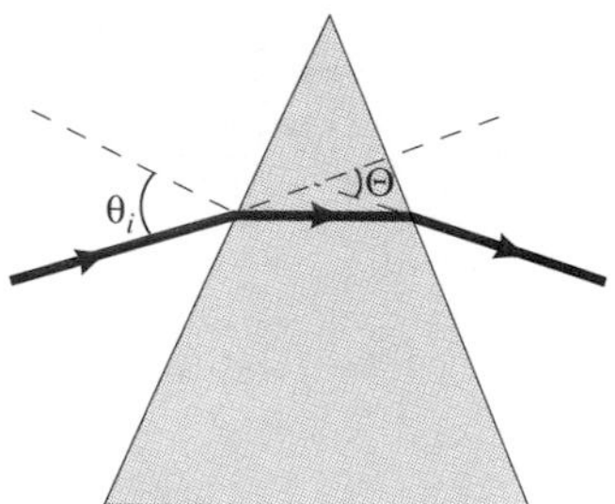

FIGURE 36–33 Problem 42.

43. (III) A ray of light incident from air onto a glass pane is partly reflected and partly refracted at the two surfaces of the pane (Fig. 36–34). The glass has an index of refraction n and a thickness d. Express in terms of n, d, and θ_i the displacement d' of the ray drawn, which enters the glass, reflects off the back surface, and exits.

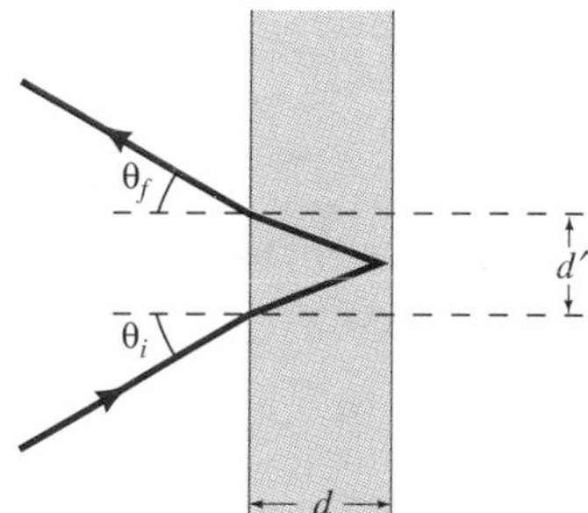

FIGURE 36–34 Problem 43.

44. (III) The first successful measurement of the speed of light, made by Olaus Roemer in 1675, was based on the following method. The mean orbital period of Io, a moon of Jupiter, is 42.5 h, but that period is measured to be about 15 s less than this value when the earth in its orbit is approaching Jupiter, and about 15 s more when the earth is receding from Jupiter. (a) Given that the earth's orbital speed around the sun is about 30 km/s, and that the earth is on a part of its orbit when it is moving toward Jupiter, how much closer will the earth have moved toward Jupiter during one orbit of Io? (b) Use the information given to estimate the speed of light.

CHAPTER 37

With the aid of a mirror, Johannes Gumpp painted an unusual self-portrait. This painting hints at the fascinating visual properties of mirrors.

MIRRORS AND LENSES AND THEIR USES

Technology often drives our ability to understand nature, because it gives us new tools and instruments that allow us to explore domains previously inaccessible. Astronomy and much of physics owe their progress to the invention of the telescope. Modern biology could not have been created without the microscope. In this chapter we shall discuss the principles that govern the construction of optical instruments. Their functioning is based on two very simple laws introduced in Chapter 36. The first law, the law of reflection, helps us to understand the images that we see in mirrors and explains the functioning of instruments such as the reflecting telescope. The second law, Snell's law, when applied to the light that passes through curved boundaries between refracting surfaces, explains the functioning of the eye, cameras, magnifying glasses, and microscopes. Because these two laws can be applied simply by tracing the geometrical paths of light rays, this aspect of the study of light is called geometric optics. Instruments whose functioning depends on, or is limited by, the wave nature of light will be discussed in Chapters 38 and 39.

The simplest reflecting surface is a flat (or plane) mirror. When you look at yourself in a mirror, you see your image. What is this image, and how is it formed? We begin by considering a point source of light in front of a plane mirror. Figure 37–1 shows several rays of light emitted by a point source (S) and reflected off a plane mirror according to the law that the angles of incidence and reflection are equal, the law of reflection. We could in fact draw an infinite number of such rays as close to one another as we like. Rays that are near one another form *bundles*, as in the bundle of rays 2. In Fig. 37–1 we have included an observing eye, which intercepts ray bundle 2.

Simple geometry allows us to see that *all the reflected rays trace back to the same point I*. To see this, consider Fig. 37–2, in which we show rays 1 and 3. We have indicated the equal angles of incidence and reflection θ_1 and θ_3 for these rays, respectively, as well as the angles α_1 and α_3. The angle formed by BP_1I is then equal to α_1, so if point B is formed by dropping a perpendicular to the mirror from point S and if point I lies along the continuation of this line, triangles BIP_1 and BSP_1 are similar triangles. By the same method, so are triangles BIP_3 and BSP_3. Because both rays 1 and 3 emanate from the same point S, the distance BS forms the base of both triangles to the left of the mirror (the *object side*), and the distance BI forms the base of both triangles to the right of the mirror (the *image side*). The (imaginary) continuations of rays 1 and 3 to the image side meet at point I, as would the continuation of *any* reflected ray. An *image point* is in fact any point other than

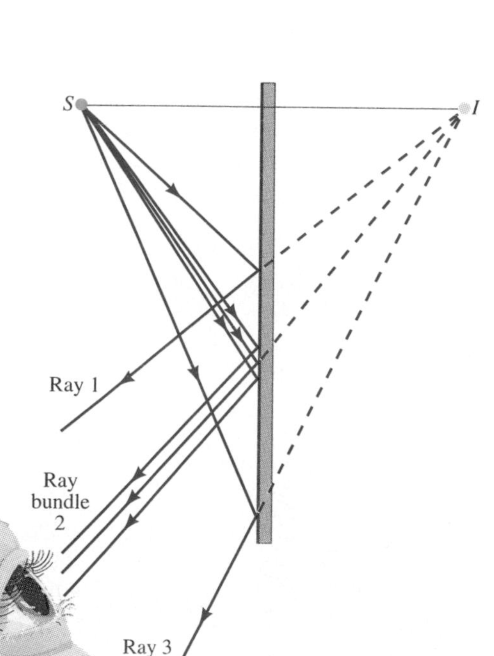

FIGURE 37–1 Rays leaving source point S reflect from a plane mirror. A bundle of such rays enters the eye, apparently from point I.

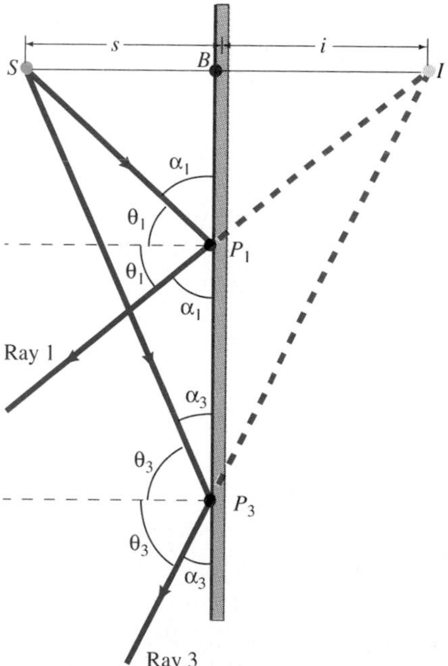

FIGURE 37–2 All the reflected rays from point S trace back to point I. The geometry implies that the perpendicular distance from the mirror to point S, s, equals the distance from the mirror to point I, i.

the object from which an unlimited number of rays emanate or appear to emanate when the rays are extended back in straight lines.

We have in fact calculated the location of point I. Because BIP_1 and BSP_1 are similar triangles, the distances BS and BI are equal. How does the eye/brain "know" where to put I? Two eyes (or one eye that moves a little) sense a bundle of rays, as in Fig. 37–1, rather than a single ray. The eye/brain can measure their degree of divergence and is capable of extrapolating this diverging bundle back to point I.

The Image of an Extended Object

If our light source is extended, a second point on the source forms a second image point. Moreover, this second image point is as close to the first image point as the second source point is to the first source point. A set of nearby source points forms nearby image points (Fig. 37–3). The entire **object**, or *source* (we use the terms interchangeably), forms a set of matching image points, which together constitute an **image**. *An image is a set of contiguous points from which reflected rays emanate or appear to emanate when the rays are extrapolated back in straight lines.* Our plane mirror forms *perfect images*, in which the image is a perfect geometrical match of the object. Not only is there no distortion of the image, but Fig. 37–3 illustrates that the source and the image have the same size. We say that a plane mirror does not produce a *magnified* image.

The definition of an image

Because no light rays actually pass through the image formed by a plane mirror, we call it a **virtual** image. We shall see below that with curved surfaces, we can form a **real** image, a continuous set of points through which any number of rays of reflected light do pass.

Virtual and real images

We have said that the image formed by a plane mirror is perfect. But it does have one striking peculiarity. If your right eye is blackened, your image, when viewed as though you were meeting yourself on a street, has a left eye that is blackened. From Fig. 37–4 we can see that the right and left eyes of the image are not reversed (the black eye of the source is closest the bottom of the page, just as the black eye of the image is). However, there *is* a front-to-back reversal (the nose of the object points in the $+x$ direction in the figure, whereas the nose of the image points in the $-x$ direction), and this is what has loosely been called a left-to-right reversal of the image.

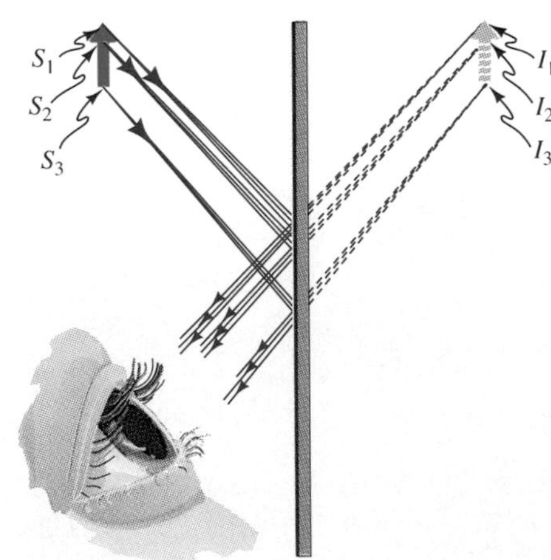

FIGURE 37–3 When the source is an extended one, there is an image point for every source point. This means that a bundle of rays will enter your eye from every image point no matter what your position before the mirror. Geometry can be used to locate the image $I_1 I_2 I_3$ of source $S_1 S_2 S_3$.

Multiple Reflections

If the image produced by one reflecting surface is reversed left-to-right, the image produced by successive reflection from two mirrors is not. What do we mean by the image produced by successive reflection? We have seen that the image produced by a source comes from the reflections of the rays emitted by the source. But these reflected rays form a set of diverging rays; this is in fact how the image is formed. The reflected rays can reflect a second time from another mirror, and we can think of the diverging set of rays incident on the second mirror as coming from a second source. In other words, *a first image can act as a source for a second image*. It makes no difference whether the first image is virtual or real. This idea allows us to calculate the effects of optical systems with many elements, whether they be plane or curved mirrors or lenses. We shall use this idea frequently in the remainder of the chapter. When there are multiple reflections, the number of images can multiply (Fig. 37–5).

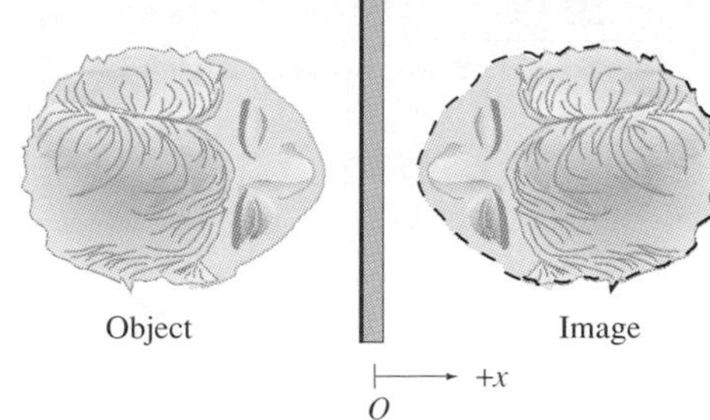

FIGURE 37–4 An image is reversed front to back. This means that if your right eye is black, your image has a black left eye.

Corner reflectors, discussed in Chapter 36, provide an example of the usefulness of multiple reflections. A corner reflector is constructed from three mirrors set at 90° with respect to each other. Any ray striking a corner reflector reflects directly back to its source (see Fig. B1–1 in Chapter 36). When it is important that all the energy of the incident light be reflected back, the incident beam can be refracted into a piece of glass cut so that the ray undergoes total internal reflection at the glass–air surface (the second surface encountered). The advantage of using total internal reflection is that there is no transmission of light into the reflecting surface. This contrasts with metallic or silvered glass mirrors, which always absorb some of the energy of the incoming light.

FIGURE 37–5 Multiple reflections from nearly parallel mirrors, in a cartoon by Charles Addams (1957).

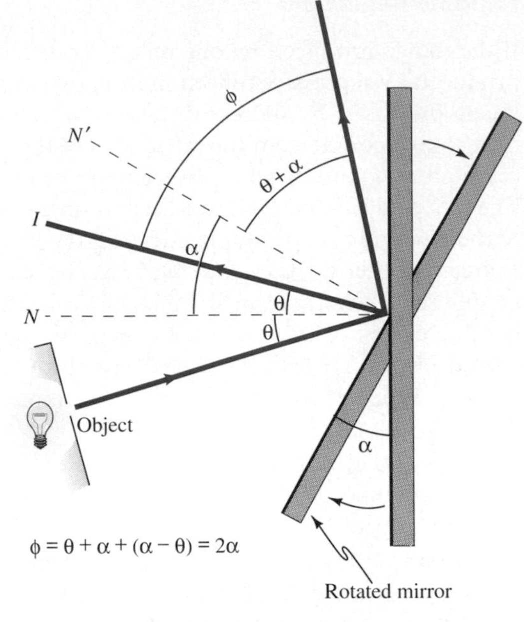

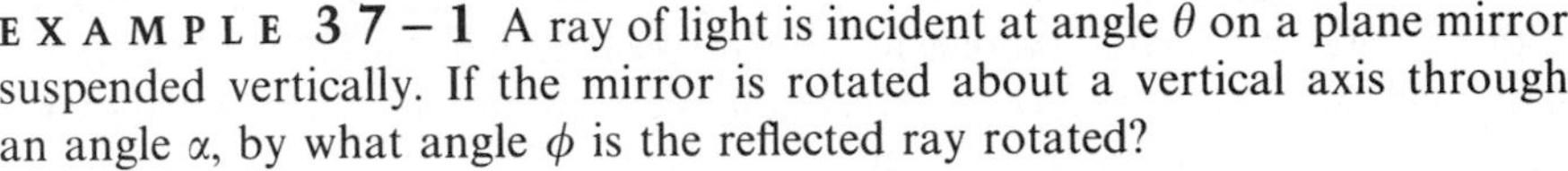

FIGURE 37–6 Example 37–1. Overhead view of a ray of light incident on a mirror that rotates about a vertical axis.

EXAMPLE 37–1 A ray of light is incident at angle θ on a plane mirror suspended vertically. If the mirror is rotated about a vertical axis through an angle α, by what angle ϕ is the reflected ray rotated?

SOLUTION: Figure 37–6 is an overhead view of the situation. Initially the incident ray makes an angle θ with the normal N to the mirror, and so does the reflected ray, labeled I. A rotation of the mirror through an angle α moves the normal through an angle α to a new normal position, N'. The new angle of incidence is $\theta + \alpha$, which is also the new angle of reflection from N'. The angle between the incident ray and the new reflected ray I' increases from 2θ to $2(\theta + \alpha)$. Because the incident ray has not moved, the reflected ray is rotated by $\phi = 2\alpha$.

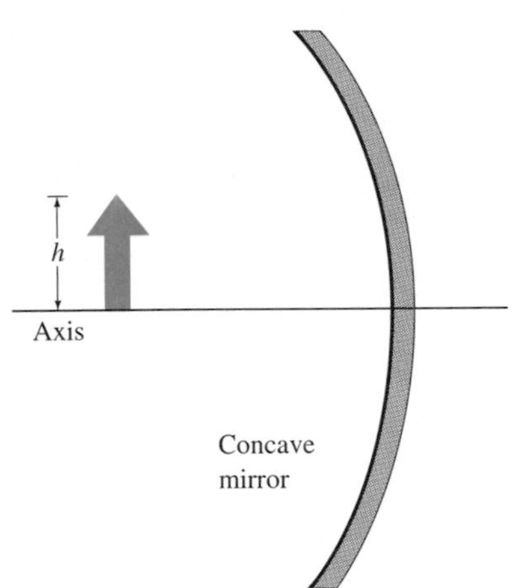

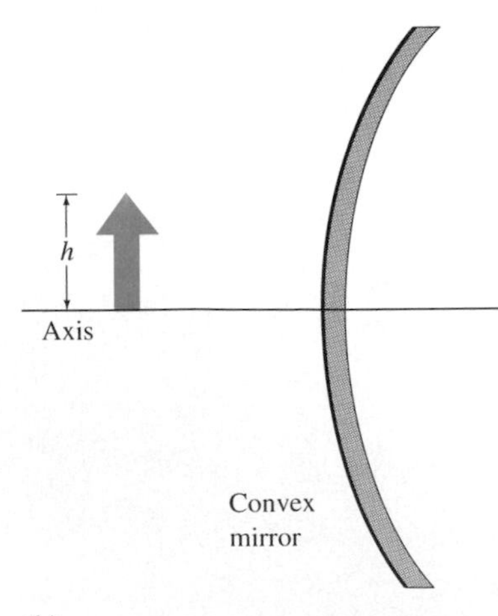

FIGURE 37–7 (a) Concave and (b) convex spherical mirrors. The object shown, an arrow, acts as an extended source.

37–2 SPHERICAL MIRRORS

We saw in Fig. 37–3 that plane mirrors produce images that are the same size as the object. We can construct mirrors that produce images of altered sizes, or real as well as virtual images, by using curved surfaces. The simplest curved surface to study, and the cheapest to manufacture, is a mirror whose surface is a segment of a sphere. As a tool to analyze the effects of reflections from such a surface, we shall extend the *ray-tracing techniques* of Section 37–1, in which we followed rays or bundles of rays that reflect from a plane mirror. In particular, you will see that ray tracing is useful because there are some particularly simple and significant rays to follow.

We can think of our mirror as a segment of a sphere with a center somewhere in space. If the light source (object) is on the same side of the surface as this center, the mirror is *concave* (Fig. 37–7a); if the source is on the other side, the mirror is *convex* (Fig. 37–7b).

In order to apply our ray-tracing techniques, we shall make some simplifying assumptions. Let us call the line perpendicular to the center point of the mirror the *axis* of the mirror. We shall study objects on or near the axis. The tips of the arrows shown in Fig. 37–7 are a distance h from the axis, and if h is small compared to the radii of curvature of the mirrors, we say that the object is *near the axis*. Finally, we shall simplify the mathematics by considering only rays sufficiently close to being parallel to the axis that we can use small-angle approximations in studying the reflections. Such rays are said to be **paraxial**.

The Concave Mirror

Figure 37–8a illustrates a concave mirror. We consider rays from a very distant point source (to the left of the figure) on the axis CB. The source is so far away that *all the rays from it arrive practically parallel to the axis.* (We say that the source is *at infinity*.) Point C indicates the position of the center of the sphere (of radius R) of which the mirror is a segment. The position of C (called the *center of curvature*) is therefore a distance R from the mirror surface, and all lines from point C to the mirror are perpendicular to the mirror.

Location of the Focal Point. We now look at ray 1, which is reflected at point A in the direction AF, in Fig. 37–8a. Angle θ is the angle of incidence (line CA is perpendicular to the mirror); hence θ is also the angle of reflection. How far is

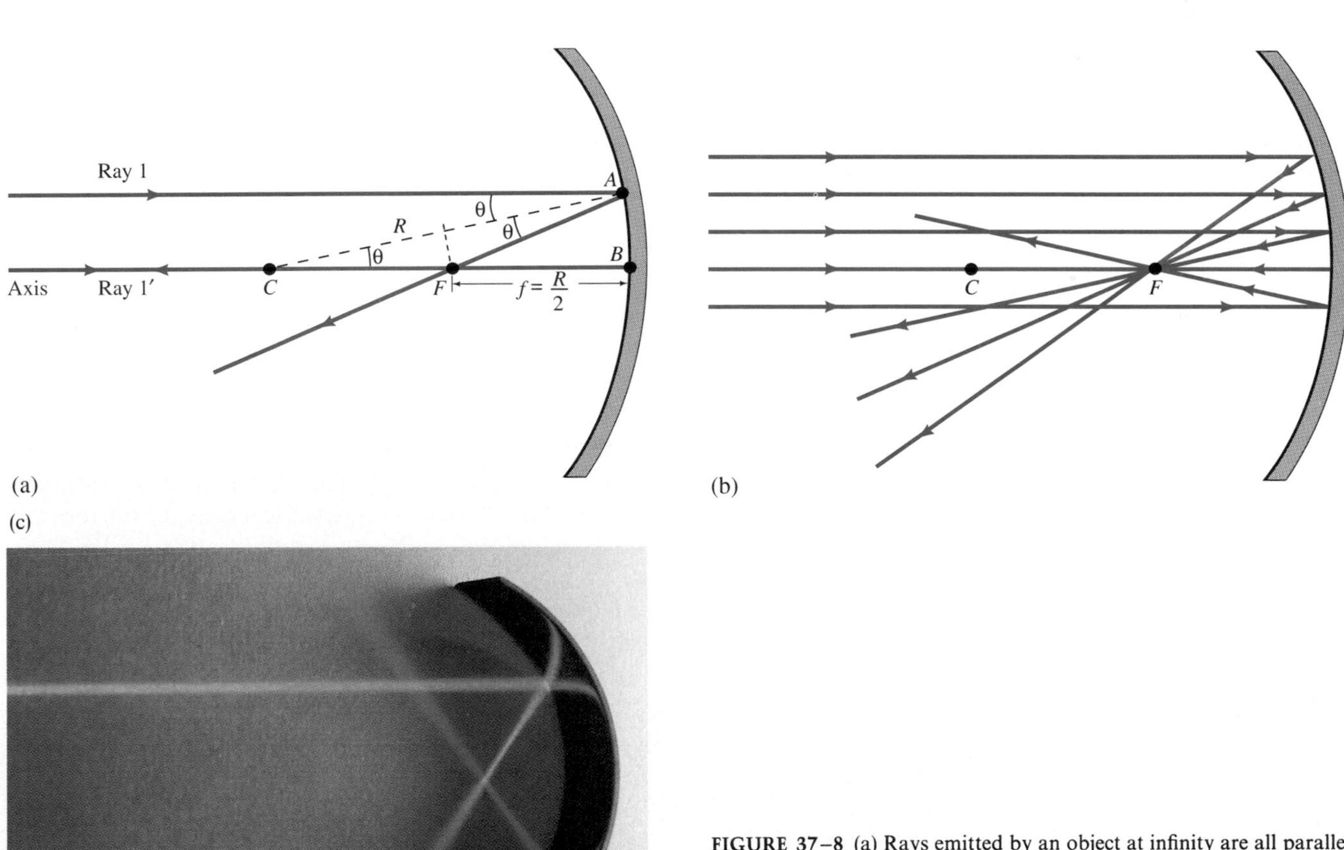

FIGURE 37–8 (a) Rays emitted by an object at infinity are all parallel to the axis. Ray 1 is reflected by the concave mirror surface and passes through the focal point, F. (b) To a good approximation, F, is independent of θ; that is, any incoming ray parallel to the axis is reflected through it. Thus all rays from infinity cross the axis at F, which is therefore an image point. (c) Photograph of paraxial rays reflecting off a concave mirror.

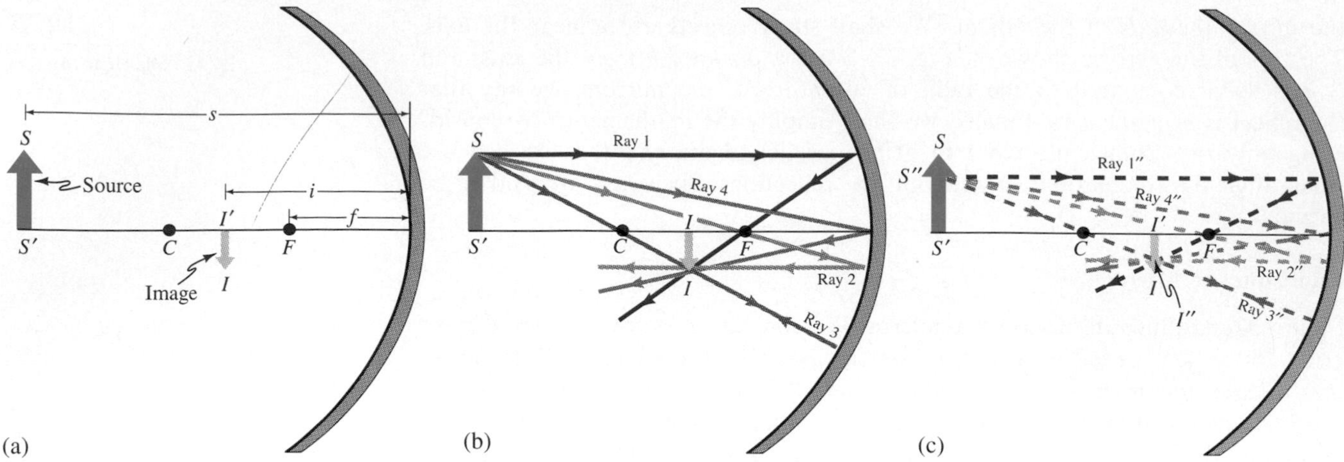

FIGURE 37–9 (a) An extended object a distance s from a mirror forms an image a distance i from the mirror. (b) Ray tracing for the concave mirror, with the principal rays for two source points. (c) By repeating this exercise, we can build up the entire image. The principal rays are a guide; *any* ray from S that reflects will cross the image point I.

point F from the mirror? Ray 1 arrives parallel to the axis, so the angle of incidence is equal to the angle ACB, which we have also labeled θ. Triangle ACF is therefore isosceles, with a base of length R. Thus by dropping a perpendicular from point F to the base AC, we see that the distance CF is $CF = (R/2)/\cos\theta$. For small θ, $\cos\theta \simeq 1$; hence $CF = R/2$, or $BF = R - CF = R/2$, *independent* of θ.[†] *All* the parallel rays near the axis reflect through point F, a distance $R/2$ from the mirror (Fig. 37–8b). An unlimited number of rays diverge from point F; hence it is the image point of the distant source point. Unlike the image points produced by a plane mirror, it is a real image point, because rays actually cross there (Fig. 37–8c).

The point F at which the rays from a source point at infinity are brought together to form an image (are *focused*) is known as the **focal point** (or *focus*), and its distance f from the mirror is called the **focal length.**[‡] This term can be applied to any optical system that produces images—including plane mirrors, whose focal length is infinite. The focal length is the distance from the image point to the optical system—the mirror or lens or whatever—when the source point is at infinity. For concave mirrors, we have shown that

The focal length of a mirror with radius of curvature R

$$f = \frac{R}{2}. \qquad (37\text{–}1)$$

If we were to reverse the directional arrow on the rays in Fig. 37–8b, we would see that a pointlike light source placed at the focus F will form an outgoing beam of light parallel to the axis of the mirror. Thus spherical concave mirrors are useful for searchlights and flashlights.

The Image of an Extended Object. Let us take an extended object, as in Fig. 37–7, that is small compared to the radius of curvature of the mirror and close enough to the axis so that the rays are paraxial. In Fig. 37–9a we label the object, which is *upright*, with the letters S and S'. As happened for the plane mirror and for the source point on the axis, entire bundles of rays from a given spot on the object pass through a corresponding spot in space after reflection, and thus an image I to I' of the entire object is formed. We want to determine the position and size of the image.

[†] This result is accurate to 1 percent for angles θ less than about 10° (see Problem 35).

[‡] Because we deal with mirrors (later, with lenses) whose radius of curvature is large, we do not need to be very careful about choosing the point along the axis from which we measure this distance. We can measure from the point where the mirror surface intersects the axis.

PROBLEM-SOLVING TECHNIQUES

For optical systems such as mirrors and lenses, it is important to be able to find the size and location of the images, given a source (or object). Because all rays cross at the image point of a source point (or behave as though they do, in the case of virtual images), we need only find the crossing points of a few rays from any point on the object, which we call *principal rays*. These rays have the virtue that it is easy to trace their paths and see where they cross. Even if the optical system is such that a ray does not actually exist,† we can pretend that it does, because for us the principal rays are only a tool to learn where the rest of the rays go.

We use here a convex mirror (Section 37–2), as in Fig. B1–1, and a converging lens (Section 37–4), as in Fig. B1–2. However, the technique applies as well to concave mirrors (Section 37–2), diverging lenses (Section 37–4), and (with the provisos listed below and in the text) to single refracting surfaces (Section 37–3). These cases are all illustrated in the text.

We count four principal rays, numbered 1 through 4, from a given source point S:

1. Rays that enter the system parallel to the optical axis. By definition, these paraxial rays are reflected or refracted to the focal point F.

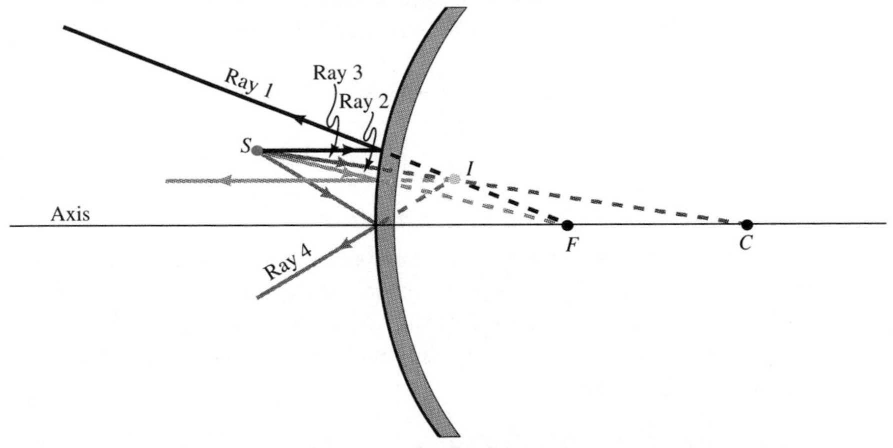

FIGURE B1–1 The four principal rays for reflection from a convex mirror.

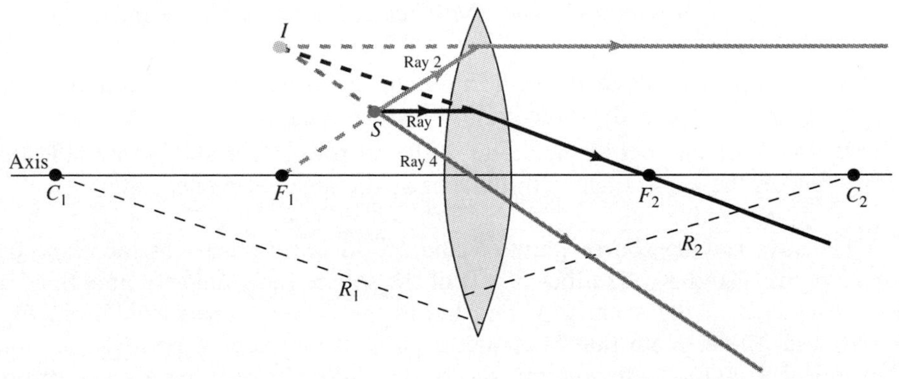

FIGURE B1–2 The three principal rays for refraction through a converging lens.

† For example, a hole may be cut out of the middle of a mirror, so the ray that approaches its center cannot reflect.

2. Rays that pass through (or are aligned as though they passed through) the focal point as they enter the system. These rays are just reversed versions of type 1 rays, and hence are reflected or refracted to leave the system parallel to the axis.
3. Rays that pass through (or are aligned as though they pass through) the center of curvature C of the sphere from which a mirror or refracting surface is formed. These rays are perpendicularly incident on the surface and will be reflected or refracted along the line of arrival. Such rays have no analogue in a thin lens, which has two surfaces and two radii of curvature.
4. Rays that strike the center of the mirror surface. The reflected rays make the same angle with the axis as the incident rays do (except for sign). For the case of thin lenses, the ray drawn directly to the center of the lens passes through it in a straight line. There is no analogous rule for single refracting surfaces.

By drawing these principal rays from any given point S on a source, we can see where their reflections or refractions cross (or appear to cross) and learn the location of the image point I of source point S. When an optical system has more than one reflecting or refracting surface (an "element"), we can apply the simple rule that the image formed by one element can serve as a new object for the next element. In that case *the principal rays must be redrawn for the new object* as they apply to the next optical element. We thereby locate the next image.

The origin of these rules is discussed more thoroughly in the text.

To do so, we use the principal-ray technique described in detail in the problem-solving techniques box. In Fig. 37–9b, the principal rays drawn to find point I of the image that corresponds to point S of the source are solid, and in Fig. 37–9c, those drawn to find the image point I'' that corresponds to source point S'' are dashed. These sets of rays are

Principal rays for a concave mirror

1. Ray 1, which approaches the mirror parallel to the axis and is reflected through the focus F (whose position we know);
2. Ray 2, which passes through the focus and reflects off the mirror parallel to the axis;
3. Ray 3, which passes through the center of curvature of the mirror and is reflected back in the direction from which it came;
4. Ray 4, which goes to the center of the mirror surface and whose reflection makes the same angle with the axis as the incident angle.

The rays leaving source points S and S'' do indeed cross at the respective image points I and I''. All other points of the source have image points that can be constructed in the same way, and the image is thus constructed as in Figs. 37–9b and 37–9c. Note that both image point I' and source point S' lie along the optical axis. Our construction shows that a vertical source gives a vertical image. This is true for all optical systems that we study and allows us to compute the location of just one image point rather than all of them. For example, the

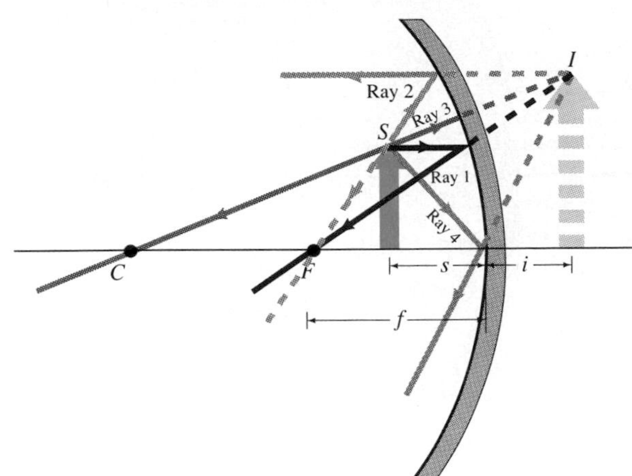

FIGURE 37–10 Ray tracing with the principal rays for a concave mirror, for a source closer to the mirror than that of Fig. 37–9b. The image becomes virtual when the source moves inside focal point F.

entire image can be constructed if we find only image point I of the top of the source (that is, of source point S), and in fact any two of the four principal rays are sufficient to determine point I. For example, rays 1 and 2 are sufficient for locating point I in Fig. 37–9b.

The image in Fig. 37–9a, located by the procedure in Figs. 37–9b and 37–9c, is *real* (real light rays pass through points I and I' and those in between), in contrast to the virtual image produced by plane mirrors. The image of the object in Fig. 37–9 is *inverted (upside-down) and reduced in size.* As the source moves farther from the mirror, the image moves closer and closer to the focal point F and also becomes smaller. When the source moves closer to the mirror, the image becomes larger and moves away from the mirror.

Let us now consider the situation depicted in Fig. 37–10, with the source inside the focal point of the mirror. The principal rays from point S are drawn according to the method of the problem-solving techniques box. Note that rays 2 and 3 only *behave* as though they pass through points F and C, respectively. In this case the reflected rays from source point S do not actually cross but are aligned as though they come from behind the mirror, at image point I. In this case, the image is virtual. It is also *upright* and enlarged. These are useful features of the concave mirrors employed for cosmetic purposes and in dentistry. As the source moves closer to the mirror, so does the image, which becomes smaller; as the source moves closer to the focal point, the image moves farther and farther away, becoming hugely magnified.

The Convex Mirror

The same ray-tracing techniques we used for concave mirrors allow us to understand convex mirrors. Point C is the center of curvature of the sphere (of radius R) of which the convex mirror is a segment. All lines from point C to the mirror are perpendicular to the mirror.

Location of the Focal Point. We start by finding the focal point, the spot where rays from a point source at infinity (that is, a set of rays parallel to the axis)

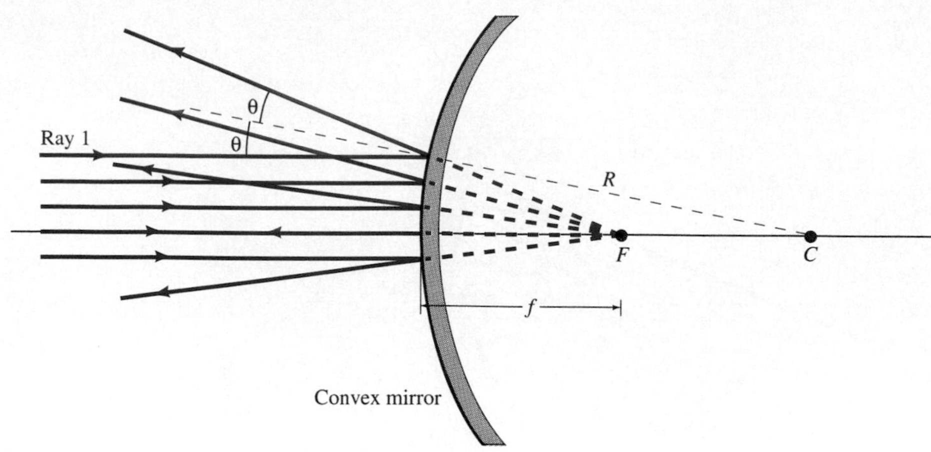

FIGURE 37–11 When a spherical mirror is convex, the focal point lies behind the mirror, as ray tracing shows. The reflected rays diverge, and their extensions all lead back to the focal point.

are focused. Figure 37–11 shows that the reflected rays diverge, so *the image is virtual*; the rays appear to originate at a common point at F behind the mirror. Of course, we must prove that there *is* a common point. To do so, we use ray 1 in Fig. 37–11. The line from its intersection with the mirror to point C is the normal to the mirror at that point. The angles marked θ are the angles of incidence and reflection for ray 1. By following the same trigonometric reasoning we used for the concave case, we can show that the distance f is again given by Eq. (37–1),

$$f = \frac{R}{2}$$

(see Problem 8), just as for the concave case. This result is *independent of* θ for small angles θ, so *all* the reflected rays appear to diverge from F. We have an image, and it is virtual. The focal point of a concave mirror is on the same side as the object, whereas the focal point of a convex mirror is on the side opposite the object.

Principal rays for a convex mirror

The Image of an Extended Object. Consider next the object shown in Fig. 37–12. We can trace the four principal rays from source point S: ray 1—parallel to the optical axis, and whose reflection extends back along the line from the mirror to point F; ray 2—drawn as though it would pass through F, and whose

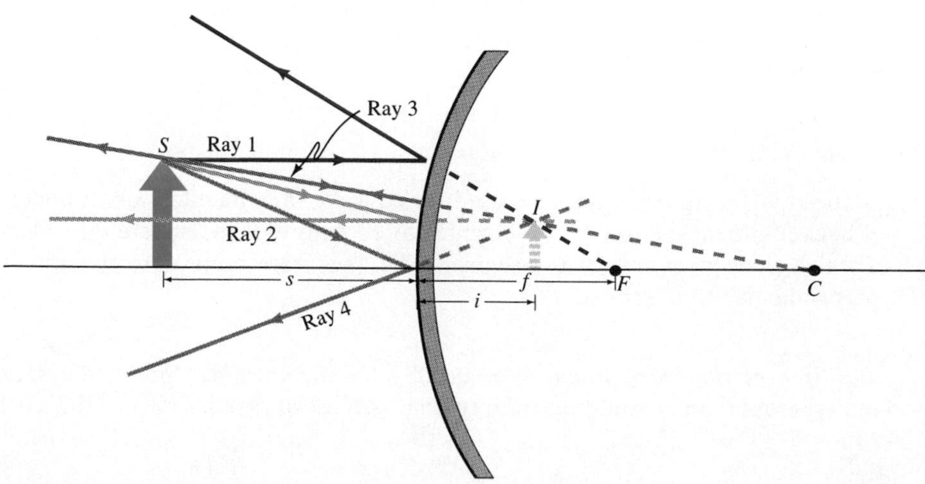

FIGURE 37–12 Ray tracing describes the formation of a virtual image by a convex spherical mirror.

reflection is parallel to the axis; ray 3—drawn as though it would pass through C, and whose reflection returns along the line of incidence; and ray 4—striking the center of the mirror surface, and whose reflection makes the same angle with the axis as the incident angle. A careful drawing shows that the reflected rays diverge from each other, but all four (indeed, *all* rays from S) appear to originate at point I. *Point I is the virtual image of point S.*

We can similarly find that there is a virtual image of the entire source; that image is upright and smaller than the source. When the source moves farther away, the image becomes smaller and remains upright, but there is no transition from virtual to real image, as there is in the concave case. These properties make convex spherical mirrors useful for rearview mirrors in motor vehicles. Figure 37–13 shows a more extreme case, an entire sphere acting as a mirror. In this case, the image is not perfect in the geometrical sense we have described.

FIGURE 37–13 Reflection by a spherical mirror that is a large portion of a sphere, seen here in the etching "Hand with Reflecting Sphere" by Maurits C. Escher.

The Relation between Source Distance and Image Distance

In Figs. 37–9a, 37–10, and 37–12 we have indicated the distance s from the mirror to the source, the distance i to the image, and the focal length f. Given f (or the radius of curvature $R = 2f$) and the source position, we can find the image position, its height, and whether or not it is inverted. Finding the quantitative relation among s, i, and f is generally a matter of straightforward geometry. The case of a concave spherical mirror is worked out in the box "Deriving Equations (37–2) and (37–8)" on p. 1086. The relation is

$$\frac{1}{s} + \frac{1}{i} = \frac{1}{f}. \qquad (37\text{–}2)$$

The relation among source distance, image distance, and focal length for mirrors

With a set of conventions about signs, we shall see how *the same relation holds for the convex mirror*. Equation (37–2) is easily understood in two limits. When the object is far away ($s \to \infty$), then $1/s \to 0$ and $i = f$ (which is the definition of f). When the object is at the focus, $s = f$, then $1/i = 0$: The image is very far away. This result is reasonable because an object that is at the focal length and produces an image at infinity is simply the ray-reversal of an object that is at infinity and produces an image at the focus.

If the object is between the concave mirror and the focus, as in Fig. 37–10, then s is smaller than f, and Eq. (37–2) implies that *i must be negative*! We associate a negative i with the image on the far side of the mirror; that is, with a virtual image. Likewise, i is positive when the image is real. Equation (37–2) may be applied to a convex mirror if we follow the convention that *the focal length f is negative when the focus is on the "virtual image" side of the mirror*. This is equivalent to saying that if the mirror's center of curvature is on the back (nonreflecting) side of the mirror, f is negative. In the application of Eq. (37–2) to a convex mirror, s is always positive, and f is always negative. Thus i must be negative (the image is always virtual). Furthermore, the image must be between the mirror and the focal point. That is because $1/s = (-1/|f|) + (1/|i|)$ is positive, so $1/|i| > -1/|f|$, and hence $|i| < |f|$. The rules for the sign of the object distance s will be discussed in Section 37–3.

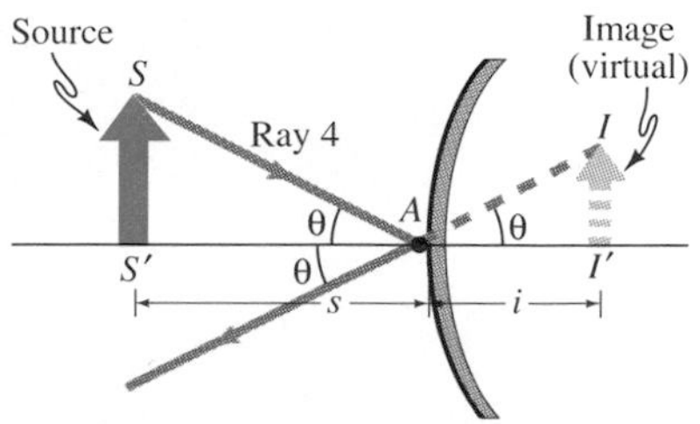

FIGURE 37–14 Geometry for the calculation of magnification.

Magnification

Our geometric constructions show that an image may not be the same size as its source. Consider the convex mirror in Fig. 37–14. Ray 4 to the center point A of the mirror is useful here, because all the angles marked θ are the same, so triangles

$AS'S$ and $AI'I$ are similar triangles. Thus the magnitude of the **magnification**, M, defined as the ratio of the heights of the source and image, is

$$|M| \equiv \frac{|II'|}{|SS'|} = \frac{|i|}{|s|}. \qquad (37\text{–}3)$$

It is usual to go a little further and specify whether the image is upright or inverted. This can be done in a very simple way, by writing

The magnification

$$M = -\frac{i}{s}. \qquad (37\text{–}4)$$

If M is negative, the image is inverted; if M is positive, the image is upright. We can verify that this form works in explicit cases:

1. When the mirror is concave and the source is outside the focal point, the image is real (i is positive). By Eq. (37–4), M is then negative and the image should be inverted, as it is in Fig. 37–9a.
2. When the mirror is concave and the source is inside the focal point, the image is virtual (i is negative). By Eq. (37–4), M is then positive and the image should be upright, as it is in Fig. 37–10.
3. When the mirror is convex, the image is virtual (i is negative). By Eq. (37–4), M is then positive and the image should be upright, as it is in Fig. 37–12.

Equation (37–2) can be rewritten as

$$\frac{1}{i} = \frac{1}{f} - \frac{1}{s} = \frac{s-f}{fs}.$$

We can thereby find M in a form in which the image distance does not appear:

$$M = -\frac{i}{s} = -\frac{f\cancel{s}/(s-f)}{\cancel{s}} = \frac{f}{f-s}. \qquad (37\text{–}5)$$

This form can be applied to both concave and convex mirrors if we recall that f is negative for convex mirrors. In the convex case, $f - s$ is always negative, so M is always positive; the image is always upright. Also, $|f - s| = |f| + |s|$ is always larger than $|f|$ for convex mirrors, so the image is always reduced in size.

EXAMPLE 37–2 A convex spherical mirror of radius of curvature R of magnitude 20.0 cm produces an upright image precisely one-quarter the size of an object, a candle. What is the separation distance between the object and its image?

SOLUTION: We can find f from Eq. (37–1), $f = R/2 = -10.0$ cm. (The negative sign indicates that the mirror is convex.) Equation (37–5) then gives

$$s = f\left(1 - \frac{1}{M}\right).$$

In this case we know that $M = \frac{1}{4}$ (it is positive because the image is upright), so

$$s = f\left(1 - \frac{1}{\frac{1}{4}}\right) = -3f = -3(-10.0\text{ cm}) = 30.0\text{ cm}.$$

The problem asks for the separation distance between the object and the

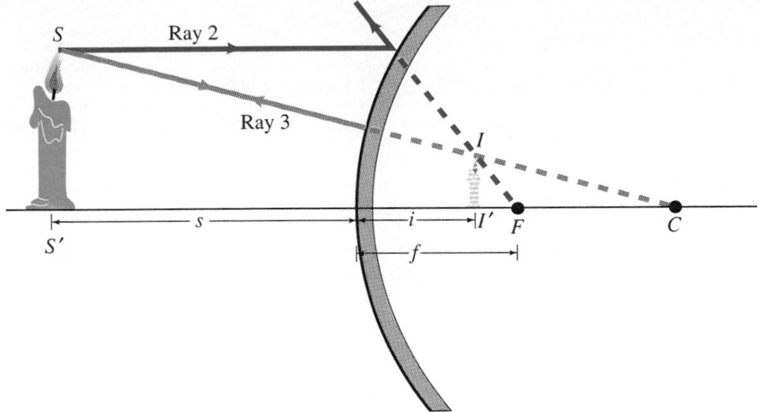

FIGURE 37–15 Example 37–2. Ray tracing to find the image.

image; hence we must also find i. It can be found in terms of s from Eq. (37–4):

$$i = -sM = -(30.0\ \text{cm})(\tfrac{1}{4}) = -7.5\ \text{cm}.$$

The minus sign is consistent with our knowledge that the image of a convex mirror is virtual (on the far side of the mirror). Then the object–image separation $s - i$ is

$$s - i = 30.0\ \text{cm} - (-7.5\ \text{cm}) = 37.5\ \text{cm}.$$

The geometric construction in Fig. 37–15 confirms these calculations.

The relation between source distance, image distance, and focal length [Eq. (37–2)], the expression for magnification [Eq. (37–4)], and the ray-tracing techniques are applicable to lenses as well as to mirrors. These three elements are sufficient for dealing with all the cases of interest to us.

The Plane Mirror as a Special Case of a Curved Surface

A plane mirror represents a concave or convex spherical mirror in the limit that $R \to \infty$, or, equivalently, $f \to \infty$. In such a case Eq. (37–2) implies that

$$i = -s,$$

which is just our earlier result that the image is virtual and as far behind the mirror as the source is in front of it. We can calculate the magnification of a plane mirror from Eq. (37–5):

$$\text{for a plane mirror:} \qquad M = \frac{f}{f - s} \xrightarrow[f \to \infty]{} \frac{f}{f} = 1.$$

The magnification is one, and the image is upright.

Ray Tracing

We have used ray-tracing techniques to learn about the images produced by certain types of mirrors (as we shall use the same techniques in Section 37–4) to learn about the images made by some simplified lenses. These techniques are the basis for the design of optical systems made from mirrors and lenses, especially the most sophisticated systems. We may want an optical system to produce a very sharp image over a very limited range of source distances; for example, there is no need for the optical systems of orbital satellites to be capable of focusing to short distances. Or we may want to sacrifice acuity in order for an optical system to operate

TABLE 37–1

SIGN CONVENTIONS FOR MIRRORS, REFRACTING SURFACES, AND LENSES

In applying the information in this table, we need to distinguish two "sides" to a reflecting or refracting surface:

Side A, the side from which light originates, and
Side B, the side to which light passes.

For mirrors, side B is identical to side A; for refracting surfaces and lenses, the two sides are opposite. Only the sign of the source position is determined by side A. All other quantities are determined by reference to side B.

Determined by side A	
Source distance s	Positive if object is on side A (real object) Negative if object is on side opposite to side A (virtual object)
Determined by side B	
Image distance i	Positive if image is on side B (real image) Negative if image is on side opposite to side B (virtual image)
Curvature R	Positive if center of curvature is on side B Negative if center of curvature is on side opposite to side B
Focal point	Positive if on side B Negative if on side opposite to side B

in dim light. Such systems may have mirrors in which the surface is nonspherical, or thick, multielement lenses in which the elements move relative to one another, as in zoom lenses. Designers of optical systems use computer programs capable of tracing large numbers of rays in a system design, of previewing the quality and placement of the image, and of creating modified designs in which the desired optical properties are attained.

We conclude this section with a comment on signs. Do not bother to try to keep track of the signs of the various quantities we have discussed. Develop your ray-tracing techniques, and you will be able to derive the signs on your own. Table 37–1 gives the signs of all the quantities necessary for mirrors, refracting surfaces, and lenses. We recommend, however, that you do not rely on the table too heavily.

37–3 REFRACTION AT SPHERICAL SURFACES

Mirrors change the direction of rays of light and create real or virtual images of objects. The same objectives can be accomplished by using refraction rather than reflection. Where the law of equal angles of incidence and reflection is enough to determine the behavior of mirrors, Snell's law is enough to determine the behavior of lenses. As for mirrors, we want only to understand the qualitative behavior of lenses, not to design real ones, and we shall assume that the surfaces are spherical sections without too much curvature and consider only paraxial rays.

Here we shall consider light rays that traverse from a medium with one index of refraction to another across a curved boundary. By repeatedly applying the rules we develop for a single such boundary, we shall be able to understand lenses. In Chapter 36 we studied Snell's law, the basic law for describing the refraction of light that crosses the boundary between one medium with index of refraction n_1 and another medium with index of refraction n_2. Figure 36–9 shows the re-

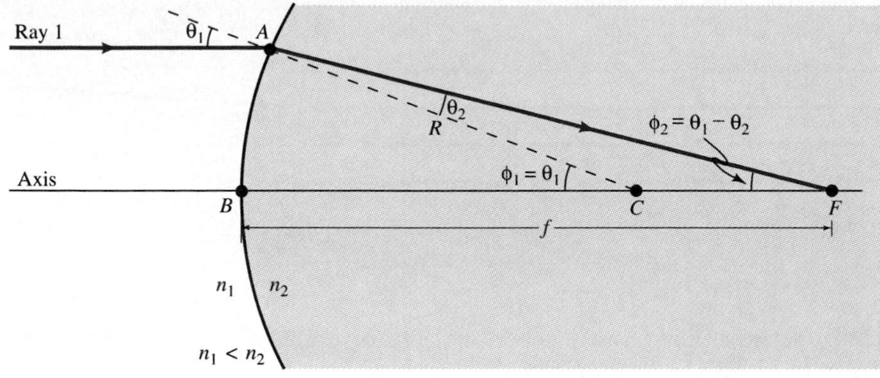

FIGURE 37–16 Ray tracing of a ray that enters a medium whose index of refraction is greater than that of the medium from which the ray came requires us to use Snell's law of refraction. Here we see refraction at a convex spherical surface.

fraction of such a ray. The angles of incidence and refraction satisfy Snell's law, Eq. (36–6):

$$n_1 \sin \theta_1 = n_2 \sin \theta_2.$$

We want to apply this law to a boundary that is not flat but forms a segment of a sphere of radius of curvature R. We take a convex surface, for which the center of curvature is point C in Fig. 37–16, in the region to which light passes. We choose $n_1 < n_2$, so that the light that passes from medium 1 to medium 2 bends toward the perpendicular to the surface. Nevertheless, the formulas we develop apply more generally.

The Focal Point of a Single Refracting Surface

As is true for a spherical mirror, a single refracting surface has a focal point F that we find by tracing rays that come from a very distant source, parallel to the axis. For the convex surface in Fig. 37–16, ray 1 bends toward the axis and crosses it at a point F. This point will be a focal point if all the incident rays that are parallel to the axis cross F—that the crossing point of a given ray is independent of the angle of incidence θ_1. For paraxial rays the angle of incidence, θ_1, and that of refraction, θ_2, are both small, so the relation $\sin \theta \simeq \theta$ is a good approximation. Thus Snell's law becomes

$$n_1\theta_1 \simeq n_2\theta_2. \tag{37–6}$$

Simple geometry shows that $\phi_1 = \theta_1$, and thus $\phi_2 = \theta_1 - \theta_2$. For small angles the relation between BF and the arc length AB in Fig. 37–16 is given by

$$BF(\theta_1 - \theta_2) \simeq AB.$$

Because $AB = R\theta_1$, this result, along with Eq. (37–6), implies that

$$BF \simeq \frac{R\theta_1}{\theta_1 - \theta_2} \simeq \frac{Rn_2}{n_2 - n_1}.$$

This distance is indeed independent of θ_1 for small angles, so *all* parallel rays near the axis pass through point F, and F is the image of a point source at infinity. The focal length f is the distance BF:

$$f = \left(\frac{n_2}{n_2 - n_1}\right)R. \tag{37–7}$$

Note that, in contrast to the case of the mirror, the focal point for a single refracting surface can be farther from the surface than the center of curvature, as shown in

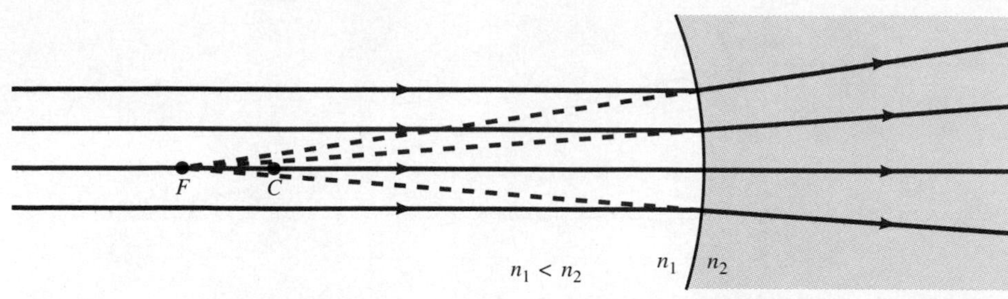

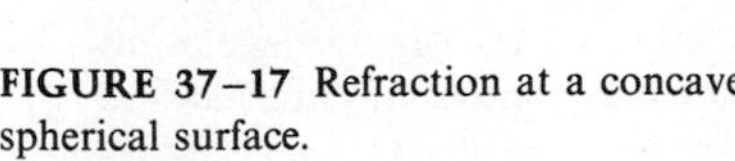
FIGURE 37–17 Refraction at a concave spherical surface.

Fig. 37–16. Whether it is or not depends on the relative sizes of n_1 and n_2. Although we have derived Eq. (37–7) for a convex surface, we can in fact derive it for a concave surface just as easily. The center of curvature C of the concave surface is on the side from which the light is incident. We find exactly the same formula, except that the focal point is to the left of the surface, on the same side as C. We see in Fig. 37–17 that the focal point F for such a surface is virtual.

The Image of an Extended Object

For the principal rays, see the Problem-Solving Techniques on pp. 1075–1076.

Convex Surface. Consider next a vertical object that stands erect on the optical axis. We already know enough about the principal rays to proceed with ray tracing. For a single refracting surface, only two of the four principal rays from any source point, such as point S, are useful. In Fig. 37–18 we show ray 1, incident parallel to the axis and refracting such that it crosses the axis at F, and ray 3, which forms the straight line through C. Ray 3 is perpendicular to the surface and hence does not deflect. The two refracted rays meet at point I. Again, these rays are only two of an unlimited number of rays that leave point S and pass through I. For example, in Fig. 37–18 we have drawn a "wild" ray. Detailed geometry (we shall not carry through this exercise) shows that the wild ray passes through I. By drawing the principal rays for any point on the object, we can reconstruct the entire image, which for the distance drawn is *real* (light rays do pass through it) and inverted.

Concave Surface. In Fig. 37–19, we offer three possibilities that depend on whether the source distance s is larger than f, smaller than R, or smaller than f but larger than R. In each case we use the same two principal rays from point S

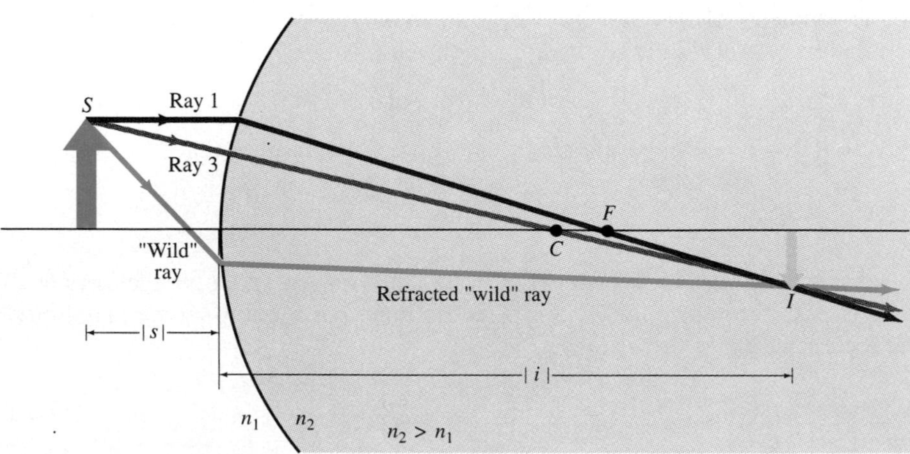

FIGURE 37–18 Ray tracing shows how a real image is formed by a convex spherical refracting surface.

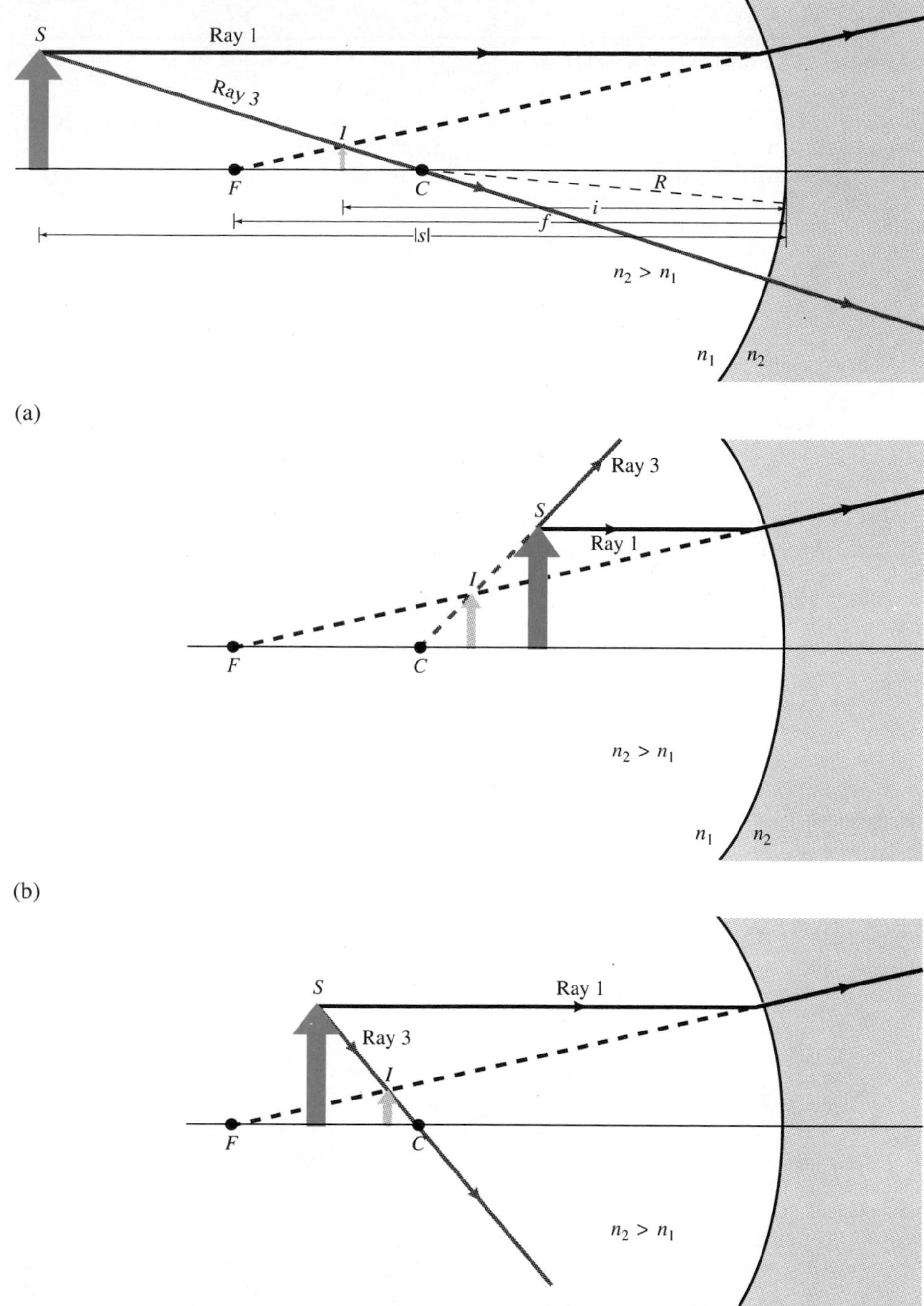

FIGURE 37–19 Ray tracing with principal rays for image formation for a concave spherical refracting surface: situations in which (a) $s > f$; (b) $R > s$; (c) $f > s > R$.

that we did for the convex case to locate the image of S at point I (Fig. 37–18). In Figs. 37–19b and 37–19c, ray 3 may not pass through the curved surface. We can extend the surface in order to pretend that ray 3 does pass through it, because the principal rays are just tools for determining where all the rays that pass through the surface cross. In each case the image is upright and virtual (the rays only appear to diverge from point I).

*DERIVATION

Deriving Equations (37–2) and (37–8)

To derive Eq. (37–2), we consider two points on an optical axis, the light source (or object), S, and its image, I, and a concave spherical surface. We can easily see from Fig. B2-1 and from the fact that the sum of the internal angles of a triangle is π that the following relationships hold:

$$\gamma = \beta + \alpha; \tag{B2–1}$$

$$\delta = \gamma + \alpha = \gamma + (\gamma - \beta) = 2\gamma - \beta. \tag{B2–2}$$

In arriving at Eq. (B2–2), we have used Eq. (B2–1). The distances of Fig. B2–1 are related to the angles by the exact relation $AB = R\gamma$ and by the *approximate* (small-angle) equations $AB = i\delta = s\beta$. These relations allow Eq. (B2–2) to be rewritten as

$$\frac{\cancel{AB}}{i} = \frac{2\cancel{AB}}{R} - \frac{\cancel{AB}}{s}.$$

We divide out the common factor AB and use the focal length $f = R/2$ for a spherical surface. We get

$$\frac{1}{i} + \frac{1}{s} = \frac{1}{f},$$

the relation we set out to establish: Eq. (37–2).

Now we derive Eq. (37–8), the relation between source distance, image distance, and focal length for a refracting surface with radius of curvature R, for the particular case of a convex surface (Fig. B2–2). We have taken $n_1 < n_2$, although the final result does not depend on this assumption. For small angles, the approximate form of Snell's law [Eq. (37–6)] holds. Simple geometry shows that $\theta_2 = \beta - \alpha$ and $\theta_1 = \beta + \gamma$, so

$$n_1(\beta + \gamma) = n_2(\beta - \alpha). \tag{B2–3}$$

The distances of Fig. B2–2 obey the exact relation $AB = R\beta$ and the small-angle approximations $AB = s\gamma = i\alpha$. Thus Eq. (B2–3) becomes

$$n_1\left(\frac{\cancel{AB}}{R} + \frac{\cancel{AB}}{s}\right) = n_2\left(\frac{\cancel{AB}}{R} - \frac{\cancel{AB}}{i}\right).$$

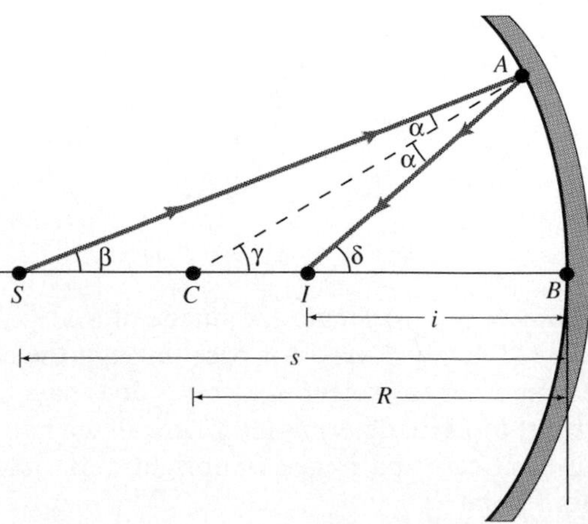

FIGURE B2–1 Geometric construction for deriving Eq. (37–2) for a spherical mirror.

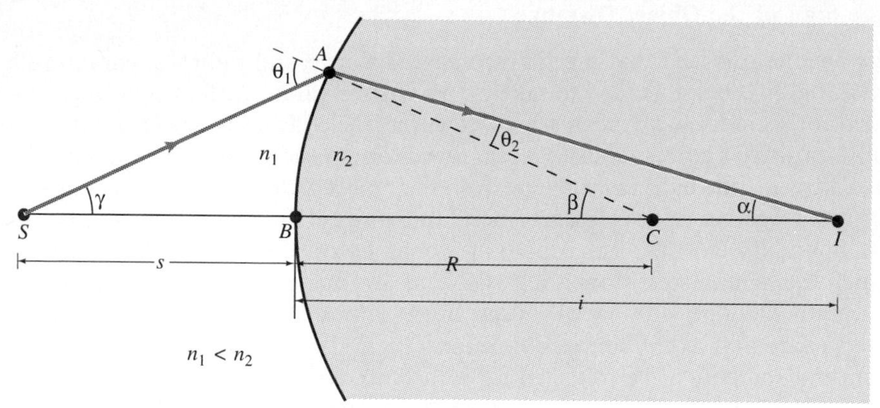

FIGURE B2–2 Geometric construction for deriving Eq. (37–8) for a spherical refracting surface.

With cancellation of the arc length AB,

$$\frac{n_1}{s} + \frac{n_2}{i} = \frac{n_2 - n_1}{R},$$

which is Eq. (37–8).

The Relation between Source Distance and Image Distance

The relationship between the positions of a source and an image for a single (concave) refracting surface is derived in the box "Deriving Equations (37–2) and (37–8)." There we find the result

$$\text{for a refracting surface:} \quad \frac{n_1}{s} + \frac{n_2}{i} = \frac{n_2 - n_1}{R}. \tag{37–8}$$

This equation is analogous to Eq. (37–2), which holds for a mirror. In deriving it, we assumed that the surface was convex; that is, that the center of curvature of the surface is on the side of the surface to which the light passes. Let us suppose that this corresponds to a positive value of R, as for the concave mirror. In addition, s was positive from the start. Whether i is positive or negative according to Eq. (37–8) is a numerical question; both are allowed. When i is positive, it is on the side of the surface to which light passes and the image is real, meaning that light does pass through it. When i is negative, it is on the side from which light is emitted and the image is virtual, meaning that light only *appears* to radiate from it when it is observed from medium 2. We can show that the result does not depend on taking $n_2 > n_1$ (see Problem 15), although a geometrical drawing would depend on it.

We can repeat this exercise with a concave spherical surface between the two media (see Problem 16). In this case the center of the spherical surface is on the side of the light source. The important result is that *Eq. (37–8) continues to hold, but with a negative value of* R. Just as when the image is on the side to which light goes, i is positive and the image is real, R is positive when it is on that side. When the image is on the side from which light radiates, i is negative and the image is virtual. Similarly, R is negative when it is on that side. See Table 37–1 for a summary of the sign conventions.

We shall not consider magnification for a single refracting surface. It is of much greater practical importance for a lens, which has two surfaces, and in Section 37–4 we shall derive it directly for that case.

The Sign of the Object Distance

We have established that a negative image distance and radius of curvature have meaning. Is it possible for s to take on a negative value? At first sight this concept does not seem to make much sense. However, the concept is both correct and useful. A positive s corresponds to a real object on the side from which the light radiates. For $+s$, the light rays diverge from the object as they approach the boundary surface. Negative values of s correspond to the rays that *converge* as they approach the boundary, so their extrapolation would be on the side of the boundary to which the light passes. How is it possible to arrange this state of affairs? Certainly it cannot be done with a real object. But it is possible if the image produced by one surface acts as the source object for a second surface (Fig. 37–20a). The rays from the source will be refracted at both boundaries 1 and 2. We can break up the problem and find first the image point I_1 produced by boundary 1 (Fig. 37–20b). In actuality, the light never forms the image point I_1, because boundary 2 intervenes. However, the image becomes the virtual object for the light refracted at boundary 2 (Fig. 37–20c). According to our convention, the source distance s_2 is negative, because the rays converge toward boundary 2, and the object is on the opposite side of the boundary from which the light radiates. The light comes from the left side of boundary 2, but the virtual object is on the right side. Equation (37–8) holds for the refraction at boundary 2, with a negative object distance s_2. The actual paths of the light rays through both boundaries are shown in Fig. 37–20d.

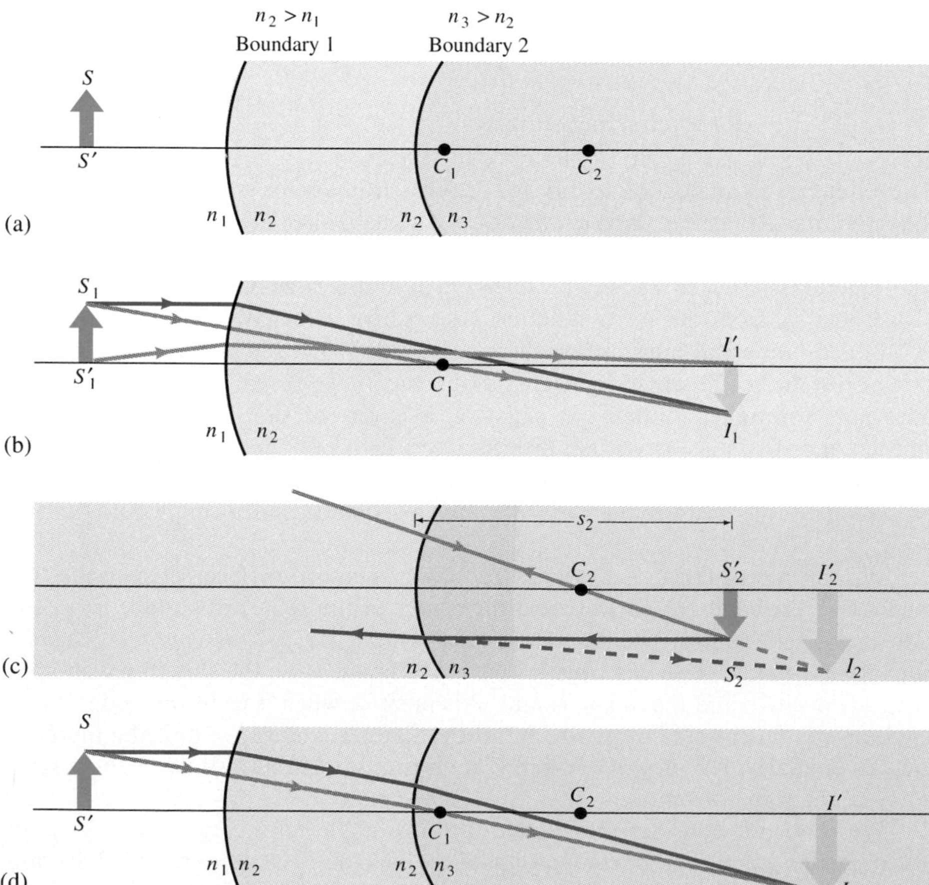

FIGURE 37–20 (a) Image construction when two refracting surfaces are involved. To simplify, (b) we can split the problem up by first finding the image point I_1 from boundary 1. (c) The resulting image serves as a virtual object; we use virtual source point S_2 for the interaction with boundary 2 to find the final image, at image point I. (d) The actual ray path.

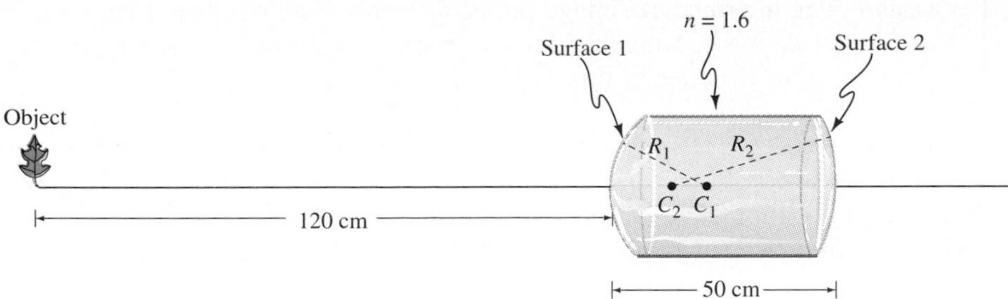

FIGURE 37–21 Example 37–3.

EXAMPLE 37–3 Consider a cylinder of glass 50 cm long, with $n = 1.6$, in air (Fig. 37–21). Surface 1 has a radius of curvature $R_1 = 0.20$ m, and surface 2 has a radius of curvature $R_2 = 0.40$ m. A small object (a leaf) is placed perpendicular to the optical axis at a distance of 120 cm from surface 1. (a) Find the location of the object's image due to refraction at surface 1. (b) Let this image be the source object for surface 2, and find the location of *its* image as light passes through surface 2.

SOLUTION: (a) We use Eq. (37–8) to calculate the distance i_1 of the image point I_1 from surface 1. We have $n_2 = 1.6$ and $n_1 = 1.0$ (air). The center of curvature is on the side to which light passes, so R_1 is positive. Finally, $s = +1.20$ m. We then have

$$\frac{\mathrm{n}_1}{s} + \frac{n_2}{i} = \frac{n_2 - n_1}{R} = \frac{1.0}{1.20\ \mathrm{m}} + \frac{1.6}{i_1} = \frac{1.6 - 1.0}{0.20\ \mathrm{m}};$$

$$i_1 = +0.74\ \mathrm{m}.$$

The image is real and located 74 cm to the right of surface 1.

(b) We now use the first image as an object for surface 2. Because the surfaces are separated by 50 cm, this new object is 24 cm to the right of surface 2. The object is on the side to which light passes, so its distance from surface 1 is negative: $s_2 = -0.24$ m. For this second step, $n_1 = 1.6$, $n_2 = 1.0$, and $R_2 = -0.4$ m (surface 2 is concave, so its center of curvature is on the side from which light radiates). Thus Eq. (37–8) now gives

$$\frac{1.6}{-0.24\ \mathrm{m}} + \frac{1}{i_2} = \frac{1.0 - 1.6}{-0.4\ \mathrm{m}};$$

$$i_2 = +0.12\ \mathrm{m}.$$

Because this is positive, the second image is real, or to the right of surface 2.

In Fig. 37–22 we have followed the principal rays 1 and 3 for surface 1 through to the production of the final image point, I_2, which is real and in-

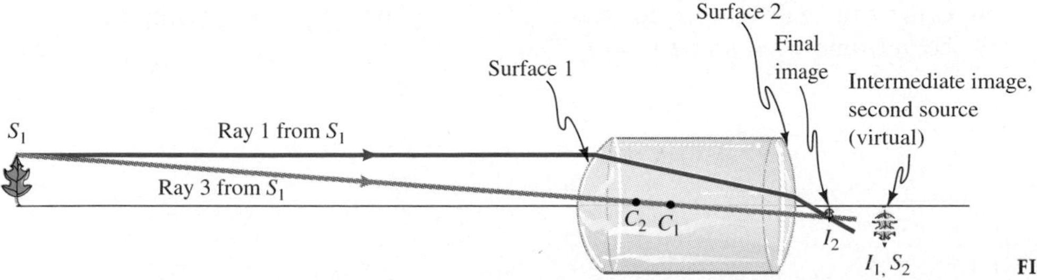

FIGURE 37–22 Two rays that form the final image in Example 37–3.

verted. The intermediate image point, I_1, is also shown. Ray tracing for this example requires too much precision in drawing to be useful. Application of the equations is better in this case.

In Example 37–3 we applied Eq. (37–8) successively. Although we have often stated that it is best to perform a calculation analytically before plugging in numbers, image location may be an exception. It is sometimes easier to locate the image of a first element numerically before using it for a second element.

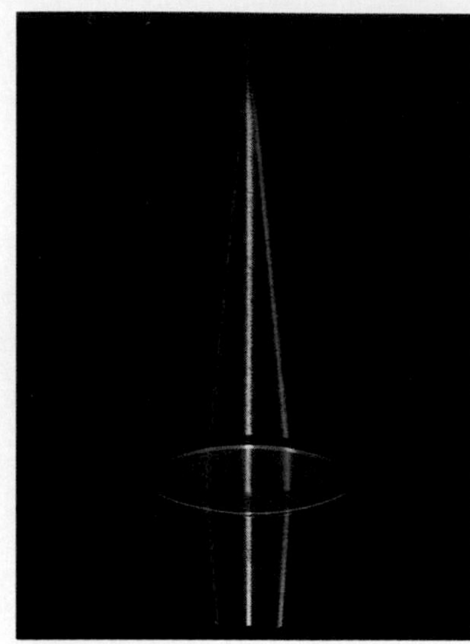

FIGURE 37–23 Light passing through a lens.

37–4 THIN LENSES

For our purposes a **lens** consists of transparent material of refractive index n embedded in a material of refractive index n_1, normally air, for which $n_1 = 1$ (Fig. 37–23). We shall assume that $n > 1$ and $n_1 = 1$. We shall also assume that lenses are thin, so that the distance from the object and the image to the lens is independent of which lens surface is involved. This simplifies the treatment considerably.† The two surface boundaries (1 and 2 in Fig. 37–24) are concave or convex spherical segments with respective radii of curvature R_1 and R_2. Whether these radii are positive or negative depends on whether the center of curvature is on the side to which light passes (positive R) or the side from which light radiates (negative R). For example, in Fig. 37–24a, R_1 is positive, whereas R_2 is negative.

Let us suppose that a real object is a distance s_1 to the left of a thin lens. We can locate the final image and identify its features by using Eq. (37–8) twice in succession for image-making at a single surface, much as we did in Example 37–3. The image produced by the first surface serves as the object for the second surface. We do not have to worry about whether the various objects and images are real or virtual, upright or inverted, and so forth, because the equation will automatically handle these questions. We have at surface 1

$$\frac{1}{s_1} + \frac{n}{i_1} = \frac{n-1}{R_1}, \tag{37–9a}$$

which we rewrite as

$$\frac{1}{i_1} = \frac{n-1}{nR_1} - \frac{1}{ns_1}. \tag{37–9b}$$

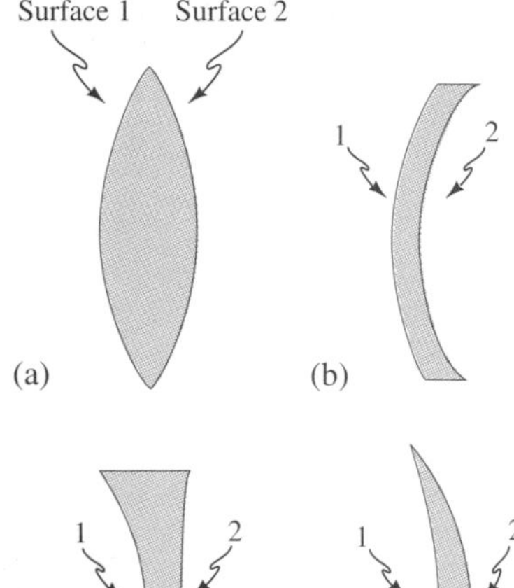

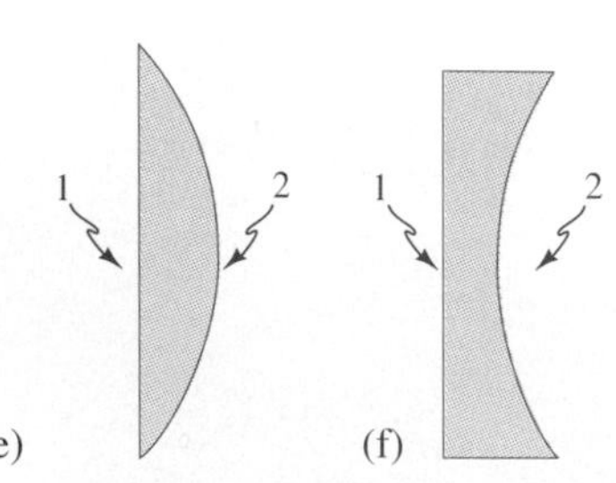

FIGURE 37–24 Six types of simple thin lenses with surfaces of different radii of curvature: (a) $R_1 > 0$, $R_2 < 0$; (b) $R_1 > 0$, $R_2 > 0$; (c) $R_1 < 0$, $R_2 > 0$; (d) $R_1 < 0$, $R_2 < 0$; (e) $R_1 = \infty$, $R_2 < 0$; (f) $R_1 = \infty$, $R_2 > 0$.

Now the image point I_1 produced by surface 1 serves as an object point S_2 for surface 2, producing a final image point at I_2. What is the sign of i_1? If i_1 is positive, the image is on the right of surface 1 and hence on the right of surface 2. This corresponds to an object distance s_2 for surface 2 that is negative. Similarly, if i_1 is negative, the image is to the left of both surfaces, corresponding to a positive object distance s_2 for surface 2. We must then reverse the sign of i_1 when we use it as the source distance s_2 for surface 2. Finally, note that in applying Eq. (37–8) a second time, $n_1 = n$ and $n_2 = 1$. Thus

$$\frac{n}{s_2} + \frac{1}{i_2} = \frac{1-n}{R_2};$$

$$-\frac{n}{i_1} + \frac{1}{i_2} = \frac{1-n}{R_2}.$$

† For instance, we draw rays for a thin lens as though they bend once, at the center of the lens.

When we substitute Eq. (37–9b) we find that

$$-n\left(\frac{n-1}{nR_1}-\frac{1}{ns_1}\right)+\frac{1}{i_2}=\frac{1-n}{R_2}.$$

If now we write $s_1 = s$ for the original object and $i_2 = i$ for the final image, we get (upon rearrangement)

for a thin lens in air: $$\frac{1}{s}+\frac{1}{i}=(n-1)\left(\frac{1}{R_1}-\frac{1}{R_2}\right). \quad (37\text{–}10)$$

Relation between image and object distances for a thin lens

Equation (37–10), which applies *only* to thin lenses in air, is the *lens-maker's equation*. By Eq. (37–10), the image can be positive or negative; that is, real or virtual. The signs are summarized in Table 37–1, and ray tracing will alternatively allow you to understand the image.

Equation (37–10) can be used to find the focal point of a lens. In the limit where $s \to \infty$, $i = f$, the distance of the focal point from the lens:

$$\frac{1}{f}=(n-1)\left(\frac{1}{R_1}-\frac{1}{R_2}\right). \quad (37\text{–}11)$$

The focal length of a thin lens

If we substitute this result into Eq. (37–10), we get Eq. (37–2), which we originally derived for mirrors. The sign of f is determined by the signs of the radii of curvature, but on the whole we can say that f is positive if the image of a point source at infinity is on the side to which light passes (real image), and negative if the image of the source at infinity is on the side from which light radiates (virtual image).

Equation (37–2) applies to lenses as well as to mirrors.

Ray 1, light coming in parallel to the optical axis of the lens, crosses (or behaves as though it crosses) the axis at f. Note that there is a certain degree of symmetry in Eq. (37–10). When light arrives from the right of the lens rather than the left, R_1 and R_2 reverse their roles, and light from infinity is focused the same distance from the lens, but on the opposite side, from the first focal point. In turn, if light radiates (or behaves as though it does) from one of the two symmetric focal points of the lens, ray 2, the light emerges as a set of parallel rays.

Principal ray 1 for a lens

Principal ray 2 for a lens

One more ray is useful for understanding the behavior of lenses. If, as in Fig. 37–25a, ray 4 (remember, principal ray 3 is not applicable to lenses) is drawn to the center of the lens, it behaves as though it passes straight through. We can see this by looking at the enlarged section (Fig. 37–25b). The two lens surfaces are to a good approximation parallel in the middle of the lens, so the ray behaves like a ray that passes through a pane of glass. (There is a *small* displacement of the ray, but it drops to zero as the lens becomes thinner.)

Principal ray 4 for a lens

We have thus found three principal rays that can be used to find the image.

For the principal rays, see the Problem-Solving Techniques on pp. 1075–1076.

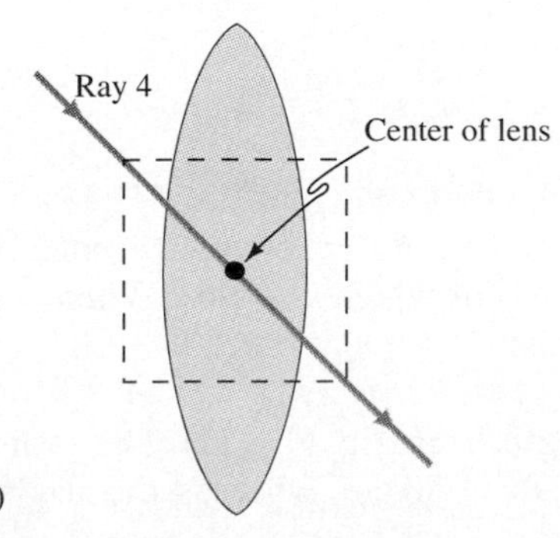

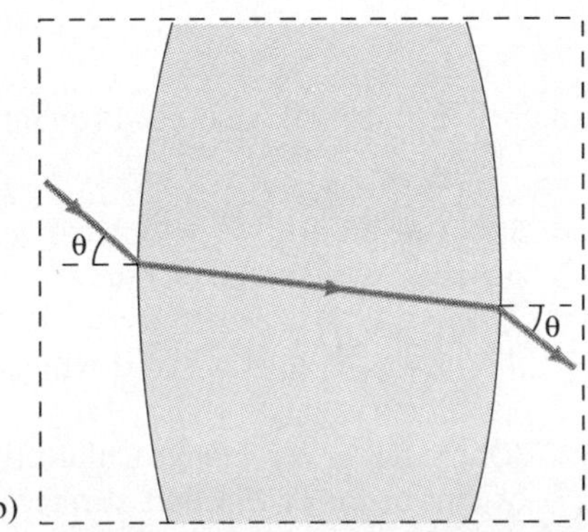

FIGURE 37–25 (a) The ray to the center of a lens passes through undeflected, because at its axis the lens is like a pane of glass. (b) An enlarged view of the same lens. If the lens is thin, the displacement of the ray is small.

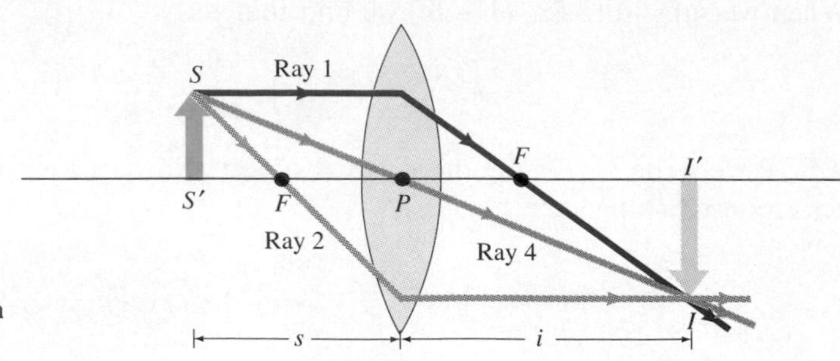

FIGURE 37–26 Ray tracing shows how a real image is formed with one type of thin lens. Point P marks the center of the lens.

Let us look at an example. Figure 37–26 shows a lens that collects light from an object. With the principal rays, we can easily find image point I of object point S. The construction works for any point on the object. The image is real and inverted, in this case. In general, if a lens causes rays that pass through it to come together, it is called a *converging lens*, and if it causes rays that pass through it to spread out, it is a *diverging lens*. Converging lenses have positive focal lengths, whereas diverging lenses have negative focal lengths. Some simple ray tracing will show that a lens like that of Fig. 37–24a is a converging lens, and one like that of Fig. 37–24c is a diverging lens.

Magnification

A thin lens produces a perfect image to the extent that the small-angle approximation is valid. Thus we can find the magnification by direct use of similar triangles. In Fig. 37–26, the magnification of the image has magnitude

$$M = \frac{II'}{SS'}.$$

From the geometry of the similar triangles $SS'P$ and $II'P$, we see that the magnitude of the magnification is $|M| = |i|/|s|$. Just as for mirrors, a systematic look at signs shows that we can decide with a single sign whether the image is upright or inverted:

$$M = -\frac{i}{s}.$$

This is Eq. (37–4), the same form we found for mirrors. If M is positive, the image is upright; if it is negative, the image is inverted. From Eq. (37-2), we have the alternate form

$$M = \frac{f}{f - s},$$

which is Eq. (37–5), also used for mirrors.

EXAMPLE 37–4 A converging lens like that shown in Fig. 37–24a has surfaces with radii of curvature $R_1 = 80$ cm and $R_2 = 36$ cm. An emerald 2.0 cm tall is placed 15 cm to the left of the lens, for which $n = 1.63$. Where will the image be located, and what will its size be?

SOLUTION: We first calculate the focal length from Eq. (37–11). The radius of curvature of the first surface is positive, $R_1 = 80$ cm, whereas the second

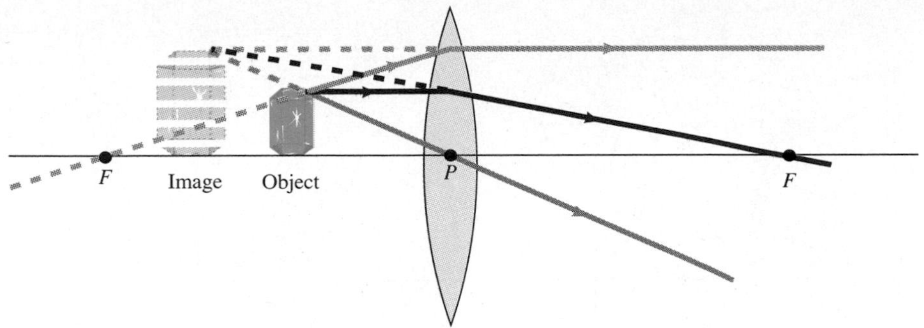

FIGURE 37–27 Example 37–4. When the object lies inside the focal point of the lens, ray tracing shows that the image formed is virtual.

surface has negative curvature, $R_2 = -36$ cm. Thus

$$\frac{1}{f} = (n-1)\left(\frac{1}{R_1} - \frac{1}{R_2}\right) = (1.63 - 1)\left(\frac{1}{80\text{ cm}} - \frac{1}{-36\text{ cm}}\right) = 0.025\text{ cm}^{-1}.$$

The object distance is positive, $s = 15$ cm, so Eq. (37–2) gives

$$\frac{1}{i} = \frac{1}{f} - \frac{1}{s} = 0.025\text{ cm}^{-1} - \frac{1}{15\text{ cm}} = -0.041\text{ cm}^{-1}.$$

Thus $i = -24$ cm. The minus sign indicates that the image is virtual and on the same side as the light source. The magnification is given by

$$M = -\frac{i}{s} = -\frac{-24\text{ cm}}{15\text{ cm}} = 1.6.$$

The positive value indicates that the image is upright. Figure 37–27 illustrates the ray paths.

We saw in Section 37–3 how the image produced by refraction at one surface acts as an object for the second surface. This principle extends to combinations of two or more lenses and lies at the heart of the design of complex optical instruments. Figure 37–28 gives an image construction for two thin converging lenses. The object SS' lies inside the focal length of lens 1 and thus gives rise to a virtual, enlarged image $I_1I'_1$. That image serves as an object $S_2S'_2$ for lens 2. Ray tracing uses the parallel ray $I_1A_2F_2I$ and I_1P_2I to determine the position of the real image, but the particular rays chosen really follow the path SA_1F_1AI and SP_1BI. This example shows that it is possible to obtain a magnified *real* image with two converging lenses, even though it is not possible with one lens.

In Example 37–5, the object for the second lens is a negative distance from the lens.

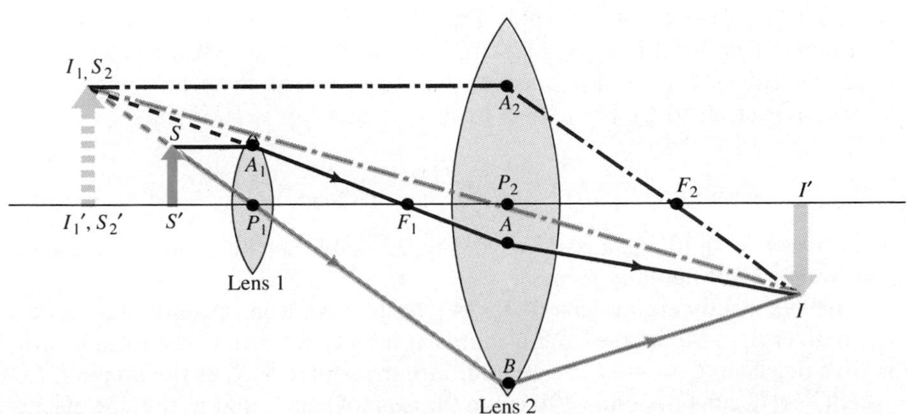

FIGURE 37–28 Ray tracing shows how two converging lenses produce a real magnified image.

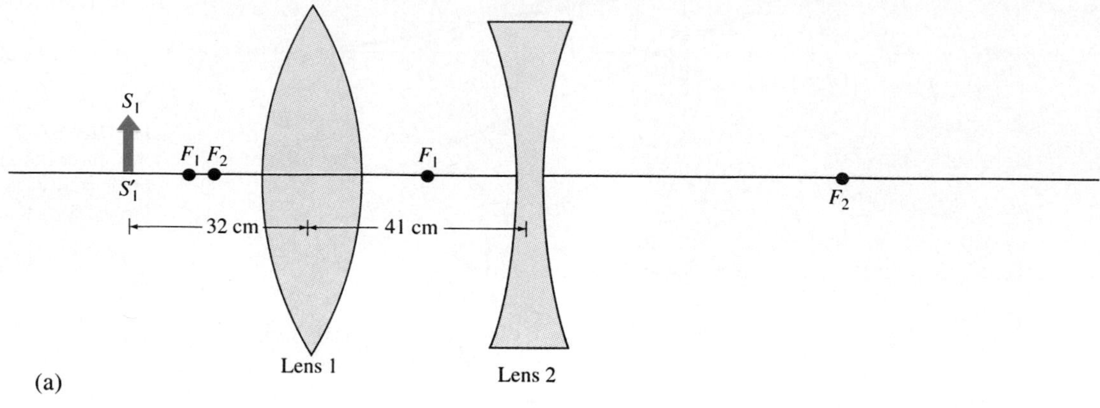

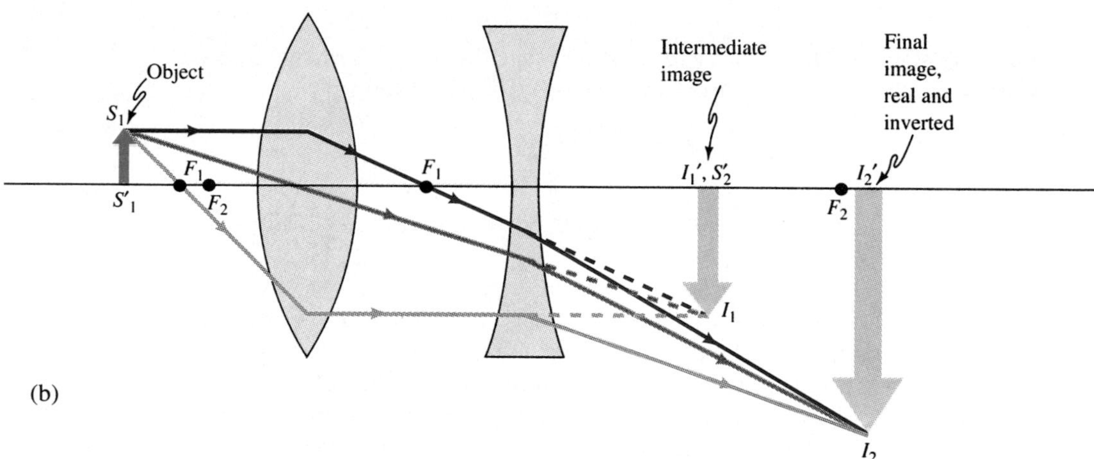

FIGURE 37–29 (a) Example 37–5.
(b) Ray tracing confirms the calculations.

EXAMPLE 37–5 Lens 1 in Fig. 37–29a is a converging lens with a focal length of 22 cm. An object is placed 32 cm to its left. Lens 2, which is a diverging lens with a focal length of 57 cm, lies 41 cm to the right of lens 1. Describe the position and other properties of the final image.

SOLUTION: We shall simply apply Eq. (37–2) twice. The focal length of lens 1 (a converging lens) is $f_1 = +22$ cm, and the real object distance $s = s_1 = +32$ cm. If lens 2 were not present, the object would form an image $I_1 I'_1$, which is determined by Eq. (37–2) to be at a position i_1 such that

$$\frac{1}{i_1} = \frac{1}{f_1} - \frac{1}{s} = \frac{1}{22\text{ cm}} - \frac{1}{32\text{ cm}}.$$

This gives $i_1 = +70$ cm, a real image (Fig. 37–29b). Because lens 2 is present, this image is not actually formed.

Lens 2 is a diverging lens. It takes parallel rays from infinity and makes them diverge, so the image from a source at infinity is virtual. The focal length is thus negative, $f_2 = -57$ cm. In addition, the source $S_2 S'_2$ is the image $I_1 I'_1$, which lies (70 cm − 41 cm) = 29 cm to the *right* of lens 2, that is, the side of lens

2 to which light passes. The source distance s_2 is thus negative, $s_2 = -29$ cm. The final image is then at distance i_2, determined by

$$\frac{1}{i_2} = \frac{1}{f_2} - \frac{1}{s_2} = \frac{1}{-57 \text{ cm}} - \frac{1}{-29 \text{ cm}} = \frac{1}{59 \text{ cm}},$$

or $i_2 = +59$ cm. The final image is real, 59 cm to the right of lens 2.

The total magnification is found by applying the magnification formula twice:

$$M_{\text{tot}} = M_{\text{lens }1} M_{\text{lens }2} = -\frac{i_1}{s_1}\left(-\frac{i_2}{s_2}\right) = \frac{-(+70 \text{ cm})}{32 \text{ cm}}\left[\frac{-(+59 \text{ cm})}{-29 \text{ cm}}\right] = -4.5.$$

The final image is inverted (M_{tot} is negative) and 4.5 times the size of the object. *The sign is automatically taken care of in the product of the two magnifications.*

Our calculations are verified qualitatively by ray tracing (Fig. 37–29b).

37–5 OPTICAL INSTRUMENTS

The Eye

The typical vertebrate eye, whose basic structure is shown in Fig. 37–30, is by itself a remarkable optical instrument. Light enters the eye proper through the *pupil*, the size of which can be changed by contraction or expansion of a membrane called the *iris* according to the intensity of the incident light. The light then passes through a convergent *crystalline lens* into a chamber filled with the *vitreous humor*, a fluid whose index of refraction is close to that of water. The light is focused onto the inner lining of the back of the eye, the *retina*, which is covered with sensitive receptor cells. The stimulation of these cells by light produces a message that is sent to the brain along the *optic nerve*, and the brain reconstructs the image.

When a normal eye is relaxed, objects at infinity form an image precisely on the retina, a distance of about 1.7 cm from the lens. When objects are brought closer, the lens is compressed by surrounding muscles and is made more convergent. The focal length is reduced, and the image continues to be focused on the retina. There is a limit to this power of *accommodation*. Objects closer than the *near point*, about 25 cm from the lens (or less for younger people), appear blurred. The near point tends to increase with age, because the lens becomes unable to compress as far as it once did, and the image of a near object is beyond the location of

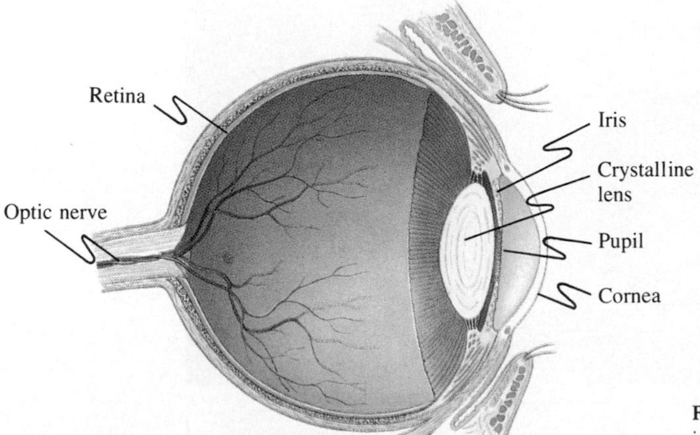

FIGURE 37–30 Schematic diagram of the human eye and some of its important features.

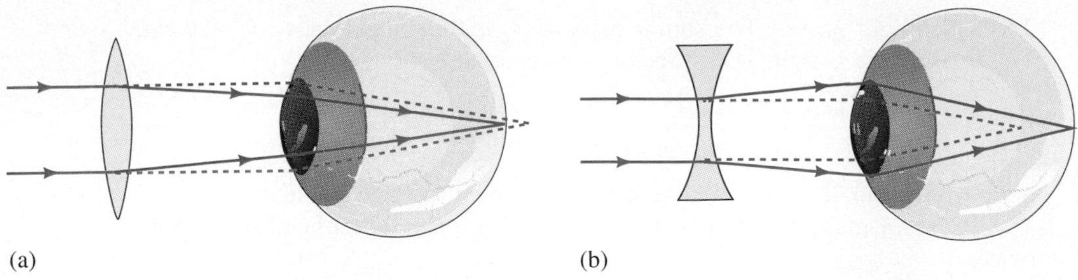

FIGURE 37–31 The dashed lines indicate the paths rays would take if no correcting lens were present. The solid lines mark the path of rays when a correcting lens is included. (a) A converging lens causes rays from an object, in this case at infinity, to focus closer to the lens of the eye. Such a lens corrects farsightedness by allowing the near point to be moved closer to the eye. (b) A diverging lens causes rays from an object, in this case at infinity, to focus farther from the lens of the eye. Such a lens corrects nearsightedness.

the retina. Converging lenses correct this problem (Fig. 37–31a). In cases of nearsightedness, the image of an object at infinity is in front of the retina. A diverging lens will provide the necessary correction (Fig. 37–31b).

The Camera

The camera is, with an important exception, optically equivalent to an eye. There is a converging lens in front, and the *film*, which plays the role of the retina, is in back. There is an *aperture*, an opening equivalent to the pupil, and a *shutter*, which provides an approximation to an instantaneous image and avoids blurring of the picture (Fig. 37–32). The difference between the simple camera shown and the eye is that in the eye, the focal length of the lens changes, whereas in a simple camera, the focal length is fixed. Instead, the camera lens moves in and out (changing the image distance) to enable objects of different source distances to produce a focused image on the film.

Angular Magnification

For those optical instruments used for observing the world closely, *angular magnification* is a critical concept, and we shall discuss it before we cover some other instruments.

From Eq. (37–5) we see that the magnification of a lens or mirror is infinite when $s = f$. This is less important than it might appear to be, because the image distance i also becomes infinite in that case. More important than the actual size of the image is the angle the image takes up in our field of vision. Given the limits of our own vision, *it is this angular coverage that determines how much detail we can see in an observed source.*

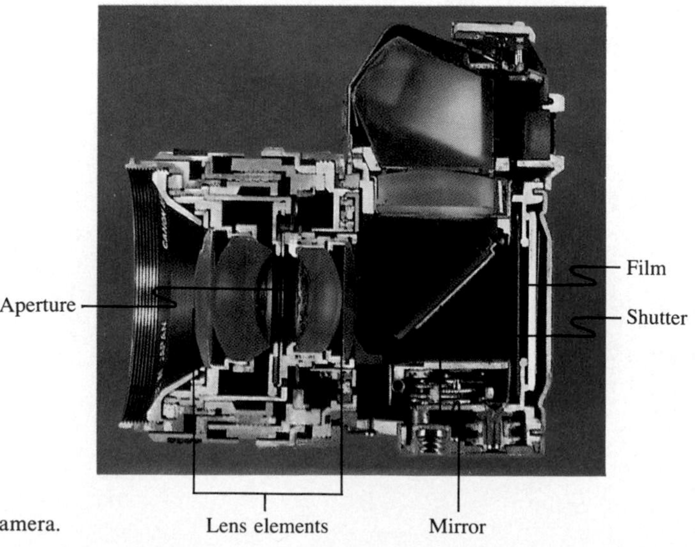

FIGURE 37–32 A cutaway view of the optical system of a camera.

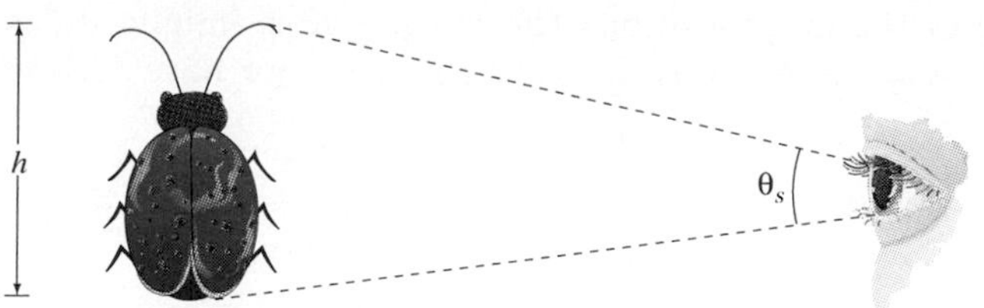

FIGURE 37–33 The angular size of an object, θ_s, is the relevant quantity for our ability to see detail in the object.

Imagine that you are a distance d from some object of height h (Fig. 37–33). For a source that does not cover an enormous part of your vision, the angular size θ_s of the source is

$$\theta_s \simeq \frac{h}{d}. \tag{37–12}$$

For normal, unaided vision, this angular size can be maximized when the object is brought to the near point of vision, around $d = d_{min} = 25$ cm, and it is $\theta_s \simeq h/(25\text{ cm})$ that is used as a reference for the angular magnification. Suppose now that we use an optical system to observe our source and that the image of the source as seen through the system has an angular size θ_i. Then the **angular magnification** of the system is

$$M_\theta \equiv \frac{\theta_i}{\theta_s}. \tag{37–13}$$

Angular magnification

We do not bother with signs here and keep track only of the magnitudes of the angular sizes. If we know the angular magnification of two elements that are superposed in an optical system, then the net angular magnification is the product of the angular magnifications of each element.

The Simple Magnifier

A converging lens has a positive focal length. By Eq. (37–2),

$$\frac{1}{i} = \frac{1}{f} - \frac{1}{s}.$$

For a real object, i passes from positive (real image) to negative (virtual image) as the object moves toward the lens through the point $s = f$. At this point, i shifts to $-\infty$. A *simple magnifier* is a converging lens with the object placed near $s = f$ (Fig. 37–34). If the object size is h, the image size is, by definition, $h_i = Mh$, where

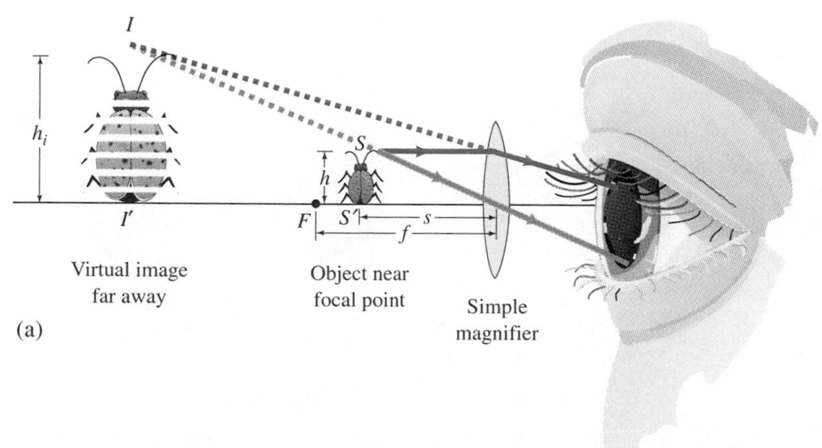

(b)

FIGURE 37–34 (a) The simple magnifier is a converging lens with an object placed near the focal point. The image is virtual and far away. (b) Such a magnifier in use.

$M = i/s$ is the magnitude of the magnification. The image size is infinite if i is infinite, but the *angular size of the image is finite.* When $s = f$, we have for the angular size

$$\theta_i = \frac{h_i}{i} = \frac{Mh}{i} = \frac{h}{s}\bigg|_{s=f} = \frac{h}{f}. \tag{37–14}$$

Note that we have no trouble seeing an image at infinity. At the near point $d_{\text{min}} =$ 25 cm, the angular size of our object is $\theta_{\text{object}} = h/d_{\text{min}}$. Thus the angular magnification of the magnifier is

$$M_\theta = \frac{\theta_i}{\theta_{\text{object}}} = \frac{\cancel{h}/f}{\cancel{h}/d_{\text{min}}} = \frac{d_{\text{min}}}{f}. \tag{37–15}$$

If we choose a converging lens with a focal length of 2 cm, we get an angular magnification of (25 cm)/(2 cm) = 12.5.

The Microscope

The **compound microscope**, invented around 1590 by Zacharias Janssen, provides high angular magnification for nearby objects (Fig. 37–35a). A two-lens system illustrates the principle of the microscope (Fig. 37–35b). The *objective* is a converging lens with a short focal length f_1, and the object to be viewed (of size h_0) is placed just outside the focal point of that lens. The image distance is given by $1/i = (1/f) - (1/s)$; i is large and positive for s just larger than f. (The image distance is roughly the distance between the two lenses, L.) This intermediate image has size $h_1 = Mh_0 \simeq Lh_0/f_1$. The second lens, the *eyepiece*, or *ocular*, acts as a simple magnifier: It is placed so that the intermediate image is at its focal length f_2 ($\ll L$). Then the angular size θ_i of the final image seen by an eye at the position of the eyepiece is given by Eq. (37–14) with $h_i \to h_1$ and $f \to f_2$:

$$\theta_i = \frac{h_1}{f_2} = \frac{Lh_0}{f_1 f_2}.$$

FIGURE 37–35 (a) A compound microscope. (b) Schematic diagram of a compound microscope. The objective produces an image close to the focal point of the eyepiece.

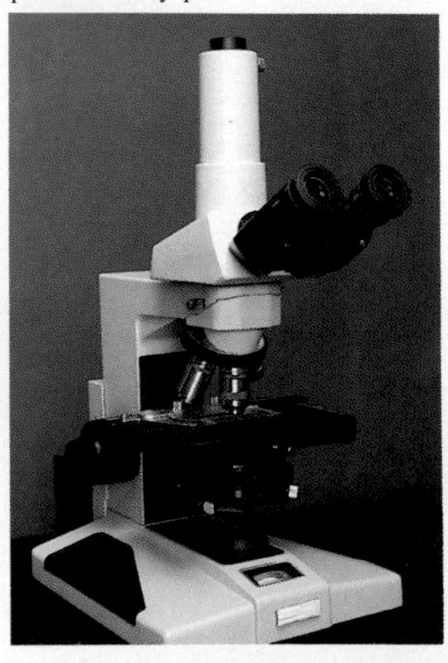

(a)

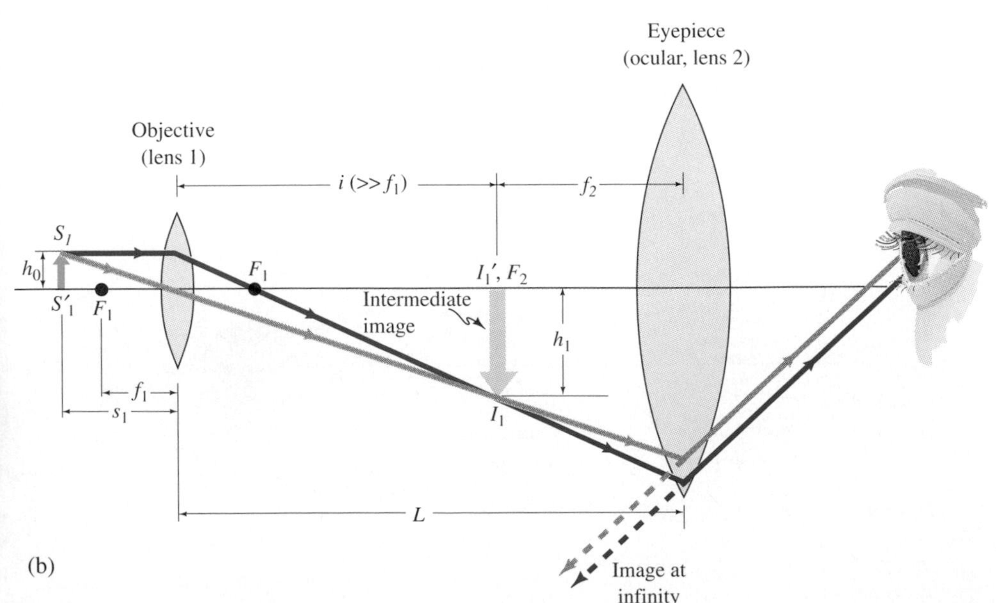

(b)

The net angular magnification is then given by Eq. (37–15):

$$M_\theta = \frac{\theta_i}{\theta_{\text{object}}} = \frac{L\not{h}_0/f_1 f_2}{\not{h}_0/d_{\text{min}}} = \frac{L d_{\text{min}}}{f_1 f_2}.$$

If we take typical values such as $L = 15$ cm, $f_1 \simeq 5$ mm, and $f_2 \simeq 2$ cm (d_{min} is always 25 cm), then the net angular magnification is 375.

The Telescope

The **telescope** magnifies very distant objects. It was invented in Holland at the beginning of the seventeenth century and made an impact on astronomy soon thereafter. Galileo built a *refracting telescope* (a telescope with strictly refracting elements) in 1609 (Fig. 37–36a). Within a year he published his first observations, which included a "bulge" around Saturn, four of the twelve moons of Jupiter, and the resolution of the Milky Way into individual, pointlike stars. Most telescopes are used as cameras rather than for direct viewing.

The refracting telescope resembles the microscope, but for the telescope the original object is in effect at infinity (Fig. 37–36b). The first lens, the objective, creates an intermediate image very close to the focal point of that lens. If that point coincides with the focal point of the eyepiece, then the eyepiece again acts as a simple magnifier. The final image is magnified. Let us calculate the angular magnification for an object that has angular size θ_s. (The moon, for example, has angular size of about 1°. We can distinguish with the naked eye stars separated by about 1′ of arc [1/60 of 1°].) If the original object has size h_0, the objective produces an image of size $h_1 = Mh_0 = ih_0/s = i\theta_s = f_1\theta_s$. The final image then has an angular

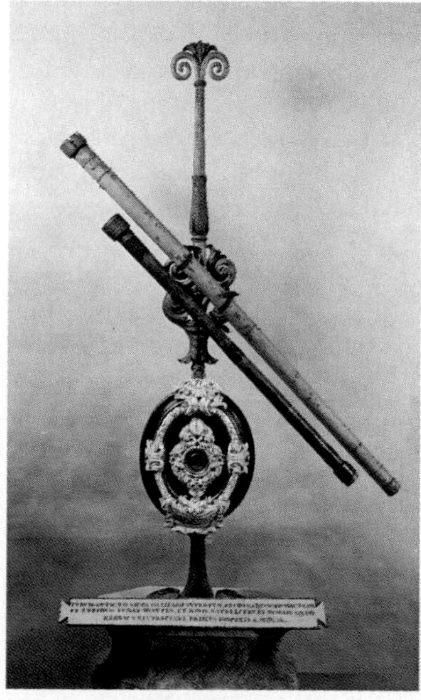

(a)

FIGURE 37–36 (a) Galileo's refracting telescope, used for viewing distant objects. (b) Schematic diagram of a refracting telescope.

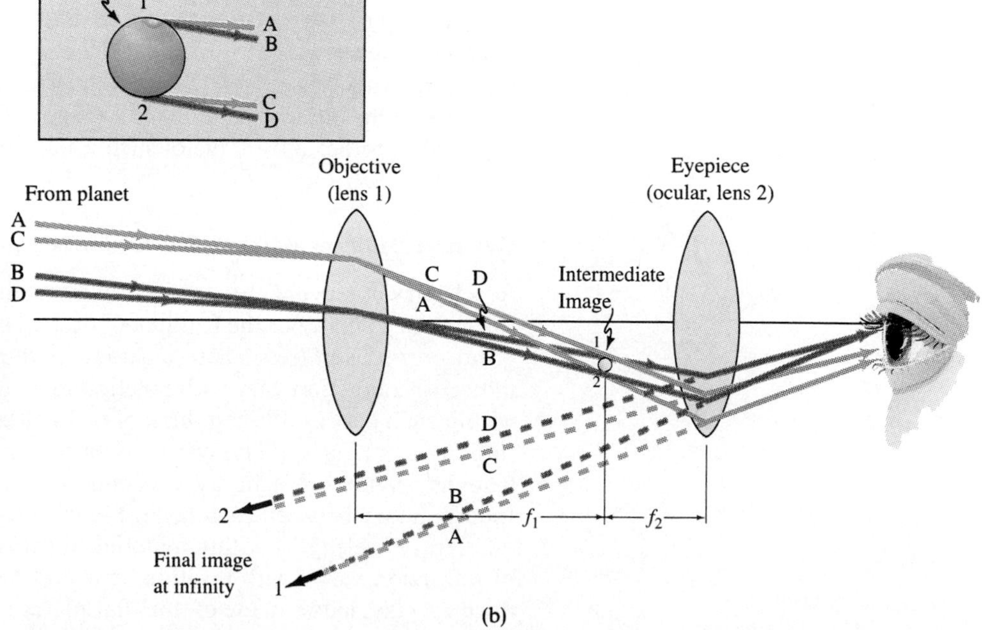

(b)

(a)

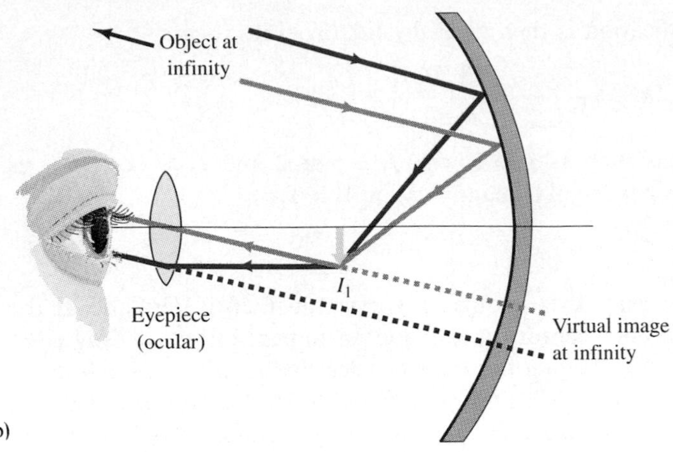

(b)

FIGURE 37–37 (a) A reflecting telescope, in which the objective of the refracting telescope is replaced by a concave mirror. (b) Schematic diagram of a reflecting telescope.

size given by Eq. (37–14) with $h \to h_1$ and $f \to f_2$, namely, $\theta_i = h_1/f_2 = \theta_s f_1/f_2$. In turn, the angular magnification is

$$M_\theta = \frac{\theta_i}{\theta_s} = \frac{f_1}{f_2}.$$

Thus, in contrast to the microscope, the objective lens of a telescope should have as long a focal length f_1 as is practical.

The study of distant galaxies depends on an examination of the spectrum of the light they emit (so that their composition and their Doppler shift can be determined) and of their energy output. The incident light from very distant objects is rather low in intensity, and more light is needed at the eyepiece in order to study spectra. To be most efficient at collecting light, the diameter of the optical system must be large. Large lenses are more difficult to construct than large mirrors, so most large telescopes are *reflecting telescopes* (Fig. 37–37a) rather than refracting telescopes. In a reflecting telescope, a mirror replaces the objective for the purpose of creating an intermediate image, which is then magnified by the eyepiece (Fig. 37–37b). Another advantage of such a telescope is that it has no chromatic aberration (see Section 37–6).

Geometric Optics without Curved Surfaces

Recent developments in manufacturing techniques hold the promise of being able to reproduce many of the features of optical systems without the need to mold or grind curved surfaces. These techniques allow the production of glasses whose indices of refraction vary with position in a controlled way. Light bends smoothly within such glasses. We can think of them as consisting of a series of infinitesimally thin layers of glass, each layer with an index of refraction infinitesimally different from its neighbor. A light ray will bend by an infinitesimally small angle as it crosses the boundary between each layer. Figure 37–38 shows the path taken by light of different wavelengths within material of this type. By tailoring the way the index of refraction varies with position, it would become possible, for example, to produce eyeglass lenses made of thin, flat plates of glass. Another possible application is exceptionally efficient collection of sunlight for solar energy.

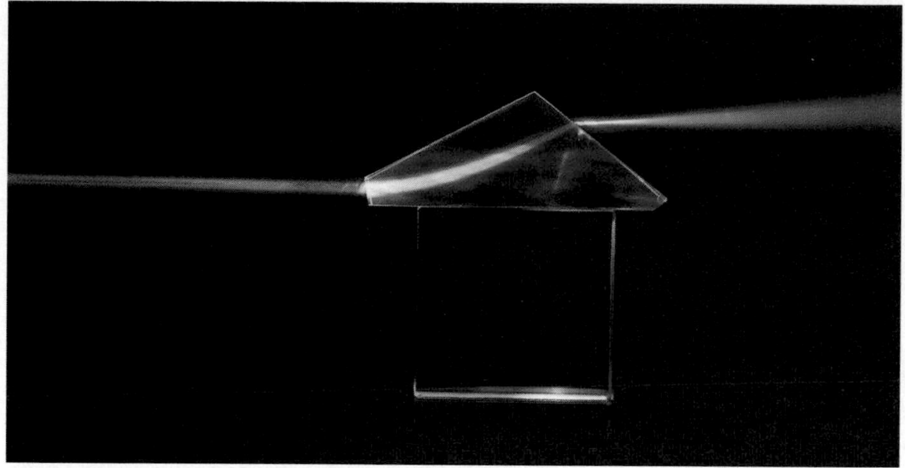

FIGURE 37–38 A sample of glass across which the index of refraction varies. As a result, light bends within the glass.

*37–6 ABERRATION

An accurate calculation would show that all rays that arrive at a spherical mirror or refracting surface from infinity cross in a small but finite region, not at a sharp point. This is but one example of **aberration** (Fig. 37–39). Aberration should be distinguished from *distortion*, in which an image is not identical in form to the object, as in Fig. 37–13. For scientific purposes, the image of Fig. 37–13 is not necessarily a bad one, because every ray from the object has its precise location in the image. Aberration concerns what we might call the quality of an image, not its geometric form. For example, if the film in a camera lies in a plane, then it is desirable that the image of a distant object or objects lies in a plane, and moreover that *all* the rays that enter the lens from a single point on an object go to precisely the same point on the film plane. When the image does not have these properties, there is aberration.

We can distinguish two important types of aberrations in geometric optics. *Monochromatic aberrations* describe the fact that in real optical systems, the rays from a given point on an object are not focused on a single image point (Fig. 37–40a). The correction for this type of aberration depends on the application. An optical system that collects images only from distant objects will have no aberration when a parabolic surface is used (Fig. 37–40b). Although such surfaces are difficult to construct from glass, a pool of mercury spinning about a vertical axis forms a parabolic surface, and such surfaces are employed in some modern telescopes. Alternatively, this type of aberration is minimized when the spherical

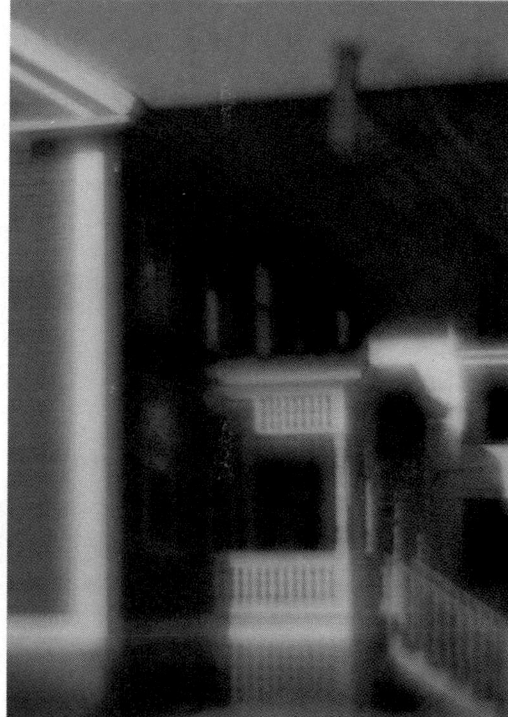

FIGURE 37–39 This poor image is the result of spherical aberration, a type of monochromatic aberration.

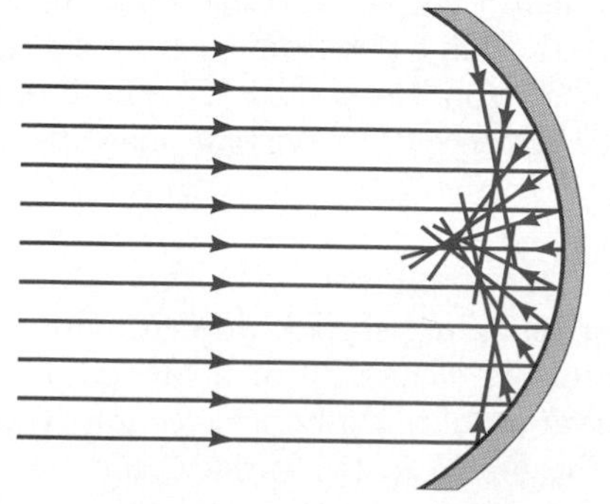

(a)

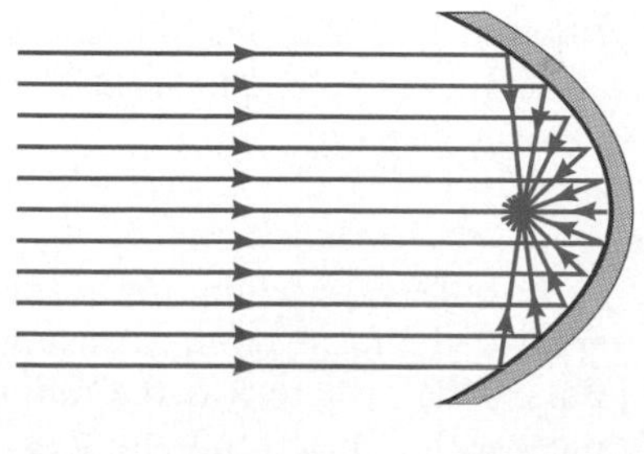

(b)

FIGURE 37–40 (a) In monochromatic aberrations, all the rays from infinity do not pass through the same point for a spherical mirror, so the focus is not sharp. (b) This type of aberration is eliminated by use of a parabolic mirror.

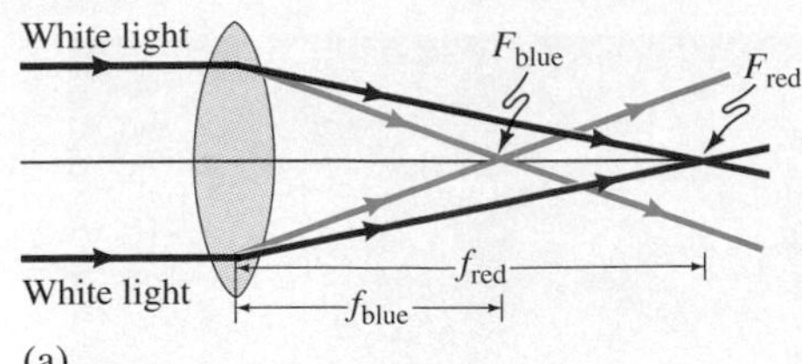

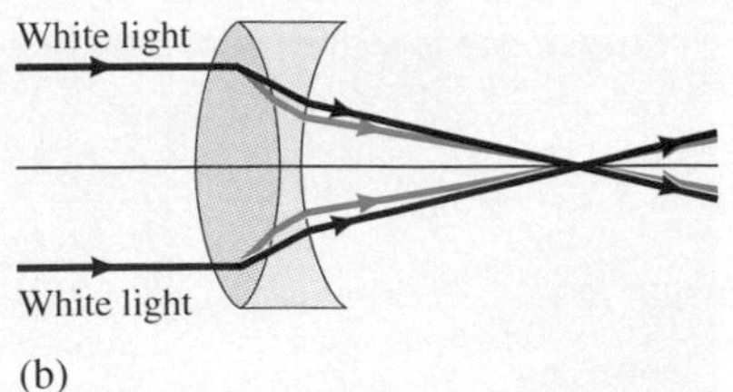

FIGURE 37–41 (a) In chromatic aberrations, the focal point of a converging lens may be different for different wavelengths. Here we show rays only for red light and blue light. (b) This type of aberration is eliminated by combining the lens with another lens with a different dispersion.

section of the lens surface or mirror is small, although then the system collects less light.

Chromatic aberrations appear in refracting systems but not in mirrors. As we remarked in Chapter 36, the index of refraction of a material depends generally on the wavelength of the incident light, the phenomenon of dispersion. Therefore the optical path of a ray at one wavelength may be different from the path followed by a ray at another wavelength (Fig. 37–41a). If a given point on an object is the source of a mixture of wavelengths (as is true of white light), then the image of this point is spread out according to the wavelength. A simple correction for chromatic aberration is to use filters that allow only a narrow band of wavelengths to pass. More commonly, several lenses are superposed (Fig. 37–41b). Different elements are designed to have canceling dispersion to minimize the net dispersion from successive passage.

Modern lenses overcome aberration by means of successive corrections with numerous refracting surfaces. A good camera lens may consist of a dozen elements of different types of glass, with complicated geometric relations, which, in the case of zoom lenses, are variable. We shall see in Chapters 38 and 39 that the wave nature of light provides a fundamental and unavoidable limitation on the ability of optical systems to produce sharp images, over and above the aberrations of geometric optics.

SUMMARY

Geometric optics is based on two basic laws of the behavior of light rays: In reflection, the angle of incidence on a reflecting surface is equal to the angle of reflection. In refraction, Snell's law, $n_1 \sin \theta_i = n_2 \sin \theta_r$, holds, where θ_i and θ_r are the angles of incidence and refraction, respectively, for a ray incident from medium 1 on medium 2. Ray tracing is a technique that allows us to locate the image of a given source.

A spherical reflecting or refracting surface forms an image of an object that is a distance s from the surface. Bundles of light rays pass through an image point (for a real image) or are directed as though they all come from such a point (for a virtual image). The image point is a distance i from the surface. One limit of such a surface is the plane mirror, for which the distance of the virtual image from the mirror is given by

$$i = -s.$$

Parallel rays falling on a reflecting or refracting surface approximately converge to the focal point, a distance f from the element. For a mirror, $f = R/2$ [Eq. (37–1)], where R is the radius of curvature of the spherical section. The distance of the object and the distance of the image to the surface, and the focal length for a spherical mirror are related by

$$\frac{1}{s} + \frac{1}{i} = \frac{1}{f}. \qquad (37\text{–}2)$$

Equation (37–2) applies to both convex and concave mirrors if proper account is taken of the signs of s, i, and R. The image size is magnified by a factor M, the magnification, times the object size, where

$$M = -\frac{i}{s} = \frac{f}{f - s}. \qquad (37\text{–}4, 37\text{–}5)$$

For a positive M, the image is upright; for a negative M, the image is inverted.

For a spherical boundary between a medium of refractive index n_1 and a medium of refractive index n_2, with light incident from medium 1, Eq. (37–2) is replaced by

$$\frac{n_1}{s} + \frac{n_2}{i} = \frac{n_2 - n_1}{R}. \qquad (37\text{–}8)$$

The same formula applies to both convex and concave spherical surfaces if proper account is taken of the signs of s, i, and R.

Thin lenses are understood by thinking of the image due to refraction at the surface nearest the object as the object for refraction at the second lens surface. For thin lenses in air, Eq. (37–2) applies, with

$$\frac{1}{f} = (n - 1)\left(\frac{1}{R_1} - \frac{1}{R_2}\right). \qquad (37\text{–}11)$$

Equations (37–2) and (37–11) combine into

$$\frac{1}{s} + \frac{1}{i} = (n - 1)\left(\frac{1}{R_1} - \frac{1}{R_2}\right). \qquad (37\text{–}10)$$

Equations (37–4) and (37–5) apply also to thin lenses.

Thin lenses can be used singly or in combination to make up optical instruments, including magnifiers, eyeglasses, cameras, mircoscopes, and telescopes. Angular magnification, which measures the ratio of the angular size of an object as seen through the instrument to the object's angular size as the naked eye sees it, is a fundamental consideration in these instruments. So is the quality of the image they produce.

QUESTIONS

1. Why is "ƎƆNA⅃UᗺMA" written on the front of an ambulance?
2. Consider a large room with walls covered with mirrors; at the center of the room is a candelabra with burning candles. Is the room brighter than a comparable room with black drapes in place of the mirrors?
3. The image of a distant candle is projected by a converging lens on a screen placed at the focal length of the lens. A piece of paper is taped over the lower half of the lens. Will only half of the image be seen?
4. Draw a right-handed coordinate system and its image in a plane mirror. Is the image a right-handed or a left-handed coordinate system? (In a right-handed system, the vector product **i** × **j** points along **k**.)
5. Would a dental mirror, the small mirror a dentist uses to examine your teeth, be concave, convex, plane, or sometimes one or another?
6. The side-view mirrors of some cars are labeled "Objects seen in this mirror may be closer than they appear." Is the mirror plane, convex, or concave?
7. For each of the simple lenses shown in Fig. 37–24, Eq. (37–4), for magnification, shows that the size of the image of a ball placed at the focal point is infinite. Can you see by ray tracing why this must be?
8. Figure 37–13 shows the reflection made by a spherical surface. Parts of all four walls of the room, even the wall behind the sphere, are visible in the image. Why?

9. Does the focal length of a lens change when the lens is in water?
10. When a magnifying glass is lined up perpendicular to the line between it and the sun, a hot spot forms on the side of the lens away from the sun. What is the relation between the distance of this hot spot from the lens and the focal length of the glass? Why does the spot become hot?
11. A camera works by forming a real image on a film plate. Can a camera take a picture of a virtual image?
12. In William Golding's novel *The Lord of the Flies* (1954), some boys rediscover fire with the aid of the sun shining through the eyeglasses of Piggy, a nearsighted boy. Has Golding made a mistake?
13. Are any principal rays useful for a point *on* the axis of an optical system?
14. When you have an eye exam, even one as simple as reading an eye chart, the examiner may dilate (open) the pupil by putting drops in your eye. Why is that useful?
15. Figure 37–42 is a photo of a lighthouse lens, which is designed to project a narrow, collimated beam of light long distances. Why is the lens glass cut into many pieces? How is it possible for the glass cut up in this way to act as a lens?

FIGURE 37–42 Question 15.

16. We mentioned in Section 37–5 the possibility of making a flat eyeglass lens with material in which the index of refraction varies with position. Sketch the profile of the index of refraction for a converging lens constructed in this way.
17. Legend has it that Archimedes, acting as an advisor to the ruler of Syracuse, devised an optical system made of shields that could concentrate sunlight sufficiently well to set enemy boats on fire from a distance. How plausible is this legend?

PROBLEMS

37–1 Images and Mirrors

1. (I) Consider two mirrors at right angles to each other. How many virtual images will a pointlike light source have?
2. (II) Consider two parallel mirrors that face each other, placed along the x-axis at $x = a$ and $x = -a$. Assume that a point source of light is placed at $x = x_0$ between the mirrors. What are the locations of the four images of the point source with the smallest values of image distance i?
3. (II) A mirror is exactly half your height, and the top of the mirror is aligned with the top of your head. (a) If your eyes were at the top of your head, how close would you need to be to the mirror in order to be able to see your feet? (b) If your height is 180 cm and your eyes are 15 cm below the top of your head, what would have to be done with the mirror so that you could see both the top of your head and your feet?
4. (III) Suppose that two plane mirrors meet at an angle of 60° (Fig. 37–43). An object is placed between the mirrors, on the line that bisects this angle. Use graphical methods or trigonometric methods to locate all the images.

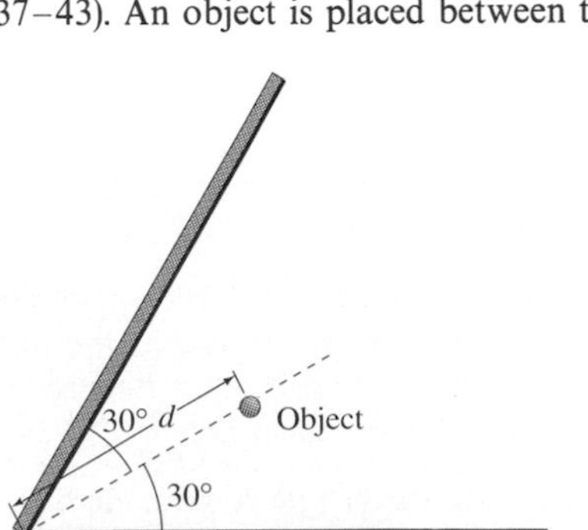

FIGURE 37–43 Problem 4.

37–2 Spherical Mirrors

5. (I) A dime 40 cm away from, and on the optical axis of, a concave spherical mirror produces an image 10 cm away from the mirror. If the dime is moved on the axis to 20 cm from the mirror, where will the image move? How large is the radius of the sphere of which the mirror is a section? Draw the system for the second case described.
6. (I) A printed page is placed 25 cm away from a convex mirror, part of a sphere of 50 cm. Where will the image be located, and what is the magnification? Make a sketch, including rays.
7. (II) Suppose that light from an object, real or virtual, falls on a convex mirror of radius of curvature R. For what value of s does i take on its largest positive value? (Consider both positive and negative values of s.) When i has its largest positive value, what is the magnification? For what positive or negative values or value of s does the magnification attain its largest absolute value?
8. (II) Show by using the same reasoning that we used in the text for the case of the concave mirror that the reflection of ray 1 in Fig. 37–11 appears to originate at point F, independent of the angle θ. Your argument shows that, at point F, there is an image of a source point at infinity.
9. (III) Use ray tracing for parallel rays far from, as well as near to, the optical axis to show that parabolic mirrors more accurately focus parallel rays than spherical mirrors do.

37–3 *Refraction at Spherical Surfaces*

10. (I) A sphere of glass ($n = 1.52$) of radius 10 cm is immersed in water ($n = 1.33$). A small flower is 5 cm outside the sphere. What are the location and nature (real or virtual) of the flower's image made by refraction at the first surface?

11. (II) Show by applying Eq. (37–8) that if light is incident on a convex refracting surface with $n_2 > n_1$ (see Fig. 37–18), there is a critical distance s_c such that the image of an object closer than s_c will be virtual. Find s_c, and show by ray tracing that the virtual image when $s < s_c$ is upright and magnified.

12. (II) Consider a convex spherical boundary between two media with an upright object whose extreme point is at S, as in Fig. 37–18. Suppose that $n_2 < n_1$ rather than $n_2 > n_1$. Find the nature of the image (inverted or upright, virtual or real, reduced or magnified) by tracing rays from S. Is there a critical distance at which the nature of the image changes, as in Problem 11?

13. (II) A convex spherical boundary produces an image whose distance from the boundary surface is governed by Eq. (37–8). Suppose that $n_2 > n_1$. (a) Show that when an object is very far from the surface, the image is a distance $i = n_2R/(n_2 - n_1)$ from the surface, and that the image is inverted, reduced, and real. (b) What is the distance s at which the image distance becomes infinite? (c) What is the position of the image for s just less than the critical value found in part (b)? Is it real? (d) As s continues to decrease, what happens to the position of the image?

14. (II) Consider a concave surface of radius of curvature R that separates two media with indices of refraction n_1 and n_2, where $n_2 > n_1$ (see Fig. 37–19). Find the distance s of an object for which the image, which is virtual, is superimposed on the object.

15. (III) In deriving Eq. (37–8), we took $n_1 < n_2$ in Fig. B2–2. Make a new drawing appropriate to $n_2 < n_1$ for a convex surface. Apply the same kind of reasoning, using small angles to show that the same algebraic formula applies whatever the relative sizes of n_2 and n_1.

16. (III) Show that Eq. (37–8) holds for a concave refracting surface with $n_2 > n_1$ by drawing a figure analogous to Fig. B2–2 and by making small-angle assumptions.

37–4 *Thin Lenses*

17. (I) The image of an object placed 24 cm away from a lens forms at a distance of 51 cm on the other side of the lens. (a) What is the focal length? (b) What type of lens is it? (c) Is the image real? upright? (d) What is the magnification?

18. (I) A double concave lens has radii of curvature of 26 cm and 20 cm. If the index of refraction of the lens is 1.53, what is the focal length?

19. (I) An apple is placed 15 cm in front of a diverging lens with a focal length of 22 cm. (a) Where is the image? (b) Is the image real? (c) upright? (d) What is the magnification?

20. (II) We want to form an image of an insect by using a converging lens with a focal length of 25 cm and a magnification of 2. (a) Where should the object be placed for the image to be real? (b) Repeat part (a) for a virtual image.

21. (II) An object 2 cm high is placed on one side of a thin converging lens of focal length 40 cm. What are the location, size, and orientation of the image when the object is (a) 80 cm from the lens, (b) 45 cm from the lens, (c) 35 cm from the lens, (d) 10 cm from the lens?

22. (II) The two surfaces of a thin lens have radii of the same sign and magnitude. Show by ray tracing that the focal length of this lens is infinite. Is the image produced by this lens real or virtual?

23. (II) A thin converging lens forms an image of a distant mountain at a distance of 25 cm from the lens. (a) What is the focal length of the lens? (b) A pine cone is placed 100 cm from the lens. Describe the resulting image: its magnification and distance from the lens, and whether it is real or virtual, upright or inverted. (c) The lens glass has an index of refraction of 1.6. The lens is immersed in a clear liquid with an index of refraction of 1.4. What is its focal length in this medium?

24. (II) Consider the thin lenses shown in Figs. 37–24a through 37–24d. Suppose that in each case the magnitudes of the radii of curvature are $R_1 = 10$ cm and $R_2 = 50$ cm, and that $n = 1.5$. (a) Find the focal lengths for each of the four lenses, and use the sign of the focal lengths to obtain the locations of the image of a distant source. (b) In each case, is the image upright or inverted, real or virtual? (c) Calculate the magnification, M, from Eq. (37–4), and check that it is consistent with your results in part (b).

25. (II) An object is placed 10 cm to the right of each of the lenses of Problem 24. For each case, locate the image, state whether it is upright or inverted and real or virtual, and give the magnification.

26. (II) Repeat Problem 25 for an object placed 4 cm to the right of each of the lenses.

27. (II) Two thin lenses of focal length f_1 and f_2, respectively, are aligned along the same axis and placed very close together. Show that the focal length f of the combination is given by

$$\frac{1}{f} = \frac{1}{f_1} + \frac{1}{f_2}.$$

37–5 *Optical Instruments*

28. (I) The eyes of an elderly person have near points of 70 cm. For such a person to be able to read a book at a distance of 30 cm with corrective lenses, what must the focal length of the lenses be?

29. (I) A nearsighted person has near and far points of 13 cm and 22 cm, respectively. (The *far point* is the farthest point at which a person can see clearly.) (a) Determine the lens required for this person to be able to see clearly at infinity. (b) What does the lens correction of part (a) do to the near point? Can the person still easily read a book?

30. (I) What is the magnification of a telescope that has an objective lens with a focal length of 60 cm and an eyepiece with a focal length of 2.9 cm?

31. (II) The two lenses of a telescope with magnification of $200\times$ are separated by 92 cm. What are the focal lengths of the lenses?

32. (II) Calculate the angular magnification of the reflecting telescope shown in Fig. 37–37b.

*37–6 *Aberration*

33. (II) Consider a spherical mirror without making the paraxial approximation (Fig. 37–44). Show that when a ray parallel to the axis makes an angle θ with the radius R at the point of contact, then f, which in this case is the distance at which the ray crosses the axis, is given by

$$f = R\left(1 - \frac{1}{2\cos\theta}\right).$$

Show that for small angles, this formula reduces to $f = R/2$. Note that F is not the focal point here (there is no sharp focus), but only the point at which some particular ray crosses the axis.

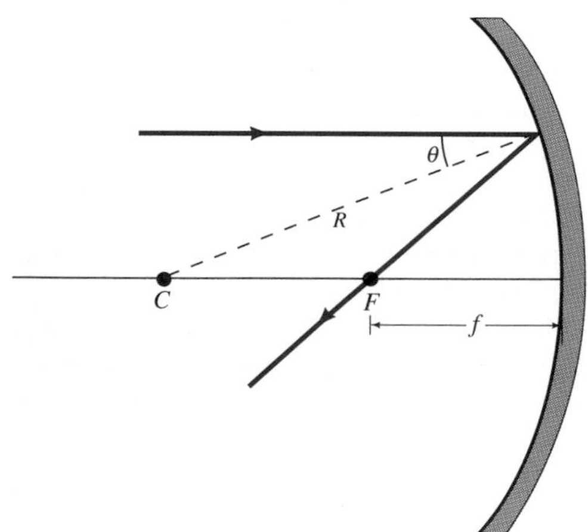

FIGURE 37–44 Problem 33.

34. (II) Use the result of Problem 33 to calculate the spread in values of f for a hemispherical mirror of radius 1 m and arc length 40 cm.

35. (II) The index of refraction of optical glass used for a thin lens with $R_1 = +20.00$ cm and $R_2 = +28.75$ cm is $n = 1.48523$ for light of wavelength $\lambda = 587.6$ nm and $n = 1.48135$ for light of wavelength $\lambda = 768.2$ nm. What is the difference in the focal length for these two wavelengths?

General Problems

36. (II) Consider a circular concave mirror of focal length f and diameter d, where $f \gg d$. This mirror's optical axis is aligned with the sun. What is the area of the spot that contains the reflected rays as a function of the distance L from the mirror if $L < f$? Sunlight has an intensity I as it arrives at the mirror. Find the intensity of the reflected rays as a function of L. Treat the sun as a point source.

37. (II) You are given a converging lens (Fig. 37–24a) with equal radii of curvature, and a diverging lens (Fig. 37–24c) with the same radii of curvature as those of the converging lens. The lenses are made of material with $n = 1.50$, and the radii of curvature are all 50 cm. They are placed at opposite ends of a tube 20 cm long, and the nearer lens is 20 cm from an object. What is the location of the image that results from the two refractions? Does it make a difference whether the converging or the diverging lens is closer to the object?

38. (II) Consider a 50-cm-long cylinder of glass in air, with $n = 1.6$, like the cylinder shown in Fig. 37–21. The two ends are shaped into sections of spheres; each has radius 20 cm. A small object is placed perpendicular to the optical axis at a distance of 15 cm from one of the spherical surfaces. (a) Find the location of the object's image due to refraction at surface 1. (b) Let this image be the object for surface 2, and find the location of *its* image as light passes through surface 2. (c) Use ray tracing to determine if the final image is upright or inverted.

39. (II) Two concave mirrors M_1 and M_2 face each other. They have respective radii of curvature of 24 cm and 40 cm and are separated by 60 cm. A light bulb is placed on the optical axis 6 cm from M_1. (a) Where is the image of the bulb formed by M_1? Draw the system. (b) The image of the light bulb formed by M_1 can in turn form an image as the result of reflections from M_2. Construct this second image by ray tracing, starting from the source light bulb.

40. (II) A lens, made of glass with $n = 1.5$, has the configuration shown in Fig. 37–45, and a candle is placed 50 cm from surface 1. The lens cannot be thought of as thin; it has a thickness of 5 cm. (a) Where is the image made by surface 1? Is it inverted? What is the magnification? (b) By using the image made by surface 1 as an object for surface 2, find the final image's location relative to the candle as well as the magnification of the image. Is it inverted or upright?

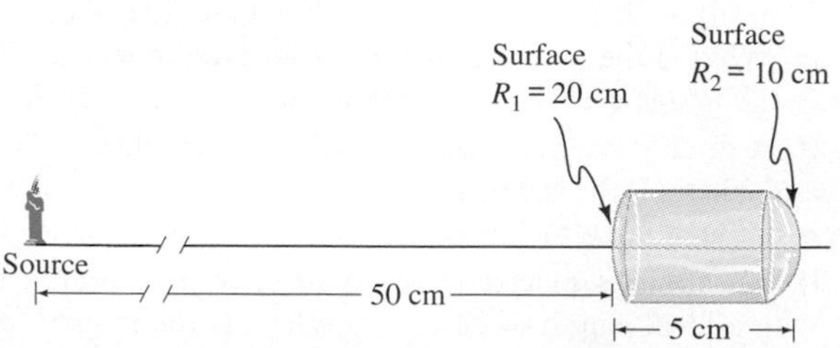

FIGURE 37–45 Problem 40.

41. (II) A thin lens with focal length f_1 is placed a distance d in front of a concave mirror with focal length f_2. What is the focal length of the combination?

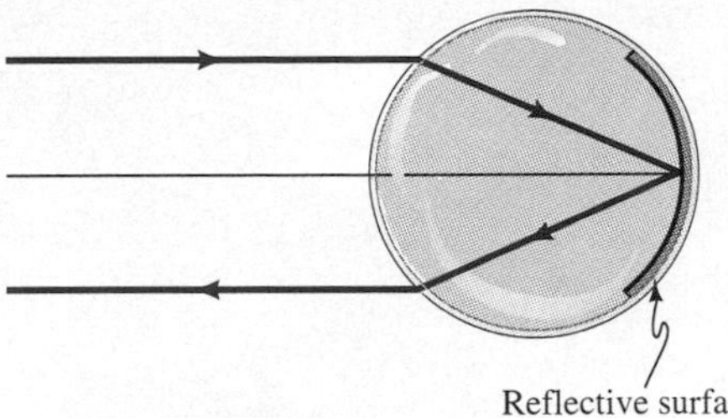

FIGURE 37–46 Problem 42.

42. (II) Consider the sphere of a glass with $n = 2$ in Fig. 37–46. Any incoming ray is parallel to an axis through the middle of the sphere and will be refracted, striking the rear surface of the sphere at the axis. Demonstrate that this is so for paraxial rays. If the back surface is painted with a reflecting material, symmetry shows that the ray will come back out in the opposite direction. Tiny spheres of this type are used for highway reflectors.

43. (III) Rays of light strike a spherical glass surface parallel to the optical axis (Fig. 37–47). The incoming ray makes an angle θ with the normal to the surface. Show that the rays will cross the optical axis at a distance $d = R/(\sqrt{n^2 - \sin^2\theta} - \cos\theta)$ beyond the center of the sphere, where n is the index of refraction of the glass. To what does this expression reduce for small angles?

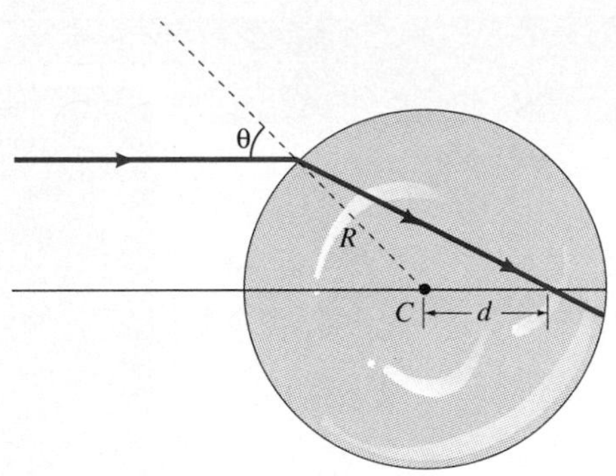

FIGURE 37–47 Problem 43.

44. (III) The index of refraction of glass varies from 1.528 (for blue light) to 1.511 (for red light). Use the result of Problem 43 to calculate the color spread on the axis for light that strikes a hemispherical cap at the end of a glass rod, at an angle $\theta = 0.4$ rad. Take the radius of curvature of the sphere to be $R = 50$ cm. What is the spread for paraxial rays?

45. (III) An optical system contains a thin lens, $n = 1.4$, with positive curvature of radius $R_1 = 25$ cm for surface 1 and negative curvature of radius $R_2 = -25$ cm for surface 2. This lens collects light from the right side. Where to the left of the lens should a flat plate of thickness t of the same glass be placed, and how thick should it be, if you want the light that radiates from a distant object to be focused on a screen 35 cm to the left of the lens?

CHAPTER 38

The brilliant colors of hummingbird feathers are due not to pigmentation, but to interference of the light reflected from them.

INTERFERENCE

In Chapters 36 and 37 we emphasized the geometrical properties of light. We were able to discuss many properties, including reflection and refraction, by treating light in terms of linear rays. We never needed to use the fact that light is a wave phenomenon. However, if we look more carefully at the behavior of light when obstacles or holes have dimensions comparable to the wavelength of the light, geometric optics is inadequate; the wave nature of light becomes important. Geometric optics cannot explain the colors observed in thin-walled soap bubbles or oil slicks. Similarly, if we look closely at shadows, we find that they are not completely sharp, in contradiction to the predictions of geometric optics. These phenomena are due to interference and diffraction, the subjects of this chapter and Chapter 39. *Physical optics*, which takes into account the wave nature of light, can explain a wider range of observations than can geometric optics.

38–1 YOUNG'S DOUBLE-SLIT EXPERIMENT

The wave phenomena discussed in Chapters 14 and 15 apply to all waves, including those of light.

When two or more harmonic waves superpose, whether they are water waves, waves on a string, or light waves, they interfere. Interference between two light waves occurs because *the electric (or magnetic) fields of the two waves add vectorially.*

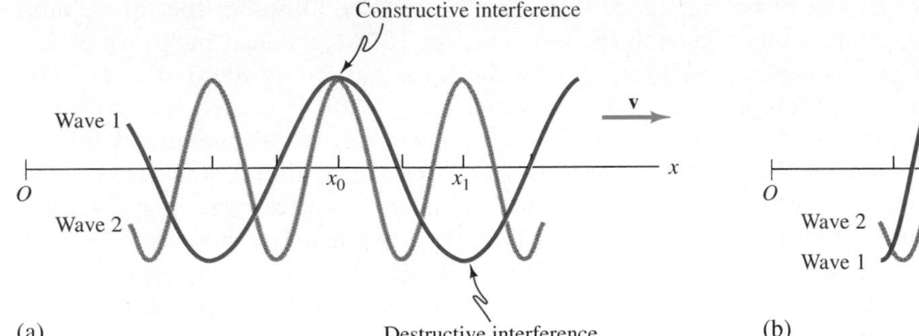

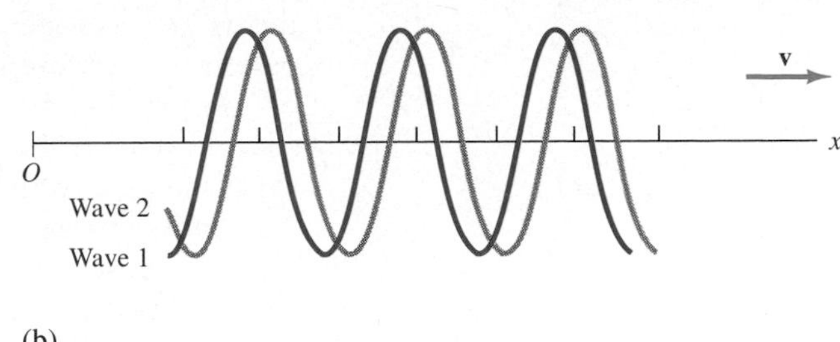

FIGURE 38–1 (a) Constructive interference between two waves occurs at point x_0 when the peaks coincide. Destructive interference between two waves occurs at point x_1 when their amplitudes cancel. (b) Two coherent waves have the same wavelength and a constant phase difference.

The resulting electromagnetic wave is a wave with a new value of the electric field. Consider the superposition of two light waves from different sources at particular points in space and at a given time, propagating along the x-axis (Fig. 38–1). Where the two waves add to a wave with a larger magnitude, we say that the waves interfere constructively, as they do at point x_0 in Fig. 38–1a. Where the two waves cancel one another, we say that they interfere destructively, as they do at point x_1. If the two waves have similar wavelengths, they may interfere either constructively or destructively over a wide region of space.

For light waves from two sources to produce an interference pattern in space, there must be a definite relation between their respective wavelengths and phases *at their respective sources*. That is, the waves must be *coherent*. The two light waves shown in Fig. 38–lb have exactly the same wavelength and a constant phase difference. These very long wave trains are the type of waves emitted by a laser, a source of coherent (and monochromatic) light, and it is easy to demonstrate in the classroom the interference pattern produced by laser light (see Chapter 39). If the waves emitted at one or both sources consisted of a mixture of waves of different wavelengths and phases, then there would be no interference pattern. An incandescent light bulb and a candle flame produce light from many independent atomic sources at different times and places within the filament or flame. Such light is said to be *incoherent*.

Waves can produce an interference pattern only if they are coherent.

Nevertheless, Thomas Young, who in the early 1800s was the first to observe interference phenomena in light, did not have a laser. How did he produce an interference pattern? Let us start with the incoherent light from a light bulb. We might first pass it through a prism and choose a single color. We would then be dealing with monochromatic light, which contains only a narrow range of wavelengths. However, monochromatic light is still incoherent, because it consists of many successive and overlapping bursts of different phases. The bursts of light from individual atomic sources within a light bulb are a quantum phenomenon. These bursts last for a time on the order of $\tau \simeq 10^{-8}$ s, and the length of the resulting individual wave trains is therefore $c\tau$, or several meters. We can now produce coherent light at two sources by illuminating a single aperture (a slit or a hole), S, with our monochromatic source. The aperture must be so small that only one burst of light enters at a time. In this way there are no phase differences among light from different bursts that may have entered different parts of the aperture. The single wave train of light that passes through forms a cylindrical wave that can illuminate two other apertures, S_1 and S_2 (Fig. 38–2). *These two apertures form sources of coherent light.* If S_1 and S_2 are equidistant from S, the light from S travels the same distance to reach S_1 and S_2, and the light is in phase as it passes through the two apertures (Fig. 38–2a). If S_1 and S_2 are not equidistant from S, the light waves that pass through them are still coherent, because they

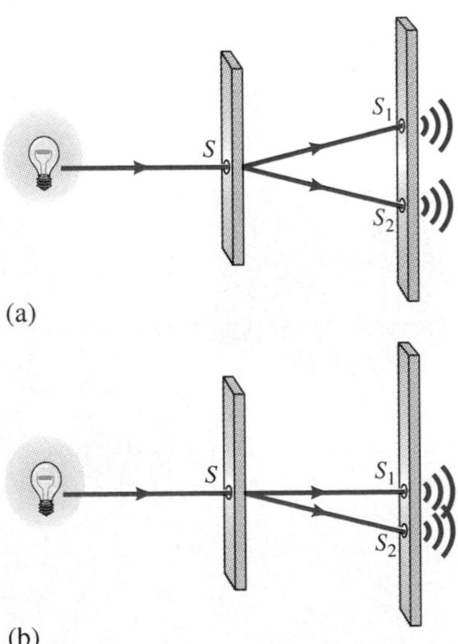

FIGURE 38–2 The light waves that pass through slits S_1 and S_2 are coherent. (a) These waves are also in phase, because the light travels the same path length from slit S to S_1 and S_2. (b) When the path lengths are different, there is a phase difference.

have a definite phase difference (Fig. 38–2b). Finally, although the interference pattern produced by a given burst lasts only for 10^{-8} s, the next burst, which has the same wavelength, produces exactly the same pattern as the first burst. The pattern is therefore a stable and observable one.

Any *single* source of light can be used in this way to produce two coherent sources. This is the method of producing coherent light that Young used in the experiments in which interference from two coherent sources was first observed. The disadvantage of Young's method is that the intensity of the pattern is very small, because the initial light has to be cut down so much. Today we can use a laser rather than a light bulb and a single aperture S. A laser is more useful because the intensity of laser light that passes through an aperture can be made quite large.

The Two-Source Interference Pattern

The interference pattern of waves from two coherent sources was treated in Chapter 15.

Let us review the spatial interference pattern produced when light from two sources of coherent waves interfere. Waves that pass through the rectangular, vertical slit S form cylindrical waves, and the same is true for the light that subsequently passes through slits S_1 and S_2 in Fig. 38–3a. Figure 38–3b is a view from above. If S_1 and S_2 are the same distance from S, then waves of the same frequency and phase emanate from S_1 and S_2 in the form of spreading circles from the source slits. The circles represent the crests of the spreading waves. Where the crests overlap, the waves interfere constructively. The pattern of these circles is apparent: As the waves progress, the positions where the circles appear advance and form lines. There is constructive interference (that is, wave motion with increased amplitude) *all along* these lines, and therefore the places where the lines intersect the screen are bright. In between the regions of constructive interference, there is destructive interference, and the screen is dark. The result is a series of bright and dark areas on the screen.

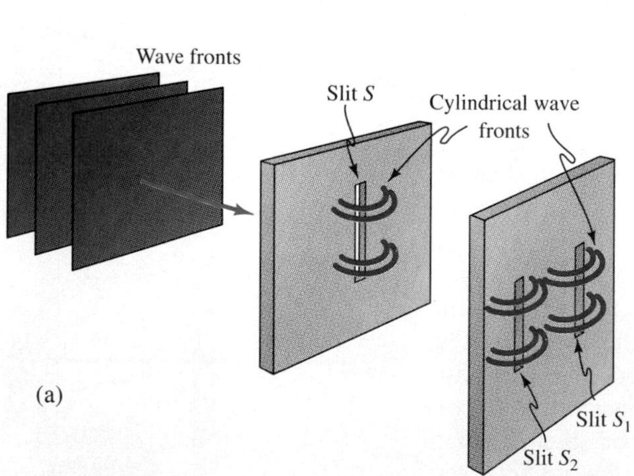

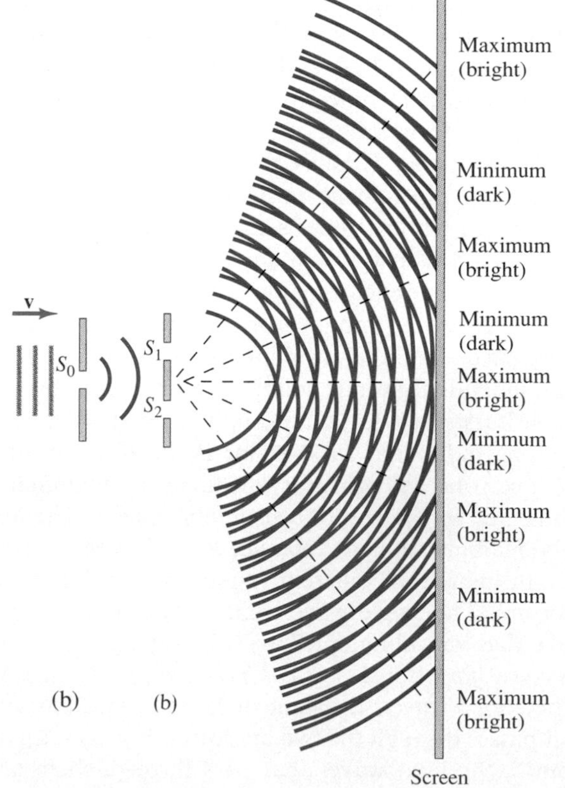

FIGURE 38–3 (a) Plane waves perpendicularly incident on thin slit S produce a series of cylindrical waves. When that series of waves reaches slits S_1 and S_2, the resulting cylindrical waves are coherent. (b) The view from above. Constructive interference occurs everywhere along the directions where the concentric circles, representing the crests of the spreading waves, overlap, because the waves are in phase along these directions. Alternating bright and dark places will be observed on a screen parallel to the walls.

We want to investigate this double-slit configuration more closely. Consider the geometry shown in Fig. 38–4. Along ray 1 and ray 2 the waves travel distances L_1 and L_2, respectively, to arrive at point P on the screen. Because the rays travel different distances, they may no longer be in phase at P, although they were in phase at the sources S_1 and S_2. Whether they are in phase or not depends on the *path-length difference* $\Delta L = L_2 - L_1$. The waves arrive in phase if ΔL is zero or if ΔL is an integral multiple of one wavelength. The electric fields can still interfere constructively even if the path lengths differ by many (integral) wavelengths. Similarly, if the peak of one wave arrives half a wavelength behind the peak of the other, the maximum electric field of one wave will occur at the same place in space as the minimum electric field of the other wave. The waves therefore cancel if ΔL is a multiple of one-half wavelength (180° out of phase). The interference is constructive when the waves add together and destructive when the waves cancel one another:

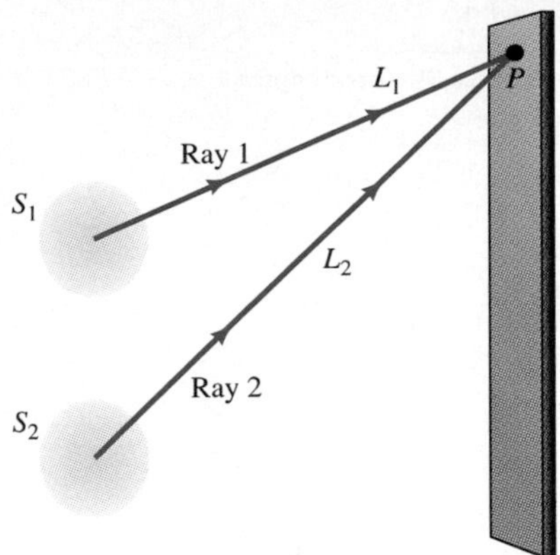

FIGURE 38–4 Light waves, rays 1 and 2, may not be in phase at point P despite being in phase at their sources, S_1 and S_2.

for constructive interference: $$\Delta L = n\lambda, \qquad n = 0, \pm 1, \pm 2, \ldots; \qquad (38\text{–}1a)$$

for destructive interference: $$\Delta L = \left(n + \frac{1}{2}\right)\lambda, \qquad n = 0, \pm 1, \pm 2, \ldots . \qquad (38\text{–}1b)$$

The result is a series of bright and dark lines (due to vertical slits) on the screen, as indicated in Fig. 38–3 and as shown in Fig. 38–5.

We can use the geometry shown in Fig. 38–6 to determine conditions for constructive and for destructive interference. We assume that the distance R to the screen is much greater than the distance d between the two slits. In this case the angle θ between the line S_1P and the line from S_1 perpendicular to the screen is approximately the same as the angle made similarly by S_2. From Fig. 38–6 we

FIGURE 38–5 The interference pattern produced by double vertical slits is a series of alternating bright and dark vertical lines on a screen.

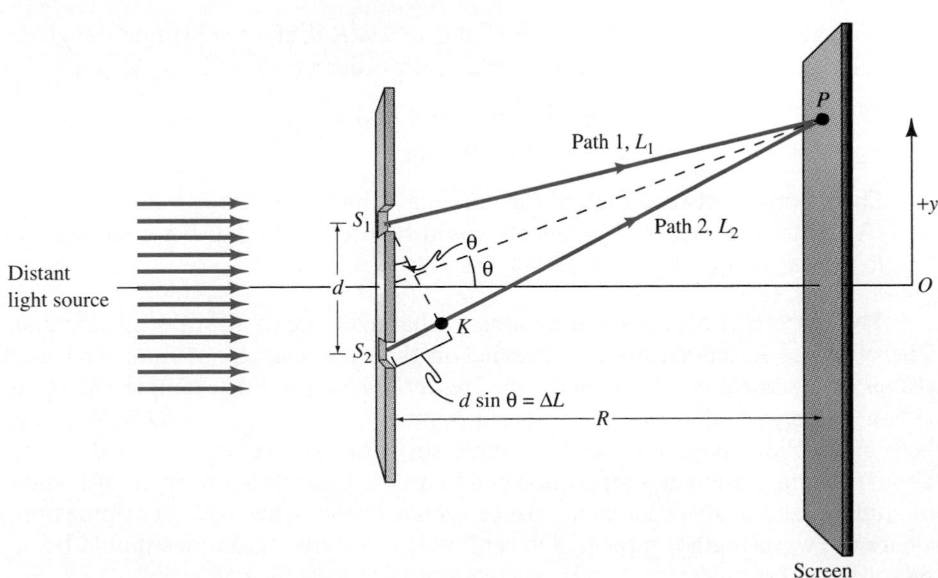

FIGURE 38–6 The geometry used to find the interference-pattern conditions for the light that reaches point P. The path-length difference $\Delta L = d \sin \theta$.

see that the angle formed by S_2S_1K is also θ. Thus

$$\Delta L = d \sin\theta. \qquad (38\text{–}2)$$

According to Eqs. (38–1a) and (38–1b), maxima (bright regions) and minima (dark regions) thus occur on the screen for angles given by

The two-source interference pattern

$$\text{for constructive interference:} \quad \sin\theta = n\frac{\lambda}{d}, \qquad n = 0, \pm 1, \pm 2, \ldots; \qquad (38\text{–}3a)$$

$$\text{for destructive interference:} \quad \sin\theta = \left(n + \frac{1}{2}\right)\frac{\lambda}{d}, \qquad n = 0, \pm 1, \pm 2, \ldots. \qquad (38\text{–}3b)$$

The point that is aligned with the sources ($\theta = 0$) is a maximum ($n = 0$). On either side of this centerline lie alternating minima and maxima. The value of n that labels the maxima is known as the *order*. The maxima on either side of the central maximum (the zeroth order) are first-order maxima ($n = \pm 1$). If θ is small, so that $\sin\theta \simeq \theta$, the maxima and minima are equally spaced in θ. Young performed the experiment described here with visible light and reported his results in 1802. His experiment shows conclusively that light behaves as a wave. Geometric optics cannot explain the result shown in Fig. 38–5.

EXAMPLE 38–1 Consider the double-slit experiment shown in Fig. 38–6, with y the distance from the center maximum along the screen. Find the positions of the maxima as a function of y. If the slit-to-screen distance $R = 3$ m, the slit separation $d = 0.2$ mm, and light comes from a helium–neon laser ($\lambda = 633$ nm) far away, determine the position y of the ninth-order maximum.

SOLUTION: Equation (38–3a) gives the angles for which maxima occur. From Fig. 38–6 the distance y is given by

$$y = R\tan\theta.$$

If $R \gg y$, $\tan\theta \simeq \sin\theta$, and we can insert the value of $\sin\theta$ from Eq. (38–3a) for the maxima. The maxima are then located at

$$y = \frac{n\lambda R}{d}. \qquad (38\text{–}4)$$

For the ninth order we have $n = 9$, and $y = 9\lambda R/d$. If we substitute the given values for λ, R, and d, we get for the ninth-order maximum

$$y = \frac{9(633 \times 10^{-9}\ \text{m})(3\ \cancel{\text{m}})}{0.2 \times 10^{-3}\ \cancel{\text{m}}} = 8.5\ \text{cm}.$$

The distance between each of the maxima is therefore about 1 cm.

Note that such measurements could be used to find λ: If we measure y, R, and d, we can solve Eq. (38–4) for λ.

The angular separation of the pattern of maxima and minima in interference phenomena increases as λ/d increases.

The pattern of maxima and minima is characteristic of all wave phenomena. The observed interference effects depend on the ratio λ/d; *as this ratio increases, the angular separation of the interference pattern increases.* For example, the separation Δy between maxima on the screen is, from Eq. (38–4), $\Delta y = R\lambda/d$. With the help of a helium–neon laser and a double slit with varying separations down to about 0.1 mm, the angular separation can be studied in a classroom demonstration or student laboratory experiment. To see similar phenomena in a water ripple tank where the wavelength is measured in centimeters, the slit separations should be on the order of centimeters.

38–2 INTENSITY IN YOUNG'S DOUBLE-SLIT EXPERIMENT

The previous discussion relied on geometrical arguments to determine the angles for which maxima and minima can be obtained. We now turn our attention to the *intensity* of the light that reaches the screen. The intensity (or brightness, for light) measures the energy delivered by a wave per unit time per unit area.

Light intensity was discussed in Chapter 36.

The energy in a given mechanical wave or superposition of waves is proportional to the displacement squared. For light, the quantity that plays the role of displacement is the electric (or magnetic) field. The intensity of a light wave (the energy delivered by the wave per unit time per unit area) is the time average of the Poynting vector (introduced in Chapter 35), which is proportional to the product of electric and magnetic field vectors in the wave. Because the magnetic field is itself proportional to the electric field in an electromagnetic wave, the intensity is proportional to the electric field squared ($I \propto E^2$). In order to find the intensity of a collection of waves, we add the electric fields of all the waves and square the sum of the net field. For example, with two sources of equal intensity I_0, the maximum electric field is twice the electric field E_0 from each source, so

$$\frac{I_{max}}{I_0} = \frac{(E_0 + E_0)^2}{E_0^2} = 4,$$

where I_{max} is the maximum intensity. The maximum intensity $I_{max} = 4I_0$. Similarly, the minimum intensity occurs when the electric fields exactly cancel, and $I_{min} = 0$.

The simple argument just given, which is based on energy, is so useful that we shall develop it further. The intensity at any point P on the screen in Fig. 38–6 is proportional to the net Poynting vector, which is in turn proportional to the square of the net electric field. The net instantaneous electric field $\mathbf{E}_{net}$ at P is the sum of the instantaneous electric fields of the light waves emitted at the two sources: $\mathbf{E}_{net} = \mathbf{E}_1 + \mathbf{E}_2$. The net Poynting vector therefore has magnitude $S \propto (\mathbf{E}_1 + \mathbf{E}_2)^2 = E_1^2 + 2\mathbf{E}_1 \cdot \mathbf{E}_2 + E_2^2$. But because light waves oscillate rapidly, it is not the Poynting vector that is of interest; it is the *time average* of the Poynting vector (that is, the intensity, I) at P. This is where the coherence of the light is important.

If we denote time averages with triangle brackets, then

$$I_{net} \propto \langle E_1^2 \rangle + 2\langle \mathbf{E}_1 \cdot \mathbf{E}_2 \rangle + \langle E_2^2 \rangle. \tag{38–5}$$

For incoherent light, there is no correlation—no definite phase relation—between the electric fields from the two sources. One moment the sources have one relative phase, the next moment the relative phase is different, and *the term* $\langle \mathbf{E}_1 \cdot \mathbf{E}_2 \rangle$ *is zero.* Thus

$$I_{incoh} = I_1 + I_2. \tag{38–6}$$

For incoherent sources, the intensities of the individual sources add.

The Intensity Pattern for Two Coherent Sources

For coherent waves, $\langle \mathbf{E}_1 \cdot \mathbf{E}_2 \rangle$ in Eq. (38–5) is not zero. If at a given time there is constructive interference at point P, where $\mathbf{E}_1 = \mathbf{E}_2$, the constructive interference will persist because the waves are coherent. Similarly, if at a given time there is destructive interference, in which $\mathbf{E}_1 = -\mathbf{E}_2$, it also persists through later times. For destructive interference, $\langle \mathbf{E}_1 \cdot \mathbf{E}_2 \rangle \propto -I_1$, and Eq. (38–5) gives $I_{net} = I_1 - 2I_1 + I_1 = 0$.

Suppose that the electric fields of the light waves from our sources S_1 and S_2

at a single point P in space are oriented in the same direction and have magnitudes[†]

$$E_1 = E_0 \sin(\omega t), \tag{38–7a}$$

$$E_2 = E_0 \sin(\omega t + \phi). \tag{38–7b}$$

The phase difference ϕ for E_2 results from the path-length difference between the waves. If $\phi = 2\pi n$, where n is an integer, the fields are identical, and there is constructive interference. This phase difference of $2\pi n$ corresponds to a path-length difference of $\Delta L = n\lambda$. The ratio of ϕ to $2\pi n$ is the same as ΔL to $n\lambda$, so we have

$$\frac{\phi}{2\pi \not{n}} = \frac{\Delta L}{\not{n}\lambda};$$

$$\frac{\phi}{2\pi} = \frac{\Delta L}{\lambda}. \tag{38–8}$$

For the distant-screen geometry of Fig. 38–6, we can use Eq. (38–2) to transform Eq. (38–8) to

$$\phi = 2\pi \frac{\Delta L}{\lambda} = \frac{2\pi}{\lambda} d \sin \theta. \tag{38–9}$$

Now, the net electric field at P has magnitude

$$E_{\text{net}} = E_1 + E_2 = E_0[\sin(\omega t) + \sin(\omega t + \phi)].$$

If we apply the equation $\sin \theta_1 + \sin \theta_2 = 2 \sin[(\theta_1 + \theta_2)/2]\sin[(\theta_1 - \theta_2)/2]$ (see Appendix IV-4), with $\theta_1 = \omega t$ and $\theta_2 = \omega t + \phi$, we have

$$E_{\text{net}} = 2E_0 \cos\left(\frac{\phi}{2}\right) \sin\left[\omega t + \left(\frac{\phi}{2}\right)\right]. \tag{38–10}$$

The Poynting vectors $\mathbf{S}_1$ and $\mathbf{S}_2$ of the light from the individual sources have magnitudes

$$S_1 \propto E_1^2 = E_0^2[\sin(\omega t)]^2 \quad \text{and} \quad S_2 \propto E_2^2 = E_0^2[\sin(\omega t + \phi)]^2, \tag{38–11}$$

respectively, whereas the net Poynting vector at P has magnitude

$$S_{\text{net}} \propto E_{\text{net}}^2 = 4E_0^2 \cos^2\left(\frac{\phi}{2}\right) \sin^2\left[\omega t + \left(\frac{\phi}{2}\right)\right]. \tag{38–12}$$

To find the intensities (the time averages of the Poynting vectors), we need know only that the time average of $\sin^2(at + b) = \frac{1}{2}$. If we write the individual intensities as $I_0 \propto E_0^2/2$, then, in terms of I_0, the net intensity from the two sources is

$$I_{\text{net}} = 4I_0^2 \cos^2\left(\frac{\phi}{2}\right). \tag{38–13}$$

When the phase ϕ in Eq. (38–7b) is related to the path-length difference by Eq. (38–8), then Eq. (38–13) for the intensity on the distant screen becomes

The intensity of the two-source interference pattern

$$I_{\text{net}} = 4I_0 \cos^2\left(\frac{\pi d}{\lambda} \sin \theta\right). \tag{38–14}$$

This is the expression for the intensity in Young's classic double-slit experiment.

[†] If two coherent sources are due to two apertures equidistant from a single source, as in Fig. 38–2a, then we know that the intensity and hence the electric field's amplitude of each source are the same.

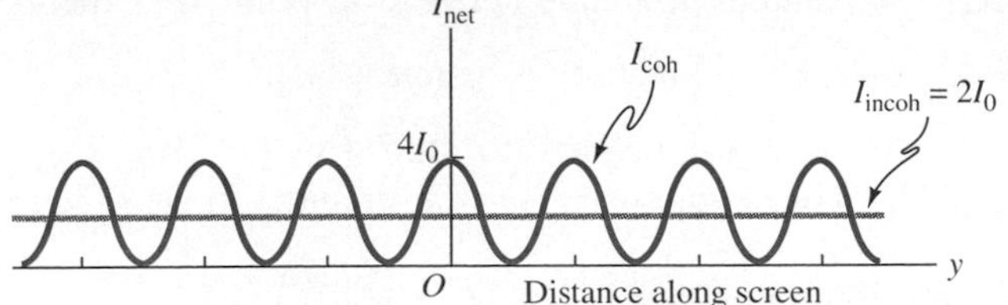

FIGURE 38–7 The net intensity of light from the double slit as a function of the distance from the center point on the screen ($y \simeq \sin\theta$). Compare the results for coherence and incoherence. The same amount of light energy reaches the screen in both cases, but in the coherent case, it occurs in peaks and valleys.

The maxima and minima occur at the angles specified by Eq. (38–3). Figure 38–7 is a plot of the intensity at the screen as a function of $\sin\theta$. This figure also serves as a plot of the intensity as a function of the distance y from the center maximum along the screen: For small θ, $y \simeq R\sin\theta$, where R is the distance from the screen. If the apertures are narrow vertical slits, then the bright maxima on the screen are vertical lines called *fringes*.

We have also plotted in Fig. 38–7 the intensity $2I_0$ that would be present on the screen if the light sources were incoherent. The result for incoherence is constant and shows no interfering maxima and minima. However, the energy reaching the screen is, *averaged over the entire screen*, exactly the same in the two cases, as required by the conservation of energy. To average the energy that reaches the screen in the case for coherence, we need only use the fact that the average of the cosine-squared factor in Eq. (38–14) is $\frac{1}{2}$, and $4I_0(\frac{1}{2}) = 2I_0$. The energy emitted at each source is the same whether the light from these sources is coherent or incoherent, and the total energy arriving at the screen must also be coherency independent. The energy is spread evenly over the screen when the sources produce incoherent light, whereas it is distributed in peaks and valleys when the light from the two sources is coherent.

EXAMPLE 38–2 You live at point H, 20 km from a vertical radio dipole antenna that broadcasts at a frequency of 1100 kHz from point B (Fig. 38–8). How well your radio picks up the signal is a direct function of the intensity of the signal. A second antenna is constructed at point A, located $d = 100$ m from the first. The second antenna broadcasts a signal identical to the first one. By calculating the new intensity at your radio in terms of the old intensity, find out if your signal is improved or not.

SOLUTION: The situation is like the double-slit situation for light, because there are two sources of coherent radiation. The only difference is that in this case, the relevant wavelengths are much longer. Because your distance from the antennas is much greater than their separation distance, the geometrical approximations we used in discussing the double-slit experiment apply. These approximations tell us that the difference in distances between you and the two antennas is $R_A - R_B = d\cos\theta$ (Fig. 38–8). We must find the net electric field at point H. The electric fields, which are parallel to the antennas, are vertical (perpendicular to the page). Suppose that the first antenna (at point B) broadcasts a signal whose z-component at point H is $E_B = E_0\cos(\omega t)$ and that the original intensity is $I_0 = CE_0^2$, where C is some constant. The second antenna's electric field takes the same form, except that there is a phase difference due to the fact that the antenna is a distance $R_A - R_B$ farther away:

$$E_A = E_0\cos(\omega t + \phi),$$

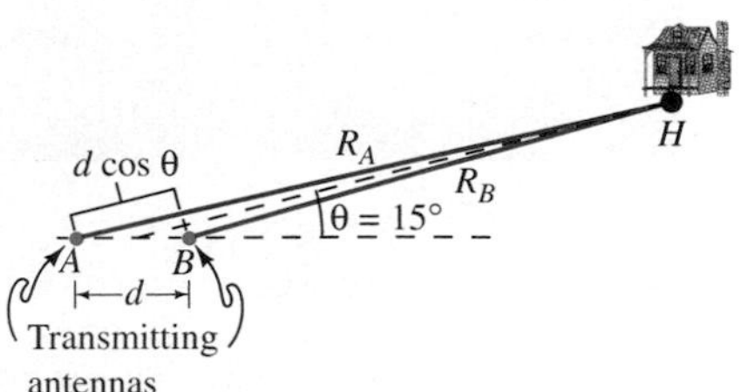

FIGURE 38–8 Example 38–2. R_A and R_B are the distances between your home and two antennas at points A and B, respectively.

where $\phi = 2\pi(R_A - R_B)/\lambda = (2\pi d \cos\theta)/\lambda$. The net field at point H is then

$$E_{\text{net}} = E_A + E_B = E_0\,[\cos(\omega t) + \cos(\omega t + \phi)].$$

This sum is similar to the one we needed to find Eq. (38–10), $E_{\text{net}} = 2E_0\cos(\phi/2)\sin[\omega t + (\phi/2)]$. The net intensity is given similarly in Eq. (38–13),

$$I_{\text{net}} = CE_{\text{net}}^2 = 4CE_0^2\cos^2\left(\frac{\phi}{2}\right) = 4I_0\cos^2\left(\frac{\phi}{2}\right).$$

It remains to calculate the factor $\cos^2(\phi/2)$ and see by what factor the intensity is changed. We require that

$$\cos\left(\frac{\phi}{2}\right) = \cos\left(\frac{2\pi d\cos\theta}{2\lambda}\right) = \cos\left(\frac{\pi d\cos\theta}{\lambda}\right).$$

We have $d = 100$ m and $\theta = 15°$. The wavelength λ comes from the frequency $f = 1100$ kHz $= 1.1\times10^6$ Hz. We get $\lambda = c/f = (3.0\times10^8\ \text{m/s})/(1.1\times10^6\ \text{s}^{-1}) =$ 273 m. Thus

$$\cos\left(\frac{\phi}{2}\right) = \cos\left[\frac{\pi(100\ \text{m})(\cos 15°)}{273\ \text{m}}\right] = \cos(1.11\ \text{rad}) = 0.44.$$

The net intensity $I_{\text{net}} = 4I_0\cos^2(\phi/2)$ is a factor $4[\cos^2(\phi/2)] = 4(0.44)^2 = 0.79$ times the original intensity I_0. The signal you receive at point H has actually become weaker, because there is partial destructive interference at your radio between the signals of the two antennas.

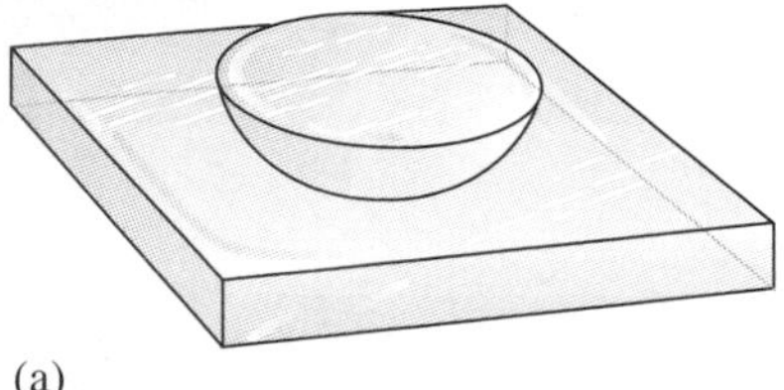

(a)

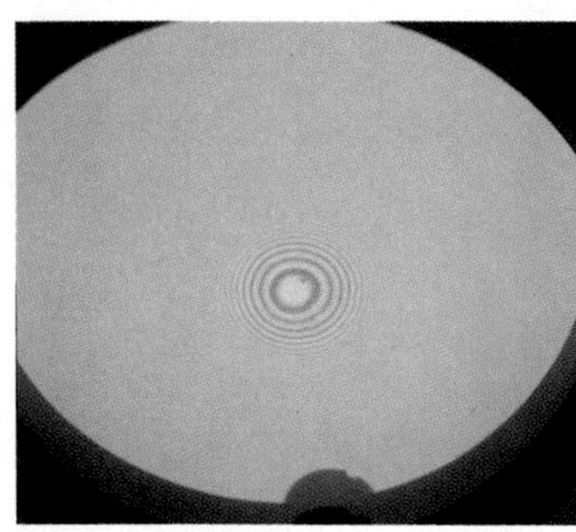

(b)

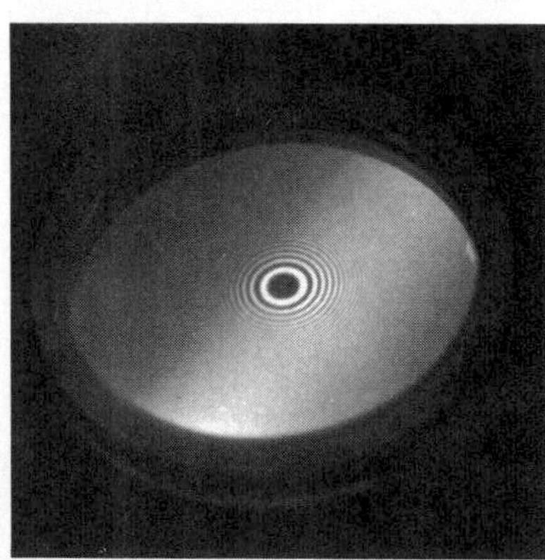

(c)

FIGURE 38–9 (a) A piece of glass whose bottom surface is curved rests on a flat glass surface. (b) The system is illuminated with white light from above, and a vertical view reveals colored rings called Newton's rings. (c) Concentric rings of alternating bright and dark appear when Newton's rings form from monochromatic light.

38–3 INTERFERENCE FROM REFLECTION

Interference phenomena occur quite commonly when a ray is partly reflected from and partly transmitted through a surface. If the transmitted ray is subsequently reflected from a second surface, then the two reflected rays may interfere with each other. The two reflected rays are automatically coherent, because they are actually pieces of the same ray.

Newton's Rings

When a curved piece of glass is placed on a flat piece of glass and illuminated from above (Fig. 38–9a), observation from above reveals rings of color (Fig. 38–9b). If monochromatic rather than white light shines down on the glass, then a series of bright and dark concentric rings appear (Fig. 38–9c). This phenomenon, called *Newton's rings*, was known to Isaac Newton in the 1700s and had been studied by his contemporaries Robert Hooke and Robert Boyle.

In order to understand this interference effect, consider the situation shown in Fig. 38–10. The curved piece of glass sits on a smooth, flat piece of glass; both pieces have index of refraction n. Air surrounds both pieces, including the space between them. It is convenient to speak of rays, which are defined to be *perpendicular to the wave fronts*, and when we speak of rays interfering, we are referring to the interference of the waves that form those fronts. The incident ray in Fig. 38–10 passes almost vertically through the top piece of glass and is partly reflected and partly refracted at the bottom, or curved, surface at point P_1. Ray 1 represents the reflected part, and ray 2 represents the refracted part, which continues through the air gap and is reflected at point P_2, at the top of the flat glass surface. Because the incident ray is almost vertical, ray 2 travels an extra distance $2P_1P_2$ in air, to a good approximation. Rays 1 and 2, exiting at the top, will interfere, and the result for monochromatic light is the series of concentric bright rings in Fig. 38–9c.

We expect the condition of bright rings for monochromatic light to depend on the extra path-length difference $2P_1P_2$ taken by ray 2. As the incident light enters the curved glass so points P_1 and P_2 are closer to the point of contact C between the two pieces of glass, the path-length difference $2P_1P_2$ decreases. As we get very close to point C, we might expect to find a bright spot at the center, caused by the constructive interference of rays 1 and 2 traveling almost the same distance. Conversely, at the center there is no gap between glass pieces, so in effect there is no boundary, and we might then expect a dark spot at the center! The second expectation is the correct one: As Fig. 38–9c shows, there is a dark spot at the center. From the point of view of rays 1 and 2, the dark spot at the center is an indication of destructive interference. *Destructive interference occurs because one of the rays undergoes a 180° phase change during reflection.* As we shall see, the light undergoes a phase change of 180° when it is reflected at P_2, whereas the light undergoes no such change when it is reflected at P_1.

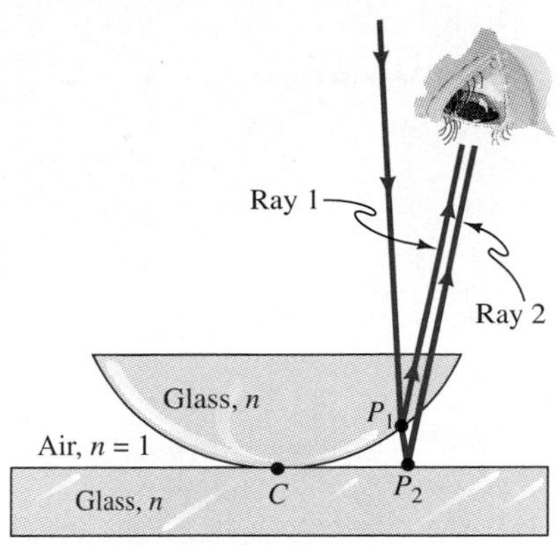

FIGURE 38–10 The geometry used to obtain the conditions for constructive interference in Newton's rings. The two rays moving toward the eye interfere after they reflect from different surfaces.

Let us look more closely at the phase change at a reflecting boundary. In Chapter 15 we discussed a corresponding phenomenon in the reflection of one-dimensional waves on strings. Suppose that two strings of different densities are connected (Fig. 38–11). The connection point forms a boundary at which there may be reflection and transmission. In Chapter 15 we discussed experiments that showed that if the wave speed on the side from which the wave comes is greater than the wave speed on the far side of the boundary, then the reflected wave is inverted, corresponding to a phase shift of 180° if the wave is harmonic. We learned also that whether the pulse is inverted or not depends on certain boundary conditions—in this case, the densities of the strings.

Maxwell's equations determine what happens to the electric and magnetic fields of an electromagnetic wave at a boundary between two dielectric media through a set of boundary conditions. For most situations of physical interest

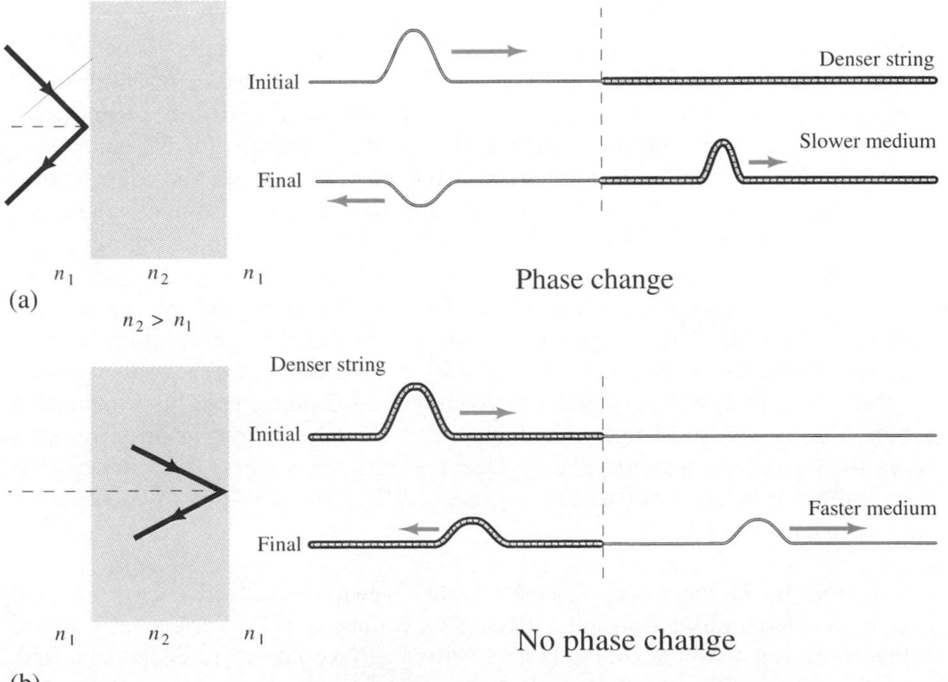

FIGURE 38–11 (a) The phase change of light upon reflection is similar to the inversion that occurs when a pulse that moves along a string meets a denser string. For light a phase change of 180° occurs when the second medium has a higher index of refraction. (b) No phase change occurs when the string encounters a less dense string or, for light, when the second medium at the reflection boundary has a lower index of refraction.

to us, these boundary conditions depend on the relative speeds of the waves in the media. The electric field changes sign or does not change sign according to whether the wave speed in the medium on the far side of the boundary is greater than or less than the wave speed in the medium from which the wave comes. The boundary conditions imply no such change of sign for the magnetic field of the wave. When the electric field changes sign but the magnetic field does not, the result is a 180° phase change in the reflected electromagnetic wave. The speed of an electromagnetic wave is inversely proportional to the index of refraction: $v = c/n$. The result for the phase change can therefore be stated as:

How phases change when light waves reflect from boundaries between different media

The phase of an electromagnetic wave that moves from a medium of index of refraction n_1 toward a medium with index of refraction n_2 will change by 180° upon reflection when $n_2 > n_1$ and will not change when $n_2 < n_1$.

Now we can understand the pattern of Newton's rings. The phase change of 180° occurs only for the reflection at P_2 because ray 2 goes from air ($n = 1$) to glass ($n > 1$). Because a phase shift of π rad (180°) corresponds to a shift of one-half wavelength, the condition for constructive interference becomes

The condition for constructive interference in Newton's rings

for constructive interference: $$\Delta L = 2P_1P_2 = \left(n + \frac{1}{2}\right)\lambda, \quad n = 0, \pm 1, \pm 2, \ldots . \tag{38–15}$$

How the path-length difference $2P_1P_2$ varies with distance from point C is a geometrical question of the curvature of the glass surface (see Problem 28). The alternate bright and dark rings correspond to radii around C in which the values of ΔL are values for which a monochromatic wave is in alternate constructive and destructive interference. In particular, the center, where $\Delta L = 0$, has destructive interference, so the center is dark. This is consistent with a much simpler model of what happens at the center: At the center is a film of air of zero thickness—that is, no film at all! When two pieces of glass with the same index of refraction touch, it is as though there is no boundary: Where there is no boundary, there is no reflection. Thus we fully expect the center to be dark. The consistency of these two models gives us confidence that our discussion of the phase change is correct.

How do we explain the colors we observe when sunlight, consisting of all wavelengths, is incident on the glass? For a particular radial distance from point C, there may be only one wavelength in the visible range, say, for the color blue, for which there is *destructive* interference. At that radius we see the color of sunlight with blue subtracted. If we look slightly farther away from C, where the distance P_1P_2 is greater, there will be destructive interference for a slightly larger wavelength, say, for green, and we see sunlight with green subtracted. Still farther out, there will be destructive interference for the color red, and we see sunlight with red subtracted. These colors are not as vivid as those of the rainbow, because they remain mixtures of different frequencies with some frequencies subtracted.

Newton's rings were explained by Young in 1802 along with his explanation for the double-slit interference phenomenon. Young's explanation in terms of a wave theory of light was not widely accepted until years later. It is curious that even though Newton observed the rings named for him, he settled on a corpuscular (particle) view of light.

Testing for Flatness with Newton's Rings. Newton's rings give us a practical method for determining how flat a given glass surface is. If the surface to be tested is placed on top of an *optical flat* (a glass with a surface known to be flat to a fraction of a wavelength of visible light), then no regions of constructive interference should appear if the tested piece is also flat and parallel to the first piece. If the second piece is flat but not parallel to the first surface, the interference fringes are *straight lines*. With this test, a surface is polished and placed against an optical flat until no curved interference fringes appear (Fig. 38–12).†

Mirrors that are flat to better than 5 percent of one wavelength, or about 25 nm, are available commercially. They are used in lasers as well as in *interferometers* (see Section 38–6), devices that use the interference of light to measure distance down to a fraction of one wavelength.

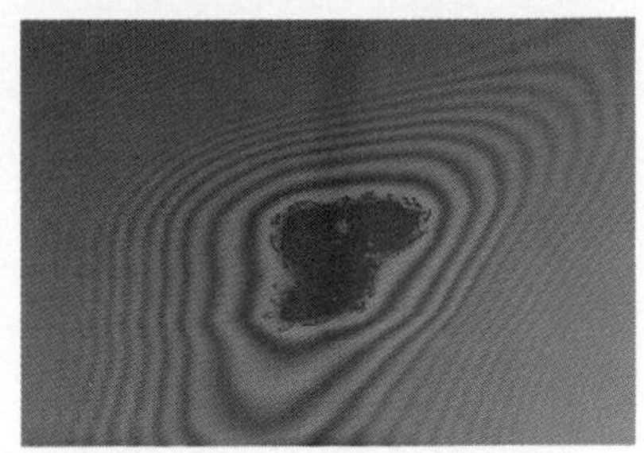

FIGURE 38–12 Fringes that occur when two nonflat pieces of glass are placed on top of one another. If each surface were perfectly flat and parallel, there would be no interference pattern. If the surfaces were flat but not parallel, the fringes would form straight lines. This is a good test of flatness for mirrors. The distortions are caused by surface irregularities.

EXAMPLE 38–3 Two flat glass plates of length $L = 10$ cm touch at one end but are separated by a wire of diameter $d = 0.01$ mm at the other end (Fig. 38–13). Light shines almost perpendicularly on the glass and is reflected into the eye as shown. What is the distance x between the observed maxima if the incident (blue) light has $\lambda = 420$ nm?

SOLUTION: This example is similar to the case of Newton's rings, but here both pieces of glass are flat. Of the two reflected light rays shown in Fig. 38–13, only the one reflected from the second glass plate undergoes a phase change of 180° ($\phi_{\text{ref}} = \pi$ rad). If the distance between the plates where the second light wave passes through the air gap is y, the phase difference between the two waves is determined by a total path-length difference $\Delta L = 2y$. From Eq. (38–9), this phase difference is

$$\phi_{\Delta L} = 2\pi \frac{2y}{\lambda} = \frac{4\pi y}{\lambda}.$$

The total phase difference becomes

$$\phi = \phi_{\text{ref}} + \phi_{\Delta L} = \pi + \frac{4\pi y}{\lambda}.$$

This phase difference must be a multiple of 2π for constructive interference to occur. The relation for maxima becomes

$$2m\pi = \pi + \frac{4\pi y}{\lambda},$$

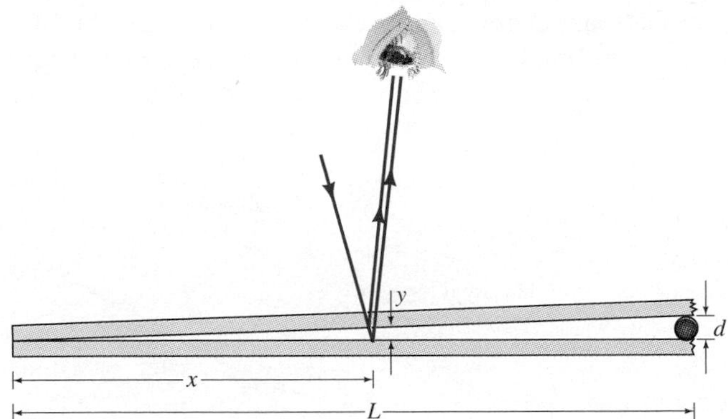

FIGURE 38–13 Example 38–3.

† A tiny spacer may be placed between the piece to be measured and the optical flat to open a wedge-shaped space between them. If the two pieces are so nearly flat and parallel that, without the spacer, *no* interference fringes are observed, a person who has spent weeks of work grinding and much money for an optical flat may have gone too far: The pieces may bond together and be virtually impossible to pull apart!

where m is an integer. This equation may be solved for y:

$$y = \frac{\lambda}{4}(2m - 1).$$

We want to determine the distance x between the bright bands. We use the geometrical relation between similar triangles in Fig. 38–13:

$$\frac{y}{d} = \frac{x}{L};$$

$$x = \frac{L}{d}y = \frac{L}{d}\frac{\lambda}{4}(2m - 1).$$

The difference in x from one maximum to another corresponds to a shift in m of 1, so

$$\Delta x = \frac{L}{d}\frac{\lambda}{4}\{(2m - 1) - [2(m - 1) - 1]\} = \frac{L}{d}\frac{\lambda}{4}2$$

$$= \frac{(10 \times 10^{-2}\ \text{m})(420 \times 10^{-9}\ \text{m})2}{(0.01 \times 10^{-3}\ \text{m})4} \simeq 2\ \text{mm}.$$

For glass plates 10 cm long, there will be about 50 bands of constructive interference.

Thin-Film Interference

FIGURE 38–14 Thin-film interference on soap bubbles.

The colors seen in soap bubbles and oil slicks are a manifestation of *thin-film interference*, which is another example of interference from reflection (Fig. 38–14). In this case the interference is between the light reflected from the two surfaces of the thin film. Consider light ray 1 that is incident on the thin film in Fig. 38–15. Part of the light is reflected at boundary I and forms ray 2. Part of ray 1 is refracted at boundary I and then reflected at boundary II. This light wave is partly refracted again at boundary I before forming ray 3.

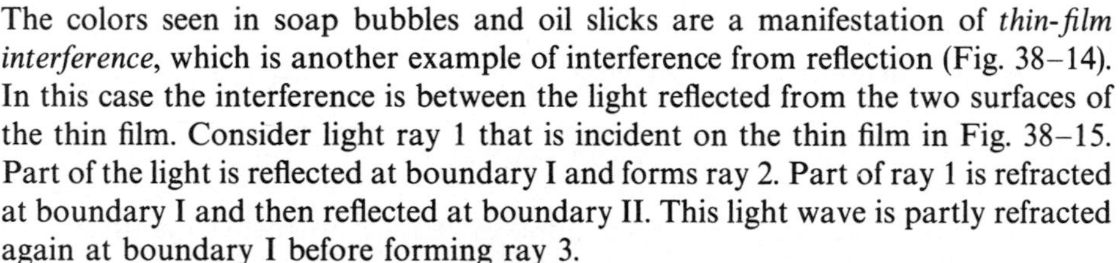

Because rays 2 and 3 both originate from ray 1 at point P_1, the conditions for constructive or destructive interference depend on the path-length difference $\Delta L = P_1P_2P_3 - P_1P_4$ as well as on any phase changes that may occur during the reflection. The rule discussed in Section 38–3 for phase changes upon reflection indicates that ray 1 undergoes a phase change upon reflection at surface I but not at II. Thin films such as soap bubbles have varying thicknesses, so different wavelengths will interfere destructively on different parts of the bubble, and the

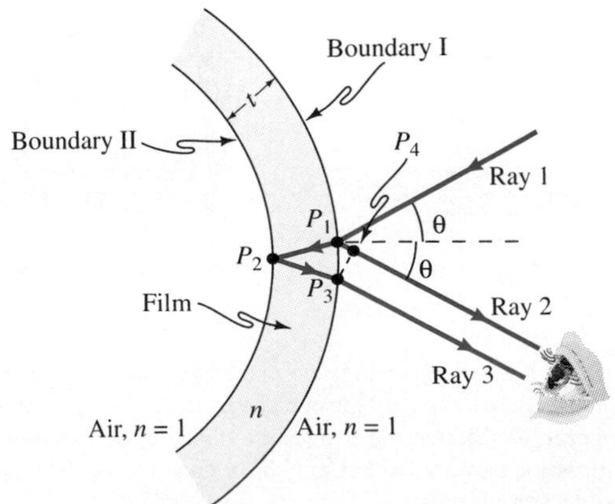

FIGURE 38–15 Geometry for thin-film interference. A light ray reflects from both the front and back boundaries of the film, and the reflected waves interfere.

colors that appear in the reflected light represent the original light minus the wavelength that interferes destructively. There is also an enhancement of those colors for which there is constructive interference. For an oil film floating on water, the oil may have an index of refraction between that of air and water, in which case there is a 180° phase change at *both* the air–oil surface and the oil–water surface. The phase changes due to reflection then cancel each other in the interference from light reflected from these two surfaces, and any phase difference is due only to a difference in the path length.

A new feature that does not arise in the case of Newton's rings appears with thin-film interference. The additional path length lies *within* the thin film of material. The phase change is then computed according to Eq. (38–9), but *the wavelength that appears is the wavelength within the material,* $\lambda_{\text{film}} = \lambda/n$, where n is the index of refraction of the material.

EXAMPLE 38–4 The soap bubble shown in cross section in Fig. 38–15 has thickness t and index of refraction n. Light of wavelength λ in air falls vertically on the bubble and is reflected back. (a) Express the condition for constructive interference for the reflected light. (b) If $t = 400$ nm and $n = 1.3$, what color or colors will interfere constructively in the reflected light?

SOLUTION: (a) When the incident light in Fig. 38–15 is vertical, distance $P_1P_4 = 0$. Light reflects at P_1 and P_2, undergoing a 180° phase change only at P_1, where the index of refraction of the medium on the far side, the soap film, is greater than that of air. The path-length difference is $\Delta L = 2t$, and the phase difference $\phi_{\Delta L}$ between the two reflected waves due to ΔL is found from Eq. (38–8) in terms of the wavelength λ_n in the soap film:

$$\frac{\phi_{\Delta L}}{2\pi} = \frac{2t}{\lambda_n} = \frac{2tn}{\lambda}.$$

The overall phase difference between the two reflected waves is then

$$\phi = \pi + \phi_{\Delta L} = \pi + \frac{4\pi tn}{\lambda},$$

where π appears as an additive factor because of the 180° phase change at P_1. There is constructive interference when $\phi = 2\pi m$, where m is an integer. Thus $2\pi m = \pi + (4\pi tn/\lambda)$, or

$$\text{for constructive interference:}\quad 4nt = (2m - 1)\lambda, \quad m = 1, 2, 3, \ldots. \tag{38–16}$$

(b) When there is constructive interference for a given wavelength, the corresponding color is strong in the reflected light. Equation (38–16) gives the condition for constructive interference in this problem, and we need only apply it to find λ, given $t = 400$ nm and $n = 1.3$:

$$\lambda = \frac{4nt}{2m-1} = \frac{4(1.3)(400\text{ nm})}{2m-1} \simeq \frac{2100\text{ nm}}{2m-1}.$$

For $m = 1$ through $m = 4$, these values of λ are

$$\lambda \simeq 2100\text{ nm}, 700\text{ nm}, 420\text{ nm}, 300\text{ nm}.$$

Only the wavelengths 700 nm (red) and 420 nm (blue) are in the visible spectrum, and these are the colors that interfere constructively in the reflected light. The color for the wavelength in between ($\lambda \simeq 560$ nm) interferes destructively and will be absent from the reflection.

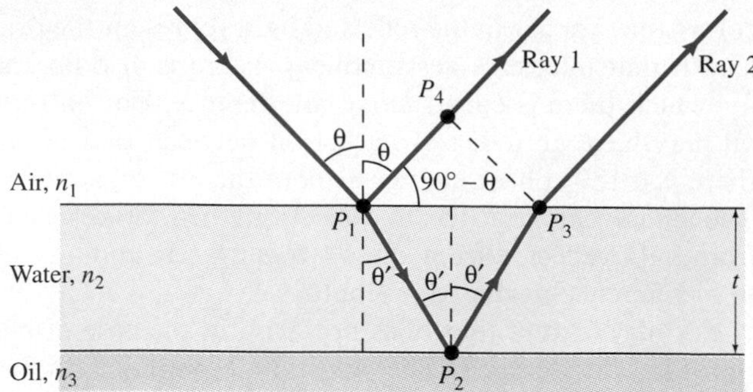

FIGURE 38–16 Example 38–5. Rays 1 and 2 arrive at a distant point where they interfere.

EXAMPLE 38–5 A thin film of water, $n_2 = 1.33$, floats on cinnamon oil, which is denser than water and has index of refraction $n_3 = 1.65$. White light reflected at 45° has a maximum intensity for a wavelength around 600 nm. What is the minimum possible thickness of the film?

SOLUTION: Figure 38–16 illustrates the paths of the two reflected rays that arrive in parallel at a distant point. *Both* reflected rays undergo a phase shift of 180°, and *no net phase difference is associated with reflection.* There is still a phase shift associated with different path lengths for the two rays—the phase shift ϕ_1 from the additional path P_1P_4 of ray 1,

$$\frac{\phi_1}{2\pi} = \frac{P_1P_4}{\lambda};$$

and the phase shift ϕ_2 for the additional path $P_1P_2 + P_2P_3$ of ray 2,

$$\frac{\phi_2}{2\pi} = \frac{P_1P_2 + P_2P_3}{\lambda_{H_2O}} = \frac{(P_1P_2 + P_2P_3)n_2}{\lambda}.$$

Note that the relevant wavelength for ray 1 is the wavelength in air ($n_1 = 1$), λ, whereas the relevant wavelength for ray 2 is $\lambda_{H_2O} = \lambda/n_2$. It is the difference $\phi_2 - \phi_1$ that enters into the net phase difference between the two waves. The net phase difference is then

$$\phi = \phi_2 - \phi_1 = (P_1P_2 + P_2P_3)\frac{2\pi n_2}{\lambda} - (P_1P_4)\frac{2\pi}{\lambda}.$$

The geometry necessary to calculate the path lengths is shown in Fig. 38–16. The incident ray enters at an angle θ perpendicular to the surface and is either reflected at angle θ or refracted into the water at angle θ'. We note that $P_1P_2 = P_2P_3 = t/(\cos\theta')$. Also, $P_1P_4 = P_1P_3\cos(90° - \theta) = (2t\tan\theta')\cos(90° - \theta)$. The phase difference is then

$$\phi = \left(\frac{2t}{\cos\theta'}\right)\frac{2\pi n_2}{\lambda} - (2t\tan\theta')\cos(90° - \theta)\frac{2\pi}{\lambda}.$$

We solve the equation above for t, the film thickness:

$$t = \frac{\lambda\phi}{4\pi}\frac{1}{(n_2/\cos\theta') - \tan\theta'\cos(90° - \theta)}.$$

From Snell's law, $\sin\theta' = (\sin\theta)/n_2$, and with $\theta = 45°$, $\theta' = 32°$. The problem states that there is a maximum in the intensity (that is, constructive interference) for $\lambda = 600$ nm. For constructive interference, $\phi = m(2\pi)$, $m = 1, 2, 3, \ldots$. (Is $m = 0$ allowed here?) Thus

$$t = \frac{\lambda m(2\pi)}{4\pi}\frac{1}{(n_2/\cos\theta) - \tan\theta'\cos(90° - \theta)}$$

$$= \frac{600\text{ nm}}{2}\, m\, \frac{1}{(1.33/\cos 32°) - \tan 32°\cos(90° - 45°)} = (1065\text{ nm})m.$$

The minimum film thickness, for $m = 1$, is 1065 nm.

APPLICATION

Antireflective Coatings

The lenses of some cameras, binoculars, and other optical devices are coated with a thin layer of material in order to reduce the intensity of the reflected light. Many optical devices have multiple lenses, and each lens typically reflects 4 percent of the energy of incident light. After several reflections a significant amount of the intensity is lost, and the final image may be degraded by the light reflected back and forth between the surfaces. Antireflective coatings are also placed on solar cells (components of solar batteries) to increase the intensity of the transmitted light. Magnesium fluoride, MgF_2, is a material commonly used to coat glass lenses. A single coating can reduce the reflected energy by over a factor of 2. A coating material must have an index of refraction between that of air and the glass to be coated. For MgF_2, $n = 1.38$ for a wavelength of 550 nm.

If light is reflected from both surface I and surface II, as shown in Fig. B1–1, then there will be a phase change in both reflections. The overall reflection is minimized when the two reflected light waves *destructively interfere*. If the thickness of the coating is t and the wavelength of light in air is λ, destructive interference will occur when the difference in optical path lengths, $2t$, is equal to $\frac{1}{2}\lambda_n, \frac{3}{2}\lambda_n, \frac{5}{2}\lambda_n, \ldots$, where λ_n is the wavelength within the coating. The condition of destructive interference is then

$$2t = \left(m + \frac{1}{2}\right)\lambda_n = \left(m + \frac{1}{2}\right)\frac{\lambda}{n}. \qquad \text{(B1–1)}$$

The coatings are applied as thinly as possible (that is, for $m = 0$), because this choice cuts down the reflection better. The thickness t for the antireflective coating then becomes

$$t = \frac{\lambda}{4n}. \qquad \text{(B1–2)}$$

A coating of *quarter-wave* thickness gives destructive interference for only one wavelength in the visible range of light. A 100-nm thickness of MgF_2 on a glass lens reduces reflectivity at 550 nm, which is in the middle of the visible range. The colors at the ends of the visible spectrum—red and violet—are still reflected, and

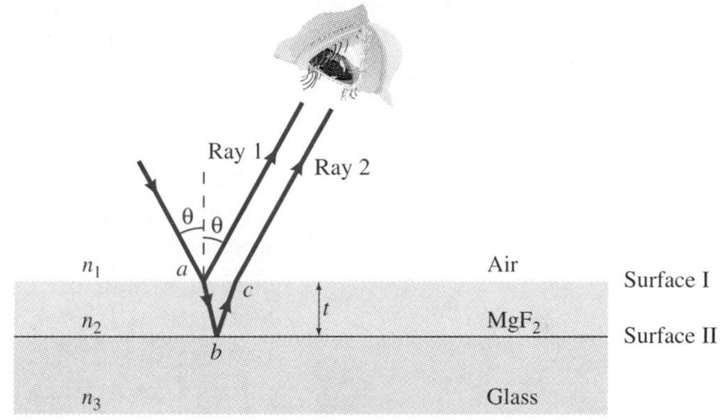

FIGURE B1–1 Thin coatings of materials such as MgF_2 can serve to reduce reflection by taking advantage of destructive interference.

the reflected light has a slight purplish hue. Reflected energies as low as 0.5 percent can be obtained with more sophisticated multiple-layer coatings. The phase condition that fixes the thickness of the coating is only part of the story. The destructive interference produced by the antireflective coating will be *total* only if the amplitudes of the two reflecting rays are equal. Because the amplitude of reflected light depends on the indices of refraction of the media involved, the condition for equal amplitudes becomes a condition of the index of refraction of the coating material. In particular, if n_2 is the index of refraction of the glass to be coated, and if the light is incident from air, then the coating material should have an index of refraction $n_1 = \sqrt{n_2}$ (see Problem 36).

Sometimes it is desirable to *increase* the intensity of reflected light. Lasers require mirrors that reflect light strongly at their operating wavelengths. For this purpose a single coating must have thickness $t = \lambda/2n$, so that there will be constructive interference between the two reflected light waves. Multiple coatings consisting of layers of various dielectric materials are even more effective in increasing the intensity of the reflected light. It is possible to reflect more than 99 percent of the energy.

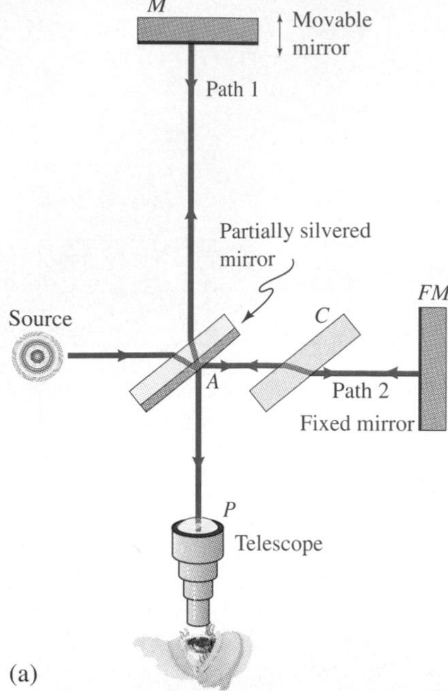

FIGURE 38–17 (a) Schematic diagram of a Michelson interferometer. Light is split by the partially silvered mirror. The resulting light travels two different paths before it returns to and interferes at point A and is subsequently observed through the telescope. (b) A modern Michelson interferometer. (c) Michelson with his interferometer.

*38–4 INTERFEROMETERS

Optical interferometers are devices that utilize the interference between light waves to measure quantities such as wavelength, small path-length differences, wave speeds, and indices of refraction. Figure 38–17a is a schematic diagram of one type of optical interferometer, the **Michelson interferometer**. In this device a light source is split by a beam splitter—for example, a partially silvered mirror—into two coherent waves that may travel different distances or through different media before they rejoin and interfere (Fig. 38–17b). The Michelson interferometer was invented in the 1880s by Albert Michelson (Fig. 38–17c).

Monochromatic light from the source in Fig. 38–17a is split at the mirror at point A. The two beams then travel along paths 1 and 2 before they rejoin at A. The recombined beam is formed from the superposition (and, therefore, interference) of the two beams that arrive at A. The element C (a *compensator*) is added to make sure that the two light waves travel through the same amount of glass. If the path lengths are exactly the same, the two light waves will constructively interfere. If mirrors M and FM are precisely perpendicular, the combined beam undergoes constructive interference and is bright. If the path lengths are not precisely the same because the mirrors are tilted slightly, the interference will produce alternating dark and bright lines, much like those discussed in Example 38–3, between two flat glass plates. The fringes will shift if the (screw-mounted) movable mirror is moved slightly. A movement of the movable mirror of only $\lambda/2$ will cause a shift from one fringe maximum to the adjacent one.

Unknown wavelengths of light can be determined by measuring very accurately the movement of the movable mirror and counting the number of maxima that pass across the telescope eyepiece. If ΔL is the distance the mirror is moved and N is the number of maxima that pass across the eyepiece for this mirror movement, then

$$N = \frac{\Delta L}{\lambda/2} = \frac{2\Delta L}{\lambda}.$$

The wavelength is then

$$\lambda = \frac{2\Delta L}{N}. \tag{38–17}$$

By performing measurements with a large number of maximum shifts, wavelengths can be accurately measured. If the wavelength is known, the same technique can be used to measure very small distances, in this case ΔL. Scientists have used interferometers creatively. For example, indices of refraction can be measured by inserting different media of known thicknesses in one of the paths. Because the wavelength in the material is λ/n, the index of refraction can be determined.

Although Michelson developed his interferometer for a single purpose, to measure the motion of the earth through the supposed *ether* (the "medium" within which light waves were thought to propagate; see Chapter 40), he used variations of his device over a period of several decades to make a number of experimental measurements. Among his accomplishments are measurements of fine and hyperfine structure in atoms and the first optical interference measurements in astronomy. His measurements by optical methods of the then-standard meter led to its redefinition in 1961 in terms of the wavelength of an orange–red spectral line of krypton-86 (^{86}Kr). A meter based on a wavelength of light is more precise and more easily reproduced than one based on two scratches on a bar in a vault.

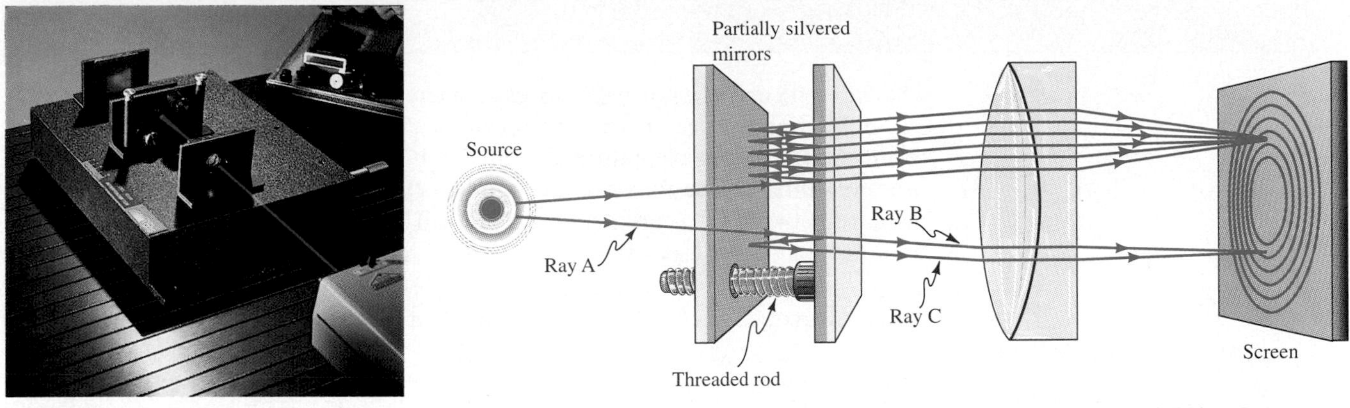

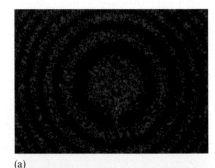

FIGURE 38–18 (a) A Fabry–Perot interferometer. (b) Schematic diagram of a Fabry–Perot interferometer. Some of the light is transmitted and some reflected at the partially silvered mirrors. Rays B and C may constructively or destructively interfere, depending on the extra path length of ray C. Multiple reflection is an important part of the operation of the Fabry–Perot interferometer and serves to make the maxima sharper and more easily locatable.

(The meter definition now depends on the defined speed of light and on an accurate time measurement.)

The Fabry–Perot Interferometer

The most widely used interferometer is the *Fabry–Perot interferometer* (Fig. 38–18a), invented by Charles Fabry and Alfred Perot and illustrated schematically in Fig. 38–18b. It contains two end plates (partially silvered mirrors) that are precisely parallel and flat and are connected by a rod that allows the distance between the plates to be changed smoothly by a screw thread. An incident laser beam (ray A) is partly transmitted (ray B) and partly reflected; the reflected ray is in turn partly reflected and partly transmitted to make ray C, which interferes with ray B. The major improvement incorporated into the Fabry–Perot interferometer is that the plates are silvered such that multiple reflections are possible, and the interference is between the multiple rays formed by these multiple reflections. The multiple reflections reinforce the regions of constructive interference, making the maxima stronger. The maxima become easier to locate, and the pattern becomes more distinct. It becomes easier to tell when these maxima have been shifted, so distances can be measured with more precision. Compare Figs. 38–19a and 38–19b, interference patterns produced by a Michelson and by a Fabry–Perot interferometer, respectively; the greater sharpness of the Fabry–Perot pattern is evident.

To measure distances with a Fabry–Perot interferometer, we begin with the plates at known positions and then count the number of interference maxima changes (fringes) as the plate separation varies by the desired distance. Because the location of maxima can be determined to great accuracy, the distance change can be measured to within an error of only a fraction of a wavelength of the laser light. The fringe counting is done automatically by electronic sensors. The number of fringes involved in distances of about 1 m is on the order of the number of wavelengths of visible light contained in 1 m, around 50 million.

The advent of lasers has made interferometers of all types even more useful. Light from ordinary sources produces waves in a series of bursts, and the phase difference between different bursts is random. This light has two disadvantages for interferometry. First, its intensity is weak, because only one burst at a time can be used (that is why the light is first passed through a small aperture). Second, the coherence length of this light, the length of the ray formed by a single sinusoidal form, is limited by the length of the bursts, on the order of 1 m. Because separate bursts do not lead to stable interference patterns, the overall size of the interferometer is limited. Laser light is intense, and its coherence length is essentially unlimited.

(a)

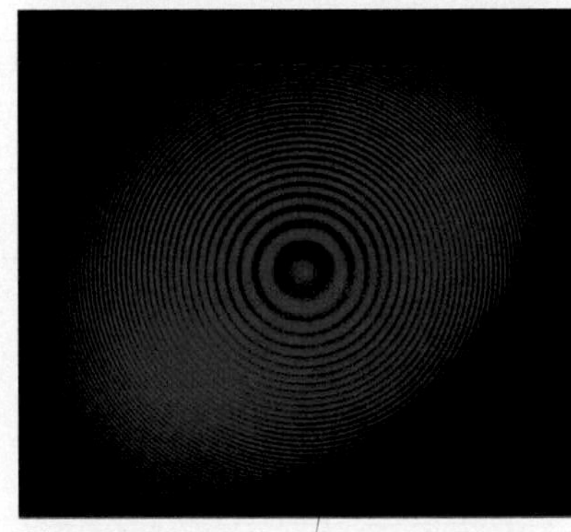

(b)

FIGURE 38–19 The interference patterns produced by (a) a Michelson interferometer and (b) a Fabry–Perot interferometer. The Fabry–Perot fringes are noticeably sharper.

SUMMARY

The wave theory of light explains experimental phenomena such as interference and diffraction that geometric optics cannot explain. Thomas Young was the first to substantiate the wave nature of light, through his double-slit experiment, which produces interference maxima and minima. If λ is the wavelength and d is the distance between the (narrow) slits, there will be maxima and minima at angles θ on a distant screen given by

for constructive interference: $$\sin\theta = n\frac{\lambda}{d}, \quad n = 0, \pm 1, \pm 2, \ldots; \tag{38–3a}$$

for destructive interference: $$\sin\theta = \left(n + \frac{1}{2}\right)\frac{\lambda}{d}, \quad n = 0, \pm 1, \pm 2, \ldots. \tag{38–3b}$$

The intensity pattern for the double slit is

$$I_{\text{net}} \propto 4I_0 \cos^2\left(\frac{\pi d}{\lambda}\sin\theta\right), \tag{38–14}$$

where I_0 is the maximum intensity for a single slit and I_{net} is the total observed intensity.

Newton's rings appear when light shines vertically down upon a curved glass that rests on a flat piece of glass. The light reflecting from two surfaces interferes. For monochromatic light, alternating rings of bright and dark are observed. For sunlight, colors are observed. In both cases the center is dark, because the phase of an electromagnetic wave will change by 180° upon reflection when the wave moves from one medium to another of higher index of refraction. The condition for constructive interference in Newton's rings is

$$\Delta L = \left(n + \frac{1}{2}\right)\lambda, \quad n = 0, \pm 1, \pm 2, \ldots, \tag{38–15}$$

where ΔL is the difference in path length of the interfering rays.

Thin-film interference is responsible for the colors observed in bubbles and oil slicks. For light falling perpendicularly from air onto the surface of the film of thickness t, the condition for constructive interference is

$$4nt = (2m - 1)\lambda, \quad m = 1, 2, 3, \ldots, \tag{38–16}$$

where n is the index of refraction of the film. Light reflected from the two surfaces of the thin film interferes destructively for some wavelengths but not for others. If the incident light is white, the wavelength undergoing destructive interference is subtracted from the reflected light. This phenomenon is used in the application of thin coatings to lenses and other optical surfaces in order to reduce the intensity of reflected light.

Optical interferometers utilize the interference of light waves to measure distances with great precision. Two interferometers with practical applications are the Michelson interferometer and the Fabry–Perot interferometer.

QUESTIONS

1. Why is a laser a more practical source of coherent light than a bright light bulb and a series of slits are?
2. In Fig. 38–3b the intersections represent places where constructive interference occurs on the crests of the advancing wave fronts. In between the circles, the troughs also reinforce each other. Do such reinforced troughs also represent a region of constructive interference? In the regions between the crests and troughs where each wave is zero, the waves add to zero. Is this a region of constructive interference?
3. In Fig. 38–3b the intersections represent places where con-

structive interference occurs. Will these bright spots be seen in air, or is a screen required for us to see the alternating bright and dark spots?

4. We refer to lines, or fringes, when we use two elongated slits for interference. If we used two holes rather than two slits, would we still observe lines, or something else?

5. Why does a single thin coating produce destructive interference in the reflection of light for only one wavelength in the visible region of light?

6. In discussing interference from thin films, we have spoken of viewing the reflected light from afar. Why? Is there no interference if the light is viewed from close to the film?

7. At an interference minimum from two sources of coherent light, there is no light intensity and therefore no energy. Yet each of the two interfering sources alone would produce energy at that point. Why does this situation not violate the conservation of energy?

8. When there is a minimum in the intensity of light reflected from thin films, there is less energy in the reflected beam. Does this mean that energy conservation is violated? If not, what happens to the energy that would otherwise have been in the reflected light? It may be helpful to think about what the light that passes all the way through a Newton's-rings apparatus must look like.

9. Is it possible to use two beams of light that travel in the z-direction—one polarized with its electric field vector aligned along the x-direction, and the other polarized with its electric field vector aligned along the y-direction—to make an interference pattern?

10. We estimated in Example 38–3 that almost 50 bands of constructive interference can occur along the glass plates. Can all these bands be observed? Will the color still appear blue?

11. In our discussion of Newton's rings, we did not consider light that reflects from the top surface of the curved piece of glass, nor that from the bottom surface of the bottom plate. Why not?

12. In Example 38–2 a second antenna has been constructed whose signal is as strong as that of the first, yet the signal you receive weakens. How is this consistent with the conservation of energy?

13. How is it possible to use interferometry to measure a small fraction of a wavelength when the distance between maxima represents a full-wavelength difference?

14. When you look into a good-quality camera lens, you will see a color tint (generally purple). What is the origin of the observed color?

15. It might appear that a given antireflective coating will work for any surface to which it is applied, provided that the thickness of the coating is given by Eq. (B1–2); this is, however, not true. Why not? You can go back to the discussion of the energy in reflection (in Chapter 36) to find out why the index of refraction of the surface to be coated must be considered.

16. Why is it important in the double-slit experiment that each slit be as narrow as possible?

17. Antireflective coatings are always applied to the front surface—the side from which light comes—of an optical element, never to the back surface. Why?

PROBLEMS

38–1 Young's Double-Slit Experiment

1. (I) A coherent source of monochromatic light of unknown wavelength shines on double slits separated by 0.20 mm. Bright spots separated by 0.70 cm appear on a screen 3.0 m away. What is the wavelength of the light?

2. (I) A double-slit interference experiment is done in a ripple tank. The slits are 4.5 cm apart, and a viewing screen is 1.5 m from the slits. The wave speed of the ripples in water is 0.15 m/s, and the frequency of the vibrator producing the ripples is 6.5 Hz. How far from the centerline of the screen will the first maximum be found?

3. (I) The source for a double-slit experiment has a wavelength of 525 nm. The slits are a distance of 120 μm apart, and a screen is 40 cm away from the wall that contains the slits. How far from the center will the third maximum occur?

4. (I) Light of wavelength 590 nm falls on a wall with two slits 0.12 mm apart. A photographic plate is placed at a distance R from the wall. The $n = 3$ maximum appears 18 cm from the central maximum on the photographic plate. How far is the plate from the wall?

5. (II) A double-slit experiment produces fringes on a distant screen. How does the linear separation between the bright maxima on the screen change when (a) the wavelength of the light doubles? (b) the separation between the slits doubles? (c) the distance between the slits and the screen doubles? (d) the intensity of the light doubles?

6. (II) A laser emitting light with $\lambda = 633$ nm shines on a double slit with a separation of 0.25 mm and produces interference fringes. If the maxima are separated by 2.0 cm, how far away is the screen on which the fringes are observed?

7. (II) In a double-slit experiment to determine an unknown wavelength of light, the measured total distance between 16 maxima (8 on each side of the central maximum) is 16.8 cm. The screen is located 3.45 m from the double slits, whose centers are 0.21 mm apart. What is the wavelength?

8. (II) Suppose that a double slit illuminates a distant screen. The light from sources S_1 and S_2 has come from a single monochromatic source S that is one-half wavelength closer to S_1 than to S_2. Use the geometry of Fig. 38–6 for the relation between the screen and the double slit to express the locations of maxima and minima. Is the point at $\theta = 0$ a maximum, a minimum, or neither?

9. (II) Two sources are 0.20 mm apart. They radiate coherently with a frequency of 5.5×10^{14} Hz but with a phase difference α between the two sources. A screen is placed as in

Fig. 38–6, 2.2 m away from the sources. How far from the center ($\theta = 0$) will the first maximum occur, as α varies from 0 to 2π?

10. (II) Light of two different wavelengths, λ_1 and λ_2, is incident on a double slit. On a distant screen, the twentieth maximum of λ_1 overlies the nineteenth minimum of λ_2. Show that the relative difference $(\lambda_1 - \lambda_2)/\lambda_1$ is small, and find a numerical value for this ratio.

11. (II) Consider two narrow slits illuminated from behind; the light impinges on a screen that is close rather than far (Fig. 38–20). Assume that the wavelength of the light, λ, is comparable to the separation d between the slits and to the screen distance R. The angle θ is measured from the point midway between the slits. (a) Show that the point corresponding to $\theta = 0$ continues to be a maximum of the pattern of light that reaches the screen. (b) At what angle θ is there a first maximum? (c) Show that your result for part (b) reduces to the distant-screen result for $R \gg d$ (and $R \gg \lambda$).

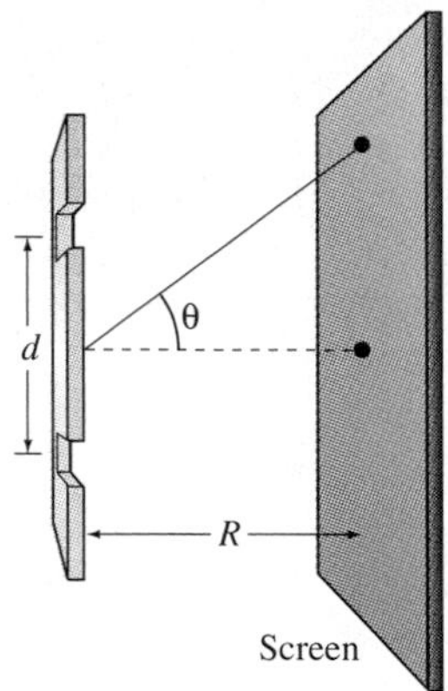

FIGURE 38–20 Problem 11.

38–2 *Intensity in Young's Double-Slit Experiment*

12. (I) Green light of wavelength 600 nm shines on two slits separated by 0.5 mm. Find the intensity ratio I/I_0 at distances 0.2 mm and -0.4 mm from the central maximum on a screen 1 m from the slits.

13. (II) Take the result of Young's classic double-slit experiment to find the average intensity on the screen by integrating the intensity over the surface of the screen. This result should be twice the average intensity from one slit alone and shows that energy conservation holds even when there is interference.

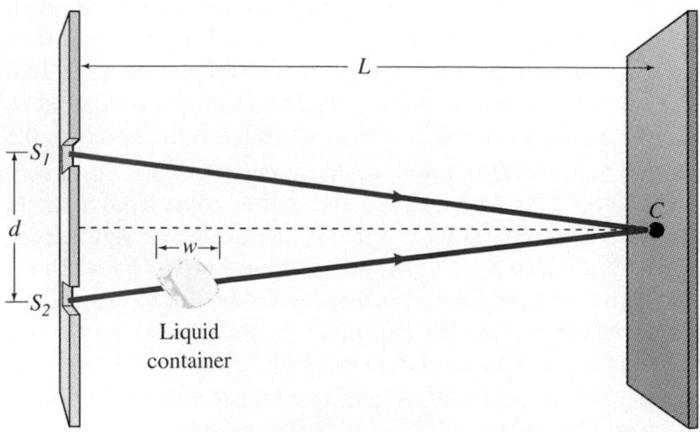

FIGURE 38–21 Problem 14.

14. (II) Consider the double-slit arrangement shown in Fig. 38–21. The center of the screen C is a point of constructive interference. A container of thickness w, holding a liquid of refractive index n, is placed in the path of the ray from slit S_2 to C. Plot, qualitatively, the intensity of light at C as a function of w, assuming that the separation between the slits is d, and that the screen is a distance L from each slit.

15. (II) The *angular width* of a maximum of the intensity pattern due to double-slit interference is defined to be the angular separation $\Delta\theta$ of the points where the intensity is half its maximum value. Express the width of the central maximum in terms of the wavelength and the slit separation. Are the widths of all the maxima the same?

16. (II) A He–Ne laser, $\lambda = 633$ nm, shines on double slits separated by 0.2 mm. At what minimum angle θ is the intensity 50 percent of the maximum? If the screen is located 3 m away, what is the distance between the two angles on either side of the maximum for which this intensity occurs? This distance is the *full width at half maximum* of the central peak.

17. (II) Two point sources of radio waves, 15 m apart, radiate in phase with a frequency of 2.4×10^7 Hz. (a) If the average intensity of each single source is 10^{-2} W/m^2, what is the direction in which the combined intensity is maximized? (b) What is the magnitude of the maximum intensity? (c) At what angle will the intensity have fallen to half its maximum value?

18. (II) Suppose that the two slits in a double-slit experiment are not exactly the same size, so the electric field from one of the slits at a particular point P on the screen is $E_1 \sin(\omega t)$, whereas the other one is $E_2 \sin(\omega t + \phi)$, where the phase ϕ is due to the path-length-difference [compare Eq. (38–7)]. Show that the intensity at P is given by $\langle I_{\text{net}} \rangle = \langle I_1 \rangle + \langle I_2 \rangle + 2\sqrt{\langle I_1 \rangle \langle I_2 \rangle} \cos\phi$, where $\langle I_1 \rangle$ and $\langle I_2 \rangle$ are the intensities due to the light from the individual slits. [*Hint:* You will need $\langle \sin^2(\omega t) \rangle = \frac{1}{2}$; $\langle \sin(\omega t) \cos(\omega t) \rangle = 0$.]

19. (II) Point sources S and S' radiate with the same intensity and the same frequency, corresponding to a wavelength of 0.020 m (Fig. 38–22). They are 45° out of phase and 2.5 m

FIGURE 38–22 Problem 19.

apart. Plot the intensity as a function of distance along the x-axis for values of x much larger than the source separation.

20. (III) Find an expression for net intensity for the situation in Problem 19 in terms of x, the wavelength (λ), and the source separation (d). Take into account that the electric field of the electromagnetic wave decreases as the inverse power of the distance between the source and the receiver, and that the distance between source S and any point on the axis is different from the distance between source S' and that point. [*Hint:* Use the results of Problem 18.]

38–3 Interference from Reflection

21. (I) Consider the two glass plates of Example 38–3 and the configuration of Fig. 38–13. The plates are 25 cm long. When light of wavelength 656 nm from hydrogen shines down perpendicularly to the glass, 102 interference fringes appear. How thick is the wire that separates the two glass plates at one end?

22. (I) Two rectangular pieces of glass are laid on top of one another on a plane surface. A thin strip of paper is inserted between them at one end, so that a wedge of air is formed. The plates are illuminated by perpendicularly incident light of wavelength 589 nm, and ten interference fringes per centimeter-length of wedge appear. What is the angle of the wedge?

23. (II) When two flat glass plates are placed on top of one another and a slip of paper is inserted between them at one edge, a thin wedge filled with air is produced between them. Interference bands form in reflection when monochromatic light falls vertically on the plates. Is the first band near the edge where the plates are in contact light or dark? Why?

24. (II) In a standard Newton's rings experiment, there is a dark spot where the convex surface touches the flat plate. Light of wavelength 500 nm is perpendicularly incident on the system. The convex lens is pulled slowly away from the flat plate until the minimum of the convex lens is 0.25 mm from the flat plate. A series of maxima and minima will appear at the center as the lens moves. How many maxima pass? Do the rings appear to move in to the center or away from the center?

25. (II) Consider the Newton's rings apparatus of Problem 24, with the minimum of the convex lens 0.25 mm from the flat plate. Water is poured into the space between the plates. Do the rings appear to move in to the center or away from the center, and how many maxima pass?

26. (II) A lens whose curved surface is part of a sphere of radius 6.0 m is placed on top of a flat plate, and the system is used to produce Newton's rings. What are the diameters of the fourth and ninth bright rings in the reflected pattern for light of wavelength $\lambda = 589$ nm?

27. (II) Light with a wavelength of 560 nm gives rise to a system of Newton's rings formed with a convex lens resting on a plane surface. The twentieth bright ring is at a radial distance of 0.98 cm. What is the thickness of the air film there, and what is the radius of curvature of the lens surface?

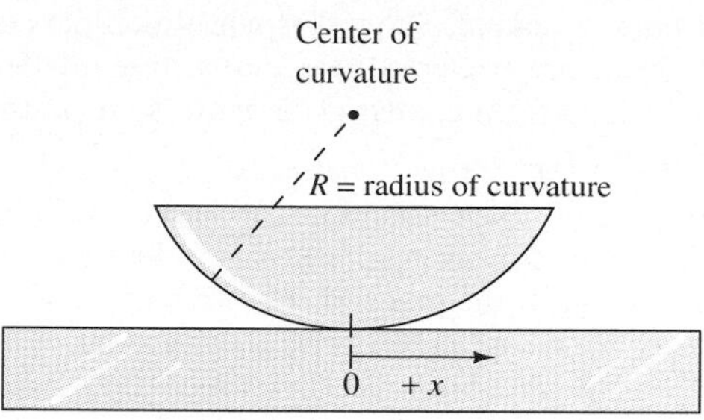

FIGURE 38–23 Problem 28.

28. (II) The radius of curvature of a convex surface used for a Newton's rings apparatus is R (Fig. 38–23). Find the position x, measured from the point where the convex surface touches the flat surface, of the nth dark ring for light of wavelength λ perpendicularly incident from above.

29. (II) Constructive interference occurs when a soap bubble reflects light of wavelength 420 nm. What is the minimum thickness of the bubble if its index of refraction is 1.38?

30. (II) What minimum thickness of antireflective coating of MgF_2 is required to minimize the reflection of blue light at 480 nm (the wavelength in air)? For MgF_2, $n = 1.38$.

31. (II) For light emitted by a He–Ne laser, $\lambda = 633$ nm, what *nonzero* minimum thickness of MgF_2 coating allows *maximum* reflectivity?

32. (II) The reflected light from an oil film floating on water shows constructive interference for light of wavelengths 495 nm and 636 nm incident along the normal. The index of refraction of the oil is $n = 1.48$. What is the film's minimum possible thickness?

33. (II) The reflected light from an oil film floating on water shows destructive interference for light of wavelengths 540 nm, 600 nm, and 675 nm incident along the normal. The index of refraction of the oil is $n = 1.60$. What is the film's minimum possible thickness?

34. (II) White light reflected at perpendicular incidence from a uniform soap film has an interference maximum at 600 nm and a minimum at 450 nm with no minima between 600 nm and 450 nm. If $n = 1.35$ for the film, what is the film thickness?

35. (II) Light is perpendicularly incident on an oil film with $n = 1.2$, suspended in air. (a) If green light ($\lambda = 550$ nm) is reflected back most strongly, what is the minimum thickness of the film? (b) If n were increased, would the maximally reflected light have a longer or shorter λ? (c) If the film were suspended on the interface between water ($n = 1.33$) and air, what would be seen?

36. (II) We mentioned in Chapter 36 that when light is perpendicularly incident from a medium of index of refraction n_1 and refracts into a medium of index of refraction n_2, then the intensity of the reflected light, I_r, is related to the incident intensity, I_0, by $I_r/I_0 = (n_2 - n_1)^2/(n_2 + n_1)^2$. In order for a coating to eliminate reflections, it is not enough that the light reflecting from the two surfaces differs in phase by 180°; the interference is totally destructive only if the amplitudes are equal. Show that when multiple reflections at interfaces are neglected, the destructive interference in light reflecting from coated glass in air is maximized when $(n_{\text{air}}/n_{\text{coat}}) = (n_{\text{coat}}/n_{\text{glass}})$.

37. (II) You would like to eliminate the reflected light from a flat glass pane for perpendicularly incident light of wavelength 600 nm. If the index of refraction of the glass is 1.55 and you have a coating material with an index of refraction of 1.25, what minimum thickness of coating material will have the desired effect?

38. (II) Blue light, $\lambda = 450$ nm, is perpendicularly incident on a vertical soap film held in a plane. A sequence of bright horizontal bands appears in the reflected light. (See Figure 38–24, in which the film is illuminated with white, not monochromatic, light.) What is the rate of change with height of the thickness of the soap film if the horizontal bright bands are 0.5 cm apart? For the soap solution, $n = 1.35$.

FIGURE 38–24 Problem 38.

39. (II) The material to be used for an antireflective coating has index of refraction of 1.25. How thick should the coating be to give the best result for $\lambda = 550$ nm and an angle of incidence of 30° with the normal?

*38–4 *Interferometers*

40. (I) Laser light with $\lambda = 575$ nm enters a Michelson interferometer. How many fringes will pass through the field of view if one of the mirrors is moved 14 mm?

41. (I) If the mirror of one arm of a Michelson interferometer is moved along the arm by 0.31 mm, 980 fringes traverse the field of view. What is the wavelength of the light used?

42. (II) A very sharp wedge of glass of index of refraction 1.60 is introduced perpendicularly in the path of one of the interfering beams of a Michelson interferometer illuminated by a narrow beam of light of wavelength 486 nm. This causes 500 dark fringes to sweep across the field of view. Calculate the thickness of the glass wedge at the point where the beam passes through it.

43. (II) A scientist wants to measure the wavelength of green light ($\lambda \simeq 550$ nm) to a precision of 0.1 percent. The freedom of motion in the screw used to adjust the movable mirror allows the change in the mirror position to be determined to no better than 0.2 mm. How many fringe shifts must be counted?

44. (II) A laser with light of wavelength 632.8 nm is used to calibrate a Fabry–Perot interferometer. As the screw controlling the position of one end plate rotates exactly 100 turns, 544 fringes are counted. Calculate the wavelength of another light source that shifts only 498 fringes for 100 turns of the screw.

General Problems

45. (II) Two ordinary light bulbs S_1 and S_2 are 1 m apart, each emitting light waves with intensity I, mainly at a frequency of 550 nm. What is the pattern (not the absolute value) of intensity on a screen 100 m away?

46. (II) AM radio waves with a wavelength of 280 m travel 10 km to your home. Halfway between the transmitting tower and your home, but off to the side, is a tower that reflects radio waves; the reflected wave has no phase shift due to the reflection. How far off the direct line is the tower if destructive interference occurs between the direct waves and the reflected waves?

47. (II) Two point sources of identical strength radiate in phase with the same frequency f. They are separated by a distance L. What is the energy density (discussed in Section 35–4) as a function of distance from one of the sources along the line that connects the sources in the central region, where the variation of the amplitudes with distance can be neglected?

48. (II) You want to hear a radio station that broadcasts at 91.3 MHz. You live on a direct line between the antenna and a large building that acts as a mirror for the radio waves broadcast by the station; you are exactly 1 km from the building. Calculate the intensity of the signal you receive in terms of the intensity you would receive if the building were not present. Assume that the reflection from the building is total, with no phase shift, and ignore any decrease of the signal with distance.

49. (II) Coherent microwave radiation reflects from two identical obstacles (Fig. 38–25). Each obstacle, of size a, is much

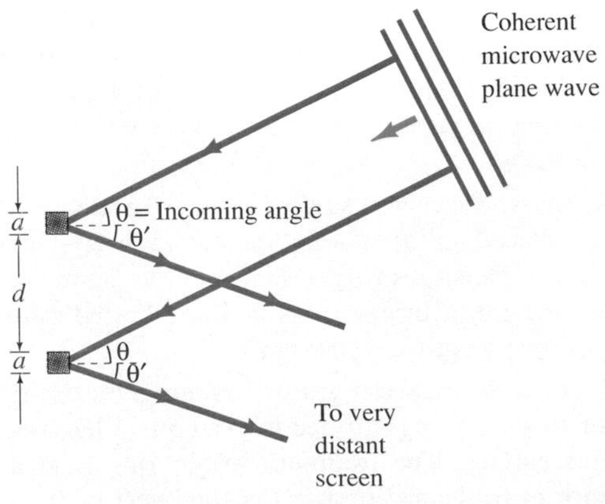

FIGURE 38–25 Problem 49.

smaller than the wavelength of the radiation, λ, and much smaller than the obstacle separation d. Find an expression for the angles θ', defined in the figure, for which there are maxima on the distant screen.

50. (II) A radio wave undergoes a phase shift of 180° when it reflects from the calm surface of the ocean. A ship in a calm port nearing a shore station receives a 200-MHz signal from the station's antenna. This antenna is located 20 m above the sea surface, as is the ship's receiving antenna. The direct and reflected signals interfere, and a succession of maxima and minima are heard in the interfering signal at the ship. How far is the ship from the station the first time the signal passes through a minimum? How fast is the ship moving if the time between this first minimum and the next one is 50 s?

51. (II) Sources S_1 and S_2 illuminate a distant screen; the distance to the screen is much larger than the separation between the sources. Each source emits light rays that are in phase at the sources, but the intensity I_1 of the light from S_1 is twice the intensity I_2 of the light from S_2. Give the ratio of the maximum to the minimum intensity of the light observed on the screen. (See Problem 18.)

52. (II) Figure 38–26 is an overhead view of two dipole radio antennas, vertical towers separated along the x-axis by a distance $\lambda/2$. This arrangement allows the radio signal to be beamed with greater intensities in some directions than in others, whereas either antenna alone would radiate its signal with the same intensity I_0 for any angle θ. (a) Find the intensity radiated by the antenna pair very far from the antennas as a function of θ, assuming that the signal of the two antennas is in phase. Describe the signal for all values of θ, from 0° to 360°. (b) The signal in the antennas is now 180° out of phase. How, if at all, does the distant intensity pattern change?

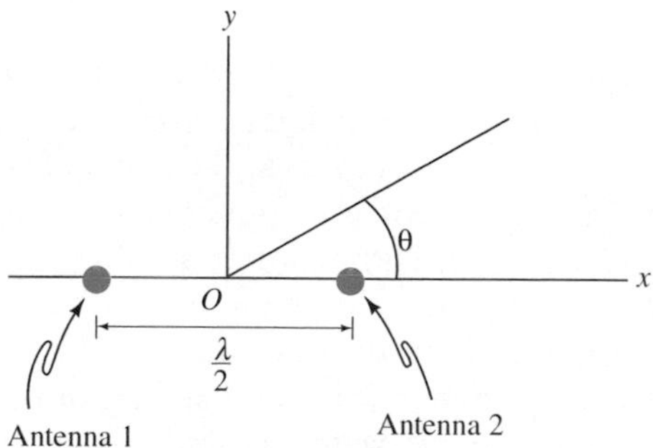

FIGURE 38–26 Problem 52.

53. (II) Consider the pair of dipole antennas in Problem 52. Suppose that the antennas are separated by one-quarter wavelength and that the signal in antenna 1 lags the signal in antenna 2 by 90° (one-quarter cycle). Show that the signal has a maximum in one direction, not two.

54. (II) Four identical loudspeakers are placed at the corners of a square with sides of length $\lambda/\sqrt{2}$. The loudspeakers emit sound coherently with wavelength λ. A listener is situated very far from the square along one of the diagonals. If the intensity with just one loudspeaker on is I_0, what are the intensities when two, three, and four speakers are on? In a table, list all combinations and the resultant intensities.

55. (III) *Lloyd's mirror* is a mirror that reflects light at large angles of incidence from a point source to a screen (Fig. 38–27). As the angle of incidence nears 90° (a grazing angle), the source and image coincide. (a) Will the direct light and reflected light interfere constructively or destructively in this limit? (b) Suppose that the (monochromatic) source is 10.0 m from the center of the mirror and 1.0 cm above the plane of the mirror and that the screen is distant. What will the angular separation be between successive maxima of the interference pattern, where the angle is measured from the source point to the screen?

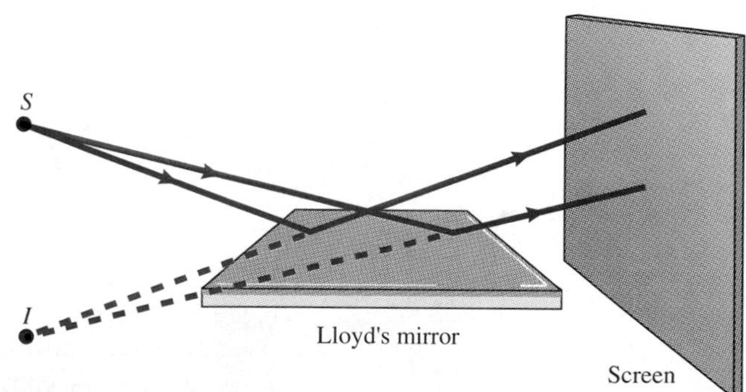

FIGURE 38–27 Problem 55.

56. (III) Light of wavelength λ is incident on a slab of glass of thickness h and index of refraction n resting on a mirror. The ray makes an angle θ with the vertical. The light reflected at the top of the glass surface and that reflected at the mirror will interfere. For what angles will the interference be totally constructive? [For angles even slightly away from grazing incidence (see Problem 55), light undergoes no phase shift upon reflection by a conductor such as a mirror.]

CHAPTER 39

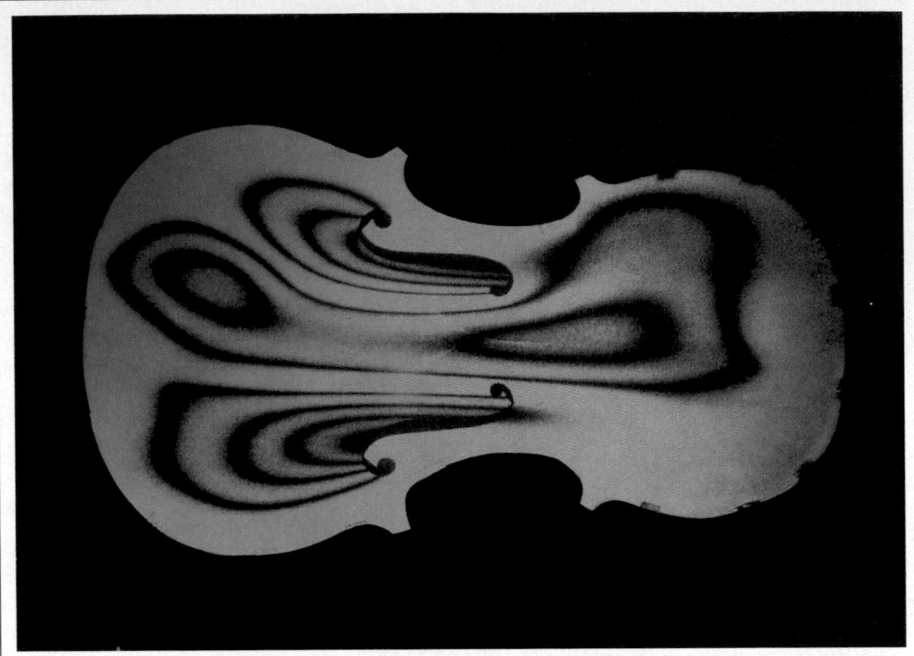

Interference and diffraction were crucial in producing this image of a violin. The lines indicate the motion of the vibrating instrument and are associated with the interference of two holographic images.

DIFFRACTION

The wave nature of light becomes apparent when there are obstacles or apertures with sizes comparable to the wavelength of light (several hundred nanometers), or when we look within comparable distances at the edges of shadows. In these cases, interference leads to effects that geometric optics cannot explain, such as the double-slit phenomena of Chapter 38. Diffraction is another manifestation of interference phenomena. The term can be used as an alternative to the term interference, but it usually refers specifically to the deviation of wave fronts from straight-line behavior. We are already familiar with the bending of ocean waves around obstacles, and the circular spread of water waves passing through an opening narrower than their wavelength. These phenomena are both diffractive effects. So is the pattern of maxima and minima that spreads across a screen in Young's double-slit experiment. The term "diffraction" also refers to interference between waves that emanate from a large number, or even a continuous set, of sources. Diffraction gratings have many slits or sources of coherent light and can be treated as a simple generalization of double slits. These gratings have important applications in the study of atomic systems and crystalline materials. We shall see that even a single, narrow slit produces characteristic and striking interference patterns that result from diffraction. Holography is a spectacular application of diffraction. Interference and diffraction effects are both consequences of the superposition of waves.

39–1 THE DIFFRACTION OF LIGHT

By the 1820s, serious attempts were underway to understand the consequences of the wavelike nature of light. Young's double-slit experiment had shown that there are clear interference effects associated with light. Although the bending of water waves around obstacles or around the edges of apertures, as in Fig. 39–1, is a familiar idea, it is not part of our everyday experience for light. This is because, as we discussed in Chapters 15 and 38, *interference effects require coherent wave sources*, and because *diffraction effects are typically most significant when the apertures or obstacles involved are comparable to the wavelength.* For light, the coherence condition is not realized in most situations, and wavelengths—several hundred nanometers—are tiny compared to the sizes of familiar objects.

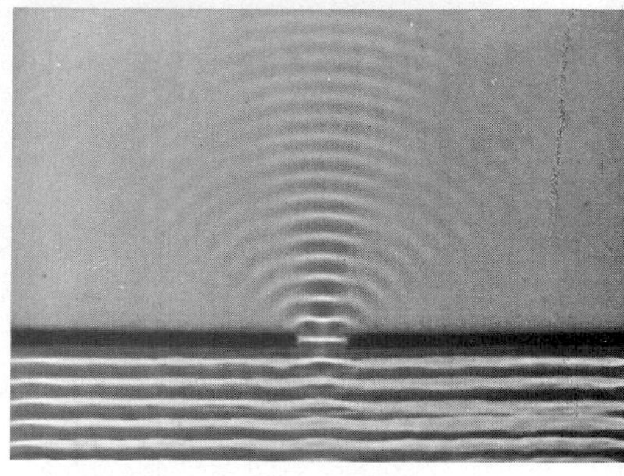

FIGURE 39–1 The way water waves bend around and through obstacles is well known. Here parallel waves pass through a hole, producing circular wave fronts on the far side.

In addition to his double-slit experiment, Young performed experiments that showed that light can pass around a single small obstacle, just as water waves can. Young firmly believed in the wave theory and publicly supported it despite considerable resistance in the scientific community. Augustin Fresnel, François Arago, and others produced further experiments and theories to establish even more firmly the reality of **diffraction phenomena** in light, which refers to situations in which light bends around obstacles by virtue of its wave properties. Fresnel carried his work sufficiently far that by 1821 he was able to use a primitive version of an interferometer to make the first quantitative measurement of the wavelength of light.

Huygens' principle had been used much earlier to show how each point on a wave front can be treated as producing spherical waves that add by superposition to a new wave front. Fresnel used the Huygens construction to understand various interference phenomena that are completely inconsistent with geometric optics. He showed that even a single aperture creates its own diffraction pattern, because waves passing through different parts of the aperture interfere with each other. Similarly, even a single obstacle creates a diffraction pattern, because parts of the original plane wave have been blocked by the obstacle and no longer participate in the downstream regeneration of the wave. With a source of coherent light, an intermediate object in the form of an obstacle or a wall with holes, and a viewing screen, we can observe diffraction effects (Fig. 39–2a). We show several such

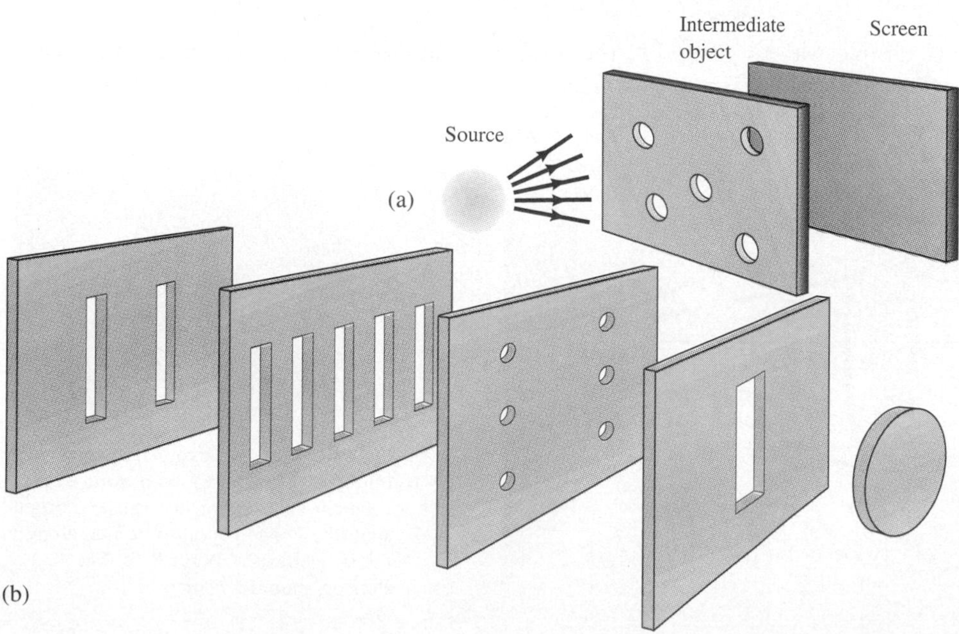

FIGURE 39–2 (a) Diffraction effects can be observed by placing an intermediate object in the path of light that passes from a source to a viewing screen. (b) Intermediate objects of various types produce different diffraction patterns.

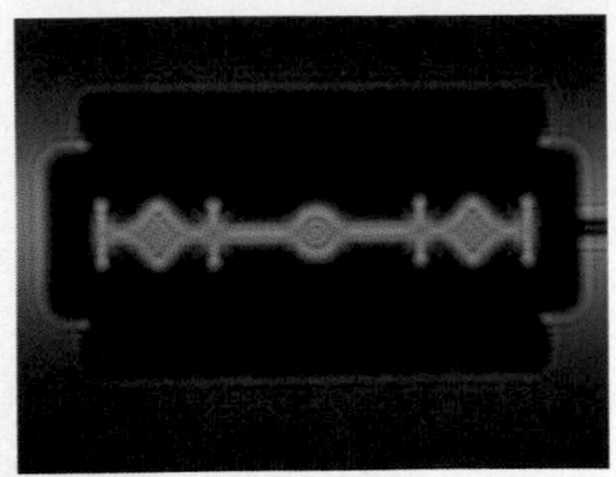

FIGURE 39–3 The diffraction of light around a razor blade.

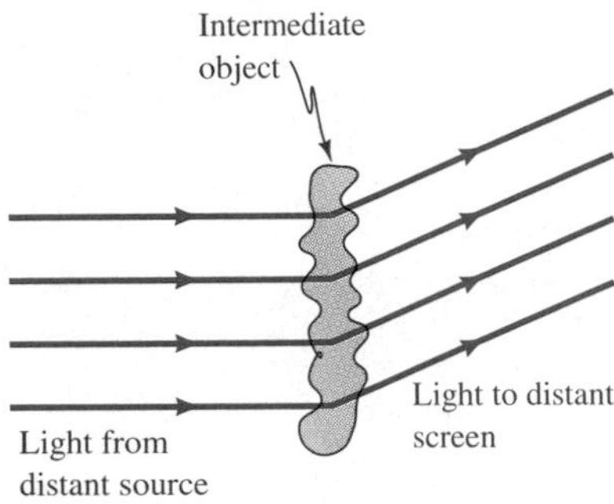

FIGURE 39–4 Parallel rays from a distant source are diffracted by an intermediate object and then viewed on a distant screen. When the screen is distant, we can treat the outgoing rays as parallel, and we have the case of Fraunhofer diffraction.

intermediate objects in Fig. 39–2b: two slits; several slits; several small apertures; one slit; and a simple disklike obstacle. Figure 39–3 is a photo of the diffraction pattern at the edge of a shadow, that of a razor blade. With a little ingenuity, the effects of diffraction are visible with the naked eye. For example, try looking at a distant mercury-vapor street lamp between the narrowest possible slit you can make between your fingers. Fresnel systematically investigated the interference patterns from apertures, edges, and small obstacles. If a source and viewing screen are each a finite distance from an intermediate object, the resulting diffraction pattern is known as *Fresnel diffraction*.

If both the source and the screen are far from the intermediate object, the mathematics is considerably simplified (Fig. 39–4). This special case of Fresnel diffraction is known as *Fraunhofer diffraction*, after Joseph von Fraunhofer. The Fraunhofer limit is easy to treat, because the waves that originate at each position of the apertures or obstacles from the source and reach a given point on the screen are nearly parallel, simplifying the calculation of path-length differences and phase differences. It was the Fraunhofer case that we treated in Chapter 38 when we studied double-slit interference patterns. Starting with Section 39–2, we shall restrict ourselves to this case.

Before considering in detail some of the cases mentioned here, let us first look at one rather spectacular demonstration of diffraction effects observed by Arago in 1818. Suppose, as in Fig. 39–5a, that we place a perfectly round obstacle in the path of a point source of coherent light. Every point on the rim of the disk is equidistant from the source, and light falling on the rim is thus perfectly in phase. According to the Huygens construction, we can think of each of these points as a new source, and they are all in phase. All the rim points are equidistant from a point P on a screen, the point that lies on the symmetry axis between the disk and the source. Because the light reemitted from the rim arrives at P in phase, there is constructive interference at P, and hence a bright spot, known as the *Poisson spot*, appears. This diffractive result is certainly inconsistent with geometric optics! Even an ordinary penny can be used to show this result (Fig. 39–5b). The central maximum is clearly visible.

39–2 DIFFRACTION GRATINGS

A simple way to generalize the double-slit interference experiment is to increase the number of narrow slits. There is a characteristic interference pattern *if the*

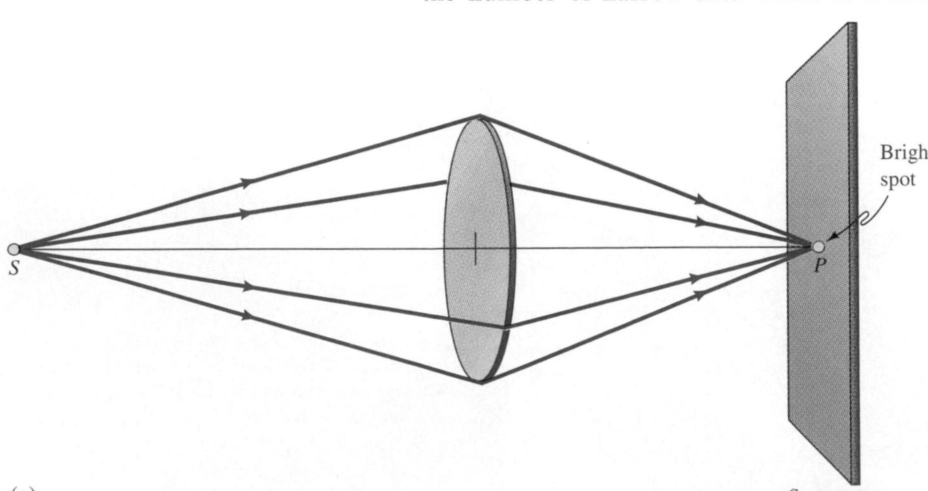

FIGURE 39–5 (a) A round, opaque object in the path of a point source of coherent light. Diffraction causes a bright spot (the Poisson spot) to be seen along the optical axis at point P. (b) The Poisson spot of a penny.

slits are regularly spaced; a screen with such an arrangement is called a **diffraction grating**. Diffraction gratings are important for two reasons: First, the multiple slits allow more light through than two slits do, thus increasing the intensity; second, the interference maxima are much sharper than they are for two slits, allowing for the wavelength of the light to be measured more precisely.

Diffraction gratings

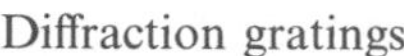

In Young's double-slit experiment, two slits were literally cut in an opaque screen. However, the presence of transparent holes in an opaque screen is not necessary. All that is required is an array of obstacles to serve as pointlike sources for the reradiation of spherical wavelets. Thus a diffraction grating can take many forms. For example, regular scratches or rulings can be inscribed on glass or metal plates. When light passes through a ruled glass plate, the scratches act as sources for the regeneration of spherical wavelets. This type of grating is called a *transmission grating.* When the scratches are made on metal plates, they act as regular point sources of reflected rather than transmitted light; such a grating is called a *reflection grating.* Even the marks on an ordinary ruler can be a reflection grating for laser light. What is important is that the light be scattered from regularly spaced centers. We shall see a vivid example of this physical property in Section 39–6 when we study X-ray diffraction, in which the atoms in the regular crystalline array of a solid act as reradiating point sources. For all these gratings, the analysis of the diffraction pattern is similar, although we shall concentrate here on transmission gratings and draw them as though there are actual slits.

FIGURE 39–6 Henry Rowland's ruling machine for cutting gratings made the use of the patterns produced by gratings a standard tool for analyzing light.

Fraunhofer made the first gratings from fine parallel wires. As we have stated, more slits for a fixed slit spacing lead to a sharper diffraction pattern and a grating more useful for the analysis of light. In the 1870s Henry Rowland invented ruling machines capable of cutting gratings with thousands of grooves per centimeter (Fig. 39–6). This process revolutionized the use of diffraction gratings, and Rowland-type gratings were very important to the discoveries and progress of modern physics. *Spectroscopes*, optical devices used to measure the wavelength of light emitted by atoms after they have been excited, are typically based on diffraction gratings. The unique set of wavelengths (or frequencies) produced by excited states of each type of atom, ion, or molecule is called its *spectrum.* The measurement of these spectra led to quantum mechanics and an understanding of the structure of the atom. Diffraction gratings are still widely used in science and technology, and simple, cheap plastic gratings are readily available. Figure 39–7a is a schematic diagram of a spectroscope, and Fig. 39–7b shows the result

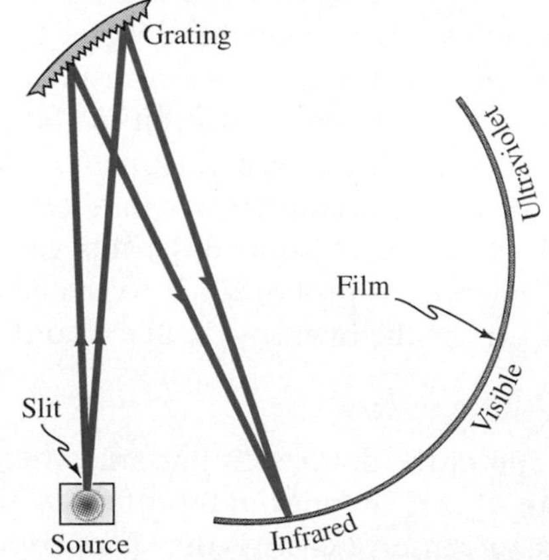

(b)

FIGURE 39–7 (a) Spectroscopes that capture an entire spectrum spread over a large sheet of film, as in this schematic diagram, are sometimes called spectrographs. The spectrograph depicted here forms part of a circle 7 m in radius, of which the film plate is a part, and employs a grating that is located along the circle. Light coming from a point on that circle is reflected back to the film. (b) The characteristic light spectrum produced by a sodium-type street lamp, as observed through a spectroscope.

of using a spectroscope to observe the light from a distant sodium-type street lamp. Arrays of antennas act to some extent like diffraction gratings. Such antennas can both broadcast and receive, as in radio telescopes or ground-control approach antennas.

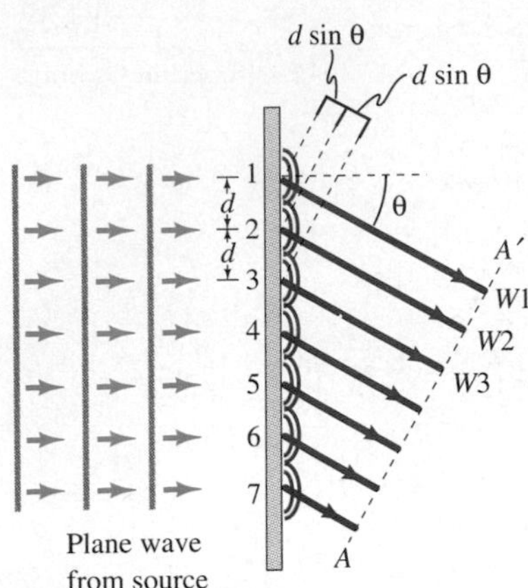

FIGURE 39–8 The geometry of a diffraction grating. Light passing through individual slits spreads in all directions. The interference of the light at a distant screen, along the directions indicated by $W1$, $W2$, $W3$, and so on, depends on the path-length differences $d \sin \theta$ between adjacent slits.

Energy Conservation and Intensity

Let us turn now to the analysis of the Fraunhofer diffraction pattern. Figure 39–8 shows a few of the slits of a grating with N slits, separated from each other by a distance d; a monochromatic plane wave of wavelength λ approaches from a distant source. The plane wave arrives with the same phase at each slit, and spherical waves are emitted in phase at each slit. We look at wave propagation along the lines labeled W. Lines $W1$, $W2$, $W3$, and so on are oriented at an angle θ to the original wave-propagation direction. Note that we allow the screen to be so distant that these lines are approximately parallel even though they point toward a particular spot on the screen (or a lens focuses them on that spot).

If the wavelets are in phase along the front AA', defined by θ, then the light that eventually reaches the distant screen at this angle will also be in phase, and there will be constructive interference—a maximum. The condition under which the waves along AA' are in phase is that each path length differs from any other by integral multiples of the wavelength λ. The path-length difference between adjacent waves is $d \sin \theta$ (Fig. 39–8). Thus there are *principal maxima*, where the light from all the slits interferes constructively, at angles such that

The location of the principal maxima of a diffraction grating

for principal maxima: $$d \sin \theta = m\lambda, \quad \text{where } m = 0, \pm 1, \pm 2, \ldots . \qquad (39\text{–}1)$$

As is the case for the two-source pattern, the integer m specifies the *order* of the principal maxima. In this result we see the universal diffraction phenomenon that *the pattern spreads in angle as the ratio λ/d increases.* Equation (39–1) is the same as Eq. (38–3a), which determines the maxima of the double-slit pattern. The intensity pattern on the screen, however, takes quite a different form, as we shall soon show.

The angular spread of the diffraction pattern increases when the ratio of the slit separation to the wavelength decreases.

Before performing the mathematical analysis for the intensity pattern, let us first examine intensity qualitatively from what we already know about light. If there were no interference whatsoever, the average intensity over the entire screen due to N slits would be NI_0, where I_0 is the average intensity for just one slit. Energy must be conserved whether or not there is interference. If the light intensity is zero in regions of destructive interference, there must be a higher light intensity in regions of constructive interference. Because intensity measures the rate at which energy arrives, energy conservation requires that even in the presence of interference, the average intensity over the entire screen must be NI_0. What is the intensity at the principal maxima, where there is constructive interference? At such points the electric or magnetic fields of the waves from all the slits add to N times the field from one slit. Because the intensity is proportional to the fields *squared*, the intensity at any maximum is N^2 times the intensity I_0 due to one slit:

The intensity of a maximum is proportional to the number of slits squared.

$$I_{\text{max}} = N^2 I_0. \qquad (39\text{–}2)$$

This result agrees with the calculation of the double-slit pattern, where we found that the heights of the maxima are $4I_0$. If the maximum intensity increases so markedly for N slits, then, in order to conserve energy, the space over which the maximum occurs must be much smaller. Suppose that the principal maxima have a *width* of $\Delta\theta$. (The width is the angular spread at the points where a principal maximum is half its maximum height of I_{max}.) The principal maxima are so strong, and there is so little light between them, that the total intensity is approximately the sum of the intensities in the principal maxima. Thus

$$I_{\text{max}}\,\Delta\theta \simeq NI_0.$$

We solve this equation for the width, and when we use Eq. (39–2), we get

$$\Delta\theta = \frac{NI_0}{I_{\text{max}}} = \frac{NI_0}{N^2 I_0} = \frac{1}{N}. \tag{39–3}$$

The width of a maximum is inversely proportional to the number of slits.

A qualitative way to understand the decrease of $\Delta\theta$ with N is to note that as N increases, it becomes less likely to have total constructive interference. If the path-length difference between adjacent slits is not precisely an integral multiple of the wavelength, over a distance of many slits, the phase differences between the light from distant slits will eventually cause some cancellation of the electric field. As N increases, the variation in the value of θ that can be tolerated to maintain a maximum becomes smaller.

To summarize, we have found that *as the number of slits N increases, the height of the principal maxima increases as* N^2. *In addition, the width decreases—that is, the principal maxima become sharper—as* $1/N$. This is the crucial feature that makes diffraction gratings so useful. A tall and narrow peak implies that the peak is easily seen, even if the source is relatively weak, and that great precision is possible in locating it. The latter feature means that the wavelengths for which the principal maxima occur can be determined with great precision from Eq. (39–1).

Intensity Pattern

In Fig. 39–8, the wave $W2$ that travels along the direction specified by θ (not necessarily the angle corresponding to a principal maximum) has a phase difference of 2β less than wave $W1$:

$$2\beta \equiv \frac{2\pi d \sin\theta}{\lambda}. \tag{39–4}$$

Similarly, wave $W3$ has the same phase 2β less than wave $W2$, and so forth. Thus, at the distant screen, the electric field due to each of the N slits is $E_1 = E_0\cos(\omega t)$, $E_2 = E_0\cos(\omega t - 2\beta)$, $E_3 = E_0\cos(\omega t - 4\beta), \ldots, E_N = E_0\cos[\omega t - 2(N-1)\beta]$. The total electric field at the screen is the sum over all N slits:

$$E = E_0 \sum_{n=0}^{N-1} \cos(\omega t - 2n\beta). \tag{39–5}$$

This sum can be carried out by advanced mathematical methods and gives

$$E = E_0\cos[\omega t - (N-1)\beta]\,\frac{\sin(N\beta)}{\sin\beta}. \tag{39–6}$$

The intensity is proportional to the time average of E^2. When we square Eq. (39–6), the time average of the cosine squared term is $\frac{1}{2}$, and we identify the factor $E_0^2/2$ in this time average as proportional to the intensity I_0 due to one slit. Thus

$$\text{for multiple slits:} \qquad I = I_0\left[\frac{\sin(N\beta)}{\sin\beta}\right]^2. \tag{39–7}$$

According to Eq. (39–1), β is zero at principal maxima. As $\beta \to 0$, $\sin\beta$ is of order β whereas $\sin(N\beta)$ is of order $N\beta$. Thus the ratio $\sin(N\beta)/\sin\beta$ approaches N in the limit that $\beta \to 0$. In this limit Eq. (39–7) agrees with Eq. (39–2), which was much simpler to derive.

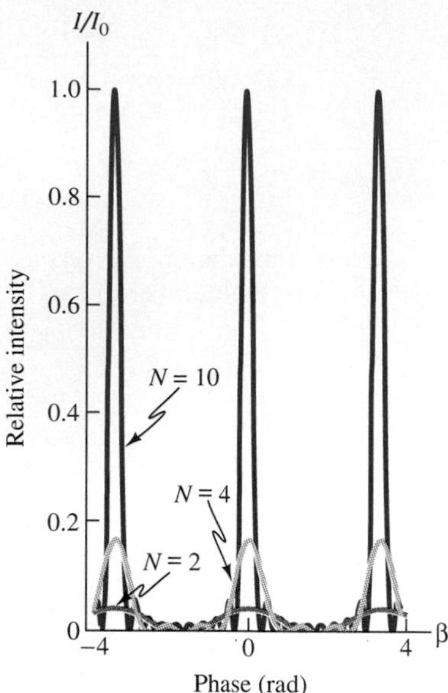

FIGURE 39–9 The ratio I/I_0 plotted against the phases $\beta = (\pi d \sin\theta)/\lambda$ for $N = 2$, 4, and 10 slits. The intensity pattern changes dramatically with N. The widths of the principal maxima decrease as $1/N$.

Equation (39–7) is most easily understood by plotting it, as is done in Fig. 39–9 for $N = 2$, 4, and 10: The ratio I/I_0 is plotted against β. The figure shows the principal maximum at $\beta = 0$ (corresponding to $\theta = 0$), the central spot on the screen, as well as the first principal maxima to the sides, corresponding to $n = \pm 1$ in Eq. (39–1). The widths of the principal maxima do indeed decrease as $1/N$. Note that there are also $N - 2$ small secondary maxima between each pair of principal maxima. They have intensities on the order of I_0 itself. For diffraction gratings in ordinary use, N is in the thousands, and the secondary maxima can safely be ignored. Physically, they occur because there is the possibility that the light from two or more of the nonneighboring slits may be in phase at a given angle, even though the light from *all* the slits is not in phase.

Demonstrations that illustrate the effect of multiple-slit diffraction are easy to arrange in the classroom. The spectral lines emitted by pure atomic sources of light are evident even with an inexpensive, handheld diffraction grating. One impressive demonstration that gives diffraction peaks is the reflection of laser light at a glancing angle from an ordinary ruler. The source separation d is large compared to the light frequency, so the principal maxima are very close together in angle, but N is large enough so that the maxima are quite distinct.

We can check our general result for N slits in the case $N = 2$. If we let $N = 2$ in Eq. (39–7), we have, with $\sin(2\beta) = 2\sin\beta\cos\beta$,

$$I = I_0\left[\frac{\sin(2\beta)}{\sin\beta}\right]^2 = I_0\,\frac{(2\cancel{\sin\beta}\cos\beta)^2}{(\cancel{\sin\beta})^2} = 4I_0\cos^2\beta.$$

This is indeed the result for two slits, given in Eq. (38–13).

Resolution of Diffraction Gratings

Angular Dispersion. Diffraction gratings were formerly used to identify the characteristic wavelengths of elements. Now that these wavelengths are known, gratings are used primarily to identify elements, ions, and compounds through the characteristic light they emit, or as a tool for understanding the structure of molecules. Because d, the distance between slits, is usually known, the angular location of a principal maximum ($m \neq 0$) gives λ according to Eq. (39–1). In this case, an important limitation is the ability to separate the spectral lines of nearly equal wavelengths λ_1 and λ_2. Two quantities determine the effectiveness of spectroscopic instruments. One is the **angular dispersion**, defined as $\Delta\theta/\Delta\lambda$, which measures the difference $\Delta\theta$ in the angles of the principal maxima due to two nearly equal wavelengths that differ by $\Delta\lambda$.† The term "dispersion" is often used in science and technology to denote a change in one variable with respect to another (remember, for example, the dispersion of light by a prism according to the wavelength). We differentiate Eq. (39–1) to determine the angular dispersion:

Angular dispersion

$$d\cos\theta\,\Delta\theta = m\,\Delta\lambda;$$

$$\frac{\Delta\theta}{\Delta\lambda} = \frac{m}{d\cos\theta}. \tag{39–8}$$

The dispersion increases for higher orders, is inversely proportional to the distance between slits, and increases away from the central maximum.

Resolution. The angular dispersion alone does not tell us whether we can visually separate two similar wavelengths. This aspect of the effectiveness of a

† Here and in the following discussion, we use Δ rather than d to refer to derivatives and differentials, in order to avoid confusion with the slit separation *d*.

grating is characterized by the **resolving power** of the grating, defined by

$$R \equiv \frac{\lambda}{\Delta\lambda}. \tag{39–9}$$

Resolving power

Here $\Delta\lambda$ is the smallest wavelength difference that can be observed with the grating. (The maxima of two wavelengths that are too close together lie so close to one another that they cannot be distinguished.) The larger the value of R, the better the grating can distinguish the relative wavelength difference of two closely spaced lines. We previously discussed the fact that the maxima peaks sharpen as the number of slits increases. It is the sharpness of the peaks that enables a grating to separate two closely spaced lines. By a detailed analysis of the peak widths of an N-slit system, it is possible to show that the resolving power of a grating is given by

$$R = mN. \tag{39–10}$$

The resolving power of a grating

The resolving power improves as the number of slits N increases and is better for larger orders; that is, for larger integers m.

EXAMPLE 39–1 Heated sodium provides an easily available source of light. It has a characteristic yellow–orange color and two intense wavelengths of 589.0 nm and 589.6 nm, called a *doublet*. (a) How many slits are required in a grating that resolves the doublet at the first-order maxima? (b) If the screen is exactly 4 m from a grating with exactly 2000 slits/cm, what are the screen positions of the two principal maxima of first order? Assume that the screen is far enough away so that the conditions of Fraunhofer diffraction apply.

SOLUTION: The solution to this problem lies in our discussion of the resolving power and the position of principal maxima. We are given the two wavelengths and must look at the first-order maxima ($m = 1$).

(a) We first find the minimum resolving power needed to resolve the two closely spaced lines. The difference in wavelengths of the doublet is only $\Delta\lambda = 589.6\text{ nm} - 589.0\text{ nm} = 0.6\text{ nm}$. The resolving power needed to separate the doublet is given by Eq. (39–9),

$$R = \frac{\lambda}{\Delta\lambda} = \frac{589.0\ \cancel{\text{nm}}}{0.6\ \cancel{\text{nm}}} = 982.$$

Equation (39–10), with $m = 1$, gives $N = R = 982$. About 1000 slits are required.

(b) The angular positions of the first-order principal maxima are given by Eq. (39–1). To use it, we need the slit separation d, which is given by $d = 1/(2000\text{ slits/cm}) = 5.000 \times 10^{-6}$ m. Then Eq. (39-1), with $m = 1$, gives

$$\text{for } \lambda_1: \quad \sin\theta_1 = \frac{\lambda_1}{\text{d}} = \frac{589.0\text{ nm}}{5.000 \times 10^{-6}\text{ m}} = 0.1178;$$

$$\text{for } \lambda_2: \quad \sin\theta_2 = \frac{\lambda_2}{d} = \frac{589.6\text{ nm}}{5.000 \times 10^{-6}\text{ m}} = 0.1179.$$

The respective angles are $\theta_1 = 0.1181$ rad and $\theta_2 = 0.1182$ rad, and the distance from the centerline of the screen is $y = L\tan\theta$. These respective distances are

$$y_1 = (4\text{ m})\tan(0.1181\text{ rad}) = 0.4745\text{ m},$$

$$y_2 = (4\text{ m})\tan(0.1182\text{ rad}) = 0.4750\text{ m}.$$

The images are separated by only 0.5 mm, but the resolution will be sufficient for the two spectral lines to be distinguishable.

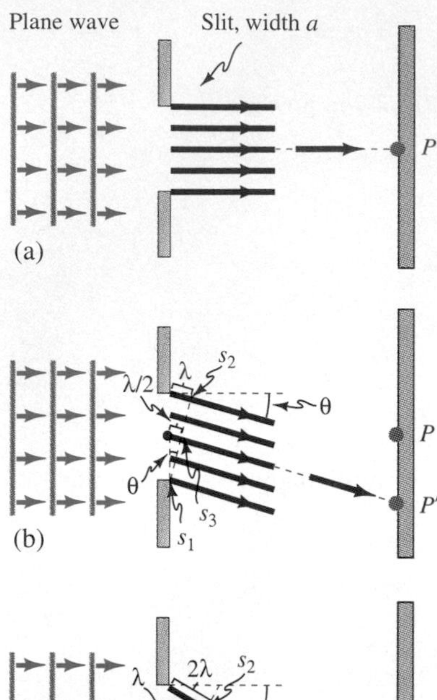

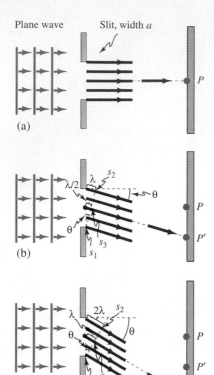

FIGURE 39–10 Monochromatic plane waves of light enter a narrow slit of width a (exaggerated here). We show three positions (P, P', and P'') on a distant screen. (a) We look along the incident direction at P. (b) We look along angle θ at P'. If the light from positions in the top half of the slit are out of phase with corresponding positions in the bottom half at s_1s_2, destructive interference occurs at P'. (c) We look along a different angle θ at P''. Destructive interference can occur also at P'', as demonstrated by breaking the slit into halves and treating each half as we did for the direction toward point P'.

39–3 SINGLE-SLIT DIFFRACTION

If the width a of a single slit is comparable to or smaller than the wavelength of coherent light that passes through it, a diffraction pattern is produced. This pattern forms because wavelets regenerated at different places across the single slit interfere with each other. To explain the diffraction pattern, we treat the single slit as an infinitely large number of infinitesimal sources of wavelets. Thus the pattern produced by a single slit has much more in common with a diffraction grating than it does with a double slit.

Consider a monochromatic parallel beam (plane waves) of light of wavelength λ that moves toward an opaque barrier with a narrow rectangular slit of width a ($a > \lambda$). A coherent wave arrives at the slit and regenerates spherical waves at each point across it. In Fig. 39–10a, we examine the light from the slit that proceeds along the initial direction toward the center point P of the distant screen. The regenerated wavelets are all in phase in this direction, and the central point of the screen is bright.

Along the direction that leads to point P' on the screen, with the angle given by $\sin\theta = \lambda/a$, there will be destructive interference under certain conditions (Fig. 39–10b). The path lengths from the line s_1s_2 to point P' are all the same because the screen is distant, so we need consider only the phase relations of the light waves at line s_1s_2. The wave emitted at the top of the slit has traveled a distance λ to point s_2, and the wave emitted from the midpoint of the slit has traveled a distance $\lambda/2$ to point s_3. Thus along the line s_1s_2, the wave emitted at the top is *out of phase* with the wave emitted at the center of the slit (and similarly at point P'). Similarly, the wave emitted just below the top of the slit is out of phase with the wave emitted from just below the center. We can follow the points *in pairs* along the slit. For every point in the top half of the slit, there is a point in the bottom half, and the waves from the two points are precisely out of phase with each other. The result is destructive interference, a minimum (or dark spot), on the screen at P' at the angle given by $\sin\theta = \lambda/a$.

Along the direction given by $\sin\theta = 2\lambda/a$, the wavelet emitted from the top of the slit travels a distance 2λ farther than the wavelet emitted from the bottom, and a distance λ farther than the wavelet emitted from the center point (Fig. 39–10c). We can think of the slit of width a as broken into two slits of width $a/2$. Along the direction chosen, we are in a situation similar to that in Fig. 39–10b. There is destructive interference at point P'' due to net destructive interference from both the top half and the bottom half of the slit. For example, for every wavelet emitted from a particular point in the top half of the slit, we can find a second wavelet emitted from another point in the top half that destructively interferes with the first wavelet, because their path-length difference is $\lambda/2$. The same happens for the bottom half. We will have another intensity minimum (dark spot) on the screen for angle $\sin\theta = 2\lambda/a$.

If we continue our analysis, we find that every time there is an additional path-length difference of λ between the top and bottom of the slit, destructive interference and a screen minimum result. Thus we have

Locations of minima in the diffraction pattern of a single slit

for destructive interference: $$\sin\theta = \frac{m\lambda}{a}, \quad \text{where } m = \pm 1, \pm 2, \pm 3, \ldots \tag{39–11}$$

The value $m = 0$ is not part of this sequence of minima: For $m = 0$, $\sin\theta = m\lambda/a = 0$, and we have seen that this central point P must always be a maximum.

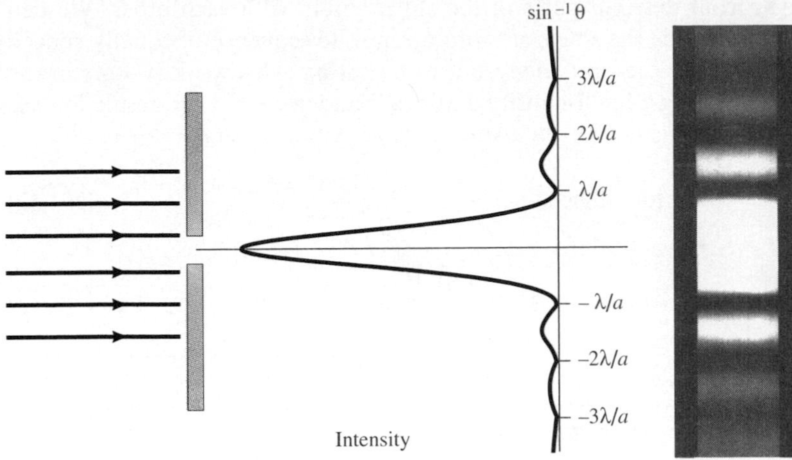

FIGURE 39–11 The interference pattern of single-slit diffraction and the relative intensities of such a pattern. Most of the light energy is in the bright central peak. The central peak is twice as wide as the secondary peaks.

The interference pattern has the typical behavior of diffractive phenomena: Larger values of a/λ give a smaller angular spread in the interference pattern, and smaller values of a/λ give a larger angular spread. In the limit that $a \gg \lambda$, the spread decreases so much that there is only a bright central spot on the screen. Do not confuse the angular spread on the screen with a projection of a slit width in geometric optics. The screen is very distant, and if geometric optics were to hold, the projection of the slit on the screen due to direct incident light would be a line with precisely the width of the slit. *The smaller the slit width compared to the wavelength, the larger the angular spread of the interference pattern.*

Approximately halfway between the successive minima, we have conditions for constructive interference and an intensity maximum (a bright region). Figure 39–11 shows the diffraction pattern of a single slit. Most of the light is in fact contained in the wide central maximum. We can understand the exact location of the maxima as well as the value of the intensity at these points, only with a more precise mathematical analysis.

EXAMPLE 39–2 Helium–neon laser light of wavelength 633 nm passes through a single slit of width 0.10 mm. The diffraction pattern is observed on a screen 3 m away, far enough so that the conditions for Fraunhofer diffraction apply. What is the distance between the two minima on either side of the central maximum?

SOLUTION: The conditions for the single-slit pattern apply here; in particular, the slit width a is some 160 times larger than the wavelength. We can use Eq. (39–11) to find the angular positions of the minima that correspond to $m = 1$ and -1. These angles are, respectively,

$$\sin\theta = \frac{m\lambda}{a} = \pm 1\,\frac{633 \times 10^{-9}\,\text{m}}{0.10 \times 10^{-3}\,\text{m}} = \pm 0.0063;$$

$$\theta = \pm 0.36^\circ.$$

At a distance of 3 m away, these minima will be (3 m) tan 0.36° = 1.9 cm from the central line. The total distance between them is thus 2(1.9 cm) = 3.8 cm.

The Intensity Pattern of Single-Slit Diffraction

We already did most of the work to find the diffraction pattern of a single slit in more detail when we analyzed the N-slit grating. As the qualitative analysis above shows, the crucial physical principle behind the diffraction pattern due to a single slit is that light from different parts of the slit interfere with each other. We can therefore formally divide the single slit into an infinite sequence of equally spaced and infinitely narrow pieces and analyze it as a grating. This work is done in the box "Deriving the Single-Slit Intensity Pattern," and we state the result for *the intensity pattern on a distant screen due to a single slit of width a*:

The intensity pattern due to a single slit

$$\text{for single slits:} \quad I = I_{max}\frac{\sin^2\alpha}{\alpha^2}, \tag{39–12}$$

where

$$\alpha = \frac{\pi a \sin\theta}{\lambda}. \tag{39–13}$$

*DERIVATION

Deriving the Single-Slit Intensity Pattern

We imagine breaking up a single slit of width a into N strips of width d (Fig. B1–1). (Each strip acts as a separate slit.) The entire slit is divided in this way, so

$$Nd = a.$$

In the limit that the number of strips $N \to \infty$, d must approach zero in order to keep a constant. Because each strip is infinitely narrow, we can treat these strips as the thin slits of a grating. We can then use the diffraction grating result of Eq. (39–7) directly to find that the intensity at the screen at angle θ is

$$I = \lim_{N\to\infty} I_0\left[\frac{\sin(N\beta)}{\sin\beta}\right]^2.$$

This expression includes a slit width $d = a/N$ that approaches zero as $N \to \infty$. According to Eq. (39–4), $\beta = (\pi d \sin\theta)/\lambda = [\pi(a/N)\sin\theta]/\lambda = \alpha/N$, where the definition of α is contained in Eq. (39–13). Thus the intensity takes the form

$$I = \lim_{N\to\infty} I_0\left[\frac{\sin\alpha}{\sin(\alpha/N)}\right]^2.$$

In the limit of large N, the factor $\sin(\alpha/N) \simeq \alpha/N$, and

$$I = N^2 I_0 \frac{\sin^2\alpha}{\alpha^2}.$$

The factor I_0 is the intensity due to one of the subslits of width d. We need only interpret the factor $N^2 I_0$ as the maximum possible intensity I_{max} of the single slit of width a to get Eq. (39–12).

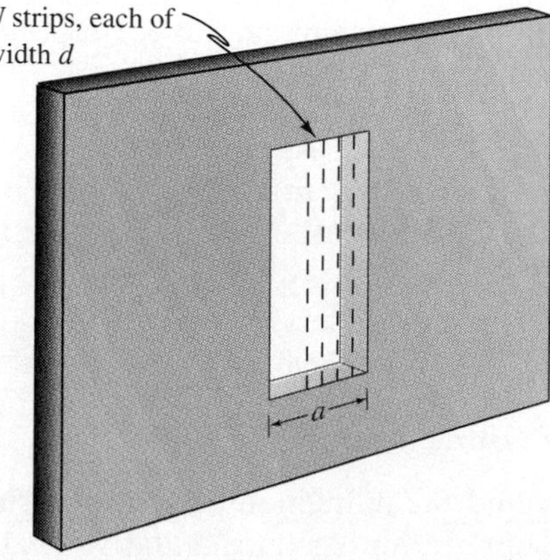

FIGURE B1–1 We can understand the intensity pattern of a single slit by supposing that the slit is composed of a large number of strips N and then use the result for diffraction gratings.

Note that there is a central maximum at $\theta = 0$: At $\theta = 0$, Eq. (39–13) shows that $\alpha = 0$. Because the limit of $(\sin \alpha)/\alpha \to 1$ as $\alpha \to 0$, the intensity at this point on the screen is just I_{max}.

Our qualitative discussion at the beginning of this section shows that angle α is simply the phase difference between the top and the middle of the slit. The intensity drops off rapidly as α increases. The intensity is zero, as we argued previously that it must be, for angles given by Eq. (39–11). We can now see that this occurs when α is an integral multiple of π:

$$\text{for minima:} \qquad \alpha = n\pi = \frac{\pi a \sin \theta}{\lambda}, \text{ where } n = \pm 1, \pm 2, \pm 3, \ldots . \qquad (39\text{–}14)$$

The intensities at the secondary maxima are estimated in Example 39–3.

EXAMPLE 39–3 Estimate the ratios of the intensities of the first and second maxima to the intensity of the central maximum for a single slit.

SOLUTION: In order to find the exact positions of the maxima, we must take the derivative of the intensity with respect to α (or θ) and set that equal to zero (see Problem 23). However, we can make a rapid estimate by realizing that the maxima are approximately halfway between the minima:

$$\text{for maxima:} \qquad \alpha \simeq \left(n + \frac{1}{2}\right)\pi, \text{ where } n = 1, 2, 3, \ldots .$$

Let us denote the intensity at these secondary maxima by I_n. We use this approximation of α in Eq. (39–12) to get

$$\frac{I_n}{I_{max}} = \left\{\frac{\sin[(n + \frac{1}{2})\pi]}{(n + \frac{1}{2})\pi}\right\}^2 = \frac{1}{(n + \frac{1}{2})^2\pi^2}.$$

[We have used $\sin^2(\pi/2) = \sin^2(3\pi/2) = \cdots = 1$.] With $n = 1$ and $n = 2$, we get

$$\frac{I_1}{I_0} = 0.045 \quad \text{and} \quad \frac{I_2}{I_0} = 0.016.$$

The intensity of the secondary maxima falls off rapidly. The central bright maximum is by far the most intense.

Figure 39–12 is a plot of the ratio of the intensity to its maximum value for the single slit as a function of θ. Note that the central maximum is *twice* as broad as the secondary maxima are, a feature that distinguishes the single-slit pattern from the double- or multiple-slit patterns. We have chosen $a = 4\lambda$ in this plot.

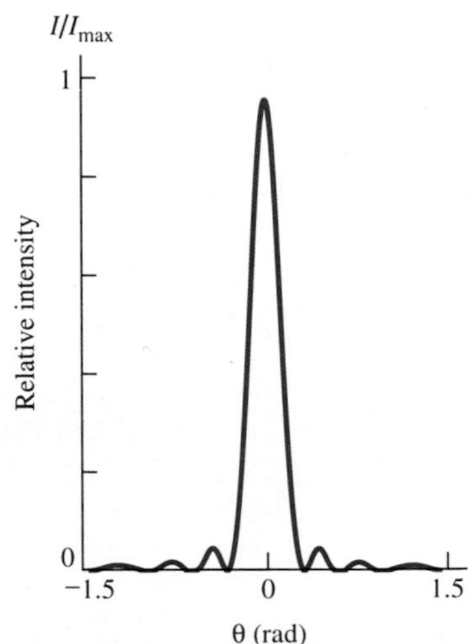

FIGURE 39–12 The intensity ratio I/I_{max} as a function of angle θ [from Eqs. (39–12) and (39–13)] for the slit width $a = 4\lambda$.

39–4 DIFFRACTION AND THE RESOLUTION OF OPTICAL INSTRUMENTS

We have already seen that the ability of a grating to resolve closely spaced lines is limited by the width of the principal maxima. Similarly, the fact that in a single aperture there is some spreading of light due to diffraction *intrinsically* limits the capacity of optical instruments to resolve objects. The resolution of optical instruments such as telescopes, binoculars, and microscopes is also limited by lens *aberration* (see Chapter 37). By careful techniques, aberration can be minimized, and there is no theoretical limit to such improvement. The limitation due to diffraction, however, is set by the aperture of the instrument. The width of a central diffraction

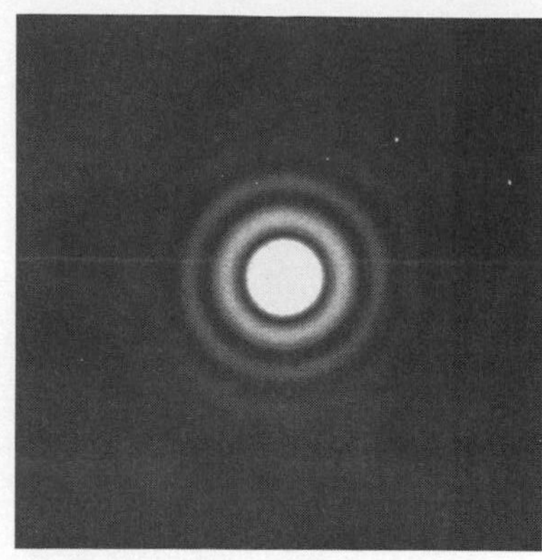

FIGURE 39–13 Diffraction pattern of light from a distant source after the light passes through a circular aperture. Some 85 percent of the light intensity is contained in the bright central maximum, called the Airy disk.

maximum is determined by the aperture size and the wavelength of light, not by lens design.

Because most optical instruments rely on circular lenses or mirrors, we shall concentrate on circular apertures. The size of the aperture of a lens system determines the diffraction pattern (first worked out by Sir George Airy in the 1830s). Figure 39–13 is a diffraction pattern of light from a distant point source that passes through a circular aperture. The bright central area, containing some 85 percent of the light intensity, is called an *Airy disk.* The rings outside the central area are the minima and secondary maxima of the diffraction pattern. The edge of the Airy disk, or, more precisely, the position of the first minimum, occurs at an angle from the central axis given by

$$\theta_{\rm min} = \frac{1.22\lambda}{D}, \tag{39–15}$$

where D is the diameter of the aperture. This result can be compared to Eq. (39–11), which gives the angle of the first minimum for a slit of width $a \gg \lambda$ as $\theta_{\rm min} = \lambda/a$. (Recall that $\sin\theta \simeq \theta$ for small θ.) The factor 1.22 arises because the width of a circular aperture varies (in effect, $a \simeq D/1.22$). The precise factor 1.22 is actually irrelevant to us, and we prefer to drop it and give an approximate $\theta_{\rm min}$ as

The approximate position of the first minimum of the diffraction pattern of an aperture

$$\theta_{\rm min} \simeq \frac{\lambda}{D}. \tag{39–16}$$

Let us look at how the presence of a diffraction pattern limits our ability to make images. We may want to observe two closely spaced objects, such as a double star, or we may need to see detail in an X-ray taken of suspected stress fractures in piping for a power plant. The presence of the Airy disk means that even a very distant star does not produce a pointlike image but rather a disklike image with an angular spread described by Eq. (39–16). The image in a telescope of two stars that are so close together that their Airy disks overlap cannot be easily recognized as an image of two stars. In Fig. 39–14a two objects have a great enough angular separation θ that they are easily resolved. In Fig. 39–14b the objects are just barely resolved. When they are even closer, as in Fig. 39–14c, the central diffraction peaks overlap too much to be resolved. It has become customary to describe the limiting case, that shown in Fig. 39–14b, as the *Rayleigh criterion*:

The Rayleigh criterion

Two point sources are just resolved if the peak of the diffraction image of the first source overlies the first minimum of the diffraction image of the second source.

The Rayleigh criterion is satisfied when the angular separation of the objects is just $\theta_{\rm min}$, defined by Eq. (39–16).† We emphasize that the limitation on the resolution of images due to the wave nature of light is fundamental.

The discussion above refers to the angular separation of the image formed by an optical system. The methods of Chapter 37 allow us to show that this angle is also the angular separation of the objects. The angles are the same because the principal rays through the center of a lens are undeviated. When we use a microscope, for example, it is useful to know the spatial separation of two objects that can just be resolved. If the objects are placed at the focal point of the lens (Fig. 39–15), the minimum separation $S_{\rm min}$ for a lens of diameter D is given by

$$S_{\rm min} = f\theta_{\rm min} \simeq \frac{f\lambda}{D}. \tag{39–17}$$

† Strictly speaking, the Rayleigh criterion contains the factor 1.22 that we dropped from Eq. (39–15). Because modern computer-fitting techniques allow us to resolve two sources even when the two maxima of their images are closer than the first minimum of one image, the precision that the factor 1.22 implies is irrelevant to the actual resolving power. This is why we dropped the factor.

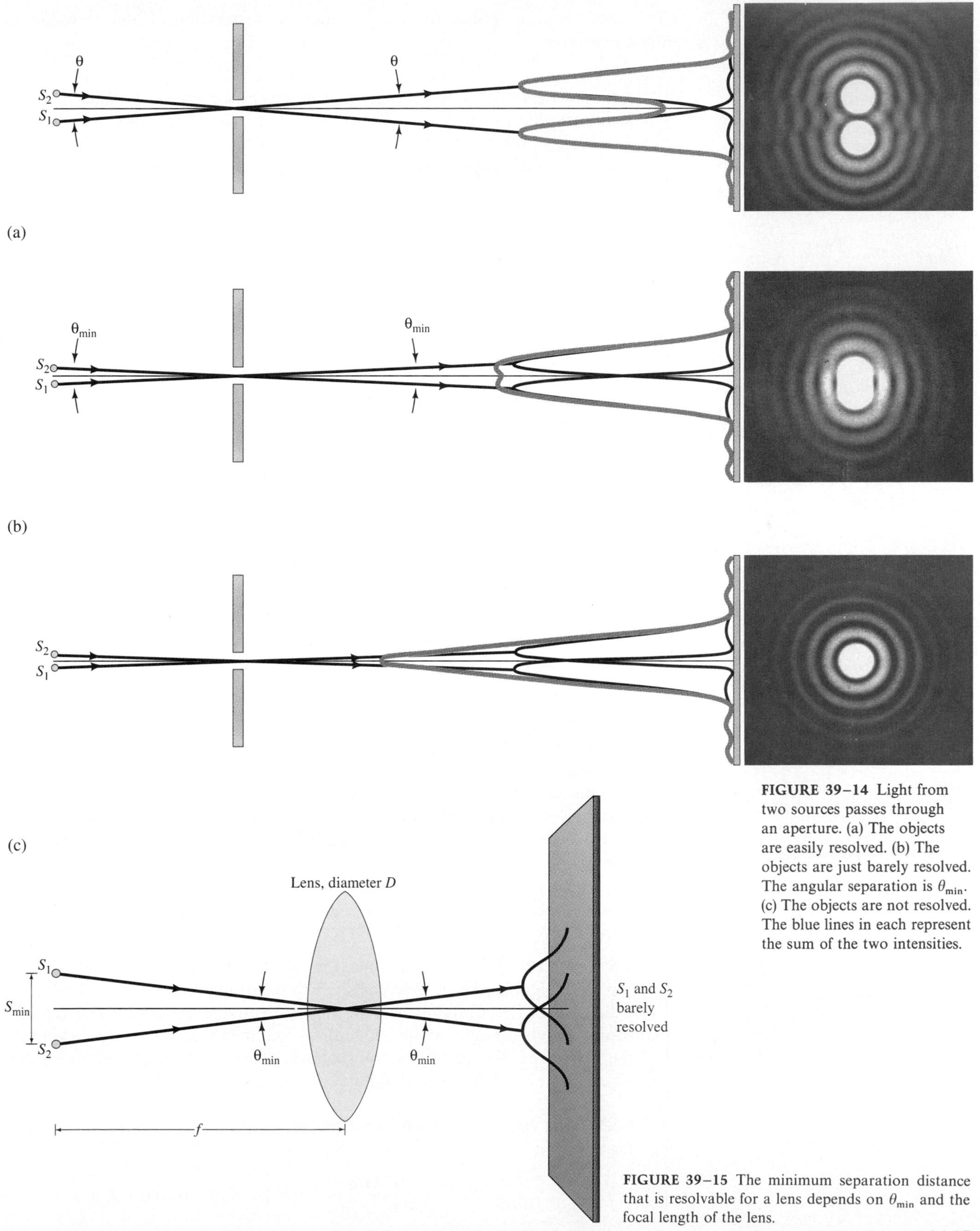

FIGURE 39–14 Light from two sources passes through an aperture. (a) The objects are easily resolved. (b) The objects are just barely resolved. The angular separation is θ_{min}. (c) The objects are not resolved. The blue lines in each represent the sum of the two intensities.

FIGURE 39–15 The minimum separation distance that is resolvable for a lens depends on θ_{min} and the focal length of the lens.

This result can be applied to find the minimum separation of objects visible to the human eye.

EXAMPLE 39–4 Estimate the minimum separation between two objects such that the human eye can still perceive them as separate (the *minimum visible object separation*) if the objects are (a) at the near-point distance (25 cm) and (b) at a distance of 5 m. Take the pupil diameter to be 2.5 mm.

SOLUTION: (a) We use the median range of visible wavelengths, or about 550 nm, in Eq. (39–17) to estimate the minimum visible object separation. For viewing an object at a given distance, the muscles of the eye adjust the focal length to that distance. Thus at 25 cm (the near point), we have

$$S_{\text{min}} = \frac{f\lambda}{D} = \frac{(0.25\ \cancel{\text{m}})(550 \times 10^{-9}\ \text{m})}{0.0025\ \cancel{\text{m}}} \simeq 0.055\ \text{mm}.$$

This is about the diameter of a small thread or a human hair. It is also, roughly, the separation between the cells of the retina. In other words, the cells that receive light and send that message to the brain are no closer than the minimum separation that could ever be resolved, an admirable example of the economy of biological systems.

(b) At 5 m, the minimum separation distance becomes

$$S_{\text{min}} = \frac{(5.0\ \cancel{\text{m}})(550 \times 10^{-9}\ \text{m})}{0.0025\ \cancel{\text{m}}} \simeq 1\ \text{mm}.$$

Thus from a few meters away we have an *intrinsic* inability to distinguish the millimeter markings on a meter stick.

The Resolution of Telescopes

Diffraction effects can limit the effectiveness of telescopes. For the visible-light region of the electromagnetic spectrum, one of the world's largest telescopes is the 200-in-diameter Hale telescope on Mt. Palomar in southern California. Its diffraction limit alone implies an angular resolution of about 0.03 s of arc. Its real resolution is some 300 times worse than this, due to the effects of atmospheric turbulence and ordinary aberration. The current record for angular resolution for an earth-based visible-light telescope is 0.36 s of arc, held by the New Technology Telescope in Chile. Radio telescopes such as that in Arecibo, Puerto Rico, operate in a wavelength range of roughly tens of centimeters. For a wavelength of 50 cm, the angular resolution of this apparatus, which has a diameter of 300 m, is about 7 min of arc.

EXAMPLE 39–5 The primary mirror on the Optical Telescope Assembly on the Hubble Space Telescope, which orbits 600 km above the earth, has a diameter of 2.4 m. (a) Calculate the minimum angular separation that it might resolve for visible light (about 550 nm). (b) Assume that the telescope is viewing the surface of the earth. What is the separation of the most closely spaced objects that it might resolve? Ignore all atmospheric effects.

SOLUTION: (a) This problem is a direct application of the resolution angle of Eq. (39–16):

$$\theta_{\text{min}} \simeq \frac{\lambda}{D} = \frac{(550 \times 10^{-9}\ \cancel{\text{m}})}{2.4\ \cancel{\text{m}}} = 2.3 \times 10^{-7}\ \text{rad} = 0.06\ \text{s of arc}.$$

This resolution is superior to the best resolution obtained on earth, because of the effects of atmospheric turbulence.

(b) The separation distance corresponding to the angle found in part (a) from a distance of $L = 600$ km is

$$d_{min} = L\theta_{min} = (600 \text{ km})(2.3 \times 10^{-7} \text{ rad}) = 0.14 \text{ m} = 14 \text{ cm}.$$

The 14-cm separation is far from the real resolution. Atmospheric turbulence sets the true limit of such a satellite. Earth-observation satellites such as Landsat and Spot have resolution capabilities of a few meters, and classified spy satellites are reported to have resolutions of less than 1 m.

An optical device need not consist of a single aperture. The mirrors of many of the largest modern reflecting telescopes are separated into several pieces, like a mosaic, in part because a single large piece of glass is difficult to form and handle. Such instruments also diminish diffractive effects, because the minimum angle that can be resolved is determined by the interference from the most widely separated pieces of the apparatus. If this maximum separation is D, then the minimum resolvable angle between objects is λ/D. (Note that D refers to a *transverse* separation; the distance between an eyepiece and an objective does not enter into diffractive effects.) One of the most intriguing suggestions for improving resolution in this way is the idea of placing on the moon an array of small optical telescopes connected electronically so that they all observe the same object and form a single large instrument. This array would be spread over a region with a diameter of 10 km and have a resolution 100,000 times better than that of the best telescope on earth. Were it not for the earth's atmosphere, such an instrument could pick up a newspaper headline on earth from the moon!

*39–5 THE EFFECTS OF SLIT WIDTH ON GRATING PATTERNS

In our discussions of double- and multiple-slit diffraction patterns, we treated each slit as a point source of light that emits a single spherical wave. The angular spread of the diffraction pattern depends on the parameter d/λ, where d is the slit separation. We have now seen that single slits of finite width have their own diffraction pattern. The angular spread of this pattern depends on a/λ, where a is the slit width. What effect does a finite slit width have on the multiple-slit pattern? The answer is that, for Fraunhofer diffraction, the overall intensity distribution is the product of the two intensity patterns. The pattern I_{mult}, corresponding to the double or multiple slit, is multiplied by the pattern I_{single}, corresponding to the single slit. The multiple-slit intensity is given by Eq. (39–7), whereas the single-slit intensity is given by Eq. (39–12). Thus their product is

$$I = I_{mult} I_{single} = I_0' \left[\frac{\sin(N\beta)}{\sin \beta}\right]^2 \left(\frac{\sin \alpha}{\alpha}\right)^2, \tag{39–18}$$

where we recall that

$$\beta = \frac{\pi d \sin \theta}{\lambda} \quad \text{and} \quad \alpha = \frac{\pi a \sin \theta}{\lambda}. \tag{39–19}$$

In Eq. (39–18) we have combined intensity factors into a single intensity factor I_0'.

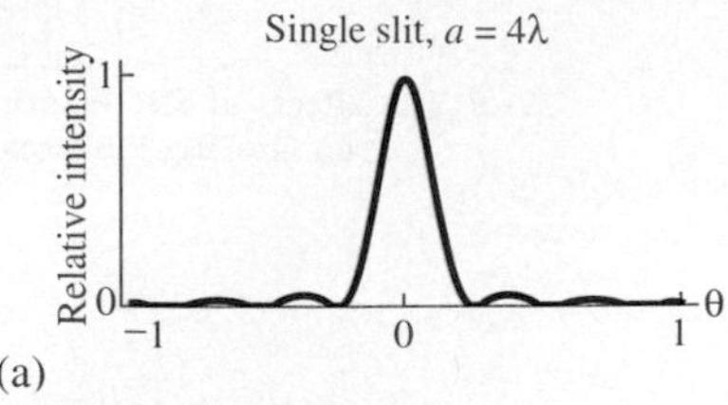

(a)

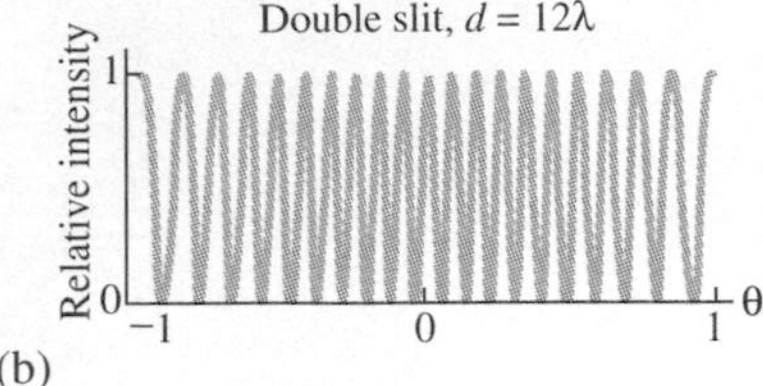

(b)

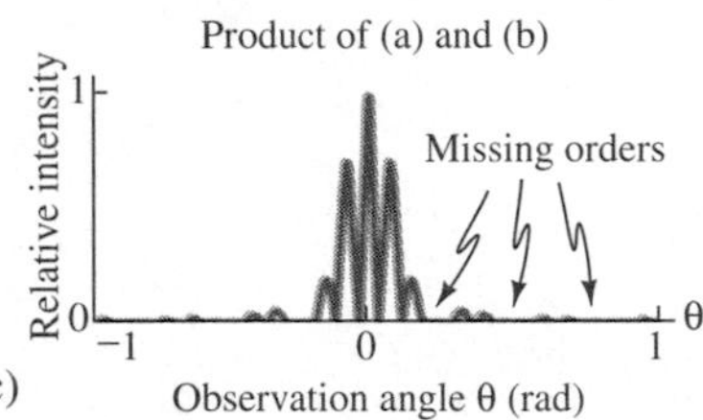

(c)

FIGURE 39–16 Intensity patterns as a function of observation angle θ for diffraction from multiple slits must include the effects of single-slit diffraction. For $a = 4\lambda$, (a) the intensity pattern for a single slit, (b) a double slit ($d = 12\lambda$), and (c) their product, which is the observed pattern. Missing orders occur; in this case, $d = 3a$, and the 3d, 6th, 9th, . . . orders are missing from the overall pattern.

FIGURE 39–17 Diffraction pattern for multiple slits where $d = 10a$. Note the missing orders.

The fact that the intensity patterns are multiplied means that the broader pattern (usually due to the single slit) acts as an envelope for the narrower pattern. For example, suppose that $d = 3a$ for a double slit ($N = 2$). In this case, the individual slit pattern is much broader than the multiple-slit pattern. At the same time, let $a = 4\lambda$, so that the single-slit pattern is easily distinguishable. Figure 39–16 shows the single-slit pattern, double-slit pattern, and combined pattern. Note that certain maxima of the double-slit pattern are absent from the combined pattern, because they fall where the minima of the single-slit diffraction pattern occur. These missing maxima are called *missing orders*. The locations of missing orders are independent of λ, as Example 39–6 illustrates. A measurement of the pattern described above, but with $d = 10a$, is shown in Fig. 39–17.

EXAMPLE 39–6 Calculate the lowest missing order of a double-slit interference pattern if the separation of the two slits is three times their individual widths, $d = 3a$.

SOLUTION: The double-slit maxima are located at

$$\beta = \frac{\pi d \sin\theta}{\lambda} = n\pi.$$

The lowest missing order will occur when a minimum in the single-slit diffraction pattern coincides with a maximum in the double-slit diffraction pattern. Minima occur at

$$\alpha = \frac{\pi a \sin\theta}{\lambda} = m\pi,$$

where m is an integer. We solve the two equations for $\sin\theta$ and equate them:

$$\sin\theta = \frac{n\lambda}{d} = \frac{m\lambda}{a}.$$

Thus we have a missing maximum when

$$\frac{d}{a} = \frac{n}{m}.$$

The wavelength has canceled and does not enter into the location of the missing orders. If d and a do not form a rational fraction, there will never be a missing order. In our case, however, $d/a = 3$, so the first missing order occurs for $m = 1$ (first-order single-slit minimum) and $n = 3$ (third-order double-slit maximum). This calculated result agrees with the results shown in Fig. 39–16.

*39–6 X-RAY DIFFRACTION

We have been emphasizing the use of gratings as a tool for the exploration of diffracted light. Light is just one form of electromagnetic radiation, and here we shall study the diffraction of X-rays. We shall see that crystalline solids form a natural kind of grating and that the diffraction of X-rays can be used to explore properties of solids. Diffraction gratings work because the apertures or obstacles serve as rescatterers. A powerful constructive interference occurs among rescattered light from *all* the apertures or obstacles, because the sources of wavelets are in a *regular* pattern. The atoms in a crystalline solid serve admirably as a grating, because they do indeed form a regular array of obstacles, even if that array is spread over three dimensions rather than two. Each atom in the array can serve as a rescatterer if the light can penetrate.

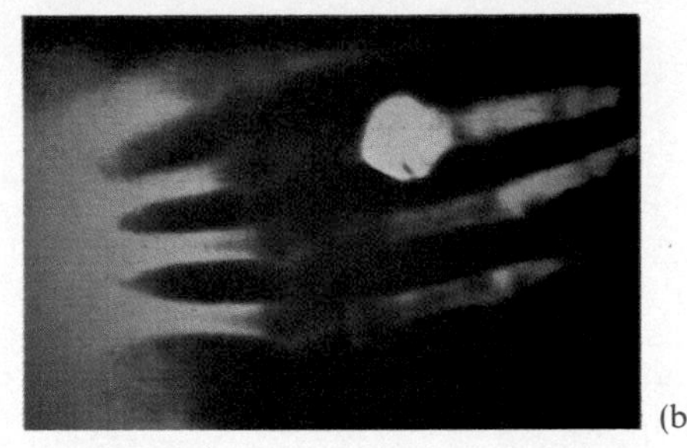

FIGURE 39–18 (a) "Beach-Party à la Roentgen" (circa 1900), a cartoon inspired by Wilhelm Roentgen's discovery of X-rays in 1895. (b) Roentgen's first X-ray photograph of a human—that of his wife's hand. X-rays have the well-known property of penetrating matter.

In 1895 Wilhelm Roentgen discovered that radiation was produced when he bombarded metal with high-energy *cathode rays* (now called electrons). This radiation was unlike any seen previously, and Roentgen called it *X-rays* (Fig. 39–18a). Shortly thereafter he produced the first X-ray picture, a human hand (Fig. 39–18b). We know now that X-rays are just electromagnetic radiation with wavelengths in the range of about 0.01 nm to 10 nm. This radiation is produced when atomic electrons change states within atoms, or when electrons are accelerated (or decelerated). X-rays are able to penetrate matter that contains a light element such as carbon (body tissue, for example) but are less able to penetrate matter that contains heavy elements such as lead—used to shield against X-rays—or the calcium in bone.

In the early 1900s it was suspected that X-rays might be some form of electromagnetic radiation. Although Roentgen concluded that X-rays could be neither reflected nor refracted, a diffraction experiment reported in 1899 vaguely suggested that X-rays might have wavelengths of about 0.1 nm, much smaller than those of visible light. At the same time, some scientists suspected that solids might be made of atoms arranged in regular arrays. In 1912 Max von Laue had the idea of scattering X-rays from solids. If X-rays had about the same wavelength as the distance between the arrays of atoms (~ 0.1 nm), then diffraction effects would be significant. Von Laue was as interested in finding a tool for the precise measurement of the wavelengths of X-rays as he was in finding a tool for the exploration of crystals. He convinced two of his colleagues, Friedrich and Knipping, to perform an experiment, and observation of the diffraction of X-rays soon followed. Von Laue's idea was a crucial step in the measurement of X-ray spectra and led to a revolution in our ability to study the nature of solids and the molecules that comprise them. The precise knowledge that table salt, NaCl, has the three-dimensional structure shown in Fig. 39–19 is a consequence of X-ray diffraction experiments, and virtually all of our knowledge of crystalline structure comes from such experiments. It was the use of X-ray diffraction on a crystallized form of DNA that led to the discovery of that molecule's double-helical structure (Fig. 39–20).

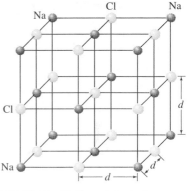

FIGURE 39–19 Crystals have three-dimensional structure with their atoms in regular arrays. This diagram shows one of the simplest, NaCl (table salt), which has a cubic structure.

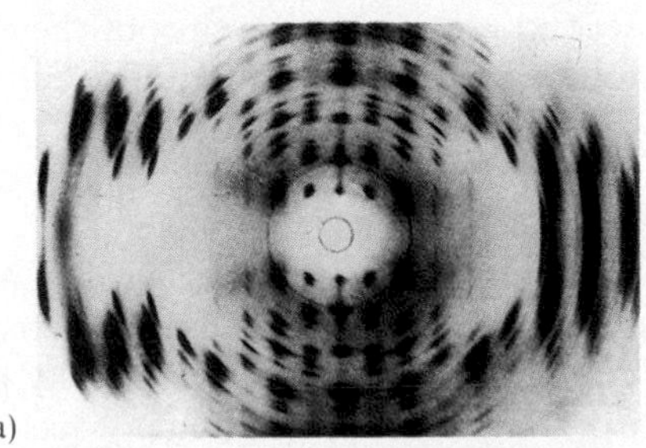

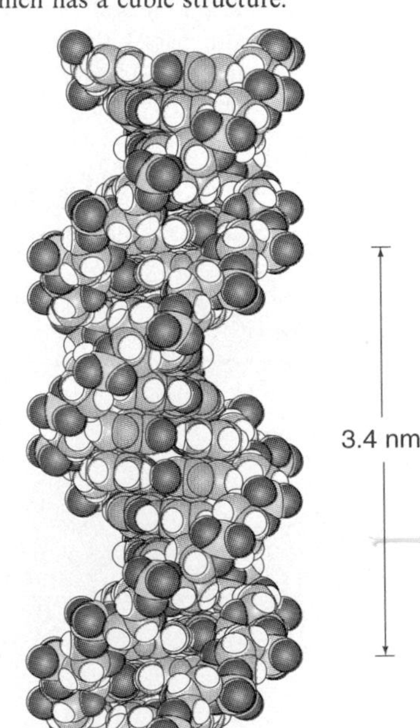

FIGURE 39–20 (a) Analysis of thousands of diffraction patterns produced by crystals of the large biological molecule deoxyribonucleic acid (DNA) showed that (b) the molecule has the shape of a double helix.

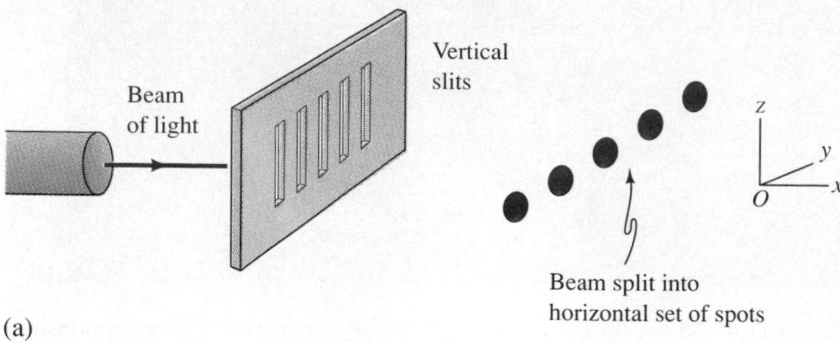

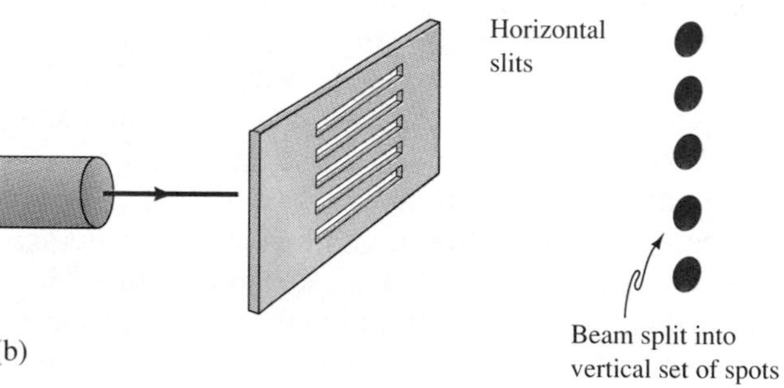

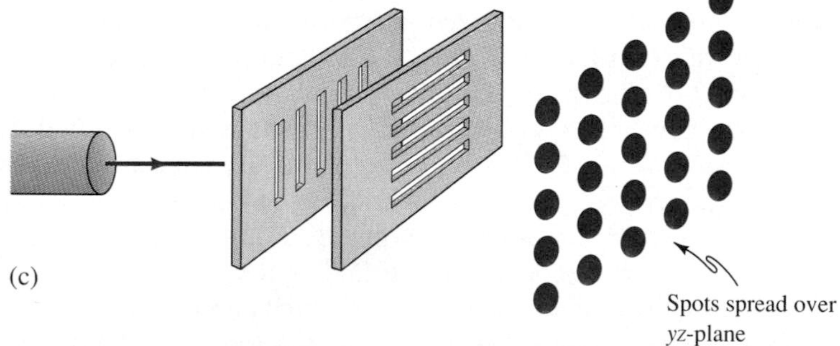

FIGURE 39–21 An analogy to show how spots are obtained when X-rays are diffracted by crystals. (a) A monochromatic beam incident on a grating with vertical slits (here, greatly exaggerated—the beam passes through many slits) produces a diffraction pattern that consists of maxima spread across a horizontal angle. (b) Similarly, vertically spread diffraction maxima are produced when the same beam passes through a series of horizontal slits (again, greatly exaggerated). (c) When the beam passes through the two gratings successively, the resulting diffraction pattern is a series of maxima spread over an entire plane.

Because the rescattering centers (the atoms of a solid) are pointlike and three-dimensional rather than slitlike, the diffraction pattern of a crystalline solid consists of a regular array of spots rather than lines. If we think of the array of points as being analogous to crossed diffraction gratings (superposed gratings with the slits on the first grating aligned along the z-axis, and the slits of the second grating aligned along the y-axis), we can understand why spots are observed. We send a well-collimated ray along the x-axis toward the first grating. The result is a set of spots spread out along the y-axis (Fig. 39–21a). These spots now approach the second grating, and each spot then produces its own set of spots aligned parallel to the z-axis (Fig. 39–21b). The combined result on a distant screen perpendicular to the x-axis is an array of spots (Fig. 39–21c). The von Laue experiment on a crystalline solid, shown schematically in Fig. 39–22a, leads to a set of spots like those shown in Fig. 39–22b.

Von Laue's idea was clarified almost immediately by W. L. Bragg, who proposed in 1912 a simple and systematic way of showing just how the positions of the spots would be determined by the solid's crystalline structure. Bragg pointed out that, in any crystal, many sets of parallel planes (called *Bragg planes*) can be

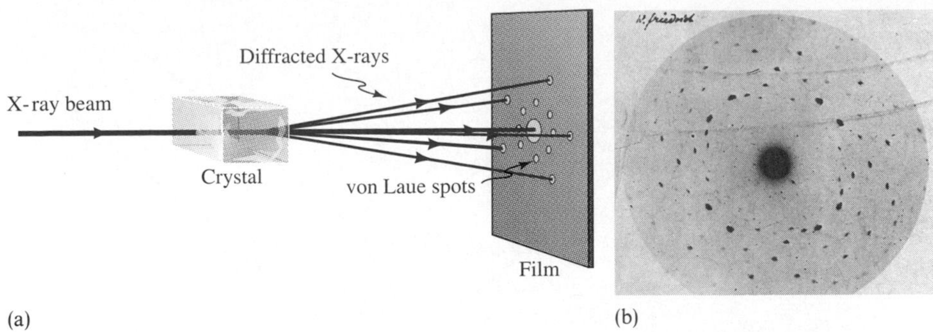

FIGURE 39–22 (a) Schematic diagram of the von Laue experiment for the diffraction of X-rays. (b) Von Laue spots in one of the first X-ray diffraction patterns. The large spot is undiffracted radiation.

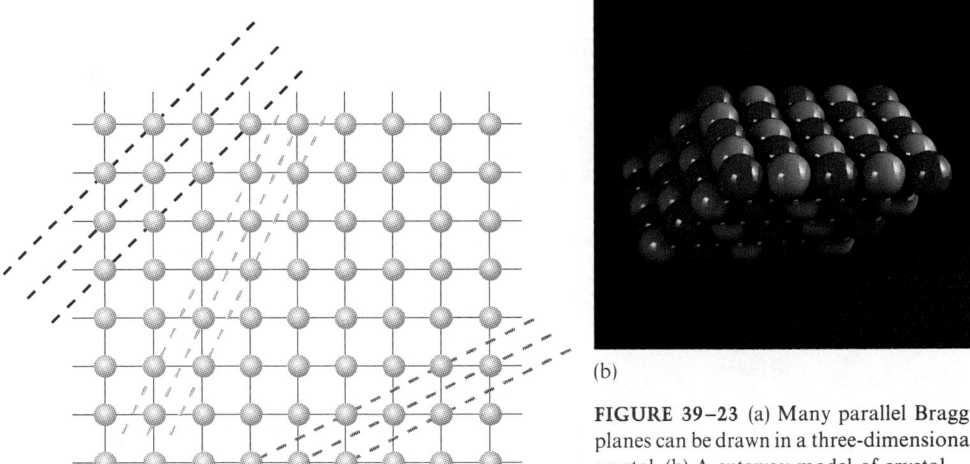

FIGURE 39–23 (a) Many parallel Bragg planes can be drawn in a three-dimensional crystal. (b) A cutaway model of crystal structure.

drawn that pass through the positions of the atoms, and that the planes of a set are separated by characteristic distances (*Bragg spacings*). Figure 39–23a shows where some of these planes cut a two-dimensional cross section of a cubic lattice similar to that formed by NaCl. One such plane is shown in a three-dimensional cutaway view of a crystal model in Fig. 39–23b, and they are all throughout the crystal. The advantage of this approach is that we can think of each family of parallel planes as a slit-type diffraction grating for the X-rays. Figure 39–24 shows two rays scattered from two parallel planes within a crystal. These rays scatter from a given plane for which the reflection angle equals the incident angle, because at that reflection angle the wavelets emitted by each atom within that plane add constructively. Now consider the interference between the scattered waves of *different* planes, which occurs because the X-rays penetrate the crystal. If the separation between the planes is d, then, from the geometry of Fig. 39–24, the difference in path lengths for the two lines is $2d \sin \theta$. Note that angle θ is measured from the plane surface rather than from the normal to the plane. Constructive interference for scattering from these two adjacent planes occurs when this path-length difference is an integer multiple of the wavelength. This relation is known as **Bragg's law**, or the **Bragg condition:**

Bragg's law: $$2d \sin \theta = n\lambda, \text{ where } n = 1, 2, 3, \ldots. \qquad (39\text{–}20)$$

The angles for which there is constructive interference in scattering from crystalline arrays of atoms

Because the planes are equally spaced, the waves that scatter from the atoms in the entire set of planes in the direction specified by the Bragg condition *all* add

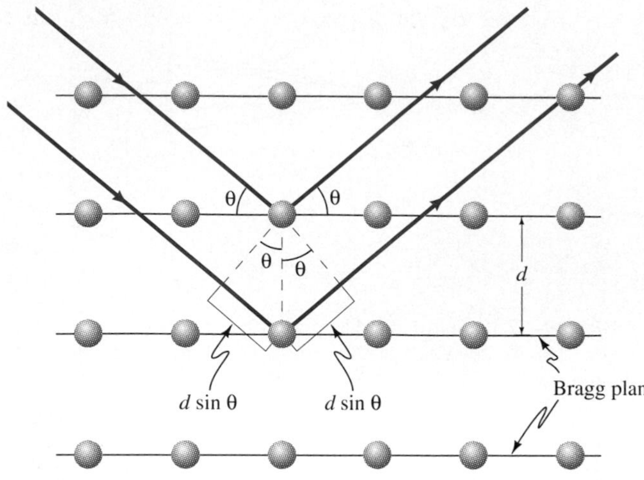

FIGURE 39–24 The geometry of X-ray diffraction between adjacent Bragg planes.

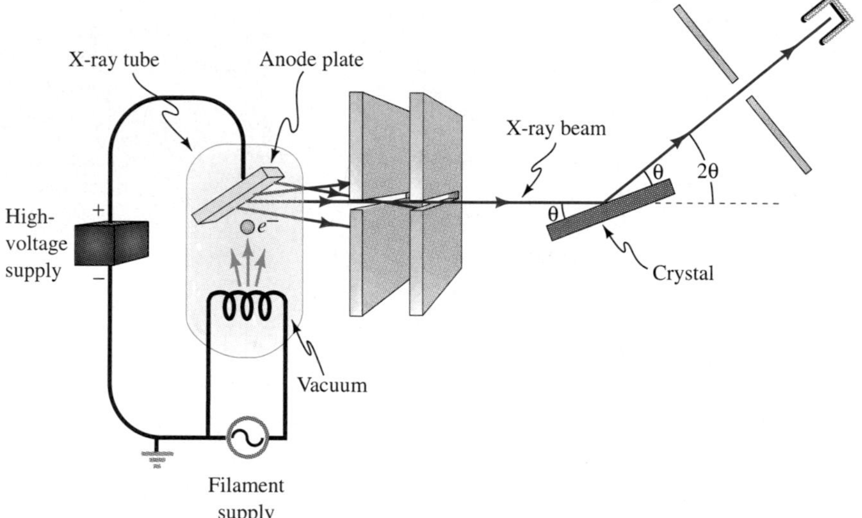

FIGURE 39–25 Schematic diagram of an X-ray spectrometer used to study properties of crystals. Electrons bombard the anode plate, producing X-rays that are collimated before being scattered by the crystal. A movable detector records X-ray intensity as a function of θ to determine where constructive interference occurs.

constructively, and, as in the case of a diffraction grating, the maximum is large and narrow.

The discussion above is inadequate to explain the *intensities* of the spots, although we may generalize that if a particular family of planes contains more atoms than another does, the maxima those planes give are more intense. Intensity information is very important in determining the crystalline structure. More advanced mathematical methods are necessary for a complete model of the diffraction pattern, which is on the whole quite complicated because of the many possible planes and orders of scattering.

In addition to clarifying the structure of von Laue scattering, W. L. Bragg collaborated with his father, W. H. Bragg, in a modified set of experiments (Fig. 39–25). Their experiment forms the basis of the X-ray spectrometer.

EXAMPLE 39–7 Figure 39–25 shows an X-ray tube, which produces a continuous distribution of wavelengths. If these wavelengths are scattered from a particular set of parallel planes of rock salt (NaCl) with a spacing $d = 0.282$ nm, what wavelengths will appear in the first and second orders at 25°?

SOLUTION: In order to take advantage of Bragg's law, either the wavelength or the atomic-plane spacing must be known. In this case, the plane spacing is known, and the Bragg law can be used to identify unknown wavelengths or to select particular wavelengths to be used for further study. We use Eq. (39–20) to determine the wavelengths:

$$\lambda = \frac{2d\sin\theta}{n} = \frac{2(0.282\text{ nm})(\sin 25^\circ)}{n} = \frac{0.238\text{ nm}}{n}.$$

The wavelengths at 25° are 0.238 nm and 0.119 nm for the first ($n = 1$) and second ($n = 2$) orders, respectively. Note that if $\theta = 25^\circ$, the overall deflection from the original beam direction is $2\theta = 50^\circ$.

*39–7 HOLOGRAPHY

In 1947 Dennis Gabor proposed that interference effects between light emitted by an object (a source) and a second coherent beam can be recorded on film, which becomes a very special diffraction grating. When light is passed through this diffraction grating, it is diffracted and forms a fully three-dimensional image of the object, an image that can be viewed from different positions and angles, just like the original object. This process is **holography**, and the film on which the interference pattern is stored is a *hologram.*

In order to understand the principles, we start with a distant point source that sends plane waves directly toward a piece of film (Fig. 39–26). At the same time we send a second beam, *coherent with the light from the source* (we shall soon see how this is done), toward the film from an angle θ_r. This second beam is known as the *reference beam.* The reference beam interferes with the light from the source. Suppose that the wave in the reference beam interferes constructively with the source wave at point P_1. There will also be constructive interference at P_2, a distance d from P_1, if the wave path along line ℓ_2 differs from the path along line ℓ_1 by λ (or an integer m times λ; we consider only the case $m = 1$). When there is constructive interference at both P_1 and P_2, the relation between θ_r, the wavelength of the light, λ, and the separation d is

$$d\sin\theta_r = \lambda. \tag{39–21}$$

On the slice of film shown in Fig. 39–26, constructive interference will occur at a series of equally spaced points, which will be recorded as dark spots on the film. (Recall that the film makes a negative.) These points are parts of continuous lines into or out of the page. The full interference pattern on the film thus consists of a set of curving lines that represent all the places where there is constructive interference. The film can record with shadings of gray places where the interference is not totally destructive.

Let us now turn to the question of how the image is viewed (or *reconstructed*). Suppose that we project a beam just like the reference beam, and at the same angle, onto the back of the film (Fig. 39–27). The dark areas on the film act as obstacles that rescatter the light. The direction indicated, *that of the original light from the source,* is a direction for which the diffracted light is a maximum, as the geometry in Fig. 39–27 shows. Thus a viewer placed at point E will see light as though it comes from a distant point I, which we may think of as an image of the original source. Note that there is no requirement that the beam that produces the image be identical to the original reference beam, as long as it is coherent across the film.

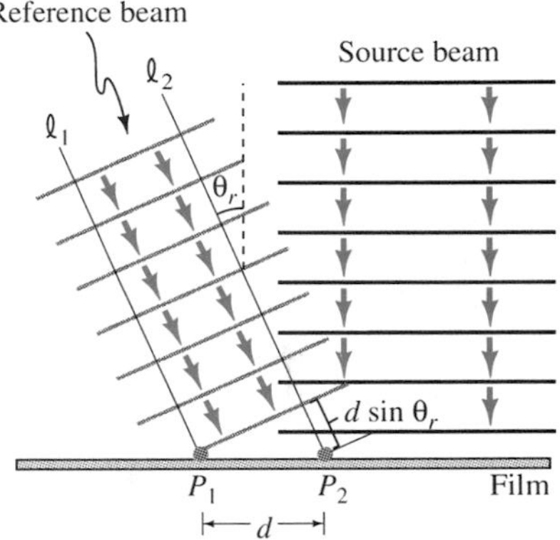

FIGURE 39–26 Coherent light from a distant source beam and from a reference beam interfere on film, producing a hologram.

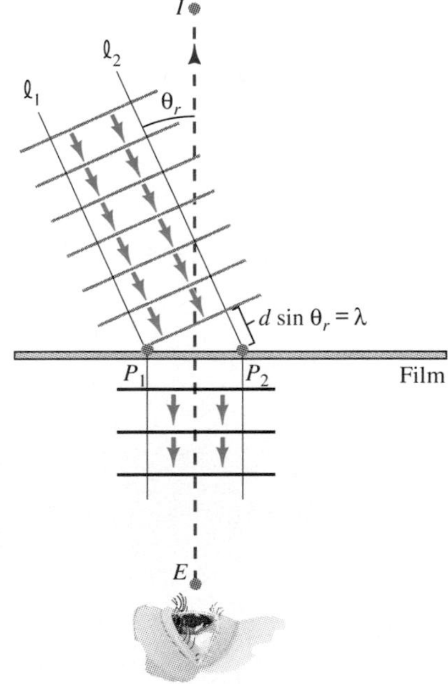

FIGURE 39–27 When a reference beam shines on holographic film, the image of the original object is reconstructed.

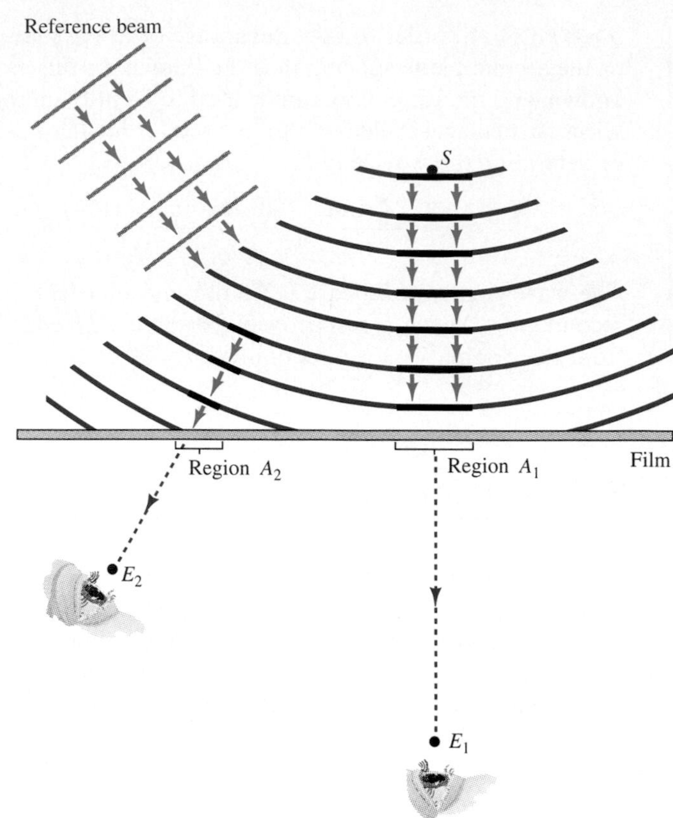

FIGURE 39–28 Holograms produce a three-dimensional image that allows the original object to be seen from different positions. A different image will be observed in each place.

If the angle of the new beam is different, the only effect is to shift the angle of the viewed image.

Suppose now that the point source is closer to the film when the image is made. In this case, the spots of constructive interference will not be spaced equally across the screen (Fig. 39–28). At region A_1 the situation described above is reproduced, but at region A_2 the points where there is constructive interference between the beams is different. When the exposed film is illuminated by a reference beam, a viewer at E_1 will see a plane wave along the direction from E_1 back to region A_1; that is, the observer will be looking back at the source from one angle. However, when the viewer is at E_2, the maximum of the diffraction pattern will indicate an image back along the direction from E_2 to A_2. *The viewer will be looking at the source from another angle.* There is a true three-dimensional image, which can be viewed from different angles, quite unlike the flat two-dimensional image made by a camera.

When the object is more complicated than a point source, light arrives at any given point on the film from many points of the object. The interference pattern that this light makes with the reference beam is far more complicated and irregular, but it is nevertheless unique to the object. Once this pattern is recorded, it serves as a diffraction grating for light from a reference beam to make a unique pattern that reproduces the light emitted by the original object, and from many angles. Although we have treated the film as a transmission grating, if the interference pattern can be recorded as scratches on a shiny surface, then it will act as a reflection grating, such as that in a hologram on a credit card.

How is the reference beam made coherent with the light from the source? If the light from an ordinary light bulb is split in two, as in Young's experiment, and

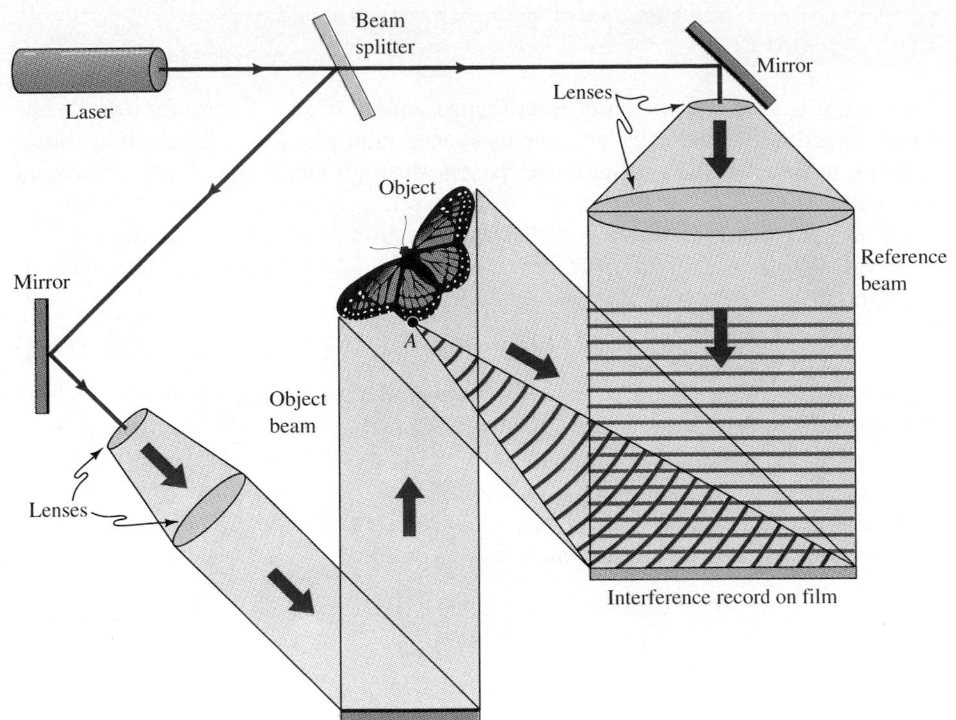

FIGURE 39–29 Schematic diagram of how laser light is split to form object and reference beams. The object beam reflects from the object, and the interference of the light is stored on a hologram.

one part illuminates the object while the other is routed to serve as a reference, then the two beams could be coherent. This would not be very satisfactory in practice, because the paths of the two beams would be limited by the short coherence length of the wave trains in a light bulb. A split laser beam makes a more practical starting point, because of the virtually unlimited coherence length of laser light (Fig. 39–29).

APPLICATIONS

The Uses of Holography

The uses of holograms go beyond the simple beauty of the image. Holography has the potential to provide an extremely compact system of information storage. Because the light from every point on, say, a printed page reaches every point of a hologram, every region of the film larger than several wavelengths across can reproduce the entire page, albeit with less detail. Moreover, successive holograms of successive pages can be made within thick photographic emulsions. If the exposure of each page is made with a reference beam oriented at a slightly different angle, then by illuminating the resulting hologram with a light beam of that particular angle, only the corresponding page appears from a particular vantage point. All the paintings in a museum could be sequentially recorded in this way with great accuracy in a very small space indeed.

Another important use of holography involves making two holograms of the same object on the same film at successive times. If the object has moved slightly between the moments when the holograms are made, then the two images interfere with each other like the light from two surfaces of a soap film does. The chapter-opening photograph shows an interference hologram of a violin in motion under the influence of a vibrating string. This interference pattern, not to be confused with the interference that makes the hologram itself, reveals detail about the motion not otherwise visible. Similarly, density variations in air are visible as the interference between two successive images of the air made on the same hologram. In this way, the mechanisms by which a candle heats the air above it or an airplane produces shock waves can be studied.

Diffraction is a manifestation of interference among waves. Examples include the pattern produced by screens with evenly spaced multiple slits (diffraction gratings) and the patterns made by light that passes through single apertures or around obstacles.

If d is the distance between slits in a diffraction grating and θ is the angle of observation from the direction of incident light, principal maxima are observed for the condition

$$d \sin \theta = m\lambda, \quad \text{where } m = 0, \pm 1, \pm 2, \ldots . \tag{39-1}$$

Here m is the order of the principal maxima. If the average intensity reaching the screen from any one slit is I_0, then the intensity of the light from the grating at the principal maxima is $N^2 I_0$, where N is the number of slits. In addition, the width of the principal maxima depends on N as $1/N$, so the diffraction peaks become sharper as N increases. This dependence on the parameters of the grating is contained in the expression for the intensity pattern:

$$I = I_0 \left[\frac{\sin(N\beta)}{\sin \beta} \right]^2, \tag{39-7}$$

where, from Eq. (39–4),

$$\beta \equiv \frac{\pi d \sin \theta}{\lambda}.$$

Angular dispersion represents the change in observation angle θ as a function of a change in wavelength and is given by

$$\frac{\Delta\theta}{\Delta\lambda} = \frac{m}{d \cos \theta}. \tag{39-8}$$

The resolving power, R, is the ability of a grating to separate closely spaced lines:

$$R = \frac{\lambda}{\Delta\lambda} = mN. \tag{39-9, 39-10}$$

A single slit produces a diffraction pattern that can be derived by considering the slit to be composed of a large number of very thin slits. The criteria for destructive interference is

$$\sin \theta = \frac{m\lambda}{a}, \quad \text{where } m = \pm 1, \pm 2, \pm 3, \ldots \tag{39-11}$$

and a is the width of the slit. The intensity pattern of a single slit is

$$I = I_{\text{max}} \frac{\sin^2 \alpha}{\alpha^2}, \tag{39-12}$$

where the angle α is given by

$$\alpha = \frac{\pi a \sin \theta}{\lambda}. \tag{39-13}$$

The minima are given in terms of α by

$$\alpha = n\pi, \quad \text{where } n = \pm 1, \pm 2, \pm 3, \ldots . \tag{39-14}$$

Most of the light from the single slit is contained in the central peak; the secondary peaks are much less intense. The narrower the slit, the broader the diffraction pattern.

The Rayleigh criterion specifies that two point sources are just resolved if the peak of the diffraction image of the first source falls on the first minimum of the diffraction image of the second source. The minimum separation angle of two closely spaced sources obtained by a circular aperture of diameter D is approximated by

$$\theta_{\text{min}} \simeq \frac{\lambda}{D}. \tag{39–16}$$

The minimum separation S_{min} of two closely spaced objects by a lens of diameter D is given by

$$S_{\text{min}} = f\theta_{\text{min}} \simeq \frac{f\lambda}{D}. \tag{39–17}$$

The practical limitation of earth-based telescopes is due to air turbulence and not to diffraction limits.

Missing orders occur in the intensity patterns of diffraction spectra due to the overlapping effects of single slits and multiple slits.

X-rays are diffracted by the atom centers of regularly spaced Bragg planes, planes formed by the regular array of atoms in a crystal. The technique is important for many aspects of modern physics, including determining the structure of crystals and atomic composition. Bragg's law gives the observation angles θ (as measured from a plane surface in a crystal lattice) for which constructive interference is obtained from planes of spacing d:

$$2d \sin\theta = n\lambda, \text{ where } n = 1, 2, 3, \ldots. \tag{39–20}$$

Holography represents a special process by which three-dimensional images are captured. A hologram is a special diffraction grating formed by the interference of two coherent beams, one a reference beam and the other an object beam reflected from a three-dimensional object. The three-dimensional image of the object can be reconstructed by projecting a reference beam on developed holographic film.

QUESTIONS

1. Would the diffraction of water waves around the timbers of a pier be reduced by decreasing the diameter of the support poles? by increasing their diameter?
2. There are tentative plans to build telescopes for waves of various wavelengths, including visible light, on the moon. What would the advantages of such facilities be?
3. Discuss how a Poisson spot might be obtained from a bowling ball. Would you want the source and screen to be close to or far away from the bowling ball? Explain.
4. Is it possible to obtain better resolution with a microscope with blue light than with red light? Why or why not?
5. Two waves are linearly polarized. The electric field of one wave is aligned with the x-axis and the other is aligned with the y-axis. In the absence of matter that might change the polarization, can these waves interfere with each other?
6. In a demonstration of diffraction peaks that involves the reflection of laser light from an ordinary ruler, does the light have to be at a glancing angle?
7. The spreading of light due to diffraction in an optical instrument is greater when the instrument uses red as opposed to blue light. Why?
8. A hologram contains information about an entire object, even in just a small portion of the film. Would you expect the image made by a small portion of the hologram to be as sharp as the image made by the entire hologram?
9. What are the differences between the interference patterns formed on a distant screen by coherent light that passes through a diffraction grating with thousands of rulings at a particular spacing and a double slit separated by the same spacing?
10. A light bulb emits light with a spectrum characteristic of blackbody radiation. What pattern will this light produce when it is observed through a grating?
11. You are standing in the ocean, and a wave passes around you. Is this an example of diffraction?
12. How do the X-rays used in X-ray diffraction "know" that

there is a given set of planes of atoms for which a diffraction pattern appears?

13. Does the fact that light bends around corners mean that, with a sensitive camera, you could read a newspaper from around a corner? (This is a serious question; try to estimate the amount of bending that the smallest obstacle would give for light, and how much information that light could contain.)

14. In so-called 3-D movies, which were introduced in the 1950s, a three-dimensional effect is achieved when different images are sent to each of your two eyes. How could you tell that they are not holographic images?

PROBLEMS

39–2 *Diffraction Gratings*

1. (I) Laser light is diffracted from a grating with 400 lines/cm. The central peak and the fourth peak are 10.34 cm apart on a screen 1.44 m away. The screen is perpendicular to the ray that makes the central peak. What is the wavelength of the light?
2. (I) A grating has a line density of 1000/cm, and a screen perpendicular to the ray that makes the central peak of the diffraction pattern is 3 m from the grating. If light of two wavelengths, 498 nm and 510 nm, passes through the grating, what is the separation on the screen between the third-order maxima for the two wavelengths?
3. (I) A student finds a diffraction grating but does not know the spacing of the ruled lines. She shines light from a laser with $\lambda = 680$ nm through the grating and examines the maxima on a screen 265 cm away. If the distance between the tenth maxima on either side of the central peak is 14.3 cm, what is the rule spacing of the grating?
4. (II) A grating with 10^4 rulings spaced uniformly over 2 cm is illuminated at normal incidence by light of wavelength 600 nm. (a) What is the dispersion of the grating in the second order? (b) What is the smallest-wavelength interval that can be resolved in the second order near $\lambda = 600$ nm?
5. (II) What is the resolving power of 3-cm-wide diffraction grating with 5000 lines/cm, for the first three orders? If light consisting of a series of discrete wavelengths around 420 nm is incident on the grating, what is the minimum wavelength separation that can be resolved in these three orders?
6. (II) Estimate the line spacing between two closely spaced lines near 580 nm if they are barely resolved in the fourth order by a grating with $N = 15{,}000$.
7. (II) A grating is to be inscribed on a 25-cm-wide glass plate so as to resolve two spectral lines with wavelengths 589.105 nm and 589.107 nm, respectively, in the first order. What is the minimum number of lines that must be ruled on the plate? What is the dispersion of the grating with this number of lines?
8. (II) A grating is made of five similar, uniformly spaced, narrow slits. For light of wavelength $\lambda = 520$ nm perpendicularly incident on the slits, the angular position of the first principal order is 25° to the normal. What is the slit separation? What is the angular position of the first principal order when the first and fifth slits are covered? when the second and fourth slits are covered?
9. (II) The resolving power of a certain grating for the first-order spectrum is 10^4. If the grating is 2 cm long, what angle separates the first- and second-order images for light with $\lambda = 580$ nm at normal incidence?
10. (II) White light shines on a diffraction grating with 5000 lines/cm. The diffracted light is observed on a screen 3 m away. Find the first- and second-order positions for blue light (420 nm), green light (550 nm), and red light (650 nm). Sketch a view of the screen.
11. (II) Visible light extends from wavelengths of 430 nm to 680 nm. If blackbody radiation, which contains all these wavelengths, is incident on a 5-cm-wide grating with 2500 slits per cm, what range of angles is covered for these wavelengths in the first-order maximum? in the second-order maximum?
12. (II) An atomic source emits two strong spectral lines, a red one of wavelength 650 nm and a blue one of wavelength 420 nm. The light falls on a diffraction grating with 4000 lines/cm that is 2 cm across, and passes to a screen 2 m away. On the screen, how far from the central maximum are the third-order maxima ($m = 3$) of the spectral lines? What is the width of these maxima?
13. (II) Light of wavelength λ is incident at an angle α to the normal of a transmission grating with spacing d between each slit (Fig. 39–30). At what angles β to the normal will diffraction maxima be located?

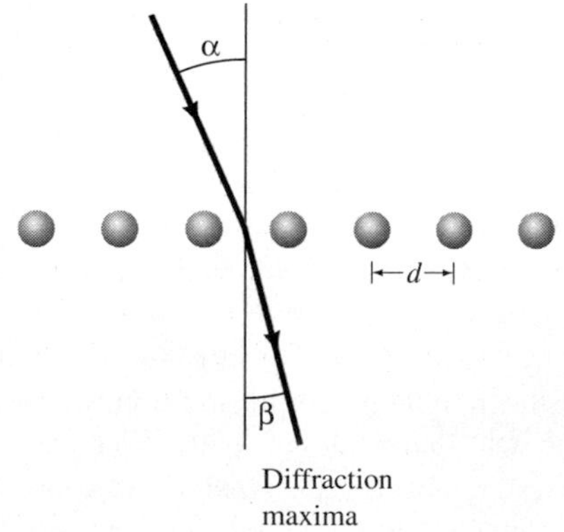

FIGURE 39–30 Problem 13.

39–3 *Single-Slit Diffraction*

14. (I) Light of wavelength $\lambda = 500.0$ nm falls on a slit of width

$a = 0.50$ mm. At what angle θ from the normal to the wall in which the slit is cut does the second dark fringe occur?

15. (I) A single slit of width 2.8×10^{-5} m diffracts light of wavelength 495 nm to a screen. The distance between the minima on either side of the central maximum is 1.8 cm. How far away is the screen?

16. (I) A single slit diffracts laser light of wavelength 612 nm onto a screen 3.0 m away. The distance between the first-order maxima on either side of the central peak is 4.0 mm. How wide is the slit?

17. (II) Blue light ($\lambda = 470$ nm) passes through a horizontal slit 10 μm wide. What is the ratio between the maximum intensity of the central peak and the maximum intensity of the next adjacent peak? At what angle θ from the horizontal will the intensity of the central peak be half its maximum value? Would θ increase or decrease if red light ($\lambda = 670$ nm) were used instead?

18. (II) Plane waves of light of wavelength 500 nm are incident on a single slit of width 20 μm. A lens focuses the plane waves on a screen 1 m away. (a) What is the width of the central maximum on the screen? (b) What is the intensity ratio between the central maximum and the first-order maxima?

19. (II) A single slit produces a diffraction pattern on a distant screen. Show that the separation distance between the two minima on either side of the central maximum is twice as large as the separation distance between all the other neighboring minima. Compare your result to the corresponding case for a double-slit pattern with very narrow slits.

20. (II) When light of wavelength 450 nm passes through a single slit of unknown width, the diffraction pattern displays a second maximum where the first minimum of light of an unknown wavelength had been observed to fall. What is the unknown wavelength?

21. (II) Light of wavelength λ arrives at a single slit of width a; the plane wave fronts arrive at the slit at an angle θ_i (Fig. 39–31). Find the angles θ for which minima appear on a very distant screen. Is there a "central maximum" in the direction defined by the incoming wave; that is, at $\theta = \theta_i$?

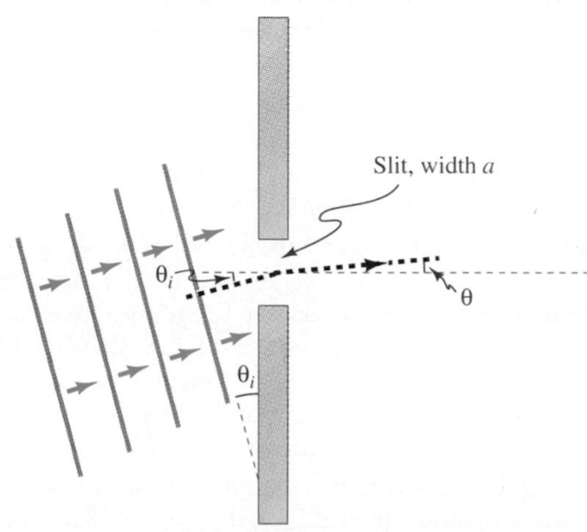

FIGURE 39–31 Problem 21.

22. (II) Suppose that light falls on a single slit at an angle ϕ with the normal to the wall that contains the slit. Show that Eq. (39–12) still holds, but $\sin\theta$ must be replaced by $(\sin\theta + \sin\phi)$ in the expression for α [Eq. (39–13)].

23. (III) When we determined the position of the minima of the Fraunhofer diffraction pattern for a single slit, we argued that the maxima are located midway between the minima. To look at the accuracy of this assumption, (a) show that the maxima of the intensity pattern $(\sin^2\alpha)/\alpha^2$ are determined by the solutions of the transcendental equation $\alpha = \tan\alpha$. (b) Compare a numerical solution of this equation for the first and second maxima with angles that are midway between the first and second, and second and third, minima (you will need trigonometric tables). (c) By plotting the intersection points of $y = \tan\alpha$ and $y = \alpha$, show that the approximation improves as the order of the maximum increases.

39–4 Diffraction and the Resolution of Optical Instruments

24. (I) A plane wave of microwave radiation, $\lambda = 2.0$ cm, passes through a circular aperture of diameter 10.0 cm. What is the angular position of the first minimum of the resulting Fraunhofer diffraction pattern?

25. (I) Astronauts leave two lunar rovers 5.00 km apart on the moon. An earth-based telescope of what minimum diameter is required to resolve laser beams ($\lambda = 650$ nm) emitted by the rovers toward the telescope? The rovers are 3.0 m long. A telescope of what diameter is required for the rovers themselves to be detected? Ignore air turbulence. (The earth-moon distance is 3.83×10^8 m.)

26. (I) An amateur astronomer uses a reflecting telescope of diameter 20 cm and focal length 200 cm to observe light of $\lambda \simeq 600$ nm from a star. (a) What minimum angular resolution can the astronomer obtain? (b) What is the diameter of the Airy disk? (c) What is the minimum separation distance of two objects on the moon that the telescope can resolve?

27. (II) An astronaut in a satellite can barely resolve two point sources on earth 200 km below. What is the separation distance between the sources, assuming ideal conditions, $\lambda = 550$ nm, and a pupil diameter of 5.0 mm?

28. (II) A telescope is used to observe two distant pointlike light sources 1.2 m apart horizontally. The light has wavelength $\lambda = 560$ nm, and the front of the telescope is covered with a screen that has a vertical slit of width 1 mm. What is the maximum distance at which the two sources may be resolved?

29. (II) The two stars of a binary star system are just resolvable when observed by a telescope with a resolution of 3 s of arc and are 75 ly from earth. Estimate their separation.

30. (II) A spy satellite is announced publicly to be capable of distinguishing detail 6 in across. If the satellite orbits at a height of 250 mi, what must the minimum size of the aperture of its lens be, assuming maximum sensitivity at $\lambda = 550$ nm? Would it be better if the film (or other sensor) were sensitive to shorter or to longer wavelengths?

31. (II) What lateral separation must there be between two objects located 1 km from a camera that must resolve them? The camera lens's aperture is 5 mm in diameter, and the film is sensitive to light of wavelength 550 nm.

32. (II) Use the Rayleigh criterion and make assumptions to estimate the distance at which the human eye should be able to resolve the headlines in a newspaper. Carry out an experiment to see how good your estimate is!

33. (II) The SR-71 Blackbird reconnaissance airplane could fly at over 70,000 ft. If the pilot's pupil has a diameter of 2 mm on a bright day, what is the distance between two objects on earth that the pilot could just resolve from 70,000 ft? Take the wavelength of light to be 550 nm.

34. (II) The headlights of a car are 1.5 m apart. At night the pupils of an oncoming driver have expanded to 4.8 mm. How close must the two cars approach before the headlights can be resolved? Take the wavelength of the light to be 550 nm.

*39–5 *The Effects of Slit Width on Grating Patterns*

35. (II) The separation distance between two narrow slits is ten times the width of either slit. What is the intensity of the tenth interference maximum, taking the center as the first, when monochromatic light passes through the two slits and falls on a distant screen?

36. (II) Light of wavelength 588 nm from helium impinges on two slits 0.80 mm apart. Each slit is 0.25 mm wide. Find the intensity ratio I/I_0 on a screen 1.0 m away at the following distances from the central maximum: 0.10 mm, −0.40 mm, 0.40 mm, and 1.7 mm.

37. (II) Light of wavelength 600 nm is perpendicularly incident on a diffraction grating. Two adjacent maxima occur at $\sin\theta = 0.30$ and $\sin\theta = 0.36$, respectively. The fifth order is missing. (a) What is the separation distance between adjacent slits? (b) What is the smallest possible individual slit width? (c) Name all orders that appear on the screen, consistent with the answers to parts (a) and (b).

38. (II) The centers of a double slit are separated by 2 mm; each slit is 0.5 mm wide. Are there missing orders? If so, at what angles are they missing on a distant screen if $\lambda =$ 550 nm?

39. (II) The slit widths of a grating with 2500 slits/cm are one-third the slit spacing. What is the ratio of the intensities of the second-order and first-order principal maxima of the grating?

40. (II) Light of wavelength 625 nm is perpendicularly incident on a screen in which double slits of width $a = 0.25$ mm have been cut. The slits are a distance $d = 0.30$ mm apart. Find the first angle away from the central axis for which the intensity on a distant screen is exactly one-half the maximum intensity.

41. (II) A grating consists of slits of width a whose centers are separated by a distance d. Sketch the diffraction pattern for (a) $d \gg a$ and (b) $d - a \ll a$ (the slits are wide compared to the strips between them).

*39–6 *X-Ray Diffraction*

42. (I) X-rays of wavelength 1.1 nm are aimed at an unknown crystal in a spectrometer. A first-order peak occurs at 13.1° (Fig. 39–32). What is the corresponding Bragg-plane spacing for the crystal?

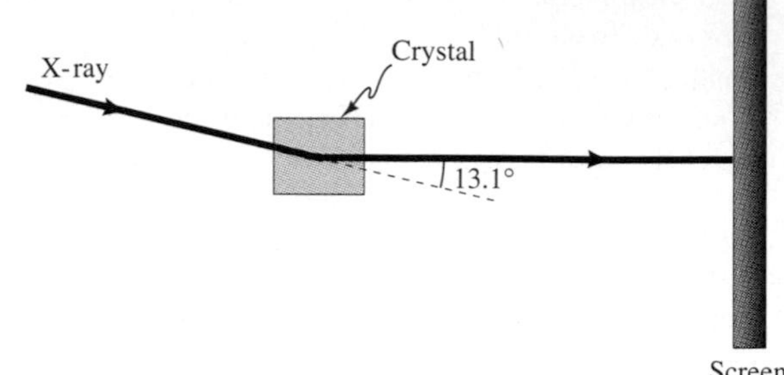

FIGURE 39–32 Problem 42.

43. (II) The distance between neighboring pairs of Bragg planes in calcite ($CaCO_3$) is 0.3 nm. At what angles to these planes will the first- and second-order diffraction peaks occur for X-rays of wavelength 0.12 nm?

44. (II) Consider a cubic crystal with a spacing between adjacent atoms of 0.25 nm. X-rays of wavelength 0.1 nm scatter elastically from the crystal. At what angle will first-order Bragg diffraction be observed?

45. (II) Mica has a Bragg-plane spacing of 1.0 nm, whereas rock salt has a spacing of 0.28 nm. For an X-ray of wavelength 0.1 nm, which material produces a diffraction pattern with the greater angular separation? What is the difference in angular separation $\Delta\theta$ for each material for the Bragg planes above when the crystals are illuminated with X-rays of wavelengths 0.096 nm and 0.104 nm?

General Problems

46. (I) A ruby laser of wavelength 690 nm with a cross-sectional area of 10^{-3} m^2 is aimed at the moon, 3.84×10^8 m away. Estimate the minimum diameter of the light beam that reaches the moon.

47. (I) A grating 3 cm long has 18,000 lines inscribed on it. A line of wavelength 640.000 nm is just resolved, in the second order, from a second line with a slightly longer wavelength. What is the wavelength of the second line?

48. (II) Radar is used to study the shapes of airplanes from as far away as 100 km. (a) Assuming that the distance scale determining a plane's shape (the size of the curves that distinguish one plane from another) is 1 m, what angular resolution is needed in the radar system? (b) Estimate the wavelength of the radar waves if the reflected radar signals are gathered in a dish of diameter 2.5 m.

49. (II) Deep-ocean waves move in linear fronts directly toward a harbor opening of width 50 m (Fig. 39–33). For what

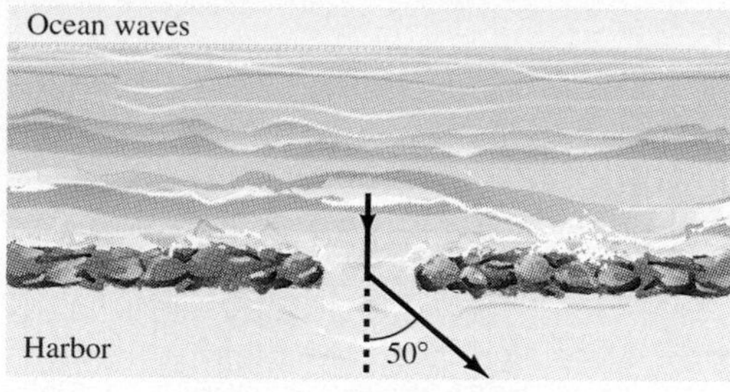

FIGURE 39–33 Problem 49.

wavelength will there be a minimum within the harbor at an angle of 50° from the axial line of the opening?

50. (II) By varying the spacing between two vertical dipole antennas as well as the phase of the signal generated by each antenna, the antennas can give signals that are stronger in some directions than in others (see Chapter 38). Suppose that N antennas are lined up along the x-axis (Fig. 39–34). The total distance between the first and last antennas is λ,

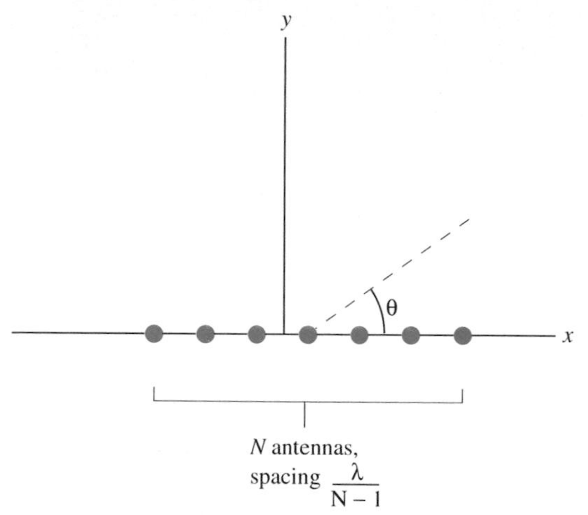

FIGURE 39–34 Problem 50.

so the spacing between the antennas is $\lambda/(N-1)$. Any one antenna would radiate its signal with the same intensity I_0 for any angle θ. (a) Find the intensity radiated very far from the array by the system of antennas as a function of angle θ in terms of I_0, assuming that the signals of all the antennas are in phase. (b) Describe the signal for all values of θ from 0° to 360°.

51. (II) In Chapter 41, we shall see that an electron behaves like a wave whose wavelength, λ, is related to its momentum, p, by $\lambda = h/p$. Here h is Planck's constant, $h = 6.63 \times 10^{-34}$ J·s. Electrons used in an electron microscope can be diffracted, and electron microscopes have a diffraction limit. If the energy of the electrons used in an electron microscope is 10 eV and if the aperture through which the electrons are channeled has a diameter of 0.01 mm, what, approximately, is the smallest angular separation the microscope can distinguish in an object?

52. (III) *Babinet's principle* is useful for the treatment of the diffraction of light by obstacles. It states that if light is incident on an opaque screen in which a hole (of any shape) is cut, then the diffraction pattern produced is the same (except at $\theta = 0$) as that obtained if the screen were removed and the hole were replaced by an obstacle. Use Babinet's principle to estimate the size of an opaque obstruction on a glass slide if a narrow laser beam (with $\lambda = 480$ nm) perpendicularly incident on the slide spreads to a spot of diameter 4.0 cm on a screen 1.6 m from the slide.

53. (III) What diffraction pattern is produced on a distant screen when light of wavelength λ is perpendicularly incident on a plane that contains N very thin hairs, each spaced a distance d apart from the next hair. [*Hint:* See Problem 52.]

54. (III) Electromagnetic radiation of frequency 1.25×10^{23} Hz is scattered by a nucleus of radius 3.2×10^{-15} m. The nucleus is totally radiation-absorbent and thus is a perfect obstacle. At what angle will the first diffraction minimum lie? [*Hint:* Use Babinet's principle (Problem 52).]

CHAPTER 40

Special relativity determines how you measure objects that move with respect to you. Distortion is significant when the relative speed between the object and you is a large fraction of the speed of light. Here we see downtown Pittsburgh from a high-speed flyby.

SPECIAL RELATIVITY

Even though electromagnetism shares with mechanics concepts such as energy and momentum, there appears to be a major difference between these two fundamental disciplines. The laws of mechanics look the same in all inertial frames—all reference frames that move with uniform velocity with respect to some standard inertial frame (for example, a frame at rest relative to distant stars). Electromagnetism appears to violate this general law. According to Maxwell's equations, electromagnetic waves propagate at the speed c, with no restrictions on the state of motion of the source or detector. This suggests the existence of an absolute frame for electromagnetism.

The special theory of relativity, proposed by Albert Einstein in a remarkable paper in 1905, extended to electromagnetism the principle that the laws that describe electromagnetism look the same in all inertial frames. This was accomplished not by altering Maxwell's equations but by modifying certain assumptions about our notions of space and time, assumptions that went unquestioned until 1905. This chapter is devoted to clarifying the nature of space and time that follows from the special theory of relativity. Exploration of the physical consequences of the theory are another aim of this chapter.

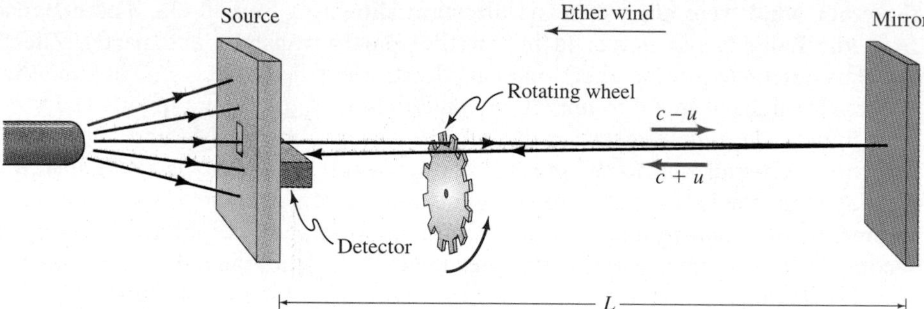

FIGURE 40–1 Schematic diagram of a light-speed measurement, with an ether wind blowing along the direction in which light travels. The ray speeds are different for the two directions of travel.

40–1 IS AN ETHER NECESSARY?

In the years following the discovery of Maxwell's equations, the absolute value of the speed of propagation of electromagnetic waves caused little concern. In an age of mechanical models, it was believed that electromagnetic waves need a medium to support them (just as sound waves need air). This presumed medium, which was thought to fill the universe, was called the **ether**. The ether was assumed to be at rest relative to the fixed stars. It would have to have an esoteric structure to support transverse waves. Many ingenious models were constructed to represent the ether. Maxwell's theory was assumed to give c for the speed of propagation of electromagnetic waves *relative to the ether's rest frame*, just as the speed of sound is given as 330 m/s relative to stationary air. In a reference frame moving at speed u relative to the ether, the speed of light emitted by a source at rest relative to the ether would be $c + u$ if the frame were moving toward the source, and $c - u$ if the frame were moving away from the source. The earth represents such a moving frame, because in its motion around the sun it travels at a speed of approximately 30 km/s relative to the fixed stars. From the point of view of a frame fixed to the earth, the ether moves past at a speed of 30 km/s. Detecting the *ether wind*, though, was likely to be very difficult.

To see why, consider a standard speed-of-light measurement, with the light beam propagating along an axis that lies in the direction of the ether wind. If the distance from the source (and detector) to a mirror is L, then in the absence of ether wind, the time for a single traversal of light from source to mirror and back is $t_0 = 2L/c$ (Fig. 40–1). With the ether wind blowing against the source, the speed of light that travels toward the mirror is $c - u$, and the speed of light that returns is $c + u$. Thus the time for a single traversal is

$$t_1 = \frac{L}{c-u} + \frac{L}{c+u} = \frac{2cL}{c^2 - u^2}$$

$$= \frac{2L/c}{1 - (u^2/c^2)}. \tag{40–1}$$

For $u = 30$ km/s and $c = 3 \times 10^5$ km/s, the factor $u^2/c^2 = 10^{-8}$. Thus a time sensitivity of better than one part in 100 million would be required to detect the ether wind.

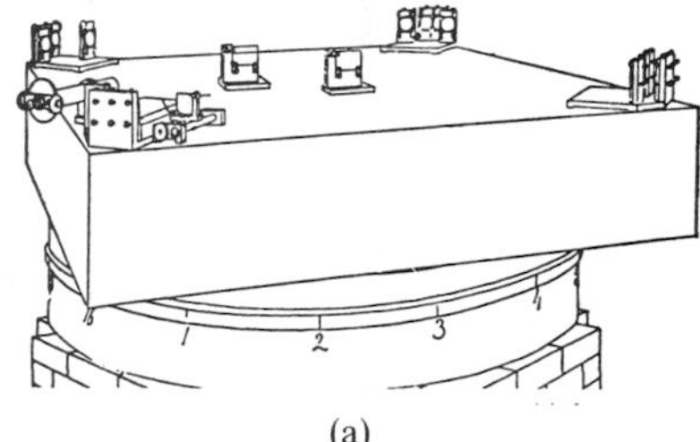

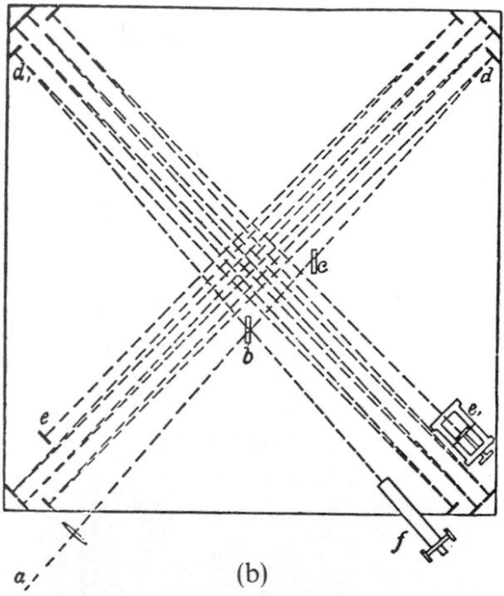

FIGURE 40–2 (a) Sketch of an interferometer built on a stone slab by Albert Michelson in 1887. (b) Schematic diagram of the workings of this interferometer, with the table rotated to the right. Follow the ray path from the source, point a, to the beam splitter, point b, and ultimately to the eyepiece, point f.

The Michelson–Morley Experiment

In 1887, Albert A. Michelson and Edward W. Morley carried out a high-precision experiment to measure the possible effect of an ether wind. They did this by using an interferometer, an instrument designed by Michelson (Fig. 40–2). Suppose that

The Michelson interferometer is described in Chapter 38.

the ether wind were aligned in the direction shown in Fig. 40–3a. The distance from the half-silvered mirror (a mirror that partly transmits and partly reflects light) to mirror M_1 in the direction aligned with the ether wind is L. The time that it takes for light to travel to mirror M_1 and back is t_1, given by Eq. (40–1). Let us next find t_2, the time it takes for the light to travel to mirror M_2 and back in the direction perpendicular to the presumed ether wind. Mirror M_2 is also a distance L away from the half-silvered mirror. (The assumption that the arms of the interferometer are equal in length is not essential, as we shall see later.) Because the second beam is perpendicular to the presumed ether wind, the light would have to travel a distance larger than $2L$; the beam would be blown "off course" by the ether wind. As Fig. 40–3b shows, it travels a distance given by the hypotenuse of a triangle in which one leg has length L and the other leg is the transverse distance the beam is blown in time $t_2/2$; that is, $ut_2/2$. The distance is $\sqrt{L^2 + (ut_2/2)^2}$. Because the speed of light is c, we get

$$\sqrt{L^2 + \left(\frac{ut_2}{2}\right)^2} = \frac{ct_2}{2}.$$

We square both sides and get

$$L^2 + \left(\frac{u^2}{4}\right)t_2^2 = \left(\frac{c^2}{4}\right)t_2^2.$$

From this equation it follows that

$$t_2^2 = \frac{4L^2}{c^2 - u^2};$$

$$t_2 = \frac{2L/c}{\sqrt{1 - (u^2/c^2)}}.$$

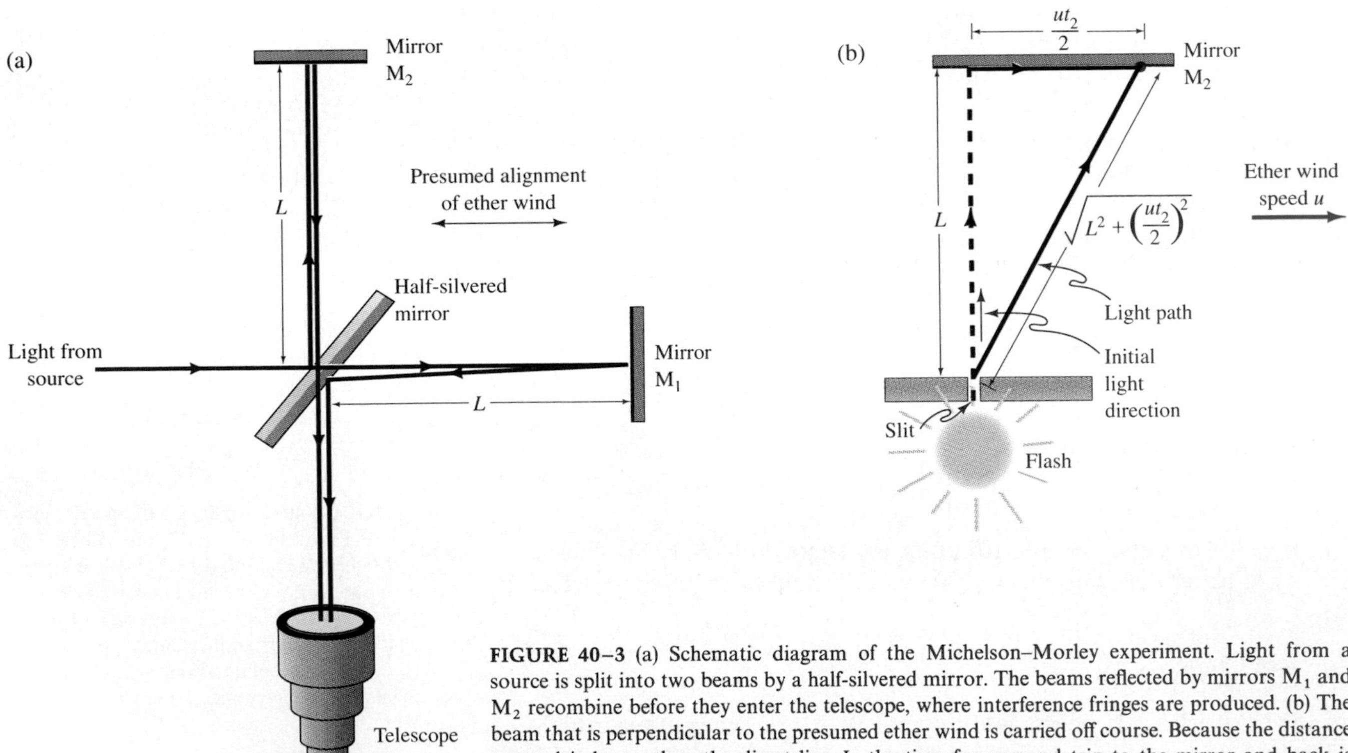

FIGURE 40–3 (a) Schematic diagram of the Michelson–Morley experiment. Light from a source is split into two beams by a half-silvered mirror. The beams reflected by mirrors M_1 and M_2 recombine before they enter the telescope, where interference fringes are produced. (b) The beam that is perpendicular to the presumed ether wind is carried off course. Because the distance covered is larger than the direct line L, the time for a round trip to the mirror and back is longer than it would be if the distance were $2L$.

We now use the fact that for small x, $1/(1-x) \simeq 1 + x$ and $1/\sqrt{1-x} \simeq 1 + (x/2)$, and apply these formulas for $x = u^2/c^2$. We then find that the time difference between the arrivals of two parts of a wave pulse is

$$\Delta t \equiv t_1 - t_2 = \frac{\cancel{2}L}{c}\frac{u^2}{\cancel{2}c^2}. \tag{40–2}$$

This corresponds to a path-length difference of $c\Delta t = Lu^2/c^2$. The two beams are combined upon their return from the two mirrors, and because they started out in phase, they will interfere according to the difference of their path lengths.

It is impossible to construct an apparatus in which the paths to the mirrors are exactly the same. In addition, because the mirrors are not exactly perpendicular to the beams, the path-length difference will vary slightly from one side of a mirror to another, and a view through the telescope yields a set of interference fringes (Fig. 40–4). Fortunately, any effects due to the apparatus itself can be accounted for by rotating the apparatus through 90°. The apparatus effects are unaltered, but the rotation effectively interchanges M_1 and M_2 and thus changes the path length to $-Lu^2/c^2$ (for unequal arm lengths, L is replaced by the average length). If there were an effect due to the ether, the fringe pattern would *shift* accordingly when the apparatus is rotated. The total path-length difference for the two orientations is $\Delta L = 2Lu^2/c^2$.

FIGURE 40–4 Interference fringes observed in a Michelson interferometer.

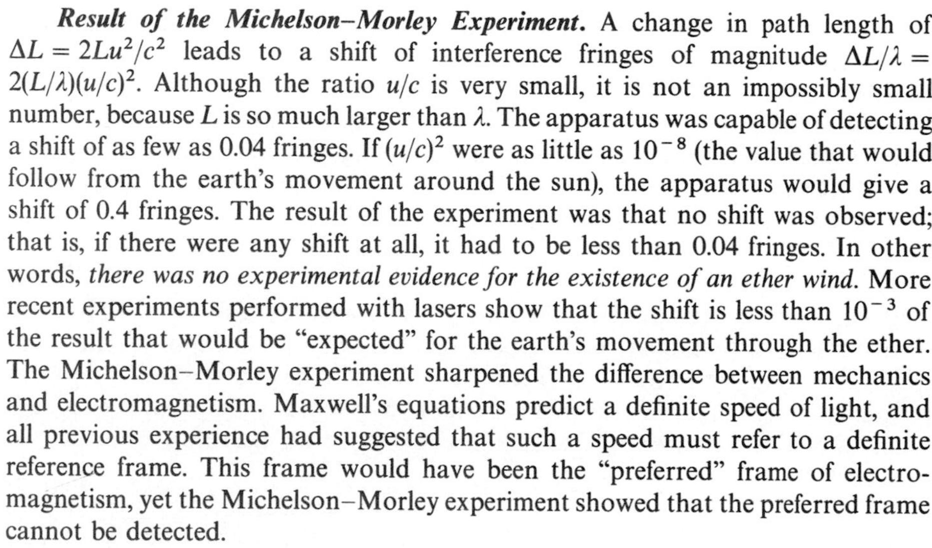

Result of the Michelson–Morley Experiment. A change in path length of $\Delta L = 2Lu^2/c^2$ leads to a shift of interference fringes of magnitude $\Delta L/\lambda = 2(L/\lambda)(u/c)^2$. Although the ratio u/c is very small, it is not an impossibly small number, because L is so much larger than λ. The apparatus was capable of detecting a shift of as few as 0.04 fringes. If $(u/c)^2$ were as little as 10^{-8} (the value that would follow from the earth's movement around the sun), the apparatus would give a shift of 0.4 fringes. The result of the experiment was that no shift was observed; that is, if there were any shift at all, it had to be less than 0.04 fringes. In other words, *there was no experimental evidence for the existence of an ether wind.* More recent experiments performed with lasers show that the shift is less than 10^{-3} of the result that would be "expected" for the earth's movement through the ether. The Michelson–Morley experiment sharpened the difference between mechanics and electromagnetism. Maxwell's equations predict a definite speed of light, and all previous experience had suggested that such a speed must refer to a definite reference frame. This frame would have been the "preferred" frame of electromagnetism, yet the Michelson–Morley experiment showed that the preferred frame cannot be detected.

FIGURE 40–5 Albert Einstein.

40–2 THE EINSTEIN POSTULATES

Albert Einstein was aware of the Michelson–Morley experiment, but its startling result apparently did not play a critical part in the creation of Einstein's theory (Fig. 40–5). He conjectured that the laws of electricity and magnetism (electrodynamics), like those of mechanics, are the same in all inertial reference frames. What results from this conjecture is the **special theory of relativity**. Einstein himself referred to the following two postulates:

The postulates of special relativity

1. **The laws of physics are the same in all inertial reference frames.**
2. **The speed of light in empty space is the same in all inertial frames.**

Strictly speaking, the second postulate is part of the first, because Maxwell's equations do not specify a frame when they predict the speed of light. As we have seen in Section 40–1, the second postulate might appear to be incompatible with the first. The combination of these apparently irreconcilable assertions led to the revolutionary insights into space and time that underlie the special theory of relativity.

The first postulate as applied to mechanics was introduced in Chapter 4, where we noted that the laws of mechanics are unchanged under the transformation

$$\mathbf{r}' = \mathbf{r} - \mathbf{u}t. \tag{40–3}$$

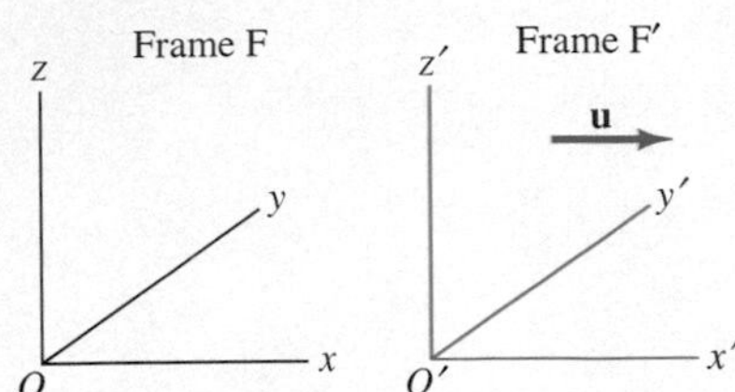

FIGURE 40–6 Reference frame F′ moves at velocity **u** with respect to frame F. An observer is in a particular frame when he or she measures the position and time of events in terms of the coordinate system and clocks of that frame.

Figure 40–6 shows the two frames to which these equations apply. The primed variables are the coordinates in a coordinate system (or reference frame) F′. Unprimed variables are the coordinates in frame F. Frames F and F′ move relative to one another. Upon differentiation with respect to t, Eq. (40–3) leads to a transformation law for the velocities:

$$\mathbf{v}' = \mathbf{v} - \mathbf{u}. \tag{40–4}$$

The first postulate seemed to Newton to imply that time is the same in all frames. Perhaps it seems so obvious to you that there is no need even to make the statement. Newton did recognize the need to make an assertion about the nature of time, and in modern language this statement is the trivial transformation law

$$t' = t. \tag{40–5}$$

This transformation law is a second one to be added to Eq. (40–3). Together these equations form the *Galilean transformations.*

The Galilean transformation laws, Eqs. (40–3) and (40–5), are incompatible with the second postulate. Suppose that we have a light source at rest at the origin of a frame F. When that source emits a flash of light in the form of a spherical wave that expands at the speed of light, then the location of the spherical wave front is given by

$$x^2 + y^2 + z^2 = \mathbf{r}^2 = c^2t^2 \tag{40–6}$$

(Fig. 40–7). Now consider how the light behaves according to an observer in a second frame, F′, that moves with respect to F, as in Fig. 40–6. Suppose that origin O' of F′ coincides with origin O of F at $t = t' = 0$, when the light flashes. The second postulate implies that *in F′ the pulse also forms a spherical wave front*, because the speed of light is the same in F′ as it is in F, whatever the direction. Thus

$$x'^2 + y'^2 + z'^2 = \mathbf{r}'^2 = c^2t'^2. \tag{40–7}$$

If we set $t' = t$, Eq. (40–7) cannot be satisfied when the relation between **r** and **r**′, given by Eq. (40–3), is used, because that would imply that $x'^2 + y'^2 + z'^2 =$

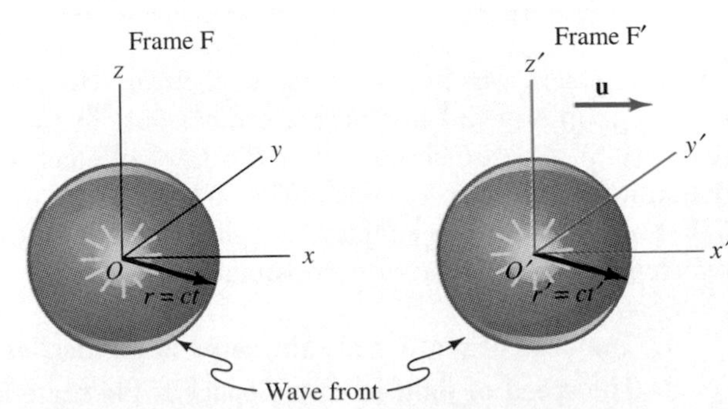

FIGURE 40–7 A spherical burst of light is emitted at the origin of frame F. To an observer in that frame, the light makes a spherical wave front centered at origin O. If origin O' of frame F′, moving with respect to frame F, is coincident with O when the burst occurs, then an observer at O' moving with frame F′ will claim that the light makes a spherical wavefront in frame F′ centered at O'.

$(x - ut)^2 + y^2 + z^2$. This conclusion shows the conflict between the Galilean transformation laws and the assertion of the absolute magnitude of the speed of light. It is apparent that we need to reexamine the very notions of space and time.

40–3 SPACE, TIME, AND SIMULTANEITY

In studying the concepts of time and space, we must first define with care how we measure time and how we establish space coordinates. In Section 40–4 we shall construct a concrete example of a clock. Now we require only that our clock be periodic: Time intervals between "ticks" are the same. We can put our clock and a light source at the origin of a coordinate system and then shine light along the x-axis. We place a mirror at some point x_1 and measure the time that it takes for the light to reach x_1 and return after reflection. If that time interval is two "ticks," we say that x_1 is one unit of length away from $x = 0$. Specifically, if the length of a tick is τ, then the distance to x_1 and back is $c\tau$. We now move the mirror farther away until the time for the light to reach there and back is four ticks. That point will be two length units ($2c\tau$) away from $x = 0$ along the x-axis. Proceeding in this way, we can in principle assign a coordinate (x, y, z) to every point in space.

To be able to discuss time at each point in the coordinate system, let us put a clock at every point for which x, y, and z are integer multiples of the unit of distance $c\tau$. We can synchronize all these clocks (that is, set all the clocks to the same time) as follows: At the origin at noon, a light signal is sent out to the point $x = 1$, $y = z = 0$. When the light ray arrives there, the clock operator at that point sets the clock to "one tick after noon," which corresponds to the point $x = 1$, $y = z = 0$. The operator at $x = 2$, $y = z = 0$ will set the clock there to "two ticks after noon" when the wave front reaches that point, and so on. In this way all the times are synchronized, and we have a reference frame in which space and time are well defined (Fig. 40–8). We have transmitted our signals in an unambiguous way, because the speed of light is, by postulate (and by experiment!), independent of any motion. Transmission of signals by means of baseballs would mean that the velocity of the baseball has to be measured, and this would involve us in complications having to do with the fact that to measure velocity, length and time must be unambiguously defined!

Although time intervals and distances within our given frame F (or another

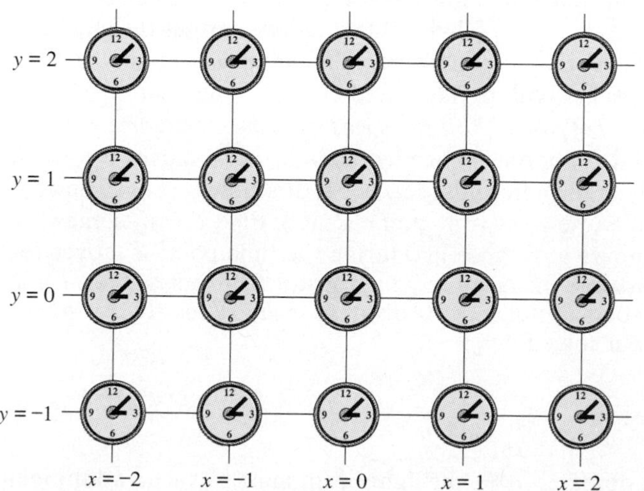

FIGURE 40–8 Clocks attached to lattice points that represent space coordinates separated by fixed distances. The location and the clock reading define the space–time coordinates of an event.

frame F′) are well defined by our setup, we must be careful about how we specify times and distances as seen from a moving frame. Consider again frames F and F′ of Fig. 40–6. An observer at O' in frame F′ who wishes to see how fast a clock ticks in frame F must receive signals from the clock attached to O every time the clock in frame F ticks. Because in Fig. 40–6 O' is moving away from O, the signals will be delayed.

Simultaneity

We define two events at different points in a given reference frame to be *simultaneous* when they occur at the same time in that frame. *This is what we mean when we say that two events are simultaneous in a given inertial frame.* The concept of simultaneity is crucial because it enters subtly into many kinds of measurements. As an example, suppose that an observer in the F′ frame wants to measure the length of a train at rest in the F frame (the observer sees the train move). To measure the length, the observer in F′ must take care that the locations x'_1 and x'_2, corresponding to the front and rear ends of the train, are marked off *at the same time*—simultaneously. Marking the position of the rear end of a moving train at midnight and the front end at 2 min past midnight and taking the difference between these two positions will not give a correct reading of the length of the train. Thus the notion of simultaneity enters into the length measurement. But as Einstein pointed out, our ordinary notion of simultaneity is strongly affected by the existence of a maximum speed for signals, the speed of light.

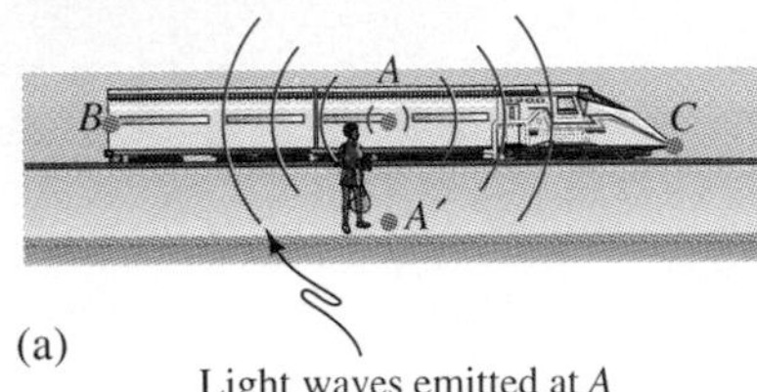

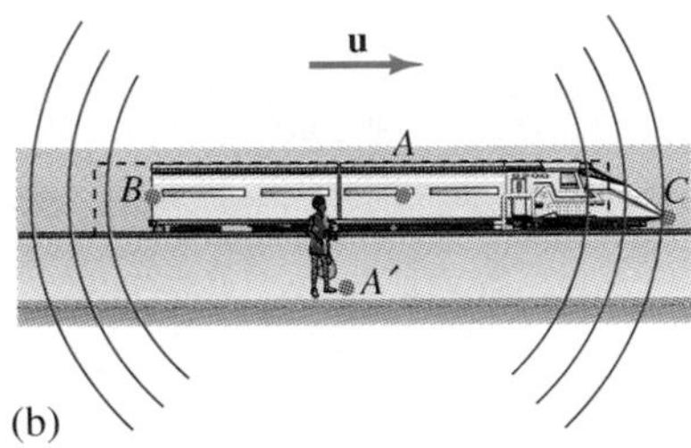

FIGURE 40–9 (a) Light emitted at point A, located midway between points B and C, reaches points B and C at the same time. (b) When B is moving toward A' and C away from A', the light reaches B before it reaches C.

Let us consider more carefully the problem of measuring a train's length. Figure 40–9a shows a train of length L initially at rest with respect to a platform. L is measured while the train is at rest, so there is no difficulty in making the measurement; we merely lay meter sticks down and count them. The rear and front of the train are labeled B and C, respectively, and two persons, A and A', are stationed at the exact midpoint of the train. Person A is *inside* the train, and person A' is *outside* on the platform. For the moment everything is measured with respect to a single inertial reference frame F. There is a set of synchronized clocks in this frame, as we described above. We label all times in frame F as t. If A sends out a spherical light pulse at $t = 0$, light reaches B and C simultaneously, at $t = L/2c$.

Now suppose that the train is moving at uniform speed u (Fig 40–9b). The frame at rest with respect to the train is F, whereas F′ is a new frame at rest with respect to the platform. Frame F′ has its own system of clocks along the railroad tracks, and times in that frame are labeled t'. At the moment person A is adjacent to person A', person A fires a light pulse, and we can set the clocks to $t = 0 = t'$. From the point of view of frame F, all is as it was in the original situation: The light pulse reaches both points B and C at $t = L/2c$. *But this cannot be true from the point of view of frame F′ if the speed of light is the same for both frames.* Person A' sees point B approach even as the light pulse moves toward B at the (fixed) speed of light, so the light pulse reaches B at a time $t' = t'_B$ somewhat earlier. Similarly, the light pulse arrives at point C at a time $t' = t'_C$ somewhat later. In fact, if the train moves a distance ut'_B during the time point B moves toward the pulse, then the distance the beam moving toward B covers is $(L/2) - ut'_B$; according to Einstein's second postulate, this distance is ct'_B. Thus $(L/2) - ut'_B = ct'_B$, an equation that we can solve for t'_B:

$$t'_B = \frac{L/2}{c + u}. \tag{40–8}$$

Similarly, according to an observer in F′, the light beam must travel an additional distance ut'_C to reach C. By the same argument,

$$t'_C = \frac{L/2}{c - u}. \qquad (40\text{–}9)$$

Time t'_B is different from time t'_C, so *events that are simultaneous in F are not simultaneous in F'*! This rather counterintuitive idea that the concept of simultaneity is not absolute is the key to all relativity, as we shall soon see. Note that the time difference between t'_B and t'_C is very small for $u \ll c$; this fact explains the origin of our nonrelativistic intuition. If we lived in a world where ordinary speeds were comparable to the speed of light, we would have developed a different intuition.

Whether or not events are simultaneous depends on the frame in which they are measured.

40–4 TIME DILATION AND LENGTH CONTRACTION

The Einstein postulates, or the fact that events that are simultaneous in one reference frame are not simultaneous in a frame that is moving with respect to the first, has two dramatic consequences. These are the slowing down of moving clocks—time dilation—and the shortening of moving rods aligned with the direction of motion—length contraction. We shall explain both consequences.

Time Dilation

To clarify our discussion, consider a very simple clock (Fig. 40–10a).† It consists of a rod with a light bulb at one end and a mirror at the other end, a distance L

† We follow the exposition of N. David Mermin in *Space and Time in Relativity*, McGraw-Hill, 1968.

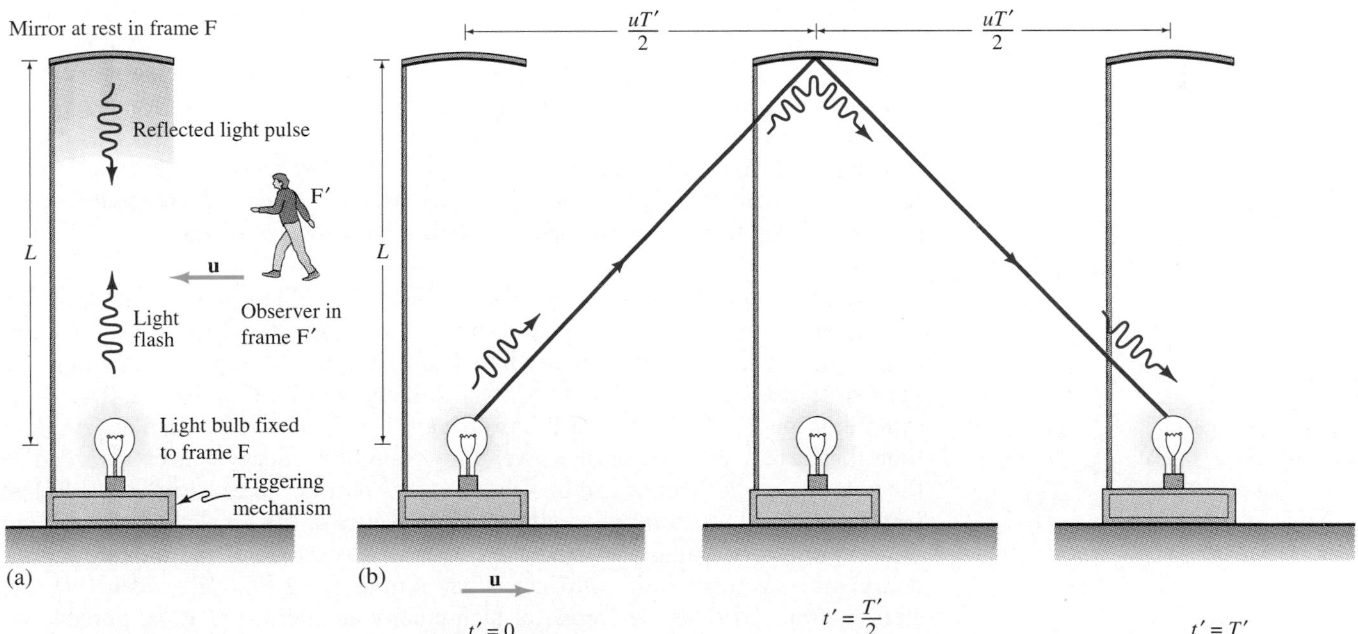

FIGURE 40–10 (a) Schematic diagram of a clock. The light bulb at one end of the rod flashes whenever it receives a light flash reflected off the mirror a distance L away from the bulb. The time interval between flashes in the frame of the clock ("ticks") is $2L/c$. (b) The path that the light ray must take according to an observer for whom the clock is moving.

apart. Attached to the light bulb is a mechanism that makes the bulb flash whenever a previous flash returns after reflecting off the mirror. According to an observer at rest relative to the clock, the bulb flashes with a period $T = 2L/c$.

The clock behaves rather differently to a moving observer. Suppose that this observer is in an inertial frame, F′, that moves at speed u to the left (Fig. 40–10a). An observer in F′ will see the clock receding to the right (Fig. 40–10b). In F′, the light still travels at speed c to the mirror, but it now has farther to go: The mirror moves during the time that the light travels to it from the light bulb. As in our discussion of the transverse light beam in the Michelson–Morley experiment, the time that it takes for the light to travel to the mirror and back is such that

$$\frac{cT'}{2} = \sqrt{L^2 + \left(\frac{uT'}{2}\right)^2};$$

that is, $T' = (2L/c)/\sqrt{1 - (u^2/c^2)}$, or

Time dilation

$$T' = \frac{T}{\sqrt{1 - (u^2/c^2)}}. \tag{40–10}$$

Time T' is greater than time T by a factor of $1/\sqrt{1 - (u^2/c^2)}$.[†] The observer in frame F′ sees longer "ticks" for the clock; in other words, *the moving clock is slower by a factor* of $\sqrt{1 - (u^2/c^2)}$. This effect is known as **time dilation:** *Moving clocks run slower than clocks at rest do.* It is not that the clocks are physically altered; rather, time is different when seen from different inertial frames.

There are two important features of the time-dilation effect. First, it is a symmetric effect. If there were a clock at rest in frame F′ identical to the clock at rest in F, then the observer in F would see the clock in F′ running slow, just as the observer in F′ sees the clock in F running slow. Were it otherwise, the observers could use this asymmetry to decide who was moving and who was standing still, an explicit violation of the original premise that only relative motion has any meaning. Second, although the particular clock we have considered (the "light clock") is an odd one, it is universal, in the sense that *every* clock imaginable must behave like it. Any additional clock in F can be synchronized with the light clock so that its ticks are directly and physically tied to the light clock's ticks. The ticks of a clock can take diverse forms, from the periodic vibrations of an atomic system, to the frequency of a light wave, to the beating of a heart. *All* these clocks run slow according to an observer who is moving with respect to them.

Experimental Tests of Time Dilation. The time-dilation effect is real. We can produce experimental evidence with measurements of half-lives of radioactive nuclei or unstable particles in motion. For example, the unstable fundamental particle called the *muon*, when at rest, has a lifetime of 2.197×10^{-6} s. (In a large sample of muons, 63 percent will have decayed in 2.197×10^{-6} s.) The length of time it takes for 63 percent of a given large sample to decay can be regarded as the tick of a clock. Muons can be produced in a particle accelerator and will then travel at a speed u determined by the characteristics of the accelerator. It is found in the laboratory frame of the accelerator that 63 percent of the moving muons decay into electrons and neutrinos after a time $t = 2.197/\sqrt{1 - (u^2/c^2)}$ μs, *not* after a time 2.197 μs. In modern high-energy accelerators, it is possible to accelerate the muons to such a high velocity that the time-dilation factor

[†] In our analysis we assumed a certain direction for the motion of frame F′. A more detailed analysis shows that our result is independent of the direction in which frame F′ moves.

$1/\sqrt{1-(u^2/c^2)}$ can be as large as 10^6, and 63 percent of these muons decay in a period $10^6 \times (2.197\ \mu s) = 2.197$ s.

The time-dilation effect was also checked in a much more pedestrian way in 1972. A very accurate cesium clock was flown in a commercial airplane around the earth and used to confirm time dilation to an accuracy of about 10 percent. The time-dilation effect has been confirmed so often that there is no doubt of its reality nor of the accuracy of Eq. (40–10).

EXAMPLE 40 – 1 Consider a clock taken on an airplane that travels at 1.0×10^3 km/h around the world along the equator. If the clock is synchronized with a stationary clock on departure, by how much will the clocks differ after 1 round trip?

SOLUTION: Equation (40–10) will provide us with the amount of time that each tick of the clock (here, 1 s) changes. We need to find the total number of seconds that it takes to fly around the world, and this is obtained by dividing the distance traveled by the speed of travel. We take the radius of the earth at the equator to be 6.38×10^3 km; the circumference is $2\pi r = 40.1 \times 10^3$ km. The speed of travel is 1.0×10^3 km/h, so the flight time is 40.1 h; that is, $(40.1\text{ h})(3600\text{ s/h}) = 1.44 \times 10^5$ s. The time-dilation factor is $\sqrt{1-(u^2/c^2)}$: The clock records $\sqrt{1-(u^2/c^2)}$ fewer seconds. With $u = (1.0 \times 10^3\text{ km/h}) \times (10^3\text{ m/km})/(3600\text{ s/h}) = 280$ m/s, we can evaluate u/c. It is given by $u/c = (280\text{ m/s})/(3.0 \times 10^8\text{ m/s}) = 9.3 \times 10^{-7}$, and thus $u^2/c^2 = 8.6 \times 10^{-13}$. For such a small value of u^2/c^2, we can write

$$\sqrt{1-\frac{u^2}{c^2}} \simeq 1-\frac{u^2}{2c^2} = 1-(4.3 \times 10^{-13}).$$

Thus N, the number of seconds lost, is

$$N = (4.3 \times 10^{-13})(1.44 \times 10^5\text{ s}) = 6.2 \times 10^{-8}\text{ s}.$$

To verify this result, we would need a clock that loses or gains no more than 6.2×10^{-8} s out of 1.44×10^5 s—a clock with an accuracy of about 1 s in 10^{12} s.

Here we have neglected corrections due to the earth's rotation and to the presence of gravity. These factors must be included when actual measurements are compared with the predictions of special relativity.

The Twin Paradox. It should be stressed that all our considerations apply to clocks that move with uniform velocities. Insufficient attention to this restriction leads to the *twin paradox*. Consider identical twins on earth. One of them takes off at high speed v on a long journey. After traveling for a long time, that twin gently comes to rest and then retraces her steps. Upon returning to earth, the stay-at-home twin will observe that the traveling twin is considerably younger, that they are no longer identical. The stay-at-home twin reasons that this is to be expected, because, relative to herself, the traveling twin was moving with uniform velocity: The traveling twin's clock, metabolism, heart rate, and so on were slowed down by a factor of $\sqrt{1-(v^2/c^2)}$. The deceleration and acceleration at the turning point are assumed to occur in such a short time that they do not affect this conclusion. The paradox appears if the traveling twin considers herself to be at rest while the stay-at-home twin, together with the earth, was traveling at speed v in the opposite direction. The traveling twin would expect the stay-at-home twin to be younger. Surely, both cannot be right!

There is no paradox: The traveling twin is not always in an inertial frame. She moves at uniform speed most of the time, but she does experience a deceleration and then an acceleration for the return. Thus she cannot make the same statements about the slowing down of clocks as her sister can. From the point of view of special relativity, only the stay-at-home twin, who is always in an inertial frame, can apply the theory to herself. Actually, with a limited use of general relativity (about which we shall say a few words in Section 40–8), the traveling twin can give a scientifically correct argument as to why and by how much the stay-at-home twin is younger at the end of the journey.

Length Contraction

The slowing down of moving clocks is accompanied by the contraction in length of moving objects along their direction of motion. We can begin by giving an argument that uses time dilation. Consider again the muon, an unstable particle with a lifetime of $\tau \simeq 2\ \mu s$. The lifetime is so small that even if a muon were moving near the speed of light, $c\tau$ would be much smaller than the atmosphere's height, and, without relativity, muons produced at the top of the atmosphere would decay before they could reach the ground. But muons are in fact copiously produced in the upper atmosphere by cosmic rays, and some muons reach the ground. Time dilation permits that. Suppose that, as seen from the ground, muons move at speed u. Because they are moving, their lifetime is increased to $\tau/\sqrt{1-(u^2/c^2)}$. According to a ground-based observer, about half the muons will cover the distance L given by the speed times the increased lifetime:

$$L = \frac{u\tau}{\sqrt{1-(u^2/c^2)}}. \tag{40–11}$$

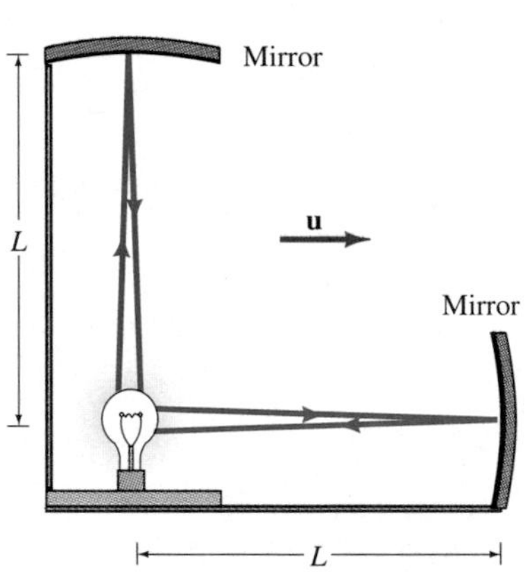

FIGURE 40–11 Schematic diagram of a two-armed clock used to exhibit length contraction. Note the similarity to the Michelson–Morley apparatus.

This length is much greater than it would be if there were no time-dilation effect, because the square-root factor approaches zero as u approaches c. The length could in particular be larger than the height of the atmosphere, and in this way muons could reach the ground before they decay.

We now suppose that an "average" muon just reaches the ground before decaying, so L is the height of the atmosphere. Let us see how this looks to an observer who is moving with the muon—that is, an observer who sees the muon at rest. He will measure the muon's lifetime to be its original value τ. However, this observer will also detect that the muon reaches the ground if the earth-based observer does; the collision with the ground is an event no observer could dispute. Thus in time τ he will see the whole atmosphere move past him at speed u. If our observer measures the atmosphere to have a height L', then the atmosphere will pass him in time L'/u. This must equal τ. Thus $L' = u\tau$, or, from Eq. (40–11),

Length contraction

$$L' = L\sqrt{1-\frac{u^2}{c^2}}. \tag{40–12}$$

The observer moving with the muon measures the atmosphere to be thinner than an earth-based observer does. To the moving observer, the atmospheric height, or any length in the direction of his motion, has undergone a **length contraction** by a factor of $\sqrt{1-(u^2/c^2)}$.

A second way to see that there must be a length contraction along the direction of motion is to modify the clock constructed at the beginning of this section. To the original rod, we add an identical rod at right angles (transverse) to it (Fig. 40–11). The mechanism is modified so that the bulb relights only when both reflected beams reach the light bulb at the same time. This can be achieved by making the length of the rods identical. Each flash of the bulb is an event, and these events are observed from any inertial frame. Suppose now that the clock moves at speed u in the direction of the added rod with respect to an observer in frame F′. According to the observer in frame F′, the round-trip time for the light on the transverse rod is

$$T' = \frac{2L}{c} \frac{1}{\sqrt{1 - (u^2/c^2)}}. \tag{40–13}$$

The round-trip time for the light that travels along the horizontal rod is the time t'_1 it takes to get to the mirror added to the time t'_2 it takes to return. The mirror is moving to the right, so the light on its outward trip has an extra distance ut'_1 to travel. If, according to the observer in frame F′, the length of the rod is L' (the quantity we want to find), then

$$ct'_1 = L' + ut'_1.$$

For the return trip, the bulb approaches the mirror at speed u, so the time t'_2 to return is determined by

$$ct'_2 = L' - ut'_2.$$

We solve these two equations for t'_1 and t'_2, respectively, and add:

$$t'_1 + t'_2 = \frac{L'}{c-u} + \frac{L'}{c+u} = \frac{2L'/c}{1 - (u^2/c^2)}. \tag{40–14}$$

This, however, must equal T', and a comparison of Eqs. (40–13) and (40–14) gives $L' = L\sqrt{1 - (u^2/c^2)}$, which is the length contraction of Eq. (40–12).

Note that length contraction occurs only *along* the direction of motion. That there is no change in directions transverse to the motion can be seen by the argument in Fig. 40–12.

(a)

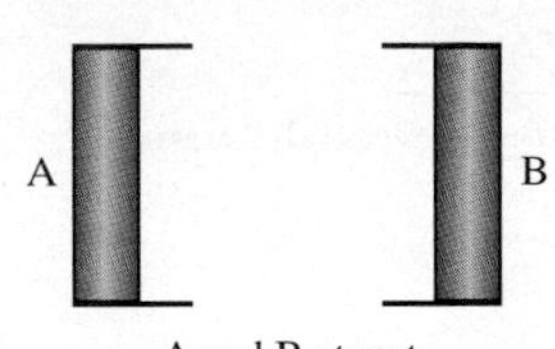

(b)

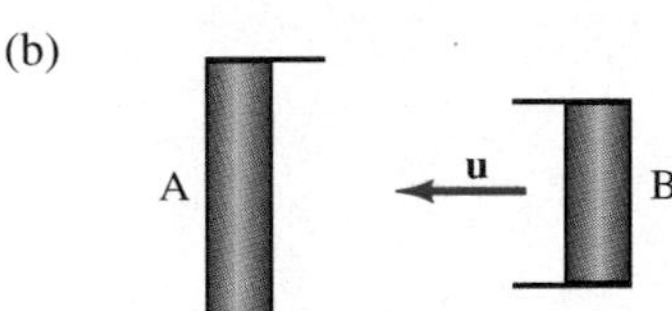

(c)

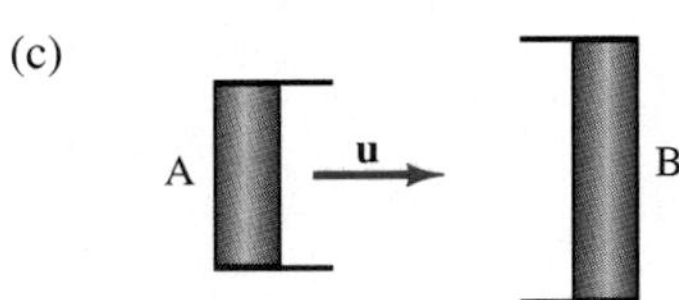

FIGURE 40–12 (a) Two rods, A and B, have the same length when they are at rest with respect to one another. (b) Now the rods approach each other. From the viewpoint of rod A, rod B might be shortened in a direction transverse to its direction of motion. Rod B could *simultaneously* scratch marks near the top and bottom of rod A as shown. (c) If the principle of relativity holds, then from the viewpoint of B, A would similarly be shortened. This shortening could be marked by scratches made by A onto B. But now bring the rods to rest and compare: A scratch at the 0.8-m mark of A, say, was made by the 1-m mark of B; a scratch at the 0.8-m mark on B was made by the 1-m mark of A. One "event," as recorded by the scratches, has been seen differently by two observers—an impossibility. The only possible resolution is that there can be no shortening in directions perpendicular to the motion.

EXAMPLE 40–2 The radius of our galaxy is 3×10^{20} m. (a) How fast would a spaceship have to travel to cross the entire galaxy in 300 yr, as measured from within the spaceship? (b) How much time would elapse on earth during the traversal?

SOLUTION: (a) Our hypothetical traveler would be at rest within the spaceship and would see the galaxy approaching at some speed v. The galaxy is contracted along the direction of motion, and it is the contracted length that must be covered in 300 yr of "spaceship time" at speed v. If L is the diameter of the galaxy, then the contracted diameter, from Eq. (40–12), is

$$L' = L\sqrt{1 - \frac{v^2}{c^2}}.$$

If the time measured in the spaceship is T, then the required speed is

$$v = \frac{L'}{T} = L\frac{\sqrt{1 - (v^2/c^2)}}{T}.$$

Thus $v^2 = (L^2/T^2)[1 - (v^2/c^2)]$, which we write as

$$x = \frac{L^2}{c^2T^2}(1 - x),$$

where $x = v^2/c^2$. When we solve for x, we get

$$x = \frac{L^2}{L^2 + c^2T^2}.$$

We know that $L = 2(3 \times 10^{20}\text{ m}) = 6 \times 10^{20}\text{ m}$ and $T = (300\text{ yr})(3.15 \times 10^7\text{ s/yr}) = 9.5 \times 10^9\text{ s}$. This gives

$$x = \frac{(6 \times 10^{20}\text{ m})^2}{(6 \times 10^{20}\text{ m})^2 + (3 \times 10^8\text{ m/s})^2(9.5 \times 10^9\text{ s})^2}$$

$$= \frac{36 \times 10^{40}}{(36 \times 10^{40}) + (0.8 \times 10^{37})} \simeq \frac{1}{1 + (2 \times 10^{-5})} \simeq 1 - (2 \times 10^{-5}).$$

Thus

$$v/c = \sqrt{x} = \sqrt{1 - (2 \times 10^{-5})} \simeq 1 - 10^{-5} = 0.99999.$$

Only if a spaceship travels at a speed extremely close to that of light can it traverse huge distances in a "reasonable" amount of time.

(b) As seen from earth, the galaxy is not contracted, and the spaceship moves at $0.99999c$. The time for the trip as seen from earth is then

$$t_{\text{earth}} = \frac{L}{v} = \frac{6 \times 10^{20}\text{ m}}{(0.99999)(3 \times 10^8\text{ m/s})} \simeq 2 \times 10^{12}\text{ s} \simeq 64{,}000\text{ yr}.$$

The earth-based observer will, however, see the spaceship's clock tick off only 300 yr as it travels from one end of the galaxy to the other.

40–5 THE RELATIVISTIC DOPPLER SHIFT

The Doppler shift for sound was covered in Chapter 14.

The Doppler shift for sound describes the changes in pitch of a train whistle as the train approaches, passes, and recedes from an observer. When a moving source that emits sound waves with frequency f travels toward an observer at rest relative to the air, the observed frequency f' is shifted from the source frequency according to

$$f' = \frac{f}{1 - (u/c)},$$

Eq. (14–50). We have changed the notation slightly, representing the source speed by u (instead of v_s) and the speed of sound by c (instead of v). If the source is at rest relative to the air and the observer is moving toward the source, then the frequency picked up by the observer is

$$f' = f\left(1 + \frac{u}{c}\right),$$

Eq. (14–54), with a similar change of notation. The frequencies f' are not the same in the two cases, so it is possible by an accurate measurement of the frequency shift and a knowledge of the relative speed to determine whether it is the source or the receiver who is moving relative to the medium (the air). The reason for the difference between the two shifts is that for sound there *is* a preferred frame, namely, the frame at rest relative to the air. The Doppler shift for electromagnetic radiation (including light) cannot, according to the relativity principle, distinguish between the two situations and must therefore have a different form.

To derive the Doppler shift for light, consider a periodically flashing light that moves at speed u toward an observer (Fig. 40–13). The source is placed at the origin of frame F′. Suppose that one pulse of light is emitted for every time interval τ', so the frequency of emission is $f_0 = 1/\tau'$, as seen by someone moving with the source. The stationary observer on the right sees frame F′ moving toward her at speed u. She measures the time between the arrival of the first wave and the

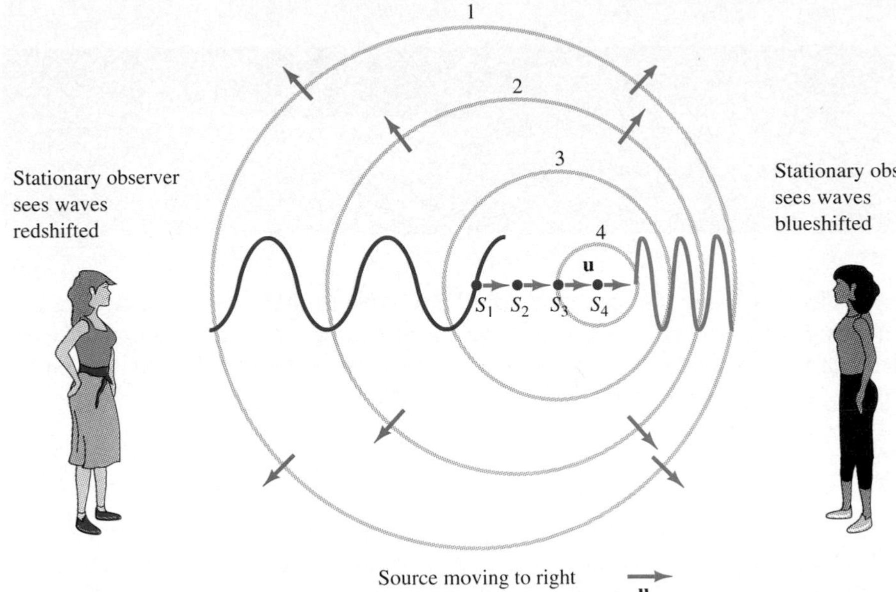

FIGURE 40–13 The Doppler effect associated with the relative movement of a light source and an observer. The observer toward whom the source moves sees the wave fronts crowded together (a decreased wavelength), and hence she sees a blueshift. The observer away from whom the source moves sees the wave fronts spread apart (an increased wavelength), and hence she sees a redshift. A time-dilation factor must be applied as well.

$(N + 1)$th wave to be t. She sees N waves fitted into a distance ct diminished by the distance ut that the source has moved in that time. Thus the wavelength is given by

$$\lambda = \frac{\text{distance}}{\text{number of waves}} = \frac{ct - ut}{N} = \frac{(c - u)t}{N},$$

and the frequency the observer measures is

$$f_1 = \frac{c}{\lambda} = \frac{cN}{(c - u)t} = \frac{N}{1 - (u/c)}\frac{1}{t}.$$

The time-dilation effect gives us a relation between t and τ. The observer measures the source clock to run slow, so the time τ she sees between pulses is

$$\tau = \frac{\tau'}{\sqrt{1 - (u^2/c^2)}}.$$

Because t is the time for N pulses to be emitted, we have

$$t = N\tau = \frac{N\tau'}{\sqrt{1 - (u^2/c^2)}}.$$

Thus

$$f_1 = \frac{\cancel{N}}{1 - (u/c)} \frac{\sqrt{1 - (u^2/c^2)}}{\cancel{N}\tau'}$$

$$= f_0\left[\frac{\sqrt{1 - (u^2/c^2)}}{1 - (u/c)}\right]. \tag{40–15}$$

If we use $1 - x^2 = (1 - x)(1 + x)$, we can write Eq. (40–15) as

$$f_1 = f_0\sqrt{\frac{1 + (u/c)}{1 - (u/c)}}, \tag{40–16a}$$

The Doppler shift for light

where f_0 is the frequency of the source in its rest frame and f_1 is the frequency

FIGURE 40–14 The spectral lines that can be attributed to a specific element are shifted to shorter wavelengths compared to spectral lines for the same element on earth. This blueshift indicates that the star Vega is moving toward our sun.

observed from a frame that moves at speed u relative to the source. The two frames are moving toward each other at relative speed u, and the frequency is increased. Instead of the frequency, we can use $\lambda = c/f$ to express the wavelength λ_1 seen by the observer in terms of the wavelength λ_0 at the source:

$$\lambda_1 = \lambda_0 \sqrt{\frac{1 - (u/c)}{1 + (u/c)}}. \tag{40–16b}$$

We call this situation, in which the source moves toward the observer and the observed wavelength decreases, a *blueshift*. The speed u can be interpreted either as the speed of the source toward a stationary receiver or as the speed of the receiver toward a stationary source. According to the principles of special relativity, these two possibilities are not distinguishable. If the source is moving away from the observer, then we must change the sign of u in our results, and

$$f_1 = f_0 \sqrt{\frac{1 - (u/c)}{1 + (u/c)}}. \tag{40–17a}$$

Equivalently, the observed wavelength is

$$\lambda_1 = \lambda_0 \sqrt{\frac{1 + (u/c)}{1 - (u/c)}}. \tag{40–17b}$$

Thus the frequency decreases (the wavelength increases) in this case. The visible spectrum is shifted toward the red colors, and Eqs. (40–17) are said to describe a relativistic *redshift*. This is the case for the stationary observer on the left in Fig. 40–13.

Cosmological Implications of the Doppler Shift for Light

Measurements of the Doppler shift of starlight have proven to be crucial in the evolution of modern astrophysics and cosmology. In one application, the Doppler shift is used to establish the velocities of stars or other radiating bodies.† Radiation emitted by atoms and molecules is characterized by *spectral lines*, discrete or very narrow frequency bands of especially intense radiation. These spectral lines provide a signature for elements and compounds. If an entire sequence of spectral lines in starlight is observed to correspond to a sequence of laboratory-observed spectral lines all shifted by the same factor, then we know that the source of the starlight is the same as the laboratory source but moves at a velocity that can be calculated from equations such as Eq. (40–16) (Fig. 40–14).

One of the most interesting uses of the Doppler shift was made by the astronomer Edwin Hubble (Fig. 40–15). In the 1920s, 1930s, and 1940s he studied the spectral lines of a large number of stars in distant galaxies, and by using the known

FIGURE 40–15 Edwin Hubble at Mt. Palomar Observatory in 1948.

† See Problem 56 for another application.

characteristic brightnesses of these stars, he estimated their distance from earth. Hubble discovered that the spectra of most of these distant stars are redshifted, which means that their galaxies are receding from us. He found that *the recession velocity of the galaxies relative to our galaxy is proportional to their distance from earth.* This result is known as **Hubble's law**, and it takes the mathematical form

Hubble's law

$$D = \frac{u}{H}. \tag{40–18}$$

Here D is the distance to a galaxy, u is the recession speed relative to us, and H is the so-called *Hubble parameter*, measured to be $H \simeq 2.5 \times 10^{-18}\ \mathrm{s}^{-1}$. The fact that the distance and speed are measured relative to earth appears to give earth a central position. This appearance is deceptive. If all stars and galaxies are moving away from each other, then an observer located at any one of them would report the same effect. This is easily visualized if we consider a simple model of dots painted uniformly on a balloon (Fig. 40–16a). As the balloon is inflated, all the dots move farther away from each other, and each dot "sees" the others moving away from it (Fig. 40–16b). The model of galaxies that move away from each other is part of the cosmological theory of the *big bang*, according to which the universe started from a point and underwent a rapid expansion. In this theory, the age of the universe is on the order of H^{-1}, roughly 13 billion yr.

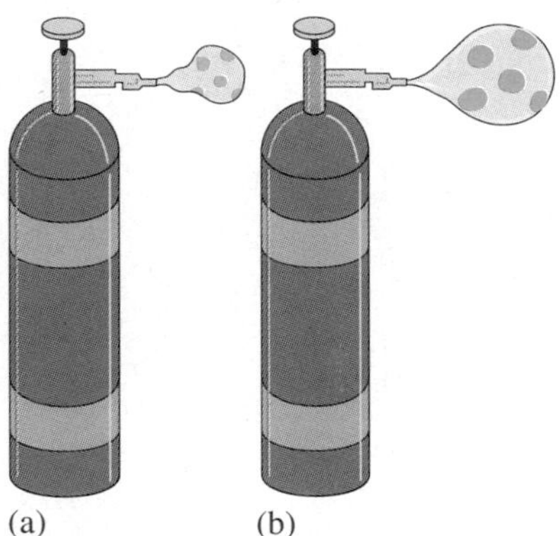

FIGURE 40–16 (a) Dots painted on the surface of a balloon represent an analogy to Hubble's expanding universe. (b) As the balloon expands, the dots move away from one another at a speed that depends on the distance between them.

Measurements of the Doppler shift, together with Hubble's law, allow us to calculate distances to galaxies. In the 1960s astronomers working with radio telescopes discovered very powerful pointlike sources of radiation in which very large redshifts ($f_1/f_0 \simeq 0.3$) were observed. These sources are *quasars*, or *quasistellar objects*. Astronomers concluded that quasars emit huge amounts of energy, which left a puzzle as to a mechanism by which that much energy is produced. An explanation that is gaining acceptance is that enormous accelerations of matter caused by the presence of black holes lead to the large amount of radiation.

EXAMPLE 40–3 Studies of a quasar show that a spectral line whose wavelength in the laboratory is 121 nm has a measured wavelength of 358 nm. With what speed is the quasar receding from earth? Assuming that Hubble's law holds, what is the distance, in light-years, of the quasar from earth?

SOLUTION: Equation (40–17b) gives the Doppler-shift formula for wavelengths. From it, we get

$$\left(\frac{\lambda_1}{\lambda_0}\right)^2\left(1 - \frac{u}{c}\right) = 1 + \frac{u}{c}.$$

When we solve this equation for u/c, we find that

$$\frac{u}{c} = \frac{(\lambda_1/\lambda_0)^2 - 1}{(\lambda_1/\lambda_0)^2 + 1}.$$

The data give $\lambda_1/\lambda_0 = (358\ \mathrm{nm})/(121\ \mathrm{nm}) = 2.96$, so $u/c = 0.79$, and

$$u = (0.79)(3.00 \times 10^8\ \mathrm{m/s}) = 2.38 \times 10^8\ \mathrm{m/s}.$$

Application of Eq. (40–18) gives

$$D = \frac{u}{H} = \left(\frac{2.38 \times 10^8\ \mathrm{m/\cancel{s}}}{2.5 \times 10^{-18}\ \cancel{\mathrm{s}^{-1}}}\right) = 0.95 \times 10^{26}\ \mathrm{m}.$$

Because 1 ly = (3.15 × 10^7 s)(3.00 × 10^8 m/s) = 0.95 × 10^{16} m, we obtain $D = 10^{10}$ ly. Such a distance is nearly at "the edge of the universe," and the light reaching earth gives information about the quasar 10 billion yr ago!†

The Relativistic Addition of Velocities

Suppose that observer A measures the velocity of an object as $\mathbf{v}_1$; in turn observer B measures observer A to move with velocity $\mathbf{u}$ with respect to him. According to the Galilean law of velocity addition [Eq. (40–4)], observer B will measure the object to move with velocity $\mathbf{v}_2 = \mathbf{v}_1 + \mathbf{u}$. As we shall soon see, this simple result cannot be consistent with special relativity.

We can use the Doppler shift for light to find the important relation that describes how velocities add. Suppose that a source emits light with the rest-frame frequency f_0 (Fig. 40–17). An observer moving away from the source at speed v_1 along the x-axis receives a redshifted frequency f_1. If that observer, O_1, immediately reradiates with frequency f_1 to another observer, O_2, who is moving away from O_1 in the same direction at speed v_2 (with respect to O_1), we would expect that the relationship between the received frequency f_2 and the original frequency f_0 would be the frequency f_0 shifted by the velocity V of the second observer relative to the source. In nonrelativistic mechanics we would expect to find the correct shift with the Galilean form $V = v_1 + v_2$. We will now use the procedure outlined above to obtain the relativistic counterpart of the Galilean form for the addition of velocities. It should come as no surprise that the formula needs modification, because for $v_1 = v_2 = 0.8c$, for example, $v_1 + v_2$ is larger than c, which our expressions for time dilation and length contraction [Eqs. (40–10) and (40–12)] do not allow.

The frequency f_1 measured by observer O_1 is, according to Eq. (40–17a),

$$f_1 = f_0\sqrt{\frac{1-(v_1/c)}{1+(v_1/c)}}.$$

If that observer immediately reradiates the light toward observer O_2, moving away at speed v_2 relative to observer O_1, then the frequency seen by observer O_2 is

$$f_2 = f_1\sqrt{\frac{1-(v_2/c)}{1+(v_2/c)}}.$$

† The study of the spectral lines of quasars thus gives us information about certain physical quantities 10 billion yr ago. In this way we know that the electron charge and the electron mass have not changed in magnitude by more than about one part in 10^{12}/yr.

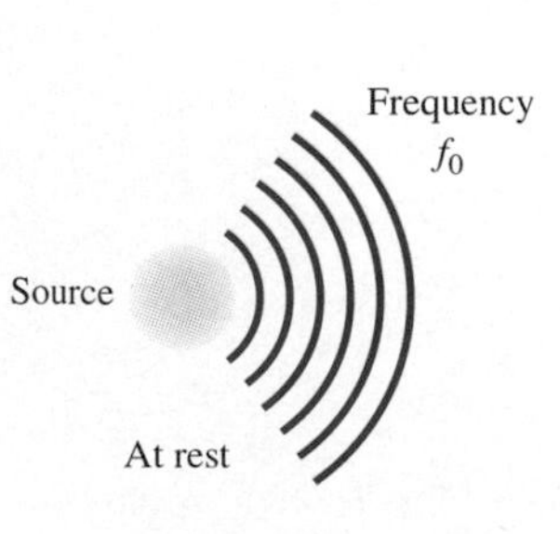

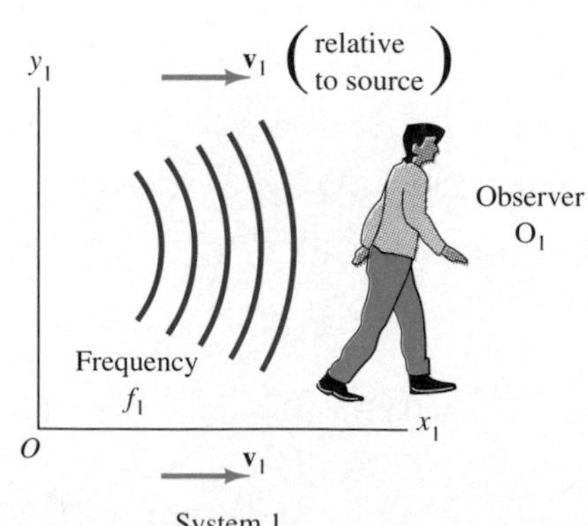

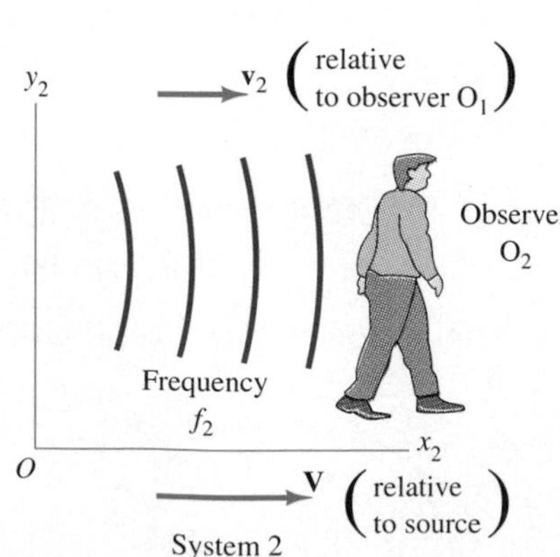

FIGURE 40–17 A light source emits a frequency f_0, seen as frequency f_1 by an observer O_1 who moves at speed v_1 to the right. Observer O_1 re-emits the light with frequency f_1, which is seen by an observer O_2 who moves to the right at speed v_2 with respect to observer O_1. (Observer O_2 moves at speed V with respect to the original source.) Observer O_2 measures a light frequency f_2, consistent with the law of addition of velocities.

We may eliminate f_1 by expressing f_2 in terms of f_0 and the (as yet unknown) velocity V of observer O_2 relative to the source. We have

$$f_2 = f_0\sqrt{\frac{1-(v_1/c)}{1+(v_1/c)}}\sqrt{\frac{1-(v_2/c)}{1+(v_2/c)}},$$

which we rewrite in the form

$$f_2 = f_0\sqrt{\frac{1-(V/c)}{1+(V/c)}}.$$

When we square (f_2/f_0) in its two forms and equate the two, we obtain

$$\frac{1-(V/c)}{1+(V/c)} = \frac{[1-(v_1/c)][1-(v_2/c)]}{[1+(v_1/c)][1+(v_2/c)]}.$$

We leave to the reader the solution of this equation for V. The result is the *law of addition of velocities*:

$$V = \frac{v_1 + v_2}{1 + (v_1 v_2/c^2)}. \qquad (40\text{–}19)$$

The relativistic addition of velocities

We see that in the nonrelativistic limit, where v_1/c and v_2/c are both small, Eq. (40–19) reduces to the Galilean law of addition of velocities: $V \to v_1 + v_2$. But if $v_1 = v_2 = 0.5c$, then Eq. (40–19) gives $V = 0.8c$, not $1.0c$. If $v_1 = c$, we get $V = c$, *independent of* v_2, so the maximum speed is c, and it is never exceeded. Equation (40–19) is a beautiful illustration of how relativistic kinematics smoothly adjoins the nonrelativistic (Galilean) form while providing a very different general result.

EXAMPLE 40–4 A light source flashes with a frequency of 1.0×10^{15} Hz. The radiation is reflected by a mirror that moves at a speed of 100 km/s away from the source. How different is the frequency of the reflected radiation, as observed at the source, from the original radiation?

SOLUTION: We shall approach this problem in the same way that we approached the derivation of the law of addition of velocities. We imagine a hypothetical observer who travels with the mirror. This observer receives the (redshifted) radiation from the source and reradiates it toward the source with the frequency that observer saw the light to have. A second observer at the source sees this radiation redshifted once more, and this radiation is completely equivalent to what results from reflection in a receding mirror.

Let us first determine the frequency as observed at the mirror. The emitted frequency is f_0. Because the source is receding from the mirror at speed u, the frequency f' observed at the mirror is redshifted:

$$f' = f_0\sqrt{\frac{1-(u/c)}{1+(u/c)}}.$$

That is also the frequency of the radiation "emitted" by the mirror in its rest frame. The fact that it is a moving source that emits to a stationary observer is of no consequence, because the same shift in frequency is obtained if the source is at rest and the observer is moving. The frequency f'' observed at the source is therefore further redshifted:

$$f'' = f'\sqrt{\frac{1-(u/c)}{1+(u/c)}}.$$

Thus

$$f'' = f_0\left[\frac{1-(u/c)}{1+(u/c)}\right].$$

The reflected radiation differs from the original radiation by the frequency difference $f_0 - f''$. We have

$$f_0 - f'' = f_0\left[1 - \frac{1-(u/c)}{1+(u/c)}\right] = \frac{f_0}{1+(u/c)}\left[\not{1} + \frac{u}{c} - \left(\not{1} - \frac{u}{c}\right)\right] = \frac{2f_0(u/c)}{1+(u/c)}.$$

Here $u/c = (10^5 \text{ m/s})/(3.0 \times 10^8 \text{ m/s}) = 0.30 \times 10^{-3}$ is very small, so we can drop it compared to 1 in the denominator. Thus

$$f_0 - f'' \simeq 2(1.0 \times 10^{15} \text{ Hz})(0.30 \times 10^{-3}) = 6.0 \times 10^{11} \text{ Hz},$$

only 6×10^{-4} times the original frequency.

Another way of obtaining this result is to think of the image of the source in the mirror as radiating directly to the observer at the source. If a mirror recedes at speed u, then the speed of recession of the image is given by Eq. (40–19), with $v_1 = v_2 = u$. When $V = 2u/[1 + (u^2/c^2)]$ is substituted into the normal redshift formula, the same result is obtained.

40–6 THE LORENTZ TRANSFORMATIONS

An observer in any frame F will describe an event by its location in space and time in that frame; that is, by its *space–time* coordinates. An observer in a second frame, F′, will describe the same event by its space–time coordinates in frame F′. Let frames F and F′ be inertial, moving at speed u along the x-axis with respect to each other. Let us also take the origins and axes of frames F and F′ to coincide at time $t = t' = 0$ (Fig. 40–18a). The origin of the moving frame F′ at a later time is given by $x = ut$ in frame F, whereas that origin in frame F′ is still given by $x' = 0$. An event, which may be an explosion, the collision of two particles, or the flash of a light bulb, is described by the variables (x, t) in F and (x', t') in F′ (Fig. 40–18b).

The assumption of a universal time would relate the times by Eq. (40–5),

$$\text{Galilean transformation:} \qquad t' = t,$$

and the position coordinates by Eq. (40–3),

$$\text{Galilean transformation:} \qquad x' = x - ut.$$

As an example of how these *transformation laws* apply, suppose that an explosion occurs at the origin of frame F ($x = 0$) at time t. It would be described in frame F′ as having occurred at $x' = x - ut = -ut$. This is physically sensible: The moving frame F′ will have left frame F behind. As we noted in Section 40–2, the Galilean transformation laws do not satisfy the condition that the speed of light is the same in all inertial frames. The correct transformation laws, known as the **Lorentz transformations**, were found by Hendrik A. Lorentz in 1890:[†]

Equations (40–20), (40–21), (40–23), and (40–24) are the Lorentz transformations.

Lorentz transformations:

$$x' = \gamma(x - ut); \tag{40–20}$$

$$t' = \gamma\left(t - \frac{ux}{c^2}\right), \tag{40–21}$$

[†] This date is well before 1905, the publication date of Einstein's work on relativity, and the date usually given as the discovery date of special relativity. Einstein's key role was perhaps less in discovering new formulas than in bringing together different results under the banner of a conceptual whole.

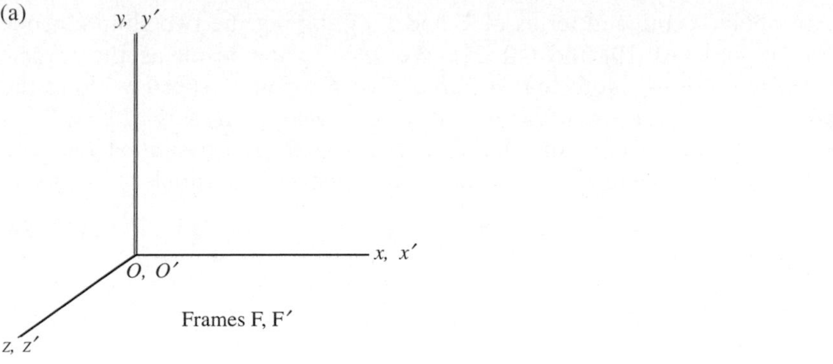

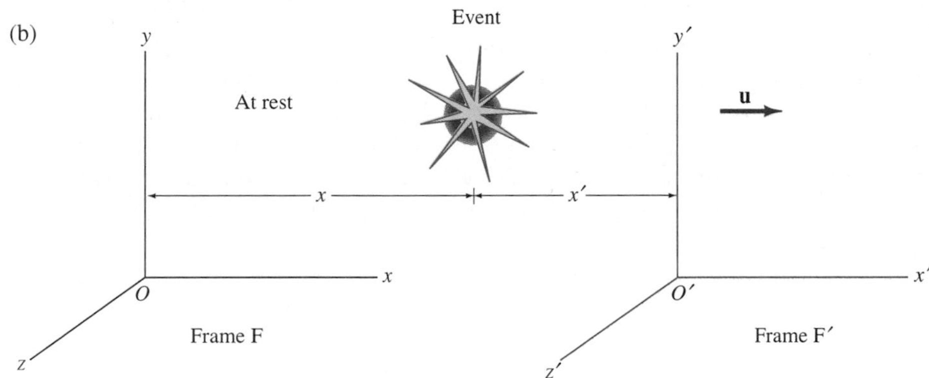

FIGURE 40–18 (a) The origins of the coordinate systems of frames F and F′ coincide at time $t = t' = 0$. (b) An event occurs some time later. An observer in frame F describes the space–time coordinates of the event differently than an observer in frame F′ would.

with γ defined by

$$\gamma \equiv \frac{1}{\sqrt{1 - (u^2/c^2)}}. \tag{40–22}$$

To these equations we should add that

$$y' = y \tag{40–23}$$

and

$$z' = z. \tag{40–24}$$

Equations (40–23) and (40–24) reflect the fact that the description of the y- and z-coordinates of an event is the same in frames F and F′.

We observe that when the relative speed of the two frames is small compared to c ($u/c \ll 1$), then $\gamma \simeq 1$, and the Lorentz transformations reduce to the Galilean transformations, Eqs. (40–3) and (40–4). This is why the necessity for the Lorentz transformations is not evident from ordinary mechanics. It can also be shown that, as in Eq. (40–6), the location of a wave front for a plane wave emitted from the joint origin of frames F and F′ at $t = t' = 0$ is given by $x^2 - c^2t^2 = 0 = x'^2 - c^2t'^2$ (see Problem 31). This result reflects the necessity that the speed of light be the same in both frames. In fact, the Lorentz transformations show more: *In general,*

$$x'^2 - c^2t'^2 = x^2 - c^2t^2. \tag{40–25}$$

An invariant quantity

Equation (40–25) states that the quantity $x^2 - c^2t^2$ is *invariant*, meaning that its value is the same in all inertial frames. There is a familiar three-dimensional analogue in geometry: For points on the surface of a sphere, $x^2 + y^2 + z^2$ is always the same.

We can obtain x and t in terms of x' and t' by solving the two simultaneous algebraic equations (40–20) and (40–21). We refer to the result as the inverse Lorentz transformations. Note that if frame F′ is moving at speed u along the x-axis away from frame F, then frame F may be viewed as moving at speed $-u$ along the x-axis away from frame F′. Thus *the inverse transformation laws are found from the original transformation laws by the simple substitution of* $-u$ *for* u:

$$x = \gamma(x' + ut'); \tag{40–26}$$

$$t = \gamma\left(t' + \frac{ux'}{c^2}\right). \tag{40–27}$$

The Lorentz Transformations, Time Dilation, and Length Contraction

We may use the Lorentz transformations of Eqs. (40–20) and (40–21) to rederive the formulas for time dilation and length contraction. Consider a clock fixed in frame F at $x = 0$. It starts at time $t = t' = 0$, and its initial location is $x = x' = 0$. The starting time and place may be viewed as the first event. If the period of the clock is τ, then the clock's first "tick" will be the second event, which will occur at $x = 0$, $t = \tau$. With the help of the Lorentz transformations given in Eqs. (40–20) and (40–21), we find that

$$x' = -\gamma u\tau, \tag{40–28a}$$

$$t' = \gamma\tau. \tag{40–28b}$$

Equation (40–28b) tells us that an observer in frame F′ sees the interval between ticks of the moving clock to be longer by a factor of γ. An observer in that frame thus sees the clock as running slow. At the first tick, the clock is seen to be at $x' = -\gamma u\tau = -ut'$, the location of the origin of frame F as seen from frame F′.

The formula for length contraction can be obtained by considering a rod of length L at rest in frame F. The length of an object in its own rest frame is called its **proper length**. The coordinates of the rod are $x_1 = 0$ and $x_2 = L$. What is the length of the rod in frame F′? As we emphasized earlier, when an object is moving in a certain reference frame, a length measurement makes sense only when the coordinates of the two ends are located simultaneously. Thus we require that $t'_1 = t'_2$. It is convenient to choose both of these to be zero; that is, make the length measurement at time $t'_1 = t'_2 = 0$. Let us see what these times are in frame F. We have

$$t'_1 = \gamma\left(t_1 - \frac{ux_1}{c^2}\right).$$

Because $x_1 = 0$ and $t'_1 = 0$, we obtain

$$t_1 = 0.$$

We also get

$$t'_2 = \gamma\left(t_2 - \frac{ux_2}{c^2}\right).$$

With $t'_2 = 0$ and $x_2 = L$, we find that

$$t_2 = \frac{uL}{c^2}.$$

Thus the length measurements are made at different times in frame F. We again see that events that are simultaneous in one frame are not simultaneous in another. We may now find the coordinates of the two ends in frame F:

$$x_1' = \gamma(x_1 - ut_1) = 0;$$

$$x_2' = \gamma(x_2 - ut_2) = \gamma\left(L - \frac{u^2L}{c^2}\right) = L\gamma\left(1 - \frac{u^2}{c^2}\right) = \frac{L}{\gamma}. \qquad (40\text{–}29)$$

This result is the length contraction as observed in frame F′.

EXAMPLE 40–5 Spaceship A of proper length L is traveling east at speed v_1, and spaceship B of proper length $2L$ is traveling west at speed v_2, as seen from earth. The pilot of spaceship A sets his clock to zero when the front of spaceship B passes him. (The spaceship pilots sit in the nose cone.) Use Lorentz transformations to calculate the time at which, according to the pilot of spaceship A, the tail of spaceship B passes him.

SOLUTION: The Lorentz transformations provide the necessary information. To proceed, we need to identify two events and describe them both in frame F, with coordinates (x, y, z), in which spaceship A is at rest, and frame F′, with coordinates (x', y', z'), in which spaceship B is at rest. We place the pilots at the origins of their respective reference frames. The Lorentz transformation formulas involve only relative speeds of frames F and F′. To calculate the relative speed u, we use the relativistic law of addition of velocities, for objects moving toward each other. This law is given by Eq. (40–19),

$$u = \frac{v_1 + v_2}{1 + (v_1v_2/c^2)}.$$

If the spaceships are aligned along their respective axes (x and x'), the front of spaceship A in its own rest frame is at $x = 0$, and its back is at $x = -L$ (Fig. 40–19a). Similarly, the front of spaceship B is at $x' = 0$, and its back is at $x' = 2L$. The time at which the two fronts just pass each other is the event that we may choose to take place at $t = t' = 0$. Our equations relating frames F and F′ are

$$x' = \gamma(x + ut)$$

and

$$t' = \gamma\left(t + \frac{ux}{c^2}\right).$$

Thus $x = 0, t = 0$ and $x' = 0, t' = 0$ are consistent.

The second event of interest to us is the alignment of the back end of spaceship B with the front end of spaceship A, when $x' = 2L$ and $x = 0$ (Fig. 40–19b). Given x' and x, we can find t, the time for the event recorded in spaceship A's rest frame. When these values for x' and x are inserted into the equations that relate x' and t' to x and t, we get

$$2L = \gamma(0 + ut) = \gamma ut;$$

$$t = \frac{1}{u}\frac{2L}{\gamma}.$$

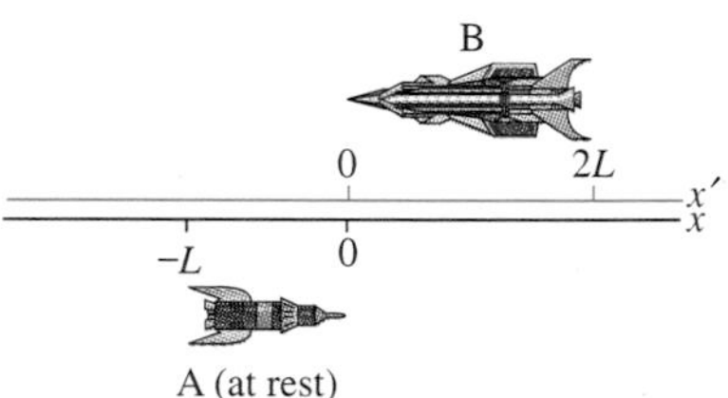

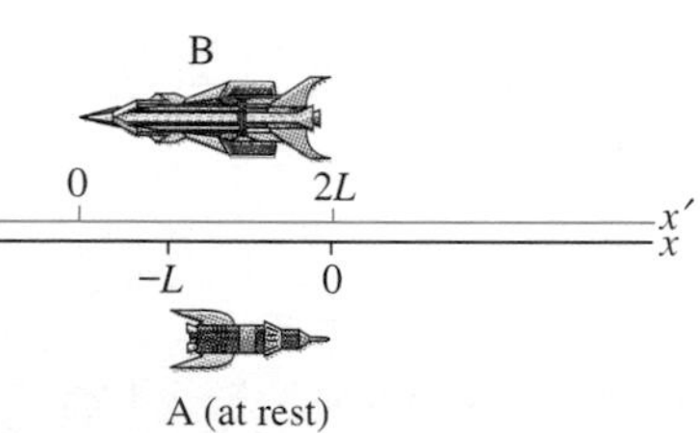

FIGURE 40–19 (a) Example 40–5. The event specified by the passing of the fronts of spaceships A and B. (b) The event specified by the passing of the rear of spaceship B by the front of spaceship A. Both cases are from the viewpoint of an observer in spaceship A, so spaceship B is shortened.

This is the time recorded by pilot A for the back of spaceship B to reach his position. This is a reasonable result: Pilot A sees the length of spaceship B contracted from $2L$ to $2L/\gamma$. Because spaceship B is moving at speed u relative to spaceship A, the time it takes to pass is its observed length $2L/\gamma$ divided by u.

EXAMPLE 40–6 A train of proper length $2L = 500$ m approaches a tunnel of proper length $L = 250$ m. The train's speed u is such that $\gamma = 1/\sqrt{1-(u^2/c^2)} = 2$. An observer at rest with respect to the tunnel measures the train's length to be contracted by a factor of 2 to 250 m and expects the whole train to fit into the tunnel. An observer on the train knows that the length of the train is 500 m, and the tunnel is contracted by a factor of 2 to 125 m. Thus the observer on the train argues that the train will not fit into the tunnel. Who is right?

SOLUTION: To analyze this problem, we start with two frames: frame F, the rest frame of the tunnel, and frame F′, the rest frame of the train. An observer in frame F would assign the position of the left side of the tunnel as $x = 0$ and the right side of the tunnel as $x = L$ (Fig. 40–20a), while an observer in frame F′ would assign the front of the train as $x' = 0$ and the rear of the train as $x' = -2L$ (Fig. 40–20b). We first need to check that, as measured in frame F, the train does indeed fit into the tunnel. Then we need to confirm that, as measured in frame F′, the train does not fit into the tunnel. In so doing, we can explain this paradox.

We identify two events, marked by arrows in Fig. 40–20, and describe them in both frames. Events must be specified by both position and time. The first event is the entry of the train into the tunnel. The clocks in the two frames are set so that at $t = t' = 0$, the front of the train ($x' = 0$) coincides with the left side of the tunnel ($x = 0$), as measured by an observer in frame F at time $t = 0$ (Fig. 40–20a) and by an observer in frame F′ at time $t' = 0$ (Fig. 40–20b). The second event is the alignment of the right end of the tunnel ($x = L$) and the front of the train ($x' = 0$), as measured by the observer in frame F (Fig. 40–20c) and the observer in frame F′ (Fig. 40–20d), respectively.

It follows from Eq. (40–20), $x' = \gamma(x - ut)$, that for the second event,

$$0 = \gamma(L - ut),$$

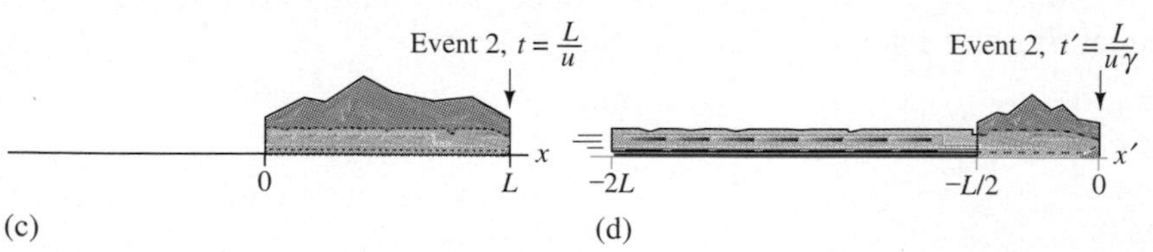

FIGURE 40–20 Example 40–6. Two events associated with the passage of a very fast train through a tunnel: the coincidence of the front of the train first with the left side of the tunnel, and second with the right side of the tunnel. Parts (a) and (c) show the events according to an observer in the rest frame of the tunnel; parts (b) and (d) show the events according to an observer in the rest frame of the train.

so $t = L/u$. This is the time of the second event as seen by the observer in frame F (Fig. 40–20c). Calculation of t' from the Lorentz transformation equation (40–21), $t' = \gamma[t - (ux/c^2)]$, yields

$$t' = \gamma\left(\frac{L}{u} - \frac{uL}{c^2}\right) = \frac{L}{u}\sqrt{1 - \frac{u^2}{c^2}} = \frac{L}{u\gamma} = \frac{t}{\gamma}.$$

This is the time of the second event as measured by the observer in frame F′ (Fig. 40–20d). Where, according to the observer in frame F, is the rear end of the train at $t = L/u$? We have

$$x' = -2L = \gamma\left[x - \cancel{u}\left(\frac{L}{\cancel{u}}\right)\right] = \gamma(x - L).$$

We solve this for x and get $x = (-2L/\gamma) + L = 0$. Thus at the time measured in frame F for the second event, the rear of the train is indeed at the tunnel entrance. As observed in frame F, the train fits into the tunnel. By this we mean that the observer in frame F sees that the front is at the exit of the tunnel and the rear is at the entrance of the tunnel *at the same time*, which in this case is $t = L/u$.

Let us check that the observer in frame F′ does not see the train fit in the tunnel. That observer considers the train to be at rest and the tunnel to be moving toward the train. What is the location in frame F′ of the tunnel entrance at the time of the event specified by the alignment of the front of the train and the tunnel exit—the value of x' when $x = 0$ at time $t' = L/\gamma u$? Substituting these values into Eq. (40–26), we get the result that $x = 0$ implies that $x' = -ut' = -L/\gamma = -L/2$ (Fig. 40–20d). Thus, from the point of view of an observer in frame F′, only one-quarter of the train is in the tunnel when its front reaches the tunnel exit!

Let us make a last check and see in frame F′ the time t' at which the rear of the train is at the tunnel entrance: What is t' when $x' = -2L$ and $x = 0$? We have

$$t' = \gamma\left[\frac{L}{u} - (0)\left(\frac{u}{c^2}\right)\right] = \frac{L/u}{\sqrt{1 - (u^2/c^2)}} = \frac{\gamma L}{u}.$$

This time is different from the time $t' = L/u\gamma$ when the front of the train reaches the end of the tunnel. Thus the observer on the train, in frame F′, measures the end of the train to pass the tunnel entrance at a later time. The observer in frame F′ will argue that the train's front passes the tunnel exit earlier than the rear passes the tunnel entrance. The differences between the interpretations of the two observers stem from their different notions of simultaneity. These differences allow both observers to be correct in their claims!

Lorentz Transformations of Electric and Magnetic Fields

Maxwell's equations predict the speed of light. It should therefore be evident that electric and magnetic fields play a distinct role in special relativity. In Chapter 31 we saw that in order to maintain Galilean invariance (invariance under the transformation laws $\mathbf{r}' = \mathbf{r} - \mathbf{u}t$ and $t' = t$) of the Lorentz force equation, the electric and magnetic fields must mix among themselves when we observe them from different reference frames. In Eqs. (31–22) and (31–23), respectively, we derived Galilean

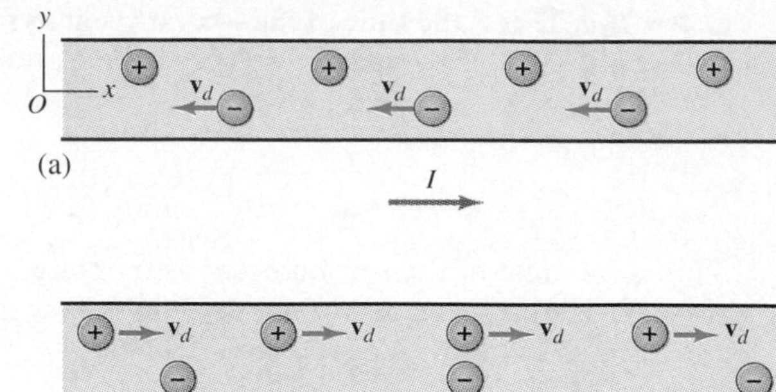

FIGURE 40–21 A current-carrying wire as observed from two frames. (a) The frame in which the wire is at rest. Here the wire is electrically neutral. (b) The frame that moves with the positive charges. Here the wire acquires a net charge density.

transformation laws for these fields:

Galilean transformation laws for electromagnetic fields

$$\mathbf{E}' = \mathbf{E} + (\mathbf{u} \times \mathbf{B});$$

$$\mathbf{B}' = \mathbf{B}.$$

(As usual, the prime refers to the field as measured in frame F′.) The replacement of the Galilean transformation law for time and position by the Lorentz transformations means that the electromagnetic fields also transform differently. Here we want to discuss briefly the physical origin of the transformation laws for these fields.

Electric fields result from the presence of electric charges, and magnetic fields result from the movement of electric charges. Let us now look at an example of how electric charge distributions are affected by Lorentz transformations. Suppose that in some frame F an electrically neutral wire carries a current in the $+x$-direction. The current consists of negative charges that move at some drift speed in the $-x$-direction against a background of stationary positive charges with the same spacing (Fig. 40–21a). This wire produces a magnetic field but no electric field.

Electric currents are described in Chapter 27.

Now consider the same wire as observed from a frame F′ that moves at the drift velocity $\mathbf{v}_d$ (Fig. 40–21b). An observer in frame F′ would see the electrons in the wire at rest and the positive charges (the ions) move in the $+x$-direction. We shall now show that special relativity implies that *the wire is not electrically neutral in frame F′*. An observer in frame F would measure the electrons to have less space between them than the F′ observer does, due to Lorentz contraction. In other words, the F′ observer measures *more* space between the electrons than the F observer does. The F′ observer measures *less* space between the positive ions than the F observer does: To the F′ observer, the positive ions are moving. Thus if the F observer sees the same spacing between electrons as between positive ions, the F′ observer will not, and the wire will no longer be electrically neutral to the F′ observer. To the F′ observer, there is an electric field.

We have shown that because of the way special relativity affects space, the presence of a magnetic field alone in one frame introduces an electric field in another. It is in this way that transformation laws between fields come about. A careful quantitative analysis of situations such as the one just described gives the full set of Lorentz transformations between electric and magnetic fields. These transformation laws, together with the Lorentz transformations for space and time, leave the physical consequences of Maxwell's equations invariant. Because the speed of light is one of the physical consequences of Maxwell's equations, that speed is the same in all frames. We have come full circle with a consistent description.

Momentum

The need to modify our notions of space and time suggests that the definitions of other kinematical quantities, which are based on measurements in space and time, also require modification. In nonrelativistic mechanics the *momentum* of a particle that moves with velocity $\mathbf{v}$ is

Momentum and its properties are treated in Chapter 8.

$$\mathbf{p} = m\mathbf{v}.$$

We sometimes call the coefficient of $\mathbf{v}$ in this expression the *rest mass m*. In the absence of external forces, the sum of the momenta of interacting particles is constant; that is, the total momentum is conserved:

$$\sum \mathbf{p}_i = \mathbf{P} = \text{a constant}.$$

Momentum conservation has its origin in Newton's third law and is a principle that holds both nonrelativistically and relativistically.

Momentum in relativity is a quantity ascribed to moving particles and has the following properties: (a) In the absence of external forces, the sum of momenta of interacting particles is conserved, and (b) in the limit that $\mathbf{v} \to 0$, $\mathbf{p} \to m\mathbf{v}$. On purely dimensional grounds, we expect that

$$\mathbf{p} = mf(v)\mathbf{v},$$

where the function $f(v)$ must be 1 for $v = 0$ and $f(v)$ is dimensionless. The function $f(v)$ depends only on the magnitude of $\mathbf{v}$, so $f(v)$ must be a function of v^2. Because f is dimensionless, it must be a function of v^2/c^2 or of the now familiar combination $\gamma = 1/\sqrt{1 - (v^2/c^2)}$.

A somewhat lengthy analysis of collisions between equal-mass particles leads to the result that $f(v) = \gamma$, so

$$\mathbf{p} = m\gamma\mathbf{v} = \frac{m\mathbf{v}}{\sqrt{1 - (v^2/c^2)}}. \quad (40\text{–}30)$$

The relativistic momentum

For $(v/c) \ll 1$ this reduces to the familiar low-velocity result $\mathbf{p} = m\mathbf{v}$. Newton's second law now reads

$$\mathbf{F} = \frac{d\mathbf{p}}{dt} = m\frac{d}{dt}(\gamma\mathbf{u}). \quad (40\text{–}31)$$

One consequence of the relativistic modification of the expression for momentum is that $\mathbf{F}$ and $d\mathbf{u}/dt$ no longer have to point in the same direction (see Problem 46).

Kinetic Energy

We may use Eq. (40–31) to derive the relativistic expression for the kinetic energy, K, from the work–energy theorem, Eq. (6–8). The work done to bring a particle of mass m from rest to speed v will be, by the work–energy theorem, the kinetic energy of the particle. In the "Deriving the Relativistic Kinetic Energy" box, we carry out the calculation of the work and find that

$$K = mc^2(\gamma - 1) = mc^2\left(\frac{1}{\sqrt{1 - (v^2/c^2)}} - 1\right). \quad (40\text{–}32)$$

The relativistic kinetic energy

For small v^2/c^2, $\sqrt{1 - (v^2/c^2)} \simeq 1 - (v^2/2c^2)$, so in the low-velocity limit, K reduces to the familiar $mv^2/2$ (as we pointed out in Section 6–6).

***DERIVATION**

Deriving the Relativistic Kinetic Energy

For simplicity we derive the expression for the kinetic energy, K, by dealing in one dimension. We define K for a particle as the work done to bring the particle from rest (at $t = 0$) to a speed v (at time t). Thus

$$K = \int F\,dx = \int_0^t F\frac{dx}{dt}\,dt = \int_0^t Fv\,dt. \tag{B1–1}$$

We now use Eq. (40–31) for the force:

$$Fv = vm\frac{d}{dt}(\gamma v) = mv^2\frac{d\gamma}{dt} + m\gamma v\frac{dv}{dt}.$$

The first term contains

$$\frac{d\gamma}{dt} = \frac{d}{dt}\frac{1}{\sqrt{1-(v^2/c^2)}} = -\frac{1}{2}\left(1-\frac{v^2}{c^2}\right)^{-3/2}\left(-\frac{2}{c^2}v\frac{dv}{dt}\right) = \left(\frac{\gamma^3}{c^2}\right)v\frac{dv}{dt}. \tag{B1–2}$$

Thus the integrand of Eq. (B1–1) takes the form

$$\left[m\gamma^3\left(\frac{v^2}{c^2}\right) + m\gamma\right]v\frac{dv}{dt}.$$

The factor in square brackets in this expression is

$$m\gamma\left[\left(\frac{v^2}{c^2}\right)\gamma^2 + 1\right] = m\gamma\left[\frac{v^2/c^2}{1-(v^2/c^2)} + 1\right] = m\gamma\left[\frac{1}{1-(v^2/c^2)}\right] = m\gamma^3.$$

Thus the integrand is $m\gamma^3 v\dfrac{dv}{dt}$, and, as Eq. (B1–2) shows, this is just $mc^2\dfrac{d\gamma}{dt}$. Thus

$$K = \int_0^t mc^2\frac{d\gamma}{dt}\,dt = mc^2\int_{\gamma'=1}^{\gamma'=\gamma} d\gamma' = mc^2(\gamma - 1) = mc^2\left[\frac{1}{\sqrt{1-(v^2/c^2)}} - 1\right],$$

which is Eq. (40–32).

Energy Associated with Mass

Einstein pointed out that energy and mass are related. It is possible to demonstrate that inertia (mass) is associated with energy, as follows. Consider a railroad car of mass M and length L standing on rails. Imagine a flash bulb attached to the left interior wall of the car (Fig. 40–22a). At a particular time the bulb emits a burst of light toward the right wall of the railroad car. As we learned in Chapter 35, if the energy of the light pulse is E, then there is momentum associated with the pulse, with magnitude E/c. Momentum conservation implies that the railroad car must move with an equal and opposite momentum toward the left (Fig. 40–22b). Because the mass of the car is M, the car will move with a velocity such that

$$Mv = \frac{E}{c}.$$

The time t that the pulse spends between the walls is given by

$$ct = L - vt,$$

because the right wall is now moving toward the light pulse. Thus

$$t = \frac{L}{c+v} = \frac{L}{c+(E/Mc)}.$$

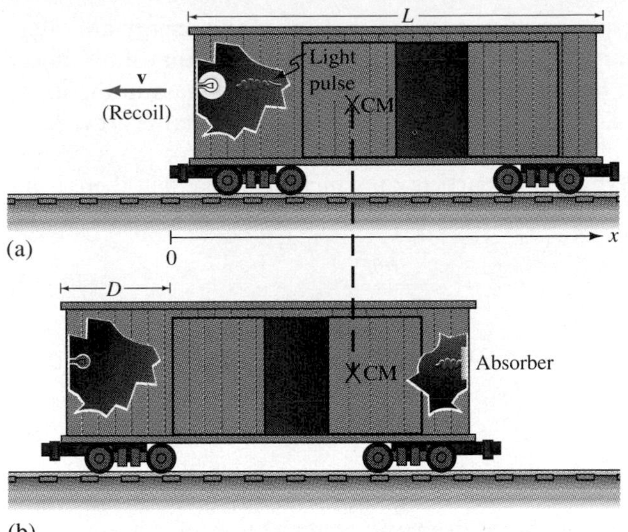

FIGURE 40–22 (a) A light source in a railroad car emits an electromagnetic pulse that carries energy E. The recoil momentum $p = E/c$ sets the car in motion. (The center of mass is exaggerated here.) (b) This motion ceases when the pulse is absorbed at the other end of the car.

The distance traveled by the car in that time is

$$D = vt = \frac{E}{Mc}\frac{L}{c + (E/Mc)} = \frac{EL}{E + Mc^2}. \qquad (40\text{–}33)$$

After the car has moved a distance D, it comes to a stop because the light, along with its momentum, is absorbed by the right wall of the car.

If momentum is to be conserved, the center of mass of the railroad car with the flashing bulb must not move. Yet the car has itself moved to the left. From this we must infer that the energy in the light is equivalent to a mass μ carried by the light flash, and that this mass moves to the right when the car moves to the left, such that the center of mass remains stationary. We can find μ by equating the position X of the overall center of mass before and after the flash event. Place the initial position of the left-hand side of the car at $x = 0$. The car can be treated as a point mass M, initially at $x = L/2$. Then, before the light is emitted,

$$X = \frac{(\mu)(0) + (M)(L/2)}{\mu + M} = \frac{M}{\mu + M}\frac{L}{2}.$$

After the light has been absorbed at the right-hand side, at position $x = L - D$,

$$X = \frac{(\mu)(L - D) + M[(L/2) - D]}{\mu + M}.$$

We equate these and solve for μ to find that

$$\mu = \frac{MD}{L - D}.$$

From Eq. (40–33), $L - D = LMc^2/(E + Mc^2)$, and when we substitute for $L - D$ as well as for D in the equation for μ, we get

$$\mu = \cancel{M}\frac{E\cancel{L}/\cancel{(E + Mc^2)}}{\cancel{L}\cancel{M}c^2/\cancel{(E + Mc^2)}} = \frac{E}{c^2}.$$

E is equivalent to a mass μ given by

$$E = \mu c^2. \qquad (40\text{–}34)$$

The relation between mass and energy

This is a very important result, because it shows that mass and energy are interchangeable concepts. We can reverse the reasoning here to say that if an object has a mass m, then it has a **rest energy** $E = mc^2$. Light has no mass in the usual sense, but any energy it has is equivalent to a mass E/c^2. The total energy, E, of any object is now the sum of the kinetic energy, the rest energy (or mass energy, E_{mass}), and the potential energy. For a particle on which no forces act, there is no potential energy, and

$$E = E_{mass} + K = \frac{mc^2}{\sqrt{1-(v^2/c^2)}}. \qquad (40\text{–}35)$$

It follows from this result and from Eq. (40–30) that

Velocity in terms of the relativistic momentum and energy

$$\mathbf{v} = \frac{c^2\mathbf{p}}{E}. \qquad (40\text{–}36)$$

The reality of the relation $E_{mass} = mc^2$ has been tested innumerable times in a large variety of nuclear reactions. More dramatic confirmation of this law came with the discovery of *antimatter*. Quantum mechanics and relativity together show that for each particle, there is a corresponding *antiparticle*. The antiparticle has a charge opposite to the particle, and an antiparticle and a particle can annihilate each other to produce electromagnetic radiation (Fig. 40–23). Similarly, out of radiation energy alone, a particle and an antiparticle can be created. The conversion of energy (in the form of radiation) into mass is thereby exhibited unambiguously.

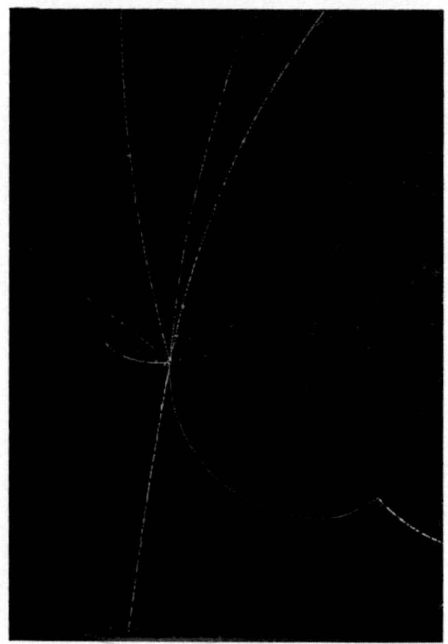

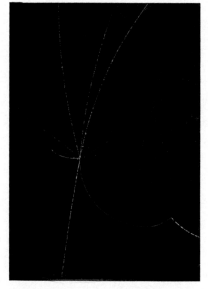

FIGURE 40–23 In this color-enhanced bubble-chamber photo, an incoming antiproton (light blue) strikes and annihilates with a proton at rest, producing 4 positive pions (red) and 4 negative pions (green), which are antiparticles of the positive pions.

EXAMPLE 40–7 The nucleus ^{8}Be is an unstable isotope of beryllium. It decays into two alpha particles, $^8\text{Be} \to {}^4\text{He} + {}^4\text{He}$. The atomic masses (in atomic mass units u) are $M(^8\text{Be}) = 8.005305$ u and $M(^4\text{He}) = 4.002603$ u.† Assume that a nucleus ^{8}Be decays while it is at rest. Find the kinetic energy of the helium nuclei (the alpha particles), in MeV.

SOLUTION: Both energy and momentum must be conserved, and these conservation laws provide the solution to the problem. The ^{8}Be is initially at rest, so the total momentum is always zero. The two helium nuclei must therefore have momenta equal in magnitude but opposite in direction, and hence the same kinetic energies. The initial energy is just the energy due to the mass of ^{8}Be. The final energy is the sum of the rest energies and the kinetic energies of the alpha particles. We equate the initial and final energies and get

$$E = M(^8\text{Be})c^2 = 2M(^4\text{He})c^2 + K_{total}.$$

The kinetic energy K_{total} is called the *Q value* in nuclear reactions. It is given by the difference between the initial and final rest energies:

$$\begin{aligned} K_{total} = Q &= [M(^8\text{Be}) - 2M(^4\text{He})]c^2 \\ &= [8.005305\text{ u} - 2(4.002603\text{ u})]c^2 = (0.000099\text{ u})c^2. \end{aligned}$$

We make use of the conversion 1 u = 931.5 MeV/c^2 to obtain

$$K_{total} = (0.000099\,\cancel{\text{u}})\left(\frac{931.5\text{ MeV}/\cancel{c^2}}{\cancel{\text{u}}}\right)\cancel{c^2} = 0.092\text{ MeV}.$$

† Atomic masses are commonly used in nuclear physics, because the electron masses are equal on both sides and the electron binding energies are so small that any differences in binding energies can be ignored.

Because we have shown above that momentum conservation implies that the kinetic energies of the two alpha particles are equal, each alpha carries kinetic energy of (0.092 MeV)/2 = 0.046 MeV.

EXAMPLE 40–8 A 1.0-kg meteorite of antimatter strikes the earth. (This is just a thought experiment; there is no solid evidence of antimatter in such large lumps.) How much energy is liberated in the annihilation process in which all the antimatter and an equal amount of matter are converted to radiant energy? Neglect the kinetic energy of the meteorite; $(v/c) \ll 1$.

SOLUTION: The rest energy of mass M is Mc^2. The antimatter interacts with an equal amount of matter in the annihilation process, so the amount of energy liberated is $2Mc^2$:

$$E = 2Mc^2 = 2(1.0\text{ kg})(3 \times 10^8\text{ m/s})^2 = 1.8 \times 10^{17}\text{ J}.$$

To get some idea of the significance of this number, the amount of (chemical) energy in 1 ton of TNT is 4.2×10^9 J. The meteorite explosion would generate the equivalent of 4×10^7 tons of TNT, or about 40 hydrogen bombs. Antimatter–matter annihilation has been suggested as a source of fuel for manned planetary journeys: It is the most efficient fuel possible.

The Relation between the Momentum and Energy of a Particle

There is an important relation between the momentum and energy of a moving particle. From Eq. (40–35) we have

$$E^2 = \frac{m^2c^4}{1 - (v^2/c^2)}.$$

Equation (40–30) gives us an expression for p^2c^2, which is a quantity with the same dimension as E^2:

$$p^2c^2 = \frac{m^2v^2c^2}{1 - (v^2/c^2)}.$$

The difference between these results is

$$E^2 - p^2c^2 = \frac{m^2c^4}{\cancel{1 - (v^2/c^2)}}\cancel{\left(1 - \frac{v^2}{c^2}\right)} = m^2c^4. \qquad (40\text{–}37)$$

This relation may also be written as

$$E = \sqrt{p^2c^2 + m^2c^4}. \qquad (40\text{–}38)$$

The relation of a particle's energy, momentum, and mass

We note that for *any* massless particle (a particle with $m = 0$),

$$E = pc. \qquad (40\text{–}39)$$

Light obeys this relation and thus behaves as a massless particle. There are other particles that are, to the accuracy that measurements allow, massless. The neutrino is such a particle. Because the velocity of a particle is measured by the ratio of p to E, as in Eq. (40–36), we see that *massless particles always move at the constant speed c*. There is no reference frame in which these particles can be brought to rest, or, for that matter, to any speed other than c. We know this to be true for light, but it is true for any other massless particle as well.

We also see from Eq. (40–37) that the combination $E^2 - p^2c^2$ is *invariant*. Invariant quantities have the same value in every inertial frame, just as $c^2t^2 - x^2$ does, as shown in Eq. (40–25). This can be a valuable result in the analysis of relativistic collision phenomena.

*40–8 BEYOND SPECIAL RELATIVITY

The Equivalence Principle

The equivalence principle of general relativity

Special relativity expresses the physical equivalence of all inertial reference frames. In noninertial, or accelerating, frames, another physical equivalence holds. It is expressed by the *equivalence principle*, formulated by Einstein in 1911:

> **It is not possible by experiment to distinguish between an accelerating frame and an inertial frame in a suitably chosen gravitational potential, provided that the observations take place in a small region of space and time.**

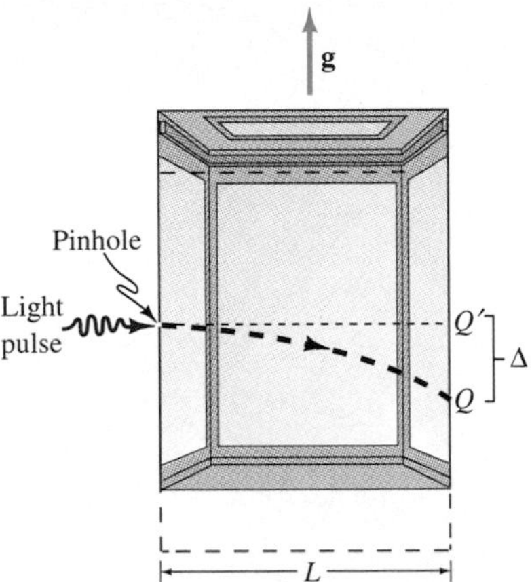

Light pulse enters elevator aimed at Q' but hits wall at Q as elevator rises

FIGURE 40–24 A pulse of light is directed into an elevator. If the elevator is at rest, the pulse would eventually arrive at point Q'. If the elevator accelerates upward with magnitude g as the light pulse crosses, an observer within the elevator would see the light pulse follow the parabolic path shown, eventually arriving at point Q on the opposite wall, a distance Δ below Q'.

A number of consequences of this principle, the principle behind *general relativity*, also known as Einstein's theory of gravitation, were described in Section 12–8. We list these again:

A. The Equality of Gravitational Mass and Inertial Mass. The *gravitational mass* m_g, the attribute of a body that appears in Newton's expression for the gravitational potential energy

$$U = -\frac{Gm_{1g}m_{2g}}{r},$$

and the *inertial mass* m_i, the attribute of a body that appears in the expression for the proportionality of force and acceleration,

$$F = m_i a,$$

must be equal. As a consequence, all bodies fall at equal rates in a given gravitational field: If a body of mass m is subject to a gravitational force due to a body of mass M, the relation $F = ma$ reads

$$\frac{Gm_gM}{r^2} = m_i a;$$

if $m_i = m_g$, the acceleration does not depend on the mass of the body. The equality of the inertial mass and gravitational mass has been verified experimentally to an accuracy of one part in 10^{11}.

B. The Gravitational Deflection of Light. If a horizontal beam of light enters a pinhole in an elevator that is accelerating upward with an acceleration g, then, in the time t that it takes for the light to cross the elevator, the horizontal level of the pinhole will have moved upward by a distance $gt^2/2$. If the width of the elevator is L, the light will hit a spot Q a distance

$$\Delta = \frac{1}{2}g\left(\frac{L}{c}\right)^2$$

below the spot Q' that is horizontally across from the pinhole (Fig. 40–24). An observer within the elevator states that the light *falls* by that amount. According to the equivalence principle, no experiment can determine whether the elevator has accelerated upward or has remained at rest within a gravitational field. Therefore, according to the equivalence principle, light must fall in a gravitational field. Note that Δ is the same distance that any other body would fall. The implications are twofold:

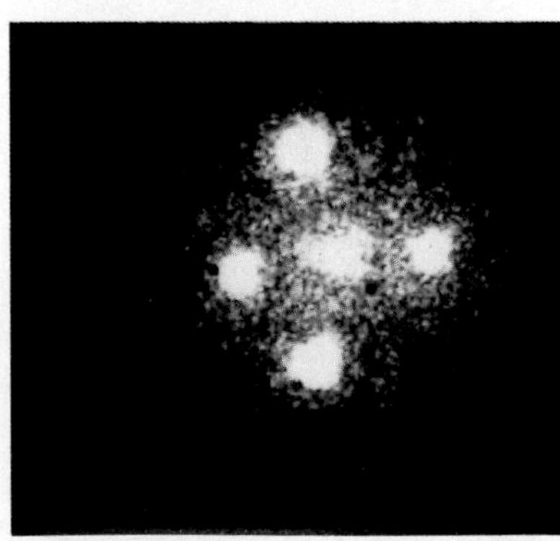

FIGURE 40–25 This astronomical image, formed by a gravitational lens, demonstrates that mass bends light. A mass (a relatively nearby galaxy) has bent the light from a quasar behind it such that four distinct images of the quasar can be observed from earth. The fuzzy spot in the center is the galaxy's core.

1. Everything that has energy falls downward with the same acceleration in a gravitational field.
2. Light is deflected as it passes stars, as we discussed in Chapter 12 (Fig. 40–25).

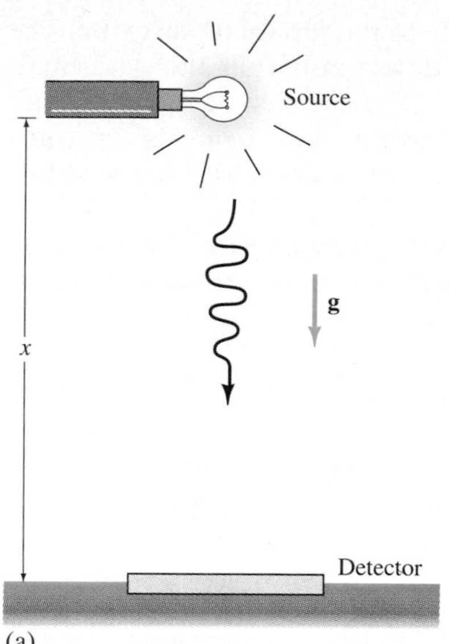

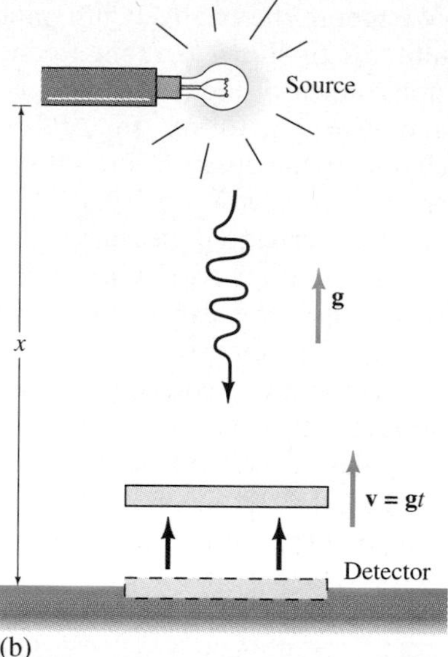

FIGURE 40–26 (a) An apparatus to measure the frequency of emitted light on earth, in the earth's gravitational field; **g** is the (downward) acceleration due to gravity. (b) The measurement of a Doppler shift due to the acceleration of the detector (as part of the whole system) is indistinguishable from what is observed when the apparatus is in a gravitational potential; **g** is the (upward) acceleration of the system.

C. The Gravitational Redshift. When light "falls," it undergoes a frequency shift. Consider a source at a height x above the ground in the presence of local gravity, emitting radiation with frequency f (Fig. 40–26a). According to the equivalence principle, the physics should be described equally well by an observer who sees the system being accelerated upward with acceleration g in empty space. If at the time of emission both the source and the detector are at rest, then according to the observer who sees the entire apparatus accelerating upward, in the time that the radiation has reached the detector ($t = x/c$), the detector will have acquired an upward velocity of magnitude $v = gt = gx/c$ (Fig. 40–26b). Thus the detector sees the radiation with a frequency f', Doppler shifted upward from the emitting frequency f. These are related by

$$\frac{f'}{f} \doteq \sqrt{\frac{1 + (v/c)}{1 - (v/c)}} \simeq 1 + \frac{v}{c} = 1 + \frac{gx}{c^2}.$$

The factor gx may be regarded as a *gravitational potential*, ϕ: We express the potential energy of a mass m in the vicinity of the earth as $mgx = m\phi$. In this way, the effect of the earth is isolated in the factor $\phi = gx$. Our result for the frequency ratio then takes the form

$$\frac{f'}{f} = 1 + \frac{\phi}{c^2}, \tag{40–40}$$

so

$$\frac{\Delta f}{f} \equiv \frac{f' - f}{f} = \frac{\phi}{c^2}. \tag{40–41}$$

Let us now apply our result to universal gravitation. The potential energy difference for a mass m and a star (of mass M and radius R) between the surface of the star and a point far from the star is $-GmM/R$. Thus we can assign to the star a corresponding gravitational potential of $-GM/R$. The frequency of light that is emitted from the star's surface and arrives at an earth-based telescope is shifted by

$$\frac{\Delta f}{f} = -\frac{GM}{Rc^2}.$$

(We ignore the relatively insignificant "fall" down to the surface of the earth.) The minus sign shows that the frequency is shifted downward. Thus the wavelength shifts upward, indicating a *gravitational redshift*. For the sun, $M = 2 \times 10^{30}$ kg and $R = 7 \times 10^8$ m, so $\Delta f/f = 2.12 \times 10^{-6}$. Recent measurements of some characteristic spectral lines of sodium in the solar spectrum have confirmed this to an accuracy of 5 percent.

In a terrestrial measurement of the gravitational redshift, carried out in 1960 by Robert Pound and Glen Rebka, light was "dropped" from a tower at Harvard University. This experiment measured a fractional shift of 3.3×10^{-15} to an accuracy of 1 percent.

When the gravitational potential is very large, the frequency shift Δf becomes large. In the extreme case that $(\Delta f/f) = 1$, the frequency of the light is shifted to zero; that is, no light can be seen. A *black hole* has been formed. This occurs when $(GM/Rc^2) > 1$. When Einstein's full theory of gravitation is taken into account, it is for the modified condition $(GM/Rc^2) > \frac{1}{2}$ that a black hole is formed.

SUMMARY

The special theory of relativity is based on the postulates that

1. The laws of physics are the same in all inertial reference frames.
2. The speed of light in empty space is the same in all inertial frames.

The first postulate generalizes the notion of Galilean invariance that we applied to the laws of mechanics. The second postulate has been justified by the result of Michelson and Morley, that there is no evidence for an absolute reference frame.

The postulates require a rethinking of the concepts of space and time. If two inertial frames F and F′ move at speed u with respect to each other, then

1. Two events that are simultaneous in frame F are not simultaneous in frame F′.
2. A time interval measured as T on a "clock" at rest in frame F is given by T' in frame F′:

$$\text{time dilation:} \qquad T' = \frac{T}{\sqrt{1-(u^2/c^2)}}. \qquad (40\text{–}10)$$

3. A length of an object at rest in frame F measured as L in frame F has length L' in frame F′:

$$\text{length contraction:} \qquad L' = L\sqrt{1-\frac{u^2}{c^2}}. \qquad (40\text{–}12)$$

4. A light source that radiates with frequency f_0 in frame F is observed to have frequency f_1 in frame F′:

$$f_1 = f_0\sqrt{\frac{1-(u/c)}{1+(u/c)}}. \qquad (40\text{–}17a)$$

5. If two objects are moving at velocities v_1 and $-v_2$, respectively, with respect to an observer, then the velocity of one object as seen in the rest frame of the other is

$$\text{law of addition of velocities:} \qquad V = \frac{v_1+v_2}{1+(v_1v_2/c^2)}. \qquad (40\text{–}19)$$

The relativistic Doppler shift for light has important astronomical implications. The measurement of the redshift of the light from a galaxy, given by Eq. (40–17), leads to a value for the speed u with which that galaxy recedes from earth. That speed is in turn related to the distance D of the galaxy by Hubble's law:

$$D = \frac{u}{H}; \qquad (40\text{–}18)$$

here H is the Hubble parameter, a number characteristic of the age of the universe.

The space–time coordinates of an event are described in reference frames F and F′ by (x, t) and (x', t'), respectively. These coordinates are related by the Lorentz transformations

$$x' = \gamma(x - ut) \qquad (40\text{–}20)$$

and

$$t' = \gamma\left(t - \frac{ux}{c^2}\right), \qquad (40\text{–}21)$$

where $\gamma \equiv 1/\sqrt{1 - (u^2/c^2)}$. An event can be specified by any two of (x, t, x', t'), and the other two coordinates can then be found from the Lorentz transformation laws, which must be supplemented by $y' = y$ and $z' = z$ when the relative motion is along the x-axis.

The relativistic momentum is defined by

$$\mathbf{p} = \frac{m\mathbf{v}}{\sqrt{1 - (v^2/c^2)}}. \qquad (40\text{–}30)$$

The general relation between energy and mass μ is $E = \mu c^2$, Eq. (40–34). The total energy of a particle is

$$E = \frac{mc^2}{\sqrt{1 - (v^2/c^2)}} \qquad (40\text{–}35)$$

and consists of a rest energy associated with the mass of the particle, $E_{\text{mass}} = mc^2$, plus the relativistic kinetic energy,

$$K = mc^2\left[\frac{1}{\sqrt{1 - (v^2/c^2)}} - 1\right]. \qquad (40\text{–}32)$$

The velocity of a particle is given in terms of the momentum and energy according to

$$\mathbf{v} = \frac{c^2\mathbf{p}}{E}. \qquad (40\text{–}36)$$

The energy and momentum are related by

$$E = \sqrt{p^2c^2 + m^2c^4}. \qquad (40\text{–}38)$$

For massless particles $E = pc$, Eq. (40–39).

Einstein extended his theory beyond inertial reference frames by proposing the equivalence principle, according to which

> It is not possible by experiment to distinguish between an accelerated frame and an inertial frame in a suitably chosen gravitational potential, provided that the observations take place in a small region of space and time.

This principle has important consequences:

1. Inertial and gravitational masses are equal, a result known to be accurate to one part in 10^{11};

2. Light falls in a gravitational field;
3. A source radiating with frequency f in a gravitational potential ϕ will be observed to radiate with frequency f' such that

$$\frac{f'}{f} = 1 + \frac{\phi}{c^2}. \tag{40–40}$$

QUESTIONS

1. Can there be such a thing as a perfectly rigid object?
2. Does the statement "moving clocks run slow" depend on the direction in which a clock is moving?
3. The rest mass of a proton is given as 937 MeV/c^2. How can a mass involve energy units?
4. As measured from the earth, what is the shortest possible travel time between earth and Alpha Centauri, the second nearest star system to us, located 4.3 ly away? Why can it not be made any shorter?
5. Suppose that the Michelson–Morley experiment were carried out over one arbitrarily short time period, much less than 1 d, and showed no sign of movement through an ether. Is this result enough to rule out the presence of an ether?
6. According to one solution proposed to make the presence of an ether consistent with the results of the Michelson–Morley experiment, there is an *ether drag*: For some reason the earth carries with it a bubble of the ether as it moves through space. Can you think of experimental consequences that could be used to rule out such an idea?
7. A golf ball is struck so well that it travels at a speed of $0.95c$ but is so durable that it is not deformed by the blow it receives. What shape does the golf ball have as it travels through the air, as measured by the golfer?
8. If a mirror recedes at speed $0.75c$ from a light source, does the image recede at speed $1.5c$ from the source?
9. Some years ago astronomers found by looking at a very distant quasar that the separation distances of certain peaks in brightness increase at a rate of 0.2 ms of arc/yr. The quasar is so far away that the separation speed translates into $v \simeq 8c$! Is this the death knell of the special theory of relativity?
10. A stick 2 m long travels toward a hole that is just short of 1 m in diameter at a speed such that $\gamma = 2$. Will the stick fall through the hole? How will things look to an observer in the rest frame of the stick?
11. Suppose that an experimenter found that some particle, such as a neutrino, travels just a bit faster than the highest-frequency radiation that has been observed so far. Would we have to give up the special theory of relativity?
12. Folklore has it that, as a teenager, Einstein worried about what would happen if somebody accelerating to a speed faster than that of light were looking in a mirror during this process. What could Einstein have been worried about?
13. Suppose that current in a wire is carried by little green men who pass negative charges along a chain from person to person while standing on positive charges, such that the wire is electrically neutral. Would the changing mechanism for the current flow change the observed behavior of a charge q that lies outside the wire? How would this look to an observer who is moving with some velocity along the wire?
14. A closed box of little mass sits on a horizontal frictionless surface. The inside walls are perfect mirrors and reflect back all radiation. A laser is inside, on the left-hand wall, and projects a very short burst of light directly at the right-hand wall. What is observed from the outside, and why?
15. According to Hubble's law, a single spectral line characteristic of a single atom is redshifted by an amount proportional to the distance of the star from earth. How do we know that the radiation of a particular color seen to come from a distant star is the redshifted radiation of a particular known spectral line?
16. The length-contraction experiment seems to imply that a meter stick accelerated to the speed of light would shrink to a point, and all the calibration markings on the meter stick would be lost. Is there something wrong with this reasoning?
17. Light falls in an accelerating elevator. But if an elevator moves upward at a constant velocity, a horizontal light beam would hit a spot *below* the horizontal projection on the opposite wall of the elevator. Does this mean that light falls in an elevator that moves at constant velocity?
18. When the supernova 1987a occurred, bursts of neutrinos—particles that are massless, according to accelerator-based experiments and to within the accuracy of the experiments—arrived at detectors at various places on the earth's surface. These neutrinos are thought to have been emitted by the supernova all at once. How could differences in the arrival times of the neutrinos be used to test whether or not neutrinos have mass?

PROBLEMS

40–1 *Is an Ether Necessary?*

1. (I) An airplane flies at an air speed of 600 mi/h. It travels east from town A to town B and returns to A without stopping. In the absence of wind, the journey takes exactly 4 h. A town C the same distance away from A is located due north of A. Suppose that a wind with ground speed

60 mi/h is blowing east to west. Calculate the times it takes for the plane to make journeys ABA and ACA.

2. (II) In one version of the Michelson–Morley experiment, light of wavelength 590 nm emitted by sodium atoms travels through a total path length of 10 m in each arm of the interferometer. To the accuracy of the apparatus, a shift of 1/40 of a fringe, no shift was seen. Estimate the greatest value possible for the speed of the earth through the ether.

40–2 *The Einstein Postulates*

3. (II) A small, powerful laser is placed on a turntable that rotates at 1200 rev/s. The laser, whose beam makes a 30° angle with the horizontal, shines on clouds 50 km away. Calculate the speed with which the light spot on the clouds moves. Does this speed violate the limitation of the speed of light? Explain.

40–4 *Time Dilation and Length Contraction*

4. (I) Two twins wave goodbye to each other. One twin, an astronaut, travels to Mars. The trip takes 3 yr in each direction, and the average speed with respect to earth is 70,000 km/h. What, approximately, will the time difference in the twin's clocks be when they are together again on earth?

5. (I) Proxima Centauri, the star nearest our own, is some 4.2 ly away. (a) If a spaceship could travel at a speed of $0.80c$, how long would it take to reach the star, according to the spaceship's pilot? (b) What would someone in the frame that moves with the spaceship measure as the distance to Proxima Centauri?

6. (I) The space shuttle orbits the earth at 17,500 mi/h in 90 min. How much time will an astronaut's atomic clock have lost during a total trip that takes 5 d?

7. (I) The diameter of our galaxy is about 10^5 ly, or 10^{21} m. Suppose that a proton moves at a speed such that $\sqrt{1-(v^2/c^2)} \simeq 10^{-7}$. (Such speeds correspond to the most energetic cosmic rays known). How long does it take the proton to cross the galaxy in (a) the galaxy's rest frame? (b) the proton's rest frame?

8. (II) A meter stick is tilted so that it makes an angle of 30° with the x-axis. How will an observer at rest in a frame F′ that moves at velocity $v = 0.80c$ in the $+x$-direction relative to the meter stick describe the stick?

9. (II) A student must complete a test in 1 h in the teacher's frame of reference F. The student puts on his rocket skates and soon is moving at a constant speed of $0.75c$ relative to the teacher. When 1 h has passed on the teacher's clock, how much time has passed on a clock that moves with the student, as measured by the teacher?

10. (II) As measured by an observer in an inertial frame, a small clock moving at a constant speed of $0.89c$ traverses a distance of 1500 m. The moving clock records 100 ticks during the passage. How many ticks pass on an identical clock at rest relative to the observer?

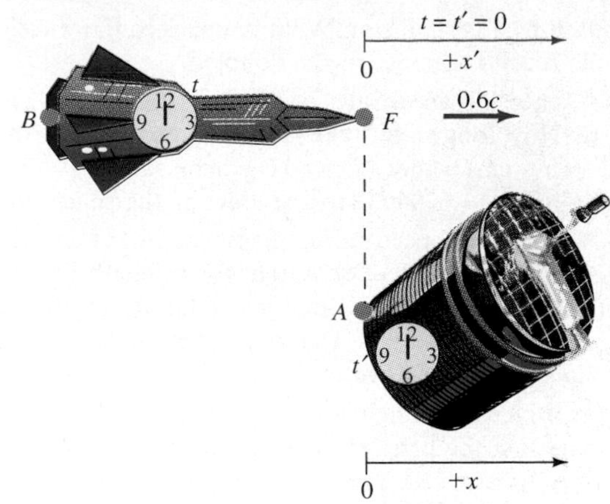

FIGURE 40–27 Problem 11.

11. (II) A spaceship of length 30 m travels at $0.6c$ past a satellite. Clocks in frame S′ of the spaceship and S of the satellite are synchronized within their respective frames of reference and are set to zero so that $t' = t = 0$ at the instant the front of the spaceship F passes point A on the satellite, located at $x' = x = 0$ (Fig. 40–27). At this time a light flashes at F. (a) What is the length of the ship as measured by an observer on the satellite? (b) What time does the observer on the satellite read from her clock when the trailing edge B of the spaceship passes her? (c) When the light flash reaches B at the rear of the spaceship, what is the reading t'_1 of a clock at B? (d) What is the reading t_1 on the clock on the satellite when, according to the observer on the satellite, the flash reaches B?

12. (III) Jessica embarks on a cosmic journey at a speed of $(24/25)c$ relative to earth. Before leaving, she tells her twin brother, Tom, who stays on earth, that she will travel outward for 25 yr of earth time, then back for another 25 yr of earth time. Tom will thus be 50 yr older when she returns. She promises to send a radio message on each of her birthdays. According to an earth-based clock, when will these messages reach Tom, and how much older than the age at which she leaves will Jessica be when she returns to earth?

13. (III) A relativistic sprinter running at speed v, near the speed of light, passes beneath a victory arch a height h above his eyes. Show that he will continue to see the arch, even though his eyes face forward, until he has run a distance $hv/[c\sqrt{1-(v^2/c^2)}] = \gamma hv/c$ *beyond* the arch. [*Hint:* Work in the rest frame of the sprinter, and think of the top of the arch as emitting pulses of light, the last of which can be seen when it travels vertically downward toward the sprinter.]

40–5 *The Relativistic Doppler Shift*

14. (I) The sodium doublet refers to light waves emitted by sodium in a closely spaced pair of frequencies. The wavelengths of this doublet are at 589.0 nm and 589.6 nm. Suppose that the lower-wavelength member of this doublet is Doppler redshifted to a wavelength of 593.5 nm in the light

emitted by a certain star. What happens to the wavelength of the second member of the doublet?

15. (I) A spaceship accelerates at a rate of 0.1 m/s^2 away from earth. How long will it take (as measured in the earth's reference system) before a green beacon on earth ($\lambda = 500$ nm) looks red ($\lambda = 600$ nm) to the crew of the spaceship?

16. (I) A particular spectral line measured in the emission of light by the star Alpha Centauri has wavelength $\lambda = 396.820$ nm. That same line measured in the laboratory has wavelength $\lambda = 396.849$ nm. Determine the radial velocity of Alpha Centauri relative to the earth.

17. (II) A driver was caught running a red light. His defense is that he saw the light as green, as a result of the Doppler shift. He is arrested. What for? Estimate the seriousness of his transgression.

18. (II) Yellow light at 587.6 nm, characteristic of helium, is found to be redshifted as it is observed in a certain star; the wavelength is measured to be 603.5 nm. (a) How fast is the star receding from earth? (b) Use Hubble's law to estimate the distance of the star from earth.

19. (II) For a particular quasar, $(\lambda - \lambda_0)/\lambda_0 = 1.95$, where λ_0 is the wavelength of the radiation emitted as measured in the quasar's rest frame. What is the speed of the quasar relative to the earth, assuming that it is traveling in a radial direction away from earth? How far away is the quasar, according to Hubble's law?

20. (II) A source radiates light with a frequency of 10^{15} Hz. The signal is reflected by a mirror that is moving at speed 100 km/s away from the source. What is the frequency of the reflected radiation, as observed at the source?

21. (II) The equation $\lambda/\lambda_0 = \sqrt{(1+\beta)/(1-\beta)}$, where $\beta = v/c$ and v is the speed of a source that is moving away from an observer or of an observer who is moving away from the source, takes a simple form if v is small compared to c. Show that if $\lambda = \lambda_0(1 + x)$, then for small β, $x \simeq \beta$.

22. (III) During the journey described in Problem 12, Tom sends a radio message to Jessica on each of his birthdays, for a total of 50 messages. With what interval in her rest frame does Jessica receive these messages during the outward part of the journey? during the return trip? Use this information to calculate how much Jessica ages during her trip, according to an earth-based clock.

23. (III) A source emits pulses with a frequency f_0. A spaceship moving at speed v_1 away from the source will receive a redshifted frequency f_1. Suppose that the spaceship immediately reemits the signals with the frequency f_1. A second spaceship, moving at speed v_2 relative to the first spaceship and in the same direction, will receive the signals with a redshifted frequency f_2. (a) Calculate f_1 and f_2. (b) If we were to eliminate the first spaceship, we could view f_2 as the redshifted frequency received by the second spaceship, which moves at some speed v relative to the source. Show that if both $v_1 \ll c$ and $v_2 \ll c$, then $v = v_1 + v_2$, as expected from the ordinary rules that govern relative motion. (c) Calculate v for arbitrary values of v_1 and v_2. This result is the relativistic law of addition of velocities, which differs from $v = v_1 + v_2$ when v_1 and v_2 are not very small compared with c.

40–6 *The Lorentz Transformations*

24. (I) Measurements of distant galaxies show that all galaxies are receding from one another at a speed proportional to their intergalactic distances. Suppose that we see galaxy 1 move away from us at a speed of $0.4c$ along the South Pole, and galaxy 2, equally far away, move away from us at the same speed along the North Pole (Fig. 40–28). What would an observer in galaxy 1 measure for the speed with which galaxy 2 moves away from him?

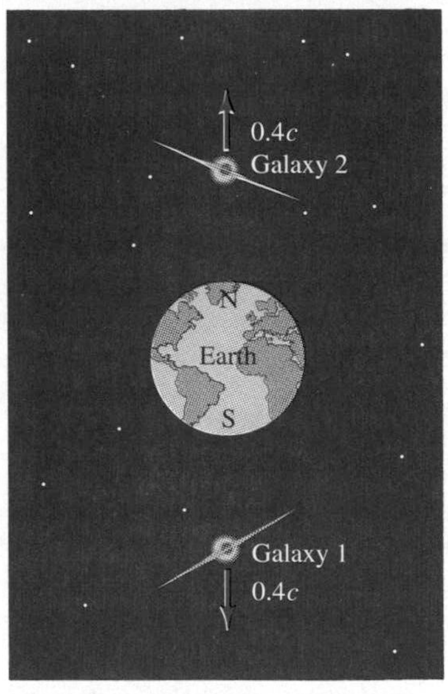

FIGURE 40–28 Problem 24.

25. (I) Events A and B are simultaneous in frame S and are 1 km apart on a line that defines the x-axis. A series of spaceships all pass at the same speed in the $+x$-direction, and they have synchronized their clocks so that together they make up a moving frame S′. They time events A and B to be separated by 10^{-6} s. What is the speed of the spaceships? How far apart in space do they measure the two events to be?

26. (II) In a given reference frame, event 1 occurs at time $t_1 = 0$ s and position $x_1 = 0$ m, while event 2 occurs at $t_2 = 5 \times 10^{-8}$ s and $x_2 = 10$ m. Is there a second frame in which these events could be at the same position but different times? If so, specify its motion with respect to the first frame. If not, what is the frame in which the events have the least possible separation in distance? [*Hint:* Use invariants.]

27. (II) A new Klingon battleship races at a top speed of $0.2c$ away from the planet XG4T. The starship *Enterprise* follows at a speed of $0.25c$ relative to the Klingon ship. With what speed does the *Enterprise* appear to catch up with the Klingon ship, according to an observer on the planet?

28. (II) An observer in frame S measures two events to occur at the same point in space and separated by a time interval

Δt. Show that in every other inertial frame, these events are separated by a larger time interval.

29. (II) A particle in frame F has velocity $v_x\mathbf{i} + v_y\mathbf{j}$. What is the velocity seen by an observer at rest in frame F′, which is moving at velocity $u\mathbf{i}$ relative to frame F?

30. (II) You shine light that moves at speed c/n through a medium of index of refraction n (Fig. 40–29). Suppose that the medium moves at speed u relative to you, parallel to the direction of the light. What is the speed of light in the medium, as seen by you? [The result for $(u/c) \ll 1$ gives a result that is different from $(c/n) + u$, and it was first obtained by Augustin Fresnel in 1818. The measurements confirming the result were made by Hippolyte Fizeau in 1851.]

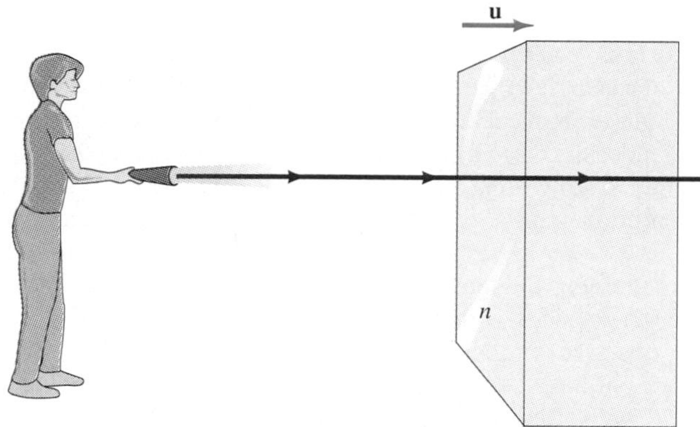

FIGURE 40–29 Problem 30.

31. (II) Use the Lorentz transformations to show that, as in Eq. (40–6), the location of a wave front for a plane wave emitted at $t = t' = 0$ from the joint origin of reference frames F and F′ is given by $x^2 - c^2t^2 = 0$ and by $x'^2 - c^2t'^2 = 0$.

32. (III) The electric field obeys the wave equation

$$\frac{\partial^2 E}{\partial t^2} - \frac{c^2(\partial^2 E)}{\partial x^2} = 0$$

when electromagnetic waves propagate along the x-axis. What is the form of the equation for a wave seen in a Lorentz-transformed frame that moves at speed u along the x-axis? (*Hint:* Use the Lorentz transformation laws to obtain expressions for $\partial/\partial t$ and $\partial/\partial x$ in terms of $\partial/\partial t'$ and $\partial/\partial x'$.)

40–7 *Momentum and Energy in Special Relativity*

33. (I) *Estimate* the mass lost when 1 million tons of TNT explodes. Assume that each chemical reaction between individual molecules involves 10 eV of energy.

34. (I) Humans generate energy at a rate of some 10^{13} W worldwide. (The United States, with less than 10 percent of the world's population, uses about 25 percent of the energy.) At what rate is mass being lost due to relativistic effects?

35. (II) Energy from the sun arrives at the earth (above the atmosphere) at a rate of about 1400 W/m^2. How fast is the sun losing mass due to energy radiation?

36. (II) What value of v/c must a particle of rest mass m have in order for its momentum to have magnitude $p = 2mv$?

37. (II) An electron accelerated from the Stanford Linear Accelerator in California has a total energy of 50 GeV. How much of this is kinetic energy? What is the momentum of the electron? What is its speed?

38. (II) A proton accelerated at Fermi National Laboratory in Illinois has a momentum of 746 GeV/c. (a) What is the proton's velocity? (b) the proton's kinetic energy?

39. (II) An electron and its antiparticle of identical mass, the positron, annihilate each other and produce two photons. Both the electron and the positron were initially at rest. What are the energy and momentum of each photon?

40. (II) A photon is the quantum unit of light. It has an energy $E = hf$, where h is Planck's constant and f is the frequency of the light. Show that when a photon is absorbed by a free electron, without anything else occurring, energy and momentum cannot be conserved simultaneously.

41. (II) You analyze the track of a particle in a photographic plate placed in a magnetic field, and find that the total energy of the particle is 870 MeV. The bending in the magnetic field gives information about momentum, and you learn that the particle's momentum is $p = 720$ MeV/c. What is the mass of the particle?

42. (II) In a generalization of Example 40–7, suppose that a ^{8}Be nucleus is moving in the x-direction with 10 MeV of kinetic energy when it decays into two alpha particles. Both alpha particles move off along the x-axis. What are their kinetic energies? [*Hint:* Note that $Q \ll M(^4\text{He})c^2$.]

43. (II) The lifetime of a particle called the neutral pion, π^0, is 0.9×10^{-16} s in the particle's rest frame. With what energy would one π^0 have to be produced so that its decay point is distinguishable from its production point in a photographic plate? Assume that a 1-mm separation is required for a measurement. The pion mass corresponds to $mc^2 =$ 135 MeV.

44. (II) The decay products of a nucleus of mass M^* include another nucleus of mass M ($M < M^*$) and radiation. If the decaying nucleus is at rest, what is the kinetic energy of the remnant nucleus of mass M? [*Hint:* Use the fact that radiation of energy E carries momentum E/c].

45. (III) Experiments have shown that for the quantum of radiation (a photon), the energy and momentum are related by $E = pc$, corresponding to a particle with mass $m = 0$. Suppose that in the observation of a supernova 170,000 ly away, the first bursts of photons with an energy range of $E = 10$ eV to 10^4 eV arrive within 10^{-8} s of each other. What limits does this set on the mass of a photon? [*Hint:* Use the fact that mc^2 is small, so $E = pc + (m^2c^3/2p)$ is a good approximation.]

46. (III) Calculate an expression for the force **F** as defined by $m(d/dt)(\gamma\mathbf{u})$, and show that the force and the acceleration $d\mathbf{u}/dt$ do not necessarily point in the same direction.

*40–8 *Beyond Special Relativity*

47. (II) A neutron star has a mass of 4×10^{30} kg and a radius of 10 km. What is the gravitational redshift of radiation emitted with a frequency of 10^{19} Hz from the star's surface?

48. (III) A clock on a disk rotating with angular speed ω, when placed at a distance R from the center of the disk, experiences an acceleration toward the center of the disk. What gravitational potential will an observer at rest relative to the clock assume that he or she is in, by the equivalence principle? Will the clock be slow or fast relative to a clock at the center of the disk?

General Problems

49. (II) Electrons and positrons (the antiparticles of electrons) of energy 50 GeV travel in opposite directions around a storage ring, a device in which the particles are held in circular orbits. What is the speed of each particle in the rest frame of the other?

50. (II) The Stanford Linear Accelerator accelerates electrons to a total energy of 50 GeV. How long does a meter stick at rest appear to a hypothetical observer at rest with respect to one such electron?

51. (II) In 1990 an SR-71 Blackbird reconnaissance airplane on its way to retirement at the Smithsonian Air and Space Museum set several speed records. The plane averaged 2153 mi/h during the 2300-mi trip from Los Angeles to Washington D.C., and 2242 mi/h during the 311-mi trip from St. Louis to Cincinnati. What would have been the difference in time elapsed for the two record-setting segments between an atomic clock placed in the airplane and another atomic clock on the ground?

52. (II) A cosmic ray is approaching the earth from outer space. A hypothetical observer in a frame that moves with the cosmic ray measures the earth as a flattened ball whose thickness is $\frac{6}{10}$ of its diameter. (a) With what speed is the cosmic ray approaching the earth? (b) The cosmic ray is identified as a proton, with mass m and mass energy given by $mc^2 = 1$ GeV. What is the energy of the approaching proton, as seen from earth?

53. (II) A new Klingon battleship has a proper length of 217 m and travels at speeds of $0.20c$ with respect to its home planet. The Klingons prepare to battle the *Enterprise*, which is moving at the same speed with respect to the same planet. If the Klingons are heading straight at the *Enterprise*, what is the length of the Klingon ship as measured by Captain Kirk?

54. (II) A particle of mass M is at rest. It decays into two identical particles, each of mass m, with $2m < M$. (a) If one of the two decay particles moves north with a momentum of magnitude p, what is the momentum of the other particle? (b) Use energy conservation to find p.

55. (II) In a quantum mechanical model, a proton and an antiproton annihilate each other and produce a pair of photons, light quanta whose frequency is related to their energy by the relation $E = hf$, where h is Planck's constant ($h \simeq 6.63 \times 10^{-34}$ J·s). The proton and the antiproton are nearly at rest when they annihilate. Find the frequencies of the emitted photons. What are these frequencies if the proton and the antiproton are approaching each other in a head-on collision in which each particle has a kinetic energy of 500 MeV? For both protons and antiprotons, $mc^2 \simeq 938$ MeV.

56. (II) According to Chapter 19, the distribution function for the z-component of the velocity of a gas at temperature T is

$$G(v_z) \propto e^{-(mv_z^2/2kT)},$$

where k is Boltzmann's constant and m is the mass of one gas molecule. If a molecule at rest emits a spectral line, light at a characteristic frequency of f_0, then what is the distribution of frequencies of the light given off by the gas of such molecules when heated to temperature T? Assume that you are looking along the z-direction, and take into account only motion in this direction. The effect described here is known as *Doppler broadening* of a spectral line. It is a tool for determining the temperature of stars and interstellar gases.

57. (II) Particles with energies as high as 10^{18} eV have been observed. Suppose that one of those particles collides with a photon of cosmic background radiation of wavelength $\lambda = 10^{-3}$ m. After a head-on collision, what will the final wavelength of the photon be? [*Hint:* You must keep the electron mass in the relativistic expression for the electron energy, E, even though $E \gg m_e c^2$. You may, however, make the approximation $\sqrt{(pc)^2 + (m_e c^2)^2} \simeq pc + (m_e c^2)^2/(2pc)$ in this case.]

58. (II) A charged pion (a particle of mass about 140 MeV/c^2) is ejected from a nuclear collision with a kinetic energy of 200 MeV. Pi mesons have a half-life (the time over which half of a given collection will decay radioactively) of about 1.5×10^{-8} s. Calculate (a) the pi meson's speed and (b) momentum. (c) How far will a collection of pi mesons travel before half of them decay?

59. (II) The *rapidity* V of a moving body is defined by $\tanh(V/c) = v/c$, where v is the body's relativistic speed. An observer in frame S′, moving at speed u in the $+x$-direction with respect to a frame S, measures a body to have speed v along the x-axis. Show that an observer in frame S measures the body to have rapidity W, given by $W = U + V$. Here U is the rapidity of frame S′ with respect to frame S. The rapidity thus adds like a Galilean velocity.

60. (III) A proton moves with a momentum of magnitude p in the $+x$-direction. It strikes a second proton, which is at rest. Three protons and one antiproton result from the collision. The four particles in the final state remain together; that is, they have no motion relative to one another. Use energy and momentum conservations to find p. For both protons and antiprotons, $mc^2 \simeq 938$ MeV.

Conferees in the 1927 Solvay Conference, including many of the founders of quantum mechanics. Front row, left to right: I. Langmuir, M. Planck, M. Curie, H. A. Lorentz, A. Einstein, P. Langevin, C. E. Guye, C. T. R. Wilson, O. W. Richardson. Second row, left to right: P. Debye, M. Knudsen, W. L. Bragg, H. A. Kramers, P. A. M. Dirac, A. H. Compton, L. V. de Broglie, M. Born, N. Bohr. Standing, left to right: A. Piccard, E. Henriot, P. Ehrenfest, E. Herzen, T. De Donder, E. Schrödinger, E. Verschaffelt, W. Pauli, W. Heisenberg, R. H. Fowler, L. Brillouin.

QUANTUM PHYSICS

It is fair to say that in the history of scientific achievement, the discovery of quantum mechanics, which explains the set of phenomena known as quantum physics, ranks with the scientific revolution wrought by Newton. The quantum physical phenomena described by quantum mechanics cannot be understood in terms of classical physics. Quantum mechanics describes the behavior of matter on a microscopic scale, far beyond what our senses can perceive. Great leaps of imagination were required to penetrate these fundamental laws of nature. The results to be discussed in this chapter include the wave properties of matter, the particle nature of radiation, and the fact that position and momentum cannot be specified simultaneously with perfect precision. According to classical physics, given a definite set of initial conditions and knowledge of the forces acting, you are able to determine the behavior of a physical system unambiguously. In contrast, quantum mechanics only predicts the probabilities of physical events and implies that this is the only information we can have about these events. In this chapter we shall describe some of the phenomena that classical physics cannot deal with and shall show how they are treated in quantum mechanics.

41–1 THE WAVE NATURE OF MATTER

One of the fundamental revelations of quantum mechanics is that *all matter exhibits wave properties.* Indeed, experiments have verified that particles such as electrons show typical wavelike properties, including interference. This possibility was conjectured by Louis de Broglie in 1924, shortly before the proper formulation of quantum mechanics (Fig. 41–1). It is possible to test the wave properties of electrons and other matter because quantum mechanics makes a clear statement about the wavelength associated with what would classically be called a particle. *When subject to experiments that test wavelike properties, particles with momentum p act like waves with a* **de Broglie wavelength** λ, *given by*

The wavelength associated with any particle of momentum p

$$\lambda = \frac{h}{p}. \tag{41–1}$$

Here $h = 6.626 \times 10^{-34}\ \text{J} \cdot \text{s}$ is *Planck's constant.* Another formulation is the following: The wave number $k = 2\pi/\lambda$ associated with a particle of momentum p is given by

The wave number and wavelength of classical waves are described in Chapter 14.

$$k = \frac{2\pi}{\lambda} = \frac{2\pi p}{h} = \frac{p}{\hbar}. \tag{41–2}$$

Here the commonly occurring combination $h/2\pi$ is written as $\hbar$:

$$\hbar \equiv \frac{h}{2\pi} = 1.05 \times 10^{-34}\ \text{J} \cdot \text{s}. \tag{41–3}$$

The smallness of h implies that the wave character of matter is evident only on a very small scale. A dust particle of mass 10^{-6} g traveling at a speed of only 10 m/s has a wavelength of

$$\lambda = \frac{h}{p} = \frac{h}{mv} = \frac{6.63 \times 10^{-34}\ \text{J} \cdot \text{s}}{(10^{-9}\ \text{kg})(10\ \text{m/s})} \simeq 6.6 \times 10^{-26}\ \text{m}.$$

This wavelength is so small that no possible diffraction experiment could ever detect it. Dust particles, baseballs, and airplanes reveal very little of their wavelike aspects. On the atomic scale, however, things are quite different. Electrons ($m_e = 9.1 \times 10^{-31}$ kg) moving at a speed of 10^6 m/s, typical of electron speeds in atoms, have a wavelength

$$\lambda = \frac{6.63 \times 10^{-34}\ \text{J} \cdot \text{s}}{(9.1 \times 10^{-31}\ \text{kg})(10^6\ \text{m/s})} = 0.7\ \text{nm}.$$

This wavelength is of the same magnitude as interatomic spacing in matter and is thus subject to tests by diffraction experiments like those described for X-rays.

In the quantum mechanical view, any particle is said to have a dual wave-particle nature. The situation is somewhat analogous to the duality of geometrical and physical optics. Experiments involving the straight-line propagation of light or its reflection and refraction tell us nothing about the wave properties of light. Conversely, interference, diffraction, and the penetration of electromagnetic radiation through materials form a second set of phenomena in which the wave properties of light emerge.

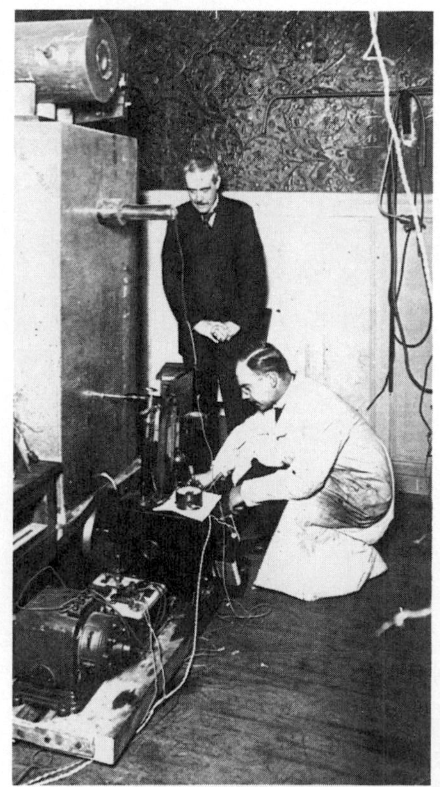

FIGURE 41–1 Louis de Broglie (standing).

EXAMPLE 41–1 What is the de Broglie wavelength of a neutron (mass $m = 1.6 \times 10^{-27}$ kg) with a speed $v = 1500$ m/s? (If v is taken as the rms speed of a gas of neutrons, the corresponding equilibrium temperature is around 35 K.)

SOLUTION: The de Broglie wavelength is

$$\lambda = \frac{h}{mv} = \frac{6.63 \times 10^{-34}\ \text{J}\cdot\text{s}}{(1.6 \times 10^{-27}\ \text{kg})(1.5 \times 10^{3}\ \text{m/s})} = 0.28\ \text{nm}.$$

This value is comparable to the typical spacing between atoms in a crystal.

Experimental Evidence for the Wavelike Behavior of Matter

The experiments that confirmed de Broglie's conjecture were first carried out in 1927 by Clinton J. Davisson and Lester H. Germer, and independently by George Paget Thomson. They found that when electrons are scattered by a crystal, certain scattering directions are preferred, as expected from constructive interference. We found the interference condition for waves, *Bragg's law*, in Chapter 39. When waves of wavelength λ are reflected from a succession of crystal planes separated by a distance d, there will be constructive interference for angles θ that satisfy Bragg's law:

$$k(2d)\sin\theta = 2\pi n, \qquad (41\text{–}4)$$

where n is an integer and $k = 2\pi/\lambda$ is the wave number. Equation (41–4) can be solved for the wavelength to get

$$\lambda = \frac{2d}{n}\sin\theta. \qquad (41\text{–}5)$$

In the Davisson–Germer experiment, the spacing between the scattering planes in the crystal was determined by X-ray diffraction experiments to be $d = 0.091$ nm (Fig. 41–2). Davisson and Germer then scattered electrons with incident energy of 86.4×10^{-19} J (54 eV) and observed a diffraction maximum—wavelike behavior—at 65°. The kinetic energy of the electron corresponds to a momentum p:

$$p = \sqrt{2m_e E} = \sqrt{2(9.1 \times 10^{-31}\ \text{kg})(86.4 \times 10^{-19}\ \text{J})} = 39.7 \times 10^{-25}\ \text{kg}\cdot\text{m/s}.$$

Now that we know the momentum, we can find the angle of constructive interference from Eq. (41–5),

$$\sin\theta = \frac{n\lambda}{2d} = \frac{nh}{2dp} = \frac{n(6.63 \times 10^{-34}\ \text{J}\cdot\text{s})}{2(9.1 \times 10^{-11}\ \text{m})(39.7 \times 10^{-25}\ \text{kg}\cdot\text{m/s})} = 0.92n.$$

For $n = 1$ this yields $\theta = 66°$, in good agreement with the measured value.

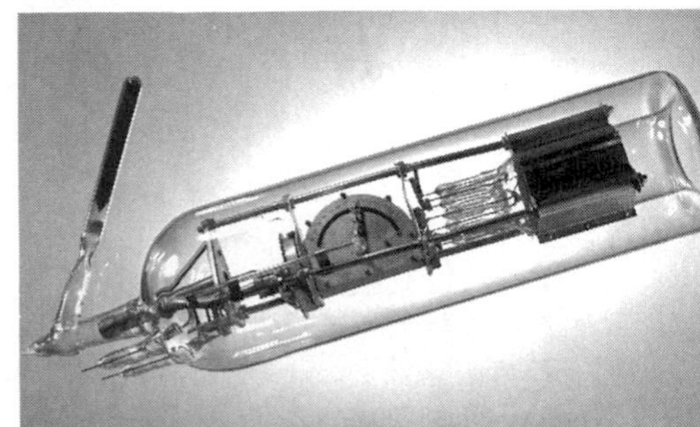

(a)

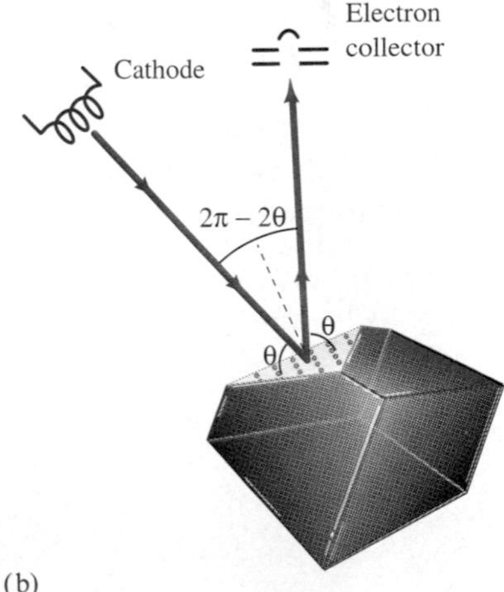

(b)

FIGURE 41–2 (a) The electron tube used in the Davisson–Germer electron diffraction experiment. (b) Experimental setup for the experiment. Electrons from a cathode strike a surface of a nickel crystal and are scattered to an electron collector.

EXAMPLE 41–2 At what angles do diffraction peaks occur for electrons of kinetic energy 120 eV incident on a crystal whose scattering planes are 0.12 nm apart?

SOLUTION: We must apply the Bragg condition. To do so, we need the de Broglie wavelength, and hence the momentum, which we calculate from $E = p^2/2m = 120$ eV. We first convert the energy to joules: 120 eV = (120 eV)(1.6×10^{-19} J/eV) = 19×10^{-17} J. Then the wavelength is given by

$$\lambda = \frac{h}{p} = \frac{h}{\sqrt{2m_e E}} = \frac{6.63 \times 10^{-34}\ \text{J}\cdot\text{s}}{\sqrt{2(9.1 \times 10^{-31}\ \text{kg})(1.9 \times 10^{-17}\ \text{J})}}$$

$$= 1.1 \times 10^{-10}\ \text{m} = 0.11\ \text{nm}.$$

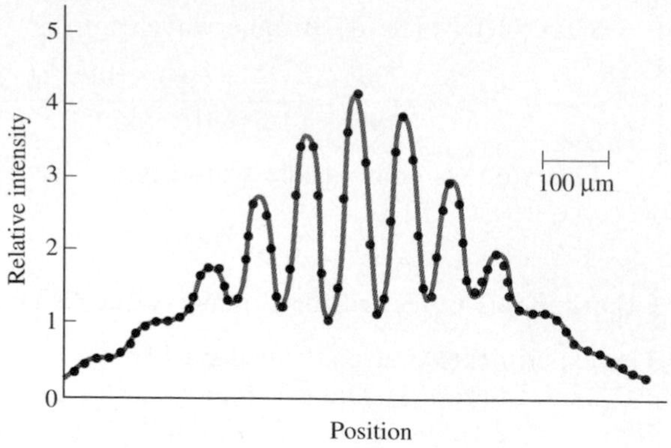

FIGURE 41–3 Diffraction pattern of neutrons passed through double slits approximately 20 μm wide and separated by 104 μm (after A. Zeilinger et al., "Single- and Double-Slit Diffraction of Neutrons," *Reviews of Modern Physics*, **60**, p. 1067 [Oct. 1988]).

With a crystal plane separation of $d = 0.12$ nm, the angles for constructive interference are

$$\sin\theta = \frac{n\lambda}{2d} = \frac{n(0.11\ \cancel{\text{nm}})}{2(0.12\ \cancel{\text{nm}})} = 0.47n.$$

Thus there will be a diffraction peak at $\theta = 28°$ for $n = 1$, and another at $70°$ for $n = 2$.

Diffraction experiments have been performed with a variety of particles. In Example 41–1, we calculated the wavelength of neutrons that move at a certain speed; the wavelengths of these neutrons satisfy the conditions for substantial diffractive effects in scattering from crystals. The diffraction of neutrons by crystal surfaces is of practical importance in the study of the surfaces of solids. Neutrons lend themselves well to such studies, because they can be slowed down by collisions in hydrogenous materials such as paraffin or water, and slower neutrons have longer wavelengths. Figure 41–3 shows the results of an experiment in which neutrons of de Broglie wavelength 2 nm were incident on a screen with two slits approximately 100 μm apart.

The simplicity of the effect should not mask the extraordinary nature of these results. In some respects electrons and neutrons behave like classical particles: They move according to Newton's second law, $F = ma$, when they are subject to forces. An ordinary water wave does not move as a classical particle. Yet electrons and neutrons deflected by regular structures produce interference patterns, just as a water wave would!

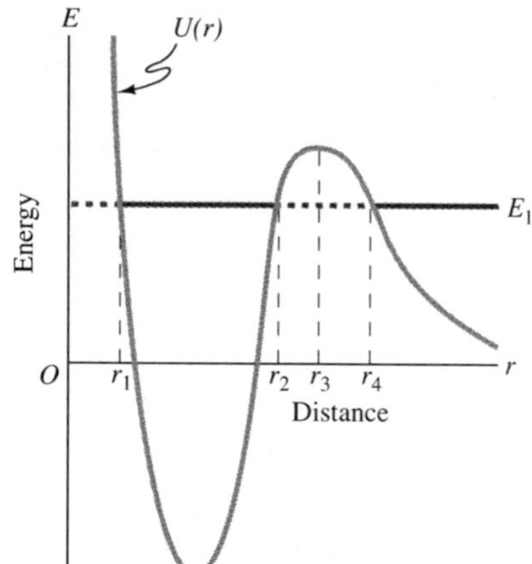

FIGURE 41–4 Energy diagram in which the potential energy forms a barrier for a classical particle with less energy than the potential energy maximum at distance r_3.

Tunneling

Potential energy barriers are discussed in Chapter 7.

There is experimental evidence that particles such as electrons, neutrons, and alpha particles (^{4}He nuclei) are able to pass through a potential energy barrier from one region in space to another. A classical particle with energy E_1 that starts within the region $r_1 < r < r_2$ in the energy diagram of Fig. 41–4 will always remain there. That is because the total energy of a particle in the region $r_2 < r < r_4$ would be less than the particle's potential energy, so its kinetic energy would be negative, which is impossible. However, quantum mechanics allows a real particle that starts in the region $r_1 < r < r_2$ to appear in the region $r > r_4$. An example of *tunneling*, or *barrier penetration*, is provided by nuclear fusion. An important application is the scanning tunneling microscope. Tunneling cannot be explained in terms of the particle aspects of matter. The wave properties of matter, however, do explain tunneling.

Tunneling was introduced in Section 7–5.

To see how wave phenomena can lead to tunneling, let us consider an

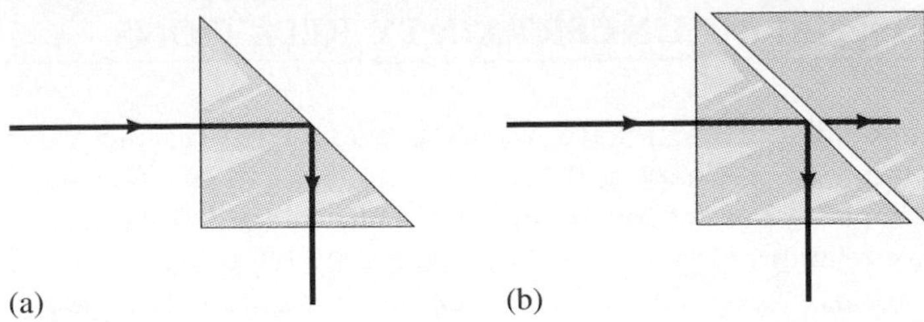

FIGURE 41–5 (a) Light incident from glass onto a glass–air interface that meets the proper geometric conditions will undergo total internal reflection. (b) If a second piece of glass is very close to the first piece, some light may cross the air gap.

analogy between quantum tunneling and the phenomenon of total internal reflection in optics. When light traveling through glass reaches a glass–air surface at an angle that exceeds the critical angle, the reflection is total (Fig. 41–5a). If, however, the air forms just a thin strip between two regions of glass, then the reflection is not total (Fig. 41–5b). The reason is as follows: Application of Maxwell's equations shows that inside the air gap, the electromagnetic field associated with the light wave decreases exponentially rather than propagating sinusoidally. If the gap is large or if there is no second piece of glass, the field will die off completely (Fig. 41–6a). But if the gap is no larger than a few wavelengths of light, then the exponential decrease of the field is not total. A reduced field remains at the second interface, and starting from there the fields can again propagate as sinusoidal waves in the second piece of glass, although greatly reduced in amplitude (Fig. 41–6b). *The light waves have tunneled through the air gap* with a calculable intensity.†

Similarly, in quantum mechanical tunneling it is possible to calculate the reduction of the intensity of quantum mechanical matter waves that tunnel through a potential barrier. This is a calculation of *the fraction of the number of particles that tunnel through the barrier*. The fraction is very small under ordinary circumstances, which is why the phenomenon is not an intuitive one. A rough estimate of this fraction shows that it depends primarily on the barrier width a and the difference $U - E$ of the barrier height and the energy:

$$\text{fraction of particles getting through} \simeq e^{[-(2/\hbar)a\sqrt{2m(\langle U\rangle - E)}]}. \qquad (41\text{–}6)$$

In this equation $\langle U\rangle$ is the average height of the potential barrier. Barrier penetration is favored when the width of the barrier is small and when the energy E is not much below the peak of the barrier. The smallness of $\hbar$ guarantees that this is ordinarily a very small fraction and classically quite negligible.

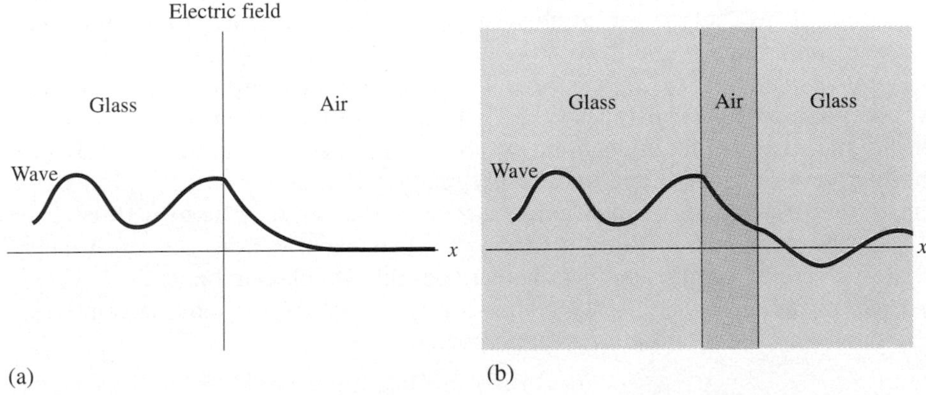

FIGURE 41–6 (a) Electric field at a glass–air interface for the case of total internal reflection. (b) Electric field at a narrow air gap between two pieces of glass, which shows the tunneling of an electromagnetic field through the air gap.

† A beautiful demonstration of the exponential falloff of light intensity is described in D. D. Coon, "Counting Photons in the Optical Barrier-Penetration Experiment," *Amer. J. Phys.* **34**, 240 (1966).

41–2 THE HEISENBERG UNCERTAINTY RELATIONS

We first saw an uncertainty relation in Chapter 8; in Chapter 15 we learned that the uncertainty relations are a general property of wave pulses.

Because electrons (and all particles) have wavelike characteristics, the wave properties already discussed in Chapters 14 and 15 apply to them. The most important of these properties, stated by Werner Heisenberg, are known as the **Heisenberg uncertainty relations** (Fig. 41–7). The uncertainty relations are stated most often in two forms. According to the *position–momentum uncertainty relation*,

> **Any attempt to localize a particle within a distance Δx necessarily limits a simultaneous determination of the x-component of that particle's momentum to an uncertainty of Δp_x, where these uncertainties are related by**

The position–momentum uncertainty relation

$$\Delta x\ \Delta p_x > \hbar. \tag{41–7}$$

There are no limitations on the simultaneous determination of x and p_y, for example. The second uncertainty relation is the *time–energy uncertainty relation*:

> **If an energy measurement is to be carried out in a time Δt, then the accuracy ΔE with which the energy can be measured during this time interval is limited by the relationship**

The time–energy uncertainty relation

$$\Delta E\ \Delta t > \hbar. \tag{41–8}$$

The value of Planck's constant sets the scale for which the uncertainty relations are important. The small size of Planck's constant ($h \simeq 6 \times 10^{-34}$ J · s, or $\hbar \simeq 10^{-34}$ J · s) guarantees that the uncertainty principle is important only on an atomic scale. For example, if we know the location of a dust particle to an accuracy of 10^{-6} m, then the uncertainty principle constrains our simultaneous knowledge of its momentum to an accuracy of 10^{-28} kg · m/s. This momentum uncertainty is so tiny that it is overwhelmed by the other experimental uncertainties associated with imperfection in a measuring apparatus. Thus the uncertainty principle has no practical role in the world of cars, baseballs, or even dust particles.

FIGURE 41–7 Werner Heisenberg.

However, when we deal with electrons in an atom (or with protons in a nucleus), the situation is quite different. The mass of an electron is about 10^{-30} kg, and its speed in an atom is in the range of 10^6 m/s. The momentum of an electron in an atom is then about 10^{-24} kg · m/s. The diameter of an atom is on the order of 10^{-10} m. If we try to pin down the location of an atomic electron to within 10 percent of the atom's size ($\Delta x \simeq 10^{-11}$ m), then the momentum becomes uncertain to about 10^{-23} kg · m/s, *10 times the value of the electron's momentum in its classical atomic orbit*. The momentum becomes so uncertain that we are not even sure that the electron will stay within the atom! The product of an atomic size and the momentum of an atomic electron is about 10^{-34} J · s, a number very close to $\hbar$. The uncertainty relation is so important for atoms and nuclei that Newtonian momentum is a concept that must be used with care.

In the remainder of this section, we shall explore how the uncertainty relations resolve certain conflicts (described below) inherent in a dual wave–particle model. The uncertainty relations are also useful for making numerical estimates when quantum effects are important.

The Double-Slit Dilemma

Diffraction experiments of the type described in Section 41–1 have been carried out with electrons, protons, neutrons, and a variety of molecular beams. The experiments have verified that these particles possess the wave properties predicted by quantum theory. This finding raises conflicts, or conceptual difficulties, that are well illustrated by the following experiment. Consider a source of electrons. The electrons are emitted at some rate, so many per second. The stream of electrons impinges on a screen after they have passed through two slits (Fig. 41–8). If we think of electrons as particles, we expect that each electron will go through one slit or the other. In fact, if slit A is open for 5 min while slit B is closed, and then B is opened while A is closed for 5 min, then the electrons will arrive at the screen in two well-defined locations (Fig. 41–8a). But if the slits are open simultaneously for a total of 10 min, an interference pattern very much like that shown in Fig. 38–5 for light forms on the screen (Fig. 41–8b).

The two sets of results seem incompatible. Suppose that an observer could tell which slit the electron was about to pass through. The observer could then, for the duration of the passage, close the other slit. The process could be repeated for each electron. The system of slits, together with the alert observer, should (and do) give the same result as the opening of one slit at a time. *Somehow, changing the experiment so that we know which of the two slits the electron passes through destroys the interference pattern.* Quantum mechanics, if it is to be a useful theory, must be able to predict the interference pattern in the double-slit case and the two-peak distribution when the slits are open one at a time or when an observer sees which of the two slits the electron passes through. We shall see that the position–momentum uncertainty relation resolves the conflict.

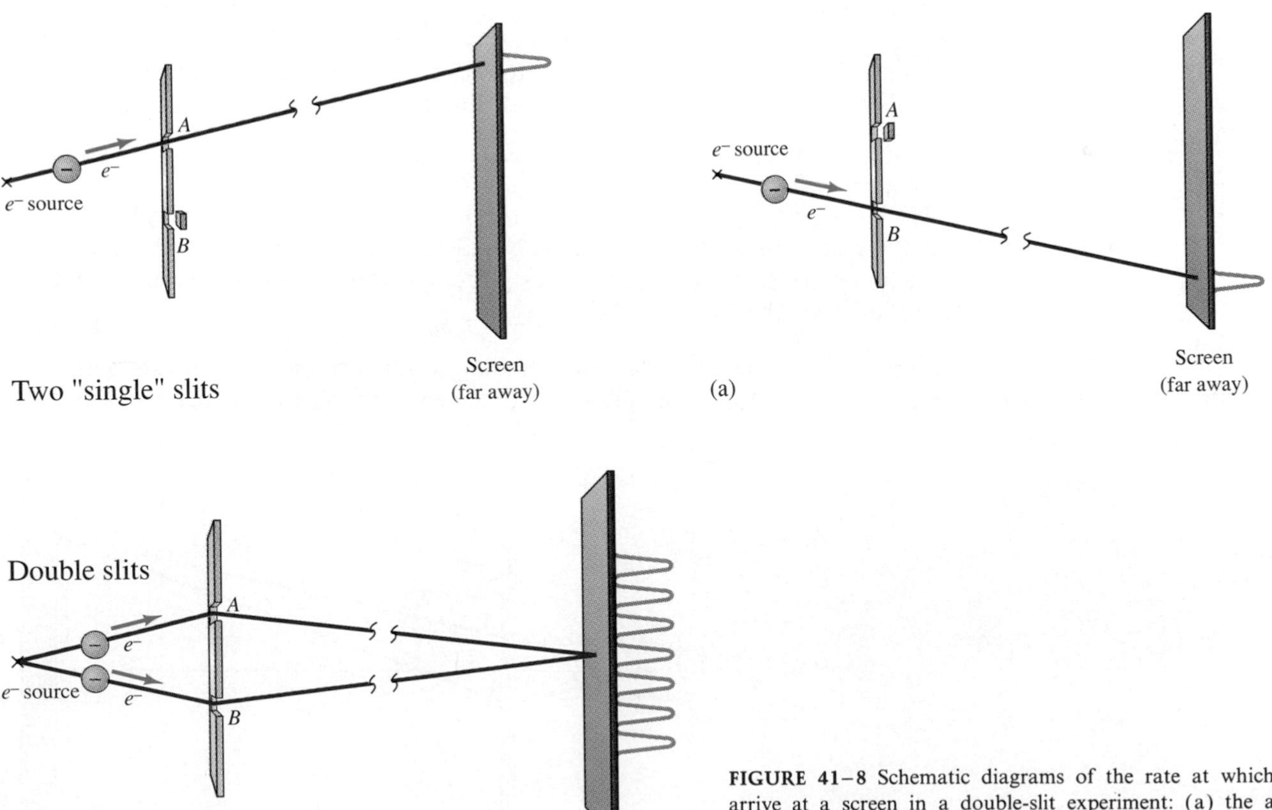

FIGURE 41–8 Schematic diagrams of the rate at which electrons arrive at a screen in a double-slit experiment: (a) the arrival-rate distribution, single peaks, when one slit is open at a time; (b) the arrival-rate distribution, an interference pattern, when both slits are open.

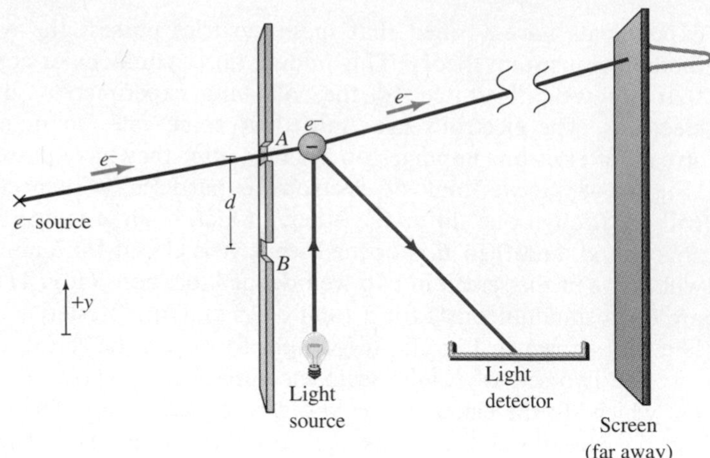

FIGURE 41–9 Schematic diagram of a monitor for a double-slit experiment designed to detect which slit an electron passes through. One mechanism: Shine light toward the slits and signal the passage of the electron by the pattern of reflected light.

Resolution of the Double-Slit Dilemma

In our double-slit electron experiment, an observer, to be able to tell which slit a given electron passes through, must use a monitor of some kind. We can now show *why*, if a monitor enables us to learn which slit the electron passes through, the interference pattern is destroyed. For this monitor to determine which slit an electron passes through, it must be able to locate the electron's y-coordinate near the wall with the slits to an accuracy $\Delta y < d/2$, where d is the separation distance between the slits and y is taken to be the direction across the slits (Fig. 41–9). Our monitor must interact with the electrons to "see" where they are going. For example, the monitor may consist of a beam of light (like the beam of a searchlight) that reflects off an electron. Any such device would transfer momentum to the electron in a direction parallel to the screen (the y-direction in Fig. 41–9). If this momentum transfer is Δp_y, then the uncertainty relation states that

$$\Delta p_y > \frac{\hbar}{\Delta y} > \frac{2\hbar}{d}. \tag{41–9}$$

This much momentum imparted to the electron is sufficient to wipe out the interference pattern. To see this, recall that for a slit separation d, the angles at which constructive interference takes place, corresponding to interference maxima, are given by

$$d \sin \theta_n = n\lambda, \tag{41–10}$$

where θ_n is the angle that the beam leading to the nth maximum makes with the x-axis (Fig. 41–10). If the distance to the viewing screen is D, then the distance

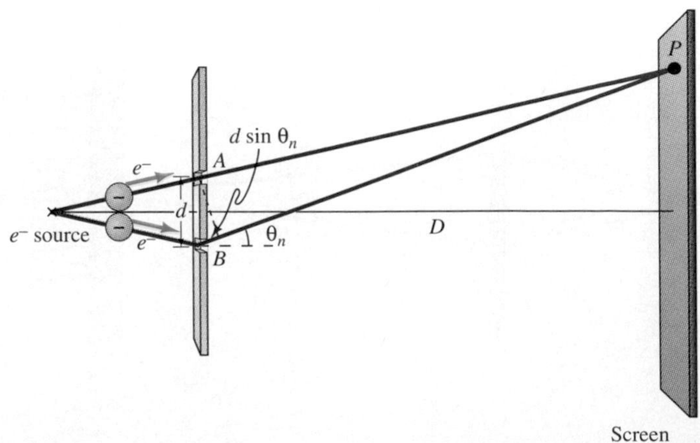

FIGURE 41–10 Geometry of pathlength differences for a double-slit experiment.

between adjacent maxima on that screen (which is also a separation in the y-direction) is

$$\Delta y_{\max} = D \sin \theta_{n+1} - D \sin \theta_n = \frac{(n+1)D\lambda}{d} - \frac{nD\lambda}{d} = \frac{D\lambda}{d}. \qquad (41\text{–}11)$$

Now suppose that the monitor gives the electron a sideways (y-direction) "kick." The electron thereby acquires an additional amount of momentum Δp_y in the $+y$-direction. This will change the angle of deflection by $\Delta p_y/p$ and thus its location at the screen by $D\,\Delta p_y/p$. It follows from the uncertainty relation that this displacement of the point of arrival at the screen, P, is

$$D\frac{\Delta p_y}{p} > \frac{D(2\hbar/d)}{p} = \frac{2D/d}{k} = \frac{D\lambda}{\pi d}. \qquad (41\text{–}12)$$

This displacement is unspecified, except that we know that it is comparable to the separation between the maxima, Eq. (41–11). Thus the interference pattern is wiped out, not just shifted.

What we have shown is that there is no paradox. A pure double-slit experiment and an experiment with an additional monitor used to determine the electrons' paths are different experiments, and different results are predicted for them. A measurement that depends on the particle nature of an electron ("Which slit does it pass through?") must, at a minimum, disturb the system just enough to remove evidence of the wavelike nature of the electron. Quite generally, the uncertainty relation is guaranteed to remove any contradictions between the particle and wave aspects of any physical system. This result is often restated as, *Any attempt to determine whether an electron (or other physical system) is "really" particlelike, or "really" wavelike, disturbs the system so much that no determination can be made.*

The Uncertainty Relations and Numerical Estimates

The uncertainty relations may be used to estimate the smallest possible energy a particle under the influence of a given force may have. This information determines the **ground-state** (or lowest) energies of atoms and molecules and is thus of great importance. Consider a particle with a potential energy $U(x)$. We choose a coordinate system such that the minimum of the potential energy is located at $x = 0$, and we change the potential energy by a constant, as we are free to do, so that $U(0) = 0$. Because the total energy of the particle is given by $E = (p^2/2m) + U(x)$, the energy is lowest when both the kinetic energy and the potential energy are lowest; that is, when $p = 0$ and $x = 0$. Because we have chosen $U(0) = 0$, $E = 0$ would be the minimum value in that case. However, quantum mechanics does not permit a perfect localization in both p and x. If we suppose that the particle is at $x = 0$ with an uncertainty Δx, then we impose an uncertainty in the momentum p of magnitude larger than $\hbar/\Delta x$. This means that p^2 cannot be zero but is unspecified to an accuracy $(\Delta p)^2 > (\hbar/\Delta x)^2$. Because p^2 cannot be zero, the energy cannot be zero, either. The energy must be at least $(\Delta p)^2/2m$, and because Δp depends on Δx, the minimum energy is also a function of Δx. We can find the value of Δx for which the energy has its lowest value, *but that minimum energy value cannot be zero.*

To understand just how the position–momentum uncertainty relation determines a minimum energy, take, for example, a particle subject to the influence of a spring. The potential energy is $U(x) = m\omega^2 x^2/2$, where ω is the angular frequency of the classical motion. The energy of the particle is given by

$$E = \frac{p^2}{2m} + \frac{m\omega^2 x^2}{2}. \qquad (41\text{–}13)$$

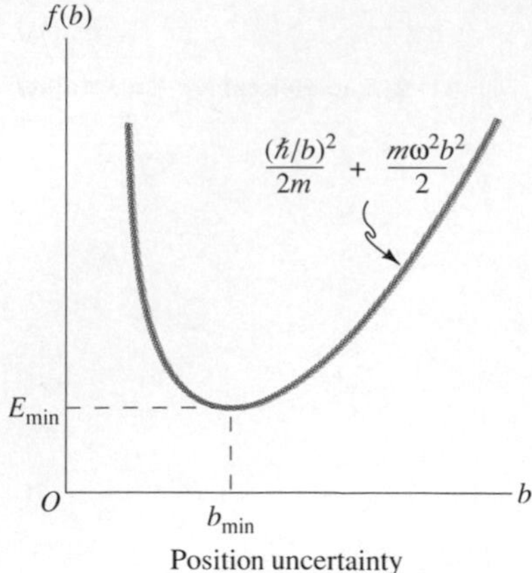

FIGURE 41–11 The right-hand side of Eq. (41–14), labeled $f(b)$, as a function of the position uncertainty b. The minimum energy is the minimum value of the curve $f(b)$.

If the position of the particle is undetermined to an accuracy b so that $\Delta x = b$, then the uncertainty in the momentum is $\Delta p > \hbar/b$. Thus the lowest value of the energy must obey the inequality

$$E > \frac{(\hbar/b)^2}{2m} + \frac{m\omega^2 b^2}{2}. \tag{41–14}$$

The right-hand side of Eq. (41–14) is plotted in Fig. 41–11. We see that it has a minimum as a function of b. Let us call the right-hand side of Eq. (41–14) $f(b)$ and find the minimum value of $f(b)$. We do so from the condition that, at the minimum, the slope of $f(b)$ is zero: $df/db = 0$. Thus

$$\frac{df}{db} = -\frac{\hbar^2}{mb^3} + m\omega^2 b = 0.$$

When we solve this equation for b^2, we get $b^2_{min} = \hbar/m\omega$. Substituting this value into the expression for $f(b)$, we obtain the minimum value of $f(b)$, which, according to Eq. (41–14), equals the minimum value of E, E_{min}:

$$E_{min} = \frac{\hbar^2}{2m(\hbar/m\omega)} + \frac{1}{2}m\omega^2\left(\frac{\hbar}{m\omega}\right) = \hbar\omega. \tag{41–15}$$

The zero-point energy

This is an estimate of the quantum mechanical *zero-point energy*. A full quantum mechanical calculation yields $E_{min} = \hbar\omega/2$. The minimum energy is never zero, so the particle on the end of the spring can never be brought completely to rest. This is certainly a nonclassical result!

EXAMPLE 41–3 Use the position–momentum uncertainty relation to estimate the lowest energy of a particle of mass m in a one-dimensional box of width L.

SOLUTION: In this case Δx is specified from the beginning: If all that is known about the particle is that it is somewhere in the box, then $\Delta x = L$. In turn $\Delta p > \hbar/L$. Thus the particle momentum is undetermined to an accuracy Δp, and the lowest energy must satisfy $E_{min} = (\Delta p)^2/2m$, or

$$E_{min} = \frac{\hbar^2}{2mL^2}.$$

EXAMPLE 41–4 Use the position–momentum uncertainty relation to estimate the lowest energy of an electron, mass m, moving in an attractive Coulomb potential of the form $U(r) = -e^2/4\pi\epsilon_0 r$. Express your result in electron-volts.

SOLUTION: Zero electric potential is at $r = \infty$, and the total energy of the particle is given by

$$E = \frac{p^2}{2m} - \frac{e^2}{4\pi\epsilon_0 r}.$$

Classically, this describes a particle in the class of orbits described in Chapter 12, and the lowest energy is −infinity, corresponding to the particle at the center of force, $r = 0$. As in our previous discussion of the use of the position–momentum uncertainty relation, we express the minimum energy in terms of the uncertainty in radius, Δr. We do not need to worry about how Δr is related to Δx, Δy, and Δz, because all the uncertainties are estimates. For simplicity, we shall let $\Delta r = R$. As the distance from the origin of an electron increases, the potential energy increases from −infinity. Thus when the uncertainty in the position is of size R, the potential energy will be as big as $-e^2/4\pi\epsilon_0 R$. The uncertainty in momentum is larger than $\hbar/R$, and therefore the kinetic energy is larger than $(\hbar/R)^2/2m$. The lowest value of the energy obeys the inequality

$$E > \frac{\hbar^2}{2mR^2} - \frac{e^2}{4\pi\epsilon_0 R}.$$

We call the function on the right-hand side of this expression $f(R)$ and find its minimum value. When the function $f(R)$ is a minimum,

$$\frac{df}{dR} = -\frac{\hbar^2}{mR^3} + \frac{e^2}{4\pi\epsilon_0 R^2} = 0.$$

This equation has the solution

$$R = \frac{4\pi\epsilon_0 \hbar^2}{me^2}.$$

When this value of R is inserted into $f(R)$, we obtain the minimum value of f—that is, the minimum energy:

$$E_{\text{min}} = \frac{\hbar^2}{2m}\frac{(me^2)^2}{(4\pi\epsilon_0\hbar^2)^2} - \frac{e^2}{4\pi\epsilon_0}\frac{me^2}{4\pi\epsilon_0\hbar^2} = -\frac{1}{2}m\left(\frac{e^2}{4\pi\epsilon_0\hbar}\right)^2.$$

The minus sign of the energy indicates that the particle is bound to the attractive Coulomb potential. To obtain a numerical value for E_{min}, we use $m = m_e = 0.9 \times 10^{-30}$ kg, $\hbar = 1.05 \times 10^{-34}$ J · s, and $e^2/4\pi\epsilon_0 = (1.6 \times 10^{-19}\text{ C})^2(8.99 \times 10^9\text{ m/F}) = 2.3 \times 10^{-28}\text{ C}^2 \cdot \text{m/F}$. The magnitude of E_{min} is then found to be 2.2×10^{-18} J. With 1 eV $= 1.6 \times 10^{-19}$ J, the energy in electron-volts is

$$E_{\text{min}} = -13.7\text{ eV}.$$

In Example 41–4 we use the aforementioned technique of determining minimum energies from the position–momentum uncertainty relation to estimate the minimum energy of an electron subject to the Coulomb force of a proton. We find a result that is extremely close to the correct value. In general, such estimates are not quite so accurate, but they can be counted on to give the correct dependence on parameters, here $\hbar$, m, and $e^2/4\pi\epsilon_0$. The principal result is that quantum mechanics provides a minimum energy even for systems that have no such classical restrictions. The fact that there is a minimum energy accounts for the stability of all atoms.

41–3 THE PARTICLE NATURE OF RADIATION

FIGURE 41–12 Max Planck.

Blackbody Radiation

Quantum mechanics treats matter and radiation on the same footing. Just as particles exhibit wave properties, so do waves exhibit particle properties. Quantum mechanics began with Max Planck's fit (1900) of the experimental data on *blackbody radiation* (Fig. 41–12). When electromagnetic radiation in any cavity is in thermodynamic equilibrium at a temperature T, the energy density of that radiation can be measured by making a small hole in the cavity and studying the intensity of the radiation emitted as a function of frequency. Using a classical application of the equipartition of energy (see Chapter 19), Lord Rayleigh and

Blackbody radiation was first discussed in Chapter 17.

James Jeans predicted that the energy density $u(f, T)$ should have the form

$$u(f, T) = \frac{8\pi f^2}{c^3} kT.$$

This result agrees with experiment for low frequencies but disagrees badly for high frequencies. Planck found that by introducing the constant h (now called Planck's constant), he could fit the observed energy density over the full range of measured frequencies with the formula given by Eq. (17–13),

$$u(f, T) = \frac{8\pi h}{c^3} \frac{f^3}{e^{hf/kT} - 1}.$$

The energy-density function has the following meaning: *The electromagnetic energy in a cavity of unit volume, for radiation with frequencies between f and $f + df$, is $u(f, T)\,df$.* Figure 17–18b shows the agreement between measured values of $u(f, T)$ and the Planck formula for a temperature of 2.7 K (see Section 17–5).

Planck's result, which was an empirical fit to data, could be derived only with a new assumption, formulated by Albert Einstein in 1905. The assumption was that *electromagnetic radiation consists of* ***quanta****, or identical, indivisible units, each carrying energy hf, where f is the frequency of the radiation.* In other words, the electromagnetic radiation comes in small "bundles" of energy. The quantum of energy is related to the frequency of the wave of radiation by

The energy of a photon is related to its frequency by $E = hf$.

$$E = hf. \tag{41–16}$$

We shall soon discuss experiments that show that quanta of radiation behave like particles. These particles have come to be called **photons**. Because the momentum and energy of any particle that travels at the speed of light are related by $p = E/c$, a photon of energy $E = hf$ carries momentum of magnitude

$$p = \frac{E}{c} = \frac{hf}{c}. \tag{41–17}$$

Note that for radiation,

$$\lambda = \frac{c}{f} = \frac{hc}{hf} = \frac{hc}{E} = \frac{h}{p}. \tag{41–18}$$

It was this formula that was later adopted for matter by de Broglie in his daring conjecture concerning the wave properties of matter [Eq. (41–1)].

EXAMPLE 41–5 An ordinary bright star easily visible to the naked eye emits radiation such that the intensity at the surface of the earth is $I = 1.6 \times 10^{-9}\ \mathrm{W/m^2}$ at a wavelength of 560 nm. Estimate the rate at which photons enter the night-adapted eye from such a star.

SOLUTION: Here we must convert the wavelength of the radiation into frequency and then into energy. The number of photons per square meter per second can then be found. The only estimate necessary is the area of the pupil for the night-adapted eye. We shall assume that the pupil is circular, with a diameter of 0.5 cm.

A wavelength λ corresponds to a frequency $f = c/\lambda$. The energy of each photon is thus $E = hf = hc/\lambda$. The intensity $I = NE$, where N is the number of photons striking the earth per square meter per second. Thus

$$N = \frac{I}{E} = \frac{I\lambda}{hc} = \frac{(1.6 \times 10^{-9}\ \cancel{\mathrm{J}}/\mathrm{m^2} \cdot \mathrm{s})(5.6 \times 10^{-9}\ \cancel{\mathrm{m}})}{(6.64 \times 10^{-34}\ \cancel{\mathrm{J}} \cdot \cancel{\mathrm{s}})(3 \times 10^8\ \cancel{\mathrm{m}}/\cancel{\mathrm{s}})}$$
$$= 0.44 \times 10^{10}\ \mathrm{photons/m^2 \cdot s}.$$

The area of the pupil is estimated to be $A = \pi r^2 = \pi(2.5 \times 10^{-3}\ \text{m})^2 = 2 \times 10^{-5}\ \text{m}^2$, so the number of photons that enter the eye in 1 s is

$$NA = (0.44 \times 10^{10}\ \text{photons}/\cancel{\text{m}^2} \cdot \text{s})(2 \times 10^{-5}\ \cancel{\text{m}^2}) = 0.9 \times 10^5\ \text{photons/s}.$$

The human eye is an extremely sensitive photon detector; it can register as few as 5 photons/s.

The Photoelectric Effect

The mechanism of blackbody radiation is so complicated that it does not provide a direct demonstration of the particle nature of radiation. Further support for the quantum nature of radiation came from the work of Albert Einstein, who in 1905 used it to explain the *photoelectric effect*. The photoelectric effect was discovered by Heinrich Hertz in 1887 in experiments shown schematically in Fig. 41–13. The 1905 data that defined the photoelectric effect were as follows:

1. When a polished metal plate is exposed to electromagnetic radiation, it may emit electrons but not positive ions. These electrons are sometimes termed *photoelectrons*.
2. Electrons will be emitted only if the frequency of the incident light exceeds a threshold value—that is, $f > f_0$. The value of f_0 may vary with the particular metal.
3. The magnitude of the emitted current of electrons is proportional to the intensity of the light source but does not depend on the frequency.
4. The energy of the emitted electrons is independent of the intensity of the light source but varies linearly with the frequency of the incident light (Fig. 41–14).
5. Subsequent to 1905, experiments showed that to an accuracy of 10^{-9} s, there is no measurable time delay between the arrival of the radiation and the appearance of the electron current.

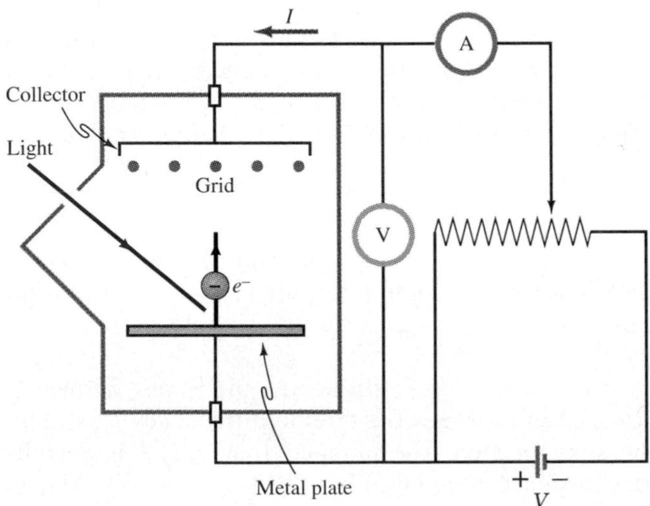

FIGURE 41–13 Schematic diagram of an experimental setup for measuring the photoelectric effect. Light strikes a metal plate in an evacuated chamber. The electron's current is measured by a collector, and the kinetic energy is determined by the grid voltage needed to slow the electrons down to rest.

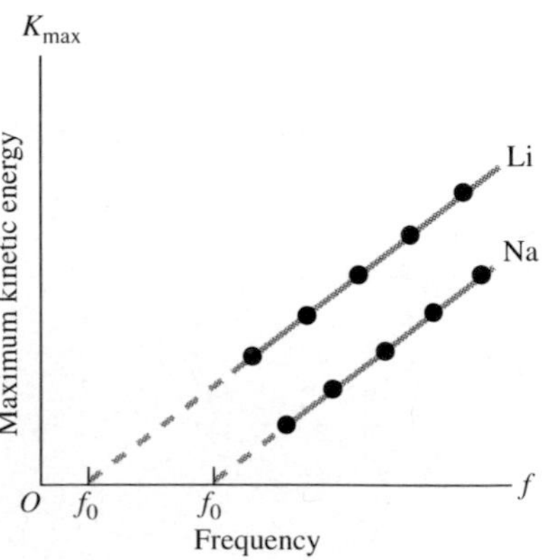

FIGURE 41–14 Data for the photoelectric effect, showing the kinetic energy of the emitted electrons (of lithium, Li, and sodium, Na) as a function of the light frequency. Note the linear relationship and the presence of a minimum frequency f_0.

The mere fact that electrons are emitted from metals subjected to electromagnetic radiation can be understood without invoking quantum ideas. Metals contain free electrons. Because these electrons do not leak out of the metal freely, it is reasonable to expect that a minimum of energy must be deposited in the metal to liberate electrons. In classical electromagnetic theory, the energy delivered to the metal by radiation is proportional to the square of the electric field, **E**; that is, to the intensity of the incoming radiation. We would therefore expect the energy carried off by the electrons to be proportional to the intensity; for example, a doubling of the intensity would double the number of electrons emitted with a given kinetic energy. What is not comprehensible in the classical view is that electrons are emitted even when the incident radiation is of very low intensity, that the energy of the emitted electrons depends linearly on the radiation frequency, and that there is a frequency threshold. Classically, energy should be delivered for all radiation frequencies. In addition, we would expect that, with low-intensity radiation, the energy required to liberate a certain number of electrons would be collected over some time, and that there would be a time delay (that increases with decreasing intensity) before the electrons would appear. Only features 1 and 3 of the experiment make any sense from the classical view.

In 1905 Einstein explained these phenomena by postulating that electrons are emitted because individual electrons absorb photons. The photons correspond to radiation with frequency f and carry energy $E = hf$. If there is a minimum energy W required to liberate an electron, then no electrons will be emitted when hf is less than W. When hf exceeds W, the excess energy can go into kinetic energy of the emitted electrons:

$$\tfrac{1}{2}mv^2 = hf - W. \tag{41–19}$$

The quantity W is a kind of potential energy that must be acquired before the electron can be liberated, and it is called the *work function*. It is a characteristic of the particular metal that emits electrons. It takes one photon to liberate one electron. Therefore, the proportionality of the current of emitted electrons to the intensity of the radiation is understandable, because the intensity is proportional to the number of photons in the electromagnetic wave. An electron absorbs a photon instantaneously, so the lack of time delay is also explained. The first accurate experiments on the photoelectric effect were carried out in 1916 by Robert Millikan, himself a disbeliever of Einstein's theory. Millikan's experiments unequivocally confirmed Eq. (41–19) and thus the quantum explanation of the photoelectric effect.

EXAMPLE 41–6 The largest wavelength of light that will induce a photoelectric effect in potassium is 564 nm. Calculate the work function for potassium, in electron-volts.

SOLUTION: The largest wavelength, or, equivalently, the lowest frequency, to induce a photoelectric effect expresses the threshold frequency f_0, and we need f_0 to calculate the work function. The minimum frequency f_0 is given by $f_0 = c/\lambda_{max}$. The work function is then given by

$$W = hf_0 = \frac{hc}{\lambda_{max}} = \frac{(6.63 \times 10^{-34}\ \text{J}\cdot\text{s})(3.00 \times 10^8\ \text{m/s})}{5.64 \times 10^{-7}\ \text{m}} = 3.53 \times 10^{-19}\ \text{J}$$

$$= (3.53 \times 10^{-19}\ \text{J})\frac{1\ \text{eV}}{1.60 \times 10^{-19}\ \text{J}} = 2.20\ \text{eV}.$$

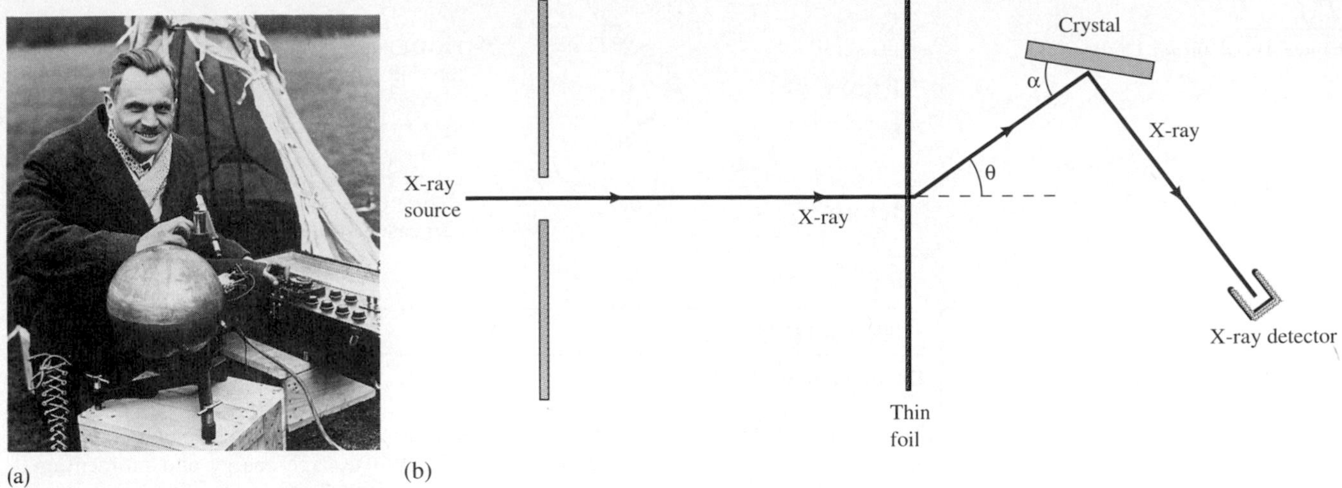

FIGURE 41–15 (a) Arthur Compton. (b) Schematic diagram of a setup for Compton's experiment. The scattered X-rays are diffracted by a crystal, with the angle α used to determine the wavelength of the scattered radiation.

The Compton Effect

Evidence for the particle properties of photons even more compelling than the photoelectric effect came from experiments carried out by Arthur Compton in 1922 (Fig. 41–15a). Compton sent X-rays through thin metallic foils (Fig. 41–15b) and discovered that the scattered X-rays emerged with one of two wavelengths. One component emerged with the same wavelength as the incident radiation. The other component emerged with a longer wavelength. This result was in contrast to what was expected from classical radiation theory, in which the electrons absorb radiation and reradiate it as dipole radiation without any change in wavelength. The experiments showed that for the second component, the wavelength varied with the scattering angle of the X-ray (Fig. 41–16). The dependence of the increased wavelength on angle fit the formula

$$\lambda' - \lambda = \frac{h}{mc}(1 - \cos\theta), \tag{41–20}$$

where λ is the wavelength of the incident X-rays and λ' that of the scattered X-rays, and m is the electron mass. The presence of h indicates that this effect must be explained by quantum mechanics, and the independence of the result on the metal used in the foils indicated to Compton that it is a property of the electrons in the metal and not of the metal's crystal structure.

Compton discovered that Eq. (41–20) can be derived by treating the photon as a particle of energy hf and momentum hf/c. The collision of this particle with an electron at rest can be treated just like a two-body collision in mechanics, by taking into account momentum conservation and energy conservation (Fig. 41–17). The initial configuration is a photon projectile incident on an electron at rest, and the final configuration has a scattered photon and an electron with momentum p. It is necessary to take the electron's energy in the relativistic form $E_e = \sqrt{p^2c^2 + m^2c^4}$. The energy of the scattered photon differs from that of the incident photon, just as the energy of a scattered billiard ball would. Because the energy of the photon is different, so is its wavelength. The calculation shown in the box "Deriving Equation (41–20)" shows that Eq. (41–20) correctly emerges. The quantity h/mc, called the *Compton wavelength* of the electron, has the dimensions of length and magnitude $h/mc = 2.4 \times 10^{-12}$ m.

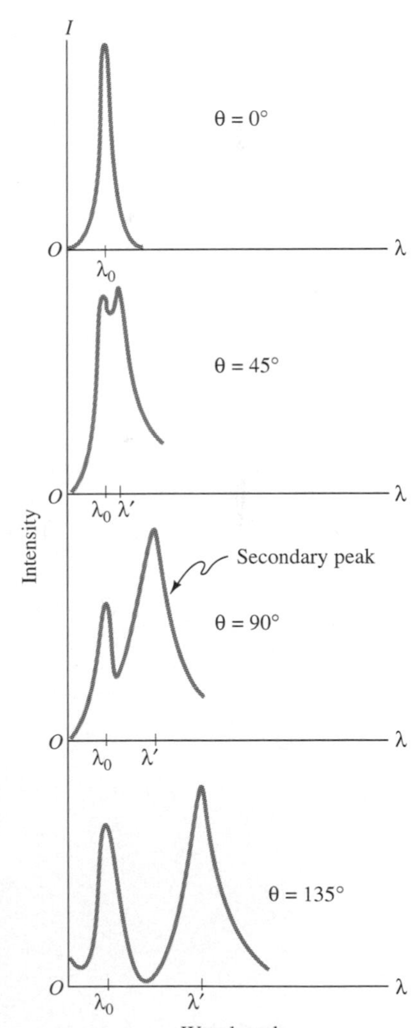

FIGURE 41–16 Experimental data for Compton's experiment. The secondary peak, due to X-ray scattering by free electrons, becomes more pronounced as the scattering angle increases.

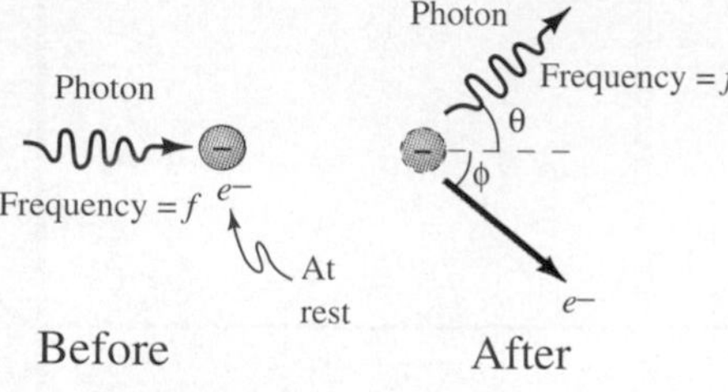

FIGURE 41–17 Kinematics for calculating the shift in frequency in the Compton effect. Light scatters from an electron as though light were a particle (photon). As in the ordinary collisions of particles, the photon's energy changes when it scatters, and by the laws of quantum mechanics, its frequency also changes to a new, calculable value.

***DERIVATION**

Deriving Equation (41–20)

To derive Eq. (41–20), we formulate the equations of conservation of momentum and energy in the collision described in Fig. 41–17. We use the energy and momentum labels of that figure. The momentum-conservation equations are

$$\text{in the } x\text{-direction:} \quad p = p' \cos\theta + p'_e \cos\phi,$$

$$\text{in the } y\text{-direction:} \quad 0 = p' \sin\theta - p'_e \sin\phi.$$

We rewrite these equations in the form

$$p - p' \cos\theta = p'_e \cos\phi,$$

$$p' \sin\theta = p'_e \sin\phi.$$

We can combine these equations by squaring both sides and adding them. After we use the trigonometric relation $\sin^2 x + \cos^2 x = 1$ for both θ and ϕ, we find that

$$p^2 - 2pp' \cos\theta + p'^2 = p'^2_e. \tag{B1–1}$$

The energy-conservation equation is

$$E + mc^2 = E' + E'_e,$$

where m is the electron mass. We substitute $E = pc$, $E' = p'c$, and $E'_e = \sqrt{p'^2_e c^2 + m^2c^4}$ and rearrange the equation to the form

$$pc - p'c + mc^2 = \sqrt{p'^2_e c^2 + m^2 c^4}.$$

We now square both sides of this equation to eliminate the square root; after doing so, we can cancel the factor c^2:

$$p^2 + p'^2 + \cancel{m^2c^2} - 2pp' + 2mpc - 2mp'c = p'^2_e + \cancel{m^2c^2}. \tag{B1–2}$$

Finally we subtract Eq. (B1–2) from Eq. (B1–1):

$$2pp'(1 - \cos\theta) - 2mpc + 2mp'c = 0.$$

We can rearrange this result as

$$\frac{1}{p'} - \frac{1}{p} = \frac{1}{mc}(1 - \cos\theta).$$

When we substitute $p = h/\lambda$ and $p' = h/\lambda'$, we get Eq. (41–20).

The particle nature of radiation manifests itself in experiments designed to study that particle nature. Experiments that probe the wave character of radiation, such as interference experiments, confirm the wave character of radiation, even at extremely high frequencies. In fact, X-ray diffraction by crystals first identified X-rays as part of the electromagnetic spectrum.

41–4 QUANTUM MECHANICS AND PROBABILITY

The phenomena we have discussed in this chapter are radically different from what our intuition would tell us. The differences are most pronounced for an electron beam that passes through a double slit, the experiment discussed in Section 41–2. Let us reconsider this experiment. When both slits are open, an interference pattern is observed on the screen, whatever the rate at which electrons go through the double-slit system. Suppose that we reduce significantly the rate at which electrons emerge from their source. Then the interference pattern is built up by a steady accumulation of electrons in regions of constructive interference, while no electrons accumulate in the regions of destructive interference. The arrival of any one electron is causally disconnected from the arrival of any other, so *each* electron must somehow carry information about the final interference pattern.

Quantum mechanics accounts for the experiment by describing the electron and the screen with the slits by means of a **wave function**, which contains all the information about the diffraction pattern. The wave function (strictly speaking, its square) gives only the *probability* that the electron will be found in one place or another on the screen. The probability is largest where the magnitude of the wave function is largest. The probability that an electron will land on the screen where the waves from the two slits interfere destructively is very small, whereas the probability that it will land where the interference is constructive is large. This implies that the outcome of the journey of any one electron is not determined; only the *probability* for a set of outcomes can be known.

An analogous situation arises when polarized light passes through an analyzer (see Chapter 35). If the polarizer makes a 45° angle with the polarization vector, then the intensity of the passed light is half that of the incident light. Classically this is easy to understand. But the classical description no longer works if light is formed from photons. The problem becomes evident when the incident light intensity is so low that the photons arrive one at a time. How will a particular photon "decide" whether to pass the analyzer or not? The resolution of this problem is that, in quantum mechanics, a wave function describes the photon, the polarizer, and the analyzer, and with this wave function we can predict *only* that a given photon has a probability of $\frac{1}{2}$ of passing the analyzer. We cannot predict whether or not a given photon will pass.

To summarize: *Quantum mechanics is different from all other theories that we have studied so far in that it does not make predictions about the outcome of single events. It makes predictions only about the probabilities of specific outcomes.*

Quantum mechanics predicts only probabilities.

A situation of this type arises in the de-excitation of an atomic electron from an excited state or in nuclear radioactivity. Radioactivity occurs when a nucleus decays to some state of lower energy by the emission, for example, of a neutron, an alpha particle (^{4}He nucleus), or a photon. Quantum mechanics predicts (and all experiments confirm) that if we start with a certain number of radioactive nuclei, N_0, then after a time t the number of nuclei left will be

$$N = N_0 e^{-t/\tau}. \tag{41–21}$$

The parameter τ has the dimensions of time and is called the *lifetime*, or *mean life*, of the radioactive nucleus. It can be calculated by using the machinery of quantum mechanics and is relatively easy to measure. After a time $t = \tau$, the number of nuclei that remain is e^{-1}, or $0.37N_0$. The value of τ is the same for a given nucleus, whether the nucleus has just been artificially produced in the laboratory or found in a rock 1 billion yr old. How does a given nucleus "know" when to decay? There is no evidence that the nucleus contains an internal clock that tells it when to decay (we say that there are no *hidden variables*). Because the quantum mechanical description of a single nucleus cannot contain information about what other nuclei are going to do, the only interpretation possible is that the *probability* that a nucleus will last a certain time t is part of the wave function of that nucleus, whereas a determination of precisely when the nucleus will decay is not.

It might appear that this is no different from the problem of life expectancy in a population. There is some probability that 100-yr-old people exist in a population, but actuarial tables make no prediction for that case for an individual. There is, however, a difference: People do have internal clocks, and an examination of the habits and jobs of individuals can give us more information about their life expectancy. With sufficiently detailed medical information, we could at least *in principle* make a prediction about a given lifetime.

A characteristic of quantum systems is that a measurement has a well-defined effect on the system. We can illustrate this with a technique known as *radiometric dating*. Consider a piece of wood discovered in an archeological dig. The method of radiometric dating depends on the fact that, in the atmosphere, the ratio of ^{14}C (whose nucleus has 6 protons and 8 neutrons) to ^{12}C (whose nucleus has 6 protons and 6 neutrons) is constant. Even though ^{14}C nuclei decay radioactively with a lifetime of 8270 yr, their number is replenished in the atmosphere through cosmic-ray bombardment of the stable nucleus ^{14}N. This bombardment has the effect of producing ^{14}C from ^{14}N. Once a tree dies and its wood stops taking in carbon from the atmosphere, the proportion of ^{14}C in the wood (relative to ^{12}C) steadily decreases according to the law given in Eq. (41–21). Suppose now that the proportion of ^{14}C atoms indicates that a sample of wood is 20,000 yr old. A measurement has been made. If the remaining (undecayed) ^{14}C atoms are now set aside, they will continue to decay in such a way that 37 percent will be left in 8270 yr. In other words, once an atom is measured not to have decayed, its clock, so to speak, has been reset to $t = 0$. This is quite different from the identification of 100-yr-old people. Once these people are identified as being alive, they are not in effect reborn!

These concepts are far removed from what we would call "common sense." It should be remembered, however, that common sense about the physical world is developed through observation, and there is no reason why the microscopic world should conform to the notions of what is sensible as developed from observation of the macroscopic world.

SUMMARY

The phenomena that comprise quantum physics are most important in microscopic systems such as atoms, molecules, and nuclei. These phenomena are rather unexpected to an intuition based on classical physics. Essentially we learn that what we think of as particles behave in some respects like waves, and what we think of as waves (electromagnetic radiation, for example) behave in some respects like particles. Quantum mechanics provides us with a unified model of these phenomena. A "particle" of momentum p will have the properties of a wave of de Broglie wavelength

$$\lambda = \frac{h}{p}, \tag{41–1}$$

where $h = 6.63 \times 10^{-34}$ J · s is Planck's constant. The frequency associated with a particle can similarly be related to its energy. The wavelike properties of particles include interference, which has indeed been observed for particles such as electrons and neutrons in diffraction experiments. We also see evidence of the wavelike properties of matter in tunneling. The smallness of h is the reason why the wave properties of matter are not evident on a macroscopic scale.

Just as classical particles have wavelike properties, electromagnetic "waves" have particlelike properties. Electromagnetic radiation of frequency f behaves as though it consists of particles (photons) with energy

$$E = hf \tag{41–16}$$

and momentum

$$p = \frac{hf}{c}. \tag{41–17}$$

Radiation exhibits its particlelike properties in the spectrum of blackbody radiation; in the photoelectric effect, in which electrons absorb incident photons and are ejected with specific energies from metals; and in the Compton effect, in which the wavelengths of photons scattered by electrons change.

There are apparent inconsistencies in the description of an "object" with both particle and wave attributes. Interference phenomena intrinsic to waves require the wave to be spread out in space and time, whereas a particle must have a definite, well-defined location. The inconsistencies are removed by the use of the Heisenberg uncertainty relations, which set limits on the use of classical variables such as position and momentum. Any attempt to specify the x-position with a precision Δx implies a limit with which the x-component of momentum can be simultaneously measured,

$$\Delta x \, \Delta p_x > \hbar, \tag{41–7}$$

where $\hbar = h/2\pi$. Similarly, there are limits on the precision of measurements of energy and time. An energy measurement is limited to a precision ΔE by the duration of time Δt that the measurement takes:

$$\Delta E \, \Delta t > \hbar. \tag{41–8}$$

These relations resolve potential inconsistencies between a simultaneous particle and wave description.

One of the consequences of the uncertainty relations is that a particle cannot be at rest at the minimum level of potential energy. The lowest energy of a quantum system (the ground-state energy) is always larger than what is expected by classical reasoning.

Quantum mechanics is special in that it cannot be used to predict the future behavior of a system, only the probability of a set of possible behaviors.

QUESTIONS

1. Would the photoelectric effect be useful in a burglar-alarm system? in a smoke alarm?
2. The shorter the wavelength of a photon, the more the photon behaves like a particle. Why?
3. The uncertainty relations provide a reason why the temperature $T = 0$ cannot be reached. What is that reason?
4. A lit cigarette can be seen at a distance of 500 m on a dark night. Outline how you would estimate the rate at which photons from the cigarette hit the retina of a night-adapted eye.
5. To probe very tiny regions of space (such as the inside of a proton) with electron beams, you need electron beams

of very high energy. Why? Can you estimate the kind of energy needed to study a region of diameter d?

6. Given that electrons behave like waves, how is a Doppler shift described in terms of momentum?
7. Before the Planck formula was discovered, Rayleigh and Jeans had obtained the expression $u(f, T) = (8\pi f^2/c^3)kT$. How could we tell that something is wrong with this expression, even with no experimental data on the subject of blackbody radiation?
8. Suppose that half of a sample of radioactive nuclei has decayed in a given time T. How long will it be before half the remaining nuclei will have decayed?
9. The lifetime τ that measures the decay rate of a sample of radioactive particles is affected by the considerations of special relativity; that is, moving radioactive particles decay more slowly than stationary ones. How do the particles "know" that they are moving and that they should decay more slowly?
10. One electron is sent through a double-slit apparatus. In what sense, if any, can we say that there is an interference pattern on the screen?
11. Does the fact that all particles, however large, have wavelike properties mean that there is some probability that a baseball can tunnel through a catcher's mitt?
12. In discussing blackbody radiation, we spoke of a cavity. What does the cavity provide? Do we mean a real cavity in bulk material?
13. Do the uncertainty relations taken together imply that there are restrictions on the simultaneous measurement of position and time?
14. An electron microscope operates by the reflection of electrons, rather than by the reflection of light from an object. Does the use of particles such as electrons eliminate the problems associated with diffraction through the viewing aperture of the microscope?
15. Does the fact that the speed of light is a definite, predictable quantity conflict with the uncertainty relations?
16. What increases the probability of tunneling the most: halving the energy difference between U and E or halving the barrier width?

PROBLEMS

41–1 *The Wave Nature of Matter*

1. (I) What is the de Broglie wavelength of an electron whose energy is (a) 1.0 eV? (b) 10 eV? (c) 100 eV? (d) 1.0×10^9 eV? (e) What size targets would you need to observe diffraction of electrons of each of these wavelengths?
2. (I) What is the kinetic energy of an electron whose de Broglie wavelength is that of visible red light, 600 nm?
3. (I) What is the de Broglie wavelength of a proton, of mass 1.7×10^{-27} kg, with kinetic energy (a) 1.0 MeV; (b) 10 MeV; (c) 300 MeV? Neglect relativistic effects.
4. (I) Consider a crystal with a planar spacing of 0.18 nm. (a) What energies would electrons need for you to be able to observe up to four interference maxima? (b) Repeat the problem for neutrons.
5. (I) The spacing between scattering planes in a crystal is 0.20 nm. What is the scattering angle from such a crystal with electrons of energy 40 eV for which a first maximum is observed?
6. (I) Express the de Broglie wavelength of a particle of mass m and kinetic energy E in terms of m and E. Suppose that the particle moves at a speed close to the speed of light. How is that relationship modified?
7. (II) Suppose that you want to carry out diffraction experiments with the protons of Problem 3. What spacing of scatterers would you need in each of the three cases of energy 1.0 MeV, 10 MeV, and 300 MeV?
8. (II) What is the de Broglie wavelength of a neutron whose kinetic energy is equal to the average kinetic energy of a gas of neutrons at temperature $T = 4.0$ K?
9. (II) Crystals are studied by means of electron and neutron diffraction (Figs. 41–18a and 41–18b, respectively) as well as by X-ray diffraction. Recall that the typical interatomic distance in a crystal is 10^{-8} cm. Estimate the energy an electron must have to be useful for diffraction experiments on crystals. Repeat the exercise for neutrons.

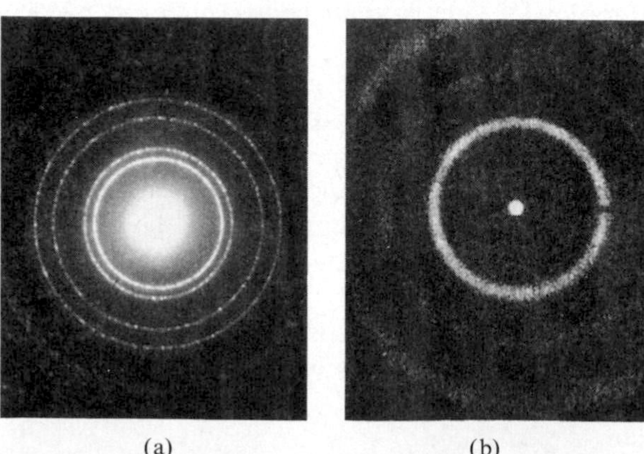

FIGURE 41–18 Problem 9.

10. (II) Although the working of an electron microscope does not depend on the wave nature of matter, the waves associated with electrons do set a limit on the resolving power of such instruments; Fig. 41–19 shows a bacterium magnified 10,000 times. (a) If the electrons in an electron microscope have an energy of 10^4 eV and the aperture of the microscope is 2.5×10^{-4} m, estimate the smallest an-

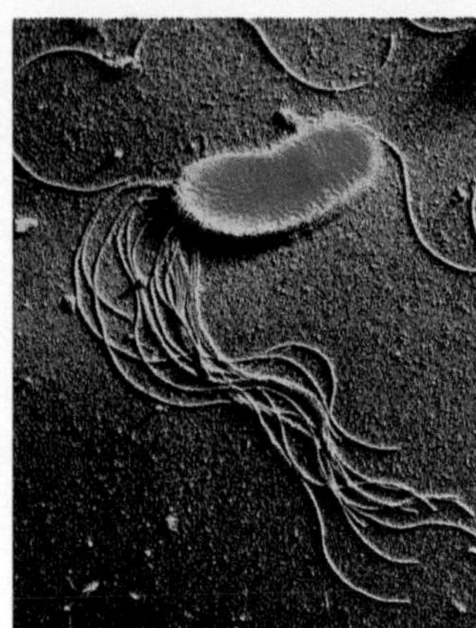

FIGURE 41–19 Problem 10.

gle that can be resolved. (b) How much energy would electrons need so that two objects separated by 0.2 nm could be resolved? Give your answer in electron-volts.

11. (II) A beam of electrons with energy 1.0 eV approaches a potential barrier with $\langle U \rangle = 2.0$ eV, whose width is 0.10 nm. Estimate the fraction of electrons that tunnel through the barrier.

12. (II) Suppose that the barrier in Problem 11 is 1.0 nm thick. What fraction of electrons tunnel through the barrier?

13. (II) A truck of mass 2000 kg travels at 1.0 m/s and approaches a smooth bump whose average height is 20 cm and whose average width is 0.5 m. Estimate the tunneling factor for the truck from Eq. (41–6).

14. (II) An electron in a semiconductor device is contained by a potential barrier of width 2.5 nm. The difference between the average potential energy and the energy of the electron is 1.4 eV. By what factor is the probability of tunneling changed if (a) the barrier width is halved? (b) the energy of the electron is increased by 1.2 eV? (c) the barrier width is halved *and* E is increased by 1.2 eV?

15. (III) The fraction F_n of particles of mass m and energy E that tunnel through a barrier whose potential energy is $U(x_n)$ and (minimal) thickness is Δx_n is given by Eq. (41–6),

$$F_n = e^{(-2/\hbar)\,\Delta x_n \sqrt{2m[U(x_n) - E]}}.$$

Suppose that a sequence of adjacent barriers at positions $x_1, x_2, \ldots$ have potential energy $U(x_1), U(x_2), \ldots$ and thickness $\Delta x_1, \Delta x_2, \ldots$, respectively. The fraction of particles that penetrate these barriers is given by the product of probabilities $F(x_1), F(x_2), \ldots$ that the particle can penetrate the individual barriers independently, namely, $F = F(x_1)F(x_2)\cdots$. Show that in the limit that each barrier is infinitesimally thin, we get

$$e^{(-2/\hbar)\int dx \sqrt{2m[U(x) - E]}}$$

for the fraction of particles that penetrate the barrier. Note that Eq. (41–6) is an approximation to this result.

41–2 *The Heisenberg Uncertainty Relations*

16. (I) An electron is localized inside a cubic region of sides 0.10 nm. What is the uncertainty in its kinetic energy?

17. (I) You are given a detector that can time the arrival of a particle to within an accuracy of 10^{-12} s. What is the accuracy of the energy determination possible with such a detector?

18. (I) The speed of an electron emitted from an atom is measured to a precision of ± 2.0 cm/s. What is the smallest uncertainty possible in the electron's position?

19. (I) A proton in a carbon atom is known to lie within a sphere whose diameter is about 6.0×10^{-15} m. What are the uncertainties in the momentum and energy of the proton?

20. (II) Monochromatic light of wavelength 720 nm passes through a fast shutter, which stays open for 1.0×10^{-9} s. What will the wavelength spread of the beam be after the light emerges through the shutter?

21. (II) The uncertainty in momentum of an electron with a kinetic energy of approximately 25 eV is 10 percent. What is the minimum uncertainty in its position?

22. (II) A completely free electron in empty space is measured to have a position within a sphere of radius $R = 1.0 \times 10^{-14}$ m, typical of an atomic nucleus. Within what volume can you say with assurance that the electron will be found after 1.0 s? Repeat the problem for an electron initially measured to lie within a sphere of radius $R = 1.0 \times 10^{-10}$ m, the radius of an atom.

23. (II) A radar gun measures the speed of a perfectly round baseball, mass 250 g, to be 93 mi/h, to an accuracy of 0.2 percent (Fig. 41–20). How well can the position of the baseball be determined in principle? Such accuracy cannot be achieved in practice.

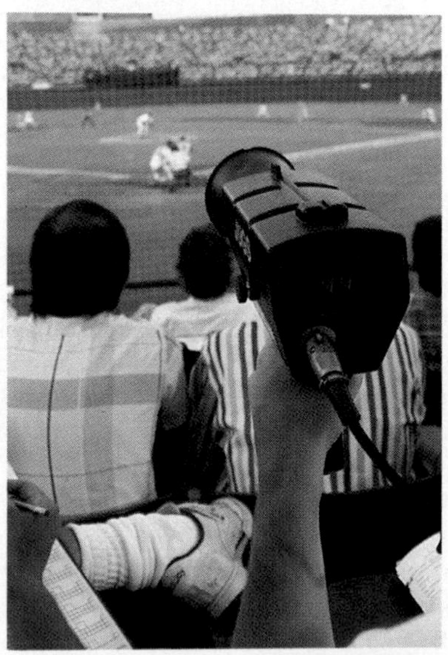

FIGURE 41–20 Problem 23.

24. (II) A beam of electrons of momentum p impinges on a slit of width a. Calculate the spread of the beam on a screen that is a distance D from the slit, taking into account both the width of the slit and the transverse spread due to the operation of the uncertainty relations.

25. (II) Consider a particle of mass m with potential energy of the form $U = -U_0(x/a)$ for $x < 0$ and $U = +U_0(x/a)$ for $x > 0$. Use the uncertainty relations to estimate the lowest energy the particle can have.

26. (II) A neutron, of mass 1.6×10^{-27} kg, is localized inside a lead nucleus, of radius 7.7 fm (1 fm $\equiv 10^{-15}$ m). Use the uncertainty relations to calculate a minimum (negative) potential energy for neutrons inside heavy nuclei.

41–3 *The Particle Nature of Radiation*

27. (I) What are the energy and momentum of a photon in He–Ne laser light of wavelength 632 nm?

28. (I) Find the energy of a photon for each of the following cases: (a) a microwave of wavelength 3.0 cm; (b) blue light of wavelength 420 nm; (c) a radio wave of frequency 1070 kHz; (d) an X-ray of wavelength 0.1 nm.

29. (I) Light of frequency 0.85×10^{15} Hz falls on a metal surface. If the maximum kinetic energy of the photoelectrons is 1.7 eV, what is the work function of the metal?

30. (I) A metal has a work function of 4.8 eV. What is the maximum kinetic energy of a photoelectron if radiation of 200 nm wavelength falls on the surface?

31. (II) Estimate the energy of a photon of each of the following radiation types: (a) visible light; (b) X-rays; (c) microwaves; (d) television signals; (e) AM radio.

32. (II) Calculate the number of photons from an AM radio station broadcasting at a frequency of 10^6 Hz that is required to equal the energy contained in one photon of visible light at a wavelength of 500 nm.

33. (II) The power generated by the sun is 4×10^{26} W. Assuming that it is emitted entirely at an average wavelength of 500 nm, calculate the number of photons emitted per second.

34. (II) The energy density of electromagnetic radiation in some region of space is 10^{-12} J/m^3. Assume that the radiation has a wavelength in the middle of the visible range, 550 nm. What is the photon density?

35. (II) Use the fact that the human eye can pick up as few as 5 photons/s in the visible range to estimate the intensity of the dimmest star that can be detected by a night-adapted eye. What is the ratio of this intensity to the intensity of noon sunlight, some 1400 W/m^2? This large intensity range means that the eye is indeed a very adaptable instrument.

36. (II) Show that the total energy in a cavity filled with blackbody radiation at temperature T in degrees kelvin—that is, of energy density $u(f, T)$ given by the Planck radiation law, Eq. (17–14)—is $U(T) = aT^4$. This result is the *Stefan–Boltzmann law*. Calculate the value of the constant a, given the integral

$$\int_0^\infty \frac{x^3}{e^x - 1}\,dx = \frac{\pi^4}{15}.$$

37. (II) Use the Stefan–Boltzmann law [see Eq. (17–14)] to calculate the temperature of the sun, given that the sun has a radius of 0.7×10^9 m, it is 1.5×10^{11} m from earth, and the rate at which the total solar radiation that falls on the earth is 1.36×10^3 J/m$^2 \cdot$ s.

38. (II) The maximum energy of photoelectrons from aluminum is 2.3 eV for radiation of wavelength 200 nm and 0.90 eV for radiation of 261 nm. Use these data to calculate Planck's constant and the work function of aluminum.

39. (II) The threshold wavelength for the photoelectric effect in tungsten is 270 nm. Calculate the work function of tungsten, and calculate the maximum kinetic energy that a photoelectron can have when radiation of 120 nm falls on tungsten.

40. (II) A photon of energy 50×10^3 eV (50 keV) collides with an electron at rest. The photon is scattered at an angle of 45°. What is its energy after the collision? What is the kinetic energy of the electron after the collision?

41. (II) The wavelength of the incoming X-rays in a Compton scattering experiment is 7.078×10^{-2} nm, and the wavelength of the outgoing X-rays is 7.314×10^{-2} nm. At what angle was the scattered radiation measured?

42. (II) Consider a case of Compton scattering in which a photon collides with a free electron and scatters backward while it gives up half its energy to the electron. (a) What are the frequency and energy of the incident photon? (b) What is the electron's velocity after the collision?

43. (III) Consider a cavity that contains blackbody radiation at 6000°C. Calculate the energy density for radiation in the wavelength range 720 nm to 750 nm, and compare it with the energy density for radiation in the range 480 nm to 510 nm. [*Hint:* To calculate energy density as a function of wavelength, use $u(f, T)\,df = u(f, T)\,d\lambda\,(df/d\lambda)$ and calculate the factor $(df/d\lambda)$. You must substitute $\lambda f = c$ in $u(f, T)$.]

44. (III) The sun's radiation peaks at a wavelength of 500 nm. How much less is the radiation intensity at 400 nm and at 700 nm? Use the results of Problem 43.

45. (III) A sodium crystal emits 6.25×10^{11} photoelectrons/m$^2 \cdot$ s. Given that the atomic weight of sodium is $A = 23$ and the mass density of the crystal is 970 kg/m^3, and assuming that the photoelectrons are supplied by the 10 layers of sodium atoms nearest the surface of the crystal, how many atoms, on average, furnish one photoelectron/s? Assume that the atoms form a cubic lattice.

41–4 *Quantum Mechanics and Probability*

46. (I) The number of atoms in an excited state whose lifetime for single-photon decay is τ is given by $N(t) = N(0)e^{-t/\tau}$. What is the number of photons emitted per second?

47. (I) A beam of light is sent along the $+z$-axis through a polarizer so that it is polarized in the direction of the $+x$-axis. This beam impinges on a second polarizer that makes a 30° angle with the x-axis. What is the probability that a photon will get through the second polarizer?

48. (II) The *half-life* of a set of radioactive nuclei is the time in which half of the nuclei decay. Express the half-life in terms of the lifetime τ that appears in Eq. (41–21).

49. (II) The half-life of ^{14}C is 5730 yr (see Problem 48). Organisms accumulate this isotope from the atmosphere while they live but cease doing so upon death. The skeleton of a mammoth is found to have a concentration of ^{14}C that is 20 percent of the atmospheric value. When did the mammoth live? Assume that the concentration of atmospheric ^{14}C does not change.

50. (II) Equation (41–21) may be written in the form $dN = -(N/\tau)\,dt$. Interpret this equation, including the sign. Does your interpretation support the assertion that the decay of any one radioactive nucleus is unaffected by the presence of others?

51. (II) A sample of radioactive material undergoing 3.7×10^{10} disintegrations/s is said to have an activity of 1 *curie* (Ci), where the activity is the rate at which decays occur. The activity of 1 g of Ra (radium) is 1 Ci. Given that $A = 226$ for radium, estimate the lifetime of ^{226}Ra.

52. (II) What is the activity of 1 g of ^{40}K (potassium), whose lifetime is 3.5×10^{8} yr? (See Problem 51.)

General Problems

53. (II) When high resolution is needed in biological microscopy, electron microscopes rather than light microscopes are used. Compare the resolution limit of an electron microscope that uses electrons of energy 30 keV with that of a light microscope.

54. (II) Find the average de Broglie wavelength of a gas of helium atoms (each of mass 6.6×10^{-27} kg) at room temperature, 293 K. Assume that the gas is ideal and is at 1 atm of pressure. Calculate the interatomic spacing. What would the temperature have to be for the same amount of helium gas in the same volume to have a de Broglie wavelength equal to the interatomic spacing?

55. (II) *Estimate* the rate at which photons in the visible range are emitted by a flashlight. Assume that 50 percent of the energy is emitted in the visible range and that the average wavelength in this range is 550 nm.

56. (II) It is a consequence of the Planck radiation law that the energy density expressed as a function of wavelength and temperature, $U(\lambda, T)$, peaks at $\lambda = \lambda_{max}$, where $\lambda_{max} = (2.9 \times 10^{-3}\ \text{m} \cdot \text{K})/T$ and T is the absolute temperature (see Problem 43). (a) Use this result to calculate λ_{max} for the radiation emitted by the sun, whose surface temperature is approximately 6000 K, and which, to a good approximation, radiates as a blackbody. (b) Use this result to estimate how hot "red-hot" is.

57. (II) Show that in the simple process of photon absorption by a free electron, without anything else occurring, energy and momentum cannot be conserved simultaneously.

58. (III) The abundances of ^{238}U and ^{235}U are assumed to have been equal when the earth formed. The mean life of ^{238}U is 0.6×10^{10} yr, and that of ^{235}U is 1.0×10^{9} yr. The present ratio of their abundances is $^{238}U/^{235}U = 140$. From these data and the assumption, estimate the age of the earth.

59. (III) The de Broglie wavelength of a free electron of energy E may be written in the form $\lambda_0 = h/p = h/\sqrt{2m_e E}$. When the electron enters a region that has an electric potential $V(x)$, the electron acquires a potential energy $U(x) = -eV(x)$, and the momentum is modified so that it satisfies $E = (p^2/2m) + U(x)$. Write an expression for the de Broglie wavelength of an electron in this region, and show that the presence of the potential $V(x)$ means that the electron behaves as though it were moving in a region of refractive index $n(x) = \sqrt{[E - U(x)]/E}$.

CHAPTER 42

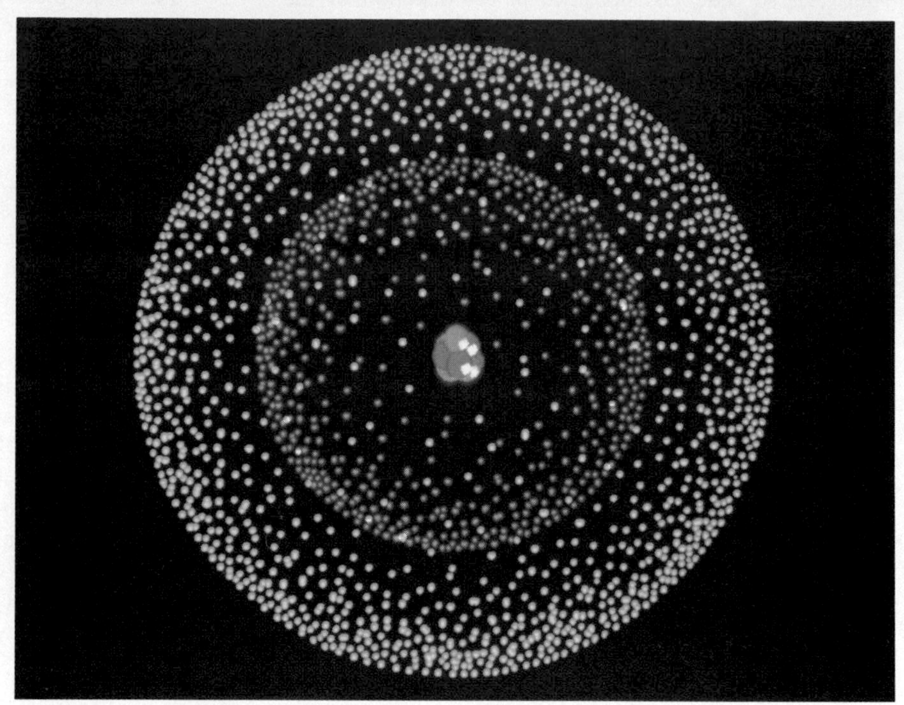

This computer-generated graphic is a representation of a beryllium atom. The electrons "orbit" in the purple and blue regions.

QUANTIZATION OF ANGULAR MOMENTUM AND OF ENERGY VALUES

In classical physics, a particle or system of particles, whether free or confined in its movements by forces, can have a continuous range of energies. Experiments have shown, however, that, when examined on an atomic scale, energy can take on only discrete values. This striking fact can be explained in terms of quantum mechanics, in particular by the wave nature of matter (see Chapter 41). The discreteness (or quantization) of energy values is closely associated with the quantization of angular momentum. The consequences of the quantizations of energy and angular momentum, the subject of this chapter, are important. When we couple the quantization principles with the so-called exclusion principle, we can explain many otherwise inexplicable features of microscopic systems, including the structure and stability of atoms and molecules. These ideas have important consequences even on a macroscopic level. For example, they explain the band structure of metals.

42–1 QUANTIZATION OF ENERGY AND ANGULAR MOMENTUM

We have stressed throughout this book that matter consists of atoms. In our study of the kinetic theory of gases, we made use of molecules and atoms with an identity defined primarily by the atomic weight. Molecules were allowed to rotate and vibrate, but in our discussion of the equipartition of energy, for example, we never asked how fast the molecules rotate or how large the amplitude of vibration is. In classical theory, an atom, made of negatively charged electrons orbiting a positively charged nucleus (the nucleus approximately 10^5 times smaller than the atom), would behave like a miniature solar system, and elliptical or circular electron orbits of any size would be allowed. Such variability in the forms of atoms is in sharp conflict with experiment. One hydrogen atom (^1H) is, *literally*, indistinguishable from another. We generally see only one form of hydrogen, one form of helium, one form of iron. Quantum mechanics explains this by a remarkable prediction: In contrast to planetary systems, in which unrestricted initial conditions determine the energy and the angular momentum, *quantum systems can have only certain—that is, quantized—values of energy and of angular momentum.* Possible energy values are always separated by gaps (although the gaps may be small). The set of allowed energy values has come to be called *energy levels*, and the first successful calculation of these levels was made by Niels Bohr in 1913.

The energies and angular momenta of systems are quantized.

Virtually all the atoms in a bottle of helium at room temperature will be in their lowest energy level, the *ground state*. Ordinary thermal collisions between these helium atoms cannot supply enough energy to change the helium atoms from their lowest state of energy to the next allowed state (an *excited state*). At high enough temperatures, some fraction of the atoms may be "kicked" up into the next, higher state, and then our bottle will contain *excited atoms* of helium. These excited atoms can "jump" back to the lowest energy level and, in the process, emit light of *discrete frequencies*. The light carries an energy that is the difference in energies between the excited state and the ground state. In this way, energy is conserved. The emitted light consists of photons with a frequency f determined by the frequency–energy relation of photons, $E = hf$. Thus atomic energy levels can be studied by looking at the radiation emitted by atoms at discrete frequencies at high temperatures (Fig. 42–1).

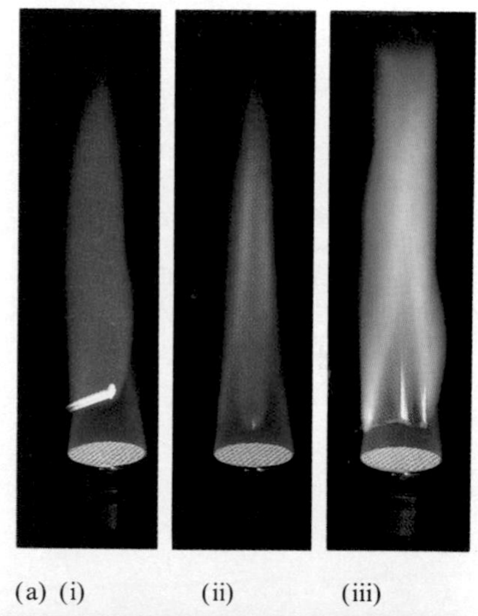

(a) (i) (ii) (iii)

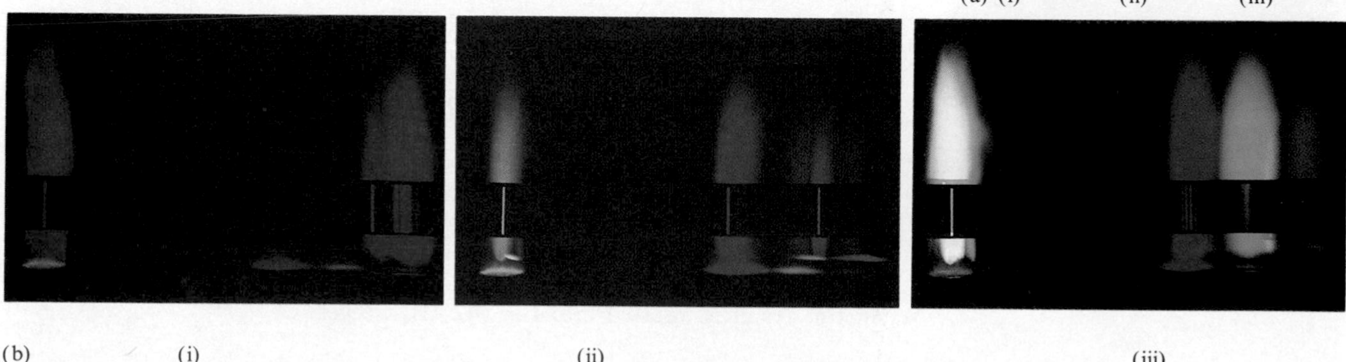

(b) (i) (ii) (iii)

FIGURE 42–1 (a) The color emitted by atoms after they have been excited by heat is characteristic of the particular element they comprise: (i) strontium, (ii) rubidium, and (iii) copper. (b) When the color of each element in part (a) is analyzed spectrally, it is seen to contain discrete frequencies (lines).

The Wave Nature of Matter and Energy Quantization

The wave theory of matter provides an explanation of the energy quantization described above. Consider, for example, a particle confined to a box with one side of finite width b. If the particle inside the box behaves like a standing wave on a string of length b fixed at its ends, then the wavelengths are constrained by the condition

$$\lambda = \frac{2b}{n}$$

[Eq. (14–8)], where $n = 1, 2, 3, \ldots$ (Fig. 42–2). The wavelength is related to the momentum, p, by the de Broglie relation, $\lambda = h/p$ (Eq. 41–1), so this condition is equivalent to

$$p = \frac{nh}{2b}.$$

The energy of the particle in the box, $E = p^2/2m$, is then constrained to the discrete values

$$E = \frac{n^2h^2}{8mb^2}. \tag{42–1}$$

The ground-state energy ($n = 1$) is therefore $h^2/8mb^2$, and the possible energy values are separated by gaps on the order of magnitude h^2/mb^2. Our more precise calculation differs from the estimate we made in Chapter 41 with the help of the position–momentum uncertainty relation [Eq. (41–7)], which is often the case with such crude estimates.

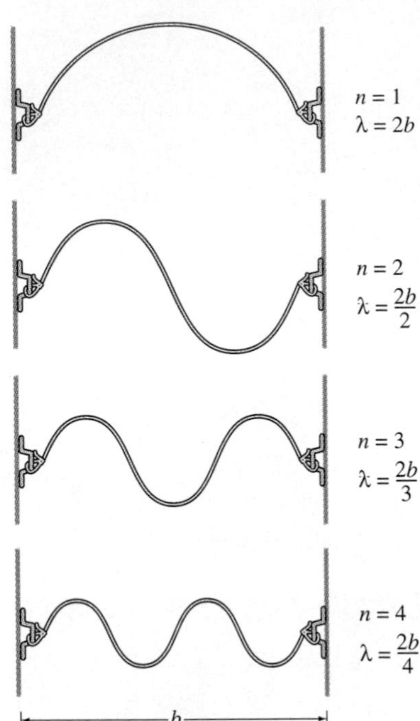

FIGURE 42–2 The wavelengths of standing waves on a string fixed at both ends are given by the length of the string divided by half-integers.

The Bohr Model of Hydrogen

Let us see how the wavelike properties of matter allow us to understand the orbits of electrons in atoms. We start with a classical description of the atom. In 1911 Ernest Rutherford discovered that atoms have a nuclear structure. The simplest atom is that of hydrogen, which has only one electron, of mass m_e and charge $-e$, and a nucleus that usually consists of a single proton, of mass $m_p \gg m_e$ and charge $+e$. If the electron moved classically, its orbits would be circular or elliptical, with the proton at the center, or focus, by analogy to planetary orbits. For circular orbits of radius r, Newton's second law, $F = ma$, becomes

$$\frac{e^2}{4\pi\epsilon_0 r^2} = m_e a = \frac{m_e v^2}{r}. \tag{42–2}$$

The left side of this equation is the Coulomb force. The total energy is

$$E = K + U = \frac{p^2}{2m_e} - \frac{e^2}{4\pi\epsilon_0 r}, \tag{42–3}$$

where zero potential energy has been chosen to be at $r = \infty$.† Newton's second law, Eq. (42–2), implies that

$$\frac{p^2}{2m_e} = \frac{1}{2}m_e v^2 = \frac{e^2}{8\pi\epsilon_0 r}.$$

When this result is inserted into Eq. (42–3), we get

$$E = \frac{e^2}{8\pi\epsilon_0 r} - \frac{e^2}{4\pi\epsilon_0 r} = -\frac{e^2}{8\pi\epsilon_0 r}. \tag{42–4}$$

† Relativity is only a minor correction to the structure of atomic orbits, which is why we use the nonrelativistic form of kinetic energy in Eq. (42–3).

The energy is negative, as is appropriate for electrons in closed orbits when zero potential energy is chosen to be at infinity.

It will be useful to find the angular momentum **L** in the circular orbit. It has magnitude $L = m_e vr$. From Eq. (42–2), $v = e/\sqrt{4\pi\epsilon_0 m_e r}$, so the angular momentum has magnitude $L = m_e vr = m_e(e/\sqrt{4\pi\epsilon_0 m_e r})r = e\sqrt{m_e r/4\pi\epsilon_0}$. We can in turn solve for r to find the radius of the orbit for a given value of angular momentum:

$$r = \frac{L^2}{m_e e^2/4\pi\epsilon_0}. \tag{42–5}$$

FIGURE 42–3 Niels Bohr.

In the classical planetary model just described, the energy, the orbital radius, and the angular momentum can take on a continuum of values. But this model has a fatal flaw. We know that a charge that moves in a circular orbit is constantly being accelerated, and we saw in Chapter 35 that an accelerating charge radiates energy. Consequently, an orbiting electron would constantly lose energy by radiation; as its energy became more negative, the radius would decrease [see Eq. (42–4)] until the proton swallowed up the electron. Detailed estimates (see Problem 14) show that this would happen in only 10^{-10} s!

The first treatment of the atom that incorporated an energy quantization condition was provided by Niels Bohr in 1913 (Fig. 42–3). We give here two of the assumptions Bohr made in his attempt to explain the hydrogen atom:

1. The classical circular orbits, with variable energy and radius, are replaced by *stationary states*, so called because the energies of those states are fixed, and hence electrons in these states cannot radiate classically. The energies of these stationary states take on only discrete values.
2. The energy values of stationary states are determined for circular orbits by imposing the condition that the angular momentum, L, is quantized in integer units of $\hbar$:

$$L = n\hbar, \qquad \text{where } n = 1, 2, 3, \ldots. \tag{42–6}$$

The Bohr quantization condition

The integer n is the *principal quantum number* for the orbit. Equation (42–6) is called the *Bohr quantization condition.*

To see how the Bohr quantization condition determines the energy values, we substitute L into Eq. (42–5), which determines the radius of a circular orbit in terms of L. We find that

$$r_n = \frac{n^2\hbar^2}{m_e e^2/4\pi\epsilon_0} = n^2 a_0, \tag{42–7}$$

The allowed radii of circular orbits for hydrogen

where the *Bohr radius* a_0 is the radius corresponding to $n = 1$:[†]

$$a_0 = \frac{\hbar^2}{m_e e^2/4\pi\epsilon_0} = 0.53 \times 10^{-10}\ \text{m}. \tag{42–8}$$

Note that the allowed radii are discrete, with an index n corresponding to the quantum number n for L. Once the radii can take on only discrete, fixed values, Eq. (42–4) shows that the energy also has discrete, fixed values.[‡] When the

[†] The fact that it is a_0, not a_1, for $n = 1$ is merely historical.

[‡] Thus saying that L is fixed (stationary) for circular orbits of single-electron atoms automatically leads to stationary energies. The first of Bohr's assumptions above is therefore redundant with the second in that case.

allowed values of r from Eq. (42–7) are inserted into Eq. (42–4) for the energy, we obtain the allowed energy values of a hydrogen atom in the Bohr model:

The allowed energies of the hydrogen atom

$$E_n = -\frac{e^2}{8\pi\epsilon_0 r_n} = -\frac{e^2}{8\pi\epsilon_0}\frac{1}{[(n^2\hbar^2)/(m_e e^2/4\pi\epsilon_0)]}$$

$$= -\frac{m_e}{2n^2}\left(\frac{e^2}{4\pi\epsilon_0\hbar}\right)^2 = -\frac{21.8\times10^{-19}\ \mathrm{J}}{n^2} = -\frac{13.6\ \mathrm{eV}}{n^2}. \qquad (42\text{–}9)$$

We have added the subscript n as a reminder of the discrete nature of the energies. The ground-state energy is $E_1 = -13.6$ eV.

Figure 42–4 shows the energy values predicted by Eq. (42–9). The excited-state energies get closer and closer together as n increases, and the energy approaches zero as $n \to \infty$. Positive energies correspond to the situation in which the electron's kinetic energy is positive when r_n approaches ∞; that is, the electron is not bound to the proton. There the kinetic energies of the electron and the proton can take on continuous values. The energy is not discrete when the system is not bound. The minimum energy required to remove an electron in the ground state from the atom is called the *ionization energy*. For hydrogen, this energy is (0 eV) − (−13.6 eV) = 13.6 eV.

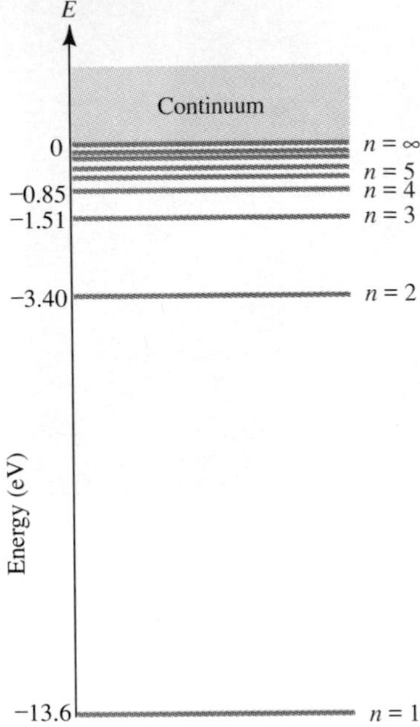

FIGURE 42–4 The energy levels in a hydrogen atom for circular orbits in the Bohr model, obtained from Eq. (42–9). The energy levels (here, not to scale) bunch up as the quantum number n increases. Above the ionization point $E = 0$, the electron and the proton are no longer bound together.

We can think of the quantization of angular momentum as a result of the requirement that *the circumference of the orbit accommodates an integral number of de Broglie waves*. This condition, which is reminiscent of the frequency-fixing conditions for standing waves, requires that for an orbital radius labeled by the integer n,

$$\frac{2\pi r_n}{\lambda} = n. \qquad (42\text{–}10)$$

Because $\lambda = h/p_n$ for that orbit, Eq. (42–10) takes the form

$$\frac{2\pi r_n}{h/p_n} = \frac{r_n p_n}{h/2\pi} = \frac{L}{\hbar} = n,$$

which is identical to Eq. (42–6).

Testing the Bohr Model

The predictions of the Bohr model for hydrogen agree well with experiment. The model has been confirmed experimentally by the observation of the discrete light frequencies that hydrogen atoms emit after they have been excited. These well-defined light frequencies are those of photons *emitted* when an electron "jumps" from one orbit to another one of lower energy (Fig. 42–5). The principle of energy conservation determines the energies and hence the frequencies of these photons:

Atomic electrons can make transitions (jumps) from one allowed level with an initial energy E_i to another allowed level with a final energy E_f. When $E_i > E_f$, energy is released. The released energy can manifest itself in the appearance of a photon that carries off the excess energy $E_i - E_f$. Because a photon of energy E has frequency f given by $E = hf$, the frequency of the emitted photon† is determined by

The frequencies of light emitted by atoms

$$hf = E_i - E_f. \qquad (42\text{–}11)$$

† Energy conservation alone allows the emission of two or more photons, the sum of whose energies is $E_i - E_f$. But quantum mechanics predicts that the emission of one photon is overwhelmingly more likely than multiple-photon emission.

Conversely, an electron starting from some initial allowed energy can *absorb* a photon of the correct frequency and "jump" to a higher-energy (excited) state (Fig. 42–5). The frequency of the absorbed photon must be given by $hf = E_f - E_i$.

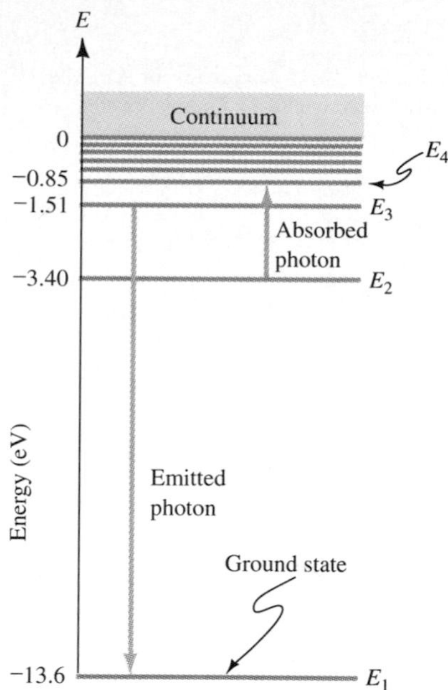

FIGURE 42–5 An atomic electron may jump down from one level to another with the emission of a photon or jump up with the absorption of a photon. The frequency of the photon is determined by the energy change.

EXAMPLE 42–1 A certain laser emits light with a wavelength of 3391 nm. What is the energy difference, in electron-volts, between the two energy levels involved in producing this light?

SOLUTION: The energy difference ΔE between the two energy levels is related to the wavelength of the emitted radiation, λ, by

$$\Delta E = hf = \frac{hc}{\lambda}.$$

Therefore

$$\Delta E = \frac{hc}{\lambda} = \frac{(6.63 \times 10^{-34}\ \text{J}\cdot\text{s})(3.0 \times 10^{8}\ \text{m/s})}{(3391 \times 10^{-9}\ \text{m})(1.6 \times 10^{-19}\ \text{J/eV})} = 0.37\ \text{eV}.$$

For hydrogen, the possible energies E_i and E_f are given by Eq. (42–9). We may rewrite Eq. (42–11) in terms of the wavelength, λ, rather than the frequency by using the relation $f = c/\lambda$. Then the wavelengths of the photons emitted when an electron jumps down from an excited state are restricted to the values

$$\frac{1}{\lambda} = \frac{E_i - E_f}{hc} = R_\infty\left(\frac{1}{n_1^2} - \frac{1}{n_2^2}\right), \qquad (42\text{–}12)$$

where n_1 and n_2 are the quantum numbers of the initial and final energies, respectively, and $R_\infty \equiv (m_e/2hc)(e^2/4\pi\epsilon_0\hbar)^2 = 1.0974 \times 10^7\ \text{m}^{-1}$ is the *Rydberg constant*, named for Johannes Rydberg. The predicted wavelengths were found to be in excellent agreement with the measured wavelengths of the spectral lines in hydrogen (Fig. 42–6). In fact, in 1890, Rydberg had already made a purely empirical fit to the frequencies of the emission spectrum of hydrogen with the formula in Eq. (42–12). Frequencies corresponding to many values of n_2 and n_1 $(n_1 < n_2)$ could be closely approximated by this formula, and new frequencies were correctly predicted on its basis. Bohr was on the right track in providing an explanation for the values of these frequencies.

FIGURE 42–6 The emission spectrum of hydrogen.

EXAMPLE 42–2 Find the magnitude of the energy difference between the lowest energy of a hydrogen atom ($n = 1$) and the first excited state ($n = 2$). Estimate the temperature at which a substantial fraction of a gas of hydrogen atoms would occupy the first excited state.

SOLUTION: To find the energy difference between the $n = 1$ and $n = 2$ levels of hydrogen, we need calculate only $E_2 - E_1$, where the energies are given by Eq. (42–9):

$$E_2 - E_1 = E_1\left[\frac{1}{(2)^2} - \frac{1}{(1)^2}\right] = (-13.6\ \text{eV})\left(\frac{1}{4} - 1\right) = 10.2\ \text{eV}.$$

To estimate the temperature T at which a substantial fraction of hydrogen atoms would occupy the first excited state, we ask how much kinetic energy is needed per atom to arrive at $10.2\,\text{eV} = (1.6 \times 10^{-19}\,\text{J/eV})(10.2\,\text{eV}) \simeq 1.6 \times 10^{-18}\,\text{J}$. If we set $\frac{3}{2}kT$ equal to this number, as the equipartition principle suggests, then $T \simeq \frac{2}{3}(1.6 \times 10^{-18}\,\text{J})/(1.38 \times 10^{-23}\,\text{J/K}) = 0.8 \times 10^5\,\text{K}$! This estimate dramatically confirms our earlier assertion that, under normal circumstances, matter will be in its ground state. A small fraction of atoms is excited at much lower temperatures, and that makes it possible to study the spectra of elements in the laboratory.

The principles on which the Bohr model is based, although originally formulated for single-electron atoms, can be applied to any classical system. For example, we shall see in Section 42–3 how we can apply them to the harmonic oscillator (which represents vibrations in molecules) and to rotating molecules. In Example 42–3, we apply Bohr-model techniques to a model of how particles such as protons are constructed.

EXAMPLE 42–3 Two massive quarks (subnuclear particles) attract each other with a constant force. Think of this situation as a single particle of mass M orbiting under the influence of a constant central force, with potential energy $U = U_0(r/r_0)$. Here r is the distance between the quarks, U_0 is a constant with the dimension of energy, and r_0 is a constant with the dimension of length. Use this potential energy function to find the quantized energies of circular orbits.

SOLUTION: To use the Bohr rules, we proceed as we did for the hydrogen atom by first expressing the energy of circular orbits in terms of angular momentum, then applying the condition that angular momentum is quantized. The energy is

$$E = \frac{1}{2}Mv^2 + U_0\frac{r}{r_0}.$$

The force corresponding to our potential energy has magnitude $F = U_0/r_0$, so for circular orbits, the dynamical equation $F = ma$ is

$$\frac{U_0}{r_0} = \frac{Mv^2}{r}.$$

The quantization of angular momentum takes the form $Mvr = n\hbar$, or $r = n\hbar/Mv$. When we substitute this into the equation of motion, we get

$$\frac{U_0}{r_0} = \frac{Mv^2}{n\hbar/Mv} = \frac{M^2v^3}{n\hbar}.$$

Thus $v = (n\hbar U_0/M^2r_0)^{1/3}$ and $r = n\hbar/Mv = (n\hbar)^{2/3}(r_0/MU_0)^{1/3}$. When we substitute these expressions into the equation for energy, we find that

$$E_n = \frac{1}{2}M\left(\frac{n\hbar U_0}{M^2r_0}\right)^{2/3} + \frac{U_0}{r_0}(n\hbar)^{2/3}(r_0/MU_0)^{1/3} = \frac{3}{2}\left(\frac{U_0^2\hbar^2}{Mr_0^2}\right)^{1/3}n^{2/3}.$$

Despite the successes of Bohr's quantization rules, which work well for single-electron atoms, the rules cannot be successfully applied to multi-electron atoms. Even though Bohr's intuition brought us to a qualitative understanding of multi-electron atoms, it was clear that what he had done was provisional.

Bohr's rules were rather artificially grafted onto classical laws, and there was no understanding of when an electron would decide to "jump" from one orbit to another, nor of where the electron was in between.

42–2 THE QUANTUM THEORY OF ANGULAR MOMENTUM AND THE TRUE SPECTRUM OF HYDROGEN

Werner Heisenberg (in 1925) and Erwin Schrödinger (in 1926) generalized their extensive studies of the "old" quantum theory to make the leap to the correct formulation of quantum mechanics. The details are beyond the scope of this course, and we shall have to quote some of the results of quantum mechanics without attempting to derive them. The complete hydrogen spectrum is one of the topics that must be treated in this way.

Heisenberg used the **Bohr correspondence principle** as a guide for his construction of a theory of quantum phenomena. This principle, which grew out of work on the original Bohr atom, states that quantum mechanical results should reduce to those that follow from a classical treatment when quantum numbers—the principal atomic quantum number n, for example—are large. Problem 15 illustrates how the correspondence principle can be applied.

In our discussion of the Bohr model, we limited ourselves for simplicity to circular orbits. A full quantum mechanical treatment deals with all types of orbits and takes into account the more complicated properties of angular momentum in quantum mechanics. When the full theory is applied to the hydrogen atom, several noteworthy features appear.

1. The concept of orbits disappears completely. An electron, when it is bound to a nucleus, as in hydrogen, is described by a *wave function* (see Chapter 41). The square of the wave function at a given point describes the probability of finding the electron at that point. The electron can exist in one of a number of *states* characterized only by energy, angular momentum, and an orientation in space (the analogue of the tilt of a planetary orbit relative to some axis). These quantities are quantized, specified by a set of *quantum numbers* taking on integral values. Once the quantum numbers are specified, the wave function (and hence all the properties of the atom) is determined.

2. The electron's angular momentum in the atom is quantized according to $L = \ell\hbar$, with ℓ taking on the values 0, 1, 2, ... (We shall see below that there is a further restriction on ℓ.) Moreover, we can speak of that angular momentum only relative to some axis (Fig. 42–7). What we mean by this is the following: The component of angular momentum along the specified axis (which henceforth we shall call the z-axis) can take on only the values $m\hbar$ [as in Eq. (42–6)], with $m = \ell, \ell - 1, \ell - 2, \ldots, 1, 0, -1, \ldots, -(\ell - 1), -\ell$. In other words, the z-component L_z of the angular momentum vector is itself quantized. The square of the angular momentum is given by $\ell(\ell+1)\hbar^2$, *not* by $\ell^2\hbar$. Figure 42–7 shows the angular momentum and its allowed projections for $\ell = 1$, $\ell = 2$, and $\ell = 3$. This behavior is totally at variance with classical mechanics, although for very large L (so that $\ell\hbar$ is some macroscopic number such as $1\ \mathrm{g \cdot cm^2/s}$), the deviations from the classical description are very small.

The quantization of angular momentum was discovered experimentally in 1921 by Otto Stern and Walter Gerlach in the following experiment: A beam of

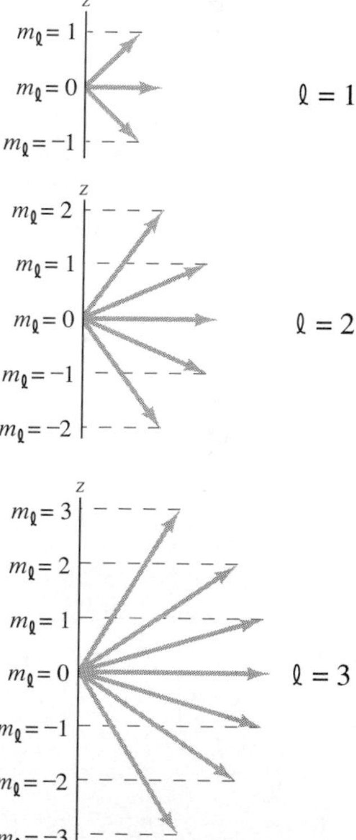

FIGURE 42–7 The directions of angular momentum vectors for $\ell = 1$, 2, and 3 allowed by the vector angular-momentum quantization conditions. The attempt to depict a quantum phenomenon with a classical model of restricted motion is successful only in that it indicates that not all spatial orientations are allowed.

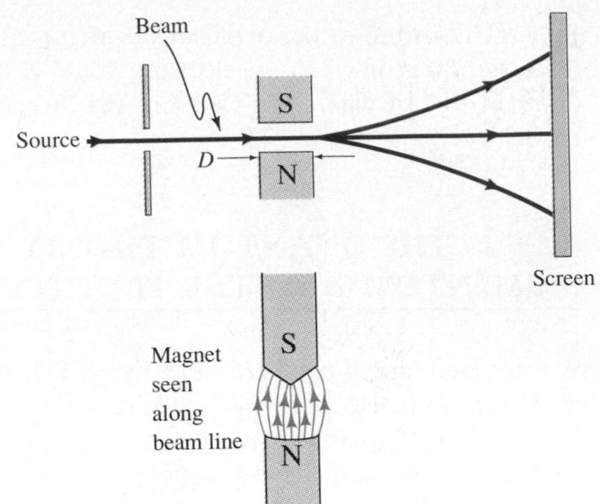

FIGURE 42–8 A Stern–Gerlach experiment. The magnetic field is not uniform. In this case, the beam is split into three components.

atoms is passed through a region of thickness D in which there is a magnetic field **B** oriented in the z-direction (Fig. 42–8).[†]

The atoms are characterized by a magnetic dipole moment $\boldsymbol{\mu}$, and the potential of the atoms in the magnetic field is given by Eq. (29–26):

$$U_{\text{mag}} = -\boldsymbol{\mu} \cdot \mathbf{B} = -\mu_z B.$$

If **B** varies with z, then there is a force on the atom proportional to its magnetic dipole moment, in the z-direction:

$$F_z = -\frac{dU_{\text{mag}}}{dz} = \mu_z \frac{dB}{dz}. \tag{42–13}$$

Classically, μ_z, which is proportional to the angular momentum component of the atom in the z-direction [see Eq. (32–12)],[‡] is not quantized. From the classical point of view, therefore, the atoms would reach the far side of the apparatus distributed in the z-direction with a uniform spread. The maximum deflection angle θ characterizing this spread would be given approximately by

$$\theta \simeq \frac{\Delta p}{p} = \frac{(\text{force})(\text{time in region containing the field})}{mv}$$

$$= \mu_z \frac{dB}{dz} \frac{1}{mv} \frac{D}{v} = \frac{D}{mv^2} \mu_z \frac{dB}{dz}, \tag{42–14}$$

where $p = mv$ is the momentum of an atom of the beam.

Rather than seeing a uniform spread, Stern and Gerlach saw that their beams were split into components at specific angles. The number of components varied with the species of atom. For helium, there was only one (undeflected) component; a beam of oxygen atoms split into five components; and a beam of silver atoms split into two components. In all cases the components were symmetrically distributed about the point of no deflection. These observations imply that the magnetic moments must be quantized, which in turn implies that

[†] In the Stern–Gerlach experiment, the magnetic field is used to deflect atoms. The study of the effects of magnetic fields on the spectra of atoms (the *Zeeman effect*) was initiated in 1896 by Pieter Zeeman. This work proved to be critical to the development of an understanding of atomic structure.

[‡] Recall that in Chapter 32 we used the notation **m** rather than $\boldsymbol{\mu}$ for the magnetic dipole moment.

the angular momentum is quantized. The initial beam of a species of atom with angular momentum quantum number ℓ consists of an equal mixture of atoms in each of the allowed $2\ell + 1$ states that correspond to quantization of its z-component. Each of these $2\ell + 1$ states has a different magnetic dipole moment z-projection, proportional to $-\ell, -(\ell - 1), \ldots, (\ell - 1), \ell$. Thus the single component of helium corresponds to $\ell = 0$, and the five components of oxygen correspond to $\ell = 2$. We shall soon discuss the case of silver, for which two components are observed.

3. The energy values of the hydrogen atom are quantized according to

$$E_n = -\frac{13.6\text{ eV}}{(n_r + \ell + 1)^2} \equiv -\frac{13.6\text{ eV}}{n^2}. \tag{42–15}$$

This is close to the prediction of the Bohr model, but in a generalized form. The original integer n of Bohr's model is replaced by $n_r + \ell + 1$, where $n_r = 0, 1, 2, 3, \ldots$ and ℓ is restricted to the integer values $0, 1, 2, 3, \ldots$. Note, however, that $n \geq \ell + 1$; in other words, $\ell \leq n - 1$ for a given n. In addition, a state with a given ℓ really refers to the collection of $2\ell + 1$ different states, all of which have the same angular momentum.

Equation (42–15) leads to a complex of energy levels that are a little more complicated than the naive Bohr model in that each energy level contains several states (Fig. 42–9). Note that the lowest state has $n_r = 0$ and $\ell = 0$ ($n = 1$), and it is unique. The next level, corresponding to $n = 2$, consists of 1 state for which $n_r = 1$ and $\ell = 0$ and $2\ell + 1 = 3$ states with $n_r = 0$ and $\ell = 1$; that is, there are 4 states for $n = 2$. The next level, corresponding to $n = 3$, consists of 1 state for which $n_r = 2$ and $\ell = 0$, 3 states with $n_r = 1$ and $\ell = 1$, and 5 states with $n_r = 0$ and $\ell = 2$—a total of 9 states. These results generalize in an obvious way: The total number of states labeled by n is n^2.

We have already noted that the conservations of energy and linear momen-

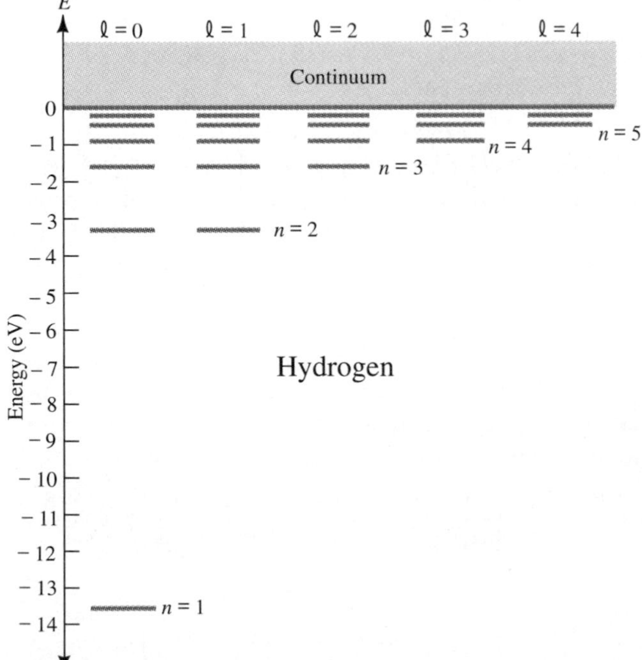

FIGURE 42–9 Spectrum of a hydrogen atom, as given by quantum mechanics. In the absence of a magnetic field, each level with a given ℓ actually consists of $2\ell + 1$ states.

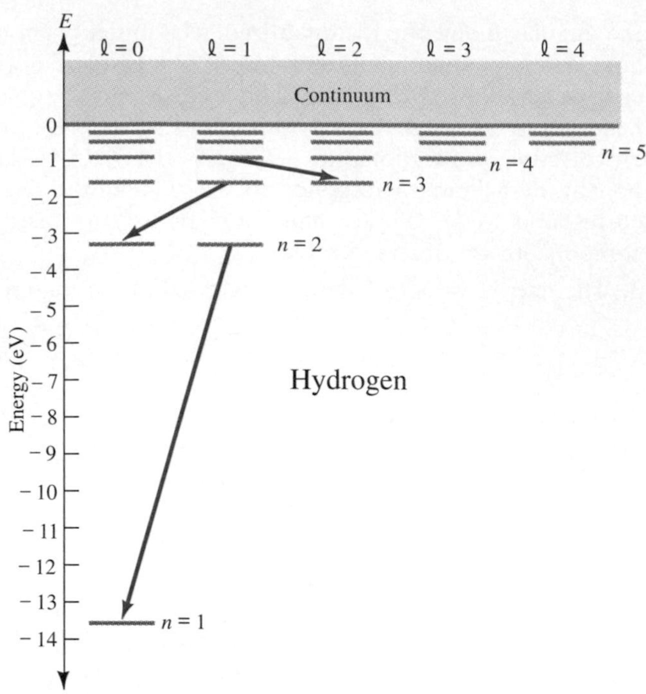

FIGURE 42–10 Some possible transitions of electrons in the quantum mechanical view of a hydrogen atom.

tum continue to hold in quantum mechanics. The same is true of angular momentum conservation in the absence of external torques. This affects the state in which a given electron can end up when it makes a jump with the emission of a photon. Electromagnetic radiation carries angular momentum, and for a photon the angular momentum is $L_{\text{photon}} = \hbar$. Thus the initial and final angular momenta of the electron states must differ by one unit of $\hbar$. Some of the possible transitions are sketched in Fig. 42–10. All the results described here are in extremely good agreement with experiment.

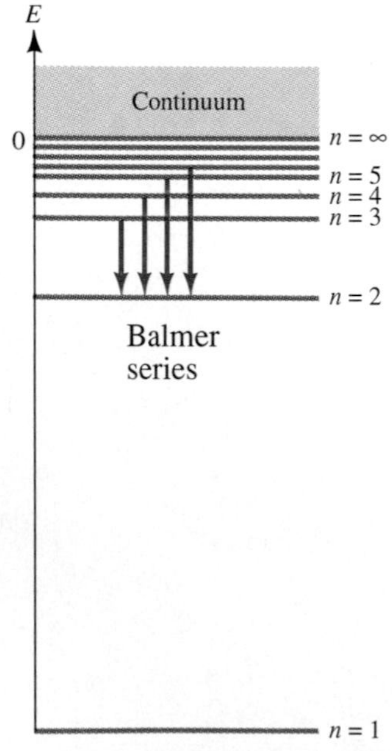

FIGURE 42–11 Example 42–4. This series of transitions is known as the Balmer series.

EXAMPLE 42–4 The *Balmer series* is a series of spectral lines that correspond to atomic transitions in hydrogen that end with a principal quantum number $n = 2$. Sketch the allowed transitions that lead to this series, and compute the longest and shortest wavelengths in this series.

SOLUTION: Figure 42–11 shows the transitions that end at $n = 2$, $\ell = 0$ (which start at $n = 3, 4, \ldots, \ell = 1$); and also at $n = 2$, $\ell = 1$ (which start at $n = 3, 4, 5, \ldots, \ell = 0, 2$). The frequencies of the transitions are given by the usual relation between the energy change of the emitting atom and the frequency of the emitted light:

$$f_2 = \frac{E_n - E_2}{2\pi\hbar} = -\frac{m_e}{2}\frac{1}{2\pi\hbar}\left(\frac{e^2}{4\pi\epsilon_0\hbar}\right)^2\left(\frac{1}{n^2} - \frac{1}{4}\right) = \frac{m_e}{16\pi\hbar^3}\left(\frac{e^2}{4\pi\epsilon_0}\right)^2\frac{n^2 - 4}{n^2}.$$

Hence

$$\lambda_2 = \frac{c}{f_2} = \frac{16\pi\hbar^3 c}{m_e(e^2/4\pi\epsilon_0)^2}\frac{n^2}{n^2 - 4} = 363 \times 10^{-9}\left(\frac{n^2}{n^2 - 4}\right)\text{m}.$$

The shortest wavelength, corresponding to $n = \infty$, is 363 nm; the longest wavelength, corresponding to $n = 3$, is 653 nm.

42–3 SPIN, THE EXCLUSION PRINCIPLE, AND THE STRUCTURE OF ATOMS

The Spin of the Electron

When an atom with a magnetic dipole moment is placed in an external magnetic field, its potential energy changes according to Eq. (29–26), $U_{\text{mag}} = -\boldsymbol{\mu} \cdot \mathbf{B}$. As we already saw in our discussion of the Stern–Gerlach experiment, each one of the $2\ell + 1$ orientations of an atom with angular-momentum quantum number ℓ has a magnetic dipole moment with a different z-component. When such an atom is placed in a magnetic field, the $2\ell + 1$ orientations no longer have the same energy. If hydrogen is placed in an external magnetic field, the energies of the first excited state with $\ell = 1$ are therefore slightly *split* (Fig. 42–12). The frequencies of the radiation emitted (or absorbed) in a transition to or from one of the three (now-split) levels are accordingly not quite the same, and this is detectable by experiment.

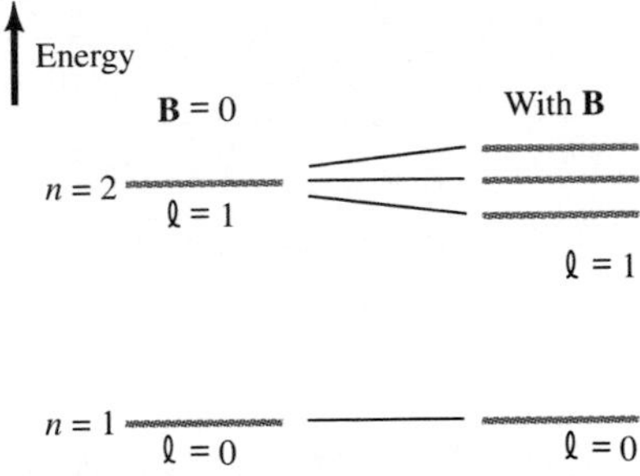

FIGURE 42–12 When a hydrogen atom is subject to an external magnetic field, the three states corresponding to $\ell = 1$, which all have the same energy in no magnetic field, have slightly different energies—the levels are split. The amount of splitting depends on the strength of the magnetic field. Note that the $\ell = 0$ state consists of only one level and is therefore not split.

States of a given ℓ that have a common energy in the absence of a magnetic field break up into $(2\ell + 1)$-member *multiplets* with slightly different energies in the presence of a magnetic field. Because $\ell = 0, 1, 2, 3, \ldots$, only odd-valued multiplets were expected. This turned out not to be the case: For some atoms—silver, for example—*doublets* appear; that is, there are two components, as we saw in our discussion of the Stern–Gerlach experiment. For a doublet, $2\ell + 1 = 2$, so $\ell = \frac{1}{2}$ for these atoms, and this was forbidden by the rules of quantum theory as they were understood in the early 1920s. In 1924 Wolfgang Pauli decided that the electron had to be described by one more quantum number, which could take only two values. A year later George Uhlenbeck and Samuel Goudsmit proposed that the electron has an *intrinsic angular momentum*, or *spin*, $\hbar/2 \equiv s\hbar$. Whereas the angular momentum $L = \ell\hbar$ discussed thus far is associated with the motion of an electron around a nucleus (a quantum mechanical version of $\mathbf{r} \times \mathbf{p}$), the spin is an *internal* property of the electron.

The electron spin

The fact that electrons have an intrinsic angular momentum $s\hbar$, with $s = \frac{1}{2}$, means that $2s + 1 = 2$, and an electron can appear in two states. For simplicity we call these states "up" and "down," corresponding to the two possible directions that the spin vector can take. In the absence of a magnetic field, the energy of an "up" electron is the same as that of a "down" electron in the hydrogen atom. However, when a magnetic field is present, the energies of these two states differ slightly. This is seen experimentally in the slightly different frequencies of photons emitted when the electrons jump from these two states. As a result of electron spin, the number of possible electron states that correspond to a given ℓ doubles from $2\ell + 1$ to $2(2\ell + 1)$. When $\ell = 0$, the number of states with the same energy is 2 (Fig. 42–13). For $\ell = 1$ there are $2 \times 3 = 6$ states, and so on. When observed closely in the presence of a magnetic field, the $\ell = 0$ state is always a doublet, the $\ell = 1$ state contains 6 levels, and so forth.

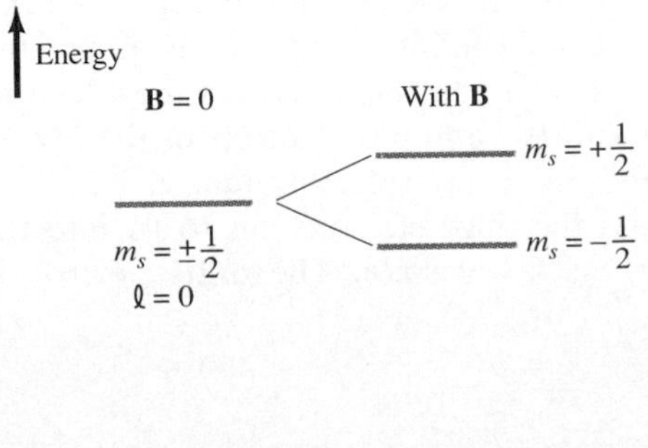

Energy
B = 0
With **B**
$m_s = +\frac{1}{2}$
$m_s = \pm\frac{1}{2}$
$\ell = 0$
$m_s = -\frac{1}{2}$

FIGURE 42–13 The existence of electron spin explains why some states are split into an even number of levels when a magnetic field is applied.

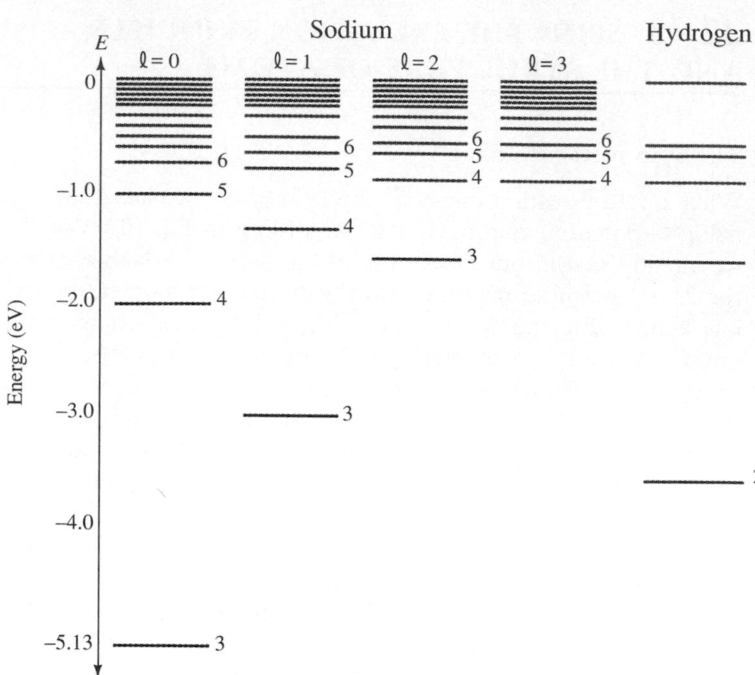

FIGURE 42–14 The schematic diagram of energy levels for an atom (here, sodium) is qualitatively similar to the hydrogen atom spectrum. The multiplets all have very nearly the same energy. For atoms with $Z > 2$, the energies for a given n value but for varying values of ℓ are no longer exactly equal. The levels with a given n and ℓ are slightly separated due to the internal magnetic field.

FIGURE 42–15 Wolfgang Pauli.

Multi-electron Atoms and the Exclusion Principle

Conceptually, the ideas on which the Bohr model is based extend to multi-electron atoms very simply. Instead of one orbit, there are many allowed orbits, and the electrons are distributed among the various orbits, filling those orbits of lowest energy first. The results predicted by the more rigorous treatment of quantum mechanics are similar. Each electron moves in the attractive Coulomb potential of the nucleus plus a repulsive potential due to the presence of the other $(Z - 1)$ electrons, where Z is the atomic number, the number of electrons in an un-ionized atom. The energy-level structure turns out to be qualitatively the same as that of the hydrogen atom. We still have an n-quantum number, which labels the total energy, and an ℓ-quantum number, which labels the angular momentum of the electron in that energy level, but it is no longer true that, for a given n, the $\ell = 0, \ell = 1, \ldots$ levels have the same energy. For a fixed value of n, the energy increases with ℓ (Fig. 42–14).

We might expect that in the ground state of any atom, all the electrons would be in the lowest energy level. The radiation spectrum from such an atom would be qualitatively the same as that of hydrogen. (The numbers would be different, because the central charge is Ze and because there would still be effects of electron–electron repulsion.) Such a spectrum would bear little resemblance to the rich structure observed throughout the periodic table of the elements.

The Pauli exclusion principle

It was again Pauli who pointed out that to understand the structure of multi-electron atoms ($Z \geq 2$), a new ingredient was needed. Pauli proposed his **exclusion principle**, according to which *each quantum state can accommodate only two electrons, one in the "up" state and one in the "down" state* (Fig. 42–15).

Let us examine what emerges when we start filling energy levels as per the exclusion principle (Fig. 42–16). Helium, $Z = 2$, has two electrons; both can fit into the $n = 1, \ell = 0$ state. There is no room for another electron in the lowest state. Helium is said to form a *closed shell*. Next consider lithium, $Z = 3$. Two electrons fit into the $n = 1, \ell = 0$ state, and the third electron has to go into the next lowest energy state, which is the $n = 2, \ell = 0$ state. The third electron is

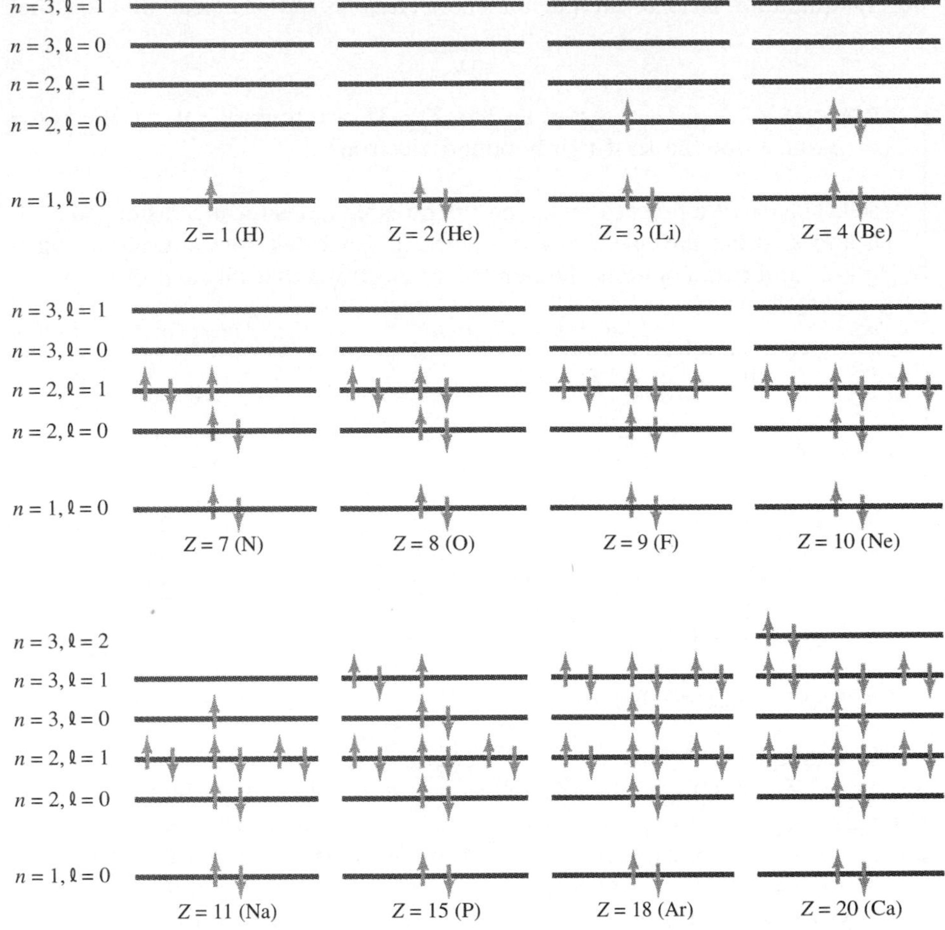

FIGURE 42–16 Pattern of electron energy-level occupation for elements from $Z = 1$ to $Z = 20$. The level splittings are not to scale.

farther from the nucleus than the other two electrons are (remember that $r \propto n^2$), and the positive charge $+3e$ of that nucleus is partly screened by the negative charge of the two electrons in the $n = 1$ orbit. As a result, the third electron is less tightly bound to the nucleus and can therefore be pulled more easily into the orbit of a nearby atom. Thus a lithium atom can bind with another atom to form a molecule and, like other atoms that have one electron outside a closed shell, is chemically very active.

For beryllium, $Z = 4$, we again fill a shell, the $n = 2, \ell = 0$ shell, and we expect beryllium to be less active than lithium. This is indeed the case. For $Z = 5$ through $Z = 10$, the $n = 2, \ell = 1$ levels are successively filled. The element $Z = 10$ is neon, which corresponds to another major closed shell, the $n = 2$ shell; it is an inert gas, one of a class of elements noted for their chemical inactivity. Fluorine, $Z = 9$, is one electron short of having a filled shell. Elements such as fluorine with a *hole* in a shell react particularly strongly with atoms such as lithium, which have one electron outside a filled shell. Fluorine has the lowest Z value of the *halogens* (atoms with a single hole in a shell), just as lithium has the lowest Z value of the *alkali metals* (atoms with one electron outside a filled shell).

The picture that we have sketched is very qualitative, and a great deal more information is needed to explain fully the chemical properties of atoms. Suffice it to say that the details of the periodic table can be understood both qualitatively and quantitatively in a quantum mechanical description. You should note that the existence of discrete energy levels, spin, and the exclusion principle are purely quantum mechanical phenomena. There is no classical hint of their existence.

EXAMPLE 42–5 An atom has $Z = 37$ electrons. What are the values of n and ℓ for the least-tightly-bound electron?

SOLUTION: We proceed by listing the possible levels for increasing values of n and ℓ, using the rule that for a given n, ℓ can take on values only up to $n - 1$, and then counting the number of electrons that fill each of the levels:

n	ℓ	*Number of Electrons*	*Cumulative Total of Electrons*
1	0	2	2
2	0	2	4
2	1	6	10
3	0	2	12
3	1	6	18
3	2	10	28
4	0	2	30
4	1	6	36
4	2	10	46

Thus the thirty-seventh electron is expected to lie in the $n = 4$, $\ell = 2$ shell.

In Example 42–5, we apply simple counting rules to determine the quantum numbers of the outermost electron in an atom with a certain number of electrons. As Z increases, these simple rules fail, because new effects enter. For example, the dynamics of the orbital motion leads to a tendency for an electron to align its spin with or against the orbital angular momentum. There is then a term in the energy that depends on the alignment of the spin and the orbital angular momentum, a so-called spin–orbit term. These effects may lead to slight differences in the order in which the levels are filled; deviations from our simple rules start at $Z = 19$.

Do All Particles Obey the Exclusion Principle?

Electrons have spin $\hbar/2$. Nuclear physics experiments have shown that protons and neutrons also have spin $\hbar/2$, and according to a very general theorem, *all particles with spins* $\hbar/2, 3\hbar/2, 5\hbar/2, \ldots$ *obey the exclusion principle*. We call particles having half-integral spin **fermions**, named after Enrico Fermi. This has an important bearing on the structure of nuclei, which are made up of protons and neutrons. By expectation and experiment, nuclei have a shell structure analogous to that of atomic electrons. In nuclei the average potential energy is the result of the mutual attraction of all the protons and neutrons by a *nuclear force*. This force is such that the energy levels tend to be spaced equally.

There are particles that do not obey the exclusion principle. Those particles with intrinsic spin of the form $s\hbar$, where $s = 0, 1, 2, \ldots$, were shown by Pauli to behave differently from particles with spin $\hbar/2$. Unlike fermions, which are unable to share the same quantum state with one another, in effect such particles "prefer" to be in the same state. Particles in this class, whose spin is an integer multiple of $\hbar$, are known as **bosons**, named after Satyendra Nath Bose. The photon provides an example of a boson; it has intrinsic angular momentum $\hbar$.

Photons, like all particles with integer spin, show a preference for congregating in the same quantum state. We shall see some physical consequences of this congregation effect when we study lasers and liquid ^4He.

42–4 THE STRUCTURE AND ENERGY STATES OF MOLECULES

The Formation of Molecules

Molecules are combinations of two or more atoms. We can calculate properties of molecules because even the lightest nucleus is some 2000 times more massive than an electron, so the nuclei in a molecule move much more slowly than their electrons. It is thus a good approximation to take the nuclei to be fixed. The electrons move rapidly around the nuclei and effectively create a smear of negative charge in which the nuclei are embedded.

Many of the important properties of simple molecules can be understood by focusing on a particular example. We shall consider the molecule H_2, which consists of two hydrogen atoms. To understand how the nuclei (protons) and electrons of the hydrogen atoms are arranged so that the two atoms bind to form the H_2 molecule, and to find the binding energy of that molecule, we study the energy as a function of r, the separation between the two hydrogen nuclei. There are several contributions to the energy. First, there is the electrostatic repulsion between the positively charged nuclei. For the H_2 molecule, this contribution to the potential energy is $U(r) = e^2/4\pi\epsilon_0 r$.

We must also include the attraction of each of the two electrons to each nucleus and the electron–electron repulsion. Although a precise calculation requires quantum mechanics, we can discuss the sum of these contributions to the energy (the "electron contribution"), and the dependence of this contribution on r, qualitatively. When r is large, the energy is lowest when one electron is close to one of the H nuclei and the other is close to the other H nucleus. In other words, for large r the H_2 molecule looks like two independent hydrogen atoms. We can also understand the limit $r \to 0$. Then the two H nuclei are directly on top of one another, and as far as the electrons are concerned, they "see" a nucleus of charge $2e$, which for atomic purposes is like a helium nucleus. Thus for $r \to 0$, the H_2 molecule looks like a helium atom.

Let us now put together some numerical values. The energy of a single hydrogen atom in its ground state is -13.6 eV, so at large r the total electron contribution is twice this value, -27.6 eV. At small r, the electron contribution is that of a helium atom. This energy can be evaluated by noting that (a) each electron sees a nucleus of charge Ze, with $Z = 2$. The electron–nucleus energy is then $-Z^2$ (13.6 eV) $= -54.4$ eV for each electron, a total of -108.8 eV. (b) The two electrons are fairly close together, on average roughly half a Bohr radius. In that case the energy associated with the electron–electron repulsion is about double 13.6 eV, and positive. When we put (a) and (b) together, the electron contribution at small r is $(-108.8\text{ eV}) + (27.2\text{ eV}) \simeq -80$ eV. This is quite a good estimate; a correct quantum mechanical calculation of the binding energy of helium (and experiment!) gives the binding energy of helium as -78.5 eV.

To estimate the electron contribution as a function of r, we draw a smooth line between the small-r result (-78.5 eV) and the large-r result (-27.6 eV).

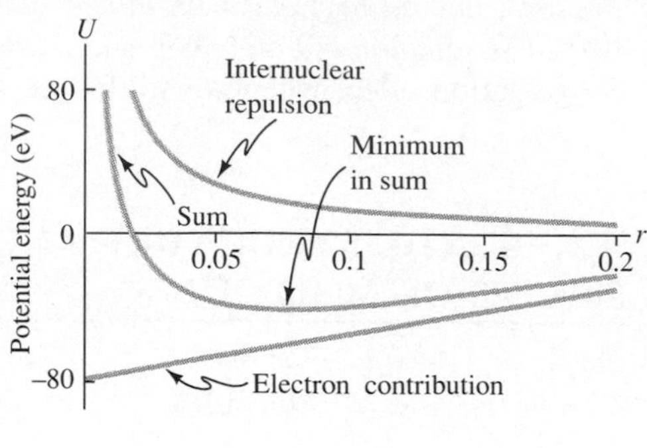

FIGURE 42–17 Energy diagram for the formation of the H_2 molecule. The upper curve is the potential energy associated with internuclear repulsion; the lower curve is an estimate of the potential energy associated with the electrons; the middle curve is the sum of these two terms. The sum has a minimum at $r \simeq 0.07$ nm, very close to the observed nuclear separation in the H_2 molecule.

Figure 42–17 shows the internuclear repulsion term, $e^2/4\pi\epsilon_0 r$; the electron contribution; and the sum of these two terms. *This sum has a minimum* at a nuclear separation of approximately 0.07 nm, and this is a stable equilibrium point. The experimental value of the separation of the H atoms in the H_2 molecule is very close to this value. This pattern holds for other stable diatomic molecules. Most simply stated, *the forces of attraction between the electrons and nuclei cancel the forces of repulsion between the electrons and between the nuclei at the position of stable equilibrium.*

The conditions under which molecules form cannot be met for all configurations of atoms, and only certain combinations of atoms can form molecules. Qualitatively, molecules can form under the following circumstances:

1. The two atoms have *paired* electrons outside closed shells, and the orbits of those electrons overlap. Here "paired" means that one electron has spin up while the other has spin down. Because the electrons have opposite spins, the Pauli exclusion principle does not prevent them from moving close to one another, into the region between the nuclei. In that region, the attraction of each to the "opposite" nucleus more than compensates for their mutual repulsion, and in effect there is a net attraction. Paired electrons of this type can form a *bond*, meaning that their interaction provides the mechanism for binding the two atoms together. The larger the number of bonds, the stronger the binding.
2. Only electrons not in closed shells can form bonds. That is because an electron in a closed shell already has a second electron acting as its "partner" within its own shell. In other words, only electrons in outer shells contribute to the chemical properties.
3. An electron may not be in a closed shell but may nevertheless already be paired within its own, partly filled shell. Such electrons cannot pair up with an electron from another atom. More precisely, whether electrons pair up with partners from their own or from another atom is a matter of which configuration has the lower energy. This is not easily calculated, but chemists have developed a set of empirical rules that govern a large number of situations.

The H_2 molecule is an example of bonding between atoms that have unfilled shells. We say that such molecules are formed by *valence bonding*. *Ionic bonding* occurs when an atom with an electron outside a closed shell combines with an atom in which the outermost shell has one vacancy (a "hole"). As an example, consider sodium fluoride (NaF). Sodium (and all alkali metals) has one electron outside a closed shell, and fluorine (and all halogens) has a hole in its outer shell. Energy must be expended for the outermost electron of a sodium atom to be freed. This energy is the ionization energy of sodium. Its value is 5.1 eV, which is compensated by energy released if the free electron drops into the hole in a fluorine atom. This released energy is the negative of the energy it would take to remove that electron from a now-filled shell of fluorine, -3.5 eV. The net energy cost to move the electron from the sodium atom to the fluorine atom is $5.1\text{ eV} - 3.5\text{ eV} = 1.6\text{ eV}$. However, we now have a positive (Na^+) ion and a negative (F^-) ion. These ions attract one another, and the energy associated with this attraction more than makes up for the 1.6 eV, so NaF has a lower energy than the separated Na and F atoms.

Van der Waals Forces

The rules listed above for conditions necessary for the formation of molecules are not without exceptions. Under special circumstances, for instance, the inert gases—whose atoms have no electrons outside of closed shells—*do* form molecules. Ar_2 (diatomic argon) is an example. The reason these molecules form is that there is an additional force between atoms, even those of inert gases. This force is the so-called *van der Waals force*. The van der Waals force is due to electromagnetism but falls off much more rapidly than the Coulomb force does with increasing distance r between the nuclei. Despite the fact that the van der Waals force is rather weak, it is always attractive and thus can give rise to binding between atoms. The van der Waals force can be understood from a classical argument, similar to the argument that explains why a comb attracts uncharged bits of paper. When one atom approaches another, its charge distribution affects the other by giving rise to a small displacement of charges, such that an electric dipole structure is created. This electric dipole has an electric field, which interacts with the electric dipole of the first atom. The two dipoles attract with a force proportional to $1/r^5$ (see Problem 33).

Van der Waals forces, although very weak, are responsible for a number of phenomena beyond the formation of exotic molecules, such as the adhesion between a liquid and the sides of its container, and the departure of a gas from ideal gas behavior (see Chapter 17).

Molecular Spectra

In the formation of molecules, the circulating electrons form an environment in which nuclei tend to be separated by a fixed distance r_0 on the order of 0.1 nm. These nuclei are nearly fixed centers of attraction for the electrons. We may expect from the general ideas about quantum mechanics in Section 42–1 that molecules will have a series of electron energy levels separated by the kinds of energies that characterizes atomic levels—that is, gaps on the order of 1 eV to 10 eV. Experiments, however, reveal an even more complicated spectrum, even for simple diatomic molecules. This complication is due to the possibility of quantized motions not allowed for the hydrogen atom: vibrational and rotational motion.

Vibrational Motion. We saw that, for diatomic molecules, there is a separation r_0 where the potential energy curve shown in Fig. 42–17 has a minimum. When such a curve has a minimum, the two nuclei can oscillate about that point. As we learned in Chapter 7, near the bottom the potential energy curve can normally be approximated by a parabolic curve; that is, by a harmonic oscillator potential. As a consequence, there will be energy levels associated with the vibrational motion of the nuclei in this harmonic oscillator potential. We show in Example 42–6 by using Bohr's techniques that these energy levels are given by $E_{\text{vib}} = n\hbar\omega$, where n is an integer and ω is a characteristic frequency given by $\sqrt{k/M}$. Here k is the "spring constant" and M is the *reduced mass* of the two nuclei in vibrational motion, given by $M = M_1 M_2/(M_1 + M_2)$, where M_1 and M_2 are the two nuclear masses.

EXAMPLE 42–6 Suppose that a mass m is attached to the end of a spring and moves in a circular orbit of radius r under the influence of the spring, which exerts an attractive centripetal force of magnitude $F = kr$. Use the condition of fitting an integral number of wavelengths into allowed circular orbits to determine the energy spectrum associated with these circular orbits. (This system is realized in diatomic molecules, in which the two atoms behave as though they were connected by a spring.)

SOLUTION: We proceed exactly as we did for the hydrogen atom: We use $F = ma$ to obtain one relation between position r and speed v, and then impose the quantization condition to obtain another relation. For $F = kr$, the equation of motion $F = ma$ for circular motion becomes

$$kr = \frac{mv^2}{r},$$

or, equivalently,

$$k\not{r} = m\omega^2\not{r}.$$

This relation can be solved for ω, yielding $\omega = \sqrt{k/m}$. The energy is the sum of the kinetic energy, K, and the potential energy, $U(r)$, of a three-dimensional harmonic oscillator:

$$E = K + U(r) = \tfrac{1}{2}mv^2 + \tfrac{1}{2}kr^2 = \tfrac{1}{2}m\omega^2r^2 + \tfrac{1}{2}m\omega^2r^2 = m\omega^2r^2.$$

The wavelength-fitting condition implies that

$$n\lambda = 2\pi r.$$

We now use $\lambda = h/p = h/mv = h/m\omega r$ in the equation $n\lambda = 2\pi r$ to obtain

$$\lambda = \frac{2\pi r}{n} = \frac{h}{m\omega r}.$$

It follows from this equation that

$$r^2 = \frac{nh}{2\pi m\omega}.$$

When this is substituted into the expression for the energy, we find for allowed energies that

$$E_n = m\omega^2r^2 = \frac{\not{m}\omega^{\not{2}}nh}{2\pi\not{m}\not{\omega}} = \frac{nh\omega}{2\pi} = n\hbar\omega, \qquad \text{where } n = 1, 2, 3, \ldots, \qquad (42\text{–}16)$$

the desired result.

Although we have found the energy levels that correspond to circular motion, we can always think of that motion as due to superimposed oscillatory motions (see Chapter 14). Thus we can think of our result as representing more generally the allowed energy levels of a three-dimensional harmonic oscillator. These levels are equally spaced in units of $\hbar\omega$.† In turn, a system of two atoms that behave as though they were connected by a spring behaves like a single (reduced) mass on the end of a three-dimensional spring.

We have seen that the minimum in the potential energy (the "spring") is created by the electron cloud in which the two nuclei are embedded, and we might therefore expect that it is parameters such as the size of the atomic orbits that determine the value of k. We can *estimate* the value of k by dimensional analysis. The dimensions of any spring constant such as k are $[k] = [EL^{-2}]$, where $[E]$ is the dimension of energy, $[ML^2T^{-2}]$. To estimate k we use this relation with typical atomic energies and distances, $E \simeq e^2/8\pi\epsilon_0 a_0$ and $L \simeq a_0$, where a_0 is the Bohr radius. Then

$$k \simeq \frac{e^2}{8\pi\epsilon_0 a_0^3} = \frac{e^2}{8\pi\epsilon_0}\left(\frac{e^2 m_e}{4\pi\epsilon_0 \hbar^2}\right)^3 = \frac{1}{2}\frac{(m_e)^3}{\hbar^2}\left(\frac{e^2}{4\pi\epsilon_0 \hbar}\right)^4. \qquad (42\text{–}17)$$

Dimensional analysis cannot specify any additional numerical factors, and to improve our estimate we must add a little additional physical reasoning. In particular, because molecules are somewhat larger than atoms, we could replace a_0 by a somewhat larger radius, say, $2a_0$. This leads to the estimate

$$k \simeq \frac{1}{16}\frac{(m_e)^3}{\hbar^2}\left(\frac{e^2}{4\pi\epsilon_0 \hbar}\right)^4. \qquad (42\text{–}18)$$

Once we know the effective spring constant, we can find the allowed energies of oscillation from Eq. (42–16):

$$E_{\text{vib}} = n\hbar\omega = n\hbar\sqrt{\frac{k}{M}} = \frac{n}{2}m_e\left(\frac{e^2}{4\pi\epsilon_0 \hbar}\right)^2\sqrt{Z_1 Z_2 \frac{m_e}{M}}, \qquad (42\text{–}19)$$

where n is an integer and M is the reduced mass of the two nuclei.

We have written Eq. (42–19) as we have because it exposes the factor $(m_e/2)(e^2/4\pi\epsilon_0\hbar)^2$, which we recall from Eq. (42–9) is the magnitude of the ground-state energy of hydrogen, 13.6 eV. Let us call this factor E_0. Then

$$E_{\text{vib}} = nE_0\sqrt{Z_1 Z_2 \frac{m_e}{M}}. \qquad (42\text{–}20)$$

Numerically, the factor by which nE_0 is multiplied is on the order of 10^{-2}; that is, the vibrational energies are on the order of 10^{-1} eV.

Rotational Motion. The fact that there is a minimum in the potential energy suggests that, independent of the vibrations, the nuclei of a diatomic molecule, separated by the distance r_0, can rotate about some common center as though they were connected by a rigid rod. Our molecule will then have energy levels associated with rotations of this dumbbell-like system. The classical energy E is given in terms of the angular momentum L by $E = L^2/2I$, and when we apply to this result the condition that the angular momentum is quantized, we find the quantized energies.

† A correct quantum mechanical treatment gives an energy spectrum for motion of a mass on the end of a spring in N dimensions that is slightly different from the Bohr-model result in Eq. (42–16); namely, $E_n = [n + (N/2)]\hbar\omega$, where again $n = 1, 2, \ldots$ and ω is the classical angular frequency.

We must still evaluate the rotational inertia. A diatomic molecule such as OH can be approximated as two masses, M_1 and M_2, connected by a massless rigid rod of length a. The molecule will rotate about an axis perpendicular to the rod and passing through the center of mass of the system. For rotations through the center of mass, the rotational inertia I is given by (see Chapter 9)

$$I = Ma^2,$$

where M is the reduced mass of the system. We now apply the quantization condition for angular momentum. The rotational energy levels are then of the form

$$E_{\text{rot}} = \frac{L^2}{2I} = \frac{\ell(\ell+1)\hbar^2}{2I} \simeq \frac{\ell(\ell+1)\hbar^2}{2Mr_0^2} = \ell(\ell+1)\frac{m_e}{2}\left(\frac{e^2}{4\pi\epsilon_0\hbar}\right)^2\frac{m_e}{4M}.$$

Here ℓ is the angular-momentum quantum number. We have substituted $r_0 = 2a_0 = 8\pi\epsilon_0\hbar^2/e^2m_e$, as we did in the derivation box. We again recognize the factor E_0, so

$$E_{\text{rot}} = \frac{\ell(\ell+1)E_0 m_e}{4M}. \quad (42\text{–}21)$$

The energy levels are suppressed by an additional factor of $\sqrt{m_e/M}$ from the factors of the vibrational levels and are on the order of 10^{-3} eV.

The depiction of molecular spectra that emerges is the following: There are electronic levels on the order of electron-volts apart. Associated with each level is a series of vibrational levels, separated by roughly 10^{-2} of the electronic levels, and associated with each of *these* levels is a series of rotational levels, with separations about 10^{-4} of the electronic levels (Fig. 42–18). The latter two sets of levels are described as vibrational bands and rotational bands. The study of rotational bands is important to chemistry, because these are most easily excited. Molecular spectra involving transitions between the rotational levels consist of wavelengths on the order of 10^4 times atomic wavelengths and thus involve infrared rather than optical spectroscopy.

The complex structure of molecular energy levels is responsible for many of the subtleties of organic chemistry and of the chemical reactions that occur in biological systems. A technological application illustrates one consequence of the complexity of molecular structure. The high electric potentials that occur within transformers can cause the electrical breakdown of pockets of air in the transformer. In breakdown, high voltage ionizes air molecules. The electrons liberated are accelerated by the large electric fields. In colliding with successive

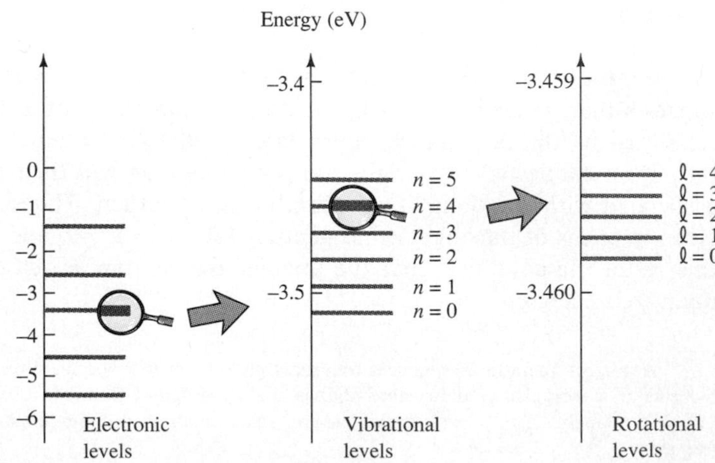

FIGURE 42–18 Molecular levels. The striking feature of these spectra is that the energy levels in each part are separated by energies 100 times finer than the energy-level separations of the previous part. The three parts represent electronic, vibrational, and rotational motions.

air molecules, they liberate more electrons, which in turn are accelerated. An avalanche of electrons results, leading to currents that short out the transformer. An excellent method to prevent this type of breakdown is to introduce a gas that allows the liberated electrons to lose energy in ways that do not lead to further ionization. Sulfur hexafluoride (SF_6), which is one gas used in this way, has a very rich molecular structure, and there are many non-ionizing ways in which it can be excited by an electron colliding with it.

42–5 BAND THEORY

Many engineering applications of quantum phenomena arise in the area of semiconductor physics. In Section 27–5 we described the *band structure* in the energy levels of crystalline solids, a structure in which there are regions of energy in which there are energy gaps—that is, no allowed energy levels. No electrons within a given solid can have energies in these forbidden regions, their energies being restricted to certain finite ranges of energy, the so-called bands of allowed energies. We discussed how the presence of energy bands and the numbers of electrons that occupy the allowed energy levels determine why some materials are conductors whereas others are insulators or semiconductors. We are now in a better position to elaborate on why bands form, a discussion of what is known as *band theory*.

Let us start by recalling from Section 42–4 our discussion of the energy levels for two hydrogen atoms. We saw that we can think about a helium atom as the result of moving together two hydrogen atoms. However, our naive expectation for the binding energy of helium, that the value of the ground-state energy of helium is a simple multiple of the binding energy of hydrogen, is different from the experimental value. The actual binding energy is different from our expectation because the electrons of the two hydrogen atoms interact with each other. In other words, a new interaction added to an existing system will generally shift the energy levels of the existing system. This statement needs to be supplemented by further remarks, about *how* the energy levels are modified as the interaction changes smoothly. When the two hydrogen atoms are far apart, there are two electrons in a pair of identical levels of energy -13.6 eV, the ground states of the two separate hydrogen atoms. We say that the two electrons are in *degenerate* energy levels. (Actually, there are four degenerate levels, because each electron can be in a spin "up" or spin "down" state, and these states have the same energy values.) What happens to these levels when the atoms are brought so close together that helium forms, and the interaction between the electrons becomes significant? We saw above that there is at least one level lower than the original -13.6 eV levels, namely, the true ground state of helium. The remaining three levels are pushed *upward* and correspond to excited states of helium. This is a general characteristic of energy levels:

The splitting of degenerate energy levels

When there are degenerate energy levels and an additional interaction comes into play, the levels are split, some moving upward and some moving downward relative to their original positions.

The observation above plays an important role in band theory. What happens when we deal with greater amounts of material rather than just two atoms? To be specific, we study N atoms of sodium, which have filled shells plus a single electron (the valence electron) in an $n = 3$, $\ell = 0$ state. There are $2N$ degenerate (equal-energy) states available to the valence electron when the atoms are far apart and there is no interaction between the atoms. (The factor of 2 is

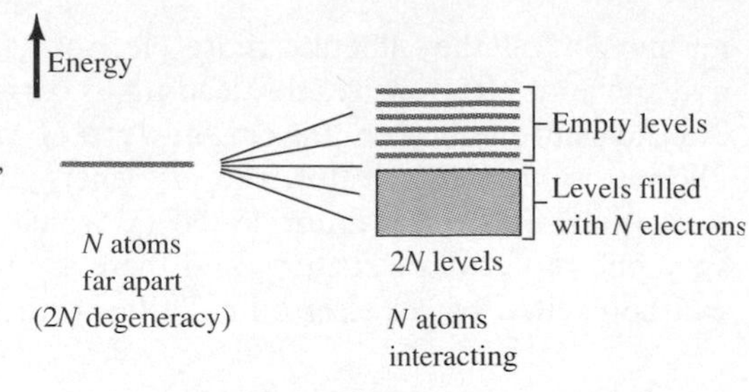

FIGURE 42–19 When many atoms are brought together to form a crystalline solid, the degenerate levels of N atoms split to form band structures, in which there are ranges of allowed electron energies and ranges (gaps) of forbidden energies. In sodium, the valence electrons fill one half of a band.

present because the electron can be in a spin "up" or "down" state, and these states have the same energy.) Now suppose that the N atoms form a crystal lattice with closely spaced lattice sites. Then the originally degenerate valence-electron states are split by new interactions between the atoms; N energy levels are pushed down and N are pushed up, each by a slightly different amount. The $2N$ energy levels then form a band of very narrowly spaced levels. The N valence electrons fill only half of this band, namely the N lowest-lying energy levels (Fig. 42–19). As we discussed in Chapter 27, an electric field can easily lift some electrons at the top of the filled levels to the empty levels, where they can move without hindrance. Metallic sodium is thus a good conductor.

For contrast, we can consider N atoms of magnesium. The Z value of magnesium is one higher than that of sodium. Magnesium has the same atomic structure as sodium, except that there are *two* electrons in the $n = 3$, $\ell = 0$ states. The band formed from the originally degenerate states when the magnesium atoms combine into a lattice is therefore completely filled. Magnesium could be an insulator. However, empty levels that originally came from the $n = 3$, $\ell = 1$ states, which were above the $n = 3$, $\ell = 0$ levels when the two sets of levels were each degenerate, fan out into a second band. For magnesium, it happens that the two bands overlap (Fig. 42–20). Thus there is room for electrons, under the influence of an electric field, to move into empty levels, and magnesium is a good conductor. There are other cases in which bands of this type do not overlap. Indeed, all the possibilities discussed in Chapter 27 can be realized.

What are the "new" interactions that occur when the atoms are assembled into a regular lattice, and that lead to the splitting of degenerate electron energy levels into a band structure? These are the interactions of electrons with ions other than the ones to which they were originally attached. In particular, if the ions were not present, the electrons would move freely in the "box" formed by the dimensions of the material, within which their wavelike aspect makes them behave analogously to waves on a string, in standing waves whose wavelengths must be integral multiples of the appropriate dimensions of the box. These waves are the waves formed by the wave function, and the electron is likely to be found

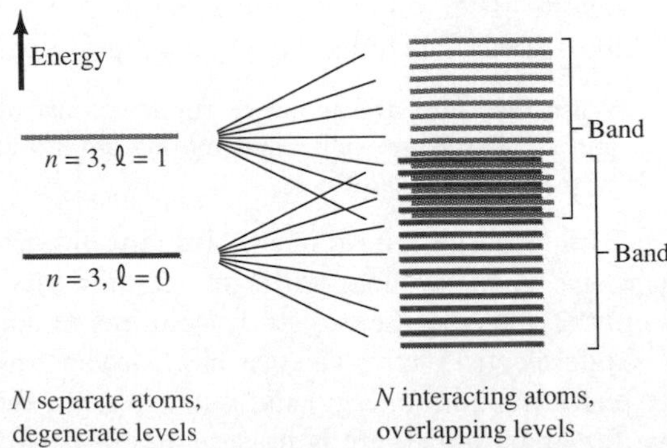

FIGURE 42–20 The origin of the band structure of N atoms of magnesium in a crystalline solid. The splitting of the degenerate levels with quantum number $n = 3$, $\ell = 0$ leads to a band that is completely filled with electrons. The degenerate levels $n = 3$, $\ell = 1$ also split, and the allowed band they form overlaps with the band formed from the $n = 3$, $\ell = 0$ levels.

where those waves have crests (or troughs). The crests and troughs are regularly spaced. At this point we can imagine adding the ions at regularly spaced sites. Two modifications to our original depiction result. First, only those electrons whose wavelengths correspond to the ionic spacing can be supported in the material, for reasons closely related to the Bragg scattering arguments we gave in Chapter 41. Second, it is possible to show that half of the allowed waves of the electrons have peaks (or troughs) where the ions are found, and half have peaks (or troughs) exactly in between the ions. The energies of the half whose peaks are where the ions are found are lowered, because these electrons are more likely to be near the ions to which they are attracted. The energies of the half whose peaks are in between the ion locations are raised, because these electrons do not feel the force of attraction as strongly as they did when they were in an isolated atom.

SUMMARY

One of the important consequences of quantum mechanics is that energies of bound systems can take on only discrete values. In the Bohr model, the quantization of atomic energy levels follows from the restriction of angular momenta to integral multiples of $\hbar$:

$$L = n\hbar, \qquad \text{where } n = 1, 2, 3, \ldots . \tag{42–6}$$

With this condition the allowed energy values are

$$E_n = -\frac{m_e}{2n^2}\left(\frac{e^2}{4\pi\epsilon_0 \hbar}\right)^2, \qquad \text{where } n = 1, 2, 3, \ldots, \tag{42–9}$$

in agreement with experiment. Energy conservation allows electrons to jump between levels with different n values while emitting photons of frequency given by

$$hf = E_i - E_f, \tag{42–11}$$

where E_i and E_f are the initial and final energies, respectively.

Quantum mechanics as developed by Heisenberg and Schrödinger shows that the structure of the possible energy levels is more complex than the Bohr model predicts. For each value of n, there are n^2 energy levels characterized by angular momentum $\ell\hbar$, where $\ell = 0, 1, 2, \ldots, (n-1)$, and $2\ell + 1$ spatial orientations are allowed for the vector angular momentum characterized by ℓ. However, the full structure of atoms cannot be understood until we add the fact that electrons carry an intrinsic angular momentum $\hbar/2$ called spin. Moreover, the Pauli exclusion principle shows that no more than two electrons can appear in any quantum state (corresponding to the $2s + 1$ states with $s = \frac{1}{2}$). With these additions, the complex structure of multi-electron atoms, as revealed in the periodic table of elements, can be explained.

Molecules form when the forces of attraction between the electrons and nuclei cancel the forces of repulsion between the electrons and between the nuclei at the position of stable equilibrium. This can occur if the atoms involved have electrons outside closed shells. Molecular spectra reveal energy levels of a molecule that are associated with vibrations of the atoms and rotations of the entire molecule. The electronic, vibrational, and rotational modes form a hierarchy.

Electrons in solids are attracted by the ions that form the crystal lattice. The effect of this attraction is to split previously degenerate levels, forming a range of very closely spaced levels called energy bands.

QUESTIONS

1. In the treatment of a particle confined to a box (in Section 42–1), why was the ground state characterized by $n = 1$ rather than $n = 0$?
2. Can we determine the atomic composition of distant objects by studying the wavelengths of their emitted photons?
3. What determines the shortest and longest wavelengths that a hydrogen atom can emit?
4. On the one hand, we say that electrons in atoms have discrete energies; on the other hand, we say that there is inherent uncertainty in our ability to measure energies. Is there a conflict here?
5. Because of the exclusion principle, only electrons with energy near the Fermi energy move under the influence of a field and create currents. If electrons did not obey the exclusion principle, how would conduction in metals differ?
6. In the Stern–Gerlach experiment, atoms of silver were observed to have only two components of angular momentum. Does that mean that all the electrons of a silver atom have angular-momentum quantum number ℓ equal to zero?
7. Is it true that all hydrogen atoms are indistinguishable from each other? How is it possible to distinguish between hydrogen with a nucleus consisting of a single proton (^{1}H), hydrogen with a nucleus consisting of a single proton and a single neutron (^{2}H, deuterium), and hydrogen with a nucleus consisting of a single proton and two neutrons (^{3}H, tritium)?
8. Why does an atom with a magnetic dipole moment need a magnetic field that varies spatially in order to deflect an atom? Would a constant magnetic field suffice?
9. The term *fine structure* is used to refer to closely spaced energy levels. Discuss the origin of such levels for some of the systems we described in this chapter.
10. Van de Graaff accelerators, which have terminals at very high voltages inside a pressurized tank, are used to accelerate nuclear particles to high energy for nuclear reactions. Would air or sulfur hexafluoride be better as an insulating gas in a van de Graaff accelerator?
11. Why does it take about twice as much energy to excite an electron from the $n = 1, \ell = 0$ state to the $n = 2, \ell = 2$ state for He^+ as it does for He (neutral helium)?
12. Would you expect the orbital radius of the lowest orbit in a helium atom to be less than, equal to, or greater than that in a hydrogen atom? Why?
13. In discussing the formation of molecules, we stated that the minimum of the net potential energy in the interaction of two hydrogen atoms is the position where the attractive forces between the electrons and the nuclei cancel the repulsive forces between each electron and between each nucleus at the position of stable equilibrium. How do we translate a statement about the potential energy minimum to a statement about the force?
14. Why do the arguments we made about the formation of the diatomic hydrogen molecule, H_2, not apply to the formation of a diatomic helium molecule, He_2, from two helium atoms?
15. Helium has two electrons. Both can be in spin "up" (or "down") states, or one can be up while the other is down. Taking into account the exclusion principle and the fact that angular-momentum states with quantum number $\ell = 1$ have a higher energy than angular-momentum states with quantum number $\ell = 0$, which spin arrangement will occur in the ground state?
16. Suppose that we add an electron to hydrogen. The second electron could be in the same orbit as the first (the spins would then have to point in opposite directions). What might prevent the existence of such a negatively charged atom? Would the existence of an atom consisting of one proton and three electrons be as likely, or unlikely?

PROBLEMS

42–1 *Quantization of Energy and Angular Momentum*

1. (I) What wavelength of radiation is necessary to ionize hydrogen? [*Hint:* Recall that to ionize an atom, it is necessary to raise the energy of the electron to be emitted from its ground state to at least $E = 0$.]
2. (I) Consider singly ionized helium (helium in which one electron has been removed). How much energy is needed to remove the second electron?
3. (I) What are the energy and wavelength of the photon emitted when a hydrogen atom jumps from its first excited state ($n = 2$) to its ground state ($n = 1$)? from $n = 4$ to $n = 3$?
4. (I) Given that, in the Bohr model, radiation is emitted only when n changes by one unit, for what range of values of n will the radiation from hydrogen lie in the visible range; that is, with wavelengths in the range 400 nm to 700 nm?
5. (II) The *Lyman series* is a series of spectral lines for hydrogen whose wavelengths correspond to Eq. (42–12) when $n_1 = 1$ (Fig. 42–21). (a) What are the quantum numbers of the states involved in the three transitions of the Lyman series with the longest wavelengths? (b) Calculate the wavelengths for the transitions of part (a). Are these wavelengths in the visible or ultraviolet regions?

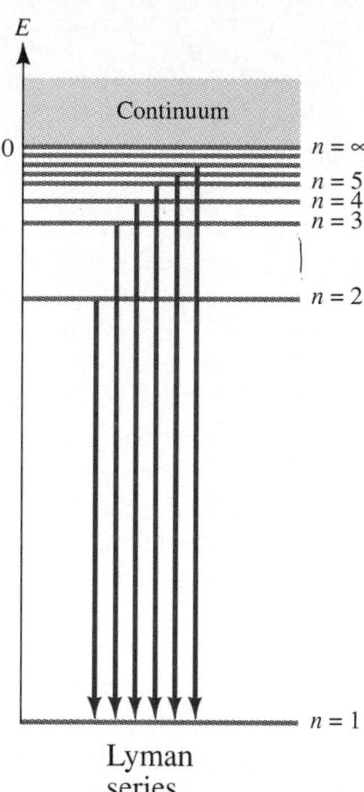

FIGURE 42–21 Problem 5.

6. (II) Singly ionized helium is a single-electron atom with $Z = 2$ and a nuclear mass four times as large as the mass of the hydrogen nucleus. Calculate the longest wavelength in the analogue of the Lyman series, those transitions that end in $n = 1$ (see Problem 5).

7. (II) What is the wavelength of the first line of the Lyman series (see Problem 5) in quadruply ionized boron ($Z = 5$)?

8. (II) The *Paschen series* is a series of spectral lines for hydrogen whose wavelengths correspond to Eq. (42–12) when $n_l = 3$ (Fig. 42–22). (a) What are the quantum numbers of the states involved in the three transitions of the Paschen series with the longest wavelengths? (b) Calculate the wavelengths for the transitions of part (a). Are these wavelengths in the visible or infrared regions?

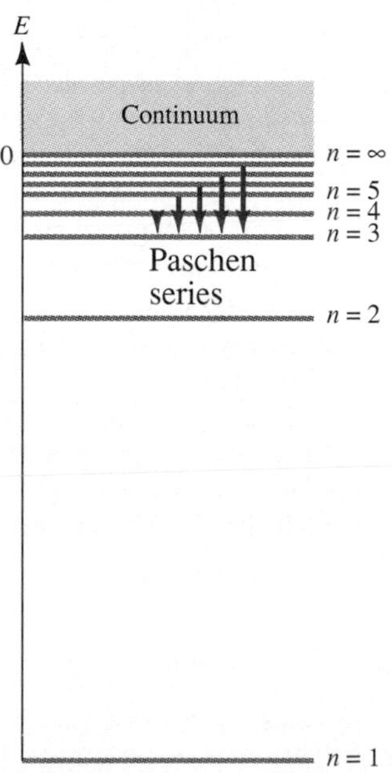

FIGURE 42–22 Problem 8.

9. (II) Derive a formula for the wavelengths of the transitions in single-electron ions whose nuclei have charge $+Ze$. For what range of Z values will radiation for $n = 2$ to $n = 1$ transitions be in the 1-nm range? in the 0.1-nm range?

10. (II) It is useful to introduce the *fine-structure constant* $\alpha \equiv e^2/4\pi\epsilon_0\hbar c$ in problems that involve atoms. (a) What are the dimensions and value of α? What is the value of $1/\alpha$, to the nearest integer? (This is a useful number to remember.) (b) Express in terms of α the energy of the nth level of the hydrogen atom, E_n. (c) Calculate in terms of α the speed of the electron in the lowest Bohr orbit of hydrogen.

11. (II) What are the orbital radius, speed, momentum, and energy of an electron in the $n = 3$ state of hydrogen? Assume a classical model to calculate the momentum and speed of the electron.

12. (II) Consider the Rutherford planetary model of circular orbits. (a) Use the expression for the radius given in Eq. (42–5) to calculate the speed of an electron in terms of $e^2/4\pi\epsilon_0$, the angular momentum L and m_e. (b) Use the result of part (a) to calculate the electron's acceleration. (c) Calculate the period of the electron's orbital motion.

13. (II) According to classical electromagnetism, the power radiated by a particle of charge e that undergoes an acceleration a is

$$P = \frac{2}{3}\frac{e^2}{4\pi\epsilon_0 c^3}a^2.$$

Use this formula and the results of Problem 12 to calculate the energy radiated by an electron per unit time for a circular orbit of angular momentum L.

14. (II) (a) Use the results of Problem 13 to show that the fraction of energy radiated by an electron in a single period T is given by

$$\frac{\Delta E}{E} = \frac{8\pi}{3}\left(\frac{e^2}{4\pi\epsilon_0 Lc}\right)^3.$$

(b) Use your result to find the number of periods over which the electron would lose all its energy, and the time over which this would take place. Evaluate this result numerically for L when the orbital radius is 0.1 nm. (Your result should be on the order of 10^{-10} s.)

15. (II) Calculate the frequency of the radiation emitted by an electron in a hydrogen atom when it jumps from a level of quantum number $n + k$ to a level of quantum number n. Find a simple expression for it when $n \gg k$. Express your result in terms of $L = n\hbar$ and compare it with the classical frequency $f_{cl} = 1/T$, where T is the period calculated in Problem 12. Show that the two will be equal only for $k = 1$, so the requirement that the results of quantum mechanics coincide with classical results for very large values of n (the Bohr correspondence principle) means that only transitions with changes in n of value $\Delta n = 1$ are allowed.

16. (III) (a) Express the power ($P = \Delta E/\Delta t$) radiated by an accelerating electron (see Problem 13) in terms of $L = n\hbar$. The energy will be radiated in the form of a photon of energy hf, where f is the frequency that corresponds to a transition $\Delta n = 1$. (b) Find $\Delta t = \Delta E/P$ in terms of n. Use this formula to calculate a numerical value for Δt when $n = 1$. This time interval may be viewed as the time constant for the $n = 2$ state of the hydrogen atom to jump to the $n = 1$ (ground) state. (c) Compare your time constant with the period of an electron in the $n = 2$ orbit. From your comparison, could you argue that the $n = 2$ orbit is almost stable? (d) Roughly, Δt is the longest time you have available for measuring the energy of the $n = 2$ state. Use the time–energy uncertainty relation to estimate the spread in energy of the $n = 2$ state, and compare this spread with the transition energy $E_2 - E_1$.

42–2 *The Quantum Theory of Angular Momentum and the True Spectrum of Hydrogen*

17. (II) A correction should be supplied to the formulas for the energy levels of hydrogenlike atoms: The electron orbits about the center of mass of the electron–proton system rather than about the proton itself. This correction, known as the *reduced-mass effect*, is small because the proton, of mass m_p, is much more massive than the electron, of mass m_e. The result is that the energy levels of hydrogen should be corrected to

$$E_n = -\frac{m_e}{1 + (m_e/m_p)} \frac{1}{2n^2} \frac{e^2}{(4\pi\epsilon_0\hbar)^2}.$$

The energy levels for deuterium, an atom with a nucleus whose charge is that of the hydrogen nucleus but whose mass is about twice that of hydrogen, obey the same formula. Find the difference in the wavelengths of radiation emitted in the transition between the $n = 2$ and $n = 1$ states for the two atoms. It was the observation of this difference that led to the discovery of deuterium by Harold Urey in 1931.

18. (II) A heavy version of the electron, called the *muon*, μ, differs from the electron only in that its mass is $m_\mu \simeq 207m_e$. In the muonic hydrogen atom (a proton–muon atom), the reduced-mass effect of Problem 17 is much more important than in ordinary hydrogen. (a) What is the radius of the muonic hydrogen atom in its ground state? (b) Calculate the wavelength of radiation emitted in the transition of a muonic hydrogen atom from the $n = 2$ state to the $n = 1$ state.

42–3 *Spin, the Exclusion Principle, and the Structure of Atoms*

19. (I) For what value of Z is the $n = 3$ level filled?

20. (I) Give the n and ℓ quantum numbers of the levels that are filled in the ground state of an atom of potassium, $Z = 19$.

21. (I) What is the lightest element with a single electron in an $n = 2$ level?

22. (II) In multi-electron atoms, the ordering of levels does not coincide with hydrogenlike atoms. In a particularly stable multi-electron atom, the following states are fully occupied: $n = 1$, $\ell = 0$; $n = 2$, $\ell = 0, 1$; $n = 3$, $\ell = 0, 1, 2$; $n = 4$, $\ell = 0, 1, 2, 3$; $n = 5$, $\ell = 0, 1, 2$; $n = 6$, $\ell = 0, 1$. What is the Z value of this atom?

23. (II) An atom with $Z = 10$ has a closed shell; an atom with $Z = 11$ (sodium) may be viewed as a "nucleus" with a net charge of $+e$ and one electron on the outside. In terms of this simplistic depiction, what would you expect sodium's ionization energy to be? [*Hint:* What levels are filled in the "nucleus"?]

24. (II) When an electron in an atomic state characterized by the quantum numbers (n, ℓ) is placed in a magnetic field, then the $(2\ell + 1)$ possible states no longer have the same energy $E_{n,\ell}$. The magnetic field *splits the degenerate levels* so the energy levels have the values $E_{n,\ell} + \kappa\ell B$, $E_{n,\ell} + \kappa(\ell - 1)B, \ldots, E_{n,\ell} - \kappa(\ell - 1)B$, $E_{n,\ell} - \kappa\ell B$, where κ is a constant. (a) In the presence of a magnetic field, how many different spectral lines are there in the transition $(n = 2, \ell = 1) \to (n = 1, \ell = 0)$? (b) in the transition $(n = 3, \ell = 2) \to (n = 2, \ell = 1)$? [*Hint:* Different transitions in which the energy change ΔE is the same give a single spectral line of frequency $f = \Delta E/h$.]

25. (II) The potential energy U of a magnetic dipole with magnetic dipole moment $\boldsymbol{\mu}$ in a magnetic field $\mathbf{B}$ is $U = -\boldsymbol{\mu} \cdot \mathbf{B}$. The magnetic dipole moment of an electron has magnitude $\mu = (e/m_e)S$. How much more (or less) energy does an electron with spin up have than an electron with spin down in the presence of an external magnetic field with magnitude $B = 1.5$ T, assuming that the field is parallel to the "up" direction?

26. (II) The inert gases are a set of elements whose outermost shells of a given n are filled. Find the Z values of all the inert gases for which $Z < 100$. [The approximation of successive filling of levels in the order $n = 1, 2, 3, \ldots$, $\ell = 0, 1, 2, \ldots, (n - 1)$ is incorrect for Z greater than about 20. Thus your numbers will not agree with the periodic table at the high end.]

27. (III) The binding energy of the $n = 1$ electron in an atom for which the nuclear electric charge is $+Ze$ is obtained by taking the hydrogen atom results and replacing e^2 by $(Ze)(e) = Ze^2$. The ionization energies of the least-tightly-bound electrons for some atoms with Z values from 1 to 29 (that is, through the $n = 3$ levels) are as follows:

Z:	2	10	11	18	19	28	29
Ionization energy (eV):	24.6	21.6	5.1	15.8	4.3	7.6	7.7

Compare these values with those that you would obtain for the removal of an $n = 1$ electron. To what do you ascribe the huge discrepancy in most cases?

28. (II) What are the energies of the three lowest levels in the vibrational spectrum of the OH molecule? Assuming that the allowed transitions correspond to $\Delta n = 1$, find the wavelengths of the allowed transitions between these levels.

29. (II) The wavelength of the $n = 1 \to n = 0$ transition in the vibrational spectrum of CO is 2.93×10^{-5} m. Use this to estimate the spring constant k in Eq. (42–18). Approximate the mass of the carbon and oxygen nuclei as 12 and 16 times the mass of a hydrogen nucleus, respectively, to compare your result with a calculation from Eq. (42–18). The discrepancy that you will find between the two numbers suggests that Eq. (42–18) is a very crude approximation to the spring constant.

30. (II) The $\ell = 1 \to \ell = 0$ rotational transition in the lowest electronic state of the CO molecule has a wavelength of 2.603 mm. Estimate the rotational inertia of the CO molecule and the equilibrium separation of the atoms.

31. (II) The energy difference between the lowest state in the CN molecule and the first excited electronic state would give rise to a single spectral line at a wavelength near 387.4 nm if there were no rotations (or vibrations). However, the lowest and first excited states actually consist of a series of rotational states with superimposed energies $\ell(\ell+1)\hbar^2/2I_0$ and $\ell(\ell+1)\hbar^2/2I_1$, respectively. Calculate I_0 and I_1 from the following data: Transition $(n = 1, \ell = 1) \leftrightarrow (n = 0, \ell = 0)$ gives $\lambda = 387.4608$ nm; transition $(n = 1, \ell = 2) \leftrightarrow (n = 0, \ell = 1)$ gives $\lambda = 387.3998$ nm; transition $(n = 1, \ell = 0) \leftrightarrow (n = 0, \ell = 1)$ gives $\lambda = 387.5763$ nm. Why, physically, should I_0 and I_1 be different?

32. (II) (a) Use the data of Problem 31 to find the internuclear separation between the C $(A = 12)$ and N $(A = 14)$ nuclei for the two electronic states $n = 1$ and $n = 0$. (b) Calculate the wavelengths for the transitions $(n = 1, \ell = 3) \to (n = 0, \ell = 2)$. Compare your results with the measured value, $\lambda = 387.3369$ nm.

33. (II) Suppose that when one atom is separated from another by a distance r, its charge distribution affects the other by giving rise to a small displacement of charges, with separation $d \ll$ the atomic radius, and by creating an electric dipole. We know from Eq. (23–14) that the electric field due to such a dipole has magnitude $E \simeq ed/4\pi\epsilon_0 r^3$. (Here we have ignored angular factors such as sines or cosines.) The charge separation created in the second atom induces a dipole moment of magnitude αE. The coefficient α has the dimensions $4\pi\epsilon_0(\text{length})^3$, as can be seen from $E \simeq ed/4\pi\epsilon_0 r^3$, and this length turns out to be the charge separation d. Now the induced dipole interacts with the electric field E. Show that the resulting interaction leads to a potential energy in the interaction of the atoms of the general form

$$U(r) = -(\text{a constant})\,\frac{e^2}{4\pi\epsilon_0}\frac{d^5}{r^6}.$$

This potential energy is what leads to the van der Waals force between the atoms.

34. (III) The rotational motion of a diatomic molecule affects the equilibrium position of the nuclei. If R_0 is the separation for zero angular momentum and R is the average separation when there is rotation, then the rotational energy is $\ell(\ell+1)\hbar^2/2MR^2$, where M is the reduced mass of the nuclei. In addition, for a given vibrational frequency ω, the vibrational potential energy in the presence of rotation is $\frac{1}{2}M\omega^2(R - R_0)^2$. Calculate the new equilibrium separation R by minimizing $E(R)$, the sum of the two new terms. Treat $R - R_0$ as small. How is the rotational inertia changed, and what effect does this have on the rotational spectrum of the molecule?

General Problems

35. (I) For what value of the angular momentum quantum number ℓ does the quantum mechanical angular momentum have the value $1.00\ \text{g}\cdot\text{cm}^2/\text{s}$?

36. (II) Suppose that two electrons are in orbit around one proton (an H^- ion), both in an $n = 1$ level. By listing all the potential energy contributions, make a crude guess of how much energy it would take to ionize one of the electrons.

37. (II) An electron in hydrogen jumps from the first excited state to the ground state. The mean duration for the transition is 2.6×10^{-10} s. What is the uncertainty in the energy value of the first excited state? Give your answer in electron-volts and as a fraction of the energy of the state.

38. (II) By using the assumptions of the Bohr model, calculate the radius of the ground-state electron orbit for doubly ionized lithium, Li^{2+}.

39. (II) A marble of mass $m = 20$ g moves in a circular orbit near the bottom of a circular bowl. The height of the sides of the bowl is given by $h = \alpha r^2$, where r is the radial distance from the bottom of the bowl. Given that $\alpha = 0.25\ \text{cm}^{-1}$, find the separation between the successive allowed energies of the marble in the bowl. It is not surprising that we have no intuitive feel for quantum mechanical phenomena.

40. (II) All integer values of the principal quantum number, n, even very large ones, are allowed in atoms. In practice, it is very hard to excite orbits that correspond to large n values in an atom unless the atom is totally isolated. Estimate the largest value of n that would be possible if you could make a gas of *atomic* hydrogen of density $\rho = 1.0 \times 10^{-8}\ \text{g/cm}^3$.

41. (II) By finding the momentum of an electron in a circular orbit with orbital angular momentum L and by using your knowledge of the magnetic dipole moment due to a current loop (Fig. 42–23), show that this orbiting electron has a magnetic dipole moment

$$M = \frac{e\hbar}{2m_e}\frac{L}{\hbar}.$$

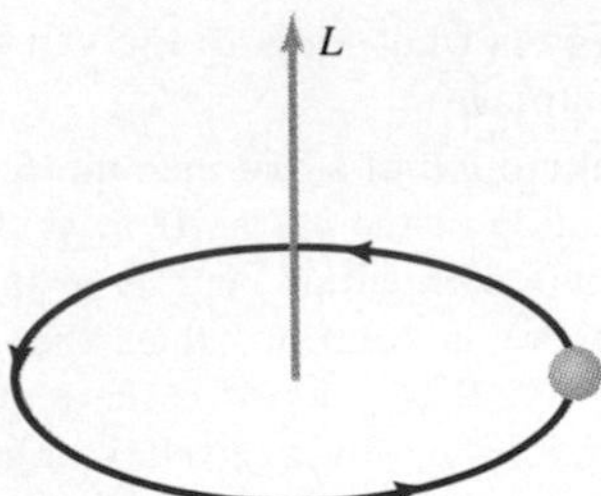

FIGURE 42–23 Problem 41.

The quantity $e\hbar/2m_e$, known as the *Bohr magneton*, is the minimum quantized value for the possible magnetic dipole moments of electrons in their orbits.

42. (III) Repeat your calculation of the energy levels of the Bohr atom, but now assume that the potential energy is given by $U(r) = -(e^2/4\pi\epsilon_0 r) + (\sigma/r^2)$, where σ is a constant. Are your energy values larger or smaller than the corresponding values for the pure Coulomb case? Is your result physically reasonable? Is the effect of the added term more important for large n or for small n, and is that result plausible?

43. (III) According to the Heisenberg uncertainty relations, it is impossible to measure to high precision the position of an atomic electron without making its momentum highly uncertain. By finding the magnitudes of the momentum for the ground state and the first excited state of an electron in a hydrogen atom, calculate just how well the position of an electron in the ground state of hydrogen can be located before you can no longer be sure whether it is in the ground state or the first excited state. It is for the reasons outlined here that it makes little sense to think of an electron as following a classical orbit like those of the planets.

44. (III) Use the position–momentum uncertainty relation to prove that it is not possible to detect orbits in hydrogen, by the following argument: (i) a measurement of the nth orbit must be such that $\Delta x \ll r_{n+1} - r_n$. Calculate Δx. (ii) This gives rise to an uncertainty in the momentum of the electron in the orbit, and hence in the energy. Calculate the uncertainty in the energy. (iii) Show that the energy is larger than the energy difference between adjacent orbits, so an orbit cannot be "photographed."

45. (III) The large-n orbits of hydrogen are more classical than quantum mechanical in the sense that uncertainties in the momenta and positions in these orbits are small compared with the momenta and positions themselves. Show that this is the case.

46. (III) Consider a particle of mass m that moves in a circular orbit of radius r around a center of attraction that exerts a force on the particle. The particle's potential energy is $U = U_0(r/a)^4$, where U_0 is a constant with dimensions of energy, and a is a constant with dimensions of length. (a) Calculate the total energy (including kinetic energy) in terms of r by using the relation $mv^2/r = |F| = |dU/dr|$. (b) Use the Bohr quantization rules to obtain an expression for the quantized energy values of the particle.

CHAPTER 43

These laser beams are made visible by smoke in the air. Lasers operate as a result of quantum phenomena that are important on a macroscopic scale.

QUANTUM EFFECTS IN LARGE SYSTEMS OF FERMIONS AND BOSONS

We have seen how the particles that compose matter are of two types: Those with angular momentum $(\ell + \frac{1}{2})\hbar$, where $\ell = 0, 1, 2, \ldots$, are fermions, and those with angular momentum $\ell\hbar$ are bosons. Identical fermions obey the Pauli exclusion principle, and the number of them allowed to be in the same quantum mechanical state—the state specifies the energy of the particle—is at most two, whereas identical bosons obey an analogue to the exclusion principle that translates into a tendency for them to congregate in the same, lowest state. The consequent behaviors of systems with many fermions or bosons are the subject of this chaper. We have already seen how the filling of the energy bands in materials is explained by the exclusion principle as it applies to electrons. Even with no dynamic forces present, identical fermions behave as though there were a repulsive force between them, just because they obey the exclusion principle. This "force" accounts for the incompressibility of matter. The congregating effects of identical bosons explain the laser and superfluidity. Fermions may congregate in pairs, forming bosons, if there are appropriate binding forces between them; when this happens, a large system of identical fermions can behave like a large system of identical bosons. This phenomenon lies behind the behavior of superconductors and leads to such properties as flux quantization and the Josephson effect, properties that go beyond the mere absence of resistivity in superconductors.

43–1 THE EXCLUSION PRINCIPLE IN METALS AND STARS

Electrons in Metals and the Fermi Energy

The classical discussion of the electrical conductivity of metals (Section 27–3) is based on the presence of free electrons in metals. These electrons move under the influence of an externally imposed electric field. Resistance to current flow is due to the existence of ions. In the classical discussion, electrons undergo collisions with ions, which leads to a retarding force and a slow, constant average speed for the electrons called the *drift speed.* The quantum mechanical description of conductivity is based on the same premise, free electrons in metals. However, the description of the motion of those electrons must be quantum mechanical, and this leads to some major differences with the classical estimates.

Picture the electrons as being confined to a one-dimensional box of macroscopic length L—several centimeters, for example. From Eq. (42–1) we get the energy levels for a single electron confined to a one-dimensional box:

$$E = \frac{\pi^2\hbar^2n^2}{2m_eL^2}.$$

(Here we have used $\hbar = h/2\pi$.) In a three-dimensional box, there are three such contributions, corresponding to motion in the x-, y-, and z-directions. The result is that

$$E = \frac{\pi^2\hbar^2(n_1^2+n_2^2+n_3^2)}{2m_eL^2}, \tag{43–1}$$

with each of the integers n_1, n_2, and n_3 allowed to take the values 1, 2, 3, ... The lowest state is the one for which $n_1 = n_2 = n_3 = 1$, and it is unique. The next energy level consists of three states: $(n_1, n_2, n_3) = (2, 1, 1)$, $(1, 2, 1)$, and $(1, 1, 2)$, and all three states have the same energy. Different states that have the same energy are degenerate; as the n value increases, the degree of degeneracy becomes high.

Suppose that we now start filling the levels described by Eq. (43–1) with electrons. According to the Pauli exclusion principle, each state can accommodate at most two electrons. There are many electrons, because in a metal one or more electrons per atom are free. In the filling process, we must respect the exclusion principle. *The lowest energy for a metal that contains N_e electrons corresponds to a situation in which the energy levels fill up from the bottom—the lowest energy level—with two electrons per energy level.* It is important to notice that the values of $n_1^2+n_2^2+n_3^2$ under consideration are very large. Suppose that an electron in the metal has energy $1\text{ eV} = 1.6\times10^{-19}$ J, a value typical of the energy of electrons in conductors. Let us also take $L = 1$ cm, so that we are dealing with a small but definitely macroscopic piece of metal. It follows from Eq. (43–1) that

$$(n_1^2+n_2^2+n_3^2) = 2m_e\frac{EL^2}{\pi^2\hbar^2} \simeq \frac{2(0.9\times10^{-30}\text{ kg})(1.6\times10^{-19}\text{ J})(0.01\text{ m})^2}{\pi^2(1.05\times10^{-34}\text{ J}\cdot\text{s})^2}$$
$$\simeq 2.6\times10^{14}.$$

Thus we are dealing with values of n_1, n_2, and n_3 that are very large. If, for example, we let $n_1 = n_2 = n_3 = 10^8$, then the energy *difference* ΔE between that state and an adjacent state, with n_1 increased by 1 (but n_2 and n_3 unchanged), is determined by

$$\frac{\Delta E}{E} = \frac{(n_1 + 1)^2 - n_1^2}{n_1^2 + n_2^2 + n_3^2} \simeq \frac{2n_1}{n_1^2 + n_2^2 + n_3^2} \simeq 10^{-8} .$$

The energy levels are so close together that we say they *almost* form a continuum.

When an electron accelerates classically under the influence of an external field, its energy increases smoothly. In quantum mechanics the electron can only jump to a state of higher energy. Here the Pauli exclusion principle plays an important role: An electron cannot move to a state of higher energy unless the state is free of electrons. Thus only electrons at the top of the filled levels can be accelerated by the electric field.

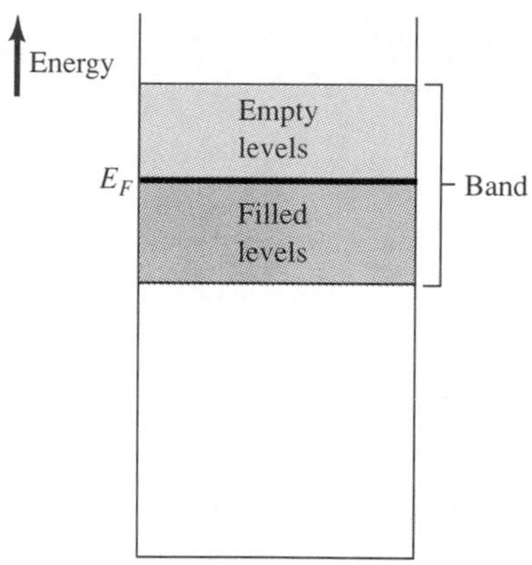

FIGURE 43–1 The energy levels in a crystalline solid form a band structure. In metals at $T = 0$, the levels in the topmost band containing electrons are filled only to the Fermi energy, E_F, of the material, leaving unoccupied levels above.

The highest-filled energy level is called the **Fermi energy**, E_F, after Enrico Fermi (Fig. 43–1). Among Fermi's many contributions to physics was the realization of the importance of the exclusion principle for many-particle systems (Fig. 43–2). In general, the de Broglie wavelength of a particle roughly corresponds to the space that it occupies. Thus the closest that two electrons with the same energy and angular momentum can get to each other is about half a de Broglie wavelength. Any closer distance would effectively superimpose the electrons, a situation forbidden by the exclusion principle.

The Fermi energy can be calculated to within an accuracy of a few percent by taking the closest possible distance between two electrons to be one-half the de Broglie wavelength that corresponds to the *Fermi momentum* $p_F = \sqrt{2m_e E_F}$. If we denote this closest distance by d, then the total number of electrons N_e in a cubical box of sides L is

$$N_e = \left(\frac{L}{d}\right)^3; \tag{43–2}$$

$$d = \left(\frac{N_e}{L^3}\right)^{-1/3} = n_e^{-1/3}. \tag{43–3}$$

Here n_e is the number density of free electrons in the metal. When we equate this closest distance to half the de Broglie wavelength at the Fermi energy, λ_F, we find that

$$d = \frac{\lambda_F}{2} = \frac{h}{2p_F} = \frac{h}{2\sqrt{2m_e E_F}} = \frac{\hbar\pi}{\sqrt{2m_e E_F}}. \tag{43–4}$$

We can combine Eqs. (43–3) and (43–4) and solve for E_F: $E_F = (\hbar^2/2m_e)(\pi^3 n_e)^{2/3}$. A more precise calculation that counts the number of electrons that can be accommodated with energy $E < E_F$, including degeneracy factors, leads to the replacement of the factor π^3 by $3\pi^2$, a very minor change:

$$E_F = \frac{\hbar^2}{2m_e}(3\pi^2 n_e)^{2/3}. \tag{43–5}$$

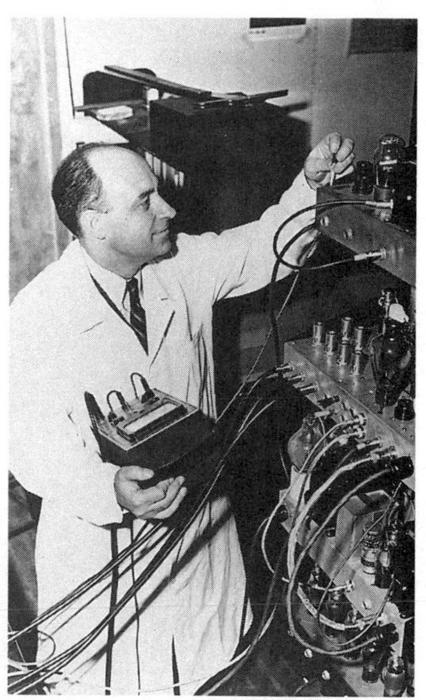

FIGURE 43–2 Enrico Fermi excelled as an experimentalist and as a theorist. Among his other accomplishments, he led the successful effort to build the first controlled nuclear fission reactor, completed in December 1942 in Chicago.

The magnitude of the Fermi energy depends on the density of free electrons. Copper, for example, has one free electron per atom. With an atomic weight of 63.5 g/mol and a mass density of 8.95 g/cm³, for copper

$$n_e = \left(\frac{1 \text{ electron}}{\cancel{\text{atom}}}\right)\left(\frac{6.02 \times 10^{23} \cancel{\text{atoms}}}{\cancel{\text{mol}}}\right)\left(\frac{1 \cancel{\text{mol}}}{63.5 \cancel{\text{g}}}\right)\left(\frac{8.95 \cancel{\text{g}}}{\cancel{\text{cm}^3}}\right)\left(\frac{10^6 \cancel{\text{cm}^3}}{\text{m}^3}\right)$$
$$= 8.5 \times 10^{28} \text{ electrons/m}^3.$$

Substituting this result into Eq. (43–5), we get $E_F = 1.12 \times 10^{-18}\text{ J} = 7.0\text{ eV}$.

EXAMPLE 43–1 Calculate the Fermi energy for N *neutrinos* (massless particles of spin $\hbar/2$, symbolized as ν) confined to a cubic volume with sides of length L. Because as far as experiment can tell, neutrinos are massless, they obey the relationship $E = pc$ between energy and momentum.

SOLUTION: Neutrinos are fermions with spin $\hbar/2$, so the exclusion principle still applies; that is, $\lambda_F = 2d$, where d is the spacing between neutrinos. The only difference between the treatment of neutrinos and our earlier treatment of electrons is that, for neutrinos, we use the massless-particle relation $E_F = p_F c$ instead of $E_F = p_F^2/2m_e$. Thus

$$d = \frac{\lambda_F}{2} = \frac{h}{2p_F} = \frac{hc}{2E_F}.$$

Hence

$$E_F = \frac{hc}{2d} = \pi\hbar c n_\nu^{1/3} = \pi\hbar c\left(\frac{N}{L^3}\right)^{1/3}.$$

A more precise counting of states leads to an E_F value that is a factor of $\pi/3$ larger than our crude calculation.

With the picture we have sketched in mind, we can ask what happens when a weak electric field is applied to a metal. Only the electrons at the top of the filled levels can accelerate and move to states of higher energies, because only they find empty levels to occupy (Fig. 43–1). In the expression for electrical conductivity, Eq. (27–22), with the collision time given by Eq. (19–50), the appropriate quantity to use is the Fermi speed v_F, defined by $E_F = \frac{1}{2}m_e v_F^2$, rather than $v_{\rm rms}$ given by the kinetic theory of gases. Recall that the use of $v_{\rm rms}$ gave an incorrect $\sqrt{T}$ temperature dependence for the resistivity [see Eq. (27–25)]. The effect of temperature changes on the behavior of electrons with a given Fermi energy is negligible. A 100-K change in temperature changes the energy of an electron by $kT = (1.4 \times 10^{-23}\ \text{J/K})\,(100\ \text{K}) = 1.4 \times 10^{-21}\ \text{J} = (1.4 \times 10^{-21}\ \text{J})/(1.6 \times 10^{-19}\ \text{J/eV}) \simeq 0.9 \times 10^{-2}\ \text{eV}$. Compared with the Fermi energy of copper, 7 eV, the factor kT is negligible. Thus the resistivity will be temperature independent for reasonable temperature changes.

The Incompressibility of Matter

The exclusion principle plays a crucial role in explaining the incompressibility of matter. A measure of this incompressibility is given by the bulk modulus, B, defined in Chapter 21 by Eq. (21–5),

$$B \equiv -\frac{p}{\Delta V/V}.$$

Here p is the pressure—actually, a pressure change—that brings about a fractional change $\Delta V/V$ in the volume of some sample of matter. Because an infinitesimal volume change $\mathrm{d}V$ is brought about by an infinitesimal pressure change, we rewrite this definition as

$$B = -V\frac{\mathrm{d}p}{\mathrm{d}V}. \qquad (43\text{–}6)$$

Suppose now that the sample of material is cylindrical with cross-sectional area A and that the pressure is applied to the ends. The work done in compressing the material along the axis of the cylinder is $\mathrm{d}W = -F\,\mathrm{d}L$, where $\mathrm{d}L$ is the length ı the cylinder is compressed. Work dW is done, so energy $dE = dW$ is the sample. Thus the change in energy of the sample is given by

$$dE = dW = -F\,dL = -\left(\frac{F}{A}\right)(A\,dL) = -p\,dV,$$

so the pressure p can be written in terms of the ratio of the energy change to the volume change:

$$p = -\frac{dE}{dV}. \tag{43–7}$$

When the volume of a metal changes, the number density changes, and so does the total energy. Let us calculate the bulk modulus under the assumption that the *only* resistance of a metal to compression comes about because of this energy change. The total energy of the free electrons in the material is equal to the number of electrons N_e multiplied by an average energy, which is a value somewhere between 0 and E_F. A calculation that we omit because of its length yields the result that the average energy factor is $\frac{3}{5}E_F$ (see Problem 46):

$$E \simeq \frac{3}{5}E_F N_e = \frac{3}{5}\frac{\hbar^2}{2m_e}\left(3\pi^2\frac{N_e}{V}\right)^{2/3} N_e.$$

Thus, from Eq. (43–7), the pressure is

$$p = -\frac{dE}{dV} = \frac{2}{5}\frac{\hbar^2}{2m_e}(3\pi^2)^{2/3}\left(\frac{N_e}{V}\right)^{5/3} \tag{43–8}$$

According to Eq. (43–6), one further derivative of the pressure is necessary to find the bulk modulus:

$$\begin{aligned} B &= -V\frac{dp}{dV} = -V\frac{d}{dV}\left[\frac{2}{5}\frac{\hbar^2}{2m_e}(3\pi^2)^{2/3}\left(\frac{N_e}{V}\right)^{5/3}\right] \\ &= -V\left[\frac{2}{\not 5}\frac{\hbar^2}{2m_e}(3\pi^2)^{2/3}(N_e)^{5/3}\right]\left(-\frac{\not 5}{3}\right)\frac{1}{V^{8/3}} \\ &= \frac{2}{3}\frac{\hbar^2}{2m_e}(3\pi^2)^{2/3}\left(\frac{N_e}{V}\right)^{5/3} = \frac{2\pi^{4/3}}{3^{1/3}}\frac{\hbar^2}{2m_e}n_e^{5/3}. \end{aligned} \tag{43–9}$$

For copper, $n_e = 8.5 \times 10^{28}\ \mathrm{m}^{-3}$, so $B = 6.4 \times 10^{10}\ \mathrm{N/m^2}$. The experimental value of B for copper is $13.4 \times 10^{10}\ \mathrm{N/m^2}$. Given the uncertainties of our estimates, the neglect of interaction between electrons, and the Coulomb repulsion between the ions, the fact that our rough approximation is within a factor of 2 of the experimental result is impressive. The effective repulsion between electrons due to the exclusion principle plays a major role in the high degree of incompressibility of matter.

White Dwarfs and Neutron Stars

The exclusion principle also plays a critical part in the evolution of stars: It is responsible for preventing a star's collapse under the mutual gravitational attraction of its mass. Stars are generally formed from the clumping together, through mutual gravitational attraction, of large clouds of gas (mainly hydrogen, which forms most of the raw material of the universe). When the hydrogen atoms fall together, gravitational potential energy is converted into kinetic energy. The density and temperature increase. The hydrogen atoms are ionized, and as further compression occurs, the protons that form the hydrogen nuclei come close enough to undergo a variety of reactions. For example, through a sequence of reactions to be described in Chapter 45, four protons can combine and form a helium nucleus and two positrons (the antimatter of electrons). A great deal of energy is released in the process, which is responsible for the luminosity of many stars and represents a kind of thermonuclear burning. While the reactions are going on, the temperature of the star stays high, and an equilibrium is established: The gravitational pressure, due to the weight of the matter, just balances the pressure in the gas of protons at the temperature of the thermonuclear burning. When the hydrogen fuel has burned to the point where the process can no longer occur, there is a substantial fraction of helium nuclei in the star. The gravitational pressure shrinks the size of the star further, until the temperature is high enough so that the helium nuclei begin to undergo reactions that produce still more energy. In what is a well-understood sequence of processes, heavier and heavier elements, up to iron, are produced. Iron does not take part in further thermonuclear reactions, and the reactions stop. However, the material continues its gravitational contraction.

The gravitational pressure that acts to compress the matter can be estimated. Suppose that the star contains N nucleons (protons and neutrons), each of mass M. (We can neglect the electron mass here.) A characteristic gravitational force at the star's surface has magnitude $G(NM)^2/R^2$, and thus a characteristic gravitational pressure (force per unit area) at the surface is

$$p_g = \frac{G(NM)^2/R^2}{4\pi R^2} = \frac{G(NM)^2}{4\pi(\frac{3}{4}V/\pi)^{4/3}} = 0.54\,\frac{G(NM)^2}{V^{4/3}},$$

where the volume $V = \frac{4}{3}\pi R^3$. Note that the gravitational pressure is directed *inward*. It actually varies through the star; a more rigorous calculation shows that the correct quantity for our calculation is three-fifths of the estimated value above. Thus

$$p_g = 0.32\,\frac{G(NM)^2}{V^{4/3}}. \qquad (43\text{–}10)$$

FIGURE 43–3 A particularly well-known white dwarf orbits the bright star Sirius as a companion star. The white dwarf can be seen here as the small bright spot just to the right of the image of Sirius.

As V decreases, the pressure grows, and without some countering pressure the star would collapse. The exclusion principle provides this countering pressure, called the *degeneracy pressure* of the electrons. This pressure, p_e, given by Eq. (43–8), points outward. (The subscript indicates that in this case the degeneracy pressure is due to electrons; the general symbol for pressure is p.) Note that the degeneracy pressure is largest when the mass of the fermion involved is smallest, which is why we use the degeneracy pressure of electrons rather than that of, say, protons. When the electron degeneracy pressure matches the gravitational pressure, equilibrium is reached. For less massive stars the end product is a *white dwarf* (Fig. 43–3). Given that N_e = number of protons $\simeq N/2$, we can calculate the radius of the resulting star. We get a radius of some 7000 km for a star the mass of the sun, 2×10^{30} kg (see Problem 49). This corresponds to a density of 7.5 metric tons/cm^3!

For a more massive star, p_g is larger, and, to maintain equilibrium, the factor N_e/V in the degeneracy pressure [Eq. (43–8)] must increase. This means, however, that the Fermi energy of the electrons grows. When the Fermi energy of an electron is comparable to the electron's rest mass, we must treat the electron relativistically. In the extreme relativistic case, the result of Example 43–1 must be used. The total energy of the electron is again N_e multiplied by the average energy, but this time p_e is proportional to $(N_e/V)^{4/3}$. This implies in turn that the factor V cancels from the equation $p_e = p_g$, which then cannot be satisfied by a judicious choice of V. The gravitational pressure always exceeds the electron degeneracy pressure when a star is more massive than about 1.4 solar masses (the *Chandrasekhar mass*), as first shown by Subrahim Chandrasekhar in the late 1930s, and gravitational collapse continues.

EXAMPLE 43–2 Show that if the electron's energy E in a star is so high that its rest mass can be ignored (see Example 43–1), then the gravitational pressure is on the order of magnitude of the degeneracy pressure for a star of one solar mass. (A star is electrically neutral, and, to a good approximation, the number of neutrons equals the number of protons when the star is formed.)

SOLUTION: The gravitational pressure is given by Eq. (43–10). We want to compare this to the degeneracy pressure, which can be calculated from the relation $p = -\mathrm{d}E/\mathrm{d}V$, Eq. (43–7). We again use $E \simeq \frac{3}{5}E_F N_e$, but we need the expression for the Fermi energy applicable to massless particles, $E_F = \pi\hbar c(N_e)^{1/3}V^{-1/3}$ (see Example 43–1). Thus

$$p = -\frac{\mathrm{d}E}{\mathrm{d}V} = -\frac{\mathrm{d}}{\mathrm{d}V}\left[\frac{3}{5}E_F N_e\right] = -0.6N_e^{4/3}\pi\hbar c\,\frac{\mathrm{d}(V^{-1/3})}{\mathrm{d}V} = -0.2\pi\hbar c N_e^{4/3}V^{-4/3}.$$

We can relate the number of electrons to the mass of the star as follows. Electrical neutrality gives us $N_e = N_p$. The total mass of the star, $M_\odot$ (one solar mass), is the total number N of neutrons and protons times the mass M of the proton (or neutron), $M_\odot = NM$. Finally, the number of protons is half the number of neutrons, so $N_e \simeq N/2 = (NM)/2M = M_\odot/2M$. Thus the ratio of the degeneracy pressure to the gravitational pressure is

$$\frac{p}{p_g} = \frac{0.2\pi\hbar c(0.5M_\odot/M)^{4/3}\cancel{V^{-4/3}}}{0.32G(NM)^2/\cancel{V^{4/3}}} = \frac{0.2\pi\hbar c(0.5M_\odot/M)^{4/3}}{0.32GM_\odot^{\,2}} \simeq 1.2.$$

This crude calculation does give the right order of magnitude. Chandrasekhar showed the more general result that if a star has a mass $sM_\odot$, then $p/p_g = 1.2s^{-2/3}$.

What happens when a star exceeds the Chandrasekhar mass and collapses can crudely be described by the statement that the gravitational pressure forces the electrons in the star to combine with protons into neutrons and neutrinos via the reaction

$$e^- + p \rightarrow n + \nu.$$

As we remarked in Example 43–1, neutrinos are apparently massless, and, more important, they interact so weakly with matter that they immediately escape from the star. The end result is a star made of N neutrons only, a *neutron star*.

Neutrons are fermions and hence also obey the exclusion principle. We have a new equilibrium condition, $p_g = p_n$, where p_n, the neutrons' degeneracy pressure, has the same form as that of p_e but replaces N_e with N and m_e with M. The relation

$$p_g = 0.32\frac{G(NM)^2}{V^{4/3}} = p_n = \frac{2}{5}\frac{\hbar^2}{2M}(3\pi^2)^{2/3}\left(\frac{N}{V}\right)^{5/3} \tag{43–11}$$

can be solved for V and thus for R. For a star of a few solar masses, numerical calculation yields R on the order of 10 km! The neutron star is as dense as a nucleus, and its compactness is responsible for the many properties that continue to intrigue astrophysicists. The existence of neutron stars was predicted in 1934 by Walter Baade and Fritz Zwicky. The pulsars first discovered by Anthony S. Hewish and collaborators in 1967 were identified by Tom Gold in 1968 as rapidly rotating neutron stars.

EXAMPLE **43–3** Calculate the radius of a neutron star whose mass is three solar masses.

SOLUTION: When the gravitational pressure p_g and the degeneracy pressure p_n are equated, we find that

$$0.32\frac{G(3M_\odot)^2}{V^{4/3}} = \frac{2}{5}\frac{\hbar^2}{2M}(3\pi^2)^{2/3}\left(\frac{3M_\odot}{MV}\right)^{5/3},$$

where we have replaced N, the number of neutrons, by $3M_\odot/M$, $M_\odot$ is the solar mass, and M is the neutron mass. From this expression, we get

$$V^{1/3} = \left(\frac{4\pi}{3}\right)^{1/3} R = \frac{\frac{2}{5}(\hbar^2/2M)(3\pi^2)^{2/3}(3M_\odot/M)^{5/3}}{0.32G(3M_\odot)^2},$$

where R is the radius of the neutron star. Thus

$$R = \left(\frac{3}{4\pi}\right)^{1/3}\frac{\frac{2}{5}(\hbar^2/2M)(3\pi^2)^{2/3}(3M_\odot/M)^{5/3}}{0.32G(3M_\odot)^2} = 1.0\times 10^4 \text{ m}.$$

If the mass of a star is so large that even the pressure due to the Pauli exclusion principle applied to the neutrons proves inadequate to resist collapse, continued gravitational contraction occurs and a black hole forms. It is clear that an understanding of matter, whether in ordinary terrestrial form or in the cores of stars, would not be possible without the clarification that the exclusion principle has provided.

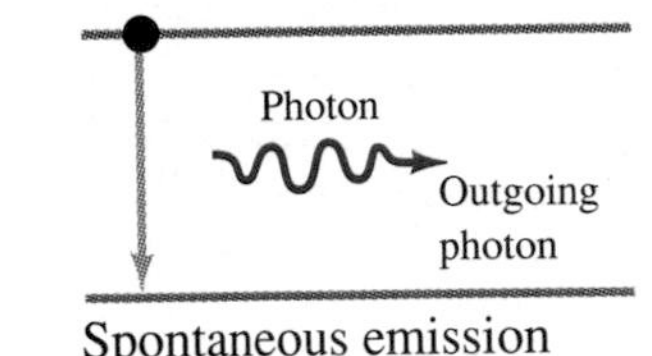

43–2 LASERS: AN APPLICATION OF THE CLUSTERING BEHAVIOR OF BOSONS

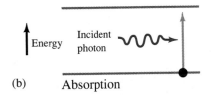

Energy
Incident photon
(b) Absorption

The fact that identical bosons tend to cluser in the same quantum mechanical states has an impact on a number of physical systems. The most direct application is the functioning of lasers, which involves the clustering of photons. We must preface our discussion of lasers with further detail about the transitions between atomic or molecular energy levels in which photons are produced.

Transitions between Energy Levels

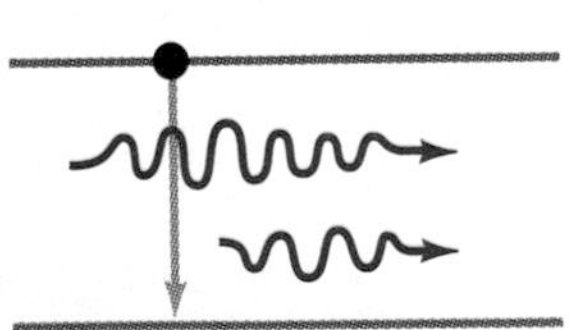

FIGURE 43–4 (a) In spontaneous emission, a photon is produced as an electron drops from an excited atomic state to a lower-lying level. (b) An atom can absorb a photon if the photon's frequency matches the energy difference between two atomic levels of the proper ℓ values. (c) In stimulated emission, the decay from an excited state occurs more readily in the presence of photons in the same state as the photon to be emitted.

As we discussed in Chapter 41, the quantum of electromagnetic radiation, the photon, is characterized by a wavelength λ or by a frequency f related to λ by $f = c/\lambda$. It is also characterized by a momentum vector $\mathbf{p}$, which gives the direction of propagation of the photon. The magnitude of this momentum is given by Eq. (41–17):

$$p = \frac{hf}{c}.$$

Photons are emitted (or radiated) when the electrons of atoms (or molecules) undergo a transition ("jump") from a higher energy level to a lower energy level, as in Fig. 43–4a. Such photons are said to be *spontaneously emitted.* We have already mentioned that transitions in which photons are emitted are subject to the law of conservation of angular momentum, and because photons act as though they have angular momentum quantum number $\ell = 1$ (with small mixtures of $\ell = 2$ and higher values), the most frequent transitions occur between atomic or molecular levels for which $|\ell_i - \ell_f| = 1$.

In order for an atom to radiate, it must first be put into a higher energy level, from which it will undergo a spontaneous transition to lower energy levels. To put the atom into a given higher energy state, it is necessary that an atomic electron be given the proper energy and angular momentum. One method of doing this is to bombard the atom with photons whose energy is the energy required for the transition. In this case, an atomic electron *absorbs* a photon and jumps into a higher energy state (Fig. 43–4b). If electrons in the ground state have $\ell = 0$, then (to high accuracy) the only states that can be reached in this way are those in which the electrons have $\ell = 1$. It is possible to reach other states by other mechanisms, such as collisions with other atoms. In this case, $\ell = 2, 3, \ldots$ states can be reached. In fact, the discovery of states with $\ell > 1$ was possible only because excitation by methods other than electromagnetic radiation exists. How does an electron in a state with $\ell = 2$ fall back to the ground state? If an $\ell = 1$ energy level lies between the $\ell = 2$ state and the $\ell = 0$ ground state, then the electron can jump first to the intermediate $\ell = 1$ state by emitting a photon, then jump to the $\ell = 0$ state by emitting a second photon (Fig. 43–5). However, what happens if there is no intermediate state through which the electron can cascade downward? A jump accompanied by photon emission is still possible, but it is on the order of 10^4 times less probable. If the change in ℓ is 3 rather than 2, then a jump accompanied by photon emission is 10^8 times less probable. These probabilities can be illustrated by thinking of a given excited energy level as a tub filled with water. A large drain that empties it in a short time corresponds to the possibility of transitions in which the angular momentum changes by one unit of $\hbar$. There are two tiny holes, 10^4 and 10^8 smaller in area than the drain hole. If the drain is open, we can ignore any leakage through the small holes, but if the drain is blocked (if there are no possible $\Delta\ell = 1$ transitions), leakage occurs through the next smallest hole. It takes much longer to empty the tub through the tiny hole. If there are no possible $\Delta\ell = 1$ transitions, then the energy level is called *metastable*, or nearly stable. Metastable states play a critical role in the operation of the laser.

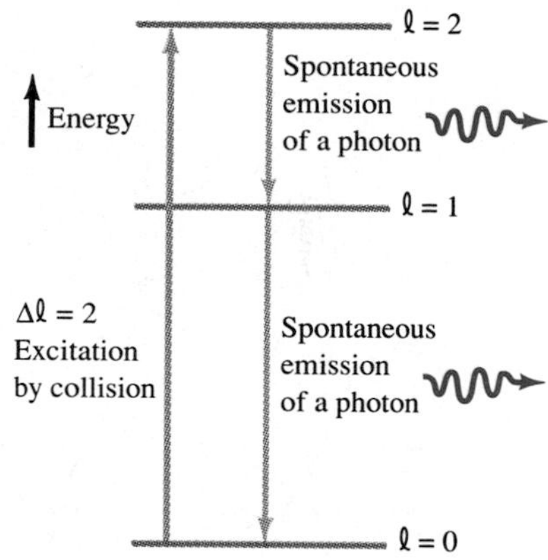

FIGURE 43–5 Excitation and deexcitation of atomic levels. An $\ell = 2$ level cannot rapidly decay by single photon emission to the ground state; a jump to an intermediate $\ell = 1$ level followed by a jump to the $\ell = 0$ level, with two photons emitted successively, is possible. Direct excitation from the $\ell = 0$ level to the $\ell = 2$ level is possible in collisions.

We have thus far treated the possibility of *spontaneous transitions* in atoms. In 1917 Albert Einstein used arguments based on thermodynamics together with the rudimentary quantum theory in existence at the time to predict the possibility of *stimulated transitions*. Consider an electron in a metastable state. It will stay in that state for a long time (long on the atomic scale, typically 10^{-8} s) before it decays with the emission of a single photon into the ground state. The energy of the photon will be $hf = E_i - E_f$, where E_i and E_f are the initial and final electron energies. If, however, photons of frequency f are present in the vicinity of the atom, then these photons will stimulate the transition, which occurs more rapidly as the number of stimulating photons grows. We say that there is *stimulated emission* (Fig. 43–4c). The more external photons, the more rapidly the stimulated emission occurs. There is another very important effect: *Because they are bosons, the photons emitted by stimulated emission will preferentially be in the same quantum state as the stimulating photons*; that is, they will have the same momentum, frequency, and phase. A *coherent* state of many photons is formed, a state in which the electromagnetic fields associated with the photons reinforce. This coherent state describes a single, intense, monochromatic plane wave, or beam. Moreover, because the photons have the same momentum (both magnitude and direction), the beam is extremely well collimated. The functioning of lasers relies on the existence of such stimulated transitions. In fact, *laser* is an acronym for "light amplification by stimulated emission of radiation."

Lasers

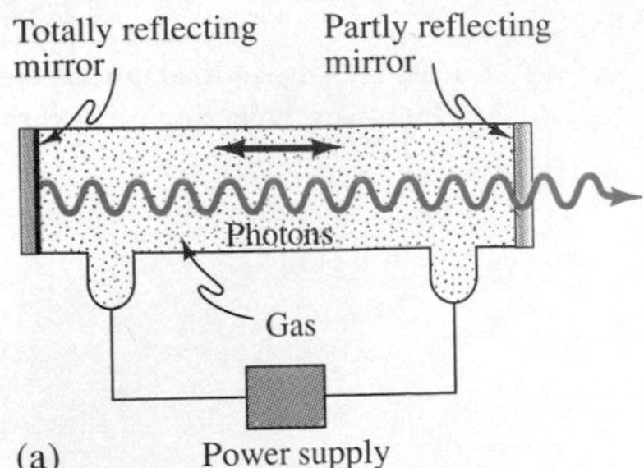

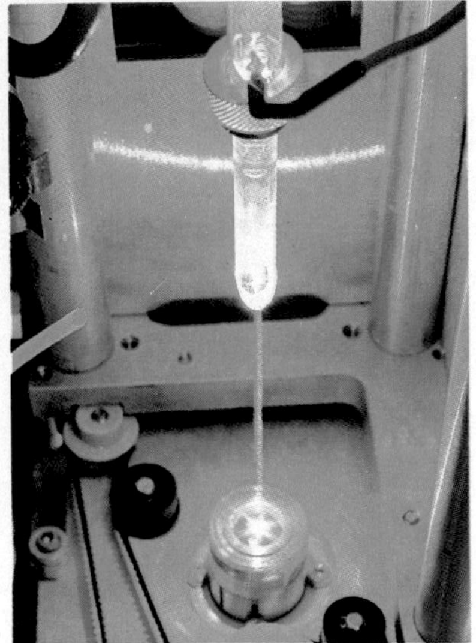

FIGURE 43–6 (a) Schematic diagram of a laser. The mirrors are present to contain photons and provide the conditions necessary for stimulated emission. (b) The laser beam formed by a krypton laser.

Suppose that we have a collection of atoms in a cavity, each of which is in a metastable state. We have seen that the presence of many photons of energy hf will stimulate a very rapid transition to the ground state, where f is the frequency corresponding to the transition to the ground state, but that otherwise the atoms will remain in the metastable state for a long time. We may then picture the following sequence: After a long time, one atom decays to the ground state and emits a photon of energy hf. The photon is constrained to stay in the cavity by means of mirrors positioned to reflect the photon back into the cavity (Fig. 43–6a). Other atoms in the metastable state, stimulated by the first photon, undergo a slightly faster transition to the ground state and produce more photons in the same state. The ever-increasing number of the "right" photons, those that induce stimulated emission, rapidly produces an avalanche of decays. These photons form a well-collimated, coherent *laser beam* (Fig. 43–6b). The laser beam can be used in a variety of ways.

The preceding description represents only the simplest type of laser. It is possible to vary the mechanisms involved to produce lasers of different types. An example of a different mechanism is provided by the helium–neon laser, in which the working material is a mixture of helium and neon gas in a tube (Fig. 43–7a). This laser illustrates one way to provide a metastable energy level. Electrons in the helium atom are raised from the ground state ($n = 1, \ell = 0$) to a pair of metastable excited states in which one of the electrons is in an $n = 2, \ell = 0$ level (Fig. 43–7b). The one state in which the electron spins are antiparallel lies about 1.2 eV higher than the one in which the electron spins are parallel. The state in which the spins are parallel is in turn about 18.6 eV above the ground state. These excitations cannot occur through absorption of a photon, because the transition has $\Delta\ell = 0$. Rather, the excitation occurs through collisions with free electrons introduced by an electrical discharge through the gas. The same reason

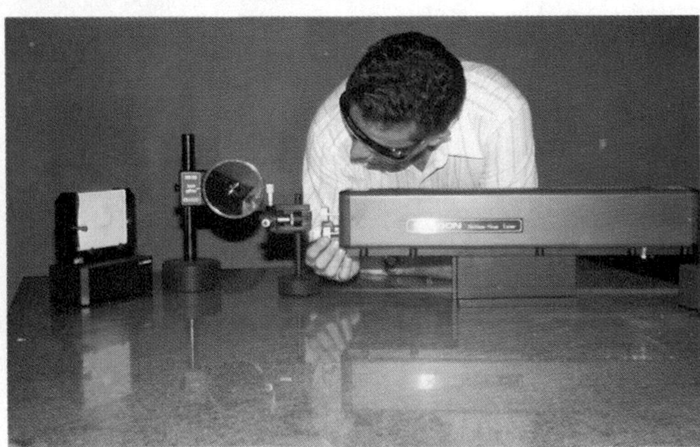

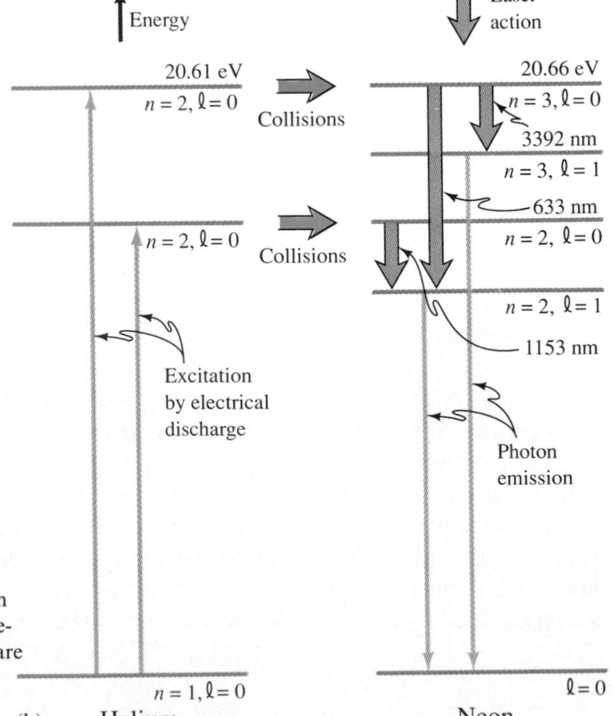

FIGURE 43–7 (a) A helium–neon laser. (b) Energy levels in helium and neon relevant to the operation of the helium–neon laser. The wide red arrows represent the transitions that produce the laser light; we say that these transitions are involved in "laser action" when photons of the proper frequency are present.

that the state cannot be excited by photon absorption ensures that deexcitation will not occur through photon emission. When many helium atoms are put into these metastable states, we say that the states are highly populated.

The metastable helium levels are *not* the levels that emit photons and thereby produce the laser beam. Instead, the helium atoms in the metastable state collide with the neon atoms of the gas and transfer the energy of the excitation to the neon. Neon has two $\ell = 0$ levels at approximately the same energies as those of the metastable states in helium, so the energy is easily transferred. In the transfer, the helium goes back to its ground state, and a large population of neon atoms in the two excited $\ell = 0$ levels is formed. The neon atoms in the two $\ell = 0$ levels can now decay to lower-lying $\ell = 1$ states.† The upper $\ell = 0$ state has a low-frequency transition with an infrared wavelength of 3392 nm. Decays to a lower-lying $\ell = 1$ level with radiation of wavelength 633 nm (red) are also possible. The lower of the two $\ell = 0$ levels decays to an $\ell = 1$ level by a laser transition in the infrared at wavelength 1153 nm. These neon atoms decay further by spontaneous (nonstimulated) photon emission to an $\ell = 0$ excited state, and the electrons in these states are deexcited by collisions with other atoms and with the cavity walls.‡ The neon atoms in their ground state are then reexcited to the higher $\ell = 0$ states by collisions with the helium atoms in their metastable states. We call this continual reexcitation *optical pumping*.

The emitted photons are contained within the cavity, which often takes the form of a narrow tube, by mirror reflectors at each end of the tube. The photons bounce back and forth between the reflectors and stimulate transitions of the metastable atoms in the tube. Because the photons also have to be extracted to form a coherent beam, the mirrors at the ends of the tube are not perfect reflectors. Instead, one of the mirrors is partly transparent. The fraction of the radiation reflected back into the tube is over 99 percent, and thus the cascade of stimulated emission occurs rapidly. The photons in the tube emerge as a coherent, monochromatic, and well-collimated beam through the partly silvered mirror. The coherent state is so intense that even though only 1 percent of the photons emerge, the beam is still useful.

In summary, the construction of a laser requires:

1. A collection of atoms that contain a metastable energy level into which many atoms can be excited.
2. A mechanism for holding emitted photons in a cavity containing excited atoms so that there is a massive stimulated deexcitation of the excited atoms (a laser transition).
3. A mechanism for repopulating the excited level after the laser transition has taken place.
4. A way for the coherent laser beam to be extracted sufficiently well to be useful.

Some Uses of Lasers

We described in Chapter 36 how the ability to measure time intervals accurately allows us to use lasers to perform *ranging*, a measurement of distance (see Fig. B1–2 on p. 1050). The high intensity and collimation of a laser beam allow

† These transitions can occur spontaneously, but they can also be stimulated and can form a coherent laser beam, in which case we refer to them as *laser transitions*. Although stimulated emission may not be considerably faster than spontaneous emission, it does produce a coherent beam.

‡ The fact that these are collisional deexcitations leads to the (expected) result that the rate of deexcitation varies inversely with the diameter of the cavity.

ranging measurements to be made over great distances, as great as the earth–moon distance.

Another application is isotope separation, whose significance was discussed in Chapter 19 (see the application box on p. 579). Laser separation of isotopes, a commercially important separation method, works as follows: The spectra of different isotopes of an element differ slightly because of tiny differences in their nuclear masses and magnetic dipole moments. Both properties affect the atomic energy levels. If atoms consisting of a mixture of isotopes are irradiated with a laser beam tuned very precisely to a transition frequency of only one of the isotopes, then the atoms of only those isotopes will be excited. For example, a suitably tuned laser beam can excite atoms of ^{235}U but not atoms of ^{238}U. A second laser beam with enough energy to ionize the already excited ^{235}U atoms but not ^{238}U atoms is then applied. The charged ions of ^{235}U can now be separated by electric fields from the un-ionized atoms of ^{238}U.

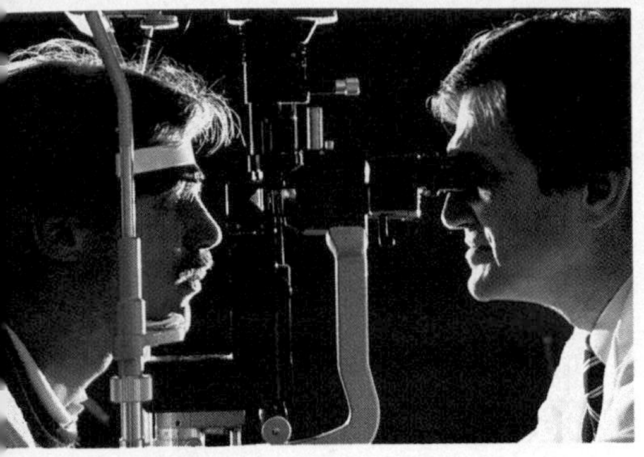

FIGURE 43–8 The reattachment of detached retinas by means of lasers is now a routine medical procedure.

Laser beams are important in *holography*. The coherence of the laser beam is crucial in this application, as we described in Chapter 39. Lasers also have important applications in medicine. One of their first uses was to reattach detached retinas. The lens of the eye focuses a laser beam onto a small area of the retina, which fuses to the tissue from which it has become detached (Fig. 43–8). Because the energy is delivered in a short time, there is no need to immobilize the eye. Infections are less likely to occur than with surgical procedures. Lasers can also be used to cauterize internal wounds and stop bleeding. We can mention also the potential application of lasers for controlled thermonuclear fusion reactions (to be discussed further in Chapter 45). Laser beams focused on lightweight nuclei may be able to provide the energy necessary for them to fuse and produce additional energy. Figure 43–9 illustrates an apparatus designed for this job.

Finally, lasers hold promise as a way to manipulate microscopic systems as small as individual atoms.† The ability to do this will be of increasing importance in quantum engineering (see Chapter 44).

FIGURE 43–9 In the OMEGA laser system at the University of Rochester, lasers are fired from many directions at pellets of material that contain nuclei suitable for thermonuclear fusion.

43–3 SUPERCONDUCTIVITY

Cooper Pairs and the BCS Theory

In our discussion of fermions and the Fermi energy in metals, we ignored all dynamic interactions among the fermions. How can this make sense, given that electrons repel each other because of their electric charges? The reason that the effects of repulsion are not very important in ordinary metals has to do with the exclusion principle. Consider two electrons in different states. Their interaction would manifest itself in some change of state, the only way we would see that something had happened. However, the vast majority of electrons cannot change states, because the states above and below their own state are already fully occupied, and the exclusion principle forbids further occupation. Thus the description of a metal as a gas of free electrons subject to the exclusion principle is better than might be expected. Interactions do lead to some modification of the predictions of the free-electron theory, and these modifications are well understood.

† See, for example, Steven Chu, "Laser Trapping of Neutral Particles," *Scientific American*, Feb. 1992, pp. 70–76.

The phenomenon of *superconductivity* relies on the interactions between the electrons and the vibrating lattices of ions in a metal (Fig. 43–10). For complicated reasons, the electron–ion interaction leads to a weak attraction between those electrons whose energies are close to the Fermi energy. Leon Cooper discovered this attraction, which leads to the formation of weakly bound pairs of electrons, known as **Cooper pairs**. These pairs are bosons, and the effective tendency of bosons to congregate in the same quantum state is carried to the point where *all* the Cooper pairs occupy the same quantum state. When this happens, we say that *the Cooper pairs have condensed into a coherent state*. The two electrons that pair up may be as far apart as 100 nm (recall that the separation between ions in the lattice may be between 0.1 nm and 1 nm), so the pairs do not form small, atomlike structures. Nevertheless, the movements of the two electrons of the pair are correlated, and, in a given region, there are many Cooper pairs whose "bonds" are entangled. Because the pairs have lower energies than free electrons do (it takes some energy to separate the pairs), a gap develops around the Fermi energy. There are no allowed energy levels within this gap, whose width (energy separation) is $2\Delta \simeq 10^{-3}$ eV (Fig. 43–11).

The properties and applications of superconductors and possible uses of high-temperature superconductivity were discussed in the optional Sections 27–6 and 32–6.

Cooper pairs

John Bardeen, Leon Cooper, and Robert Schrieffer pointed out in 1957 that because a Cooper pair is a boson (the two electrons have a net angular momentum of 0 or $\hbar$), the pairs can, and indeed will, all go into the same state.† Their theory is known as the BCS theory. The pairs condense when the material's temperature drops below a critical temperature T_c. The abrupt change of state at a critical temperature is analogous to the phase change from liquid to solid.

FIGURE 43–10 Superconducting wire; the filaments are the actual superconducting material. The applications of such wires are discussed in Chapters 27 and 32.

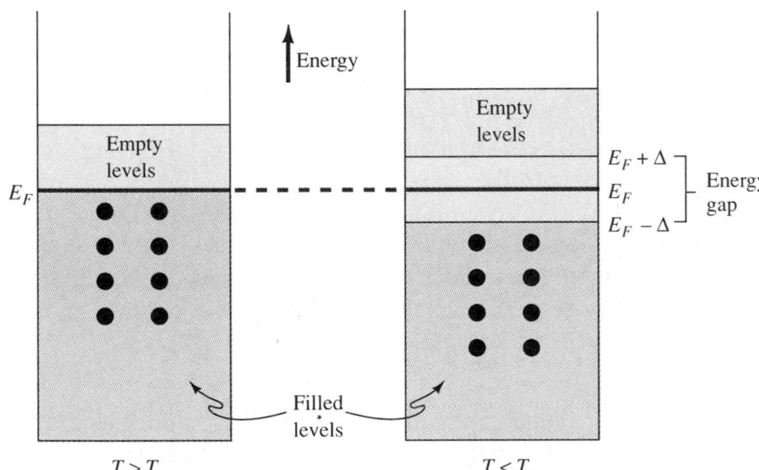

FIGURE 43–11 Formation of the energy gap in the transition of a metal at T_c from a normal state to a superconducting state. The vertical scale is grossly distorted.

Electromagnetic Properties of Superconductors

The most striking property of superconductors is that below T_c, their resistivity is zero. This phenomenon is explained by the coherence of the superconducting state. In a normal metal, when a current starts and there is no potential difference to maintain it, the electrons scatter from the lattice and give up energy. The energy appears as ohmic heating, and the current rapidly decays. However, in a superconductor, a large number of electron pairs move together in a

† In fact, the pairing occurs only because of the presence of other pairs in the same state. This is analogous to the stimulated emission of photons that takes place in a laser.

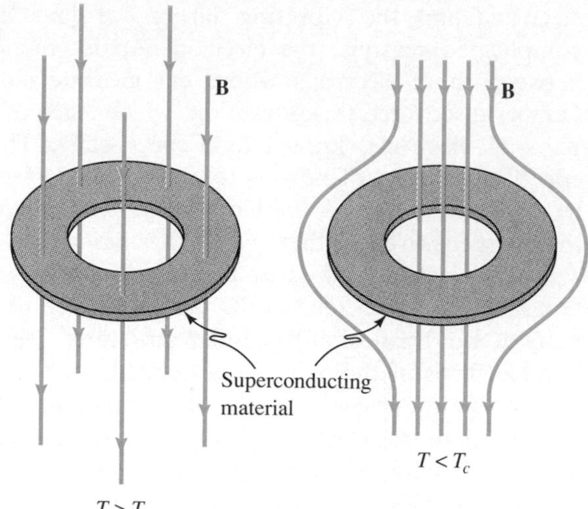

FIGURE 43–12 Magnetic field lines expelled by a ring of material that makes a transition to the superconducting state when the temperature drops below the critical temperature T_c. Flux is trapped by the hole.

coordinated way. For the superconductor to lose energy, *the entire coherent state would have to be broken up*, and this would require a large amount of energy to be supplied at once. The mechanisms of ordinary electrical resistance cannot work if a macroscopic number of electron–electron pairs (say, 10^{22}) must be slowed down *all at once*. An analogy is the difference between pushing your hand through liquid water (a "normal" metal) and pushing your hand against an icicle (a "superconductor"). In the first case, there are easy ways to lose just a little energy: You can form arbitrarily small waves and eddies. In the second case, you can lose a large amount of energy only all at once, when you break the icicle.

Another property, the *Meissner effect*, is that no magnetic field can penetrate a superconductor, and that a magnetic field already present is expelled when a metal is cooled below T_c (Fig. 43–12). The inability to establish a magnetic field within a superconductor is also a consequence of the coherence of the superconducting state. If you were to try to increase a magnetic field inside a superconductor, you would, by Faraday's law, induce an emf within the material. A current would be induced to oppose the change in magnetic flux. But because all the Cooper pairs act together, even the tiniest change in magnetic flux can generate a current sufficiently large to cancel the inducing field entirely.

The magnetic properties of superconductors were first described in Chapter 32.

Flux Quantization

Suppose that we place a ring of superconducting material that is above its critical temperature in a magnetic field. The ring lies in a plane perpendicular to the field direction. We now lower the temperature of the material below T_c so that the material becomes superconducting. Because superconductors expel magnetic fields, no field lines will pass through the material. All the field lines will be either outside the ring or confined within the ring.

The BCS theory can be used to show that the coherence of the superconducting state implies that *the magnetic flux through the ring is quantized in units of* $h/2e$:

Flux quantization

$$\Phi_B = \frac{nh}{2e} \equiv n\Phi_0, \quad \text{where } n = 0, 1, 2, \ldots . \tag{43–12}$$

The quantity Φ_0 is the *magnetic flux quantum*, with value 2.07×10^{-15} Wb = $2.07 \times 10^{-15}\ \text{T} \cdot \text{m}^2$. The charge $2e$ that appears here is the charge of a Cooper pair. The prediction that flux is quantized has been successfully tested by experiment. An explanation of flux quantization requires the full machinery of quantum mechanics.

EXAMPLE 43–4 A single unit of the quantized flux Φ_0 is confined within a cylinder of copper wire of diameter 1.3×10^{-5} m. What is the magnitude of the magnetic field in the wire, assuming that it is uniform throughout the wire?

SOLUTION: The unit of flux is $\Phi_0 = 2.07 \times 10^{-15}\ \mathrm{T \cdot m^2}$. If the magnetic field is uniform within a wire of radius R, its magnitude is given by

$$B = \frac{\Phi_0}{\pi R^2} = \frac{2.07 \times 10^{-15}\ \mathrm{T \cdot m^2}}{\pi(0.65 \times 10^{-5}\ \mathrm{m})^2} = 1.6 \times 10^{-5}\ \mathrm{T}.$$

This is only about a factor of 6 less than the earth's magnetic field and is therefore quite measurable.

Tunneling of Pairs and the Josephson Effects

Tunneling, which we studied in Chapters 7 and 41, can occur across junctions between normal metals, between normal metals and superconductors, and between superconductors. Here "junction" refers to a very thin insulating strip between two pieces of metal. The insulating strip is formed from a material that can be evaporated and then deposited in thin layers. The ability to create extremely thin and uniform films such as an insulating strip is one of the major technological advances of recent times. If the strip is thick, no current can flow between the metals. If it is thin enough (10 nm to 20 nm), an electron can tunnel across the strip from one metal to the other. Consider two metals, labeled 1 and 2, with Fermi energies E_{F1} and E_{F2}, respectively, for which $E_{F2} > E_{F1}$ (Fig. 43–13a). If the metals are joined at a junction, the electrons in filled energy levels just below E_{F2} in metal 2 can tunnel into empty energy levels in metal 1 (Fig. 43–13b). A current will flow, even in the absence of an external electric potential. If the two metals are identical, however, there will be no current unless an external potential V_{ext} is imposed. This potential would lower the Fermi energy

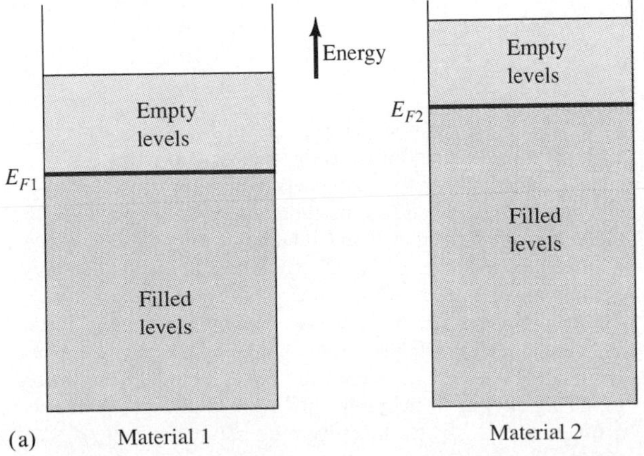

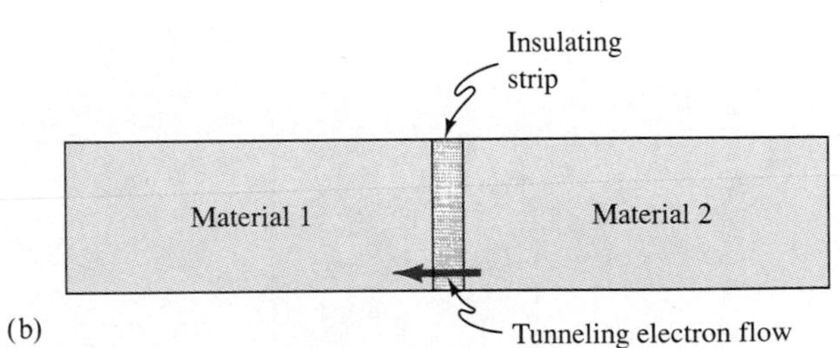

FIGURE 43–13 (a) Two metals with different Fermi energy levels. (b) The metals are joined with an insulating strip between them. Even in the absence of an external electric potential, current flows from metal 2 to metal 1 as electrons tunnel through the insulating strip.

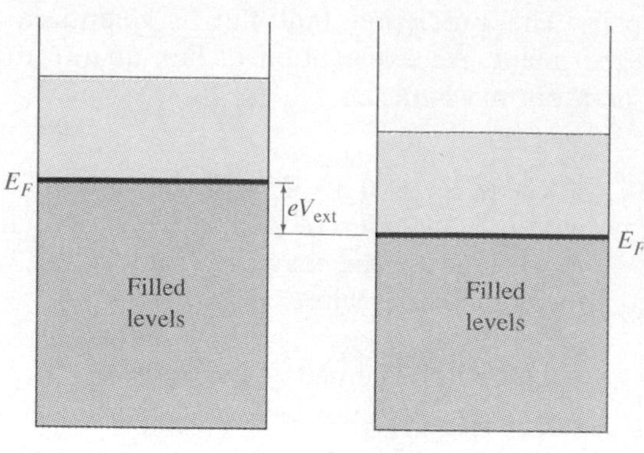

FIGURE 43–14 When an electric potential that lowers the levels on the right-hand side compared to those of the left-hand side is applied, a tunneling current flows from left to right across the insulating strip. In this case, the metals on both sides of the strip are the same.

on one side of the insulating strip (or raise it on the other side), so tunneling is again possible (Fig. 43–14).

Now suppose that we have a junction between a metal in its normal state and a metal in a superconducting state (Fig. 43–15a) and that each piece has the same Fermi energy. With no external potential, no electrons flow. Suppose that an external potential V_{ext} that lowers the Fermi energy of the superconductor is applied. A current will flow between normal metals, but because of the energy gap in the superconductor, the external potential must exceed a minimum value before empty levels become available for tunneling into (Fig. 43–15b). This minimum value is given by $eV_{min} = \Delta$, and by varying the external potential and observing the onset of tunneling, it is possible to measure the gap size. Figure 43–15c illustrates the relationship between current and potential in this case.

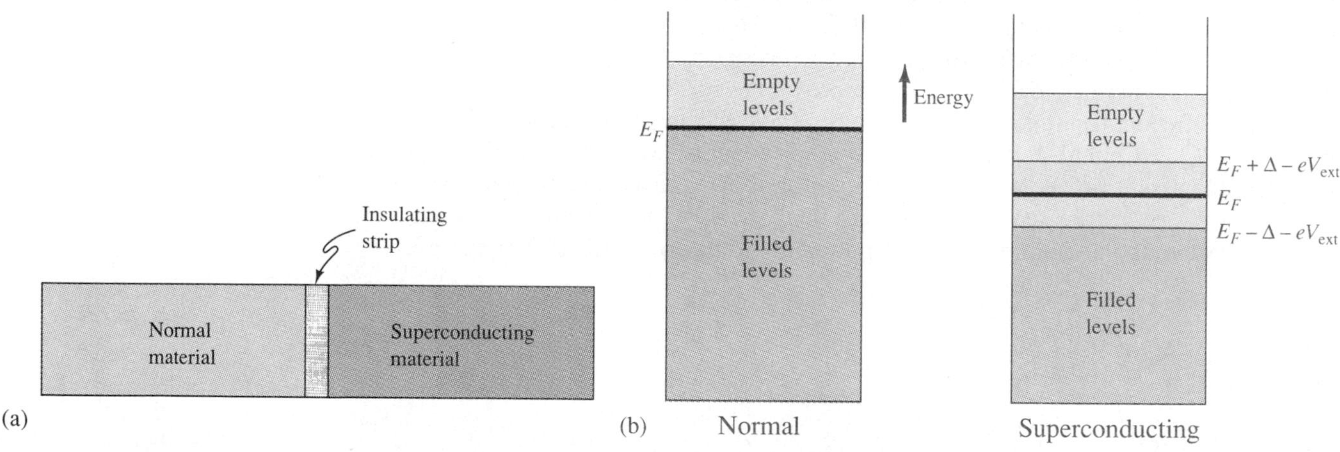

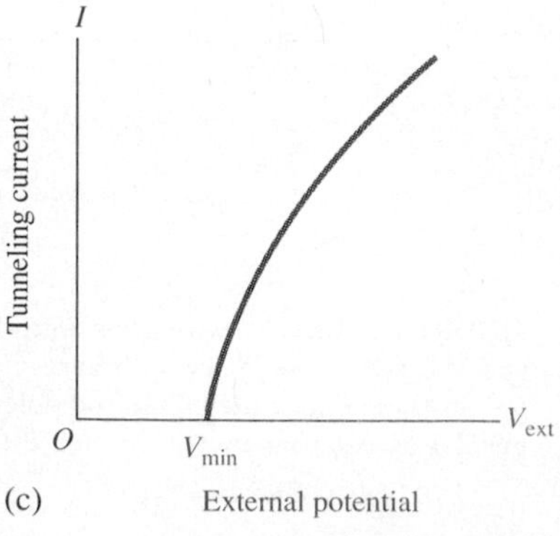

FIGURE 43–15 (a) A metal in its normal state separated by a thin insulating strip from the same metal in its superconducting state. (b) An energy diagram of this situation, including an external potential, V_{ext}, applied so that the energy levels in the superconducting metal are lowered with respect to those in the normal metal. Because an energy gap is present in the superconductor, the potential must be a certain minimum size before electrons from the normal metal can tunnel into the empty levels of the superconductor. This minimum value of V_{ext} is given by $eV_{min} = \Delta$. (c) The flow of tunneling current in the junction described in Fig. 43–15a as a function of V_{ext}. V_{ext} must have the minimum value V_{min} before current can flow.

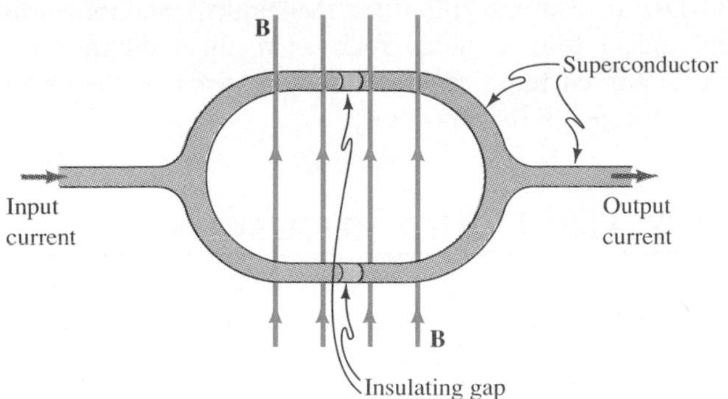

FIGURE 43–16 The current that passes through a pair of Josephson junctions enclosing a magnetic flux varies with that flux.

Now let us consider two superconductors separated by a thin insulating strip, an arrangement known as a *Josephson junction.* At first there might appear to be little current across the barrier. If the probability of a single electron tunneling through is very small, then, because the probability of two independent electrons tunneling through is the product of the tunneling probabilities of the individual electrons, there would be very little tunneling for an electron pair. In 1962, Brian Josephson, who at the time was still a student, noted that if the barrier is less than 1 nm thick, the two superconductors form a single, coherent quantum system. As a result, what tunnels through the barrier is not two individual electrons, but a pair of electrons as a unit. This has two consequences (we shall prove neither here): First, *even in the absence of a potential difference between two identical superconductors, a tunneling current will flow across a Josephson junction.* This is called the *DC Josephson effect.* Second, *when there is a constant potential difference V across the barrier between two superconductors, the current that flows between them oscillates with angular frequency*

$$\omega = \frac{2e}{\hbar} V. \tag{43–13}$$

This phenomenon is called the *AC Josephson effect.* Because frequencies can be measured with great precision, the AC Josephson effect provides physicists and engineers with the most accurate way known of standardizing voltage measurements. Equivalently, it allows the most precise measurement possible of the fundamental ratio $e/\hbar$.

Note that Eq. (43–13) can be rewritten as $\hbar\omega = (2e)V$, or, because $\hbar = h/2\pi$ and $\omega = 2\pi f$, as $hf = (2e)V$. The left side gives the energy of a photon of frequency f, whereas the right side gives the work done to move a Cooper pair through the potential V.

Consider two Josephson junctions connected in parallel (Fig. 43–16). Like the interference pattern produced by an electron passing through a double slit, an interference pattern appears in the net current. The interference pattern is affected by a magnetic flux enclosed by the two arms of the pair of junctions. The flux plays the same role in the pair of Josephson junctions that the slit separation plays in a double-slit experiment. The dependence of the current on the enclosed magnetic flux Φ_B is given by

$$J = J_{\max} \cos\left(\frac{e\Phi_B}{\hbar}\right). \tag{43–14}$$

This phenomenon allows us to measure magnetic flux precisely. The SQUID (Superconducting QUantum Interference Device), which is based on the phenomenon described above, is widely used for high-precision measurements of very tiny magnetic fluxes (Fig. 43–17). Magnetic fields as small as 10^{-13} T have

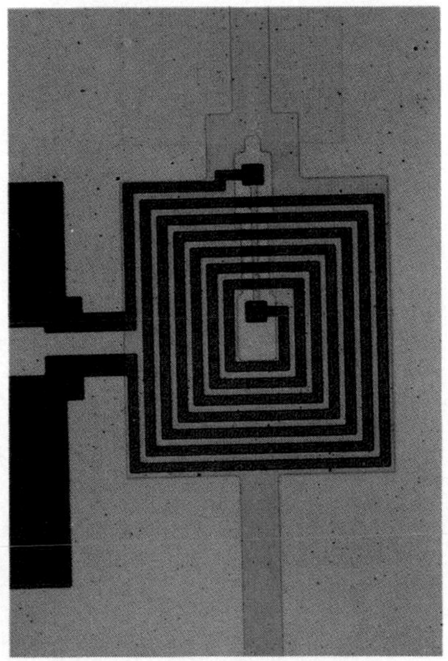

FIGURE 43–17 This SQUID (Superconducting QUantum Interference Device), manufactured from high-T_c superconducting material, is used as a high-sensitivity magnetometer, a device that measures magnetic fields.

been measured. With SQUIDs, an esoteric quantum mechanical phenomenon has become a major technological tool in fields such as medical diagnostics, where tiny variations in the electric currents generated by the heart or the brain can be measured by the magnetic fields they produce.

43–4 SUPERFLUIDITY AND LIQUID HELIUM

In addition to superconductivity, there is another technologically important low-temperature phenomenon that involves the condensation of bosons into a macroscopic coherent state. Helium atoms, ^{4}He, with a nucleus consisting of two protons and two neutrons, are bosons; their nucleus has an angular momentum of zero, and the two electrons of helium atoms have an angular momentum that is an integral multiple of $\hbar$. Helium forms a liquid at low temperatures, but in contrast to other liquids, it does not solidify when the temperature is lowered further (unless it is simultaneously compressed under a pressure of more than 25 atm). At a temperature of 2.17 K, a change of state commences. As the temperature is lowered below 2.17 K, an increasing fraction of the liquid helium flows as though it has no internal friction (viscosity). This property is called **superfluidity**. Superfluid helium can flow freely through the narrowest of channels. It creeps up the sides of a beaker containing it and flows to the outside (Fig. 43–18). It can be used to make the most nearly perfect (frictionless) heat engine, one that converts thermal energy to kinetic energy with no moving parts save that of the liquid helium itself. Finally, and most important technologically, because a superfluid material can flow as though there were no internal friction, it makes a nearly perfect heat conductor.

Superfluidity

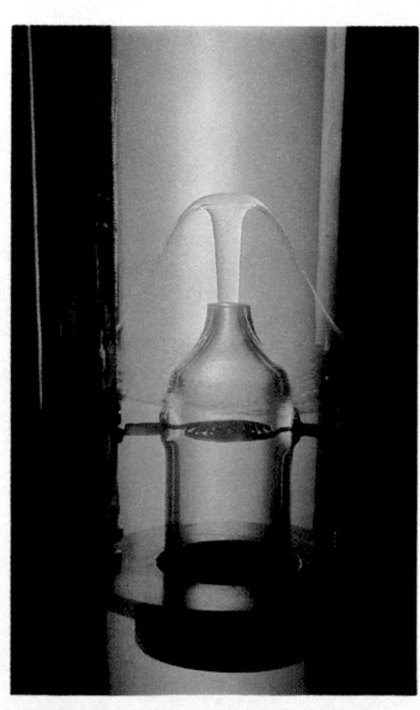

FIGURE 43–18 The fountain effect of superfluid helium, which acts as though it has no viscosity.

The fact that atoms of ^{4}He are bosons allows us to explain these properties. At sufficiently low temperatures, the helium atoms congregate in the same (coherent) ground state. When energy is added to the system, excitations are produced, a process analogous to the movement of electrons from the ground state to the first rotational state in molecules. In atoms and molecules, the levels are discrete, and a minimum energy is needed to produce an excitation. The same is true in the system of liquid helium, but the separation between the energy levels is tiny in that system. In particular, the lowest excitations are compressional, or sound, waves. (These waves are of the same type that form sound waves in solids.) Very little energy needs to be added to the system to produce these excitations. However, the energy of the excitation is not the whole story. Even though sound waves are the excitations of lowest energy, it is difficult to excite them mechanically, and this class of excitations is not useful for carrying off mechanical energy; that is, they produce no internal friction. To understand the significance of this remark, let us consider the process by which friction is produced.

Suppose that we drag a sphere of mass M through liquid helium. If the sphere gives up some energy and momentum to the liquid, it will create an excitation. Such excitations are discrete quantum mechanical states that correspond to motion of the entire system of liquid helium, and the excitations have a definite momentum $\mathbf{p}$ and energy E. Momentum conservation states that

$$M\mathbf{v}_i = \mathbf{p} + M\mathbf{v}_f,$$

where $\mathbf{v}_i$ is the initial velocity of the sphere and $\mathbf{v}_f$ is its velocity after it has given up some energy. Then $\mathbf{v}_f = \mathbf{v}_i - (\mathbf{p}/M)$. Similarly, energy conservation gives

$$\tfrac{1}{2}Mv_i^2 = E + \tfrac{1}{2}Mv_f^2.$$

We substitute $\mathbf{v}_f$ into the energy conservation equation:

$$\frac{1}{2}Mv_i^2 = E + \frac{1}{2}\left(Mv_i^2 - (2\mathbf{v}_i \cdot \mathbf{p}) + \frac{p^2}{M}\right).$$

If we drop the last term on the right because M is large (macroscopic), we can then solve for E:

$$E = \mathbf{v}_i \cdot \mathbf{p}. \tag{43–15}$$

As long as this relation can be satisfied, it will be possible for the sphere to lose kinetic energy by causing excitations in the liquid, and there will therefore be resistance to the motion of the sphere.

Let us see whether it is possible for sound waves to be excited in this way. For sound waves, the relation between energy and momentum is given by

$$E = v_s p,$$

where v_s is a constant, the velocity of a sound wave. Thus for $v_i < v_s$, Eq. (43–15) cannot be satisfied, and *there is no way for the sphere to lose energy*. It therefore travels as though the liquid exerts no friction on it; that is, as if there were no viscosity. The same conclusion follows in a reference frame in which the sphere is at rest and the liquid flows past it. Superfluid ^{4}He therefore moves with no viscosity.

The creeping of liquid helium up the sides of a beaker can be understood in terms of the absence of viscosity, together with the fact that the van der Waals forces between neighboring helium atoms and between the atoms of helium and of the beaker walls are stronger than the force of gravity on a helium atom. These forces then pull helium atoms up against gravity. The strange properties of liquid ^{4}He are another demonstration of quantum effects on a large scale.

The technological importance of these properties is closely linked to the importance of superconductivity. Superconducting systems, such as the large arrays of superconducting magnets used in particle accelerators, must be kept at temperatures below their critical temperature. The only practical superconductors that can be used for such arrays are those for which liquid helium is sufficiently cold to form a coolant. Here the fact that the liquid ^{4}He is a nearly perfect heat conductor—its coefficient of thermal conductivity is hundreds of thousands of times greater than that of copper—is critical. The entire array of magnets, which, for the Fermilab accelerator, forms a ring 6 km long, can be cooled by a completely connected system of helium, and no "hot spots" can form. In particular, boiling cannot occur at any one location, and this offers crucial insurance against any of the magnet coils heating beyond its critical temperature.

SUMMARY

In large systems for which quantum mechanics governs the dynamics, the Pauli exclusion principle plays a crucial role if the system is composed of fermions. This principle states that only one fermion—for our purposes, a particle with spin $\hbar/2$—of a given type can occupy a particular state such as an energy level. In metals, this means that very closely spaced energy levels are filled with electrons up to the Fermi energy, E_F. The Fermi energy of electrons is

$$E_F = \frac{\hbar^2}{2m_e}(3\pi^2 n_e)^{2/3}, \tag{43–5}$$

where n_e is the electron number density in the metal. Only electrons near the top of the filled levels take part in dynamic processes such as current flow.

The exclusion principle is responsible for the incompressibility of matter. In ordinary matter the incompressibility can be estimated to within a factor of 2 by assuming that the electrons in the matter are free. The exclusion principle prevents stars that are not too massive from collapsing under the mutual gravitational attraction of their constituents.

A system composed of bosons—particles whose spins are integral multiples of $\hbar$, such as photons and ^{4}He atoms—manifests quantum mechanical effects quite different from those of fermionic systems. Identical bosons have an enhanced probability of aggregating in the same quantum state. As a consequence there can be stimulated emission of radiation (photons) into states already occupied by other photons. The operation of the laser is based on stimulated emission and on the bosonic properties of photons. To construct a laser, we need a way to populate an excited energy level in a collection of atoms; a way to build up many photons of the "proper" energy to provoke a massive stimulated transition from the excited level; and a way to extract the coherent laser beam formed by the stimulated emission.

Fermions can pair up to form bosons, and those bound pairs tend to congregate in the same quantum state. This is the mechanism behind superconductivity. Superconductors form when a collection of electron–electron pairs known as Cooper pairs occupy a single, coherent quantum state at low temperatures. Cooper pairs can tunnel, and this phenomenon has advanced our ability to observe very small magnetic fields. Superfluidity, characterized by flow without viscosity in superfluid ^{4}He, is another phenomenon that follows from the tendency of bosons to condense in the same quantum state at low temperatures.

QUESTIONS

1. How would the behavior of electrons in materials change if the Pauli exclusion principle were not applicable?
2. Most heavenly bodies rotate and have some angular momentum. Astrophysicists believe that pulsars are rapidly rotating neutron stars, some with periods as short as 0.005 s. Explain how such a small, massive object could rotate so fast.
3. Where does the fact that photons tend to congregate in the same quantum state play a role in our discussion of lasers?
4. Chemical compounds in metastable excited states can be formed in chemical reactions. Is this a suitable way to arrive at the conditions necessary for laser action?
5. The *Fermi temperature*, T_F, is defined by $E_F \equiv kT_F$. What does it mean if T_F in a given material is much larger or much smaller than the material's temperature?
6. Describe the steps in the evolution of a star that becomes a black hole.
7. A He–Ne laser emits light that is not coherent as well as coherent laser light. Why?
8. An electron on the moon and an electron on earth both have spin up and are in the ground state of hydrogen. Given that both electrons are in the same state, how is the Pauli exclusion principle bypassed?
9. Suppose that a crystal could be made from the atoms of the inert gases (by going to low temperatures, for example). Would such a crystal be likely to act as a metal?
10. The spins of the two electrons in the ground state of helium can be either parallel or antiparallel. The Pauli exclusion principle does not apply to antiparallel spins, but electrons with parallel spins must avoid each other. Would you expect the two possible spin arrangements to have the same energy? If not, which energy would be lower?
11. How could you tell if electrons in very distant galaxies obey the Pauli exclusion principle?
12. Given our discussion of stellar evolution, why is Jupiter not a star? (Assume that Jupiter is an aggregate of hydrogen atoms held together by gravity.)
13. An electron traveling through a collection of protons in a plasma is slowed down by collisions and captured into an orbit with a large principal quantum number, far from the proton it orbits. Given that radiative transitions are largely those for which $\Delta\ell = \pm 1$, would the electron end up in the ground state through one transition or through a sequence of transitions?
14. The filament of a light bulb has many atoms that become

thermally excited into the same excited states. Why does the bulb not act as a laser?

15. What difficulties might you encounter in an attempt to make an X-ray laser?

16. Use the argument as to why the flow of superconducting helium around an obstacle is nonviscous when the flow occurs at a speed less than that of sound to explain why electron waves propagate without energy loss through a rigid crystal lattice at $T = 0$. Why will the electrons experience resistance at finite temperatures?

17. In our treatment of the incompressibility of ordinary matter, we spoke about metals rather than a more general class of solids. Does our discussion apply to insulators? (See Chapter 27 for the difference between conductors and insulators.)

PROBLEMS

43–1 The Exclusion Principle in Metals and Stars

1. (I) What is the value of the Fermi energy of beryllium, which has two valence electrons per atom? The gram-atomic weight of beryllium is 9.01 g/mol, and its mass density is 1.85 g/cm^3.

2. (I) Find the Fermi temperature (see Question 5) of copper, for which the Fermi energy is 7.1 eV.

3. (I) Calculate the Fermi energies of sodium ($n_e = 2.65 \times 10^{28}\ \text{m}^{-3}$), potassium ($n_e = 1.40 \times 10^{28}\ \text{m}^{-3}$), and aluminum ($n_e = 18.1 \times 10^{28}\ \text{m}^{-3}$). All these metals have a single valence electron per atom.

4. (I) Calculate the bulk moduli, B, of sodium, potassium, and aluminum, using the data given in Problem 3. (The experimental values of B for these materials are 0.64, 0.28, and 7.6, respectively, in units of $10^{10}\ \text{N/m}^2$.)

5. (I) Sodium, a metal with one free electron per atom, has a gram-atomic weight of 23 g/mol and a density of 0.97 g/cm^3. What is the density of electrons in sodium?

6. (I) The density of electrons in copper is 8.47×10^{28} electrons/m^3. Calculate the bulk modulus of copper. (The experimental value is $13.4 \times 10^{10}\ \text{N/m}^2$.)

7. (I) What values of $n_1 (= n_2 = n_3)$ give an energy close to 1 eV for an electron in a cubical box whose sides are precisely 100 nm long? What is the actual energy of this state? of the state in which n_1 and n_2 are unchanged but n_3 is increased by 1?

8. (I) The Fermi energy of electrons in a metal, E_F, may be used to define the Fermi momentum, p_F, defined by $E_F \equiv p_F^2/2_{m_e}$. Express the Fermi momentum in terms of n_e. Use your results to calculate the Fermi speed, v_F, the speed of an electron with the Fermi energy.

9. (I) What is the speed of an electron with the Fermi energy in sodium, given that $n_e = 2.65 \times 10^{28}$ electrons/m^3?

10. (II) When the Fermi speed is $0.25c$, where c is the speed of light, it is more appropriate to use a relativistic connection between E_F and p_F rather than the nonrelativistic connection that we used in Problem 8. Estimate the value of n_e for which the relativistic connection is necessary.

11. (II) Three electrons are in an infinitely deep, one-dimensional potential energy well of width L. (a) What is the lowest possible energy value for this system? (b) What is the next highest energy value? In both parts, ignore Coulomb interactions between the electrons.

12. (II) Consider N identical, noninteracting fermions of mass m in an infinitely deep one-dimensional potential energy well of width L. Assume that $N \gg 1$. Find (a) the Fermi energy and (b) the total energy of this system of particles in its lowest energy state. (c) What is the smallest amount of energy that can be absorbed by the system; that is, what is the minimum excitation energy of the system from its ground state?

13. (II) Assume that the nucleus is composed of free neutrons and protons. What is the equation analogous to Eq. (43–1) for the case of a nucleus? Assume that the nucleus is a cubic box with sides L. L is such that the box volume is the same as a sphere of radius $R = r_0 A^{1/3}$, where A is the total number of neutrons and protons and $r_0 \simeq 1.2 \times 10^{-15}$ m.

14. (II) Calculate the Fermi energy for the neutrons confined to the nucleus ^{56}Fe, which roughly forms a sphere of radius 4.6×10^{-15} m. (There are 30 neutrons in the ^{56}Fe nucleus.)

15. (II) Show that resistivity will be temperature independent in the quantum description of conducting materials for temperatures much less than the Fermi temperature (see Question 5).

16. (II) The general definition of the pressure in a gas with internal energy U is $p = -(\partial U/\partial V)$, where V is the volume of gas. (This follows from the first law of thermodynamics and from the fact that the infinitesimal work done when a system changes volume by dV is $dW = p\,dV$.) Given the fact that for a gas of identical fermions at $T = 0$, $U = \frac{3}{5}NE_F$, show that at $T = 0$, $p = \frac{2}{3}U/V$ for the fermion gas. This result is the *same* as our result in Chapter 19, where we studied the atomic origin of pressure. (That result did not depend on the precise distribution, so it is not surprising that we recover it here.)

17. (II) Using the result of Problem 16, (a) calculate the pressure of an ideal gas of N identical fermions at $T = 0$ in terms of the volume, and (b) evaluate the pressure of the degenerate electron gas at $T = 0$ in a sample of copper. The density of free electrons in copper is 8.5×10^{22} electrons/cm^3.

18. (II) Calculate the ratio of the degeneracy pressure of electrons to that of protons for a star such as our sun. Assume that the star is electrically neutral.

19. (II) Suppose that electrons are confined to a plane at a density of n_e electrons per unit area. Follow the steps that lead to Eq. (43–5) to calculate the Fermi energy for such a system.

20. (II) A lead nucleus consists of 82 protons and 126 neutrons in a sphere of radius 7×10^{-15} m. Assume that none of the particles interact with each other. Calculate the Fermi energies of the protons and the neutrons.

21. (II) (a) Calculate the radius of a neutron star (such as the Crab pulsar in Fig. 43–19) as a function of its mass. (b) Find the Fermi energy and Fermi momentum of a neutron star in terms of its radius and its mass. (c) Estimate the mass for which it is necessary to use relativistic expressions in part (b). Express your results in solar masses.

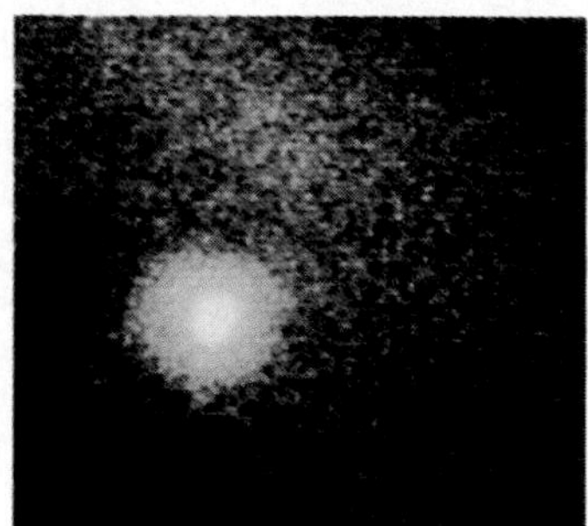

FIGURE 43–19 Problem 21.

22. (III) A proper calculation of the Fermi energy in a cubical box with sides L amounts to counting the number of electrons that fill all the states up to an energy E_F. There are two states for each set of positive integers $\{n_1, n_2, n_3\}$ that satisfy the condition $n_1^2 + n_2^2 + n_3^2 = 2(m_e E_F L^2/\pi^2\hbar^2) \equiv R^2$. In a "continuum approximation" to the problem of counting the number of states, appropriate to large values of n_1, n_2, and n_3, the number of states is twice the volume of an octant (one-eighth) of a sphere of radius R; that is, $2(\frac{1}{8})(4\pi/3)R^3$. Compare the answer in terms of the density with the approximate result that we gave in the text, $E_F \simeq (\hbar^2/2m_e)(\pi^3 n_e)^{2/3}$.

23. (III) By counting the number of states (that is, by using the techniques described in Problem 22), calculate the Fermi energy for a two-dimensional gas of free, noninteracting electrons. Show that $d = \lambda_F/2$, an approximation we made in our text discussion, is not as good as it is for a three-dimensional gas.

43–2 Lasers: An Application of the Clustering Behavior of Bosons

24. (I) A laser emits radiation of wavelength 880 nm and delivers power of 12 W. How many photons are emitted per second?

25. (I) The characteristic red light of a He–Ne gas laser (Fig. 43–7a) is due to stimulated emission between neon levels at 20.66 eV and 18.70 eV. Calculate the wavelength and frequency of the photon emission.

26. (I) A He–Ne laser emits radiation with a power of 1/100 W at a wavelength of 632.8 nm. How many photons are emitted per second?

27. (I) A laser emits radiation at a wavelength of 660 nm. Photons are emitted at a rate of $1.5 \times 10^{17}\ \text{s}^{-1}$. What is the power of the laser?

28. (II) Consider N identical bosons of mass m in an infinitely deep, one-dimensional potential energy well of width L. (a) What is the total energy of this system of particles in its lowest energy state? (b) What is the smallest amount of energy that can be absorbed by the system; that is, what is the minimum excitation energy of the system from its ground state?

29. (II) The collimation of a laser beam is limited by diffraction; that is, the angular beam spread $\Delta\theta$ is at least as large as $\Delta\theta \simeq \lambda/D$, where D is the diameter of the laser's aperture. What is the diameter of the beam spot projected on a wall 1 km from a laser of aperture diameter 0.35 cm emitting light of wavelength 675 nm?

30. (II) A laser beam projects light of wavelength 550 nm at the rate of 5×10^{18} photons/s. (a) What is the power of the laser? (b) What is the radiation pressure exerted by the beam on a shiny surface if the laser beam projects a circle 2.2 mm in diameter?

31. (III) In 1917, Einstein wrote a paper that describes the stimulated-emission process. We can summarize his results as follows: Suppose that an atom has two states. State 1 is the ground state; state 2 is an excited state. The atom is in an enclosure filled with blackbody radiation, which we recall from Chapter 17 has the energy-density function $u(f, T)$. The transition rate for excitation from state 1 to state 2, $\Gamma_{1\to2}$, is proportional to the number of atoms in state 1, N_1, times the density of photons present that have the proper frequency (such that the photons have the transition energy): $\Gamma_{1\to2} = N_1 u(f, T)B_{12}$, where B_{12} is a proportionality constant. The transition rate for deexcitation from state 2 to state 1, $\Gamma_{2\to1}$, has two terms, a stimulated-emission term proportional to the number of atoms in state 2 times the density of photons present with the proper frequency, $N_2 u(f, T)B_{21}$, and a spontaneous-emission term proportional to the number of atoms in state 2. Thus $\Gamma_{2\to1} = N_2[u(f, T)B_{21} + A]$, where A is a constant. (a) In thermodynamic equilibrium, the transition rate in the two directions must be equal. Use this fact, together with the Boltzmann relation for the populations, $(N_1/N_2) = e^{-E_1/kT}/e^{-E_2/kT}$, to show that

$$u(f, T) = \frac{A}{B_{12}e^{hf/kT} - B_{21}}.$$

(b) In the classical limit, $u(f, T) = 8\pi f^2 kT/c^3$. Show that the quantum mechanical expression for u can match this form for large T only if $B_{12} = B_{21}$, and find the value of the ratio A/B_{12} from the matching requirement.

32. (III) From Problem 31, the rate per atom at which atoms

in equilibrium with radiation at temperature T emit photons of frequency f has the form

$$\frac{\Gamma_{2\to 1}}{N_2} = C\left(1 + \frac{1}{e^{hf/kT} - 1}\right),$$

where C is a constant. (a) Using the hint below, show that the average number of photons of frequency f per unit volume, $\langle n \rangle$, is given by

$$\langle n \rangle = 1/(e^{hf/kT} - 1).$$

(b) Use this result to show that the deexcitation transition rate per atom is given by

$$\frac{\Gamma_{2\to 1}}{N_2} = A(1 + \langle n \rangle).$$

This establishes the result that the transition rate for deexcitation increases linearly with the number of photons of the proper frequency already present, as is required for laser operation. [*Hint:* The distribution function to find n photons of energy hf is given according to Planck's work on blackbody radiation as $P_n = e^{-nhf/kT}(1 - e^{-hf/kT})$. The sum representing the average number is $\Sigma P_n n$, and this sum can be performed by noticing that $\sum_n x^n$ is a geometric series, and that $\sum_n nx^n = x(d/dx) \sum_n x^n$].

43–3 *Superconductivity*

33. (I) The critical temperature of a superconductor, T_c, varies with the isotopic mass of the element making up the superconductor, M, according to the relation $T_c\sqrt{M}$ = a constant. In mercury, $T_c = 4.185$ K for the isotopic molar mass 199.5 g. What is the critical temperature for the isotopic molar mass 203.4 g?

34. (I) An unknown constant potential difference V is placed across a Josephson junction, and an alternating current of frequency 12.5 GHz is produced. (Figure 43–20 shows a circuit that contains some 12,000 Josephson junctions.) What is the value of V?

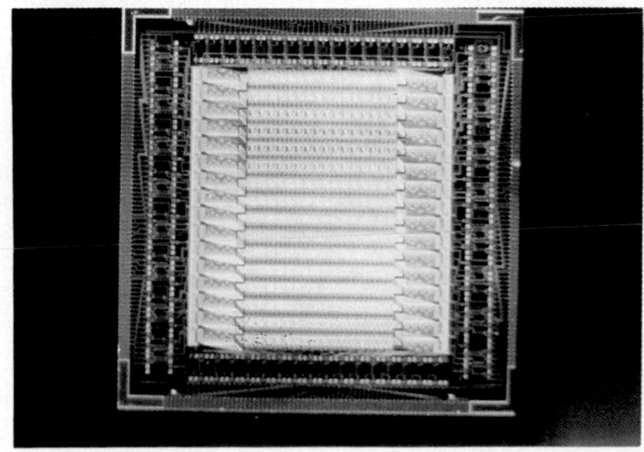

FIGURE 43–20 Problem 34.

35. (I) What is the frequency of the AC component of the current when a DC voltage of 2.75 V is placed across a Josephson junction? a DC voltage of 10.00 μV?

36. (II) Verify that the units of magnetic flux are consistent with Eq. (43–14).

37. (II) When a current passes through a Josephson junction, the resulting AC current can be regarded as corresponding to photons whose frequency is that of the AC current. For a particular Josephson junction, if these photons have an energy of 10^{-6} eV, what is the DC potential difference across the junction?

43–4 *Superfluidity and Liquid Helium*

38. (II) According to one approach that takes into account the quantum mechanical "clustering" behavior of the bosons that make up a superfluid, the temperature at which the superfluid forms is given by $T_c = [N/(2.612)(0.886)C]^{2/3}$, where $C = 2\pi V(2mk/h^2)^{3/2}$, V is the volume in which the N bosons are confined, and m is the mass of the boson. Apply this result to liquid ^{4}He; take the density of the liquid to be 0.147 g/cm^3, and calculate T_c. (The experimental value of T_c for this system is 2.2 K.)

General Problems

39. (II) (a) The energy-distribution function of electrons at $T = 0$ is $n(E) = (V/2\pi)(2m/\hbar^2)^{3/2}\sqrt{E}$. Use this to calculate the speed-distribution function. (b) Calculate $\langle v^4 \rangle$ by using the result of part (a).

40. (II) What is the radius of a star with a mass of 10^{-3} solar masses, assuming that the star has the density of the sun? Compare your answer with the radius of Jupiter.

41. (II) Assume that the Fermi energy depends only on $\hbar$, the electron density (n_e), and the mass of the electron. Then use dimensional analysis to get the dependence.

42. (II) Assume that the Fermi energy depends only on $\hbar$, the electron density (n_e), and the speed of light (c). Then use dimensional analysis to get the dependence.

43. (II) You want to verify the isotopic effect described in Problem 33 with a set of data you collected. Show that if an isotope equation of the type $T_c M^{\gamma}$ = a constant were correct, the data on a plot of $\ln(T_c)$ versus $\ln(M)$ would be a straight line whose slope determines γ. What would your curve look like if $\gamma = \frac{1}{2}$?

44. (II) Show that the isotopic-effect equation $T_c\sqrt{M}$ described in Problem 33 implies that T_c is proportional to v, where v is the speed of sound in a material with isotopic mass M.

45. (II) (a) Using the methods in the text, show that in a gas of free electrons in a box with sides of length L at $T = 0$ K, the total number of electrons with energy less than E is

$$N = \left(E \frac{2m_e}{\hbar^2 \pi^2} L^2\right)^{3/2} \frac{\pi}{3},$$

where $E < E_F$. (b) Use the results of part (a) to show that the number of electrons with energies between E and $E + dE$ is given by

$$n(E)\, dE = \frac{V}{2\pi^2}\left(\frac{2m}{\hbar^2}\right)^{3/2} \sqrt{E}\, dE,$$

where $n(E)$ is the energy-distribution function of the electrons. [*Hint:* The expression you want is given by $n(E)\, dE \equiv N(E + dE) - N(E)$.]

46. (II) Use the energy-distribution function of Problem 45 to show that the total energy of a gas of electrons in a box with sides of length L at $T = 0$ K is given by $E_{tot} = \frac{3}{5} N E_F$.

47. (III) The behavior of the atoms of a crystal at temperature T under the influence of the springlike forces between the atoms is analogous to the behavior of light in a cavity at temperature T (blackbody radiation). The behavior of the lattice can be described by *phonons*, collective oscillations of the lattice that can be treated as particles analogous to the photons in a hot cavity. The phonons are identical bosons that obey the quantum rules for such particles. There is one important difference between the lattice vibrations and blackbody radiation: The number of possible modes of a lattice made up of N atoms is limited to $3N$ (corresponding to the three-dimensional motion), and there is a maximum frequency f_{max}. By following the kind of reasoning Planck used in deriving his blackbody spectrum, it is possible to show that the total lattice vibration energy, U, of the crystal is given by

$$U = \frac{9NkT}{\Lambda^3} \int_0^{\Lambda} \frac{x^3}{e^x - 1}\, dx,$$

where $\Lambda \equiv T_D/T$; T_D, called the *Debye temperature*, is given by $T_D = hf_{max}/k$. (a) Show that in the limit of $T \gg T_D$, the energy of the solid is the classical expression $3NkT$. (b) Show that in the limit $T \ll T_D$, $U \propto T^4$. (c) What is the temperature dependence of the specific heat at low temperatures? The solid curve in Fig. 43–21, a plot of the expression above for U, matches well the experimental data over a large temperature range in many solids. (After G. Burns, *Solid State Physics*, Academic Press Inc., 1985, p. 369).

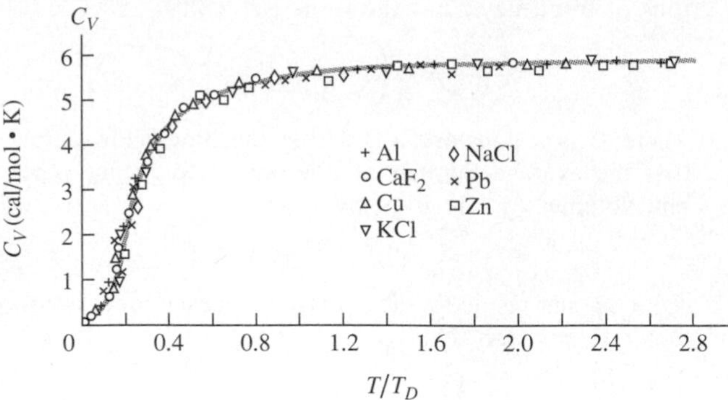

FIGURE 43–21 Problem 47.

48. (III) For stars whose mass is roughly one solar mass (2×10^{30} kg) and whose thermonuclear reactions have stopped because the fuel has been used up, the repulsive effects due to the Pauli exclusion principle *for electrons* are strong enough to keep the star from collapsing gravitationally. In that case, white dwarfs are formed. (a) Calculate the Fermi energy of the electrons in a white dwarf of mass M. (b) Calculate the radius of a white dwarf of mass M. If $M = M_\odot$ is one solar mass, what is the ratio of your result to the present radius of the sun? [*Hint:* Follow our procedure for a neutron star. The gravitational pressure is the same as before, because only the protons and neutrons account for the mass. The number of electrons is the same as the number of protons, which is roughly half the total of protons plus neutrons. The radius of the sun is 6.96×10^8 m.]

49. (III) When all the hydrogen in a star has burned, helium nuclei and electrons remain. Suppose that there are N electrons, and that the star is a sphere of radius R and mass $M \simeq (N/2)m_{He}$. (a) Use Eq. (43–8) to estimate the degeneracy pressure. (b) The gravitational force of the helium nuclei exerts an inward pressure, which is approximately $p_{grav} = E_{grav}/V \simeq (GM^2/R)/V$. Find p_{grav} in terms of R. (c) The star will collapse until the two pressures cancel, when it has a radius R_f. Calculate R_f in terms of m_e, m_{He}, and N. (d) Evaluate R_f for a star of one solar mass. (The result will be surprisingly small.)

50. (III) Use the methods of Problem 22 to calculate the Fermi energy of a gas of electrons, assuming that the mass of electrons can be neglected, so that $E = pc$. Under what conditions will the star of Problem 49 *not* collapse?

51. (III) What is the bulk modulus of a nucleus that consists of N neutrons and Z protons and has a radius $R = r_0 A^{1/3}$, where $r_0 \simeq 1.2 \times 10^{-15}$ m?

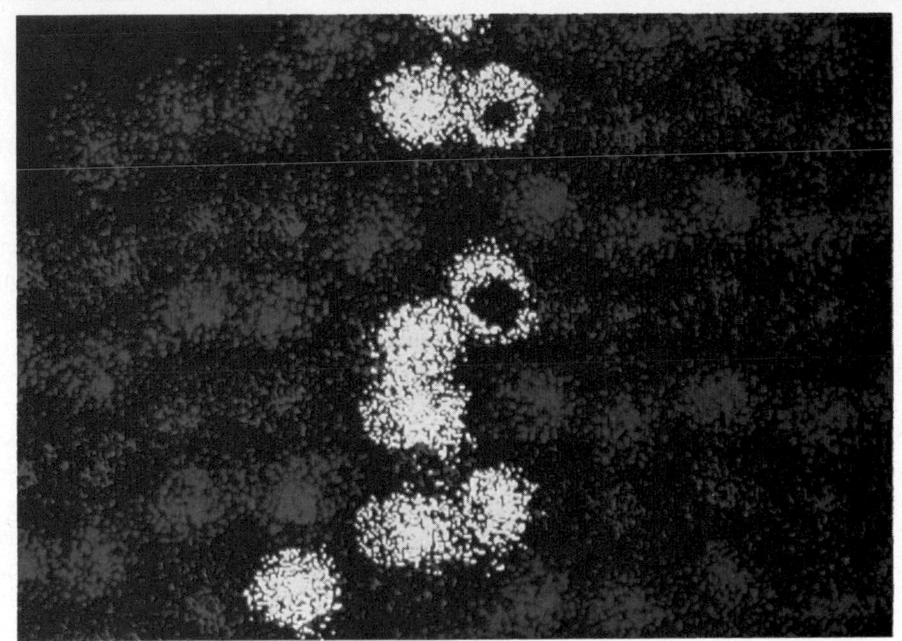

This scanning tunneling micrograph shows atoms in a solar cell, a device that operates through the quantum behavior of carefully engineered semiconductor materials. We can tailor the structure of such materials to produce the electronic properties that we want.

QUANTUM ENGINEERING

Science and technology have always been closely intertwined. Sometimes technology has led to new scientific advances, as in the case of thermal physics, which was stimulated by the need to understand the engines that propelled the Industrial Revolution. At other times, science has led technology, as was true in the development of the electric power industry out of the discoveries of Faraday and Maxwell. The second half of the twentieth century has seen simultaneous advances in technology and physics. The discovery of quantum mechanics has led to an ever-increasing understanding of the detailed properties of materials, and this understanding has been accompanied by rapid developments in the fabrication of materials required to satisfy the needs of technology. In this chapter we shall begin with a study of semiconductors and their applications and shall then make a qualitative and rapid tour of more recent technological developments that rely on an understanding of quantum physics. We conclude with a brief discussion of some of the devices whose function relies on quantum phenomena and whose potential for making major contributions to future technology has already been demonstrated.

44–1 SEMICONDUCTORS

Conduction is discussed in Chapter 27.

The electrical conduction properties of solids start with two overriding features: the Pauli exclusion principle, which is obeyed by all electrons (see Chapter 43), and the presence of a band structure, with gaps, that follows from the lattice of ions in the material (see Chapters 27 and 42). The Pauli exclusion principle restricts the number of electrons that can exist in each energy level to no more than two. The lowest of a set of electron energy levels are filled first. The energy of the highest filled level is the *Fermi energy*, E_F. The band structure expresses the presence of energy gaps between allowed bands; there are no levels within such gaps, so no electrons can have energies there. In conductors, levels are filled to the Fermi energy, but there are empty energy levels in the band within which the Fermi energy is found. In semiconductors and insulators, the *valence band*, which is the topmost band containing electrons, is filled right up to a gap. Materials with a totally filled valence band will in general act as insulators, *unless the energy gap between the valence band and the next band is small.*† In that case it is possible to induce conductivity by thermally exciting some electrons across the gap. The next band above the valence band is therefore referred to as the *conduction band* (see Fig. 27–18a). Typical gap widths—the energy difference from the top to the bottom of the gap—at room temperature are listed in Table 44–1 for some semiconductors. In contrast, the gap width of the diamond form of carbon, which is regarded as an insulator, is 5.5 eV.

TABLE 44–1

GAP WIDTHS OF SOME SEMICONDUCTORS†

Material	*Gap Width (in eV)*
Tin (gray form)	0.08
Tellurium	0.35
Lead sulfide	0.37
Germanium	0.67
Silicon	1.12
Gallium arsenide	1.43
Boron	1.5
Selenium	1.8

† At room temperature.

The Effects of Temperature

What we have said about the Fermi energy applies to systems at zero temperature, $T = 0$ K. Let us now study the effects of a nonzero temperature. We must do so because the behavior of semiconductors depends crucially on temperature. It is simplest to look first at finite temperature effects in conductors. At $T = 0$ K, we are equally likely to find an electron in any energy level up to E_F. If we suppose that the possible levels are so closely spaced as to represent a continuum, then at $T = 0$ K the probability $f(E)$ that the state with energy E is occupied by an electron (actually up to two electrons, due to spin) is

$$\begin{aligned} &\text{for } E < E_F: \quad f(E) = 1, \\ &\text{for } E > E_F: \quad f(E) = 0. \end{aligned} \tag{44–1}$$

When $T > 0$ K, some of the electrons will be thermally excited into previously unoccupied levels, leaving vacant energy levels behind them. For temperatures that are not too large, this can happen only for electrons with energies near E_F, because electrons with energies far below E_F are unlikely to get a sufficiently large thermal "kick." If electrons behaved according to the classical rules we described in Chapter 19, they would obey a Maxwell–Boltzmann distribution. But because electrons obey the Pauli exclusion principle, their energy distribution is instead described by the **Fermi–Dirac distribution**, whose form is

The Fermi–Dirac distribution

$$f(E) = \frac{1}{1 + e^{(E - E_F)/kT}}. \tag{44–2}$$

We can check that we recover Eq. (44–1) from this distribution in the limit $T \to 0$. For $E > E_F$, the exponent gets very large as T approaches zero, so $f(E)$ has a limiting value of zero; for $E < E_F$, the exponent is very large but negative, so the exponential factor goes to zero, and $f(E) \to 1$. Figure 44–1 shows the

† Note that the distinction between a semiconductor and an insulator is not a sharp one.

distribution for the increasing temperatures $T = 0$, T_1, and T_2. Notice that for $(E - E_F) \gg kT$, the exponential factor dominates in the denominator, and we get the approximation

$$f(E) \simeq e^{-(E-E_F)/kT} \qquad (44\text{–}3)$$

Equation (44–3) has the form of the Maxwell–Boltzmann distribution discussed in Chapter 19, so we have something like a classical distribution when the energy difference $E - E_F$ is large compared to kT. Just what is a typical value of kT? A useful number to remember is that room temperature, $T = 300$ K, corresponds to 1/40 eV.[†]

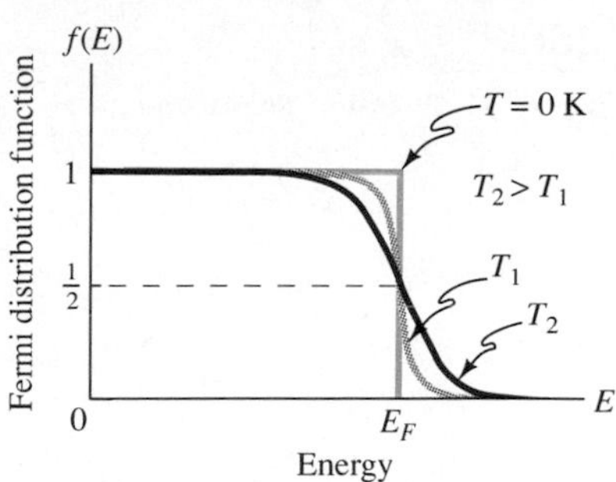

FIGURE 44–1 The Fermi–Dirac distribution, $f(E)$, which describes the probability of finding one of a set of identical fermions with an energy E. The distribution is sketched for three successively higher temperatures, $T = 0$, T_1, and T_2.

EXAMPLE 44–1 Assume that electrons in copper are distributed according to the Fermi–Dirac distribution. The Fermi energy in copper is $E_F = 7.04$ eV. (a) The *Fermi speed*, v_F, is the speed of an electron that moves with a kinetic energy equal to E_F. What is v_F for copper? (b) From the Fermi–Dirac distribution, $f(E_F) = \frac{1}{2}$. What is the value of $f(E)$ at a temperature of 300 K for an electron with a speed 1 percent higher than v_F? (c) Estimate the temperature at which the probability $f(E)$ is 10^{-10} when E is 1.2 eV above E_F.

SOLUTION: (a) We assume that the electrons are nonrelativistic, so $E_F = \frac{1}{2}m_e v_F^2$, or

$$v_F = \sqrt{\frac{2E_F}{m_e}} = \sqrt{\frac{2(7.04\ \text{eV})(1.60 \times 10^{-19}\ \text{J/eV})}{0.911 \times 10^{-30}\ \text{kg}}} = 1.57 \times 10^6\ \text{m/s}.$$

(b) When the speed of an electron increases by 1 percent, the energy, proportional to the speed squared, increases by approximately 2 percent. Thus $E = (1.02)(7.04\ \text{eV}) = 7.18$ eV. With $kT = 2.6 \times 10^{-2}$ eV at $T = 300$ K, $(E - E_F)/kT = 5.4$, and hence

$$f(E) = \frac{1}{1 + e^{5.4}} \simeq 4.6 \times 10^{-3}.$$

(c) We set $f(E)$ equal to 10^{-10} and solve for kT, given $E - E_F = 1.2$ eV. When $f(E) = 10^{-10}$, the exponential term in the denominator of $f(E)$ in Eq. (44–2) is much larger than 1. Thus

$$f(E) \simeq e^{-(E-E_F)/kT} = 10^{-10},$$

and $-(1.2\ \text{eV})/kT = \ln(10^{-10}) = -10\ln(10) = -23$. Thus $kT = (1.2\ \text{eV})/23 = 0.052$ eV. This is twice the value of kT at $T = 300$ K [see part (b)], so the desired estimate is 600 K.

Semiconductors, Electrons, and Holes

To apply our ideas about finite temperatures to semiconductors, we must reconsider the role of the energy gap. Suppose that the band structure is that of a semiconductor; that is, at $T = 0$ the valence band is filled, the conduction band is empty, and the gap energy $E_g \equiv E_c - E_v$ is relatively small. Here E_v is the maximum energy of the valence band (the *edge of the valence band*) and E_c is the lowest energy of the conduction band (the *edge of the conduction band*) (Fig. 44–2).

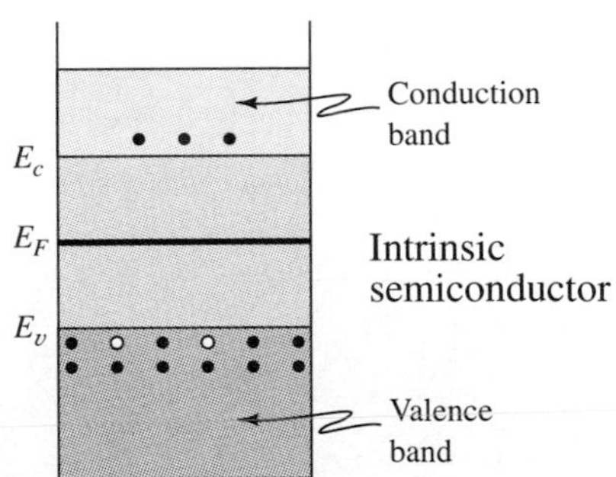

FIGURE 44–2 The band structure of an intrinsic semiconductor is that of a filled valence band, a relatively narrow band gap, and, at $T = 0$, an empty conduction band. The solid circles represent electrons, and the open circles represent holes. At $T = 0$, there are no electrons in the conduction band and no holes in the valence band.

[†] For $T = 300$ K, $kT = (1.38 \times 10^{-23}\ \text{J/K})(300\ \text{K})/(1.6 \times 10^{-19}\ \text{J/eV}) = 2.6 \times 10^{-2}\ \text{eV} = 1/40$ eV.

To determine what happens when T is not zero, we must clarify what we mean by the Fermi energy for this situation. If we think of the Fermi energy as the energy above which there are no electrons (at zero temperature), then *any* energy value within the energy gap would serve. This is important because we must find a correct distribution for electron energies analogous to Eq. (44–2), and that expression contains the Fermi energy. Our procedure is as follows: We use Eq. (44–2) but with an unknown quantity μ replacing the Fermi energy:

$$f(E) = \frac{1}{1 + e^{(E-\mu)/kT}}. \tag{44–4}$$

The quantity μ is in fact a thermodynamic variable, the *chemical potential*, but the detailed properties of this variable, which are consequences of the laws of thermodynamics, need not concern us further. For the moment, μ is unknown, but we shall *determine* its value from a condition of the conservation of charge. *In the following discussion, we shall conform to established practice for semiconductor engineering by writing E_F rather than μ throughout.* When you see formulas for semiconductors with the parameter E_F, keep in mind that this is not the Fermi energy in the sense in which we first defined it.

When $T \neq 0$, some electrons go into the conduction band, and the valence band is correspondingly depleted of electrons (Fig. 44–3). At room temperature, $E_g \gg kT$ (about $1\text{ eV} \gg \frac{1}{40}\text{ eV}$). We can then very roughly estimate that the relative number of electrons able to jump the gap in, say, silicon, for which $E_g \simeq 1$ eV, is $e^{-E_g/kT} \simeq e^{-25} \simeq 10^{-11}$. Thus only a very small fraction of electrons will reach the conduction band. This same reasoning would hold if, instead of E_g, we used the smaller quantity $E_c - E_F$, as long as $E_c - E_F$ were of the same order of magnitude as E_g, as we shall see it typically is. We would thus have a situation in which an energy distribution of the same form as Eq. (44–3) holds (although we have yet to determine E_F). Because the function in Eq. (44–3) decreases so rapidly as energy increases, the number of electrons per unit volume (the number density n) in the conduction band is well approximated by replacing E in Eq. (44–3) with the lowest energy in the conduction band, E_c:

$$n = N_c e^{-(E_c - E_F)/kT}. \tag{44–5}$$

The proportionality constant N_c will be treated later. The electrons in the conduction band are free to move when an electric field is applied. The electrons that are free to move are called *n-carriers* (n for negative).

Electrons that are thermally "promoted" into the conduction band leave

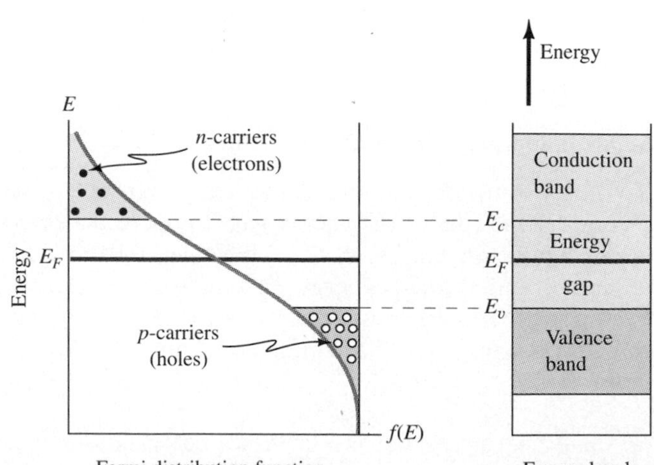

FIGURE 44–3 Electrons near the top of the valence band can jump the gap in a semiconductor at finite temperature. The Fermi–Dirac distribution describes the probability that an electron will have enough energy to be promoted to the conduction band. For every electron that can be promoted, a hole is left behind in the valence band.

vacancies in the valence band. Only the electrons with energies near E_v are promoted, so the vacancies are located at those energies. Such a vacancy is called a **hole**, as in the study of atomic and nuclear shells. Holes propagate when one hole is filled by an electron, which in turn leaves a hole in the electron's original position, which in turn is filled by an electron, and so on. If there is an electric field pointing in the $+x$-direction, then electrons will flow in the $-x$-direction, and thus the hole will flow in the $+x$-direction, leading to an additional current. (A helpful analogy might be an air bubble floating upward in a tank of water. The "vacancy" in the water, the air bubble, is constantly replaced by water falling down under the force of gravity; this water leaves its own vacancy, which thus continually moves upward.) The hole thus acts as a positively charged particle, which explains why a hole is alternatively called a *p-carrier* (p for positive). The probability of finding a hole with energy E can be determined by the Fermi–Dirac distribution. If $f(E)$ is the probability of finding an electron with energy E, then the probability of not finding an electron with that energy is

$$1-f(E)=1-\frac{1}{1+e^{(E-E_F)/kT}}=\frac{e^{(E-E_F)/kT}}{1+e^{(E-E_F)/kT}}$$

$$=\frac{e^{-(E_F-E)/kT}}{1+e^{-(E_F-E)/kT}}. \qquad (44\text{–}6)$$

Here E is the energy of the empty electron state (the hole), which is an energy in the valence band, so the factor E_F-E is positive. For $E_F-E \gg kT$, as we shall see is typically the case, the exponential in the denominator is very small compared with unity, and the right-hand side of Eq. (44–6) reduces to $e^{-(E_F-E)/kT}$. This exponential factor decreases so rapidly as the factor E_F-E increases that to a good approximation we can replace E_F-E by E_F-E_v. Thus the number density of holes is accurately given by

$$p=N_v e^{-(E_F-E_v)/kT}. \qquad (44\text{–}7)$$

We shall give the proportionality constant N_v later. Note that the product of the n-carrier concentration and the p-carrier concentration is given by

$$np=N_c e^{-(E_c-E_F)/kT}N_v e^{-(E_F-E_v)/kT}=N_cN_v e^{-(E_c-E_v)/kT}=N_cN_v e^{-E_g/kT}. \qquad (44\text{–}8)$$

We have assumed thus far that each hole in the valence band was created by the promotion of an electron into the conduction band, so the number of p-carriers equals the number of n-carriers. Semiconductors for which this is the case are called *intrinsic semiconductors.* We can now use this condition to determine E_F. We have

Intrinsic semiconductors

$$n_i=p_i,$$

where the subscript i indicates that we are dealing with an intrinsic semiconductor. It follows from Eq. (44–8) that

$$n_i=p_i=\sqrt{n_ip_i}=\sqrt{N_cN_v}e^{-E_g/2kT}. \qquad (44\text{–}9)$$

Comparison of the exponential factor in this result with the exponential factors of Eqs. (44–5) and (44–7)—which is valid as long as N_c and N_v do not vary too rapidly with temperature—shows that

$$E_c-E_F=E_F-E_v=\tfrac{1}{2}E_g. \qquad (44\text{–}10)$$

In other words, E_F *(or more properly, the chemical potential) is at the halfway point of the gap.* This result justifies our expectation that E_c-E_F and E_F-E_v are much larger than kT if E_g is, too.

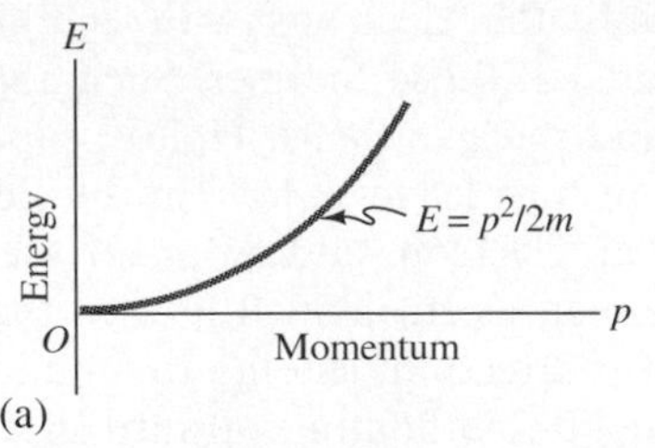

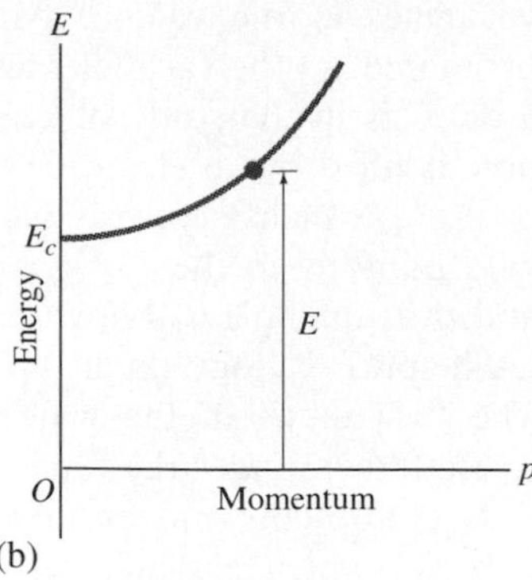

FIGURE 44–4 (a) Energy versus momentum for a free electron. (b) Energy versus momentum for an electron in the conduction band. The lowest possible energy for the electron is E_c.

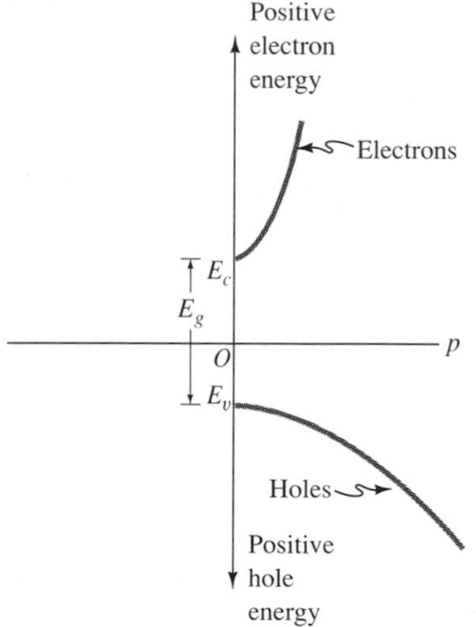

FIGURE 44–5 The energy–momentum relation for an electron in the conduction band (Fig. 44–4b) is modified because the electron behaves as though it has a modified mass m^*. Also shown is the energy–momentum relation for a hole, which in general has a different effective mass than does an electron.

An additional effect is important to semiconductors. The energy of a free electron with momentum p is $E = p^2/2m_e$, as in Fig. 44–4a. Figure 44–4b shows the energy–momentum relation that would hold for the electrons in the conduction band if those electrons were free. However, because of electron interactions with the lattice, this relationship changes. For small momentum, the principal effect is to change the steepness of the parabola, and this is equivalent to changing the electron mass to an *effective electron mass*, denoted by m^*. For semiconductors, m^* is typically an order of magnitude *smaller* than the true electron mass; for gallium arsenide (GaAs), $m^*/m_e \simeq 0.067$, and for indium antimonide (InSb), this ratio is only 0.015. Furthermore, the effective mass of n-carriers need not be the same as that of p-carriers; generally m^* is much smaller for n-carriers (Fig. 44–5). The most important consequence of a small effective mass is that the drift velocity of the carriers is increased (see Chapter 27).

An understanding of why effective masses take on their values is beyond the scope of our discussion. Roughly speaking, the ratio m^*/m_e is small if E_g is small compared to the width of the conduction band. E_g is large compared to the width of the conduction band when atomic energy levels are not disturbed very much by the presence of neighboring atoms. In such a "tight binding" situation, the effective masses are typically much larger than those of free electrons, because it is hard to "move" electrons in such a material.

Doping

Extrinsic semiconductors

We have thus far concentrated on intrinsic semiconductors. Now we shall consider *extrinsic*, or *impurity*, *semiconductors*, in which additional n-carriers or p-carriers may be supplied by the addition of impurity atoms to the crystal lattice. The addition of impurities (whether artificially or naturally) is known as *doping*. It works as follows: Suppose that we add some arsenic atoms to germanium. The crystal structure will remain unchanged (each arsenic ion replaces one germanium ion in the lattice), but the electronic structure will change. Germanium has four valence electrons, whereas arsenic has five; because one electron that can affect conduction has been added per replacement atom, the arsenic atom is said to be a *donor impurity*. When n-carriers are added in this way, the doped material is said to be an *n-type semiconductor*.

If the additional electron were not bound at all to the arsenic ion, it could fill an energy level only in the conduction band and therefore could easily carry current. The energy such an electron would have is on the order of E_c. Although the additional electron is indeed bound to the arsenic ion, it is bound only very weakly. If the arsenic were not part of the semiconductor lattice, we might expect the electron to be bound to the arsenic ion with an energy like the Bohr energy

[Eq. (42–9) with $n = 1$]:

$$E = -\frac{e^4 m_e}{2(4\pi\epsilon_0 \hbar)^2} \simeq -13.6 \text{ eV}.$$

Within the semiconductor, this binding energy is weakened for two reasons. First, the mass of the electron is effectively reduced, typically by more than a factor of 10, from m_e to m^*. The radius of the electron's Bohr orbit is now so large that the electron is circulating within the larger environment of the semiconductor lattice. Second, the semiconductor medium has a dielectric constant κ (defined by $\epsilon = \kappa\epsilon_0$) that is typically more than 10 times larger than the value of 1 for a vacuum. These two factors change the binding energy to

$$E = -\frac{e^4 m^*}{2(4\pi\epsilon\hbar)^2}. \tag{44–11}$$

Our numerical values for m^* and ϵ show that the magnitude of this energy is less than $\frac{1}{1000}$ that of the typical Bohr energy. This small binding energy, on the order of $\frac{1}{100}$ eV or less, manifests itself in the presence of additional energy levels very close to the bottom of the conduction band (Fig. 44–6a). The amount by which these energies differ from E_c is on the order of kT or less at room temperature. Thus the additional electron, which is bound at $T = 0$, is very easily excited to the conduction band at room temperature. Not only do the additional electrons in an n-type semiconductor provide more electrons for carrying current, but the ease with which these electrons are excited to the conduction band may make them far more important for conduction than the electrons associated with the intrinsic semiconductor.

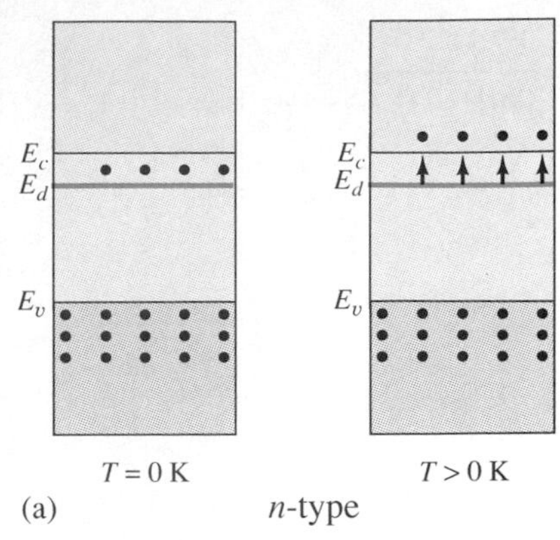

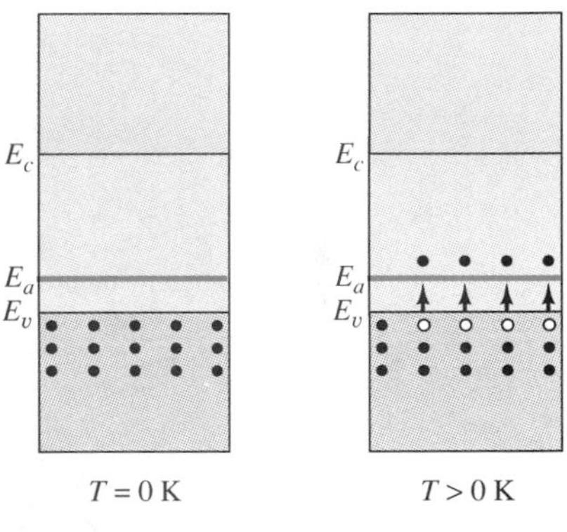

FIGURE 44–6 Energy-band structure for doped semiconductors. (a) In n-type semiconductors, E_F (labeled E_d) lies close to E_c. The levels of electrons (solid circles) from donor impurities lie above E_F and are easily excited at $T > 0$ into the conduction band. (b) In p-type semiconductors, E_F (labeled E_a) lies close to E_v. There are levels associated with the acceptor impurities that are quite close to E_v, so at finite temperatures, electrons from the valence band are easily promoted to these levels, leaving holes (open circles) in the valence band.

EXAMPLE 44–2 Calculate the ionization energy for a donor electron in doped indium antimonide, a semiconductor in which $\epsilon/\epsilon_0 = 17.9$ and $m^*/m_e = 0.015$.

SOLUTION: The ionization energy E is the negative of the binding energy of the ground state of the donor atom [Eq. (44–11)], so

$$E = \frac{e^4 m^*}{2(4\pi\epsilon\hbar)^2} = (13.6 \text{ eV})\left(\frac{m^*}{m_e}\right)\left(\frac{\epsilon_0}{\epsilon}\right)^2$$

$$= (13.6 \text{ eV})(0.015)\left(\frac{1}{17.9}\right)^2 = 6.4 \times 10^{-4} \text{ eV}.$$

This much energy corresponds to a temperature of $T = E/k = 0.025$ K.

If an impurity has a deficit of valence electrons, a surplus of p-carriers is created, because holes form when electrons from the intrinsic semiconductor atoms become strongly attached to the impurity atoms. Such impurities are known as *acceptor impurities*, because they accept electrons. Boron, for example, has three valence electrons and is an acceptor impurity for germanium. Semiconductors doped to have an excess of holes (p-carriers) are known as *p-type semiconductors*. By the same sort of reasoning that we used for n-type semiconductors, p-type semiconductors have the level structure shown in Fig. 44–6b, with some additional levels close to E_v. The n-carriers from the valence band are easily promoted to the acceptor levels, leaving holes in the valence band that are very effective at carrying charge.

As is true of intrinsic semiconductors, the effective Fermi energy of doped semiconductors determines the probability of promoting electrons and holes to levels where they can conduct. The Fermi energy is found by correctly relating the densities of n-carriers and p-carriers. Because the Fermi energy basically describes a jumping-off point for determining how the carrier densities increase as a function of temperature, as in Eq. (44–3), we can generalize that the Fermi energy nears E_c (increases) in n-type semiconductors and nears E_v (decreases) in p-type semiconductors. Indeed, if the donor in an n-type semiconductor provides the bulk of the n-carriers, as is typically the case, then the Fermi energy will be above the donor levels, quite close to E_c; similarly, in a p-type semiconductor that is doped sufficiently, the Fermi energy will lie close to E_v, below the acceptor levels. These results can be formally derived by starting with Eq. (44–8), which was established without using the conditions special to an intrinsic semiconductor and continues to hold whether or not the semiconductor is doped. The constants N_c and N_v in Eq. (44–8) can be obtained by summing the number of electrons in the conduction band and the number of holes in the valence band, respectively:

$$N_c = 2\left(\frac{m_n^* kT}{2\pi\hbar^2}\right)^{3/2} \quad \text{and} \quad N_v = 2\left(\frac{m_p^* kT}{2\pi\hbar^2}\right)^{3/2}, \tag{44–12}$$

where m_n^* and m_p^* are the effective masses of the n-carriers and p-carriers, respectively. The fact that Eq. (44–8) continues to hold even when there is doping means that†

$$np = n_i^2 = N_c N_v e^{-(E_c - E_v)/kT} = N_c N_v e^{-E_g/kT}. \tag{44–13}$$

Note that Eq. (44–13) is *not* identical to Eq. (44–8), even though it has the same form. Equation (44–13) relates the n-carrier and p-carrier densities in a *doped* semiconductor to the intrinsic carrier densities, whereas Eq. (44–8) relates the carrier densities in the undoped semiconductor.

Equation (44–13) has important consequences. For example, if we dope an intrinsic semiconductor with acceptor impurities to increase the density of holes, the density of electrons must decrease so that Eq. (44–13) continues to be satisfied. It is typical that the n-carriers dominate in an n-type extrinsic semiconductor, whereas p-carriers dominate in a p-type extrinsic semiconductor.

EXAMPLE 44–3 The effective masses in the semiconductor gallium arsenide are $m_n^*/m_e = 0.067$ and $m_p^*/m_e = 0.48$; the band gap is $E_g = 1.43$ eV. Find the intrinsic concentration of n-carriers or p-carriers in gallium arsenide at room temperature, where $kT = 0.026$ eV.

SOLUTION: The concentration of intrinsic carriers in a semiconductor (doped or not) is given by Eq. (44–12). With the help of Eq. (44–13), we get

$$n_i = \sqrt{N_c N_v e^{-E_g/2kT}} = 2\left(\frac{mkT}{2\pi\hbar^2}\right)^{3/2}\left(\frac{m_n^* m_p^*}{m^2}\right)^{3/4} e^{-E_g/2kT}$$

$$= 2\left[\frac{(0.91 \times 10^{-30}\ \text{kg})(1.38 \times 10^{-23}\ \text{J/}\cancel{\text{K}})(300\ \cancel{\text{K}})}{2\pi(1.05 \times 10^{-34}\ \text{J} \cdot \text{s})^2}\right]^{3/2}$$

$$\times [(0.067)(0.48)]^{3/4} e^{-(1.43\ \cancel{\text{eV}})/2(0.026\ \cancel{\text{eV}})}$$

$$= 2.2 \times 10^{12}\ \text{carriers/m}^3.$$

† The only important assumption that goes into Eq. (44–8) is that energy differences are large compared to kT.

Optical Effects in Semiconductors

The probability that a photon striking a semiconductor will be absorbed has a threshold as a function of the photon's frequency, f (Fig. 44–7). Only if the energy of the photon, hf, is greater than the gap energy can an electron in the material absorb the photon, because only at such energies are there empty levels available for electron occupation in the conduction band. Thus some materials are transparent (a photon passes through without being absorbed) in certain frequency ranges but cease to be so for higher frequencies. If a photon for which $hf > E_g$ is absorbed by an electron in the valence band of a semiconductor, the electron is kicked up into the conduction band, often to an energy well above E_c. It will lose energy by colliding with lattice atoms until its energy is near E_c, when further energy loss is inhibited by the exclusion principle and/or the absence of energy levels below E_c. At that stage, the only way in which the electron can lose energy is to fall back into a hole in the valence band, a process called *recombination*. The electron–hole recombination is accompanied by the emission of radiation whose frequency is such that $hf_g = E_g$. The process is shown schematically in Fig. 44–8, and the radiation is called *photoluminescence*. The electrons can alternatively be initially excited into the conduction band by the bombardment of the material by high-energy electrons or by the action of an electric current in the material. In these cases, the resulting radiation is called *cathodoluminescence* and *electroluminescence*, respectively.

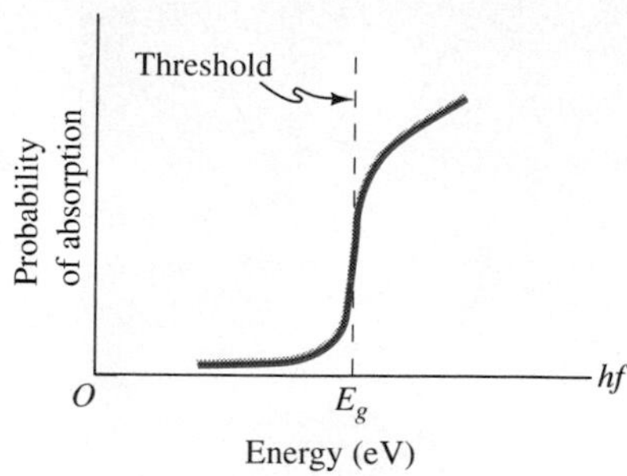

FIGURE 44–7 The probability of photon absorption in a semiconductor has a threshold when the photon energy equals the gap energy, because no energy levels are available for the absorbing electron at lower energies. Because a photon's energy is proportional to its frequency, the energy threshold can be expressed as a threshold in frequency.

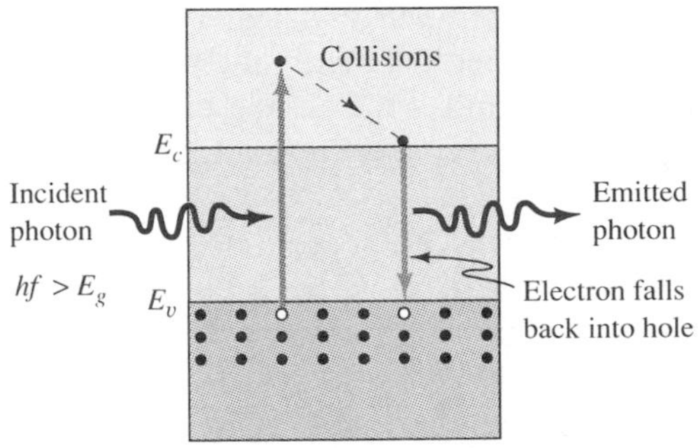

FIGURE 44–8 An electron excited to the conduction band by photon absorption loses energy by collision until its energy reaches E_c; at that point it recombines with a hole in the valence band, with photon emission.

When an electron–hole pair recombines rapidly, on the order of 10^{-8} s, the process is called *fluorescence*. (The time 10^{-8} s is also typical of atomic transitions, and the term fluorescence is also used for atomic processes.) A different situation may arise when there are impurity levels in the gap, and the electron undergoes a transition to those levels. If an impurity level is metastable, in that a direct recombination transition from it to the valence band is forbidden by conservation laws, then the electron is said to be *trapped*. It will ultimately undergo a transition to a vacancy in the valence band (electron–hole recombination), but this may take as much as seconds or even minutes to occur. This process is termed *phosphorescence*, and the materials in which this occurs are called *phosphors*. The frequency of the phosphorescent radiation is determined by the energy difference between the metastable level and the top of the valence band. Thus the color of light emitted by a phosphor is characteristic of the impurity levels in the band gap, and we can choose the impurities such that radiation will be emitted in a variety of colors. In a color television tube the screen is coated with phosphor dots that emit radiation in the primary colors (Fig. 44–9). The phosphors are excited by three incoming electron beams (cathodoluminescence), which are deflected to the proper spot on the screen by electric and magnetic fields determined by the incoming signal to the television's

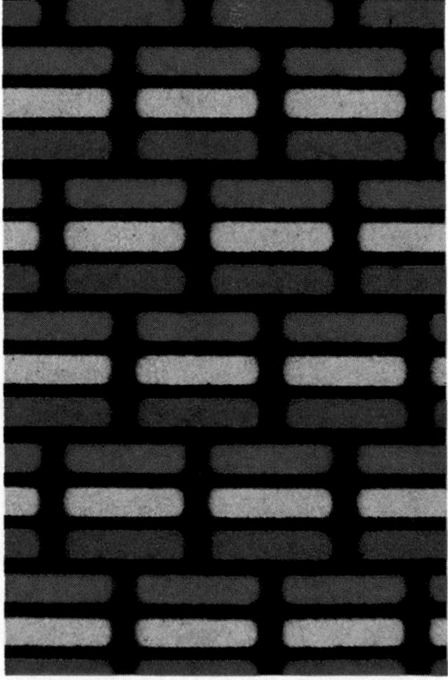

FIGURE 44–9 Excited red, green, and blue phosphors are visible in this photomicrograph of a color television screen.

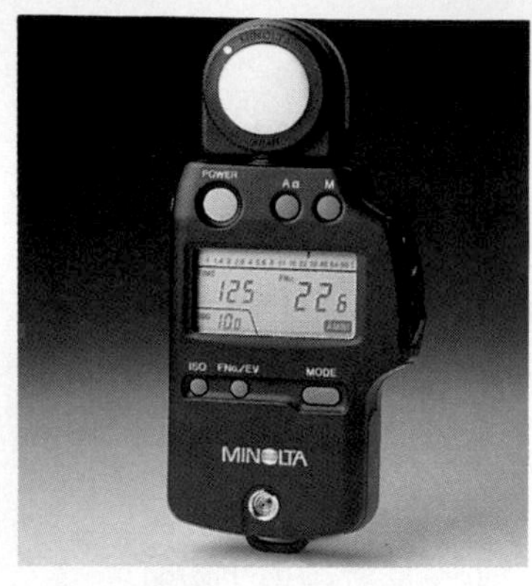

FIGURE 44–10 A light meter aids a photographer in finding the correct light settings for the camera. At the heart of the device is a material whose electrical response depends on the intensity of the light it receives.

antenna. Each of the beams excites a different color, and the incoming signal determines the relative intensity of the colors.

A semiconductor exposed to radiation for which $hf > E_g$ has an increased number of n-carriers and p-carriers and thus an increased conductivity. Devices whose conductivity changes when they are exposed to light—*photoconductive* devices—are used for lights that automatically turn on at dusk and off at dawn, or for measuring light intensity, as in exposure meters for cameras (Fig. 44–10). More sophisticated applications, including light-emitting diodes and semiconductor lasers, involve semiconductor structures, as we shall see in Section 44–2.

44–2 SEMICONDUCTOR STRUCTURES

The Fermi energies equalize across materials that are in electrical contact.

In applying our knowledge of energy distributions in single semiconductors to semiconductor structures (combinations of semiconductors in contact so that charge carriers can pass between them), one principle determines essentially all the important features: Like temperature, *the chemical potential—that is, what we have agreed to call the Fermi energy for semiconductors—has the same value throughout a system in equilibrium.* This fact follows from the laws of thermodynamics, but we do not need thermodynamic language to derive it. We need only apply the same ideas we used for single semiconductors, namely, that in the absence of external potentials and in thermodynamic equilibrium, there is no net flow of charge or of energy across the boundary between two materials. The materials may be the same type of semiconductor doped differently, two different intrinsic semiconductors, or even a semiconductor and a metal. We prove this important result in the box "Why the Fermi Energy Is Constant across a Boundary."

***DERIVATION**

Why the Fermi Energy Is Constant across a Boundary

When two materials are in contact so that charge carriers can pass between them, then in thermal equilibrium, with no external potentials, the Fermi energy must be the same in the two materials. To establish this principle, we note that, at thermal equilibrium, there is no net energy flow and no net charge flow across the boundary. Let us denote the two materials by A and B; number densities of possible energy states for electron occupation at energy E are $n_A(E)$ and $n_B(E)$, respectively. (It is sufficient to consider the electrons alone, because the behavior of the holes can be derived from the behavior of the electrons.) The conditions of no energy flow and no charge flow mean that the flow of electrons from A to B at any E value equals the flow of electrons from B to A at that same E value. The flow of electrons from A to B at energy E is proportional to the number of electrons present in A times the number of empty states present in B; that is, to the density of occupied states in A times the density of empty states in B. To express this mathematically, we need the densities of occupied and unoccupied states:

$$\text{density of occupied states} = (\text{density of available states})(\text{probability of occupation})$$
$$= n(E)f(E),$$

where $f(E)$ is the probability of occupation, given by the Fermi–Dirac distribution, Eq. (44–2). If we use the result that the probability of a state being vacant is one minus the probability of that state being occupied, we also have

$$\text{density of unoccupied states} = (\text{density of available states})(\text{probability of vacancy})$$
$$= n(E)[1 - f(E)].$$

We apply these results to our situation:

$$(\text{flow of electrons from } A \text{ to } B \text{ at energy } E) \propto [n_A(E)f_A(E)]\{n_B(E)[1 - f_B(E)]\};$$
$$(\text{flow of electrons from } B \text{ to } A \text{ at energy } E) \propto [n_B(E)f_B(E)]\{n_A(E)[1 - f_A(E)]\}.$$

We set these equal to each other:

$$[\cancel{n_A(E)}f_A(E)]\{\cancel{n_B(E)}[1 - f_B(E)]\} = [\cancel{n_B(E)}f_B(E)]\{\cancel{n_A(E)}[1 - f_A(E)]\};$$
$$f_A(E) - [\cancel{f_A(E)f_B(E)}] = f_B(E) - [\cancel{f_B(E)f_A(E)}];$$
$$f_A(E) = f_B(E).$$

If we now compare this equation with Eq. (44–2), we see that the only way it can be satisfied is for the parameters E_F in the two materials to be equal.

The *p-n* Junction

When an n-type and a p-type semiconductor are brought into contact, a *p-n junction* is formed. Such a junction is of interest because it acts as a diode, as we shall see. We consider p-n junctions in which both semiconductors are made by doping the same intrinsic semiconductor. In Fig. 44–11a we have used p_p and p_n to denote the hole densities and n_p and n_n to denote the electron densities in p-type and n-type semiconductors, respectively. The sketches are highly distorted, because the ratio of the dominant carriers to the carriers of opposite sign on each side of the junction is typically on the order of 10^{11}; drawn to scale, the n_p and p_n levels would be indistinguishable from the horizontal axis. We shall henceforth refer to the regions of n-type and p-type semiconductors as the n-side and p-side, respectively.

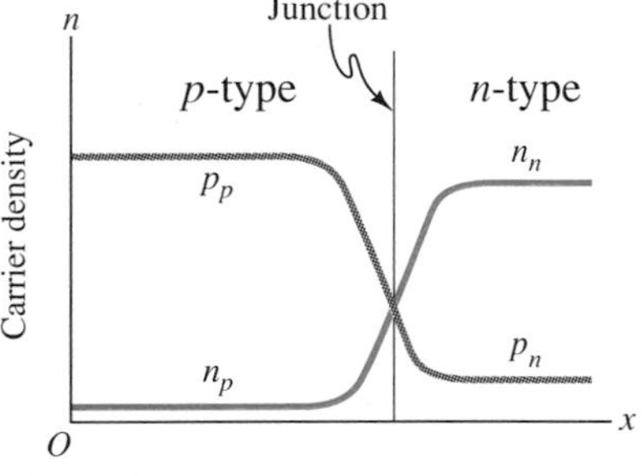

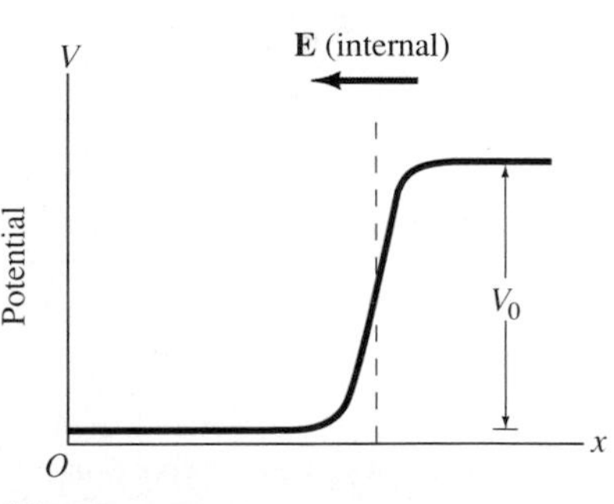

FIGURE 44–11 A p-n junction is formed by placing an n-type and a p-type semiconductor in contact. (a) The number densities of p-carriers (p_p and p_n, respectively), and the number densities of n-carriers (n_p and n_n, respectively) in p-type and n-type semiconductors. These densities are not drawn to scale. (b) As described in the text, the diffusion of carriers causes a potential difference to build up across the boundary. The direction of the resulting electric field is shown.

The p-side has an excess of mobile holes, and the n-side has an excess of mobile electrons, not necessarily with the same density. Electrical neutrality is preserved on the p-side by the (immobile) negative acceptor ions in the crystal lattice and on the n-side by the (immobile) positive donor ions. When the two materials are put in contact, the mobile holes on the p-side tend to diffuse into the n-side, and the mobile electrons on the n-side tend to diffuse into the p-side. This mixing continues until the Fermi energies in the two materials reach the same level, by the following mechanism: Charges flow until the net positive charge remaining on the n-side (the right) and the net negative charge remaining on the p-side (the left) set up an electric field that stops the diffusion (Fig. 44–11b). The n-side will be at a higher potential, V_0, than the p-side. But electrons have a *negative* charge, so the energy of electrons on the right is *reduced* by eV_0.

Let us look at the p-n junction with the aid of energy diagrams. Figure 44–12a shows the energy levels of the semiconductors before they are brought together. The gap energies are the same, because both semiconductors are made by doping the same intrinsic semiconductor. When the junction is formed, the conduction and valence bands will be distorted (Fig. 44–12b). The internal potential V_0 is what brings the Fermi energies to the same level. From Figs. 44–8 and 44–9, we see this statement is equivalent to the relation

$$E^0_{Fn} - E^0_{Fp} = eV_0,$$

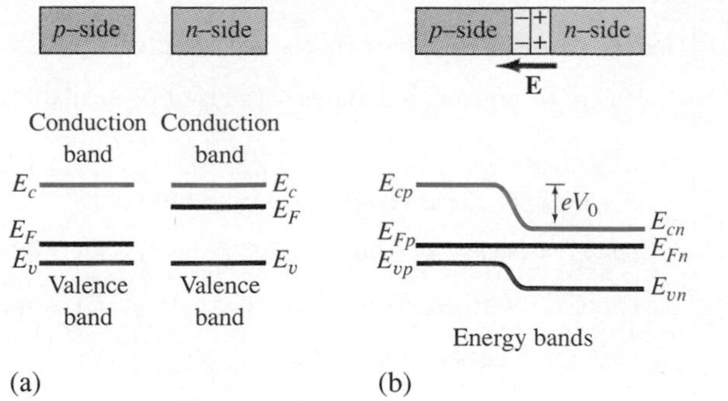

FIGURE 44–12 (a) The energy-level structure of n-type and p-type semiconductors before they are joined in a junction. (b) After the junction is formed and equilibrium is established, the energy levels adjust in such a way that the Fermi levels of the two materials equalize. This lowers the electron energy levels on the n-side and raises the electron energy levels on the p-side. (After B. G. Streetman, *Solid State Electronic Devices*, Prentice Hall, 1980, p. 141.)

where E^0_{Fn} and E^0_{Fp} are the Fermi energies in the n-type and p-type semiconductors, respectively, *before* the materials are brought into contact. *After* the materials are brought into contact,

$$E_{cp} - E_{cn} = eV_0, \tag{44–14}$$

where E_{cp} and E_{cn} are the edges of the conduction band in the p-type and n-type semiconductors, respectively. This relation follows directly from Fig. 44–12b. A similar expression holds for the edges of the valence bands.

There is an internally generated potential difference across the boundary in a p-n junction.

Note that V_0, which is known as the *contact potential*, is *not* an externally imposed potential. It is a property of the junction itself.

Let us suppose that the potential jump across the boundary is abrupt and study the consequences. (The change cannot be abrupt in real materials. The distance across which this potential change occurs is called the *transition region*, or *depletion region*, as in Fig. 44–13. However, we shall ignore effects due to the finite width of the transition region.) The densities of n-carriers on the two sides are, according to Eq. (44–5),

$$n_n = N_c e^{-(E_{cn} - E_F)/kT} \quad \text{and} \quad n_p = N_c e^{-(E_{cp} - E_F)/kT}$$

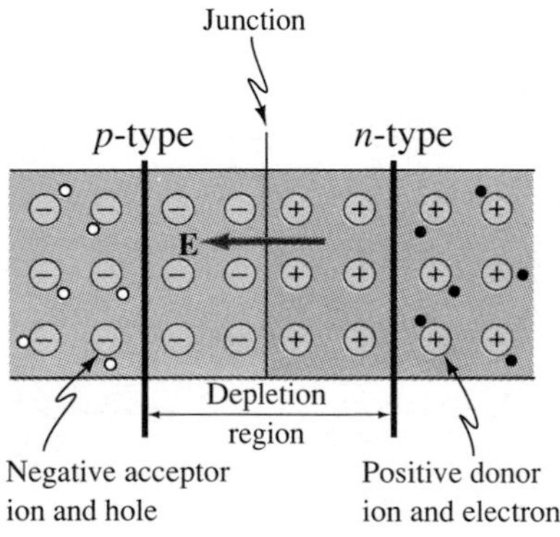

FIGURE 44–13 The potential in a p-n junction takes place over a region of finite width, called the depletion region. There is an internally generated electric field across the depletion region. (After Narciso Garcia and Arthur C. Damask, *Physics for Computer Science Students*, John Wiley & Sons, 1986, p. 457.)

Thus

$$\frac{n_p}{n_n} = \frac{\cancel{N_c} e^{-(E_{cp} - E_F)/kT}}{\cancel{N_c} e^{-(E_{cn} - E_F)/kT}} = e^{-(E_{cp} - E_{cn})/kT}$$
$$= e^{-eV_0/kT}, \tag{44–15}$$

where in the last step we have used Eq. (44–14). Similarly, we can show that, in equilibrium, the ratio of the hole densities on the two sides is

$$\frac{p_n}{p_p} = e^{-eV_0/kT}. \tag{44–16}$$

The ratios in Eqs. (44–15) and (44–16) help us to understand the physical mechanism of the balance of currents in the two directions. Let us consider the electrons (the flow of holes would work the same way). From Eq. (44–15), the concentration of electrons on the p-side is less than the concentration on the n-side by the factor $e^{-eV_0/kT}$, but these electrons have no potential barrier to climb to get to the n-side. There is a large density of electrons on the n-side, but to get to the p-side they need enough energy to climb a potential barrier of height eV_0, and only a number reduced by the Maxwell–Boltzmann factor $e^{-eV_0/kT}$ will have that much energy. In this way the flow of electrons in the two directions matches in equilibrium.

It follows from Eq. (44–14) that each of the products $p_n n_n$ and $n_p p_p$, which from Eqs. (44–15) and (44–16) are equal in equilibrium, is also equal to n_i^2. Thus

$$p_n n_p = N_c N_v e^{-(E_g + eV_0)/kT} \quad \text{and} \quad p_p n_n = N_c N_v e^{-(E_g - eV_0)/kT}. \tag{44–17}$$

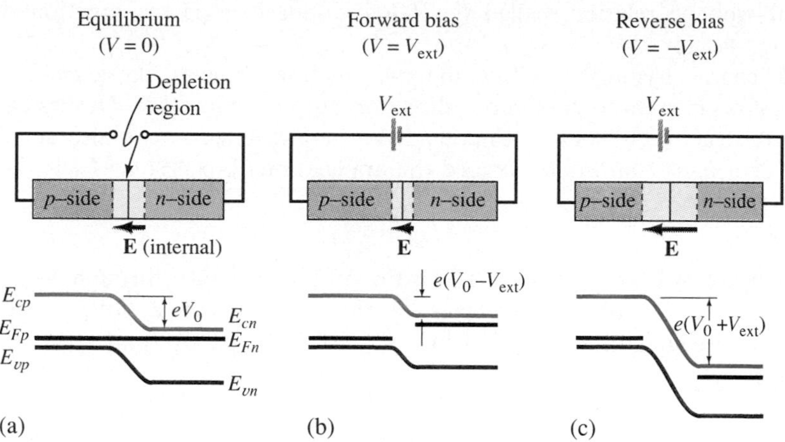

FIGURE 44–14 (a) In the unbiased *p-n* junction, the intrinsic contact potential leads to equilibrium. (b) In forward biasing of a *p-n* junction, the energy levels on the *n*-side are raised relative to those on the *p*-side. (c) In reverse biasing of the same junction, the energy levels on the *n*-side are lowered relative to those on the *p*-side. (After B. G. Streetman, *Solid State Electronic Devices*, Prentice Hall, 1980, p. 151.)

Biasing

Let us now see what happens when we apply an external potential, V_{ext}, across a *p-n* junction, which is then said to be *biased*. In that case there can be a net flow of charge and energy, and the Fermi energies no longer need be the same, unlike the unbiased case of Fig. 44–14a. Let us first assume that the external potential is such that the energies on the *p*-side are brought closer to the energies on the *n*-side (Fig. 44–14b). This external potential, by which the magnitude of the potential difference between the *n*-side and the *p*-side is *reduced* from $|V_0|$ to $|V_0| - |V_{\text{ext}}|$, is described as a *forward bias* voltage. The electrons that diffuse from the *n*-side to the *p*-side have less of a potential barrier to overcome. The same is true for the holes that diffuse from the *p*-side to the *n*-side (the potential difference is the same, but the charges of the carriers have opposite signs, so both have the same hill to climb). There is a net current from the *p*-side to the *n*-side, which rapidly increases as $|V_0| - |V_{\text{ext}}|$ decreases in magnitude. Note that there is also a small fixed current from the *n*-side to the *p*-side, associated with movement of electrons and holes without a potential barrier to overcome.

If the external potential is applied in such a way that the potential difference between the *n*-side and the *p*-side is *increased* from $|V_0|$ to $|V_0| + |V_{\text{ext}}|$, we have a *reverse bias* voltage (Fig. 44–14c). Both electrons from the *n*-side and holes from the *p*-side have an even harder time overcoming the additional barrier, and there is little current due to those charge carriers. However, the small fixed current from the *n*-side to the *p*-side is unaffected.

To make our argument more quantitative, we suppose that a positive V_{ext} is a forward bias and a negative V_{ext} is a reverse bias. Consider first the current due to the motion of the holes, with positive current corresponding to positive charges moving to the right. The potential barrier $e(V_0 - V_{\text{ext}})$ inhibits positive carriers from the left (of density p_p) from flowing, and only the fraction $p_p e^{-e(V_0 - V_{\text{ext}})/kT}$ get through. Positive carriers from the right do not encounter such a barrier, but their density is smaller by a factor $e^{-eV_0/kT}$, independent of V_{ext}. (The same is true separately for the *n*-carriers.) Take I_0, which is proportional to $e^{-eV_0/kT}$, to be the current of *p*-carriers from the right and *n*-carriers from the left. Without the external potential, I_0 would be canceled by the current of *p*-carriers from the left and *n*-carriers from the right, but those currents have now been modified by the factor $e^{eV_{\text{ext}}/kT}$. Thus the *net* current flowing to the right is given by

$$I_{\text{net}} = I_0(e^{eV_{\text{ext}}/kT} - 1). \qquad (44\text{–}18)$$

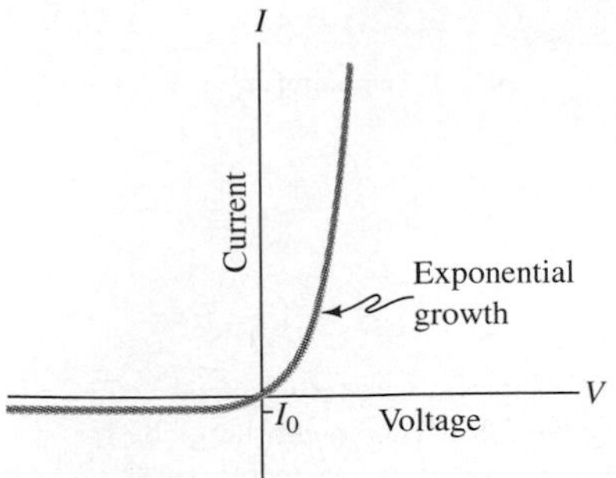

FIGURE 44–15 I-V characteristic of a p-n junction. The junction behaves as a diode under an external potential V_{ext}.

This current–voltage relation, called the *I-V characteristic* of the junction, is shown in Fig. 44–15.

The I-V characteristic shows that the p-n junction behaves like a *diode*, a device that allows current to pass in one direction but not in another. Diodes can be used to construct rectifiers (see Chapter 34). The p-n junction can also act as a *solar cell* or a *light-emitting diode* (see the applications box "Solar Cells and LEDs").

EXAMPLE 44–4 Calculate the ratio of the currents through a p-n junction diode for positive voltages to those for negative voltages for voltages whose magnitude is given by $eV_{ext} = 0.1$ eV, 0.2 eV, and 0.3 eV at $T = 300$ K.

SOLUTION: The current through a junction diode biased by a voltage is given by Eq. (44–18). Thus the ratio of the currents for positive to negative bias, R, is given by

$$R = \left| \frac{e^K - 1}{e^{-K} - 1} \right| = \frac{e^K - 1}{1 - e^{-K}},$$

APPLICATIONS

Solar Cells and LEDs

If a photon of light with frequency $f > E_g/h$ shines on a p-n junction, an electron from the valence band that absorbs that photon will be excited to a state in the conduction band. The electron now becomes an n-carrier and leaves behind a hole, a p-carrier. The intrinsic contact potential V_0 separates the electrons and the holes: The p-carriers on the p-side will move to the junction in the direction of the intrinsic electric field, while the n-carriers on the n-side will also move to the junction. The movement of these new carriers produces a potential difference opposite to V_0. It thus acts as a forward bias voltage across the junction, even in the absence of V_{ext}. A current is produced by this *photovoltaic effect*, and power can be delivered to an external circuit. All that is needed is a conductor attached to the two components of the junction. In silicon, photons of wavelength shorter than 1200 nm will give rise to a photovoltaic effect.

A solar cell, a device that produces electricity from sunlight, is a p-n junction constructed in such a way that the solar photons can be absorbed in the area closest to the built-in contact potential—that is, near the junction (Fig. B1–1). To maximize that area, the sunlight is allowed to fall onto the top semiconductor, covered with an antireflective coating. The light penetrates the semiconductor and is absorbed in the junction region below. The materials are chosen

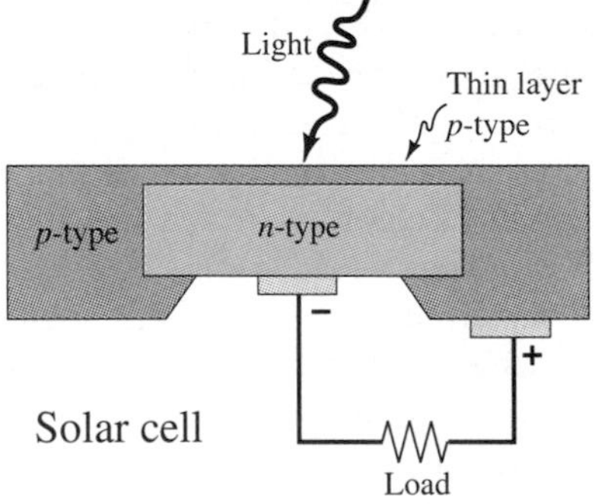

FIGURE B1–1 Schematic diagram of a solar cell, which generates current across the load (resistor) when light shines on the p-side of a p-n junction.

with a band gap small enough to absorb even the long-wavelength component of solar light. It is also important to have a large absorption coefficient for the light. In order to conserve energy and momentum, an electron must give up some momentum when it absorbs a photon. In crystalline silicon, an electron can do this only by giving some momentum (and energy) to the lattice, and this reduces by some two orders of magnitude the probability that a transition will occur. Thus in crystalline silicon the absorption is not very good. Amorphous silicon has 50 times the absorptive power of crystalline silicon, and copper indium where $K = eV_{ext}/kT$. At $T = 300$ K, $kT = 2.6 \times 10^{-2}$ eV, so the parameter K has the values 3.85, 7.69, and 11.5 for $eV_{ext} = 0.1$ eV, 0.2 eV, and 0.3 eV, respectively. The respective values of R are then 48, 2.2×10^3, and 10.2×10^4. The rapid growth of these values as eV_{ext} increases illustrates the dramatic biasing power of relatively small potential differences.

The Bipolar Junction Transistor

The **transistor** is a three-terminal device that can be used for sensitive control of currents flowing through circuits, particularly for amplifying or switching currents. One of the first transistors, the *bipolar junction transistor* (BJT), started a revolution in electronics that is still going on. That transistor and the theory of its operation were developed in 1948 and 1949 by John Bardeen, Walter Brattain, and William Shockley (Fig. 44–16). Bipolar junction transistors can be described schematically as *n-p-n* or *p-n-p* three-terminal devices. In an *n-p-n* transistor, there are two junctions in which a narrow *p*-type *base* is sandwiched between two wide *n*-type semiconductors, called the *emitter* and *collector* (Fig. 44–17a). In a *p-n-p* transistor, a narrow *n*-type base is sandwiched between a *p*-type emitter and collector (Fig. 44–17b). Figure 44–18a shows the band structure for an *n-p-n* BJT where there is no external (bias) voltage.

FIGURE 44–16 "Bardeen's box," constructed in 1949, contains a circuit with transistors as elements. The instant turn-on of the circuit impressed audiences who had been used to waiting a long time for circuits containing tubes to warm up.

diselenide is another factor of 10 more efficient for absorption, because in these materials momentum can be conserved without the involvement of the ions.

Because of limitations on collecting light and on absorbing photons, there is a theoretical maximum of 28 percent for conversion of solar energy into electric energy in silicon. Practical silicon solar cells operate with efficiencies of around 10 percent. Perhaps the most serious barrier to large-scale utilization of solar cells for power generation is cost. In 1991, the cost per kilowatt hour for energy from solar cells was 30 to 60 times more than the cost of energy produced from fossil fuels. The expectation is that, with improvements in the technology of known materials, solar cells will become competitive with conventional power sources before the year 2000. Power generation by solar cells is already indispensable in satellites and for areas on earth that are remote from other power sources (Fig. B1–2).

A light-emitting diode (or LED), which is a *p-n* junction acting as a signal light, is essentially a solar cell that operates in reverse. When a forward bias is applied across the junction, electrons flow from the *n*-side to the *p*-side and holes from the *p*-side to the *n*-side. When the electrons arrive at the *p*-side, they combine with available holes and emit light in the process. The holes that arrive at the *n*-side combine with available electrons and also emit light. LEDs are widely used as electronic signaling lights because they are long-lasting, use little energy, can be switched on and off rapidly, and are compact. LEDs are used also for digital readouts in some watchfaces and calculator displays.

FIGURE B1–2 The high initial cost of solar cells is compensated when they operate in situations in which alternatives are few, as in a remote application such as this satellite.

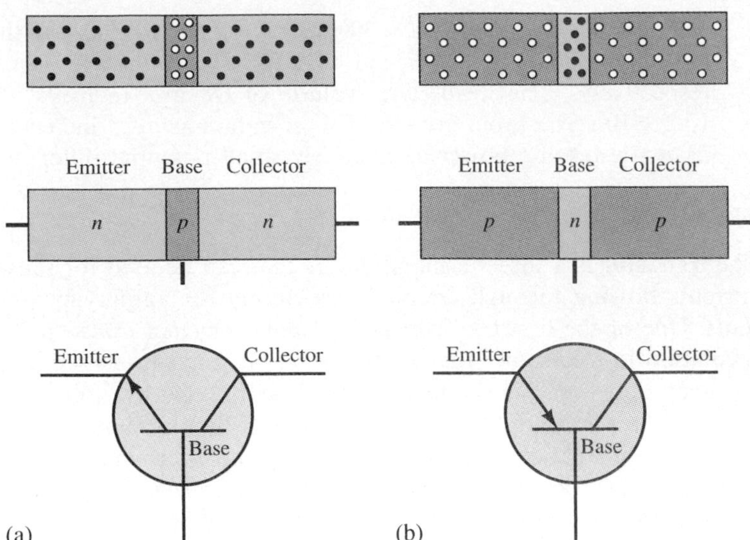

FIGURE 44–17 Diagrams of (a) *n-p-n* and (b) *p-n-p* bipolar junction transistors. The circuit-diagram symbols for these devices are shown at the bottom of the figures.

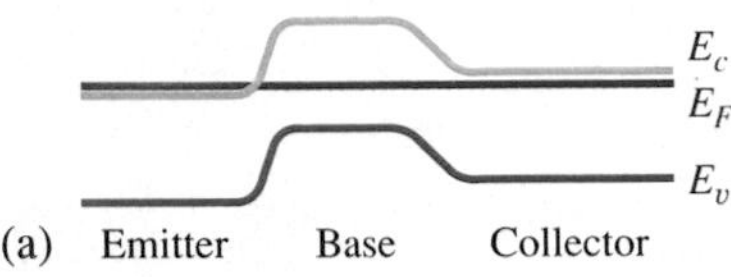

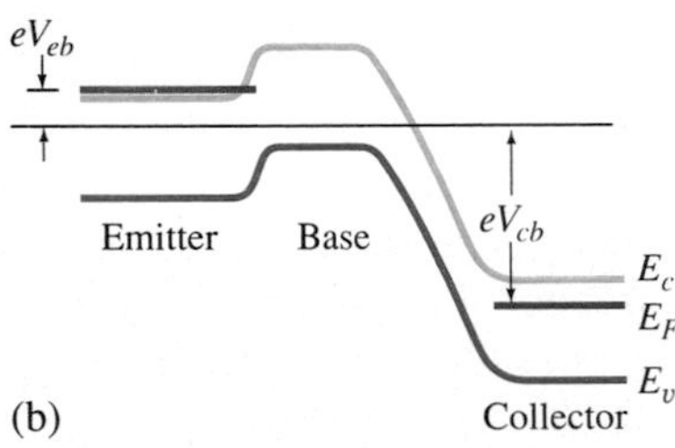

FIGURE 44–18 The band structure of an *n-p-n* bipolar junction transistor (a) in the absence of an external biasing voltage and (b) with forward biasing on the emitter–base junction and reverse biasing on the collector–base junction. (After M. Shur, *Physics of Semiconductor Devices*, Prentice Hall, 1990, p. 242.)

In Fig. 44–18b we show how the band structure is altered when the emitter–base junction is forward biased and the collector–base junction is reverse biased, as can be arranged with the circuit of Fig. 44–19a. Because the base is so narrow, the transistor is more than just two *n-p* junctions back to back. Electrons that enter the base from the emitter can diffuse all the way across the base, and no electric field of the type encountered in the *n-p* junction described earlier will build up. Once electrons from the emitter reach the base–collector junction, they flow easily into and through the collector (Fig. 44–19b). Note that the current flowing through the collector is controlled by the forward bias voltage across the emitter–base junction.

Let us define a *current transfer ratio*, α, by

$$\alpha \equiv \frac{I_c}{I_e}, \tag{44–19}$$

the ratio of the change, I_c, in the collector current with the change, I_e, in the emitter current. As can be seen in Fig. 44–19a, the emitter current is the sum of the collector current and base current (I_b):

$$I_e = I_c + I_b. \tag{44–20}$$

The *current gain*, β, is a measure of the change in collector current for a change,

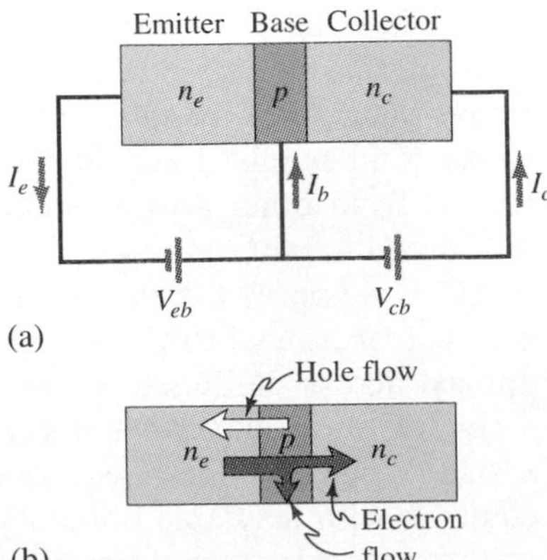

FIGURE 44–19 (a) The biasing of the *n-p-n* bipolar junction transistor of Fig. 44–18b is accomplished by a circuit such as this one. (b) Schematic diagram of the resulting flows of charged particles for this situation.

I_b, in base current, defined by

$$\beta \equiv \frac{I_c}{I_b}. \quad (44\text{–}21)$$

It follows from Eqs. (44–19) through (44–21) that

$$\frac{1}{\alpha} = \frac{I_e}{I_c} = \frac{I_c + I_b}{I_c} = 1 + \frac{1}{\beta}.$$

Thus

$$\beta = \frac{\alpha}{1 - \alpha}. \quad (44\text{–}22)$$

When α is close to unity, β can be large. For typical values for a transistor whose base is about 1 μm wide, forward bias is about 0.2 V, and backward bias is 5 to 10 V; α is in the range 0.97 to 0.99, corresponding, according to Eq. (44–21), to gains of from 32 to 99. Thus a small variation of the current into the base is *amplified* into a very large variation of the current into the collector. Note that the transistor sensitively fulfills its amplification role by manipulating bias voltages.

The Field-Effect Transistor

The *field-effect transistor* is a semiconductor transistor structure widely used in modern electronics. A semiconductor plate called a *channel* is connected to two terminals, called the *source* and the *drain*. A charge is induced in the channel by a metal plate known as a *gate*, which is separated from the semiconductor by insulating material. The channel and the gate together act as a capacitor. In the metal-oxide semiconductor field-effect transistor (MOSFET), the insulator separating the gate from the semiconducting channel is silicon dioxide (Fig. 44–20). When a positive voltage is applied to the gate, a negative charge is induced in the channel; that is, n-carriers fill the channel. As the gate voltage increases, so does the concentration of n-carriers in the channel, and hence so does its conductance. If a drain-to-source voltage is applied, a current flows across the channel. This current can be changed by changing the gate voltage as just described. Thus the field-effect transistor is a three-terminal device in which the current between two terminals is sensitively controlled by changing the voltage at the third terminal, the one connected to the gate.

MOSFETs illustrate the principles behind the building blocks of today's electronics. By the techniques to be described in Section 44–3, thousands or even millions of tiny MOSFETs and other semiconductor structures making up complicated circuits are fabricated as one unit; we call these circuits *integrated circuits*. The layers of material that make up the devices in integrated circuits are as thin as 0.2 nm, and the area of a single MOSFET is as small as 5 μm × 5 μm (Fig. 44–21). Typical currents in these circuits are on the order of 20 mA. The major disadvantage of MOSFETs is that there are limits to the speed at which they operate, because currents flow through them in the same region that the doping impurities are located, so collisions with impurities restrict the drift speed of current carriers. A different geometry, to be discussed in Section 44–3, improves the mobility of the electrons.

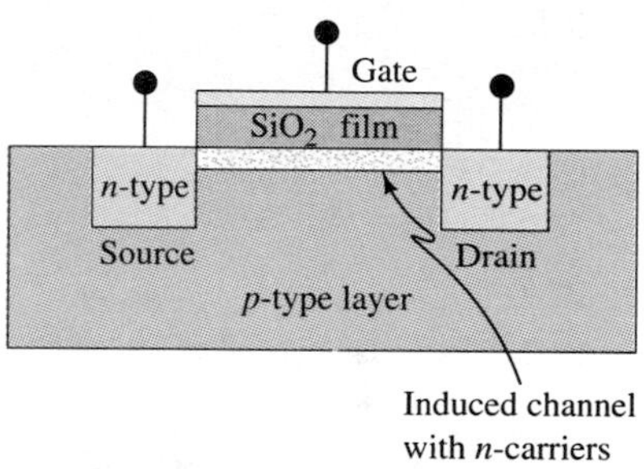

FIGURE 44–20 Schematic diagram of a MOSFET, a type of field-effect transistor.

FIGURE 44–21 In 1992, IBM engineers produced this MOSFET, so tiny—its length is several hundred nanometers—that memory chips containing 4 gigabits of memory are made possible. (The largest memory chip available until then had a capacity of 16 megabits.) This transistor is so small that many of its properties are quantum-mechanical in nature.

44–3 BAND-GAP ENGINEERING

Technological advances in the fabrication of semiconductor structures have created a new tool chest for the electronics engineer and computer designer. One

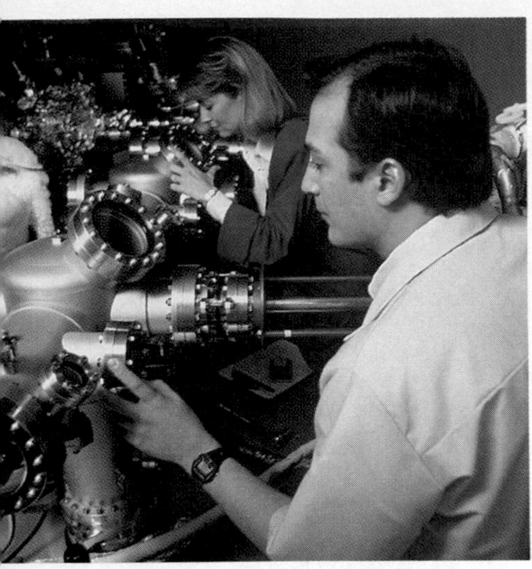

FIGURE 44–22 The molecular-beam epitaxy machine shown here is used to construct microscopic structures molecular layer by molecular layer.

advance involves the ability to place thin layers of one or several materials on a single underlying layer (a *substrate*) of another material with very similar crystal structure. (Such techniques are the basis of integrated-circuit construction.) This procedure, known as *epitaxy*, has reached its highest refinement with *molecular-beam epitaxy* (Fig. 44–22). Examples are provided in structures in which single crystals of gallium arsenide form a substrate and compounds of aluminum and gallium arsenide ($Ga_{1-x}Al_xAs$, which are alloys of variable content; x represents a fraction) form junctions with the substrate. Gallium, aluminum, and arsenic are evaporated in a vacuum chamber and projected as beams onto the gallium arsenide substrate. The beams form layers on the substrate whose thickness can be controlled. The components of the deposited layers can be varied by exposing the substrate to different beams in different proportions. The growth of the layers is slow, about one atomic layer per second. In addition, donor impurities (such as silicon) and acceptor impurities (such as beryllium) are easily added in desired proportions.

Lithography is another technology indispensable for the fabrication of integrated circuits. With this process, we can create detailed patterns on a crystal surface (Fig. 44–23). One type is scanning-electron-beam lithography, in which an opaque film on a transparent base is removed in a precise pattern by a moving electron beam. The resulting transparent features are as small as 20 nm. Light passes through this "negative" onto a semiconductor blank that has been coated with a polymer that disintegrates when light hits it. The coated semiconductor can then be etched by chemicals such as chlorine gas that act where the polymer has disintegrated. Alternatively, other materials can be deposited on the areas where the polymer has disintegrated, or doping ions can be introduced in those areas. A process such as this can be repeated to produce elaborate structures with many layers. The resulting integrated circuits have enabled us to produce desktop computers many times more powerful than yesterday's mainframes.

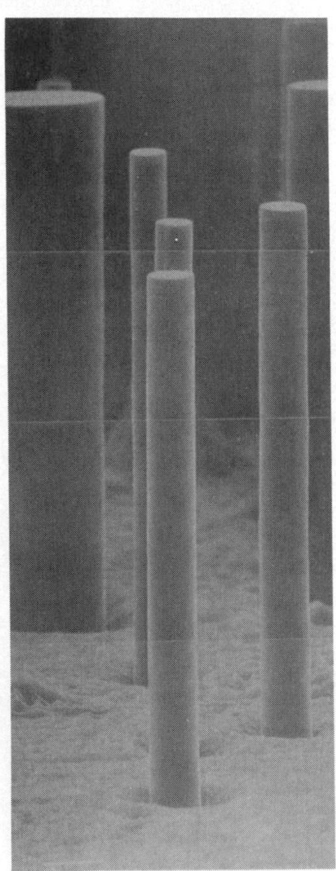

FIGURE 44–23 Some of the elaborate microscopic structures produced by lithography. They are less than a half micron high.

Heterojunctions

A *heterostructure* is a single crystal in which the occupancy of the ion lattice sites changes at an interface, or boundary. Such crystals are created by epitaxy, as described above. Examples are GeAs/GaAs and $Al_xGa_{1-x}As$/GaAs, where the slash indicates the interface. When we use two different semiconductors, we say that we have a *heterojunction*. The fact that different semiconductors have different band gaps leads to some new behaviors. Figure 44–24a shows the band-gap structure of two different semiconductors, each doped as an n-type material. The semiconductor on the right has a wider gap than the one on the left, and in each case the Fermi energy, E_F, lies closer to E_c than to E_v. (The doping in the semiconductors need not be uniform. In Fig. 44–24a the donor impurities are located some distance away from what will become a junction.)

When the materials are joined to form a heterojunction, the flow of charges causes the Fermi energies to equalize, as in Fig. 44–24b. The form of the resulting potential difference depends on the materials at the junction. In a heterostructure, the edge of the conduction band (or, depending on the materials, the edge of the valence band) is distorted to contain the kink shown in Fig. 44–24b. We can understand this kink as follows. When the heterojunction is formed, the energies shift so as to equalize the Fermi energies of the two semiconductors. The gap separations on either side of the junction, however, must remain as they were before the junction was formed, because those are intrinsic to the semiconductors. The only way to satisfy both constraints is to develop a kink such as the one shown in Fig. 44–24b.

As a result of the kink, *electrons in the conduction band at the interface find themselves in a potential energy well* for motion across the junction. Thus, as an example, the potential "step up" in energy between GaAs on the left and $Al_{0.45}Ga_{0.55}As$ on the right is 0.35 eV, so the electrons are confined to the bottom of the kink, on the GaAs side. The potential well is far from the donor ions, which in this example are on the $Al_{0.45}Ga_{0.55}As$ side, so if an electric field is applied in the direction perpendicular to the plane of Fig. 44–24b, electrons will flow along the channel formed by the well without finding many positive impurity ions in their path. These electrons can therefore move very rapidly. Typical drift speeds are on the order of 1.8×10^5 m/s, some 1000 times faster than in MOSFETs. The drift speed of electrons is an important factor in devices that act as switches in high-speed computers.

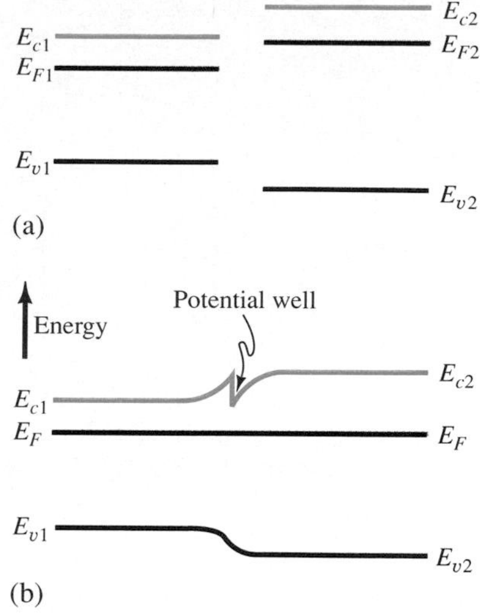

FIGURE 44–24 (a) Energy diagrams for two different semiconductors, each doped to n-type, before they are joined in a heterojunction. (b) When the semiconductors are joined to form a heterojunction, the edge of the conduction band forms a kink that can act as a potential energy well for electrons. There is no restriction to movement into or out of the page for an electron in this well.

Quantum Wells, Quantum Wires, and Quantum Dots

When two or more heterojunctions are put back to back, the peculiar shapes taken by the edge of the conduction band may form a region in which electrons have "walls" on two sides. Figure 44–25a shows a thin GaAs layer sandwiched between two AlGaAs layers, surrounded by GaAs that contains the donor ions; Fig. 44–25b is an energy diagram for this structure. The electrons in the GaAs layer move in a *quantum well.* Such potential wells are characterized by quantized energy levels. We can find the values of these energies by thinking of standing waves set up in the direction of the (small) distance a between the "walls" that confine the electron, just as atomic energies in the Bohr model follow from thinking about fitting standing waves into atomic orbits. The allowed energies of the quantum well are then given by

$$E_n = \frac{n^2\hbar^2\pi^2}{2m^*a^2}, \tag{44–23}$$

where n is any integer. The confining distance a is typically 10 to 100 nm. Although a is large compared with atomic dimensions, the fact that m^* is small partly compensates for this effect. Only the energy of lateral motion is quantized: the motion in the direction perpendicular to the plane of the page for Fig. 44–25a is not quantized.

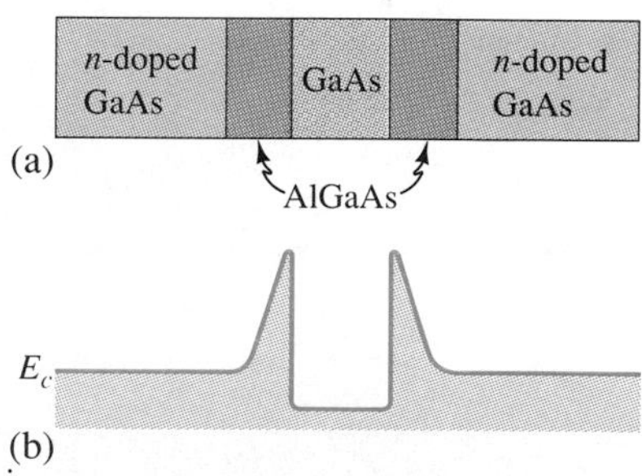

FIGURE 44–25 (a) A heterostructure within which (b) a potential energy well forms and confines electrons.

EXAMPLE 44–5 Calculate the energy difference between the ground state and the first excited state in a quantum well of width 20 nm, given that the effective mass is $m^* = 0.07m_e$. What is the wavelength of a photon emitted when an electron undergoes a transition between the two states?

SOLUTION: We use Eq. (44–22) to calculate the energy difference between levels characterized by $n = 2$ (the first excited state) and $n = 1$ (the ground state). It follows that

$$\Delta E = E_2 - E_1 = \frac{(2^2 - 1^2)\hbar^2\pi^2}{2m^*a^2} = \frac{3\hbar^2\pi^2}{2m^*a^2} = \frac{3(1.05 \times 10^{-34}\ \text{J}\cdot\text{s})^2(3.14)^2}{2(0.07)(0.9 \times 10^{-30}\ \text{kg})(2 \times 10^{-8}\ \text{m})^2}$$

$$= 6.2 \times 10^{-21}\ \text{J} = \frac{6.2 \times 10^{-21}\ \cancel{\text{J}}}{1.6 \times 10^{-19}\ \cancel{\text{J}}/\text{eV}} = 3.9 \times 10^{-2}\ \text{eV}.$$

The frequency of a photon emitted in a transition is given by $f = \Delta E/2\pi\hbar$, and the wavelength is given by

$$\lambda = \frac{c}{f} = \frac{2\pi\hbar c}{\Delta E} = \frac{2(3.14)(1.05 \times 10^{-34}\ \cancel{\text{J}}\cdot\cancel{\text{s}})(3 \times 10^8\ \text{m}/\cancel{\text{s}})}{6.2 \times 10^{-21}\ \cancel{\text{J}}} = 3.2 \times 10^{-5}\ \text{m}.$$

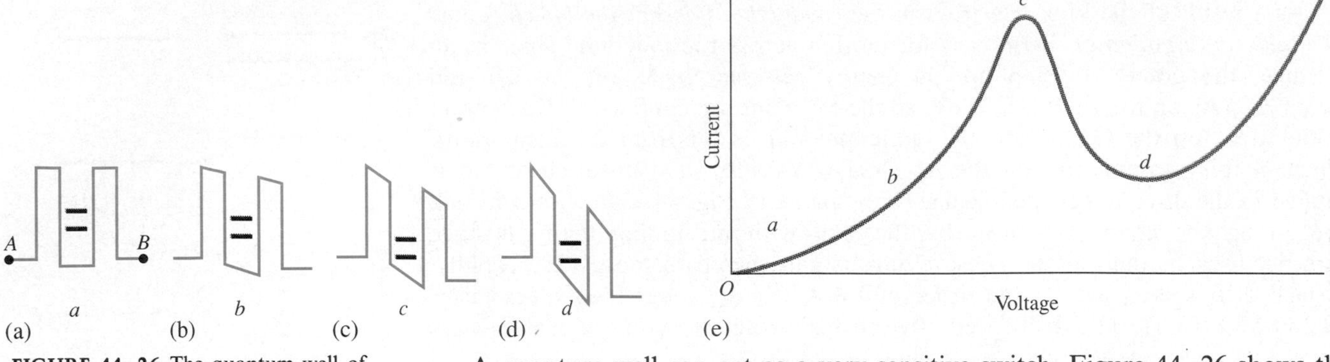

FIGURE 44–26 The quantum well of Fig. 44–25 subjected to an external potential. (a through d) As the potential is increased, there is an enhanced probability that an electron will tunnel from one side of the well to another when the location of allowed levels in the well matches an occupied level in the conduction band on the left. (e) I-V characteristic for the situation.

A quantum well can act as a very sensitive switch. Figure 44–26 shows the I-V characteristic for a quantum well device. With zero voltage between points A and B (Fig. 44–26a), the quantum well presents a barrier to the n-carriers. There is no tunneling, because the energy levels on both sides of the quantum well are occupied equally. When a voltage is applied, the n-carriers on the left are at a higher energy, and they can tunnel into empty levels. Thus the current rises with V, as shown in Fig. 44–26b. When the bottom of the conduction band on the left is lifted high enough to match one of the low-lying quantized levels inside the well, there is a drastic increase in the flow of the current (Fig. 44–26c). A further voltage rise decreases the current, because the bottom of the conduction band on the left no longer matches a well level (Fig. 44–26d).

The effect is very similar to the *resonant behavior* that occurs when the frequency of a harmonic driving force matches the characteristic frequency of a slightly damped harmonic oscillator. As the voltage increases further, the current decreases and then increases again as the second energy level is approached, as we see in Fig. 44–26e. The resonance makes for a sensitive switch.

The existence of discrete energy levels in the well can be confirmed by the selective absorption of laser light at frequencies that correspond to energy differences between the well levels, according to the relation $\Delta E = hf$. By making a long string of AlGaAs–GaAs–AlGaAs sandwiches, we can create a series of quantum wells that resemble a crystal lattice; the arrangement is in fact called a *superlattice*. The energy-level structure of a superlattice is just like an ordinary band structure, with an important difference: Crystal lattice spacings are on the order of 0.1 nm, whereas quantum well spacings are on the order of 10 to 100 nm. The superlattice then has band gaps in the range of milli-electron-volts.

The motion in the (two-dimensional) plane perpendicular to the width of the well is not quantized. The presence of the well walls restricts the electron motion to be in effect two-dimensional (Fig. 44–27). The physics of electron motion in two dimensions is a particularly interesting field of research, especially because it appears that high-temperature superconductivity takes place in materials that have a planar structure, with electrons moving preferentially in two dimensions.

Quantum wires and *quantum dots* represent further steps in the realization of systems in which quantum behavior is dominant. They correspond to electron-confining "pipes" and "boxes," respectively. The creation of such structures is more complicated technically than that of quantum wells, and it may be some time before they are incorporated into quantum devices. Electrons in a quantum dot, the equivalent of a tiny three-dimensional box, have widely separated energy levels (see Problem 31). Electrons with these energies passing through the box display resonant behavior and provide a current only when well-defined voltages are applied to them. Such structures are promising as very precise switching devices.

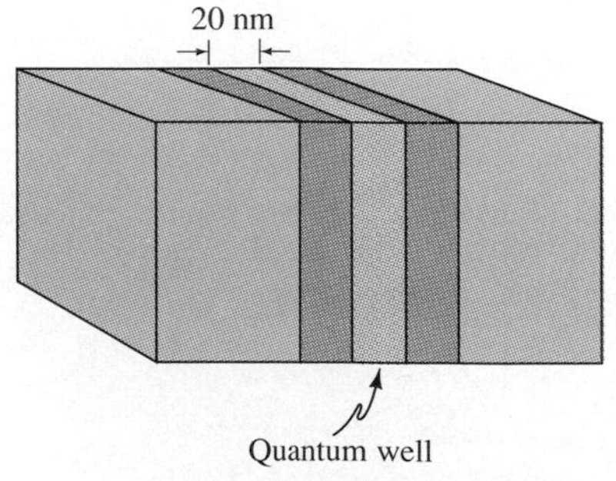

FIGURE 44–27 The quantum wells we have described allow electrons to move in what is in effect a two-dimensional region.

Semiconductor Lasers

Semiconducting materials are well suited to the construction of microscopic lasers. Such lasers have already found uses in compact-disk players, amplifiers in satellite receivers, and communications with fiber optics. An energy diagram for a heterostructure illustrates how these lasers may be constructed. The semiconductor within which the laser light will be produced (for example, GaAs) is sandwiched between two layers of appropriately doped AlGaAs. The right side of the energy diagram in Fig. 44–28 is doped to n-type, and the left side is doped to p-type. A potential barrier keeps the electrons from flowing to the left near the bottom of the conduction band, and a counterpart barrier keeps the holes from flowing to the right near the top of the valence band. If a voltage is applied that causes more electrons to flow into the GaAs region and more holes to flow into the laser region, and the barrier prevents both n-carriers and p-carriers from flowing back, a population inversion is built up. Note that unlike the lasers described in Chapter 43, the "pumping" is done automatically in these materials. The electrons and holes can recombine by the emission of photons, and these photons stimulate further, rapid recombination. The faces of the single crystal that makes up the heterostructure act as mirrors, because about 30 percent of the light is reflected at the crystal–air interface. This describes the process we treated in Chapter 43 for the production of laser light.

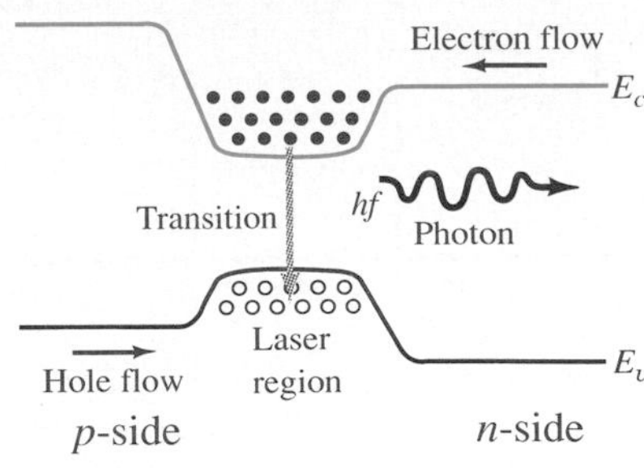

FIGURE 44–28 Energy diagram for a semiconductor laser. The laser light is produced in the recombination of n-carriers and p-carriers at the central region.

Figure 44–29 is an image of a set of *microlasers* that consist of a thin InGaAs quantum well, with stacks of AlAs, GaAs and AlGaAs layers both above and below it. The laser action takes place in the quantum well, and because the well is narrow, the power required to make the laser work is very small. Many reflections of the light are required to get suitable amplification, and the large number of layers provide a certain amount of reflection at each interface. Even though the reflectivity at each interface is less than 1 percent, the large number of layers gives a total reflectivity back to the lasing material of 99 percent. Many lasers, each with cylindrical geometry, can be formed on a single substrate by molecular-beam epitaxy and subsequent etching. Figure 44–30 shows a portion of a 7 mm × 8 mm surface on which two million microlasers were created! Such laser arrays are useful for optical communications in fiber optics.

FIGURE 44–29 Scanning tunneling microscope (STM) image of a microlaser. The multilayer structure is evident.

FIGURE 44–30 Part of a surface on which about two million microlasers have been fabricated.

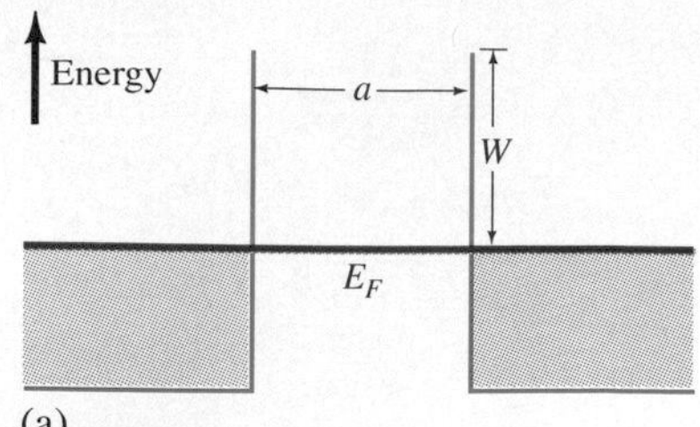

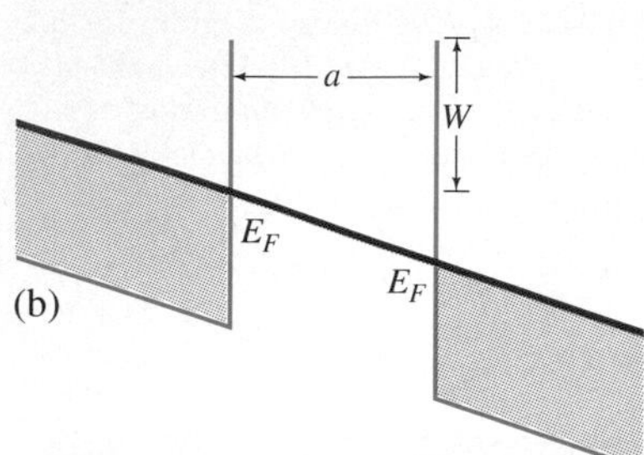

FIGURE 44–31 (a) When two metals are separated by a gap, there is no tunneling in equilibrium. (b) If an external potential is applied, the Fermi energy within one of the metals is lowered, leaving empty levels into which electrons from the side with the higher electron energies can tunnel. The height of the barrier to overcome is the metal's work function, W.

44–4 SCANNING MICROSCOPY

We have learned that, because of diffractive effects, the resolution with which an object can be studied is limited by the wavelength of the radiation used to study the object. Thus to resolve atoms, whose size is roughly 0.1 nm, the radiation must have a wavelength on the order of 0.1 nm or less. For electromagnetic radiation, this corresponds to a frequency of $f = c/\lambda = (3 \times 10^8\ \text{m/s})/(10^{-10}\ \text{m}) = 3 \times 10^{18}\ \text{Hz}$. A photon of this frequency has an energy $hf = (6.62 \times 10^{-34}\ \text{J} \cdot \text{s})\ (3 \times 10^{18}\ \text{s}^{-1}) = 2.0 \times 10^{-15}\ \text{J} = (2.0 \times 10^{-15}\ \text{J})/(1.6 \times 10^{-19}\ \text{J/eV}) = 12\ \text{keV}$. This energy is 1000 times larger than an atomic ionization energy, and such photons are not readily available. The situation is improved somewhat when electrons are used, but the required electron energy is about 150 eV, which is still very large. J. A. O'Keefe pointed out in 1956 that it would be possible to bypass the diffraction limitation with a new type of microscope. The basic idea is simple: Light passes through a tiny hole in a screen and illuminates the object to be viewed, which is on the other side of the hole. The transmitted light (or reflected light) is recorded as the object is moved back and forth across the hole. *The size of the hole, not the wavelength of the light, determines the resolution.* The use of *piezoelectric* materials has provided the technology for controlling the position of an object with the kind of precision needed to realize O'Keefe's idea (see Fig. 25–26). Piezoelectrics are ceramic materials that expand or contract when an electric field is applied to them. The ability to move them with an accuracy of 10^{-5} nm (!) has made possible the development of scanning microscopy into a superb tool for the study of surfaces.

Scanning Tunneling Microscopy

STMs were first described in Chapter 7, and tunneling is treated in more detail in Chapter 41.

The scanning tunneling microscope (STM) is based on the quantum mechanical result that electrons can tunnel from one region to another through domains that are classically inaccessible. Figure 44–31a shows the energy levels in two metallic samples separated by a vacuum of width a. Electrons at the Fermi energy can cross over from one sample to the other in significant numbers only if (1) there are empty energy levels of the same energy, so that the exclusion principle does not prevent tunneling, and (2) the tunneling probability is sufficiently large. If an electric field is applied, thereby lowering the Fermi energy on one side as in Fig. 44–31b, then condition (1) is satisfied and tunneling can occur. An application of Eq. (41–6) shows that the fraction of electrons that cross such a barrier is

$$f = e^{-(2/\hbar)a\sqrt{2mW}}, \tag{44–24}$$

where m is the electron mass and W is the work function of the metal (the height of the potential barrier between the metals). The important feature of Eq. (44–24) is the extreme sensitivity of the tunneling fraction (that is, the current) to the magnitude of the separation.

Figure 44–32 shows how an STM is built. A tungsten tip acts as a "hole in the screen." When such a tip is poised above a surface at a potential different from that of the tip, then according to Eq. (44–24) the current passing through the tip is a sensitive indicator of its distance to the surface. The resolution depends on the tip size. It is possible to make tips that end in a *single* atom by heating a crude tungsten tip and applying a strong electric field to it. The field pulls atoms away layer by layer, leaving a single-atom tip. With such tips it is now possible to resolve features 0.1 nm across: *We can see single atoms on a surface.*

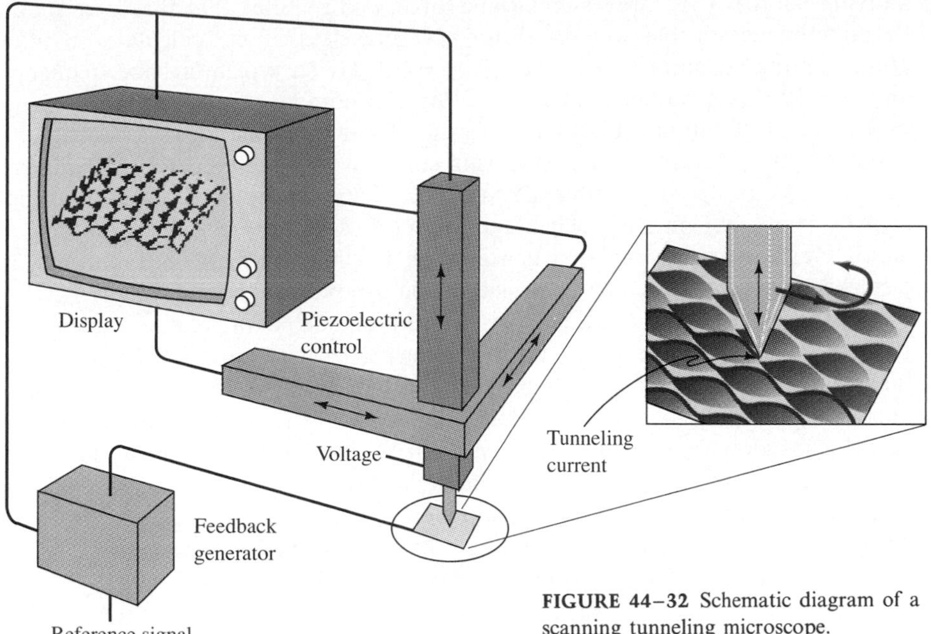

FIGURE 44–32 Schematic diagram of a scanning tunneling microscope.

There are two ways of scanning a surface. In one method, piezoelectric supports that hold the tip are arranged so as to sense the tunneling current and move the tip up or down so as to maintain a constant current or, equivalently, a constant distance from the surface. The voltages applied to the piezoelectric supports during this process provide a record of the surface topography. A second way of scanning the surface is to move the tip horizontally across the surface and measure the dependence of the tunneling current at any given point on the voltage applied. The I-V characteristics can be used to identify the atoms and the type and strength of their bonding to the surface of the substrate. STMs are especially effective because miniaturization of the apparatus helps in the reduction of "noise": The smaller the apparatus, the higher the frequency of the random thermal vibrations, and this noise is relatively easy to filter out. In addition, STMs work better in air than in a vacuum, because the random bombardment by air molecules also tends to average out noise.

STMs are used in a variety of surface studies. Because the presence of occupied states reduces the current, by measuring the dependence of the current on the applied voltage at a given point, it is possible to study the energy-band structure at the surface of crystals. Electron charge distributions at surfaces can be measured, as can the way the distributions are affected by the deposition of very thin films on the surfaces. Another application involves the use of the STM tip to manipulate individual atoms or groups of atoms and move them around on the surface of the sample (Fig. 44–33; see also Fig. 25–28). There is potential for the creation of tailor-made molecular structures for specific tasks in microelectronic devices.

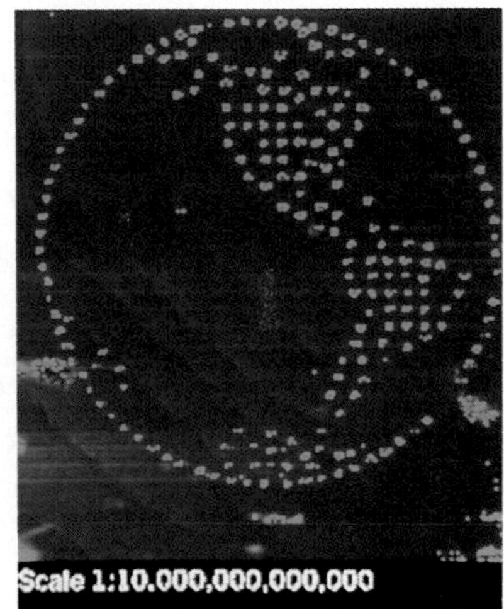

FIGURE 44–33 A map made of tiny dots that consist of just a few thousand atoms of gold. The map, about 1 micron across, was made by manipulating the gold dots with the tip of a scanning tunneling microscope.

Atomic Force Microscopy

Because currents do not flow through insulators, the STM cannot make images of insulating surfaces. To do so we can use another device, the *atomic force microscope* (AFM), which instead of probing surfaces with a current, images a surface by measuring the effect of the force it exerts on a small cantilever (like a diving board). This force is an atomic force, and the idea that this force should be directly measurable sounds absurd. Nevertheless, if we calculate an interatomic spring constant $k = m\omega^2$, with $\omega = 10^{13}$ Hz (a typical atomic frequency) and $m = 10^{-25}$ kg, we find that $k = 10$ N/m. For comparison, the spring constant of a piece of aluminum foil 4 mm long and 1 mm wide acting as a cantilever is about 1 N/m. Because the atomic spring constant and the mechanical cantilever spring constant are comparable, atomic forces can be measured by sensing small displacements (0.1 nm) of such a cantilever. It is thereby possible to study an atomic surface in the way that a needle on a long-playing record, through its displacement, senses the structure within the grooves of the record.

In some AFMs, the cantilever is a piece of silicon or silicon oxide about 100 μm long and 1 μm thick. Such cantilevers have a spring constant in the range of 0.1 to 1.0 N/m. The sensor that detects the motion of the cantilever must detect displacements of less than 0.1 nm. One way of detecting the motion of the cantilever is to shine laser light on it and record the position of the reflected beam with a light detector. The light detector activates the piezoelectric base on which the sample is placed, and a current controls the base such that the distance between the sample and tip is held constant. The movement of the sample is translated into an image of the surface (Fig. 44–34).

The effect of external vibrations of the building containing the AFM is quite negligible. Typical frequencies of such vibrations are about 10 to 20 Hz, whereas the resonant frequencies of the cantilever are about 10 to 100 kHz. Amplitude distortions due to external vibrations of frequency f are on the order of $(f/f_{res})^2$. Building vibrations thus contribute a correction of less than 10^{-3} nm to the displacement.

AFMs can be operated in a variety of environments: in air, in water, at low temperatures, or in a vacuum. They have been applied to the imaging of biological molecules such as amino acids, DNA, and proteins and have even

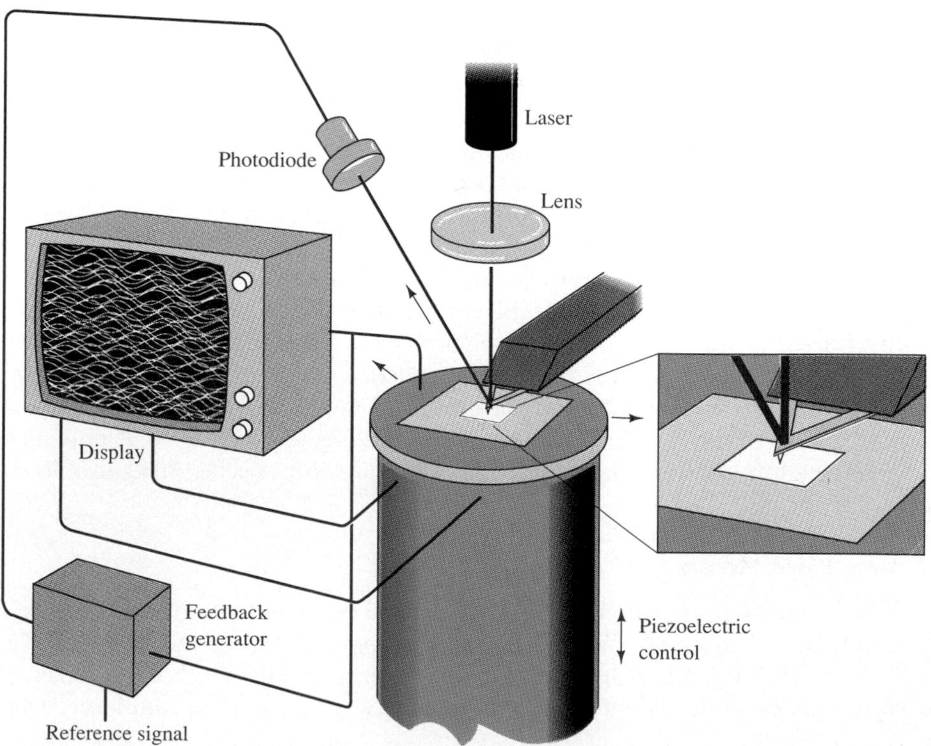

FIGURE 44–34 Schematic diagram of an atomic force microscope (AFM).

imaged certain chemical processes *as they occur*. The AFM has also been used to study atomic-scale structure on the surfaces of graphite, mica, sodium chloride, and many other materials. Another promising application is in *tribology*, the study of friction. For example, when the nickel tip of a cantilever is brought to within about 0.4 nm of a surface of gold, some of the gold atoms jump up to the tip and bind to it. As the tip is pulled away, a vertical column of gold atoms, like the funnel of a tornado, leads up from the gold surface to the nickel. This intermetallic junction manifests itself at the macroscopic level as friction. It is hoped that these studies will lead to improvements in lubrication; layers of material that inhibit the formation of such metallic "necks" between two surfaces will make excellent lubricants.

The Ultimate in Quantum Engineering

In this chapter we have emphasized the interplay of physics and modern technology. We have touched on only a few topics, concentrating on those for which quantum physics is important. The ultimate aim of the kind of work we have described here is the production of devices in which single atoms or single electrons play the major role. For example, a computer memory that stores information by the presence or absence of an individual atom in a given surface site could conceivably store all the knowledge of the human race on a surface of area 1 cm^2! And transistors that operate by allowing single electrons to tunnel could compose the smallest possible circuit. The realization of these ideas may not be impossibly distant. The link between physics and technology in the area of materials is so strong that the distinction between the two fields has all but disappeared.

SUMMARY

Because electrons obey the Pauli exclusion principle, their energies within crystalline solids are distributed according to the Fermi–Dirac distribution,

$$f(E) = \frac{1}{1 + e^{(E - E_F)/kT}}, \tag{44–2}$$

where E_F is the Fermi energy. Within semiconductors, whose band structure is that of a valence band separated from a conduction band by a small energy gap (with the Fermi energy in the middle of the gap), this distribution leads to strongly temperature-dependent electrical conduction. Both electrons promoted to the conduction band and the holes they leave behind can conduct electricity. Within semiconductors, electrons (and holes) have effective masses that can be quite different from the mass of free electrons. The conduction properties of semiconductors can also be affected by doping, the introduction of impurities that increase the number of negatively or positively charged carriers (n-carriers and p-carriers, respectively). Doping shifts the position of the Fermi energy but leaves the densities of the n-carriers and p-carriers linked in a way that depends on the behavior of the undoped semiconductor. Semiconductors may emit light when n-carriers and p-carriers recombine, leading to fluorescence and phosphorescence, and conductivity may depend on whether or not light strikes the material.

A pair of semiconductors placed in contact with one another, whether the same semiconductor doped differently or different semiconductors, forms a junction. Junctions exhibit a variety of behaviors and are the basis for much of the electronic technology in use today. The conservation of energy and of charge leads to the principle that the Fermi energy across a junction is constant. This principle dictates that an internally generated electric potential, the contact potential, is generated across the boundary of a *p-n* junction, a semiconductor structure consisting of a semiconductor doped with *p*-carriers in contact with the same semiconductor doped with *n*-carriers. The contact potential explains why *p-n* junctions act as diodes, devices for which current I_{net} in effect passes in one direction only when an external potential V_{ext} is applied:

$$I_{net} = I_0(e^{eV_{ext}/kT} - 1), \qquad (44\text{–}18)$$

where I_0 is a current associated with the contact potential. A *p-n* junction can also act as a light-emitting diode or as a solar cell.

The transistor is a three-terminal semiconductor device in which small changes in potential across one pair of terminals can lead to large amplification of the current that passes through another pair of terminals. Such devices are therefore useful for the sensitive control of currents and of voltages in circuits. We studied in particular both the bipolar junction transistor and the field-effect transistor.

Modern techniques of material manipulation permit the elaboration of heterostructures, single-crystal structures built in a precisely controlled way from different materials. The electrical behavior of heterostructures depends intimately on quantum phenomena, even beyond the collective quantum effects that lead to any band structure. Microscopic semiconductor lasers and quantum wells are manifestations of this new technology whose limits are not yet known.

Scanning tunneling microscopes employ quantum tunneling to explore conductors, and atomic force microscopes allow us to explore nonconductors. Both instruments avoid resolution difficulties associated with diffraction and permit the study of matter at sizes around one Bohr radius, atomic size.

QUESTIONS

1. You have just calculated the number density of *n*-carriers in undoped silicon, a semiconductor, at room temperature. Is it necessary to make a second calculation to find the number density of holes?

2. When a hole moves from the *p*-side of a junction to the *n*-side, why does its energy increase?

3. Does the Fermi energy still lie in the middle of the energy gap when the effective masses of the *n*-carriers and *p*-carriers are different?

4. Is the mechanism that makes metals opaque to visible light but transparent to X-rays similar to the mechanism that makes semiconductors opaque to visible light but transparent to infrared light?

5. There are several methods by which electrons can be removed from a solid's surface. How can holes be removed?

6. What happens to the Fermi energy at low temperatures when a semiconductor is so heavily doped that the energy levels of the donor states overlap the conduction band?

7. How can the energy of a hole be raised as it travels from the *p*-side of a *p-n* junction to the *n*-side?

8. Why, physically, is it true that the smaller the effective mass of an *n*-carrier, the larger the drift velocity of the carrier?

9. We described why new states for electrons are available near the bottom of the conduction band in an *n*-type extrinsic semiconductor. Why are new states for holes available just above the top of the valence band in a *p*-type extrinsic semiconductor?

10. Is there a type of viscosity that acts on holes in a semiconductor and gives them a terminal velocity in an electric field, or do holes just accelerate under the influence of such a field? If there is a terminal velocity, is it the same as that of the electrons?

11. We referred to a current amplification in our discussion of transistors. Can this mean that the power is amplified? If so, where does the necessary energy come from?

12. The gap widths of the semiconductors in Table 44–1 are given at room temperature. How could a gap width depend on temperature?

44–1 *Semiconductors*

1. (I) In silicon, $E_g = 1.12$ eV, and the effective mass of the n-carriers is $m^* = 0.31m_e$, where m_e is the electron mass. Find the number densities of n-carriers at 100 K and at 300 K. Compare your result with the number density of "free" electrons in copper, 8.5×10^{22} electrons/cm^3.

2. (I) What is the density of electrons in the conduction band of germanium, for which $m_n^* = 0.55m_e$, at temperatures of (a) 30 K, (b) 100 K, and (c) 300 K?

3. (I) For copper ($E_F = 6.95$ eV) at room temperature (300 K), calculate the occupation probability of an electron state of energy (a) 5 eV, (b) 6 eV, (c) 6.9 eV, (d) 7 eV, and (e) 8 eV.

4. (I) For copper ($E_F = 6.95$ eV), at what temperature is the probability that an energy state at 7 eV will be populated equal to 33 percent?

5. (I) Find the energy gap of an intrinsic semiconductor in which 1.02×10^{-30} of the available energy levels near the bottom of the conduction band are occupied at $T = 126$ K.

6. (I) What is the probability of finding an electron at the bottom of the conduction band in germanium, for which $E_g = 0.67$ eV, at $T = 300$ K?

7. (I) The Fermi temperature, T_F, is defined by E_F/k, where k is Boltzmann's constant. What is T_F for (a) silver ($E_F = 5.1$ eV), (b) copper ($E_F = 6.95$ eV), and (c) a white dwarf star ($E_F \simeq 2 \times 10^5$ eV)?

8. (I) What is the maximum wavelength of light that will excite an electron from the valence band to the conduction band in (a) silicon ($E_g = 1.12$ eV), (b) germanium ($E_g = 0.67$ eV), and (c) carbon ($E_g = 5.5$ eV)?

9. (I) The following table gives the energy above the edge of the valence band, ΔE, for the energy levels of various acceptor impurities in silicon:

Impurity:	Boron	Gallium	Indium	Nickel	Zinc
ΔE (in eV):	0.045	0.065	0.16	0.22	0.31

Find the maximum wavelength of the radiation required to excite electrons from the top of the valence band to these acceptor levels.

10. (I) Calculate the intrinsic carrier concentration in germanium at 380 K, given that $m_n^* = 0.55m_e$ and $m_p^* = 0.37m_e$.

11. (I) At room temperature (300 K), undoped GaSb has an energy gap of 0.68 eV. (a) What is the probability that an electron occupies an energy state at the bottom of the conduction band? (b) What is the probability that there is a hole state at the top of the valence band?

12. (II) The effective masses of n- and p-carriers in germanium are given by $m_n^* = 0.55m_e$ and $m_p^* = 0.37m_e$, respectively. The concentration n_i of n-carriers at 300 K is 2.5×10^{19} m^{-3}. Use this datum to calculate the gap width of germanium.

13. (II) The effective masses of n- and p-carriers in germanium are given by $m_n^* = 0.55m_e$ and $m_p^* = 0.37m_e$, respectively. The gap width of germanium is 0.67 eV. Use this information to find the intrinsic carrier density at 260 K.

14. (II) A sample of germanium is doped so that the donor density is $n = 1.0 \times 10^{22}$ m^{-3}. Find $E_c - E_F$ at $T = 300$ K, given that the effective n-carrier mass in germanium is $m_n^* = 0.55m_e$. Compare your result with E_g for germanium.

15. (II) A sample of germanium is doped so that the hole density is $p = 2.8 \times 10^{24}$ m^{-3}. Find $E_F - E_v$ at $T = 300$ K, given that the effective p-carrier mass in germanium is $m_p^* = 0.37m_e$.

16. (II) The intrinsic semiconductor InSb has a gap energy of 0.18 eV, and the effective masses of the electrons and holes are $0.015m_e$ and $0.39m_e$, respectively. Find the carrier densities at $T = 100$ K and 300 K.

17. (II) The *mobility*, μ, of a charge carrier in a material is defined as the drift speed of the carrier, v_d, divided by the magnitude of the electric field driving the current, $\mu \equiv v_d/E$. For silicon at 300 K, the mobilities of electrons and holes are about 0.14 m^2/V · s and 0.05 m^2/V · s, respectively. (a) Use the definition of μ and the relation between v_d and the current to calculate the free-electron density, n, if the conductivity is 3.8×10^{-4} $(\Omega \cdot \text{m})^{-1}$ and $n \gg p$. (b) Repeat part (a) for the hole density, p, assuming doping such that $p \gg n$.

18. (II) Calculate the electrical conductivity at room temperature of silicon doped to n-type if the density of n-carriers and p-carriers are 1.1×10^{19} m^{-3} and 1.1×10^{13} m^{-3}, respectively. [*Hint:* Use the results of Problem 17.]

19. (II) Use the energy distribution for a collection of identical fermions at finite temperature T to (a) show that the distribution of speeds (the probability that a fermion has a speed between v and $v + dv$) is given by

$$n(v)\,dv = \frac{8\pi m^3}{h^3}\frac{v^2\,dv}{1 + e^{[(mv^2/2) - E_F]/kT}}.$$

(b) What is the distribution for $kT \ll E_F$? [*Hint:* See Problem 45 in Chapter 43.]

20. (II) The Fermi energy can be defined as the energy at which the probability of finding an electron is $\frac{1}{2}$. (a) Show that this is the case at any temperature. (b) Refine this idea by showing that if the probability that an electron has an energy E *above* E_F is P, then the probability that an electron has an energy E *below* E_F is $1 - P$.

21. (III) Using a computer, plot the Fermi distribution as a function of energy for a material with a Fermi energy of 11 eV at temperatures 20 K, 100 K, 300 K, and 1000 K.

44–2 *Semiconductor Structures*

22. (I) A GaAs solar cell produces electric current when the frequency of the light that falls on it is above 3.4×10^{14} Hz. What is the width of the band gap?

23. (I) Plot the factor $e^{(eV_{\text{ext}})/kT} - 1$ [from Eq. (44–18)] as a function of temperature for a sample in which $eV_{\text{ext}} = 0.02$ eV over the range $100\text{ K} < T < 350\text{ K}$. Plot the same factor as a function of eV_{ext} (positive and negative) for $T = 300$ K.

24. (II) The Fermi energy of a heavily doped n-type semiconductor is close to E_c, whereas the Fermi energy of a heavily doped p-type semiconductor is close to E_v. (a) By using an energy diagram, show that when a p-n junction is formed from these semiconductors, a sufficient amount of reverse biasing will allow electrons to tunnel from the p-side to the n-side, a phenomenon known as the *Zener effect*. (b) The tunneling current is a sensitive function of the width of the barrier across which the electrons must tunnel. Show by means of energy diagrams that this width systematically decreases when the reverse bias increases, leading to a quickly building current. [*Hint:* Recall that the transition region has a finite width.] The Zener effect can be used in a *Zener diode*, in which a current flows when a reverse bias potential reaches a certain value.

25. (II) Consider a p-n junction formed from two samples of doped germanium. The two samples are as described in Problems 14 and 15. What is the value of eV_0 for this junction?

26. (II) A silicon p-n junction has $n = 4.5 \times 10^{21}\ \text{m}^{-3}$ on the n-side and $p = 1.2 \times 10^{24}\ \text{m}^{-3}$ on the p-side. For silicon, $E_g = 1.1$ eV, $m_n^* = 1.1m_e$, and $m_p^* = 0.56m_e$. Find the locations of the Fermi energies before the junction is formed. Use your results to find eV_0 at (a) $T = 177$ K and (b) $T = 300$ K.

27. (II) Many LEDs give off red light (Fig. 44–35). Given this fact, estimate the width of the band gap for the materials from which such LEDs are constructed.

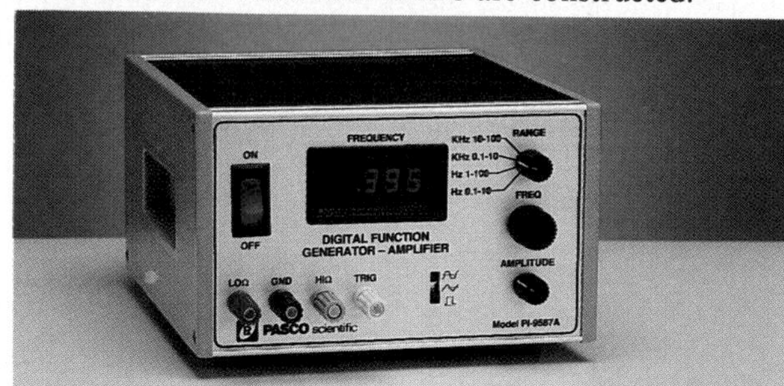

FIGURE 44–35 Problem 27.

28. (II) A current of 4 mA flows through an LED with a band-gap energy of 1.8 eV. Assume that each current-carrying electron drops into a hole, and a single photon is thereby emitted. (a) What is the power emitted in the light? (b) How many photons/s are emitted?

29. (II) Consider a heterostructure in which an intrinsic semiconductor with a narrow band gap is sandwiched between an intrinsic semiconductor with a larger energy gap. (a) Draw an energy diagram for this system. (b) What happens to any n-carriers in this system in the absence of external potentials?

44–3 *Band-Gap Engineering*

30. (II) A quantum well with a width of 12 nm is produced, and spectroscopic measurements show that the energy difference between the ground state and the first excited state within this well is 1.64×10^{-2} eV. What is the effective mass of electrons within the well?

31. (II) For a quantum dot, in which an electron is confined within three dimensions (Fig. 44–36), the allowed energy values have the form

$$E = \frac{\hbar^2\pi^2 n_1^2}{2m^* a_1^2} + \frac{\hbar^2\pi^2 n_2^2}{2m^* a_2^2} + \frac{\hbar^2\pi^2 n_3^2}{2m^* a_3^2},$$

where m^* is the effective mass of an electron; a_1, a_2, and a_3 are the three confining dimensions of the box; and the n_i values are integers. Suppose that a particular quantum dot has dimensions $20\text{ nm} \times 20\text{ nm} \times 40\text{ nm}$ and that $m^* = 0.065m_e$. (a) Calculate the energy gap between the first and second excited states. (b) What is the wavelength of the radiation emitted in the transition between these two states?

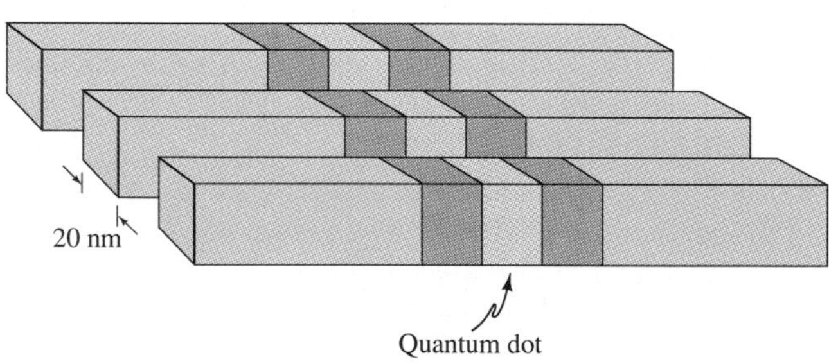

FIGURE 44–36 Problem 31.

44–4 *Scanning Microscopy*

32. (I) The current measured in a scanning tunneling microscope whose tip height is held fixed increases by a factor of 15 when the tip moves from point A to point B on the surface. By what factor is point B closer to the tip than point A?

33. (II) The cantilever of an atomic force microscope has a spring constant of 0.6 N/m, and a laser sensor detects that it has been deflected by 0.2 nm. What is the ratio of the force acting on the cantilever to that of the force of gravity on a fly of mass 0.1 g?

General Problems

34. (I) An electric field of magnitude 10^6 V/m is applied in the $+x$-direction in a piece of intrinsic Si, for which the mobilities of electrons and holes at room temperature are $0.14\ m^2/V \cdot s$ and $0.05\ m^2/V \cdot s$, respectively (see Problem 17). What are the drift velocities (direction and magnitudes) of electrons and holes at room temperature?

35. (I) A silicon sample is doped with 0.2×10^{23} atoms/m^3 of arsenic. Given that the intrinsic n-carrier density is $n_i = 1.5 \times 10^{16}$ carriers/m^3, what is the hole concentration at equilibrium?

36. (II) The effective electron mass in silicon is given by $m_n^* = 1.1m_e$. Calculate the value of N_c [from Eq. (44–12)] at 300 K, and use your result to find the location of E_F relative to the edge of the conduction band for the doped silicon sample of Problem 35.

37. (II) A donor electron moves in doped indium antimonide, for which $\epsilon/\epsilon_0 = 17.9$ and $m^* = 0.015m_e$. Find the radius of a circular atomic orbit of such an electron in terms of the Bohr radius a_0. The effective nuclear charge for such a loosely bound electron is $Z = 1$.

38. (II) A magnetic field of 1.80 T is applied to doped indium antimonide. What is the angular frequency of the donor electrons? [*Hint:* See Problem 37 for the parameters of this semiconductor, and Chapter 29 for a discussion of the cyclotron frequency.]

39. (II) The number of electrons with energy in the range E to $E + \Delta E$ contained in a volume V is given by $f(E)\,\Delta N(E)$, where $f(E)$ is the Fermi–Dirac distribution function and $\Delta N(E)$ is the number of states available to the electrons in the range E to $E + \Delta E$. Given that

$$\Delta N(E) = \frac{V}{2\pi^2}\left(\frac{2m_e}{\hbar^2}\right)^{3/2}\sqrt{E}\,\Delta E,$$

calculate the densities of electrons that have energies in the range 10.45 eV to 10.47 eV for gallium ($E_F =$ 10.40 eV) at the temperatures 90 K and 300 K.

40. (II) Repeat the calculation of Problem 39 for the energy interval between 12.65 eV and 12.67 eV.

41. (II) For a set of identical fermions at high temperature, the average occupation number of any given energy level $f(E) \ll 1$. Show that the Fermi–Dirac distribution approaches the Maxwell–Boltzmann distribution [Eq. (19–40)] in the limit of high temperatures. [*Hint:* The factor $e^{-E_F/kT}$ is determined from the normalization condition $\Sigma_i n_i = N$, where the index i labels the allowed energy levels.]

42. (II) The laser that reads a compact disk is a semiconductor laser of power in the milliwatt range with light of wavelength around 800 nm. (a) Estimate the rate at which photons are emitted. (b) If the compact disk turns beneath the (narrow) laser beam at several hundred revolutions per minute, how many photons will strike a region 0.1 mm in length of the disk?

43. (II) Consider a group of N atoms that obey Maxwell–Boltzmann statistics. The atoms have only two energy states, a ground state at energy E_0 and an excited state at energy E_1. The energy difference $\Delta E = E_1 - E_0 \gg kT$. Show that the average energy of the atoms at temperature T is $E_0 + \Delta E e^{-\Delta E/kT}$.

44. (II) Monochromatic light shines on a thin film of GaAs ($E_g = 1.521$ eV), and the electrical resistance is measured across the film. (a) What happens to the resistance as the wavelength of the light increases, starting with a wavelength in the ultraviolet range? (b) At what critical wavelength does a change in resistance occur? (c) Does the resistance increase or decrease at the critical wavelength?

45. (III) Use the expression for $\Delta N(E)$ given in Problem 39 to express (a) the total number of electrons in a volume V and (b) the average energy of an electron in a volume V at temperature T. (c) Find the limits of your results for parts (a) and (b) for $T \to 0$.

46. (III) A degenerate semiconductor is a semiconductor that is so heavily doped that the doping ions participate in the formation of their own band structure. The Fermi energy may lie within the original conduction band in a degenerate n-type semiconductor, and within the original valence band in a degenerate p-type semiconductor. When a p-n junction is formed from a semiconductor that is degenerate on both the n- and p-sides, a *tunnel diode* is formed. Under these circumstances tunneling is possible when there is an external voltage. Draw an energy diagram (a) for the equilibrium state (no external voltage across the junction); (b) for a reverse bias. How will the tunneling current vary as the reverse bias increases from zero? (c) Draw an energy diagram for a *small* forward bias. How will the tunneling current vary as the small forward bias increases from zero? (d) Show that when the forward bias increases beyond a critical value, the tunneling current begins to *decrease*. This phenomenon is called *negative resistance*.

CHAPTER 45

The nucleus of every atom is composed of neutrons and protons, whose interactions determine the properties of the nucleus. As we shall learn in Chapter 46, even neutrons and protons have structure.

NUCLEAR PHYSICS

Once it became known that the atom has a nucleus that contains all the positive charge and almost all the atomic mass, the structure and properties of the nucleus itself became objects of study. Early on, many of the scientists involved in this research were convinced that nuclear physics would have no practical applications. Of course we now know that this is not true: The understanding of the properties of the nucleus has had enormous consequences for technology and for society in general. Here we shall describe the nuclear constituents, the static properties of nuclei, and the radioactive decays of certain nuclei. We shall see how nuclei react with one another. The practical applications of nuclear physics are widespread and include power generation, radiometric dating, microscopic crack and void testing, food conservation, and cancer treatment.

45–1 STATIC PROPERTIES OF NUCLEI

Nuclear Constituents

The idea that an atom contains a *nucleus* at its center emerged through the research of Ernest Rutherford, who in 1911 discovered the nuclear structure of the atom through scattering experiments. Virtually all of an atom's mass is concentrated in the nucleus, a sphere with a radius that is some 10^5 times smaller than the radius of the atom as a whole, defined by the outermost orbiting electrons. Let us review these results, first described in Chapter 8, in some detail. Even before Rutherford made his discovery, Joseph J. Thomson noted the presence of electrons in atoms. Further experiment established that the number of electrons in an atom is roughly half its atomic weight. The number of electrons is equal to the *atomic number*, Z, a quantity defined by chemical properties. Thomson had measured the charge-to-mass ratio of the electron, and, together with H. A. Wilson's and Robert Millikan's measurements of the electron charge, the electron's mass can be deduced to be much less than the atomic mass. Because the atom is neutral, it must contain a positive charge $+Ze$ in addition to the electron charge $-Ze$. What was *not* known was how the positive and negative charges were distributed within the atom or what carried the positive charge.

The experiments that led to the discovery of the nuclear structure involved scattering. We described in Section 8–8 how we could learn of the existence of a steel pellet in a volume of cotton candy by firing BBs at the candy. Similar reasoning applies to the exploration of atoms by scattering. The atomic model favored at the turn of the twentieth century—proposed by Thomson and known as the Thomson model—had pointlike electrons embedded within a uniform distribution of positive charge. Rutherford undertook the study of atoms by scattering *alpha* (α) *particles* from atoms in a thin gold foil (Fig. 45–1). Alpha particles are produced when certain heavy elements decay, as we shall see in Section 45–3. Rutherford had analyzed the charge and mass of α particles and knew that they consist of doubly ionized helium atoms (charge $+2e$).[†] An α particle is some 8000 times more massive than an electron and would scarcely be deflected by a collision with an electron. A smooth distribution of positive charge, such as the positively charged matrix of the Thomson model, would hardly deflect α particles that passed through the middle of the charge distribution, because the effects of repulsion from each side of the distribution would tend to cancel.

Rutherford's experiment showed that far too many α particles scattered through large angles than could be explained by the Thomson model.[‡] However,

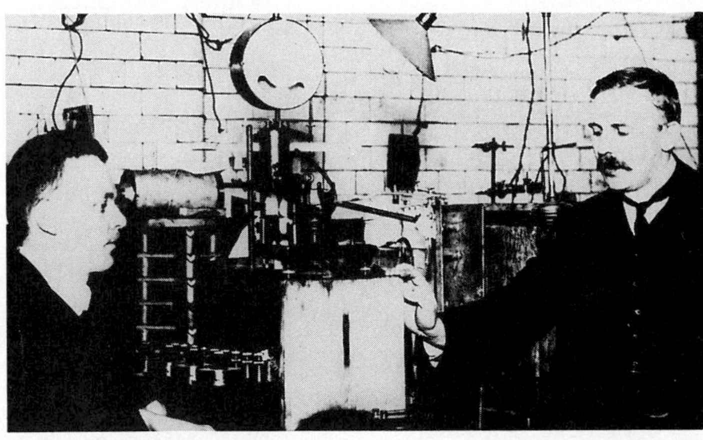

(a)

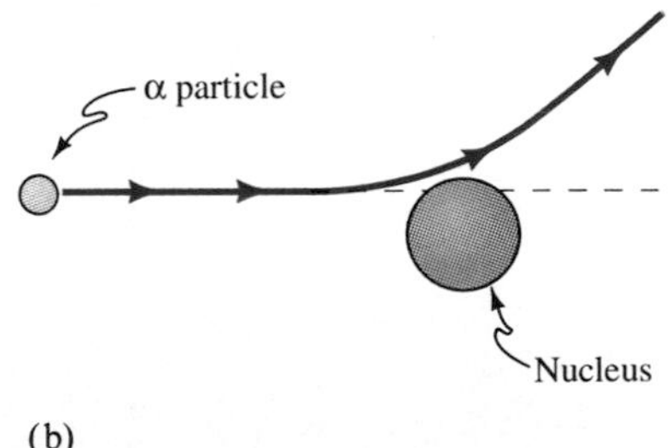

(b)

FIGURE 45–1 (a) Ernest Rutherford (here, on the right, with Hans Geiger) directed the scattering experiments that revealed the atom's nuclear structure. (b) Alpha-particle scattering from a nucleus. A typical deflection occurs when the α particle passes the edge of the nucleus.

[†] The Z value of helium is 2. Thus α particles are themselves helium nuclei, consisting of 2 protons and 2 neutrons.

[‡] Thomson's model predicts a factor of some 10^{10} fewer α particles scattered through angles larger than 90° than were observed.

if the positive charge and the mass of the atom were concentrated in a central structure, the scattering angle could be large. We can estimate the typical scattering angle, θ, of an incident α particle deflected by its electrical interaction with a positively charged sphere of radius R as follows: $\theta \simeq \Delta p/p$, where $p = m_\alpha v$ is the momentum of the α particle, v is its speed, and Δp is the momentum transfer that the α particle receives as it passes the spherical charge distribution. The typical deflection occurs when the α particle passes close to the edge of the positive charge distribution (Fig. 45–1b). In this case, we can approximate Δp by $F\,\Delta t$, where $F = (Ze)(2e)/4\pi\epsilon_0 R^2$ is the Coulomb force between the charge distribution and the α particle that reaches the edge of the distribution, and $\Delta t \simeq R/v$ is the time period over which the α particle is deflected. Thus

$$\theta \simeq \frac{F\,\Delta t}{p} = \frac{(Ze)(2e)}{4\pi\epsilon_0 R^{\cancel{2}}}\frac{\cancel{R}/v}{m_\alpha v} = \frac{2Ze^2}{4\pi\epsilon_0 m_\alpha v^2 R}. \tag{45–1}$$

The factor $1/R$ in this result indicates that the smaller the radius of the nucleus, the larger the typical scattering angle will be. For R equal to the atomic radius (about 0.1 nm), the deflection is negligible. The amount of large-angle scattering that Rutherford discovered can be explained if R is on the order of 10^{-5} of the atomic radius.

EXAMPLE 45–1 An α particle with kinetic energy of 5 MeV heads directly toward a gold nucleus ($Z = 79$) at rest. The gold nucleus is much heavier than the α particle. Find how close the α particle comes to the gold nucleus before it turns around.

SOLUTION: Coulomb repulsion between the gold nucleus and the α particle will cause the α particle to slow down, stop, and then turn around. Let us assume the gold nucleus is so heavy that it remains fixed and use the conservation of energy to solve this problem. The initial total energy of the system is $E = K + U$. We take the potential energy, U, to be zero when the colliding particles are far apart. E is then equal to K, or 5 MeV, and it remains that value throughout. When the α particle reaches the turn-around point, its kinetic energy is zero, and the energy is entirely potential energy. At this point the separation is R (Fig. 45–2). The Coulomb potential energy of two objects with charges $Z_1 e$ and $Z_2 e$ is given by

$$U(r) = \frac{Z_1 Z_2 e^2}{4\pi\epsilon_0}\frac{1}{r}.$$

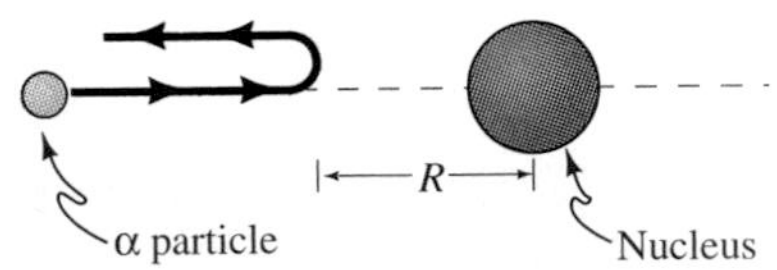

FIGURE 45–2 Example 45–1. A head-on collision in which an α particle is deflected straight back from a nucleus.

We apply the conservation of energy by equating the energies at $r = \infty$ and at $r = R$, setting $Z_1 = 2$ (the α-particle charge) and $Z_2 = 79$ (the gold nucleus charge):

$$K|_{r=\infty} + U|_{r=\infty} = K|_{r=R} + U|_{r=R};$$

$$K + 0 = 0 + \frac{(2)(79)e^2}{4\pi\epsilon_0}\frac{1}{R}.$$

We solve for R:

$$R = \frac{(2)(79)e^2}{4\pi\epsilon_0}\frac{1}{K} = (2)(79)(1.6 \times 10^{-19}\,\cancel{\mathrm{C}})^2(9 \times 10^9\ \mathrm{N \cdot m^2 \cdot \cancel{C^{-2}}})$$

$$\times \frac{1}{5 \times 10^6\,\cancel{\mathrm{eV}}}\frac{1\,\cancel{\mathrm{eV}}}{1.6 \times 10^{-19}\ \mathrm{J}} = 4.55 \times 10^{-14}\ \mathrm{m}.$$

This distance is somewhat outside the combined nuclear radii of the two particles involved, so the nuclear force will be negligible. However, the distance is well inside even the innermost Bohr radius.

Scattering Distributions

Rutherford's observation of large scattering angles led him to propose the planetary orbit model discussed in Chapter 42. He was also able to make a more sophisticated calculation of what happens when a collimated beam of α particles is fired at a set of atoms constructed on the basis of his model. Some of the α particles follow a line directly toward a nucleus, and some follow a line off to the side (Fig. 45–3). For a given line, the force anywhere along that line is the Coulomb force, and the trajectory of the α particle can be calculated. If the beam of α particles is spread uniformly across a region with a collection of target nuclei, then the number of α particles that arrive at any distance off to the side of any nucleus can be calculated. Rutherford used this technique to predict a precise *distribution* of the number of α particles deflected at any given angle.†

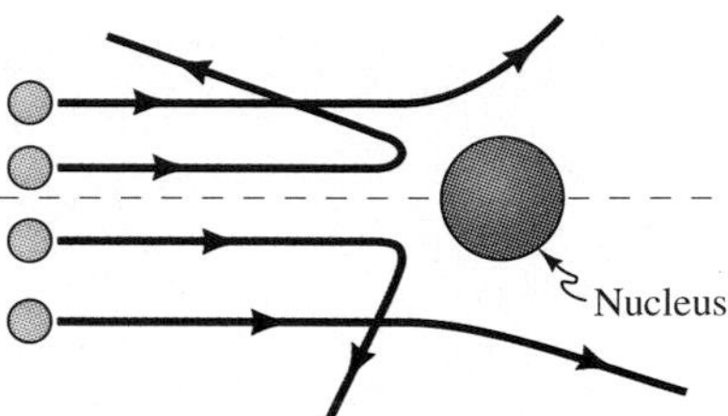

FIGURE 45–3 A beam of α particles sent into a sample of material will scatter from a nucleus at many different angles.

Rutherford considered the Coulomb scattering of nuclear charges only, because he suspected that the region containing the positive charge was at the center of the atom and contained *all* the positive charge. He assumed that the target was at rest and, in his initial calculations, that the target was much more massive than the projectile. If M is the target mass (of charge Z_2e) at rest and m is the projectile particle (of charge Z_1e) moving with kinetic energy K, then if $M \gg m$, the target particle will not recoil. We will consider only this case, because of its simplicity. The probability that the particle will be scattered at an angle θ is proportional to the *collision cross section*, $\sigma(\theta)$. Rutherford showed that, for $M \gg m$, this cross section is given by

The collision cross section was described in Chapter 19.

$$\sigma(\theta) = \frac{Z_1 Z_2 e^2}{16(4\pi\epsilon_0)^2} \frac{1}{K^2} \frac{1}{\sin^4(\theta/2)}. \qquad (45\text{–}2)$$

The critical features are the inverse-square dependence on K and the strong dependence on θ.

Rutherford's assistants Hans Geiger and Ernest Marsden painstakingly counted the scattering of α particles from a thin gold foil by observing the particles that arrived on a fluorescent screen (Fig. 45–4). These experiments

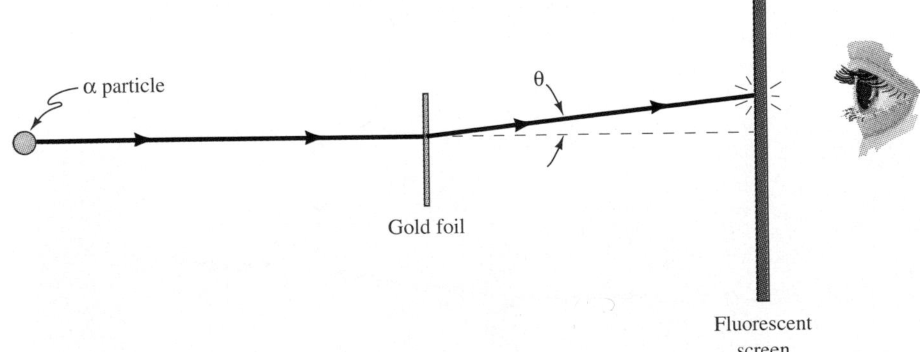

FIGURE 45–4 Schematic diagram of the experiment in which Geiger and Marsden observed the scattering of α particles from gold atoms in gold foil.

† Rutherford could make only a classical calculation, but we now know that quantum physics dominates at these scales. By an extraordinary stroke of luck, the inverse-square law obeyed by the Coulomb force is the *only* force law for which a quantum mechanical calculation gives a distribution identical to that given by a classical calculation.

verified Eq. (45–2) and hence also the existence of the nucleus and Rutherford's concept of the atom. Rutherford also noticed a deviation from the prediction of Eq. (45–2). When an α particle is on a path directly toward a nucleus, it backscatters at 180° (see Example 45–1). If the Z value of the target is small enough and the energy of the incident α particle is high enough (greater than 5 MeV in Rutherford's experiment), the α particle has enough energy to overcome the Coulomb force and penetrate into the nucleus. The force that the α particle experiences is thereby modified and Eq. (45–2) is no longer correct, which accounts for the deviation from Eq. (45–2) that Rutherford observed. The penetrating α particle experiences a nuclear force, and Rutherford's observation was the first time the nuclear force was seen in scattering.

Neutrons. Because almost all the mass of an atom is in its nucleus and because the atomic weight is roughly twice the atomic number, the nucleus cannot be composed of protons alone. Rutherford speculated in 1920 that a nucleus could consist of Z protons and $N = A - Z$ particles of about the same mass as the proton but with no electric charge, particles later called **neutrons**.† Here A is the *mass number*, the number of protons plus neutrons in the nucleus. Experimental verification was difficult, because neutral particles are hard to detect. When charged particles pass through matter, Coulomb forces cause electrons to be ejected from the atoms of the matter, leaving a path of ionized atoms that is easy to observe by various detection techniques. But when neutral particles such as photons or neutrons pass through matter, they do not ionize the matter very easily.

When atoms of boron or beryllium are bombarded with α particles, observation of the recoiling nuclei shows that neutral particles are produced. These neutral particles are, however, not necessarily neutrons; they could, for example, be photons. James Chadwick showed in 1932 that the neutral particles produced by the α-particle bombardment of boron and beryllium are actually neutrons. Chadwick allowed the neutral component to pass into material that contains a good deal of hydrogen, such as paraffin, and observed that a rather energetic proton was occasionally produced (Fig. 45–5). He interpreted this as the result of a collision between the neutral component and a hydrogen nucleus, which consists of just a single proton. By measuring the momentum of the observed proton and by using momentum conservation, Chadwick verified that the neutral component could not be a photon. Rather, he concluded that the particle, which is now called the *neutron*, must have about the same mass as the proton. Thus, some 20 yr after Rutherford's discovery of the nucleus, the nature of the nuclear components was firmly established.

The proton and the neutron collectively are known as **nucleons**, because they

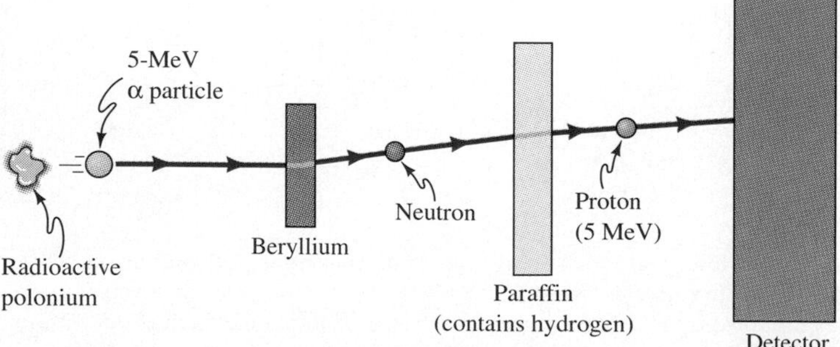

FIGURE 45–5 Schematic diagram of Chadwick's experiment to discover neutrons. Paraffin contains a high percentage of hydrogen and hence of nuclei consisting of single protons that act as targets for any incoming projectiles. A kinematic analysis shows that only neutrons scattering from hydrogen nuclei could have produced the energetic protons in the paraffin.

† For reasons such as those described in Question 4 and Problem 56, the presence of electrons in the nucleus was ruled out.

are both found in the nucleus and they share many of the same properties. An important difference between protons and neutrons is that, unlike protons, free neutrons are not stable, but decay with a lifetime $\tau \simeq 890$ s. Neutrons do *not*, however, decay within stable nuclei; we shall soon see why.

The lifetime is discussed in Chapter 41.

The properties of the nucleus depend on the force that holds it together, called the *nuclear*, or *strong*, force. We know that such a force must exist, because without it the repulsive Coulomb force between protons would cause the nucleus to disintegrate.

Some Terminology

A given nucleus contains Z protons and N neutrons, making in all $A = Z + N$ nucleons.

Recall that the number of nucleons in a single nucleus is the mass number, A, a quantity that very nearly determines the mass of the nucleus; the number of protons is the atomic number, Z; and the number of neutrons is $N = A - Z$. A **nuclide** is a nucleus of an element with a given Z and A and is written as

$$^{A}_{Z}X,$$

where X is the chemical symbol of the element. For example, the nuclide $^{4}_{2}\text{He}$ is a helium nucleus ($Z = 2$), with 2 protons and 2 neutrons. The nuclide $^{3}_{2}\text{He}$ also occurs naturally. This terminology is redundant, because the chemical symbol alone specifies Z, so we normally write the simpler form ^{4}He. The proton is a hydrogen nucleus, $^{1}_{1}\text{H}$, but we shall more often use the standard symbol p for the hydrogen nucleus. As we have just seen for helium, the nucleus of a given chemical element has a particular Z value, but a range of N values are possible. For example, neon has three stable nuclides (^{20}Ne, ^{21}Ne, and ^{22}Ne) and several unstable ones. Nuclides with the same Z value are known as **isotopes** of the given element (see Chapter 19). The gram-atomic weight of a chemical element—the mass of 1 mol of that element—is actually an average over the different naturally occurring isotopes of that element weighted by the relative abundance of those isotopes.

Nuclear Masses and Binding Energies

The masses of nuclei are equal, to within less than 0.5 percent, to the masses of the corresponding atoms, and these masses can be determined at this accuracy by purely chemical methods. The mass of a nucleus, which we label $m(^{A}_{Z}X)$, can be determined more accurately, however, by first stripping its atom of one or more electrons, and then sending the nucleus through a mass spectrometer (see Chapter 29). This spectrometer may, for example, first use crossed electric and magnetic fields to find the speed of the ion, which has been accelerated in another electric field. The ion then passes through a region of fixed magnetic field; there the radius of curvature of the ion's path accurately determines the momentum. Knowing the speed and the momentum, we can determine the mass. With such devices, the masses of ions can be determined to great accuracy: Atomic masses can be measured to an accuracy of better than 1 part in 10^8. That atomic masses and not nuclear masses are obtained is not a handicap because, as we shall see, the electron masses normally cancel in calculations with nuclei.

We remark in passing that nuclear abundances, the relative amounts of isotopes that occur in nature, can also be established from mass spectrometry. Ionized atoms from a naturally occurring sample of the element pass through the spectrometer. Because their masses are different, each isotope follows a different path, and the relative number that follow each path is a measure of their abundance.

TABLE 45–1

SOME ATOMIC MASSES	
Element	*Atomic Mass (u)*
$^{1}_{1}H$	1.007825
$^{2}_{1}H$	2.014102
$^{3}_{2}He$	3.016029
$^{4}_{2}He$	4.002603
$^{6}_{3}Li$	6.015121
$^{7}_{3}Li$	7.016003
$^{10}_{5}B$	10.012937
$^{12}_{6}C$	12.000000
$^{14}_{6}C$	14.003242
$^{14}_{7}N$	14.003074
$^{15}_{7}N$	15.000109
$^{16}_{8}O$	15.994915
$^{23}_{11}Na$	22.989768
$^{24}_{12}Mg$	23.985042
$^{27}_{13}Al$	26.981539
$^{35}_{17}Cl$	34.968853
$^{40}_{20}Ca$	39.962591
$^{55}_{25}Mn$	54.938047
$^{56}_{26}Fe$	55.934939
$^{60}_{28}Ni$	59.930788
$^{60}_{27}Co$	59.933820
$^{74}_{32}Ge$	73.921177
$^{90}_{40}Zr$	89.904703
$^{138}_{56}Ba$	137.905232
$^{181}_{73}Ta$	180.947992
$^{208}_{82}Pb$	207.976627
$^{209}_{82}Pb$	208.981065
$^{210}_{84}Po$	209.982848
$^{232}_{90}Th$	232.038051
$^{235}_{92}U$	235.043924
$^{238}_{92}U$	238.050785

A useful mass unit for nuclear and atomic physics is the *atomic mass unit* (u), which is defined as $\frac{1}{12}$ the mass of a ^{12}C atom. Such a mass standard is useful because the *relative* masses of atoms (and nuclei) can be measured with high precision, even if we can express with much less precision the atomic mass unit in terms of the kilogram:

$$1\ \text{u} = 1.66057 \times 10^{-27}\ \text{kg}. \qquad (45\text{–}3)$$

Another useful mass unit follows from the relativistic relation between mass and energy, which allows us to express masses in eV/c^2, or MeV/c^2:

$$1\ \text{u} = 931.494\ \text{MeV}/c^2. \qquad (45\text{–}4)$$

Energy is also conveniently expressed in terms of uc^2, because of the relation in Eq. (45–4). In atomic mass units, the masses of the nucleons are

$$m_p = 1.00728\ \text{u} = 938.272\ \text{MeV}/c^2 = 1.67262 \times 10^{-27}\ \text{kg} \quad \text{and} \qquad (45\text{–}5a)$$

$$m_n = 1.00866\ \text{u} = 939.566\ \text{MeV}/c^2 = 1.67493 \times 10^{-27}\ \text{kg}, \qquad (45\text{–}5b)$$

whereas the mass of the electron is

$$m_e = 5.48578 \times 10^{-4}\ \text{u} = 0.510995\ \text{MeV}/c^2 = 0.910939 \times 10^{-30}\ \text{kg} \qquad (45\text{–}5c)$$

and the mass of the hydrogen atom is

$$M(^1\text{H}) = 1.007825\ \text{u} = 938.783\ \text{MeV}/c^2 = 1.67356 \times 10^{-27}\ \text{kg}. \qquad (45\text{–}5d)$$

Note in Eqs. (45–5) the standard notation of capital em (M) for atomic mass and lower case em (m) for nuclear and particle masses; we shall use this notation throughout. Table 45–1 gives the values of some atomic masses in u. The corresponding nuclear mass can be found by subtracting Zm_e from the atomic mass.

Binding Energy. From Eqs. (45–5), the sum of the masses of 6 protons, 6 neutrons, and 12 electrons is 12.0989 u, about 1 percent greater than the mass of a ^{12}C atom, exactly 12 u. A closer look shows that this discrepancy is due entirely to the nuclear mass: The mass of a carbon nucleus is less than the sum of the masses of 6 protons and 6 nucleons by this same amount. *In general the mass of a stable nucleus is less than* $Zm_p + Nm_n$, due to the energies of binding the nucleons into the nucleus and to special relativity, as we shall soon see. To separate a nucleon (or collection of nucleons) from a stable nucleus takes energy.

The *binding energy* E_b of a nucleus is the energy released when the nucleus is formed from A independent constituent nucleons. To understand the relation between binding energy and nuclear mass, we must use mass–energy equivalence (see Chapter 40). Because energy must be added to a nucleus to transform it to free nucleons, the energy of the nucleus is less than the sum of the energies of free nucleons, and by the equivalence of mass and energy, the mass of the nucleus is less than the sum of the masses of the constituents. These considerations hold for *any* bound system, but for atoms and the other bound systems we have considered previously, this effect is a small one. For the nucleus, the attractive forces between nucleons are so strong that the binding energy is a significant fraction of the total mass. Thus, by the relativistic relation between mass and energy, *the mass of a nucleus is on the order of 1 percent less than the sum of its nucleon masses.*

The nuclear mass is about 1 percent less than the total masses of the nucleus's constituents.

If we had added together the masses of 6 hydrogen *atoms* and 6 neutrons instead of comparing the masses of the carbon nucleus plus 6 electrons with all the constituent particles, we would still have obtained 12.0989 u, as compared

TABLE 45–2

SOME NUCLEAR BINDING ENERGIES

Nucleus	*Total Binding Energy (MeV)*	*Binding Energy per Nucleon (MeV/nucleon)*
$^{2}_{1}H$	2.23	1.1
$^{4}_{2}He$	28.3	7.1
$^{12}_{6}C$	92.1	7.7
$^{16}_{8}O$	127.5	8.0
$^{56}_{26}Fe$	492.3	8.8
$^{63}_{29}Cu$	552.1	8.8
$^{238}_{92}U$	1803	7.6

with exactly 12 u for the ^{12}C *atom*: The 6 electron masses cancel. We can neglect the effects of atomic electron binding energies on masses because these binding energies are so much smaller than the total mass–energy. The binding energy is thus c^2 times the difference between the atomic mass and the sum of the masses of Z hydrogen atoms and N neutrons:

$$E_b \equiv [ZM(^1\text{H}) + Nm_n - M(^AX)]c^2. \tag{45–6}$$

This quantity is positive for stable nuclei. Table 45–2 gives the total binding energies of some nuclei. In the case of ^{12}C, we get $E_b = (12.0989\ \text{u} - 12.0\ \text{u})\ c^2(931.494\ \text{MeV/u}c^2) = 92.12\ \text{MeV}$, from Eq. (45–6).

Table 45–2 also gives the *binding energy per nucleon*, E_b/A, in addition to the total binding energy. Once A is 12 or larger, the binding energy per nucleon is fairly constant, between 7 and 9 MeV/nucleon. We plot this quantity as a function of the mass number, A, in Fig. 45–6. Note the position on this curve of ^{4}He, which has an exceptionally large E_b/A value and is therefore strongly bound, and the region around ^{56}Fe, for which the curve exhibits a broad maximum. The fact that the binding energy per nucleon increases from zero for small nuclei and decreases as A increases for $A > 60$ tells us a lot about nuclear stability. We see that the total mass–energy per nucleon *decreases* (the total binding energy increases) when two small nuclei combine into an intermediate one (for instance, E_b/A for ^{2}H is 1.1 MeV, whereas E_b/A for ^{4}He is 7.1 MeV). Similarly, the total mass–energy per nucleon decreases when one large nucleus, such as ^{235}U, for which $E_b/A \simeq 7.5$ MeV, forms two intermediate nuclei, for which E_b/A may be around 8.5 MeV, or when a large nucleus forms a remnant nucleus plus a ^{4}He nucleus (an α particle). Thus decays of large nuclei into smaller nuclei are to be expected. Whether these decays actually occur or not, and the rate at which they occur, depend on details of the structure of the decaying nucleus. For example, a decay may occur only when there is tunneling through a potential barrier (see Chapter 41). We shall discuss these questions further.

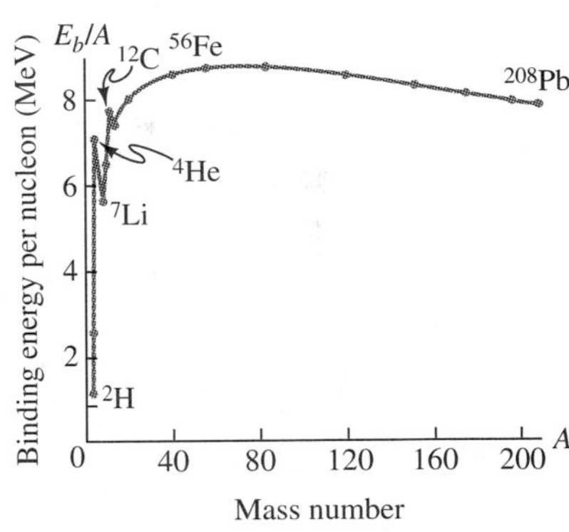

FIGURE 45–6 The binding energy per nucleon, E_b/A, as a function of the mass number A.

Other Properties

Size and Internal Distribution of Mass and Charge. Rutherford's experiments have been refined. When the energy of the bombarding particle is increased to the point at which the particle can penetrate the nucleus, we can study the *internal* structure of the nucleus by analyzing the distribution of the scattering. If the nucleus is bombarded with electrons, which interact *only* through the Coulomb force with the nucleus, we learn about the distribution of charge within the nucleus. If the nucleus is bombarded with α particles, protons, or neutrons, we learn instead about the distribution of nucleons within the nucleus. Both types of experiments give roughly the same picture of both the radius and the internal makeup of the nucleus; in other words, the protons and neutrons are distributed in much the same way within the nucleus. Scattering experiments show that both the mass and the charge of a nucleus of mass number A is distributed *uniformly* out to a radius R_A given by

$$R_A = r_0 A^{1/3}, \tag{45–7}$$

where $r_0 \simeq 1.2 \times 10^{-15}$ m. The mass (and charge) density within this volume is very nearly the same for all nuclei. Beyond R_A, the mass and charge densities drop rapidly to zero. The fact that R_A is proportional to $A^{1/3}$ means that the volume, approximately equal to $\frac{4}{3}\pi R_A^3$, is proportional to A: volume $= \frac{4}{3}\pi R_A^3 = \frac{4}{3}\pi(r_0 A^{1/3})^3 = \frac{4}{3}\pi r_0^3 A$. This is an indication that the nucleons are spread uniformly throughout the volume of the nucleus and is confirmed by the fact that the internal distributions of mass and charge have more or less the same shape for all nuclei. In the classical model, the nucleus resembles a set of closely packed marbles (Fig. 45–7). The nucleons inside the nucleus have a very strong attraction to their immediate neighbors, whereas the surface nucleons are less strongly bound.

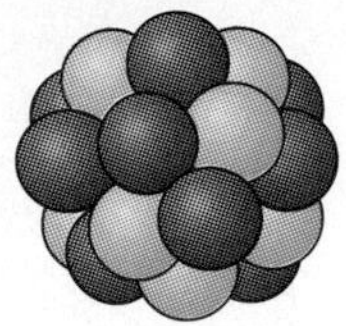

FIGURE 45–7 Nucleons are closely packed within nuclei. We have used red to indicate neutrons and blue to indicate protons.

Stability. Although nuclides with Z values up to $Z = 92$ (uranium) occur naturally, not all these nuclides are stable. The nuclide $^{209}_{83}$Bi is the heaviest stable nucleus. Even though uranium is not stable, however, its longest-lived isotope, ^{238}U, has a half-life of some 4 billion yr.† At low Z values, nuclei tend to have the same number of neutrons and protons (that is, $N \simeq Z$), but as Z increases, N tends to exceed Z, because there is a repulsive Coulomb force between protons that is not present for neutrons. The nuclide ^{40}Ca is the heaviest stable nuclide for which $N = Z$. Figure 45–8, a plot of the stable nuclides, shows a region called the *line of stability* on which the stable nuclides lie. This curve has the rough form $N \simeq Z$ for small Z and $N \simeq 1.6Z$ for large Z.

There are approximately 250 stable isotopes. When the number of stable isotopes is expressed in terms of whether Z and N are even or odd, a striking pattern emerges. About 60 percent of the stable isotopes have both Z and N even; about 20 percent have Z even, N odd; and 20 percent have Z odd, N even. Only 5 stable isotopes have both Z and N odd. The strong preference for nuclei to have both Z and N even can be understood by a pairing force of the kind that leads to Cooper pairs in superconductivity (see Chapter 43).

We shall discuss further in Sections 45–3 and 45–4 the energetics of decay, the various ways in which unstable nuclei decay, and the rates at which those decays occur.

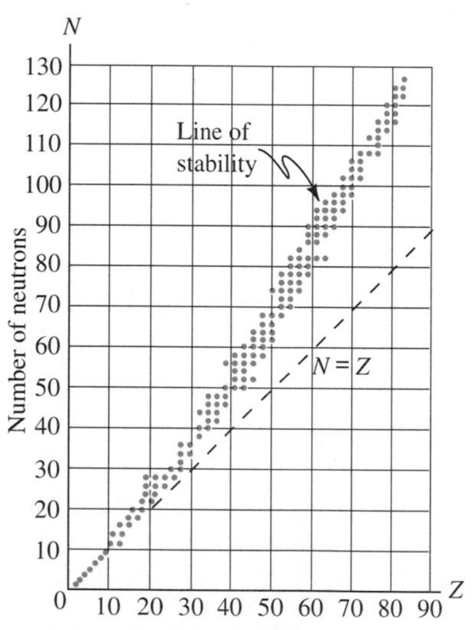

FIGURE 45–8 The stable nuclides plotted on a graph of neutron number, N, versus proton number, Z. Note that for heavier nuclides, N is larger relative to Z. The stable nuclides group along a curve called the line of stability.

Spins and Magnetic Dipole Moments. Nuclei have angular momentum, just as atoms do. This angular momentum is usually called the *spin* of the nucleus, because, without knowing the internal structure of the nucleus, we can regard it as intrinsic. Nucleons have a spin of $\hbar/2$. And as is true of electrons in atoms, the nucleons that compose the nucleus can have an orbital angular momentum. The total angular momentum of a nucleus results from adding the orbital angular momentum and the spins according to the rules of quantum mechanics, supplemented by the exclusion principle. (The exclusion principle states, for example, that two protons cannot move in the same orbit and both have spin up.) One of

† Recall from Chapter 41 that the half-life is the time over which half of a collection of unstable particles decay. It is equal to 0.69τ, where τ is the lifetime.

these rules states that when the angular momenta and spins of an odd number of particles of spin $\hbar/2$ are added, the result is a net angular momentum that is an odd multiple of $\hbar/2$, whereas when the angular momenta and spins of an even number of particles of spin $\hbar/2$ are added, the result is a net angular momentum that is an integral multiple of $\hbar$. Thus, in their respective ground states, ${}^3_2\text{He}$ has spin $\hbar/2$, but ${}^4_2\text{He}$ has spin 0; ${}^{107}_{49}\text{In}$ has spin $9\hbar/2$, whereas ${}^{158}_{67}\text{Ho}$ has spin $5\hbar$.

Associated with the angular momentum of a nucleus is a magnetic dipole moment, and this magnetic movement provides the means of detecting the angular momentum of the nucleus. We described *nuclear magnetic resonance*, in which the nuclear magnetic dipole moment is detected because of a resonance phenomenon, in Chapter 32. Another means of detecting the nuclear magnetic dipole moment is to note that a magnetic dipole produces a magnetic field, and because atomic electrons have both spin and orbital angular momenta, the energy of these electrons depends on this nuclear magnetic field. The small resulting difference in electron energy levels is referred to as *hyperfine splitting* of the atomic spectra, a splitting some thousand times smaller than the fine-structure splitting discussed in Chapter 42.

Although a universal explanation of all nuclear magnetic dipole moments is lacking, the orders of magnitude of these magnetic moments are those of the magnetic dipole moments of the proton and neutron themselves. We expect these magnetic dipole moments in turn to be on the order of $e\hbar/2m_pc$ (a quantity known as the *nuclear magneton*) for the proton and zero for the (neutral) neutron, similar to the magnetic dipole moment of $e\hbar/2m_ec$ for the electron.† This order-of-magnitude estimate is correct. Note that, because of the masses involved, the nuclear magnetic moment is some 2000 times smaller than the electron magnetic moment. That is why the hyperfine splitting of atomic spectra is some 1000 times smaller than the fine-structure splitting.

EXAMPLE 45–2 (a) From data on atomic masses, determine the total binding energy and binding energy per nucleon of ^{56}Fe. (b) How much energy does it take to remove the least-tightly-bound proton from ^{56}Fe?

SOLUTION: (a) Table 45–1 and Eqs. (45–5) contain the masses necessary for us to be able to use Eq. (45–6) to determine the binding energy. When Eq. (45–6) is applied to ^{56}Fe, we find that

$$\begin{aligned} E_b(^{56}\text{Fe}) &= [Z_{^{56}\text{Fe}}M(^1\text{H}) + N_{^{56}\text{Fe}}m_n - M(^{56}\text{Fe})]c^2 \\ &= [26M(^1\text{H}) + 30m_n - M(^{56}\text{Fe})]c^2 \\ &= [26(1.007825\text{ u}) + 30(1.008665\text{ u}) - 55.934939\text{ u}]c^2 \\ &= (0.52846\,\text{u}c^2)(931.494\text{ MeV}/\text{u}c^2) = 492.3\text{ MeV}. \end{aligned}$$

The binding energy per nucleon is (492.3 MeV)/(56 nucleons) = (8.8 MeV)/nucleon.

(b) To determine how much energy binds the least-tightly-bound proton, which we label S_p, we use a relation like Eq. (45–6) but instead compare the mass of an atom of ^{56}Fe to that of a hydrogen atom plus an atom of ^{55}Mn (the ^{55}Mn nucleus is what remains when a proton is removed from an ^{56}Fe

† The electron is, as far as we know, "elementary" (without internal structure), whereas the proton and the neutron are themselves particles with internal structure. As a result, the magnetic dipole moments are not $e\hbar/2m_pc$ and zero but $e\hbar/2m_pc$ times 2.8 and −1.9 for the proton and neutron, respectively.

nucleus):

$$S_p(^{56}\text{Fe}) = [M(^1\text{H}) + M(^{55}\text{Mn}) - M(^{56}\text{Fe})]c^2$$
$$= [1.007825 \text{ u} + 54.938046 \text{ u} - 55.934939 \text{ u}]c^2$$
$$= (0.0109 \cancel{\text{u}c^2})(931.494 \text{ MeV}/\cancel{\text{u}c^2}) = 10.2 \text{ MeV}. \qquad (45\text{–}8)$$

The energy required to remove just one nucleon from a nucleus is called the *separation energy*. The proton separation energy for ^{56}Fe is particularly large, 10.2 MeV, because ^{56}Fe is a particularly strongly bound nucleus.

45–2 NUCLEAR FORCES AND NUCLEAR MODELS

Nuclear Forces

One of the major tasks of nuclear physics is to discover the nature of the nuclear force. It must be strong enough to overcome the electrical repulsion between protons. An energy of 5 to 10 MeV is required to free a nucleon from a nucleus. This value can be compared with the 13.6 eV of energy needed to ionize hydrogen or the 3 to 5 eV needed to free an electron from a metal (the work function). By this measure, the nuclear forces are about a million times stronger than the electric forces that bind atoms.

The nuclear force is a short-range force.

The fact that the nucleon separation energy is on the order of a few MeV even for nuclei as heavy as lead ($A \simeq 208$) implies that the number of "bonds" that attach a nucleon to a nucleus does not grow as the nucleus grows; we say that the nuclear force becomes *saturated* when a nucleon is surrounded by other nucleons. *The nuclear force has a short range, comparable to the spacing between nucleons in a nucleus.* By the *range*, we mean the distance beyond which the nuclear force rapidly decreases; we shall soon give a more quantitative definition. From our formula for the nuclear radius, Eq. (45–7), we can deduce that the spacing between nucleons is on the order of 1 **fermi** (fm), or 10^{-15} m (also known as a *femtometer*). The conclusion that the nuclear force has a short range is supported by scattering experiments, which also allow us to measure the range quantitatively.

We can summarize the quantitative behavior of nuclear interactions as follows. Except at distances much less than 1 fm, a reasonable representation of the nucleon–nucleon potential energy is†

$$U(r) = -g^2 \frac{e^{-r/R}}{r}, \qquad (45\text{–}9)$$

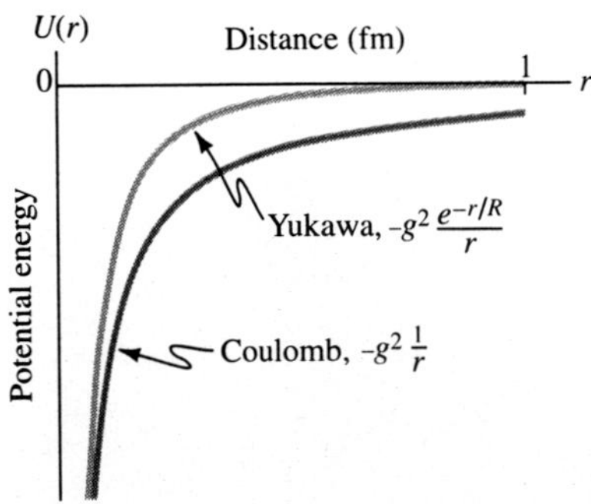

FIGURE 45–9 The Yukawa potential $U(r) = -g^2e^{-r/R}/r$, and the Coulomb form $-g^2/r$ for comparison. We have chosen $g^2 = 1$ and $R = 0.3$ fm.

where r is the distance between the two nucleons, and the parameter R, which we define as the range, is approximately 2×10^{-15} m. Figure 45–9 illustrates this potential energy as well as a potential energy of the same strength but with a Coulomb form, $U_C = -g^2/r$, for comparison; these two potential energies have the same behavior for small r. The potential energy of Eq. (45–9) is known as the *Yukawa potential*, named for its inventor, Hideki Yukawa. It can be justified in the context of quantum mechanics (as we shall see in Chapter 46).

In our discussion, we have referred to the nucleon–nucleon force, not the proton–proton force, the proton–neutron force, or the neutron–neutron force. Numerous scattering experiments, as well as the study of nuclear binding

† Because the two nucleons have the same mass, it is necessary to treat the system in terms of a reduced mass acting under the influence of a central force.

energies for nuclei with the same A value but different Z values and N values, show that, basically, the only differences in the interactions of neutrons and protons are due to electromagnetism.

A full description of nuclear potentials is much more complicated than the Yukawa potential. First, there is a repulsion (a *core repulsion*) between the nucleons at separation distances less than about 1 fm. Second, nuclear forces have a substantial dependence on the orientation of the spin of the nucleons. Third, and perhaps most important, nuclear forces normally involve more than just two interacting particles. In atomic physics, this does not present much of a problem, because the forces are relatively weak, and mathematical tools exist by which we can handle such forces when many bodies are involved. In nuclear physics, the forces are *strong*, as measured by the parameter g^2 in Eq. (45–9). Analytic mathematical techniques for the systematic treatment of such forces do not exist. The problems posed by nuclear forces are too difficult to be solved from a fundamental theory. Despite this difficulty, scientists know much about nuclei. As in the case of materials, models and approximations that explain different, limited features of nuclei have been developed for describing the nucleus. We discuss two of these models, the shell model and the liquid-drop model.

The Shell Model

In Chapter 42, we learned that the energy levels of electrons in an atom form a shell structure. The nucleus attracts the electrons through the Coulomb force, which decreases as $1/r^2$ and affects even distant electrons.† Electrons also interact with each other, but this interaction leads only to relatively minor modifications of the atomic structure dictated by the pure Coulomb force from the nucleus. We have seen in this chapter that nuclear forces are quite different from Coulomb forces, especially in that they are short-ranged. Despite this important difference, individual nucleons in a nucleus also act as though they are in an external potential. In effect, the interactions of one nucleon and surrounding nucleons lead to an average force, or average potential energy, that has well-defined energy levels. The description of the nucleus based on this idea is known as the **shell model** (or as an *independent-particle model*).

What seems at first incomprehensible in the description above is why nucleons would stay within any "orbit" for any length of time. The exclusion principle plays an important role here: Nucleons can undergo transitions to other orbits only if, in colliding with other nucleons, they and their partners have empty energy levels to move into. However, most of the available energy levels are filled, and this is what leads to the stability of orbits and the existence of a shell structure.

To examine what such a structure would look like, we assume that the force experienced by a given nucleon is a three-dimensional spring force.‡ In three dimensions the energy levels of a harmonic oscillator are a simple generalization of the result of Example 42–4, corresponding to the possibility of motion in three dimensions:

$$E = \hbar\omega(2n_r + \ell + \tfrac{3}{2}). \qquad (45\text{–}10)$$

Here the radial quantum number $n_r = 0, 1, 2, 3, \ldots$ and ℓ, the angular momentum quantum number, takes on the values $\ell = 0, 1, 2, 3, \ldots$. As usual, there are

† *Screening*, in which intervening electrons partly "block" the electric field of the nucleus at the position of a more distant electron, must be taken into account.

‡ The form of the potential energy is less important here than the fact that a single nucleon behaves as though it experiences a central force.

$2\ell + 1$ states with the same energy for each value of ℓ. We separately fill energy levels with protons and neutrons. The lowest-energy state of Z protons, for example, is obtained by filling successive levels with two (spin $\hbar/2$) protons each—that is, $2(2\ell + 1)$ protons could fit into the levels for a given ℓ value. Thus the order of filling levels for protons is the following:

			Total protons
$E = \frac{3}{2}\hbar\omega$;	$n_r = 0, \ell = 0$;	2 protons	2
$E = \frac{5}{2}\hbar\omega$;	$n_r = 0, \ell = 1$;	6 protons	8
$E = \frac{7}{2}\hbar\omega$;	$n_r = 0, \ell = 2$;	10 protons	18
$E = \frac{7}{2}\hbar\omega$;	$n_r = 1, \ell = 0$;	2 protons	20
$E = \frac{9}{2}\hbar\omega$;	$n_r = 0, \ell = 3$;	14 protons	34
$E = \frac{9}{2}\hbar\omega$;	$n_r = 1, \ell = 1$;	6 protons	40

and so on; the same sequence holds for neutrons. Nuclei in which the states that correspond to given quantum numbers (shells) are filled (closed shells) are expected to be particularly stable. According to the list above, closed shells occur for $Z = 2$, $Z = 2 + 6 = 8$, $Z = 8 + (10 + 2) = 20$, $Z = 20 + (14 + 6) = 40$, and so on. Because the same list holds for neutrons, particularly stable nuclei ought to occur in this simple model for Z values or N values of 2, 8, 20, 40, 70, ... These values are called *magic numbers*.

The *experimental* values of Z for which the binding energy is particularly large are 2, 8, 20, 28, 50, 82, 126. These numbers do not represent a perfect match to the predicted magic numbers and indicate that the order of filling levels is different in reality than the simple harmonic oscillator potential would suggest. In 1949, Hans Jensen and Maria Goeppert-Mayer independently pointed out that the energy should contain terms not included earlier that involve both the spin of a single nucleon and its orbital angular momentum. This *spin–orbit coupling* is the same type of coupling that causes splitting of atomic levels (see Section 42–3). By including this type of coupling, splitting of levels occurs, and the ordering of levels is upset. The observed magic numbers are thereby more correctly predicted, as is the complicated pattern of level splittings seen in nuclei. This discussion confirms that the energy levels of nuclei, like those of atoms and molecules, are filled in a manner dictated by the exclusion principle.

The Liquid-Drop Model

The short range of nuclear forces and the small amount of space between the constituent nucleons suggested to Niels Bohr that a nucleus should behave in some ways like a continuous fluid, such as water. The model based on this idea, the **liquid-drop model** (also referred to as a *collective model*), contains no reference to the behavior of individual nucleons and in this sense is complementary to the shell model. It is based on the resemblance between the following two types of data on nuclear physics and the behavior of incompressible fluids. First, the density of all nuclear matter is roughly constant, as is the density of an incompressible fluid. (Indeed, that is what we mean by saying that the fluid is incompressible.) Second, just as the binding energy per nucleon, E_b/A, is roughly constant from nucleus to nucleus, the heat of vaporization per unit mass is constant for different size drops of an incompressible fluid. (The heat of vaporization is the energy required to separate a drop of fluid into its component molecules, and the total mass of a nucleus is proportional to A. Thus E_b/A is analogous to the heat of vaporization per unit mass.)

Following the analogy above, we can write an *empirical* formula for the mass of a nucleus, $M(^AX)$, by including terms of the type that would be included in computing the energy of a liquid drop, as follows:

1. A zeroth-order term that is just the sum of the rest masses of the nucleons:
$$ZM_H + Nm_n = ZM_H + (A - Z)m_n. \tag{45–11}$$
2. A term proportional to the volume of the nucleus (or $\propto A$, as we have shown), which takes into account the nearly constant binding energy per nucleon:
$$-a_V A, \tag{45–12}$$
where a_V is positive, because a binding energy reduces the mass of the nucleus.
3. A term proportional to $A^{2/3}$, or, equivalently, to the surface area of the nucleus:
$$+a_S A^{2/3}. \tag{45–13}$$
A nucleon near the surface is not bound as strongly as one in the interior, so a_S takes into account a reduced binding energy and is positive. This term is the analogue of a surface-tension term in a liquid drop.

The remaining three terms take into account some simple observations specific to the nucleus:

4. A term for the Coulomb repulsion between all the protons in the nucleus. This term tends to increase the mass of the nucleus. If we suppose that the protons are distributed uniformly throughout a sphere of radius $R_A = r_0 A^{1/3}$ [Eq. (45–7)], then the energy required to assemble that charge is
$$\frac{3}{5}\frac{Z^2e^2}{4\pi\epsilon_0 r_0 A^{1/3}} = 0.72\frac{Z^2}{A^{1/3}}\,\text{MeV} \tag{45–14}$$
(see Problem 59 of Chapter 12). The contribution of the energy to the mass is this quantity divided by c^2.
5. A term that has a minimum for $N = Z$ accounts for the tendency for the number of neutrons to equal the number of protons. If terms 1 through 4 were the only terms present, then the energy could be lowered by taking Z to zero; that is, nuclei would consist exclusively of neutrons. In fact, there are no nuclei with many more neutrons than protons, and nuclei are particularly stable when $N \simeq Z$. This term is written as
$$+a_A\frac{(A - 2Z)^2}{A}, \tag{45–15}$$
where a_A is positive.
6. Several terms that describe the tendency for nuclei with even numbers of protons and/or neutrons to be more deeply bound than nuclei with odd numbers of protons and/or neutrons. This tendency occurs because the

spins of two nucleons in each "shell" are antiparallel. Such terms have the empirical form

for Z, N even: $$\Delta = -\frac{a_p}{\sqrt{A}} \quad (45\text{–}16a)$$

for Z even, N odd; or Z odd, N even: $$\Delta = 0 \quad (45\text{–}16b)$$

for Z, N odd: $$\Delta = +\frac{a_p}{\sqrt{A}} \quad (45\text{–}16c)$$

The sum of all these terms is the **semiempirical mass formula:**†

$$M_A = ZM_H + (A - Z)m_n - a_V A + a_S A^{2/3} + \frac{a_C Z^2}{A^{1/3}} + a_A \frac{(A - 2Z)^2}{A} + \Delta. \quad (45\text{–}17)$$

The parameters are determined empirically by fitting this relation with many different nuclear masses. One set of parameters is given in Table 45–3. For these values, the curve in Fig. 45–6 for $A = 20$ and above is in good agreement with Eq. (45–17). For very light nuclei, the notion of a nucleus that looks like a droplet with a volume and a surface loses meaning, and the mass formula does not work very well. The semiempirical mass formula does not take into account minor variations from nucleus to nucleus having to do with the shell structure. Otherwise, the fit is excellent.

TABLE 45–3

PARAMETERS OF THE SEMIEMPIRICAL MASS FORMULA

Parameter	*Value* (u)
a_V	1.7×10^{-2}
a_S	1.8×10^{-2}
a_C	7.5×10^{-4}
a_A	2.5×10^{-2}
a_p	1.3×10^{-2}

A droplet can deform in shape. This changes the energy, because the surface area changes. In fact, droplets oscillate because of this effect, even when the fluid is incompressible. Surface forces tend to return the droplet to a spherical shape, but the Coulomb energy term favors a larger deformation. When we compare these two effects for heavy nuclei, we find that the liquid drop is unstable and breaks up into two smaller droplets. The liquid-drop model thus provides an explanation of the fission process.

EXAMPLE 45–3 (a) Use the semiempirical mass formula to calculate the atomic masses of ^{208}Pb and ^{209}Pb. (b) Calculate the neutron separation energy for ^{209}Pb.

SOLUTION: (a) We use the semiempirical mass formula with the parameters of Table 45–3:

$$\begin{aligned} M(^{208}\text{Pb}) &= (82)(1.007825\text{ u}) + (208 - 82)(1.008665\text{ u}) - (1.7 \times 10^{-2}\text{ u})(208) \\ &\quad + (1.8 \times 10^{-2}\text{ u})(208)^{2/3} + (7.5 \times 10^{-4}\text{ u})\left[\frac{(82)^2}{(208)^{1/3}}\right] \\ &\quad + (2.5 \times 10^{-2}\text{ u})\left\{\frac{[208 - 2(82)]^2}{208}\right\} - \left(\frac{1.3 \times 10^{-2}\text{ u}}{\sqrt{208}}\right) \\ &= 207.912\text{ u}; \end{aligned}$$

$$\begin{aligned} M(^{209}\text{Pb}) &= (82)(1.007825\text{ u}) + (209 - 82)(1.008665\text{ u}) - (1.7 \times 10^{-2}\text{ u})(209) \\ &\quad + (1.8 \times 10^{-2}\text{ u})(209)^{2/3} + (7.5 \times 10^{-4}\text{ u})\left[\frac{(82)^2}{(209)^{1/3}}\right] \end{aligned}$$

† In some instances, the calculated coefficient $0.72/c^2$ of the factor $Z^2/A^{1/3}$ in Eq. (45–14) is replaced by the empirical parameter $a_C \simeq 0.60/c^2$. This is a very small difference in the contribution of the energy to the masses of nuclei. In Table 45–3, we use the empirical form.

$$+(2.5\times 10^{-2}\ \mathrm{u})\left\{\frac{[209-2(82)]^2}{209}\right\}+0$$

$$= 208.915\ \mathrm{u}.$$

(b) To find the neutron separation energy, we write an equation similar to Eq. (45–8):

$$S_n(^{209}\mathrm{Pb}) = [m_n + M(^{208}\mathrm{Pb}) - M(^{209}\mathrm{Pb})]c^2$$
$$= [1.008665\ \mathrm{u} + 207.912\ \mathrm{u} - 208.915\ \mathrm{u}]c^2$$
$$= (0.0057\ \mathrm{u}c^2)(931.494\ \mathrm{MeV}/\mathrm{u}c^2) = 5.3\ \mathrm{MeV}.$$

The experimental values of the masses are $M(^{208}\mathrm{Pb}) = 207.97663$ u and $M(^{209}\mathrm{Pb}) = 208.98107$ u, and the experimental value of the neutron separation energy is 3.9 MeV. The 5.3-MeV separation energy that we calculated from the semiempirical mass equation does not take into account the shell-model magic numbers, for which nuclei are particularly strongly bound. The nuclide ^{208}Pb is an example of a "doubly magic" nucleus ($Z = 82$, $N = 126$), so it should be *very* strongly bound. The nuclide ^{209}Pb gives up its neutron rather easily in order to reach this special configuration. Therefore, the actual value, 3.9 MeV, is below our calculated value.

45–3 ENERGETICS OF NUCLEAR REACTIONS

By nuclear reactions, we generally mean the processes that occur when nuclei interact through collisions. All the possibilities for different types of collisions outlined in Chapter 8 occur. Two nuclei may have an elastic collision or various types of inelastic collisions. A nucleus may absorb energy and then decay by one of several modes, which we shall discuss in Section 45–4. Mass may be transferred from one nucleus to another by the exchange of nucleons. Colliding nuclei may coalesce—a process called **nuclear fusion**. The decay of a single unstable nucleus may be regarded as a nuclear reaction. Examples are α decay, in which a nucleus decays to a helium nucleus and another nucleus; and **nuclear fission**, in which the decay products include two nuclei of more or less equal size. In all these interactions, the conservation laws we have developed throughout this book apply, including the conservations of energy, momentum, angular momentum, and charge.

At the simplest level, collisions between nuclei allow us to measure kinematic quantities such as masses or angular momenta. For studying nuclear reactions, a terminology special to their kinematics has been developed. Consider a nuclear collision in which both the initial and final states consist of two nuclei, a reaction that we write as

$$A + B \to C + D.$$

Energy conservation applies, but because relativity is important in nuclear interactions, we must use relativistic relations. For any one nucleus (or particle) of mass M, this energy takes the form

$$E = K + Mc^2,$$

where K is the kinetic energy of the nucleus. The value of K depends on the frame of reference, but the conservation of energy, $E_A + E_B = E_C + E_D$, applies in any reference frame. If we suppose that nucleus B is the target nucleus, at rest in the laboratory frame of reference, then the conservation of energy takes the form

$$(K_A + M_A c^2) + M_B c^2 = (K_C + M_C c^2) + (K_D + M_D c^2) \qquad (45\text{–}18)$$

in that reference frame. Depending on the masses of the nuclei involved, the reaction can be *exothermic* (having more kinetic energy in the final state than in the initial state) or *endothermic* (having less kinetic energy in the final state). A kinematic parameter that describes such properties is the *Q value* of the reaction:

$$Q \equiv K_C + K_D - K_A - K_B. \qquad (45\text{–}19)$$

Note that the Q value is defined with respect to the laboratory frame of reference. By including the possibility that nucleus B has kinetic energy, we have written a general relation. A positive Q value corresponds to an exothermic reaction, and a negative Q value corresponds to an endothermic reaction. By moving all the kinetic energy terms in Eq. (45–18) to one side of the equation and all the mass terms to the other, we can express the Q value in terms of atomic masses:

$$Q = (M_A + M_B - M_C - M_D)c^2. \qquad (45\text{–}20)$$

In this way, the Q value can confirm measurements made by mass spectrometry. By measuring kinetic energies in a reaction and comparing Eqs. (45–19) and (45–20), we can use the Q value for mass measurements that cannot otherwise be made.

The variety of measurements of different nuclear reactions—including their dependence on different variables of the collision process, such as the projectile energy and angle—gives us much of our knowledge of nuclear structure and nuclear forces. The most useful concept for collisions is the *collision cross section*, which, as we described in Chapter 19, measures the effective area taken up by the colliding nuclei. The cross section is a measure of the probability of interaction; the larger the cross section, the more probable the collision and hence the reaction. We can thus define a cross section *for each possible reaction* that measures the relative probability of each of these processes.

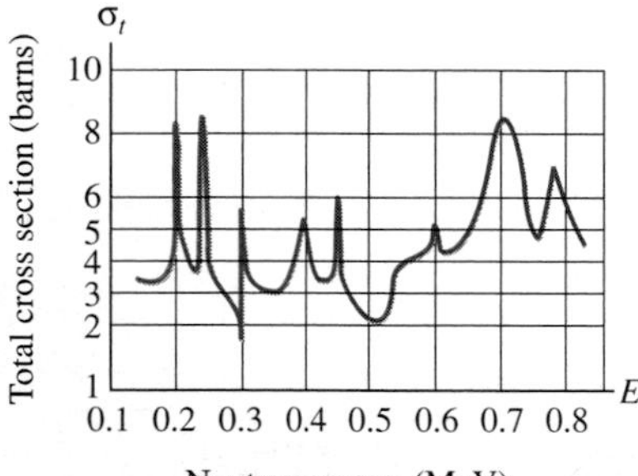

FIGURE 45–10 The total cross section for the absorption of neutrons by ^{23}Na. Energy is measured in the laboratory frame of reference. The peaks, also known as resonances, correspond to excited states of ^{24}Na. (After S. F. Mughabghab et al., *Neutron Cross Sections*, Academic Press, 1984.)

One of the important features of the cross section is the presence of *resonances*, which we define here as bumps in the curve of cross section versus collision energy of the same form as the resonance curve of Fig. 13–21. We see in Fig. 45–10 resonance shapes in the probability (total cross section) for neutron absorption by ^{23}Na. Each peak corresponds to an excited state of the nuclide ^{24}Na, which then decays. *The presence of a resonance peak in a cross section signals the presence of an excited state at that energy.* Indeed, these excited states are often referred to as resonances. The *width* of a resonance peak, ΔE, has a simple interpretation. When a resonance decays with a lifetime τ, the time we have to attempt to measure the energy of the state is limited to τ. Then the Heisenberg uncertainty principle suggests that the position of the resonance in energy is unknown (and unknowable) to a precision ΔE given by

$$\Delta E \cdot \tau \simeq \hbar. \qquad (45\text{–}21)$$

The width of a resonance is inversely proportional to its lifetime. See Chapters 13, 15, and 41.

Thus *the width in energy of the resonance is* $\hbar$ *divided by the lifetime of the resonance*. The shorter-lived the resonance is, the broader the resonance peak is. In nuclear physics, resonances may have widths ranging from about 0.1 eV, corresponding to a lifetime of 10^{-14} s, to as much as 1 MeV, corresponding to a lifetime of 10^{-21} s. In some interactions of nucleons and other particles that participate in the nuclear force, resonances occur with widths as large as 100 MeV. This corresponds to a lifetime of 10^{-23} s, similar to the time it takes for light to cross the nucleus!

(a)

(b)

FIGURE 45–11 (a) Henri Becquerel, who discovered radioactivity after he had left a photographic plate under uranium salts in a closed drawer for several days in 1896. (b) The blurred images that formed on Becquerel's photographic plate—the first evidence of radioactivity.

45–4 RADIOACTIVITY

Henri Becquerel discovered radioactivity in 1896, even before the existence of the nucleus had been established (Fig. 45–11). **Radioactivity**—the name was coined by Marie Curie—is the phenomenon of nuclear decay (Fig. 45–12). Nuclei can decay in a variety of *modes* or *channels*—for example, by emitting photons or α particles. Different physical laws may govern different decay modes, so each mode may have a different probability of occurring. The lifetime, τ, of a collection of radioactive nuclei is a constant with dimensions of time that determines the rate at which a set of such nuclei decay: If N_0 is the number of nuclei present at $t = 0$, then $N(t)$, the number of nuclei that remain after a time t, is given by Eq. (41–21), $N(t) = N_0 e^{-t/\tau}$. An alternate way to write this equation is to define the **decay constant**, λ, as $\lambda \equiv 1/\tau$, so

$$N(t) = N_0 e^{-\lambda t}. \qquad (45\text{–}22)$$

The rate at which nuclei decay, the *decay rate*, is $-\mathrm{d}N/\mathrm{d}t$. From Eq. (45–22),

$$\frac{\mathrm{d}N}{\mathrm{d}t} = -\lambda N_0 e^{-\lambda t} = -\lambda N. \qquad (45\text{–}23)$$

The decay rate λN, also called the *activity*, is high when the number of unstable nuclei is large and when the decay constant is large (the lifetime is small). We also employ the *half-life*, $t_{1/2}$, the time for half of a given sample of unstable nuclei to decay. The half-life is related to the lifetime by $t_{1/2} = 0.693\tau$ (see Problem 48 in Chapter 41). The SI unit of activity is the *becquerel* (Bq): 1 Bq = 1 decay/s.

If a nucleus can decay in two or more ways, then the total decay rate is a sum of decay rates for the various decay modes. Think of several holes in a leaky can: The total rate of flow from the can is the sum of the rates from each of the holes. The decay rate from the mode labeled by the subscript j (for example, j may label the α-decay mode) is proportional to the decay constant λ_j for that mode, so the total decay rate is proportional to the sum of all the decay constants. In other words, there is a total decay constant λ given by

$$\lambda = \sum_j \lambda_j. \qquad (45\text{–}24)$$

FIGURE 45–12 Marie Curie in her laboratory at the Sorbonne, Paris, circa 1908. Curie and her husband, Pierre Curie, performed important early work on radioactivity for which she earned two Nobel prizes.

If the decay in one mode, say, mode 1, is much more rapid than that in all the other modes, then λ_1 is much larger than the other decay constants, and the λ_1 term dominates the sum in Eq. (45–24). In that case, $\lambda \simeq \lambda_1$. It is as though one hole in our leaky can is much larger than the others. We say that the nucleus *decays predominantly through the channel* labeled by λ_1. For example, ^{214}Bi decays 0.021 percent of the time through the α-decay channel and the rest of the time by the emission of an electron.

With the exception of fission reactions (to be discussed in Section 45–5), a typical radioactive decay involves a nucleus (the *parent nucleus*) that emits a particle and is thereby converted into a *daughter* nucleus. Three different types of decays can occur, labeled α, β, or γ according to the particles emitted.

EXAMPLE **45–4** A sample of uranium ore emits α radiation characteristic of ^{235}U at a rate of 9.3×10^5 decays/s. What mass of chemical uranium is present in the ore? The half-life of ^{235}U is 7.04×10^8 yr, and the abundance of naturally occurring ^{235}U in a sample of uranium is 0.72 percent.

SOLUTION: We are given the decay activity and half-life for ^{235}U. From the half-life we can find the decay constant, λ; with λ and the activity we can find the number of ^{235}U nuclei. Then we can use the abundance of ^{235}U to find the total quantity of uranium. We have $\lambda = 1/\tau = 0.693/t_{1/2}$. From the relation for the activity, Eq. (45–23), we have for the number of ^{235}U atoms

$$N = -\frac{1}{\lambda}\frac{dN}{dt} = -\frac{t_{1/2}}{0.693}\frac{dN}{dt}$$

$$= -\frac{(7.04 \times 10^8 \cancel{\text{yr}})(3.16 \times 10^7 \cancel{\text{s}}/\cancel{\text{yr}})}{0.693}(-9.3 \times 10^5 \text{ decays}/\cancel{\text{s}}) = 3.0 \times 10^{22}.$$

Given this number of ^{235}U atoms, we can use Avogadro's number to find the mass m of ^{235}U:

$$m = (3.0 \times 10^{22} \cancel{\text{atoms}})\left(\frac{1 \cancel{\text{mol}}}{6.02 \times 10^{23} \cancel{\text{atoms}}}\right)(235 \text{ g}/\cancel{\text{mol}}) \simeq 12 \text{ g}.$$

Because ^{235}U makes up 0.72 percent of naturally occurring uranium, the total amount of uranium is

$$(1/0.0072)(12 \text{ g}) = 140(12 \text{ g}) = 1.6 \text{ kg}.$$

Alpha Decay

Decay in the **α-decay mode** is written as

$$^A_ZX \rightarrow {}^{A-4}_{Z-2}X' + {}^4_2\text{He}, \qquad (45\text{–}25)$$

where X and X' are the chemical symbols of the parent and daughter nuclei, respectively. Here we have recognized that the emitted α particle is actually a ^{4}He nucleus. An example of this type of decay is $^{238}_{92}\text{U} \rightarrow {}^{234}_{90}\text{Th} + {}^4_2\text{He}$.

Tunneling is discussed in Chapter 41.

Alpha decay involves the quantum mechanical tunneling of an α particle through a potential barrier. For α decay to occur, the mass of the parent nucleus must be greater than the sum of the masses of an α particle and the daughter nucleus. (Any leftover difference in energy is converted to kinetic energy of the α particle and the daughter nucleus.) The systematics of nuclear masses implies that α decay is increasingly more likely to occur as the size of a nucleus increases. Many nuclei above $Z = 83$ are unstable in this mode.

Nuclear forces are understood well enough to allow us to predict the probabilities of α decay. Because the rate at which tunneling occurs is extremely sensitive to the width and height of the potential barrier, the range of lifetimes of nuclei in the α-decay mode is huge, from 3×10^{-7} s to 1.4×10^{17} s!

EXAMPLE 45–5 ^{241}Am decays in the α-decay mode: $^{241}\text{Am} \rightarrow \alpha + {}^{237}\text{Np}$. Calculate the kinetic energy of the α particle if the ^{241}Am nucleus decays at rest and the atomic masses are $M(^{241}\text{Am}) = 241.05682$ u, $M(^{4}\text{He}) = 4.002603$ u, and $M(^{237}\text{Np}) = 237.04817$ u.

SOLUTION: Both conservation of energy and conservation of momentum must be applied here. From the conservation of energy, we have

$$M(^{241}\text{Am}) = M(^{4}\text{He}) + M(^{237}\text{Np}) + \frac{K_f}{c^2},$$

where K_f is the kinetic energy of the two-body final state. Thus

$$\begin{aligned} K_f &= [M(^{241}\text{Am}) - M(^{4}\text{He}) - M(^{237}\text{Np})]c^2 \\ &= (241.05682\text{ u} - 4.002603\text{ u} - 237.04817\text{ u})c^2 \\ &= 0.006047\text{ u}c^2 = (0.006047\,\cancel{\text{u}c^2})(931.474\text{ MeV}/\cancel{\text{u}c^2}) = 5.63\text{ MeV}. \end{aligned}$$

This kinetic energy is much less than the rest mass of either final nucleus, so the motion can be treated nonrelativistically. $K_f = K(^{4}\text{He}) + K(^{237}\text{Np})$, where $K = p^2/2M$ of the respective nuclei. The conservation of momentum applied to the final state gives $0 = p(^{4}\text{He}) + p(^{237}\text{Np})$, or $p(^{237}\text{Np}) = -p(^{4}\text{He})$. Thus

$$\begin{aligned} K(^{4}\text{He}) &= K_f - K(^{237}\text{Np}) = K_f - \frac{p(^{237}\text{Np})^2}{2M(^{237}\text{Np})} = K_f - \frac{[-p(^{4}\text{He})]^2}{2M(^{237}\text{Np})} \\ &= K_f - \frac{p(^{4}\text{He})^2}{2M(^{4}\text{He})}\frac{M(^{4}\text{He})}{M(^{237}\text{Np})} = K_f - K(^{4}\text{He})\frac{M(^{4}\text{He})}{M(^{237}\text{Np})}. \end{aligned}$$

We can solve this relation for $K(^{4}\text{He})$:

$$K(^{4}\text{He}) = \frac{K_f}{1 + \left[\dfrac{M(^{4}\text{He})}{M(^{237}\text{Np})}\right]} = \frac{5.63\text{ MeV}}{1 + (4.00\text{ u}/237\text{ u})} = 5.53\text{ MeV}.$$

The ^{237}Np nucleus is so much more massive than the α particle that it moves very slowly, and the α particle carries off most of the kinetic energy.

Beta Decay

In the **β-decay mode** of a nucleus, an electron and an *antineutrino* ($\bar{\nu}$) (or a positron and a neutrino; see below) are emitted, leaving a daughter nucleus.† The antineutrino is the antiparticle of the neutrino (ν), just as the positron is the antiparticle of the electron.‡ The properties of the antineutrino and the neutrino are, for our purposes, identical.

The neutrino was introduced in Chapter 6.

Wolfgang Pauli postulated the existence of the neutrino in 1930 on the basis of experimental data on β decay. An important feature of the neutrino, which

† In early usage, the emitted electron was called a beta particle.

‡ Strictly speaking, the neutrino of β decay is a special variety of neutrino, the electron neutrino. But this distinction plays no role in our discussion.

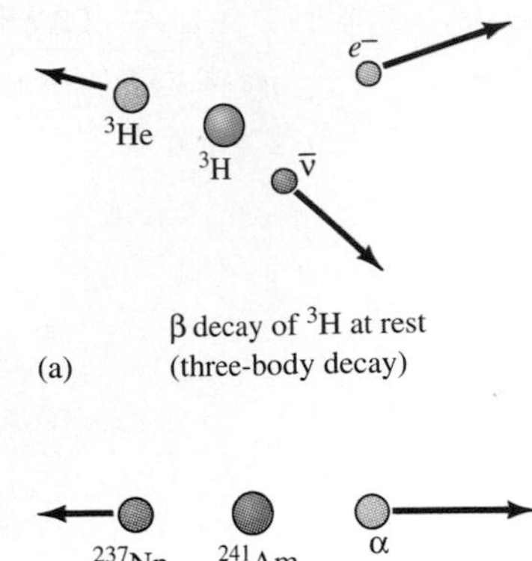

FIGURE 45–13 (a) Beta decay is a three-body decay, as opposed to (b) α decay, a two-body decay. In each case the parent nucleus is at rest. In the three-body decay, a variety of directions and energies are possible for the decay products, subject to overall energy and momentum conservations. In the two-body decay, the decay products emerge back-to-back and with fixed energies.

has zero charge and spin $\hbar/2$, is that it interacts so weakly with matter that any one neutrino will traverse *light-years* of ordinary matter before the probability of interacting with the matter is significant. It is therefore not surprising that the neutrino was not *independently* observed to initiate a collision process until 1956, 60 yr after β decay was first observed. As a result of the neutrino's weak interaction, what is actually observed in β decay is the electron and (sometimes) the daughter nucleus but never the neutrino. Finally, the neutrino has one other important property: It is virtually massless.[†]

When an electron is emitted in nuclear β decay, the daughter nucleus has a charge $+e$ more than that of the parent nucleus; that is, *Z increases by 1 in this β-decay process.* As we did in Eq. (45–25), we write this β-decay process symbolically as[‡]

$$^A_ZX \to {}_{Z+1}^{\ \ A}X' + e^- + \bar{\nu}. \tag{45–26}$$

An example of this type of decay is $^3\text{H} \to {}^3\text{He} + e^- + \bar{\nu}$. In this form of β decay, a neutron has in effect been converted to a proton, an electron, and an antineutrino within the nucleus. Indeed, the most primitive β decay is that of the neutron itself:

$$n \to p + e^- + \bar{\nu}. \tag{45–27}$$

For *free* neutrons, the lifetime for this process is approximately 890 s. Thus neutrons by themselves are unstable. In contrast, neutrons within nuclei that do not undergo β decay are stable. The primary β-decay mode, Eq. (45–26), is possible only for the mass condition

$$M_P > M_D \tag{45–28}$$

(see Problem 27). We have assumed here that the neutrino is indeed massless; M_P and M_D are the atomic masses of the parent and daughter nuclei, respectively. For nuclei that undergo β decay, the kinetic energy of the electrons ranges up to 10 MeV, with 1 MeV being typical.

If the neutrons in β decay cannot be observed directly, how was it possible for Pauli to predict their existence and properties as early as 1930? The existence of the neutrino was inferred on the basis of several types of conservation laws. First, conservations of momentum and of energy have rather different consequences for three-body decay (one in which the parent nucleus decays to three particles; Fig. 45–13a) than for two-body decay (Fig. 45–13b). In two-body decay of a nucleus at rest, the energies of each of the two decay products are fixed: The conservations of momentum and of energy *uniquely* determine the magnitudes of the momentum and the energy of the decay products. This is not true in three-body decay, in which different configurations of the three bodies allow for a range of energies for all three bodies. Various configurations allow the electron's kinetic energy to range from zero up to a maximum value, as shown in Fig. 45–14. A similar plot for a two-body decay would show a single energy value at the energy maximum K_{max}.

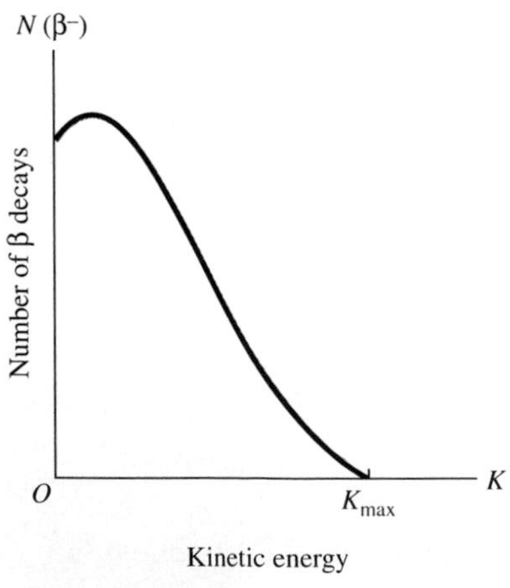

FIGURE 45–14 The energy of the electron for a large sample of β decays. The energy is not fixed, but rather spread, a consequence of the kinematics of three-body decay. The maximum possible electron energy, K_{max}, is the energy the electron would have if there were no neutrino in the decay.

Pauli surmised from the broad curve observed in Fig. 45–14 that three particles—including the otherwise invisible neutrino—must be present.[§] Angular

[†] More precisely, the best experiments indicate that the neutrino's mass is at most 17 eV/c^2, a value only 3×10^{-5} times the electron's mass.

[‡] The daughter nucleus produced in the various forms of β decay is typically left in an excited nuclear state and will then itself decay in some mode.

[§] When the broad electron spectrum was first observed, some physicists suggested that it would be necessary to abandon the principle of the conservation of energy.

momentum conservation leads to the same conclusion. If the decay of the neutron were the two-body decay $n \rightarrow p + e^-$ rather than a three-body decay, we would have a discrepancy. Neutrons, like protons and electrons, have a spin of $\hbar/2$, but the rules of angular momentum addition do not allow two $\hbar/2$ spins to add to a total spin of $\hbar/2$. The addition of a third particle with a spin of $\hbar/2$ can resolve the discrepancy; in this way Pauli was able to predict that the invisible, massless third particle, the neutrino, was a fermion.

Lifetimes of β decays vary from about 1 s to more than 10^{20} s. The energy dependence and low rate of β decay of some nuclei suggest that the β-decay process is not a quantum mechanical tunneling process but rather a primary manifestation of the *weak force* (one aspect of the electroweak force; see Chapter 5). By using the methods of quantum mechanics, we can compute to good accuracy the lifetimes and other properties of β decays in terms of postulated properties of the weak force.

We can mention two other forms of β decay here, both of them consequences of the weak force. The first is *positron emission*, which takes the form

$$ {}_{Z+1}^{A}X \rightarrow {}_{Z}^{A}X' + e^+ + \nu. \qquad (45\text{–}29)$$

This process is allowed under the mass condition

$$M_X > M_{X'} + 2m_e. \qquad (45\text{–}30)$$

The second variant is *electron capture*, in which an electron from an atomic orbit is absorbed by the nucleus:

$$ {}_{Z+1}^{A}X + e^- \rightarrow {}_{Z}^{A}X' + \nu. \qquad (45\text{–}31)$$

Electron capture is possible under the mass condition

$$M_X > M_{X'}. \qquad (45\text{–}32)$$

Note that the mass conditions for the primary decay mode and electron capture are identical. However, β decay tends to occur for neutron-rich nuclei, and electron capture for neutron-poor nuclei. Practically all the kinetic energy in electron capture goes to the neutrino, because of the conservation of energy and linear momentum (see Example 45–5).

Gamma Decay

A nucleus in an excited state can decay in the **γ-decay mode**: It emits photons as it decays to its ground state or to lower-lying excited states (Fig. 45–15), just as an atom in an excited electronic state can. Because the typical energy differences between nuclear excited states and the ground state are some 10^6 times greater than the corresponding energy differences in atoms, lighting a match is not enough to excite the nucleus, as it is to excite an atom. However, daughter nuclei are often produced in their excited states in α, β, or even γ decay, and these residual states can then decay by the γ-decay mode. Photons with energies as high as a few MeV were not at first recognized as electromagnetic radiation and were therefore given a special name, γ rays; we speak of the γ-decay mode of the nucleus. The lifetimes of γ decay range from 10^{-17} s to as much as 10^8 s.

The electromagnetic forces responsible for the properties of the γ-decay modes are well known. Therefore we can use observations of lifetimes and the energies and angular distributions of the emitted photons as sensitive indicators of the quantum numbers and nature of the decaying excited states. In this way models (such as the shell model) and their predictions for the energies and quantum numbers of the excited states can be tested or developed.

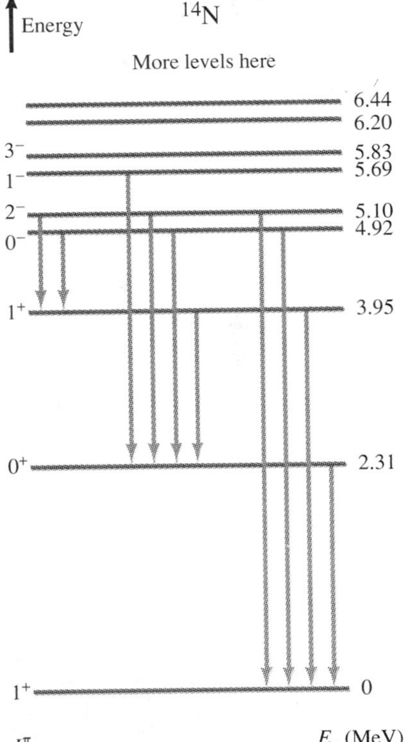

FIGURE 45–15 Some γ-ray decay transitions in ^{14}N. The notation J^π describes the respective quantum states labeled by quantum numbers not discussed here. E_x labels the energy of the states relative to the ground state. Not all possible transitions are shown. Many more levels exist above 6.44 MeV in ^{14}N.

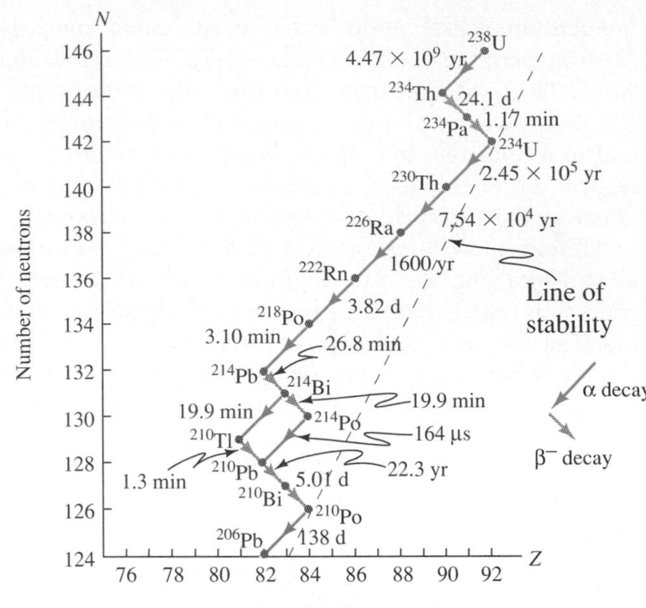

FIGURE 45–16 The decay series of ^{238}U plotted on a graph of N versus Z. Half-lives of each nuclide are also shown. Only one of the four different branchings of this series is shown.

Natural Radioactivity

It is often true that the daughter nucleus produced in a radioactive decay is itself unstable. The result is a *radioactive series*, in which a succession of nuclei decay in a cascade to some stable nucleus or nuclei. Along the steps of the cascade, some nuclei may have two competitive decay modes, such as an α-decay mode and a β-decay mode, in which case we say that there is *branching*. Figure 45–16 shows the uranium series, which starts with ^{238}U. It generally exists on the neutron-rich side of the line of stability (discussed in Section 45–1), also visible in the figure. Beta decays pull the branching line closer to the line of stability. Of the three possible decay modes, only α decay changes A, and by 4 units. Therefore if a series starts with a parent whose A value has the form $A = 4n$, where n is an integer, then all the nuclides with $A = 4n - 4$, $4n - 8$, and so forth will be encountered, until a stable nucleus is obtained. Similarly, if the starting nucleus has $A = 4n + 1$, then nuclei with $A = 4n - 3, 4n - 7, \ldots$ are reached. A third series with $A = 4n + 2$ and a fourth with $A = 4n + 3$ are also possible, but the "fifth" series, with $A = 4n + 4$, is identical to the $A = 4n$ series. Thus *there are only four different radioactive series*. The series shown in Fig. 45–16 for $^{238}_{92}$U is the $A = 4n + 2$ series; a series starting with $^{232}_{90}$Th is the $A = 4n$ series.

The lifetimes of the original parent nuclei are 6.4×10^9 yr for $^{238}_{92}$U and 2.0×10^{10} yr for $^{232}_{90}$Th. The fact that these numbers are comparable to the age of the Solar System explains why we can still find these nuclides in nature: If the lifetimes of the parent nuclei were shorter, all the parents would have decayed by now. Moreover, the fact that the daughter nuclei are continuously being replenished through decays of the parent nuclei explains why we find extremely short-lived nuclei in nature. For example, in the decay series starting with $^{238}_{92}$U, the sixth step produces $^{222}_{86}$Rn, and this nucleus undergoes α decay to $^{218}_{84}$Po with a half-life of 3.8 d, certainly much less than the age of the Solar System. Yet $^{222}_{86}$Rn can be found by chemical means in uranium ore. The general problem of how much of some daughter nucleus exists at any given time for a certain amount of parent nucleus in a known decay chain is solvable; it involves the solution of differential equations (see Problem 38).

The end product of a radioactive series is a stable nucleus. The series starting with $^{238}_{92}U$ includes 14 steps with 4 different branchings. In the series starting with $^{232}_{90}Th$, there are 10 steps with 1 branching. Despite the branchings, there is a unique end product in each case: The stable nuclides $^{206}_{82}Pb$ and $^{208}_{92}Pb$, respectively.

Radioactivity and Life†

Radioactivity has significant effects on biological systems. Some effects are desirable—medical diagnosis and treatment, for example—and some are just the opposite. Undesirable effects occur when one of the forms of radiation deposits energy in a normal living cell. That energy is typically absorbed when ionization occurs. Chemical bonds are broken, destroying needed substances in the cell. New ions may induce abnormal chemical reactions. We can distinguish three levels of damage: The chemical functioning of cellular processes may be so badly interrupted that the cell dies or is unable to reproduce; internal controls on the cell's behavior may be lost, so that the cell becomes cancerous; or the gene structure of the cell may be changed, resulting in abnormal offspring.

The danger of a given radioactive substance is associated with the activity, the energy of the radioactive decay products, and the amount of that energy deposited in the body. One measure of possible damage is the energy of the radiation absorbed per mass of the organism. The corresponding SI unit is the *gray* (Gy): $1\ \text{Gy} \equiv 1\ \text{J/kg}$. A dose of several Gy over the entire human body may be lethal, and 10 Gy or more is lethal within a short period. Because damage depends on just how the energy is deposited, a more precise measure takes into account the type of radioactive decay products that produce a given dose. For example, because α particles ionize more effectively, they are more likely to produce severe damage than β rays or γ rays are. (Conversely, they are less likely to penetrate tissue.) These considerations are incorporated (in a complex way) in a modification of the gray called the *sievert* (Sv), also an SI unit.‡ However, for practical purposes we can think of the gray and the sievert as roughly equivalent.

These measures are tempered by still other factors; for example, some cells, especially those that reproduce frequently, are more susceptible to radiation damage than others are. Even a small dose can lead to long-term damage. In deciding whether a risk from radioactivity is worth a possible benefit, recall that we are constantly bombarded by cosmic rays, and that both the earth and elements in our own bodies have some natural radioactivity. We receive a dose of some 10^{-3} Sv/yr from such sources. Exposure to radioactivity from human-made sources must be compared to this natural background. For example, an inhabitant of the United States receives on average less than half the natural dose in diagnostic X-rays, and the background radioactivity associated with nuclear power generation is only about 3×10^{-5} of this value. Nevertheless, a nuclear disaster would create a much greater risk from radiation exposure. Because we know little about the long-term effects of low-level radiation, we are conservative about the doses that are allowable; the recommended limit for whole-body exposure is 0.05 Sv/yr.

† For a brief article on this subject, see S. C. Bushong, "Radiation Exposure in Our Daily Lives," *The Physics Teacher* (March 1977); A. C. Upton, "Health Effects of Low-Level Ionizing Radiation," *Physics Today* (August 1991). For more detail, see E. L. Alpen, *Radiation Biophysics*, Prentice Hall, 1990.

‡ Two non-SI units analogous to the gray and the sievert in customary usage are the *rad* and the *rem*, respectively: 1 rad = 0.01 Gy; 1 rem = 0.01 Sv.

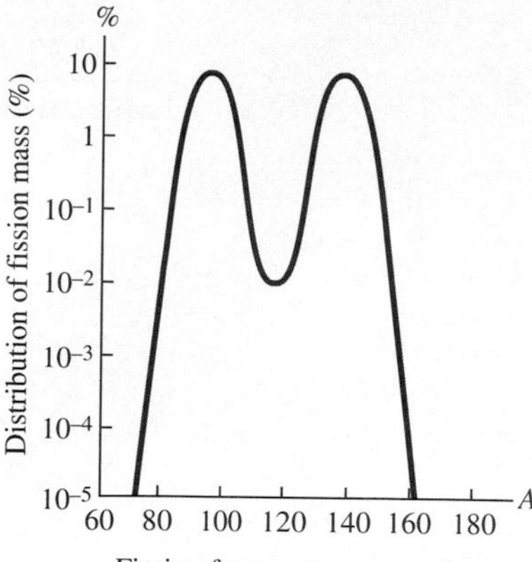

FIGURE 45–17 The relative probability of producing a fragment of atomic number A when a ^{236}U nucleus undergoes fission. When slow neutrons induce fission, a variety of daughter nuclides (fragments) are produced. Note that it is *less* likely that fragments with exactly half the size of the parent, in this case ^{236}U, are produced than that the fragments are of unequal size. (After W. E. Burcham, *Nuclear Physics*, McGraw-Hill, 1963.)

45–5 FISSION AND FUSION

Fission

We have mentioned that α decay is a tunneling process. It is a type of **fission**, a process in which a nucleus breaks into two pieces of more or less equal size. Fission occurs in large nuclei because the energetics favor it. Two terms in the liquid-drop expression for the energy of a nucleus help us to understand why fission occurs. When a large nucleus breaks into two fragments, the sum of the Coulomb repulsion terms from Eq. (45–14) for the two pieces is considerably less than that of the corresponding single term of the parent nucleus. The Coulomb term alone implies that the most energetically favorable fission process is one in which the parent nucleus breaks into two equal pieces. Other terms in the energy take into account detailed differences. Some nuclei are bound more strongly than others. In real fission processes, the two fragments are typically of different sizes (Fig. 45–17).

EXAMPLE **45–6** Find the difference in energies due to the Coulomb term alone between a $^{238}_{92}$U nucleus and two $^{119}_{46}$Pd (palladium) nuclei.

SOLUTION: The difference in energies due to the Coulomb energy term, Eq. (45–14), for a parent nucleus with Z protons and A nucleons and the sum of the same terms for each of the two daughter nuclei with $Z/2$ protons and $A/2$ nucleons is

$$\Delta E = \frac{3}{5}\frac{Z^2e^2}{4\pi\epsilon_0 r_0 A^{1/3}} - 2\left[\frac{3}{5}\frac{(Z/2)^2e^2}{4\pi\epsilon_0 r_0 (A/2)^{1/3}}\right] = (1-2^{-2/3})\frac{3}{5}\frac{Z^2e^2}{4\pi\epsilon_0 r_0 A^{1/3}}.$$

We are interested in the case $Z = 92$ and $A = 238$. Numerical evaluation gives

$$\Delta E \simeq (1-2^{-2/3})(9.8\times10^8\ \text{eV}) = 3.6\times10^8\ \text{eV} = 360\ \text{MeV}.$$

This is a substantial amount of energy even for a nuclear reaction. But it is not the whole story, because other terms contribute to energy differences of this type. It does, however, give a reasonable order-of-magnitude estimate.

If the Coulomb energy destabilizes large nuclei, how can such nuclei exist? The presence of an energy proportional to the surface of a liquid drop provides a barrier against breakup. The energy term is positive and is minimized when the nucleus is spherical [Eq. (45–14)]. In Fig. 45–18 we show a sequence in which a large liquid-drop nucleus formed by neutron absorption in ^{235}U breaks into two fragments. As the fragments start to separate, the surface, and hence the surface energy, *increases* without the Coulomb energy changing very much. As the fragments separate further, the surface-energy term makes it energetically favorable for the separating fragments to "neck off." It is only when the nuclei have separated into two spheres that the surface energy no longer changes. The fact that the sum of surface and Coulomb energies initially increases as the fragments start to separate means that there is a potential barrier. This barrier is on the order of 5 MeV for nuclei such as uranium.

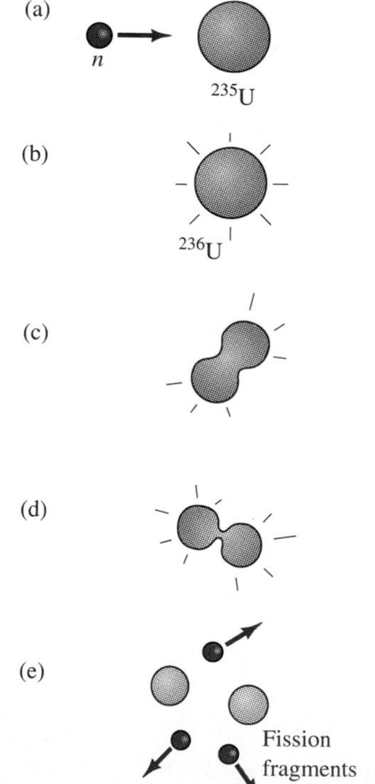

FIGURE 45–18 The fission process. (a), (b) The ^{236}U nucleus produced by neutron bombardment of ^{235}U is in unstable equilibrium, and (c)–(e) breaks into two daughter nuclei. Several free neutrons are also produced.

The presence of a potential barrier implies the possibility of fission through quantum mechanical tunneling. This process, called *spontaneous fission*, does indeed occur, although at a slow rate. Another process, called *induced fission*, occurs when neutrons are projected toward and captured by heavy nuclei such as uranium. The resulting nucleus is an excited state, with the added neutron having a binding energy of about 5 MeV. This is about the height of the potential barrier against fission, and the new nucleus thus has enough energy to undergo fission easily. For ^{233}U and ^{235}U, the energy is carried *above* the potential barrier, so fission is guaranteed; for ^{238}U, the neutron must also supply at least 1 MeV of kinetic energy. Induced fission is important in the possibility of using sustained fission for power production (we shall explore this possibility in Section 45–6).

Fusion

Heavy elements release energy during fission because the curve of binding energy per nucleon decreases as A increases for large A values (see Fig. 45–6). Conversely, this curve increases as A increases for *small* A values; light elements thereby release energy during **fusion**, the combining of small nuclei into larger nuclei. An example is the combining of free nucleons into nuclei: 2.23 MeV of energy is released when a free proton and a free neutron combine and form a deuteron, ^{2}H. More typically, the combining nuclei are each charged, and because the Coulomb force has long range whereas the nuclear force does not, each of the combining nuclei must have considerable energy in order to surmount the potential barrier due to the Coulomb force. Once the potential barrier is passed and the nuclei come close enough to fuse, the energy released by their fusion is much more than the total kinetic energy of the nuclei.

Fusion reactions are central to the "burning" processes of stars. These reactions occur in cycles of reactions or decays, the primary example of which is the *proton cycle*:

$$p + p \rightarrow {}^2\text{H} + e^+ + \nu, \quad \text{with 0.4 MeV of energy released;} \qquad (45\text{–}33a)$$

$$^2\text{H} + p \rightarrow {}^3\text{He} + \gamma, \quad \text{with 5.5 MeV of energy released;} \qquad (45\text{–}33b)$$

$$^3\text{He} + {}^3\text{He} \rightarrow {}^4\text{He} + 2p + \gamma, \quad \text{with 13.0 MeV of energy released.} \qquad (45\text{–}33c)$$

The internal temperature of stars is high enough to give some colliding nuclei sufficient energy to overcome the Coulomb repulsion involved in all three of these reactions. The *net* effect of this cycle is to convert four protons into an α particle along with the emission of energy in the form of photons and a neutrino. Although the energy given to the neutrino is lost (in the sense that the neutrino does not interact further), the photons continue to heat the interior. This is the source of the energy that eventually arrives on earth from our own star, the sun.

EXAMPLE 45–7 Calculate the total energy released each time a ^{4}He nucleus is produced in the proton cycle. To the energies described in Eqs. (45–33), add the energy released when each of the 2 positrons produced annihilates with 2 electrons (already present in the star), producing 4 photons and a kinetic energy equivalent to four times the electron mass—the positron and electron have the same mass, 0.51 MeV/c^2—namely, 2.0 MeV: $2e^+ + 2e^- \rightarrow 4\gamma + 2.0$ MeV.

SOLUTION: For the third reaction of the cycle, Eq. (45–33c), two ^{3}He nuclei are present, so the reaction of Eq. (45–33b) must occur twice. Similarly, the reaction of Eq. (45–33a) must also occur twice. In effect, the chain of reactions that produces one ^{4}He nucleus is

$$
\begin{aligned}
&2(p+p)+p+p+2e^- \to 2(^2\text{H}+e^+ +\nu+0.4\text{ MeV})+p+p+2e^-\\
&\to 2(^3\text{He}+\gamma+5.5\text{ MeV})+(2e^+ +2\nu+0.8\text{ MeV})+2e^-\\
&\to (^4\text{He}+2p+\gamma+13.0\text{ MeV})+(2\gamma+11.0\text{ MeV})\\
&\quad +(2e^+ +2\nu+0.8\text{ MeV})+2e^-\\
&\to (^4\text{He}+2p+\gamma+13.0\text{ MeV})+(2\gamma+11.0\text{ MeV})+(2\nu+0.8\text{ MeV})\\
&\quad +(4\gamma+2.0\text{ MeV}) = {}^4\text{He}+2p+7\gamma+2\nu+26.8\text{ MeV}.
\end{aligned}
$$

Notice that 2 protons are left; these protons can contribute anew to the proton cycle.

EXAMPLE 45–8 The total power output of the sun, its *luminosity*, is about 3.9×10^{26} W. Assume that the sun is made entirely of protons, that its luminosity remains constant, and that it "burns" protons via the proton cycle until all the protons have been converted into α particles. How long can the sun burn in this way? You may use the results of Example 45–7.

SOLUTION: From the results of Example 45–7, 4 protons are required to produce one α particle, so the energy released *per proton* is 26.8 MeV/4 = 6.7 MeV. We know the total mass of the sun, 2.0×10^{30} kg, so the number of protons in the sun is $N_p = (\text{mass of sun})/m_p = (2.0 \times 10^{30}\text{ kg})/(1.67 \times 10^{-27}\text{ kg/proton}) = 1.2 \times 10^{57}$ protons. We can find the number of protons converted per second, N, from the sun's total power:

$$N = \frac{3.9 \times 10^{26}\ \cancel{\text{J}}/\text{s}}{(6.7\ \cancel{\text{MeV}}/\text{proton})(1.602 \times 10^{-13}\ \cancel{\text{J}}/\cancel{\text{MeV}})} = 3.6 \times 10^{38}\text{ protons/s}.$$

The time it would take for all of the sun's protons to be burned can now be found by dividing the number of protons by their rate of use in the proton cycle:

$$\frac{1.2 \times 10^{57}\ \cancel{\text{protons}}}{3.6 \times 10^{38}\ \cancel{\text{protons}}/\text{s}} = 3.3 \times 10^{18}\text{ s}.$$

This amounts to about 10^{11} yr, or 100 billion yr. The actual burning time for the sun is more like 10 billion yr (half of which has passed), because only the innermost 10 percent of the sun's mass is hot enough to burn through the proton cycle.

45–6 APPLICATIONS OF NUCLEAR PHYSICS

Radiometric Dating

In Chapter 41, we discussed how the radioactive nucleus ^{14}C can be used to date biological materials. The lifetime involved is 8268 yr, and that is the rough order of magnitude of ages that can be measured with this method (Fig. 45–19). A second set of radioactive decays, with much longer lifetimes, can be used to date longer geological times, in the range 10^8 yr to 10^{10} yr. Rocks form when liquid magma solidifies; after that point, chemically induced separation of elements is more difficult. As long as there is no lead in a sample of rock when that rock forms, any ^{206}Pb found in the rock must be the product of ^{238}U decays (for which $\tau = 4.5$ billion yr). Therefore the ratio of ^{238}U to ^{206}Pb in the rock allows us to find the rock's age. To use such a technique, we must have confidence that there was no lead in the liquid magma that formed the rock initially. We can

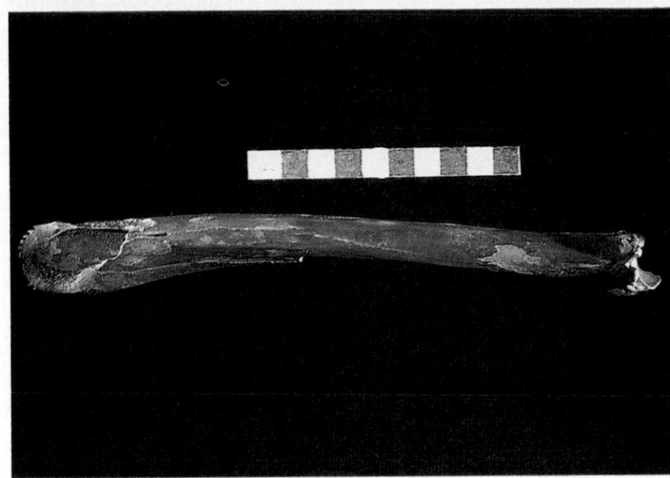

FIGURE 45–19 A caribou bone used in the Yukon Territory for fleshing hides. The carbon in this implement was radiometrically dated to be 1350 ± 150 yr old.

similarly measure the percentages of radioactive intermediate products of the ^{238}U decay chain.

A rather different technique measures the α particles of α decay in the rocks whose age we want. These rocks form with no helium, which would have escaped as a gas from the liquid phase of the rock, and with a certain percentage of heavy elements such as ^{238}U. However, once the rock solidifies, any helium that might appear as a result of one or several sequential α decays of the uranium may be trapped. When a ^{238}U nucleus decays sequentially to ^{206}Pb, eight α particles are produced. By measuring the ratio of the percentage of uranium to that of helium in a rock, we can calculate the date at which the helium started to accumulate. To use this technique, we must be confident that helium produced by radioactive decay *remains* in the rock.

Radioisotopes

Radioisotopes, unstable isotopes produced in nuclear reactions, have characteristic lifetimes for decay. Atoms with these nuclides behave chemically like stable isotopes of the same atom. Observation of the location of the decay of a radioisotope helps us to understand where a chemical or physical process has taken the radioisotope. The applications of radioisotopes include the use of ^{153}Gd, ^{67}Ga, ^{201}Pb, and ^{123}I in medical diagnosis—a cancerous tumor, for example, may concentrate one of these elements, and the observed location of α decays indicates the tumor site; ^{60}Co in cancer treatment—the nuclide can be placed in a tumor and, upon decaying, deposits large amounts of energy that kills cells there; and ^{85}Kr in leak detection—krypton can penetrate very small cracks, and decays indicate the location of those cracks.

Nuclear Power Generation

Here we discuss two ways in which nuclei can be used for electric power generation. The first, using the fission process, is a proven, well-established technology. The second, using the fusion process, is not yet commercially feasible. In both cases, the energy released in the respective nuclear reactions is converted through collisions to thermal energy, which is then used to turn turbines and generate electric power. We shall not concern ourselves further with this aspect of the process.

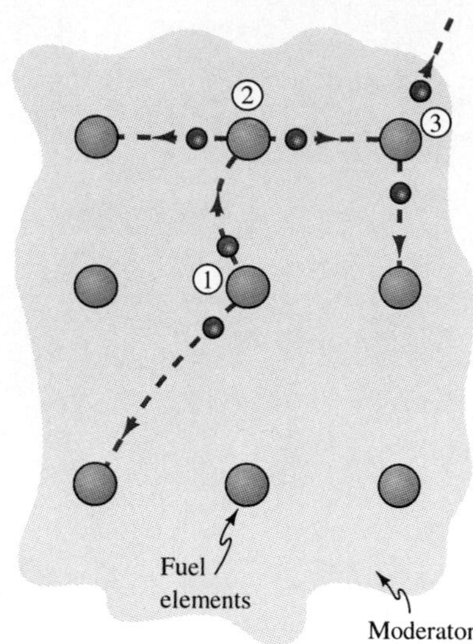

FIGURE 45–20 A chain reaction, in which the free neutrons produced in a fission process induce the fission of still other nuclei, indicated in the sequence ①, ②, ③. The moderator slows down the released neutrons and increases the likelihood that they will produce another fission.

Controlled Fission. It is possible to generate energy from fission because the proportion of mass in neutrons is larger for heavy nuclei such as those of uranium than for light nuclei. (In Example 45–6, we treated the decay of ^{238}U into two ^{119}Pd nuclei, but in fact the heaviest *stable* isotope of palladium has $A = 110$.) This fact, a consequence of the Coulomb repulsion of protons, implies that when a parent nucleus undergoes induced fission into two fragments, the daughter nuclei have relatively too many neutrons to be stable. They may undergo β decay, or, more typically, a few neutrons may be produced along with the two daughter fragments in fission (Fig. 45–18e).

The induced fission process tends to occur more readily when the bombarding neutrons are slow. Although the neutrons that are *produced* in the fission process generally have too much kinetic energy to induce efficiently the fission of other uranium nuclei, their kinetic energy can be reduced by allowing them to collide within a *moderator*, a material whose atoms have light nuclei: The recoiling neutrons lose a substantial fraction of their kinetic energy in such collisions. Once this has happened, these neutrons can induce other fission processes. Of course, some neutrons do not do so, but in each fission reaction as many as 2 to 3 neutrons are produced, on average. If enough fissionable nuclei (^{233}U or ^{235}U, for example) are present, so that at least one released neutron induces another fission, the process can be sustained. We thereby have a *chain reaction* (Fig. 45–20), which takes place within a *reactor*.

By increasing or decreasing the relative amounts of either fissionable nuclei or moderating material in the form of "control rods," the reaction process can be either speeded up or slowed down and thus satisfactorily controlled. The problem of disposing of the waste products has proven to be more difficult to resolve. These waste products, which include the fission fragments themselves, are typically radioactive and have lifetimes that are long compared to a human lifespan. Because radioactivity can cause genetic damage and cancer, we must dispose of such waste material in such a way that it is isolated from humans and other living things for many generations.

A Natural Reactor. There is evidence that sufficiently rich deposits of uranium occur in nature to have sustained a "natural" chain reaction. Such ore deposits must have not only a large concentration of uranium but also deuterium—which is contained in water—to act as a moderator. The ^{235}U-to-^{238}U ratio that exists today is not large enough for a natural reactor to operate. But because ^{235}U decays with a shorter lifetime than ^{238}U does, we know that ^{235}U was relatively more abundant in the past, and ore bodies could have sustained a chain reaction at earlier times. The remnants of a possible *natural reactor* that shut itself down after a certain amount of ^{235}U was burned have been discovered in Africa.[†] The evidence includes a depleted supply of ^{235}U and a characteristic distribution of daughter nuclei descended from fission products.

Controlled Nuclear Fusion. Controlling the processes that power the sun for direct power generation on earth is an old idea. Deuterium, a hydrogen atom with a ^{2}H nucleus, which is abundant in sea water, could supply fuel for the fusion reaction, and the radioactive waste problem for nuclear fusion appears to be less acute than that for fission. But the technology of controlling fusion reactions has proved to be a difficult one to master. The essence of the problem is not simply to make nuclei collide with enough energy to overcome the

[†] See G. A. Cowan, "A Natural Fission Reactor," *Scientific American*, **235**: 36 (July 1976).

Coulomb repulsion barrier and fuse, but to make such collisions occur in large numbers. A collection of nuclei could in principle be brought to energies high enough to fuse by heating them, but the temperatures required to do so correspond to energies on the nuclear scale rather than the atomic scale. We can estimate this energy to be $E \simeq e^2/4\pi\epsilon_0 R$, the Coulomb energy between two singly charged nuclei separated by a nuclear radius R, where R is on the order of 10^{-14} m. Upon substituting, we find that $E \simeq 150$ keV. This energy corresponds to a temperature of about 10^9 K, far higher than any ordinary container could withstand. Indeed, at such temperatures, atoms lose their electrons and matter breaks down to a fully ionized gas of electrons and nuclei, a *plasma*. Although in the Maxwell distribution there are enough particles with high speeds so that a temperature of some 10^8 K is still sufficient to produce reactions, this temperature is still much too high to allow for conventional treatment.

Before we discuss current ideas of how to overcome the problem of handling hot plasma, let us discuss the reactions themselves. The most promising of these reactions in the so-called D–T reaction. The deuteron, D, is the ^{2}H nuclide, and the triton, T, is the ^{3}H nuclide. (The atom with a triton nucleus is tritium.) The D–T fusion reaction is

$$D + T \rightarrow {}^4\text{He} + n.$$

The Q value of this reaction is 17.6 MeV. A second reaction of possible interest is the D–D reaction,

$$D + D \rightarrow {}^3\text{He} + n,$$

in which a total of 4 MeV of kinetic energy is produced. In addition to the fact that more energy is produced in the D–T reaction, the cross section for the D–T reaction is almost 10 times higher than that of the D–D reaction when the energies of the colliding particles are in the range of 100 keV. These positive features largely override the negative features of tritium, which is highly radioactive and must be produced in other nuclear reactions. If the D–D reaction could be mastered, however, a virtually unlimited supply of energy would be available, because deuterium occurs so abundantly in water.

If the problem of producing and containing the hot plasma could be solved, the fusion reaction would be self-sustaining: The kinetic energy of the neutrons produced is more than enough to maintain the plasma at high temperatures even though a large fraction of that kinetic energy is used up in energy production. Two rather different approaches have been taken to address the problem of confinement of the hot plasma. The first is *inertial confinement*, a scheme in which tiny pellets of material that contains nuclei suitable for fusions are compressed by powerful laser beams or ion beams until the necessary temperatures and densities are reached (Fig. 45–21). In this scheme there is no effort to hold the plasma confined for long periods; the beams are fired in pulses, and the pellets are replaced with each pulse. The principal difficulties of this approach are associated with understanding just how the pellets will behave when the energy-supplying beams stike them, with the construction and handling of the pellets, and with the guidance of the beams.

In the second scheme, *magnetic confinement*, the plasma, which is given additional thermal energy by the absorption of electromagnetic waves, is held in place by magnetic forces. The confinement can last for long periods. A difficulty faced in this approach is that magnetic fields do not affect the component of a charged particle's motion that is along the direction of the magnetic field lines, so the plasma tends to "squirt out" in one direction or another. To avoid this

(a)

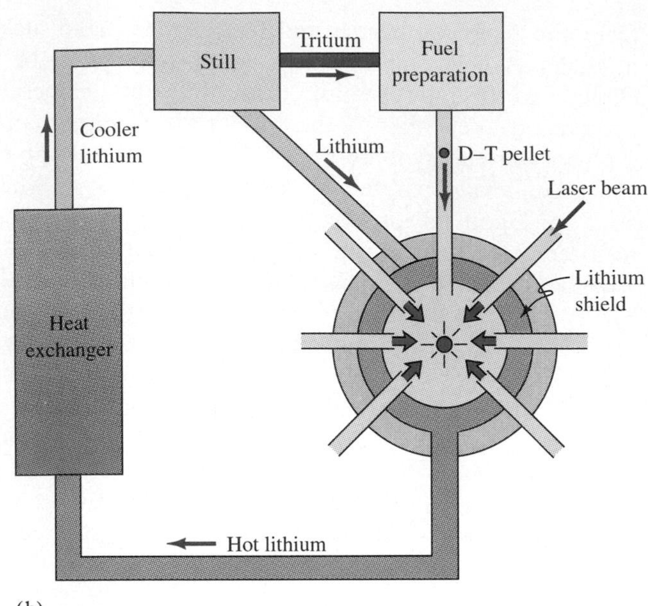

(b)

FIGURE 45–21 (a) The Particle-Beam Fusion Accelerator II at Sandia National Laboratories in Albuquerque, New Mexico, used for inertial-confinement experiments. (b) Schematic diagram of fusion by inertial confinement with laser beams or particle beams. D–T pellets are sent to a site where momentum from many laser or particle beams implodes them, bringing them to high enough temperatures and densities for significant amounts of fusion to occur. The neutrons from the fusion process produce tritium in the lithium shield, which also becomes hot. The tritium can be distilled out of the hot lithium and used to produce more fuel pellets. Deuterons are abundant in water. Electricity could be made from the thermal energy removed at the heat exchanger.

problem, complicated configurations of magnetic fields must be used. One field configuration employs magnetic mirrors like those that contain the earth's Van Allen belts (see Chapter 29). The *tokamak* is a device that attempts to hold the plasma with magnetic fields contained within a finite volume (Fig. 45–22; see the opening photograph of Chapter 30). (An example of a contained magnetic field is that of the toroidal solenoid.) In the early 1990s, large tokamaks came close to achieving self-sustaining reactions. Although both inertial confinement and magnetic confinement schemes show promise, we are still far from making fusion a commercial technology.

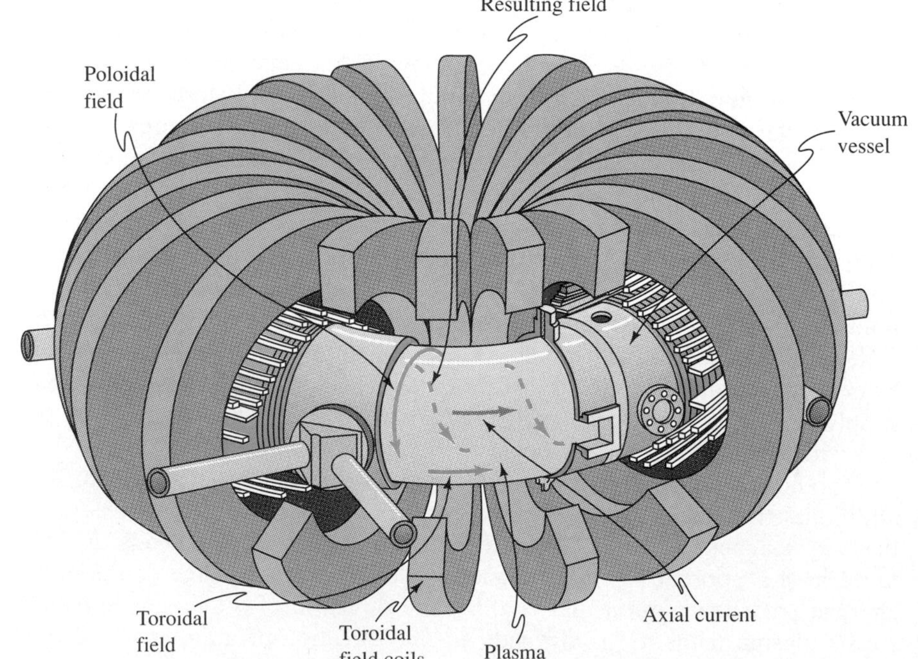

FIGURE 45–22 Schematic diagram of fusion by magnetic confinement, containment in the magnetic fields of a tokamak. The so-called poloidal field forms circles around the toroidal field. [After W. M. Stacey, Jr., "Fusion Reactor Development: A Review," *Advanced Nuclear Science and Technology*, **15:** 131 (1983).]

SUMMARY

The nucleus was discovered by means of scattering experiments that showed significant numbers of incident α particles scattered through large angles, which follows if the atom has a small, massive, positive core structure. The nucleus, held together by the nuclear force, is composed of Z protons and $N = A - Z$ neutrons, where A is the mass number. Nuclei with the same Z value but different N values are isotopes of a given atomic species. The masses of nuclei and of atoms are conveniently measured in atomic mass units, u, defined so that the mass of a ^{12}C atom is exactly 12 u. The binding energy, the energy released when a nucleus is formed from its A constituents (nucleons), is significant; a binding energy per nucleon of 7 to 9 MeV is typical of larger nuclei. This binding energy is so large that, due to the relativistic equivalence of mass and energy, the mass of a nucleus is on the order of 1 percent less than the sum of its constituent masses. To a good approximation, the constituents of the nucleus are closely packed into a sphere of radius

$$R_A = r_0 A^{1/3}, \tag{45–7}$$

with $r_0 \simeq 1.2$ fm. The lifetimes of unstable nuclei, the spectrum of states, spins, and magnetic dipole moments all provide useful ways to understand nuclear behavior.

For distances greater than about 1 fm, the nuclear force between nucleons is well described in terms of the Yukawa potential,

$$U(r) = -g^2 \frac{e^{-r/R}}{r}, \tag{45–9}$$

where g^2 measures the strength of the force and R is the range, approximately 2 fm. Because we are not sure of the nuclear force at short distances, and because the strength of the nuclear force is so great that our mathematical tools for calculating its consequences are inadequate, we cannot make a full set of predictions from fundamental physical laws. However, a useful set of models has been developed to explain nuclear behavior. The shell model is an approach in which individual nucleons move in a central nuclear potential. The resulting spectrum, including the presence of magic numbers—Z and N values for which a nucleus is particularly strongly bound—can be explained by this model. The liquid-drop model treats the nucleus as a continuous medium and is able to explain behavior such as fission, the splitting of large nuclei. The semiempirical mass formula, which is inspired by the liquid-drop model, successfully fits the masses of nuclei.

The strong binding of nuclei has important implications for whether nuclear reactions, including scattering and decay, can occur. Short-lived resonances, excited states of nuclei, are visible in cross sections for scattering, and their widths are inversely proportional to their lifetimes according to the Heisenberg uncertainty principle.

Nuclei can decay through three principal processes: β decay, α decay, and γ decay, depending on whether the decaying nucleus emits an electron, an α particle, or a γ ray (a highly energetic photon), respectively. Collectively, these processes comprise the phenomenon of radioactivity. The rate of a single decay process is described by the equation

$$N(t) = N_0 e^{-\lambda t}. \tag{45–22}$$

Here $N(t)$ is the number of undecayed nuclei that remain after time t, N_0 is the initial number at $t = 0$, and λ is the decay constant, which is different for each type of decay. Tunneling is an essential feature of α decay. Natural radioactivity is associated with decay chains, processes in which the products of radioactive decay are themselves radioactive and decay. The presence of such processes supplies us with useful radiometric dating techniques. Certain artificially produced radioactive nuclides have a variety of uses, including radioactive and medical applications.

The systematics of nuclear masses shows that fission, the breakup of large nuclei into smaller ones, and fusion, the combining of small nuclei into larger ones, are energetically favorable. When these processes occur, energy is released. However, both can occur only if a potential energy barrier is overcome. Fusion processes are the means by which the stars obtain energy. Fission induced by neutrons can occur in controlled chain reactions, in which a fissioning nucleus releases neutrons that can initiate other fission reactions. Such chain reactions are a commercially important source of energy.

QUESTIONS

1. Alpha particles have a charge of $+2e$ and a mass of roughly four nucleon masses. Describe experiments that would allow you to measure these quantities.

2. Neutron stars are stars that have collapsed under the influence of the gravitational force into what are essentially enormous nuclei composed mostly of neutrons. Why do these neutrons not decay?

3. In our discussion of why some nuclei undergo fission, we did not mention the effect of the energy term proportional to A in the liquid-drop model. Why not?

4. Protons and electrons each have intrinsic spins $\hbar/2$. Use the fact that the spin of ^{6}Li is $\hbar$ to show that the ^{6}Li nucleus cannot consist exclusively of protons and electrons.

5. Why is it easier to produce ions with a positive charge rather than with a negative charge from neutral atoms?

6. How can a mass spectrometer be used to determine the isotopic abundances of nuclides of an element such as oxygen?

7. Experiments to measure the inner structures of the nucleus typically use electrons as incident projectiles. Why might electrons be more useful than α particles for such experiments?

8. Of the calcium isotopes ^{39}Ca, ^{40}Ca, ^{41}Ca, and ^{42}Ca, which would you expect to have the smallest neutron separation energy, and why?

9. Of the nuclides ^{15}N, ^{16}O, and ^{17}F, which would you expect to have the largest proton separation energy, and why?

10. The mass conditions for β decay and electron capture are the same. Under what conditions is each one more likely to occur for a given nuclide?

11. Certain chemical elements tend to concentrate in bone marrow. Why should you avoid any such elements that are radioactive?

12. Why are light nuclei more efficient moderators for controlled nuclear fission than heavy nuclei are?

13. Neutrons that have been slowed down by a moderator are unable to induce rapid fission in ^{238}U nuclei. Why?

14. When we gave the (final) kinetic energies of the ^{4}He and neutron products in the D–T fusion reaction, we assumed that the collision takes place at rest. Yet we know there is a Coulomb energy potential barrier to overcome. Why was it possible for us to ignore the initial kinetic energy?

15. Most of the daughter nuclei that result from fission are radioactive. Why?

16. Why might you expect a neutron to penetrate farther into a sample of matter than an α particle of the same kinetic energy would?

PROBLEMS

45–1 Static Properties of Nuclei

1. (I) How many neutrons and how many protons do the following nuclides have: ^{7}Li, ^{15}N, ^{36}Cl, ^{53}Cr, ^{74}Ge, ^{101}Ru, ^{135}Cs, ^{179}Hf, ^{224}Ra, and ^{260}Lr?

2. (I) What is the Rutherford collision cross section of a 5.0-MeV α particle ($Z = 2$) on a lead nucleus ($Z = 82$) at a scattering angle of 90°? Express your answer in units of *barns* (1 barn $\equiv 10^{-24}$ cm^2).

3. (I) Consider the deflection of an α particle of kinetic energy 5 MeV by a gold atom, for which $Z = 79$. Estimate the typical angular deflection (a) for $R = 0.1$ nm; (b) for $R = 10$ fm.

4. (II) An α particle with (nonrelativistic) speed v is approaching the center of a nucleus of radius R and charge $+Ze$. Calculate the distance of closest approach. Assume that the nucleus is much heavier than the α particle. For what value of v will the distance of closest approach equal R?

5. (II) Calculate the distance of closest approach that a 7.00-MeV α particle makes when it scatters head-on with an aluminum nucleus. Neglect recoil of the aluminum nucleus. Compare this with the sum of the nuclear radii for helium and aluminum.

6. (II) An α particle with 4.0 MeV of energy scatters at a 45° angle from a lead atom. (a) How close did the α particle come to the center of the nucleus? (b) to the surface of the lead nucleus? (Assume that the α particle is pointlike.)

7. (II) The masses of the stable isotopes of silicon are 27.976928 u for ^{28}Si, 28.976496 u for ^{29}Si, and 29.973772 u for ^{30}Si. Calculate the binding energies of each nuclide.

8. (II) Express the Rutherford collision cross section in terms of the momentum transfer, $\Delta\mathbf{p}$, of a scattered particle, which has an initial momentum $mv\mathbf{i}$ and a final momentum $(mv\cos\theta)\mathbf{i} + (mv\sin\theta)\mathbf{j}$ (Fig. 45–23).

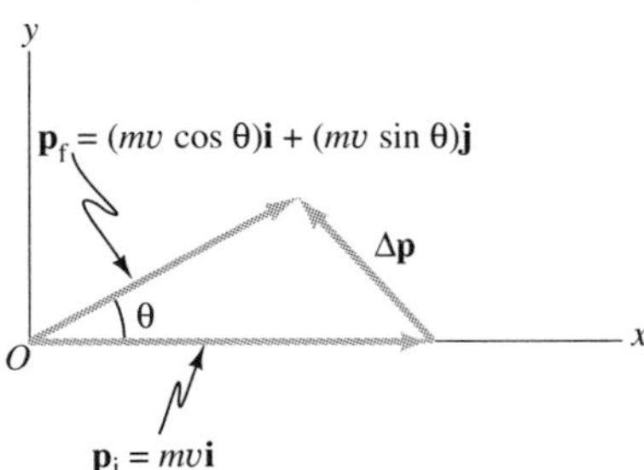

FIGURE 45–23 Problem 8.

9. (II) If we want to use an electron to probe distances to the size of a uranium nucleus, what energy should the electron have? What if we want to probe distances to the size of a nucleon, about 1 fm?

10. (II) Use the work–energy theorem to solve Example 45–1. Integrate the work over the distance from infinity to the turnaround distance R, the point at which the Coulomb force has done work equal to the initial kinetic energy of the α particle.

11. (III) *Mirror nuclei* are pairs of nuclei for which Z and N of each nucleus are opposites: For example, $^{23}_{11}$Na and $^{23}_{12}$Mg are mirror nuclei with $Z = 11$, $N = 12$ and $Z = 12$, $N = 11$, respectively. The idea that the *only* difference in the energies of these nuclei is due to a difference in electrostatic energy gives us another way to measure nuclear radii and in particular the constant r_0 in Eq. (45–7). If you solved Problem 69 in Chapter 25, you showed that when a charge Q is distributed uniformly throughout a spherical volume $\frac{4}{3}\pi R^3$, the electrostatic energy is $\frac{3}{5}Q^2/4\pi\epsilon_0 R$. Use this result to express the *difference* ΔE in the electrostatic energies of the pair of mirror nuclei above, and use that expression to find r_0, as defined by $R = r_0 A^{1/3}$, in terms of ΔE.

45–2 *Nuclear Forces and Nuclear Models*

12. (I) (a) The energy of the first excited state of a nucleus of ^{4}He is 20.1 MeV. Calculate the frequency and wavelength of a photon needed to excite this state from helium atoms at rest. (b) Do the same for the nuclide ^{187}Os, whose first excited state has an energy of 0.0098 MeV.

13. (I) Use the shell model to determine whether the following nuclides have a closed shell (magic number) for either neutrons or protons: ^{17}O, ^{38}Ar, ^{48}Ca, ^{115}Sn, ^{144}Sm, ^{208}Pb.

14. (II) Assume that the two protons inside a ^{4}He nucleus are approximately 4 fm apart. (a) Calculate the Coulomb force between them. (b) Use the binding energy of ^{4}He to *estimate* the nuclear force. Compare the two forces.

15. (II) Find the force that corresponds to the Yukawa potential energy, Eq. (45–9). Sketch it as a function of the separation distance; on the same plot, sketch an inverse-square force, assuming that that force corresponds to the same potential energy as Eq. (45–9) for small r.

16. (II) The energy required to assemble Z protons in a uniform distribution throughout the volume of a sphere of radius $r_0 A^{1/3}$ is $\frac{3}{5}(Z^2e^2/4\pi\epsilon_0 r_0 A^{1/3})$. The parameter r_0 is roughly 1.2×10^{-15} m. What fraction of the mass of a carbon nucleus does this energy represent?

17. (II) Use the semiempirical mass formula to determine the neutron, proton, and α-particle separation energies for the nuclide ^{48}Ca.

18. (II) Use the semiempirical mass formula to find the Z value that minimizes the total mass $M(A, Z)$ for fixed A by solving $\partial M/\partial Z = 0$.

19. (III) Use the semiempirical mass formula to calculate the A values for which spontaneous fission into two fragments, each with $A/2$ nucleons, occurs. In other words, find the A values for which $M(A, Z) - 2M(A/2, Z/2) > 0$.

45–3 *Energetics of Nuclear Reactions*

20. (II) The first nuclear reaction ever observed, by Ernest Rutherford, was the reaction $\alpha + {}^{14}\text{N} \rightarrow p + {}^{17}\text{O}$. If 5-MeV α particles are used to bombard ^{14}N at rest, determine the sum of the kinetic energies of the two outgoing particles from the atomic masses in Table 45–2 [$M(^{17}\text{O}) = 16.999131$ u].

21. (II) The α particle Rutherford used to initiate the first observed nuclear reaction (see Problem 20) had to overcome the Coulomb barrier between the protons in ^{4}He

and those in ^{14}N. Assume that the α particle must come within 1 fm of a proton in the target nucleus for a nuclear reaction to occur. Calculate the minimum energy the α particle must have to initiate the reaction. Compare your result with the 5.26-MeV energy for the α particle in Rutherford's experiment. Was the reaction likely?

22. (II) In the fusion reaction $^3\text{H} + {}^2\text{H} \to {}^4\text{He} + n$, the energy released is 17.6 MeV. Taking the He nucleus to have four times the mass of the neutron, and assuming that the initial particles are at rest when the reaction takes place, what is the kinetic energy of the neutron?

23. (II) Nucleons are not static, but move within nuclei. The approximate energy of a nucleon inside a nucleus is about 20 MeV. Use this kinetic energy and the uncertainty principle to *estimate* the dimensions of a sphere that might contain the nucleon. Is this size approximately equal to that of a nucleus? Interpret your result.

24. (II) The nuclear reaction James Chadwick used in his experiment to identify the neutron was $\alpha + {}^{11}\text{B} \to n + {}^{14}\text{N}$. Assume that the α particle had energy 5.3 MeV and that the ^{11}B was at rest; $M(^{11}\text{B}) = 11.009305$ u. If the kinetic energy of the ^{14}N is 0.8 MeV, what energy would the neutron have?

25. (II) Which of the following reactions are not possible, and why not: (a) $\alpha + {}^{40}\text{Ar} \to n + {}^{43}\text{K}$; (b) $p + {}^{58}\text{Ni} \to {}^3\text{He} + {}^{56}\text{Co}$; (c) $n + {}^{157}\text{Gd} \to d + {}^{156}\text{Dy}$; (d) $^{210}\text{Po} \to \alpha + {}^{206}\text{Ti}$?

26. (III) A foil made of 5 mg of ^{113}Cd (metallic cadmium) is exposed for 1 h to a beam of neutrons with a flux of 10^{14} neutrons/cm$^2 \cdot$ s. If the *capture cross section* (the reaction $^{113}\text{Cd} + n \to {}^{114}\text{Cd}$) for such neutrons is 2×10^4 barns (see Problem 2), how many ^{114}Cd nuclei are formed? [*Hint:* Find the effective area taken up by the target cadmium atoms.]

45–4 *Radioactivity*

27. (I) Show that the primary β-decay process, Eq. (45–26), is possible only when the atomic mass of the parent nucleus is greater than the atomic mass of the daughter nucleus.

28. (I) Show that positron emission, Eq. (45–29), is possible only when $M_P > M_D + 2m_e$, where M_P is the atomic mass of the parent nucleus and M_D is the atomic mass of the daughter nucleus.

29. (I) Show that electron capture is possible only when the atomic mass of the parent nucleus is greater than the atomic mass of the daughter nucleus.

30. (I) The number of unstable nuclei $N(t)$ present at time t diminishes with time according to the differential equation

$$\frac{dN(t)}{dt} = -\frac{1}{\tau} N(t),$$

where τ is the lifetime. Show by direct substitution that the solution to this equation is indeed $N(t) = N_0 e^{-t/\tau}$.

31. (II) The theory of α decay as a tunneling process allows us to predict an experimental relation between the half-life, $t_{1/2}$, and the kinetic energy, K, of α particles emitted during radioactivity. An approximate version of this prediction is

$$\ln t_{1/2} = \frac{Z(3.97\ \text{MeV}^{1/2})}{\sqrt{K}} - 123,$$

where K is in MeV. (a) Fill in the missing values in the following table:

Nucleus	*K (MeV)*	$t_{1/2}$
^{210}Po	5.30	?
^{210}Ra	?	2.4 h
^{215}Fr	9.36	?
^{232}Th	?	1.41×10^{10} yr

(b) In order to perform an experiment, we need a source of α particles with a half-life of at least 90 min. What is the highest-energy source we are likely to find?

32. (II) By looking up the masses of the neutron, proton, and electron in Appendix II, find the maximum energy, in MeV, of an electron produced in the decay of a neutron.

33. (II) In a sample of $^{212}_{83}$Bi nuclei, 33.7 percent decay by α decay to $^{208}_{81}$Tl and the rest decay by β decay to $^{212}_{84}$Po. What is the ratio of the lifetime for α decay, τ_α, to the lifetime for β decay, τ_β?

34. (II) The half-life of ^{14}C is 5730 yr, and the tissues of organisms accumulate this isotope from the atmosphere while the organisms are living. The skeleton of a mammoth is found to have 20 percent as much ^{14}C as the atmosphere has. When did the mammoth live? Assume that the concentration of atmospheric ^{14}C does not change.

35. (II) ^{80}Br is one of a small number of nuclides that is able to decay by all three β-decay processes. Show that this is possible. $M(^{80}\text{Br}) = 79.918528$ u, $M(^{80}\text{Kr}) = 79.916376$ u, $M(^{80}\text{Se}) = 79.916521$ u.

36. (II) Suppose that the lifetime for γ decay of a nucleus at rest is so long that, for all practical purposes, the natural width of the spectral line can be ignored (that is, the γ energy takes on a single, fixed value). When the nucleus is part of a collection of atoms at finite temperature, it moves randomly, sometimes toward and sometimes away from a γ-ray detector, and the spectral line width is determined by the Doppler shift of the emitted γ rays. This phenomenon is known as *Doppler broadening*. Estimate the width due to Doppler broadening for these atoms at a temperature $T = 800$ K.

37. (III) A radioactive nucleus a decays with a lifetime τ_1 into nucleus b, which decays with a lifetime τ_2 into nucleus c. The change in the number of nuclei a obeys the relation $dN_a = -(N_a/\tau_1)\,dt$ (see Problem 41–49). The number of nuclei b does not obey a simple equation of this type, because those nuclei are continuously being depleted through decays into nuclei c while they are replenished through decays of nuclei a. Show that the correct expression for the change in the number of nuclei b is $dN_b =$

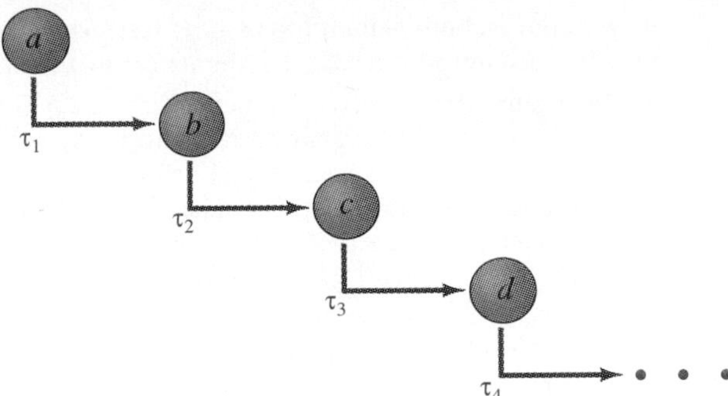

FIGURE 45–24 Problem 37.

$-(N_b/\tau_2)\,dt + dN_a$. How will this expression generalize to a chain of decays $a \to b \to c \to d \ldots$ with lifetimes $\tau_1, \tau_2, \tau_3, \tau_4, \ldots$, respectively (Fig. 45–24)?

38. (III) We have a chain of decays in a three-level system of nuclei a, b, c with $M_a > M_b > M_c$. Level a decays both to level b (with lifetime τ_{ab}) and level c (with lifetime τ_{ac}), while level b decays to level c (with lifetime τ_{bc}). Thus the number of nuclei at level a at time t obeys the differential equation

$$\frac{dN_a(t)}{dt} = -\frac{1}{\tau_{ab}}N_a(t) - \frac{1}{\tau_{ac}}N_a(t) \equiv -\frac{1}{\tau_a}N_a(t).$$

(a) Show that the solution to this equation is $N_a(t) = N_{0a}e^{-t/\tau_a}$. (b) Show that the number of nuclei of level b at time t obeys the differential equation

$$\frac{dN_b(t)}{dt} = \frac{1}{\tau_{ab}}N_a(t) - \frac{1}{\tau_{bc}}N_b(t).$$

(c) Show by substitution that the number of level b nuclei is

$$N_b(t) = N_{0a}\frac{\tau_a\tau_{bc}/\tau_{ab}}{\tau_{bc} - \tau_a}(e^{-t/\tau_{bc}} - e^{-t/\tau_a}).$$

(d) Show that there is no need to write an equation for N_c once N_a and N_b are known as functions of time, because $N_c(t) = N_{0a} - N_a(t) - N_b(t)$ if we start with only a collection of N_{0a} nuclei at time $t = 0$.

39. (III) Show that when the rate of decay from nucleus a in Problem 38 is much smaller than the rate of decay from nucleus b to nucleus c, then after a time that is long compared to τ_{bc}, $N_b(t) \simeq (\tau_{bc}/\tau_{ab})N_a(t)$. (You may use the results derived in Problem 38.)

45–5 *Fission and Fusion*

40. (I) The energy of nuclear weapons is expressed in terms of megatons of TNT, where 1 megaton of TNT $= 4.3 \times 10^{15}$ J. What minimum mass of ^{235}U is required to produce a nuclear weapon of 100 kilotons? [*Hint:* Each fission of ^{235}U produces about 200 MeV of energy.]

41. (II) The temperature of the interior of stars is about 10^7 K. (a) Estimate the typical kinetic energy of the colliding protons that participate in the proton cycle. (b) Compare your answer to part (a) with an estimate of the kinetic energy that protons initially distant from one another must have to come within 1 fm of each other, a distance at which the strong forces can lead to fusion.

42. (II) Calculate the difference between the Coulomb energy for a parent nucleus (Z, A) and the sum of the Coulomb energies of two daughter nuclei (fZ, fA) and $[(1-f)Z, (1-f)A]$, respectively, where f is some fraction between 0 and 1. Show that this energy difference is maximized when $f = \frac{1}{2}$; that is, from the Coulomb energy terms alone it is energetically favorable for fission to produce two equal-sized fragments.

45–6 *Applications of Nuclear Physics*

43. (II) A 300-Mwatt (electrical output) nuclear power plant uses ^{235}U as its fuel. The uranium fuel contains 4 percent ^{235}U, and each fission reaction produces 200 MeV of energy. Assume that the power plant is 30 percent efficient in producing electric energy. (a) Calculate the amount of uranium fuel used in 1 yr. (b) Determine the amount of thermal energy released to the environment in that time. (c) How many fission events occur per second?

44. (II) One of the reactions in the proton cycle is Eq. (45–33b), $^2\text{H} + p \to {}^3\text{He} + \gamma$. Calculate the energy released in this reaction. Use the data given in Table 45–1, but take into account that atoms inside burning stars are completely stripped of their electrons.

General Problems

45. (I) The *carbon cycle*, a stellar burning process secondary to the proton cycle, consists of the following sequential series of nuclear reactions and decays:

$^{12}\text{C} + p \to {}^{13}\text{N} + \gamma$, with 1.9 MeV of energy released;

$^{13}\text{N} \to {}^{13}\text{C} + e^+ + \nu$, with 1.2 MeV of energy released;

$^{13}\text{C} + p \to {}^{14}\text{N} + \gamma$, with 7.6 MeV of energy released;

$^{14}\text{N} + p \to {}^{15}\text{O} + \gamma$, with 7.4 MeV of energy released;

$^{15}\text{O} \to {}^{15}\text{N} + e^+ + \nu$, with 1.7 MeV of energy released;

$^{15}\text{N} + p \to {}^{12}\text{C} + {}^4\text{He}$, with 5.0 MeV of energy released.

(a) What is the net effect of one cycle, and how much energy is produced per cycle? (b) Consider a star of mass 3×10^{30} kg, consisting mainly of hydrogen, with 0.1 percent of its mass in ^{12}C. The characteristic duration of a carbon cycle in the star is 5×10^6 yr. Estimate the energy produced each year by the carbon cycle, assuming that each carbon nucleus in the star acts as a catalyst for the cycle.

46. (II) (a) Use the semiempirical mass formula to determine an expression for the binding energy of a nucleus. (b) Apply your result to calculate the binding energy of ^{12}C.

47. (II) A fissioning ^{235}U nucleus produces two fission fragments, ^{146}La and ^{87}Br (Fig. 45–25). Assume that the two fragments are spherical in shape and that the charge is uniformly distributed in both of them. What is the Coulomb potential energy of the system if the two fragments are just touching?

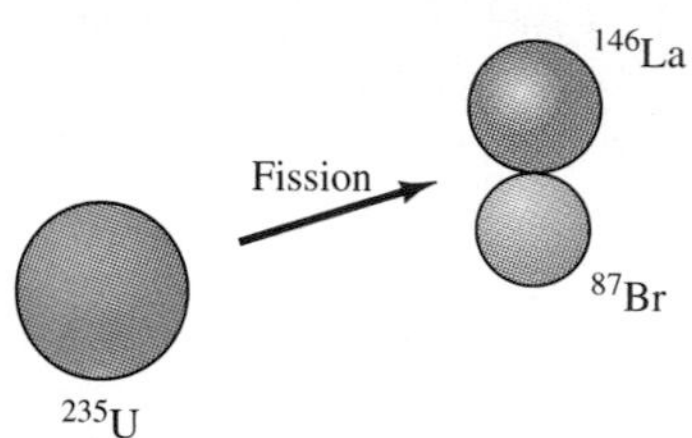

FIGURE 45–25 Problem 47.

48. (II) Use the semiempirical mass formula to find the neutron separation energy for the nuclides ^{16}O and ^{17}O. Explain the difference.

49. (II) Two protons within a star, each with energy 5 keV, collide head-on. What is the distance of their closest approach?

50. (II) A bone containing 10 g of carbon emits β particles from ^{14}C at the rate of 1.3 per second. At the time the ^{14}C is taken from the air and incorporated into bone, the activity of natural carbon due to its ^{14}C content is 15 Bq/mol. The lifetime of ^{14}C is 8268 yr. Estimate the bone's age.

51. (II) Two identical and neutral helium atoms approach each other. Because there is no Coulomb repulsion between them, they should be able to get very close together, close enough for the nuclei to touch. (a) What is wrong with this argument? (b) Given that the nuclear radius of helium is about 2 fm and that the center of mass of two approaching helium atoms is at rest, what energy must the helium atoms have for the nuclei to touch?

52. (II) A beam of neutrons of kinetic energy 2.5 GeV is produced through the bombardment of nuclei by protons. What distance does this beam travel before 90 percent of the neutrons have decayed? [*Hint:* The neutrons are relativistic, and time-dilation effects are important.]

53. (II) One of the early suggestions about the composition of the nucleus was that it contains A protons and $A - Z$ electrons. Use the uncertainty principle to estimate that if an electron is confined to a sphere with a radius of about 10^{-14} m, the size of a nucleus, then its momentum must be larger than about 10^2 MeV/c. Given that β rays emitted by the nucleus have energy on the order of 1 MeV, this estimate rules out the possibility that there are electrons in the nucleus.

54. (II) Calculate the overall kinetic energy produced in the carbon cycle (see Problem 45) after the two positrons produced in the cycle have annihilated against electrons.

55. (II) The energy of a collection of noninteracting fermions of mass m with number density n in a volume V is given by $E = V(\hbar^2\pi^3/10m)(3n/\pi)^{5/3}$. Use this expression to calculate the energy of the protons, E_p, in a nucleus in terms of A, Z, and r_0, where r_0 is the nuclear radius scale in Eq. (45–7). Repeat your calculation for the energy of the neutrons, E_n.

56. (III) Starting with the results of Problem 55, sum the proton and neutron energy terms, using the variables A and $\tau \equiv N - Z$. You will want to make the substitutions $Z = (A - \tau)/2$ and $N = (A + \tau)/2$. Approximate your result for $\tau \ll A$ by using $(1 + x)^p \simeq 1 + px + \frac{1}{2}p(p - 1)x^2$, an approximation good for $x \ll 1$. Show that this result has one term proportional to A and another term proportional to $\tau^2/2A$. Compare the coefficients of these two terms with the corresponding terms in the semiempirical mass formula. What does this tell you about the model of a nucleus as a collection of noninteracting protons and neutrons in a box?

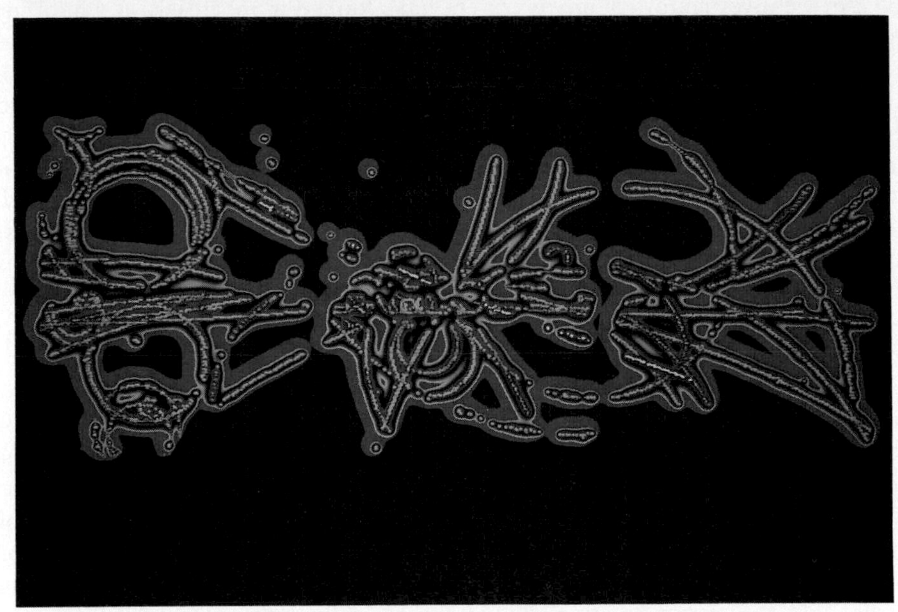

This enhanced image records the discovery of a fundamental carrier of the electroweak force, the Z particle, at CERN (Geneva) in 1983. The Z particle is produced by the collision of a proton–antiproton pair. Its tell-tale signature is an electron–positron pair (in gold) moving back-to-back away from the collision point. The remaining tracks represent other debris from the collision.

PARTICLES AND COSMOLOGY

Although the realms of the interior of a proton and the far reaches of the universe might appear to have little to do with each other, we believe that the same physical principles govern them. Because both realms are at the edges of our knowledge of the fundamental laws of physics, they act as proving grounds for our current ideas about those laws. But these realms converge in a more profound way. Particle physicists study the rules that govern the behavior of matter at the shortest distances we can examine. Astronomers attempt to see the behavior of the cosmos at the farthest distances. Because the speed of light is finite, when we look at a very distant object, we see the radiation that object emitted long ago: Looking at distant objects is equivalent to looking back in time. There is compelling evidence for the big-bang theory, according to which the universe expanded from a tiny volume some 15 billion years ago. Thus our efforts to see back to the era when the universe began return us to the questions of the behavior of matter at short distances.

46–1 RUSSIAN DOLLS, SUBNUCLEAR FORCES, AND SPECIAL RELATIVITY

A "Russian doll" is a set of nested dolls; each but the smallest can be opened to reveal a smaller version of itself. The investigation of the structure of matter and of the forces that hold it together can be compared to the opening of a Russian doll. As the investigators of any given era have looked more closely at what they regarded as the fundamental constituents and interactions of matter, they have found, at smaller scales within that matter, other constituents interacting through other forces. The constituents and interactions at larger scales can be explained in terms of the physics at smaller scales. The first real step in "opening the Russian doll" was John Dalton's realization (1808), based on a body of late-eighteenth-century work, that when chemical elements combine chemically, they do so in precise proportions. Dalton realized also that the simplest explanation for this phenomenon is that a variety of different types of atoms exist, one for each chemical element. The atoms of the chemical elements supply the building blocks for the construction of molecules. The interatomic forces that supply the "glue" for this construction are electrical in origin.

But atoms are not truly indivisible. We saw in Chapter 42 how an atom is constructed from electrons and a nucleus. The structure of the nucleus and of electrons represents the Russian doll within an atom. The atom is held together by electromagnetic forces between the electrons and the nuclei.

As we know from Chapter 45, even nuclei have structure. They consist of nucleons—protons and neutrons. The nucleus is held together by the nuclear, or strong, force. This force differs from the electromagnetic force in that it has finite range. And the energy required to remove a neutron or proton from a given nucleus is so large that relativistic effects—those that relate energy to mass—are significant for the first time in our brief overview of the structure of the matter. The origin of those large energies is to be found in the strength of the nuclear force.

Do electrons and nucleons have structure? If so, what forces hold those structures together? As far as we know today, there is no Russian doll within the electron: We have no evidence to suppose that electrons are composed of more fundamental constituents. This is not the case for nucleons and other strongly interacting particles, known collectively as **hadrons**. Scattering experiments carried out with particle accelerators have provided us with evidence for their structure as well as with information about the nature of the interactions involved.

Hadrons

Scattering

How do we determine the shape and size of an object in everyday life? We look at it. This means that we observe light (photons) bouncing off the object. Our eyes serve as a detector for observing a pattern of reflected light, which tells us about the properties of the object.

What prevents us from seeing atoms in this way? Light diffraction limits our ability to discern details of small objects, as we learned in Chapter 39. Resolution of details in an object of size d requires light of wavelength $\lambda \leq d$. Thus to study atoms in this way, we would need light of short wavelength. Because of the relation $\lambda = h/p$ [Eq. (41–1)], our light consists of photons with large momentum. The fact that matter has wavelike properties extends this limitation to the case in which we scatter projectiles of matter (electrons, for example).

In Chapter 8 we discussed the connection between the momentum (or energy) of a projectile and the size of the projectile's target that can be studied. In order for a projectile to "see" an object of size d, the projectile must have a minimum momentum (or energy). Let us review that relation, and this time use our knowledge of diffraction patterns. The characteristic angular size of the diffraction pattern made by a projectile of wavelength λ scattering from an object of size d is $\sin\theta \simeq \lambda/d$. The size of the momentum transfer in a scattering experiment is characteristically $\Delta p \simeq p \sin\theta$. Therefore

$$\Delta p \simeq p \sin\theta \simeq p\frac{\lambda}{d} = \frac{h}{d}, \tag{46–1}$$

where in the last step we have used the de Broglie momentum–wavelength relation, $p = h/\lambda$. Thus an increasing momentum is needed to study ever smaller objects. We stress that in the laboratory frame (a moving projectile and a fixed target particle), much of the energy of the system is contained in the motion of the center of mass, which, because of momentum conservation, continues unaltered after the collision. This energy is unavailable for probing the target. *Thus Eq. (46–1) actually applies to the center-of-mass momentum and momentum transfer.*

Equation (46–1), which expresses the limits on our ability to see the details of a system due to the wavelike aspects of the incident "projectile," is essentially contained in the Heisenberg uncertainty relations. Recall that the momentum–position uncertainty relation is $\Delta x\,\Delta p > \hbar$, Eq. (41–7), where $\hbar = h/2\pi$. If we interpret d as the position uncertainty Δx, then the connection is evident.

The Heisenberg uncertainty relations are discussed in Chapter 41.

EXAMPLE 46–1 The proton has a radius of approximately 1.2 fm. You want to use electrons as projectiles to study the structure of protons on a scale of 0.5 to 1.2 fm. Assuming that the maximum observable electron-scattering angle in the center-of-mass system is 30°, what center-of-mass momentum must the incident electrons have? Is it necessary to use relativistic kinematics?

SOLUTION: Equation (46–1) gives us a relation between the incident momentum and the scale d that can be probed, namely,

$$p > \frac{h}{d \sin\theta}.$$

The largest value of p is determined by the smallest value of d, namely, $d = 0.5$ fm:

$$p \simeq \frac{6.63\times10^{-34}\,\text{J}\cdot\text{s}}{(0.5\,\text{fm})(\sin 30^\circ)} = \frac{6.63\times10^{-34}\,\text{J}\cdot\text{s}}{(0.5\,\cancel{\text{fm}})(0.5)}\,\frac{1\,\cancel{\text{fm}}}{10^{-15}\,\text{m}} = 2.7\times10^{-18}\,\text{kg}\cdot\text{m/s}.$$

Thus

$$pc = (2.7\times10^{-18}\,\text{kg}\cdot\text{m/s})(3.0\times10^{8}\,\text{m/s})$$
$$= (8.0\times10^{-10}\,\cancel{\text{J}})\frac{1\,\text{eV}}{1.6\times10^{-19}\,\cancel{\text{J}}} = 5.0\times10^{9}\,\text{eV},$$

or 5.0 GeV. Because $m_e c^2 \simeq 0.5$ MeV for an electron, the value of pc required is some 10,000 times the rest mass, and the motion is highly relativistic.

Note that an electron with 5 GeV of energy in the center-of-mass reference frame is an electron with much higher energy in the laboratory reference frame, namely about 53 GeV, or 53,000 MeV (see Problem 6).

Features at Different Levels

The scattering experiments we have described are the tools by which we probe the different levels of the structure of matter. These levels have not turned out to be monotonous repetitions of one another—the same Russian doll does not emerge at each level. Two important features appear for the first time at the structural level of subnuclear matter (the level of the structure of protons): Relativity dominates, and the constituents are confined. Let us look at these features in turn.

FIGURE 46–1 The result of a simulated collision between protons in the beams of the Superconducting Supercollider (SSC), an accelerator under construction in Texas. The tracks are formed by particles that leave the collision site or are produced in subsequent collisions. The tracks curve because a magnetic field is present. Many particles are produced, due to the highly relativistic energies involved. Such energies are required to probe the interior of the proton with the detail possible at the SSC.

Relativity. Even at the level of nuclear structure, the effects of relativity cannot be ignored. In Chapter 45, we saw evidence for this in the fact that the masses of nuclei in excited states are slightly but measurably different from the masses of the same nuclei in their ground states. Here special relativity enters via the energy–mass equivalency, $E = mc^2$. At the subnuclear level, relativity is crucial: The mass of a proton in its first excited state (a "proton resonance") is some 30 percent higher than that of the proton.

Similarly, the center-of-mass energies of the projectiles required to study proton structure can be estimated from the uncertainty relation for momentum and position, namely, $\Delta p\ \Delta r \geq \hbar$. A proton's radius is roughly 10^{-15} m. Suppose that we want to look at a region whose radius is $\frac{1}{100}$ of this value. Then, with $\Delta r \simeq 10^{-17}$ m, we estimate a momentum transfer $\Delta p = \hbar/\Delta r$. If we use an electron as a projectile and assume that the momentum transfer is of the same order of magnitude as the initial momentum, the electron projectile must have a (relativistic) center-of-mass energy $E \simeq (\Delta p)c \simeq (\hbar/\Delta r)c$. With $\hbar c \simeq$ 200 MeV · fm (see Appendix II), $E \simeq 20$ GeV. By comparison with the electron's mass, whose energy equivalent is about 0.5 MeV, our projectile is ultrarelativistic. An energy of 20 GeV is also much larger than the energy that corresponds to the proton's mass, on the order of 1 GeV.

Once the energies required in our "microscope" become so large, there is an immediate complication: Additional particles can be created from the incident energy (consistent with conservation laws, such as that of charge). In particular, we recognize the presence of antiparticles such as antiprotons, which are like their counterpart particles, except that all the intrinsic quantum numbers—electric charge, for example—have the opposite sign. Particle–antiparticle pairs can always be created if the excess energy is sufficiently large. The electron–proton scattering experiment described above thus becomes much more complicated, because the final state may contain many particles (Fig. 46–1).

EXAMPLE 46–2 We wish to collide a pair of electrons head-on in a *symmetric collider* (an accelerator in which beams of equal energy meet head-on) to produce a proton–antiproton ($p\bar{p}$) pair. Given that the proton mass is 938 MeV/c^2, find the magnitude of the minimum momentum of either of the colliding electrons. The reaction takes the form $e^- + e^- \rightarrow e^- + e^- + p + \bar{p}$. Ignore the electron mass throughout.

SOLUTION: We must satisfy conservation of both energy and momentum. The minimum momentum corresponds to a minimum energy, just enough energy to account for the rest mass of all the final particles, with nothing left over for kinetic energy. The outcome is a pair of electrons and a $p\bar{p}$ pair at rest in the laboratory reference frame, which for the case of a collider is also the center-of-mass frame. At the same time, we automatically satisfy momentum conservation, because the initial momentum of the pair of electrons is zero for a symmetric collider, and because all the final-state particles are at rest, this is the momentum of our minimum-energy final state as well.

Let the magnitude of the minimum incoming momentum of each of the two electrons be k_{min}. The total initial energy is then $2\sqrt{k_{\text{min}}^2c^2 + m_e^2c^4}$. This energy must equal the energy of a pair of electrons and a $p\bar{p}$ pair, all at rest:

$$\cancel{2}\sqrt{k_{\text{min}}^2c^2 + m_e^2c^4} = \cancel{2}m_ec^2 + \cancel{2}m_pc^2,$$

where we have used the fact that the mass of an antiproton is equal to that of a proton. We square both sides and subtract $m_e^2c^4$ to get

$$k_{\text{min}}^2c^2 = (m_ec^2 + m_pc^2)^2 - m_e^2c^4 = 2m_em_pc^4 + m_p^2c^4.$$

The proton mass is some 2000 times larger than the electron mass, so we can safely neglect the $2m_em_pc^4$ term, containing m_e. We are left with

$$k_{\text{min}} \simeq m_pc = \frac{m_pc^2}{c}.$$

Numerically, $k_{\text{min}} \simeq (938 \text{ MeV})/c = 0.938 \text{ GeV}/c$. (Note the use of the convenient unit GeV/c for momentum.) Equivalently, the minimum total initial energy, E_{min}, is the energy corresponding to the rest mass of a $p\bar{p}$ pair:

$$E_{\text{min}} = 2m_pc^2 = 2[(0.938 \text{ GeV})/\cancel{c^2}]\cancel{c^2} = 1.88 \text{ GeV}.$$

The minimum energy required for an electron incident on another electron at rest to produce a $p\bar{p}$ pair is much larger (see Problem 7).

Confinement. If we excite an atom, the energies of successive excitations increase until at some point the energy exceeds the binding energy of the atom. One or more electrons may escape from the atom, leaving behind a positive ion. With enough energy, all the electrons can be separated from the nucleus. Similarly, when we excite the nucleus, the energies of the excitations increase until the nucleus breaks apart into constituent daughter nuclei or nucleons. In this way the constituents of the atom, or of a nucleus, can be isolated, or observed as free particles. *The constituents of the nucleons or other hadrons cannot, however, be separated and observed individually, no matter how much energy is added.* This phenomenon is known as **confinement**. In particular, the excitations of the nucleons, whose masses are significantly greater than that of their ground-state counterparts, do not decay into the constituents of the nucleon. This does not mean that these excitations do not decay. They decay most readily into other (lower-lying) excitations or into the nucleons themselves and into other strongly interacting particles such as *mesons* (see Section 46–2). We shall briefly discuss in Section 46–3 the reasons for confinement.

Confinement

46–2 NEW PARTICLES AND CONSERVATION LAWS

How the Nuclear Force Is Transmitted

One of the original goals of what is now called high-energy physics was the determination of the potential energy associated with nuclear forces. Once the nuclear potential was found, the application of the principles of quantum mechanics allowed nuclei to be treated in a manner analogous to the treatment of atoms. In 1935 this goal underwent a deep modification as a result of an insight by Hideki Yukawa. Yukawa proposed that at the most fundamental level, forces between interacting particles result from the exchange of a special quantum that we shall call a "bonding" particle. One of two interacting particles

The range of the nuclear force is described in Chapter 45.

emits a bonding particle, and the other interacting particle absorbs it. Yukawa was able to predict the mass of the bonding particle responsible for the nuclear force—he called that particle the *meson*—from knowledge of the range of nuclear forces.

The idea that a force is mediated by an exchange is not a complicated one; a baseball pitcher and catcher exert a (repulsive) force on one another through the exchange of a baseball. What is new here is that the exchanged object contains a substantial amount of energy. Take the mass of the meson to be m. When a meson has been emitted, and before it is reabsorbed, the system of interacting particles has an energy deficit on the order of mc^2. Energy conservation can be consistent with such a deficit only if the time Δt for which the deficit exists satisfies the relation $\Delta E\,\Delta t > \hbar$, which is the limit allowed by the time–energy Heisenberg uncertainty relation. If we suppose that this system is dominated by quantum mechanical effects so that the uncertainty relations are satisfied as equalities, then $\Delta E\,\Delta t \simeq \hbar$. With $\Delta E = mc^2$,

$$\Delta t \simeq \frac{\hbar}{\Delta E} = \frac{\hbar}{mc^2}. \qquad (46\text{–}2)$$

Energy nonconservation in the exchange of quanta is unmeasurable because of an uncertainty relation.

This relation allows us to retain the principle of the conservation of energy, in that we cannot *measure* the nonconservation of energy within time Δt. When a particle is emitted and reabsorbed in this way, we say that it exists in a *virtual* state and that the process is virtual.

Range. Let us see how far a meson can travel, a measure of the distance over which the nuclear force extends. During the time Δt, the distance the meson can travel is $v\,\Delta t$, and because v must be less than c, the greatest distance the particle can travel is $R = c\,\Delta t$. From Eq. (46–2),

$$R = c\,\Delta t \simeq \frac{c\hbar}{mc^2} = \frac{\hbar}{mc}. \qquad (46\text{–}3)$$

We refer to this maximum distance as the *range* of the force; we have already described the range of nuclear forces in Chapter 45.

Yukawa's meson, the bonding particle of the nuclear force, is called the pi-meson, or *pion*, denoted by the symbol π. We can estimate its mass, m_π, if we take $R \simeq 1.4$ fm, an experimental value for the range of nuclear forces:

$$m_\pi = \frac{\hbar}{Rc} = \frac{1.05 \times 10^{-34}\ \text{J}\cdot\text{s}}{(1.4 \times 10^{-15}\ \text{m})(3.0 \times 10^{8}\ \text{m/s})} = 2.5 \times 10^{-28}\ \text{kg} = 0.15 m_p, \qquad (46\text{–}4)$$

where m_p is the proton mass. Thus $m_\pi c^2 \simeq (0.15)(938\ \text{MeV}) = 140$ MeV. Yukawa also calculated the potential energy as a function of particle separation r for the force associated with meson exchange:

$$U(r) = g^2 \frac{e^{-m_\pi rc/\hbar}}{r}. \qquad (46\text{–}5)$$

Here g^2 is a measure of the strength of the force between the interacting particles, analogous to $e^2/4\pi\epsilon_0$ for the Coulomb force. The Coulomb force does not have the exponentially decreasing factor characteristic of a short-ranged force. However, the Coulomb force fits Yukawa's framework perfectly well if it is regarded as the result of the exchange of a massless particle—the photon.

Several comments and qualifications about Yukawa's idea are worth making:

1. The potential energy of Eq. (46–5) is oversimplified. Depending on the nature of the exchanged meson, the potential energy may have a complicated dependence on the spins of the interacting particles.
2. The meson must be a boson; that is, its intrinsic angular momentum (spin) must be an integral multiple of $\hbar$. That is because the particles that emit and absorb the meson—the nucleons—are fermions, and if angular momentum is to be conserved, they cannot themselves emit or absorb a particle with half-integral spin and remain fermions.
3. The meson can be electrically neutral or electrically charged, but any electric charge must be conserved by an emission or absorption. Nuclear forces occur with nearly equal range between neutrons and neutrons, protons and protons, and neutrons and protons, and this means that Yukawa's meson occurs in the three charge states π^0, π^+, and π^-, all of which must have roughly the same mass.

Yukawa's idea implies that more particles are involved with the nuclear force than had been suspected. His idea is eminently testable: Pions, with predicted masses, should be produced in collisions of sufficient energy. In the late 1940s, accelerators with energies sufficient for this purpose were built, and all three pions, with masses close to the predicted values, were discovered.

EXAMPLE 46–3 A proton projectile collides with a stationary proton target. What minimum kinetic energy must the projectile have so that the final state contains two protons and a neutral pion?

SOLUTION: In the laboratory reference frame, the incoming proton has momentum of magnitude p and energy E, related by $E = \sqrt{p^2c^2 + m_p{}^2c^4}$. The target proton has zero momentum and energy m_pc^2. The minimum energy is the energy that accounts for the final particle rest masses, with no leftover kinetic energy for the final particles. Therefore the three final particles—two protons and one π^0—have a total momentum p but no motion relative to each other. Thus the final state can be viewed as a single particle of momentum p and mass $2m_p + m_\pi$. Energy conservation gives the relation

$$E_i = E + m_pc^2 = \sqrt{p^2c^2 + m_p{}^2c^4} + m_pc^2 = E_f = \sqrt{p^2c^2 + (2m_p + m_\pi)^2c^4}\,.$$

We square both sides of this equation to find that

$$\cancel{p^2c^2} + m_p{}^2c^4 + 2Em_pc^2 + m_p{}^2c^4 = \cancel{p^2c^2} + (2m_p + m_\pi)^2c^4,$$

or, solving for E,

$$E = \frac{2m_p{}^2c^4 + 4m_pm_\pi c^4 + m_\pi{}^2c^4}{2m_pc^2}\,.$$

Thus the minimum kinetic energy needed is

$$K = E - m_pc^2 = 2m_\pi c^2 + \frac{m_\pi{}^2}{2m_p}c^2 = 2(140\text{ MeV}) + \frac{(140\text{ MeV})^2}{2(938\text{ MeV})} = 290\text{ MeV}.$$

An intensive program of experimentation produced pions in such large numbers that they could themselves be scattered from nuclear targets. One of the most striking discoveries of this program was the existence of many new particles. These particles are all hadrons, because the ease with which they are produced suggests that they all interact with a strength equal to that of the nuclear force. There are large numbers of new bosons, all referred to as mesons, and there are large numbers of new fermions. The new hadrons differ by electric charge and by new quantum numbers. All these particles are as "elementary" as the nucleons and the pions. They had not been discovered earlier because they (and the pions) are all unstable, decaying with lifetimes between 10^{-8} s and as little as 10^{-23} s (the time it takes for light to cross two proton diameters!). The mesons can in principle be exchanged by two nucleons or, for that matter, by any two fermions. Nuclear forces are far from simple. The underlying dynamics of the particles that interact through nuclear forces has been better understood by studying the forces between *constituents* of the nucleons.

Baryon Number

In Chapter 40 we described the existence of antiparticles. Relativity and quantum mechanics together predict that each particle should have a counterpart with the opposite charge but the same mass and lifetime for decay. The fact that the charge is opposite makes it easy to identify the positron (the antielectron) or the antiproton. But what is the antiparticle of the neutron, which is electrically neutral? The antineutron is identifiable through a new quantum number with some similarities to the electric charge. It is called the **baryon number**, B, and nucleons are each assigned the baryon number $B = 1$. The antineutron and the antiproton have the baryon number $B = -1$, and all the fermionic cousins of the nucleons discovered in scattering experiments have a nonzero baryon number. These particles—that is, *all strongly interacting fermions*—collectively are referred to as **baryons**. The pion and all the other mesons have a baryon number $B = 0$.

Baryons

The existence of the baryon number, together with a conservation law for it, was postulated when the remarkable stability of matter was recognized. Charge conservation and angular momentum conservation would allow the decay

$$p \to e^{+} + \pi^{0}. \tag{46–6}$$

Unless this decay and others like it were somehow hindered, atoms would disappear. But experiment shows that *if protons have any lifetime at all, that lifetime is greater than 10^{30} yr*. How can we say this if the universe has existed for only some 10^{10} yr? We can measure such a lifetime by assembling 10^{30} protons and watching them carefully; if the lifetime were 10^{30} yr, we would expect one decay per year. An extraordinary experiment in which a tank filled with 8000 tons of water (about 10^{32} protons) was surrounded by electron counters showed that no decays like those in Eq. (46–6) occurred over a period of several years (Fig. 46–2). The tank was placed in a salt mine deep underground to avoid interference from high-energy particles coming from outer space. Because electrons and pions are assigned a baryon number $B = 0$, and because we postulate that *baryon number is conserved in any reaction*, the decay of Eq. (46–6) is forbidden.

Baryon number is a conserved and quantized quantum number.

Although baryon number is conserved like electric charge, baryon number must be rather different from electric charge. We know that when an electric charge is present, there is an electromagnetic field, and forces are associated with that field. There is, however, no comparable field associated with baryon number. If there *were* Coulomblike forces associated with baryon number, of the form kB_1B_2/r^2, then the force attracting a mass of uranium to the earth would differ from the force attracting the same mass of hydrogen, because uranium has

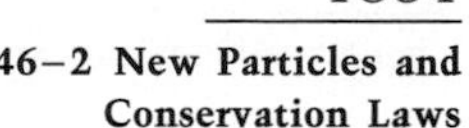

FIGURE 46–2 The very pure water in this tank is surrounded by phototubes, detectors sensitive to the tiniest trace of light. Here the apparatus is being checked by a scientist with scuba equipment. The phototubes are activated when the lights are turned out; they are capable of capturing the tiny flash of light associated with a proton decay somewhere in the tank.

a larger fraction of its mass in nucleons. But tests of the equivalence principle show to very high accuracy that the forces are the same and therefore show that there is no detectable field associated with baryon number.

Another way in which baryon number differs from electric charge is that, under certain circumstances, the conservation of baryon number *can* be violated. Conditions in the early stages of the universe appear to have provided these circumstances and in fact may be necessary to explain why the universe today contains an excess of matter over antimatter.

The Quark Model

The large number of "elementary" particles discovered in scattering experiments is reminiscent of atoms and their excited states. The Bohr model of hydrogen allows us to explain all the states of hydrogen in terms of an electron, a proton, and the forces between them. An analogous explanation of the numerous mesons and baryons observed is provided by the *quark model*, an economical description of baryons and mesons in terms of a few constituents and their interactions. This model was proposed independently by Murray Gell-Mann and George Zweig in 1964.

In the quark model, neutrons, protons, and pions are composed of two fundamental constituents called **quarks**, each with spin $\hbar/2$: the *u-quark* ("up" quark) and the *d-quark* ("down" quark). Both types of quarks have an antiparticle, $\bar{u}$ and $\bar{d}$, respectively. The quantum numbers of these quarks are listed in Table 46–1. The electric charge of the quarks is a fraction of the electron charge.

TABLE 46–1

PROPERTIES OF QUARKS†

Quark	*Electric Charge*	*Baryon Number*	*Spin*
u	$+\frac{2}{3}e$	$+\frac{1}{3}$	$\frac{\hbar}{2}$
d	$-\frac{1}{3}e$	$+\frac{1}{3}$	$\frac{\hbar}{2}$
$\bar{u}$	$-\frac{2}{3}e$	$-\frac{1}{3}$	$\frac{\hbar}{2}$
$\bar{d}$	$+\frac{1}{3}e$	$-\frac{1}{3}$	$\frac{\hbar}{2}$

†Quarks actually come in three distinct sets, or *families*, totaling six quarks. These lead to the prediction of still more hadrons, most of them already observed.

This does not mean that charge is no longer quantized, but that the true quantization is in units of $\frac{1}{3}e$.† The fact that quarks and antiquarks have spin $\hbar/2$ means that the spin of an even number of quarks plus antiquarks is an integer times $\hbar$, whereas the spin of an odd number of quarks plus antiquarks is a half-integer times $\hbar$.

The proton is a composite of three quarks, *uud*; the neutron has the structure *udd*. The antiproton is made of the combination $\bar{u}\bar{u}\bar{d}$, whereas the antineutron is $\bar{u}\bar{d}\bar{d}$. A combination of three quarks can have a variety of angular momenta, and these different combinations successfully explain all the baryons thus far discovered. For example, among the observed baryons is a quartet of states with similar masses, Δ^{++}, Δ^{+}, Δ^{0}, Δ^{-}, all with spin $3\hbar/2$ and baryon number $B = 1$. These are formed from the combinations *uuu*, *uud*, *udd*, and *ddd*, respectively. The quark model also accounts for mesons. Mesons have $B = 0$ and are bosons (their spin is an integral multiple of $\hbar$), so they must be formed from combinations of quarks and antiquarks. In particular, $\pi^{+} = (u\bar{d})$ and $\pi^{-} = (d\bar{u})$, whereas π^{0} is a combination of $u\bar{u}$ and $d\bar{d}$.

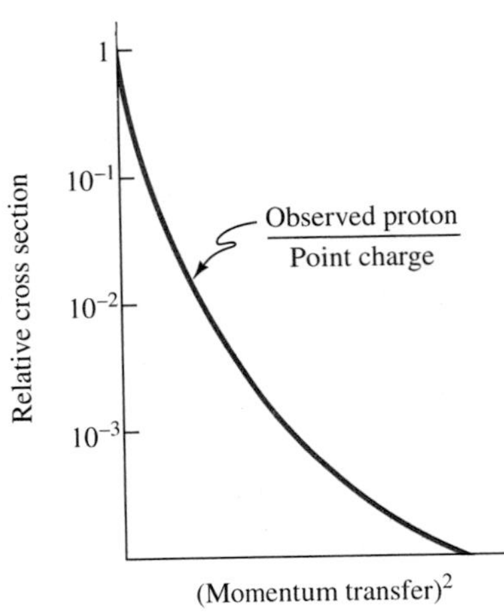

FIGURE 46–3 Elastic electron–proton scattering experiments show that the proton is not pointlike. The vertical axis is the ratio of the measured cross section to the cross section that would follow if the proton were pointlike. This curve would be flat, not decreasing, if the proton were pointlike. We know from other scattering experiments that the corresponding curve for quarks is flat, to the accuracy with which we can measure it.

We mentioned in Section 46–1 that quarks are *confined* inside hadrons; that is, that they cannot be separated from a nucleon or other strongly interacting particle and exist independently. The evidence for the existence of quarks does, however, go beyond the fact that quarks in combination account for the observed hadrons. Complementary evidence is provided by scattering experiments. When electrons with high energy were first used to probe protons in a hydrogen target in the 1950s, the results quite clearly revealed a distribution of charge within the proton (that is, the proton is not pointlike). Figure 46–3 shows the observed distribution, compared to the distribution a pointlike proton would give. There was wide expectation that further experiments would continue to verify that the proton's charge is distributed smoothly within it, much as was expected for the α-particle scattering from the atom before Rutherford's work proved otherwise. If the proton charge were distributed uniformly, electrons passing through the proton (and possibly breaking it apart) would be deflected very little. To the surprise of the physics community, experiments of this type performed in 1968 showed a very different behavior: Electrons were scattered in large numbers at large angles and in a pattern characteristic of a proton constructed of three pointlike particles, each carrying a fraction of the proton charge. These particles are the quarks that we used to describe the observed hadrons.

In addition to classifying the hadrons and explaining some of the results of scattering experiments, the quark model has led us to new questions that could not be asked in the framework of protons, neutrons, and pions. In particular, if quarks are the constituents of hadrons, how is the force between two quarks transmitted? Is there an exchanged particle analogous to the pion? We shall discuss such questions in Section 46–3.

Lepton Number

We have seen that a proper understanding of the forces that act in nature cannot be achieved until we have some knowledge of the constituents of matter. The electron is not a hadron, but it certainly plays an important role in the structure of matter. In fact, both the electron and the neutrino (ν) are fundamental constituents of matter. These particles, and other particles that do *not* participate in the strong force, are known as **leptons**.

Leptons

†Charge can be *isolated* only in units of e because the quarks are confined (see Section 46–1).

We recall from our discussion in Chapter 45 that the neutrino has no charge and little or no mass. But, along with the electron, it does carry a conserved **lepton number** analogous to the baryon number. Both the electron and the neutrino, each with spin $\hbar/2$, have the lepton number $+1$; their antiparticles e^+ and $\bar{\nu}$ have the lepton number -1. Electrons interact through the forces of electromagnetism, whereas neutrinos do not. But both participate in the weak force associated with β decay. We shall see in Section 46–3 that these forces are closely related.

46–3 FUNDAMENTAL FORCES REVISITED

We introduced fundamental forces in Chapter 5 and are now in a position to add to that discussion. At present physicists believe that three fundamental forces act in nature (Table 46–2): gravitation, the electroweak force, and the strong force. (The electroweak force is the result of a mid-1970s unification of electromagnetism and the weak force, responsible for β decay.) Each of the fundamental forces is thought to act through the exchange of one or more particles. In Table 46–2 we distinguish the "constituent" particles that interact by means of these forces and the exchange particles, or bonding particles, through which the force is transmitted. Note that the nuclear force is not listed; today we believe that it is ultimately due to the strong force, the force between the quarks. Gravitation plays no practical role in our understanding of the behavior of matter at energies that are currently accessible. Indeed, the bonding particle for gravitation, the *graviton*, has never been observed. We shall discuss the other two forces in more detail.

Strong Force. The strong force involves the quarks, described earlier, and the **gluons**, a set of particles with spin $\hbar$. The gluons are bonding particles, like photons for electromagnetism or pions for nuclear forces. The quarks carry a quantum number analogous to the electric charge that, like electric charge, is both quantized and conserved. This quantum number is called the *color*, but it should not be confused with the colors of the visible-light spectrum. The electric charge occurs in two different types; the color quantum number occurs in three different types. Gluons are emitted and absorbed by quarks or by any particles with a color quantum number, just as photons are emitted and absorbed by particles with electric charge. But there is a crucial difference between electromagnetism and the strong force: Photons do not *carry* electric charge, *but gluons carry the color quantum number and thus can themselves interact via the strong*

TABLE 46–2

FUNDAMENTAL FORCES

Force	*Particles on which Force Acts*	*Particles that Transmit Force*
Gravitation	Any particle with mass	Gravitons[†]
Electroweak force	Particles with electric charge or "weak" charge	Photons, weak bosons
Strong force	Quarks, gluons	Gluons

[†]Gravitons have never been observed in any sense, nor has the law of gravitation (general relativity) been satisfactorily integrated with quantum ideas.

force through the exchange of other gluons. This difference is crucial. The theory that describes these interactions is **quantum chromodynamics**.

The fact that gluons interact among themselves through the strong force leads to a force between quarks and gluons that *increases* with distance. (The restoring force due to a spring has this property, but the spring force is not a fundamental one, and a spring will distort and eventually break if it is stretched too far.) The farther apart the quarks and gluons are, the more strongly they attract one another. Any effort to pull quarks and gluons out of the nucleon by supplying energy is therefore bound to fail. *This is the basis of the confinement mechanism discussed earlier*.

Note that because strong forces increase as distance increases, the concept of range and its connection to the mass of the transmitting meson has less meaning for the strong force than it does for the nuclear force or the electroweak force. Whether the exchange particles for the strong force, the gluons, have mass is an open question.

Electroweak Force. At the energies typical of the everyday world, the two aspects of the electroweak force, electromagnetism and the "weak" force responsible for β decay, appear to be rather different. The photon is an exchange particle that transmits the electromagnetic aspect of the electroweak force. The weak force manifested in β decay appears to have a very different strength than the electromagnetic force, but this is not the case: The difference is that, in addition to photons, there is another set of force transmitters, the *weak bosons*, W^+, W^-, and Z^0. These bosons, whose superscripts indicate their electric charge, have very substantial masses, some 86 and 97 times the mass of a proton for the W (W^+ and W^- both) and Z^0, respectively. This fact implies that the weak force has not only a limited range (through Yukawa's reasoning), but also a certain energy dependence. If the energies of the particles that interact via the weak force are greater than the mass of the weak bosons, the strength of the "weak" interaction becomes comparable to that of electromagnetism. Thus the apparent difference of the electromagnetic and weak aspects of the electroweak force disappears at high energies.

Quarks as well as leptons can emit or absorb the carriers of the electroweak force and are subject to the conservation of electric charge, of baryon number, and of lepton number. Processes such as β decay, $n \to p + e^- + \bar{\nu}$, can be visualized as a two-step process in which a d-quark within a neutron is transformed to a u-quark:

$$d \to u + W^-, \qquad W^- \to e^- + \bar{\nu}.$$

The unification of the electromagnetic and weak forces is an important step in our understanding of the fundamental forces. In addition to simple economy, a number of predictions follow from the unification. For example, we have mentioned that the "weak" force becomes as strong as the electromagnetic force at high energies, a fact that manifests itself in the dramatically increased rate at which reactions such as $e^- + p \to n + \nu$ occur. To this point, all the predictions of the electroweak unification that have been tested have confirmed that this unification is correct. Further unification is a continuing goal of current research.

A Summary Chart

Figure 46–4 is a chart that summarizes much of the current state of knowledge of fundamental particles and forces.

FUNDAMENTAL PARTICLES AND INTERACTIONS

FERMIONS

matter constituents
spin = 1/2, 3/2, 5/2,...

Leptons spin = 1/2

Flavor	Mass GeV/c²	Electric charge
ν_e electron neutrino	$<2 \times 10^{-8}$	0
e electron	5.1×10^{-4}	−1
ν_μ muon neutrino	$<3 \times 10^{-4}$	0
μ muon	0.106	−1
ν_τ tau neutrino	$<4 \times 10^{-2}$	0
τ tau	1.784	−1

Quarks spin = 1/2

Flavor	Approx. Mass GeV/c²	Electric charge
u up	4×10^{-3}	2/3
d down	7×10^{-3}	−1/3
c charm	1.5	2/3
s strange	0.15	−1/3
t top (not yet observed)	>89	2/3
b bottom	4.7	−1/3

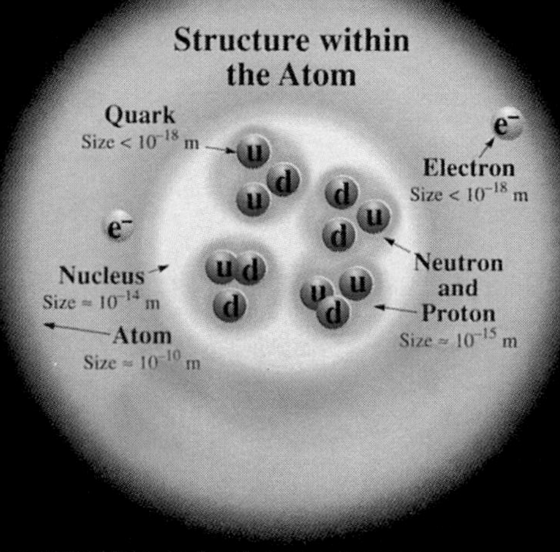

Structure within the Atom

Quark Size $< 10^{-18}$ m
Electron Size $< 10^{-18}$ m
Nucleus Size $\approx 10^{-14}$ m
Neutron and Proton Size $\approx 10^{-15}$ m
Atom Size $\approx 10^{-10}$ m

BOSONS

force carriers
spin = 0, 1, 2,...

Unified Electroweak spin = 1	Mass GeV/c²	Electric charge
γ photon	0	0
W^-	80.6	−1
W^+	80.6	+1
Z^0	91.16	0

Strong or color spin = 1	Mass GeV/c²	Electric charge
g gluon	0	0

PROPERTIES OF THE INTERACTIONS

Property \ Interaction	Gravitational	Weak (Electroweak)	Electromagnetic (Electroweak)	Strong: Fundamental	Strong: Residual
Acts on:	Mass – Energy	Flavor	Electric Charge	Color charge	
Particles experiencing:	All	Quarks, Leptons	Electrically charged	Quarks, Gluons	Hadrons
Particles mediating:	Graviton (not yet observed)	W^+ W^- Z^0	γ	Gluons	Mesons
Strength for two u quarks at: (relative to electromagnetic) 10^{-18} m	10^{-41}	0.8	1	25	Not applicable to quarks
Strength for two u quarks at: (relative to electromagnetic) 3×10^{-17} m	10^{-41}	10^{-4}	1	60	Not applicable to quarks
for two protons in nucleus	10^{-36}	10^{-7}	1	Not applicable to hadrons	20

Sample Fermionic Hadrons

Baryons qqq and Antibaryons $\bar{q}\bar{q}\bar{q}$

Symbol	Name	Quark content	Electric charge	Mass GeV/c²	Spin
p	proton	uud	1	0.938	1/2
$\bar{p}$	anti-proton	$\bar{u}\bar{u}\bar{d}$	−1	0.938	1/2
n	neutron	udd	0	0.940	1/2
Λ	lambda	uds	0	1.116	1/2
Ω^-	omega	sss	−1	1.672	3/2

Sample Bosonic Hadrons

Mesons $q\bar{q}$

Symbol	Name	Quark content	Electric charge	Mass GeV/c²	Spin
π^+	pion	$u\bar{d}$	+1	0.140	0
K^-	kaon	$s\bar{u}$	−1	0.494	0
ρ^+	rho	$u\bar{d}$	+1	0.770	1
D^+	D+	$c\bar{d}$	+1	1.869	0
η_c	eta-c	$c\bar{c}$	0	2.980	0

FIGURE 46–4 Electric charges of the particles listed here are given in units of the proton's charge, and spins are given in units of $\hbar$. There are two more neutrinos, two more leptons, and four more quarks than we describe in the text (the "top" quark was not discovered as of 1992, but its presence has been inferred); we do not now understand why these additional particles exist. There are many more fermionic hadrons and bosonic hadrons than are described here, all of which can be accounted for by quantum chromodynamics. More complete copies of this chart can be obtained from CPEP, MS 50-308, Lawrence Berkeley Laboratory, Berkeley, CA 94720.

46–4 TOOLS OF PARTICLE PHYSICS

Richard Feynman compared the process of understanding the behavior of elementary particles to understanding how a watch works by throwing two watches at each other and studying the array of gears and springs that come out. Actually, the situation is somewhat worse. We have seen that we require high energies to see things more sharply (that is, at short distances), but high energies bring with them relativistic effects such as pair production (pairs of "watches and antiwatches" are produced). In addition, because of confinement, the gears and wheels cannot themselves be final products, just more watches, perhaps of a different design but made from those same gears and wheels. Many of the particles that result are unstable and decay so quickly that we can see only their decay products. Finally, the collisions being studied are relatively rare, so a high flux of colliding particles (many watches) is necessary.

The problem of designing experiments to study the behavior of particles at short distances is twofold: First, we require particle accelerators that have the best possible combination of high energies and high beam currents, appropriate to the particles and processes being studied. Second, we need detectors that enable us to see and analyze the results of collisions. In constructing accelerators and detectors, physicists have pushed technologies to their limits. The results have been scientifically and technologically rewarding.

Accelerators

The principle of modern particle accelerators is the repeated acceleration of a stable charged particle by an electric field. A magnetic field controls the path the beam of accelerated particles follows. Beyond these basic ideas, a variety of techniques are used in different accelerators.

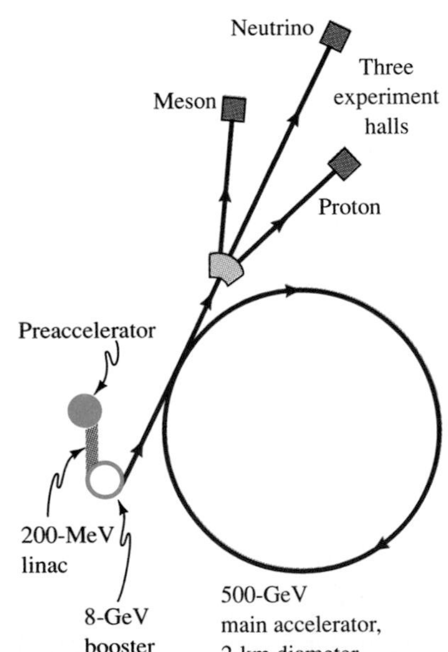

FIGURE 46–5 Schematic diagram of a fixed-target machine, in this case an early version of the accelerator at the Fermi National Accelerator Laboratory. Note the several stages of acceleration (beginning with the preaccelerator and a linear accelerator section, or linac) and the extensive target areas (experiment halls), where beams extracted from the ring of the main accelerator collide with fixed targets.

Electrons or Protons. Electrons are analogous to quarks in the sense that neither has been observed to have a fundamental substructure: Electrons and quarks are elementary. In addition, it appears that electrons and quarks are *both* necessary in quantum mechanical models with calculable predictions. These facts may suggest a possible unification between strongly interacting particles and particles that interact by means of the electroweak interactions. Thus there is as much interest in the observation of electron collisions as in the observation of proton collisions; accelerators exist that allow us to study pp, $p\bar{p}$, ep, and $e\bar{e}$ collisions. Proton collisions are of interest in the sense that they represent collisions involving quarks. Given that a proton shares its energy among three constituent quarks and that our ability to accelerate a particle depends on its total charge, we can accelerate quarks up to only one-third the energy to which we can accelerate electrons.

Fixed-Target Machines or Colliders. Until the 1960s, virtually all accelerators were *fixed-target machines*, in which an accelerated beam is incident on a target (Fig. 46–5). In a fixed-target machine, the beam is typically extracted and directed to a fixed target such as liquid hydrogen. The advantages of such an accelerator are that a huge number of target particles can be spread throughout any volume and that there is only one beam to manage. These machines have systematically been replaced by *colliders*, in which two moving beams of particles with equal but opposite momenta interact in head-on collisions with one another

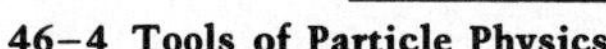

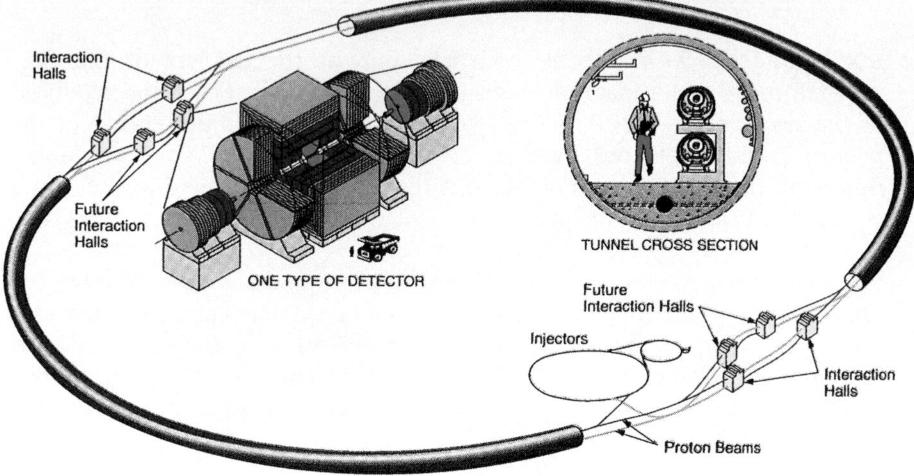

FIGURE 46–6 Schematic diagram of a collider, in this case the Superconducting Supercollider. The entire ring has a circumference of some 90 km.

TABLE **46–3**

CHARACTERISTICS OF SOME COLLIDERS

Name	*Laboratory, Location*	*Particles*	*Maximum Beam Energy (GeV)*
Sp$\bar{p}$S	European Laboratory for Particle Physics (CERN), Geneva, Switzerland	$p\bar{p}$	315
TEVATRON	Fermi National Accelerator Laboratory (FNAL), Batavia, Illinois	$p\bar{p}$	1000
HERA	Deutsches Elektronen–Synchrotron (DESY), Hamburg, Germany	ep	e: 26; p: 820
Large Electron–Positron Collider (LEP)	European Laboratory for Particle Physics (CERN), Geneva, Switzerland	$e\bar{e}$	60
Stanford Linear Collider (SLC)	Stanford Linear Accelerator Center (SLAC), Stanford, California	$e\bar{e}$	50
Superconducting Supercollider (SSC)†	SSC Laboratory, Waxahatchie, Texas	pp	20,000

†Planned operation date: late 1990s

(Fig. 46–6).† Table 46–3 lists the characteristics of some advanced colliders. We know that the momentum of the center of mass of a system of interacting particles is constant. When a moving particle collides with a stationary target, the center of mass has nonzero momentum, which is maintained after the collision. Thus a large fraction of the incident energy remains in the motion of the center of mass rather than going into the production of particles, the most interesting part of the reaction. In colliders, the system's center of mass is at rest, and all the energy goes into the production of particles. Given that the cost of an accelerator increases rapidly with the energy of the beam or beams it accelerates, a collider buys more center-of-mass energy per dollar.

†Actually, in some colliders the particles of the two beams do not have equal momenta. We assume throughout that we deal with the symmetric case.

EXAMPLE 46–4 The W^+ particle, one of the carriers of the electroweak force, has a mass of about 80 GeV/c^2. (a) For W^+ to be produced in the reaction $p + \bar{p} \to W^+ + \pi^-$, what would the minimum energy of the proton (and antiproton) have to be in a collider? (b) What would the minimum energy of the $\bar{p}$ have to be if the same reaction were to occur in a collision with a stationary proton?

SOLUTION: Figure 46–7 shows the $p\bar{p}$ collision in the frames of reference appropriate to the collider and to the fixed-target machine, with the momenta and energies labeled. (a) Energy conservation in the center-of-mass reference frame gives $2E = E_W + E_\pi$. The minimum energy, $E = E_{min}$, corresponds to the W^+ and π^- both produced at rest, so $E_W = M_W c^2$ and $E_\pi = m_\pi c^2$. With $m_\pi \ll M_W$, we have $E_\pi \ll E_W$, so

$$E_{min} \simeq \tfrac{1}{2} M_W c^2 \simeq \tfrac{1}{2}(80\ \text{GeV}) = 40\ \text{GeV}.$$

(b) In the laboratory reference frame, the total momentum is k, the momentum of the initial $\bar{p}$. At the minimum energy, the W and π are at rest relative to each other and act like a single particle of mass $M_W + m_\pi \simeq M_W$. Momentum conservation ensures that the momentum of the W–π pair is k. Energy conservation then gives

$$E_{lab,min} + m_p c^2 = \sqrt{k^2c^2 + M_W{}^2 c^4},$$

where $E_{lab,min}$ is the minimum energy of the $\bar{p}$. We square both sides of this equation to get

$$E^2_{lab,min} + 2E_{lab,min} m_p c^2 + m_p{}^2 c^4 = k^2c^2 + M_W{}^2 c^4.$$

Finally, we use the fact that in the laboratory reference frame, the energy of the initial $\bar{p}$ is related to its momentum by $E_{lab,min} = \sqrt{k^2c^2 + m_p{}^2c^4}$. We square this relation and substitute for the term $E^2_{lab,min}$ in the preceding equation:

$$k^2c^2 + m_p{}^2c^4 + 2E_{lab,min} m_p c^2 + m_p{}^2 c^4 = k^2c^2 + M_W{}^2 c^4,$$

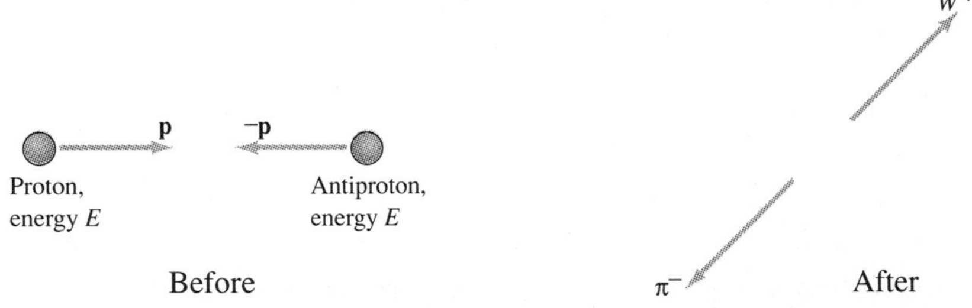

FIGURE 46–7 Example 46–4. (a) The reaction as seen in the center-of-mass reference frame. (b) The reaction as seen in the laboratory reference frame.

an equation we can solve for $E_{\text{lab,min}}$. With $m_p \ll M_W$, we get

$$E_{\text{lab,min}} = \frac{M_W{}^2c^4}{2m_pc^2} \simeq \frac{(80\text{ GeV})^2}{2(0.94\text{ GeV})} = 3400\text{ GeV}.$$

The huge difference between the answers to parts (a) and (b) illustrates the advantage of colliders. It is far cheaper to accelerate two beams to 40 GeV each than to accelerate one beam to 3400 GeV.

The difficulty of controlling the two beams of a collider is not as great as you might imagine. In Fig. 46–8, we can deduce that the same electric field that accelerates particles in one direction accelerates their antiparticles in the other direction, and the same magnetic field that forces charged particles to follow a clockwise path also forces their antiparticles to follow a counterclockwise path. Thus the same fields can be used for a beam of both electrons and positrons, or protons and antiprotons. The major complication is that the two beams must be concentrated and steered so that they cross at isolated points where the collisions occur; at those points, elaborate detectors monitor the results of the collisions.

We have shown how the fact that the center of mass of a system of particles in a collider is at rest is advantageous. A second advantage of a collider concerns the distribution of the many particles that are produced in a collision. In the center-of-mass reference frame, the particles produced are spread over all angles (Fig. 46–9), and this is the way the reaction appears in a collider. The fact that

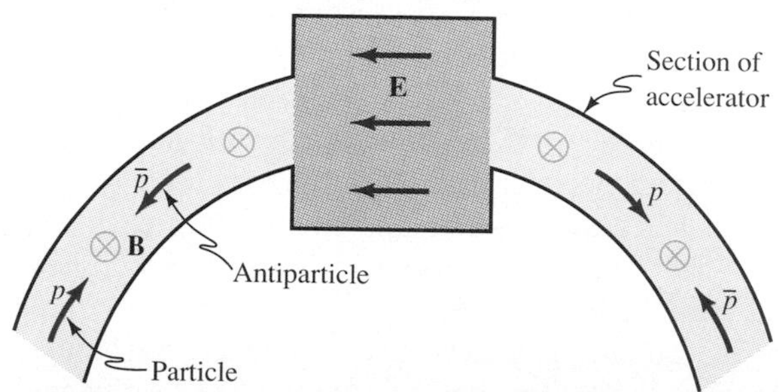

FIGURE 46–8 In a collider, the same magnetic fields guide both particle p and antiparticle $\bar{p}$, and the same electric fields accelerate both p and $\bar{p}$.

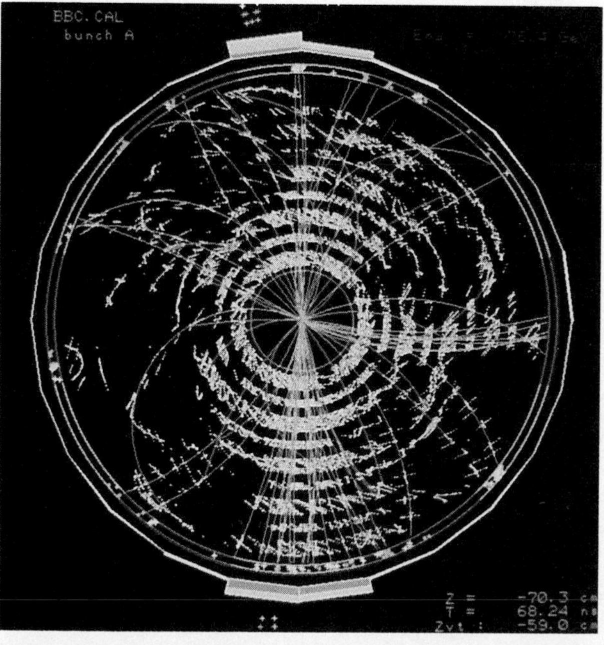

FIGURE 46–9 A collision at the TEVATRON collider at FNAL, as recorded in the Collider Detector Facility. The view is an end-on one—that is, along the direction of the colliding beams. A substantial number of particles are produced at right angles to the beams.

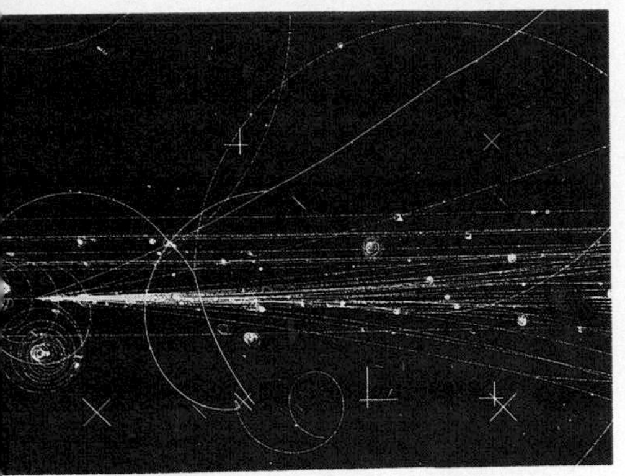

FIGURE 46–10 A proton of energy 300 GeV strikes a stationary proton in an experiment performed at FNAL. In this fixed-target machine, most of the particles produced are thrown forward in a forward cone.

the center of mass moves in a fixed-target machine means that the particles produced are carried forward with it and hence concentrated in a narrow cone in the forward direction. As Fig. 46–10 shows, this concentration makes it difficult to separate the products.

Each beam of a collider provides a target for the other beam that contains far fewer target particles than does a fixed-target machine. Thus the rate at which collisions occur is less in a collider than in a fixed-target machine. This is in fact not always a disadvantage, because the reactions we observe at very high energies produce so much information that even the best computers cannot keep up with data that arrive too rapidly. But when rare reactions are to be studied, a collider may be at a real disadvantage compared to a fixed-target machine.

Circular versus Linear Colliders. The problem of beams that are not very dense (compared to target densities in a fixed-target machine) is in part overcome in colliders that are circular, so that the beams cross and recross frequently. However, recall from Chapter 36 that energy is radiated in the form of electromagnetic waves (or photons) when charged particles accelerate. For the beams of an accelerator, this is undesirable in that we want the beam to retain as much energy as possible. This is unavoidable for the initial acceleration that gives the particle its energy. However, the circular motion in a circular collider means that the beam undergoes a centripetal acceleration that is proportional to the momentum squared. Thus the higher the energy of the beam, the greater the radiation of energy. Moreover, the particles' acceleration is inversely proportional to their mass, so lighter particles, such as electrons, radiate energy at a higher rate than do heavier particles, such as protons. LEP, a circular machine that accelerates electrons and positrons to 60 GeV each in a ring of radius 5 km (Table 46–3), is at about the limit for an electron collider in the form of a ring—at higher energies, the electron beam loses too much of its energy with each turn. The alternative is to construct a linear accelerator, in which a new electron beam is repeatedly generated along a straight beam tube without following a circular trajectory. Because a collider has advantages over a fixed-target machine, as described above, two linear machines with their beams meeting head-on provide the best solution. The major disadvantage of such a machine, which has indeed been constructed on a demonstration basis (SLC in Table 46–3), is that the beams cross only once, so there is relatively little chance for collision. To overcome this difficulty, the beams are concentrated to only a few microns in size, a tiny fraction of the size of the beams of other accelerators; these beams must then cross with a remarkable precision. Such machines appear to be the only way to construct electron colliders with energies much greater than those of LEP.

FIGURE 46–11 Photographic emulsions can record the passage of charged particles: Such particles produce tracks in the developed emulsion. In this 1948 image, a charged pion is produced at point A and decays at point B. Emulsions are now rarely used, because they are slow and cannot be "triggered."

Detectors

The apparatus that observe and analyze collisions in particle accelerators are called *detectors*. Detectors have undergone continual evolution since Joseph J. Thomson first detected the electron in 1897. Subnuclear particles can be detected because they leave a trace of energy as they pass through or decay within a detector. For example, a charged particle that passes through a *photographic emulsion*, a suspension used to coat photographic plates, film, and paper, leaves the same kind of trace that light does: The particle may dislodge electrons from atoms along its path and thereby cause a chemical reaction within the material and the consequent deposit of silver along the track. Stacks of photographic emulsion formed the earliest detectors (Fig. 46–11)—they were in use until the 1950s—and allowed scientists to learn much about the charged particles that enter the earth's atmosphere (*cosmic rays*). The resolution of emulsions—their ability to pinpoint a particle track—is less than a micron (10^{-6} m). A similar detection technique uses the fact that a liquid near its boiling point forms bubbles of vapor along the path of a passing charged particle. In use until the 1980s, *bubble chambers* filled with liquid hydrogen utilized this technique (Fig. 46–12). Other, more modern types of detectors register the passage of a charged particle or of neutral particles, such as photons, by their characteristic effects on the electronic properties of the detector. The resolution of these modern detectors can be quite high, although not yet as high as the resolution of emulsions.

A detector may be sensitive to the total amount of energy a particle deposits as it is brought to a stop within the detector. This type of detector is called a *calorimeter*. Alternatively, a particle can be identified by the rate at which it loses energy as it passes through parts of the detector. In a supplementary technique, by measuring the amount of curvature in the path of a passing particle in a magnetic field of large, known magnitude within a detection region, the particle's momentum is determined. Knowing both the momentum and energy, we can calculate the mass, so the particle can be identified. Particle identification is important and not so simple at high energies.

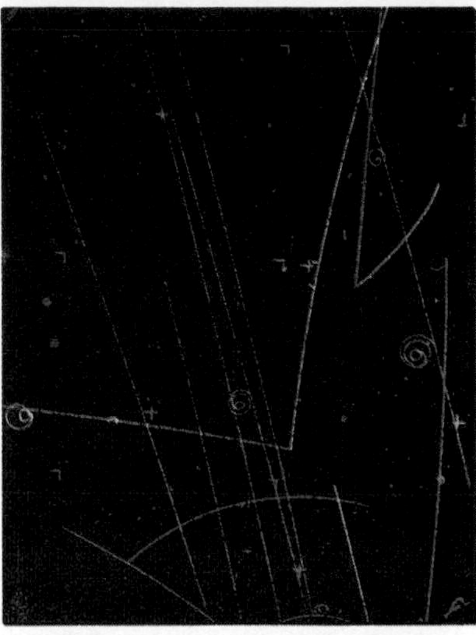

FIGURE 46–12 Bubble chambers can record the passage of charged particles: A chain of bubbles forms around the particle tracks. In this false-color image, a π^- (green) collides with a stationary proton in the chamber and produces two neutral particles that travel invisibly through the chamber until they decay into charged particles. One neutral particle decays to a proton (red) and a π^- (green); the other neutral particle decays to a π^- (green) and a π^+ (yellow). The charged tracks curve because a magnetic field is present. Analysis of the decay tracks reveals properties of the neutral particles. The blue tracks were made by particles not involved in this interaction.

Events cannot be recorded selectively with photographic emulsions and bubble chambers. Techniques that use electronic *triggers* avoid this difficulty. For example, suppose that you are interested in viewing only those events in which a muon (see Chapter 40) of high energy is produced. Muons pass through matter more easily than do protons or other strongly interacting particles. Then it is possible to surround the interaction region with a great deal of matter—such as many tons of steel—and outside this matter to place a detector that registers electronically the passage of charged particles. This detector emits a signal that what has passed is likely to be a muon and thereby acts as a trigger so that the rest of the event will be registered. Modern detectors are huge and complex (Fig. 46–13). The electronic registration of the passage of dozens or even hundreds of particle tracks requires vast information storage facilities and powerful computers to extract the most useful information. High-energy physicists have been pioneers in the use of large scientific computers. The computers and analysis techniques that triggered detectors require have filtered down to more general use.

FIGURE 46–13 The detectors used in colliders are large and complex and may require the work of hundreds of physicists, engineers, and technicians over several years to operate. Here we show a view of the wiring of the H1 detector, which is to be used at HERA, in Hamburg, Germany (see Table 46–3).

46–5 THE EXPANDING UNIVERSE

Giordano Bruno was burned at the stake in 1600 in good part for claiming that the universe is infinite. Although passions have calmed somewhat, people have always had a lively interest in the nature of the cosmos as a whole (*cosmology*). Powerful telescopes that increase our ability to see into the depths of the universe give us a glimpse into the ancient past as well, because the finite speed of light implies that the light we observe from distant sources was emitted long ago. Our instruments have allowed us to see so far back into the past that we now have, for the first time, concrete data about the very early stages of our universe.

Olbers's Paradox. *Olbers's paradox* was stated in 1823, when telescopic observations had made it clear that the universe contains many, many more stars than are visible with the naked eye. Heinrich Olbers argued that if the universe were infinite and unchanging, then no matter in what direction we looked, we would see a star and star light would cover the entire sky.† We cannot dispute this argument by arguing that distance makes stars too faint to see. Suppose that N stars are contained in a thin spherical shell a distance R away from us. Then, because the area of a sphere increases as the radius-squared increases, there would be $4N$ stars in a shell a distance $2R$ away from us. The intensity of the light decreases as the distance-squared increases, so each star in the more distant shell is only one-fourth as intense as those in the nearer shell, but the total intensity is *independent* of the distance of the shell. The net intensity of an infinite number of concentric shells would be infinite. Yet—and this is what constitutes Olbers's paradox—the night sky is not bright. The situation is actually worse than Olbers imagined: He did not realize that in an infinite and unchanging (static) universe, all the stars would come to thermal equilibrium with one another (see Chapter 17) at the temperature of the surface of a typical star such as the sun, some 6000 K. A night sky as bright and hot as the surface of the sun would not be very comfortable! Before we can resolve Olbers's paradox, we must learn about more modern astronomical observations.

Hubble's Law

In Chapter 40 we discussed Hubble's law, which was discovered in 1929 by Edwin Hubble through the use of the relativistic Doppler shift. The spectral lines of the stars in distant galaxies showed, on average, a redshift. Such a shift implies that those galaxies are receding from us. Hubble had at his disposal the Mount Wilson telescope in southern California, the first telescope sufficiently powerful to distinguish individual stars in what were then termed nebulae. Before 1923, nebulae were thought to be clouds of gas within our own galaxy. (It was only a few years before this that Harlow Shapley had determined that what we now know to be our galaxy has a definite shape and that our sun is far from the galaxy's center.) Hubble identified the nebulae as galaxies like our own and thereby took a giant step toward the understanding of the universe. Continued work led to Hubble's law, which relates the average speed u with which another galaxy retreats from our own to a distance D away from us [Eq. (40–18)]:

$$D = \frac{u}{H}.$$

H is the *Hubble parameter*, measured to be $H \simeq 2.5 \times 10^{-18}\ \mathrm{s}^{-1}$. The redshift gives the speed of recession directly, but how did Hubble know the distance of

† Johannes Kepler had also realized in 1610 that there was a possible problem.

a given galaxy from our own? He used the cepheids, a class of powerful stars whose systematic waxing and waning make them easily identifiable. The intensity of these stars within our galaxy is characteristic, and Hubble supposed that any cepheid, even one in a distant galaxy, would emit light with that characteristic intensity. To assign a distance to a given cepheid, Hubble used the fact that the intensity at any distance decreases as the inverse square of the distance from the star.

Astronomers have refined Hubble's procedure by identifying other characteristic objects whose distance can be measured by their intensity. There is also *independent* evidence that objects with large redshifts, such as quasars (referred to in Chapter 40), are very distant: We see some of them through gravitational lenses, which we discussed in Chapter 12. This is possible only if these quasars are much farther than the galaxies that produce their images through the bending of the quasar light. The distances calculated from the lensing effect are consistent with those provided by Hubble's law.

The Big-Bang Model

An immediate conclusion follows from the fact that other galaxies recede from ours at a speed proportional to their distance from us: *At some time in the distant past, all galaxies must have been very close together*. This fact defines what is conventionally called the *big-bang* model of the history of the universe. Note that Hubble's law does not mean that the earth (or our galaxy) occupies a privileged position in the universe. In fact, it means just the opposite, as the following one-dimensional analogue shows. Figures 46–14a and 46–14b show a line of galaxies at times T and $2T$, respectively, after the moment when the big bang occurred, $t = 0$. Note that at time T the galaxies are evenly spaced; there is no preferred point in this arrangement. We have identified our galaxy (galaxy A) and supposed that Hubble's law holds. Therefore galaxy C, which is twice as far from us as galaxy B, recedes from us at twice the speed with which galaxy B recedes. Thus at time $2T$, galaxy C is still twice as far from us as galaxy B. The galaxies continue to be uniformly spaced, and no point along the line is preferred. Hubble's law can be contrasted with a law in which other galaxies retreat from ours at constant velocity. Even though the galaxies are evenly spaced at time T, at time $2T$ this is no longer the case, and galaxy A occupies a special point (Fig. 46–14c). In fact, Hubble's law is the unique law for galactic speeds by which no one galaxy occupies a special point within the cosmos. This

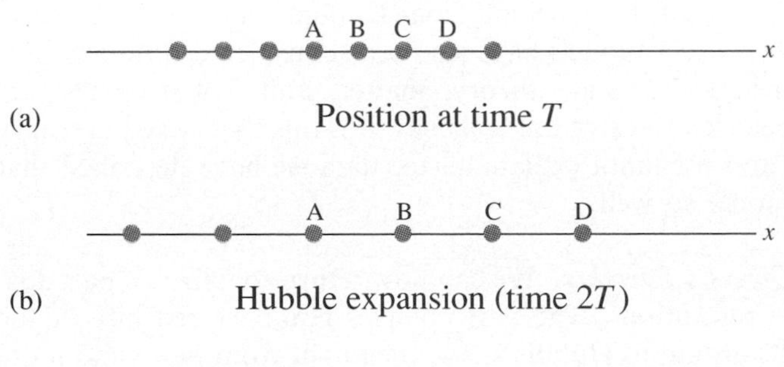

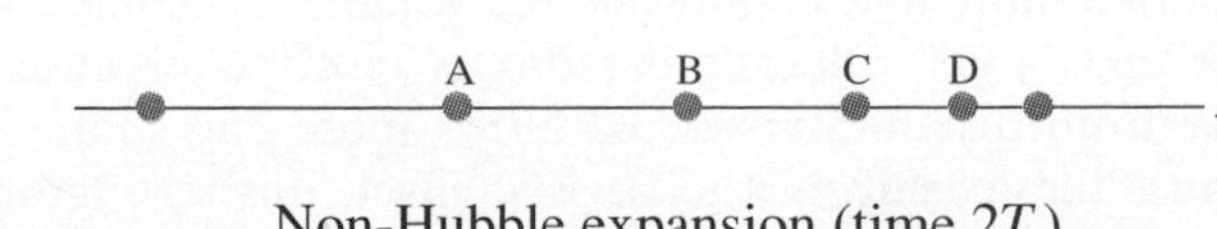

FIGURE 46–14 (a), (b) Two successive times in a one-dimensional version of Hubble's law. The points (galaxy positions, for example) remain equally spaced along the line. (a), (c) The same two successive times in a (non-Hubble) expansion law in which all the points recede from point A at the same speed, independent of their distance from A. Point A is a special point in the expansion law.

can be confirmed by studying Figs. 46–14a and 46–14b from the point of view of galaxy B (see Problem 29). An inhabitant of galaxy B would agree that Hubble's law holds for him or her as well. A two-dimensional analogue of the situation described by Hubble's law is provided by dots painted on a spherical balloon (Fig. 40–16). When the balloon is blown up, *all* the dots recede from each other, with no dot occupying any special position, and a law like Hubble's law holds. If we lived on the surface of such a balloon, any voyage we took would eventually take us back home, and we would conclude that there is no "outside."

That the earth's galaxy occupies no privileged position is the end of centuries of discussion in which, gradually, humankind has understood that it is not at the universe's physical center. There *is* no center. It is not as though we are now living in a space in which an explosion once occurred; such a description would imply a privileged point, the position of the explosion itself. Rather, all of space began as a small volume *with no outside*.

Hubble's law with a constant value of H can be used to find the time required for galaxies to separate by given amounts (see Problem 32). Unfortunately, if the law is used to find the time at which all matter was together, that time is infinite (see Problem 33)! Thus if there was an initial moment when the universe started to expand, Hubble's law with a constant value of H cannot hold all the way back to that time. We can, however, use the law to make an estimate: If we assume that the recession speed is constant, Eq. (40–18) states that the expansion time (the *Hubble time*) is just H^{-1}, some 13 billion years. Other information (for example, theoretical knowledge of stellar evolution and experimental identification of the characteristics of the oldest stars) confirms this estimate to within a factor of 2.

We discussed background radiation and its discovery in Chapter 17.

We can cite three important pieces of evidence that support the idea that the entire visible universe has expanded away from a small volume. The first and most striking piece of evidence is the presence and character of the observed background blackbody radiation. According to the big-bang model, this radiation is a relic of the universe at the tender age of 300,000 years. The second piece of evidence is that the relative amounts of light elements, such as hydrogen and helium, that are distributed throughout the visible universe are successfully predicted by the big-bang model (see Section 46–6). The third piece of evidence is that stars and galaxies exhibit signs of having evolved over time; for example, there appears to have been a time when quasars were more abundant than they are now, and the optical emissions of older radio galaxies (galaxies that emit radio waves strongly) are different from those of younger radio galaxies.

Although it may be natural to conclude from Hubble's law that all matter was once located together, other models have also been considered seriously. For instance, in the so-called steady-state theory, matter, and new galaxies, are created in the regions between the receding galaxies. But this theory requires new physical laws to work, and it cannot explain all the data we have described that support the big-bang model so well.

The Resolution of Olbers's Paradox. We can now return to Olbers's paradox, which has two possible resolutions. The first involves Hubble's redshift. If the universe is expanding according to Hubble's law, then light from more and more distant stars will be redshifted more and more while the volume to be filled with radiation becomes larger and larger. Olbers's paradox is resolved because the light reaching our galaxy from distant stars is redshifted more and more with increasing distance. Because the redshift is a relativistic effect, this resolution of the paradox uses the theory of general relativity and is often cited as evidence for that theory. The second resolution is not at all relativistic but does depend on a big bang: If the universe has existed for only a finite time, then any one star has been radiating for a finite time, and there has not been enough time for the radiation from all stars to have filled the universe and created a bright night sky. Either resolution recognizes a time dependence in the evolution of the universe that is incompatible with a static model.

What does it mean to say that the universe has a finite "size?" Or, to rephrase, what would happen if we tried to throw something beyond the limits of the universe? This troubling question can be answered in the context of general relativity, which allows space to have a curvature in the presence of matter. We can best describe the curvature of three-dimensional space by analogy with a two-dimensional space such as the surface of the earth. If we were restricted to move *only* on the surface of the earth, we would find that a long voyage would have no limits yet would still lead us back, after a long circumferential trip, to our starting point. The space on the surface of a sphere is closed—that is, finite—yet has no real boundary, even though the "size" of this surface is well defined. The surface of the earth, as opposed to the surface of an infinite plane, has *curvature*. Whether or not our three-dimensional space has curvature is a question that can be answered, at least in principle, without making a round trip, which in the case of our universe would be a very long one indeed (see Question 1).

46–6 THE EARLY MOMENTS OF THE UNIVERSE

Special relativity tells us that energy cannot be distinguished from mass. When we move back in time to the early universe, the mass density of the universe increases, which means that the energy density increases. And by the principles of thermodynamics, the temperature increases as well.[†] Only one mathematical model, the *Friedmann model*, seems to describe the overall dynamic structure of an expanding universe in a way consistent with the formalism of general relativity—indeed, this model can be used to define the big-bang theory. When Alexander Friedmann proposed his model in 1922, he had no idea that he was correctly predicting the expanding universe that Hubble observed. Friedmann was merely trying to find the simplest model predicted by Einstein's general relativity, and that model implies a dynamic, expanding universe.[‡]

The Friedmann model follows from two simple assumptions that together are referred to as the *cosmological principle*: The universe looks the same in all directions (the universe is *isotropic*), and it looks the same from any observation point (the universe is *homogeneous*). Both assumptions are roughly consistent with observation. We say "roughly" because the presence of clumps of matter such as planets, stars, galaxies, and clusters of galaxies appear to be inconsistent with homogeneity. However, if we take a distant enough look, the distribution of matter is relatively uniform, as in Fig. 46–15 (but see "The Uniformity of the Universe" on p. 1370). The Friedmann model has a definite prediction for

[†] Actually, thermodynamics applies only to situations in which there is thermal equilibrium, and a dynamic, expanding universe is not necessarily in thermal equilibrium at each stage. We shall assume, however, that we can speak about a well-defined temperature at each stage of the expansion.

[‡] Einstein later deeply regretted that, influenced by the astronomers' belief that their observations revealed a static universe, he had spent much time and effort on a model consistent with general relativity and a static universe. Such a model is not only more artificial than Friedmann's model, but it was eventually ruled out by Hubble's observations.

FIGURE 46–15 Structure is evident in this very large scale composite image of some one million galaxies. That structure repeats uniformly across the entire view. On this very large scale, the universe is homogeneous.

temperature of the universe as a function of its size: For all but the earliest times in the history of the universe, temperature is inversely proportional to "size," or *distance scale*, R:

$$T \propto \frac{1}{R}. \qquad (46\text{–}7)$$

Thus as we move forward in time and the universe expands (R increases), the temperature decreases. The Friedmann model also describes how R, and hence T, changes as a function of time.

Temperature and Energy Scales

Equipartition is described in Chapter 19.

The equipartition theorem of statistical mechanics states that the energy of an individual particle in a system at temperature T is on the order of kT. As the temperature increases, the increasing energy can bring us to the realm where atoms or nuclei can be broken up or where the collisions between nucleons are in the energy regime described in the first two sections of this chapter. The thermodynamics of systems of particles in this regime is complicated, because particles and antiparticles can be created or annihilated, but statistical techniques are sufficiently well developed to treat these complicated systems. At the high temperatures of the early universe, the interests of physicists studying the behavior of matter at the shortest distances and those studying cosmology converge. The further back in time we go, the higher the temperatures we encounter, from energies where the fundamental physics is well understood to energies where we are still groping for answers about the behavior of matter.

It is helpful to view the changes in the early universe in terms of a logarithmic time scale, because we can "stretch" smaller and smaller fractions of a second into finite segments on the logarithmic scale. Such a scale never arrives at time zero, because $t = 0$ is the point $-\infty$ on the logarithmic scale. However, we do not in any case know how to go all the way to $t = 0$, because of our difficulty in reconciling quantum physics with general relativity, in which the presence of matter affects the nature of space and time. All that our knowledge really allows us to say is that, thus far, we have a satisfactory model of our dynamic universe back in time to when it occupied a very small but finite volume. To go further back in time is pure speculation. We shall now take a brief tour, starting with the earliest time our present understanding of the physics allows us to treat.†

A Time Chart of the Early Universe

Figure 46–16 illustrates the change in various parameters of the universe, in particular temperature, as a function of time. The temperatures in the chart are

† For more information on this fascinating subject, there are several accessible treatments. We cite in particular Steven Weinberg, *The First Three Minutes: A Modern View of the Origin of the Universe*, Basic, 1988.

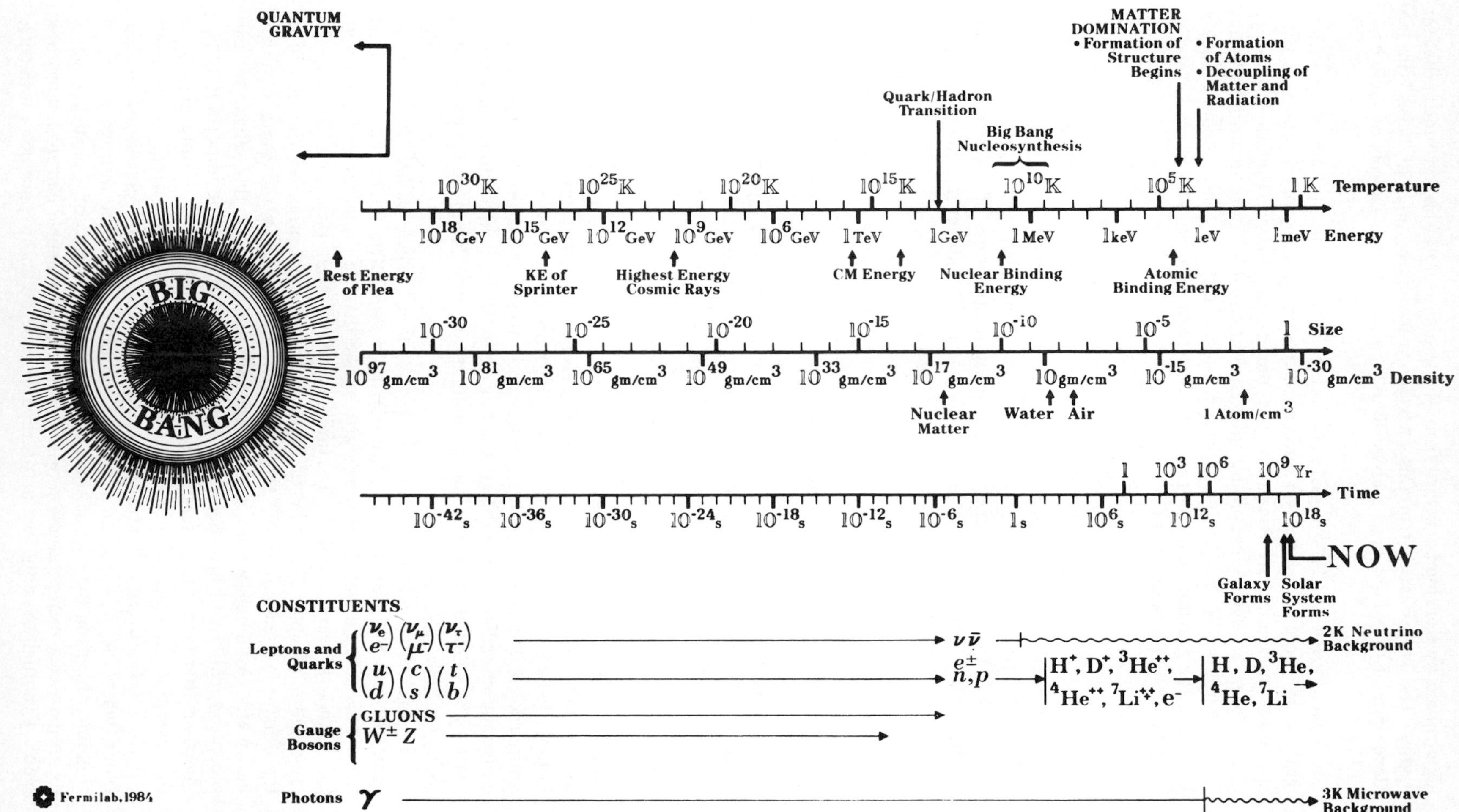

FIGURE 46–16 Stages in the early universe. One of the horizontal axes represents time increasing on a logarithmic scale. An increase in time means an increase in size and a decrease in temperature, characteristic energy, and density. In addition to signaling various "events" in the history of the universe, the chart indicates the changing composition of the universe.

determined by working backward from the temperature of background radiation today. The stages labeled in this chart correspond to different "eras" in the development of the universe, and we shall comment on key aspects of these eras.

Time Period $t < 10^{-43}$ **s.** This period can be called the era of quantum gravity. Physicists have not yet succeeded in combining general relativity with quantum mechanics, so this era is not understood in any fundamental way. Dimensional analysis suggests that this era must involve both quantum physics and gravitation: We can form a mass (or, equivalently, an energy) from Planck's constant, the gravitational constant, and the speed of light. This mass, known as the *Planck mass*, is given by

$$M_P \equiv \sqrt{\frac{\hbar c}{G}} \simeq 2.2 \times 10^{-8} \text{ kg}. \tag{46–8}$$

(Keep in mind that although this mass may appear small, it is roughly equal to 1.3×10^{19} proton masses.) The energy equivalent to this mass is $M_P c^2 \simeq 1.2 \times 10^{19}$ GeV, corresponding to a temperature of 1.4×10^{32} K (see Problem 34). This energy is higher than any we shall ever achieve with an accelerator built with our current technology.

Time Period 10^{-43} **s** $< t < 10^{-12}$ **s.** This period also lies beyond what we can study with current accelerators, whose highest energy corresponds to $t \simeq 10^{-12}$ s. However, if our ideas about strong and electroweak interactions are correct, we can say that nucleons have not yet formed and that the universe is a soup of all the fundamental particles we know about. Particles and antiparticles, including quarks and antiquarks, are continuously being formed in pairs and annihilated.

One feature of this period, having to do with the apparent domination of matter over antimatter in the universe today, deserves special mention. Although we have no direct evidence that distant galaxies are made of matter rather than antimatter (antimatter would form antiatoms with exactly the same emission spectra as atoms), there is much indirect evidence that the amount of antimatter in the universe is limited. For example, even though the earth is continually bombarded with cosmic rays, these particles from distant space are *particles*, not antiparticles. The collision of antimatter and matter would produce spectacular emissions of radiant energy. Yet if a hot early universe had equal amounts of matter and antimatter, which is a fair assumption, how could we have arrived at today's situation?

Experimental evidence from the decays of a particle called the K-meson indicates that physical processes are not quite symmetric under the reversal of time. This is not a statistical time's arrow, like that we discussed in Chapter 20, but rather a real, if small, directionality of physical processes. To use the language of Chapter 20, a film of fundamental physical processes that is run in the forward direction would look slightly different than the same film run backward. Although we do not understand the origin of this effect, we can accept its reality and investigate the consequences. For us the important consequence is that matter is preferentially created over antimatter, if only slightly, in the ensemble of possible reactions that could have occurred up to time $t = 10^{-12}$ s. For every 10^9 antiquarks created, roughly $10^9 + 1$ quarks are created. Later, when the universe had cooled to the point where the average energy of colliding objects in the cosmic soup was less than the mass of a particle (and of its antiparticle), then production of that particle–antiparticle pair no longer took place. Eventually, as the temperature dropped, no more particle–antiparticle production took place. The existing particles and antiparticles could then annihilate each other without being re-created in other collisions. (This occurred when the universe was about 1 s old.) At this point, the slight predominance of matter over antimatter was all-important, because the remaining bit of matter, the matter that exists around us today, had no antimatter to annihilate against, and it persisted.

***Time Period* $\mathbf{10^{-12}\,s < \mathit{t} < 1\,s}$.** The period from 10^{-12} s on covers a realm where we have investigated at least the basic interactions with earth-based experiments. But we have no direct experiments on the behavior of large quantities of quarks and antiquarks. Thus our ability to trace the behavior of the universe is a combination of direct knowledge and deduction. A particularly significant moment in the evolution of the universe was that of the *quark–hadron transition*, when the energy density decreased to the point that quarks, antiquarks, and the gluons formed nucleons and other particles that participate in the nuclear force. This occurred at about 10^{-6} s, when the temperature was some 10^{13} K, corresponding to an energy of around 1 GeV. Of the strongly interacting particles, only the nucleons are stable on the time scale of 1 s, and even that fact must be tempered by our knowledge that neutrons decay due to the weak interaction. Even though up and down quarks occurred in equal numbers in the cosmic soup, so initially the number of neutrons equaled the number of protons, the neutrons decayed such that, at around 1 s, there was one neutron for every five protons. Between 1 s and 100 s, the time at which some nuclei formed and held the neutrons permanently, many more neutrons decayed. At 100 s, neutrons and protons existed in about a 1:7 ratio.

***Time Period* $\mathbf{1\,s < \mathit{t} < 300{,}000\,yr}$.** This period is sometimes referred to as the *radiation era*, because during this time the energy density of electromagnetic radiation was greater than that of matter. The Friedmann model shows that, for this period, the distance scale was proportional to the square root of time,

$$R \propto \sqrt{t}. \tag{46–9}$$

Both the energy density of radiation and the energy density of matter decreased as the universe expanded, but the energy density of radiation decreased more rapidly, because the energy density in mass itself (as opposed to the kinetic energy of matter) was preserved. At roughly 300,000 yr, the two energy densities were equal. We may pick out a significant moment within this era, $t \simeq 100$ s, when the temperature had dropped to the point where neutrons and protons combined to form primarily the nuclei H^+ and $^4He^{2+}$, with some $^2H^+$, $^3He^{2+}$, and $^7Li^{3+}$. (Only these nuclei formed, because they are so strongly bound that potential barriers exist against the formation of others.) All the positive charge in this set of nuclei was matched by an equal quantity of negative charge in the form of free electrons, which continuously absorbed and reradiated the electromagnetic radiation.

***Time Period* $\mathbf{\mathit{t} > 300{,}000\,yr}$.** The energy density of matter was greater than the energy density of electromagnetic radiation once the universe was older than about 300,000 yr, a period often called the *matter-dominated era*. Today, the energy density of matter is several thousand times greater than that of radiation. Since some 300,000 yr ago, the expansion has followed the law

$$R \propto t^{2/3}. \tag{46–10}$$

At about the moment that the universe became matter dominated, which was the moment when the temperature was some 3000 K, the temperature was no longer high enough to keep electrons and ions from combining into stable neutral atoms. The ratio cited above of one neutron for every seven protons determined the number of nuclei formed at 100 s and hence the ratios of the stable elements, primarily hydrogen and helium. With the 1:7 ratio of neutrons to protons, the number of moles of hydrogen formed was 12 times the number of moles of helium formed, and hence the mass of hydrogen was roughly 3 times that of the helium (see Problem 37). All the stellar evolution that has occurred since that time has not significantly changed this difference (only 2 percent of the mass of today's universe is in elements heavier than helium), so this ratio of 3 in the mass abundance can be regarded as a prediction and has been confirmed by observation. This measurement is an important confirmation of the big-bang theory.

All the free electrons are used to form stable atoms, and without free electrons, electromagnetic radiation no longer interacts as strongly with matter. The universe is said to have become *transparent* because radiation *decoupled* from matter. It is this radiation that we spoke of in Chapter 17 when we described the discovery of the 3-K background blackbody radiation. The wavelengths of this radiation increase as the universe expands; in fact, they are a measure of R. Blackbody radiation with larger wavelengths corresponds to blackbody radiation at a lower temperature. What was blackbody radiation at 3000 K at the moment of decoupling has become, through the expansion, radiation at 3 K. The fact that we observe this radiation, which is the oldest directly observable relic of the big bang, supports the ideas discussed here.

In the period that followed decoupling, stars formed, as did galaxies. Just how stellar evolution proceeded is a well-studied subject. The stellar evolution process, which involves series of nuclear reactions that occur in the hearts of stars, is on the whole well understood and confirmed by observation. According to this process, heavy elements, and hence the matter that makes up the earth and its inhabitants, are produced in supernovas, violent explosions of stars. We are thus part of a second generation of stars.

The Uniformity of the Universe

Matter distribution in the universe is granular, or clumped. Galaxies, themselves clumps of stars, are grouped into clusters, and there are large regions with no visible matter (Fig. 46–17). The granular structure is thought to have developed from small, random fluctuations in a uniform energy density, forming small clumps of higher density. Such clumps act as seeds, gathering matter into them under the influence of the gravitational force. The process is much like the formation of clouds, whose droplets require "nuclei" of some kind. The initial

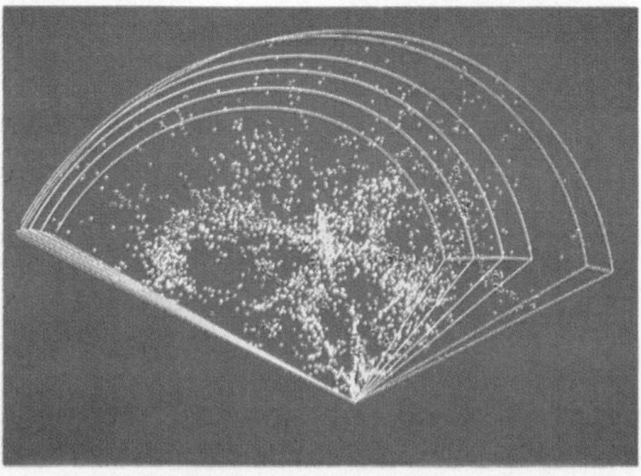

FIGURE 46–17 This composite view of galaxies, which covers a small region of the view in Fig. 46–15, reveals a well-developed structure of clumping and voids. The central structure has come to be known as "stickman."

density fluctuations that started the process must have begun at a time not too late in the evolution of the universe. The 3-K background radiation should contain some "memory" of these fluctuations. For the big-bang theory to remain viable, the observed background radiation must show a departure from uniformity at a certain level of accuracy. In 1992, the COBE satellite (see Chapter 17) discovered nonuniformities in the background radiation consistent with the observed granularity of the matter distribution.

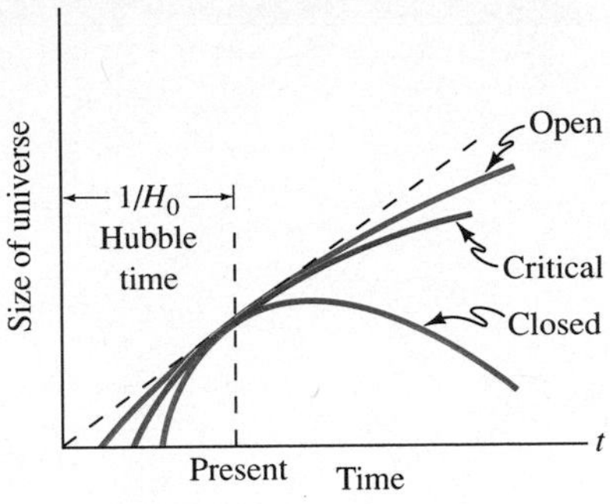

FIGURE 46–18 The evolution with time of open, closed, and critical universes. The vertical axis is some measure of the size of the universe.

The Future

Will the universe continue to expand forever? The answer to that question lies in a parameter of the Friedmann model. At large distances, the gravitational force controls the destiny of the universe. If there is more than a *critical density* of matter, the expansion would slow, stop, and reverse itself (resulting in a *closed* universe). An era of collapse would inevitably follow the expansion era. With less than the critical density, the expanding universe would continue to expand (an *open* universe). In between these cases, there is a *critical* universe, with the critical density of matter (Fig. 46–18). Current observations can account directly for enough matter for only about 10 percent of the critical density. But indirect evidence—details about the rotation of galaxies, for example—indicates the presence of matter that we cannot observe directly (*dark matter*). And on the cosmic scale, 10 percent is close enough to 100 percent to make it tempting to think that the actual density may equal the critical density! In that case, the universe would take forever to reach a point where it is neither expanding nor contracting. We shall surely learn much about physics in trying to untangle the answer to our question.

46–7 A FEW LAST WORDS

In the introduction to this text, we spoke of macroscopic and microscopic realms. For centuries engineers (or the inventors of technology) were limited to manipulating matter in the macroscopic realm on the basis of empirical knowledge. That matter was known to consist of atoms, and that the behavior of atoms was governed by quantum mechanics, initially had little impact on the way engineers worked. Scientists, by contrast, were interested from the start in the fundamental questions for their own time, those that appeared to be rather remote from practical application. The interest of engineers, with certain exceptions, came later and in unforeseeable ways. Thus the bulk of today's technology is built on quantum principles and on an understanding of the quantum nature of matter. It is unimaginable that our society could operate without the quantum-physics-based circuits that make computers or electronic communication possible or that a chemical engineer could construct new molecules without knowledge of their microscopic structure. While engineers continue to build bridges, those bridges are build on a solid understanding of the microscopic behavior of the materials that comprise the bridges and with the aid of computers that depend on the microscopic realm.

We are still learning how to apply quantum mechanical ideas to bulk materials. We are still in the process of evaluating how to make the best use of our scientific understanding of atoms and nuclei. And we have not yet seen how we can directly apply the knowledge we have gained in the areas of particle physics and cosmology. But we can say with confidence that if our society survives in anything like its present form, we shall use our new knowledge.

SUMMARY

At the microscopic scale, matter exhibits a series of levels, one within the other. Bulk matter is made of atoms and molecules (themselves made of atoms); atoms are built from electrons and nuclei; nuclei are constructed from nucleons (protons and neutrons); and nucleons and other particles that participate in the nuclear force are comprised of quarks. These different levels, which correspond to smaller and smaller distances, are probed by microscopes that must, because of the Heisenberg uncertainty relation between momentum and distance, use beams of higher and higher momenta. This fact implies that special relativity becomes important when we want to probe distances of nuclear size and less.

At the level of the structure of nucleons, a fundamentally new feature emerges: confinement. Quarks, unlike the constituents of matter at larger scales, are confined; that is, they have never been observed as isolated particles. It is believed that, by the nature of the strong force, they can never be isolated.

Three forces can be regarded as fundamental: the gravitational force, the electroweak force, and the strong force. The gravitational force, unlike the electroweak and strong forces, has not yet been successfully reconciled with the ideas of quantum mechanics. Particles exert these forces on one another through the exchange of still other particles. The electroweak force is a unified version of the electromagnetic force and the weak force, which nevertheless appear to be rather different from one another at low energies. The exchanged particles of the electroweak force are the photon, which is massless, and the W and Z bosons, whose masses are some 100 times that of the proton. The strong force is mediated by the exchange of the gluon. Because the gluon is itself subject to the strong force, this force increases with increasing distance, a fact that guarantees that not only quarks, but gluons as well, are confined.

Various conservation laws apply to the fundamental forces and manifest themselves in certain rules for how reactions can proceed. The conservation of electric charge, of baryon number, and of lepton number applies in strong and electroweak interactions.

The reactions that occur in high-energy collisions form the most important pieces of evidence for the deduction of the fundamental laws. Colliders, accelerators in which beams of electrons or protons collide head-on, allow us to produce reactions with the highest energies.

Experiments on high-energy reactions share a common ground with astronomical observations. Hubble's law describes an expanding universe that started from a tiny initial volume, at the so-called big bang. When the expansion started, the universe was hot and dense, conditions appropriate to the high-energy reactions where the fundamental forces act. As the universe expanded, the temperature dropped, and a succession of energy scales resulted. By applying their knowledge of the behavior of matter at these different energy scales, physicists have successfully accounted for the background blackbody radiation measured to be present as well as the observed amounts of hydrogen and helium in today's universe. The early evolution of the universe must have involved gravity in a form consistent with quantum ideas, as we can see from the Planck mass,

$$M_P = \sqrt{\frac{\hbar c}{G}}. \tag{46–5}$$

The value of M_P is characteristic of very high energies, hence of very early times. Accounting for these earliest times is a formidable challenge for scientists.

QUESTIONS

1. Whether a surface has curvature can be determined from certain geometrical properties of that surface. The sum of the angles of a triangle drawn in a flat plane is 180°. Show that this is not the case for the surface of the earth by summing the angles of a triangle that links the North Pole with any two points on the equator (Fig. 46–19).

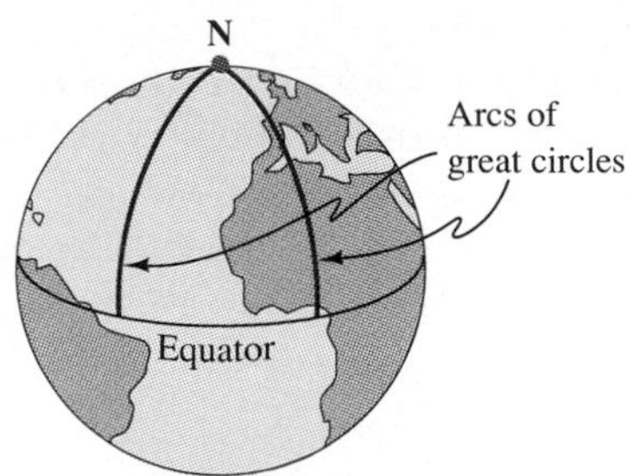

FIGURE 46–19 Question 1

2. There are at least three known "families" of leptons. For example, in addition to the electron there is a particle known as the muon (μ), which behaves in all respects like the electron, except that it is more massive. By virtue of the extra mass, the muon is unstable. We also believe that the muon has a muon–lepton number, which, like the lepton number associated with the electron, is conserved. What experiment might tell you that the lepton numbers of the muon and the electron are conserved separately?
3. The lifetime of the proton is at least 10^{30} s, much greater than the age of the universe. Yet it is possible to create and destroy protons in particle reactions. How?
4. Why are electrons rather than protons used to study the charge distribution in the nucleus?
5. In a reaction between colliding high-energy electrons and positrons, why is the excess energy likely to create particles and antiparticles?
6. What fundamental forces most nearly account for the following phenomena: (a) walking; (b) the structure of gasoline molecules; (c) moonrise; (d) the source of sunlight; (e) the passage of sunlight from the sun to the earth; (f) neutron stars (stars made mainly of neutrons packed to a density on the order of that of the nucleus)?
7. Can a particle be a lepton and hadron simultaneously? meson and hadron? hadron and baryon?
8. Suppose that electrons could not be removed from atoms, isolated, and studied independently. Would we still have experimental evidence of their existence?
9. Cosmic rays, consisting of protons and nuclei, arrive from outer space with energies that sometimes surpass those of the largest particle accelerators. What are the advantages and disadvantages of using such particles for beams for the study of high-energy reactions?
10. The earth orbits the sun, the sun moves within our galaxy, and our galaxy moves within its own galactic group. With all this movement, how could Hubble have discovered Hubble's law?
11. If the universe is closed, then a beam of light emitted from one point will eventually make a round trip and come back to that point, just as Magellan went all the way around the world without ever making a 180° turn. Does this mean that you would be able to see your own back if you looked through a sufficiently powerful telescope?
12. By energy and momentum conservations, an isolated electron cannot emit a photon and remain an electron. Why, then, can this happen in a virtual process?
13. Suppose that an accelerator can accelerate either electrons or protons to a given energy and that these projectiles are to collide with protons at rest. Which type of projectile should be chosen to obtain the greatest center-of-mass energy in the collision?
14. We know that the early universe was denser than it is now. What are some pieces of evidence that indicate that the early universe was hotter?
15. If the universe is closed, do we live inside a black hole?

PROBLEMS

46–1 Russian Dolls, Subnuclear Forces, and Special Relativity

1. (I) Robert Hofstadter received a 1961 Nobel Prize for experiments in which 500-MeV electrons scattered from stationary nuclei. He found that the nucleus has a fairly constant density, but over a thickness of about 2.4 fm at its surface, the density drops to zero. With what spatial resolution could Hofstadter probe the nucleus with 500-MeV electrons?
2. (I) The Stanford Linear Accelerator accelerates electrons up to an energy of 52 GeV. These electrons can collide with stationary targets. (a) Determine the spatial accuracy with which such a probe can discern details of the nucleus. (b) In the 1960s, experiments performed there with 20-GeV electrons showed that the proton contains pointlike particles, the quarks. What is the best spatial resolution obtainable with such an electron beam?

3. (II) We want to use electrons as projectiles to study the structure of protons by means of the scattering reaction $e + p \to e + p$. The target proton is initially at rest in the

laboratory; we measure those electrons that recoil directly backward. Suppose that all the motion is ultrarelativistic (you can ignore all rest masses except that of the particle at rest). Calculate the momentum transferred to the proton target. If the energy of the incident electron is doubled, how much does the momentum transferred to the proton increase?

4. (II) (a) Repeat Problem 3, but keep the rest mass of the proton, so that $E = \sqrt{(m_p c^2)^2 + (qc)^2}$ for a proton of momentum q. Continue to treat the ratio of the mass of the electron to that of the proton as zero. (b) Is there an initial electron energy for which the electron motion must be treated relativistically while the motion of the proton can be treated nonrelativistically?

5. (II) Equation (45–2) gives the collision cross section of a *pointlike* projectile of charge $Z_1 e$, mass m, and kinetic energy K scattering from a *pointlike* target particle of charge $Z_2 e$ and mass M ($M \gg m$) at rest, as a function of the scattering angle θ:

$$\sigma(\theta) = Z_1^2 Z_2^2 \left(\frac{e^2}{4\pi\epsilon_0}\right)^2 \frac{1}{16} \frac{1}{K^2 \sin^4(\theta/2)}.$$

Suppose that the initial and final momenta of the projectile are $\mathbf{p}_\mathrm{i}$ and $\mathbf{p}_\mathrm{f}$, respectively, with $|\mathbf{p}_\mathrm{i}| = |\mathbf{p}_\mathrm{f}| = p$. Then for large M the magnitude of the momentum transfer $\mathbf{\Delta} \equiv \mathbf{p}_\mathrm{f} - \mathbf{p}_\mathrm{i}$ in the collision is given by $\Delta^2 = (\mathbf{p}_\mathrm{f} - \mathbf{p}_\mathrm{i})^2 = 2p^2(1 - \cos\theta)$. (a) Express the cross section in terms of Δ^2, and plot it as a function of Δ^2. (b) Suppose that the target is not pointlike but spread out over a sphere of radius r_0. Give some qualitative arguments based on the position–momentum uncertainty relation of how the dependence of the cross section on Δ^2 would change.

6. (III) Consider a colliding electron and proton. Show that if in the center-of-mass reference frame the electron has energy 5000 MeV, then in the laboratory reference frame (in which the proton is at rest), the electron has an energy of about 53 GeV, or 53,000 MeV. [*Hint:* You may suppose that in the center-of-mass frame, the electron and proton collide and produce some object with zero total momentum, and with a mass M that can be calculated, given the electron's momentum. In the laboratory frame, that same object is created, and, by the conservation of momentum, moves with a momentum p_lab equal to that of the incoming electron. The energy of the object in the laboratory frame is then given by $E_\mathrm{object} = \sqrt{(Mc^2)^2 + (p_\mathrm{lab}c)^2}$. You may then use energy conservation, $p_\mathrm{lab}c + m_p c^2 = E_\mathrm{object}$, and your calculation of the mass M to find $p_\mathrm{lab}c$.]

7. (III) An electron collides with a second electron at rest. What is the minimum energy the incident electron must have to produce a proton–antiproton pair? The reaction takes the form $e + e \to e + e + (p\bar{p})$. Ignore the electron mass. (See the hint to Problem 6.)

46–2 *New Particles and Conservation Laws*

8. (I) The K^+ meson (the *kaon*) decays by the process $K^+ \to \pi^+ + \pi^0$. Assume that in this decay the conservation laws of charge, baryon number, and lepton number are obeyed. What are the values of these quantum numbers for the K^+ meson?

9. (I) Show that the conservation laws of charge, baryon number, and lepton number are satisfied in neutron decay, $n \to p + e^- + \bar{\nu}$.

10. (I) The reaction $p + e^- \to \bar{p} + e^+$ does not occur in sensitive experiments designed to observe it. Why not?

11. (II) The π^+ meson decays as $\pi^+ \to \mu + \nu$. What is the baryon number of the μ? What is its lepton number? Is it a boson or a fermion? What is its electric charge?

46–3 *Fundamental Forces Revisited*

12. (I) The Z meson has a mass of 91 GeV/c^2. Calculate the range of the force transmitted by the exchange of the Z meson.

13. (I) Describe the quark content of the antiproton.

14. (II) There exist mesons like the pion that can be emitted and absorbed by nucleons, but with mass some 5.5 times that of the pion. The exchange of these particles gives rise to a repulsive nucleon–nucleon force. (a) Estimate the range of this repulsive force. (b) Estimate at what energy, in the center-of-mass reference frame of a proton–proton collision, the repulsive force will begin to affect the scattering.

15. (II) You have 1 kg each of hydrogen (gram-atomic weight 1 g/mol) and uranium 235 (gram-atomic weight 235 g/mol). What are the baryon numbers of the two samples of material?

16. (II) The pion is the exchange boson for nuclear forces: It is exchanged between two protons or two neutrons and leads to the nuclear force. Given that the transition $p \to p + \pi^0$ is possible (virtually), where π^0 is the uncharged pion, describe the minimum quark content for this particle.

17. (II) The quark contents of some strongly interacting particles are (uud) for protons, (udd) for neutrons, $(u\bar{d})$ for π^+, $(d\bar{u})$ for π^-, and $(u\bar{u} - d\bar{d})/\sqrt{2}$ for π^0. Figure 46–20 is a sketch of the reaction $p + \pi^- \to n$ that uses this

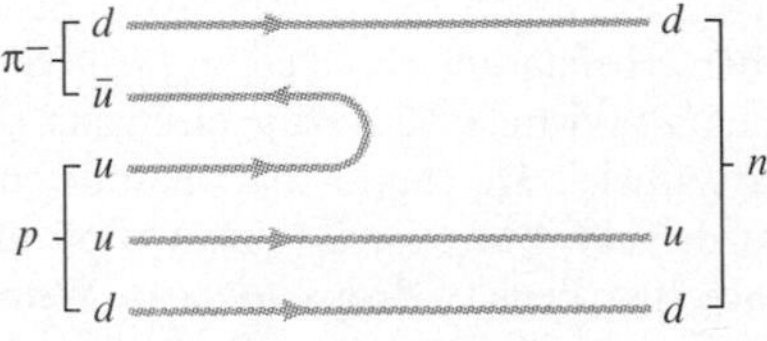

FIGURE 46–20 Problem 17

information. Here all lines represent particles moving to the right, with right-pointing arrows for quarks and left-pointing arrows for antiquarks. Moreover, quark lines never end in such a graph. Use this technique to draw graphs for the reactions (a) $p+\pi^- \to n+\pi^0$, (b) $p+\bar{p} \to \pi^+ + \pi^- + \pi^0$, and (c) $p+\pi^- \to p+n+\bar{p}$.

18. (II) Combine the quark-line rules of Problem 17 with the additional graphical rules for emission of a W^- boson by a d-quark, emission of a W^+ boson by a u-quark, and the decay $W^- \to e^- + \bar{\nu}$, as drawn in Fig. 46–21. Use these rules to give the graphical representation of the processes (a) $n \to p + e^- + \bar{\nu}$, (b) $\pi^+ \to \pi^0 + e^+ + \nu$, and (c) $\pi^- \to e^- + \bar{\nu}$, all of which proceed with a W boson present at an intermediate stage.

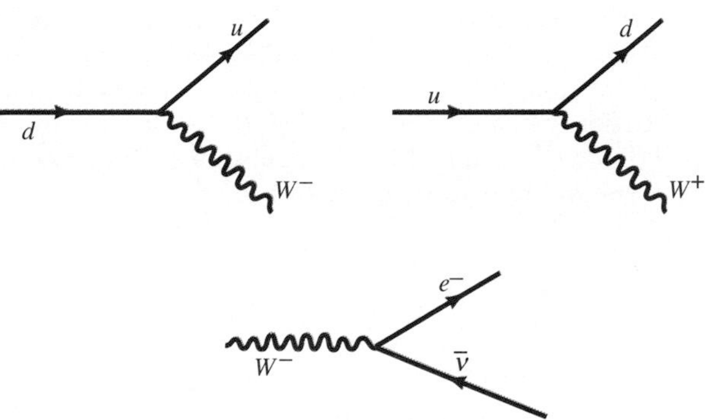

FIGURE 46–21 Problem 18

46–4 *Tools of Particle Physics*

19. (I) According to some theories, a proton at rest can decay (with very long lifetime!) in the mode $p \to e^+ + \pi^0$. Use $m_p \simeq 940\ \text{MeV}/c^2$ and $m_\pi \simeq 135\ \text{MeV}/c^2$ to calculate the momentum, q, of the π^0 in the decay; neglect the electron mass. Express your answer in terms of the combination qc, in units of MeV. Detectors looking for the decay would use this characteristic momentum as a signal.

20. (II) A proton–antiproton collider is actually a quark–antiquark collider, if we think of each quark in a proton (or each antiquark in an antiproton) as carrying one-third the momentum of the proton (or antiproton). What is the proton (and antiproton) energy in a symmetric $p\bar{p}$ collider that produces quark–antiquark collisions with the same total energy as the collisions produced by an electron–positron collider in which each electron carries a momentum of $8\ \text{GeV}/c$? Ignore the masses of the electron, positron, quark, and antiquark but not those of the proton and antiproton.

21. (II) Calculate the momentum in the center of mass of each incident proton for an accelerator in which 450-GeV protons are incident on stationary proton targets. (See the hint for Problem 6.)

22. (II) Antiprotons of energy 900 GeV collide with protons of the same energy but opposite momentum in the TEVATRON collider (Table 46–3). What is the momentum of antiprotons incident on stationary protons if the center-of-mass energy of the $p\bar{p}$ system is the same as that of the TEVATRON?

23. (II) An experimenter wants to study the reaction $e^- + e^+ \to Z^0$, where the Z^0, of mass $91.1\ \text{GeV}/c^2$, is the neutral intermediate boson of the electroweak interactions. The experimenter has a fixed-target machine that accelerates a beam of positrons that can be directed at atoms, which provide a stationary target of electrons. What is the minimum energy of the positron beam for which the reaction can occur?

24. (II) The *luminosity* of a colliding beam machine is the rate at which particles pass in one beam times the number of collisions per unit area of a second beam. The design luminosity of the SSC (see Table 46–3) is $10^{37}/\text{m}^2 \cdot \text{s}$. If the total collision cross section for proton–proton interactions is 100 millibarns ($1\ \text{barn} = 10^{-28}\ \text{m}^2$), how many interactions per second will there be at the SSC? (The value of the total cross section is one of the quantities to be measured at the SSC.)

25. (II) The TEVATRON collider (see Table 46–3) has proton and antiproton beams that follow a circular path of circumference 6.3 km. What is the frequency with which a given proton in the proton beam crosses a given antiproton in the antiproton beam, as seen in the laboratory reference frame?

26. (II) A proton–antiproton pair can be produced in the reaction $\gamma + p \to p + \bar{p} + p$. If the target proton is at rest, what is the minimum energy that the photon must have to produce this reaction?

27. (III) Table 46–3 shows one collider, HERA, that is asymmetric in that its two beams, one of electrons and the other of protons, are of different momenta. This collider is thus a kind of hybrid between symmetric colliders and fixed-target machines. In the following, neglect the baryon-number and lepton-number conservation laws that must be obeyed in proton–electron collisions. (a) What is the maximum number of neutral pions that can be produced in HERA collisions? (b) What momentum would beams of protons and electrons of equal momentum have in order to be able to produce the number of pions in part (a)? (c) What momentum would a beam of electrons incident on protons at rest have to produce the number of pions in part (a)?

28. (III) Electron accelerators that produce copious quantities of X-rays for research on atomic and condensed-matter physics employ electron beams of several GeV. The X-rays are produced when the electrons scatter from photons of lesser energy. (For example, the National Synchrotron Light Source at Brookhaven National Laboratory accelerates electrons to an energy of 2.5 GeV and produces photons with energies up to 310 MeV.) Consider laser photons (with wavelength of, say, 350 nm) that scatter head-on from 6-GeV electrons. Calculate the maximum energy of the recoil photon produced in such a (Compton scattering) reaction. At what angle would this photon be produced? [*Hint:* You must keep the electron mass in the relativistic expression for the electron energy, E, even though $E \gg m_e c^2$. You may, however, make the approximation $\sqrt{(pc)^2 + (m_e c^2)^2} \simeq pc + (m_e c^2)^2/2pc$ in this case.]

46–5 *The Expanding Universe*

29. (II) Consider the one-dimensional line of galaxies in Figs. 46–14a and 46–14b. Show that an inhabitant of galaxy B would agree that Hubble's law holds, but with the distance being the distance from galaxy B. In this way you confirm that Hubble's law is the unique law consistent with no special central point.

30. (II) Assume that the size of the universe is determined by a distance scale factor R. This factor is increasing with time t at the rate $R = kt^n$, where k and n are constants. Hubble's law then becomes $dR/dt = HR$. Show that the Hubble parameter H is not constant, and find its time dependence.

31. (II) The universe is expanding, but the expansion may be slowing down. (a) If R varies with time as in Problem 30, find the deceleration of R. (b) The deceleration parameter, q, is defined by $q \equiv -R(d^2R/dt^2)/(dR/dt)^2$. Calculate the value of q.

32. (III) Use Hubble's law to find the time it would take for the distance from earth to some galaxy to double from what it is today. Assume that for the period over which you make your calculation, H is constant. [*Hint:* First find the time interval for the distance to change from x to $x + dx$, assuming that dx is an infinitesimal quantity. Then integrate your result.]

33. (III) By using the techniques of Problem 32, show that if Hubble's law with a constant value of H has held ever since the time when there was no separation between galaxies, it would have taken an infinite amount of time for the universe to have expanded to its present situation.

46–6 *The Early Moments of the Universe*

34. (I) Show that the Planck mass, 2.2×10^{-8} kg, is equivalent to an energy of 1.2×10^{19} GeV and to a temperature of 1.4×10^{32} K.

35. (I) In addition to the Planck mass, there is a distance referred to as the *Planck length.* Use dimensional analysis to form a length from the constants $\hbar$, G, and c.

36. (I) In addition to the Planck mass, there is a time referred to as the *Planck time.* Use dimensional analysis to form a time from the constants $\hbar$, G, and c.

37. (II) Suppose that at the time when stable atoms formed, neutrons and protons were present in the ratio of 1:7. (a) Show that if hydrogen and helium formed with these nucleons (plus an appropriate number of electrons), then the number of moles of hydrogen formed was 12 times the number of moles of helium formed. (b) Show that the total mass of hydrogen formed is roughly 3 times that of helium formed then.

38. (III) Under conditions that apply to our universe, Einstein's theory of gravitation relates the density of matter to the Hubble parameter through the relation $\rho = 3H^2/8\pi G$. (a) Given the present value of the Hubble parameter, $H_0 = 2.5 \times 10^{-18}\ \mathrm{s}^{-1}$, calculate the present value of the density of matter, ρ_0. Assuming that matter consists (almost) entirely of hydrogen, with $M_H = 1.7 \times 10^{-27}$ kg, find the number density of hydrogen. (b) Show that the conservation of baryon number implies that the time dependence of the density is given by $\rho(t) = \rho_0[R_0/R(t)]^3$, where R_0 is the current distance scale of the universe and $R(t)$ is the distance scale at a general time t. (c) Combine the result of part (b) with the general relation between ρ and $H \equiv (1/R)\, dR/dt$ to show that $R(t) = kt^{2/3}$, and find k. [*Hint:* Substitute the general form $R(t) = kt^n$ into your equations.]

General Problems

39. (II) Below what temperature will stable hydrogen atoms form? helium atoms? The ionization energies of H and He are 13.6 eV and 24.6 eV, respectively.

40. (II) What is the maximum number of antiprotons that can be created in a scattering reaction with 10-GeV protons incident on a stationary hydrogen target?

41. (II) Consider a fixed-target machine with protons of momentum 1.7×10^6 GeV/c. If the largest magnetic field we could use to guide the beam in a circular path has magnitude 9.5 T, what is the radius of such a machine?

42. (II) *Ĉerenkov counters* are detectors that measure the presence and speed of charged particles by observing the angle of the "wake" of light (Ĉerenkov radiation; see Fig. 14–24) produced by those particles when their speed through a medium is greater than the speed of light in that medium. What is the minimum energy of electrons that produce Ĉerenkov radiation in water?

43. (II) In a reaction in which 52-GeV electrons hit 52-GeV positrons head-on (as has been done at the Stanford Linear Accelerator Center), what is the maximum number of π mesons (and antimesons) that can be created for each colliding pair? The mass of a π meson is about 140 MeV/c^2.

44. (II) Find a relationship between the de Broglie wavelength and the kinetic energy, K, of ultrarelativistic particles (particles for which $K \gg mc^2$).

45. (II) The muon, symbol μ, is for all practical purposes just like the electron, except that it is some 207 times more massive. The muon can be destroyed by a proton in the reaction $\mu + p \rightarrow n + \nu$. A muon sent into hydrogen gas slows and is finally captured in a hydrogenlike Bohr orbit. Once it is in the lowest orbit (radius R_0), it will be destroyed by the reaction above if it comes within a distance $L = 0.2$ fm of the proton. The probability that a muon in the lowest orbit will come that close to the proton is estimated from quantum mechanics to be the ratio of the volume of a sphere of radius L to that of a sphere of radius R_0. Suppose that the lifetime for destruction from the lowest Bohr orbit in hydrogen is τ_c. Show that the lifetime for destruction from the ground state of a Bohrlike atom in which the muon orbits a nucleus with Z protons is τ_c/Z^4.

46. (II) Consider muon destruction from the lowest state in a hydrogenlike Bohr atom (see Problem 45). What is the energy of the neutron that emerges from the reaction? If the process occurs when the muon is in orbit around a heavy nucleus, will the reaction look the same as in the hydrogenlike atom?

47. (II) Using energy conservation and momentum conservation, prove that an isolated electron cannot emit a photon and remain an electron. (See Question 12.)

48. (II) Estimate the repulsive force between protons in a uranium nucleus. Compare that force to the attraction between the uranium nucleus and the innermost electron in a uranium atom.

49. (II) Free neutrons have a lifetime of approximately 890 s. Given that the ratio of neutrons to protons was 1:5 when the universe was 1 s old, estimate the ratio at 100 s. Take into account only the loss of neutrons through their decay.

50. (II) A pi meson (of mass approximately 140 MeV/c^2) is ejected from a nuclear collision with a kinetic energy of 200 MeV. Pi mesons at rest have a lifetime of about 2.6×10^{-8} s. Calculate (a) the speed of the pi meson, (b) its momentum, and (c) how far, on average, such a pi meson will travel before it decays.

51. (III) When an antiproton slows down by interacting with matter and approaches a proton, it can be captured and form a hydrogenlike atom called *protonium* through its Coulomb attraction to the proton. Find (a) the binding energy of the atom in electron-volts and (b) the radius of the atom in its ground state. (Do not forget the reduced-mass effect.)

52. (III) The probability that a proton and an antiproton in the ground state of protonium (see Problem 51) will come within a distance R_1 of each other is estimated from quantum mechanics to be the ratio of the volume of a sphere of radius R_1 to that of a sphere of radius $2R_0$, where R_0 is the ground-state radius of protonium. If the lifetime for annihilation of a proton and an antiproton is 10^{-22} s when they are within 0.2 fm of each other and is infinite beyond this distance, estimate how long the protonium atom lasts in its ground state.

APPENDIX I

THE SYSTÈME INTERNATIONALE (SI) OF UNITS

I–1 SOME SI BASE UNITS

Physical Quantity	*Name of Unit*	*Symbol*
length	meter	m
mass	kilogram	kg
time	second	s
electric current	ampere	A
thermodynamic temperature	kelvin	K
amount of substance	mole	mol

I–2 SOME SI DERIVED UNITS

Physical Quantity	*Name of Unit*	*Symbol*	*SI Unit*
frequency	hertz	Hz	s^{-1}
energy	joule	J	$kg \cdot m^2/s^2$
force	newton	N	$kg \cdot m/s^2$
pressure	pascal	Pa	$kg/m \cdot s^2$
power	watt	W	$kg \cdot m^2/s^3$
electric charge	coulomb	C	$A \cdot s$
electric potential	volt	V	$kg \cdot m^2/A \cdot s^3$
electric resistance	ohm	Ω	$kg \cdot m^2/A^2 \cdot s^3$
capacitance	farad	F	$A^2 \cdot s^4/kg \cdot m^2$
inductance	henry	H	$kg \cdot m^2/A^2 \cdot s^2$
magnetic flux	weber	Wb	$kg \cdot m^2/A \cdot s^2$
magnetic flux density	tesla	T	$kg/A \cdot s^2$

I–3 SI UNITS OF SOME OTHER PHYSICAL QUANTITIES

Physical Quantity	*SI Unit*
speed	m/s
acceleration	m/s^2
angular speed	rad/s
angular acceleration	rad/s^2
torque	$kg \cdot m^2/s^2$, or $N \cdot m$
heat flow	J, or $kg \cdot m^2/s^2$, or $N \cdot m$
entropy	J/K, or $kg \cdot m^2/K \cdot s^2$, or $N \cdot m/K$
thermal conductivity	$W/m \cdot K$

I–4 SOME CONVERSIONS OF NON-SI UNITS TO SI UNITS

Energy:
1 electron-volt (eV) = 1.6022×10^{-19} J
1 erg = 10^{-7} J
1 British thermal unit (BTU) = 1055 J
1 calorie (cal) = 4.186 J
1 kilowatt-hour (kWh) = 3.6×10^6 J

Mass:
1 gram (g) = 10^{-3} kg
1 atomic mass unit (u) = 931.5 MeV/c^2 = 1.661×10^{-27} kg
1 MeV/c^2 = 1.783×10^{-30} kg

Force:
1 dyne = 10^{-5} N
1 pound (lb or #) = 4.448 N

Length:
1 centimeter (cm) = 10^{-2} m
1 kilometer (km) = 10^3 m
1 fermi = 10^{-15} m
1 Angstrom (Å) = 10^{-10} m
1 inch (in or ″) = 0.0254 m
1 foot (ft) = 0.3048 m
1 mile (mi) = 1609.3 m
1 astronomical unit (AU) = 1.496×10^{11} m
1 light-year (ly) = 9.46×10^{15} m
1 parsec (ps) = 3.09×10^{16} m

Angle:
1 degree (°) = 1.745×10^{-2} rad
1 min (′) = 2.909×10^{-4} rad
1 second (″) = 4.848×10^{-6} rad

Volume:
1 liter (L) = 10^{-3} m^3

Power:
1 kilowatt (kW) = 10^3 W
1 horsepower (hp) = 745.7 W

Pressure:
1 bar = 10^5 Pa
1 atmosphere (atm) = 1.013×10^5 Pa
1 pound per square inch (lb/in^2) = 6.895×10^3 Pa

Time:
1 year (yr) = 3.156×10^7 s
1 day (d) = 8.640×10^4 s
1 hour (h) = 3600 s
1 minute (min) = 60 s

Speed:
1 mile per hour (mi/h) = 0.447 m/s

Magnetic field:
1 gauss = 10^{-4} T

APPENDIX II

SOME FUNDAMENTAL PHYSICAL CONSTANTS†

Constant	*Symbol*	*Value*	*Error*
speed of light in a vacuum	c	2.99792458×10^8 m/s	exact
gravitational constant	G	6.67259×10^{-11} m³/kg·s²	128
Avogadro's number	N_A	6.02214×10^{23} mol^{-1}	0.6
universal gas constant	R	8.31451 J/mol·K	8.4
Boltzmann's constant	k	1.38066×10^{-23} J/K	8.5
elementary charge	e	1.60218×10^{-19} C	0.3
permittivity of free space	ϵ_0	$8.85418781762 \times 10^{-12}$ C²/N·m²	exact
	$1/4\pi\epsilon_0$	8.987552×10^9 kg·m³·s^{-2}·C^{-2}	
permeability of free space	μ_0	$4\pi \times 10^{-7}$ T·m/A	exact
electron mass	m_e	9.10939×10^{-31} kg	0.6
proton mass	m_p	1.67262×10^{-27} kg	0.6
neutron mass	m_n	1.67493×10^{-27} kg	0.6
Planck's constant	h	6.62608×10^{-34} J·s	0.6
$h/2\pi$	$\hbar$	1.05457×10^{-34} J·s	0.6
		$= 6.58212 \times 10^{-22}$ MeV·s	0.3
	$\hbar c$	197.327 Mev·fm	0.3
electron charge-to-mass ratio	$-e/m_e$	-1.75882×10^{11} C/kg	0.3
proton-electron mass ratio	m_p/m_e	1836.15	0.15
molar volume of ideal gas at STP		22414.1 cm³/mol	8.4
Bohr magneton	μ_B	9.27402×10^{-24} J/T	0.3
magnetic flux quantum	$\Phi_0 = h/2e$	2.067783×10^{-15} Wb	0.3
Bohr radius	a_0	0.529177×10^{-10} m	0.05
Rydberg constant	R_∞	1.09737×10^7 m^{-1}	0.001

† From E. R. Cohen and B. N. Taylor, *The 1986 Adjustment of the Fundamental Constants*, Report of the CODATA Task Group on Fundamental Constants, CODATA Bulletin 63, Pergamon, Elmsford, N.Y. (1986). The next scheduled reassessment of these values is due in 1994–1995.

We have given values of the measured constants to six significant figures, even though they may be known to greater accuracy. We quote the error, which expresses the uncertainty in the values of these constants, in parts per million. Defined constants have no error, and we give their full definition; they are indicated by the notation "exact" in the error column.

APPENDIX III

OTHER PHYSICAL QUANTITIES

III–1.1 SOME ASTRONOMICAL CONSTANTS

Constant	*Symbol*	*Value*
standard gravity at earth's surface	g	$9.80665\ m/s^2$
equatorial radius of earth	R_e	6.374×10^6 m
mass of earth	M_e	5.976×10^{24} kg
mass of moon		7.350×10^{22} kg $= 0.0123\ M_e$
mean radius of moon's orbit around earth		3.844×10^8 m
mass of sun	$M_\odot$	1.989×10^{30} kg
radius of sun	$R_\odot$	6.96×10^8 m
mean radius of earth's orbit around sun	AU	1.496×10^{11} m
period of earth's orbit around sun	yr	3.156×10^7 s
diameter of our galaxy		7.5×10^{20} m
mass of our galaxy		2.7×10^{41} kg $= (1.4 \times 10^{11})\ M_\odot$
Hubble parameter	H	$2.5 \times 10^{-18}\ s^{-1}$

III–1.2 PLANETARY DATA

Planet	*Diameter (in km)*	*Diameter Relative*[†]	*Relative Mass*[†]	*Average Density (in g/cm³)*	*Period of Rotation*	*Surface Gravity*[†] *(in g)*	*Escape Speed (in km/s)*	*Semimajor Axis (AU)*	*Period of Solar Orbit*	*Average Orbital Speed (in km/s)*
Mercury	4,800	0.38	0.05	5.2	58 d 15 h	0.39	4.3	0.387	87.96 d	47.8
Venus	12,100	0.95	0.82	5.3	243 d 4 h	0.90	10.3	0.723	224.7 d	35.0
Earth	12,750	1.00	1.00	5.5	23 h 56 min	1.00	11.2	1.000	365.26 d	29.8
Mars	6,800	0.53	0.11	3.8	24 h 37 min	0.38	5.1	1.524	687.0 d or 1.88 yr	24.2
Jupiter	142,800	11.23	317.9	1.3	9 h 50 min	2.58	59.5	5.20	11.86 yr	13.1
Saturn	120,660	9.41	95.2	0.7	10 h 39 min	1.11	35.6	9.54	29.46 yr	9.7
Uranus	51,800	3.98	14.6	1.3	17 h	1.07	21.4	19.18	84.01 yr	6.8
Neptune	49,500	3.88	17.2	1.7	18 to 22 h	1.40	23.6	30.06	164.79 yr	5.4
Pluto	4,000	0.23	0.002	0.4	6 d 9 h 17 min	0.02	1.2	39.44	248 yr	4.7

[†] Relative to the earth.

III–2 ENERGY SUPPLY AND DEMAND†

† From the *Physics Vade Mecum*, Ed. Herbert L. Anderson, American Institute of Physics (New York, 1981); and U.S. Congress, Office of Technology Assessment, *Changing by Degrees: Steps to Reduce Greenhouse Gases*, OTA-O-482 (Washington, D.C.: U.S. Government Printing Office, February 1991).

III–2.1 Fuel Resources (1980, Estimated)

Resource	*U.S. Resources*	*World Resources*
coal (recoverable)	5×10^{21} J	2×10^{22} J
oil (not including oil shales)	10^{21} J	10^{22} J
natural gas	2×10^{21} J	10^{22} J
hydroelectric	10^{22} J/yr (North America)	6×10^{22} J/yr

III–2.2 Annual Usage of Resource (1989, Percentage of total)

Resource	*U.S. Usage (total = 9×10^{19} J)*	*World Usage (total = 4×10^{20} J)*
coal	40	24
oil	23	35
natural gas	23	18
nuclear	7	5
hydroelectric	3	5
biomass	3	13

III–2.3 Energy Content of Fuels

Fuel	*Energy Content (in J/kg)*
bread	10×10^6
glucose ($C_6H_{12}O_6$)	16×10^6
white pine wood	20×10^6
methyl alcohol (CH_4O)	20×10^6
anthracite coal	32×10^6
domestic heating oil	45×10^6
propane (C_3H_8)	50×10^6
natural gas (96% CH_4)	51×10^6
fission of U^{235}	5.8×10^{11}
perfect mass-energy conversion	9×10^{16}

III–2.4 Solar Energy Output

total radiated power from the sun	4×10^{26} W
power per unit area at the top of earth's atmosphere	1.4 kW/m^2
average power per unit area delivered to an average horizontal surface in the United States in 1 yr	0.2 W/m^2

III–2.5 Energy Consumption in Transportation

Mode	*Energy Consumption (J/passenger·km)*
bicycle	5×10^4
foot travel	1.5×10^5
intercity bus	3×10^5
intercity train	9×10^5
automobile	1.5×10^6
747 jet airplane	2×10^6
snowmobile	6×10^6

III–2.6 Energy Consumption of Electrical Appliances

Appliance	*Power (in W)*	*Energy Use Per Year (in kWh)*
window air conditioner	1565	1390
clock	2	17
dishwasher	1200	363
window fan	200	170
hair dryer	380	14
iron	1000	144
microwave oven	1450	190
radio	71	86
refrigerator-freezer	615	1830
stove	12,200	1175
color television	200	440
vacuum cleaner	630	46
washing machine	512	103

APPENDIX IV

MATHEMATICS

IV–1 SOME MATHEMATICAL CONSTANTS†

Constant	*Value*
π	3.14159
e (Euler's constant)	2.71828
$\sqrt{2}$	1.41421
$1/\sqrt{2}$	0.707107
ln(10)	2.30259
ln(2)	0.693147
1 rad	57.2958°
1°	0.0174533 rad

† To six significant figures.

IV–2 SOLUTION OF QUADRATIC EQUATIONS

Quadratic equation:

$$ax^2 + bx + c = 0$$

Two solutions:

$$x = \frac{-b \pm \sqrt{b^2 - 4ac}}{2a}$$

IV–3 BINOMIAL THEOREM

$$(x + y)^n = \sum_{k=0}^{n} \binom{n}{k} x^{n-k} y^k,$$

where

$$\binom{n}{k} = \frac{n!}{(n-k)!\,k!}.$$

The factorial $m! \equiv 1 \cdot 2 \cdot 3 \ldots \cdot m$; $0! \equiv 1$. Some particular cases of the binomial theorem:

(1) $(x \pm y)^2 = x^2 \pm 2xy + y^2$;

(2) $(x \pm y)^3 = x^3 \pm 3x^2y + 3xy^2 \pm y^3$;

(3) $(x \pm y)^4 = x^4 \pm 4x^3y + 6x^2y^2 \pm 4xy^3 + y^4$.

IV–4 TRIGONOMETRY

1. For a right triangle with sides a, b, and c (the hypotenuse), where the angle opposite side a is θ_a (Fig. A–1),

$$\text{sine of } \theta_a = \sin\theta_a \equiv \frac{a}{c};$$

$$\text{cosine of } \theta_a = \cos\theta_a \equiv \frac{b}{c};$$

$$\text{tangent of } \theta_a = \tan\theta_a = \frac{a}{b}.$$

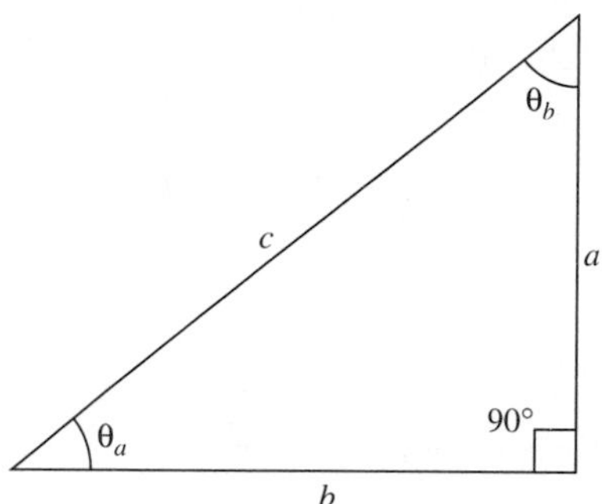

2. The cosine function is even, $\cos(-x) = \cos x$; the sine function is odd, $\sin(-x) = -\sin x$.

3. (1) $\tan\theta = \dfrac{\sin\theta}{\cos\theta}$

(2) $\sec\theta = \dfrac{1}{\cos\theta}$

(3) $\operatorname{cosec}\theta = \dfrac{1}{\sin\theta}$

(4) $\cot\theta = \dfrac{1}{\tan\theta}$

4. (1) $\sin^2\theta + \cos^2\theta = 1$

(2) $\sec^2\theta - \tan^2\theta = 1$

(3) $\operatorname{cosec}^2\theta - \cot^2\theta = 1$

5. (1) $\sin(\theta_1 \pm \theta_2) = \sin\theta_1\cos\theta_2 \pm \cos\theta_1\sin\theta_2$

(2) $\cos(\theta_1 \pm \theta_2) = \cos\theta_1\cos\theta_2 \mp \sin\theta_1\sin\theta_2$

(3) $\sin\theta_1 \pm \sin\theta_2 = 2\sin\left(\dfrac{\theta_1 \pm \theta_2}{2}\right)\cos\left(\dfrac{\theta_1 \mp \theta_2}{2}\right)$

(4) $\cos\theta_1 + \cos\theta_2 = 2\cos\left(\dfrac{\theta_1 + \theta_2}{2}\right)\cos\left(\dfrac{\theta_1 - \theta_2}{2}\right)$

(5) $\cos\theta_1 - \cos\theta_2 = -2\sin\left(\dfrac{\theta_1 + \theta_2}{2}\right)\sin\left(\dfrac{\theta_1 - \theta_2}{2}\right)$

(6) $\tan(\theta_1 + \theta_2) = \dfrac{\tan\theta_1 + \tan\theta_2}{1 - (\tan\theta_1)(\tan\theta_2)}$

(7) $\cos\left(\theta \pm \dfrac{\pi}{2}\right) = \mp\sin\theta$

(8) $\sin\left(\theta \pm \dfrac{\pi}{2}\right) = \pm\cos\theta$

6. (1) $\sin(2\theta) = 2\sin\theta\cos\theta = \dfrac{2\tan\theta}{1 + \tan^2\theta}$

(2) $\cos(2\theta) = \cos^2\theta - \sin^2\theta = 2\cos^2\theta - 1 = 1 - 2\sin^2\theta$

(3) $\tan(2\theta) = \dfrac{2\tan\theta}{1-\tan^2\theta}$

(4) $\sin\left(\dfrac{\theta}{2}\right) = \pm\sqrt{\dfrac{1-\cos\theta}{2}}$

(5) $\cos\left(\dfrac{\theta}{2}\right) = \pm\sqrt{\dfrac{1+\cos\theta}{2}}$

7. Expansions of Trigonometric Functions (θ in rad):

(1) $\sin\theta = \theta - \dfrac{\theta^3}{3!} + \dfrac{\theta^5}{5!} - \dfrac{\theta^7}{7!} + \cdots \qquad (\theta^2 < 1)$

(2) $\cos\theta = 1 - \dfrac{\theta^2}{2!} + \dfrac{\theta^4}{4!} - \dfrac{\theta^6}{6!} + \cdots \qquad (\theta^2 < 1)$

(3) $\tan\theta = \theta + \dfrac{1}{3}\theta^3 + \dfrac{2}{15}\theta^5 + \dfrac{17}{315}\theta^7 + \cdots \qquad \left(\theta^2 < \dfrac{\pi^2}{4}\right)$

IV–5 GEOMETRICAL FORMULAS

1. (circumference of a circle of radius r) = $2\pi r$
2. (area of a circle of radius r) = πr^2
3. (area of a sphere of radius r) = $4\pi r^2$
4. (volume of a sphere of radius r) = $\frac{4}{3}\pi r^3$
5. (area of a rectangle with sides of lengths L_1 and L_2) = $L_1 L_2$
6. For a right triangle with sides a, b, and c and angles θ_a and θ_b opposite the sides a and b, respectively (Fig. A–1):
(1) $a^2 + b^2 = c^2$ (the Pythagorean theorem)
(2) area = $\frac{1}{2}$(base)(height) = $\frac{1}{2}ab$
7. For a triangle with sides a, b, and c opposite the angles θ_a, θ_b, and θ_c, respectively (Fig. A–2):
(1) $\theta_a + \theta_b + \theta_c = 180° = \pi$ rad
(2) $a^2 = b^2 + c^2 - 2bc\cos\theta_a$
(3) $\dfrac{a}{\sin\theta_a} = \dfrac{b}{\sin\theta_b} = \dfrac{c}{\sin\theta_c}$
(4) $a = b\cos\theta_c + c\cos\theta_b$
(5) area $= \dfrac{1}{2}$(base)(height) $= \dfrac{1}{2}ab\sin\theta_c = \dfrac{1}{2}a^2\dfrac{\sin\theta_b\sin\theta_c}{\sin\theta_a}$

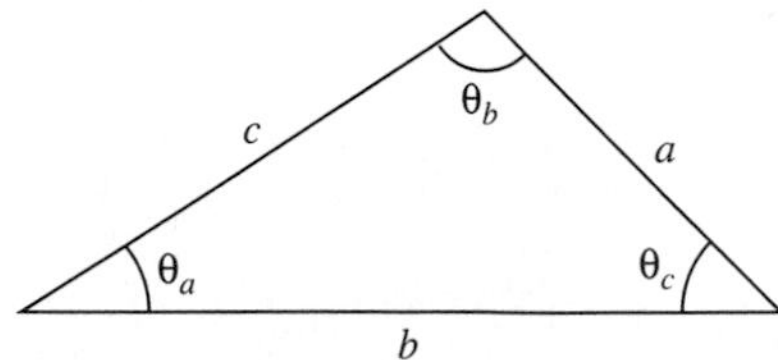

8. (volume of a right cylinder of height h and radius r) = $\pi r^2 h$

IV–6 SOME PROPERTIES OF ALGEBRAIC FUNCTIONS

1. General properties
(1) $a^x a^y = a^{x+y}$
(2) $a^0 = 1$
(3) $(ab)^x = a^x b^x$

2. Properties of exponential of x, e^x:
(1) $e^{\ln(x)} = x$
(2) $e^{x_1}e^{x_2} = e^{x_1+x_2}$
(3) $e^0 = 1$
(4) expansion: $e^x = 1 - x + \dfrac{x^2}{2!} - \dfrac{x^3}{3!} + \cdots$

3. Properties of the natural logarithm of x, $\ln(x)$:
(1) $\ln(e^x) = x$
(2) $\ln(x_1x_2) = \ln(x_1) + \ln(x_2)$
(3) $\ln(x_1/x_2) = \ln(x_1) - \ln(x_2)$
(4) $\ln(1) = 0$
(5) expansion: $\ln(1+x) = x - \dfrac{x^2}{2} + \dfrac{x^3}{3} - \dfrac{x^4}{4} + \cdots \quad (x^2 < 1)$

IV–7 DERIVATIVES

In the following, b and p are constants and u and v are functions of x:

1. $\dfrac{db}{dx} = 0$

2. $\dfrac{d}{dx}(bu) = b\dfrac{du}{dx}$

3. $\dfrac{d}{dx}(u+v) = \dfrac{du}{dx} + \dfrac{dv}{dx}$

4. $\dfrac{d}{dx}(uv) = v\dfrac{du}{dx} + u\dfrac{dv}{dx}$

5. $\dfrac{dx^p}{dx} = px^{p-1}$

6. Chain rule: If u is a function of y and y is in turn a function of x, then $\dfrac{du}{dx} = \dfrac{du}{dy}\dfrac{dy}{dx}$

7. $\dfrac{d}{dx}(\sin x) = \cos x$

8. $\dfrac{d}{dx}(\cos x) = -\sin x$

9. $\dfrac{d}{dx}(\tan x) = \dfrac{1}{\cos^2 x}$

10. $\dfrac{d}{dx}(e^{bx}) = be^{bx}$

11. $\dfrac{d}{dx}\ln(x) = \dfrac{1}{x}$

IV–8 INTEGRALS

In the following, b and p are constants and u and v are functions of x:

1. $\displaystyle\int \frac{du}{dx}\,dx = u$

2. $\displaystyle\int_{x_1}^{x_2} \frac{du}{dx}\,dx = u(x_2) - u(x_1)$

3. $\displaystyle\int bu\,dx = b\int u\,dx$

4. $\int (u + v) = \int u\,dx + \int v\,dx$

5. $\int u \frac{dv}{dx}\,dx = uv - \int v \frac{du}{dx}\,dx$ (integration by parts)

6. If u is a function of y and y is in turn a function of x, then

$$\int u\,dy = \int u \frac{dy}{dx}\,dx$$

7. $\int x^p\,dx = \frac{x^{p+1}}{p+1} \quad (p \neq -1)$

8. $\int \frac{dx}{x} = \ln(|x|)$

9. $\int \sin x\,dx = -\cos x$

10. $\int \cos x\,dx = \sin x$

11. $\int e^{bx}\,dx = \frac{1}{b} e^{bx}$

12. $\int x e^{bx}\,dx = e^{bx}\left(\frac{x}{b} - \frac{1}{b^2}\right)$

13. Some definite integrals

(1) $\int_0^\infty x^n e^{-x}\,dx = n!$

(2) $\int_0^\pi \sin^2 x\,dx = \int_0^\pi \cos^2 x\,dx = \frac{\pi}{2}$

(3) $\int_0^\infty e^{-b^2x^2}\,dx = \frac{\sqrt{\pi}}{2b} \quad (b > 0)$

(4) $\int_0^\infty x e^{-x^2}\,dx = \frac{1}{2}$

(5) $\int_0^\infty x^2 e^{-x^2}\,dx = \frac{\sqrt{\pi}}{4}$

(6) $\int_0^\infty \frac{b}{b^2 + x^2}\,dx = \begin{Bmatrix} \frac{\pi}{2} & (b > 0) \\ 0 & (b = 0) \\ -\frac{\pi}{2} & (b < 0) \end{Bmatrix}$

IV–9 SOME EXPANSIONS APPROPRIATE FOR $x^2 < 1$

1. The following expression is good for any n, positive or negative, integer or noninteger:

$$(1 + x)^n = 1 + nx + \frac{n(n-1)}{2!}x^2 + \frac{n(n-1)(n-2)}{3!}x^3 + \cdots$$

2. $\sin x = x - \frac{x^3}{3!} + \frac{x^5}{5!} + \cdots$

3. $\cos x = 1 - \frac{x^2}{2!} + \frac{x^4}{4!} + \cdots$

4. $\tan x = x + \frac{x^3}{3} + \frac{2}{15}x^2 + \cdots$

5. $e^{ax} = 1 + ax + \frac{(ax)^2}{2!} + \frac{ax^3}{3!} + \cdots$

IV–10 SOME MATHEMATICAL NOTATION

1.	$=$	is equal to
2.	$\simeq$	is approximately equal to
3.	$\propto$	is proportional to
4.	$\equiv$	is defined to be
5.	$\neq$	is unequal to
6.	$>$	is greater than
7.	$\geq$	is greater than or equal to
8.	$<$	is less than
9.	$\leq$	is less than or equal to
10.	Δx	the change in x
11.	$\lvert x \rvert$	the absolute value of x
12.	$O(N)$	on the order of the magnitude of N
13.	$\pm$	plus or minus
14.	$\mp$	minus or plus
15.	$\langle x \rangle$	average of x
16.	$\sum_{i=i_1}^{i_2} f_i$	the sum of all f_i over the integers i from a smallest integer i_1 to a largest integer i_2
17.	$\ln(x)$	natural logarithm of x
18.	$\log(x)$	logarithm to the base 10 of x
19.	$\int$	one-dimensional integral, line integral
20.	$\oint$	line integral around a loop
21.	$\iint$	two-dimensional integral, integral over a surface
22.	$\oiint$	integral over a closed surface

APPENDIX V

PERIODIC TABLE OF THE ELEMENTS

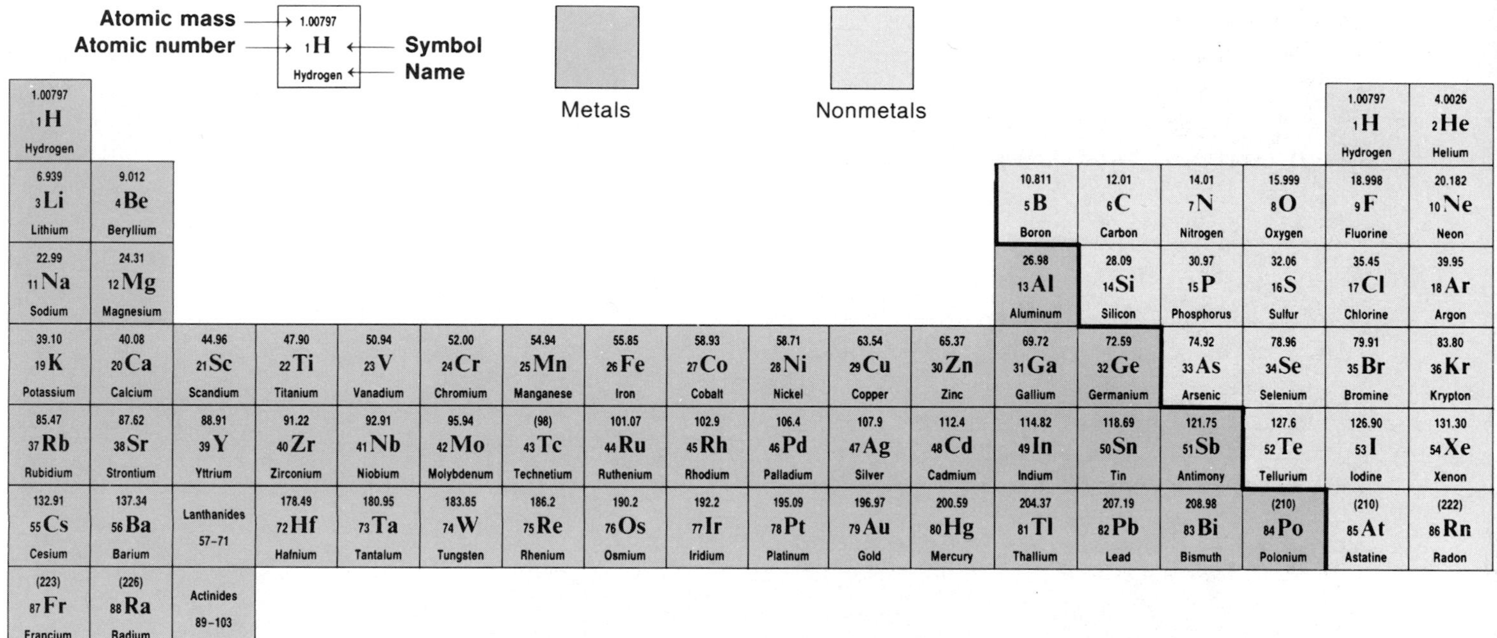

1.00797 1 H Hydrogen																1.00797 1 H Hydrogen	4.0026 2 He Helium
6.939 3 Li Lithium	9.012 4 Be Beryllium											10.811 5 B Boron	12.01 6 C Carbon	14.01 7 N Nitrogen	15.999 8 O Oxygen	18.998 9 F Fluorine	20.182 10 Ne Neon
22.99 11 Na Sodium	24.31 12 Mg Magnesium											26.98 13 Al Aluminum	28.09 14 Si Silicon	30.97 15 P Phosphorus	32.06 16 S Sulfur	35.45 17 Cl Chlorine	39.95 18 Ar Argon
39.10 19 K Potassium	40.08 20 Ca Calcium	44.96 21 Sc Scandium	47.90 22 Ti Titanium	50.94 23 V Vanadium	52.00 24 Cr Chromium	54.94 25 Mn Manganese	55.85 26 Fe Iron	58.93 27 Co Cobalt	58.71 28 Ni Nickel	63.54 29 Cu Copper	65.37 30 Zn Zinc	69.72 31 Ga Gallium	72.59 32 Ge Germanium	74.92 33 As Arsenic	78.96 34 Se Selenium	79.91 35 Br Bromine	83.80 36 Kr Krypton
85.47 37 Rb Rubidium	87.62 38 Sr Strontium	88.91 39 Y Yttrium	91.22 40 Zr Zirconium	92.91 41 Nb Niobium	95.94 42 Mo Molybdenum	(98) 43 Tc Technetium	101.07 44 Ru Ruthenium	102.9 45 Rh Rhodium	106.4 46 Pd Palladium	107.9 47 Ag Silver	112.4 48 Cd Cadmium	114.82 49 In Indium	118.69 50 Sn Tin	121.75 51 Sb Antimony	127.6 52 Te Tellurium	126.90 53 I Iodine	131.30 54 Xe Xenon
132.91 55 Cs Cesium	137.34 56 Ba Barium	Lanthanides 57–71	178.49 72 Hf Hafnium	180.95 73 Ta Tantalum	183.85 74 W Tungsten	186.2 75 Re Rhenium	190.2 76 Os Osmium	192.2 77 Ir Iridium	195.09 78 Pt Platinum	196.97 79 Au Gold	200.59 80 Hg Mercury	204.37 81 Tl Thallium	207.19 82 Pb Lead	208.98 83 Bi Bismuth	(210) 84 Po Polonium	(210) 85 At Astatine	(222) 86 Rn Radon
(223) 87 Fr Francium	(226) 88 Ra Radium	Actinides 89–103															

Lanthanides →	138.91 57 La Lanthanum	140.12 58 Ce Cerium	140.91 59 Pr Praseodymium	144.24 60 Nd Neodymium	(145) 61 Pm Promethium	150.4 62 Sm Samarium	151.96 63 Eu Europium	157.25 64 Gd Gadolinium	158.9 65 Tb Terbium	162.5 66 Dy Dysprosium	164.9 67 Ho Holmium	167.3 68 Er Erbium	168.9 69 Tm Thulium	173.0 70 Yb Ytterbium	175.0 71 Lu Lutetium
Actinides →	(227) 89 Ac Actinium	232.04 90 Th Thorium	(231) 91 Pa Protactinium	238.03 92 U Uranium	(237) 93 Np Neptunium	(242) 94 Pu Plutonium	(243) 95 Am Americium	(247) 96 Cm Curium	(247) 97 Bk Berkelium	(251) 98 Cf Californium	(254) 99 Es Einsteinium	(253) 100 Fm Fermium	(256) 101 Md Mendelevium	(253) 102 No Nobelium	(257) 103 Lr Lawrencium

†Atomic masses given in parentheses refer to the most stable isotope of an unstable element.

APPENDIX VI

SIGNIFICANT DATES IN THE DEVELOPMENT OF PHYSICS

History can rarely be stated as a simple series of dates, and the history of science is no exception. Throughout the text we have alluded to important discoveries in physics. The list below is a personal choice and should be thought of as a guide. It oversimplifies some of the history, including stories that are covered more thoroughly in the text (for example, Coulomb's law). Some of the dates are to be taken with a grain of salt, because discoveries are rarely made in a single identifiable moment. Our list includes some names (and discoveries) not mentioned in the text. Far more numerous are the names not listed, the names of those who built the experimental foundations, those who explored the false paths and cleared the way for those whose names we remember today, or those who verified the speculations that we call laws.

1583	Galileo	Pendulum motion
1600	Gilbert	Study of magnets
1602	Galileo	Early statement of Newton's first law
1602	Galileo	Laws of falling bodies
1609	Kepler	First two laws of planetary motion
1619	Kepler	Third law of planetary motion
1620	Snell	Law of refraction
1648	Pascal	Atmospheric pressure
1650	Grimaldi	Diffraction of light
1661	Boyle	Chemical elements
1669	Newton	Light dispersion in prisms
1678	Huygens	Wave propagation
1687	Newton	Laws of motion; universal gravitation
1760	Black	Calorimetry
1785	Coulomb	Coulomb's law
1789	Lavoisier	Conservation of mass
1798	Cavendish	Measurement of G
1800	Volta	Electric battery
1801	Young	Interference of light
1801	Dalton	Laws of chemical combination
~1802	Charles; Gay-Lussac	Ideal gases
1807	Dalton	Atomic theory
1812	Fourier	Decomposition of waves
1815	Fraunhofer	Discrete spectral lines
1819	Fresnel	Wave picture of light
1820	Oersted	Magnetic fields from currents
1820	Biot; Savart	Law of magnetic field produced by current
1824	Carnot	Second law of thermodynamics
1827	Ohm	Ohm's law
1827	Ampère	Ampère's law
1831	Faraday; Henry	Induction

1842	Joule	Mechanical equivalent of heat
1847	Helmholtz	Conservation of energy
1849	Fizeau	Direct measurement of the speed of light
1865	Maxwell	The laws of electricity and magnetism; light waves
1877	Boltzmann; Gibbs	Statistical mechanics
1879	Stefan	Blackbody radiation
1885	Osmond	Crystalline structure of metals
1887	Hertz	Electromagnetic waves
1887	Michelson and Morley	Constancy of the speed of light
1896	Becquerel	Radioactivity
1897	Thomson	Charge-to-mass ratio of the electron
1900	Planck	Quanta in blackbody radiation
1903	Rutherford; Soddy	Isotopes
1905	Einstein	Special relativity; quanta in photoelectric effect
1908	Kammerlingh Onnes	Superfluidity
1911	Kammerlingh Onnes	Superconductivity
1911	Rutherford	Nuclear structure of atom
1911	Millikan	Quantization of charge
1912	von Laue	X-ray diffraction in crystals
1913	Bohr	Atomic structure
1916	Einstein	General relativity
1923	Hubble	Discovery of galaxies
1924	de Broglie	Wave nature of particles
1925	Pauli	Exclusion principle
1925	Heisenberg	Formulation of quantum mechanics
1925	Goudsmit and Uhlenbeck	Electron spin
1926	Davisson and Germer; Thomson	Diffraction of electrons by crystals
1926	Schrodinger	Alternate formulation of quantum mechanics
1926	Born	Probabilistic interpretation of quantum theory
1927	Heisenberg	Uncertainty relations
1929	Hubble	Hubble's law
1930	Dirac	Antiparticles
1932	Anderson	The positron
1932	Lawrence and Livingston	The cyclotron
1932	Chadwick	The neutron
1934	Yukawa	Nuclear forces and the pi meson
1948	Feynman; Schwinger; Tomonaga	Electromagnetism as a quantum theory
1954	Townes	The maser
1957	Lee and Yang	Nonconservation of parity
1957	Bardeen, Cooper, and Schrieffer	Theory of superconductivity
1962	Josephson	Josephson junction
1964	Gell-Mann; Zweig	Quarks
1964	Penzias and Wilson	Background radiation of the universe
1967–1970	Glashow; Salam; Weinberg	Unification of electromagnetic and weak forces

APPENDIX VII

TABLES IN THE TEXT

APPENDIX VIII

SELECTED TEXT BOXES

PROBLEM-SOLVING TECHNIQUES

APPLICATION

A CLOSER LOOK

ANSWERS TO ODD-NUMBERED PROBLEMS

CHAPTER 1

1. 2.500×10^3 green jelly beans
3. 10^7, 10^{14}
5. 170 cm
7. 32.2 ft/s^2
9. 5.5 g·cm^{-3}
11. 5.3 L/100 km, 34 L/100 km
13. $[ML^2T^{-2}]$
15. $[L]$
17. (a) 0.4 mi, 1 significant figure; (b) 0.414 mi, 3 significant figures, which is 0.4 to 1 significant figure; (c) 0.414 ± 0.004 mi
19. 2×10^{-2} kg·m·s^{-2}
21. 1.1×10^3 kg·m^{-3} $\pm$ 17%
23. 2.5%
25. $\simeq 2 \times 10^9$ beats
27. $\simeq 10^4$ peas
29. $\simeq 40$ lb, 10 gal
31. $\simeq 9.3 \times 10^5$ mi
33. $\simeq 4.6 \times 10^{-4}$
35. $\simeq 900$ cars
37. $\simeq 5 \times 10^4$ apples
39. ______; $\simeq 2 \times 10^{-8}$ cm
41. 0.83 paces east and 5.83 paces north, or 5.9 paces 82° north of east
43. $V_x = -V \sin \alpha$
45. $V_x = V \cos(\alpha + 45°)$
49. (a) $[M^{1/2}L^{3/2}T^{-1}]$; (b) $[ML^2T^{-2}]$
51. (a) 10^{-6} metric tons; (b) 10^{-6} m^3/cm^3
53. 500 s (8.3 min)
55. 3.0×10^{-26} kg
57. $\simeq 3 \times 10^{-8}$ cm
59. 2×10^{41} kg, 1.2×10^{68} H atoms
63. $\simeq 1000$ trucks/day, 2000 trucks/day
65. 7.8×10^8 km, 2.5×10^2 yr
67. $\simeq 10^{44}$ molecules
69.

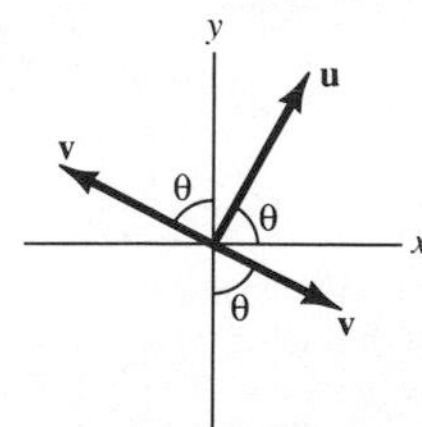

(a) $\mathbf{v} = -v \sin \theta \, \mathbf{i} + v \cos \theta \, \mathbf{j}$ or $\mathbf{v} = v \sin \theta \, \mathbf{i} - v \cos \theta \, \mathbf{j}$

71. (c) $r_y = r \sin \theta \sin \phi$
73. $\simeq 50$ cm

CHAPTER 2

1. 40**i**

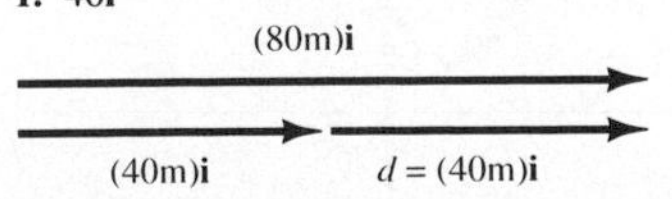

3. 8.9**i** m/s, 11.8**i** m/s, 10.1**i** m/s
5.

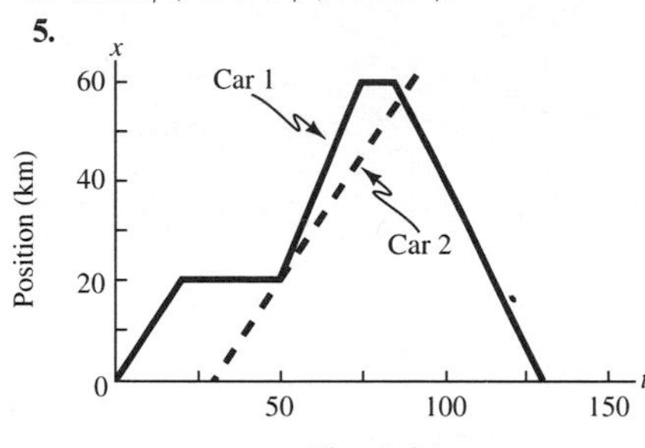

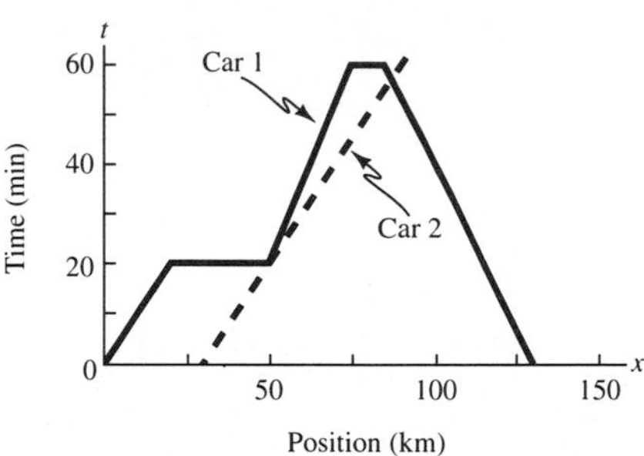

With the axes reversed, Car 1 travels infinitely fast at $t = 20$ s and at $t = 60$ s and then travels backward in time; Car 2 travels at constant speed.

7. 11.0 m/s, 10.5 m/s, 11.4 m/s
9.

Time (s)	*Position (m)*
0.0	0.0
0.5	0.19
1.5	1.44
2.5	7.31
3.5	22.56
4.5	53.31
5.5	108.06
6.5	184.81
7.5	291.56

11. $\mathbf{v}_1 = -0.1\mathbf{i}$ cm/s, $\mathbf{v}_5 = 4.7\mathbf{i}$ cm/s, $\mathbf{v}_{10} = 24.2\mathbf{i}$ cm/s; $\mathbf{v}_{av} = 7.2\mathbf{i}$ cm/s; v will get very large due to its t^2 term
13.
15. 2.8 m/s^2
17.

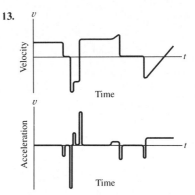

Time (s)	*Acceleration (m/s^2)*
1.0	1.0
2.0	7.0
3.0	13.0
4.0	18.0
5.0	23.0
6.0	28.0
7.0	32.0

19. 8.8×10^2 m/s^2
21. 12.3 s
23. 38.6 s
25. (a) $v = 8.0 - 0.50t$ easterly (t in s and v in m/s); (b) 34 m to the east
27. 312 m
29. (a) 2.6 s; (b) 5.8 s; (c) 2.4 s; (d) 8.2 s; (e) the early times are longer, but the later times are shorter, because runners cannot maintain high acceleration
31. 6.33 m/s
33. No, 107 ft
35. (a) Car A: 50.0 m/s, 10.0 s; Car B: 47.4 m/s, 10.5 s; (b) Car A: 14.5 s, 61.2 m/s; Car B: 15.1 s, 61.2m/s
37. 150 mi/h, 0.14 mi, 1.3 mi
39. 54 m
41. -5.0×10^5 m/s^2; 8.0×10^{-4} s
43. 40 cm, 90 cm, 160 cm
45. 1.1 s
47. 40 m
49. 10 m/s^2
51.

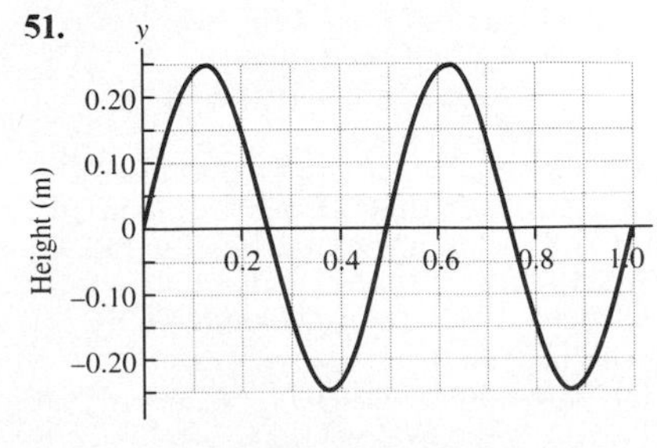

(a)

(b) -0.8 m/s; (c) 2.5 m/s; (d) -30 m/s^2

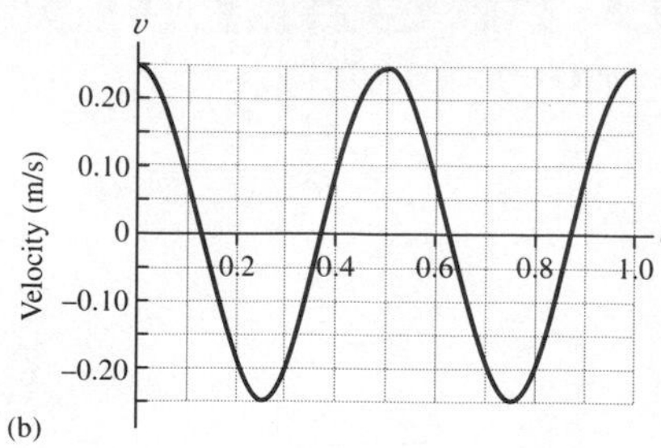

53. 0.16 s
55. -9.8 (8 ft/Δ ft) m/s^2, where Δ is the distance your legs bend
57. 11 m
59. 31 m above ground

CHAPTER 3

1. 1st turn: $\mathbf{r}_1 = 15\mathbf{i} + 15\mathbf{j}$ km; 2nd turn: $\mathbf{r}_2 = 30\mathbf{i} + 15\mathbf{j}$ km; total: $\mathbf{r} = 30\mathbf{i} + 43\mathbf{j}$ km
3. $\mathbf{r}_A = 0$, $\mathbf{r}_B = (25\text{ m})\mathbf{i}$; $\mathbf{r}_C = (25\text{ m})\mathbf{i} + (35\text{ m})\mathbf{j}$; $\mathbf{r}_D = (35\text{ m})\mathbf{j}$
5.

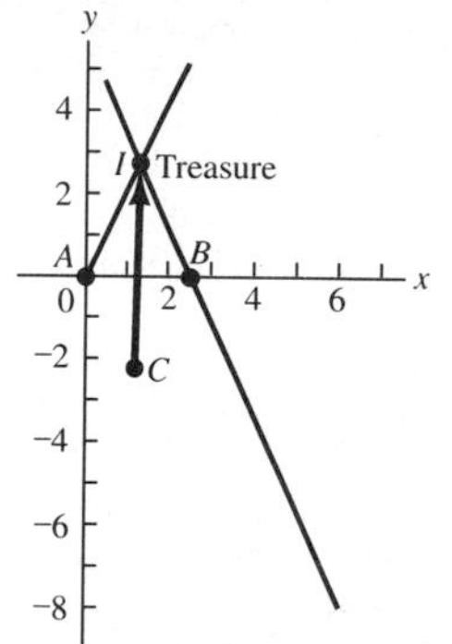

Intersection: $\mathbf{r}_I = (1.3\text{ km})\mathbf{i} + (2.7\text{ km})\mathbf{j}$;
$\mathbf{r}_I - \mathbf{r}_C = (0.1\text{ km})\mathbf{i} + (4.9\text{ km})\mathbf{j}$
7. $\mathbf{r}_{T/3} = (2.0\text{ m})\mathbf{i} - (3.5\text{ m})\mathbf{j}$, 4.0 m to origin; $\mathbf{r}_{T/2} = (-4.0\text{ m})\mathbf{j}$, 4.0 m to origin; $\mathbf{r}_{2T} = (4.0\text{ m})\mathbf{i}$, 4.0 m to origin; $\theta(t) = -\pi t/T$
9. $\mathbf{r} = [(4 + 3t + t^2)\mathbf{i} + (6 - 4t + 0.5t^2)\mathbf{j}]$ m, $\mathbf{v} = [(3 + 2t)\mathbf{i} + (-4 + t)\mathbf{j}]$ m/s, $\mathbf{a} = [2\mathbf{i} + \mathbf{j}]$ m/s^2; $v_x = v_y$ at $t = -7$ s
11. (a) 1.02 m/s; (b) 1.17 m/s
13. $\mathbf{v} = -(4\pi/T)\sin(\pi t/T)\,\mathbf{i} - (4\pi/T)\cos(\pi t/T)\,\mathbf{j}$ m/s, $(\pi/2) - (\pi t/T)$
15. (a) $30\mathbf{i} - 48\mathbf{j}$ m, $22\mathbf{i} - 22\mathbf{j}$ m/s; (b) $-58°$ (below the x-axis)
17. $\mathbf{a}_1 = -[4R(\pi f)^2\cos(2\pi ft)]\mathbf{i} - [4R(\pi f)^2\sin(2\pi ft)]\mathbf{j} = -4\pi^2 f^2\mathbf{r}_1$,
$\mathbf{a}_2 = -\{[R(\pi f)^2/4]\cos(\pi ft/4)\}\mathbf{i} - \{[R(\pi f)^2/4]\sin(\pi ft/4)\}\mathbf{j} = -(\pi^2 f^2/16)\mathbf{r}_2$,
$\mathbf{a}_2 - \mathbf{a}_1 = -R(\pi f)^2\{[\tfrac{1}{4}\cos(\pi ft/4) - 4\cos(2\pi ft)]\mathbf{i} + [(\tfrac{1}{4}\sin(\pi ft/4) - 4\sin(2\pi ft)]\mathbf{j}\}$
19. **(a) (889 m)i − (160 m)j (903 m at 10.2° below horizontal); (b) (111 m/s)i − (40 m/s)j m/s at 19.8° below horizontal); (c) (889 m)i + (2.34 × 10³ m)j**
21. $(6.6\text{ km})\mathbf{i} + (11.5\text{ km})\mathbf{j}$, $(72\text{ km/h})\mathbf{i} + (125\text{ km/h})\mathbf{j}$
23. (a) 5.6 m/s; (b) 5.4 m/s; (c) 1.7 s
25. 63°
27. (a) 7.1 m/s; (b) 58 m; (c) 34 m/s
29. 30°
31. (a) 1.25 s; (b) no, 0.4 m beneath the bar
33. 15° and 75°; two angles arise from the combination of time of flight (shorter for the smaller angle) and horizontal speed (higher for the smaller angle); 0° and 90°
35. (a) 7.8×10^3 m/s; (b) 9.1 m/s^2
37. 2.2 m/s^2
39. 9.5 s
41. (a) (0.43 m, 0.25 m); (b) $-79\mathbf{i}$ m/s^2; (c) $-79\mathbf{j}$ m/s^2
43. $1.21R$
45. 10.8 km/h, 22° south of east
47. (a) 95 km/h; (b) 88 km/h, 25° west of south; **(c) −(185 km)i − (397 km)j**
49. **(a)** $t_0 = 2L/v$; **(b)** $t_1 = 2Lv/(v^2 - v_w^2)$

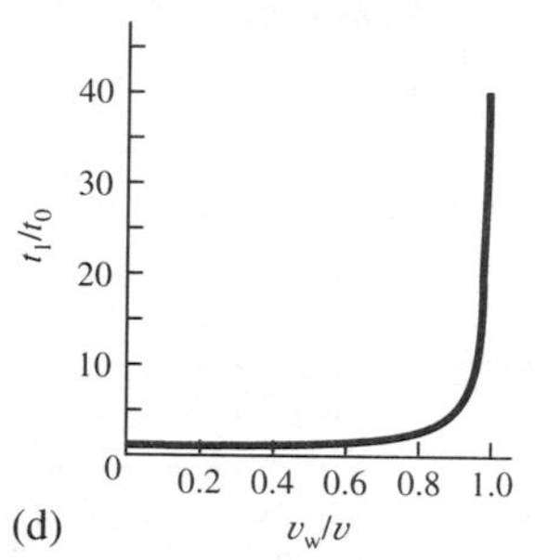

51. 9.4×10^3 m/s, tangential; 3.0×10^3 m/s^2 = 300g, centripetal
53. 27 m/s
55. **With the current from left to right, $\mathbf{r} = (10\cos\theta + 6)t\mathbf{i} + (10\sin\theta)t\mathbf{j}$ for r in km and t in h; 127°; 68 s**
57. It will fall onto the deck 8.3 m from the mast, if the deck is horizontal at that moment
59. (a) 7.35 m/s; (b) 0 (thrown), 2.45 m going up, 2.45 m coming down, 0 (caught); (c) 4.9 m

CHAPTER 4

1. (a) The force of gravity (toward earth); (b) the force of gravity (down), the normal force from the ice (up), and a small friction force (opposite the motion); (c) essentially none
3. 200 N in the $-y$ direction
5. No, because the z-component must be balanced.
9.

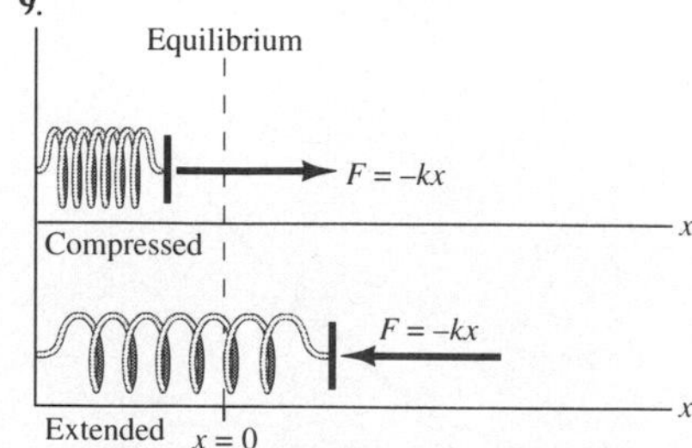

11. **2.39 × 10³ N; 3.63 × 10³ N forward; 1.48 m/s² forward**
13. **Away from the body, along the rope; (a) 22 N, 7.2 N, both in the direction of pull; (b) 30 N up, balanced by the downward force at the axle**
15. 10^5 m/s; 10^6 m/s; 10^8 m/s
17. (a) The earth; (b) the ice and the earth; (c) none
19. Consider only horizontal forces (vertical normal forces balance gravity forces):

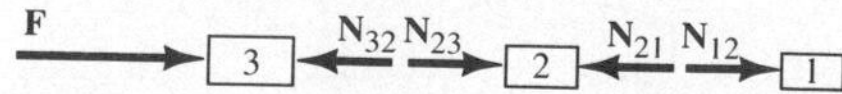

(a) 2.0 m/s^2; (b) $F = 12$ N to the right, $N_{32} = 6.0$ N to the left, $F_{\text{net }3} = 6.0$ N to the right; (c) $N_{23} = 6.0$ N to the right, $N_{21} = 2.0$ N to the left, $F_{\text{net }2} = 4.0$ N to the right; (d) $N_{12} = 2.0$ N to the right, $F_{\text{net }1} = 2.0$ N to the right
21. If the observer does not realize that he is in free fall, he will think that there is a force opposing but equal to gravity.
23. (a) Parallel to the edge; (b) angled toward the back of the plane such that $\tan\theta = a'/g$; (c) parallel to the edge
25.

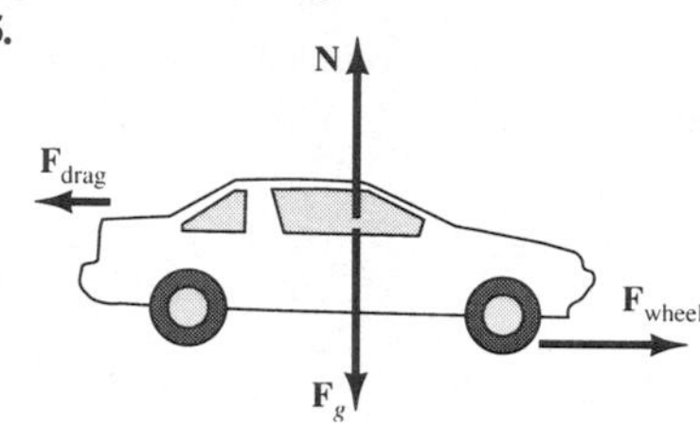

27.

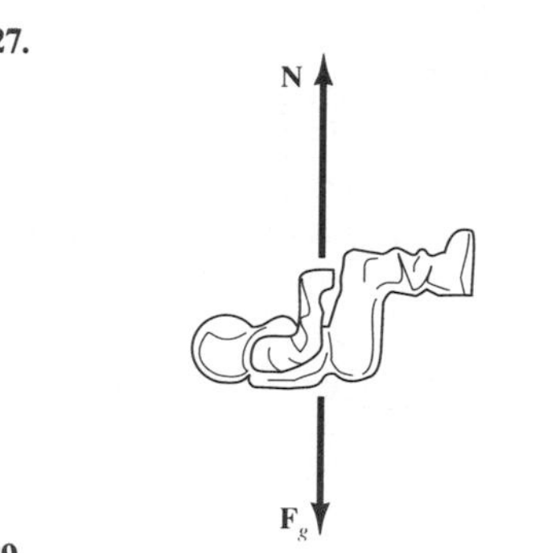

29.

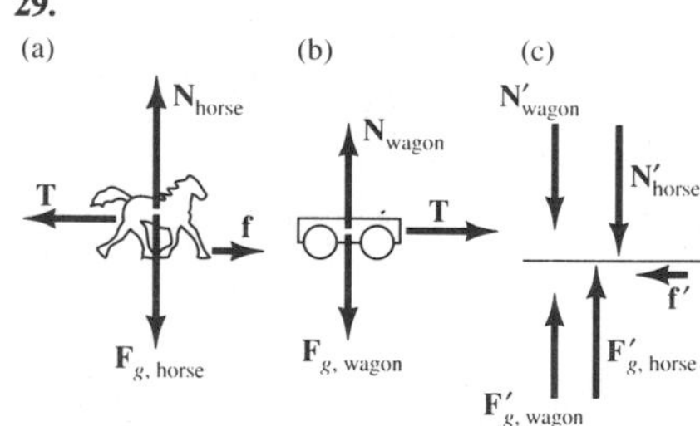

(d) *Net* force determines acceleration. There is a forward force on the horse's hooves.
31. $\mathbf{F}_g = F_g\sin\theta\,\mathbf{i} - F_g\cos\theta\,\mathbf{j}$; $\mathbf{N} = N\mathbf{j}$; $\mathbf{f} = -f\mathbf{i}$

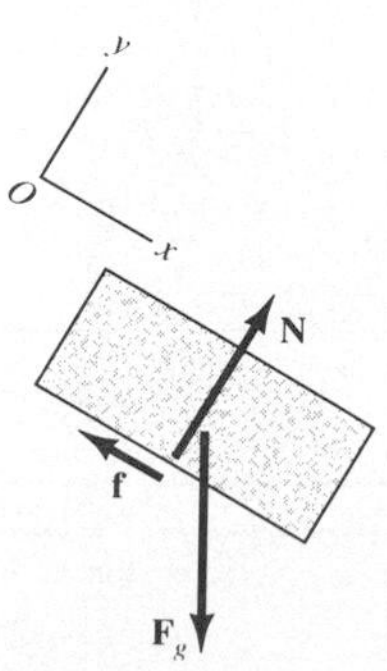

33.

The ladder might slip because there is a net horizontal force from the wall.

35.

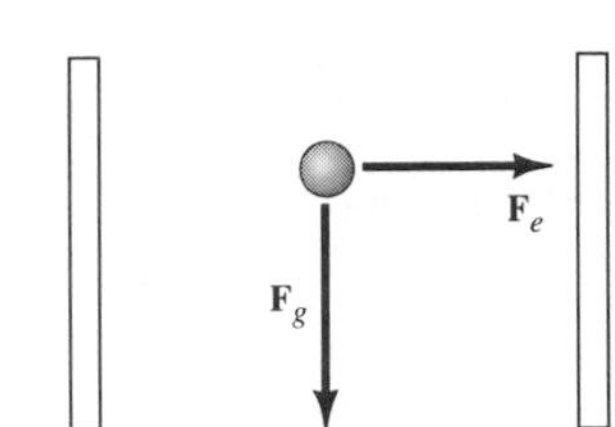

37. (a)

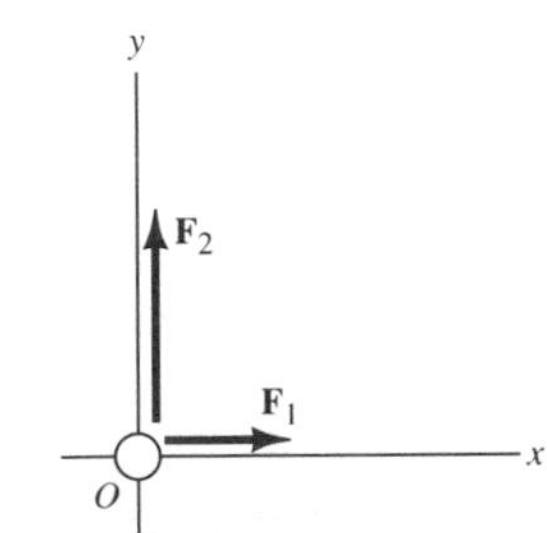

(b) $\mathbf{r}_O = (0, 0, 0)$; $\mathbf{v}_O = (0, 0, 0)$;
(c) $\mathbf{r}_{2.000} = (18.98, 32.46, 0)$ m, $\mathbf{v}_{2.000} = (18.98, 32.46, 0)$ m/s;
(d) $\mathbf{r}_{3.000} = (42.70, 73.04, 0)$ m, $\mathbf{v}_{3.000} = (28.47, 48.69, 0)$ m/s

39. (a) -1.096 N, -1.875 N; (b) 0.059 N, -0.720 N

41. (a) ______, $\sqrt{k/m}$; (b) 0, $v_0\sqrt{m/k}$

43. (a)

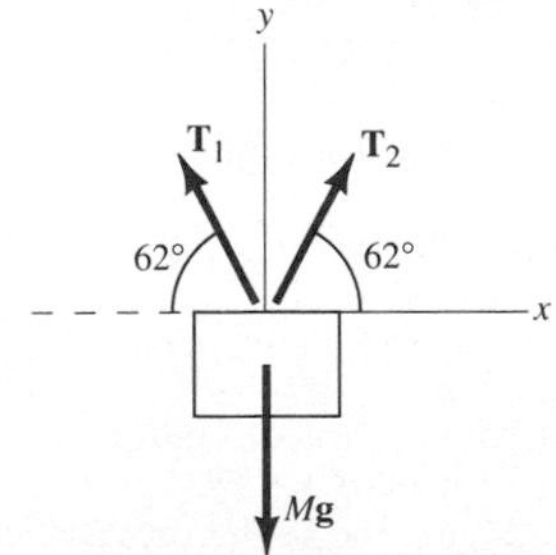

(b) 1.85×10^3 N up; (c) 5.69 m/s^2 up

45. $F_g \simeq 0.20$ N down; $F_{\text{floor}} \simeq 1.7 \text{ N} + mg \simeq 1.9$ N up during impact

47. (a)

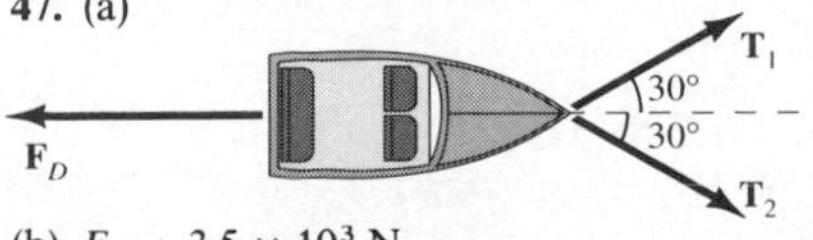

(b) $F_D = 3.5 \times 10^3$ N

49.

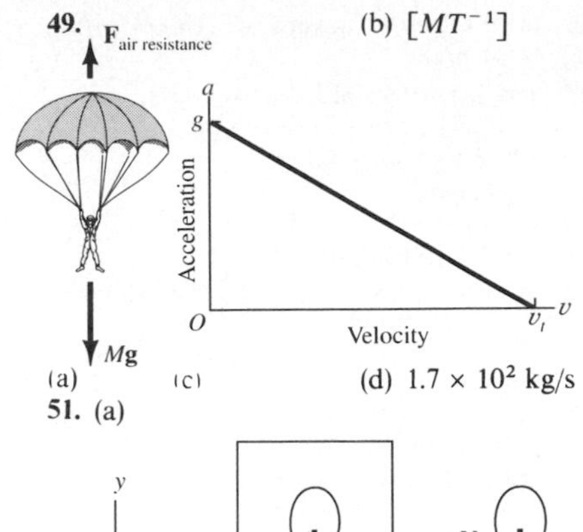

(b) $[MT^{-1}]$

(d) 1.7×10^2 kg/s

51. (a)

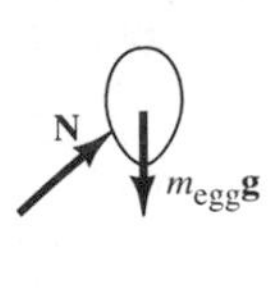

(b) $\mathbf{F} - (m_{\text{egg}} + m_{\text{box}})g\mathbf{j} = (m_{\text{egg}} + m_{\text{box}})\mathbf{a}$, $\mathbf{N} - m_{\text{egg}}g\mathbf{j} = m_{\text{egg}}\mathbf{a}$;
(c) $\mathbf{a} = (13.2 \text{ m/s}^2)\mathbf{i} - (1.80 \text{ m/s}^2)\mathbf{j}$, $\mathbf{F}_{\text{egg(net)}} = (0.66 \text{ N})\mathbf{i} - (0.90 \text{ N})\mathbf{j}$

53. (a) 1×10^{-8} s; (b) 1×10^4 m/s

55. (a) $g_0 - g_{\text{eq}} = 4\pi^2Rf^2$; (b) 9.77 m/s^2; (c) $g = g_0 - 0.034\cos^2\phi$; (d) the earth is not a perfect sphere, and the elevation may be above sea level

CHAPTER 5

1. 1.35×10^4 N
3. 980 N
5. 0.095 s
7. (a) 17 N; (b) 8.9 N
9. (a) 2.15 m; (b) 1.13 s; (c) 1.13 s; (d) 3.8 m/s down the plane
11. (a) 2.5 m/s^2 forward; (b) 3.5×10^3 N forward
13. For each mass: $a = [(m_1 \sin\theta_1 - m_2 \sin\theta_2)g]/(m_1 + m_2)$, $v = v_0 + at$, $x = x_0 + v_0t + (at^2/2)$, for m_2 up, m_1 down
15. $a_M = (mg \sin\theta \cos\theta)/(M + m\sin^2\theta)$, horizontal
17. $T(h) = \lambda gh$
19. (a) $a_1 = -1.21$ m/s^2, $a_2 = +1.99$ m/s^2, $a_3 = +0.43$ m/s^2 (down is positive); (b) $T_1 = 11.0$ N, $T_2 = 5.5$ N
21. 3.3×10^2 N
23. 0.40
25. (a) 52 m; (b) 3.9 s
27. (a) 4.5 s; (b) 0.65 N; (c) 0.35
29. (a) $370/(\cos\theta + 0.75\sin\theta)$ N; (b) ______, $\theta_{\min} = 37°$, $F_{\min} = 2.9 \times 10^2$ N; at small angles, the normal force, N, and thus the friction force, f, are large, so F must be large. Near 90°, N is small but F has a small horizontal component, so F must be large. Thus there is an angle where the decrease in N is balanced by the decrease in horizontal component, and therefore there is a minimum F.
31. 0.077
33. $a = [m_1g(\sin\theta_1 + \mu_k\cos\theta_1) - m_2g(\sin\theta_2 - \mu_k\cos\theta_2)]/(m_1 + m_2)$, opposite to the motion
35. 47 kg/m
37. 98 cm^2
39. 0.46 m/s
41. $C_{D,\text{ max speed}}/C_{D,\text{ min speed}} = 0.67$
43. 1.4×10^2 N toward the center
45. (a) 2.3°; (b) 6.9°; (c) 11.4°
47. 1.0×10^3 m/s tangent to the orbit
49. (a) 4.43 m/s; (b) 3.35 m/s
51. (a)

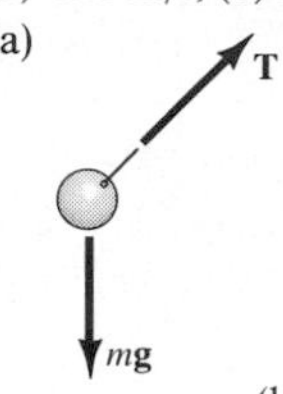

(b) 0.4 m/s

53. 2.1 m
55. 55 m/s ($\simeq$120 mi/h); there are many slower speeds, because friction can be less than μ_sN
57. $\mathbf{F}_{\text{net}} = -(14 \times 10^4 \text{ N})\mathbf{i} - (5 \times 10^4 \text{ N})\mathbf{j}$, $|\mathbf{F}_{\text{net}}| = 1.5 \times 10^5$ N
59. α(angle from radial direction) = $\theta - \tan^{-1}\{[1 - (R\omega^2/g)]\tan\theta\} \simeq (R\omega^2/g)\sin\theta\cos\theta$
61. (d) To the right of the direction of throw
63. (a) 60 N; (b) 96 N
65. (a) $2\,\Delta x_1$ in the opposite direction; (b) $a_1 = 2.8$ m/s^2 up, $a_2 = 5.6$ m/s^2 down; (c) 7.6 N
67. $1.5m_2$
69. 12 m/s ($\simeq$27 mi/h)
71. (a) 2.6×10^3 N forward; (b) 2.6×10^3 N; (c) 2.8×10^3 N forward
73. 28 m (91 ft)

CHAPTER 6

1. (a) 39 J; (b) 1.0 m/s and 89 m/s; (c) 0.40 m/s
3. (a) 98 N (gravity) down and 98 N (pull) up, $F_{\text{net}} = 0$; (b) 0; (c) 98 J
5. 1.0×10^4 J
7. (a) 2.4×10^2 J; (b) friction; (c) 0
9. (a) 9.8 J; (b) -7.8 J
11. 1.8×10^3 J
13. (a) 1.6×10^4 J; (b) 1.6×10^4 J
15. $\frac{1}{2}MgL$
17. 3.9×10^3 J
19. 9.7×10^3 J
21.

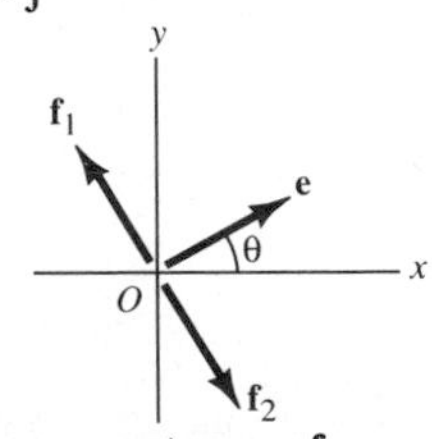

______; two vectors: $\mathbf{f}_1 = -\sin\theta\,\mathbf{i} + \cos\theta\,\mathbf{j}$, $\mathbf{f}_2 = +\sin\theta\,\mathbf{i} - \cos\theta\,\mathbf{j}$
23. 3.2
25. 12 J
27. 0.20, -4.3×10^2 J
29. $2g_1 - 4g_2$
31. $330s + 2.12s^2$ J (s in meters)
33. 8.1×10^{-2} J
35. 1.3×10^2 J
37. (a) 6.3×10^{-2} J; (b) 0; (c) 5.9×10^{-2} J; (d) -5.9×10^{-2} J
39. (a) $0.16C$ J, 0 J; (b) $0.32C$ J for both
41. 30 bulbs

43. 1 min, $\simeq 0.7$ kW
45. 52 kW
47. (a) 5.0 kW; (b) 7.1 kW
49. 4.7×10^7 W
51. (a) 16 m/s; (b) 0.5 s; (c) $\simeq 12$ s
53. 5.8×10^{-12} J, 5.8×10^{-9} J
55. 0.62, 0.31
57. -0.2 J
59. 0.034
61. (a) 2.2×10^7 J; (b) -2.2×10^7 J
63. (a) $mg = 26$ N (down), $N = 22$ N normal to the plane (up), $f_k = 5.4$ N parallel to the plane (down); (b) $W_g = -18$ J, $W_N = 0$, $W_f = -7.0$ J; (c) 4.4 m/s
65. 5.3 J
67. (a) 0; (b) -0.20 J; (c) 9.8 J
69. (b) $mgH(1 + \mu_k \cot\theta)$
71. $W_g = mgL(\cos\theta_f - \cos\theta_i)$, $v = \sqrt{2gL(1 - \cos\theta_i)}$, $W_T = 0$ because T is normal to the displacement
73. $0.01mK/x$

CHAPTER 7

1. 2.1×10^3 J either way
3. -15 J, -5 J, yes
7. (a) 196 J; (b) 29 J, 225 J
11. (a) $-7x$ J; (b) 10 J; (c) 5.5 m/s
13. 2.5 m, 4.5 m/s
15. 0.7248 m, 6.596 kg
17.

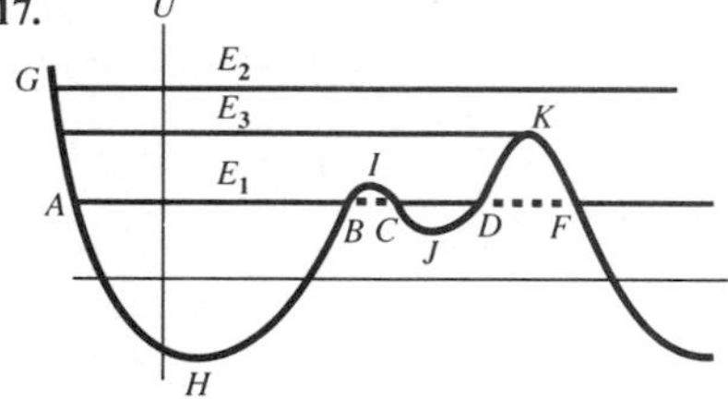

(a) E_1: $A < x < B$, $C < x < D$, or $F < x$; E_2: $G < x$; (b) when the particle has energy of positions H, I, J, or K; (c) $E < E_3$, position must be in the valleys; (d) H and J are stable equilibrium positions, I and K are unstable
19.

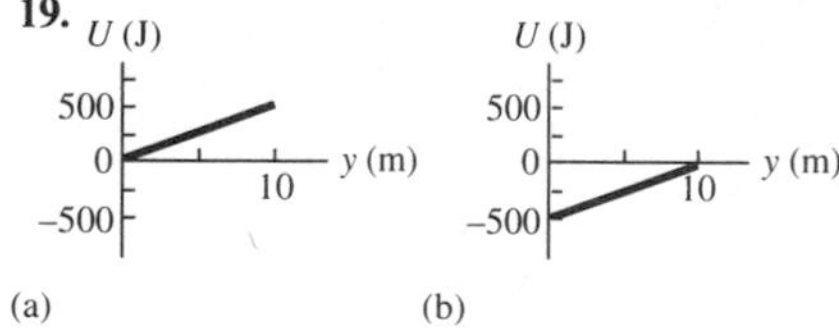

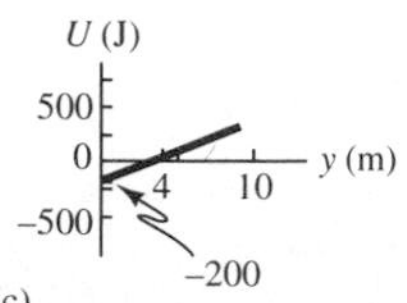

21. $x = 0$, unstable; $x = \pm 1.15$ m, stable
23. 0.529 m, 0.135 m
25. 0.50 m, 0.55 m
27. (a) 7.1 m/s; (b) -0.25 J; (c) 0.73
29. (b) $\sqrt{2gh}$; (c) $2\sqrt{gh}$
33. -1.2×10^5 J
35. (a) 1.4 m/s; (b) 1.2 m/s; (c) 6.7 cm from point b
37. $\frac{1}{2}k_1(\sqrt{L^2 + y'^2} - L)^2 + \frac{1}{2}k_2 y'^2$, $-(k_1 + k_2)y' + k_1 L y'/\sqrt{L^2 + y'^2}$
39. (a) $+GMm/r^2$ toward M; (b) $GMm/2r$; (c) $-GMm/2r$
43. $x = \frac{1}{4}v_0\sqrt{m/A}\sin(2t\sqrt{A/m}) + \frac{1}{2}v_0 t$, $y = \frac{1}{4}v_0\sqrt{m/A}\sin(2t\sqrt{A/m}) - \frac{1}{2}v_0 t$
45. 4.9×10^{12} particles
47. $\simeq 20$ cents
49. 16 cm
51. -9.0 J for both paths
53. (a) 1.4 m/s, 0.63 m/s, v_d not possible; (b) reaches 10 cm high
55. 33 strokes, possible for gravity, not for resistance force
57. (a) Yes; (b) $U(r) = -C/r$, $U(\infty) = 0$; (c) 2.1×10^7 m/s
59. (a) 1.13 J; (b) -0.083 J; (c) -0.30 J; (d) 0.70 m/s; (e) 8.8 cm; (f) 5.0 cm
61.

(a)

$r \times 10^{-15}$ (m)

$U(r)$ ($A \times 10^{15}$)

(b) $r = 0$; (c) $-Ae^{-kr}[(1 + kr)/r^2]$; (d) $(1.0 \times 10^{+32})A$, $(5.0 \times 10^{+24})A$
65. $L(\sin\alpha + 2\sin\alpha\cos^2\alpha + 2\sqrt{\cos^3\alpha - \cos^6\alpha})$

CHAPTER 8

1. (a) 0.63 kg·m/s; (b) 5.8 kg·m/s; (c) 7.3×10^2 kg·m/s; (d) 7.7×10^2 kg·m/s
3. (a) 4.2×10^2 kg·m/s; (b) 6.0×10^6 kg·m/s; (c) 1.2×10^4 kg·m/s; (d) 3.3×10^{-22} kg·m/s; (e) 1.4×10^{-2} kg·m/s
5. (a) $\sqrt{mRF}$; (b) $R = p/K$
7. (b) $r/R = M/m$
9. (a) The first and second are at rest; the third moves with the velocity of the original ball; (b) the first ball recoils straight back at speed $0.20v_0$; the other two move at speed $0.69v_0$—one at 30° above the original line, the other 30° below.
11. 0.63 kg·m/s, 0.13 m/s, in the direction of the arrow's initial motion for both
13. 2.3 kg·m/s, 180 N (both up)
15. (a) $\simeq 1$ kg·m/s; (b) $\simeq 22$ kg·m/s
17. (a) 1.2×10^3 kg·m/s; (b) 6×10^3 N
19. (a) 5 kg·m/s; (b) 250 N; (c) 500 N
21. (a) 5.7×10^2 N; (b) 1.4×10^2 J; (c) 0.011 s
23. $0.40\sqrt{2g(0.85 + 0.15N)}$ kg·m/s
25. $-m/(m_0 + m)$
27. (a) $M/(m + M)$; (b) $v = [(m + M)/m]\sqrt{2gh}$
29. (a) $v[m_2^2 + M(m_1 + m_2)]/M(m_2 + M)$ forward; (b) $v[m_1^2 + M(m_1 + m_2)]/M(m_1 + M)$ forward; (c) $v(m_1 + m_2)/M$ forward
31. (a) -0.2 m/s; (b) inelastic, 64%
33. 2.59 m/s
35. 15°
37. (a) $\sqrt{2gh}$; (b) $\sqrt{2gh}$, each;
(c) $[(m - 3M)/(m + M)]\sqrt{2gh}$ up;
(d) $[(m - 3M)/(m + M)]^2 h$; (e) $9h$
39. $-v_1\mathbf{i} - v_2\mathbf{j}$
41. (3.9 m/s)**i** + (0.28 m/s)**j**
43. (a) 71.4°; (c) $K_p = 0.935K$, $K_\alpha = 0.065K$

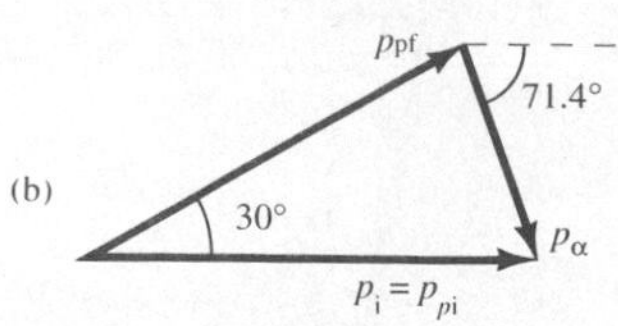

45. (0.78 m)**i** + (2.11 m)**j**
47. In both cases, $v_{CM} = v/3$ before and after, in the initial direction of the first ball
49. 1.0 m from the bottom of the handle, ignoring the dimensions of the mallet head
53. $4R/3\pi$ from the center of the circular arc, along the bisector
55. $a\ln[(L^2 + a^2)/a^2]/[2\tan^{-1}(L/a)]$
57. 1.3×10^4 kg
59. 904 m/s
61. $v = u_{ex}\ln\{m_0(m_0 - m_1 - m_1')/[(m_0 - m_1 - m_1' - m_2)(m_0 - m_1)]\} - gt$, where up is positive
63. 2.2 m
65. (a) $[8M(M - m)/(M + m)^2]h$;
(b) $[4mM/(M + m)^2]h$
67. 158 m
69. (a) O; (b) $(-3.5$ m/s$)$**i** $-$ (5.0 m/s)**j**; (c) (0.43 m)**i** + (0.18 m)**j**; (d) CM has moved due to the initial velocity and to friction forces from the floor
71. (a) -0.070 m/s; (b) 0; (c) 0.084 m; (d) 0

CHAPTER 9

1. 0.68 rad/s²
3. 28 rad/s²
5. 7.3×10^{-5} rad/s from the South Pole to the North Pole, 2.0×10^{-7} rad/s perpendicular to the orbital plane
7. (a) -2.62×10^2 rad/s²; (b) 1.67×10^2 m
9. (a) 0.10 rad/s²; (b) 11 rev; (c) 2.1×10^2 m
11. (a) $x = (x_0 + 0.05t^2)$ m, 0.10 m/s²; (b) 5 rad/s²; (c) $\omega = (5\text{ rad/s}^2)t$
13. 1.35×10^{-4} m
19. $I_{earth} = 9.8 \times 10^{37}$ kg·m², $I_{neutron\ star} = 1.0 \times 10^{39}$ kg·m² $\simeq 10I_{earth}$
23. $MR^2/2$, where $M = \rho R^2 h\theta_0/2$
25. $(77/30)\pi h\rho_0 a^4$
27. $(2m/3\mu^2)[K_4(\mu R)/K_2(\mu R)]$
29. 22.5 lb
31. 1.2×10^2 kg
33. $2.9\cos\phi$
35. $\tau_{net\,A} = 17$ N·m counterclockwise, $\tau_{net\,B} = 19$ N·m clockwise
37. (a) 11.5 kg·m²; (b) 89.3 kg·m²/s; (c) 1.14 rev; (d) 349 J
39. 1.9 rad/s up, the student rotates left
41. (a) It increases to 8.3 rad/s; (b) the platform and the child have the original angular speed of 6.2 rad/s.
43. (a) 0; (b) 26.8 m/s; (c) 19.0 m/s; (d) 19.0 m/s
45. (a) 1/3; (b) 1/2; (c) 2/7
47. 2.3 rev
49. 4.0×10^3 J, 10%
51. $t_{rolling} = \sqrt{2.8\ell/g\sin\theta}$, $t_{slipping} = \sqrt{2\ell/g\sin\theta}$, $t_{rolling}/t_{slipping} = 1.18$

53. (a) $0.040\ \text{kg}\cdot\text{m}^2/\text{s}$; (b) 0.016 J; (c) 0.032 J

55.

F_N T (a) Mg T mg

(b) $TR = \frac{2}{5}MR^2\alpha$, $mg - T = ma$, $a = R\alpha$; (c) $4.9\ \text{m/s}^2$; (d) $98\ \text{rad/s}^2$ into the page; (e) 0.98 N; (f) 63 rad/s into the page, $3.1 \times 10^{-2}\ \text{kg}\cdot\text{m}^2/\text{s}$ into the page, 0.98 J

57. $MgR/(R + 2r)$

CHAPTER 10

1. (a) $1.1 \times 10^{10}\ \text{kg}\cdot\text{m}^2/\text{s}$ down; (b) $7.9 \times 10^9\ \text{kg}\cdot\text{m}^2/\text{s}$ down

3. $5 \times 10^4\ \text{kg}\cdot\text{m}^2/\text{s}$ down, $5 \times 10^4\ \text{kg}\cdot\text{m}^2/\text{s}$ down

5. $-(mbv/\sqrt{1 + a^2})\mathbf{k}$

7. $15.4\sqrt{\cos\theta - 0.882}$ into the plane of the swing

9. $(mbvt^2/2)\mathbf{i} - (mawt^2/2)\mathbf{j} - (mavt^2/2)\mathbf{k}$

13. (a) $m\omega d^2$; (b) $\frac{3}{4}m\omega d^2$; (c) $\frac{1}{4}m\omega d^2$, along the rotation axis for each

15. $3.7 \times 10^3\,(1 - 0.09t)\ \text{kg}\cdot\text{m}^2/\text{s}^2$

17. $MgR\sin(\omega t)$

19. $L_i = mvd\sin(2\theta)$ up, $L_f = 0$, the wall exerts an impulsive force on the ball and thus an impulsive torque that changes the angular momentum

21. $-20\mathbf{i} - 25\mathbf{j} - 30\mathbf{k}$

23. 1.9 rad/s

25. 0.20 rad/s counterclockwise; the torque on the turntable due to the force of friction between the man and the turntable

27. $r_n = (n^2\hbar^2/mk)^{1/4}$, $v_n = (n^2\hbar^2k/m^3)^{1/4}$, $K_n = \sqrt{k/m}\,n\hbar/2$

29. $E_1 = -13.6$ eV and $E_2 = -3.4$ eV, so there is no energy level 2.0 eV above the lowest-energy state; possible energies are $[13.6 - (13.6/n^2)]$ eV, or 10.2 eV, 12.1 eV, 12.75 eV, ...

31. $1.2 \times 10^4\ \text{N}\cdot\text{m}$ along the axis

33. $\sqrt{\frac{4}{3}g\ell}\,\sin\theta$

35. (a) $1.6 \times 10^3\ \text{kg}\cdot\text{m}^2$; (b) 3.4×10^7 J

39. 2.8 rad/s

41. 0.38

43. 2.6 rad/s, $0.083\ \text{kg}\cdot\text{m}^2/\text{s}$ (both clockwise), 0.105 J; no

45. (a) $1.2\ \text{m/s}^2$; (b) 2.5 s

47. (a) $6.67\ \text{kg}\cdot\text{m}^2$; (b) 0.70 rad/s; (c) 0.070

49. 4.57 min

51. ______; $K = (L^2/2mr^2) + (p_r^2/2m)$

53. (a) $\sqrt{3g/\ell}$ clockwise; (b) $\sqrt{M^2g\ell^3/3}$ clockwise; (c) $Mg\ell/2$

55. $\sqrt{(v^2 + 10g\ell)}/2\ell$, $(M\ell/3)\sqrt{v^2 + 10g\ell}$ (both clockwise), $(M/12)(v^2 + 10g\ell)$

57. $Mg\sqrt{h(2R - h)}/(R - h)$

CHAPTER 11

1. 5.8×10^2 N down, 4.0×10^2 N down

3. (a) $L/3$; (b) no

5. Lighter child: 0.97 m from the center; heavier child: 1.53 m from the center

7. $1.6417L$, $(1 + \frac{1}{2} + \frac{1}{3} + \cdots + \frac{1}{N}) = 2(1 + \frac{1}{2} + \frac{1}{3} + \frac{1}{4} + \frac{1}{5})$—could use computer to do the summation

9. $F_{N\text{bottom}} = 9.8 \times 10^2$ N up, $f_{\text{bottom}} = 6.4 \times 10^2$ N toward the wall, $F_{N\text{top}} = 6.4 \times 10^2$ N away from the wall, $F_{\text{man}} = 7.8 \times 10^2$ N down, $W_{\text{ladder}} = 2.0 \times 10^2$ N down

11. $F = 9.4 \times 10^2$ N, $F_V = 1.0 \times 10^3$ N down, $F_H = 8.5 \times 10^2$ N to the left

13. $0.58Mg$ outward

15. $T = 490$ N, $\mathbf{F}_{\text{wall}} = (424\ \text{N})\mathbf{i} + (245\ \text{N})\mathbf{j}$

17. $f_{\text{front}} = 32.6$ N/leg up slope, $f_{\text{rear}} = 15.2$ N/leg up slope; $F = 69$ N down slope

19. 100 kg, no

21. $T = 28$ N, $\mathbf{F} = (-28\ \text{N})\mathbf{i} + (98\ \text{N})\mathbf{j}$

23. 1.33×10^4 N

25. (a) $N_1 = 0$, $N_2 = N_3 = \frac{1}{2}(M + m)g - (mgx/L)$, $N_4 = 2mgx/L$; (b) $N_1 = -2mgx/L$, $N_2 = N_3 = \frac{1}{2}(M + m)g + (mgx/L)$, $N_4 = 0$

27. (a) $\sum F_x = 0$, $\sum F_y = 0$, $\sum F_z = 0$; (b) yes—after symmetry considerations, 3 unknowns, no equations; (c) yes—after symmetry considerations, 1 unknown, no equations

29. (a) Pull in; (b) 1.98×10^4 N outward

31.

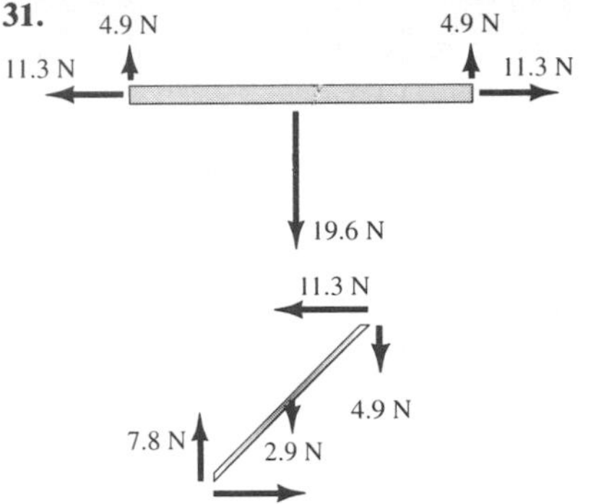

We assume for simplicity that only the crosspiece is involved.

33. 85 N

35. 1.36×10^3 N

37. (a) $Mg(\frac{1}{4} + \frac{1}{2}\mu_k)$ up and $\mu_k Mg(\frac{1}{4} + \frac{1}{2}\mu_k)$ to the left for each leg, $Mg(\frac{1}{4} - \frac{1}{2}\mu_k)$ up and $\mu_k Mg(\frac{1}{4} - \frac{1}{2}\mu_k)$ to the left for each leg; (b) 0.5

39. 1.3 cm (below the center of mass)

41.

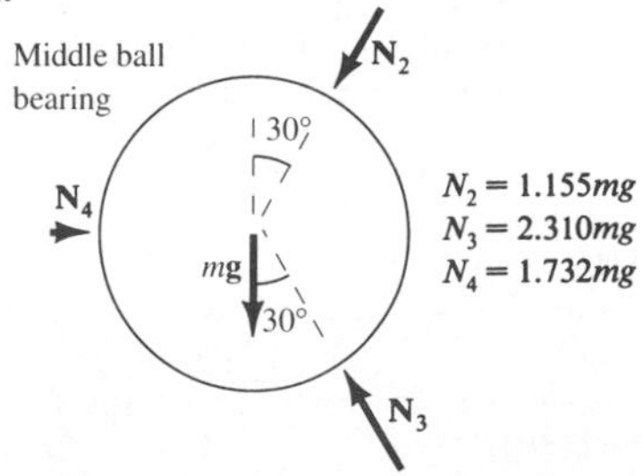

CHAPTER 12

1. (a) $[ML^4T^{-2}]$; (c) T/R^2 = a constant

5. 2.8×10^{-47} N

7. (a) $1.62\ \text{m/s}^2$; (b) $3.70\ \text{m/s}^2$; (c) $24.9\ \text{m/s}^2$; (d) $274\ \text{m/s}^2$

9. 1.51 h

11. (a) 696 kg; (b) 4.1 kg

13. (a) 2.59×10^8 m from earth's center; (b) 3.46×10^8 m from earth's center

15. (a) 2.37×10^3 m/s; (b) 5.02×10^3 m/s; (c) 5.96×10^4 m/s

17. 9.2×10^{-4} rad/s

19. 3.38×10^9 J

21. (a) 2.42×10^8 m; (b) 6.09×10^8 m; (c) 5.2×10^7 m

23. 7.68 km/s tangent to the orbit, 2.95×10^{10} J, $5.2 \times 10^{13}\ \text{kg}\cdot\text{m}^2/\text{s}$ perpendicular to the orbit

25. 1.96 h

27. 1.49×10^{11} m

29. (a) 12×10^3 km; (b) 18×10^3 km (c) 16×10^3 km

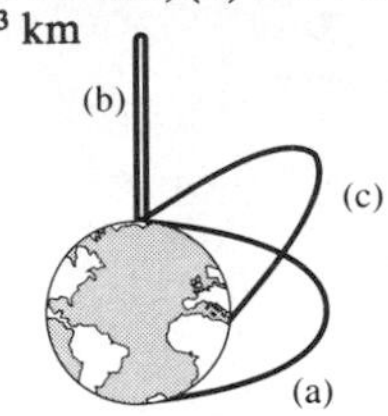

31. 3.2×10^{30} kg

33. $0.134R_1$, $1.86R_1$

35. Elliptical; (a) 2.67×10^{12} m; (b) 0.967; (c) 5.25×10^{12} m

37. $2.4 \times 10^{-3}\,g$

39. (a) Zero; (b) $4Gmm'/L^2$ toward the center; (c) $8\sqrt{2}\,Gmm'L/(d^2 + 2L^2)^{3/2}$ toward the center

41. (a) $U = GMmr^2/2R^3$, $U = 0$ at $r = 0$; (b) $U = GMmx^2/2R^3$, $U = 0$ at $x = 0$

43. $(2GMm/R^2)\{1 - [1 + (R^2/H^2)]^{-1/2}\}$ toward the center of the disk

45. 4.66×10^3 km from the earth's center (below the surface)

47. No

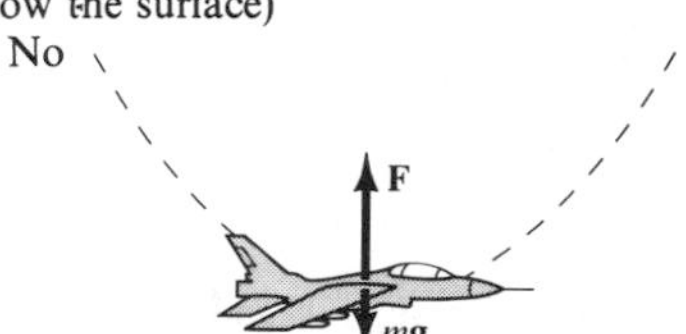

49. $\tan\theta = gw/2c^2$

51. (a) $19.6\ \text{m/s}^2$; (b) 1.79 d; (c) $R_A = 4R_B$

53. (a) $E = \frac{1}{2}Mv^2 - \frac{4}{3}\pi\rho GMR^2$; (b) $4.5 \times 10^{-27}\ \text{kg/m}^3$

55. $\Delta r = -4\pi fr^3/GMm$, $\Delta U = -4\pi fr$, $\Delta E = -2\pi fr$, $\Delta K = 2\pi fr$

57. (a) 5.0 m/s relative to himself (5.1 m/s relative to the ship); (b) 2.0 h

CHAPTER 13

1. $\pi/2$ rad

3.

0.8, 0.4, 0, −0.4, −0.8; 2, 4, 6, 8, 10; x, t, v, a; Time

5. 4.09 s

7. −0.038 m

9. 7.5452×10^{-4} m

11. 5 m, $\pi/2$ rad, 20 s, 0.050 Hz, $\pi/10$ rad/s

13. 6.3 m/s

15. (a) 4.08 cm/s; (b) 6.41 cm/s, $23.6\ \text{cm/s}^2$

17. (b) still clockwise

19. (a)

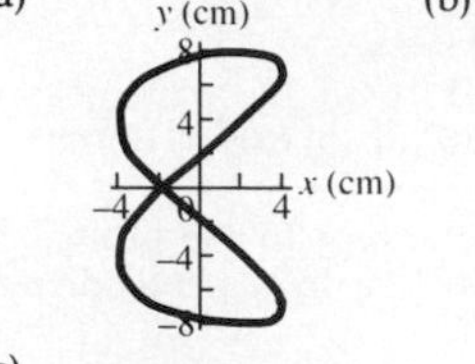

(b) 1.6×10^{-2} N

(c)

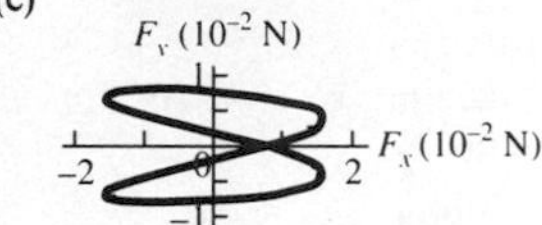

No

21. (a) 4.9 cm; (b) 0.81 cm
23. 20 N
25. 1.2 s
27. 0.35 J
29. 0.85 m
31. 0.66 m, it would increase by $\sqrt{1.5}$ to 0.81 m
35. 0.74 m, 1.67 m/s, 0.27 J
37. 2.20 s
39. $\theta = 0.19\cos(1.085t)$ rad, with t in s
41. 2.2 s
43. 1.11 Hz
45. $2\pi\sqrt{2(L^2 - 3Ly + 3y^2)/3(L - 2y)g}$
47. 1.03 Hz
49. (a) $\frac{1}{2}MR^2 + M\ell^2$;
(b) $-gL\theta = (\frac{1}{2}R^2 + \ell^2)\, d^2\theta/dt^2$;
(c) $2\pi\sqrt{(R^2 + 2\ell^2)/2g\ell}$;
(d) as $\ell \to 0$, $T \to \infty$ (no torque)
51. (a) 6.3 s; (b) 1.25×10^{-6} J;
(c) 5.0×10^{-3} m/s
55. (a) $2\pi\sqrt{R/g}$
57. 37.999995 rad/s
59. 1.5003 s
61. −4.85 cm, 1.48 cm; 5.21 cm
63. $\frac{1}{2}kA^2e^{-bt/m}$, $-2\pi b/m\omega_0$
65. 2.5 Hz
67. (a) 0.38 N·s/m; (b) 5.9 s; (c) $\Delta\omega =$ 0.34 rad/s, $Q = 5.4$
69. 0.36 m
71. $\sqrt{2}\omega$
73. (a) $0.553H$; (b) $0.347H$
75. 6.6 s
77. (a) $2A/e^2$; (b) $U = (-e^4/4A) + (e^4/4A)x^2$; $e^4/\sqrt{4A^3m}$
79. $(1/2\pi)\sqrt{g/2R}$
81. $(1/2\pi)\sqrt{2k/3m}$

CHAPTER 14

1. 1.1 m, 0.54 m, 0.36m

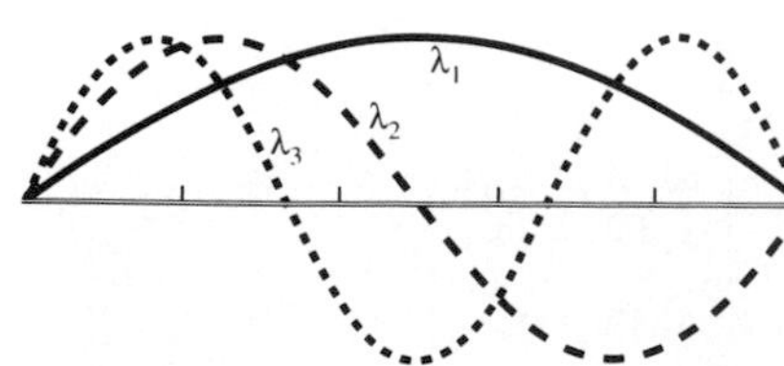

3. $z(x, t) = 0.064\cos(3.9x)\sin(0.63t + 0.67)$, with t in s, x and z in m
5. (a) 24 Hz; (b) 8.0 Hz
7. 1.61×10^3 N
9. (a) 2.1 m; (b) 0.32 Hz; (c) 0.4 m; (d) 3.1 s; (e) 3.0 m^{-1}
11. $z = z_0\sin[(3.3 \times 10^6)\pi x - (1.0 \times 10^{15})\pi t + \phi]$, with t in s, x in m; need z and dz/dt at $t = 0$
13.
(a)

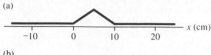

x (cm)
−10 0 10 20

(b)

15. 0.31 s
17. $v =$ (a constant)$\sqrt{gD}$
19. 1.7×10^2 m/s (600 km/h)
21. 1.9 m/s
23. $K = U = \pi^2 z_0^2 T/\lambda$
25. 5.2×10^{-9} J/m^3
27. 3.32×10^{24} J, 2.65×10^8 power plants
29. 330 m
31. 3.2×10^3
33. $\sqrt{2}$ increase
35. 1.3 cm
37. 1.97×10^9 N/m^2
39. 254 Hz
41. 1.5×10^7 m/s
43. 1.88×10^3 Hz
45. 32.6 m/s, 24.9 m/s
47. 965 m/s (3.5×10^3 km/h)
49. $\lambda_n = 2.40/(2n - 1)$ m, where $n = 1, 2, 3, \ldots$
51. (a) Reduced by a factor of $1/\sqrt{2}$; (b), (c) no change; (d) increased by a factor of $4/\sqrt{2}$
53. (a) 1.2×10^2 Hz; (b) 36 m/s; (c) 2.6×10^4 m/s^2
55. 40 cm, 20 cm, 13 cm
57. ______; $k = n\pi/L$, where $n = 1, 2, 3, \ldots$, $q = m\pi/L$, where $m = 1, 2, 3, \ldots$

CHAPTER 15

1. ______; $\sqrt{2}$, $\pi/4$
3. 1, 30° = $\pi/6$ rad
5.

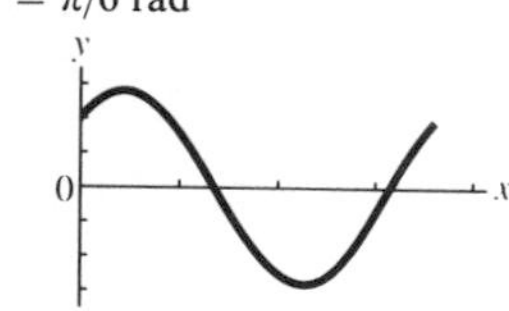

All waves have the same f, λ, and v; the resulting sinusoidal wave has the same f, λ, and v
7. Yes; $z = 2z_0\sin[kx + (\pi/4)]\cos[\omega t + (\pi/4)]$, x(nodes) $= (n - \frac{1}{4})\,\pi/k$, where $n = 0, \pm 1, \pm 2, \ldots$
11. 2 m
13. $f_{pulse} = 2.0$ Hz ($f_{beat} = 1.0$ Hz)
15. −1.6%
17. 0.59 cm
19. 3.3×10^{-6} m
21.

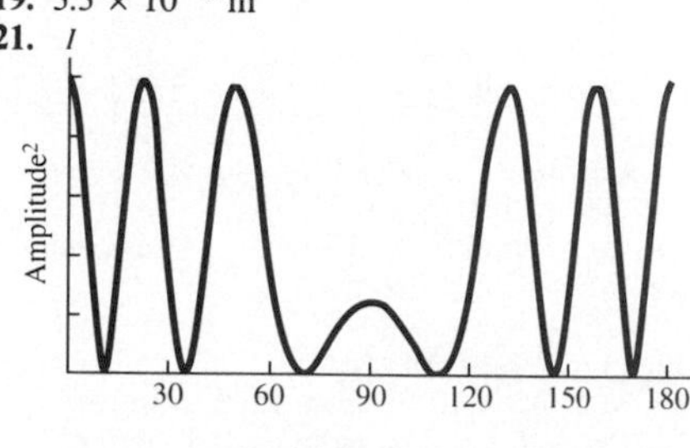

23. (a) $x = \pm n$, where $n = 0, 1, 2, \ldots$;
(b) $\psi = (4\text{ cm})\sin(\pi y - \pi v t)$
25. (d) For parts (a) and (b), there are regions where the intensity is $4I$ of one source but other regions where $I = 0$. Average intensity over a circle will be $2I$ of one source. For part (c), a doubling of intensity I means $2I$ of one source. Total energy is conserved.
27. (a)

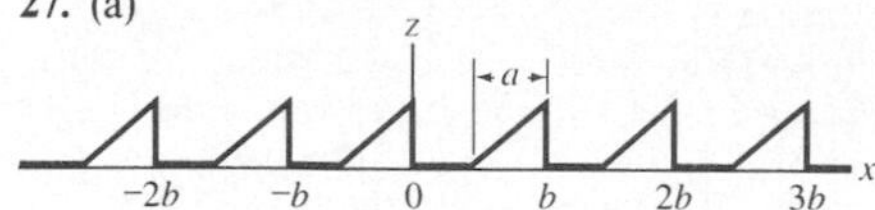

(b) $z(x, t) = 0$ for $x < (-a + nb + vt)$ and $x > (nb + vt)$, $z(x, t) = k(x - vt + a - nb)$ for $(-a + nb + vt) < x < (nb + vt)$; (c) b/v, b
29. $\approx \mu z_0^2 v^2/\alpha$, $E \to \infty$
31.

35. No; when the rope is flat, there is no potential energy, but the instantaneous transverse velocity is not zero, so there is only kinetic energy.
37.

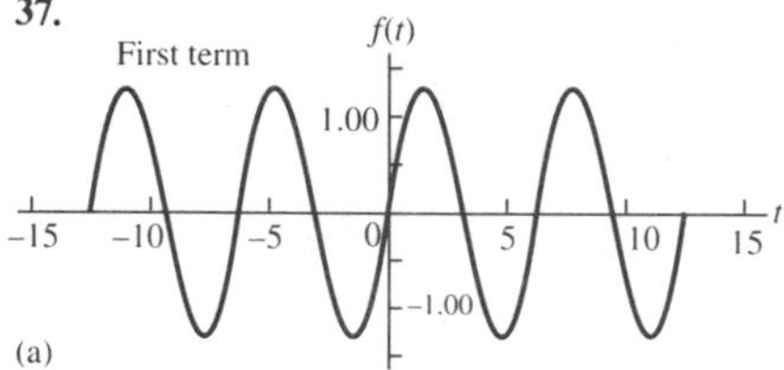

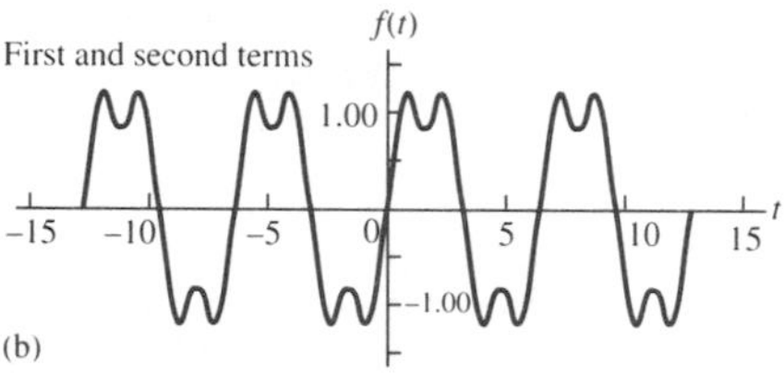

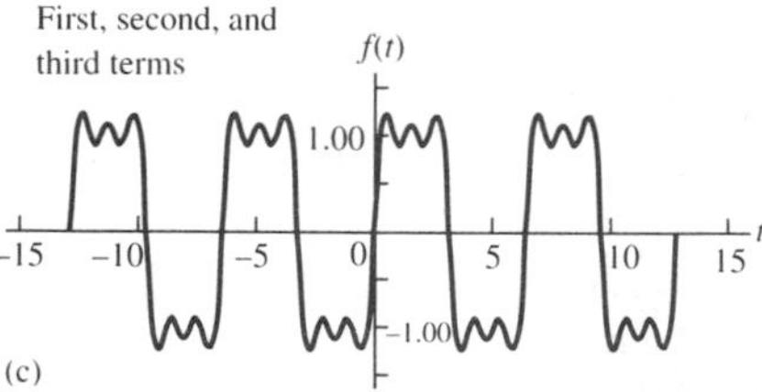

39. (a) $f_1 = f_2 = 400$ Hz, $\lambda_1 = \lambda_2 = 0.4$ m, $v_1 = v_2 = 160$ m/s; (b) 7.7 cm
41. (a) 1387 Hz; (b) 1633 Hz
43. (a) 390 Hz, 395 Hz, 399 Hz; (b) 0.5%
45. $d = (2n - 1)$ 0.5 cm, where $n = 1, 2, 3, \ldots$; yes, there is a standing wave with nodes at $x = (n - 1)$ 0.5 cm, where $n = 1, 2, 3, \ldots$
47. (a) $\pm\sqrt{2\{1 + \cos[(k_1 - k_2)L]\}}$, where L is the length of each pipe; (c) yes: Waves start in phase at the entrance to the pipes (two crests leave at the same time); when one crest has reached the other end, the second wave may not be at a crest (over the same distance, there is a different number of wavelengths in each pipe, because λ is different).

49. (a) and (b) are linear: Each term has ψ, and $\psi_2 = b\psi_1$ is a solution; (c) and (d) are nonlinear: ψ^2 or $\psi\, d\psi/dt$ are present, and $\psi_2 = b\psi_1$ is not a solution.
51. $z(\text{envelope}) \simeq \sin[\Delta k(x/2) - \Delta\omega(t/2) + (\pi/4)]$; $f = (\omega_1 - \omega_2)/2\pi$, $\lambda = 2\pi/(k_1 - k_2)$
53. (a) $y_1(x, t) = y_0 \sin(k_1 x + \omega_1 t + \delta)$, $y_2(x, t) = y_0 \sin(k_2 x - \omega_2 t + \delta)$;
(b) $\omega_2 = k_2\omega_1/k_1$; (c) $\delta\omega = v\, dk$; (d) ______; $\lambda = 2\pi/k_1$, $\omega = \omega_1$

CHAPTER 16

1. 1.15×10^3 kg/m³
3. 0.0855, 0.973
5. (a) 5.1×10^4 N; (b) 1.3×10^2 N (equivalent to lifting 13 kg)
7. 3.7×10^3 N/m², decrease
9. 0.027 m, 2.0 m
11. Sink
13. 7.5×10^2 kg/m³
15. 12.7 m²
17. 2.8 m
19. Bathroom scale: 17.7 N, spring scale: 9.3 N
21. $0.65R$
23. 2.5 m/s
25. 0.54-cm diameter
27. 1.6×10^2 W (0.22 hp)
29. -4.7×10^2 Pa (gauge), $\Delta P = 4.7 \times 10^2$ Pa
31. (a) 3.1 m/s, 0.32 cm; (b) 4.9×10^3 Pa (gauge); (c) 2.2 m/s, 0.38 cm
33. $v_1 = 6.2$ m/s, $v_2 = 3.1$ m/s (both along the pipe)
35. (b) $k = 2\rho g A$, $f = (1/2\pi)\sqrt{2Ag/V}$
39. (a) $v_t = \frac{2}{9} g R^2(\rho_m - \rho)/\eta$; (b) $v_t \propto R^2$; (c) 1.2 m/s, 1.9×10^{-3} m/s
41. 3.9 atm; less air, because pressure on the chest and lungs decreases the volume change
43. (a) 27 m/s; (b) $v_{\text{Grand Jet}} \simeq 200 v_{\text{hose}}$; (c) 1.9×10^{-3} m²; (d) 3.65×10^5 Pa (gauge)
45. The work of insertion is greater than the energy acquired from the buoyant force by at least the potential energy gain during the rise.
47. (a) 32 Pa; (b) 7 m/s
51. (a) $\sigma\sqrt{2gx}\,\Delta t$; (b) $-\sigma\sqrt{2gx}\,\Delta t/A$; (c) $dx/dt = -\sigma\sqrt{2gx}/A$; (e) $(A/\sigma)\sqrt{2x_0/g}$

CHAPTER 17

1. (b), (e), and (f)
3. Extensive: (a) and (d); intensive: (b) and (c)
5. 506 K
7. (a) 5727°C, 10,340°F; (b) −270°C, −454°F; (c) 294 K, 21°C; (d) 4.4°C, 277 K
9. $t_R = \frac{4}{5}T - 218$, $t_R = \frac{4}{9}t_F - 14.2$
11. 492°R, 139°R, 491.69°R, 672°R, 538°R, 551°R
13. 22°C (72°F)
15. 1.8 cm³
17. 0.59 kg/m³
19. 0.445 mol, 2.68×10^{23} molecules
21. 3.1×10^{-10} m
23. (a) 6.06×10^4 Pa, 308 K; (b) 3.5×10^{-3} m³; (c) 0.77 atm
25. 22.4×10^{-3} m³, 1.29 kg/m³
27. 6.07 kg/m³
29. 2.4×10^9 molecules
31. (a) 0.53 atm; (b) 0.53 atm
33. (a) 5.1 atm; (b) 0.21; (c) 3.1 atm
35. 4.0×10^{-2} m³
37. 12°C, 2.8×10^4 Pa
39. 342°C (615 K)
41. (a) 305.4 K (32.3°C); (b) 303.7 K (30.5°C), 0.55%
43. $T_c = 8a/27bR$, $p_c = a/27b^2$, $V_c = 3bn$
45. 0.27 m
47. 9.4×10^{-6} m
49. 4.47×10^{26} W
51. (b) $5 - (hc/\lambda_{\max}kT) = 5e^{-hc/\lambda_{\max}kt}$; (c) $\lambda_{\max}T = 2.90 \times 10^{-3}$ m·K; between 4500 K and 6400 K
53. 2×10^{-16} Pa $= 2 \times 10^{-21}$ atm $= 1.5 \times 10^{-18}$ torr
55. 9.1°C
57. 83 cm³
59. (a) $0.20/(101 + 9.8D)$; (b) $(2.0 \times 10^3)/(101 + 9.8D)$; (c) 31 m; (d) unstable
61. 46 s

CHAPTER 18

1. Reversible: (c), (e), and (h); Irreversible: (a), (b), (d), (f), and (g)

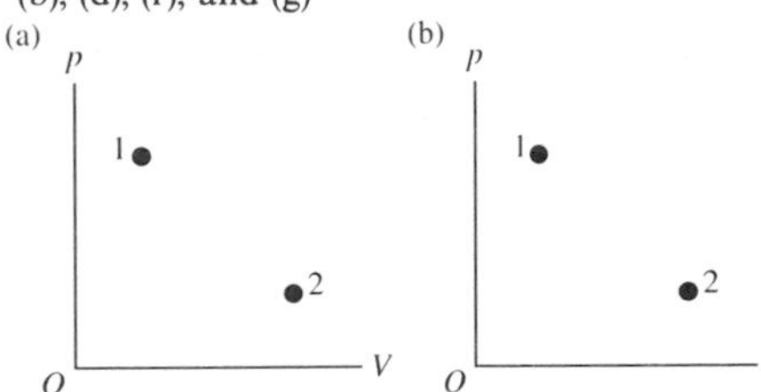

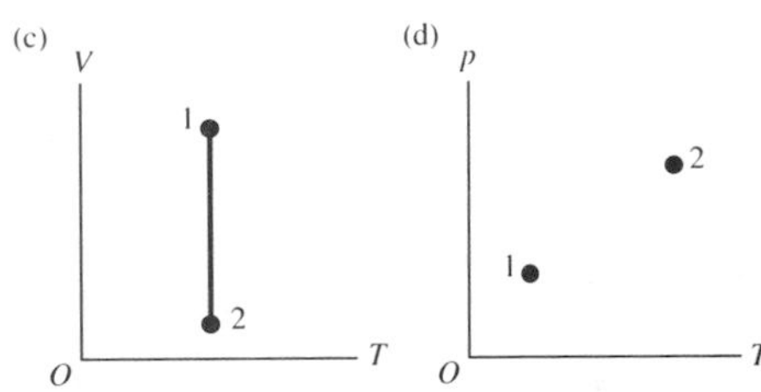

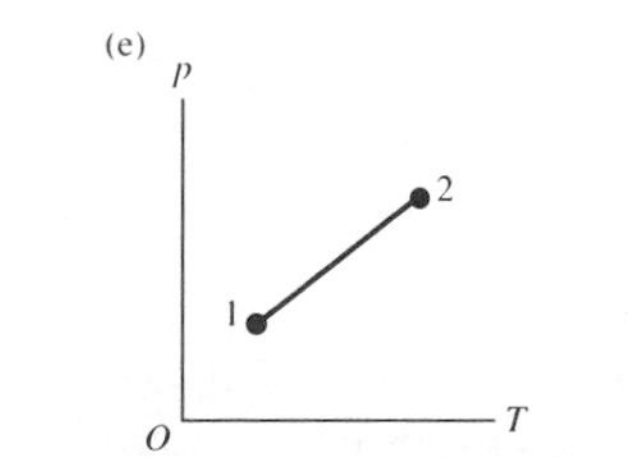

3. 2.1×10^5 cal/s
5. 9.4×10^4 cal; no, the latent heat of fusion must be absorbed
7. (a) 0.095 cal/g·K, 6.1 cal/mol·K; (b) 0.118 cal/g·K, 6.6 cal/mol·K; (c) 0.054 cal/g·K, 5.8 cal/mol·K
9. 308 K
11. 167 K
13. 18 times
15. 49.5 L
17. 69 min
19. $W = nRT_0 \ln(p_i/p_f)$
21. 1.6×10^2 J
23. (a) 1.8 L, 1.9×10^5 Pa; (b) 1.1×10^2 J
25. (a) $[ML^{-4}T^{-2}]$ (units of N/m⁵); (b) $(V_2 - V_1)[p_0 - \frac{1}{2}\beta(V_2 + V_1)]$; (c) same as (b)
27. (a) $p(V_B - V_A)$

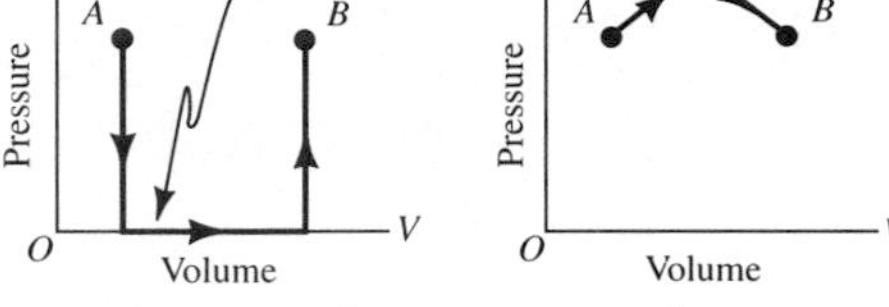

29. (a) $+2.7 \times 10^4$ cal; (b) 3.5×10^4 cal liberated
31. 41 cal/g (7.6%)
33.

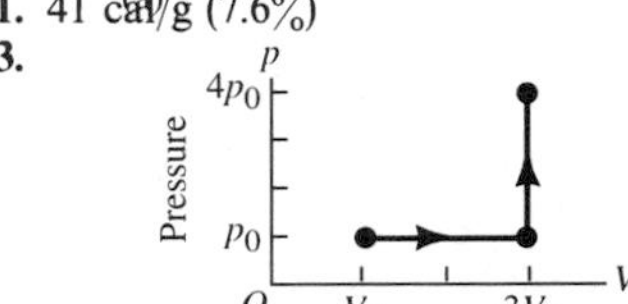

(b) $(33/2)p_0V_0$; (c) $2p_0V_0$; (d) $(37/2)p_0V_0$
35. (a) 2.88×10^{-3} m³, 1.15×10^{-3} m³; (b) 2.7×10^3 J; (c) 0; (d) -2.7×10^3 J (heat is liberated)
37. (a) 3400 J, 3700 J; (b) +200 J; (c) 3400 J, 3900 J, 0
39. 7.2°C
41. (a) 0.40; (b) 0.040
43. 1.28×10^5 J
45. 830 K (560°C)
49. (a)

Volume (L)

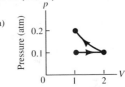

(b) 0.20 atm; (c) −3.9 J
51. (a) $+7.1 \times 10^3$ J; (b) $+2.5 \times 10^3$ J
53. The adiabatic process has the greater (negative) slope; $\text{slope}_{\text{adiabatic}}/\text{slope}_{\text{isothermal}} = \gamma$
55. (a) 4.9×10^{-3} m³; (b) 12.3×10^{-3} m³; (c) $+2.3 \times 10^3$ J; (d) $+2.3 \times 10^3$ J; (e) 120 K; (f) 8.0×10^2 J
57. (a) 1.30×10^6 J; (b) 8.6×10^4 J; (c) $+1.21 \times 10^6$ J
59. 1.1×10^3 cal
61. 5.2 min, 36 min
63.

p	T	V
5.0 atm	673 K	1.11×10^{-2} m³
2.5 atm	673 K	2.22×10^{-2} m³
1.25 atm	336 K	2.22×10^{-2} m³
1.25 atm	168 K	1.11×10^{-2} m³

$W = +2.48 \times 10^3$ J
65. (a) 4.0×10^5 Pa; (b) 1.98×10^{-2} m³; (c) 2.94×10^3 J; (d) 162 K; (e) $+4.93 \times 10^3$ J; (f) $+9.6 \times 10^2$ J

CHAPTER 19

1. (a) $R_V = 3.8 \times 10^{-8}$, $R_L = 4.2 \times 10^{-3}$; (b) $R_V = 3.8 \times 10^{-7}$, $R_L = 9.1 \times 10^{-3}$;

(c) $R_V = 3.8 \times 10^{-12}$, $R_L = 2.0 \times 10^{-4}$;
(d) $R_V = 3.8 \times 10^{-11}$, $R_L = 4.2 \times 10^{-4}$

3. 3.9×10^{24} components, 3.9×10^5 m^2

5. (a) 152 J; (b) 12 K; (c) 7.9×10^4 (m/s)^2; (d) 2.8×10^2 m/s

7. 6.1×10^{23} molecules/mol

9. (a) $p = U/V$; (b) $U = nRT$

11. $v_{\text{rms }2} = 1.17 v_{\text{rms }1}$

13. (a) 257 K; (b) 3.2×10^3 J; (c) 2.1×10^4 Pa

15. 5×10^{-8} K

17. 3.6×10^8 K; yes, because $\langle K \rangle = 7.5 \times 10^{-15}$ J, which is much greater than the binding energy

19. (a) 61.8 mi/h; (b) 62.4 mi/h

21. (a) 1/1078; (b) 10/1078; (c) 9/98

23. $13!(39!/52!) = 1/(6.35 \times 10^{11})$

25. (a) 37,500; (b) 41.7

27. 157 m/s

29. ______; no

31. (a) $\sqrt{3kT/m}$; (b) $(2/\sqrt{\pi})\sqrt{2kT/m}$

35. $$\frac{N_5}{N_{10}} = \frac{\int_7^\infty u^2 e^{-u^2}\, du}{\int_{4.9}^\infty u^2 e^{-u^2}\, du}$$

37. 1.2×10^{-31} kg·m^2/s

39. $3R$

41. (a) 0.034 Pa; (b) 1.0×10^4 Pa

43. $\simeq 100$ m

45. 2×10^{21} atoms, 2×10^{17} atoms, 4 Pa

47. 3×10^{-7}

49. $\simeq 1.5 \times 10^5$ K, N_2 could exist at T = 6000 K

51. (a) 2.4×10^3 m/s; (b) 470 K; as faster molecules escape, v_{rms} and T decrease; (c) yes, because the speed of some H_2 molecules would be greater than the escape speed, but the fraction would be less, so H_2 molecules would escape more slowly; (d) the lighter component will escape more rapidly and will make up a smaller fraction of the atmosphere

53. (a) 7.0×10^5 m; (b) decreases by a factor of 1000 to 7.0×10^2 m

55. 1.1×10^{-2} m/s

59. (a) 3.2×10^{12} molecules/m^3; (b) 1.6×10^3 s; (c) 7.8×10^5 m

61. $$P = \frac{4}{\sqrt{\pi}} \int_{\sqrt{3/2}\, v_{\text{escape}}/v_{\text{rms}}}^{\infty} u^2 e^{-u^2}\, du$$

CHAPTER 20

1. 28%

3. $10^{-3 \times 10^{22}}$

5. 445 K

7. 8.6%

9. 42.6%, 1.12×10^3 cal, 6.45×10^2 cal

11. 1170 MW, 62%

15. 127 W

19. 0.46 dQ/dt

21. 7.2×10^2 W

23. (a) $Q_\text{I} = 6.9 \times 10^3$ J, $Q_\text{II} = -6.2 \times 10^3$ J, $Q_\text{III} = -3.5 \times 10^3$ J, $Q_\text{IV} = 6.2 \times 10^3$ J, Q_I and Q_IV; (b) $W_\text{I} = 6.9 \times 10^3$ J, $W_\text{III} = -3.5 \times 10^3$ J, W_I; (c) 26%, $\eta_{\text{Carnot}} = 50\% = 1.9\eta_{\text{Stirling}}$

27. (a) $+Q/T_1$, $-Q/T_2$; (b) $\Delta S = (Q/T_1)[1 - (T_1/T_2)]$; ______

29. (a) +1.2 J/K; (b) +6.1 J/K

31. (a) +0.11 J/K·s; (b) $\Delta S_{\text{heating element}} < 0$, $\Delta S_{\text{room}} > 0$, $\Delta S_{\text{oven}} = 0$

33. $+1.5 \times 10^2$ J/K

35. $A \to B$: $Q = \frac{5}{2}nR(T_B - T_A)$, $W = nR(T_B - T_A)$, $\Delta U = \frac{3}{2}nR(T_B - T_A)$, $\Delta T = T_B - T_A$, $\Delta S = \frac{5}{2}nR \ln(T_B/T_A)$; $B \to C$: $Q = 0$, $W = \frac{3}{2}nR(T_B - T_C)$, $\Delta U = \frac{3}{2}nR(T_C - T_B)$, $\Delta T = T_C - T_B$, $\Delta S = 0$; $C \to D$: $Q = \frac{3}{2}nRT_C \times \ln(T_A^{2/3}T_C/T_B^{5/3})$, $W = \frac{3}{2}nRT_C \ln(T_A^{2/3}T_C/T_B^{5/3})$, $\Delta U = 0$, $\Delta T = 0$, $\Delta S = \frac{3}{2}nR \ln(T_A^{2/3}T_C/T_B^{5/3})$; $D \to A$: $Q = \frac{3}{2}nR(T_A - T_C)$, $W = 0$, $\Delta U = \frac{3}{2}nR(T_A - T_C)$, $\Delta T = T_A - T_C$, $\Delta S = \frac{3}{2}nR \times \ln(T_A/T_C)$

37. 0, +32.5 J/K

39.

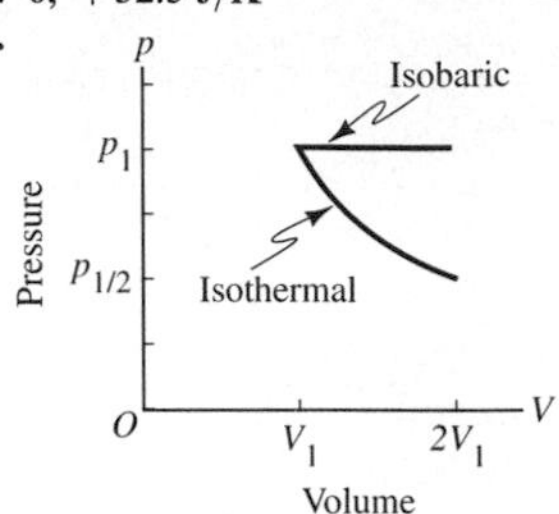

(b) $W_{\text{isobaric}} = p_1V_1$, $W_{\text{isothermal}} = 0.693p_1V_1$; (c) $\Delta S_{\text{isobaric}} = 0.693nc_p'$, $\Delta S_{\text{isothermal}} = 0.693nR$

41. Decreases; increases; the total entropy change is zero, which agrees with the second law, because the process is reversible

43. (a) T_0; (b) $+nR \ln(3)$; (c) $+nR \ln(3)$; (d) 0

45. (a) +87 J; (b) 2.7 K; (c) 0.38 J/K

47. (a) 2.25×10^{-3} m^3; (b) 273 K; (c) −11.5 J/K

49. (a) 3.07×10^{-2} m^3; (b) 373 K; (c) 0

51. $1 - (T_2/T_1)$

53. (a) $\frac{1}{2}(r - 1)RT_a$; (b) $Q_\text{i} = \frac{5}{2}R(r - 1)T_a$, $Q_\text{ii} = -\frac{3}{4}RrT_a$, $Q_\text{iii} = -\frac{5}{4}R(r - 1)T_a$, $Q_\text{iv} = \frac{3}{4}RT_a$; (c) $(2r - 2)/(10r - 7)$

55. area $= Q_{\text{net}} = W_{\text{net}}$

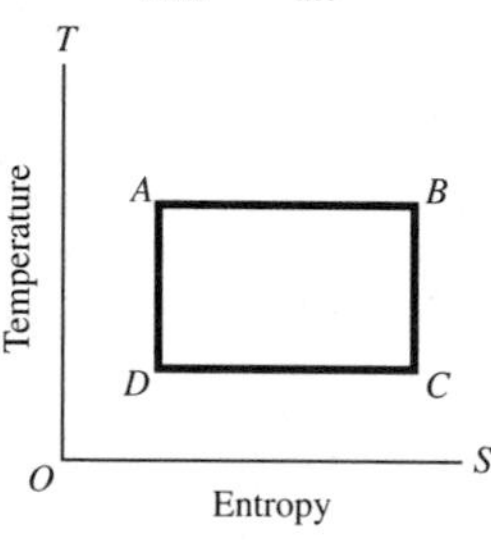

59. (a) $W_\text{i} = p_2(V_2 - V_1)$, $W_\text{ii} = p_1(V_2 - V_1)$, $W_\text{iii} = \frac{1}{2}(p_1 + p_2)(V_2 - V_1)$, $Q_\text{i} = (c_V/R) \times (p_2V_2 - p_1V_1) + p_2(V_2 - V_1)$, $Q_\text{ii} = (c_V/R) \times (p_2V_2 - p_1V_1) + p_1(V_2 - V_1)$, $Q_\text{iii} = (c_V/R) \times (p_2V_2 - p_1V_1) + \frac{1}{2}(p_1 + p_2)(V_2 - V_1)$; (b) $\Delta S_\text{i} = \Delta S_\text{ii} = \Delta S_\text{iii} = nc_V \ln(p_2/p_1) + nc_p \ln(V_2/V_1)$

CHAPTER 21

1. 3.9 mm

3. 6.9 mm

5. 2.4×10^3 N

7. 2.0×10^{11} N/m^2, 2.0×10^3 kg

9. No, because the maximum stress (210 MN/m^2) is less than the tensile strength

11. 1.99928 cm

15. (b) 1.9×10^{-11} m^2/N

17. (a) 5.1×10^3 m/s; (b) 3.5×10^3 m/s; (c) 5.0×10^3 m/s; (d) 1.2×10^3 m/s

19. $Y = 3 \times 10^5$ MN/m^2, $G = 8 \times 10^4$ MN/m^2, 1.3×10^4 km

21. Gains 6.5 s

23. (a) Steel; (b) 0.99971 Hz $\le f_{\text{aluminum}} \le$ 1.00029 Hz, 0.99979 Hz $\le f_{\text{copper}} \le$ 1.00021 Hz, 0.99987 Hz $\le f_{\text{steel}} \le$ 1.00013 Hz

25. 22 m

27. 16 kW

29. $\Delta Q/\Delta t_{\text{pine}} = 144$ Btu/h, $\Delta Q/\Delta t_{\text{Fiberglas}} = 62$ Btu/h; $R_{\text{eff}} = 16$ ft·h·°F/Btu

31. (a) $(L_1 + L_2)\kappa_1\kappa_2/(\kappa_1L_2 + \kappa_2L_1)$; (b) $(\kappa_1L_2T_1 + \kappa_2L_1T_2)/(\kappa_1L_2 + \kappa_2L_1)$

37. $(1/L)\, dQ/dt = 2\pi\kappa(T_2 - T_1)/\ln[1 + (\alpha/R_1)]$

39. 6.7×10^5 N, 6.8×10^4 kg

41. 0.0073 m^2 (8.5 cm × 8.5 cm)

43. 2.0×10^5 N/m^2

CHAPTER 22

1. 6.2×10^9 fewer electrons

3. First: -1×10^{-11} C, 6.2×10^7 electrons; second: -0.5×10^{-11} C, 3.1×10^7 electrons; third: -0.5×10^{-11} C, 3.1×10^7 electrons

5.

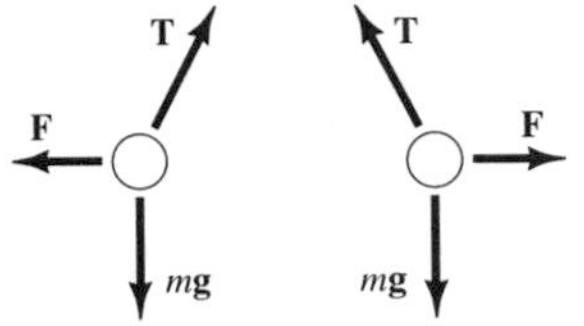

(b) 4.9×10^{-3} N, 6.8×10^{-3} N; (c) 44°

7. (a) and (c)

9. -8.5×10^{-24} C

11. 46 N repulsion between the two up quarks, 23 N attraction between up and down quarks

13. 1.4×10^{-9}; because the masses are so much larger than they are at the atomic scale

15. (a) 3.6×10^3 m; (b) 1.0×10^{-10} m

17. 4.2 N repulsion

19. (a) 11.3 g; (b) 2.27×10^{-7} C

21. (a) 2.0×10^{20} N; (b) 1.0; (c) 5.4×10^{-8} C/m^3; (d) 3.4×10^{11} protons/m^3; (e) 1.65×10^{30} protons/m^3, 2.1×10^{-19} times 1.65×10^{30} protons/m^3

23. (a) -2.5×10^{-7} C; (b) 8×10^{-40} of electron charge

25. (−1.17 cm, 1.17 cm); yes for displacements along the line joining the charges, no for displacements away from the line

27. 8.0 N toward the opposite corner

29. (a) $\sqrt{3}kq^2/2L^2$, 9.7° above the $-x$-axis; (b) $3.2kqQ/L^2$, 6.3° below the $-x$-axis

31. $\mathbf{F} = (2kq\lambda/x_0)\mathbf{i}$

33. $F = kqQ/d(L + d)$ away from the rod

35. 53 N away from the center

37. (a) It hangs 30° from the vertical away from the sheet; (b) it hangs 16° from the vertical toward the sheet

39. (a) $F = kQ \sum_{m=0}^{n} \frac{q_m(-1)^m}{(D - md)^2}$; (b) $F = \frac{kQ}{D^2} \sum_{m=0}^{n} q_m(-1)^m + \frac{2kQd}{D^3} \sum_{m=0}^{n} mq_m(-1)^m$

41. 1.7×10^{-7} C, 1.7×10^{-7} C; Q would be larger

43. 3.6 N repulsion

45. (a) 5×10^{-15}; (b) weight reduced by 500 N

47. (a) $F_{net} = kq^2\{(1/x^2) - [1/(\ell - x)^2]\}$ away from the closer charge, $F_{net} = 0$ at $x = \ell/2$; (b) $\mathbf{F}_{net} = -2kq^2\ell\, \Delta\mathbf{x}/[(\ell/2)^2 - (\Delta x)^2]^2$, where Δx is the displacement from equilibrium; (c) $(1/2\pi)\sqrt{32kq^2/\ell^3 m}$

49. $F = \dfrac{kQ^2R}{4L^2}\displaystyle\int_{-L}^{L}\int_{-L}^{L}\frac{dy\, dy'}{[(y - y')^2 + R^2]^{3/2}}$ repulsion; $F = kQ^2/R^2$ repulsion, equivalent to two point charges

CHAPTER 23

1. $\mathbf{E} = 5.2 \times 10^6\ (0.55\mathbf{i} + 0.83\mathbf{j})$ N/C

3. (a) $-(1.08 \times 10^5$ N/C$)\mathbf{i}$; (b) $-(8.9 \times 10^6$ N/C$)\mathbf{i}$; (c) $+(4.5 \times 10^{10}$ N/C$)\mathbf{i}$; (d) $-(4.5 \times 10^{10}$ N/C$)\mathbf{i}$

5. $\mathbf{E} = 1.67 \times 10^8$ N/C 4.3° above the horizontal

7. (a) $\mathbf{E} = 0$; (b) unstable

9. $\mathbf{E} = (-1)^n\ (qnd/4\pi\epsilon_0 Y^3)\ \mathbf{i}$

11.

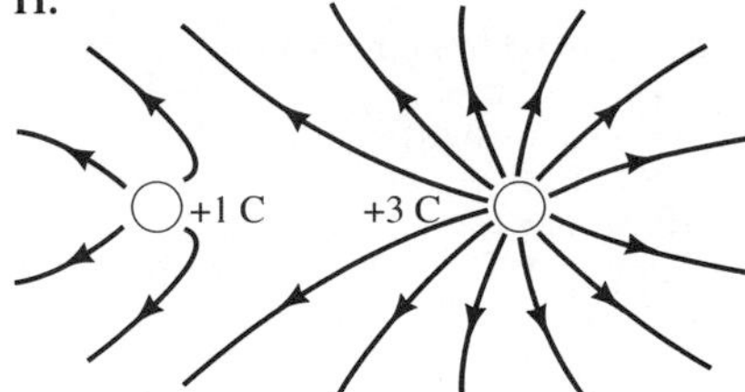

13.

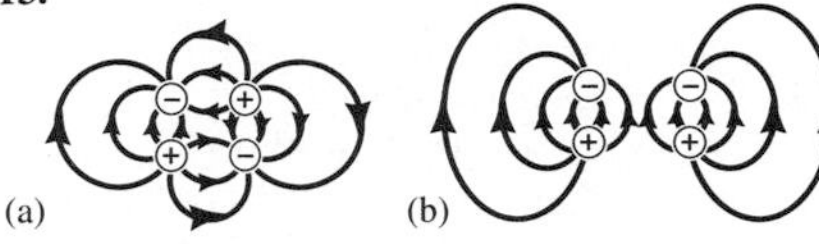

(a) (b)

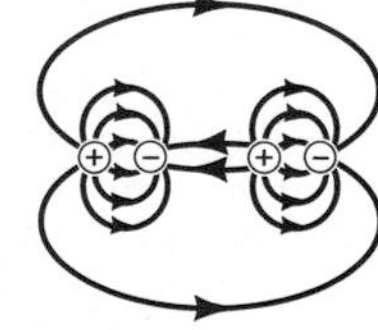

(c) (d)

15. $\mathbf{E}_1 = (1.27 \times 10^7$ N/C$)\mathbf{i}$; $\mathbf{E}_2 = (1.27 \times 10^7$ N/C$)\mathbf{j}$

17.

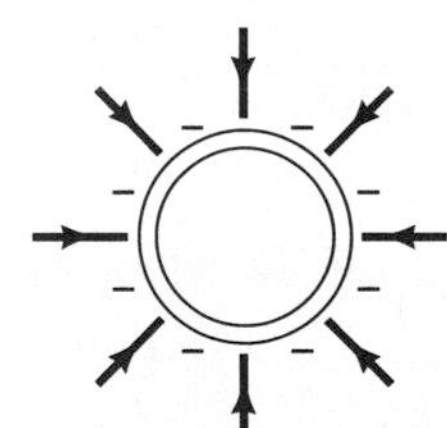

19.

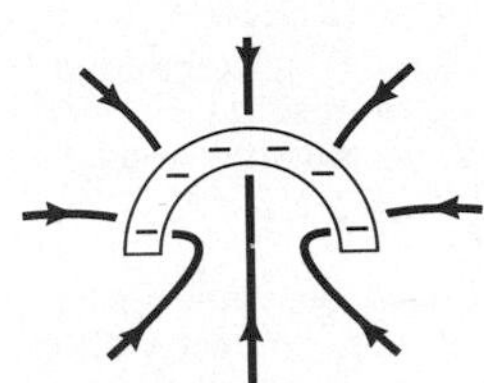

21. Around the plates: $E = \sigma/\epsilon_0$ perpendicular to them and away from them; between the plates: $E = 0$

23. $\mathbf{E}(0, D) = (Q/4\pi\epsilon_0 LD) \times [(D/\sqrt{D^2 + L^2} - 1)\mathbf{i} + (L/\sqrt{D^2 + L^2})\mathbf{j}]$, $\mathbf{E}(L/2, D) = (Q/4\pi\epsilon_0 D)(2/\sqrt{4D^2 + L^2})\mathbf{j}$

25. $E = 4.52 \times 10^5$ N/C away from the semicircle

27. (a) $\mathbf{E} = \dfrac{\sigma L z_0}{2\pi\epsilon_0}\mathbf{k} \times$

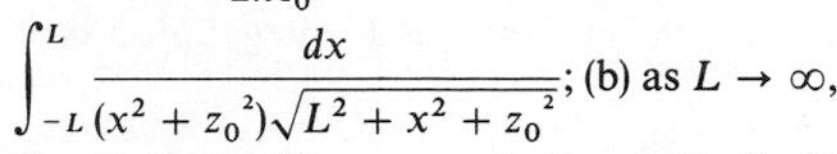

$\displaystyle\int_{-L}^{L}\frac{dx}{(x^2 + z_0^2)\sqrt{L^2 + x^2 + z_0^2}}$; (b) as $L \to \infty$, $\mathbf{E} \to (\sigma/2\epsilon_0)\mathbf{k}$; (c) as $z_0 \to 0$, $\mathbf{E} \to (\sigma/2\epsilon_0)\mathbf{k}$ ($\mathbf{E}$ for an infinite plane)

29. 6.0×10^{-7} C/m^2

31. $m\, d^2y/dt^2 = -q\lambda/2\pi\epsilon_0 y$, $m\, d^2x/dt^2 = 0$

33. 1.2×10^{-2} C/m^2

35. 9.6 cm

37. 5.1×10^{-6} s

39. Decreased by a factor of 20

41. (a) 6.2×10^{-11} m; (b) 9.8×10^{-11} m

43.

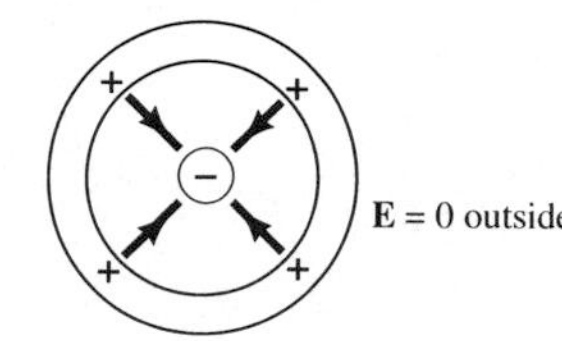

45.

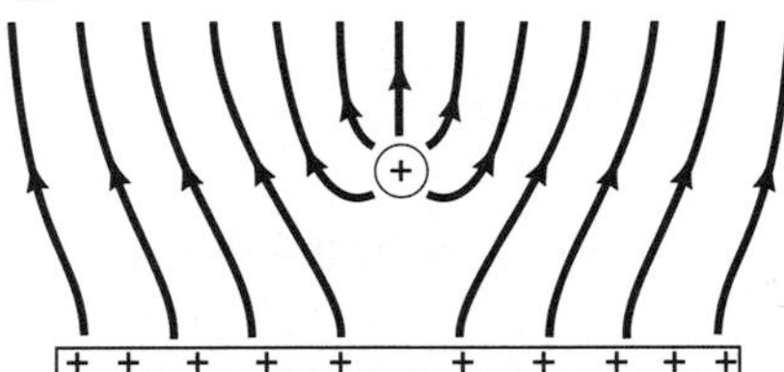

47. (a) $\mathbf{E}_1 = (-4.03 \times 10^6$ N/C$)\mathbf{i} + (+8.05 \times 10^6$ N/C$)\mathbf{j}$; $\mathbf{E}_2 = (-6.48 \times 10^6$ N/C$)\mathbf{i} + (-8.64 \times 10^6$ N/C$)\mathbf{j}$; (b) $\mathbf{F}_1 = (-4.03 \times 10^{-3}$ N$)\mathbf{i} + (+8.05 \times 10^{-3}$ N$)\mathbf{j}$; $\mathbf{F}_2 = (-6.48 \times 10^{-3}$ N$)\mathbf{i} + (-8.64 \times 10^{-3}$ N$)\mathbf{j}$; (c) $(-10.5 \times 10^{-3}$ N$)\mathbf{i} + (-0.59 \times 10^{-3}$ N$)\mathbf{j}$; (d) $(-10.5 \times 10^6$ N/C$)\mathbf{i} + (-0.59 \times 10^6$ N/C$)\mathbf{j}$

49. $\lambda^2/2\pi\epsilon_0 R$ (attraction)

51. (a) 0; (b) 0; (c) $\mathbf{F} = (2.43$ N$)\mathbf{i} - (1.89$ N$)\mathbf{j}$

53. (c) 1.74×10^6 N/C

55. $t \propto R\sqrt{m\epsilon_0/q\lambda}$

57.

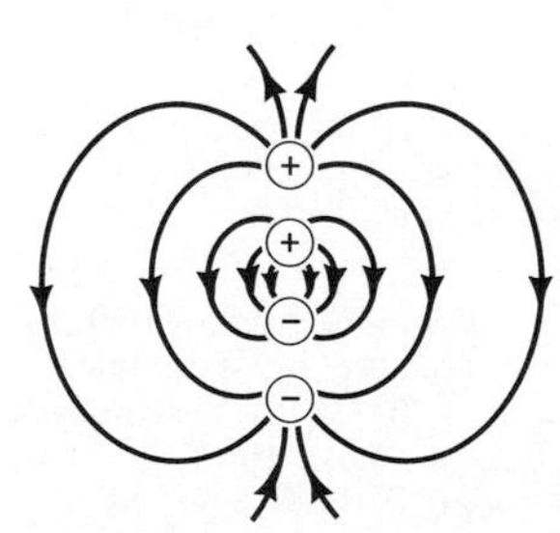

CHAPTER 24

1. (a) $\sigma\pi R^2/2\epsilon_0$; (b) $0.866\sigma\pi R^2/2\epsilon_0$

3. 18 N·m^2/C

5. $2\pi bR$, where b is a constant

7. q/ϵ_0

9. (a) 0; (b) 1.13×10^8 N·m^2/C

11. (a) -3.54×10^{-9} C; (b) -3.54×10^{-9} C; (c) -3.54×10^{-9} C

13. 5.65 N·m^2/C out of the sides parallel to the xy- or xz-planes, 5.15 N·m^2/C out of the side perpendicular to the $+x$-axis, 6.15 N·m^2/C out of the side perpendicular to the $-x$-axis

15. $q/\sqrt{3}\epsilon_0 L^2$

19. -1.3×10^{-4} C

21. $\mathbf{E} = (1.69 \times 10^3)\hat{\mathbf{r}}$ N/C; $\mathbf{E} = (1.69 \times 10^3)\hat{\mathbf{r}}$ N/C

23. $r < r_1$: $\mathbf{E} = 0$; $r_1 < r < r_2$: $\mathbf{E} = [\rho(r^2 - r_1^2)/2\epsilon_0 r]\hat{\mathbf{r}}$; $r_2 < r$: $\mathbf{E} = [\rho(r_2^2 - r_1^2)/2\epsilon_0 r]\hat{\mathbf{r}}$

25. $r < R_1$: $\mathbf{E} = 0$; $R_1 < r < R_2$: $\mathbf{E} = [Q(r^3 - R_1^3)/4\pi\epsilon_0(R_2^3 - R_1^3)r^2]\hat{\mathbf{r}}$; $R_2 < r$: $\mathbf{E} = (Q/4\pi\epsilon_0 r^2)\hat{\mathbf{r}}$

27. For $E = 8.2 \times 10^5$ N/C and $\theta = 34°$, 1st quadrant: E at $-\theta$; 2nd quadrant: E at $180° + \theta$; 3rd quadrant: E at $180° - \theta$; 4th quadrant: E at θ

29. $r < 4$ cm: $\mathbf{E} = (-5.6 \times 10^9)r\hat{\mathbf{r}}$ N/C; 4 cm $< r <$ 10 cm: $\mathbf{E} = [(4.8 \times 10^8)r - (3.9 \times 10^5)/r^2]\hat{\mathbf{r}}$ N/C; 10 cm $< r$: $\mathbf{E} = [(9.0 \times 10^4)/r^2]\hat{\mathbf{r}}$ N/C

31. $r < R$: $\mathbf{E} = 0$; $R < r < 2R$: $\mathbf{E} = (q/4\pi\epsilon_0 r^2)\hat{\mathbf{r}}$; $2R < r$: $\mathbf{E} = -(q/4\pi\epsilon_0 r^2)\hat{\mathbf{r}}$

33. 5.3×10^{-5} C/m^2

35. -4.5×10^5 C, on the surface, -8.9×10^{-10} C/m^2

37.

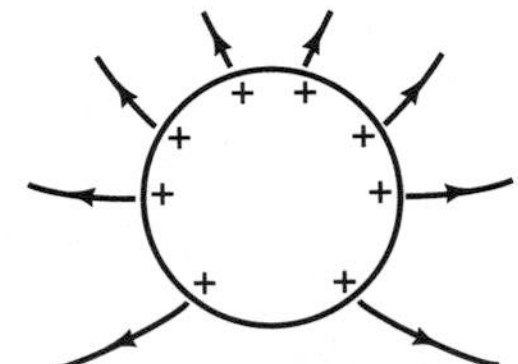

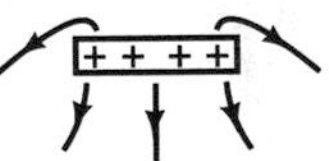

39. $\Phi_{x=0} = 0$, $\Phi_{x=a} = ba^4$, $\Phi_{y=0} = 0$, $\Phi_{y=a} = 0$, $\Phi_{z=0} = 0$, $\Phi_{z=a} = ca^4/2$; $q = \epsilon_0[b + (c/2)]a^4$

41. $r \propto R\cos(\omega t)$, motion is simple harmonic; $\tau = 2\pi\sqrt{4\pi\epsilon_0 mR^3/qQ}$, $U = qQ/8\pi\epsilon_0 R$

43. (a) $(7.9 \times 10^6)\hat{\mathbf{r}}$ N/C; (b) $(2.8 \times 10^6)\hat{\mathbf{r}}$ N/C; (c) no change; (d) $(1.3 \times 10^7)\hat{\mathbf{r}}$ N/C

45. $\rho \propto 1/r$; at the center $\rho \to \infty$, because the volume of a sphere $\to 0$ faster than the area $\to 0$

47. $\rho = -\frac{500}{3}\epsilon_0$ C/m^3 (constant); a field of 500 N/C entering at $x = 0$ is produced by some external source

CHAPTER 25

1. 8.0 J

3. (a) 0; (b) -1.35 J; (c) $+1.35$ J

5. $+0.92$ J, 0

7. -1.77 J

9. At ∞

11. $+4.7$ V

13. $+3.0 \times 10^{-2}$ J

15.

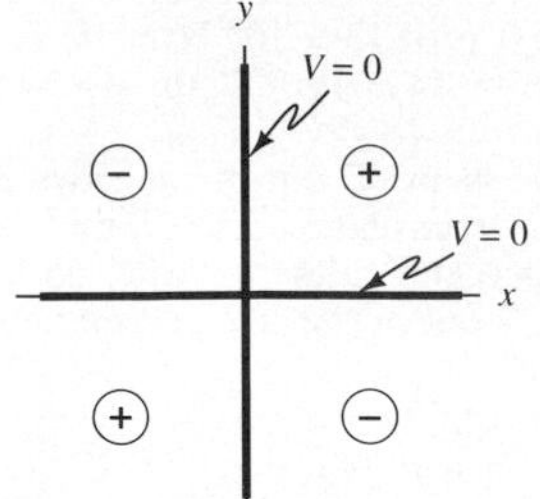

17. $+3.7 \times 10^5$ V
19. -11.6 J, no
21. $\frac{1}{2}Q^2/4\pi\epsilon_0 R$
23.

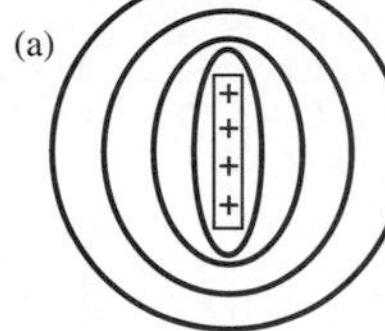

(b)

25.

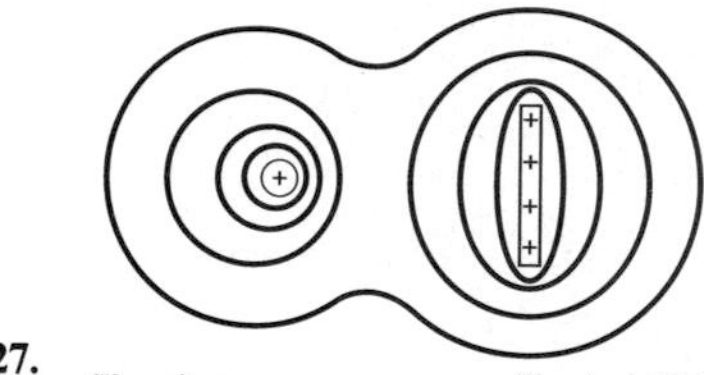

27.

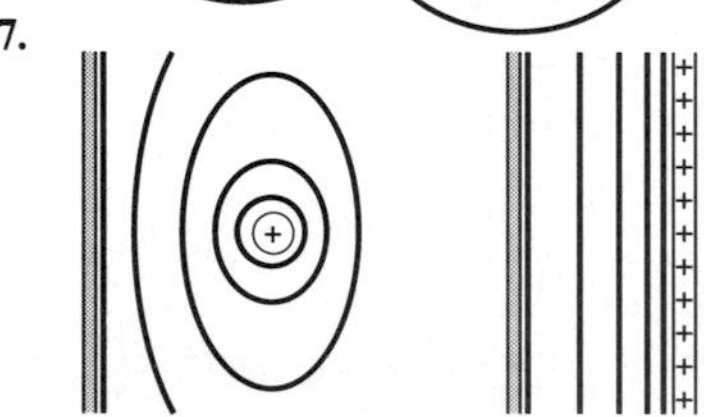

29.

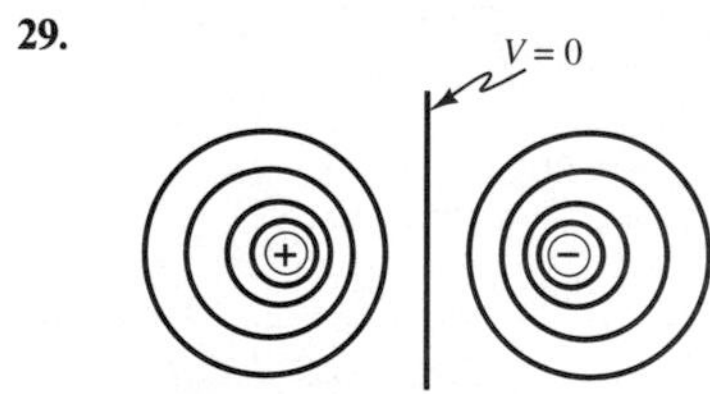

31. $\mathbf{E} = Q/(4\pi\epsilon_0\, x^2)\mathbf{i}$
33. $\mathbf{E} = Qx/[4\pi\epsilon_0(R^2 + x^2)^{3/2}]\mathbf{i}$
35. $\mathbf{E}(0, 0, 0) = 0$, $\mathbf{E}(0, 0, 1) = -a_2\mathbf{i} - 2a_3\mathbf{k}$
37. (a) $r < R$: $\mathbf{E} = (Q/4\pi\epsilon_0)(8r/R^3)\hat{\mathbf{r}}$; $r > R$: $\mathbf{E} = (Q/4\pi\epsilon_0\, r^2)\hat{\mathbf{r}}$; (b) $r < R$: $\rho = 6Q/\pi R^3$, $r = R$: $q = -7Q$, $r > R$: $\rho = 0$
39. 0.5 m
41. $V = -(2\lambda/4\pi\epsilon_0)\ln(R) +$ (a constant)
43. $V = (\lambda/4\pi\epsilon_0) \times \ln(\{z + (L/2) + \sqrt{x^2 + y^2 + [z + (L/2)]^2}\}/\{z - (L/2) + \sqrt{x^2 + y^2 + [z - (L/2)]^2}\})$;
45. 3.4×10^{-5} J
47. 1430 V
49. 3×10^5 V
51. (a) 2.2 μC, 11 μC; (b) $\mathbf{E}_1 = (4.95 \times 10^7)\hat{\mathbf{r}}_1$ V/m, $\mathbf{E}_2 = (9.90 \times 10^6)\hat{\mathbf{r}}_2$ V/m
53. (a) 5.5×10^6 eV, 8.8×10^{-13} J; (b) 3.3×10^7 m/s
55. (a) 3×10^6 V; (b) 3 MeV (4.8×10^{-13} J); (c) 3.3×10^{-4} C
57. Total energy $= -qQ/4\pi\epsilon_0 r$, total energy $= -qQ/8\pi\epsilon_0 r$
59. Put the first electron at -2 mm; $U_1 = 0$; put the next at $+2$ mm; $U_2 = 0.58 \times 10^{-25}$ J; put the third at the origin; $U_3 = 2.30 \times 10^{-25}$ J; the order will affect the individual terms but not the total energy of 2.88×10^{-25} J
61. $\mathbf{E} = [Qx/4\pi\epsilon_0(R^2 + x^2)^{3/2}]\mathbf{i}$; direct integration requires the selection of differential elements and the use of symmetry to handle vector components, and because potential is a scalar, that is generally easier.
63. -7.68×10^{-19} J, $+3.84 \times 10^{-19}$ J, -3.84×10^{-19} J
65. (a) 9.9×10^{-3} kg; (b) $(0.023/\sin\theta) + 0.078(1 - \cos\theta)$
67. $r < R$: $V = (-\rho r^2/4\epsilon_0) - (\rho R^2/2\epsilon_0) \times [\ln(R/a) - \frac{1}{2}]$; $r > R$: $V = -(\rho R^2/2\epsilon_0)\ln(r/a)$
69. $4\pi\rho^2 R^5/15\epsilon_0$

CHAPTER 26

1. (a) 17.7 pF; (b) 15.9 cm
3. (a) 2 μC; (b) 12 μC
5. 13 pF
7. 89 kV
9. 4.1×10^{-7} C/m^2, 6.5×10^{-8} C
11. (a) 10 μF; (b) 4.1×10^{10} J
13. (a) 5.67×10^{-10} F; (b) 2.83×10^{-4} J, 2.83×10^{-2} J
15. 6.91×10^{-2} J, 6.91×10^7 J
17. 1.1×10^{-2} J
19. 1.9 J
21. (a) 4.00×10^5 V/m; (b) 2.12×10^{-7} C; (c) 4.25×10^{-2} N; (d) 8.5×10^{-6} J, yes
23. $\epsilon_0 A(\ell_1 + \ell_2)/2\ell_1\ell_2$; yes, in parallel
25. $C = \epsilon_0 A/(D - d)$, independent of x
27. 1.4 μF
29. (a) 3.8×10^{-3} J; (b) nothing will change
31. The 4-μF capacitor is in parallel with a series combination of the 2-μF, 3-μF, and 5-μF capacitors
33. 2.1×10^{12} m^2
35. $(\kappa - 1)4\pi\epsilon_0 R$; $\sigma_{ind}/\sigma = (\kappa - 1)/\kappa$
37. $2.3q_0$
39. $\kappa\epsilon_0 A/[d + \kappa(D - d)]$
41. 5.3×10^{-10} C, 2.0×10^{-9} C, 3.6×10^{-9} C, 2.6×10^{-9} C, 1.8×10^{-7} C
43. $\kappa 4\pi\epsilon_0 r_1 r_2/(r_2 - r_1)$, $(1 - \kappa)Q^2/2C$ (a decrease)
45. 3.9 nF, 9.7 nF
47. 2.8×10^{-6} C, 9.8×10^5 V/m, 2.5×10^{-3} J
49. (a) Two plates, each of area 50 cm^2, 1 cm apart with Bakelite between them; 1.08×10^{-8} C, 2.71×10^{-6} J; (b) the same system but with no Bakelite; 5.0×10^4 V/m
51. (a) 1.1×10^{-10} F; (b) $4Q$ C/m^2; (c) $U_2/U_1 = \frac{2}{3}$ (a decrease); (d) $C_{dielectric}/C_{metal} = 2\kappa/(2\kappa + 1)$
53. $\simeq 10^{-6}$ F, $\simeq 10^{10}$ J
55. $V_A = 1.8 \times 10^2$ V, $V_B = V_D = 57$ V, $V_C = 3.6 \times 10^2$ V
57. 0.72 J, -0.36 J
59. $(\kappa_1 - \kappa_0)[(\epsilon_0 L^2/D)\ln(\kappa_1/\kappa_0)]$
61. Three in series: $C_{min} = 0.57$ μF, three in parallel: $C_{max} = 7.0$ μF

CHAPTER 27

1. 1.2×10^5 A/m^2, 0.46 C
3. 5.8×10^{17} electrons
5. 8.7×10^{-6} m/s, 2.0×10^{-5} m/s
7. 4.5×10^{21} electrons, 4.2×10^9 electrons
9. $1.001nev_e$ to the right
11. 2.8 A, 5.6×10^4 A/m^2, 5.8×10^{-6} m/s
13. $v_{larger} = 1.6 \times 10^{-5}$ m/s, $v_{smaller} = 6.4 \times 10^{-4}$ m/s; charge conservation is equivalent to mass conservation in water flow, so the smaller area requires a greater speed
15. $\pi qR^2[n_0 v_0 - (2n'v_0 R/3) - (n_0 v' R^2/2) + (2n'v'R^3/5)]$
17. (a) 6.37 Ω; (b) 361 ft
19. 1.1 Ω, 2.0×10^2 A
21. 19.6 A, 17.1 A
23. $+2\%$
25. 120 turns
27. 2.2×10^{-2} Ω, 2.1×10^2 kg, 4.2×10^2 kg
29. 33.4 kg
31. 1.82×10^{-3} Ω, 0.17 cm, the masses are the same
33. 0.723, 0.003 from thermal change
35. 8.4×10^{-6} m/s
37. $\rho = \rho_0/[1 - (r/R)]$
39. $\simeq 0.24$ m/s
41. 190 Ω
43. 35 mA, 50 mA, 71 mA, 0.10 A, 0.14 A
45. $\sqrt{PR}$
47. 2.1×10^2 m
49. 1.04×10^6 J (0.288 kWh)
51. $\simeq 110$ V, $\simeq 11$ Ω
53. Power decreases by $\frac{3}{4}$
55. $R = \rho L/\pi r_0(r_0 + \alpha L)$, $R/R_m = 1 + [\alpha^2 L^2/4r_0(r_0 + \alpha L)]$
57. 6.6 kW, 51 s, 6.4 min
59. (a) $I = I_0/(1 + \alpha kt^2)$; (b) $P = VI_0/(1 + \alpha kt^2)$; (c) $dP/dt < 0$, so thermal equilibrium would be reached
61. $R_2/R_1 = 0.128$
63. $dT/dt = k/[1 + \alpha(T - T_0)]$, where $k = V^2/mcR_0$; $I(t) = V/R_0\{1 + \alpha[T(t) - T_0]\}$

CHAPTER 28

1. 0.11 Ω
3. 3.2×10^6 J
5. 0.030 Ω, 300 W
7. R_1, ∞
9. 120 W
11. (a) $220 - 1.2I$; (b) $120 + I$; (c) 0.30 Ω; (d) 4.0×10^3 W
13. $+0.5$ A, $+1.0$ A, -1.0 A
15. 0.039 A (up)
17. No, $I_1 = R_3\mathscr{E}/(R_1R_2 + R_1R_3 + R_2R_3)$, $I_2 = (R_1 + R_3)\mathscr{E}/(R_1R_2 + R_1R_3 + R_2R_3)$, $I_3 = R_1\mathscr{E}/(R_1R_2 + R_1R_3 + R_2R_3)$
19. (a) 9.6 Ω; (b) in series: $R_2 = 8.64$ Ω, $P_2 = 5.4 \times 10^3$ W, not possible in parallel
21. $V_{CD} = [R_2/(R_1 + R_2)]V_{AB}$
23. 0.15 kΩ
25. $R_{eq} = R$; yes, but only because the numbers are such that symmetry allows removal of the center resistor
27. $I_4 = 3.98(1 + 0.26R_x)/(1 + 0.21R_x)$, $P_4 = 63[1 + 0.52R_x + (6.66 \times 10^{-2})R_x^2]/[1 + 0.43R_x + (4.58 \times 10^{-2})R_x^2]$

29. 8 Ω, not possible
31. $\frac{5}{6}R$ Ω, $\frac{2}{5}V/R$ in resistors attached to A and B, $\frac{1}{5}V/R$ in others
33. (a) $2R$; (b) $\frac{5}{3}R$; (c) $\frac{13}{8}R$; (d) $\frac{1}{2}(1 + \sqrt{5})R = 1.618R$
35. 0.010 Ω
37. 2.0 MΩ
39. 9.38 V, 9.60 V
41. $R_x = (V/I) - R_A$; when $R_A \ll V/I$ (when $R_A \ll R_x$)
43. 100 Ω
45. ______; 20 s, 6 μs, 3 ns
49. $(\mathscr{E}/R_1)e^{-t/R_1C} + (\mathscr{E}/R_2)$
51. Two 2-Ω resistors connected in parallel with each other and in series with a 4-Ω resistor and with two capacitors
53. 1.30×10^4 s (3.6 h)
55. Four 80-Ω resistors connected in parallel or four 5-Ω resistors connected in series
57. $P_1P_2/(P_1 + P_2)$, $P_1 + P_2$
59. $I_{mixer} = 10$ A, $I_{vacuum} = 4$ A, $I_{lamp} = 1.25$ A; 120 W
61. $\mathscr{E}/R$ in the battery, 0 in the bottom leg, $\mathscr{E}/2R$ in all the others (down on the left, up on the right)
63. (a) $J_{Al}/J_{Cu} = 1$, $E_{Al}/E_{Cu} = 1.64$, $P_{Al}/P_{Cu} = 1.64$; (b) $J_{Al}/J_{Cu} = 0.61$, $E_{Al}/E_{Cu} = 1$, $P_{Al}/P_{Cu} = 0.61$; (c) all have the same J, silver has the lowest E and P
65. (a) V_0/d; (b) $\sigma\pi r^2 V_0/d = V_0/R$

CHAPTER 29

1.

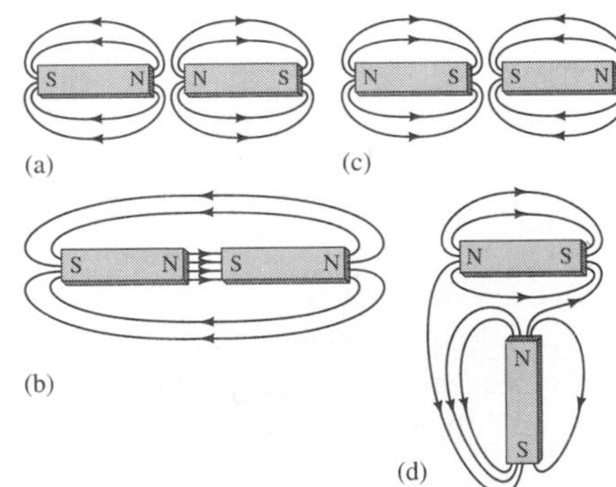

3. $-(7.2 \times 10^{-5}$ T)**k**
5. (a) $\theta = (eB\,\Delta t)/m$; (b) 0.15 T
7. $-(1.5$ mm)**j**
9. (a) 2.25×10^{-3} T; (b) 2.5×10^{-2} T
11. 1.7×10^{-13} m, 4.8×10^{-5} N
13. 3.6×10^{-6} T
15. (a) 5.9×10^7 m/s, 3.4° from the original direction; (b) 1.2 cm
17. (a) 24 J (15×10^{19} eV); (b) 16×10^{-8} kg·m/s, 48 J; (c) 8×10^{-8} kg·m/s, 24 J
19. 1.1×10^{-4} m
21. (a) $R_p/R_\alpha = 1/\sqrt{2}$; (b) $f_p/f_\alpha = 2$
23. 2.1×10^6 V/m $< E < 6.5 \times 10^6$ V/m; 1.5×10^{-4} T $< B < 4.9 \times 10^{-4}$ T
25. (a) 3.0×10^7 Hz; (b) 1.5×10^{-11} J (94 MeV); (c) 1.5×10^7 Hz, 94 MeV
27. 2.0
29. (a)

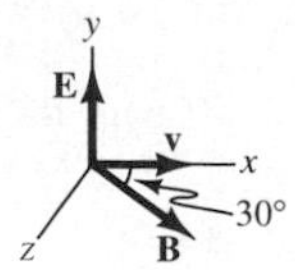

(b) $v_{min} = 2 \times 10^3$ m/s, $v_{max} = \infty$
31. 8.7×10^{-10} N perpendicular to the wire and to **B**
33. d**F** $= -IRB\cos\theta\, d\theta$ **k**
35. 0.2 T; use a variable current and measure I as a function of y
37. 5.8 A·m² perpendicular to the loop, 0.29 N·m parallel to the loop and perpendicular to **B**
39. $IN\pi R^2 B(1 - \cos\theta)$
41. $[(R_2 - R_1)/(R_2 + R_1)]\frac{1}{2}ILBd$ in the direction of the current
43. 1.1×10^{-3} N·m

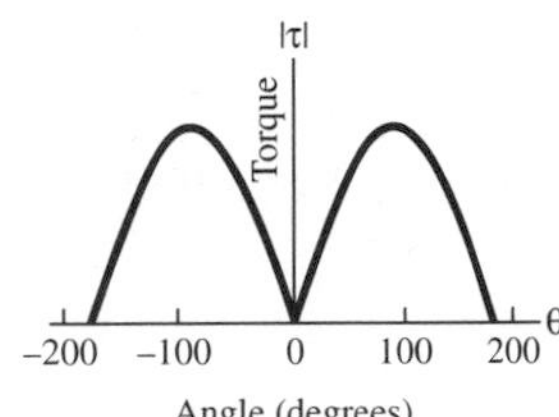

45. (a) $2IRB$**j**; (b) $-2IRB$**j**; (c) 0; (d) 0
47. $K = \frac{1}{2}\mu B\theta_0^2 \sin^2(\omega t)$; ______
49. ______; $v = \Delta V/wB$
51. 1 electron/atom
53. 3.9 mA
55. 5.8×10^{11} N/C, 2.6×10^5 T
57. 3.4 T
59. 0.05 (decrease)
61. $2V/B^2R^2$
63. (a) $NmvR$ perpendicular to the orbit; (b) $\frac{1}{2}NevR$; (c) $2m/e$
65. E = 0, B = (6.6×10^{-2} T)j

CHAPTER 30

1. 1.0×10^{-3} N/m
3. 0
5.

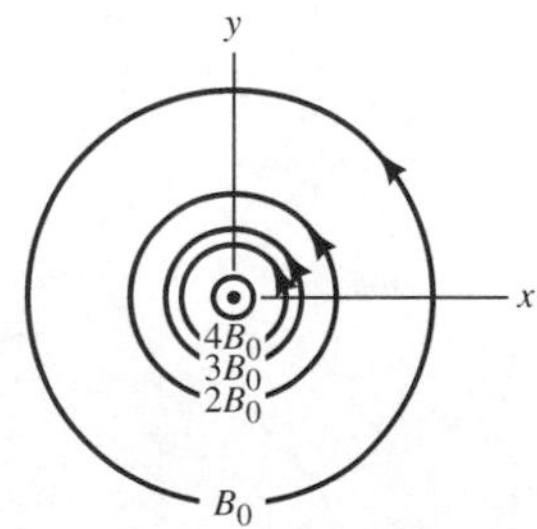

7. $B = \mu_0 h/2$ parallel to the sheet and perpendicular to the current (opposite directions on the two sides)

9. (a) $\mathbf{B} = \dfrac{\mu_0 I}{2\pi(x^2 + y^2)}(-y\mathbf{i} + x\mathbf{j})$;

(b) $$\mathbf{B} = \frac{\mu_0 I}{2\pi}\left\{-\left[\frac{y}{(x-a)^2+y^2} + \frac{y}{(x+a)^2+y^2}\right]\mathbf{i} + \left[\frac{x-a}{(x-a)^2+y^2} + \frac{x+a}{(x+a)^2+y^2}\right]\mathbf{j}\right\};$$

(c) $$\mathbf{B} = \frac{\mu_0 I}{2\pi}\left\{\left[\frac{-y}{(x-a)^2+y^2} + \frac{y}{(x+a)^2+y^2}\right]\mathbf{i} + \left[\frac{x-a}{(x-a)^2+y^2} - \frac{x+a}{(x+a)^2+y^2}\right]\mathbf{j}\right\}$$

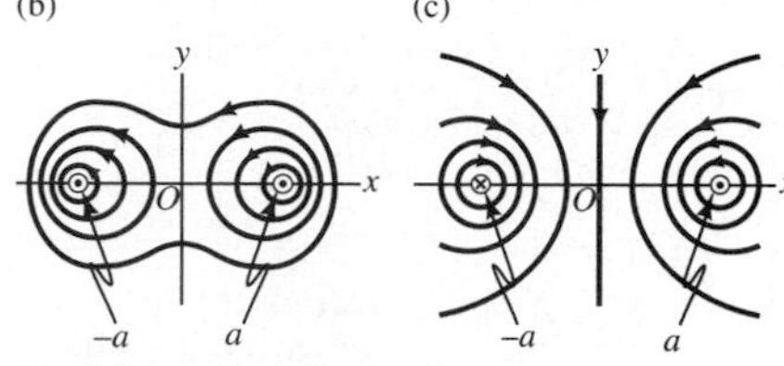

11. 1.54×10^{-4} T, circular
13.

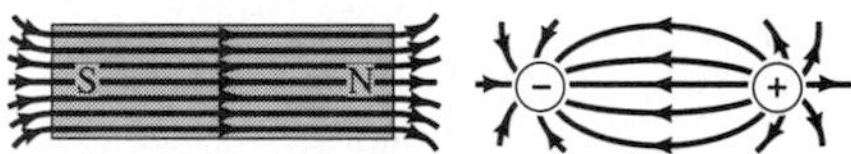

The internal fields are opposite in direction.
19. 1.9 T along the axis
21. 2.5×10^{-3} Wb
23. $(\mu_0 NIL/2\pi)\ln[(R + L)/R]$
25. $B = (\mu_0 I/4\pi D)[L/\sqrt{L^2 + D^2}]$ **circular**
27. 0
29. 1.6×10^{-7} T along the axis
31. $4\mu_0 I/\pi L\sqrt{2}$ out of the page; $B_{square} = 0.900B_{circle}$
33. $(\mu_0 I/4\pi)[(x + y + \sqrt{x^2 + y^2})/xy]$**k**
35. 3.62×10^{-4} T along the axis
37. 0.766R, 4.53R
39. $\frac{1}{3}Q\omega R^2$**k**
41. $(\xi/R)e^{-t/RC}$; 0
43. (a) 2.0×10^4 V/s; (b) 1.1×10^5 V·m/s, 1.0×10^{-6} A
45. (a) 3.8×10^6 V/m·s; (b) 9.0×10^{-2} A
47. $\frac{1}{2}\mu_0 h^2$ (attractive)
49. $12.4/n^5$ T perpendicular to the orbit, $(9.3 \times 10^{-24})\, n$ A·m² perpendicular to the orbit
51. (a) $$B(x) = \frac{\mu_0 I}{2R}\left\{\frac{1}{[1 + (x/R)^2]^{3/2}} + \frac{1}{[2 - 2(x/R) + (x/R)^2]^{3/2}}\right\},$$ 0.677 $\mu_0 I/R$, 0.713$\mu_0 I/R$, 0.716$\mu_0 I/R$
53. $(\mu_0/2\pi)I_1 I_2 L$**k** (out of the page)
55. Maximize the number of turns (length of wire) subject to a resistance that will produce a maximum current. The wire is 2.23×10^3 m long with diameter 0.44 mm. $B_{max} = 3.3 \times 10^{-2}$ T; 57 layers with 273 turns/layer.

CHAPTER 31

1. 7.5×10^{-2} V
3. (0.78 T/s)**i**
5. 1.7×10^{-10} V **counterclockwise**
7. 37 Hz (2.3×10^2 rad/s)
9. $\mathscr{E} = 0$, $\mathscr{E}$ is clockwise, $\mathscr{E} = 0$, $\mathscr{E}$ is clockwise
11. 0
13. $-B_0 r_0 \alpha/R_0(1 + \beta t)$ clockwise
17. 0.18 V, it would not change
19. 2.1×10^{-4} V (the bottom end is higher)
21. $-2Bv\sqrt{2Rvt - v^2t^2}$ for $0 < t < 2R/v$
23. $B = \pi I_{av} R/2NA\omega$

25. If $t = 0$ when the front tip of the loop is just leaving **B**, $P = (2Bv^2t)^2/R$ for $0 < t < L/v\sqrt{2}$ and $P = [2Bv(\sqrt{2}L - vt)]^2/R$ for $L/v\sqrt{2} < t < \sqrt{2}L/v$
27. (a) $F = [\mu_0 I_0 L^2/2\pi D(L + D)]^2(v/R)$; (b) $P = [\mu_0 I_0 L^2 v/2\pi D(L + D)]^2(1/R)$; (c) the same
31. $-(\mu_0 n I_0 \omega R/4)\cos(\omega t)$ circular, $-(\mu_0 n I_0 \omega R/4)\cos(\omega t)$ circular
33. -20 V counterclockwise
35. (a) 1.6 mA; (b) 62 μW
37. 0.8 V
39. 1.3 m/s
41. **(a) $(\pi r^2 \mu_0 N I \omega/2R)\sin(\omega t)$; (b) 90°**
43. Counterclockwise, away from the wire
45. 0.24 C
47. (a) $4.7\sin(120\pi t)$ V; (b) $0.95\sin(120\pi t)$ A
49. (a) 1.8×10^5 m/s; (b) 1.4×10^{-20} J; (d) -10%

CHAPTER 32

1. 1.0×10^{-2} A·m² in the direction of the magnetic field
3. 3.4×10^3 A/m
5. -46 A/m
7. $(9.5 \times 10^{-6})°$
9. 9.3×10^{-24} A·m²
11. (a) 0; (b) $(1.76 \times 10^{11})\,\hbar$
15. 1.1×10^6 A/m, 5.6×10^2
17. 16 A
19. 0.36 A
21. $(4.2 \times 10^3)\mu_0$
25. (a) $-\frac{1}{2}R\,dB/dt$ clockwise; (b) $(eR/2m_e)\,dB/dt$; (c) $v_f = v_i + (eR/2m_e)B_f$; (d) $\Delta\mathbf{L} = \frac{1}{2}eR^2\mathbf{B}_f$; (e) $\Delta\mathbf{m} = -(e/2m_e)\,\Delta\mathbf{L} = -e^2R^2\mathbf{B}_f/4m_e$
27. (a) 0; (b) $\mathbf{m} = -e^2R^2\mathbf{B}_f/2m_e$
29.

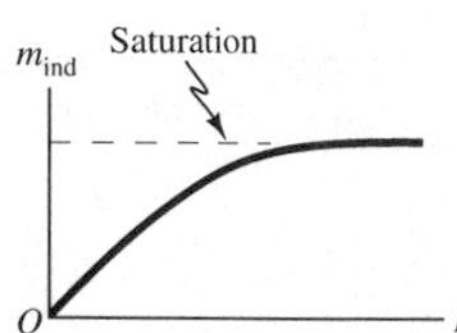

31. $\simeq 1.5 \times 10^{-3}$
33. 0.19 A/m, 2.4×10^{-7} T
37. $\simeq -5.2 \times 10^{-5}$ $(0.24\chi_{mCv})$
39. 8×10^{22} A·m², 6×10^2 A/m, 2×10^9 A
41. $m_{ring}/m_{disk} = 2$
43. (a) $C = (NmB/2\pi kT)/(e^{mB/kT} - e^{-mB/kT})$; (b) $\langle\cos\theta\rangle = [(e^{mB/kT} + e^{-mB/kT})/(e^{mB/kT} - e^{-mB/kT})] - (kT/mB)$

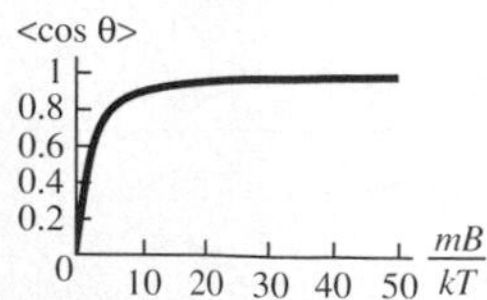

CHAPTER 33

1. 6.0×10^{-5} Wb
3. 1.5 μH
5. 30 mH
7. (a) 1.0 mH; (b) 3×10^{-3} Wb
9. $L_1L_2/(L_1 + L_2)$
11. 0 for $t < 0$; -6.6 nA for $0 < t < 0.1$ s; $+6.6$ nA for 0.1 s $< t < 0.2$ s; 0 for $t > 0.2$ s
13.

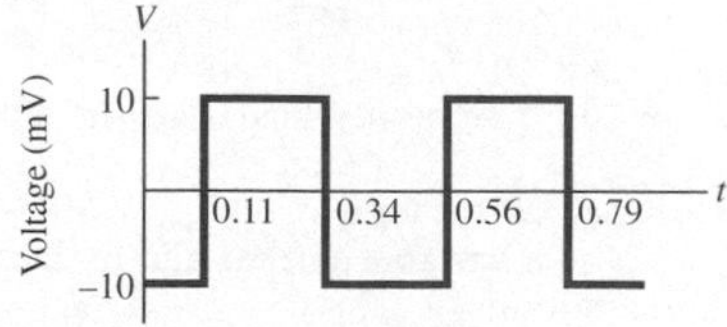

17. (a) 9.0×10^{-4} H; (b) 1.8 H
19. $(\mu_0 a/2\pi)\ln[1 + (a/d)]$
21. 24 A
23. (a) 0.60 W; (b) 1.8 W; (c) 3.0 W
25. **(a) 1.0 H; (b) 5×10^{-3} J**
27. 2.6×10^2 J
29. (a) $\mu_0 I^2/8\pi^2 r^2$; (b) $(\mu_0 I^2/4\pi)\ln(R/a)$
31. (a) $\mu_0 I^2 r^2/8\pi^2 a^4$; (b) $\mu_0 I^2/16\pi$
33. (a) 1.7×10^3 s⁻¹, 5.5×10^3 rad/s; (b) 35 Ω
35. $2\omega_0$
39. $q/\sqrt{LC}$, at $t = \frac{1}{2}(2n - 1)\,\pi\sqrt{LC}$, where $n = 1, 2, 3, \ldots$
41. $Q_0^2/2C$; ______
43. 2.0×10^{-4} C, 0.22 A; 0, 0
47. (a) $I_3 = I_{\mathscr{E}} = I_L = 2/3$ A, $I_4 = 0$; (b) 18 mH; (c) $I_3 = I_{\mathscr{E}} = 0$, $I_L = I_4 = 74$ mA
49. 1.2×10^{-4}
51. $I_3 = I_4 = I_6 = 1.7$ A; $I_3 = 2.2$ A, $I_4 = 1.3$ A $I_6 = 0.9$ A
53. 5.0×10^{-2} H; the iron core increases B and thus also the linked flux and M

CHAPTER 34

1. (a) \$53; (b) \$68 × 10⁶
3. 0.55 A
5. 0.8 A
7. 50 turns
9. **$1.36\sin(120\pi t)$ A**
11. 1.50×10^2 V
13. 5.0 Ω, 16 mH
15. 15 Ω, 95 Hz
17. 66 pF to 580 pF
19. 2.7×10^{-4} C, 4.0 V
21. 1.3 Ω, 57 Ω, 207 Ω, 1.5 mC, $-74.4°$, 0.58 A
23. $I_{max} = 92$ mA, $V_{Rmax} = 55.5$ V, $V_{Cmax} = 123$ V, $V_{Lmax} = 27.9$ V
25. It would double
27.

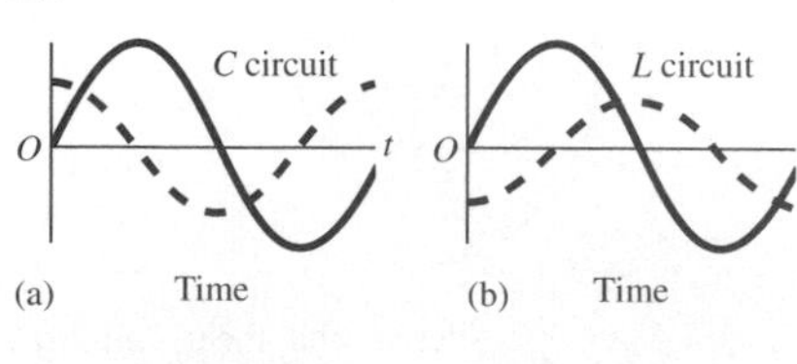

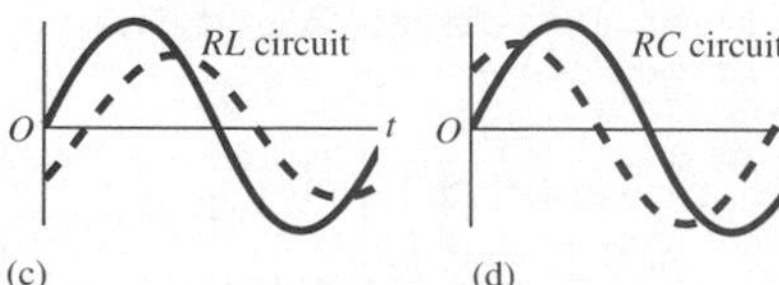

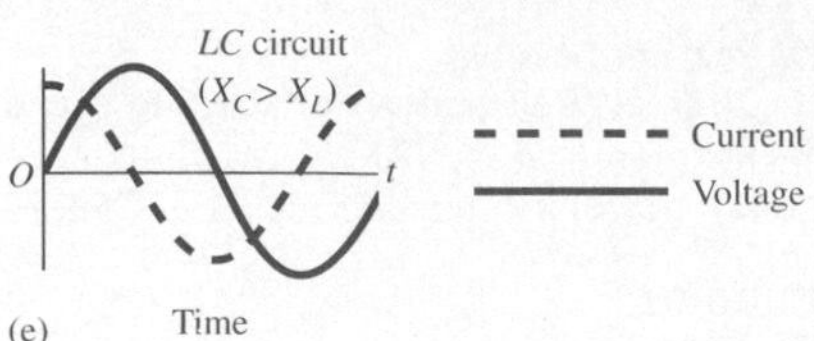

29. 66 V, 67 V
31. $L = 0.18$ H, there is no change in ω_0 when R changes
33. (a) **D** (b) 90°

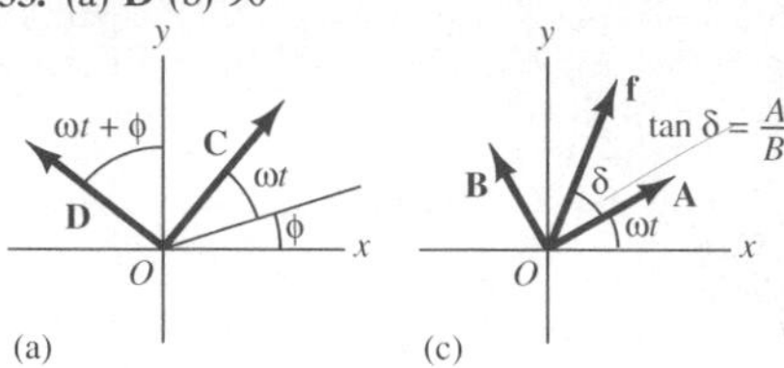

35. 111 W
37. $\omega C V_0^2\sin(\omega t)\cos(\omega t)$, 0
39. (a) 0; (b) 0; (c) 1
41. (a) $\cos\phi = R/\sqrt{X_L^2 + R^2}$; (b) $\cos\phi = R/\sqrt{X_C^2 + R^2}$; (c) 0
43. **27 μF, 46 μF**
45. 3.9×10^3 W
47. 2.1 A, 180 W, 0.77; they would become 1.9 A, 140 W, 0.69
53. (a) 0.25 V; (b) 0.16 V; (c) 0.20×10^{-3} V, low-pass (filters high frequencies)
55. $(\mathscr{E}_0/Z_1Z_2)\sqrt{Z_1^2 + Z_2^2 + 2(R_1R_2 + X_1X_2)}$, where $X = \omega L$ and $Z = \sqrt{X^2 + R^2}$, $\cos\phi = (R_1Z_2^2 + R_2Z_1^2)/Z_1Z_2\sqrt{Z_1^2 + Z_2^2 + 2(R_1R_2 + X_1X_2)}$
57. I_0/π, $I_0/2$
59.

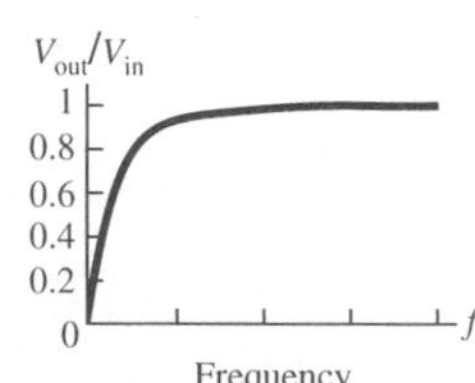

At low frequencies, X_C is large, so the current and $V_{out} = IR$ are small.
63. **(a) 0.10 A; (b) 280 Hz; (c) $V_{C1} = 0.64$ V, $V_{C2} = 0.35$ V; (d) 13 V**
65. 4.5 W
67. $\mathscr{E} - L_1(dI_1/dt) - M(dI_2/dt) = 0$; $-M(dI_1dt) - L_2dI_2/dt) - I_2R = 0$
69. 57 A
71. (a) 12.1 A; (b) 144 mH

CHAPTER 35

3. $\oint \mathbf{E}\cdot d\mathbf{s} = \mu_0(dM/dt) - [d(\iint \mathbf{B}\cdot d\mathbf{A})/dt]$; the dM/dt term is the "current" of monopoles
5. $\mathbf{B} = (E_0/c)\cos(kz + \omega t)\mathbf{i}$, in the $-z$-direction
7. (a) 3×10^2 m, 2×10^{-2} m⁻¹, 1×10^6 Hz, **6×10^6 rad/s; (b) 3 m, 0.2 m⁻¹, 1×10^8 Hz,** 6×10^8 rad/s; (c) 3×10^{-2} m, 2×10^2 m⁻¹, 1×10^{10} Hz, 6×10^{10} rad/s; (d) 6×10^{-7} m, 1×10^7 m⁻¹, 5×10^{14} Hz, 3×10^{15} rad/s;

(e) 6×10^{-10} m, 1×10^{10} m^{-1}, 5×10^{17} Hz, 3×10^{18} rad/s
9. $\partial B_y/\partial x = \mu_0\epsilon_0 \partial E_z/\partial t$, $\partial B_y/\partial t = \partial E_z/\partial x$
11. ______; in the plane formed by the z-axis and the line $y = -(\tan\theta)x$
15. $E_0 = 1.7 \times 10^{-1}$ V/m, $B_0 = 5.8 \times 10^{-10}$ T at 10 km; $E_0 = 1.7 \times 10^{-3}$ V/m, $B_0 = 5.8 \times 10^{-12}$ T at 1000 km
17. 8.9×10^{-10} N/m^2
19. 2.4×10^{-2} W/m^2, 0.72 J
21. $E_0 = 77.4$ V/m, $E_{rms} = 54.7$ V/m, $B_0 = 2.58 \times 10^{-7}$ T, $B_{rms} = 1.82 \times 10^{-7}$ T
23. (a) 4.0×10^{-7} W/m^2; (b) 1.2×10^{-2} V/m
25. $[MT^{-3}]$, kg/s^3 = W/m^2
27. 2.9×10^6 W, the paper would burn or vaporize
29. 3.1×10^{10} V/m, 1.0×10^2 T
31. (a) $+y$-direction; (b) $-x$-direction; (c) $+z$-direction; (d) $-y$-, $+x$-, $+z$-directions
33. (a) $\mathbf{S} = 2c\epsilon_0 E_0^2 \cos^2(kz - \omega t)\mathbf{k}$, only where E_0 is the field from one arm; (b) $\mathbf{S} = 2c\epsilon_0 E_0^2 \times \cos^2(kz - \omega t)\mathbf{k}$; (c) $\mathbf{S} = c\epsilon_0 E_0^2 \cos^2(kz - \omega t)\mathbf{k}$
35. 0.277
37. $0.096 I_0$
39. Linearly polarized at 30° to the x-axis
41. $(\sin^2\theta \cos^2\theta) I_0 = \frac{1}{4} I_0 \sin^2(2\theta)$
43. 6×10^8 photons
45. (a) $\mathbf{B} = \mu_0 n I_0 \cos(\omega t)\,\mathbf{k}$ (along the axis); (b) $\mathbf{E} = \frac{1}{2}\mu_0 n I_0 \omega r \sin(\omega t)$ (circular); (c) $\mathbf{S} = \frac{1}{4}\mu_0 n^2 I_0^2 \omega r \sin(2\omega t)\hat{\mathbf{r}}$, if $T = 2\pi/\omega$, for $0 < t < \frac{1}{4}T$: $\mathbf{S}$ is $+\hat{\mathbf{r}}$, for $\frac{1}{4}T < t < \frac{1}{2}T$: $\mathbf{S}$ is $-\hat{\mathbf{r}}$, for $\frac{1}{2}T < t < \frac{3}{4}T$, $\mathbf{S}$ is $+\hat{\mathbf{r}}$, for $\frac{3}{4}T < t < T$, $\mathbf{S}$ is $-\hat{\mathbf{r}}$
49. $\mathbf{E} = (7.09 \times 10^2$ **V/m)** $\cos[(2.2 \times 10^4)x + (2.2 \times 10^4)z - (2\pi \times 10^{12}\,t)]\mathbf{j}$
51. 3×10^6 J
53. 4.5×10^{13} photons/m^3
55. $[(c - v)/(c + v)]f$
57. (a) $\mathbf{B} = (\mu_0 r/2\pi R^2)\,dQ/dt$, circular; (b) $\mathbf{S} = -(Qr/2\pi^2 R^4 \epsilon_0)(dQ/dt)\,\hat{\mathbf{r}}$

CHAPTER 36

1. 4.0×10^{16} m
3. 600 nm, 400 nm; 2.0
5. 16.7 min
7. $\simeq 1.5$ s, 2.7×10^{-5} s
9. 45°
11. 1.3×10^2 m/s
13. 40.6°, 32.3°
15. $\theta = 19.4°$ (at a latitude of 8.8°), no

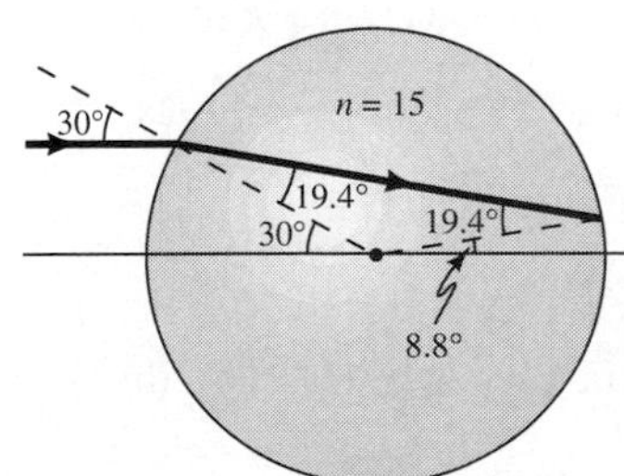

17. 3.9 m
19. 74° from the normal
21. 2.7 mm along the glass surface
23. $n_1/n_2 = 1.414$, $n_2/n_3 = 1.355$, $n_1/n_3 = 1.916$
25. 2.99°
31. $D(\sin\phi - \cos\phi \tan\{\sin^{-1}[(\sin\phi)/n]\})$
33. 0.009
35.

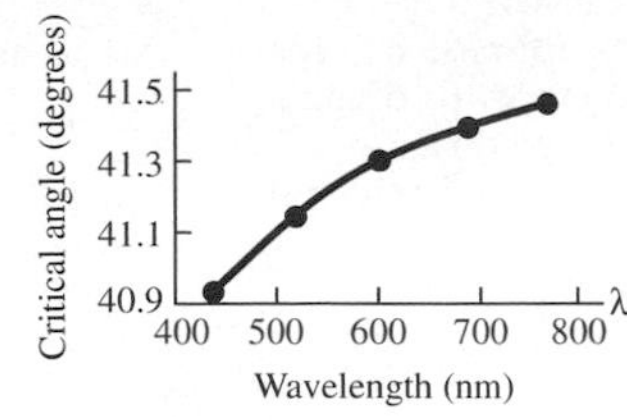

37. 434 nm, 1.97×10^8 m/s
39. 1.75 cm
41. 2.8×10^2 m
43. $d' = 2d \sin\theta_i/\sqrt{n^2 - \sin^2\theta_i}$

CHAPTER 37

1. 3
3. (a) Any distance; (b) lower the top of the mirror by 7.5 cm
5. $i_2 = 13.3$ cm, $R = 16$ cm

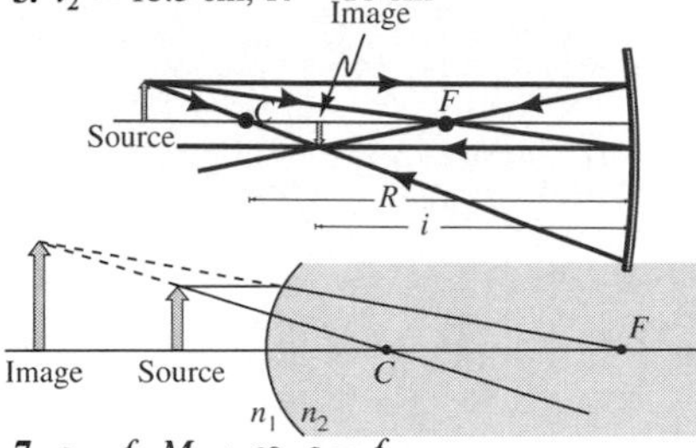

7. $s = f$, $M \to \infty$, $s = f$
11. ______; $s_c = n_1 R/(n_2 - n_1)$
13. (b) $n_1 R/(n_2 - n_1)$; (c) very far in front of the boundary, no (virtual); (d) it approaches the boundary
15.

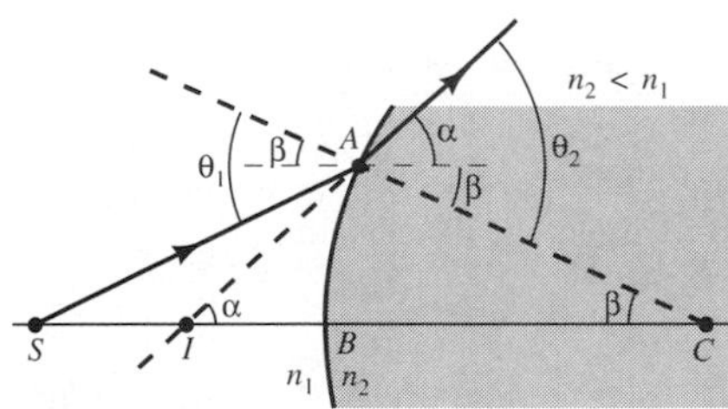

17. (a) 16 cm; (b) converging; (c) real, inverted; (d) -2.1
19. (a) 8.9 cm in front of the lens; (b) no (virtual), yes; (c) $+0.60$
21. (a) 80 cm behind the lens (real), 2 cm, inverted; (b) 360 cm behind the lens (real), 16 cm, inverted; (c) 280 cm in front of the lens (virtual), 16 cm, upright; (d) 13 cm in front of the lens (virtual), 2.7 cm, upright
23. (a) $+25$ cm; (b) $-1/3$, 33.3 cm, real, inverted; (c) $+105$ cm
25. (a) 25 cm to the right of the lens, upright, **virtual, +2.5; (b) 17 cm to the right of the lens, upright, virtual, +1.7; (c) 6.3 cm to the right of the lens, upright, virtual, +0.63; (d) 7.1 cm to the right of the lens, upright, virtual, +0.71**
29. (a) A diverging lens, $f = -22$ cm; (b) moves it to 31.8 cm, probably still readable
31. $f_1 = 91.5$ cm, $f_2 = 0.46$ cm
35. 1.09 cm
37. Positive lens closer to the object: Image is 5.8 cm from that lens on the object side; negative lens closer to the object: image 88 cm from that lens
39. (a) 12 cm behind M_1; (b) 28 cm in front of M_2

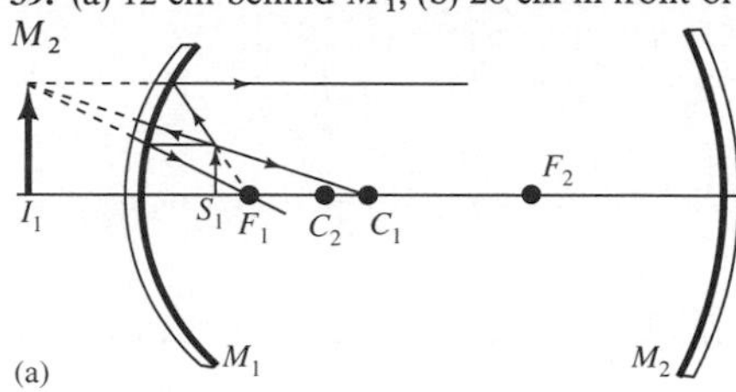

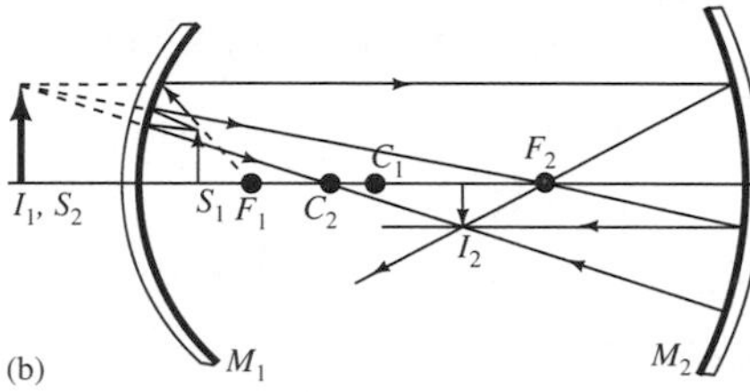

41. $[f_1(d^2 - f_1 d - 2f_2 d + f_1 f_2)]/(d^2 - 2f_1 d - 2f_2 d + 2f_1 f_2 + f_1^2)$
43. $d = R/(n - 1)$
45. Anywhere between the lens and the screen, 13.1 cm

CHAPTER 38

1. 4.7×10^2 nm
3. 5.3 mm
5. (a) It doubles; (b) it is reduced by $\frac{1}{2}$; (c) it doubles; (d) no change
7. 6.4×10^2 nm
9. $6.0[1 - (\alpha/2\pi)]$ mm
11. (b) $\theta = \tan^{-1}[\lambda\sqrt{1/(d^2 - \lambda^2) + (1/4R^2)}]$
13. $I = 2I_0$
15. $\Delta\theta = 2\sin^{-1}(\lambda/4d)$; in general, no, except for maxima near the central maximum (small angles)
17. (a) Along the perpendicular bisector; (b) 4×10^{-2} W/m^2; (c) 12.0°
19.

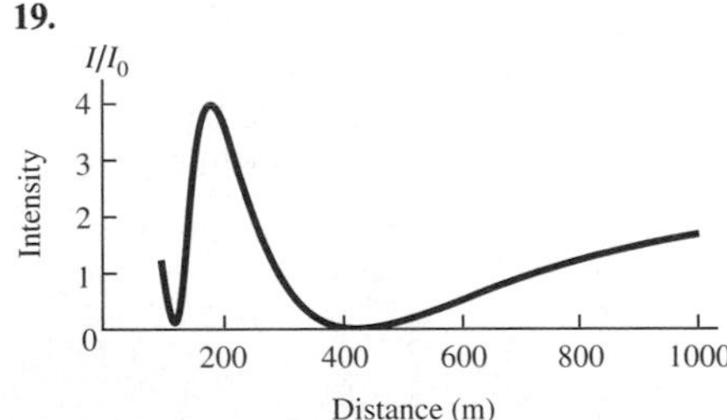

21. 33.6 μm
23. Dark, due to a phase shift π from reflection off the bottom glass
25. In to the center, 330 maxima
27. 5.46 μm, 8.8 m
29. 76.1 nm
31. 229 nm
33. 1.69 μm
35. (a) 115 nm; (b) longer; (c) no reflection of green light but still some reflection of red and blue, giving a purple hue

37. 120 nm
39. 120 nm
41. 633 nm
43. 7.273×10^5 fringes
45. Uniform intensity, because the sources are not coherent
47. $u \propto \sin^2\{2\pi f[x-(L/2)]/c\}$
49. $\sin\theta' = \sin\theta - (m\lambda/d)$, where $m = 0, \pm 1, \pm 2, \ldots$
51. $I_{max}/I_{min} = 34.0$
55. Destructively; 50λ rad, with λ in m

CHAPTER 39

1. 450 nm
3. 39.7 lines/cm
5. 15,000, 30,000, 45,000; 0.028 nm, 0.014 nm, 0.0093 nm
7. 2.945×10^5 lines, 1.64×10^6 rad/m
9. 18.6°
11. 6.17° to 9.79°; 12.42° to 19.88°
13. $\sin\beta = \sin\alpha - (m\lambda/d)$, where $m = 0, \pm 1, \pm 2, \ldots$
15. 0.51 m
17. $I_0/I_1 = 22.2$, 1.19°, increase
19. ______; the separation between neighboring minima is the same for the double-slit case
21. $\sin\theta = \sin\theta_i - (m\lambda/a)$, where $m = \pm 1, \pm 2, \ldots$; yes
23. (b) 1st maximum: 257.43°; midway between 1st and 2nd minima: 270°; 2nd maximum: 442.61°; midway between 2nd and 3rd minima: 450°;
(c)

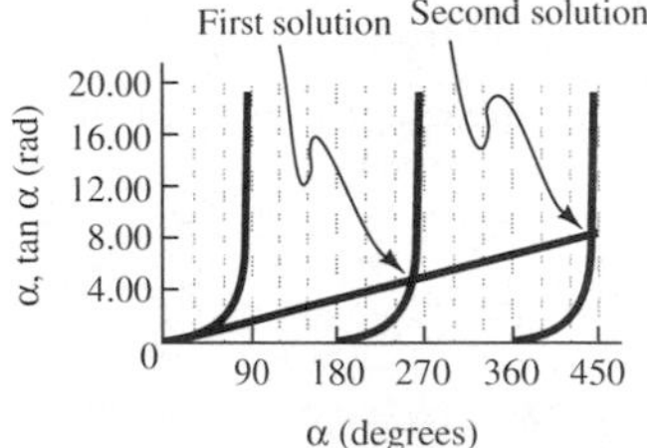

25. 5.0 cm, 83 m
27. 22.0 m
29. 10^{10} km
31. 11.0 cm
33. 19 ft
35. $I/I_{max} = 0.048$
37. (a) 10 μm; (b) 2.0 μm; (c) 1, 2, 3, 4, 6, 7, 8, 9, 11, 12, 13, 14, 16, . . . (no multiples of 5)
39. 1/4
41.

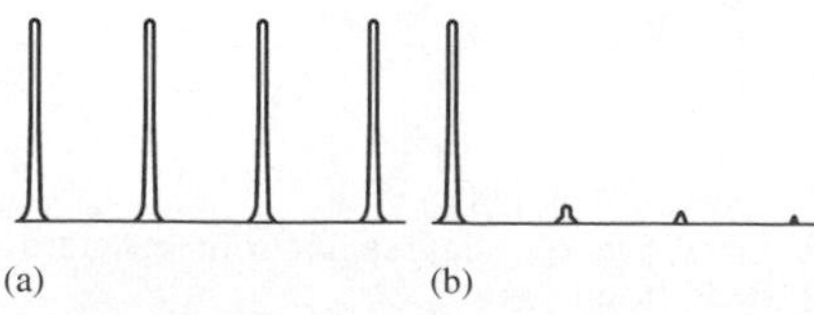

43. 11.5°, 23.6°
45. Rock salt; $\Delta\theta_{mica} = 0.23°$, $\Delta\theta_{rock\ salt} = 0.83°$
47. 640.018 nm
49. 38 m
51. 3.9×10^{-5} rad $= (2.2 \times 10^{-3})°$
53. The same as the diffraction grating, with maxima at $\sin\theta = m\lambda/d$

CHAPTER 40

1. 4.04 h; 4.02 h
3. 3.3×10^8 m/s; no, because this is not a speed of energy or of mass
5. (a) 3.1 yr; (b) 2.5 ly
7. (a) 3.3×10^{12} s (10^5 yr); (b) 3.3×10^5 s (~4 days)
9. 0.66 h
11. (a) 24 m; (b) $+13.3 \times 10^{-8}$ s; (c) 10×10^{-8} s; (d) 5.0×10^{-8} s
15. 5.4×10^8 s (17 yr)
17. Speeding: $u \simeq 0.4c$
19. 2.38×10^8 m/s, 9.5×10^{25} m (1.0×10^9 ly)
23. (a) $f_0\sqrt{[1-(v_1/c)]/[1+(v_1/c)]}$, $f_0\sqrt{[1-(v_1/c)][1-(v_2/c)]/[1+(v_1/c)][1+(v_2/c)]}$; (c) $v = (v_1+v_2)/[1+(v_1v_2/c^2)]$
25. $0.287c = 8.62 \times 10^7$ m/s, 1.044 km
27. $0.23c$
29. $\mathbf{v}' = (v_x - u)/[1-(uv_x/c^2)]\mathbf{i} + v_y/\gamma[1-(uv_x/c^2)]\mathbf{j}$
33. $\simeq 50$ g
35. 4.4×10^9 kg/s
37. $0.99999E$, 2.7×10^{-17} kg·m/s, $[1-(5 \times 10^{-11})]c$
39. 0.511 MeV, 2.73×10^{-22} kg·m/s (0.511 MeV/c)
41. 488 MeV/c^2
43. 5.0 TeV
45. $m = 1.1 \times 10^{-45}$ kg ($mc^2 = 6.1 \times 10^{-10}$ eV)
47. $\Delta f = 3 \times 10^{18}$ Hz
49. c
51. -19.6 ns, -2.7 ns
53. 200 m
55. 2.26×10^{23} Hz; 3.47×10^{23} Hz
57. 1.2×10^{-24} m

CHAPTER 41

1. (a) 1.23 nm; (b) 0.39 nm; (c) 0.12 nm; (d) 1.24×10^{-15} m; (e) approximately the same size as the wavelengths
3. (a) 2.9×10^{-14} m; (b) 9.2×10^{-15} m; (c) 1.7×10^{-15} m
5. 58°
7. (a) 1.4×10^{-14} m; (b) 4.6×10^{-15} m; (c) 8.5×10^{-16} m
9. 2.4×10^{-17} J (150 eV), 1.3×10^{-20} J (0.082 eV)
11. 0.36
13. $\simeq e^{-3.2 \times 10^{37}} \simeq 0$
17. 10^{-22} J (6×10^{-4} eV)
19. $\simeq 1.8 \times 10^{-20}$ kg·m/s, 9.7×10^{-14} J (0.61 MeV)
21. 0.39 nm
23. $\Delta x = 5.0 \times 10^{-33}$ m
25. $\frac{3}{2}(\hbar^2 U_0^2/ma^2)^{1/3}$
27. 3.15×10^{-19} J (1.97 eV), 1.05×10^{-27} kg·m/s
29. 1.8 eV
31. (a) 2.4 eV; (b) 1.2×10^4 eV; (c) 1.2×10^{-4} eV; (d) 1.2×10^{-7} eV; (e) 1.2×10^{-8} eV
33. 1.0×10^{45} photon/s
35. 6×10^{-14} W/m², 4×10^{-17}
37. 5.8×10^3 K
39. 7.37×10^{-19} J (4.60 eV), 9.21×10^{-19} J (5.75 eV)
41. 88.4°
43. 720–750 nm: 0.028 J/m³, 480–510 nm: 0.040 J/m³, ratio = 0.71.
45. 1.38×10^8
47. 0.75
49. 1.33×10^4 yr ago (≃11,300 BC)
51. 7.2×10^{10} s $= 2.3 \times 10^3$ yr
53. $RP_{elect}/RP_{light} = 8 \times 10^4$
55. $\simeq 10^{17}$ photons/s

CHAPTER 42

1. 91.4 nm
3. 10.2 eV, 122 nm; 0.66 eV, 1880 nm
5. (a) 2, 3, 4; (b) 122 nm, 103 nm, 97 nm; all in ultraviolet
7. 4.88 nm
9. $(8\epsilon_0^2 h^3 c/m_e Z^2 e^4)[n_2^2 n_1^2/(n_2^2 - n_1^2)]$, 11, 35
11. 47.7 nm, 7.28×10^5 m/s, 6.63×10^{-25} kg·m/s, -1.51 eV
13. $\frac{2}{3}(e^2/4\pi\epsilon_0)^7 m_e^2/c^3 L^8$
15. $f = (m_e/4\pi\hbar^3)(e^2/4\pi\epsilon_0)^2 \times \{(1/n^2) - [1/(n+k)^2]\}$; $f(n \gg k) \simeq (m_e/2\pi L^3)(e^2/4\pi\epsilon_0)^2 k$, which is k greater than $1/T$; ______
17. 0.033 nm
19. 28
21. Li ($Z = 3$)
23. 1.5 eV
25. 1.73×10^{-4} eV less than an electron with spin down
27. 54.4 eV, 1.36×10^3 eV, 1.65×10^3 eV, 4.41×10^3 eV, 4.91×10^3 eV, 1.07×10^4 eV, 1.14×10^4 eV; for outer electrons, the nucleus is screened by inner electrons: Effective charge $< Ze$; and $n > 1$.
29. 47 kg/s², smaller than 97 kg/s² from Eq. (42–18)
31. $I_0 = 1.48 \times 10^{-46}$ kg·m², $I_1 = 1.43 \times 10^{-46}$ kg·m²; at different vibrational energies, nuclei are in different electron potentials, which are not harmonic oscillator potentials.
35. 9.48×10^{26}
37. 2.5×10^{-6} eV, 0.74×10^{-6}
39. 2.3×10^{-33} J (1.4×10^{-14} eV)
43. $2a_0 = 1 \times 10^{-10}$ m

CHAPTER 43

1. 14.4 eV
3. 3.25 eV, 2.13 eV, 11.7 eV
5. 2.5×10^{28} electrons/m³
7. 94; 0.999 eV; $\Delta E = 7.1 \times 10^{-3}$ eV
9. 1.1×10^6 m/s
11. (a) $6(\pi^2\hbar^2/2m_eL^2)$; (b) $9(\pi^2\hbar^2/2m_eL^2)$
13. $(\pi^2\hbar^2/2r_0^2)\left(\frac{4\pi A}{3}\right)^{2/3}$ $\{[(n_{p1}^2 + n_{p2}^2 + n_{p3}^2)/m_p] + [(n_{n1}^2 + n_{n2}^2 + n_{n3}^2)/m_n]\}$
17. (a) $(\hbar^2/5m)(3\pi^2)^{2/3}(N/V)^{5/3}$; (b) 3.8×10^{10} N/m²
19. $(\hbar^2\pi^2/2m_e)n_e$
21. (a) $12/N_{sm}^{1/3}$ km, where N_{sm} = number of solar masses; (b) $E_F = 60N_{sm}^{4/3}$ MeV, $p_F = 335N_{sm}^{2/3}$ MeV/c; (c) $4.7M_\odot$
23. $E_F = (\hbar^2\pi/m_e)n_e$; ______
25. 634 nm, 4.73×10^{14} Hz
27. 45 mW
29. 19 cm
31. (b) ______, $8\pi hf^3/c^3$

33. 4.144 K
35. 1.33×10^{15} Hz, 4.83×10^{9} Hz
37. 0.5 μV
39. (a) $n(v) = (V/\pi)(m/\hbar)^3 v^2$; (b) $\frac{12}{7}E_F^2/m^2$
41. $(\hbar^2/m_e)n_e^{2/3}$
43. ______; slope would be $-\frac{1}{2}$
47. (c) $c \propto T^3$
49. (a) $(9.15 \times 10^{55})N_{\text{sm}}^{4/3}/R^5$ N/m², where N_{sm} = number of solar masses; (b) $(6.37 \times 10^{49})N_{\text{sm}}^2/R^4$ N/m²; (c) $(\hbar^2/5Gm_e m_{\text{He}}^2) \times (324\pi^2/N)^{1/3}$; (d) 1.4×10^6 m
51. $(7.78 \times 10^{32}\ \text{N/m}^2)[(N/A)^{5/3} + (Z/A)^{5/3}]$; if $N = Z$, 4.90×10^{32} N/m²

CHAPTER 44

1. 5.5×10^{-11} *n*-carriers/cm³; (b) 1.7×10^9 *n*-carriers/cm³; $\simeq 10^{-14}$ that of copper
3. (a) 1; (b) 1; (c) 0.87; (d) 0.13; (e) 2.4×10^{-18}
5. 1.5 eV
7. (a) 5.9×10^4 K; (b) 8.1×10^4 K; (c) 2.3×10^9 K
9. 28 μm, 19 μm, 7.8 μm, 5.6 μm, 4.0 μm
11. (a) 2.0×10^{-6}; (b) 2.0×10^{-6}
13. 2.0×10^{18} carriers/m³
15. 0.018 eV $< \frac{1}{2}E_{\text{Ge}}$
17. (a) 1.7×10^{16} carriers/m³, (b) 4.8×10^{16} carriers/m³.
19. (b) $n(v)\,dv = \left(\frac{8\pi m_e^3}{h^3}\right) \times e^{-(\frac{1}{2}m_e v^2 - E_F)/kT} v^2\,dv$
21.

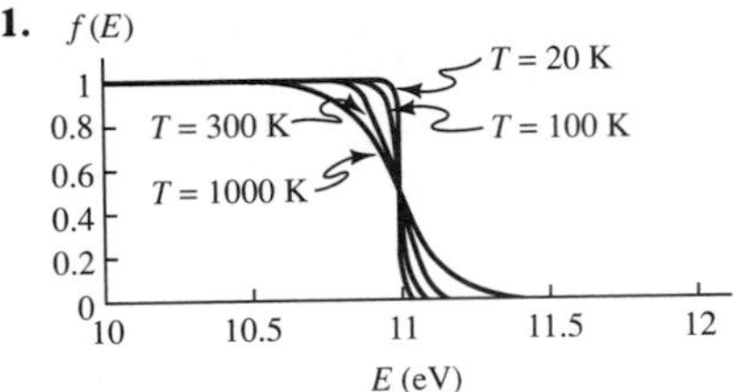

23.

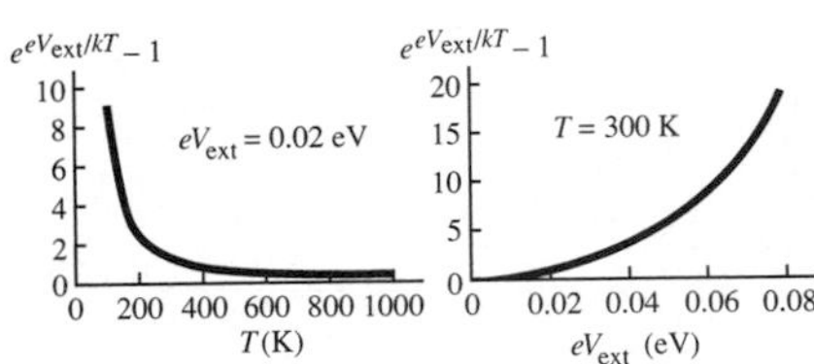

25. 0.47 eV
27. 2.1 eV
29. (a)

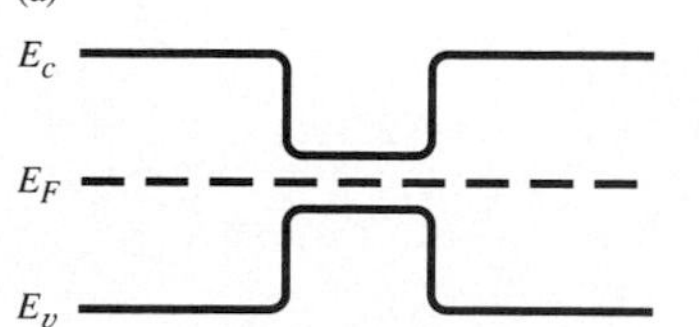

(b) Any *n*-carriers will accumulate in the semiconductor with the narrow gap.

31. (a) 18 meV; (b) 68 μm
33. 1.2×10^{-7}
35. 1.1×10^{10} carriers/m³
37. $(1.2 \times 10^3)a_0$
39. 1.9×10^{23} electrons/m³, 3.9×10^{25} electrons/m³
45. (a) $\left(\frac{V}{2\pi^2}\right)\left(\frac{2m_e}{\hbar^2}\right)^{3/2} \int_0^\infty \frac{E^{1/2}\,dE}{1 + e^{(E-E_F)/kT}}$;

(b) $E_{\text{av}} = \dfrac{\int_0^\infty \frac{E^{3/2}\,dE}{1 + e^{(E-E_F)/kT}}}{\int_0^\infty \frac{E^{1/2}\,dE}{1 + e^{(E-E_F)/kT}}}$;

(c) $N \to \frac{2}{3}\left(\frac{V}{2\pi^2}\right)\left(\frac{2m_e}{\hbar^2}\right)^{3/2} E_F^{3/2}$, $E \to \frac{3}{5}E_F$

CHAPTER 45

1. (n, p): (4, 3); (8, 7); (19, 17); (29, 24); (42, 32); (57, 44); (80, 55); (107, 72); (136, 88); (157, 103)
3. (a) 2.3×10^{-4} rad; (b) 2.3 rad
5. 5.35 fm; sum of radii = 5.50 fm
7. 236 MeV, 245 MeV, 255 MeV
9. 170 MeV; 1.2×10^3 MeV
11. $\Delta E = \frac{3}{5}(e^2/4\pi\epsilon_0 R)(Z_2^2 - Z_1^2)$, $r_0 = 6.99/\Delta E$, with ΔE in MeV and r_0 in fm
13. ^{17}O: $Z = 8$; ^{38}Ar: $N = 20$; ^{48}Ca: $Z = 20$, $N = 28$; ^{115}Sn: $Z = 50$; ^{144}Sm: $N = 82$; ^{208}Pb: $Z = 82$, $N = 126$
15. $F = g^2[(1/r^2) + (1/Rr)]e^{-r/R}$, attractive

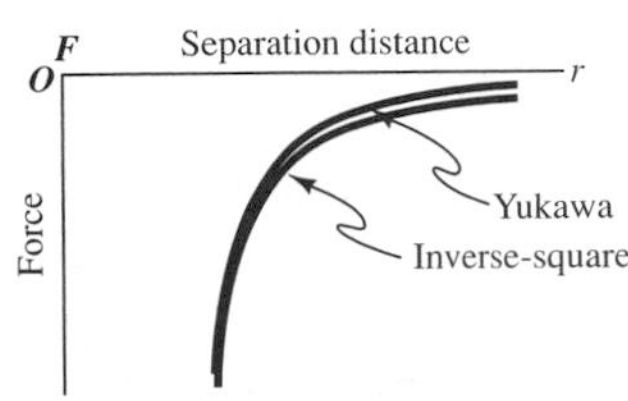

17. $S_n(^{48}\text{Ca}) = 8.27$ MeV; $S_p(^{48}\text{Ca}) = 16.57$ MeV; $S_\alpha(^{48}\text{Ca}) = 12.53$ MeV
19. $Z^2/A > 0.703(a_s/a_c) \simeq 17$
21. 20.1 MeV; this is greater than the α-particle energy, so tunneling must occur
23. $\simeq 1$ fm; this size represents a single nucleon, not a nucleus.
25. (a), (c), and (d) are not possible, because charge is not conserved.
31. (a)

Nucleus	K (MeV)	$t_{1/2}$
^{210}Po	5.30	98.2 yr
^{210}Ra	7.00	2.4 h
^{215}Fr	9.36	40.8 ns
^{232}Th	4.77	1.4×10^{10} yr

(b) 7.7 MeV
33. 1.97
37. ______; $dN_a/dt = -N_a/\tau_1$; $dN_b/dt = (-N_b/\tau_2) + (N_a/\tau_1)$; $dN_c/dt = (-N_c/\tau_3) + (N_b/\tau_2)$; $dN_d/dt = (-N_d/\tau_4) + (N_c/\tau_3)$; . . .
41. (a) 1.3 keV; (b) 0.7 MeV, much greater than thermal energy
43. (a) 9.7 metric tons; (b) 7.3×10^{16} J; (c) 3.1×10^{19} fissions

45. (a) $4p \to {}^4\text{He} + 2e^+ + 2\nu + 3\gamma$, 24.8 MeV released; (b) 7.5×10^{47} MeV $= 1.2 \times 10^{34}$ J
47. 2.5×10^2 MeV
49. 1.4×10^2 fm
51. (a) When the nuclei get inside the electron cloud, there will be a repulsive Coulomb force; (b) 0.7 MeV each.
55. $E_p = (\hbar^2/10m_p r_0^2)(3^7\pi^2 Z^5/16A^2)^{1/3}$; $E_n = (\hbar^2/10m_n r_0^2)[3^7\pi^2(A - Z)^5/16A^2]^{1/3}$

CHAPTER 46

1. 2.5 fm
3. $p_i + \frac{1}{2}Mc$; increases by a factor of $(4p_i + Mc)/(p_i + Mc)$
5. (a) $\sigma(\theta) = Z_1^2 Z_2^2 (e^2/4\pi\epsilon_0)^2[4m^2/(\Delta^2)^2]$

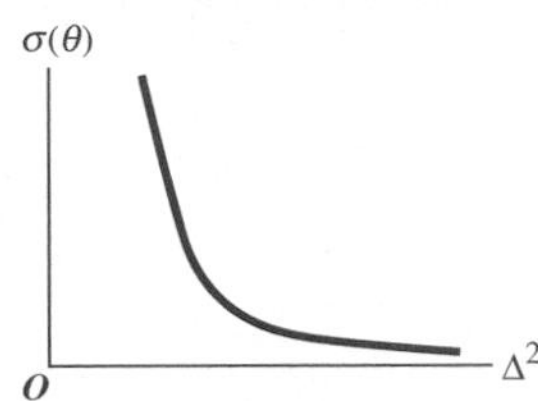

(b) with $d \simeq \hbar/\Delta$, large Δ corresponds to interaction over a short distance, where extended structure is not important, so Δ would not change. For small Δ, $\Delta < \hbar/r_0$, collision takes place over a large distance where spread-out charge would appear smaller and σ would be smaller than for a point charge.
7. 3.44 TeV
11. 0, -1, fermion, $+1e$
13. $\overline{uud}$
15. 6.02×10^{26} baryons for each
17.

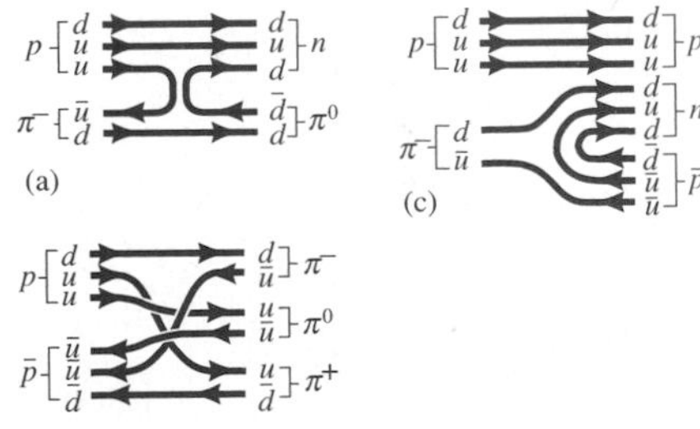

19. 460 MeV
21. 14.5 GeV/c
23. 8.1×10^9 MeV
25. 9.5×10^4 s^{-1}
27. (a) 2.16×10^3; (b) 146 GeV/c; (c) 4.6×10^4 GeV/c
31. (a) $n(n - 1)kt^{n-2}$; (b) $(1 - n)/n$
35. $(\hbar G/c^3)^{1/2}$
39. 1.6×10^5 K; 3.0×10^5 K
41. 6.0×10^2 km
43. 740
49. 0.175
51. (a) 1.25×10^4 eV; (b) 58 fm

INDEX

Note: Italics indicate a definition or primary entry for multiple entries, where applicable.

PHOTO CREDITS

Chapter 1 **CO** The Granger Collection **1-1** The Granger Collection **1-2a** Frederica Georgia/Photo Researchers **1-2b** From Ramsey, Norman F., *Precise Measurement of Time,* AMERICAN SCIENTIST, Vol. 76, Jan./Feb. 1988, p. 43 **1-2c** Patricia McDonough **1-3** John G. Ross/Photo Researchers **1-20** (no credit) **1-21** Photofest **1-22** Courtesy Franklin Institute **1-23** Sidney Harris **1-24** Thomas J. Cutitta, International Imaging, Inc.

Chapter 2 **CO** James Sugar/Black Star **2-1** Gerard Vandystadt/Photo Researchers **2-5** Audrey Gottlieb/Monkmeyer Press **2-15** Pasco Scientific **2-22a** Richard Megna/Fundamental Photographs **2-25** The Image Bank **2-28** JAL/PPS/Photo Researchers **2-30** Richard Megna/Fundamental Photographs

Chapter 3 **CO** Kraft/Explorer/Photo Researchers **3-3a** D'Lynn Waldron/The Image Bank **3-8** Scala/Art Resource **3-11** Gary S. Settles/Photo Researchers **3-18c** C. Cuny/Rapho/Photo Researchers **3-24a** Kent Wood/Photo Researchers **3-24b** Estate of André Kertész **3-28** Kenneth Murray/Photo Researchers **3-30** The Granger Collection

Chapter 4 **CO** The Granger Collection **4-4** Bernard Asset/Agence Vandystadt/Photo Researchers **4-6a** Focus on Sports **4-6b** Susan McCartney/Photo Researchers **4-9** D. Guravich/Photo Researchers **4-12a** Michael Freeman **4-12b** Michael Freeman **4-13** NASA **4-16** Bernard Asset/Agence Vandystadt/Photo Researchers **B1-1** NASA **4-23** NASA **4-35** Bill Liske/Photo Researchers

Chapter 5 **CO** Bettmann **5-10** Richard Megna/Fundamental Photographs **5-12** The Granger Collection **5-15** Scala/Art Resource (Hirshhorn Museum) **5-17a** Paul Silverman/Fundamental Photographs **5-17b** Paul Silverman/Fundamental Photographs **5-18** Peter Johansky/FPG **5-19** Eunice Harris/Photo Researchers **5-25a** Photofest **5-28** Berenice Abbott/Photo Researchers **5-30** Robert A. Isaacs/Photo Researchers

Chapter 6 **CO** The Granger Collection **6-1** Focus on Sports **6-2** Movie Star News **6-8a** Jean-Marc Barey/Agence Vandystadt/Photo Researchers **6-8b** Focus on Sports **6-8c** Gale Zucker/Stock, Boston **6-8d** Michael A. Keller/The Stock Market **6-16** Richard Mackson/Sports Illustrated **6-17** Theodore Anderson/Stockphotos **6-20** The Granger Collection

Chapter 7 **CO** Rafael Macia/Photo Researchers **7-4** Guido Benetton/Agence Vandystadt/Photo Researchers **7-5** Anthony A. Boccaccio/The Image Bank **7-6** R. & M. Magruder/The Image Bank **7-14** Andy Levin/Photo Researchers **7-16** Courtesy Digital Instruments, Inc., Santa Barbara, CA **7-17** Richard Megna/Fundamental Photographs

Chapter 8 **CO** Richard Megna/Fundamental Photographs **8-3** Joe Strunk **8-10a** David Drapkin **8-10b** David Drapkin **8-22** Richard Megna/Fundamental Photographs **8-28** NASA **8-30** Cenco **8-32b** Cenco

Chapter 9 **CO** Gene Ahrens/FPG **9-9** NASA **9-20** Richard Megna/Fundamental Photographs **9-29a** Thomas J. Cutitta, International Imaging, Inc. **9-31** Richard Megna/Fundamental Photographs

Chapter 10 **CO** R. J. Dufour, Rice University, Hansen Planetarium **10-16a** Stephen Dalton, Agence Nature/NHPA **10-16b** Focus on Sports **10-17c** Thomas J. Cutitta, International Imaging, Inc. **10-17d** Thomas J. Cutitta, International Imaging, Inc. **10-19** Courtesy Bausch & Lomb Inc. **10-23** Thomas J. Cutitta, International Imaging, Inc. **10-24** Thomas J. Cutitta, International Imaging, Inc.

Chapter 11 **CO** National Museum of American Art, Smithsonian Institution, Transfer from General Services Administration **11-1** Addison Geary/Stock, Boston **11-6** Richard Megna/Fundamental Photographs **11-8** Thomas J. Cutitta, International Imaging, Inc. **11-11** Jeffrey Muir Hamilton/Stock, Boston **11-14** Richard Megna/Fundamental Photographs **11-17** Lincoln Russell/Stock, Boston

Chapter 12 **CO** NASA **12-2** Erich Lessing, Culture and Fine Arts Archive **12-4** The Granger Collection **12-9** National Museum of American History, Smithsonian Institution, Washington, D.C. **12-10a** NASA **12-11b** Kitt Peak Observatory **12-15** Farley Lewis/Photo Researchers **12-20** National Radio Astronomy Observatory **12-22** Courtesy AIP Niels Bohr Library

Chapter 13 **CO** Barry L. Runk/Grant Heilman **13-8** Berenice Abbott/Photo Researchers **13-15** W & B Productions **13-17a** Special Collections Division, University of Washington Libraries **13-17b** Special Collections Division, University of Washington Libraries **13-18** Richard Hutchings/Photo Researchers **13-19** David Brownell/The Image Bank **13-32** D. Gleiter/FPG

Chapter 14 **CO** © M. C. Escher Heirs, Cordon Art, Baarn, Holland **14-5** Cenco **14-7** By permission of National Film Board of Canada **14-8** Dohrn/Science Photo Library/Photo Researchers **14-16** Martin Bough/Fundamental Photographs **14-23** J. Kim Vandiver and Harold E. Edgerton, 1973 **14-24** Bertrand/Explorer/Photo Researchers **14-25** Jim W. Grace/Photo Researchers

Chapter 15 **CO** John S. Shelton **15-4** Physics, 2nd ed., Physical Science Study Committee; D.C. Heath and Company, Boston, 1965 **15-6** Thomas J. Cutitta, International Imaging, Inc. **15-9** Dr. E. R. Degginger, FPSA **15-12a** Thomas J. Cutitta, International Imaging, Inc. **15-12b** Thomas J. Cutitta, International Imaging, Inc. **15-25** Stephen Dalton/NHPA **B1-1** David Halpern/Photo Researchers **B1-2** J. Lees & P. Malin, University of California, Santa Barbara/Science Photo Library/Photo Researchers **B1-3** Harry J. Mzkop, Jr./Medichrome

Chapter 16 **CO** Randall Hyman/Stock, Boston **16-6** Michael Grecco/Stock, Boston **16-7** Cenco **16-12** Balloon Excelsior, Inc. **16-16** Tom McHugh/Photo Researchers **16-18** From *An Introduction to Fluid Dynamics,* by G. K. Batchelor, F.R.S., 1970, plate 10 **16-19** NASA **16-24a** Cenco **16-26b** Cenco **16-30** Diane Schiumo/Fundamental Photographs **16-32a** Dr. Gary Settles/Photo Researchers **16-32b** NASA/Science Source/Photo Researchers **16-32c** Philippe Plailly/Science Photo Library/Photo Researchers **16-34** The Granger Collection **16-39** Giraudon/Art Resource

Chapter 17 **CO** Dawson Jones/Stock, Boston **17-1** Diane Schiumo/Fundamental Photographs **17-2** Thomas J. Cutitta/International Imaging, Inc. **17-5a, b, c** Cenco **17-7** Courtesy AIP Niels Bohr Library. Photo by A. G. Webster **17-15** NASA/Science Source/Photo Researchers **17-16** A.T. & T. Bell Laboratories **17-18a** NASA

Chapter 18 **CO** William Felger/Grant Heilman **18-7a** Grant Heilman **18-7b** Richard Folwell/Science Photo Library/Photo Researchers **18-7c** David Cavagnaro/Peter Arnold, Inc. **18-9** Richard Megna/Fundamental Photographs **18-10** The Granger Collection **18-11a** The Science Museum, London **18-12** The Granger Collection

Chapter 19 **CO** J. Zalon/FPG International **19-1** The Granger Collection **19-4** (no credit) **19-5** FPG International **B1-1** U.S. Dept. of Energy **19-16** FPG International **19-21** Beitel Lottery Products **19-23** Media Services, CERN

Chapter 20 **CO** Private Collection/Superstock **20-3** Scientific American, Jan. 1968, Vol. 218, No. 1 **20-7** Brock May/Photo Researchers **20-8b** Cenco **20-14** Carrier Corporation **20-18** Richard Megna/Fundamental Photographs **20-19** Thomas J. Cutitta, International Imaging, Inc.

Chapter 21 **CO** J & L Weber/Peter Arnold, Inc. **21-4** Tom McHugh/Photo Researchers **21-5a** Richard P. Feynman, Robert B. Leighton, and Matthew Sands, *The Feynman Lectures on Physics,* © 1964, Addison-Wesley Publishing Company, Inc., figure 2, p. 30-13. **21-5b** *The Feynman Lectures on Physics,* figure 6a, p. 30-15. **21-6a** *The Feynman Lectures on Physics,* figure 5b, p. 30-14. **21-6b** *The Feynman Lectures on Physics,* figure 12, p. 30-20. **21-9** Earth Satellite Corporation **21-10** Haling/Science Source/Photo Researchers **21-13c** Thomas J. Cutitta, International Imaging, Inc. **21-15** Richard Megna/Fundamental Photographs **21-16** NASA/Science Source/Photo Researchers

Chapter 22 **CO** The Granger Collection **22-1** Courtesy of the Bundy Library, Norwalk, CT **22-4** The British Library **22-7** Science Photo Library/Photo Researchers **22-8** Courtesy of the Bundy Library, Norwalk, CT

Chapter 23 CO Fundamental Photographs **23–3a** Harold M. Waage **23–9a** (no credit) **23–11a** Harold M. Waage **23–12a** Harold M. Waage **23–13d** Harold M. Waage **B1–1** Thomas J. Cutitta, International Imaging, Inc. **23–22** Phil Jude/Science Photo Library/Photo Researchers

Chapter 24 CO Collection, The Museum of Modern Art, New York. Gift of the Advisory Committee. **24–3** Harold M. Waage **24–5** John Colwell/Grant Heilman Photography **24–20** Thomas J. Cutitta, International Imaging, Inc. **24–21a** Cenco

Chapter 25 CO FPG International **25–9** U.S. Geological Survey **25–23b** Lester V. Bergman & Associates **B1–2a** (after B. Long and H. Nordan) **B1–3** Prof. Erwin Mueller/Science Photo Library/Photo Researchers **25–26a** Philippe Plailly/Science Photo Library/Photo Researchers **25–27** Philippe Plailly/Science Photo Library/Photo Researchers **25–28** Courtesy IBM Corp. **25–29** Berenice Abbott/Photo Researchers

Chapter 26 CO Hailes-Hamann/Peter Arnold **26–4a** The Image Works **B1–1** The Image Works **26–6** The Image Works **26–7** Lawrence Livermore National Laboratory **26–8** Harold M. Waage **26–15** By courtesy of the Royal Institution, London

Chapter 27 CO Dan McCoy/Rainbow **27–9** Dan McCoy/Rainbow **27–11** The Image Works **27–13** The Image Works **27–14** Los Angeles Department of Water & Power **27–21** Brookhaven National Laboratory **27–22** T. J. Florian/Rainbow

Chapter 28 CO George Haling/Photo Researchers **28–1** The Image Works **28–3** J. L. Charmet/Science Photo Library/Photo Researchers **28–4** Ray Pfortner/Peter Arnold, Inc. **28–16** Cenco **28–17a** Thomas J. Cutitta, International Imaging, Inc. **28–19a** Thomas J. Cutitta, International Imaging, Inc. **28–22a** The Image Works **28–25c** Thomas J. Cutitta, International Imaging, Inc.

Chapter 29 CO NASA **29–2a** Richard Megna/Fundamental Photographs **29–2b** Werner H. Muller/Peter Arnold, Inc. **29–2c** Richard Megna/Fundamental Photographs **29–3a** Thomas J. Cutitta, International Imaging, Inc. **29–3b** Thomas J. Cutitta, International Imaging, Inc. **29–11** Graphics Art Department, Lawrence Berkeley Laboratory, University of California **29–12** Brookhaven/Science Photo Library/Photo Researchers **29–14b** University of California, Lawrence Berkeley Laboratory **29–17a** The Granger Collection **29–24a** Thomas J. Cutitta, International Imaging, Inc. **29–25a** Thomas J. Cutitta, International Imaging, Inc.

Chapter 30 CO Plasma Physics Laboratory, Princeton University **30–1a** Thomas J. Cutitta, International Imaging, Inc. **30–1b** Thomas J. Cutitta, International Imaging, Inc. **30–2a** Thomas J. Cutitta, International Imaging, Inc. **30–2b** Thomas J. Cutitta, International Imaging, Inc. **30–9** The Granger Collection **30–15a** Cenco **30–16a** General Electric Engineering Co. **30–18a** Richard Megna/Fundamental Photographs **30–28a** Richard Megna/Fundamental Photographs

Chapter 31 CO Royal Institution of London **31–2** From *Experimental Researches in Electricity* by Michael Faraday (1839) **31–12** Thomas J. Cutitta, International Imaging, Inc. **31–20** IBM Research/Peter Arnold, Inc. **31–24** Thomas J. Cutitta, International Imaging, Inc. **31–28** Howard Liverance/The Image Works

Chapter 32 CO Leonard Lessin/Peter Arnold, Inc. **32–6** Dr. John Unguris **32–12a** Courtesy of the IBM Corporation **32–12b** IBM Research **32–14** D. Clegg & Roxy Wilson **B1–1** Courtesy General Electric Medical Systems **B1–2** Mallinckrodt Institute of Radiology

Chapter 33 CO Thomas J. Cutitta, International Imaging, Inc. **33–2b** Cenco **33–4** Michael Gallitelli/Metroland Photo **33–5** The Image Works **33–8a** Smithsonian Institution, Washington, D.C. **33–8b** Janet Kroboth-Weber

Chapter 34 CO Hank Morgan/Rainbow **34–1a** Central Scientific **34–1b** Westinghouse Historical Collection, Westinghouse Electric **34–4a** Westinghouse Electric Corp. **34–4b** Steve Allen/Peter Arnold, Inc. **34–10** The Image Works **34–13** The Image Works

Chapter 35 CO Raja Guhathakurta/NRAO/AUI **35–8a** Ann Ronan Picture Library **35–10a** FPG International **35–10b** Leonard Lessin/Peter Arnold, Inc. **35–12** Bruce Iverson/Science Photo Library/Photo Researchers **35–16** Diane Schiumo/Fundamental Photographs **35–17a** Thomas J. Cutitta, International Imaging, Inc. **35–17b** Thomas J. Cutitta, International Imaging, Inc. **35–17c** Thomas J. Cutitta, International Imaging, Inc. **35–21a** Nina Barnett **35–21b** Nina Barnett **35–24** NASA

Chapter 36 CO Scala/Art Resource **36–1a, b** *Scientific American,* Sept. 1968, p. 55 36–4b Courtesy of the Bancroft Library, University of California, Berkeley **36–6** Courtesy A.T. & T. **36–7a** Richard Megna/Fundamental Photographs **B1–2a** Frank Armstrong/News & Information Service, University of Texas at Austin **B1–2b** NASA **36–10a** Fundamental Photographs **36–14b** Thomas J. Cutitta, International Imaging, Inc. **B2–1** Courtesy A.T. & T. **B2–2** *Scientific American,* May 1989, p. 122 **36–16a** Cenco **36–20** The Granger Collection **36–21a** Courtesy Bausch & Lomb **36–24b** Douglas W. Johnson

Chapter 37 CO Scala/Art Resource **37–5** Charles Addams **37–8c** Richard Megna/Fundamental Photographs **37–13** © M.C. Escher Heirs, Cordon Art, Baarn, Holland **37–23** Richard Megna/Fundamental Photographs **37–30** Frederic Martini, *Fundamentals of Anatomy and Physiology,* 2nd ed., Prentice Hall, 1992, illustration by William Ober **37–32** Leonard Lessin/Peter Arnold, Inc. **37–34** Richard Megna/Fundamental Photographs **37–35a** Leonard Lessin/Peter Arnold, Inc. **37–36a** Scala/Art Resource **37–37a** Dr. E.I. Robson/Science Photo Library/Photo Researchers **37–38** Light Path Technologies **37–39** Tom Pantages **37–42** Tom Pantages

Chapter 38 CO Luiz C. Marigo/Peter Arnold, Inc. **38–5** From the Atlas of Optical Phenomena, Michel Cagnet, Maurice Francon, Jean Claude Thrierr. © by Springer-Verlag OHG, Berlin, 1962. Published by Prentice Hall, Inc., Englewood Cliffs, N.J. **38–9b** Cenco **38–9c** Courtesy Bausch & Lomb **38–12** Dr. E.R. Degginger **38–14** Dr. Jeremy Burgess/Science Photo Library/Photo Researchers **38–17b** Courtesy of Pasco Scientific **38–17c** Courtesy AIP Niels Bohr Library **38–18a** Cenco **38–19a** Cenco **38–19b** Cenco **38–24** Richard Megna/Fundamental Photographs

Chapter 39 CO Michael Freeman **39–1** (no credit) **39–3** Ken Kay/Fundamental Photographs **39–5** P.M. Rinard, from *Am. J. Phys.,* 44, 1976, p. 70 **39–6** The Ferdinand Hamburger, Jr. Archives of the Johns Hopkins University **39–7b** Bausch & Lomb **39–11** From the Atlas of Optical Phenomena, Michel Cagnet, Maurice Francon, Jean Claude Thrierr. © by Springer-Verlag OHG, Berlin, 1962. Published by Prentice Hall, Inc., Englewood Cliffs, N.J. **39–13** From the Atlas of Optical Phenomena **39–14a, b, c,** (no credit) **39–17** From the Atlas of Optical Phenomena **39–18a** Science Photo Library/Photo Researchers **39–18b** From "The Particle Explosion," p. 25; Special Collections University of Virginia Library **39–20a** Photo Researchers **39–22b** Courtesy of the Burndy Library, Norwalk, CT **39–23b** Thomas J. Cutitta, International Imaging, Inc.

Chapter 40 CO Ping-Kang Hsiung, PhD. **40–2a** *Invention and Technology,* Fall 1987, p. 47 **40–2b** *Invention and Technology,* Fall 1987, p. 47 **40–4** Tom Pantages **40–5** The Bettmann Archive **40–14** National Optical Astronomy Observatories **40–15** UPI/Bettmann Newsphotos **40–23** Lawrence Berkeley Laboratory/Science Photo Library/Photo Researchers **40–25** NASA

Chapter 41 CO Courtesy AIP Niels Bohr Library **41–1** Culver Pictures, Inc. **41–2a** Courtesy Bell Laboratories **41–7** Courtesy AIP Niels Bohr Library **41–12** Maison Albert Schweitzer, Courtesy AIP Niels Bohr Library **41–15a** Courtesy AIP Niels Bohr Library **41–18a** Courtesy Educational Development Center & Prof. C. G. Shull **41–18b** Courtesy Educational Development Center & Prof. C. G. Shull **41–19** Dr. Tony Brain/Science Photo Library/Photo Researchers **41–20** Phil Huber/Sports Illustrated

Chapter 42 CO Ken Edward/Science Source/Photo Researchers **42–1a, b** Wolfgang Rueckner, Harvard Science Center **42–3** Princeton University, Courtesy AIP Niels Bohr Library **42–6** Courtesy Bausch & Lomb Inc. **42–15** Courtesy AIP Niels Bohr Library, Gondsmit Collection

Chapter 43 CO Dan McCoy/Rainbow **43–2** University of Chicago, Courtesy AIP Niels Bohr Library **43–3** Astronomical Society of the Pacific, Leaflet No. 390, Dec. 1961. *Sirius and Its Companion* by Sarah Lee Lippincott, Sproul Observatory, Swarthmore College, Fig. 1 **43–6b** Dan McCoy/Rainbow **43–7a** Courtesy Jodon **43–8** Andy Levin/Photo Researchers **43–9** James Kilkelly, Scientific American, August 1986, Volume 255, Number 2, p. 69 **43–10** Lawrence Berkeley Laboratory, University of California **43–17** Courtesy Conductus Inc. **43–18** (no credit) **43–19** F. R. Harnden, Jr., Harvard-Smithsonian Center for Astrophysics **43–20** ©

Chapter 44 CO D.O.E./Science Source/Photo Researchers **44–B1–2** NASA **44–9** Courtesy Thomson Consumer Electronics **44–10** Courtesy of Minolta **44–16** Nick Holonyak, Jr., University of Illinois **44–21** Courtesy IBM Research **44–22** Courtesy Texas Instruments **44–23** Courtesy of Fraunhoffer Institute, IMT-Berlin **44–29** © Bellcore, Axel Scherer **44–30** © Bellcore, Axel Scherer **44–33** Courtesy IBM **44–35** Pasco

Chapter 45 CO Michael Gilbert/Science Photo Library/Photo Researchers **45–1a** AIP Niels Bohr Library, Physics Today Collection **45–11a** The Mansell Collection, London **45–11b** AIP Niels Bohr Library, William G. Myers Collection **45–12** The Granger Collection **45–19** Dr. Bonnichsen **45–21a** Sandia National Laboratories, photo by Walter Dickenman

Chapter 46 CO Patrice Loiez, Cern/Science Photo Library/Photo Researchers **46–1** (no credit) **46–2** Joe Stancampiano/National Geographic Society **46–4** Courtesy Contemporary Physics Education Project **46–6** Courtesy T. Myhrer and P. F. Dahl; **46–9** Courtesy Fermilab Photo Dept. **46–10** Courtesy Fermi National Accelerator Laboratory **46–11** C. Powell, P. Fowler & D. Perkins/Science Photo Library/Photo Researchers **46–12** Lawrence Berkeley Laboratory/Science Photo Library/Photo Researchers **46–13** Pedro Waloschek, DESY, Notkestr. 85, 2000 Hamburg 52, Germany **46–15** S. J. Maddox, W. J. Sutherland, G. P. Efstathiou, and J. Loveday, Oxford Astrophysics **46–16** Fermilab **46–17** Harvard-Smithsonian Center for Astrophysics

Interactive Physics II™

A Complete Motion Lab on the Computer!

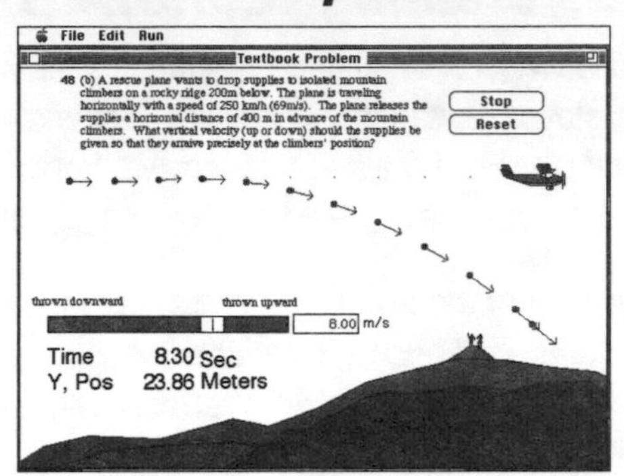

Draw It.

Any motion experiment you can imagine. Create this air drop experiment, for example, to investigate projectile motion.

Click RUN.

Your experiment moves acording to real-world physical laws. The plane moves across the screen and releases a survival packet in smooth movie-like animation.

The 'Word Processor' of Motion Simulation.

Interactive Physics II is a complete motion lab on the computer. Choose an experiment — **any experiment you can think of** — at any level of sophistication. Select it from the library of ready-to-run experiments or build it yourself using the simple Macintosh® palette and pull-down menus. Roll a ball down an inclined plane, launch a rocket between orbiting planets, send a stream of electrons through a varying magnetic field, build a working model of a combustion engine. Model virtually any experiment, visualize any principle, all in smooth animation.

Model with Powerful Tools.

Mass objects include circles, squares, rectangles, and polygons. Simulation elements include ropes, motors, actuators, pulleys, pin joints, springs, dampers, and forces. Change object mass, elasticity, friction, and charge. Vary and modify gravity and electrostatics, or create your own custom forces. Use **pulleys** and **pin joints** to connect one object to another. Attach **motors** and **actuators** to create advanced models such as dump trucks and motorcycles. Paste in your own graphics and watch them work in your simulations. Modify and control your simulations while they are running.

What You Get:

- Three 3.5-inch disks (one program disk and two experiment disks).
- A complete curriculum guide on disk with experiments.
- Interactive Physics II User's Guide with step-by-step tutorials.

System Requirements:

- Apple Macintosh
- System 6.0.5 or greater
- System 7.0 Savvy
- 2 MB RAM (2.5 MB RAM for System 7)

15 Brush Place
San Francisco, CA 94103 USA
800-766-6615

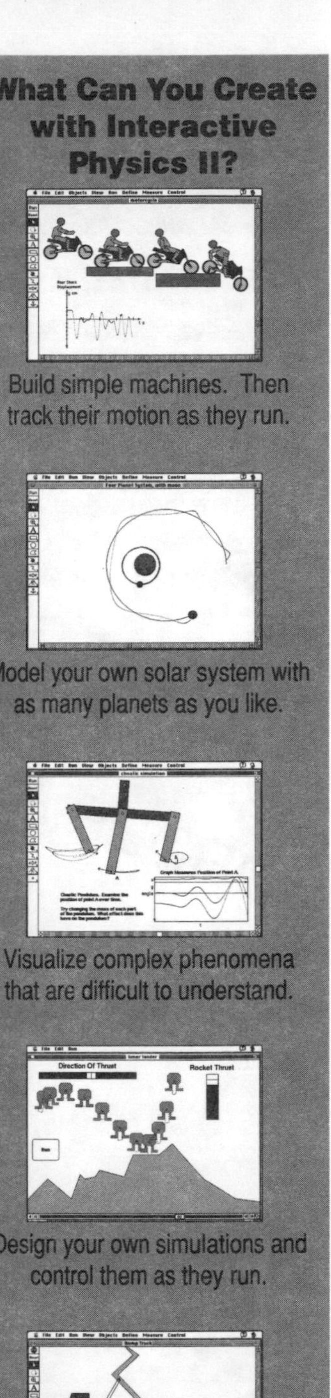

What Can You Create with Interactive Physics II?
Build simple machines. Then track their motion as they run.
Model your own solar system with as many planets as you like.
Visualize complex phenomena that are difficult to understand.
Design your own simulations and control them as they run.
Make QuickTime™ movies of any simulation. Export your data.

TABLES IN THE TEXT